PROPERTIES OF VARIOUS HOMOGENEOUS SOLIDS

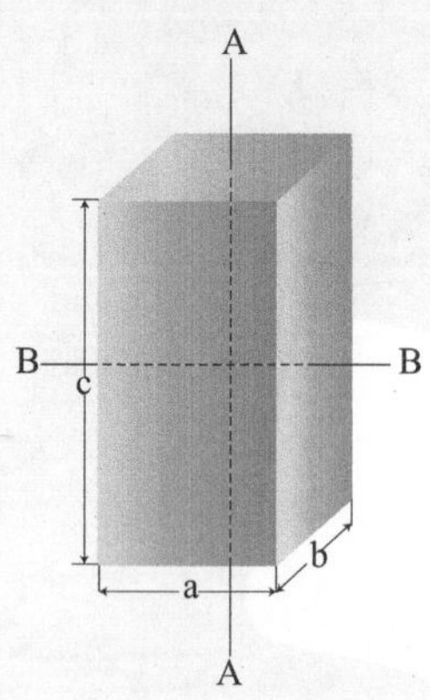

Rectangular prism

$$V = abc$$

$$I_{AA} = \frac{1}{12}M(a^2 + b^2)$$

$$k_{AA} = \sqrt{\frac{a^2 + b^2}{12}}$$

$$I_{BB} = \frac{1}{12}M(b^2 + c^2)$$

$$k_{BB} = \sqrt{\frac{b^2 + c^2}{12}}$$

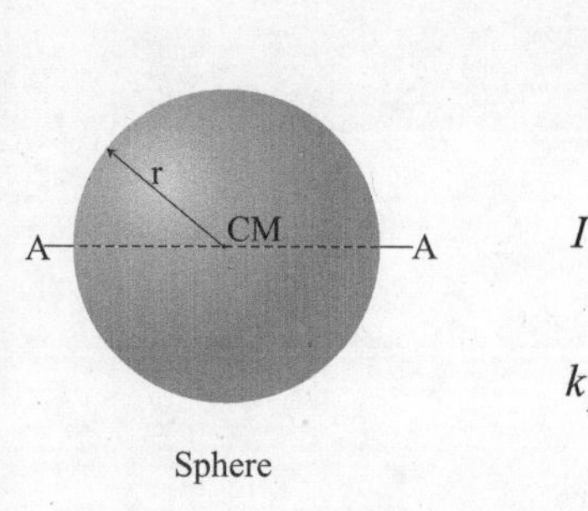

Sphere

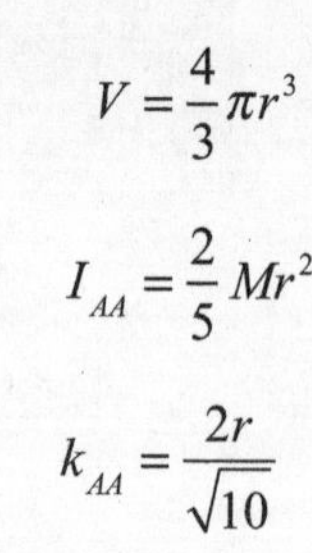

$$V = \frac{4}{3}\pi r^3$$

$$I_{AA} = \frac{2}{5}Mr^2$$

$$k_{AA} = \frac{2r}{\sqrt{10}}$$

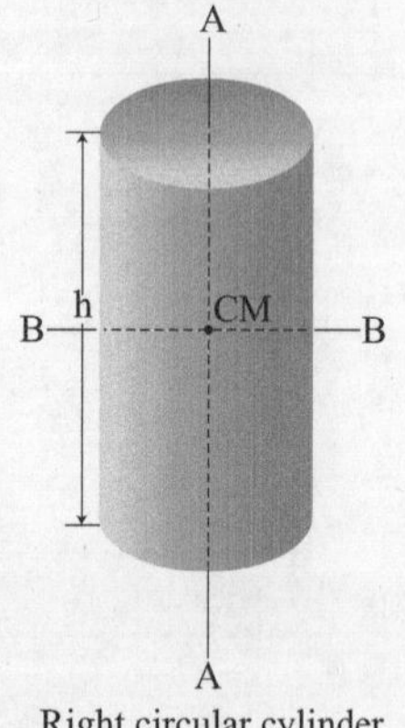

Right circular cylinder

$$V = \pi r^2 h$$

$$I_{AA} = \frac{1}{2}Mr^2$$

$$k_{AA} = \frac{r}{\sqrt{2}}$$

$$I_{BB} = \frac{1}{12}M(3r^2 + h^2)$$

$$k_{BB} = \sqrt{\frac{3r^2 + h^2}{12}}$$

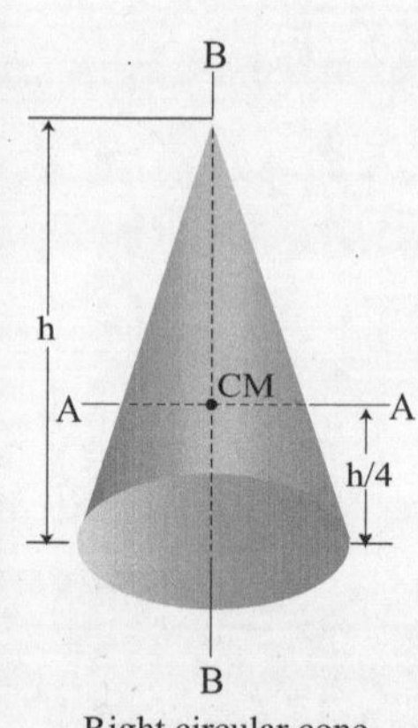

Right circular cone

$$V = \frac{1}{3}\pi r^2 h$$

$$I_{AA} = \frac{3}{20}M\left(r^2 + \frac{h^2}{4}\right)$$

$$k_{AA} = \sqrt{\frac{3}{80}(4r^2 + h^2)}$$

$$I_{BB} = \frac{3}{10}Mr^2$$

$$k_{BB} = \frac{3r}{\sqrt{30}}$$

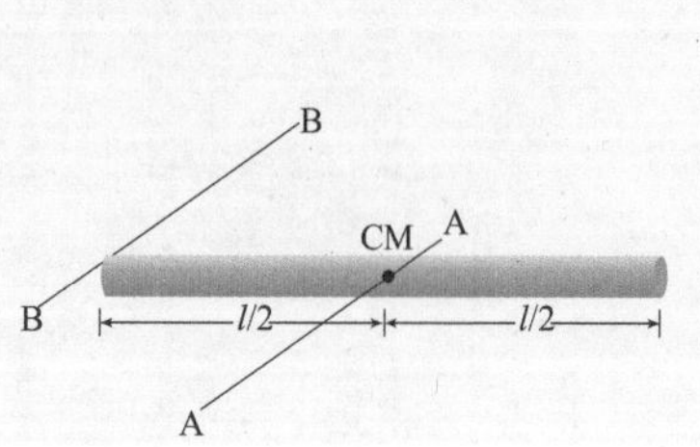

Slender rod

$$I_{AA} = \frac{1}{12}Ml^2$$

$$k_{AA} = \frac{l}{\sqrt{12}}$$

$$I_{BB} = \frac{1}{3}Ml^2$$

$$k_{BB} = \frac{1}{\sqrt{3}}$$

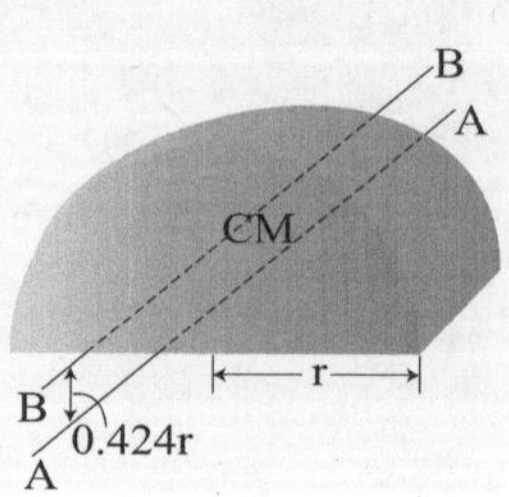

Semi-circular cylinder

$$I_{AA} = \frac{1}{2}Mr^2$$

$$I_{BB} = .320\,Mr^2$$

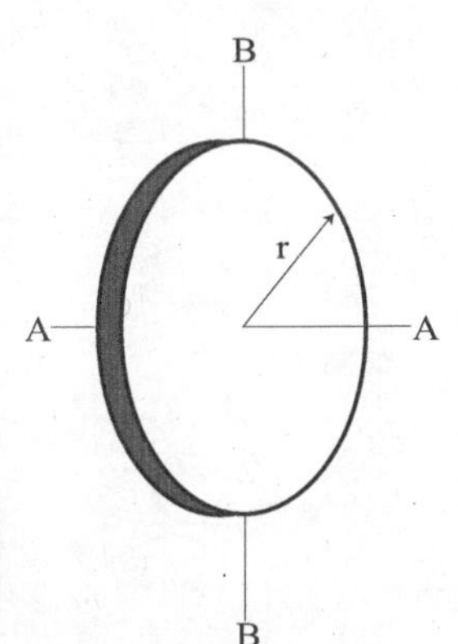

Thin disc

$$I_{AA} = \frac{1}{2} Mr^2$$

$$I_{BB} = \frac{1}{4} Mr^2$$

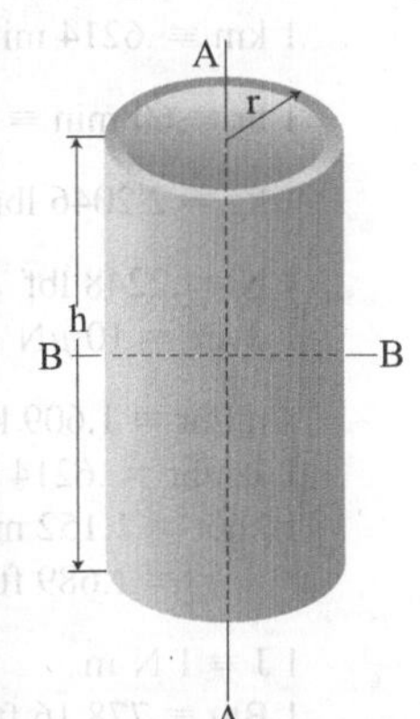
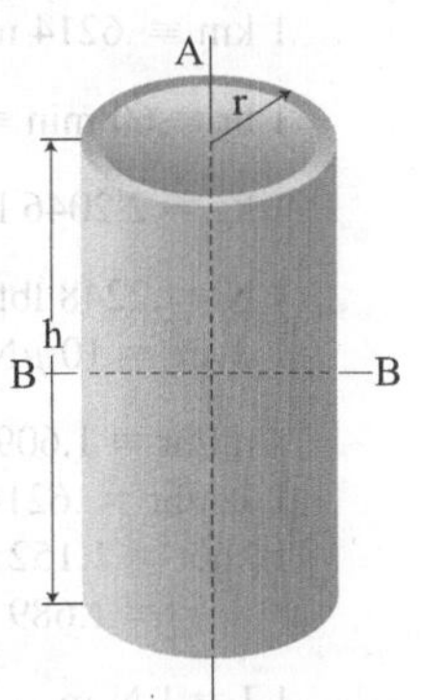

Thin-walled cylinder

$$I_{AA} = Mr^2$$

$$I_{BB} = \frac{M}{2}\left(r^2 + \frac{h^2}{6}\right)$$

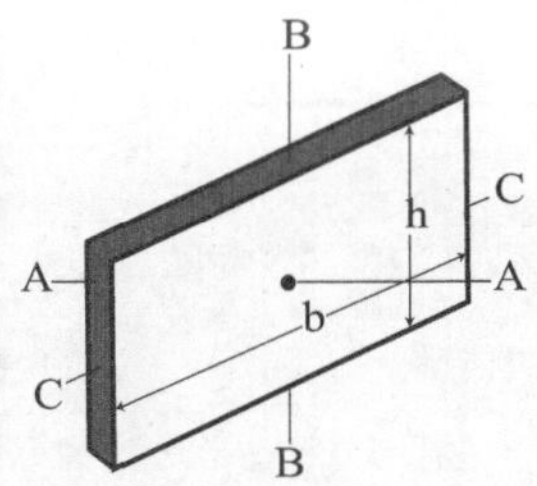

Thin rectangular plate

$$I_{AA} = \frac{1}{12} M(b^2 + h^2)$$

$$I_{BB} = \frac{1}{12} Mb^2$$

$$I_{CC} = \frac{1}{12} Mh^2$$

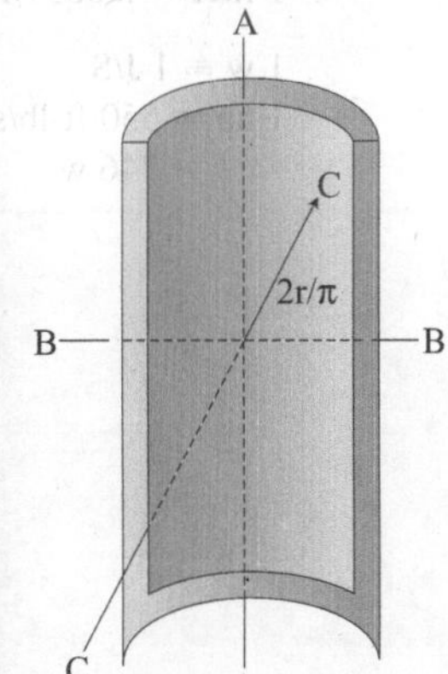

Thin half cylinder

$$I_{BB} = \frac{1}{2} M\left(\frac{r^2 + h^2}{6}\right)$$

$$I_{AA} = Mr^2$$

$$I_{CC} = \frac{1}{2} M\left(\frac{r^2 + h^2}{6}\right)$$

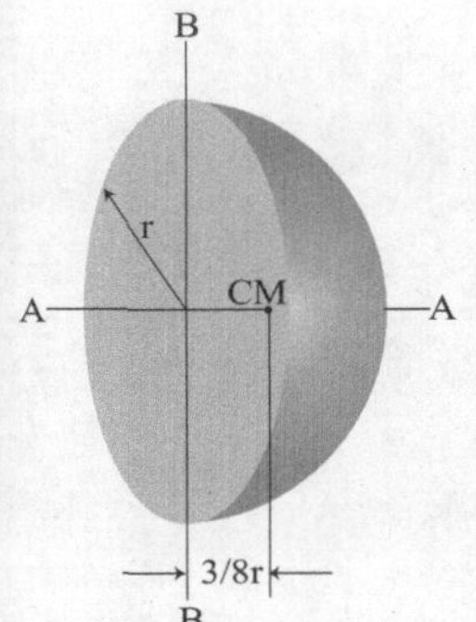

Hemisphere

$$V = \frac{2}{3}\pi r^3$$

$$I_{AA} = \frac{2}{5} Mr^2$$

$$I_{BB} = \frac{2}{5} Mr^2$$

SELECTED DIMENSIONAL EQUIVALENTS

LENGTH	1 m ≡ 3.281 ft ≡ 39.37 in. 1 mi ≡ 5280 ft ≡ 1.609 km 1 km ≡ .6214 mi
TIME	1 hr ≡ 60 min ≡ 3600 sec
MASS	1 kg ≡ 2.2046 lbm ≡ .068521 slug
FORCE	1 N ≡ .2248 lbf 1 dyne ≡ 10 μN
SPEED	1 mi/hr ≡ 1.609 km/hr ≡ 1.467 ft/sec 1 km/hr = .6214 mi/hr 1 knot = 1.152 mi/hr ≡ 1.853 km/hr ≡ 1.689 ft/sec
ENERGY	1 J ≡ 1 N-m 1 Btu ≡ 778.16 ft-lbf ≡ 1.055 kJ 1 watt-hour ≡ 2.778 × 10^{-4} J
VOLUME	1 gal ≡ .16054 ft^3 ≡ .0045461 m^3 1 liter ≡ .03531 ft^3 = .2642 gal
POWER	1 w ≡ 1 J/S 1 hp ≡ 550 ft-lb/sec ≡ .7068 Btu/sec ≡ 746 w

Engineering Mechanics

Statics and Dynamics

Fourth Edition

Irving H. Shames

Professor
Dept. of Civil, Mechanical and Environmental Engineering
The George Washington University

G. Krishna Mohana Rao

Assistant Professor
Dept. of Mechanical Engineering
JNTU College of Engineering, Kukatpally
India

ISBN 9788177581232
First Impression, 2006
Current Impression, 2026

Published by Pearson India Education Services Pvt. Ltd, CIN: U72200TN2005PTC057128.

Head Office: 1st Floor, Berger Tower, Plot No. C-001A/2, Sector 16B, Noida - 201 301, Uttar Pradesh, India.
Registered Office: 6th Floor, Tower A, Unit A and B,International Tech Park, CapitaLand Chennai,

200 Feet Radial Road, Zamin Pallavaram, Old Pallavaram,Chennai – 600117, Tamil Nadu, India

Website: in.pearson.com; Email: companysecretary.india@pearson.com

Printed in India at: Manipal Technologies Limited, Manipal

Contents

Part B—Dynamics

16 Vibrations 771

APPENDIX I

APPENDIX II

Preface

Engineering Mechanics has been adapted to suit the curriculum of various Indian universities and engineering colleges. The book provides comprehensive coverage in a clear, straightforward and well-illustrated manner with excellent style of presentation. Both SI and FPS systems of units have been used.

The goal is aimed towards working problems as soon as possible from first principles. Thus, examples are carefully chosen during the development of a series of related areas to instill continuity in the evolving theory and then, after these areas have been carefully discussed with rigor, come the problems. Furthermore at the end of each chapter, there are many problems that have not been arranged by text section. The instructor is encouraged as soon as he/she is well along in the chapter to use these problems. The text is not chopped up into many methodologies each with an abbreviated discussion followed by many examples for using the specific methodology and finally a set of problems carefully tailored for the methodology. The nature of the format is to discourage excessive mapping of homework problems from the examples. And second, it is to lessen the memorization of specific, specialized methodologies in lieu of absorbing basic principles.

Salient features

- Spatial systems of forces are covered.
- Friction chapter is enriched with topics on wedge, screw jack and differential screw jack.
- A new topic on transmission of power through belt drives has been introduced.
- Separate chapter on mechanical vibrations has been introduced.

Pedagogical features

- Chapter objectives at the begining of each chapter.
- Review questions and exercises at the end of each section in a chapter.
- Supplementary problems after each worked example.
- Chapter summaries at the end of all chapters.
- Extensive sectionalized problem sets at the end of each chapter.

Topical coverage

Part A—Statics

Chapter 1 provides detailed introduction to engineering mechanics with the help of illustrations and photographs coupled with brief on units, vector quantities, and scalar quantities. Fundamental equations and theorems are explained to create a base for the remaining chapters.

Chapter 2 deals with most of the problems that are solved and explained using vector concept. Fundamentals of vectors and scalars and their analysis are explained in this chapter.

Chapter 3 explains about various force systems and the result of these forces on the objects. Creation of moments and couples with the help of different force systems is explained with clarity.

Chapter 4 deals with the analysis of resultant force of spatial and distributed force systems.

Chapter 5 explains the fundamental step in solving the problems by introducing the concept of free body diagram.

Chapter 6 is enriched with additional topics on screw jack, differential screw jack, wedge, etc. Transmission of power through belt drives has been included as another application of friction in real life.

Chapter 7 deals with the properties of surfaces like center of gravity of plane area and center of gravity of volumes and masses.

Chapter 8 clearly defines and explains the concept of moment of inertia of plane areas, and volumes and solids about their centroidal axes or any other axis chosen.

Part B—Dynamics

Chapter 9 in general provides introduction to various quantities, which decide the dynamic behavior of a particle. Velocity and acceleration are explained for both rectangular and cylindrical coordinates.

Chapter 10 deals with rectangular translation, central force motion, and a system of particles.

Chapter 11 includes a different technique of analyzing the particle. Concept of energy is utilized to analyze individual particles and also systems of particle.

Chapter 12 explains the concept of linear momentum and moment of momentum.

Chapter 13 deals with concept of relative motion.

Chapter 14 deals with rolling bodies under different conditions.

Chapter 15 deals with work energy relations of rigid bodies.

Chapter 16 deals with mechanical vibrations. Topics like simple pendulum, compound pendulum, and simple harmonic motion have been explained in detail.

In overall summary, two main goals have been pursued. They are

1. To encourage working problems from first principles and thus to minimize excessive mapping from examples and to discourage rote learning of specific methodologies for solving various and sundry kinds of specific problems.

2. To "open-end" the material to later course work in other engineering sciences with the view towards making smoother transitions and to provide for greater continuity. Also, the purpose is to engage the interest and curiosity of students for further study of mechanics.

Acknowledgments

At this stage of my career, I will risk impropriety by presenting now an extended section of acknowledgments. I want to give thanks to SUNY at Buffalo where I spent 31 happy years and where I wrote many of my books. And I want to salute the thousands (about 5000) of fine students who took my courses during this long stretch. I wish to thank my eminent friend and colleague Professor Shahid Ahmad who among other things taught the sophomore mechanics sequence with me and who continues to teach it. He gave me a very thorough review of the fourth edition with many valuable suggestions. I thank him profusely. I want particularly to thank Professor Michael Symans, from Washington State University, Pullman for his superb contributions to the entire manuscript. I came to The George Washington University at the invitation of my longtime friend and former Buffalo colleague Dean Gideon Frieder and the faculty in the Civil, Mechanical and Environmental Engineering Department. Here, I came back into contact with two well-known scholars that I knew from the early days of my career, namely Professor Hal Liebowitz (president-elect of the National Academy of Engineering) and Professor Ali Cambel (author of recent well-received book on chaos). I must give profound thanks to the chairman of my new department at G.W., Professor Sharam Sarkani. He has allowed me to play a vital role in the academic program of the department. I will be able to continue my writing at full speed as a result. I shall always be grateful to him. Let me not forget the two dear ladies in the front office of the department. Mrs. Zephra Coles in her decisive efficient way took care of all my needs even before I was aware of them. And Ms. Joyce Jeffress was no less helpful and always had a humorous comment to make.

I was extremely fortunate in having the following professors as reviewers.

Professor Shahid Ahmad, SUNY at Buffalo
Professor Ravinder Chona, Texas A&M University
Professor Bruce H. Karnopp. University of Michigan
Professor Richard F. Keltie, North Carolina State University

Professor Stephen Malkin, University of Massachusetts
Professor Sudhakar Nair, Illinois Institute of Technology
Professor Jonathan Wickert, Carnegie Mellon University

I wish to thank these gentlemen for their valuable assistance and encouragement.

I have two people left. One is my good friend Professor Bob Jones from V.P.I. who assisted me in the third edition with several hundred excellent statics problems and who went over the entire manuscript with me with able assistance and advice. I continue to benefit in the new edition from his input of the third edition. And now, finally, the most important person of all, my dear wife Sheila. She has put up all these years with the author of this book, an absent-minded, hopeless workaholic. Whatever I have accomplished of any value in a long and ongoing career, I owe to her.

To my Dear, Wonderful Wife Sheila

About the Author

Irving Shames presently serves as a Professor in the Department of Civil, Mechanical, and Environmental Engineering at The George Washington University. Prior to this appointment Professor Shames was a Distinguished Teaching Professor and Faculty Professor at The State University of New York—Buffalo, where he spent 31 years.

Professor Shames has written up to this point in time 10 textbooks. His first book *Engineering Mechanics, Statics and Dynamics* was originally published in 1958, and it is going into its fourth edition in 1996. All of the books written by Professor Shames have been characterized by innovations that have become mainstays of how engineering principles are taught to students. *Engineering Mechanics, Statics and Dynamics* was the first widely used Mechanics book based on vector principles. It ushered in the almost universal use of vector principles in teaching engineering mechanics courses today.

Other textbooks written by Professor Shames include:

- *Mechanics of Deformable Solids*, Prentice-Hall, Inc.
- *Mechanics of Fluids*, McGraw-Hill
- *Introduction to Solid Mechanics*, Prentice-Hall, Inc.
- *Introduction to Statics*, Prentice-Hall, Inc.
- *Solid Mechanics—A Variational Approach* (with C.L. Dym), McGraw-Hill
- *Energy and Finite Element Methods in Structural Mechanics* (with C.L. Dym), Hemisphere Corp., of Taylor and Francis
- *Elastic and Inelastic Stress Analysis* (with F. Cozzarelli), Prentice-Hall, Inc.

In recent years, Professor Shames has expanded his teaching activities and has held two summer faculty workshops in mechanics sponsored by the State of New York, and one national workshop sponsored by the National Science Foundation. The programs involved the integration both conceptually and pedagogically of mechanics from the sophomore year on through graduate school.

Statics

CHAPTER 1

REVIEW I*

Fundamentals of Mechanics

†1.1 Introduction

Mechanics is the physical science concerned with the dynamical behavior (as opposed to chemical and thermal behavior) of bodies that are acted on by mechanical disturbances. Since such behavior is involved in virtually all the situations that confront an engineer, mechanics lies at the core of much engineering analysis. In fact, no physical science plays a greater role in engineering than does mechanics, and it is the oldest of all the physical sciences. The writings of Archimedes covering buoyancy and the lever were recorded before 200 B.C. Our modern knowledge of gravity and motion was established by Isaac Newton (1642–1727), whose laws founded Newtonian mechanics, the subject matter of this text.

In 1905, Einstein placed limitations on Newton's formulations with his theory of relativity and thus set the stage for the development of relativistic mechanics. The newer theories, however, give results that depart from those of Newton's formulations only when the speed of a body approaches the speed of light (186,000 miles/sec). These speeds are encountered in the large-scale phenomena of dynamical astronomy. Furthermore for small-scale phenomena involving subatomic particles, quantum mechanics must be used rather than Newtonian mechanics. Despite these limitations, it remains nevertheless true that, in the great bulk of engineering problems, Newtonian mechanics still applies.

†1.2 Basic Dimensions and Units of Mechanics

To study mechanics, we must establish abstractions to describe those characteristics of a body that interest us. These abstractions are called *dimensions*. The dimensions that we pick, which are independent of all other dimensions, are termed *primary* or *basic dimensions*, and

*The reader is urged to be sure that Section 1.9 is thoroughly understood since this section is vital for a good understanding of statics in particular and mechanics in general.

Also, the notation † before the titles of certain sections indicates that specific questions concerning the contents of these sections requiring verbal answers are presented at the end of the chapter. The instructor may wish to assign these sections as a reading assignment along with the requirement to answer the aforestated associated questions as the author routinely does himself.

the ones that are then developed in terms of the basic dimensions we call *secondary dimensions*. Of the many possible sets of basic dimensions that we could use, we will confine ourselves at present to the set that includes the dimensions of length, time, and mass. Another convenient set will be examined later.

Length—A Concept for Describing Size Quantitatively. In order to determine the size of an object, we must place a second object of known size next to it. Thus, in pictures of machinery, a man often appears standing disinterestedly beside the apparatus. Without him, it would be difficult to gage the size of the unfamiliar machine. Although the man has served as some sort of standard measure, we can, of course, only get an approximate idea of the machine's size. Men's heights vary, and, what is even worse, the shape of a man is too complicated to be of much help in acquiring a precise measurement of the machine's size. What we need, obviously, is an object that is constant in shape and, moreover, simple in concept. Thus, instead of a three-dimensional object, we choose a one-dimensional object.[1] Then, we can use the known mathematical concepts of geometry to extend the measure of size in one dimension to the three dimensions necessary to characterize a general body. A straight line scratched on a metal bar that is kept at uniform thermal and physical conditions (as, e.g., the meter bar kept at Sèvres, France) serves as this simple invariant standard in one dimension. We can now readily calculate and communicate the distance along a certain direction of an object by counting the number of standards and fractions thereof that can be marked off along this direction. We commonly refer to this distance as length, although the term "length" could also apply to the more general concept of size. Other aspects of size, such as volume and area, can then be formulated in terms of the standard by the methods of plane, spherical, and solid geometry.

A *unit* is the name we give an accepted measure of a dimension. Many systems of units are actually employed around the world, but we shall only use the two major systems, the American system and the SI system. The basic unit of length in the American system is the foot, whereas the basic unit of length in the SI system is the meter.

Time—A Concept for Ordering the Flow of Events. In observing the picture of the machine with the man standing close by, we can sometimes tell approximately when the picture was taken by the style of clothes the man is wearing. But how do we determine this? We may say to ourselves: "During the thirties, people wore the type of straw hat that the fellow in the picture is wearing." In other words, the "when" is tied to certain events that are experienced by, or otherwise known to, the observer. For a more accurate description of "when," we must find an action that appears to be completely repeatable. Then, we can order the events under study by counting the number of these repeatable actions and fractions thereof that occur while the events transpire. The rotation of the earth gives rise to an event that serves as a good measure of time—the day. But we need smaller units in most of our work in engineering, and thus, generally, we tie events to the second, which is an interval repeatable 86,400 times a day.

Mass—A Property of Matter. The student ordinarily has no trouble understanding the concepts of length and time because he/she is constantly aware of the size of things through his/her senses of sight and touch, and is always conscious of time by observing the flow of events in his/her daily life. The concept of mass, however, is not as easily grasped since it does not impinge as directly on our daily experience.

[1]We are using the word "dimensional" here in its everyday sense and not as defined above.

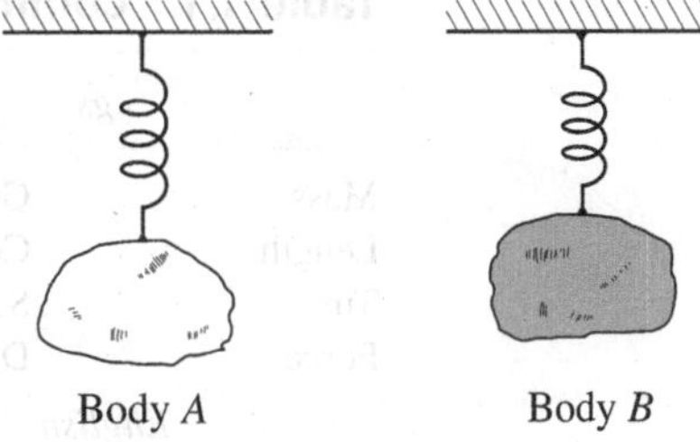

Figure 1.1. Bodies restrained by identical springs.

Mass is a property of matter that can be determined from *two* different actions on bodies. To study the first action, suppose that we consider two hard bodies of entirely different composition, size, shape, color, and so on. If we attach the bodies to identical springs, as shown in Fig. 1.1, each spring will extend some distance as a result of the attraction of gravity for the bodies. By grinding off some of the material on the body that causes the greater extension, we can make the deflections that are induced on both springs equal. Even if we raise the springs to a new height above the earth's surface, thus lessening the deformation of the springs, the extensions induced by the pull of gravity will be the same for both bodies. And since they are, we can conclude that the bodies have an equivalent innate property. This property of each body that manifests itself in the *amount of gravitational attraction* we call *mass*.

The equivalence of these bodies, after the aforementioned grinding operation, can be indicated in yet a second action. If we move both bodies an equal distance downward, by stretching each spring, and then release them at the same time, they will begin to move in an identical manner (except for small variations due to differences in wind friction and local deformations of the bodies). We have imposed, in effect, the same mechanical disturbance on each body and we have elicited the same dynamical response. Hence, despite many obvious differences, the two bodies again show an equivalence.

The property of mass, then, characterizes a body both in the action of gravitational attraction and in the response to a mechanical disturbance.

To communicate this property quantitatively, we may choose some convenient body and compare other bodies to it in either of the two abovementioned actions. The two basic units commonly used in much American engineering practice to measure mass are the *pound mass*, which is defined in terms of the attraction of gravity for a standard body at a standard location, and the *slug*, which is defined in terms of the dynamical response of a standard body to a standard mechanical disturbance. A similar duality of mass units does not exist in the SI system. There only the *kilogram* is used as the basic measure of mass. The kilogram is measured in terms of response of a body to a mechanical disturbance. Both systems of units will be discussed further in a subsequent section.

We have now established three basic independent dimensions to describe certain physical phenomena. It is convenient to identify these dimensions in the following manner:

$$\begin{array}{ll} \text{length} & [L] \\ \text{time} & [t] \\ \text{mass} & [M] \end{array}$$

These formal expressions of identification for basic dimensions and the more complicated groupings to be presented in Section 1.3 for secondary dimensions are called "dimensional representations."

Often, there are occasions when we want to change units during computations. For instance, we may wish to change feet into inches or millimeters. In such a case, we must replace the unit in question by a *physically equivalent* number of new units. Thus, a foot

Table 1.1 Common Systems of Units

cgs		*SI*	
Mass	Gram	Mass	Kilogram
Length	Centimeter	Length	Meter
Time	Second	Time	Second
Force	Dyne	Force	Newton
English		*American Practice*	
Mass	Pound mass	Mass	Slug or pound mass
Length	Foot	Length	Foot
Time	Second	Time	Second
Force	Poundal	Force	Pound force

is replaced by 12 inches or 305 millimeters. A listing of common systems of units is given in Table 1.1, and a table of equivalences between these and other units is given on the inside covers. Such relations between units will be expressed in this way:

$$1 \text{ ft} \equiv 12 \text{ in.} \equiv 305 \text{ mm}$$

The three horizontal bars are not used to denote *algebraic* equivalence; instead, they are used to indicate physical equivalence. Here is another way of expressing the relations above:

$$\begin{aligned} \left(\frac{1 \text{ ft}}{12 \text{ in.}}\right) &\equiv 1, & \left(\frac{1 \text{ ft}}{305 \text{ mm}}\right) &\equiv 1 \\ \left(\frac{12 \text{ in.}}{1 \text{ ft}}\right) &\equiv 1, & \left(\frac{305 \text{ mm}}{1 \text{ ft}}\right) &\equiv 1 \end{aligned} \tag{1.1}$$

The unity on the right side of these relations indicates that the numerator and denominator on the left side are physically equivalent, and thus have a 1:1 relation. This notation will prove convenient when we consider the change of units for secondary dimensions in the next section.

†1.3 Secondary Dimensional Quantities

When physical characteristics are described in terms of basic dimensions by the use of suitable definitions (e.g., velocity is defined[2] as a distance divided by a time interval), such quantities are called *secondary dimensional quantities*. In Section 1.4, we will see that these quantities may also be established as a consequence of natural laws. The dimensional representation of secondary quantities is given in terms of the basic dimensions that enter into the formulation of the concept. For example, the dimensional representation of velocity is

$$[\text{velocity}] \equiv \frac{[L]}{[t]}$$

That is, the dimensional representation of velocity is the dimension length divided by the dimension time. The units for a secondary quantity are then given in terms of the units of the constituent basic dimensions. Thus,

[2]A more precise definition will be given in the chapters on dynamics.

$$[\text{velocity units}] \equiv \frac{[\text{ft}]}{[\text{sec}]}$$

A *change* of units from one system into another usually involves a change in the scale of measure of the secondary quantities involved in the problem. Thus, one scale unit of velocity in the American system is 1 foot per second, while in the SI system it is 1 meter per second. How may these scale units be correctly related for complicated secondary quantities? That is, for our simple case, how many meters per second are equivalent to 1 foot per second? The formal expressions of dimensional representation may be put to good use for such an evaluation. The procedure is as follows. Express the dependent quantity dimensionally; substitute existing units for the basic dimensions; and finally, change these units to the equivalent numbers of units in the new system. The result gives the number of scale units of the quantity in the new system of units that is equivalent to 1 scale unit of the quantity in the old system. Performing these operations for velocity, we would thus have

$$1\left(\frac{\text{ft}}{\text{sec}}\right) \equiv 1\left(\frac{.305\text{ m}}{\text{sec}}\right) \equiv .305\left(\frac{\text{m}}{\text{sec}}\right)$$

which means that .305 scale unit of velocity in the SI system is equivalent to 1 scale unit in the American system.

Another way of changing units when secondary dimensions are present is to make use of the formalism illustrated in relations 1.1. To change a unit in an expression, multiply this unit by a ratio physically equivalent to unity, as we discussed earlier, so that the old unit is canceled out, leaving the desired unit with the proper numerical coefficient. In the example of velocity used above, we may replace ft/sec by m/sec in the following manner:

$$1\left(\frac{\text{ft}}{\text{sec}}\right) \equiv \left(\frac{1\,\cancel{\text{ft}}}{\text{sec}}\right) \cdot \left(\frac{.305\text{m}}{1\,\cancel{\text{ft}}}\right) \equiv .305\left(\frac{\text{m}}{\text{sec}}\right)$$

It should be clear that, when we multiply by such ratios to accomplish a change of units as shown above, we do not alter the magnitude of the *actual physical quantity* represented by the expression. Students are strongly urged to employ the above technique in their work, for the use of less formal methods is generally an invitation to error.

†1.4 Law of Dimensional Homogeneity

Now that we can describe certain aspects of nature in a quantitative manner through basic and secondary dimensions, we can by careful observation and experimentation learn to relate certain of the quantities in the form of equations. In this regard, there is an important law, the law of *dimensional homogeneity*, which imposes a restriction on the formulation of such equations. This law states that, because natural phenomena proceed with no regard for man-made units, *basic equations representing physical phenomena must be valid for all systems of units*. Thus, the equation for the period of a pendulum, $t = 2\pi\sqrt{L/g}$, must be valid for all systems of units, and is accordingly said to be *dimensionally homogeneous*. It then follows that the fundamental equations of physics are dimensionally homogeneous; and all equations derived analytically from these fundamental laws must also be dimensionally homogeneous.

What restriction does this condition place on an equation? To answer this, let us examine the following arbitrary equation:

$$x = ygd + k$$

For this equation to be dimensionally homogeneous, the numerical equality between both sides of the equation must be maintained for all systems of units. To accomplish this, the change in the scale of measure of each group of terms must be the same when there is a change of units. That is, if the numerical measure of one group such as *ygd* is doubled for a new system of units, so must that of the quantities *x* and *k*. *For this to occur under all systems of units, it is necessary that every grouping in the equation have the same dimensional representation.*

In this regard, consider the dimensional representation of the above equation expressed in the following manner:

$$[x] = [ygd] + [k]$$

From the previous conclusion for dimensional homogeneity, we require that

$$[x] \equiv [ygd] \equiv [k]$$

As a further illustration, consider the dimensional representation of an equation that is *not* dimensionally homogeneous:

$$[L] = [t]^2 + [t]$$

When we change units from the American to the SI system, the units of feet give way to units of meters, but there is no change in the unit of time, and it becomes clear that the numerical value of the left side of the equation changes while that of the right side does not. The equation, then, becomes invalid in the new system of units and hence is not derived from the basic laws of physics. Throughout this book, we shall invariably be concerned with dimensionally homogeneous equations. Therefore, we should dimensionally analyze our equations to help spot errors.

†1.5 Dimensional Relation Between Force and Mass

We shall now employ the law of dimensional homogeneity to establish a new secondary dimension—namely *force*. A superficial use of Newton's law will be employed for this purpose. In a later section, this law will be presented in greater detail, but it will suffice at this time to state that the acceleration of a particle[3] is inversely proportional to its mass for a given disturbance. Mathematically, this becomes

$$a \propto \frac{1}{m} \tag{1.2}$$

where $\propto$ is the proportionality symbol. Inserting the constant of proportionality, F, we have, on rearranging the equation,

$$F = ma \tag{1.3}$$

[3]We shall define particles in Section 1.7.

The mechanical disturbance, represented by F and called *force*, must have the following dimensional representation, according to the law of dimensional homogeneity:

$$[F] \equiv [M]\frac{[L]}{[t]^2} \tag{1.4}$$

The type of disturbance for which relation 1.2 is valid is usually the action of one body on another by direct contact. However, other actions, such as magnetic, electrostatic, and gravitational actions of one body on another involving no contact, also create mechanical effects that are valid in Newton's equation.

We could have initiated the study of mechanics by considering *force* as a basic dimension, the manifestation of which can be measured by the elongation of a standard spring at a prescribed temperature. Experiment would then indicate that for a given body the acceleration is directly proportional to the applied force. Mathematically,

$$F \propto a; \text{ therefore, } F = ma$$

from which we see that the proportionality constant now represents the property of mass. Here, mass is now a secondary quantity whose dimensional representation is determined from Newton's law:

$$M \equiv [F]\frac{[t]^2}{[L]} \tag{1.5}$$

As was mentioned earlier, we now have a choice between two systems of basic dimensions—the *MLt* or the *FLt* system of basic dimensions. Physicists prefer the former, whereas engineers usually prefer the latter.

1.6 Units of Mass

As we have already seen, the concept of mass arose from two types of actions—those of motion and gravitational attraction. In American engineering practice, units of mass are based on both actions, and this sometimes leads to confusion. Let us consider the *FLt* system of basic dimensions for the following discussion. The unit of force may be taken to be the pound-force (lbf), which is defined as a force that extends a standard spring a certain distance at a given temperature. Using Newton's law, we then define the *slug* as the amount of mass that a 1-pound force will cause to accelerate at the rate of 1 foot per second per second.

On the other hand, another unit of mass can be stipulated if we use the gravitational effect as a criterion. Here, the *pound mass* (lbm) is defined as the amount of matter that is drawn by gravity toward the earth by a force of 1 pound-force (lbf) at a specified position on the earth's surface.

We have formulated two units of mass by two different actions, and to relate these units we must subject them to the *same* action. Thus, we can take 1 pound mass and see what fraction or multiple of it will be accelerated 1 ft/sec^2 under the action of 1 pound of force. This fraction or multiple will then represent the number of units of pound mass that are equivalent to 1 slug. It turns out that this coefficient is g_0, where g_0 has the value corresponding to the acceleration of gravity at a position on the earth's surface where the pound mass was standardized. To three significant figures, the value of g_0 is 32.2. We may then make the statement of equivalence that

$$1 \text{ slug} \equiv 32.2 \text{ pounds mass}$$

To use the pound-mass unit in Newton's law, it is necessary to divide by g_0 to form units of mass, that have been derived from Newton's law. Thus,

$$F = \frac{m}{g_0} a \tag{1.6}$$

where m has the units of pound mass and m/g_0 has units of slugs. Having properly introduced into Newton's law the pound-mass unit from the viewpoint of physical equivalence, let us now consider the dimensional homogeneity of the resulting equation. The right side of Eq. 1.6 must have the dimensional representation of F and, since the unit here for F is the pound force, the right side must then have this unit. Examination of the units on the right side of the equation then indicates that the units of g_0 must be

$$[g_0] \equiv \frac{[\text{lbm}][\text{ft}]}{[\text{lbf}][\text{sec}]^2} \tag{1.7}$$

How does *weight* fit into this picture? Weight is defined as *the force of gravity on a body*. Its value will depend on the position of the body relative to the earth's surface. At a location on the earth's surface where the pound mass is standardized, a mass of 1 pound (lbm) has the weight of 1 pound (lbf), but with increasing altitude the weight will become smaller than 1 pound (lbf). The mass, however, remains at all times a 1-pound mass (lbm). If the altitude is not exceedingly large, the measure of weight, in lbf, will practically equal the measure of mass, in lbm. Therefore, it is unfortunately the practice in engineering to think erroneously of weight at positions other than on the earth's surface as the measure of mass, and consequently to use the symbol W to represent either lbm or lbf. In this age of rockets and missiles, it behooves us to be careful about the proper usage of units of mass and weight throughout the entire text.

If we know the weight of a body at some point, we can determine its mass in slugs very easily, provided that we know the acceleration of gravity, g, at that point. Thus, according to Newton's law,

$$W(\text{lbf}) = m(\text{slugs}) \times g(\text{ft/sec}^2)$$

Therefore,

$$m(\text{slugs}) = \frac{W(\text{lbf})}{g(\text{ft/sec}^2)} \tag{1.8}$$

Up to this point, we have only considered the American system of units. In the SI system of units, a *kilogram* is the amount of mass that will accelerate 1 m/sec^2 under the action of a force of 1 newton. Here we do not have the problem of 2 units of mass; the kilogram is the basic unit of mass. However, we do have another kind of problem—that the kilogram is unfortunately also used as a measure of force, as is the newton. One kilogram of force is the weight of 1 kilogram of mass at the earth's surface, where the acceleration of gravity (i.e., the acceleration due to the force of gravity) is 9.81 m/sec^2. A newton, on the other hand, is the force that causes 1 kilogram of mass to have an acceleration of 1 m/sec^2. Hence, 9.81 newtons are equivalent to 1 kilogram of force. That is,

$$9.81 \text{ newtons} \equiv 1 \text{ kilogram(force)} \equiv 2.205 \text{ lbf}$$

Note from the above that the newton is a comparatively small force, equaling approximately one-fifth of a pound. A kilonewton (1000 newtons), which will be used often, is about 200 lb. In this text, we shall *not* use the kilogram as a unit of force. However, you should be aware that many people do.[4]

Note that at the earth's surface the weight W of a mass M is:

$$W(\text{newtons}) = [M(\text{kilograms})](9.81)(\text{m/s}^2) \tag{1.9}$$

Hence:

$$M(\text{kilograms}) = \frac{W(\text{newtons})}{9.81\ (\text{m/s}^2)} \tag{1.10}$$

Away from the earth's surface, use the acceleration of gravity g rather than 9.81 in the above equations.

1.7 Idealizations of Mechanics

As we have pointed out, basic and secondary dimensions may sometimes be related in equations to represent a physical action that we are interested in. We want to represent an action using the known laws of physics, and also to be able to form equations simple enough to be susceptible to mathematical computational techniques. Invariably in our deliberations, we must replace the actual physical action and the participating bodies with hypothetical, highly simplified substitutes. We must be sure, of course, that the results of our substitutions have some reasonable correlation with reality. All analytical physical sciences must resort to this technique, and, consequently, their computations are not cut-and-dried but involve a considerable amount of imagination, ingenuity, and insight into physical behavior. We shall, at this time, set forth the most fundamental idealizations of mechanics and a bit of the philosophy involved in scientific analysis.

Continuum. Even the simplification of matter into molecules, atoms, electrons, and so on, is too complex a picture for many problems of engineering mechanics. In most problems, we are interested only in the average measurable manifestations of these elementary bodies. Pressure, density, and temperature are actually the gross effects of the actions of the many molecules and atoms, and they can be conveniently assumed to arise from a hypothetically continuous distribution of matter, which we shall call the *continuum*, instead of from a conglomeration of discrete, tiny bodies. Without such an artifice, we would have to consider the action of each of these elementary bodies—a virtual impossibility for most problems.

Rigid Body. In many cases involving the action on a body by a force, we simplify the continuum concept even further. The most elemental case is that of a rigid body, which is

[4]This is particularly true in the marketplace where the word "kilos" is often heard.

a continuum that undergoes theoretically no deformation whatever. Actually, every body must deform to a certain degree under the actions of forces, but in many cases the deformation is too small to affect the desired analysis. It is then preferable to consider the body as rigid, and proceed with simplified computations. For example, assume that we are to determine the forces transmitted by a beam to the earth as the result of a load P (Fig. 1.2). If P is small enough, the beam will undergo little deflection, and we can carry out a straightforward simple analysis using the *undeformed geometry* as if the body were indeed rigid. If we were to attempt a more accurate analysis—even though a slight increase in accuracy is not required—we would then need to know the exact position that the load assumes relative to the beam *after* the beam has ceased to deform, as shown in an exaggerated manner in Fig. 1.3. To do this accurately is a hopelessly difficult task, especially when we consider that the support must also "give" in a certain way. Although the alternative to a rigid-body analysis here leads us to a virtually impossible calculation, situations do arise in which more realistic models must be employed to yield the required accuracy. For example, when determining the internal force distribution in a body, we must often take the deformation into account, however small it might be. Other cases will be presented later. *The guiding principle is to make such simplifications as are consistent with the required accuracy of the results.*

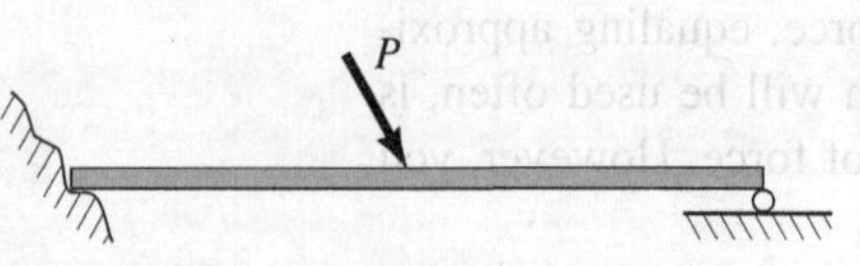

Figure 1.2. Rigid-body assumption—use original geometry.

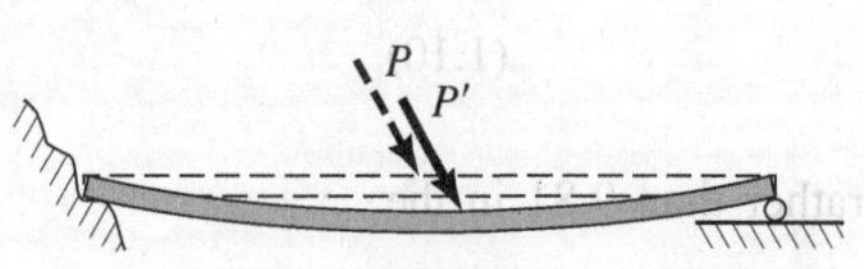

Figure 1.3. Deformable body.

Point Force. A finite force exerted on one body by another must cause a finite amount of local deformation, and always creates a finite area of contact between the bodies through which the force is transmitted. However, since we have formulated the concept of the rigid body, we should also be able to imagine a finite force to be transmitted through an infinitesimal area or point. This simplification of a force distribution is called a *point force*. In many cases where the actual area of contact in a problem is very small but is not known exactly, the use of the concept of the point force results in little sacrifice in accuracy. In Figs. 1.2 and 1.3, we actually employed the graphical representation of the point force.

Particle. The *particle* is defined as an object that has no size but that has a mass. Perhaps this does not sound like a very helpful definition for engineers to employ, but it is actually one of the most useful in mechanics. For the trajectory of a planet, for example, it is the mass of the planet and not its size that is significant. Hence, we can consider planets as particles for such computations. On the other hand, take a figure skater spinning on the ice. Her revolutions are controlled beautifully by the orientation of the body. In this motion, the size and distribution of the body are significant, and since a particle, by definition, can have no distribution, it is patently clear that a particle cannot represent the skater in this case. If, however, the skater should be billed as the "human cannonball on skates" and be shot out of a large air gun, it would be possible to consider her as a single particle in ascertaining her trajectory, since arm and leg movements that were significant while she was spinning on the ice would have little effect on the arc traversed by the main portion of her body.

You will learn later that the *center of mass* or *mass center* is a hypothetical point at which one can concentrate the mass of the body for certain dynamics calculations.

Actually in the previous examples of the planet and the "human cannonball on skates," the particle we refer to is actually the mass center whose motion is sufficient for the desired information. Thus, when the motion of the mass center of a body suffices for the information desired, we can replace the body by a particle, namely the mass center.

Many other simplifications pervade mechanics. The perfectly elastic body, the frictionless fluid, and so on, will become familiar as you study various phases of mechanics.

†1.8 Vector and Scalar Quantities

We have now proposed sets of basic dimensions and secondary dimensions to describe certain aspects of nature. However, more than just the dimensional identification and the number of units are often needed to convey adequately the desired information. For instance, to specify fully the motion of a car, which we may represent as a particle at this time, we must answer the following questions:

1. How fast?
2. Which way?

The concept of velocity entails the information desired in questions 1 and 2. The first question, "How fast?", is answered by the speedometer reading, which gives the value of the velocity in miles per hour or kilometers per hour. The second question, "Which way?", is more complicated, because two separate factors are involved. First, we must specify the angular orientation of the velocity relative to a reference frame. Second, we must specify the sense of the velocity, which tells us whether we are moving *toward or away from* a given point. The concepts of angular orientation of the velocity and sense of the velocity are often collectively denoted as the *direction* of the velocity. Graphically, we may use a *directed line segment* (an arrow) to describe the velocity of the car. The *length* of the directed line segment gives information as to "how fast" and is the *magnitude* of the velocity. The angular orientation of the directed line segment and the position of the arrowhead give information as to "which way"—that is, as to the *direction* of the velocity. The directed line segment itself is called the *velocity*, whereas the length of the directed line segment—that is, the magnitude—is called the *speed*.

There are many physical quantities that are represented by a directed line segment and thus are describable by specifying a magnitude and a direction. The most common example is force, where the magnitude is a measure of the intensity of the force and the direction is evident from how the force is applied. Another example is the *displacement vector* between two points on the path of a particle. The magnitude of the displacement vector corresponds to the distance moved along a *straight line* between two points, and the direction is defined by the orientation of this line relative to a reference, with the sense corresponding to which point is being approached. Thus, $\boldsymbol{\rho}_{AB}$ (see Fig. 1.4) is the displacement vector from A to B (while $\boldsymbol{\rho}_{BA}$ goes from B to A).

Certain quantities having magnitude and direction combine their effects in a special way. Thus, the combined effect of two forces acting on a particle, as shown in Fig. 1.5, corresponds to a single force that may be shown by experiment to be equal to the diagonal of a parallelogram formed by the graphical representation of the forces. That is, the quantities add according to the *parallelogram law*. All quantities that have magnitude and direction and that add according to the parallelogram law are called *vector quantities*. Other quantities that have only magnitude, such as temperature and work, are called *scalar quantities*.

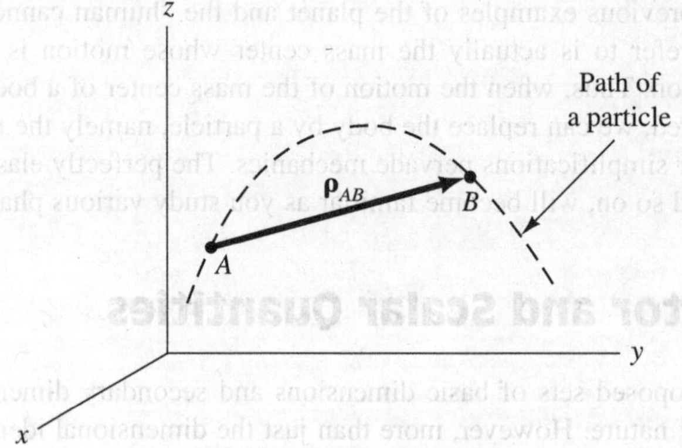

Figure 1.4. Displacement vector $\boldsymbol{\rho}_{AB}$.

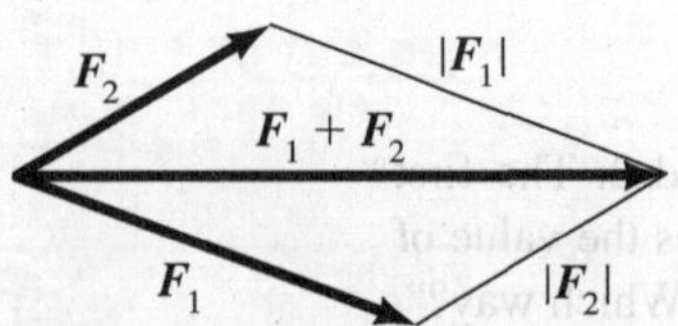

Figure 1.5. Parallelogram law.

A vector quantity will be denoted with a boldface italic letter, which in the case of force becomes $\boldsymbol{F}$.[5]

The reader may ask: Don't all quantities having magnitude and direction combine according to the parallelogram law and, therefore, become vector quantities? No, not all of them do. One very important example will be pointed out after we reconsider Fig. 1.5. In the construction of the parallelogram it matters not which force is laid out first. In other words, "$\boldsymbol{F}_1$ combined with $\boldsymbol{F}_2$" gives the same result as "$\boldsymbol{F}_2$ combined with $\boldsymbol{F}_1$." In short, the combination is *commutative*. If a combination is not commutative, it cannot in general be represented by a parallelogram operation and is thus not a vector. With this in mind, consider the *finite* angle of rotation of a body about an axis. We can associate a magnitude (degrees or radians) and a direction (the axis and a stipulation of clockwise or counterclockwise) with this quantity. However, the finite angle of rotation cannot be considered a vector because in general two finite rotations about different axes cannot be replaced by a single finite rotation consistent with the parallelogram law. The easiest way to show this is to demonstrate that the combination of such rotations is not commutative. In Fig. 1.6(a) a book is to be given two rotations—a 90° counterclockwise rotation about the x axis and a 90° clockwise rotation about the z axis, both looking in toward the origin. This is carried out in Figs. 1.6(b) and (c). In Fig. 1.6(c), the sequence of combination is reversed from that in Fig. 1.6(b), and you can see how it alters the final orientation of the book. Finite angular rotation, therefore, is not a vector quantity, since the parallelogram law is not valid for such a combination.[6]

You may now wonder why we tacked on the parallelogram law for the definition of a vector and thereby excluded finite rotations from this category. The answer to this query is as follows. In the next chapter, we will present certain sets of very useful operations termed *vector algebra*. These operations are valid in general *only* if the parallelogram law is satisfied

[5]Your instructor on the blackboard and you in your homework will not be able to use boldface notation for vectors. Accordingly, you may choose to use a superscript arrow or bar, e.g., $\vec{F}$ or $\bar{F}$ ($\underline{F}$ or $\underset{\sim}{F}$ are other possibilities).

[6]However, *vanishingly small* rotations can be considered as vectors since the commutative law applies for the combination of such rotations.

as you will see when we get to Chapter 2. Therefore, we had to restrict the definition of a vector in order to be able to use this kind of algebra for these quantities. Also, it is to be pointed out that later in the text we will present yet a third definition consistent with our latest definition. This next definition will have certain advantages as we will see later.

Before closing the section, we will set forth one more definition. The *line of action* of a vector is a hypothetical infinite straight line collinear with the vector (see Fig. 1.7). Thus, the velocities of two cars moving on different lanes of a straight highway have different lines of action. Keep in mind that the line of action involves no connotation as to sense. Thus, a vector $\mathbf{V}'$ collinear with $\mathbf{V}$ in Fig. 1.7 and with opposite sense would nevertheless have the same line of action.

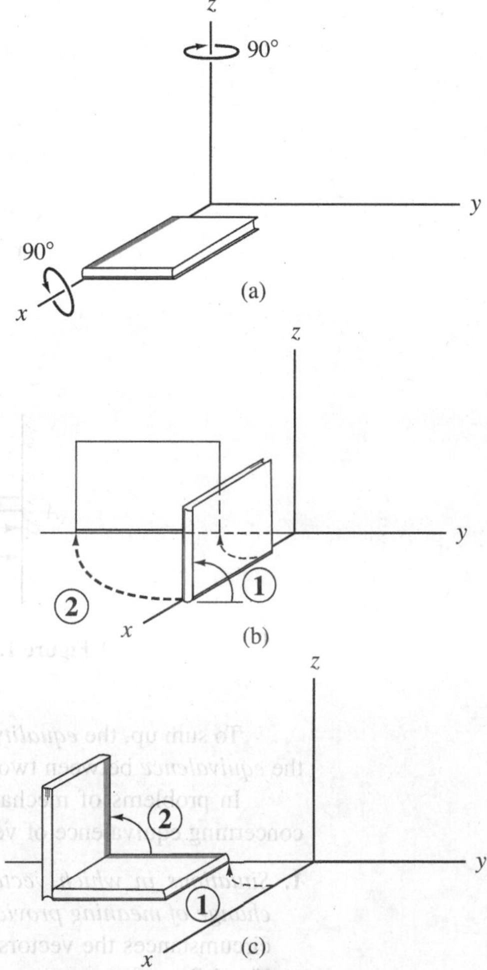

Figure 1.6. Successive rotations are not commutative.

1.9 Equality and Equivalence of Vectors

We shall avoid many pitfalls in the study of mechanics if we clearly make a distinction between the equality and the equivalence of vectors.

Two vectors are equal if they have the same dimensions, magnitude, and direction. In Fig. 1.8, the velocity vectors of three particles have equal length, are identically inclined toward the reference *xyz*, and have the same sense. Although they have different lines of action, they are nevertheless equal according to the definition.

Two vectors are equivalent in a certain capacity if each produces the very same effect in this capacity. If the criterion in Fig. 1.8 is change of elevation of the particles or total distance traveled by the particles, all three vectors give the same result. They are, in addition to being equal, also equivalent for these capacities. If the absolute height of the particles above the *xy* plane is the question in point, these vectors will not be equivalent despite their equality. Thus, it must be emphasized that *equal vectors need not always be equivalent; it depends entirely on the situation at hand.* Furthermore, vectors that are not equal may still be equivalent in some capacity. Thus, in the beam in Fig. 1.9, forces F_1 and F_2 are unequal, since their magnitudes are 10 lb and 20 lb, respectively. However, it is clear from elementary physics that their moments about the base of the beam are equal, and so the forces have the same "turning" action at the fixed end of the beam. In that capacity, the forces are equivalent. If, however, we are interested in the deflection of the free end of the beam resulting from each force, there is no longer an equivalence between the forces, since each will give a different deflection.

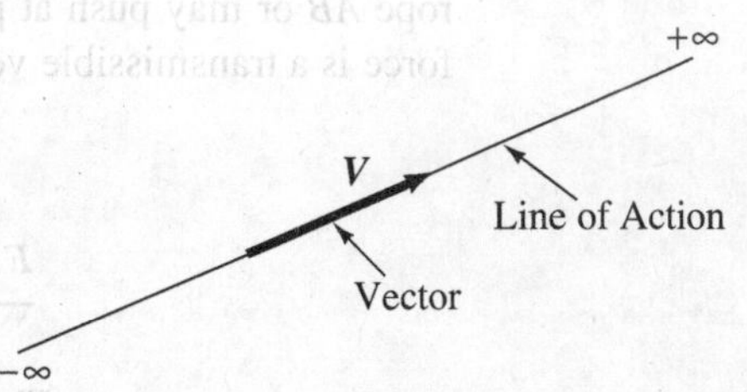

Figure 1.7. Line of action of a vector.

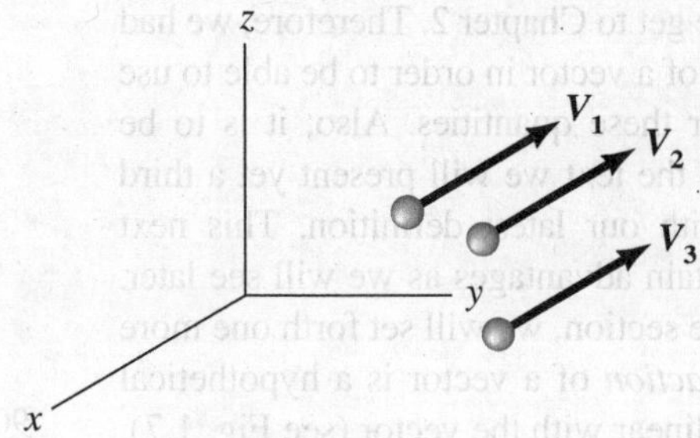

Figure 1.8. Equal-velocity vectors.

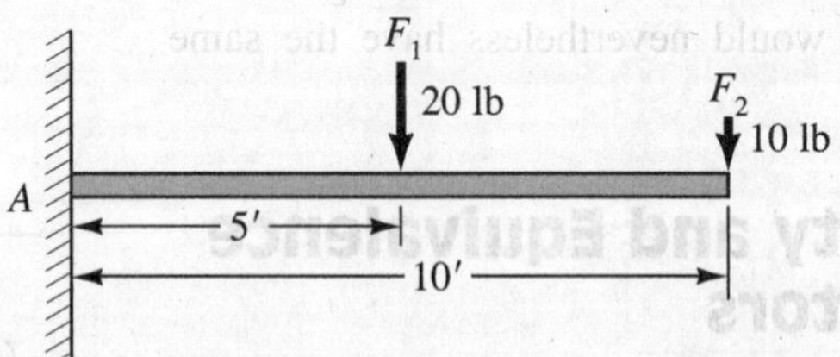

Figure 1.9. F_1 and F_2 equivalent for moment about A.

To sum up, the *equality* of two vectors is determined by the vectors themselves, and the *equivalence* between two vectors is determined by the task involving the vectors.

In problems of mechanics, we can profitably delineate three classes of situations concerning equivalence of vectors:

1. *Situations in which vectors may be positioned anywhere in space without loss or change of meaning provided that magnitude and direction are kept intact*. Under such circumstances the vectors are called *free vectors*. For example, the velocity vectors in Fig. 1.8 are free vectors as far as total distance traveled is concerned.
2. *Situations in which vectors may be moved along their lines of action without change of meaning*. Under such circumstances the vectors are called *transmissible vectors*. For example, in towing the object in Fig. 1.10, we may apply the force anywhere along the rope AB or may push at point C. The resulting motion is the same in all cases, so the force is a transmissible vector for this purpose.

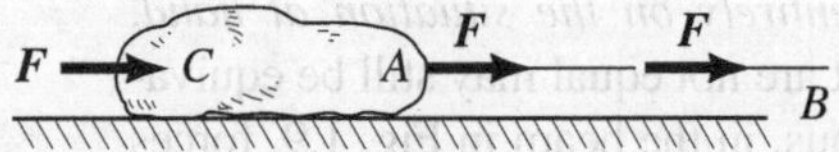

Figure 1.10. F is transmissible for towing.

3. *Situations in which the vectors must be applied at definite points*. The point may be represented as the tail or head of the arrow in the graphical representation. For this case, no other position of application leads to equivalence. Under such circumstances, the vector is called a *bound vector*. For example, if we are interested in the deformation induced

by forces in the body in Fig. 1.10, we must be more selective in our actions than we were when all we wanted to know was the motion of the body. Clearly, force $\boldsymbol{F}$ will cause a different deformation when applied at point C than it will when applied at point A. The force is thus a bound vector for this problem.

We shall be concerned throughout this text with considerations of equivalence.

†1.10 Laws of Mechanics

The entire structure of mechanics rests on relatively few basic laws. Nevertheless, for the student to comprehend these laws sufficiently to undertake novel and varied problems, much study will be required.

We shall now discuss briefly the following laws, which are considered to be the foundation of mechanics:

1. Newton's first and second laws of motion.
2. Newton's third law.
3. The gravitational law of attraction.
4. The parallelogram law.

Newton's First and Second Laws of Motion. These laws were first stated by Newton as

Every particle continues in a state of rest or uniform motion in a straight line unless it is compelled to change that state by forces imposed on it.

The change of motion is proportional to the natural force impressed and is made in a direction of the straight line in which the force is impressed.

Notice that the words "rest," "uniform motion," and "change of motion" appear in the statements above. For such information to be meaningful, we must have some frame of reference relative to which these states of motion can be described. We may then ask: relative to what reference in space does every particle remain at "rest" or "move uniformly along a straight line" in the absence of any forces? Or, in the case of a force acting on the particle, relative to what reference in space is the "change in motion proportional to the force"? Experiment indicates that the "fixed" stars act as a reference for which the first and second laws of Newton are highly accurate. Later, we will see that any other system that moves uniformly and without rotation relative to the fixed stars may be used as a reference with equal accuracy. All such references are called *inertial references*. The earth's surface is usually employed as a reference in engineering work. Because of the rotation of the earth and the variations in its motion around the sun, it is not, strictly speaking, an inertial reference. However, the departure is so small for most situations (exceptions are the motion of guided missiles and spacecraft) that the error incurred is very slight. We shall, therefore, usually consider the earth's surface as an inertial reference, but will keep in mind the somewhat approximate nature of this step.

As a result of the preceding discussion, we may define *equilibrium* as *that state of a body in which all its constituent particles are at rest or moving uniformly along a straight line relative to an inertial reference*. The converse of Newton's first law, then, stipulates for the equilibrium state that there must be no force (or equivalent action of no force) acting on the body. Many situations fall into this category. The study of bodies in equilibrium is called *statics*, and it will be an important consideration in this text.

In addition to the reference limitations explained above, a serious limitation was brought to light at the turn of this century. As pointed out earlier, the pioneering work of Einstein revealed that the laws of Newton become increasingly more approximate as the speed of a body increases. Near the speed of light, they are untenable. In the vast majority of engineering computations, the speed of a body is so small compared to the speed of light that these departures from Newtonian mechanics, called *relativistic effects*, may be entirely disregarded with little sacrifice in accuracy. In considering the motion of high-energy elementary particles occurring in nuclear phenomena, however, we cannot ignore relativistic effects. Finally, when we get down to very small distances, such as those between the protons and neutrons in the nucleus of an atom, we find that Newtonian mechanics cannot explain many observed phenomena. In this case, we must resort to quantum mechanics, and then Newton's laws give way to the Schrödinger equation as the key equation.

Newton's Third Law. Newton stated in his third law:

> *To every action there is always opposed an equal reaction, or the mutual actions of two bodies upon each other are always equal and directed to contrary points.*

This is illustrated graphically in Fig. 1.11, where the action and reaction between two bodies arise from direct contact. Other important actions in which Newton's third law holds are gravitational attractions (to be discussed next) and electrostatic forces between charged particles. It should be pointed out that there are actions that do not follow this law, notably the electromagnetic forces between charged moving bodies.[7]

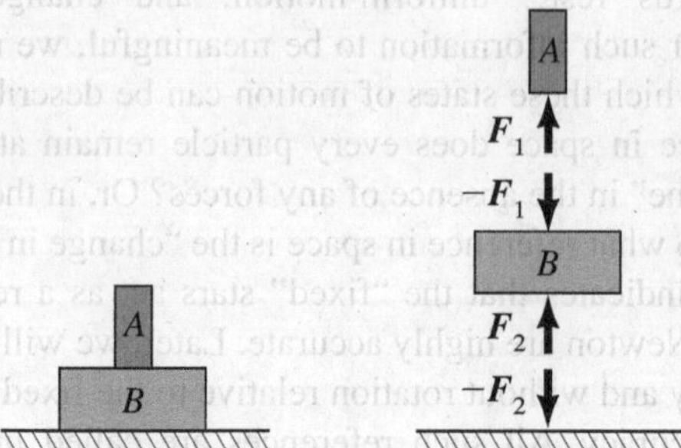

Figure 1.11. Newton's third law.

[7]Electromagnetic forces between charged moving particles are equal and opposite but are not collinear and hence are not "directed to contrary points."

Law of Gravitational Attraction. It has already been pointed out that there is an attraction between the earth and the bodies at its surface, such as A and B in Fig. 1.11. This attraction is mutual and Newton's third law applies. There is also an attraction between the two bodies A and B themselves, but this force because of the small size of both bodies is extremely weak. However, the mechanism for the mutual attraction between the earth and each body is the same as that for the mutual attraction between the bodies. These forces of attraction may be given by the *law of gravitational attraction*:

Two particles will be attracted toward each other along their connecting line with a force whose magnitude is directly proportional to the product of the masses and inversely proportional to the distance squared between the particles.

Avoiding vector notation for now, we may thus say that

$$F = G\frac{m_1 m_2}{r^2} \tag{1.11}$$

where G is called the *universal gravitational constant*. In the actions involving the earth and the bodies discussed above, we may consider each body as a particle, with its entire mass concentrated at its center of gravity.[8] Hence, if we know the various constants in formula 1.11, we can compute the weight of a given mass at different altitudes above the earth.

Parallelogram Law. Stevinius (1548–1620) was the first to demonstrate that forces could be combined by representing them by arrows to some suitable scale, and then forming a parallelogram in which the diagonal represents the sum of the two forces. As we pointed out, all vectors must combine in this manner.

1.11 Closure

In this chapter, we have introduced the basic dimensions by which we can describe in a quantitative manner certain aspects of nature. These basic, and from them secondary, dimensions may be related by dimensionally homogeneous equations which, with suitable idealizations, can represent certain actions in nature. The basic laws of mechanics were thus introduced. Since the equations of these laws relate vector quantities, we shall introduce a useful and highly descriptive set of vector operations in Chapter 2 in order to learn to handle these laws effectively and to gain more insight into mechanics in general. These operations are generally called *vector algebra*.

[8]To be studied in detail in Chapter 4.

Check-Out for Sections with †

1.1. What are two kinds of limitations on Newtonian mechanics?

1.2. What are the two phenomena wherein mass plays a key role?

1.3. If a pound force is defined by the extension of a standard spring, define the pound mass and the slug.

1.4. Express mass density dimensionally. How many scale units of mass density (mass per unit volume) in the SI units are equivalent to 1 scale unit in the American system using (a) slugs, ft, sec and (b) lbm, ft, sec?

1.5. **(a)** What is a necessary condition for *dimensional homogeneity* in an equation?

(b) In the Newtonian viscosity law, the frictional resistance τ (force per unit area) in a fluid is proportional to the distance rate of change of velocity dV/dy. The proportionality constant μ is called the *coefficient of viscosity.* What is its dimensional representation?

1.6. Define a vector and a scalar.

1.7. What is meant by *line of action* of a vector?

1.8. What is a *displacement* vector?

1.9. What is an *inertial reference*?

CHAPTER 2

REVIEW II*

Elements of Vector Algebra

†2.1 Introduction

In Chapter 1, we saw that a scalar quantity is adequately given by a magnitude, while a vector quantity requires the additional specification of a direction. The basic algebraic operations for the handling of scalar quantities are those familiar ones studied in grade school, so familiar that you now wonder even that you had to be "introduced" to them. For vector quantities, these methods may be cumbersome since the directional aspects must be taken into account. Therefore, an algebra has evolved that clearly and concisely allows for certain very useful manipulations of vectors. It is not merely for elegance or sophistication that we employ vector algebra. Indeed, we can achieve greater insight into the subject matter—particularly into dynamics—by employing the more powerful and descriptive methods introduced in this chapter.

†2.2 Magnitude and Multiplication of a Vector by a Scalar

The magnitude of a quantity, in strict mathematical parlance, is always a *positive* number of units whose value corresponds to the numerical measure of the quantity. Thus, the magnitude of a quantity of measure –50 units is +50 units. Note that the magnitude of a quantity is its absolute value. The mathematical symbol for indicating the magnitude of a quantity is a set of vertical lines enclosing the quantity. That is,

$$|-50 \text{ units}| = \text{absolute value } (-50 \text{ units}) = +50 \text{ units}$$

Similarly, the magnitude of a vector quantity is a positive number of units corresponding to the length of the vector in those units. Using our vector symbols, we can say that

$$\text{magnitude of vector } A = |A| \equiv A$$

*The reader is urged to pay particular attention to Section 2.4 on **Resolution of Vectors** and Section 2.6 on **Useful Ways of Representing Vectors**.

†Again, as in Chapter 1, we have used the symbol † for certain section headings to indicate that at the end of the chapter there are questions to be answered in writing pertaining to these sections. The instructer may wish to assign the reading of these sections along with the aforementioned questions.

Thus, A is a positive scalar quantity. We may now discuss the multiplication of a vector by a scalar.

The definition of the product of vector $\mathbf{A}$ by scalar m, written simply as $m\mathbf{A}$, is given in the following manner:

$m\mathbf{A}$ is a vector having the same direction as $\mathbf{A}$ and a magnitude equal to the ordinary scalar product between the magnitudes of m and $\mathbf{A}$. If m is negative, it means simply that the vector $m\mathbf{A}$ has a direction directly opposite to that of $\mathbf{A}$.

The vector $-\mathbf{A}$ may be considered as the product of the scalar -1 and the vector $\mathbf{A}$. Thus, from the statement above we see that $-\mathbf{A}$ differs from $\mathbf{A}$ in that it has an opposite sense. Furthermore, these operations have nothing to do with the line of action of a vector, so $\mathbf{A}$ and $-\mathbf{A}$ may have different lines of action. This will be the case of the couple to be studied in Chapter 3.

†2.3 Addition and Subtraction of Vectors

In adding a number of vectors, we may repeatedly employ the parallelogram construction. We can do this graphically by scaling the lengths of the arrows according to the magnitudes of the vector quantities they represent. The magnitude of the final arrow can then be interpreted in terms of its length by employing the chosen scale factor. As an example, consider the coplanar[1] vectors $\mathbf{A}$, $\mathbf{B}$, and $\mathbf{C}$ shown in Fig. 2.1(a). The addition of the vectors $\mathbf{A}$, $\mathbf{B}$, and $\mathbf{C}$ has been accomplished in two ways. In Fig. 2.1(b) we first add $\mathbf{B}$ and $\mathbf{C}$ and then add the resulting vector (shown dashed) to $\mathbf{A}$. This combination can be represented by the notation $\mathbf{A} + (\mathbf{B} + \mathbf{C})$. In Fig. 2.1(c), we add $\mathbf{A}$ and $\mathbf{B}$, and then add the resulting vector (shown dashed) to $\mathbf{C}$. The representation of this combination is given as $(\mathbf{A} + \mathbf{B}) + \mathbf{C}$. Note that the final vector is identical for both procedures. Thus,

$$\mathbf{A} + (\mathbf{B} + \mathbf{C}) = (\mathbf{A} + \mathbf{B}) + \mathbf{C} \tag{2.1}$$

When the quantities involved in an algebraic operation can be grouped without restriction, the operation is said to be *associative*. Thus, the addition of vectors is both commutative, as explained earlier, and associative.

To determine a summation of, let us say, two vectors without recourse to graphics, we need only make a simple sketch of the vectors approximately to scale. By using familiar trigonometric relations, we can get a direct evaluation of the result. This is illustrated in the following examples.

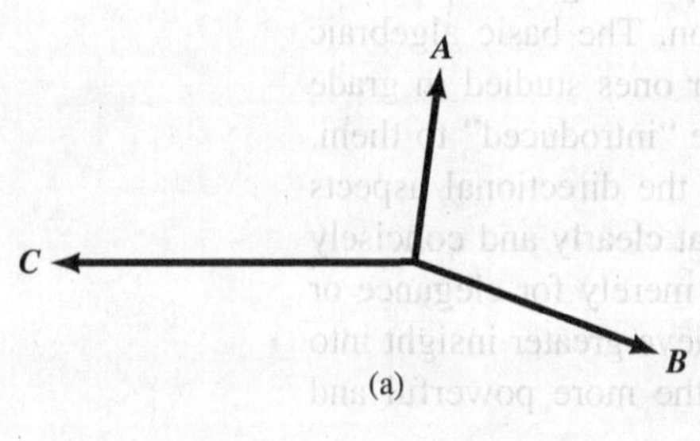

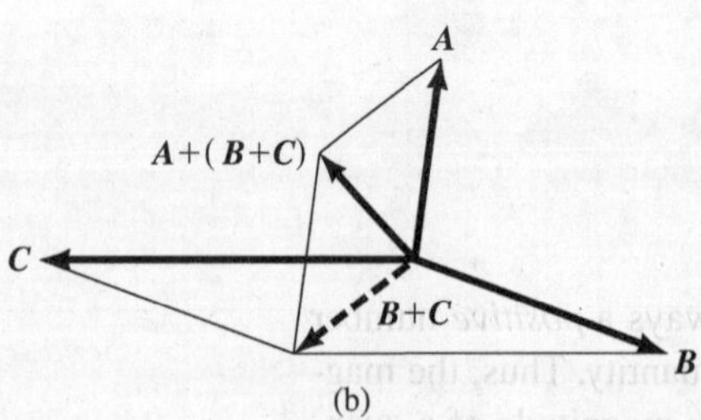

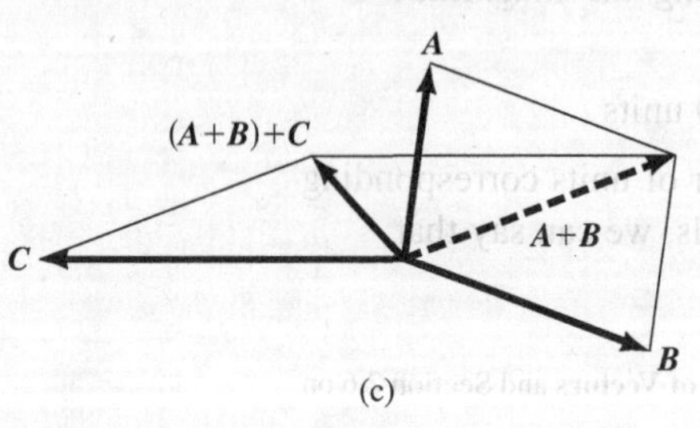

Figure 2.1. Addition by parallelogram law.

[1]Coplanar, meaning "same plane," is a word used often in mechanics.

Example 2.1

Add the forces acting on a particle situated at the origin of a two-dimensional reference frame (Fig. 2.2). One force has a magnitude of 10 lb acting in the positive x direction, whereas the other has a magnitude of 5 lb acting at an angle of 135° with a sense directed away from the origin.

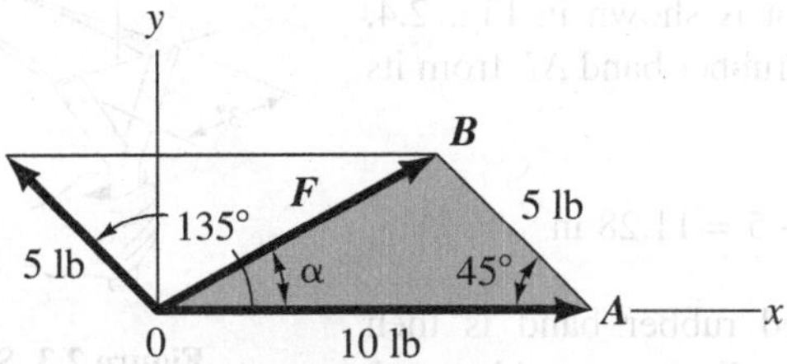

Figure 2.2. Find F and α using trigonometry.

To get the sum (shown as $\boldsymbol{F}$), we may use the law of cosines[2] for one of the triangular portions of the sketched parallelogram. Thus, using triangle OBA,

$$|\boldsymbol{F}| = \left[10^2 + 5^2 - (2)(10)(5)\cos 45°\right]^{1/2}$$
$$= (100 + 25 - 70.7)^{1/2} = \sqrt{54.3} = 7.37 \text{ lb}$$

The direction of the vector may be described by giving the angle and the sense. The angle is determined by employing the law of sines for triangle OBA.[3]

$$\frac{5}{\sin\alpha} = \frac{7.37}{\sin 45°}$$

$$\sin\alpha = \frac{(5)(0.707)}{7.37} = 0.480$$

Therefore,

$$F = 7.37 \text{ lb}$$
$$\alpha = 28.6°$$

The sense is shown using the directed line segment.

[2]
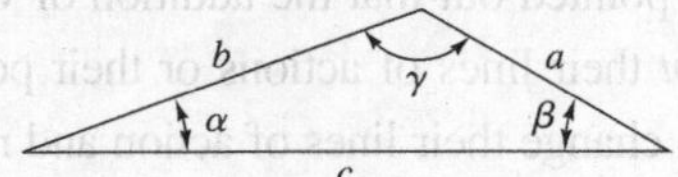

You will recall from trigonometry that the *law of cosines* for side b of a triangle is given as

$$b^2 = a^2 + c^2 - 2ac\cos\beta$$

[3]The *law of sines* is given as follows for a triangle:

$$\frac{a}{\sin\alpha} = \frac{b}{\sin\beta} = \frac{c}{\sin\gamma}$$

Example 2.2

A simple slingshot (see Fig. 2.3) is about to be "fired." If the entire rubber band requires 3 lb per inch of elongation, what force does the band exert on the hand? The total unstretched length of the rubber band is 5 in.

The top view of the slingshot is shown in Fig. 2.4. The change in overall length of the rubber band ΔL from its unstretched length is

$$\Delta L = 2\left(1.5^2 + 8^2\right)^{1/2} - 5 = 11.28 \text{ in.}$$

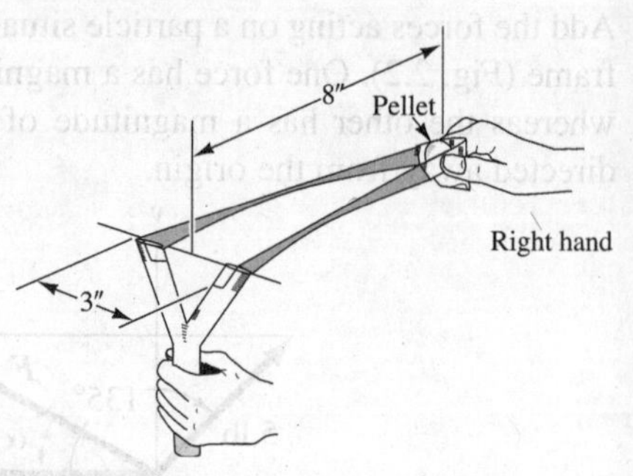

Figure 2.3. Simple slingshot.

The tension in the entire extended rubber band is then (11.28)(3) lb. Consequently, the force F transmitted by *each leg* of the slingshot is

$$F = (11.28)(3) = \boxed{33.84 \text{ lb}}$$

and the value of θ

$$\theta = \tan^{-1}\frac{1.5}{8} = 10.62°$$

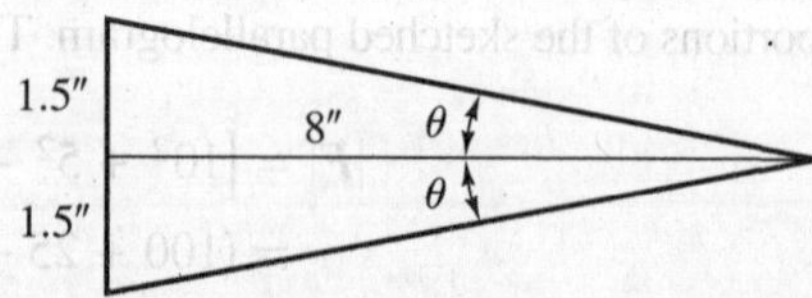

Figure 2.4. Top view of the slingshot.

In Fig. 2.5, we show a parallelogram involving the forces F and their sum R where R is the force that the band exerts on the hand. We can use the law of cosines on either of the triangles to get R. Thus

$$R^2 = 33.84^2 + 33.84^2 - (2)(33.84)(33.84)\cos\alpha$$

Noting that $\alpha = 180° - (2)(10.62°) = 158.8°$ we have

$$R = \left[(2)(33.84)^2(1 - \cos 158.8°)\right]^{1/2} = \boxed{66.52 \text{ lb}}$$

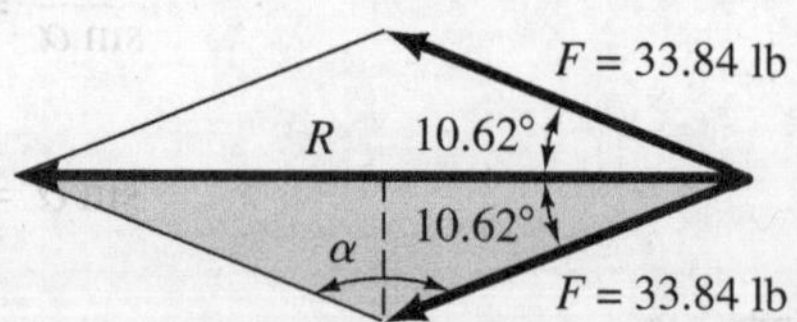

Figure 2.5. Parallelogram of forces.

A more direct calculation can be used by considering two right triangles within the chosen triangle. Then using elementary trigonometry we have

$$R = (2)(33.84)\cos(10.62)° = 66.52 \text{ lb}$$

It must be emphatically pointed out that the addition of vectors **A** and **B** only involves the vectors themselves and *not* their lines of actions or their positions along their respective lines of action. That is, we can change their lines of action and move them along their respective lines of action so as to form two sides of a parallelogram. For the additional vector algebra that we will develop in this chapter, we can take similar liberties with the vectors involved.

We may also add the vectors by moving them successively to parallel positions so that the head of one vector connects to the tail of the next vector, and so on. The sum of the vectors will then be a vector whose tail connects to the tail of the first vector and whose head connects to the head of the last vector. This last step will form a polygon from the vectors, and we say that the vector sum then "closes the polygon." Thus, adding the 10-lb vector to the 5-lb vector in Fig. 2.2, we would form the sides *OA* and *AB* of a triangle. The sum ***F*** then closes the triangle and is *OB*. Also, in Fig. 2.6(a), we have shown three coplanar vectors $\boldsymbol{F}_1$, $\boldsymbol{F}_2$, and $\boldsymbol{F}_3$. The vectors are connected in Fig. 2.6(b) as described. The sum of the vectors then is the dashed vector that closes the polygon. In Fig. 2.6(c), we have laid off the vectors $\boldsymbol{F}_1$, $\boldsymbol{F}_2$, and $\boldsymbol{F}_3$ in a different sequence. Nevertheless, it is seen that the sum is the same vector as in Fig. 2.6(b). Clearly, the *order* of laying off the vectors is not significant.

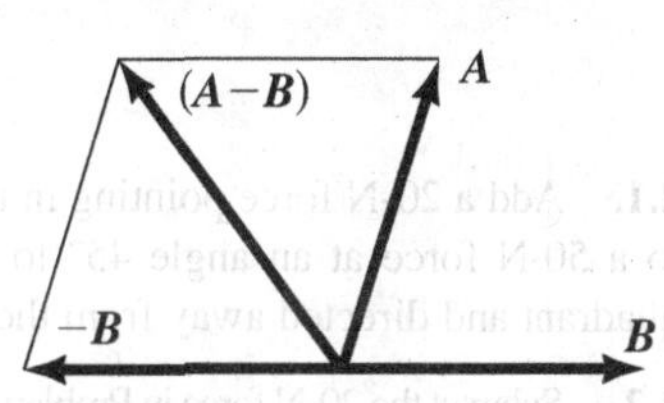

Figure 2.7. Subtraction of vectors.

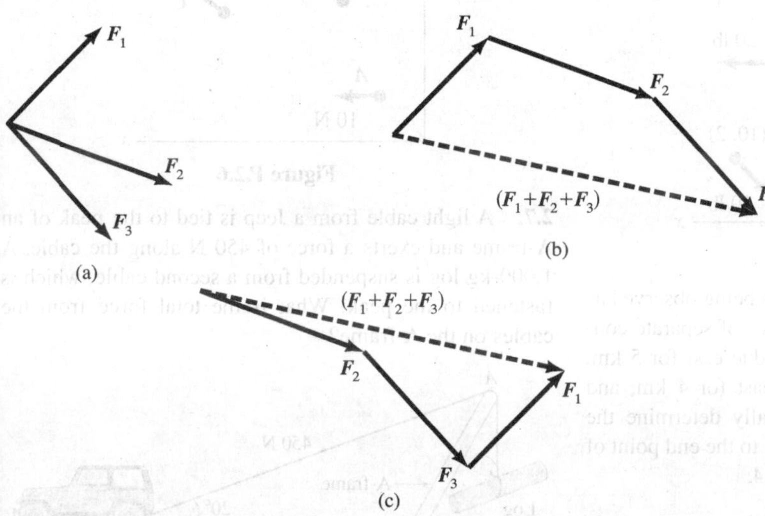

Figure 2.6. Addition by "closing the polygon."

A simple physical interpretation of the above vector sum can be formed for vectors each of which represents a movement of a certain distance and direction (i.e., a *displacement* vector). Then, traveling along the system of given vectors you start from one point (the tail of the first vector) and end at another point (the head of the last vector). The vector *sum* that closes the polygon is equivalent to the system of given vectors, in that it takes you from the same initial to the same final point.

The polygon summation process, like the parallelogram of addition, can be used as a graphical process, or, still better, can be used to generate analytical computations with the aid of trigonometry. The extension of this procedure to any number of vectors is obvious.

The process of *subtraction* of vectors is defined in the following manner: to subtract vector ***B*** from vector ***A***, we reverse the direction of ***B*** (i.e., multiply by −1) and then add this new vector to ***A*** (Fig. 2.7).

This process may also be used in the polygon construction. Thus, consider coplanar vectors ***A***, ***B***, ***C***, and ***D*** in Fig. 2.8(a). To form ***A***+***B***−***C***−***D***, we proceed as shown in Fig. 2.8(b). Again, the order of the process is not significant, as can be seen in Fig. 2.8(c).

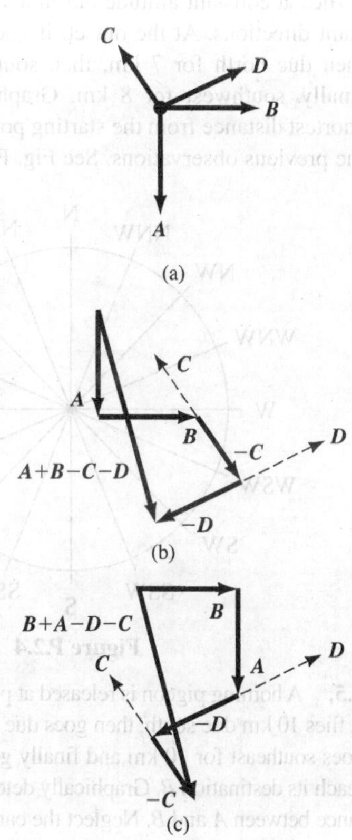

Figure 2.8. Addition and subtraction using polygon construction.

PROBLEMS

2.1. Add a 20-N force pointing in the positive x direction to a 50-N force at an angle 45° to the x axis in the first quadrant and directed away from the origin.

2.2. Subtract the 20-N force in Problem 2.1 from the 50-N force.

2.3. Add the vectors in the xy plane. Do this first graphically, using the force polygon, and then do it analytically.

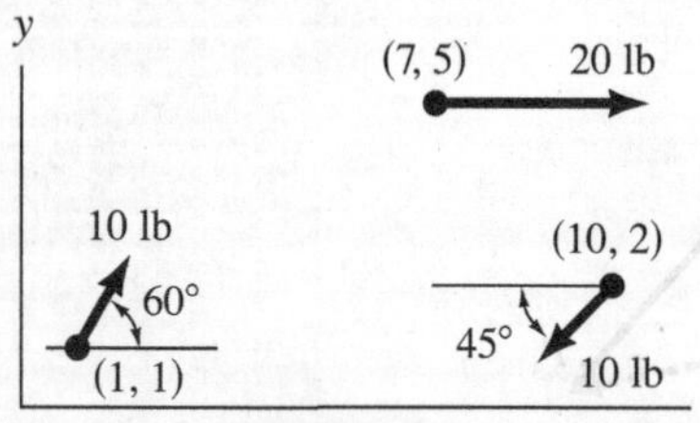

Figure P.2.3

2.4. A lightweight homemade plane is being observed as it flies at constant altitude but in a series of separate constant directions. At the outset, it goes due east for 5 km, then due north for 7 km, then southeast for 4 km, and finally, southwest for 8 km. Graphically determine the shortest distance from the starting point to the end point of the previous observations. See Fig. P.2.4.

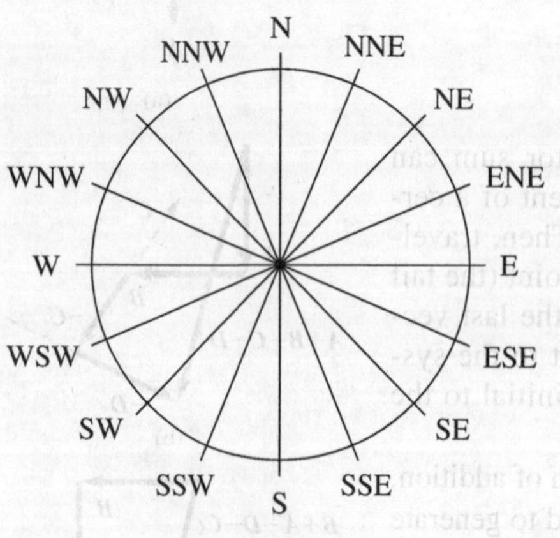

Figure P.2.4

2.5. A homing pigeon is released at point A and is observed. It flies 10 km due south, then goes due east for 15 km. Next it goes southeast for 10 km and finally goes due south 5 km to reach its destination B. Graphically determine the shortest distance between A and B. Neglect the earth's curvature.

2.6. Force $\boldsymbol{A}$ (given as a horizontal 10-N force) and $\boldsymbol{B}$ (vertical) add up to a force $\boldsymbol{C}$ that has a magnitude of 20 N. What is the magnitude of force $\boldsymbol{B}$ and the direction of force $\boldsymbol{C}$? (For the simplest results, use the force polygon, which for this case is a right triangle, and perform analytical computations.)

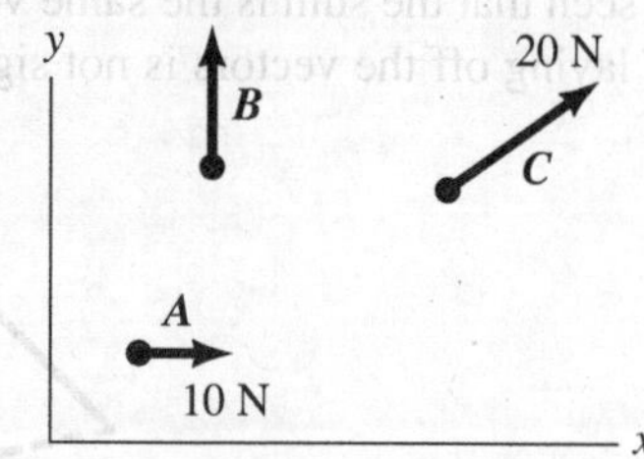

Figure P.2.6

2.7. A light cable from a Jeep is tied to the peak of an A-frame and exerts a force of 450 N along the cable. A 1,000-kg log is suspended from a second cable, which is fastened to the peak. What is the total force from the cables on the A-frame?

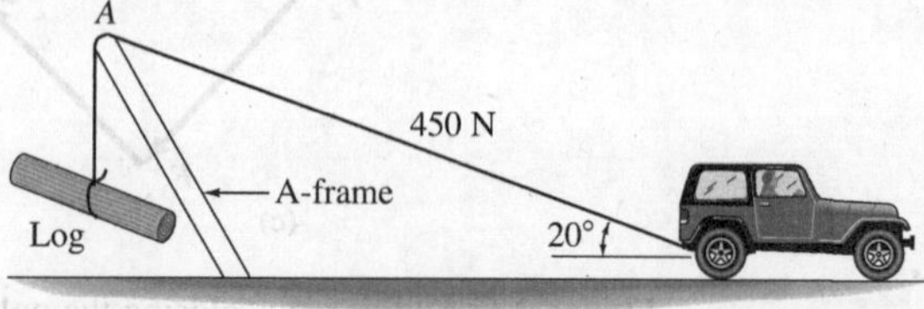

Figure P.2.7

2.8. Find the total force and its direction from the cable acting on each of the three pulleys, each of which is free to turn. The 100-N weight is stationary.

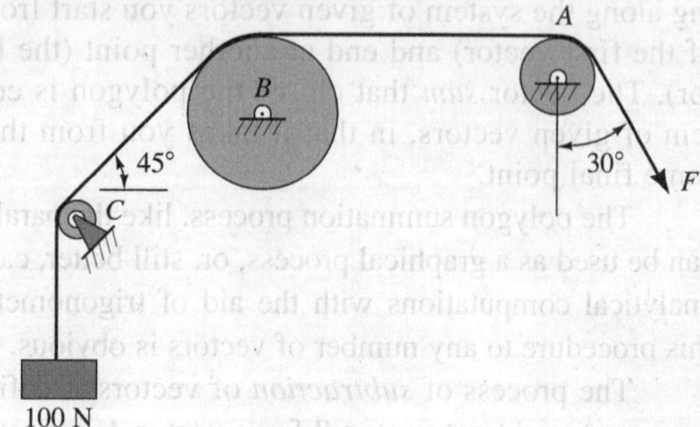

Figure P.2.8

2.9. If the difference between forces $\boldsymbol{B}$ and $\boldsymbol{A}$ in Fig. P.2.6 is a force $\boldsymbol{D}$ having a magnitude of 25 N, what is the magnitude of $\boldsymbol{B}$ and the direction of $\boldsymbol{D}$?

2.10. What is the sum of the forces transmitted by the structural rods to the pin at A?

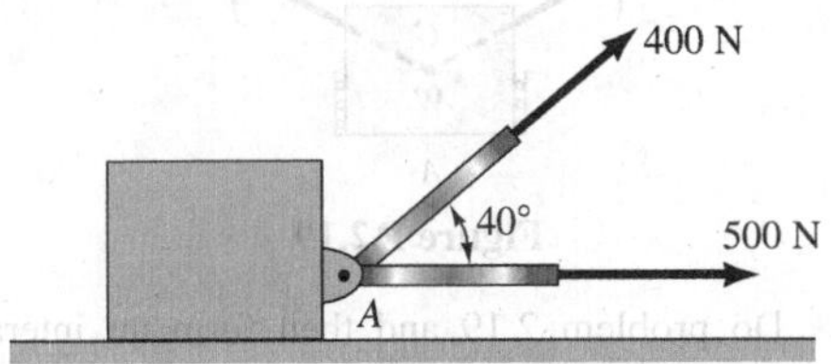

Figure P.2.10

2.11. Suppose in Problem 2.10 we require that the total force transmitted by the members to pin A be inclined 12° to the horizontal. If we do not change the force transmitted by the horizontal member, what must be the new force for the other member whose direction remains at 40°? What is the total force?

2.12. Using the parallelogram law, find the tensile force in cable AC, T_{AC}, and the angle α. (We will do this problem differently in Example 5.4.)

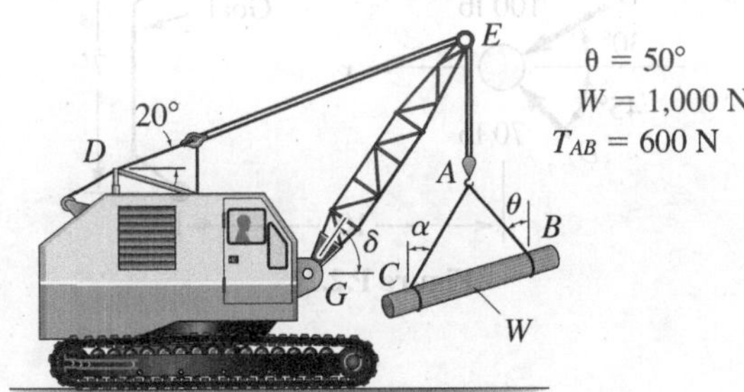

Figure P.2.12

2.13. In the preceding problem, what should the angle δ be so that the sum of the forces from cable DE and cable EA is colinear with the boom GE? Verify that $\delta = 55°$.

2.14. Three forces act on the block. The 500-N and the 600-N forces act, respectively, on the upper and lower faces of the block, while the 1,000-N force acts along the edge. Give the magnitude of the sum of these forces using the parallelogram law twice.

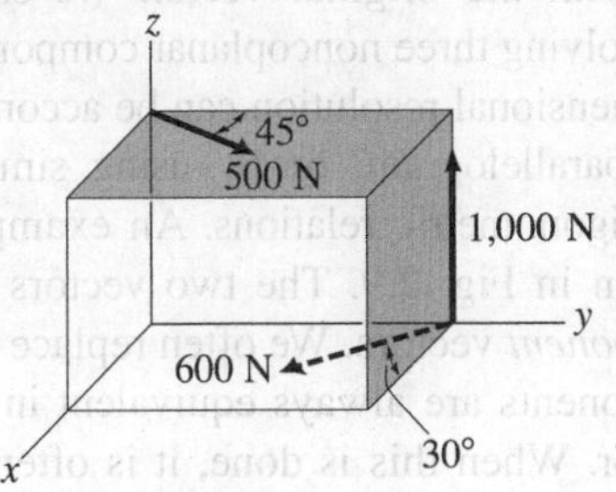

Figure P.2.14

2.15. A man pulls with force W on a rope through a simple frictionless pulley to raise a weight W. What total force is exerted on the pulley?

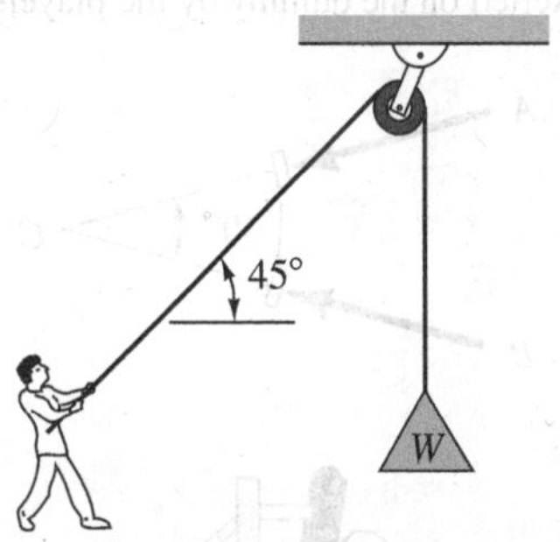

Figure P.2.15

2.16. Add the three vectors using the parallelogram law twice. The 100-N force is in the xz plane, while the other two forces are parallel to the yz plane and do not intersect. Give the magnitude of the sum and the angle it forms with the x axis.

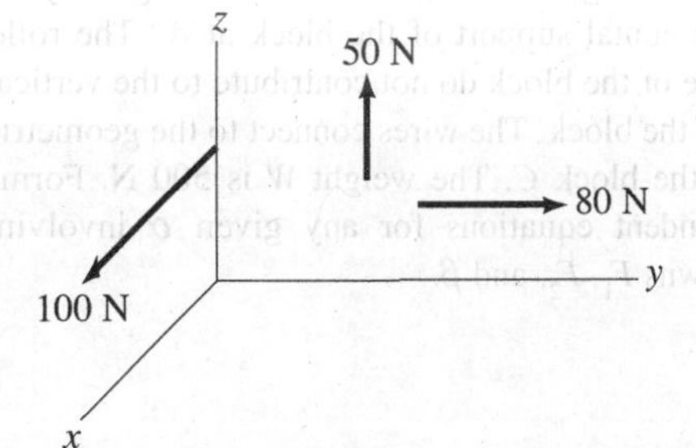

Figure P.2.16

2.17. A mass M is supported by cables (1) and (2). The tension in cable (1) is 200 N, whereas the tension in (2) is such as to maintain the configuration shown. What is the mass of M in kilograms? (You will learn very shortly that the weight of M must be equal and opposite to the vector sum of the supporting forces for equilibrium.)

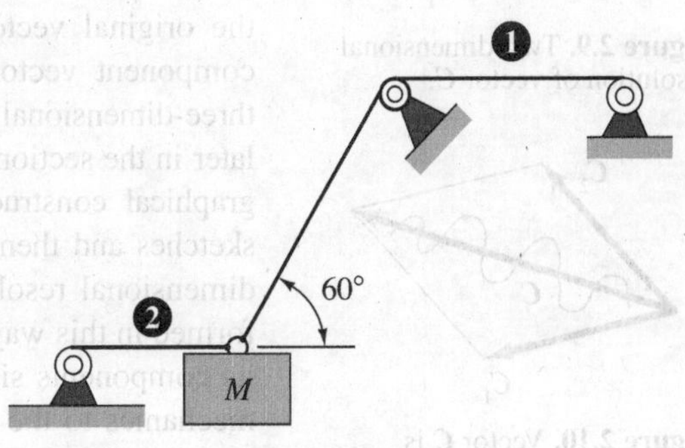

Figure P.2.17

2.18. Two football players are pushing a blocking dummy. Player A pushes with 100-lb force while player B pushes with 150-lb force toward bow C of the dummy. What is the total force exerted on the dummy by the players?

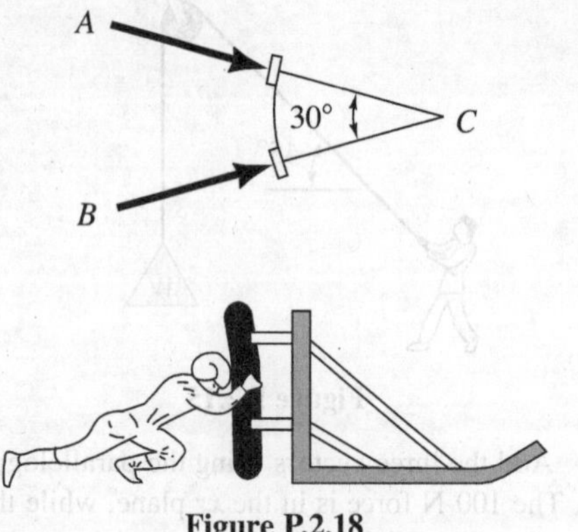

Figure P.2.18

2.19. What are the forces F_1 and F_2 and the angle β for any given angle α to relieve the force of gravity W from the horizontal support of the block at A? The rollers on the side of the block do not contribute to the vertical support of the block. The wires connect to the geometric center of the block C. The weight W is 500 N. Form three independent equations for any given α involving the unknowns F_1, F_2, and β.

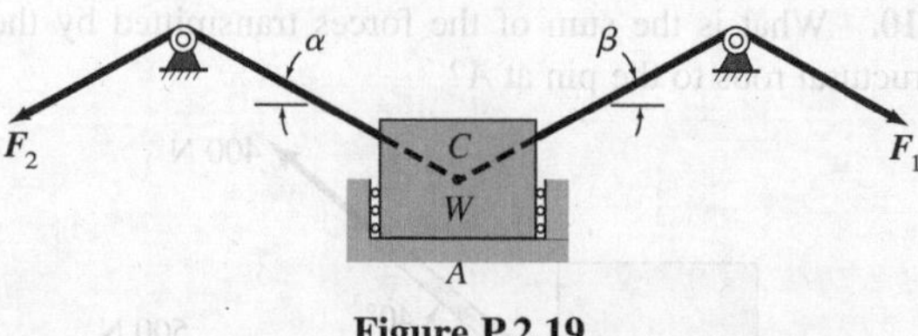

Figure P.2.19

***2.20.** Do problem 2.19 and then form an interactive computer program so that the user at a prompt is asked to insert an angle α in radians for which the program will deliver the correct values of F_1, F_2, and β.

2.21. Two soccer players approach a stationary ball 10 ft away from the goal. Simultaneously, a player on team O (offense) kicks the ball with force 100 lb for a split second while a player on team D (defense) kicks with force 70 lb during the same time interval. Does the offense score (assuming that the goalie is asleep)?

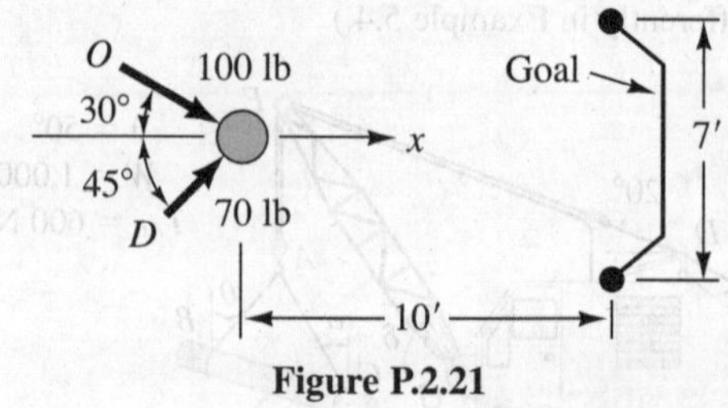

Figure P.2.21

2.4 Resolution of Vectors; Scalar Components

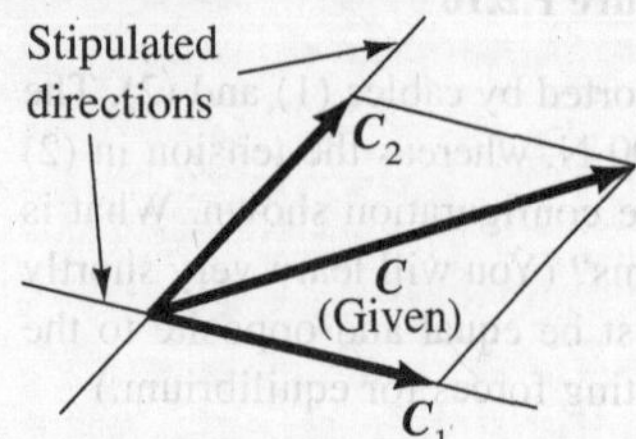

Figure 2.9. Two-dimensional resolution of vector C.

The opposite action of addition of vectors is called *resolution*. Thus, for a given vector $\boldsymbol{C}$, we may find a pair of vectors in any two stipulated directions coplanar with $\boldsymbol{C}$ such that the two vectors, called *components*, sum to the original vector. This is a *two-dimensional* resolution involving two component vectors *coplanar* with the original vector. We shall discuss three-dimensional resolution involving three noncoplanar component vectors later in the section. The two-dimensional resolution can be accomplished by graphical construction of the parallelogram, or by using simple helpful sketches and then employing trigonometric relations. An example of two-dimensional resolution is shown in Fig. 2.9. The two vectors $\boldsymbol{C}_1$ and $\boldsymbol{C}_2$ formed in this way are the *component* vectors. We often replace a vector by its components since the components are always equivalent in rigid-body mechanics to the original vector. When this is done, it is often helpful to indicate that the original vector is no longer operative by drawing a wavy line through the original vector as shown in Fig. 2.10.

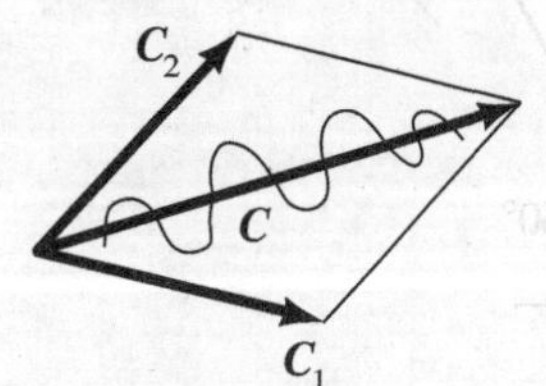

Figure 2.10. Vector C is replaced by its components and is no longer operative.

Example 2.3

A sailboat cannot go directly into the wind, but must *tack* from side to side as shown in Fig. 2.11 wherein a sailboat is going from marker *A* to marker *B* 5,000 meters apart. What is the additional distance ΔL beyond 5,000 m that the sailboat must travel to get from *A* to *B*?

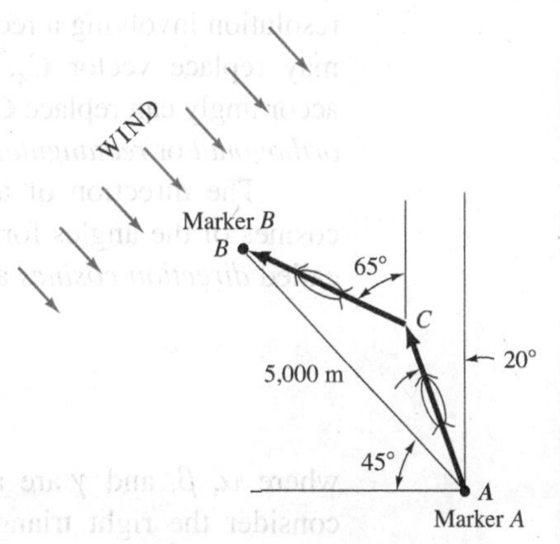

Figure 2.11. Sailboat tacking.

Clearly the displacement vector[4] $\boldsymbol{\rho}_{AB}$ is equivalent to the vector sum of displacement vectors $\boldsymbol{\rho}_{AC}$ plus $\boldsymbol{\rho}_{CB}$ in that the same starting points *A*, and the same destination points *B*, are involved in each case. Thus, vectors $\boldsymbol{\rho}_{AC}$ and $\boldsymbol{\rho}_{CB}$ are two-dimensional components of vector $\boldsymbol{\rho}_{AB}$. Accordingly, we can show a parallelogram for those vectors for which triangle *ABC* forms half of the parallelogram (see Fig. 2.12). We leave it for you to justify the various angles indicated in the diagram. Now we first use the *law of sines*.

$$\frac{AC}{\sin\beta} = \frac{5{,}000}{\sin\alpha}$$

$$\therefore\ AC = \frac{5{,}000\sin 20^\circ}{\sin 135^\circ} = 2{,}418.4\ \text{m}$$

And

$$\frac{BC}{\sin\gamma} = \frac{5{,}000}{\sin\alpha}$$

$$\therefore\ BC = \frac{5{,}000\sin 25^\circ}{\sin 135^\circ} = 2{,}988.4\ \text{m}$$

Hence the increase in distance ΔL is

$$\Delta L = (2{,}418.4 + 2{,}988.4) - 5{,}000 = \boxed{406.8\ \text{m}}$$

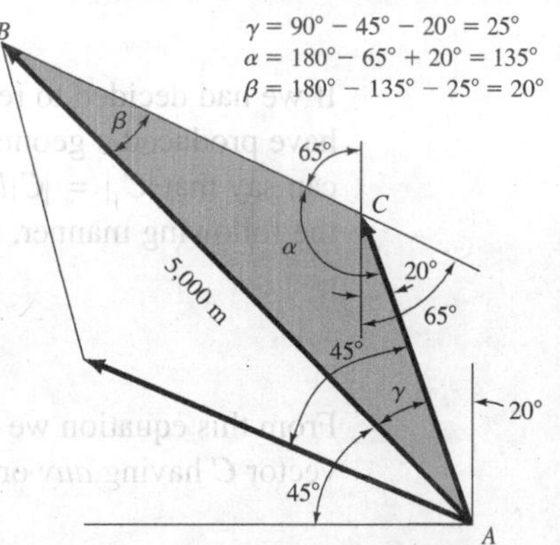

Figure 2.12. Enlarged parallelogram.

[4]A *displacement vector*, we remind you, connects two points *A* and *B* in space and is often denoted as $\boldsymbol{\rho}_{AB}$. The order of the subscripts gives the sense of the vector—here going from *A* to *B*.

It is also readily possible to find *three* components *not in the same plane as* ***C*** that add up to ***C***. This is the aforementioned three-dimensional resolution. Consider the specification of three *orthogonal* directions[5] for the resolution of ***C*** positioned in the first quadrant, as is shown in Fig. 2.13. The resolution may be accomplished in two steps. Resolve ***C*** along the *z* direction, and along the intersection of the *xy* plane and the plane formed by ***C*** and the *z* axis. This is a two-dimensional resolution with the parallelogram becoming a rectangle because of the normalcy of

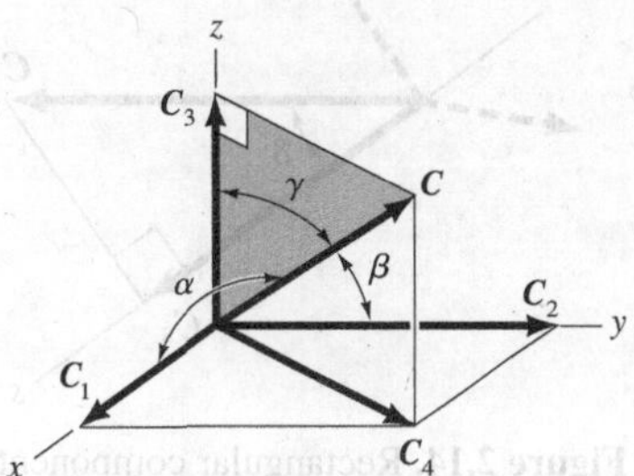

Figure 2.13. Orthogonal or rectangular components.

[5]Although the vector can be resolved along three *skew* directions (hence nonorthogonal), the orthogonal directions are used most often in engineering practice.

the z axis to the xy plane. This gives orthogonal vectors $\boldsymbol{C}_3$ and $\boldsymbol{C}_4$ that replace original vector $\boldsymbol{C}$. Next take vector $\boldsymbol{C}_4$ and resolve it along axes x and y by a second two-dimensional resolution involving a rectangle once again thus forming orthogonal vectors $\boldsymbol{C}_1$ and $\boldsymbol{C}_2$ that may replace vector $\boldsymbol{C}_4$. Clearly orthogonal vectors $\boldsymbol{C}_1$, $\boldsymbol{C}_2$, and $\boldsymbol{C}_3$ add up to $\boldsymbol{C}$ and accordingly can replace $\boldsymbol{C}$ under any and all circumstances. Hence $\boldsymbol{C}_1$, $\boldsymbol{C}_2$, and $\boldsymbol{C}_3$ are called *orthogonal* or *rectangular component vectors* of vector $\boldsymbol{C}$.

The direction of a vector $\boldsymbol{C}$ relative to an orthogonal reference is given by the cosines of the angles formed by the vector and the respective coordinate axes. These are called *direction cosines* and are denoted as

$$\begin{aligned} \cos(\boldsymbol{C}, x) &= \cos\alpha \equiv l \\ \cos(\boldsymbol{C}, y) &= \cos\beta \equiv m \\ \cos(\boldsymbol{C}, z) &= \cos\gamma \equiv n \end{aligned} \tag{2.2}$$

where α, β, and γ are associated with the x, y, and z axes, respectively. Now let us consider the right triangle, whose sides are $\boldsymbol{C}$ and the component vector, $\boldsymbol{C}_3$, shown shaded in Fig. 2.13. It then becomes clear, from trigonometric considerations of the right triangle, that for the first quadrant

$$|\boldsymbol{C}_3| = |\boldsymbol{C}|\cos\gamma = |\boldsymbol{C}|\,n \tag{2.3}$$

If we had decided to resolve $\boldsymbol{C}$ first in the y direction instead of the z direction, we would have produced a geometry from which we could conclude that $|\boldsymbol{C}_2| = |\boldsymbol{C}|m$. Similarly, we can say that $|\boldsymbol{C}_1| = |\boldsymbol{C}|l$. We can then express $|\boldsymbol{C}|$ in terms of its orthogonal components in the following manner, using the Pythagorean theorem.[6]

$$|\boldsymbol{C}| = \left[(|\boldsymbol{C}|l)^2 + (|\boldsymbol{C}|m)^2 + (|\boldsymbol{C}|n)^2\right]^{1/2} \tag{2.4}$$

From this equation we can define the *orthogonal* or *rectangular scalar components* of the vector $\boldsymbol{C}$ having *any* orientation as

$$C_x = |\boldsymbol{C}|l, \qquad C_y = |\boldsymbol{C}|m, \qquad C_z = |\boldsymbol{C}|n \tag{2.5}$$

Note that C_x, C_y, and C_z may be negative, depending on the sign of the direction cosines. Finally, it must be pointed out that *although C_x, C_y, and C_z are associated with certain axes and hence certain directions, they have been developed as scalars and must be handled as scalars*. Thus, an equation such as $10\boldsymbol{V} = V_x \cos\beta$ is not correct, because the left side is a vector and the right side is a scalar. This should spur you to observe care in your notation.

Sometimes only *one* of the scalar orthogonal components of a vector (often called a rectangular scalar component) is desired. Then, just one direction is prescribed, as shown in Fig. 2.14. Thus, the scalar rectangular component C_s is $|\boldsymbol{C}|\cos\delta$. Note we have shown a pair of other rectangular components as dashed vectors in Fig. 2.14. However, it is only the single component C_s that we often use, disregarding other rectangular components. It is always the case that the triangle formed by the vector and its scalar rectangular component is a right triangle. In establishing C_s we therefore speak of "dropping a perpendicular from $\boldsymbol{C}$ to s" or of "projecting along s."

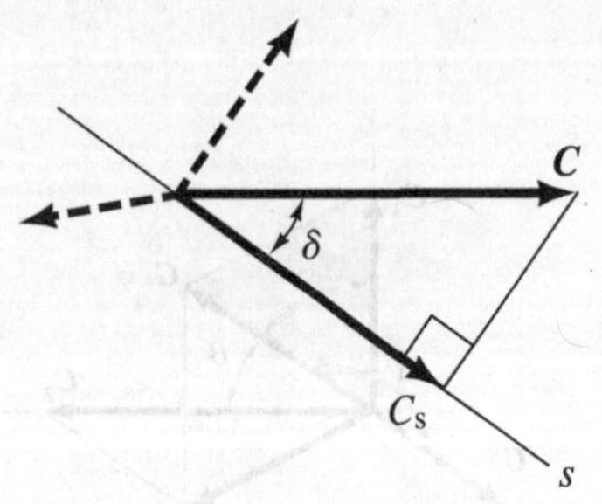

Figure 2.14. Rectangular component of $\boldsymbol{C}$.

[6]From Eq. 2.4 one readily can conclude that $l^2 + m^2 + n^2 = 1$, which is a well-known geometric relation.

The scalar rectangular component C_s could also be the result of a *two-dimensional* orthogonal resolution wherein the other component is in the plane of $\boldsymbol{C}$ and C_s and is *normal* to C_s. It is important to remember, however, that a component of a *nonorthogonal* two-dimensional resolution is *not* a rectangular component.

As a final consideration, let us examine vectors $\boldsymbol{A}$ and $\boldsymbol{B}$, which, along with direction s, form a plane as is shown in Fig. 2.15. The sum of the vectors $\boldsymbol{A}$ and $\boldsymbol{B}$ is found by the parallelogram law to be $\boldsymbol{C}$. We shall now show that the projection of $\boldsymbol{C}$ along s is the same as the sum of the projections of the two-dimensional components of $\boldsymbol{A}$ and $\boldsymbol{B}$, taken along s. That is,

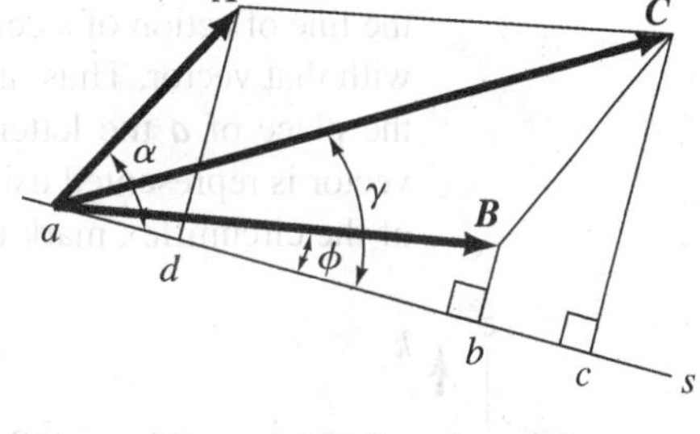

Figure 2.15. $C_s = A_s + B_s$.

$$C_s = A_s + B_s$$

On the diagram, then, the following relation must be verified:

$$ac = ad + ab \tag{a}$$

But

$$ac = ab + bc \tag{b}$$

Also, it is clear that

$$ad = bc \tag{c}$$

By substituting from Eqs. (b) and (c) into Eq. (a), we reduce Eq. (a) to an identity which shows that the projection of the sum of two vectors is the same as the sum of the projections of the two vectors.

2.5 Unit Vectors

It is sometimes convenient to express a vector $\boldsymbol{C}$ as the product of its magnitude and a vector $\boldsymbol{a}$ of unit magnitude and having direction corresponding to the vector $\boldsymbol{C}$. The vector $\boldsymbol{a}$ is called a *unit vector*. The unit vector is also at times denoted as $\hat{\boldsymbol{a}}$. (You will write it as $\hat{a}$.) It has no dimensions. We formulate this vector as follows:

$$\boldsymbol{a}(\text{unit vector in direction } \boldsymbol{C}) = \frac{\boldsymbol{C}}{|\boldsymbol{C}|} \tag{2.6}$$

Clearly, this development fulfills the requirements that have been set forth for this vector. We can then express the vector $\boldsymbol{C}$ in the form

$$\boldsymbol{C} = |\boldsymbol{C}|\boldsymbol{a} \tag{2.7}$$

The unit vector, once established, does not have, per se, an inherent line of action. This will be determined entirely by its use. In the preceding equation, the unit vector $\boldsymbol{a}$ might be considered collinear with the vector $\boldsymbol{C}$. However, we can represent the vector $\boldsymbol{D}$, shown in Fig. 2.16 parallel to $\boldsymbol{C}$, by using the unit vector $\boldsymbol{a}$ as follows:

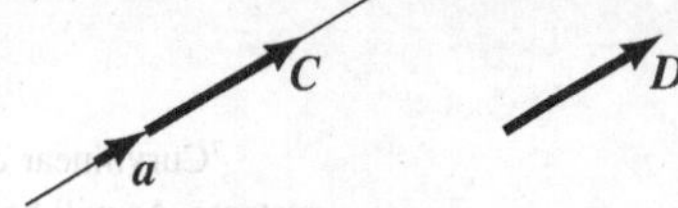

Figure 2.16. Unit vector $\boldsymbol{a}$.

$$\boldsymbol{D} = |\boldsymbol{D}|\boldsymbol{a} \tag{2.7a}$$

It thus acts as a free vector. Occasionally, it is useful to label a unit vector meant to have the line of action of a certain vector with the lowercase letter of the capital letter associated with that vector. Thus, in Eqs. 2.7 and 2.7(a) for this purpose we might have employed in the place of $\boldsymbol{a}$ the letters $\boldsymbol{c}$ and $\boldsymbol{d}$ (in your case $\hat{c}$ and $\hat{d}$), respectively. Next, if a given vector is represented using a lowercase letter, such as the vector $\boldsymbol{r}$, then we often make use of the circumflex mark to indicate the associated unit vector. Thus,

$$\boldsymbol{r} = |\boldsymbol{r}|\hat{\boldsymbol{r}} \tag{2.7b}$$

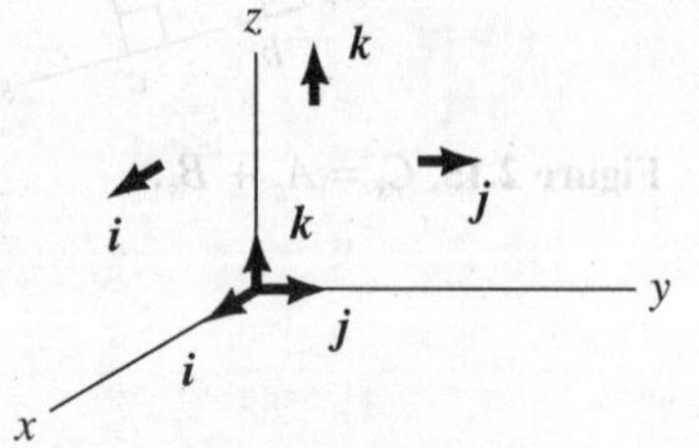

Figure 2.17. Unit vectors for xyz axes.

Unit vectors that are of particular use are those directed along the directions of coordinate axes of a rectangular reference, where $\boldsymbol{i}$, $\boldsymbol{j}$, and $\boldsymbol{k}$ (your instructor will probably use the notation $\hat{\imath}$, $\hat{\jmath}$, and $\hat{k}$) correspond to the x, y, and z directions, as shown in Fig. 2.17.[7]

Since the sum of a set of concurrent vectors is equivalent in all situations to the original vector, we can always replace the vector $\boldsymbol{C}$ by its rectangular scalar components in the following manner:

$$\boldsymbol{C} = C_x\boldsymbol{i} + C_y\boldsymbol{j} + C_z\boldsymbol{k} \tag{2.8}$$

In Chapter 1, we saw that vectors that are equal have the same magnitude and direction. Hence, if $\boldsymbol{A} = \boldsymbol{B}$, we can say that

$$A_x\boldsymbol{i} + A_y\boldsymbol{j} + A_z\boldsymbol{k} = B_x\boldsymbol{i} + B_y\boldsymbol{j} + B_z\boldsymbol{k} \tag{2.9}$$

Then, since the unit vectors have mutually different directions, we conclude from above that

$$A_x\boldsymbol{i} = B_x\boldsymbol{i}$$
$$A_y\boldsymbol{j} = B_y\boldsymbol{j}$$
$$A_z\boldsymbol{k} = B_z\boldsymbol{k}$$

It then follows that

$$A_x = B_x$$
$$A_y = B_y$$
$$A_z = B_z$$

Hence, the vector equation, $\boldsymbol{A} = \boldsymbol{B}$, has resulted in three scalar equations that in totality are equivalent in every way to the vector statement of equality. Thus, in Newton's law we would have

$$\boldsymbol{F} = m\boldsymbol{a} \tag{2.10a}$$

as the vector equation, and

$$F_x = ma_x, \quad F_y = ma_y, \quad F_z = ma_z \tag{2.10b}$$

as the corresponding scalar equations.

[7]Curvilinear coordinate systems have associated sets of unit vectors just as do the rectangular coordinate systems. As will be seen later, however, certain of these unit vectors do not all have fixed directions in space for a given reference as do the vectors $\boldsymbol{i}$, $\boldsymbol{j}$, and $\boldsymbol{k}$.

2.6 Useful Ways of Representing Vectors

Quite often, we show a rectangular parallelepiped with sides oriented parallel to the coordinate axes and positioned somewhere along the line of action of a vector (see Fig. 2.18) such that this line of action coincides with an inside diagonal of the rectangular parallelepiped. The purpose of this rectangular parallelepiped and diagonal is to allow for the easy determination of the orientation of the line of action and hence the orientation of a vector. *AB* in the diagram is such a diagonal used for the determination of the line of action of vector $\boldsymbol{F}$. Numbers for this purpose are shown along the sides of the rectangular parallelepiped without units. *Any* set of numbers can be used as long as the *ratios* of these numbers remain the ones required for the proper determination of the orientation of the vector. That determination proceeds by first replacing the *displacement vector* $\boldsymbol{\rho}_{AB}$ from corner *A* to corner *B* by a set of three vector displacements going from *A* to *B* along the sides of the rectangular parallelepiped. We thereby can replace the vector $\boldsymbol{\rho}_{AB}$ by the sum of its rectangular components. Thus, for the case shown in Fig. 2.18 we can say[8]

$$\boldsymbol{\rho}_{AB} = 10\boldsymbol{j} - 4\boldsymbol{i} + 6\boldsymbol{k} = -4\boldsymbol{i} + 10\boldsymbol{j} + 6\boldsymbol{k}$$

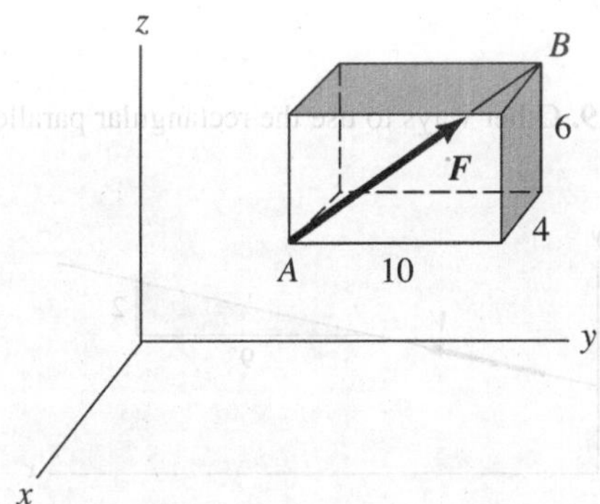

Figure 2.18. Rectangular parallelepiped used for specifying the direction of a vector.

Now using the Pythagorean theorem, divide $\boldsymbol{\rho}_{AB}$ by its magnitude, namely: $\sqrt{4^2+10^2+6^2}$. We thus form the unit vector $\hat{\boldsymbol{\rho}}_{AB}$. That is,

$$\hat{\boldsymbol{\rho}}_{AB} = \frac{\boldsymbol{\rho}_{AB}}{|\boldsymbol{\rho}_{AB}|} = \frac{-4\boldsymbol{i} + 10\boldsymbol{j} + 6\boldsymbol{k}}{\sqrt{4^2+10^2+6^2}} = -.3244\boldsymbol{i} + .8111\boldsymbol{j} + .4867\boldsymbol{k}$$

As a final step we can give vector $\boldsymbol{F}$ as follows:

$$\boldsymbol{F} = F(-.3244\boldsymbol{i} + .8111\boldsymbol{j} + .4867\boldsymbol{k})$$

If $F = 100$ N we then can say:

$$\boldsymbol{F} = -32.44\boldsymbol{i} + 81.11\boldsymbol{j} + 48.67\boldsymbol{k} \text{ N}$$

[8]Imagine you are "walking" from *A* to *B* but restricting your movements to be along the coordinate directions. This movement is equivalent to going directly from *A* to *B* in that the same endpoints result.

Note that the rectangular parallelepiped can be anywhere along the line of action of F including cases where F is not inside of the parallelepiped or extends beyond the parallelepiped (see Fig. 2.19). In the two-dimensional case, a right triangle serves the same purpose as the rectangular parallelepiped in three dimensions. This is shown in Fig. 2.20 where vector V is in the xy plane. Here we can say,

$$V = V\left[\frac{9}{\sqrt{2^2+9^2}}\,i + \frac{2}{\sqrt{2^2+9^2}}\,j\right] = V(.9762\,i + .2169\,j)$$

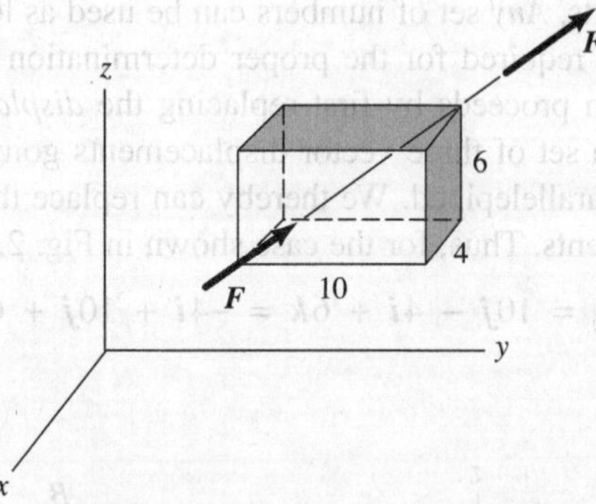

Figure 2.19. Other ways to use the rectangular parallelepiped.

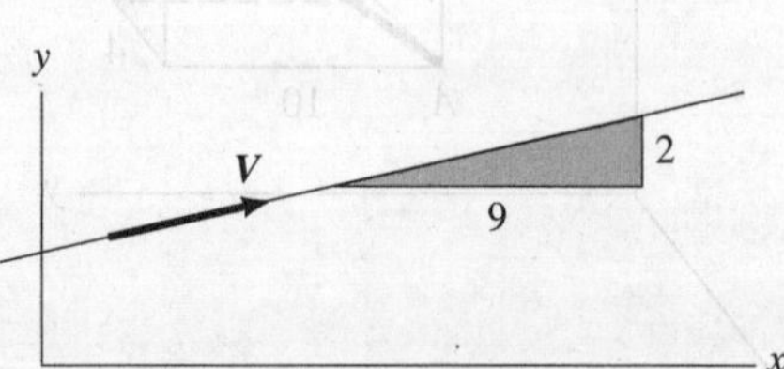

Figure 2.20. Right triangle used for specifying the direction of a vector in two dimensions.

There are times when the rectangular parallelpiped is not shown explicitly. However, the length of the sides of one having the proper diagonal may be available so that the replacement of the diagonal displacement vector into rectangular components can be readily achieved. The simplest procedure is to mentally move from the beginning point of the diagonal to the final point always moving along coordinate directions, or, in other words, always moving along the sides of the hypothetical rectangular parallelepiped. Thus, in Fig. 2.21, for the vector F_1 we can consider AB to be the diagonal and in going from A to B we could first move in the minus x direction by an amount –1, then move in the plus y direction by the amount 1.5, and finally in the z direction by an amount 3. This would take us from initial point A to final point B. The corresponding displacement vector would then be

$$\rho_{AB} = -1i + 1.5j + 3j$$

The following example will illustrate the use of orthogonal resolution as well as the use of rectangular components of vectors.

Example 2.4

A crane (not shown) is supporting a 2,000-N crate (see Fig. 2.21) through three cables: *AB*, *CB*, and *DB*. Note that *D* is at the center of the outer edge of the crate; *C* is 1.6 m from the corner of this edge; and *B* is directly above the center of the crate. What are the forces F_1, F_2, and F_3 transmitted by the cables?

We will soon learn formally what our common sense tells us, namely that the vector sum of force F_1, force F_2, and force F_3 must equal 2,000$\boldsymbol{k}$ N. We first express these three forces in terms of rectangular components. Thus,

$$\boldsymbol{F}_{AB} = F_1\left(\frac{\boldsymbol{\rho}_{AB}}{|\boldsymbol{\rho}_{AB}|}\right) = F_1\left[\frac{-1\boldsymbol{i} + 1.5\boldsymbol{j} + 3\boldsymbol{k}}{\sqrt{1^2 + 1.5^2 + 3^2}}\right]$$

$$= F_1(-.2857\boldsymbol{i} + .4286\boldsymbol{j} + .8571\boldsymbol{k})\text{ N}$$

$$\boldsymbol{F}_{DB} = F_2\left(\frac{\boldsymbol{\rho}_{DB}}{|\boldsymbol{\rho}_{DB}|}\right) = F_2\left[\frac{1\boldsymbol{i} + 0\boldsymbol{j} + 3\boldsymbol{k}}{\sqrt{1^2 + 3^2}}\right] = F_2(.3162\boldsymbol{i} + .9487\boldsymbol{k})\text{ N}$$

$$\boldsymbol{F}_{CB} = F_3\left(\frac{\boldsymbol{\rho}_{CB}}{|\boldsymbol{\rho}_{CB}|}\right) = F_3\left[\frac{-.6\boldsymbol{i} - 1.5\boldsymbol{j} + 3\boldsymbol{k}}{\sqrt{.6^2 + 1.5^2 + 3^2}}\right]$$

$$= F_3(-.1761\boldsymbol{i} - .4402\boldsymbol{j} + .8805\boldsymbol{k})\text{ N}$$

We now sum the three forces to equal 2,000$\boldsymbol{k}$ N.

$$F_1(-.2857\boldsymbol{i} + .4286\boldsymbol{j} + .8571\boldsymbol{k}) + F_2(.3162\boldsymbol{i} + .9487\boldsymbol{k}) + F_3(-.1761\boldsymbol{i} - .4402\boldsymbol{j} + .8805\boldsymbol{k}) = 2{,}000\boldsymbol{k}$$

We have three scalar equations from the previous equation.

$$-.2857\,F_1 + .3162\,F_2 - .1761\,F_3 = 0$$

$$.4286\,F_1 + 0 - .4402\,F_3 = 0$$

$$.8571\,F_1 + .9487\,F_2 + .8805\,F_3 = 2{,}000$$

Solving simultaneously, we get the following results:

F1 = 648.1 N
F2 = 937.1 N
F3 = 631.1 N

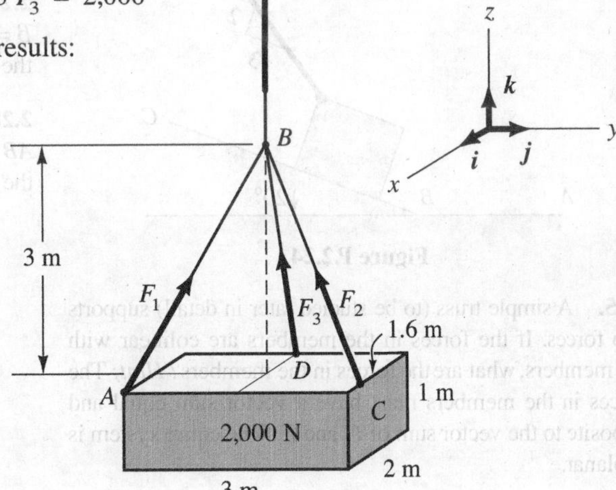

Figure 2.21. A crate is supported by three forces.

PROBLEMS

2.22. Resolve the 100-lb force into a set of components along the slot shown and in the vertical direction.

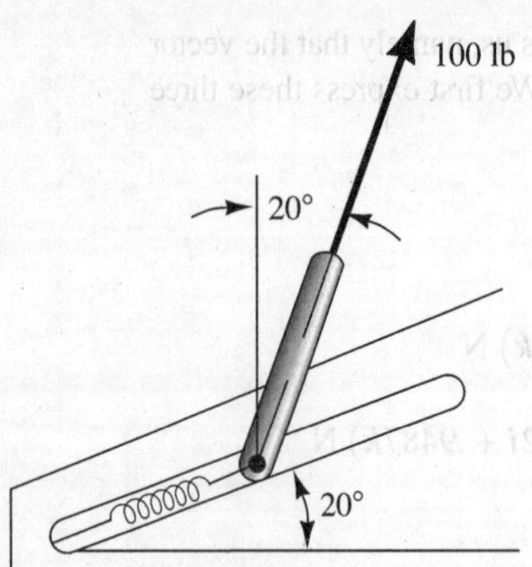

Figure P.2.22

2.23. A farmer needs to build a fence from the corner of his barn to the corner of his chicken house 30 m away in the NE direction. However, he wants to enclose as much of the barnyard as possible. Thus, he runs the fence east, from the corner of his barn to the property line and then NNE to the corner of his chicken house. How long is the fence?

2.24. Resolve the force $\boldsymbol{F}$ into a component perpendicular to AB and a component parallel to BC.

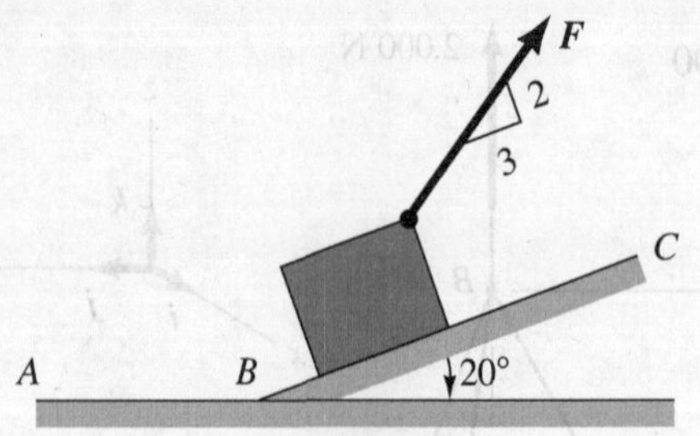

Figure P.2.24

2.25. A simple truss (to be studied later in detail) supports two forces. If the forces in the members are colinear with the members, what are the forces in the members? *Hint:* The forces in the members must have a vector sum equal and opposite to the vector sum of F_1 and F_2. The entire system is coplanar.

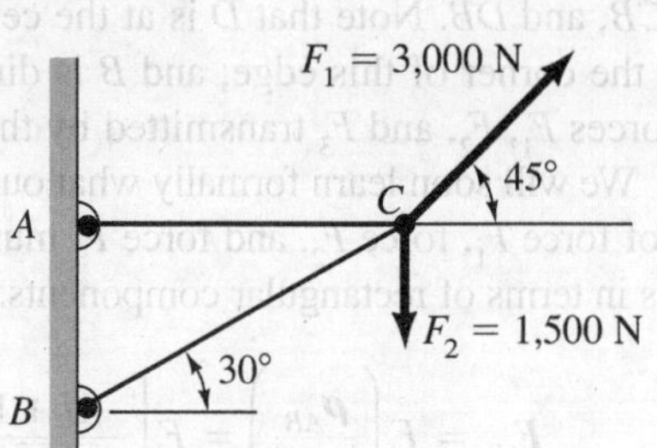

Figure P.2.25

2.26. Two tugboats are maneuvering an ocean liner. The desired total force is 3,000 lb at an angle of 15° as shown in the diagram. If the tugboat forces have directions as shown, what must the forces F_1 and F_2 be?

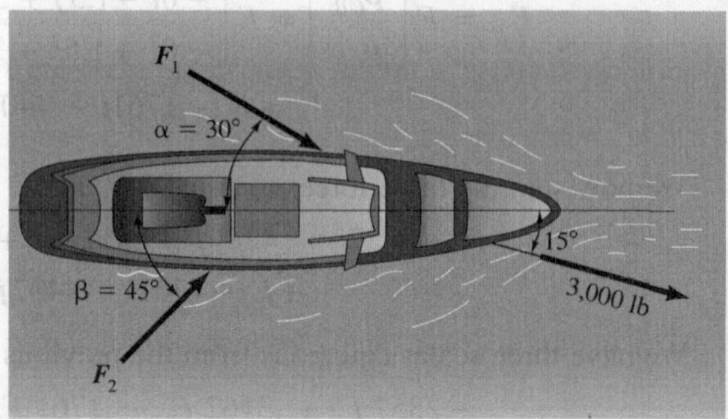

Figure P.2.26

2.27. In the previous problem, if $F_2 = 1{,}000$ lb and $\beta = 40°$, what should F_1 and α be so that $F_1 + F_2$ yields the indicated 3,000-lb force?

2.28. A 1,000-N force is resolved into components along AB and AC. If the component along AB is 700 N, determine the angle α and the value of the component along AC.

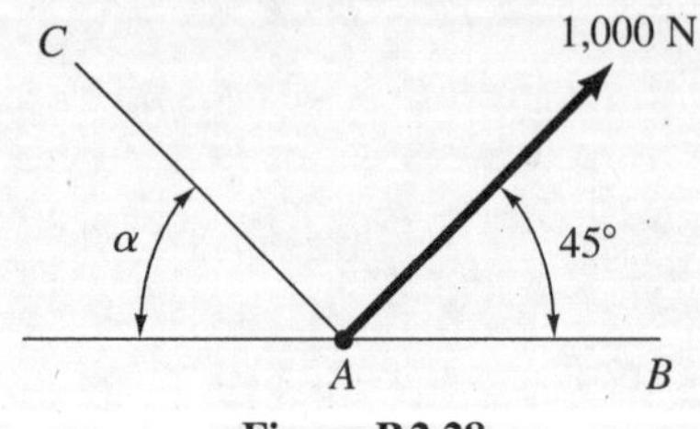

Figure P.2.28

2.29. Two men are trying to pull a crate which will not move until a 150-lb total force is applied in any one direction. Man A can pull only at 45° to the desired direction of crate motion, whereas man B can pull only at 60° to the desired motion. What force must each man exert to start the box moving as shown?

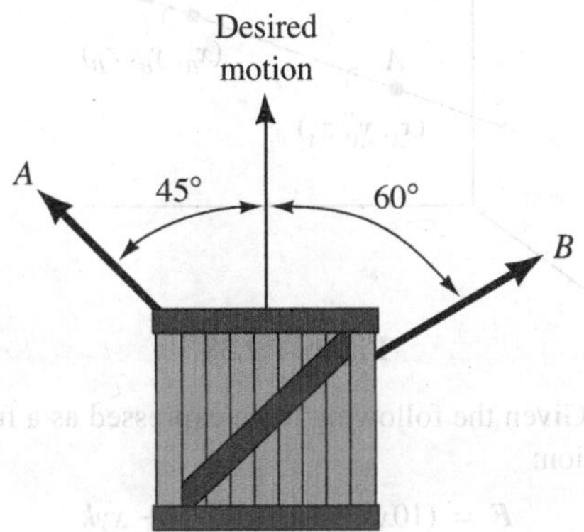

Figure P.2.29

2.30. What is the sum of the three forces? The 2,000-N force is in the yz plane.

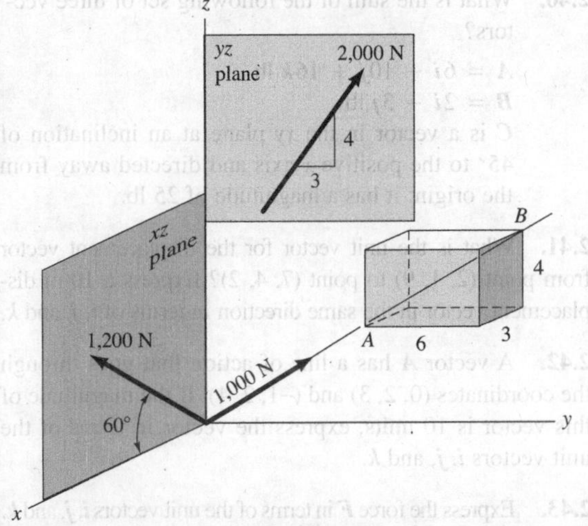

Figure P.2.30

2.31. The 500-N force is to be resolved into components along the AC and AB directions in the xy plane measured by the angles α and β. If the component along AC is to be 1,000 N and the component along AB is to be 800 N, compute α and β.

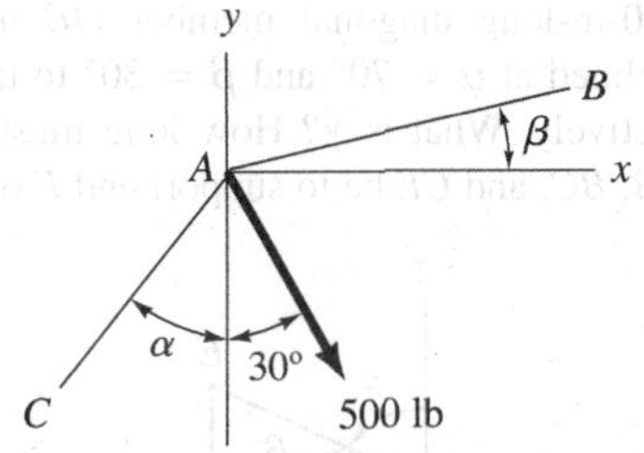

Figure P.2.31

2.32. The orthogonal components of a force are:
x component 10 lb in positive x direction
y component 20 lb in positive y direction
z component 30 lb in negative z direction
(a) What is the magnitude of the force itself?
(b) What are the direction cosines of the force?

2.33. What are the rectangular components of the 100-lb force? What are the direction cosines for this force?

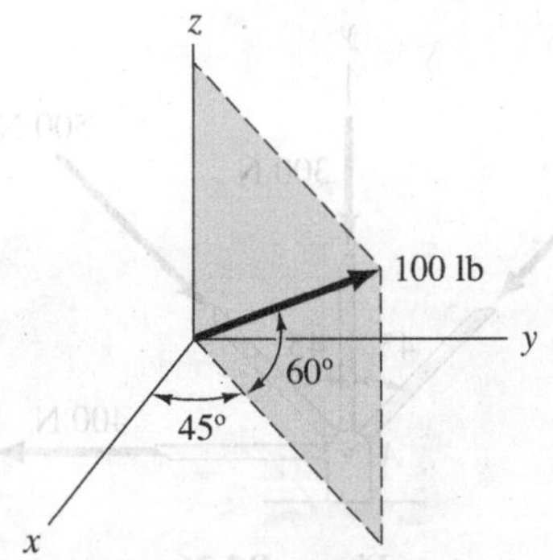

Figure P.2.33

2.34. The 1,000-N force is parallel to the displacement vector $\overrightarrow{OA}$ while the 2,000-N force is parallel to the displacement vector $\overrightarrow{CB}$. What is the vector sum of these forces?

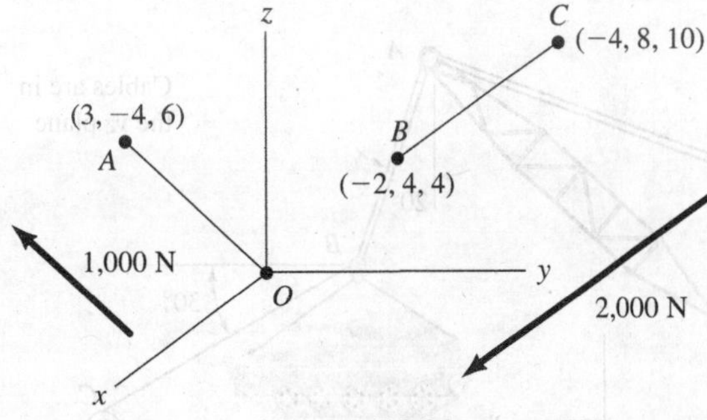

Figure P.2.34

2.35. A 50-m-long diagonal member OE in a space frame is inclined at $\alpha = 70°$ and $\beta = 30°$ to the x and y axes, respectively. What is γ? How long must members OA, AC, OB, BC, and CE be to support end E of OE?

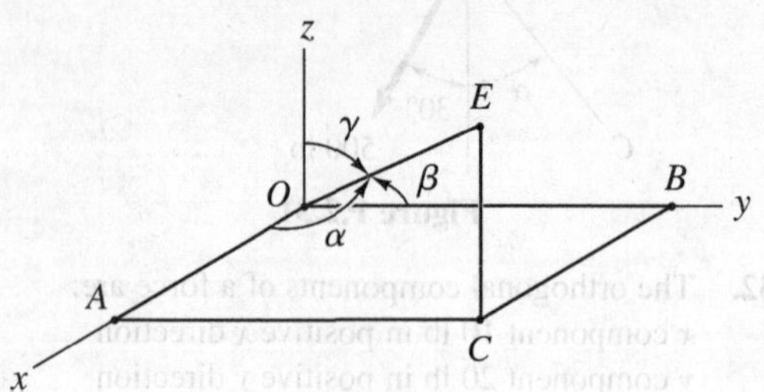

Figure P.2.35

2.36. What is the orthogonal total force component in the x direction of the force transmitted to pin A of a roof truss by the four members? What is the total component in the y direction?

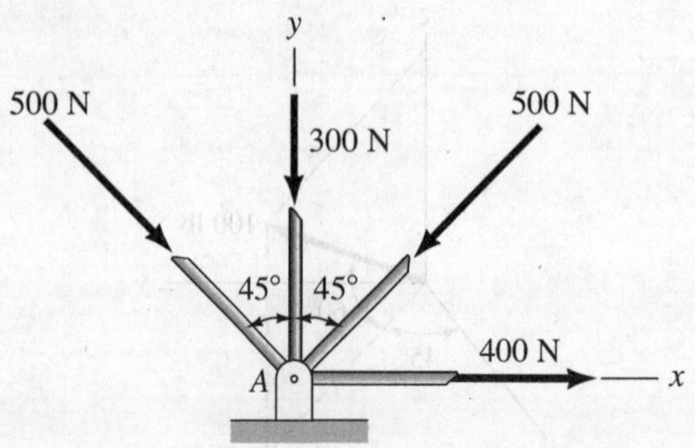

Figure P.2.36

2.37. A 30-ton light tank is being lowered from a ship using two cables AB and BC. What are the forces in these cables? Use a parallelogram sketch. Also, do this problem using rectangular components.

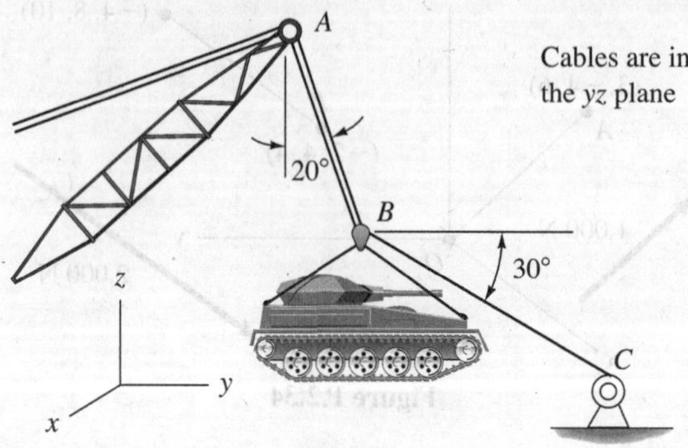

Figure P.2.37

2.38. Suppose two points A and B in space are designated and a velocity vector $\boldsymbol{V}$ is colinear with the line of action of the displacement vector $\boldsymbol{\rho}_{AB}$ as has been shown. Express $\boldsymbol{V}$ in terms of its rectangular components.

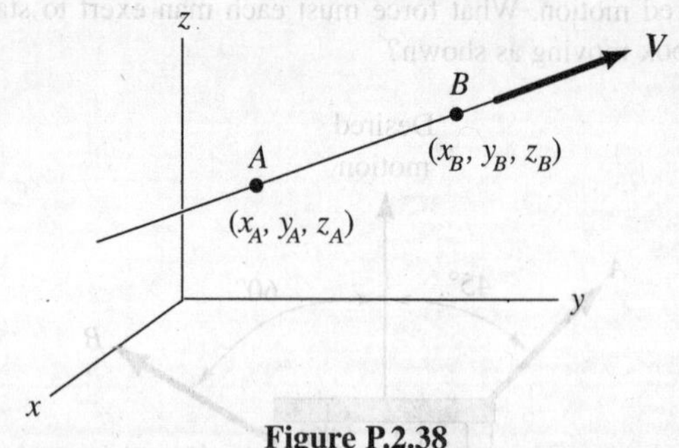

Figure P.2.38

2.39. Given the following force expressed as a function of position:

$$\boldsymbol{F} = (10x - 6)\boldsymbol{i} + x^2 z\boldsymbol{j} + xy\boldsymbol{k}$$

What are the direction cosines of the force at position (1, 2, 2)? What is the position along the x coordinate where $F_x = 0$? Plot F_y versus the x coordinate for an elevation $z = 1$.

2.40. What is the sum of the following set of three vectors?

$\boldsymbol{A} = 6\boldsymbol{i} + 10\boldsymbol{j} + 16\boldsymbol{k}$ lb

$\boldsymbol{B} = 2\boldsymbol{i} - 3\boldsymbol{j}$ lb

$\boldsymbol{C}$ is a vector in the xy plane at an inclination of 45° to the positive x axis and directed away from the origin; it has a magnitude of 25 lb.

2.41. What is the unit vector for the displacement vector from point (2, 1, 9) to point (7, 4, 2)? Express a 10-m displacement vector in the same direction in terms of $\boldsymbol{i}$, $\boldsymbol{j}$, and $\boldsymbol{k}$.

2.42. A vector $\boldsymbol{A}$ has a line of action that goes through the coordinates (0, 2, 3) and (−1, 2, 4). If the magnitude of this vector is 10 units, express the vector in terms of the unit vectors $\boldsymbol{i}$, $\boldsymbol{j}$, and $\boldsymbol{k}$.

2.43. Express the force $\boldsymbol{F}$ in terms of the unit vectors $\boldsymbol{i}$, $\boldsymbol{j}$, and $\boldsymbol{k}$.

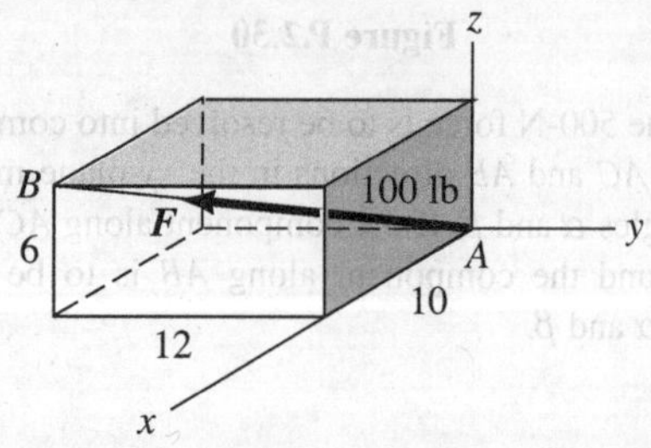

Figure P.2.43

2.44. Express the 100-N force in terms of the unit vectors $\boldsymbol{i}, \boldsymbol{j}$, and $\boldsymbol{k}$. What is the unit vector in the direction of the 100-N force? The force lies along diagonal AB.

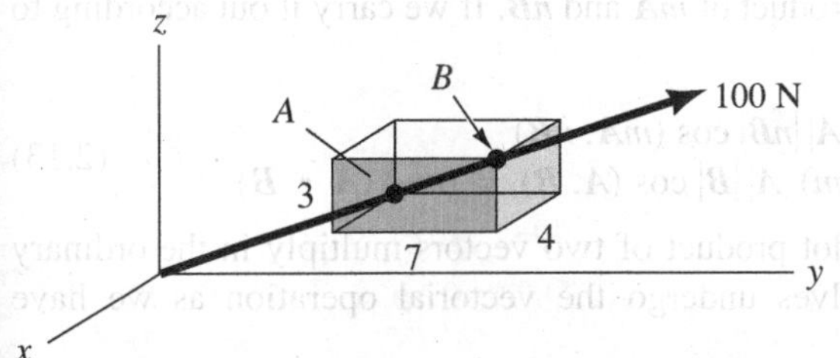

Figure P.2.44

2.45. Express the unit vectors $\boldsymbol{i}, \boldsymbol{j}$, and $\boldsymbol{k}$ in terms of unit vectors $\boldsymbol{\epsilon}_r$, $\boldsymbol{\epsilon}_\theta$, and $\boldsymbol{\epsilon}_z$. (These are unit vectors for *cylindrical coordinates*.) Express the 1,000-lb force going through the origin and through point (2, 4, 4) in terms of the unit vectors $\boldsymbol{i}, \boldsymbol{j}, \boldsymbol{k}$ and $\boldsymbol{\epsilon}_r$, $\boldsymbol{\epsilon}_\theta$, $\boldsymbol{\epsilon}_z$ with $\theta = 60°$. (See the footnote on p. 34.)

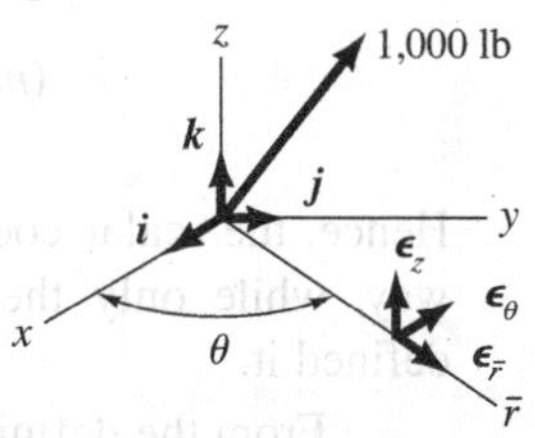

Figure P.2.44

2.7 Scalar or Dot Product of Two Vectors

In elementary physics, work was defined as the product of the force component, in the direction of a displacement, times the displacement. In effect, two vectors, force and displacement, are employed to give a scalar, work. In other physical problems, vectors are associated in this same manner so as to result in a scalar quantity. A vector operation that represents such operations concisely is the scalar product (or dot product), which, for the vector $\boldsymbol{A}$ and $\boldsymbol{B}$ in Fig. 2.22, is defined as[9]

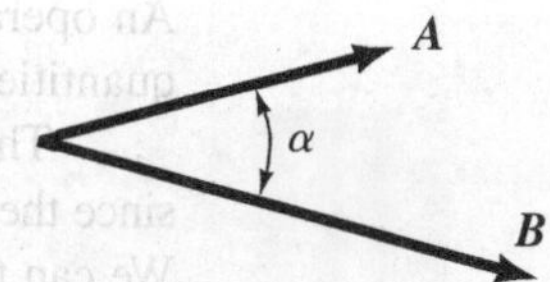

Figure 2.22. α is smallest angle between $\boldsymbol{A}$ and $\boldsymbol{B}$.

$$\boldsymbol{A} \bullet \boldsymbol{B} = |\boldsymbol{A}||\boldsymbol{B}| \cos \alpha \tag{2.11}$$

where α is the smaller angle between the two vectors. Note that the dot product may involve vectors of different dimensional representation, and may be positive or negative, depending on whether the smaller included angle α is less than or greater than 90°. Note also that $\boldsymbol{A} \bullet \boldsymbol{B}$ is equivalent to first projecting vector $\boldsymbol{A}$ onto the line of action of vector $\boldsymbol{B}$ (this gives us $|\boldsymbol{A}| \cos \alpha$), and then multiplying by the magnitude of vector $\boldsymbol{B}$ (or vice versa). The appropriate sign must, of course, be assigned positive if the projected component of vector $\boldsymbol{A}$ and vector $\boldsymbol{B}$ point in the same direction; negative, if not.

The work concept for a force $\boldsymbol{F}$ acting on a particle moving along a path described by s can now be given as

$$W = \int \boldsymbol{F} \bullet d\boldsymbol{s}$$

where $d\boldsymbol{s}$ is a displacement on the path along which the particle is moved.

[9]To ensure that there is no confusion between the dot product of two vectors and the ordinary product of two scalars that you have used up to now, we urge you to read $\boldsymbol{A} \bullet \boldsymbol{B} = \boldsymbol{C}$ as "$\boldsymbol{A}$ dotted into $\boldsymbol{B}$ yields $\boldsymbol{C}$."

As with addition and subtraction of vectors, the dot product operation involves only the vectors themselves and not their respective lines of action. Accordingly, for a dot product of two vectors, we can move the vectors so as to intersect at their tails as in Fig. 2.22. Remember in so doing we must not alter the magnitudes and directions of the vectors.

Let us next consider the scalar product of $m\boldsymbol{A}$ and $n\boldsymbol{B}$. If we carry it out according to our definitions:

$$\begin{aligned}(m\boldsymbol{A}) \bullet (n\boldsymbol{B}) &= |m\boldsymbol{A}|\,|n\boldsymbol{B}| \cos(m\boldsymbol{A}, n\boldsymbol{B}) \\ &= (mn)\,|\boldsymbol{A}|\,|\boldsymbol{B}| \cos(\boldsymbol{A}, \boldsymbol{B}) = (mn)(\boldsymbol{A} \bullet \boldsymbol{B})\end{aligned} \tag{2.13}$$

Hence, the scalar coefficients in the dot product of two vectors multiply in the ordinary way, while only the vectors themselves undergo the vectorial operation as we have defined it.

From the definition, clearly the dot product is *commutative,* since the number $|\boldsymbol{A}|\,|\boldsymbol{B}|$ cos $(\boldsymbol{A}, \boldsymbol{B})$ is independent of the order of multiplication of its terms. Thus,

$$\boldsymbol{A} \bullet \boldsymbol{B} = \boldsymbol{B} \bullet \boldsymbol{A} \tag{2.14}$$

Let us now consider $\boldsymbol{A} \bullet (\boldsymbol{B} + \boldsymbol{C})$. By definition, we may project the vector $(\boldsymbol{B} + \boldsymbol{C})$ onto the line of action of $\boldsymbol{A}$ and then, assigning the appropriate sign, multiply the magnitude of $\boldsymbol{A}$ times the projection of $\boldsymbol{B} + \boldsymbol{C}$. However, in Section 2.4 we showed that the projection of the sum of two vectors is the same as the sum of the projections of the vectors, which means that

$$\boldsymbol{A} \bullet (\boldsymbol{B} + \boldsymbol{C}) = \boldsymbol{A} \bullet \boldsymbol{B} + \boldsymbol{A} \bullet \boldsymbol{C} \tag{2.15}$$

An operation on a sum of quantities that is the same as the sum of the operations on the quantities is called a *distributive operation.* Thus, the dot product is distributive.

The scalar product between unit vectors will now be carried out. The product $\boldsymbol{i} \bullet \boldsymbol{j}$ is 0, since the angle α in Eq. 2.11 is 90°, which makes $\cos \alpha = 0$. On the other hand, $\boldsymbol{i} \bullet \boldsymbol{i} = 1$. We can thus conclude that the dot product of equal orthogonal unit vectors for a given reference is unity and that of unequal orthogonal unit vectors is zero.

If we express the vectors $\boldsymbol{A}$ and $\boldsymbol{B}$ in Cartesian components when taking the dot product, we get

$$\begin{aligned}\boldsymbol{A} \bullet \boldsymbol{B} &= (A_x\boldsymbol{i} + A_y\boldsymbol{j} + A_z\boldsymbol{k}) \bullet (B_x\boldsymbol{i} + B_y\boldsymbol{j} + B_z\boldsymbol{k}) \\ &= A_xB_x + A_yB_y + A_zB_z\end{aligned} \tag{2.16}$$

Thus, we see that a scalar product of two vectors is the sum of the ordinary products of the respective components.[10]

If a vector is multiplied by itself as a dot product, the result is the square of the magnitude of the vector. That is,

$$\boldsymbol{A} \bullet \boldsymbol{A} = |\boldsymbol{A}|\,|\boldsymbol{A}| = A^2 \tag{2.17}$$

Conversely, the square of a number may be considered to be the dot product of two equal vectors having a magnitude equal to the number. Note also that

$$\boldsymbol{A} \bullet \boldsymbol{A} = A_x^2 + A_y^2 + A_z^2 = A^2 \tag{2.18}$$

We can conclude from Eq. 2.18 that

[10]Thus the ordinary grade school product of two numbers, i.e. $(a)(b)$, is a special case of the dot product $\boldsymbol{a} \bullet \boldsymbol{b}$ where the the vectors have the same direction. Thus

$$a\boldsymbol{i} \bullet b\boldsymbol{i} = (a)(b)$$

$$A = \sqrt{A_x^2 + A_y^2 + A_z^2}$$

which checks with the Pythagorean theorem.

The dot product may be of immediate use in expressing the scalar rectangular component of a vector along a given direction as discussed in Section 2.4. If you refer back to Fig. 2.14, you will recall that the component of $\boldsymbol{C}$ along the direction s is given as

$$C_s = |\boldsymbol{C}| \cos \delta$$

Now let us consider a unit vector $\boldsymbol{s}$ along the direction of the line s. If we carry out the dot product of $\boldsymbol{C}$ and $\boldsymbol{s}$ according to our fundamental definition, the result is

$$\boldsymbol{C} \bullet \boldsymbol{s} = |\boldsymbol{C}|\,|\boldsymbol{s}| \cos \delta$$

But since $|\boldsymbol{s}|$ is unity, when we compare the preceding two equations, it is apparent that

$$C_s = \boldsymbol{C} \bullet \boldsymbol{s}$$

Similarly, the following useful relations are valid:

$$C_x = \boldsymbol{C} \bullet \boldsymbol{i}, \quad C_y = \boldsymbol{C} \bullet \boldsymbol{j}, \quad C_z = \boldsymbol{C} \bullet \boldsymbol{k}$$

Finally, express the unit vector $\hat{\boldsymbol{r}}$ directed out from the origin (see Fig. 2.23) in terms of the orthogonal scalar components:

$$\hat{\boldsymbol{r}} = (\hat{\boldsymbol{r}} \bullet \boldsymbol{i})\boldsymbol{i} + (\hat{\boldsymbol{r}} \bullet \boldsymbol{j})\boldsymbol{j} + (\hat{\boldsymbol{r}} \bullet \boldsymbol{k})\boldsymbol{k}$$

But

$$\hat{\boldsymbol{r}} \bullet \boldsymbol{i} = |\hat{\boldsymbol{r}}|\,|\boldsymbol{i}| \cos(\hat{\boldsymbol{r}}, x) = l$$

Similarly, $\hat{\boldsymbol{r}} \bullet \boldsymbol{j} = m$ and $\hat{\boldsymbol{r}} \bullet \boldsymbol{k} = n$. Hence, we can say that

$$\hat{\boldsymbol{r}} = l\boldsymbol{i} + m\boldsymbol{j} + n\boldsymbol{k} \tag{2.19}$$

Thus, *the orthogonal scalar components of a unit vector are the direction cosines of the direction of the unit vector.* Now, computing the square of the magnitude of $\hat{\boldsymbol{r}}$, we have

$$|\hat{\boldsymbol{r}}|^2 = 1 = l^2 + m^2 + n^2 \tag{2.20}$$

We thus arrive at the familiar geometrical relation that the sum of the squares of the direction cosines of a vector is unity.

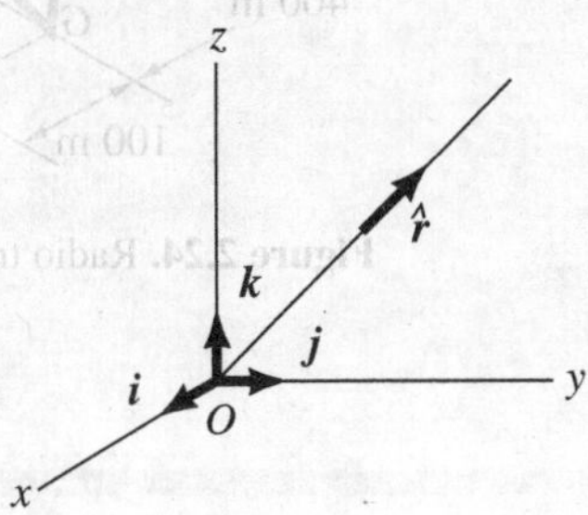

Figure 2.23. Unit vector $\hat{\boldsymbol{r}}$ directed from O.

Example 2.5

Cables *GA* and *GB* (see Fig. 2.24) are part of a guy-wire system supporting two radio transmission towers. What are the lengths of *GA* and *GB* and the angle α between them?

We may directly set up the vectors $\vec{GA}$ and $\vec{GB}$ by inspecting the diagram. Thus on moving along the coordinate directions, it is easy to see that

$$\vec{GA} = 300\boldsymbol{j} - 400\boldsymbol{i} + 500\boldsymbol{k} \text{ m}$$

$$\vec{GB} = 300\boldsymbol{j} + 100\boldsymbol{i} + 500\boldsymbol{k} \text{ m}$$

Using the Pythagorean theorem, we can say for the lengths of $\vec{GA}$ and $\vec{GB}$:

$$GA = (300^2 + 400^2 + 500^2)^{1/2} = 707 \text{ m}$$
$$GB = (300^2 + 100^2 + 500^2)^{1/2} = 592 \text{ m}$$

Now we use the dot product definition to find the angle.

$$\vec{GA} \bullet \vec{GB} = (GA)(GB) \cos \alpha$$

Therefore,

$$\cos \alpha = \frac{\vec{GA} \bullet \vec{GB}}{(GA)(GB)} = \frac{90{,}000 - 40{,}000 + 250{,}000}{(707)(592)}$$
$$= .717$$

Hence,

$$\alpha = 44.18°$$

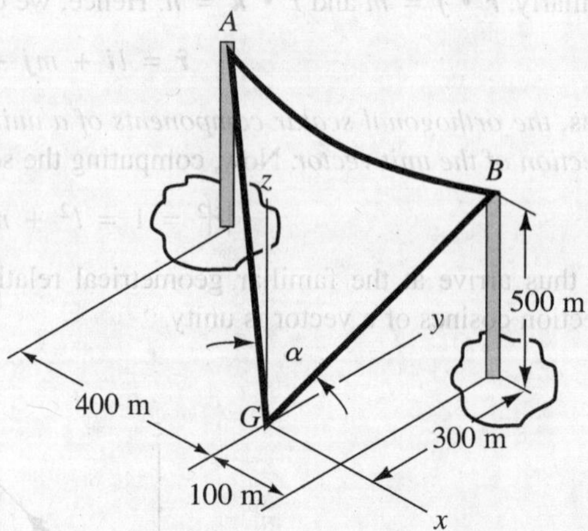

Figure 2.24. Radio transmission towers.

PROBLEMS

2.46. Given the vectors

$$A = 10i + 20j + 3k$$
$$B = -10j + 12k$$

what is $A \cdot B$? What is $\cos(A, B)$? What is the projection of A along B?

2.47. Given the vectors

$$A = 16i + 3j, \quad B = 10k - 6i, \quad C = 4j$$

compute

(a) $C(A \cdot C) + B$
(b) $-C + [B \cdot (-A)]C$

2.48. Given the vectors

$$A = 6i + 3j + 10k$$
$$B = 2i - 5j + 5k$$
$$C = 5i - 2j + 7k$$

what vector D gives the following results?

$$D \cdot A = 20$$
$$D \cdot B = 5$$
$$D \cdot i = 10$$

2.49. A sailboat is tacking into a 20-knot wind. The boat has a velocity component along its axis of 6 kn but because of side slip and water currents, it has a speed a speed of .2 kn at right angles to its axis. What are the x and y components of the wind velocity and the boat velocity? What is the angle between the wind velocity and the sailboat velocity?

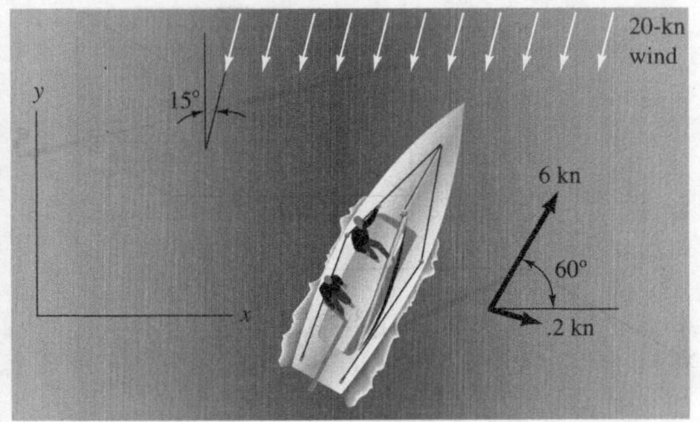

Figure P.2.49

2.50. Show that

$$\cos(A, B) = ll' + mm' + nn'$$

where l, m, n and l', m', n' are direction cosines of A and B, respectively, with respect to the given xyz reference.

2.51. Explain why the following operations are meaningless:

(a) $(A \cdot B) \cdot C$
(b) $(A \cdot B) + C$

2.52. A block A is constrained to move along a 20° incline in the yz plane. How far does the block have to move if the force F is to do 10 ft-lb of work?

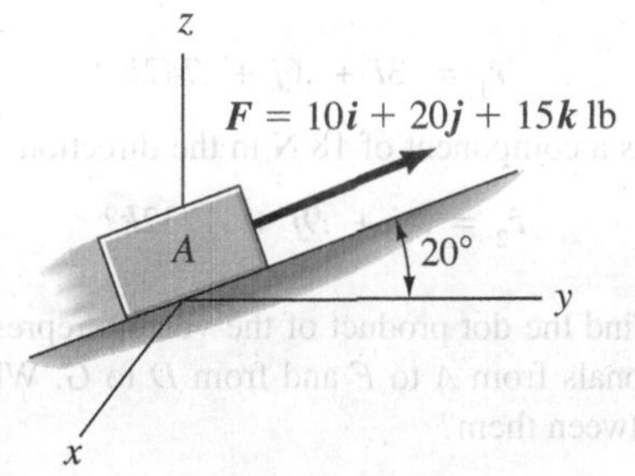

Figure P.2.52

2.53. An electrostatic field E exerts a force on a charged particle of qE, where q is the charge of the particle. If we have for E:

$$E = 6i + 3j + 2k \text{ dynes/coulomb}$$

what work is done by the field on a particle with a unit charge moving along a straight line from the origin to position $x = 20$ mm, $y = 40$ mm, $z = -40$ mm?

2.54. A force vector of magnitude 100 N has a line of action with direction cosines $l = .7$, $m = .2$, $n = .59$ relative to a reference xyz. The vector points away from the origin. What is the component of the force vector along a direction a having direction cosines $l = -.3$, $m = .1$, and $n = .95$ for the xyz reference? (*Hint:* Whenever simply a component is asked for, it is virtually always the *rectangular* component that is desired.)

2.55. What is the angle between the 1,000-N force and the axis AB? The force is in the diagonal plane $GCDE$.

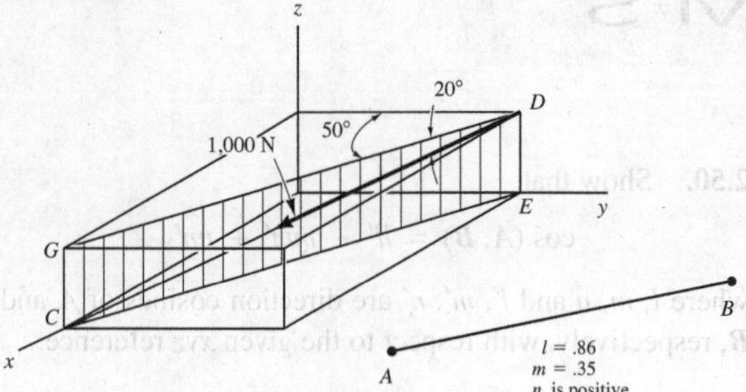

Figure P.2.55

2.56. Given a force $\boldsymbol{F} = 10\boldsymbol{i} + 5\boldsymbol{j} + A\boldsymbol{k}$ N. If this force is to have a rectangular component of 8 N along a line having a unit vector $\hat{\boldsymbol{r}} = .6\boldsymbol{i} + .8\boldsymbol{k}$, what should A be? What is the angle between $\boldsymbol{F}$ and $\hat{\boldsymbol{r}}$?

2.57. Given a force $A\boldsymbol{i} + B\boldsymbol{j} + 20\boldsymbol{k}$ N, what must A and B be to give a rectangular component of 10 N in the direction

$$\hat{\boldsymbol{r}}_1 = .3\boldsymbol{i} + .6\boldsymbol{j} + .742\boldsymbol{k}$$

as well as a component of 18 N in the direction

$$\hat{\boldsymbol{r}}_2 = .4\boldsymbol{i} + .9\boldsymbol{j} + .1732\boldsymbol{k}?$$

2.58. Find the dot product of the vectors represented by the diagonals from A to F and from D to G. What is the angle between them?

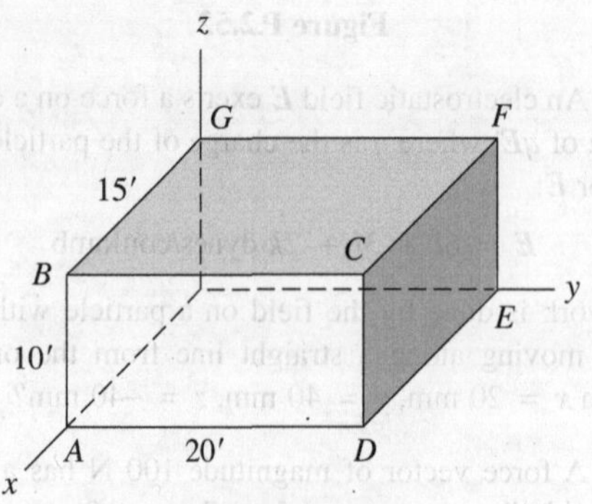

Figure P.2.58

2.59. A force $\boldsymbol{F}$ is given as

$$\boldsymbol{F} = 800\boldsymbol{i} + 600\boldsymbol{j} - 1{,}000\boldsymbol{k} \text{ N}$$

What is the *rectangular component* along an axis A-A equally inclined to the positive x, y, and z axes?

2.60. What is the rectangular component of the 500-N force along the diagonal from B to A?

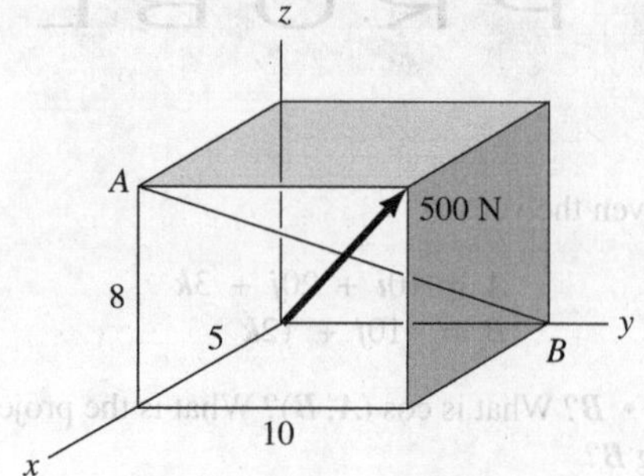

Figure P.2.60

2.61. A radio tower is held by guy wires. If AB were to be moved to intersect CD while remaining parallel to its original position, what is the angle between AB and CD?

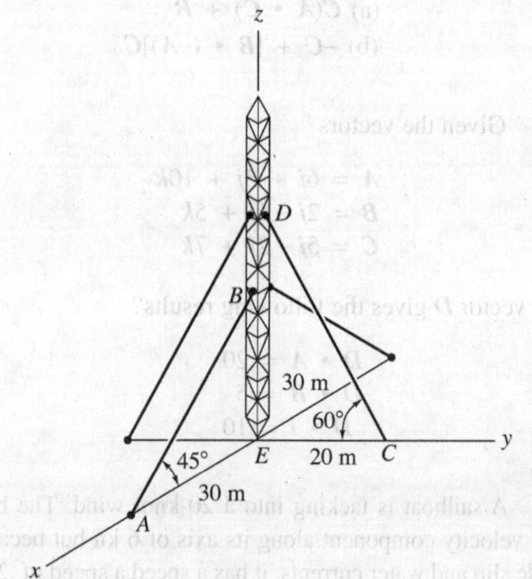

Figure P.2.61

2.62. What is the angle between the 1,000-N force and the position vector $\boldsymbol{r}$?

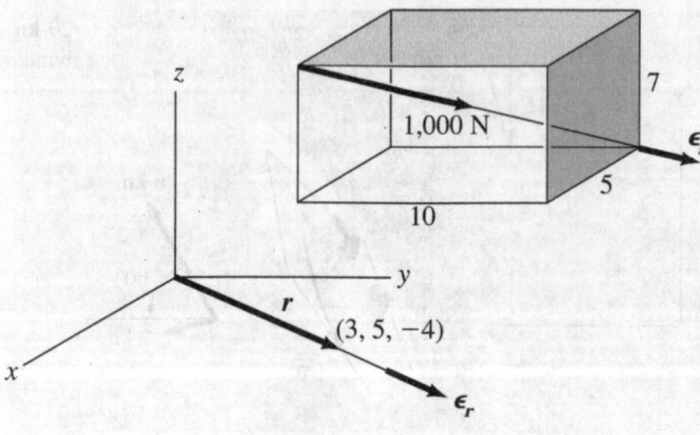

Figure P.2.62

2.8 Cross Product of Two Vectors

There are interactions between vector quantities that result in vector quantities. One such interaction is the moment of a force, which involves a special product of the force and a position vector (to be studied in Chapter 3). To set up a convenient operation for these situations, the *vector cross product* has been established. For the two vectors (having possibly different dimensions) shown in Fig. 2.25 as $\boldsymbol{A}$ and $\boldsymbol{B}$, the operation[11] is defined as

$$\boldsymbol{A} \times \boldsymbol{B} = \boldsymbol{C} \tag{2.21}$$

where $\boldsymbol{C}$ has a magnitude that is given as

$$|\boldsymbol{C}| = |\boldsymbol{A}|\,|\boldsymbol{B}| \sin \alpha \tag{2.22}$$

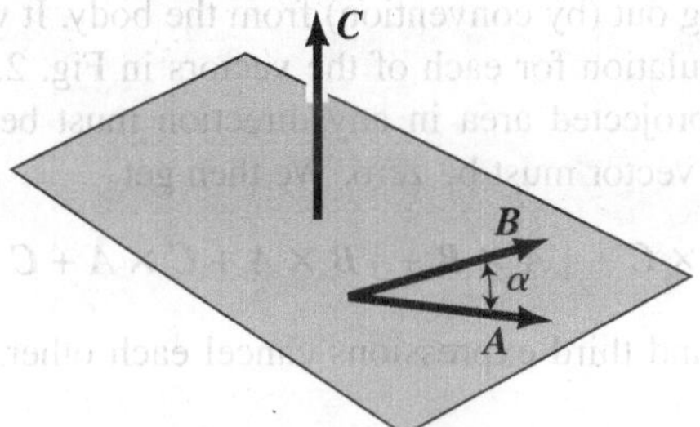

Figure 2.25. $\boldsymbol{A} \times \boldsymbol{B} = \boldsymbol{C}$.

The angle α is the smaller of the two angles between the vectors, thus making $\sin \alpha$ always positive. The vector $\boldsymbol{C}$ has an orientation normal to the plane of the vectors $\boldsymbol{A}$ and $\boldsymbol{B}$. The sense, furthermore, corresponds to the advance of a right-hand screw rotated about $\boldsymbol{C}$ as an axis while turning from $\boldsymbol{A}$ to $\boldsymbol{B}$ through α—that is, from the first stated vector to the second stated vector through the smaller angle between them. In Fig. 2.25, the screw would advance upward in rotating from $\boldsymbol{A}$ to $\boldsymbol{B}$, whether the procedure is viewed from above or below the plane formed by $\boldsymbol{A}$ and $\boldsymbol{B}$. The reader can easily verify this. The description of vector $\boldsymbol{C}$ is now complete, since the magnitude and direction are fully established. The line of action of $\boldsymbol{C}$ is not determined by the cross product; it depends on the use of the vector $\boldsymbol{C}$.

Again we remind you that the cross product, like the other vector algebraic operations, does not involve lines of action, so in taking a cross product we can move the vectors so as to come together at their tails as in Fig. 2.25.

As in the previous case, the coefficients of the vectors will multiply as ordinary scalars. This may be deduced from the nature of the definition. However, the *commutative* law breaks down for this product. We can verify, by carefully considering the definition of the cross product, that

$$(\boldsymbol{A} \times \boldsymbol{B}) = -(\boldsymbol{B} \times \boldsymbol{A}) \tag{2.23}$$

We can readily show that the cross product, like the dot product, is a distributive operation. To do this, consider in Fig. 2.26 a prism *mnopqr* with edges coinciding with the vectors $\boldsymbol{A}$, $\boldsymbol{B}$, $\boldsymbol{C}$, and $(\boldsymbol{A} + \boldsymbol{B})$. We can represent the area of each face of the prism as a vector whose magnitude equals the area of the face and whose direction is normal to the

[11]Again, we urge you to read $\boldsymbol{A} \times \boldsymbol{B} = \boldsymbol{C}$ as "$\boldsymbol{A}$ crossed into $\boldsymbol{B}$ yields $\boldsymbol{C}$."

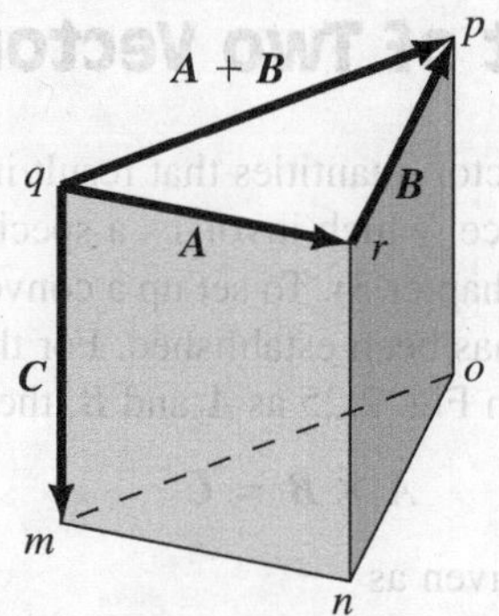

Figure 2.26. Prism using *A*, *B*, and *C*.

face with a sense pointing out (by convention) from the body. It will be left to the student to justify the given formulation for each of the vectors in Fig. 2.27. Since the prism is a closed surface, the net projected area in any direction must be zero, and this, in turn, means that the total area vector must be zero. We then get

$$(\boldsymbol{A}+\boldsymbol{B})\times\boldsymbol{C}+\tfrac{1}{2}\boldsymbol{A}\times\boldsymbol{B}+\tfrac{1}{2}\boldsymbol{B}\times\boldsymbol{A}+\boldsymbol{C}\times\boldsymbol{A}+\boldsymbol{C}\times\boldsymbol{B}=\boldsymbol{0}$$

Noting that the second and third expressions cancel each other, we get, on rearranging the terms,

$$\boldsymbol{C}\times(\boldsymbol{A}+\boldsymbol{B})=\boldsymbol{C}\times\boldsymbol{A}+\boldsymbol{C}\times\boldsymbol{B} \tag{2.24}$$

We have thus demonstrated the *distributive* property of the cross product.

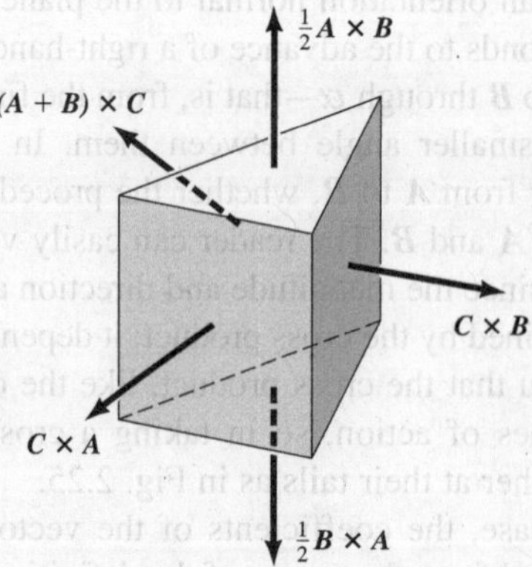

Figure 2.27. Area vectors for prism faces.

Next, consider the cross product of rectangular unit vectors. Here, the product of equal vectors is zero because α and, consequently, $\sin\alpha$ are zero. The product $\boldsymbol{i}\times\boldsymbol{j}$ is unity in magnitude, and because of the right-hand-screw rule must be parallel to the z axis. If the z axis has been erected in a sense consistent with the right-hand-screw rule when rotating from the x to the y direction, the reference is called a *right-hand triad* [see Fig. 2.28(a)] and we can write

$$\boldsymbol{i}\times\boldsymbol{j}=\boldsymbol{k}$$

If a left-hand triad is used, the result is a $-\boldsymbol{k}$ for the cross product above [see Fig. 2.28(b)]. In this text, we will use a right-hand triad as a reference. For ease in evaluation of unit cross products for

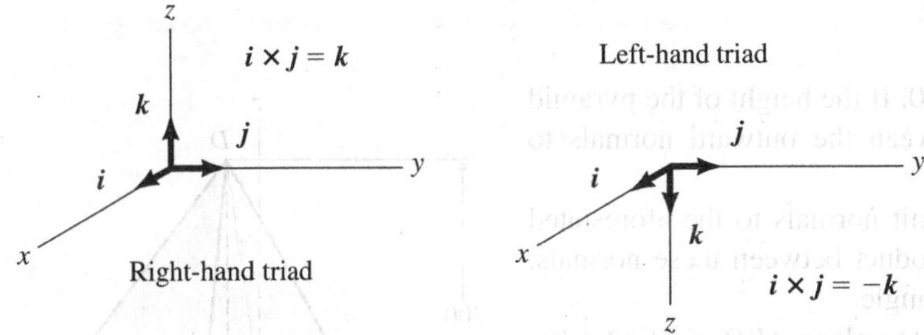

Figure 2.28. Different kinds of references.

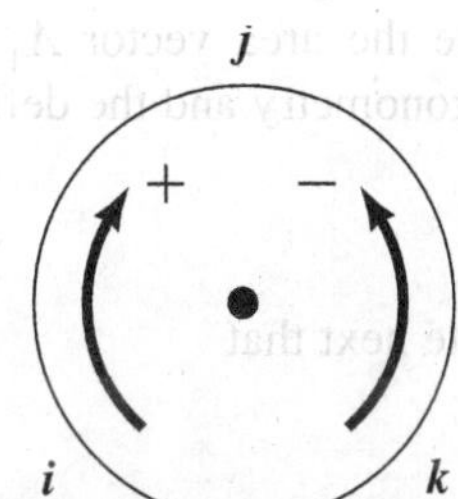

Figure 2.29. Permutation scheme.

such references, a simple permutation scheme is helpful. In Fig. 2.29, the unit vectors $\boldsymbol{i}$, $\boldsymbol{j}$, and $\boldsymbol{k}$ are indicated on a circle in a clockwise sequence. Any cross product of a pair of unit vectors results in a positive third unit vector if going from the first vector to the second vector involves a clockwise motion on this circle. Otherwise, the vector is negative. Thus,

$$\boldsymbol{k} \times \boldsymbol{j} = -\boldsymbol{i}, \quad \boldsymbol{k} \times \boldsymbol{i} = \boldsymbol{j}, \quad \text{etc.}$$

Next, the cross product of two vectors in terms of their rectangular components is

$$\begin{aligned} \boldsymbol{A} \times \boldsymbol{B} &= (A_x\boldsymbol{i} + A_y\boldsymbol{j} + A_z\boldsymbol{k}) \times (B_x\boldsymbol{i} + B_y\boldsymbol{j} + B_z\boldsymbol{k}) \\ &= (A_yB_z - A_zB_y)\boldsymbol{i} + (A_zB_x - A_xB_z)\boldsymbol{j} + (A_xB_y - A_yB_x)\boldsymbol{k} \end{aligned} \tag{2.25}$$

Another method of carrying out this long computation is to evaluate the following determinant:

$$\begin{vmatrix} A_x & A_y & A_z \\ B_x & B_y & B_z \\ \boldsymbol{i} & \boldsymbol{j} & \boldsymbol{k} \end{vmatrix} \tag{2.26}$$

The determinant may easily be evaluated in the following manner. Repeat the first two rows below the determinant, and then form products along diagonals.

$$\begin{array}{ccc} A_x & A_y & A_z \\ B_x & B_y & B_z \\ \boldsymbol{i} & \boldsymbol{j} & \boldsymbol{k} \\ A_x & A_y & A_z \\ B_x & B_y & B_z \end{array} \tag{2.27}$$

For the products along the dashed diagonals, we must remember in this method to multiply by −1. We then add all six products as follows:

$$\begin{aligned} A_xB_y\boldsymbol{k} + B_xA_z\boldsymbol{j} + A_yB_z\boldsymbol{i} - A_zB_y\boldsymbol{i} - B_zA_x\boldsymbol{j} - A_yB_x\boldsymbol{k} \\ = (A_yB_z - A_zB_y)\boldsymbol{i} + (A_zB_x - A_xB_z)\boldsymbol{j} + (A_xB_y - A_yB_x)\boldsymbol{k} \end{aligned}$$

Clearly, this is the same result as in Eq. 2.25. It must be cautioned that this method of evaluating a determinant is correct only for 3 × 3 determinants. If the cross product of two vectors involves less than six nonzero components, such as in the cross product

$$(6\boldsymbol{i} + 10\boldsymbol{j}) \times (5\boldsymbol{j} - 3\boldsymbol{k})$$

then it is advisable to multiply the components directly and collect terms, as in Eq. 2.25.

Example 2.6

A pyramid is shown in Fig. 2.30. If the height of the pyramid is 300 ft, find the angle between the outward normals to planes *ADB* and *BDC*.[12]

We shall first find the unit normals to the aforestated planes. Then, using the dot product between these normals, we can easily find the desired angle.

To get the unit normal $\boldsymbol{n}_1$ to plane *ABD*, we first compute the area vector $\boldsymbol{A}_1$ for this plane. Thus, from simple trigonometry and the definition of the cross product,

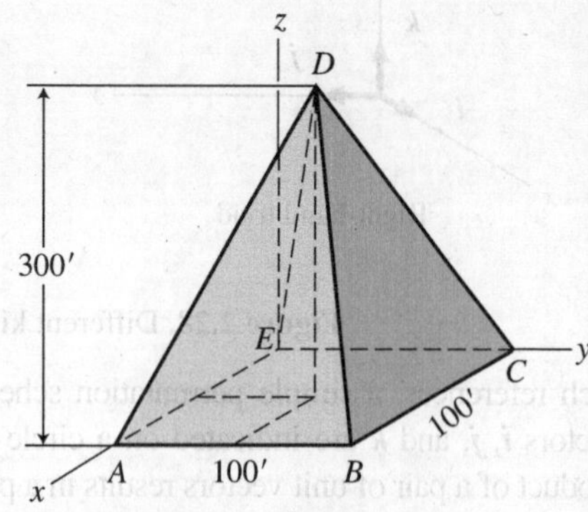

Figure 2.30. Pyramid.

$$\boldsymbol{A}_1 = \tfrac{1}{2}\,\overrightarrow{AB} \times \overrightarrow{AD}$$

Note next that

$$\overrightarrow{AB} = 100\boldsymbol{j}\text{ ft}$$

Furthermore, we can express $\overrightarrow{AD}$ in rectangular components by moving from *A* to *D* along coordinate directions as follows:

$$\overrightarrow{AD} = 50\boldsymbol{j} - 50\boldsymbol{i} + 300\boldsymbol{k}\text{ ft}$$

Hence,

$$\begin{aligned}\boldsymbol{A}_1 &= \tfrac{1}{2}(100\boldsymbol{j}) \times (-50\boldsymbol{i} + 50\boldsymbol{j} + 300\boldsymbol{k}) \\ &= 15{,}000\boldsymbol{i} + 2{,}500\boldsymbol{k}\text{ ft}^2\end{aligned}$$

Accordingly,

$$\begin{aligned}\boldsymbol{n}_1 &= \frac{\boldsymbol{A}_1}{|\boldsymbol{A}_1|} = \frac{15{,}000\boldsymbol{i} + 2{,}500\boldsymbol{k}}{\sqrt{15{,}000^2 + 2{,}500^2}} \qquad \text{(a)} \\ &= .9864\boldsymbol{i} + .1644\boldsymbol{k}\end{aligned}$$

As for unit normal $\boldsymbol{n}_2$ corresponding to plane *BDC*, whose area vector we denote as $\boldsymbol{A}_2$, we have

$$\boldsymbol{A}_2 = \tfrac{1}{2}\,\overrightarrow{BC} \times \overrightarrow{BD}$$

Note that

$$\overrightarrow{BC} = -100\boldsymbol{i}\text{ ft}$$

And once again, moving along coordinate directions, we have for $\overrightarrow{BD}$

$$\overrightarrow{BD} = -50\boldsymbol{j} - 50\boldsymbol{i} + 300\boldsymbol{k}\text{ ft}$$

[12] π minus this angle is the angle between these planes.

Example 2.6 (Continued)

Hence,

$$A_2 = \tfrac{1}{2}(-100\boldsymbol{i}) \times (-50\boldsymbol{i} - 50\boldsymbol{j} + 300\boldsymbol{k})$$
$$= 15{,}000\boldsymbol{j} + 2{,}500\boldsymbol{k}\ \text{ft}^2$$

Accordingly,

$$\boldsymbol{n}_2 = \frac{A_2}{|A_2|} = \frac{15{,}000\boldsymbol{j} + 2{,}500\boldsymbol{k}}{\sqrt{15{,}000^2 + 2{,}500^2}}$$
$$= .9864\boldsymbol{j} + .1644\boldsymbol{k} \qquad \text{(b)}$$

Now, we use the dot product of $\boldsymbol{n}_1$ and $\boldsymbol{n}_2$. Thus,

$$\boldsymbol{n}_1 \cdot \boldsymbol{n}_2 = \cos\beta \qquad \text{(c)}$$

where β is the angle between the normals to the planes. Substituting from Eqs. (a) and (b) into (c), we get

$$\cos\beta = .0270$$

Therefore,

$$\beta = 88.5°$$

We see from this example that a plane surface can be represented as a vector, and if that plane surface is part of a closed surface, by convention the area vector is in the direction of the outward normal.

2.9 Scalar Triple Product

A very useful quantity is the *scalar triple product,* which for a set of vectors $\boldsymbol{A}$, $\boldsymbol{B}$, and $\boldsymbol{C}$ is defined as

$$(\boldsymbol{A} \times \boldsymbol{B}) \cdot \boldsymbol{C} \qquad (2.28)$$

This clearly is a scalar quantity.

A simple geometric meaning can be associated with this operation. In Fig. 2.31, we have shown $\boldsymbol{A}$, $\boldsymbol{B}$, and $\boldsymbol{C}$ as an arbitrary set of concurrent vectors. We have set up an *xyz* reference such that the $\boldsymbol{A}$ and $\boldsymbol{B}$ vectors are in the *xy* plane. Further, a parallelogram *abcd* in the *xy* plane is shown in the diagram. We can say that

$$|\boldsymbol{A} \times \boldsymbol{B}| = |\boldsymbol{A}|\,|\boldsymbol{B}| \sin\alpha = \text{area of } abcd$$

Furthermore, the direction of $\boldsymbol{A} \times \boldsymbol{B}$ is in the z direction. Clearly, when we carry out Eq. 2.28, we are thus multiplying the scalar component of $\boldsymbol{C}$ in the z direction by the area of the aforementioned parallelogram. Thus, we have, for Eq. 2.28:

$$(\boldsymbol{A} \times \boldsymbol{B}) \bullet \boldsymbol{C} = (\text{area of } abcd)(C_z)$$

But C_z is the altitude of the parallelepiped formed by vectors $\boldsymbol{A}$, $\boldsymbol{B}$, and $\boldsymbol{C}$ (see Fig. 2.31). We then conclude from solid geometry that *the scalar triple product is the volume of the parallelepiped formed by the concurrent vectors of the scalar triple product.*

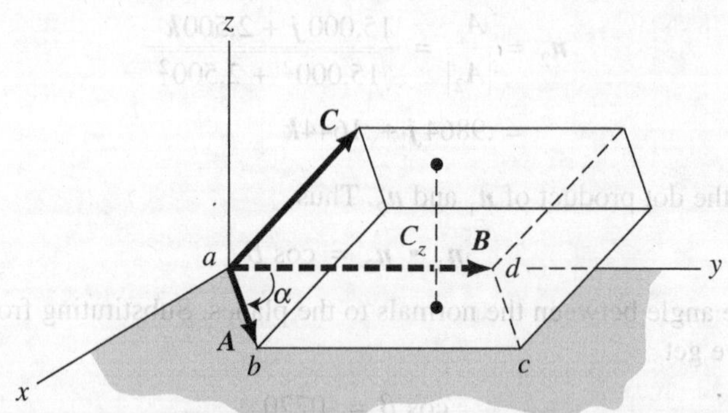

Figure 2.31. $\boldsymbol{A}$ and $\boldsymbol{B}$ in xy plane.

Using this geometrical interpretation of the scalar triple product, the reader can easily conclude that

$$(\boldsymbol{A} \times \boldsymbol{B}) \bullet \boldsymbol{C} = -(\boldsymbol{A} \times \boldsymbol{C}) \bullet \boldsymbol{B} = -(\boldsymbol{C} \times \boldsymbol{B}) \bullet \boldsymbol{A} \tag{2.29}$$

The computation of the scalar triple product is a very straightforward process. It will be left as an exercise (Problem 2.72) for you to demonstrate that

$$(\boldsymbol{A} \times \boldsymbol{B}) \bullet \boldsymbol{C} = \begin{vmatrix} A_x & A_y & A_z \\ B_x & B_y & B_z \\ C_x & C_y & C_z \end{vmatrix} \tag{2.30}$$

In later chapters, we shall employ the scalar triple product, although we shall not always want to associate the preceding geometric interpretation of this product.

Another operation involving three vectors is the *vector triple product* defined for vectors $\boldsymbol{A}$, $\boldsymbol{B}$, and $\boldsymbol{C}$ as $\boldsymbol{A} \times (\boldsymbol{B} \times \boldsymbol{C})$. The vector triple product is a vector quantity and will appear quite often in studies of dynamics. It will be left for you to demonstrate that

$$\boldsymbol{A} \times (\boldsymbol{B} \times \boldsymbol{C}) = \boldsymbol{B}(\boldsymbol{A} \bullet \boldsymbol{C}) - \boldsymbol{C}(\boldsymbol{A} \bullet \boldsymbol{B}) \tag{2.31}$$

Notice here that the vector triple product can be carried out by using only dot products.

Example 2.7

In Example 2.6, what is the area projected by plane *ADE* onto an infinite plane that is inclined equally to the *x*, *y*, and *z* axes?

The normal $\boldsymbol{n}$ to the infinite plane must have three equal direction cosines. Hence, noting Eq. 2.20 for the sum of the squares of a set of direction cosines, we can say that

$$l^2 = m^2 = n^2 = \tfrac{1}{3}$$

Therefore,

$$l = m = n = \frac{1}{\sqrt{3}}$$

Hence,

$$\boldsymbol{n} = \frac{1}{\sqrt{3}}\boldsymbol{i} + \frac{1}{\sqrt{3}}\boldsymbol{j} + \frac{1}{\sqrt{3}}\boldsymbol{k}$$

The projected area then is given as

$$\begin{aligned} A_n &= \left(\tfrac{1}{2}\,\overrightarrow{AD} \times \overrightarrow{AE}\right) \bullet \boldsymbol{n} \\ &= \left[\tfrac{1}{2}(-50\boldsymbol{i} + 50\boldsymbol{j} + 300\boldsymbol{k}) \times (-100\boldsymbol{i})\right] \bullet \frac{1}{\sqrt{3}}(\boldsymbol{i} + \boldsymbol{j} + \boldsymbol{k}) \end{aligned}$$

The preceding result is a scalar triple product that can readily be solved as follows (disregarding the final sign):

$$A_n = \frac{1}{2\sqrt{3}} \begin{vmatrix} -50 & 50 & 300 \\ -100 & 0 & 0 \\ 1 & 1 & 1 \end{vmatrix} \begin{matrix} -50 & 50 & 300 \\ -100 & 0 & 0 \end{matrix} = \boxed{7{,}217 \text{ ft}^2}$$

2.10 A Note on Vector Notation

When expressing *equations,* we must at all times clearly denote scalar and vector quantities and handle them accordingly. When we are simply identifying quantities in a *discussion* or in a *diagram,* however, instead of using the vector representation, $\boldsymbol{F}$, we can just use F. On the other hand, F will be understood to represent in an equation the magnitude of the vector $\boldsymbol{F}$. Thus, using $\boldsymbol{f}$ as the unit vector in the direction of $\boldsymbol{F}$, we can then say:

$$\begin{aligned}\boldsymbol{F} &= F\boldsymbol{f} \\ &= F[\cos(\boldsymbol{F}, x)\boldsymbol{i} + \cos(\boldsymbol{F}, y)\boldsymbol{j} + \cos(\boldsymbol{F}, z)\boldsymbol{k}]\end{aligned}$$

As another example, we might want to employ the force F, which is shown in the coplanar diagram of Fig. 2.32(a) at a known inclination and acting at a point a. A correct representation of this force in a vector equation would be $F(-\cos\alpha\boldsymbol{i} + \sin\alpha\boldsymbol{j})$.

As for scalar components of any vector $\boldsymbol{F}$, we shall adopt the following understanding. The notation F_x, F_y, or F_z labeling some vector component in a *diagram* will be understood to represent the *magnitude* of that particular component. Thus, in Fig. 2.32(b) the two components shown are equal in magnitude but opposite in sense. Nevertheless, they are both labeled F_x. However, in an equation involving these quantities, the sense must properly be accounted for by the appropriate use of signs.

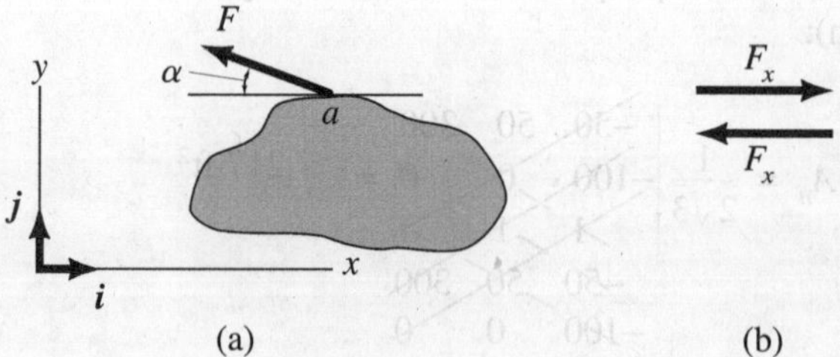

Figure 2.32. Notation in diagrams.

PROBLEMS

2.63. If $A = 10i + 6j - 3k$ and $B = 6i$, find $A \times B$ and $B \times A$. What is the magnitude of the resulting vector? What are its direction cosines relative to the xyz reference in which A and B are expressed?

2.64. What are the cross and dot products for the vectors A and B given as:

$$A = 6i + 3j + 4k$$
$$B = 8i - 3j + 2k?$$

2.65. If vectors A and B in the xy plane have a dot product of 50 units, and if the magnitudes of these vectors are 10 units and 8 units, respectively, what is $A \times B$?

2.66. (a) If $A \cdot B = A \cdot B'$, does B necessarily equal B'? Explain.
(b) If $A \times B = A \times B'$, does B necessarily equal B'? Explain.

2.67. What is the cross product of the displacement vector from A to B times the displacement vector from C to D?

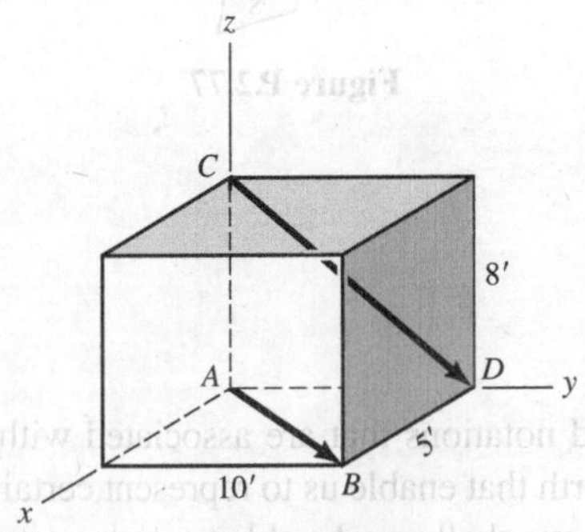

Figure P.2.67

2.68. Making use of the cross product, give the unit vector n normal to the inclined surface ABC.

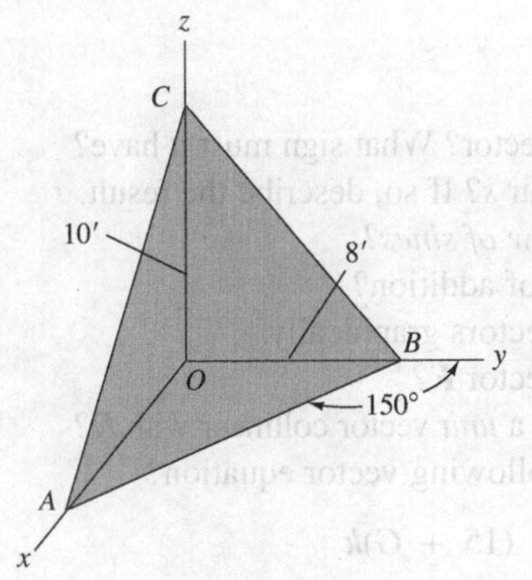

Figure P.2.68

2.69. If the coordinates of vertex E of the inclined pyramid are (5, 50, 80) m, what is the angle between outward normals to faces ADE and BCE?

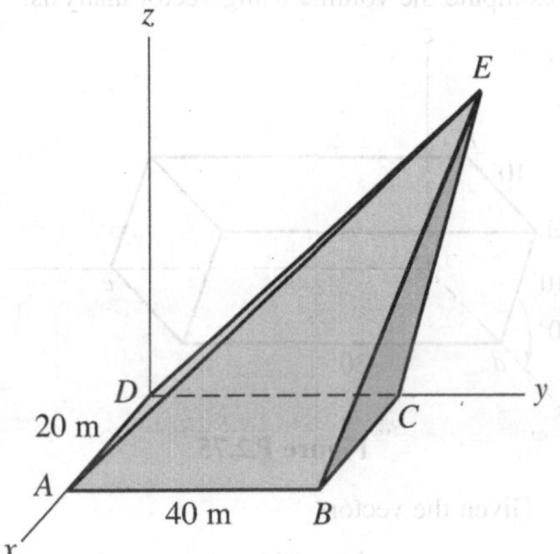

Figure P.2.69

2.70. In Problem 2.69, what is the area of face ADE of the pyramid? What is the projection of the area of face ADE onto a plane whose normal is along the direction ϵ where:

$$\epsilon = 0.6i - 0.8j$$

2.71. (a) Compute the product

$$(A \times B) \cdot C$$

in terms of orthogonal components.
(b) Compute $(C \times A) \cdot B$ and compare with the result in part (a).

2.72. Compute the determinant

$$\begin{vmatrix} A_x & A_y & A_z \\ B_x & B_y & B_z \\ C_x & C_y & C_z \end{vmatrix}$$

where each row represents, respectively, the scalar components of A, B, and C. Compare the result with the computation of $(A \times B) \cdot C$ by using the dot-product and cross-product operations.

2.73. In Example 2.5, what is the area vector for GAB assuming a straight line connects points A and B? Give the results in kilometers squared.

2.74. What is the component of the cross product $\boldsymbol{A} \times \boldsymbol{B}$ along the direction $\boldsymbol{n}$, where

$$\boldsymbol{A} = 10\boldsymbol{i} + 16\boldsymbol{j} + 3\boldsymbol{k}$$
$$\boldsymbol{B} = 5\boldsymbol{i} - 2\boldsymbol{j} + 2\boldsymbol{k}$$
$$\boldsymbol{n} = 0.8\boldsymbol{i} + 0.6\boldsymbol{k}$$

2.75. The surface *abcd* of the parallelepiped is in the *xz* plane. Compute the volume using vector analysis.

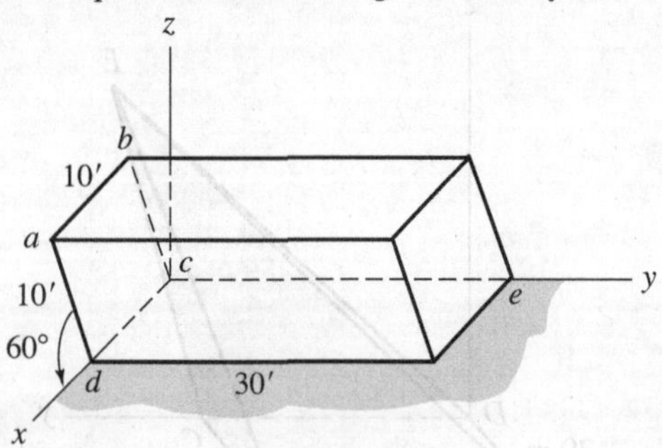

Figure P.2.75

2.76. Given the vectors

$$\boldsymbol{A} = 10\boldsymbol{i} + 6\boldsymbol{j}$$
$$\boldsymbol{B} = 3\boldsymbol{i} + 5\boldsymbol{j} + 10\boldsymbol{k}$$
$$\boldsymbol{C} = \boldsymbol{i} + \boldsymbol{j} - 3\boldsymbol{k}$$

find

(a) $(\boldsymbol{A} + \boldsymbol{B}) \times \boldsymbol{C}$
(b) $(\boldsymbol{A} \times \boldsymbol{B}) \cdot \boldsymbol{C}$
(c) $\boldsymbol{A} \cdot (\boldsymbol{B} \times \boldsymbol{C})$

***2.77.** A mirror system is used to relay a laser-beam signal from mountain *M* to hill *H*. The mountain is 5,000 m high and 20,000 m NW from the mirror site *S*, while the hill is 200 m high and 15,000 m ENE from the mirror site *S*. Set up, but do not necessarily solve, the equations to find the direction of the mirror to properly relay the signal. Recall that the angle of reflection for a mirror equals the angle of incidence and that the incident ray, reflecting ray, and normal to the mirror are coplanar.

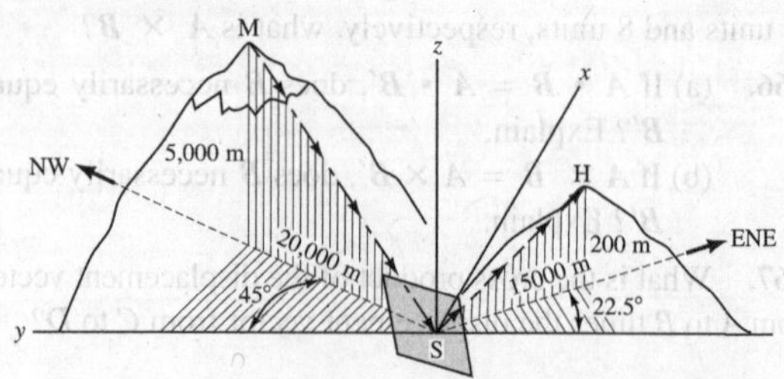

Figure P.2.77

2.11 Closure

In this chapter, we have presented symbols and notations that are associated with vectors. Also, various vector operations have been set forth that enable us to represent certain actions in nature mathematically. With this background, we shall now be able to study certain vector quantities that are of essential importance in mechanics. Some of these vectors will be formulated in terms of the operations contained in this chapter.

Check-Out for Sections with †

2.1 What is meant by the *magnitude* of a vector? What sign must it have?
2.2 Can you multiply a vector $\boldsymbol{C}$ by a scalar s? If so, describe the result.
2.3 What are the *law of cosines* and the *law of sines*?
2.4 What is meant by the *associative* law of addition?
2.5 Describe two ways to add any three vectors graphically.
2.6 How do you subtract vector $\boldsymbol{D}$ from vector $\boldsymbol{F}$?
2.7 Given a vector $\boldsymbol{D}$, how would you form a *unit* vector collinear with $\boldsymbol{D}$?
2.8 What are the scalar equations of the following vector equation?

$$D\boldsymbol{i} + E\boldsymbol{j} - 16\boldsymbol{k} = 20\boldsymbol{i} + (15 + G)\boldsymbol{k}$$

PROBLEMS

2.78. Flight 304 from Dallas is flying NE to Chicago 900 miles away. To avoid a massive storm front, the pilot decides instead to fly due north to Topeka, Kansas, and then ENE (see Fig. P.2.4 for compass settings) to Chicago. What are the distances that he must travel from Dallas to Topeka and from Topeka to Chicago?

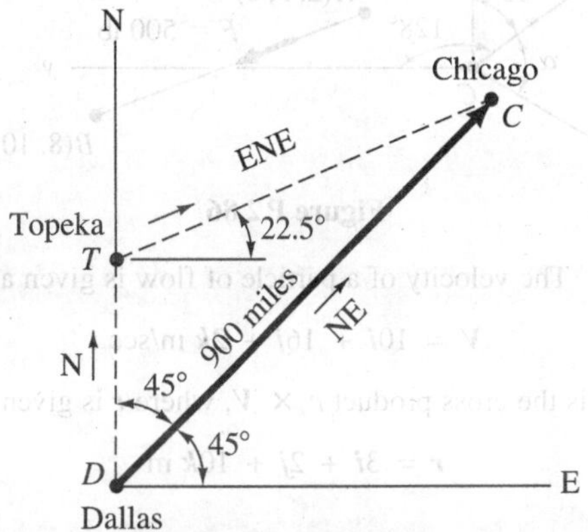

Figure P.2.78

2.79. What is the cross product between the 1,000-N force and the displacement vector $\boldsymbol{\rho}_{AB}$?

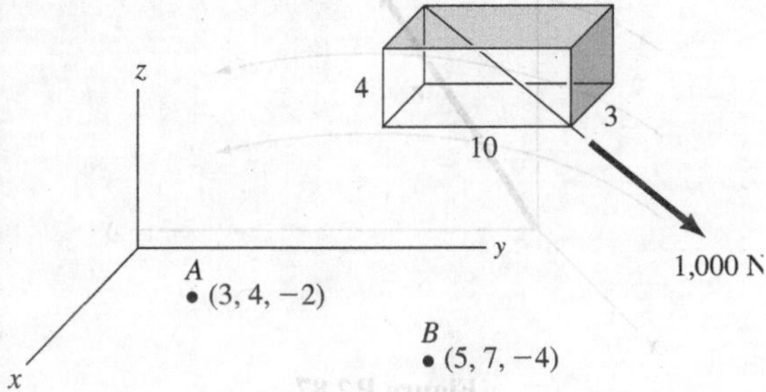

Figure P.2.79

2.80. Four members of a space frame are loaded as shown. What are the orthogonal scalar components of the forces on the ball joint at O? The 1,000-N force goes through points D and E of the rectangular parallelepiped.

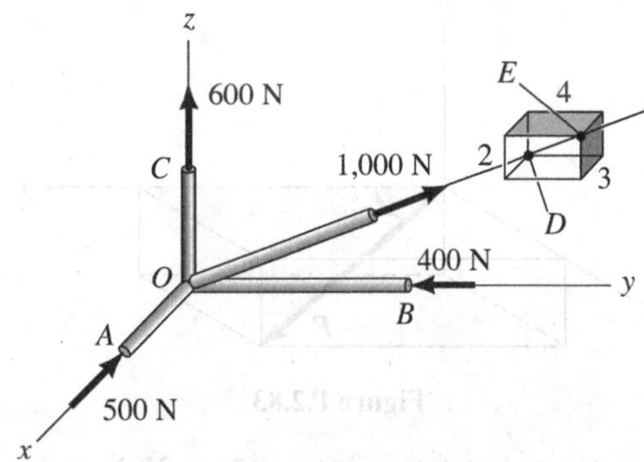

Figure P.2.80

2.81. Contractors encountered an impassable swamp while building a road from town T to city C 50 km SE. To avoid the swamp, they built the road SSW from T and then ENE to C. How long is the road? (*Hint:* See the compass-settings diagram, Fig. P.2.4.)

2.82. Sum all forces acting on the block. Plane A is parallel to the xy plane. We will later study the special properties of two parallel forces (called a *couple*) that are opposite in direction and equal in magnitude.

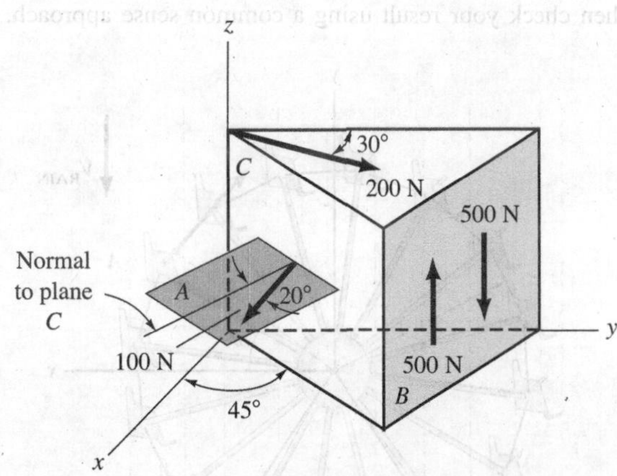

Figure P.2.82

2.83. The x and z components of the force F are known to be 100 lb and −30 lb, respectively. What is the force F and what are its direction cosines?

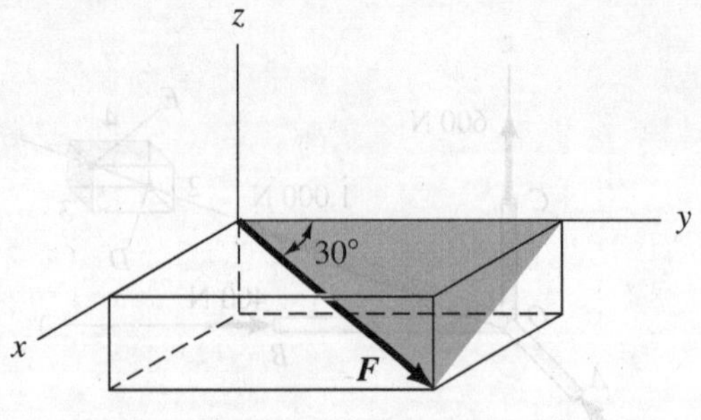

Figure P.2.83

2.84. A constant force given as $2\boldsymbol{i} + 3\boldsymbol{k}$ N moves a particle along a straight line from position $x = 10$, $y = 20$, $z = 0$ to position $x = 3$, $y = 0$, $z = -10$. If the coordinates of the xyz reference are given in meter units, how much work does the force do in ft-lb?

2.85. Rain is falling on a ferris wheel. The velocity of the rain is constant having reached a terminal velocity of 5 ft/sec. The angular speed of the ferris wheel is constant at 0.5 RPM. At what position θ will the angle between the velocity of the occupant at A and that of the rain drops be 158°? You will recall from freshman physics that the speed of a particle in circular motion is $r\omega$ with ω in radians per unit time. First use the dot-product approach and then check your result using a common sense approach.

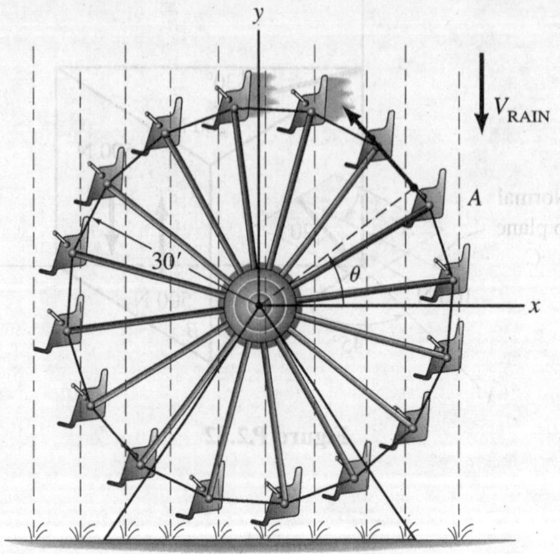

Figure P.2.85

2.86. A force $F = 500$ lb has a line of action that goes through points A and B. What is the angle δ between $\boldsymbol{F}$ and the displacement vector $\boldsymbol{\rho}_{CD}$. What is the scalar *rectangular component* of the force along axis CD?

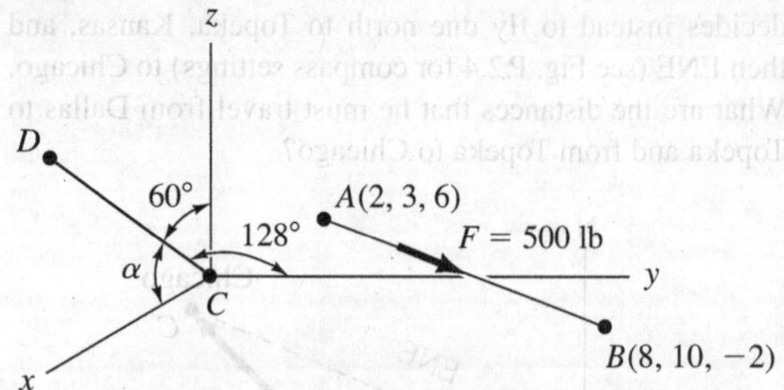

Figure P.2.86

2.87. The velocity of a particle of flow is given as

$$V = 10\boldsymbol{i} + 16\boldsymbol{j} + 2\boldsymbol{k} \text{ m/sec}$$

What is the cross product $\boldsymbol{r} \times \boldsymbol{V}$, where $\boldsymbol{r}$ is given as

$$\boldsymbol{r} = 3\boldsymbol{i} + 2\boldsymbol{j} + 10\boldsymbol{k} \text{ m}$$

Give the proper units.

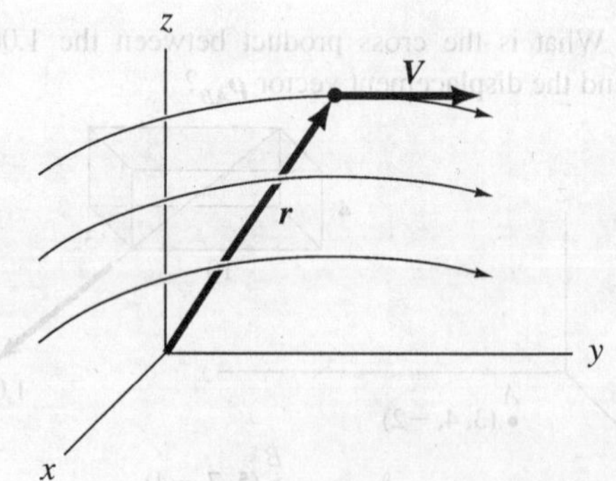

Figure P.2.87

2.88. A bridge truss has bar forces as shown in the cutaway sketch. What is the total force on the supporting pin at point A from the members?

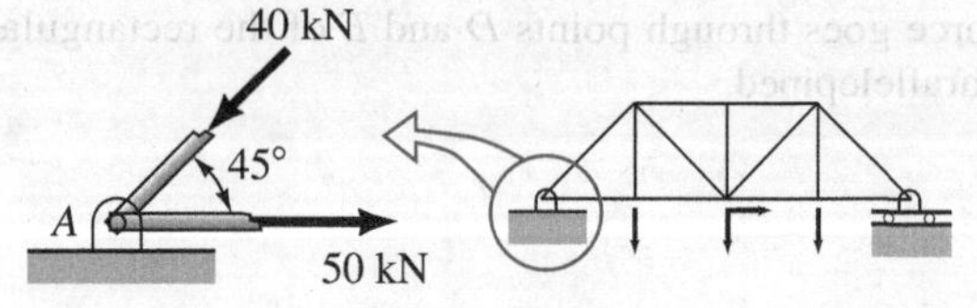

Figure P.2.88

2.89. Forces are transmitted by two members to pin A. If the sum of these forces is 700 lb directed vertically, what are the angles α and β?

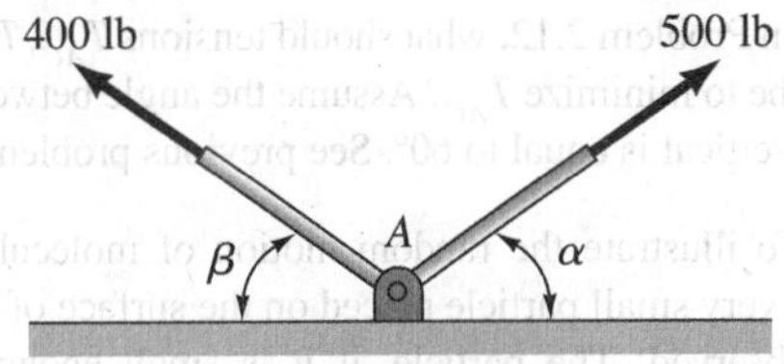

Figure P.2.89

2.90. A skeet shooter is aiming his gun at point A. What is the height z of point A?

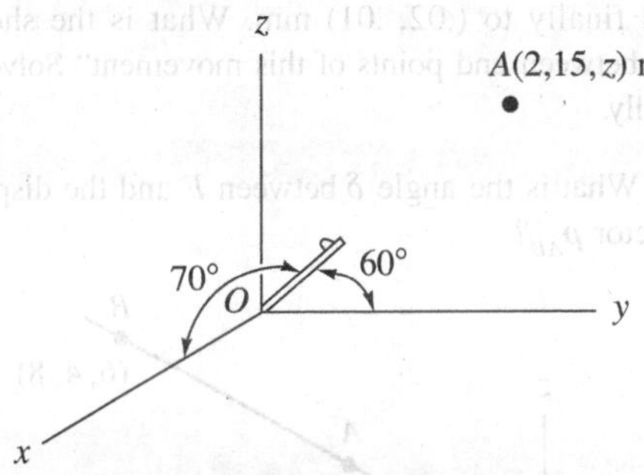

Figure P.2.90

2.91. If F_1 and the 500-N force sum vectorially to F_T, determine F_1 and F_T.

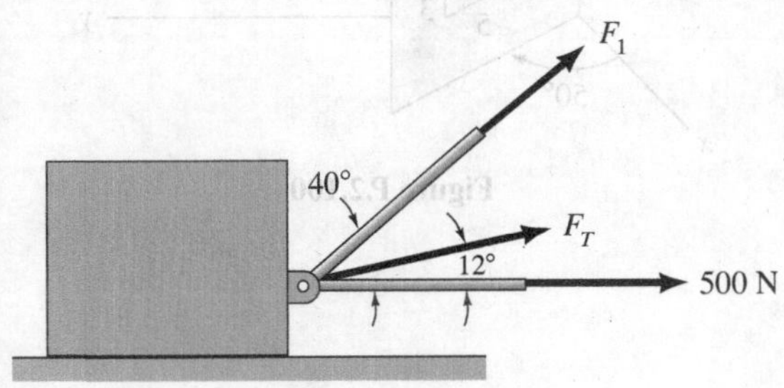

Figure P.2.91

2.92. The force on a charge moving through a magnetic field $\boldsymbol{B}$ is given as

$$\boldsymbol{F} = q\boldsymbol{V} \times \boldsymbol{B}$$

where q = magnitude of the charge, coulombs
$\boldsymbol{F}$ = force on the body, newtons
$\boldsymbol{V}$ = velocity vector of the particle, meters per second
$\boldsymbol{B}$ = magnetic flux density, webers per meter^2

Suppose that an electron moves through a uniform magnetic field of 10^6 Wb/m^2 in a direction inclined 30° to the field, as shown, with a speed of 100 m/sec. What are the force components on the electron? The charge of the electron is 1.6018×10^{-19} coulombs.

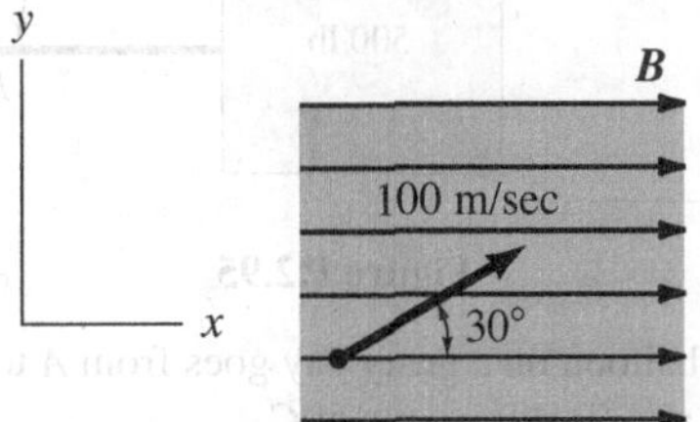

Figure P.2.92

2.93. For the line segment AB, determine z_B and direction cosines m and n.

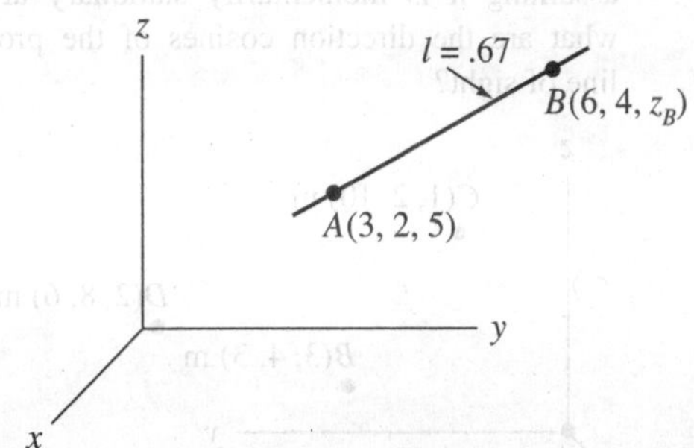

Figure P.2.93

2.94. Using the scalar triple product, find the area projected onto the plane N from the surface ABC. Plane N is infinite and is normal to the vector

$$\boldsymbol{r} = 50\boldsymbol{i} + 40\boldsymbol{j} + 30\boldsymbol{k} \text{ ft}$$

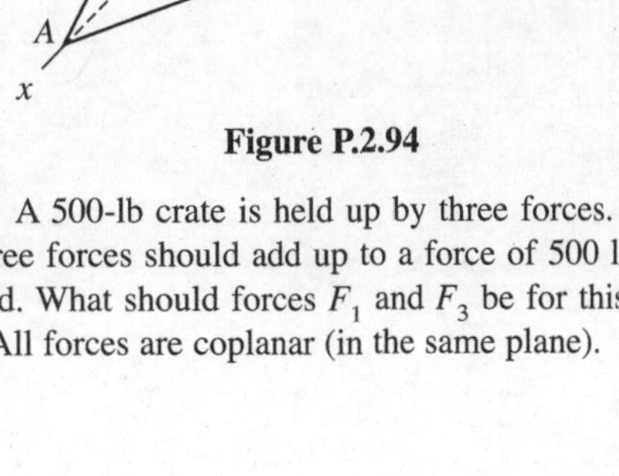

Figure P.2.94

2.95. A 500-lb crate is held up by three forces. Clearly the three forces should add up to a force of 500 lb going upward. What should forces F_1 and F_3 be for this condition? All forces are coplanar (in the same plane).

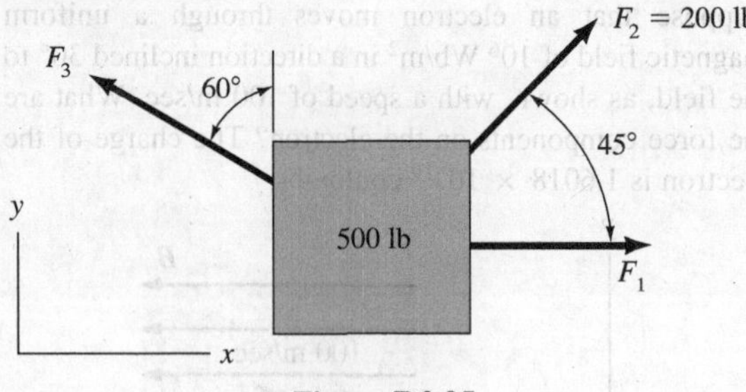

Figure P.2.95

2.96. A balloon on a gusty day goes from A to B, then is seen at D and finally is seen at C.

(a) If the balloon moves along straight lines, how far has it moved?

(b) If a bee-bee gun is to shoot down the balloon, assuming it is momentarily stationary at C, what are the direction cosines of the proper line of sight?

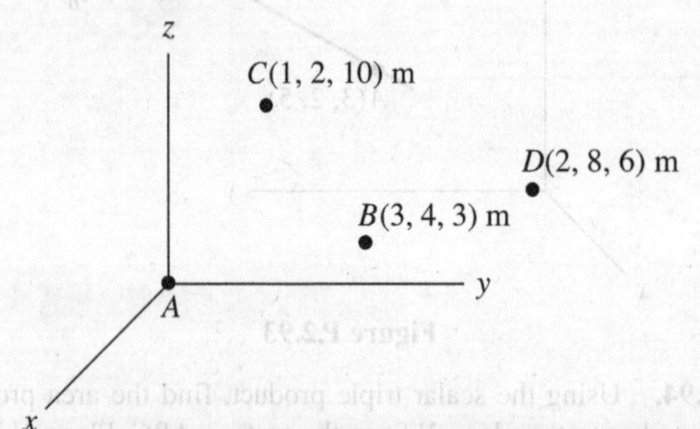

Figure P.2.96

2.97. In Problem 2.12 what should T_{AC}, T_{AB}, and the acute angle θ between T_{AB} and the vertical to minimize the tension T_{AB}? Take $\alpha = 36.5°$ and $W = 1{,}000$ N. *Hint:* Consider various possible force triangles and decide on inspection how to minimize T_{AB}.

2.98. In Problem 2.12, what should tensions T_{AC}, T_{AB}, and angle α be to minimize T_{AC}? Assume the angle between T_{AB} and the vertical is equal to 60°. See previous problem.

2.99. To illustrate the random motion of molecules in a liquid, a very small particle placed on the surface of the liquid is observed. The particle, if it is small enough, will jump around in a random manner giving rise to the *random walk* phenomenon described in your physics course. Suppose such a particle jumps from position (0,0) mm in the xy plane to (.01, −.03) mm to (−.01, −.02) mm to (.01, −.01) mm and finally to (.02, .01) mm. What is the shortest distance between end points of this movement? Solve this graphically.

2.100. What is the angle δ between F and the displacement vector $\boldsymbol{\rho}_{AB}$?

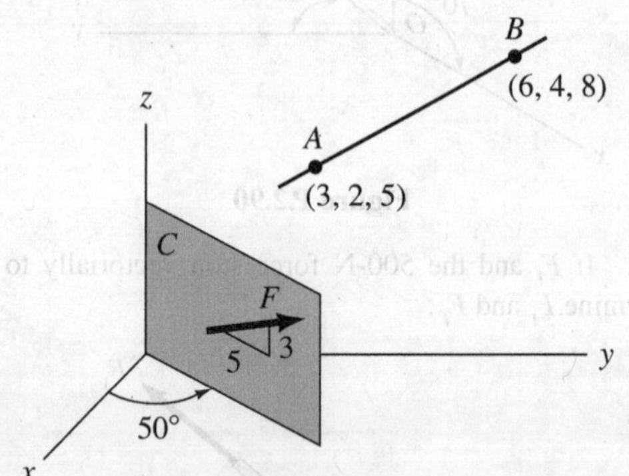

Figure P.2.100

CHAPTER 3

Systems of Forces

3.1 Position Vector

In this chapter, we shall discuss a number of useful vector quantities. Consider first the path of motion of a particle shown dashed in Fig. 3.1. As indicated in Chapter 1, the *displacement vector* $\boldsymbol{\rho}$ is a directed line segment connecting any two points on the path of motion, such as points 1 and 2 in Fig. 3.1. The displacement vector thus represents the shortest movement of the particle to get from one position on the path of motion to another. The purpose of the rectangular parallelepiped shown in the diagram is to convey the magnitude and direction of $\boldsymbol{\rho}$ as explained earlier. We can readily express $\boldsymbol{\rho}$ between points 1 and 2 in terms of rectangular components by noting the distance in the coordinate directions needed to go from 1 to 2. Thus, in Fig. 3.1, $\boldsymbol{\rho}_{12} = -2\boldsymbol{i} + 6\boldsymbol{j} + 3\boldsymbol{k}$ m.

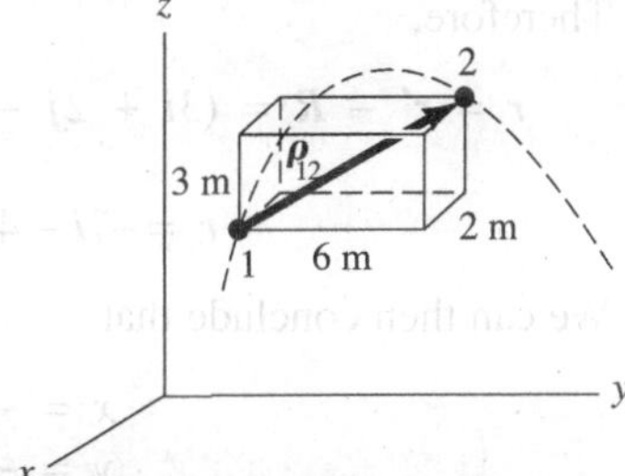

Figure 3.1. Displacement vector $\boldsymbol{\rho}$ between points 1 and 2.

The directed line segment $\boldsymbol{r}$ from the origin of a coordinate system to a point P in space (Fig. 3.2) is called the *position vector.* The notations $\boldsymbol{R}$ and $\boldsymbol{\rho}$ are also used for position vectors. You can conclude from Chapter 2 that the magnitude of the position vector is the distance between the origin O and point P. The scalar components of a position vector are simply the coordinates of the point P. To express $\boldsymbol{r}$ in Cartesian components, we then have

$$\boldsymbol{r} = x\boldsymbol{i} + y\boldsymbol{j} + z\boldsymbol{k} \tag{3.1}$$

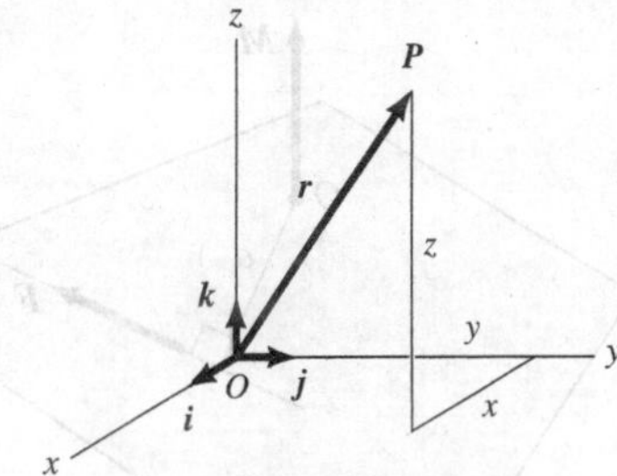

Figure 3.2. Position vector.

We can obviously express a displacement vector $\boldsymbol{\rho}$ between points 1 and 2 (see Fig. 3.3) in terms of position vectors for points 1 and 2 (i.e., $\boldsymbol{r}_1$ and $\boldsymbol{r}_2$) as follows:

$$\boldsymbol{\rho} = \boldsymbol{r}_2 - \boldsymbol{r}_1 = (x_2 - x_1)\boldsymbol{i} + (y_2 - y_1)\boldsymbol{j} + (z_2 - z_1)\boldsymbol{k} \tag{3.2}$$

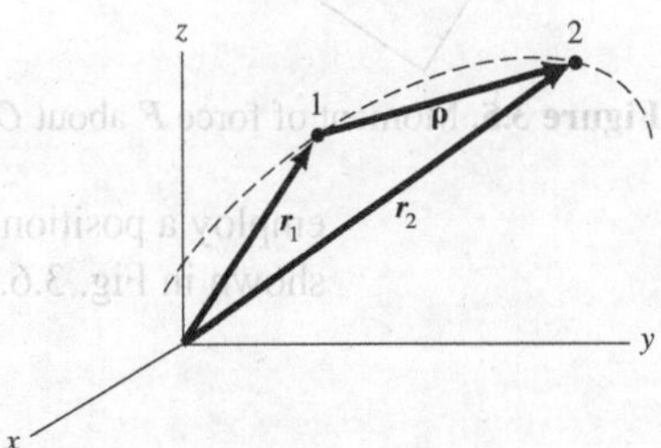

Figure 3.3. Relation between a displacement vector and position vectors.

Example 3.1

Two sets of references, xyz and XYZ, are shown in Fig. 3.4. The position vector of the origin O of xyz relative to XYZ is given as

$$\boldsymbol{R} = 10\boldsymbol{i} + 6\boldsymbol{j} + 5\boldsymbol{k} \text{ m} \qquad \text{(a)}$$

The position vector, $\boldsymbol{r}'$, of a point P relative to XYZ is

$$\boldsymbol{r}' = 3\boldsymbol{i} + 2\boldsymbol{j} - 6\boldsymbol{k} \text{ m} \qquad \text{(b)}$$

What is the position vector $\boldsymbol{r}$ of point P relative to xyz? What are the coordinates x, y, and z of P?

From Fig. 3.4, it is clear that

$$\boldsymbol{r}' = \boldsymbol{R} + \boldsymbol{r} \qquad \text{(c)}$$

Therefore,

$$\boldsymbol{r} = \boldsymbol{r}' - \boldsymbol{R} = (3\boldsymbol{i} + 2\boldsymbol{j} - 6\boldsymbol{k}) - (10\boldsymbol{i} + 6\boldsymbol{j} + 5\boldsymbol{k})$$

$$\boldsymbol{r} = -7\boldsymbol{i} - 4\boldsymbol{j} - 11\boldsymbol{k} \text{ m} \qquad \text{(d)}$$

We can then conclude that

$$x = -7 \text{ m}, \quad y = -4 \text{ m}, \quad z = -11 \text{ m} \qquad \text{(e)}$$

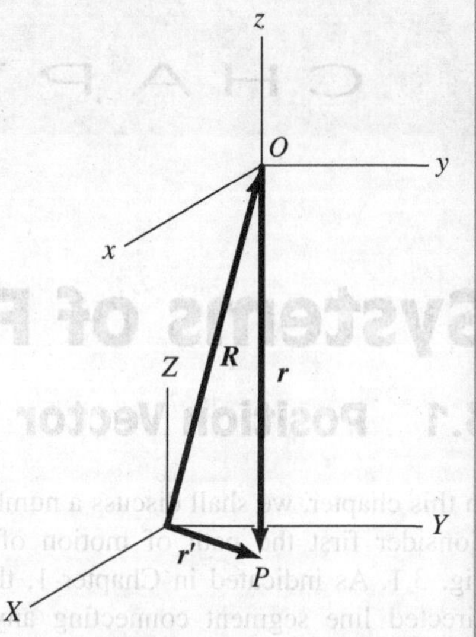

Figure 3.4. References xyz and XYZ separated by position vector $\boldsymbol{R}$.

3.2 Moment of a Force About a Point

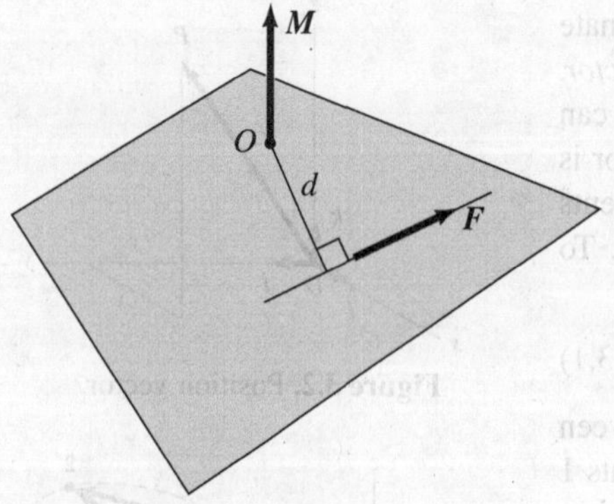

Figure 3.5. Moment of force $\boldsymbol{F}$ about O is Fd.

Case A. For Simple Cases. The moment of a force about a point O (see Fig. 3.5), you will recall from physics, is a vector $\boldsymbol{M}$ whose magnitude equals the product of the force magnitude times the perpendicular distance d from O to the line of action of the force. And the direction of this vector is perpendicular to the plane of the point and the force, with a sense determined from the familiar right-hand-screw rule.[1] The line of action of $\boldsymbol{M}$ is determined by the problem at hand. In Fig. 3.5, the line of action of $\boldsymbol{M}$ is taken for simplicity through point O.

Case B. For Complex Cases. Another approach is to employ a position vector $\boldsymbol{r}$ from point O to *any point* P along the line of action of force $\boldsymbol{F}$ as shown in Fig. 3.6. The moment $\boldsymbol{M}$ of $\boldsymbol{F}$ about point O will be shown to be given as[2]

$$\boldsymbol{M} = \boldsymbol{r} \times \boldsymbol{F} \qquad (3.3)$$

[1]The sense of $\boldsymbol{M}$ would be that of the direction of advance of an ordinary right-hand screw at O, oriented normal to the plane of O and $\boldsymbol{F}$, when this screw is turned with a sense of rotation corresponding to that of $\boldsymbol{F}$ around O.

[2]It is worth pointing out again that in determining the moment of force $\boldsymbol{F}$ about any point O, we must always take the position vector $\boldsymbol{r}$ going *from* point O to *any point along the line of action of the force* $\boldsymbol{F}$. It is easy to make errors here.

For the purpose of forming the cross product, the vectors in Fig. 3.6 can be moved to the configuration shown in Fig. 3.7. Then the cross product between $\boldsymbol{r}$ and $\boldsymbol{F}$ obviously has the magnitude

$$|\boldsymbol{r} \times \boldsymbol{F}| = |\boldsymbol{r}|\,|\boldsymbol{F}| \sin \alpha = |\boldsymbol{F}|\,|\boldsymbol{r}| \sin \beta = |\boldsymbol{F}|\, r \sin \beta = Fd$$

where $r \sin \beta = d$, the perpendicular distance from O to the line of action of $\boldsymbol{F}$, as can readily be seen in Fig. 3.7. Thus, we get the same *magnitude* of $\boldsymbol{M}$ as with the elementary definition. Also, note that the *direction* of $\boldsymbol{M}$ here is identical to that of the elementary definition. Thus we have the same result as for the elementary definition in all pertinent respects. We shall use either of these formulations depending on the situation at hand.

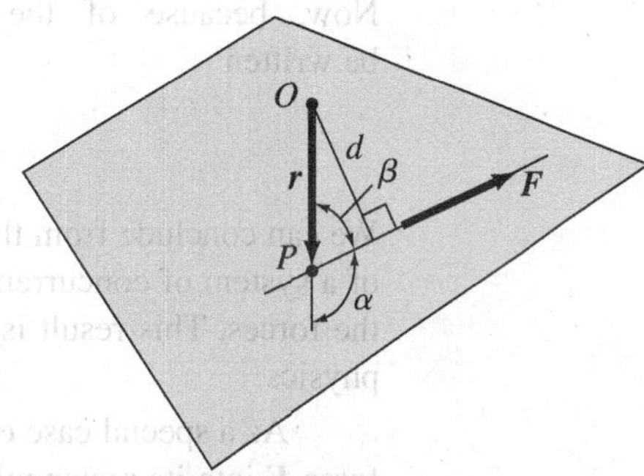

Figure 3.6. Put $\boldsymbol{r}$ from O to any point along the line of action of $\boldsymbol{F}$.

The first of these formulations will be used generally for cases where the force and point are in a convenient plane, and where the perpendicular distance between the point and the line of action of the force is easily measured. As an example, we have shown in Fig. 3.8 a system of coplanar forces acting on a beam. The moment of the forces about point A is then[3]

$$\begin{aligned} \boldsymbol{M}_A &= -(5)(1{,}000)\boldsymbol{k} - (4)(600)\boldsymbol{k} + (11)R_B\boldsymbol{k} \text{ ft-lb} \\ &= (11R_B - 7{,}400)\boldsymbol{k} \text{ ft-lb} \end{aligned}$$

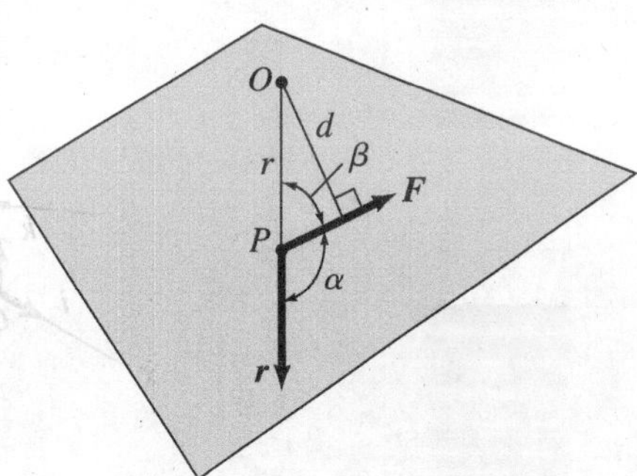

Figure 3.7. Move vector $\boldsymbol{r}$ end $\boldsymbol{F}$.

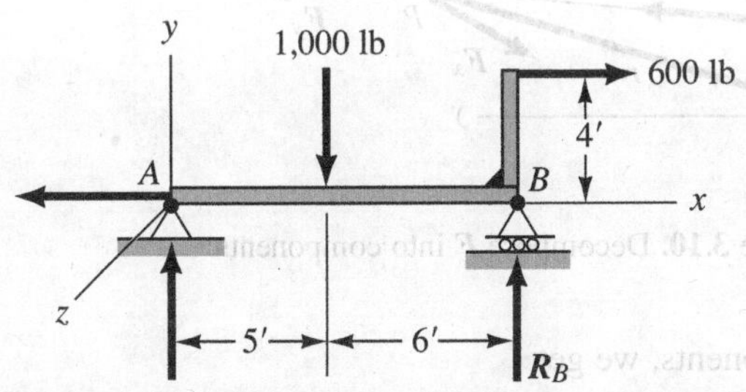

Figure 3.8. Coplanar forces on a beam.

For a coplanar force system such as this, we may simply give the scalar form of the equation above, as follows:

$$M_A = 11R_B - 7{,}400 \text{ ft-lb}$$

The second formulation of the moment about a point, namely $\boldsymbol{r} \times \boldsymbol{F}$, is used for complicated coplanar cases and for three-dimensional cases. We shall illustrate such a case in Example 3.2 after we discuss the rectangular components of $\boldsymbol{M}$.

Consider next a system of n concurrent forces in Fig. 3.9 whose total moment about point O (where we have established reference xyz) is desired. We can say that

$$\begin{aligned} \boldsymbol{M} &= \boldsymbol{M}_1 + \boldsymbol{M}_2 + \boldsymbol{M}_3 + \ldots + \boldsymbol{M}_n \\ &= \boldsymbol{r} \times \boldsymbol{F}_1 + \boldsymbol{r} \times \boldsymbol{F}_2 + \boldsymbol{r} \times \boldsymbol{F}_3 \qquad (3.5) \\ &\quad + \ldots + \boldsymbol{r} \times \boldsymbol{F}_n \end{aligned}$$

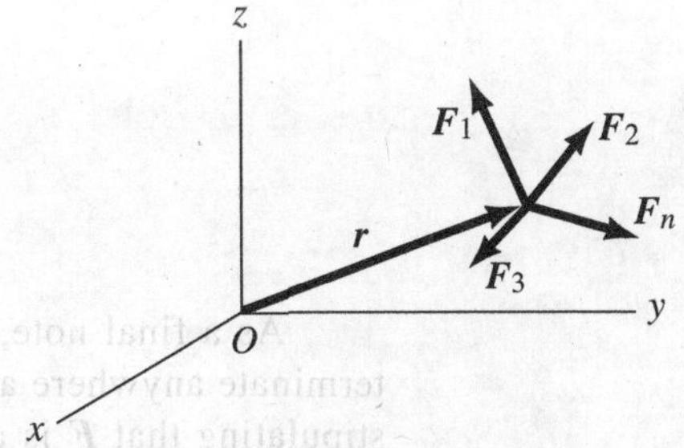

Figure 3.9. Concurrent forces.

[3]Please note that we still use the right-hand-screw rule in determining the signs of the respective moments.

Now, because of the distributive property of the cross product, Eq. 3.5 can be written

$$M = r \times (F_1 + F_2 + F_3 + \ldots + F_n) \tag{3.6}$$

We can conclude from the preceding equations that the sum of the moments about a point of a system of concurrent forces is the same as the moment about the point of the sum of the forces. This result is known as *Varignon's theorem,* which you may well recall from physics.

As a special case of Varignon's theorem, we may find it convenient to decompose a force F into its rectangular components (Fig. 3.10), and then to use these components for taking moments about a point. We can then say that

$$M = r \times F = r \times (F_x i + F_y j + F_z k) \tag{3.7}$$

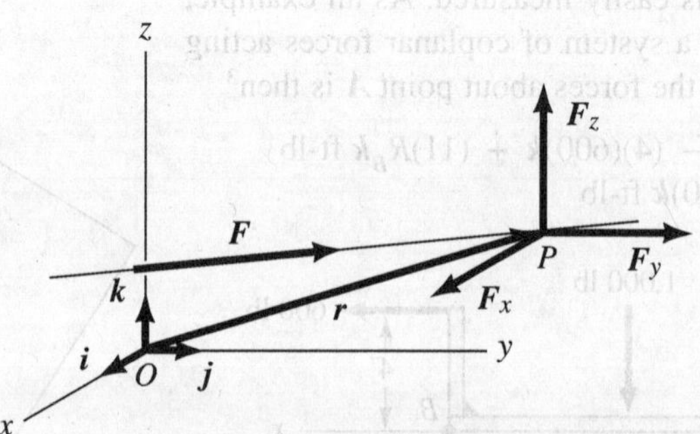

Figure 3.10. Decompose F into components.

Now, replacing r by its components, we get

$$\begin{aligned} M &= (xi + yj + zk) \times (F_x i + F_y j + F_z k) \\ &= (yF_z - zF_y)i + (zF_x - xF_z)j + (xF_y - yF_x)k \end{aligned} \tag{3.8}$$

The scalar rectangular components of M are then

$$M_x = yF_z - zF_y \tag{3.9a}$$

$$M_y = zF_x - xF_z \tag{3.9b}$$

$$M_z = xF_y - yF_x \tag{3.9c}$$

As a final note, it should be apparent that, because we can choose r so as to terminate anywhere along the line of action of F in computing M, we are, in effect, stipulating that F is a *transmissible* vector (defined in Chapter 1) in the computation of M. However, it must be emphasized that we cannot change the line of action of F here.

We are now ready for the example involving the direct vectorial computation of M promised just before the above discussion.

Example 3.2

Determine the moment of the 100-lb force F, shown in Fig. 3.11, about points A and B, respectively.

As a first step, let us express force F vectorially. Note that the force is collinear with the vector $\boldsymbol{\rho}_{DE}$ from D_{DE} to E, where

$$\boldsymbol{\rho}_{DE} = 8\boldsymbol{i} + 4\boldsymbol{j} - 4\boldsymbol{k} \qquad \text{(a)}$$

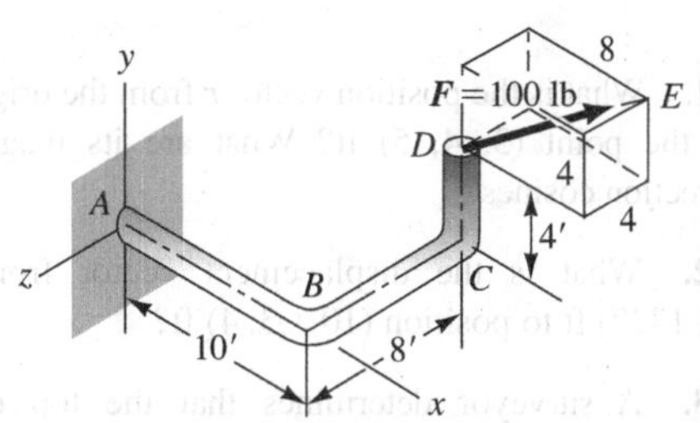

Figure 3.11. Find moments at A and B.

To get a unit vector $\hat{\boldsymbol{\rho}}$ in the direction of $\boldsymbol{\rho}_{DE}$, we proceed as follows:

$$\hat{\boldsymbol{\rho}}_{DE} = \frac{\boldsymbol{\rho}_{DE}}{|\boldsymbol{\rho}_{DE}|} = \frac{8\boldsymbol{i} + 4\boldsymbol{j} - 4\boldsymbol{k}}{\sqrt{8^2 + 4^2 + 4^2}}$$

$$= .816\boldsymbol{i} + .408\boldsymbol{j} - .408\boldsymbol{k} \qquad \text{(b)}$$

We can then express the force $\boldsymbol{F}$ in the following manner:

$$\boldsymbol{F} = F\hat{\boldsymbol{\rho}}_{DE} = (100)(.816\boldsymbol{i} + .408\boldsymbol{j} - .408\boldsymbol{k})$$
$$= 81.6\boldsymbol{i} + 40.8\boldsymbol{j} - 40.8\boldsymbol{k} \qquad \text{(c)}$$

To get the moment $\boldsymbol{M}_A$ about point A, we choose a position vector from point A to point D which is on the line of action of force F. Thus, we have, for $\boldsymbol{r}_{AD}$,

$$\boldsymbol{r}_{AD} = 10\boldsymbol{i} + 4\boldsymbol{j} - 8\boldsymbol{k} \text{ ft} \qquad \text{(d)}$$

and for $\boldsymbol{M}_A$, we than get

$$\boldsymbol{M}_A = \boldsymbol{r}_{AD} \times \boldsymbol{F} = (10\boldsymbol{i} + 4\boldsymbol{j} - 8\boldsymbol{k}) \times (81.6\boldsymbol{i} + 40.8\boldsymbol{j} - 40.8\boldsymbol{k})$$

$$= \left|\begin{array}{ccc} 10 & 4 & -8 \\ 81.6 & 40.8 & -40.8 \\ \boldsymbol{i} & \boldsymbol{j} & \boldsymbol{k} \\ 10 & 4 & -8 \\ 81.6 & 40.8 & -40.8 \end{array}\right.$$

$$= (10)(40.8)\boldsymbol{k} + (81.6)(-8)\boldsymbol{j} + (4)(-40.8)\boldsymbol{i}$$
$$-(-8)(40.8)\boldsymbol{i} - (-40.8)(10)\boldsymbol{j} - (4)(81.6)\boldsymbol{k}$$

Therefore,

$$\boldsymbol{M}_A = 163.2\boldsymbol{i} - 245\boldsymbol{j} + 81.6\boldsymbol{k} \text{ ft-lb} \qquad \text{(e)}$$

As for the moment about reference point B, we employ the position vector $\boldsymbol{r}_{BD}$ from B to position D, again on the line of action of force $\boldsymbol{F}$. Thus, we have $\boldsymbol{r}_{BD} = 4\boldsymbol{j} - 8\boldsymbol{k}$ ft

Accordingly,

$$\boldsymbol{M}_B = \boldsymbol{r}_{BD} \times \boldsymbol{F} = (4\boldsymbol{j} - 8\boldsymbol{k}) \times (81.6\boldsymbol{i} + 40.8\boldsymbol{j} - 40.8\boldsymbol{k})$$
$$= (4)(81.6)(-\boldsymbol{k}) + (4)(-40.8)(\boldsymbol{i}) + (-8)(81.6)(\boldsymbol{j}) + (-8)(40.8)(-\boldsymbol{i}) \qquad \text{(f)}$$

$$\boldsymbol{M}_B = 163.2\boldsymbol{i} - 653\boldsymbol{j} - 326\boldsymbol{k} \text{ ft-lb}$$

PROBLEMS

3.1. What is the position vector $\boldsymbol{r}$ from the origin (0, 0, 0) to the point (3, 4, 5) ft? What are its magnitude and direction cosines?

3.2. What is the displacement vector from position (6, 13, 7) ft to position (10, –3, 4) ft?

3.3. A surveyor determines that the top of a radio transmission tower is at position $\boldsymbol{r}_1 = (1{,}000\boldsymbol{i} + 1{,}000\boldsymbol{j} + 1{,}000\boldsymbol{k})$ m relative to her position. Similarly, the top of a second tower is located by $\boldsymbol{r}_2 = (2{,}000\boldsymbol{i} + 500\boldsymbol{j} + 700\boldsymbol{k})$ m. What is the distance between the two tower tops?

3.4. Reference xyz is rotated 30° counterclockwise about its x axis to form reference XYZ. What is the position vector $\boldsymbol{r}$ for reference xyz of a point having a position vector $\boldsymbol{r}'$ for reference XYZ given as

$$\boldsymbol{r}' = 6\boldsymbol{i}' + 10\boldsymbol{j}' + 3\boldsymbol{k}' \text{ m?}$$

Use $\boldsymbol{i}, \boldsymbol{j}$, and $\boldsymbol{k}$ (no primes) for unit vectors associated with reference xyz.

3.5. Find the moment of the 50-lb force about the support at A and about support B of the simply supported beam.

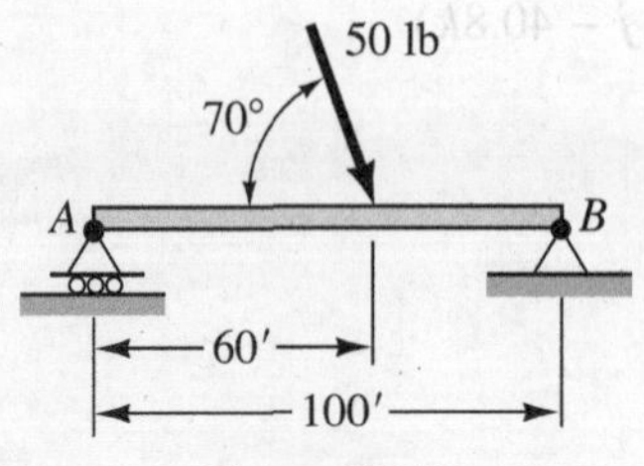

Figure P.3.5.

3.6. Find the moment of the two forces first about point A and then about point B.

(a) Do not use $\boldsymbol{r} \times \boldsymbol{F}$ format—only scalar products.

(b) Use vector approach.

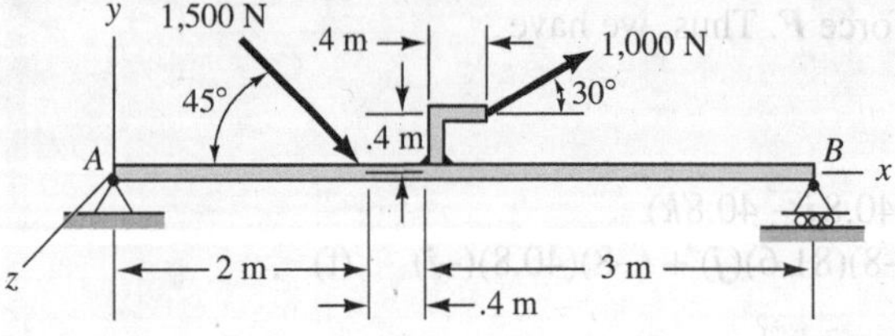

Figure P.3.6.

3.7. A particle moves along a circular path in the xy plane. What is the position vector $\boldsymbol{r}$ of this particle as a function of the coordinate x?

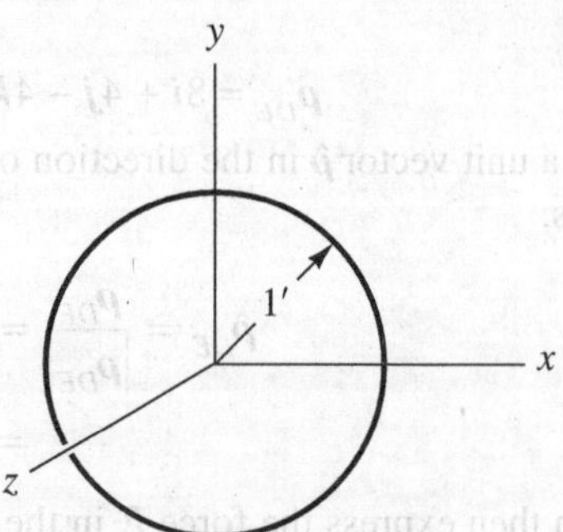

Figure P.3.7.

3.8. A particle moves along a parabolic path in the yz plane. If the particle has at one point a position vector $\boldsymbol{r} = 4\boldsymbol{j} + 2\boldsymbol{k}$, give the position vector at any point on the path as a function of the z coordinate.

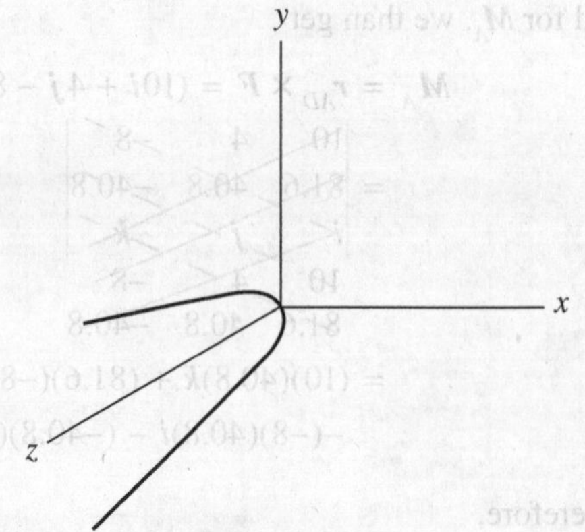

Figure P.3.8.

3.9. An artillery spotter on Hill 350 (350 m high) estimates the position of an enemy tank as 3,000 m NE of him at an elevation 200 m below his position. A 105-mm howitzer unit with a range of 11,000 m is 10,000 m due south of the spotter, and a 155-mm howitzer unit with a range of 15,000 m is 13,000 m SSE of the spotter (see Fig. P.2.4). Both gun units are located at an elevation of 150 m. Can either or both gun units hit the tank, or must an air strike be called in?

3.10. Find the moment of the forces about points A and B.
(a) Use scalar approach.
(b) Use vector approach.

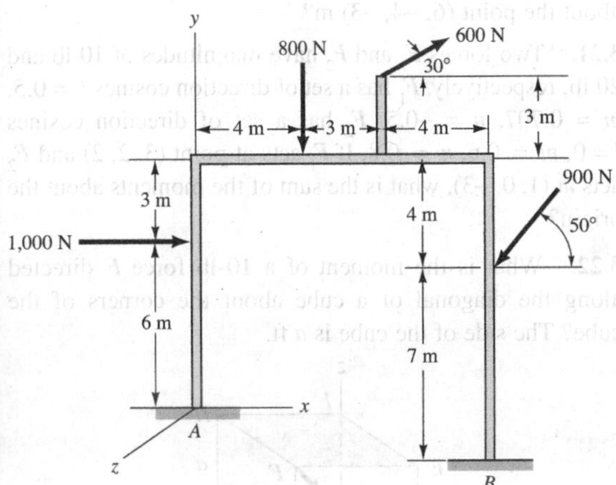

Figure P.3.10.

3.11. The crew of a submarine patrol plane, with three-dimensional radar, sights a surfaced submarine 10,000 yards north and 5,000 yards east while flying at an elevation of 3,000 ft above sea level. Where should the pilot instruct a second patrol plane flying at an elevation of 4,000 ft at a position 40,000 yards east of the first plane to look for confirmation of the sighting?

3.12. A power company lineman can comfortably trim branches 1 m from his waist at an angle of 45° above the horizontal. His waist coincides with the pivot of the work capsule. How high a branch can he trim if the maximum elevation angle of the arm is 75° and the maximum extended length is 12 m?

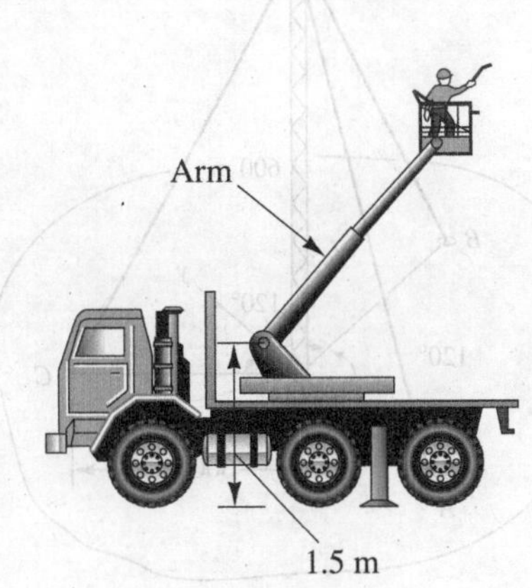

Figure P.3.12.

3.13. The total equivalent forces from water and gravity are shown on the dam. (We will soon be able to compute such equivalents.) Compute the moment of these forces about the toe of the dam in the right-hand corner.

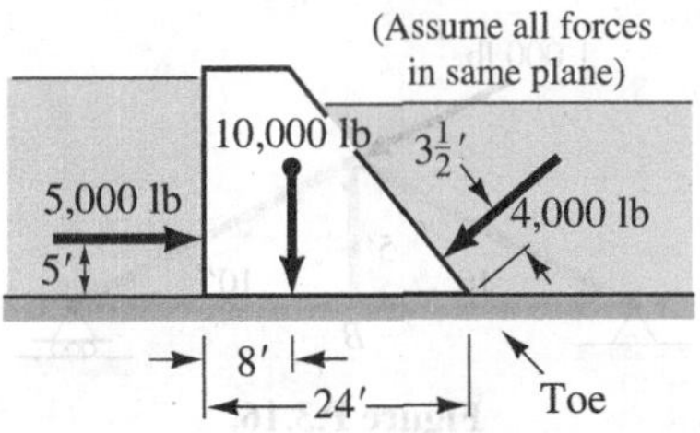

Figure P.3.13.

3.14. In an underwater "village" for research, an American flag is in place as shown. It is of plastic material and can rotate so as to be oriented parallel to the flow of water. A uniform friction force distribution from the flow is present on both faces of the flag having the value of 10 N per square meter. Also the flagpole has a uniform force from the flow of 20 N per meter of length of the flagpole. Finally there is an upward buoyant force on the flag of 30 N and on the flagpole of 8 N. What is the moment vector of these forces at the base of the flagpole?

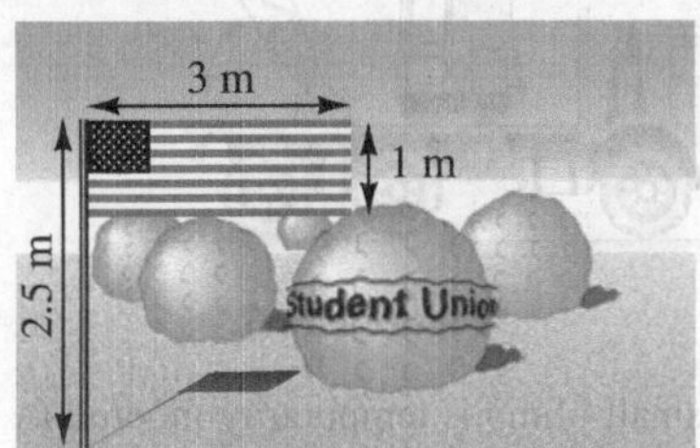

Figure P.3.14.

3.15. Three transmission lines are placed unsymmetrically on a power-line pole. For each pole, the weight of a single line when covered with ice is 2,000 N. What is the moment at the base of a pole?

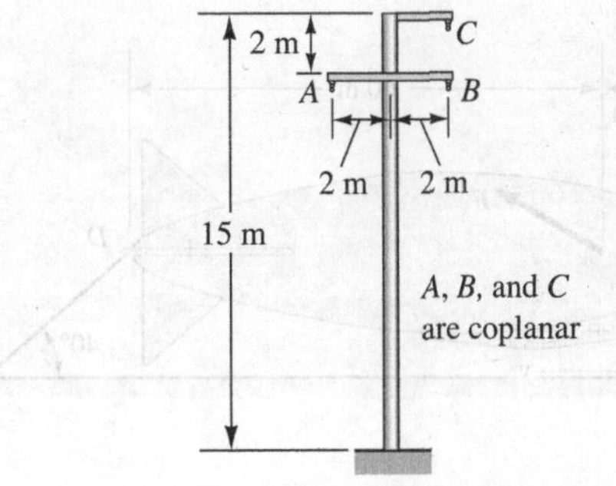

Figure P.3.15.

3.16. Compute the moment of the 1,000-lb force about points A, B, and C. Use the transmissibility property of force and rectangular components to make the computations simplest.

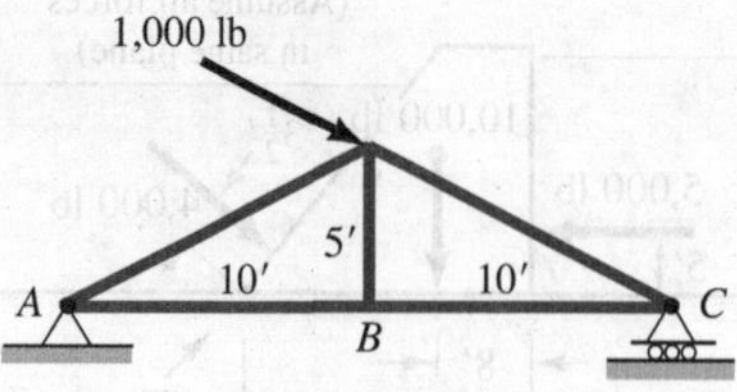

Figure P.3.16.

3.17. A truck-mounted crane has a 20-m boom inclined at 60° to the horizontal. What is the moment about the boom pivot due to a lifted weight of 30 kN? Do by vector and by scalar methods.

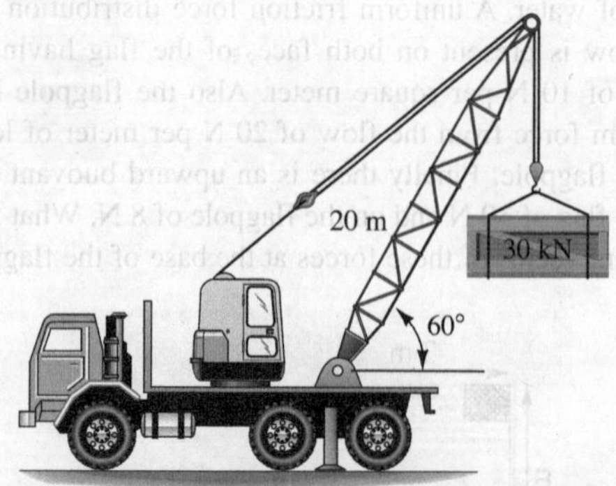

Figure P.3.17.

3.18. A small blimp is temporarily moored as shown in the diagram wherein DC and the centerline of AB are coplanar. A force $\boldsymbol{F}$ from wind, weight, and buoyancy is shown acting at the centerline of the blimp. If

$$\boldsymbol{F} = 5\boldsymbol{i} + 10\boldsymbol{j} + 18\boldsymbol{k} \text{ kN}$$

what are the moment vectors from $\boldsymbol{F}$ about A, B, and C?

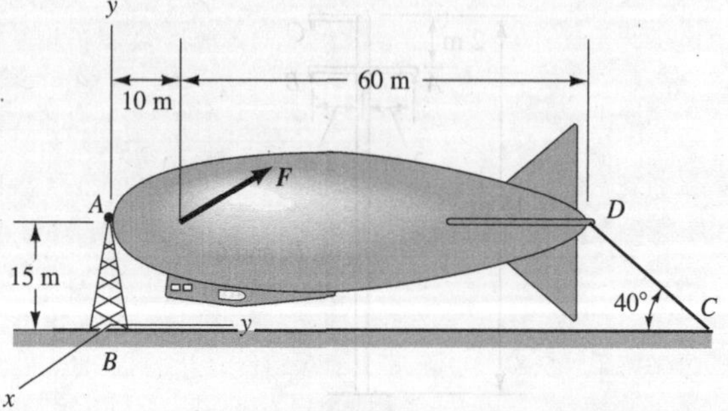

Figure P.3.18.

3.19. A force $\boldsymbol{F} = 10\boldsymbol{i} + 6\boldsymbol{j} - 6\boldsymbol{k}$ N acts at position (10, 3, 4) m relative to a coordinate system. What is the moment of the force about the origin?

3.20. What is the moment of the force in Problem 3.19 about the point (6, –4, –3) m?

3.21. Two forces $\boldsymbol{F}_1$ and $\boldsymbol{F}_2$ have magnitudes of 10 lb and 20 lb, respectively. $\boldsymbol{F}_1$ has a set of direction cosines $l = 0.5$, $m = 0.707$, $n = -0.5$. $\boldsymbol{F}_2$ has a set of direction cosines $l = 0$, $m = 0.6$, $n = 0.8$. If $\boldsymbol{F}_1$ acts at point (3, 2, 2) and $\boldsymbol{F}_2$ acts at (1, 0, –3), what is the sum of the moments about the origin?

3.22. What is the moment of a 10-lb force $\boldsymbol{F}$ directed along the diagonal of a cube about the corners of the cube? The side of the cube is a ft.

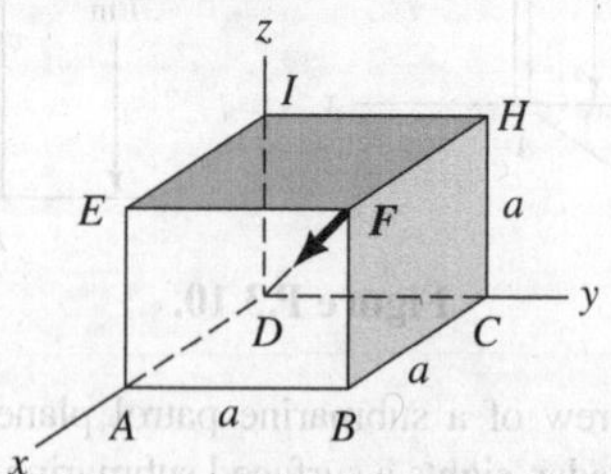

Figure P.3.22.

3.23. Three guy wires are used in the support system for a television transmission tower that is 600 m tall. Wires A and B are tightened to a tension of 60 kN, whereas wire C has only 30 kN of tension. What is the moment of the wire forces about the base O of the tower? The y axis is collinear with AO.

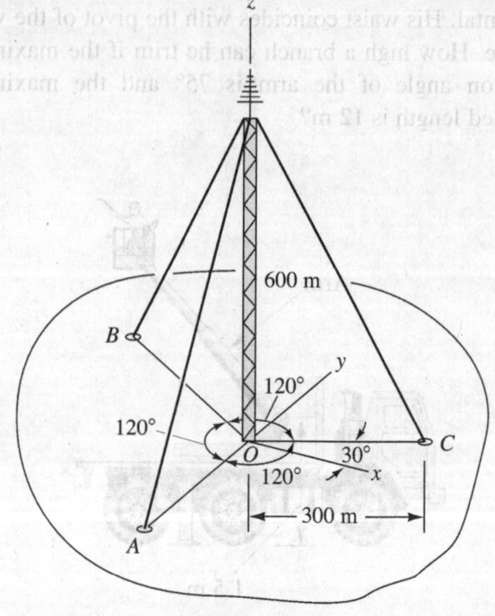

Figure P.3.23.

3.24. Cables CD and AB help support member ED and the 1,000-lb load at D. At E there is a ball-and-socket joint which also supports the member. Denoting the forces from the cables as F_{CD} and F_{AB}, respectively, compute moments of the three forces about point E. Plane EGD is perpendicular to the wall. Get results in terms of F_{CD} and F_{AB}.

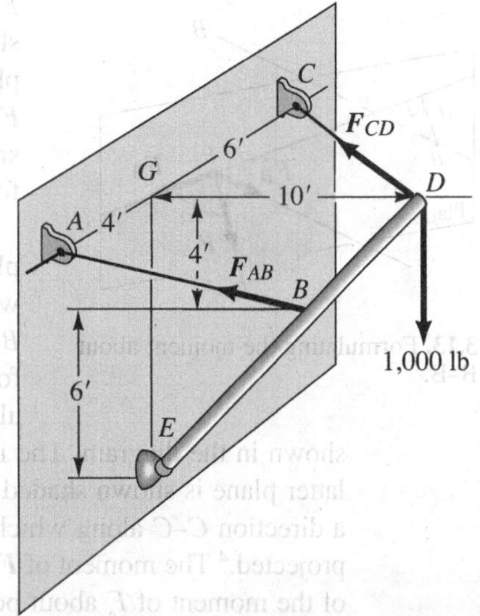

Figure P.3.24.

3.3 Moment of a Force About an Axis

Case A. For Simple Cases. By means of a simple situation, we shall set forth a definition of the moment of a force about an axis. Suppose that a disc is mounted on a shaft that is free to rotate in a set of bearings, as shown in Fig. 3.12. A force F, inclined to the plane A of the disc, acts on the disc. We decompose the force into two coplanar rectangular components, one normal to plane A of the disc and one tangent to plane A of the disc, that is, into forces F_B and F_A, respectively, so as to form a plane shown tinted, normal to plane A.

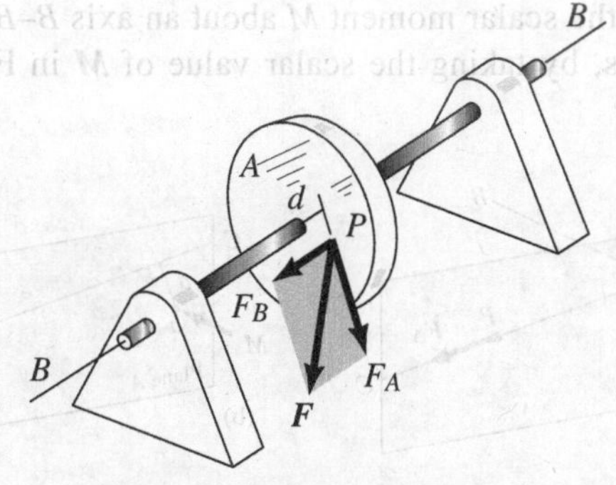

Figure 3.12. F_A turns disc.

We know from experience that F_B does not cause the disc to rotate. And we know from physics and intuition that the rotational motion of the disc is determined by the product of

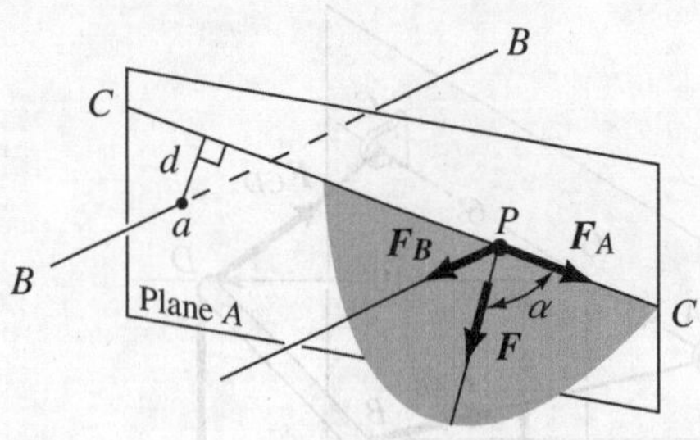

Figure 3.13. Formulating the moment about an axis B–B.

F_A and the perpendicular distance d from the centerline of the shaft to the line of action of F_A. You will remember from physics, this product is nothing more than the moment of force $\boldsymbol{F}$ about the axis of the shaft. We shall next generalize from this simple case to the general case of taking the moment of *any* force $\boldsymbol{F}$ about *any* axis.

To compute the moment (or torque) of a force $\boldsymbol{F}$ in a plane perpendicular to plane A about an axis B–B (Fig. 3.13), we pass any plane A perpendicular to the axis. This plane cuts B–B at a and the line of action of force $\boldsymbol{F}$ at some point P. The force $\boldsymbol{F}$ is then projected to form a rectangular component F_B along a line at P normal to plane A and thus parallel to B–B, as shown in the diagram. The intersection of plane A with the plane of forces F_B and $\boldsymbol{F}$ (the latter plane is shown shaded and is a plane through $\boldsymbol{F}$ and perpendicular to plane A) gives a direction C–C along which the other rectangular component of $\boldsymbol{F}$, denoted as F_A, can be projected.[4] The moment of $\boldsymbol{F}$ about the line B–B is then defined as the scalar representation of the moment of F_A about point a with a magnitude equal to $F_A d$—a problem discussed at the beginning of the previous section (Case A). Thus in accordance with the definition, the component F_B, which is parallel to the axis B–B, contributes no moment about the axis, and we may say:

$$\text{Moment about axis } B\text{–}B = (F_A)(d) = |\boldsymbol{F}|(\cos\alpha)(d)$$

with an appropriate sign. The moment about an axis clearly is a scalar, even though this moment is associated with a particular axis that has a distinct direction. The situation is the same as it is with the scalar components V_x, V_y, V_z, etc., which are associated with certain directions but which are scalars. The reader will be quick to note that Fig. 3.13 represents a generalization of Fig. 3.12 having an axis B–B, a plane A normal to this axis and finally an arbitrary force $\boldsymbol{F}$. To explain further, we have redrawn Fig. 3.13 [see Fig. 3.14(a)] showing only plane A, axes B–B and C–C, and the force F_A. In Fig. 3.14(b), we have also included the moment vector $\boldsymbol{M}$. This latter diagram then takes us back to Fig. 3.5 where we first defined the moment vector of a force about a point in a most simple manner. Accordingly, we note, on the one hand, for the moment about point a [see Fig. 3.14(b)], we can get the vector $\boldsymbol{M}$, whereas on the other hand in Fig. 3.14(a) we can get the scalar moment M about an axis B–B at point a and perpendicular to plane A. Thus, by taking the scalar value of $\boldsymbol{M}$ in Fig. 3.14(b), we get the

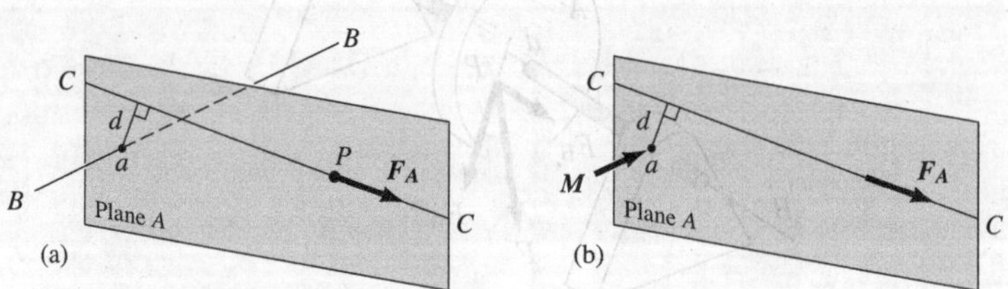

Figure 3.14. Comparison of Figs. 3.13 and 3.5.

[4]Notice that we are decomposing $\boldsymbol{F}$ into only *two* rectangular components, which, to replace $\boldsymbol{F}$, must be coplanar with $\boldsymbol{F}$.

moment about the axis at point a normal to plane A as formulated in the development of Fig. 3.13.

We can thus conclude, on considering Fig. 3.15, that the moment of $\boldsymbol{F}$ about point C can be considered in two ways as follows:

Moment vector about point C = $-Fd\,\mathbf{k}$
Torque (or moment) about z axis at point C = $-Fd$

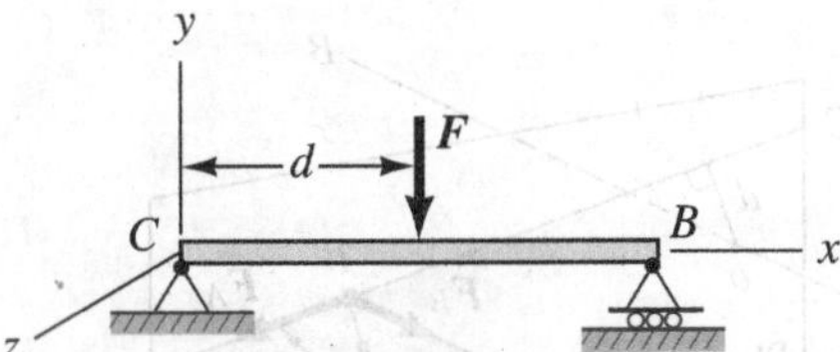

Figure 3.15. Consideration of the moment of $\boldsymbol{F}$ about point A.

Before continuing, we wish to point out that F_A in Fig. 3.13 can be decomposed into pairs of components in plane A. From Varignon's theorem we can employ these components instead of F_A in computing the moment about the B–B axis. For each force component, we multiply the force times the perpendicular distance from a to the line of action of the force component using the right-hand-screw rule to determine the sense and thus the sign.

Case B. For Complex Cases. When we discussed the moment of a force about a *point*, we presented a formulation useful for simple cases (i.e., a vector of magnitude Fd) as well as a more powerful formulation that would be needed for more complex situations (i.e., $\boldsymbol{r} \times \boldsymbol{F}$). Thus far, for moments about an *axis*, we have presented a formulation Fd that is useful for simple cases,[5] and now we shall present a formulation that is needed for more complex cases. For this purpose, we have redrawn Fig. 13.13 as Fig. 13.16(a). In Fig. 13.16(b), we have shown axis B–B of Fig. 13.16(a) as an x axis and have set up coordinate axes y and z at *any* point O anywhere on axis B–B. The coordinate distances x, y, and z for the point P are shown for this reference. The position vector $\boldsymbol{r}$ to P is also shown. The force component F_B of Fig. 3.16(a) now becomes force component F_x. And, instead of using F_A, we shall decompose it into components F_y and F_z in plane A as shown in Fig. 3.16(b). We now compute the moment about the x axis for force $\boldsymbol{F}$ using this new arrangement which does not require $\boldsymbol{F}$ to be in a plane perpendicular to plane A. Clearly, F_x contributes no moment, as before. The force components F_y and F_z are in plane A that is perpendicular to the axis of interest and so, as before in the case of F_A, we multiply each of these forces by the perpendicular distance of point a to the respective lines of action of these forces. For force F_z, this perpendicular distance is clearly y, as can readily be seen from the diagram, and, for force F_y, this perpendicular

[5]That is, for cases where the *force* and *point* in question are in a plane easily seen to be normal to the *axis* in question, thus allowing for an easy determination of the perpendicular distance between the point and the line of action of the force.

distance is z. Using the right-hand-screw rule for ascertaining the sense of each of the moments, we can say:

$$\text{moment about } x \text{ axis} = (yF_z - zF_y) \tag{3.10}$$

Were we to take moments of $\boldsymbol{F}$ about the origin O, we would get (see Eq. 3.8)

$$\begin{aligned} \boldsymbol{M} &= M_x\boldsymbol{i} + M_y\boldsymbol{j} + M_z\boldsymbol{k} = \boldsymbol{r} \times \boldsymbol{F} \\ &= (yF_z - zF_y)\boldsymbol{i} + (zF_x - xF_z)\boldsymbol{j} + (xF_y - yF_x)\boldsymbol{k} \end{aligned} \tag{3.11}$$

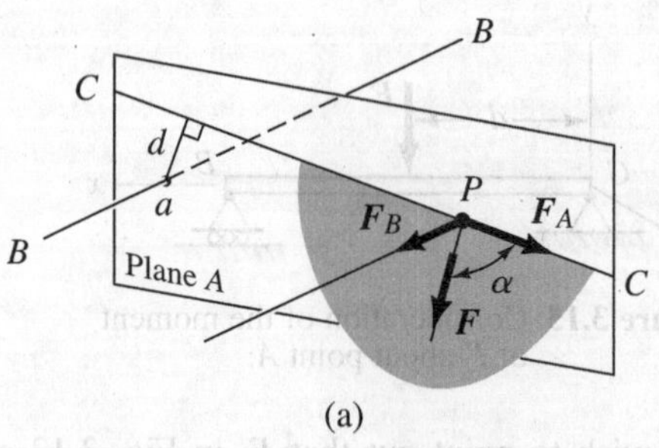

(a)

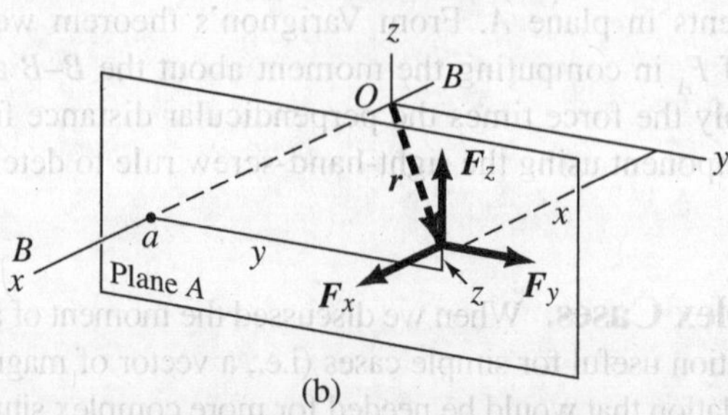

(b)

Figure 3.16. Moment about an axis.

Comparing Eqs. 3.10 and 3.11, we can conclude that the moment about the x axis is simply M_x, the x component of $\boldsymbol{M}$ about O. We can thus conclude that the moment about the x axis of the force $\boldsymbol{F}$ is the component in the x direction of the moment of $\boldsymbol{F}$ about a point O positioned *anywhere* along the x axis. That is,

$$\text{moment about } x \text{ axis} = M_x = \boldsymbol{M}_o \cdot \boldsymbol{i} = (\boldsymbol{r} \times \boldsymbol{F}) \cdot \boldsymbol{i} \tag{3.12}$$

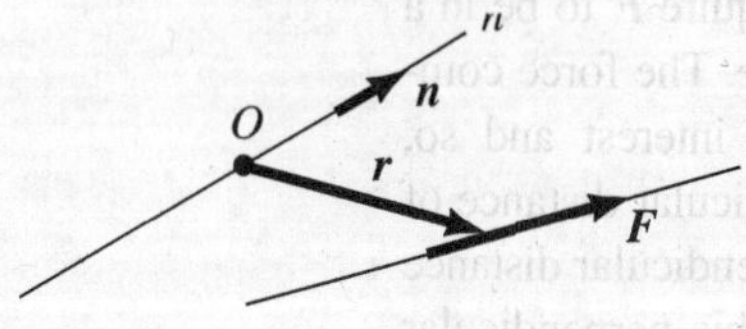

Figure 3.17. $M_n = (\boldsymbol{r} \times \boldsymbol{F}) \cdot \boldsymbol{n}$.

We may generalize the preceding discussion as follows. Consider an arbitrary axis n–n to which we have assigned a unit vector $\boldsymbol{n}$ (Fig. 3.17). An arbitrary force $\boldsymbol{F}$ is also shown. To get the moment M_n of force $\boldsymbol{F}$ about axis n–n, we choose any point O along n–n. Then draw a position vector $\boldsymbol{r}$ from point O to any point along the line of action of $\boldsymbol{F}$. This has been shown in the diagram. We can then say, from our previous discussion,

$$M_n = (\boldsymbol{r} \times \boldsymbol{F}) \cdot \boldsymbol{n} \tag{3.13}$$

(Notice from Eqs. 3.12 and 3.13 that the moment of a force about an axis involves a scalar triple product.) Equation 3.13 stipulates in words that:

The moment of a force about an axis equals the scalar component in the direction of the axis of the moment vector taken about any point along the axis.

This is the more powerful formulation that can be used for complex cases.

Note that the unit vector $\boldsymbol{n}$ can have two opposite senses along the axis n, in contrast to the usual unit vectors $\boldsymbol{i}$, $\boldsymbol{j}$, and $\boldsymbol{k}$ associated with the coordinate axes. A moment M_n about the n axis determined from $\boldsymbol{M} \bullet \boldsymbol{n}$ has a sense consistent with the sense chosen for $\boldsymbol{n}$. That is, a positive moment M_n has a sense corresponding to that of $\boldsymbol{n}$, and a negative moment M_n has a sense opposite to that of $\boldsymbol{n}$. If the opposite sense had been chosen for $\boldsymbol{n}$, the sign of $\boldsymbol{M} \bullet \boldsymbol{n}$ would be opposite to that found in the first case. However, the same physical moment is obtained in both cases.

If we specify the moments of a force about *three* orthogonal concurrent axes, we then single out *one* possible point in space for O along the axes. Point O, of course, is the origin of the axes. These three moments about orthogonal axes then become the orthogonal scalar components of the moment of $\boldsymbol{F}$ about point O, and we can say:

$$\begin{aligned}\boldsymbol{M} = {} & (\text{moment about the } x \text{ axis})\boldsymbol{i} + \\ & (\text{moment about the } y \text{ axis})\boldsymbol{j} + \\ & (\text{moment about the } z \text{ axis})\boldsymbol{k} = M_x\boldsymbol{i} + M_y\boldsymbol{j} + M_z\boldsymbol{k}\end{aligned} \tag{3.14}$$

From this relation, we can conclude that:

The three orthogonal components of the moment of a force about a point are the moments of this force about the three orthogonal axes that have the point as an origin.

You may now ask what the physical differences are in applications of moments about an axis and moments about a point. The simplest example is in the dynamics of rigid bodies. If a body is constrained so it can only spin about its axis, as in Fig. 3.12, the rotary motion will depend on the moment of the forces about the axis of rotation, as related by a scalar equation. The less familiar concept of moment about a point is illustrated in the motion of bodies that have no constraints, such as missiles and rockets. In these cases, the rotational motion of the body is related by a vector equation to the moment of forces acting on the body about a point called the *center of mass*. (The center of mass will be defined completely later.)

Example 3.3

Compute the moment of a force $\boldsymbol{F} = 10\boldsymbol{i} + 6\boldsymbol{j}$ N, which goes through position $\boldsymbol{r}_a = 2\boldsymbol{i} + 6\boldsymbol{j}$ m (see Fig. 3.18), about a line going through points 1 and 2 having the respective position vectors

$$\boldsymbol{r}_1 = 6\boldsymbol{i} + 10\boldsymbol{j} - 3\boldsymbol{k} \text{ m}$$
$$\boldsymbol{r}_2 = -3\boldsymbol{i} - 12\boldsymbol{j} + 6\boldsymbol{k} \text{ m}$$

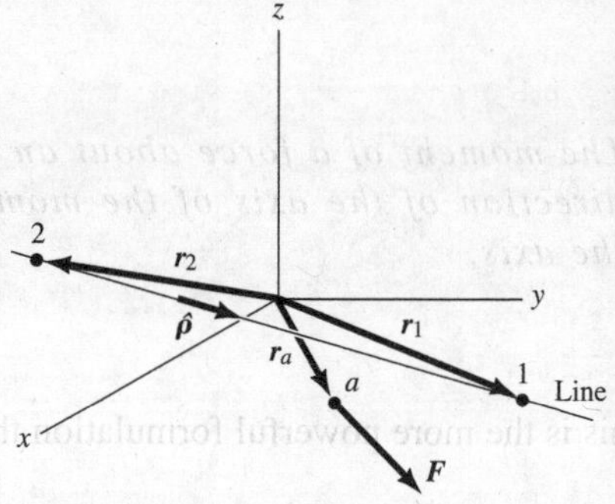

Figure 3.18. Find moment of $\boldsymbol{F}$ about line.

To compute this moment, we can take the moment of $\boldsymbol{F}$ about either point 1 or point 2, and then find the component of this vector along the direction of the displacement vector between 1 and 2 or between 2 and 1. Mathematically, we have, using a displacement vector from point 1 to point a, namely $(\boldsymbol{r}_a - \boldsymbol{r}_1)$,

$$M_\rho = [(\boldsymbol{r}_a - \boldsymbol{r}_1) \times \boldsymbol{F}] \bullet \hat{\boldsymbol{\rho}} \qquad \text{(a)}$$

where $\hat{\boldsymbol{\rho}}$ is the unit vector along the line chosen to have a sense going from point 2 to point 1. The formulation above is the scalar triple product examined in Chapter 2 and we can use the determinant approach for the calculation once the components of the vectors $(\boldsymbol{r}_a - \boldsymbol{r}_1)$, $\boldsymbol{F}$, and $\hat{\boldsymbol{\rho}}$ have been determined. Thus, we have

$$\boldsymbol{r}_a - \boldsymbol{r}_1 = (2\boldsymbol{i} + 6\boldsymbol{j}) - (6\boldsymbol{i} + 10\boldsymbol{j} - 3\boldsymbol{k})$$
$$= -4\boldsymbol{i} - 4\boldsymbol{j} + 3\boldsymbol{k} \text{ m}$$
$$\boldsymbol{F} = 10\boldsymbol{i} + 6\boldsymbol{j} \text{ N}$$
$$\hat{\boldsymbol{\rho}} = \frac{\boldsymbol{r}_1 - \boldsymbol{r}_2}{|\boldsymbol{r}_1 - \boldsymbol{r}_2|} = \frac{9\boldsymbol{i} + 22\boldsymbol{j} - 9\boldsymbol{k}}{\sqrt{81 + 484 + 81}}$$
$$= .354\boldsymbol{i} + .866\boldsymbol{j} - .354\boldsymbol{k}$$

We then have, for M_ρ:

$$M_\rho = \begin{vmatrix} -4 & -4 & 3 \\ 10 & 6 & 0 \\ .354 & .866 & -.354 \end{vmatrix} = \boxed{13.94 \text{ N-m}} \qquad \text{(b)}$$

Because M_ρ is positive, we have a clockwise moment about the line as we look from point 2 to point 1. If we had chosen $\hat{\boldsymbol{\rho}}$ to have an opposite sense, then M_ρ would have been computed as –13.94 N-m. Then, we would conclude that M_ρ is a counterclockwise moment about the line as one looks from point 1 to point 2. Note that the same physical moment is determined in both cases.

Example 3.4

A deep submergence vessel is connected to its mother ship by a cable (Fig. 3.19). The vessel becomes snagged on some rocks and the mother ship steams ahead in a forward direction in an attempt to free the submerged vessel. The connecting cable is suspended from a crane directed up over the water 20 m above the center of mass of the mother ship and 15 m out from the longitudinal axis of the mother ship. The cable transmits a force of 200 kN. It is inclined 50° from the vertical in a vertical plane which, in turn, is oriented 20° from the longitudinal axis of the ship. What is the moment tending to cause the mother ship to roll about its longitudinal axis (i.e., the *x*-axis)?

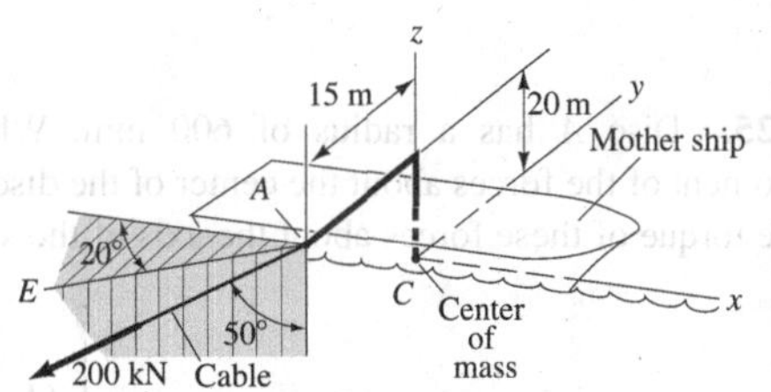

Figure 3.19. Torque on a ship about a horizontal axis through the center of mass.

The position vector from the center of mass C to point A is

$$\boldsymbol{r} = -15\boldsymbol{j} + 20\boldsymbol{k} \text{ m}$$

Next consider the 200-kN force on the cable. Notice that the cable is in a plane which was rotated 20° counterclockwise from an orientation parallel to the xz plane and in this rotated plane, the cable is at an angle of 50° from the vertical. We first decompose the 200-kN force into two coplanar rectangular components, one along the vertical and one along AE. We have then

$$\boldsymbol{F}_{CABLE} = -200 \cos 50° \,\boldsymbol{k} + \boldsymbol{F}_{AE} = -128.6\boldsymbol{k} + \boldsymbol{F}_{AE} \text{ kN}$$

Next, decompose $\boldsymbol{F}_{AE}$ into two coplanar rectangular components, one parallel to the y axis, and the other parallel to the x axis. We then get, noting that $|\boldsymbol{F}_{AE}| = 200 \sin 50°$ kN,

$$\boldsymbol{F}_{AE} = (200 \sin 50°)[-\cos 20°\boldsymbol{i} - \sin 20°\boldsymbol{j}] = -144\boldsymbol{i} - 52.4\boldsymbol{j} \text{ kN}$$

Hence,

$$\boldsymbol{F}_{CABLE} = -144\boldsymbol{i} - 52.4\boldsymbol{j} - 128.6\boldsymbol{k} \text{ kN}$$

To get the desired moment, we have the following formulation:

$$M_x = \left[\mathbf{r} \times \boldsymbol{F}_{CABLE}\right] \bullet \mathbf{i}$$

$$M_x = [(-15\mathbf{j} + 20\mathbf{k}) \times (-144\mathbf{i} - 52.4\mathbf{j} - 128.6\mathbf{k})] \bullet \mathbf{i}$$

Using the determinant format for the triple scalar product we get

$$M_x = \begin{vmatrix} 0 & -15 & 20 \\ -144 & -52.4 & -128.6 \\ 1 & 0 & 0 \end{vmatrix} \text{ kN-m}$$

We get on carrying out the determinant

$$M_x = 2{,}977 \text{ kN–m}$$

PROBLEMS

3.25. Disc *A* has a radius of 600 mm. What is the moment of the forces about the center of the disc? What is the torque of these forces about the axis of the shaft?

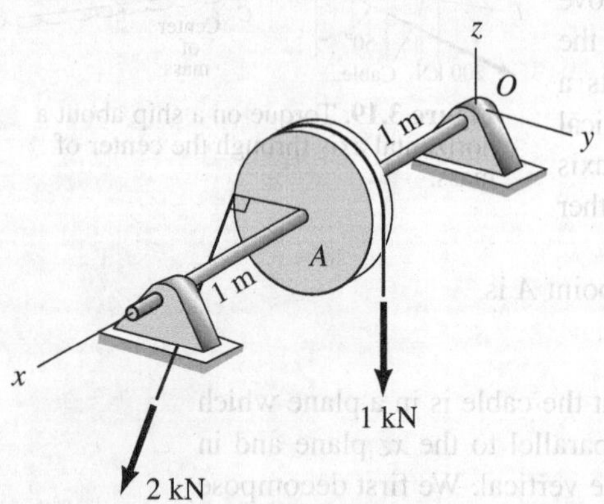

Figure P.3.25.

3.26. A force $\boldsymbol{F}$ acts at position (3, 2, 0) ft. It is in the *xy* plane and is inclined at 30° from the *x* axis with a sense directed away from the origin. What is the moment of this force about an axis going through the points (6, 2, 5) ft and (0, −2, −3) ft?

3.27. A force $\boldsymbol{F} = 10\boldsymbol{i} + 6\boldsymbol{j}$ N goes through the origin of the coordinate system. What is the moment of this force $\boldsymbol{F}$ about an axis going through points 1 and 2 with position vectors?

$$\boldsymbol{r}_1 = 6\boldsymbol{i} + 3\boldsymbol{k} \text{ m}$$
$$\boldsymbol{r}_2 = 16\boldsymbol{j} - 4\boldsymbol{k} \text{ m}$$

3.28. Given a force $\boldsymbol{F} = 10\boldsymbol{i} + 3\boldsymbol{j}$ N acting at position $\boldsymbol{r} = 5\boldsymbol{j} + 10\boldsymbol{k}$ m, what is the torque about the diagonal shown in the diagram? What is the moment about point *E*?

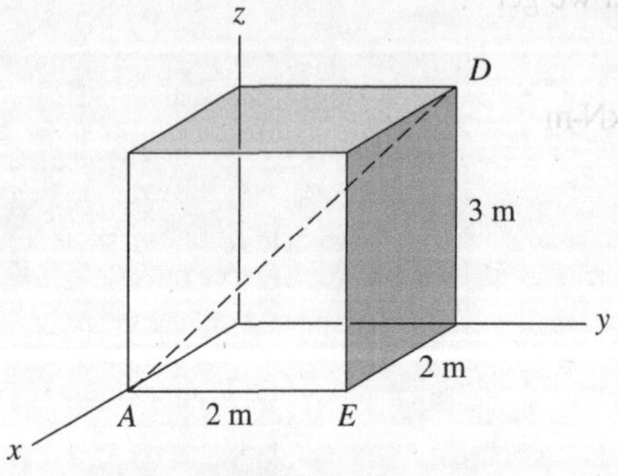

Figure P.3.28.

3.29. A blimp is moored to a tower at *A*. A force on *A* from this blimp is

$$\boldsymbol{F} = 5\boldsymbol{i} + 3\boldsymbol{j} + 1.8\boldsymbol{k} \text{ kN}$$

What is the moment about axis *C* on the ground? Knowledge of this moment and other moments at the base is needed to properly design the foundation of the tower.

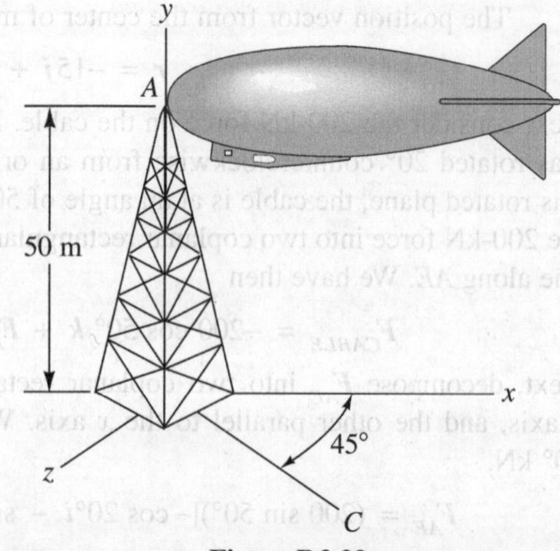

Figure P.3.29.

3.30. Compute the thrust of the applied forces shown along the axis of the shaft and the torque of the forces about the axis of the shaft.

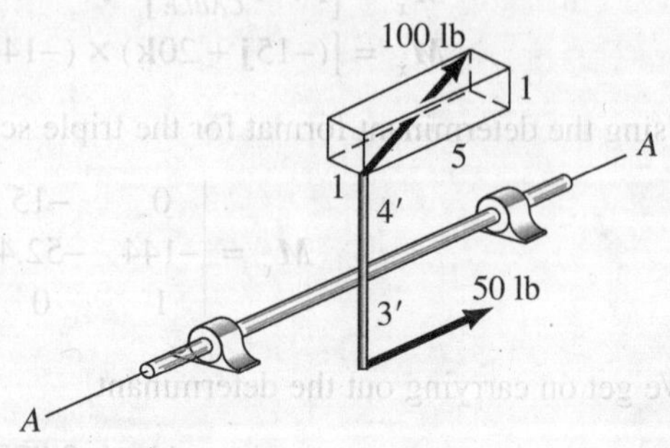

Figure P.3.30.

3.31. What is the maximum load *W* that the crane can lift without tipping about *A*? *Hint:* when tipping is impending, what is the supporting force at the wheel at *B*?

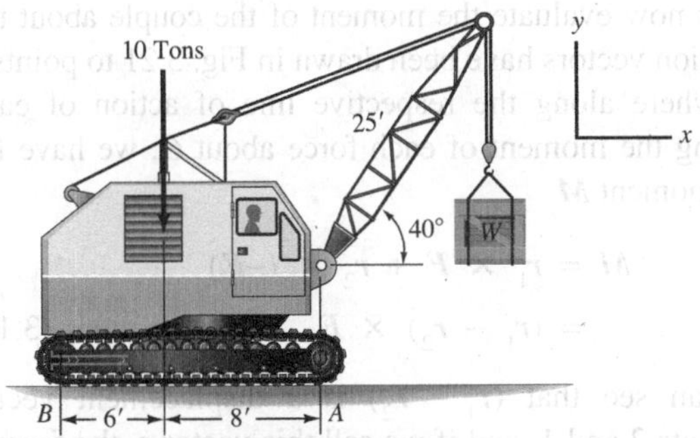

Figure P.3.31.

3.32. Find the moment of the 1,000-lb force about an axis going between points D and C.

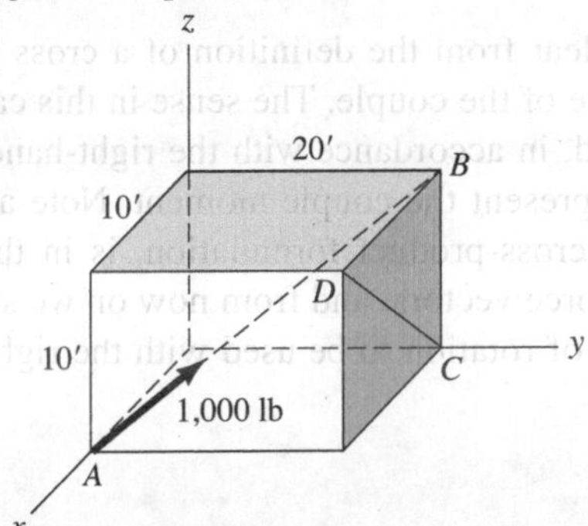

Figure P.3.32.

3.33. In Problem 3.24, what is the moment of the three indicated forces about axis GD?

3.34. The base of a fire truck extension ladder is rotated 75° counter-clockwise. The 25-m ladder is elevated 60° from the horizontal. The ladder weight is 20 kN and is regarded as concentrated at a point 10 m up from the base (the lower part of the ladder weighs much more than the upper part). A 900-N fireman and the 500-N young lady he is rescuing are at the top of the ladder. (a) What is the moment at the base of the ladder tending to tip over the fire truck? (b) What is the moment about the horizontal axis having unit vector $\hat{\boldsymbol{\rho}}$ shown in the diagram?

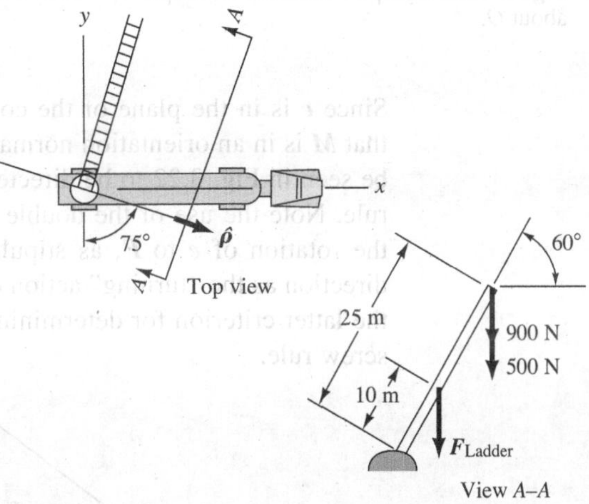

Figure P.3.34.

3.4 The Couple and Couple Moment

A special arrangement of forces that is of great importance is the *couple. The couple is formed by any two equal parallel forces that have opposite senses* (Fig. 3.20). On a rigid body, a couple has only *one* effect, a "turning" action. Individual forces or combinations of forces that do not constitute couples may "push" or "pull" as well as "turn" a body. The turning action is given quantitatively by the moment of forces about a point or an axis. We shall, accordingly, be most concerned with the moment of a couple, or what we shall call the *couple moment.*

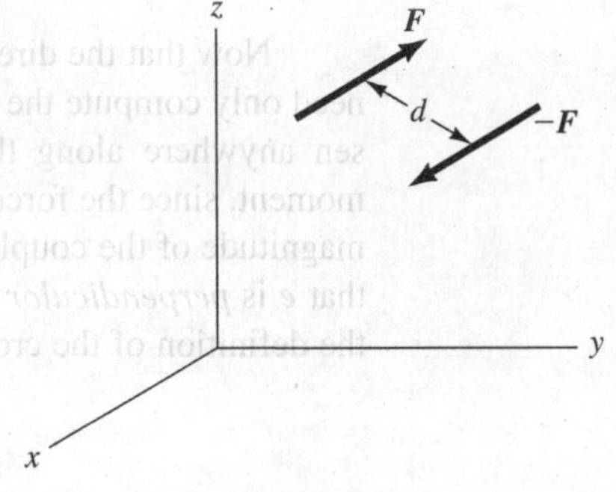

Figure 3.20. A couple.

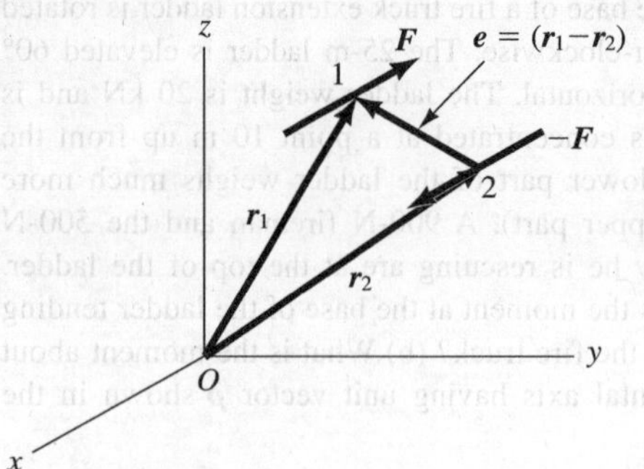

Figure 3.21. Compute moment of couple about O.

Let us now evaluate the moment of the couple about the origin. Position vectors have been drawn in Fig. 3.21 to points 1 and 2 anywhere along the respective line of action of each force. Adding the moment of each force about O, we have for the couple moment $\boldsymbol{M}$

$$\boldsymbol{M} = \boldsymbol{r}_1 \times \boldsymbol{F} + \boldsymbol{r}_2 \times (-\boldsymbol{F})$$
$$= (\boldsymbol{r}_1 - \boldsymbol{r}_2) \times \boldsymbol{F} \tag{3.15}$$

We can see that $(\boldsymbol{r}_1 - \boldsymbol{r}_2)$ is a displacement vector between points 2 and 1, and if we call this vector $\boldsymbol{e}$, the formulation above becomes

$$\boldsymbol{M} = \boldsymbol{e} \times \boldsymbol{F} \tag{3.16}$$

Since $\boldsymbol{e}$ is in the plane of the couple, it is clear from the definition of a cross product that $\boldsymbol{M}$ is in an orientation normal to the plane of the couple. The sense in this case may be seen in Fig. 3.22 to be directed downward, in accordance with the right-hand-screw rule. Note the use of the double arrow to represent the couple moment. Note also that the rotation of $\boldsymbol{e}$ to $\boldsymbol{F}$, as stipulated in the cross-product formulation, is in the same direction as the "turning" action of the two force vectors, and from now on we shall use the latter criterion for determining the sense of rotation to be used with the right-hand-screw rule.

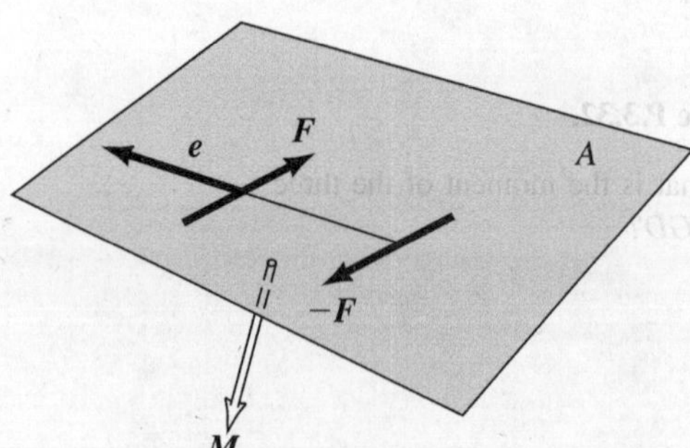

Figure 3.22. The couple moment $\boldsymbol{M}$.

Now that the direction of couple moment $\boldsymbol{M}$ has been established for the couple, we need only compute the magnitude for a complete description. Points 1 and 2 may be chosen anywhere along the lines of action of the forces without changing the resulting moment, since the forces are transmissible for taking moments. Therefore, to compute the magnitude of the couple moment vector it will be simplest to choose positions 1 and 2 so that $\boldsymbol{e}$ is *perpendicular* to the lines of action of the forces ($\boldsymbol{e}$ is then denoted as $\boldsymbol{e}_\perp$). From the definition of the cross product, we can then say:

$$|\boldsymbol{M}| = |\boldsymbol{e}_\perp||\boldsymbol{F}| \sin 90° = |\boldsymbol{e}_\perp||\boldsymbol{F}| = |\boldsymbol{F}|d \tag{3.17}$$

where the more familiar notation, d, has been used in place of $|\boldsymbol{e}_\perp|$ as the perpendicular distance between the lines of action of the forces.

To summarize the preceding discussions, we may say that: The moment of a couple is a vector whose orientation is normal to the plane of the couple and whose sense is determined in accordance with the right-hand-screw rule, using the "turning" action of the forces to give the proper rotation. The magnitude of the couple moment equals the product of either force magnitude comprising the couple times the perpendicular distance between the forces.

Note that in the computation of the moment of the couple about origin O, the final result in no way involved the position of point O. Thus, we can assume immediately that the couple has the *same* moment about every point in space. More about this in the next section.

3.5 The Couple Moment as a Free Vector

Had we chosen any other position in space as the origin, and had we computed the moment of the couple about it, we would have formed the same moment vector. To understand this, note that although the position vectors to points 1 and 2 will change for a new origin O', the *difference* between these vectors (which has been termed $\boldsymbol{e}$) does *not* change, as can readily be observed in Fig. 3.23. Since $\boldsymbol{M} = \boldsymbol{e} \times \boldsymbol{F}$, we can conclude that *the couple has the same moment about every point in space*. The particular line of action of the vector representation of the couple moment that is illustrated in Fig. 3.24 is then of little significance and can be moved anywhere. In short, the *couple moment is a free vector.* That is, we may move this vector anywhere in space without changing its meaning, provided that we keep the direction and magnitude intact. Consequently, *for the purpose of taking moments,* we may move the couple itself anywhere in its own or a parallel plane, provided that the direction of turning is not altered—i.e., we cannot "flip" the couple over. In any of these possible planes, we can also change the magnitude of the forces of the couple to other equal values, provided that the distance d is simultaneously changed so that the product $|\boldsymbol{F}|d$ remains the same. Since none of these steps changes the direction or magnitude of the couple moment, all of them are permissible.

As we pointed out earlier, the only effect of a couple on a rigid body is its turning action, which is represented quantitatively by the moment of the couple—i.e., the couple moment. Since this is its sole effect, it is only natural to represent the couple by

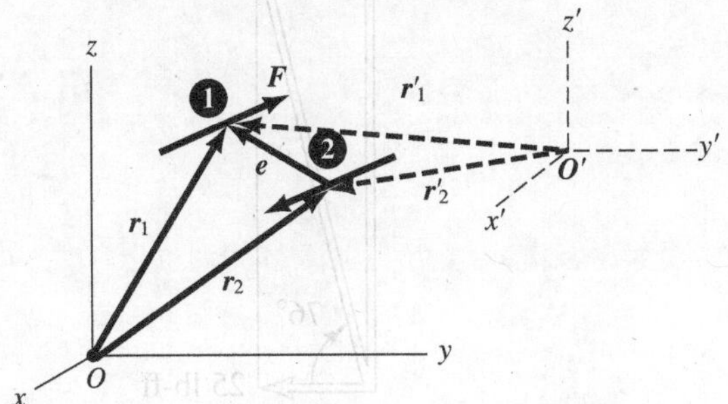

Figure 3.23. Vector $\boldsymbol{e}$ is the same for both references.

specifications of its moment; its magnitude, then, becomes $|F|d$ and its direction that of its moment. This is the same as identifying a person by her/his job (i.e., as a teacher, plumber, etc.). Thus, in Fig. 3.24, the couple moment C may be used to represent the indicated couple.

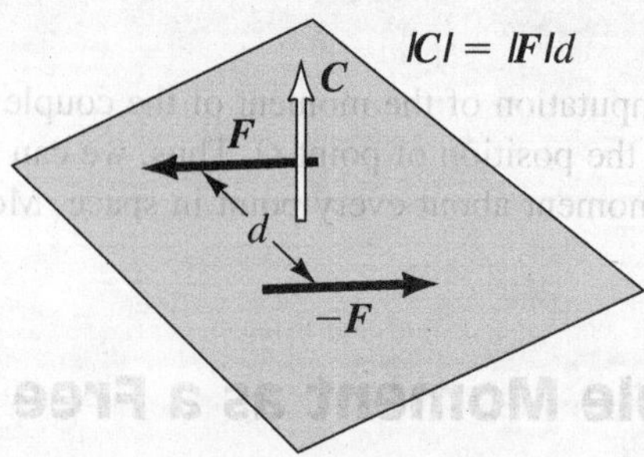

Figure 3.24. C represents couple.

3.6 Addition and Subtraction of Couples

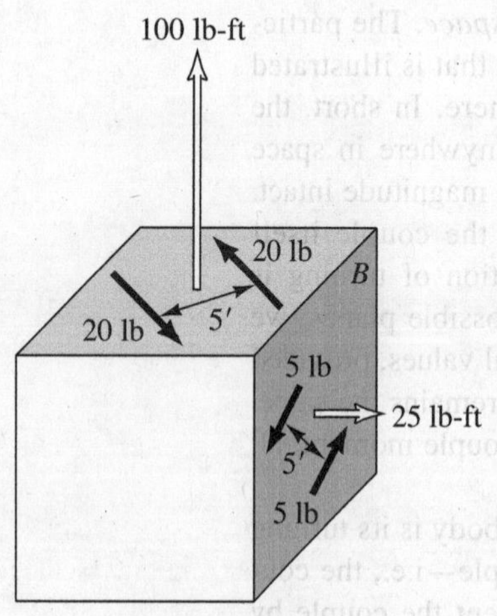

Figure 3.25. Add couples.

Since couples themselves have zero net forces, addition per se of couples always yields zero force. For this reason, the addition and subtraction of couples is interpreted to mean addition and subtraction of the *moments* of the couples. Since couple moments are free vectors, we can always arrange to have a concurrent system of vectors. We shall now take the opportunity to illustrate many of the earlier remarks about couples by adding the two couples shown on the face of the cube in Fig. 3.25. Notice that the couple moment vectors of the couples have been drawn. Since these vectors are free, they may be moved to a convenient position and then added. The total couple moment then becomes 103.2 lb-ft at an angle of 76° with the horizontal, as shown in Fig. 3.26. The couple that creates this turning action is in a plane at right angles to this orientation with a clockwise sense as observed from below.

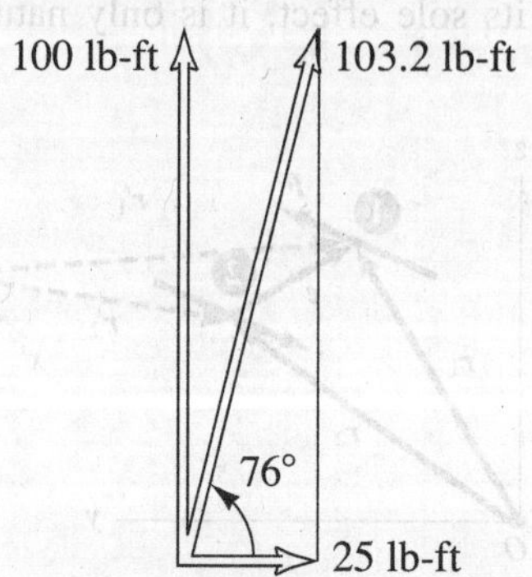

Figure 3.26. Add couple moments.

This addition may be shown to be valid by the following more elementary procedure. The couples of the cube are moved in their respective planes to the positions shown in Fig. 3.27, which does not alter the moment of the couples, as pointed out in Section 3.5. If the couple on plane B is adjusted to have a force magnitude of 20 lb and if the separating distance is decreased to 5/4 ft, the couple moment is not changed (Fig. 3.28). We thus form a system of forces in which two of the forces are equal, opposite, and collinear and, since these two forces together cannot contribute moment, they may be deleted, leaving a single couple on a plane inclined to the original planes (Fig. 3.29). The distance between the remaining forces is

$$\sqrt{25 + \tfrac{25}{16}}\ \text{ft} = 5.16\ \text{ft}$$

and so the magnitude of the couple moment may then be computed to be 103.2 lb-ft. The orientation of the normal to the plane of the couple is readily evaluated as 76° with the horizontal, making the total couple moment identical to our preceding result.

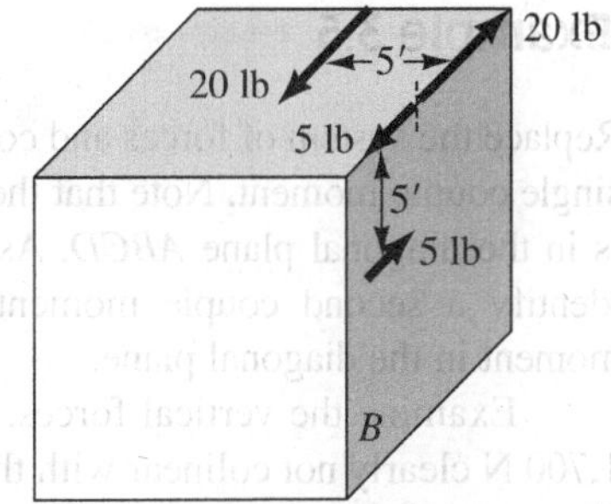

Figure 3.27. Move couples.

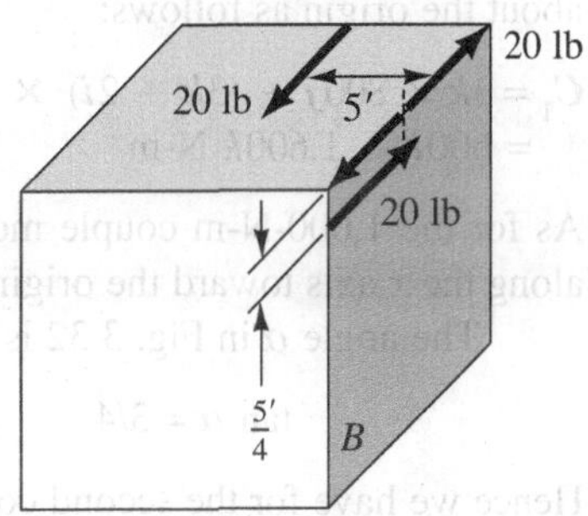

Figure 3.28. Change values of two forces.

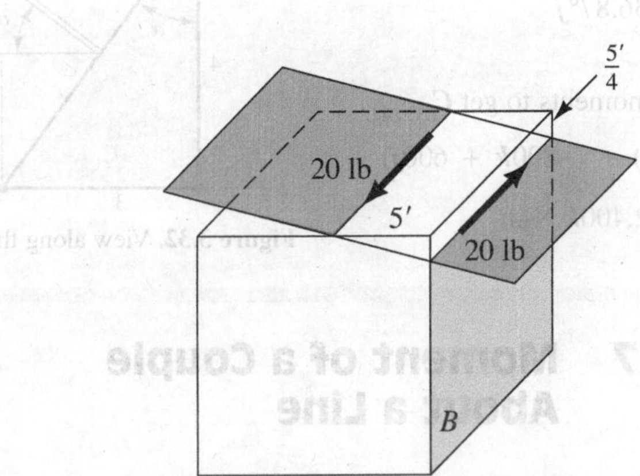

Figure 3.29. Eliminate collinear 20-lb forces.

A common notation for couples in a plane is shown in Fig. 3.30. The values given will be that of the couple moments.

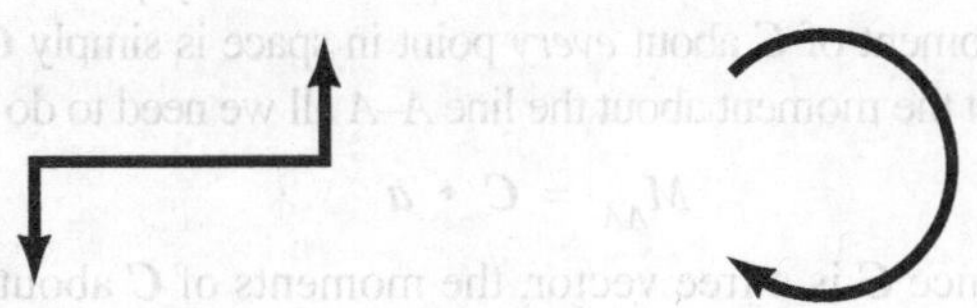
Figure 3.30. Representation of couple moments in a plane.

Example 3.5

Replace the system of forces and couple shown in Fig. 3.31 by a single couple moment. Note that the 1,000-N-m couple moment is in the diagonal plane *ABCD*. As a first step in the problem, identify a second couple moment in addition to the couple moment in the diagonal plane.

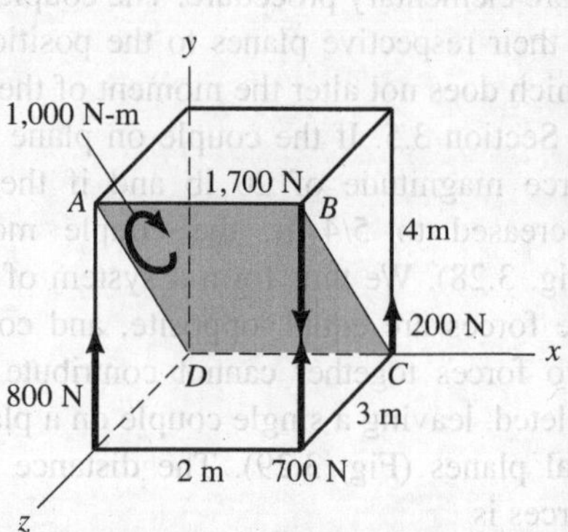

Figure 3.31. Replace system of forces by a single couple moment.

Examine the vertical forces. There is an upward sum of 1,700 N clearly not colinear with the downward 1,700-N force. These forces form the second couple. To get the couple moment for these forces, we take moments of these forces about the origin as follows:

$$C_1 = 3\boldsymbol{k} \times 800\boldsymbol{j} + (3\boldsymbol{k} + 2\boldsymbol{i}) \times (700 - 1700)\boldsymbol{j} + 2\boldsymbol{i} \times 200\boldsymbol{j}$$
$$= 600\boldsymbol{i} - 1{,}600\boldsymbol{k} \text{ N-m}$$

As for the 1,000-N-m couple moment, we look in a direction along the *x* axis toward the origin as shown in Fig. 3.32.

The angle α in Fig. 3.32 is given as follows:

$$\tan\alpha = 3/4 \qquad \therefore \alpha = 36.87°$$

Hence we have for the second couple moment $\boldsymbol{C}_2$.

$$\boldsymbol{C}_2 = -1{,}000 \cos 36.87° \boldsymbol{k} + 1{,}000 \sin 36.87° \boldsymbol{j}$$
$$= -800\boldsymbol{k} + 600\boldsymbol{j} \text{ N-m}$$

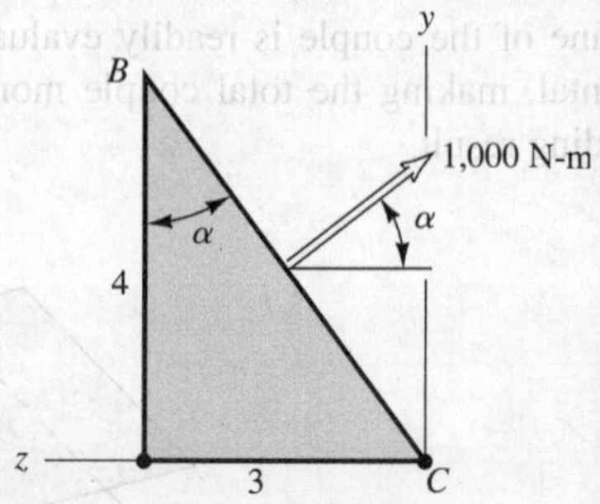

Figure 3.32. View along the *x* axis.

Now we can add the two couple moments to get $\boldsymbol{C}_{TOTAL}$.

$$\boldsymbol{C}_{TOTAL} = \boldsymbol{C}_1 + \boldsymbol{C}_2 = (600\boldsymbol{i} - 1{,}600\boldsymbol{k}) + (-800\boldsymbol{k} + 600\boldsymbol{j})$$

$$\boldsymbol{C}_{TOTAL} = 600\boldsymbol{i} + 600\boldsymbol{j} - 2{,}400\boldsymbol{k} \text{ N-m}$$

3.7 Moment of a Couple About a Line

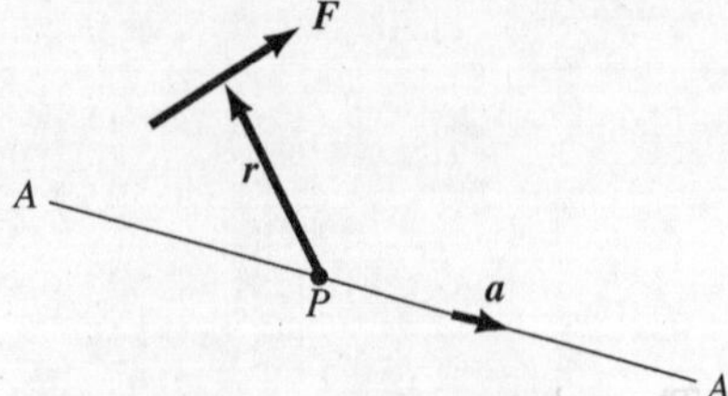

Figure 3.33. To find moment of *F* about *A–A*.

In Section 3.3, we pointed out that the moment of a force ***F*** about a line *A–A* (see Fig. 3.33) is found by first taking the moment of ***F*** about *any* point *P* on *A-A* and then dotting this vector into ***a***, the unit vector along the line. That is,

$$M_{AA} = (\boldsymbol{r} \times \boldsymbol{F}) \cdot \boldsymbol{a} \tag{3.18}$$

Consider now the moment of a couple about a line. For this purpose, we show a couple moment ***C*** and line *A–A* in Fig. 3.34. As before, we first want the moment of the couple about any point *P* along *A–A*. But the moment of ***C*** about *every* point in space is simply ***C*** itself. Therefore, to get the moment about the line *A–A* all we need to do is dot ***C*** into ***a***. Thus,

$$M_{AA} = \boldsymbol{C} \cdot \boldsymbol{a} \tag{3.19}$$

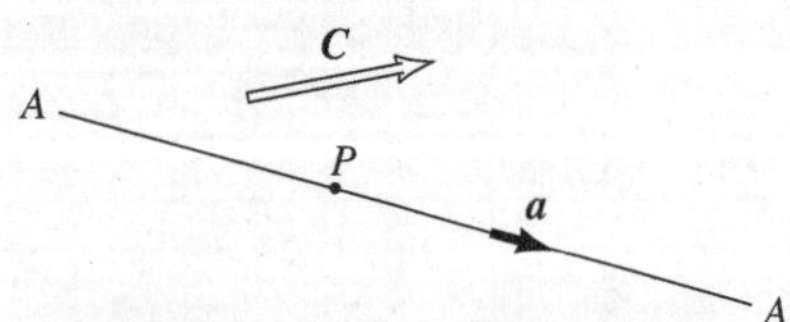

Figure 3.34. To find moment of couple about *A–A*.

Since ***C*** is a free vector, the moments of ***C*** about all lines parallel to *A–A* must have the same value.

Example 3.6

Consider the steering mechanism for a go-cart in Fig. 3.35. The linkages are all in a plane oriented at 45° to the horizontal. This plane is perpendicular to the steering column. In a hard turn, the driver exerts oppositely directed forces of 30 lb with each hand in order to turn the 12-in. diameter steering wheel clockwise as the driver looks at the steering wheel. What is the moment applied to each wheel about an axis normal to the ground? Assume half the transmitted torque goes to each wheel.

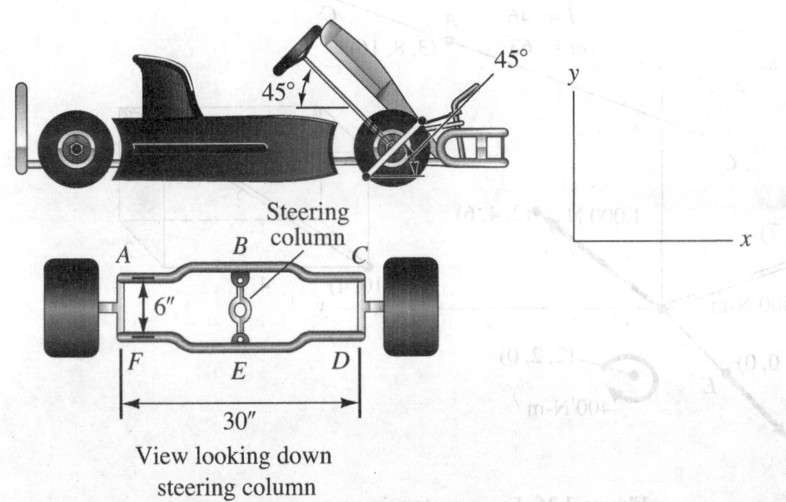

Figure 3.35. Steering mechanism of a go-cart.

The couple moment applied to the steering wheel is readily determined as

$$\boldsymbol{C} = (30)(12)(.707\boldsymbol{i} - .707\boldsymbol{j}) = 254.5\boldsymbol{i} - 254.5\boldsymbol{j} \text{ in-lb}$$

The torque about the vertical axis for each wheel is now easily evaluated. Thus

$$Torque = \frac{1}{2}(254.5\boldsymbol{i} - 254.5\boldsymbol{j}) \cdot \boldsymbol{j} = \boxed{-127.3 \text{ in-lb}}$$

Example 3.7

In Fig. 3.36, find

(a) the sum of the forces
(b) the sum of the couples
(c) the torque of the entire system about axis C–C having direction cosines $l = .46$ and $m = .63$ and going through point A.

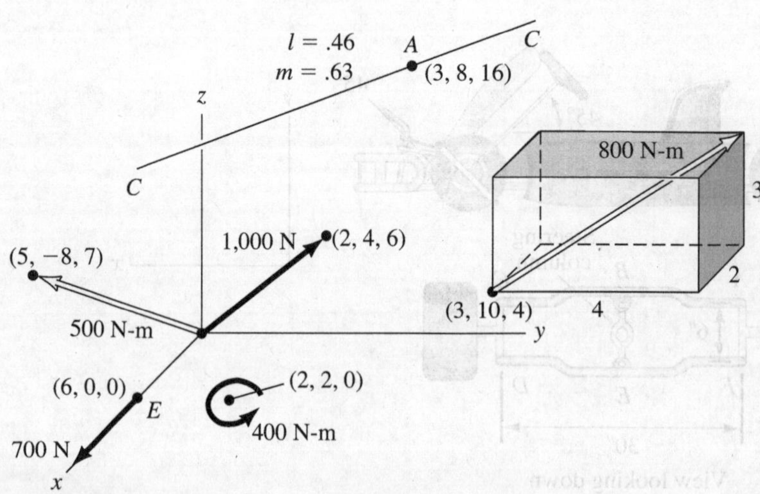

Figure 3.36. Force system in space.

(a)
$$\sum \boldsymbol{F} = 700\boldsymbol{i} + 1{,}000\left[\frac{2\boldsymbol{i} + 4\boldsymbol{j} + 6\boldsymbol{k}}{\sqrt{2^2 + 4^2 + 6^2}}\right]$$

$$= 700\boldsymbol{i} + 267.3\boldsymbol{i} + 534.5\boldsymbol{j} + 801.8\boldsymbol{k}$$

$$\sum \boldsymbol{F} = 967.3\boldsymbol{i} + 534.5\boldsymbol{j} + 801.8\boldsymbol{k}\ \text{N}$$

(b)
$$\sum \boldsymbol{C} = 400\boldsymbol{k} + 500\left[\frac{5\boldsymbol{i} - 8\boldsymbol{j} + 7\boldsymbol{k}}{\sqrt{5^2 + 8^2 + 7^2}}\right]$$

$$+ 800\,\frac{4\boldsymbol{j} - 2\boldsymbol{i} + 3\boldsymbol{k}}{\sqrt{4^2 + 2^2 + 3^2}}\ \text{N-m}$$

Example 3.7 (Continued)

$$\sum C = 400\boldsymbol{k} + 212.8\boldsymbol{i} - 340.5\boldsymbol{j} + 297.9\boldsymbol{k} + 594.2\boldsymbol{j} - 297.1\boldsymbol{i} + 445.7\boldsymbol{k}$$

$$\sum C = -84.3\boldsymbol{i} + 253.7\boldsymbol{j} + 1{,}144\boldsymbol{k} \text{ N-m}$$

(c) To find M_{CC} we proceed by first finding the unit vector along C–C, which we denote as $\hat{\boldsymbol{c}}$. Thus from geometry

$$l^2 + m^2 + n^2 = 1$$
$$\therefore\ .46^2 + .63^2 + n^2 = 1$$
$$n = .6257$$

Hence,

$$\hat{\boldsymbol{c}} = .46\boldsymbol{i} + .63\boldsymbol{j} + .6257\boldsymbol{k}$$

We now get M_{CC}.

$$M_{CC} = \left[(\boldsymbol{r}_E - \boldsymbol{r}_A) \times (700\boldsymbol{i})\right] \bullet \hat{\boldsymbol{c}} + \left[(\boldsymbol{O} - \boldsymbol{r}_A) \times (1{,}000)\frac{2\boldsymbol{i} + 4\boldsymbol{j} + 6\boldsymbol{k}}{\sqrt{2^2 + 4^2 + 6^2}}\right] \bullet \hat{\boldsymbol{c}}$$
$$+ \left[500\frac{5\boldsymbol{i} - 8\boldsymbol{j} + 7\boldsymbol{k}}{\sqrt{5^2 + 8^2 + 7^2}}\right] \bullet \hat{\boldsymbol{c}} + \left[800\frac{4\boldsymbol{j} - 2\boldsymbol{i} + 3\boldsymbol{k}}{\sqrt{4^2 + 2^2 + 3^2}}\right] \bullet \hat{\boldsymbol{c}} + 400\boldsymbol{k} \bullet \hat{\boldsymbol{c}}$$

$$M_{CC} = \{(6\boldsymbol{i} - 3\boldsymbol{i} - 8\boldsymbol{j} - 16\boldsymbol{k}) \times (700\boldsymbol{i}) + (-3\boldsymbol{i} - 8\boldsymbol{j} - 16\boldsymbol{k})$$
$$\times (267.3\boldsymbol{i} + 534.5\boldsymbol{j} + 801.8\boldsymbol{k}) + (212.8\boldsymbol{i} - 340.5\boldsymbol{j} + 297.9\boldsymbol{k})$$
$$+ (594.2\boldsymbol{j} - 297.1\boldsymbol{i} + 445.7\boldsymbol{k}) + 400\boldsymbol{k}\} \bullet (.46\boldsymbol{i} + .63\boldsymbol{j} + .6257\boldsymbol{k})$$

Carrying out calculations in the large bracket, we get:

$$M_{CC} = \{(5{,}600\boldsymbol{k} - 11{,}200\boldsymbol{j}) + (2{,}138\boldsymbol{i} - 1{,}872\boldsymbol{j} + 534.9\boldsymbol{k})$$
$$+ (212.8\boldsymbol{i} - 340.5\boldsymbol{j} + 297.9\boldsymbol{k}) + (594.2\boldsymbol{j} - 297.1\boldsymbol{i} + 445.7\boldsymbol{k})$$
$$+ 400\boldsymbol{k}\} \bullet (.46\boldsymbol{i} + .63\boldsymbol{j} + .6257\boldsymbol{k})$$

$$M_{CC} = 944.7 - 8{,}075 + 4{,}554 = -2{,}576 \text{ N-m}$$

PROBLEMS

3.35. A truck driver, while changing a tire, must tighten the nuts holding on the wheel using a torque of 80 lb-ft. If his tire tool has a length such that the forces from his hand are 22 in. apart, how much force must he exert with each hand? To remove the nuts, he exerted 70 lb with each hand. What torque did he apply?

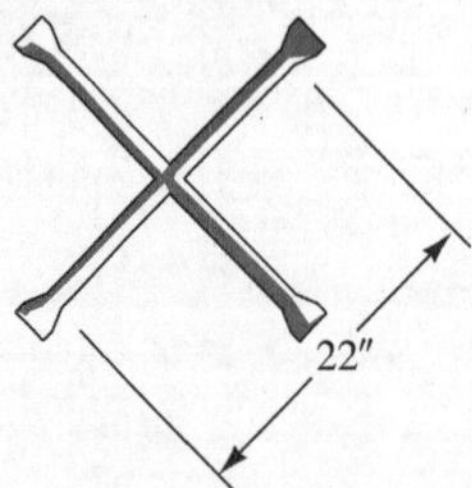

Figure P.3.35.

3.36. Equal couples in the plane of a wheel are shown in Fig. P.3.36. Explain why they are equivalent for the purpose of turning the wheel. Are they equivalent from the viewpoint of the deformation of the wheel? Explain.

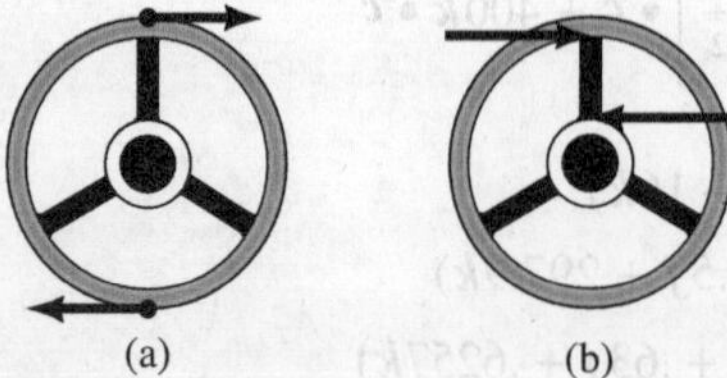

Figure P.3.36.

3.37. Oil-field workers can exert between 50 lb and 125 lb with each hand on a valve wheel (one hand on each side). If a couple moment of 100 lb-ft is required to close the valve, what diameter d must the wheel have?

Figure P.3.37.

3.38. Two children push with 30 lb of force each on the rail of a 10-ft-diameter merry-go-round. What couple moment do they produce? They fasten a 20-ft-long 2-in. by 4-in. board to the merry-go-round so that the middle of the board is at the middle of the merry-go-round. What is the resulting couple moment if they push on the ends of the board? What moment about the merry-go-round axis would they generate by fastening one end of the board to the middle of the merry-go-round and both pushing in the same direction on the other end?

3.39. A posthole digger has a 2-ft-long handle on a 5-ft-long shaft fastened to the digging (scraping) base. From tests, we know that a couple moment of 100 lb-ft is required to dig a posthole in clay, but only 65 lb-ft is needed in sandy soil. What force $\boldsymbol{F}$ must be applied in each case to dig a hole if the distance between the forces from a person's hands is 20 in.?

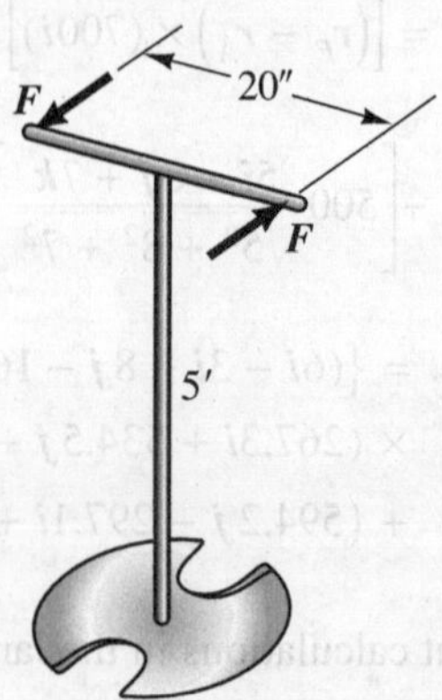

Figure P.3.39.

3.40. While stopping, a truck develops 350 N-m of torque at the rear axle due to the action of the brake drum on the axle. What forces are generated at the front and rear supports of the springs to which the axle is attached?

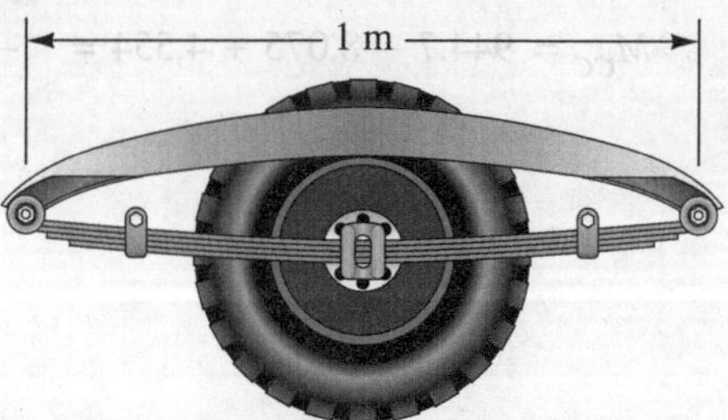

Figure P.3.40.

3.41. A couple is shown in the yz plane. What is the moment of this couple about the origin? About point (6, 3, 4) m? What is the moment of the couple about a line through the origin with direction cosines $l = 0$, $m = .8$, $n = -.6$? If this line is shifted to a parallel position so that it goes through point (6, 3, 4) m, what is the moment of the couple about this line?

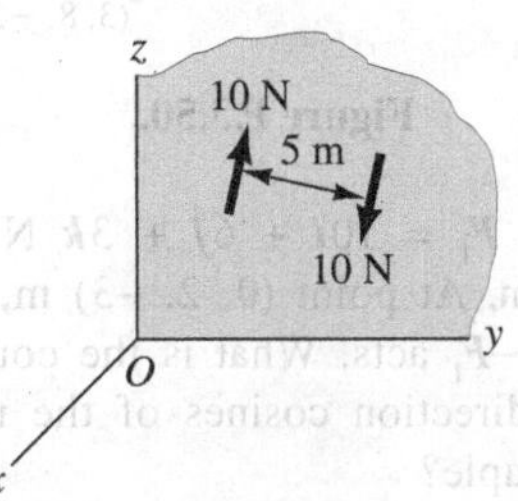

Figure P.3.41.

3.42. Given the indicated forces, what is the moment of these forces about points A and B?

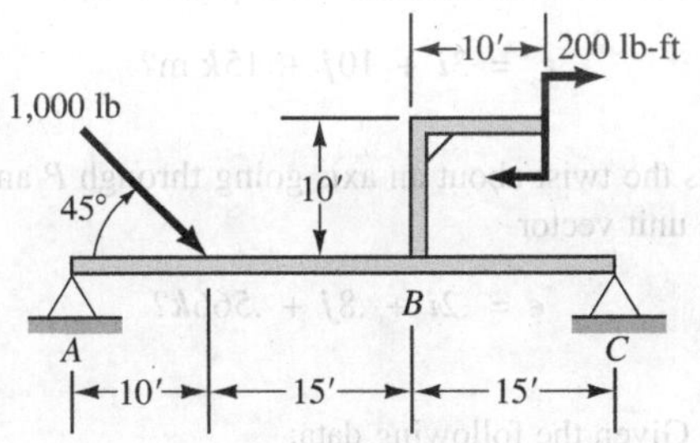

Figure P.3.42.

3.43. An eight-bladed windmill used for power generation and pumping water stops turning because a bearing on the blade shaft has "frozen up." However, the wind still blows, so each blade is subjected to a 25-lb force perpendicular to the (flat) blade surface. The force effectively acts at 2 ft from the centerline of the shaft to which the blades are attached. The blades are inclined at 60° to the axis of rotation. What is the total thrust of all the blade forces on the windmill shaft? What is the moment on the stalled shaft?

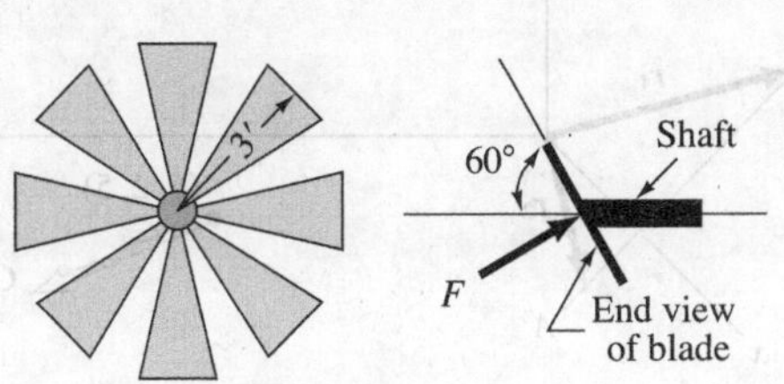

Figure P.3.43.

3.44. What is the moment of the forces shown about point A and about a point P having a position vector

$$r_p = 10i + 7j + 15k \text{ m?}$$

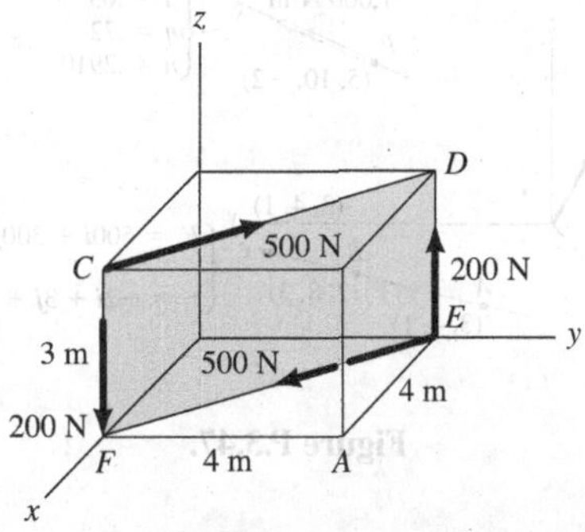

Figure P.3.44.

3.45. Find the torque about axis A–A developed by the 100-lb force and the 3,000-ft-lb couple moment. The position vector r_1 is:

$$r_1 = 10i + 8j + 12k \text{ ft}$$

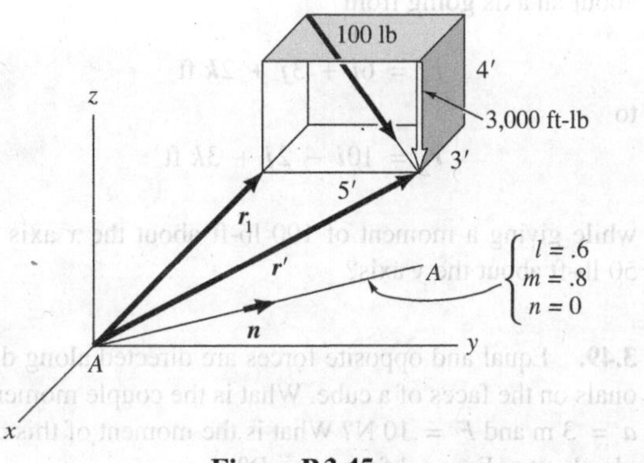

Figure P.3.45.

3.46. Find M_{CD}. Note that the 400-ft-lb couple moment is along the diagonal from point A to point B.

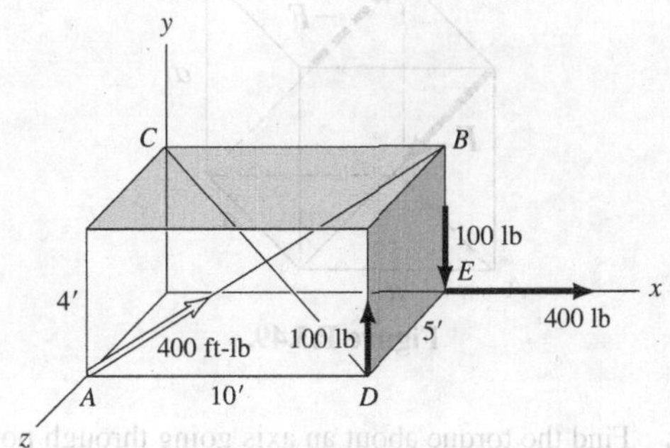

Figure P.3.46.

3.47. Find the torque about an axis going from A to B.

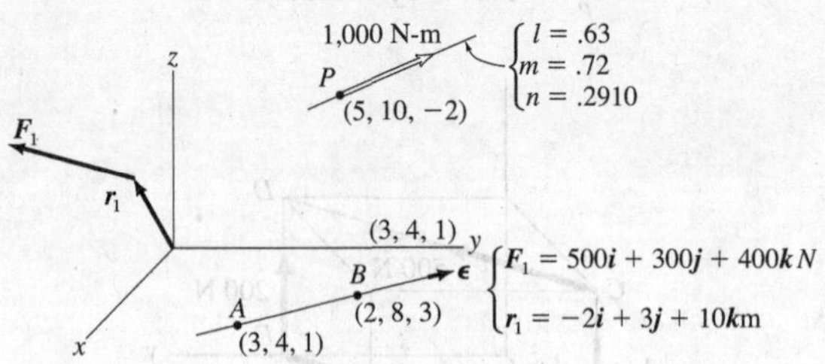

Figure P.3.47.

3.48. Given the couple moments

$$\begin{aligned} C_1 &= 100i + 30j + 82k \text{ lb-ft} \\ C_2 &= -16i + 42j \text{ lb-ft} \\ C_3 &= 15k \text{ lb-ft} \end{aligned}$$

what couple will restrain the twisting action of this system about an axis going from

$$r_1 = 6i + 3j + 2k \text{ ft}$$

to

$$r_2 = 10i - 2j + 3k \text{ ft}$$

while giving a moment of 100 lb-ft about the x axis and 50 lb-ft about the y axis?

3.49. Equal and opposite forces are directed along diagonals on the faces of a cube. What is the couple moment if $a = 3$ m and $F = 10$ N? What is the moment of this couple about a diagonal from A to D?

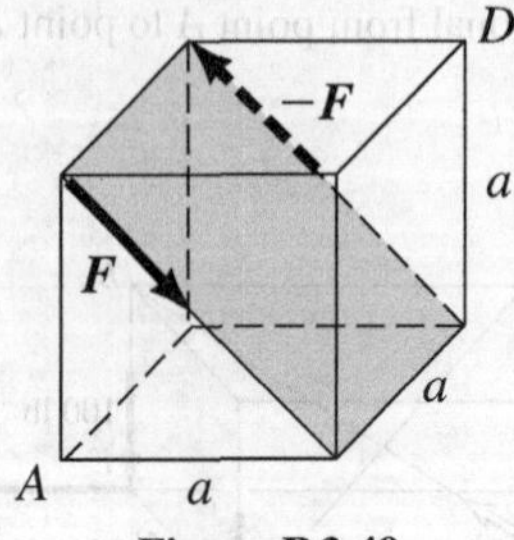

Figure P.3.49.

3.50. Find the torque about an axis going through points A and B.

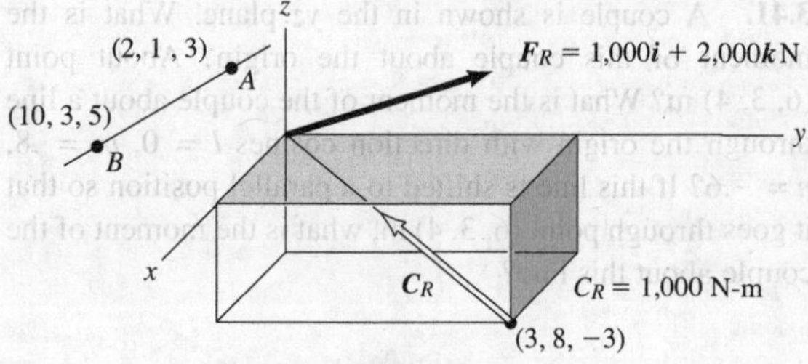

Figure P.3.50.

3.51. A force $F_1 = 10i + 6j + 3k$ N acts at position (3, 0, 2) m. At point (0, 2, −3) m, an equal but opposite force $-F_1$ acts. What is the couple moment? What are the direction cosines of the normal to the plane of the couple?

3.52. Force $F_1 = -16i + 10j - 5k$ N acts at the origin while $F_2 = -F_1$ acts at the end of a rod of length 12 m protruding from the origin with direction cosines $l = .6$, $m = .8$. What is the moment about point P at

$$r_p = 3i + 10j + 15k \text{ m?}$$

What is the twist about an axis going through P and having the unit vector

$$\epsilon = .2i + .8j + .566k?$$

3.53. Given the following data:

$$\begin{aligned} F &= 100i + 300j \text{ N} \\ C &= 200j + 300k \text{ N-m} \\ r_1 &= 3i - 6j + 4k \text{ m} \\ r_2 &= 8i + 3j \text{ m} \end{aligned}$$

Find the torque about axis A–A from F and C.

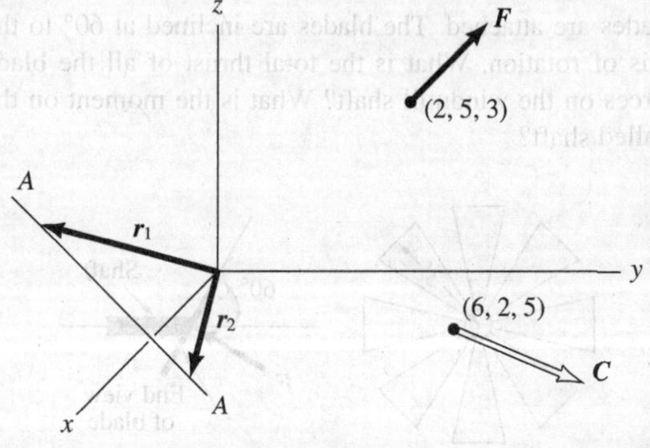

Figure P.3.53.

3.54. An oil-field pump has two valves, one on top and one on the side, that must be closed simultaneously. The valve wheels are each 27 in. in diameter and are turned with both hands by workers who can exert between 50 lb and 125 lb with each hand. If a weak worker turns the side wheel and a strong worker turns the top wheel, what is the total twisting moment (couple moment) on the pump?

3.55. What is the total moment about the origin of the force system shown?

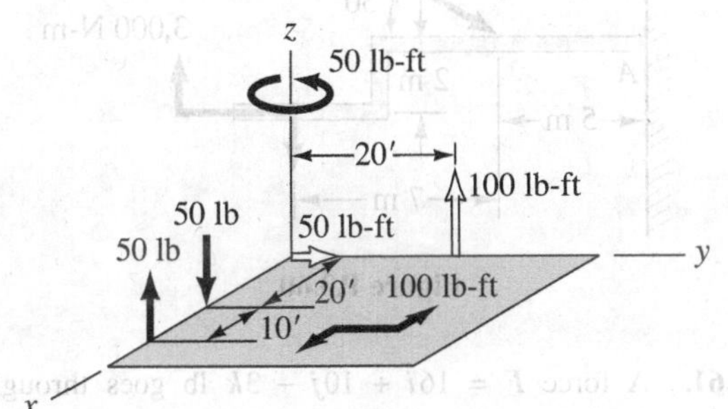

Figure P.3.55.

3.56. Add the couples whose forces act along diagonals of the sides of the rectangular parallelepiped.

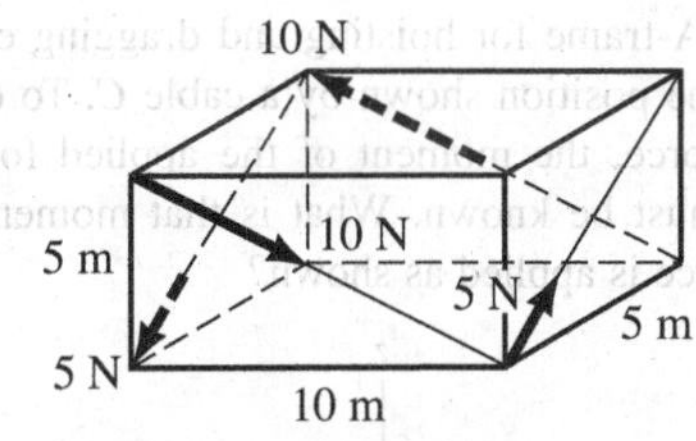

Figure P.3.56.

3.8 Closure

In this chapter, we have considered several important vector quantities and their properties. In particular, for rigid bodies we found we could take certain liberties with a couple without invalidating the results.

Note in particular that in the chapter on vector algebra, the line of action was of no significance. However, it should now be abundantly clear that in taking moments we *cannot* change the line of action of the force. *Only* the line of action of a *couple moment* can be changed to any parallel position for rigid bodies. Moreover, it is important to remember that we can always move a force along its line of action any time we are computing moments.

We are now ready to pursue in greater detail the important subject of equivalence of force systems for rigid body considerations. We will see that in equivalence considerations of rigid bodies, we again must be careful about what to do with lines of action. They will play a vital role in our deliberations.

PROBLEMS

3.57. An A-frame for hoisting and dragging equipment is held in the position shown by a cable C. To determine the cable force, the moment of the applied force about axis B–B must be known. What is that moment when a 1,000-N force is applied as shown?

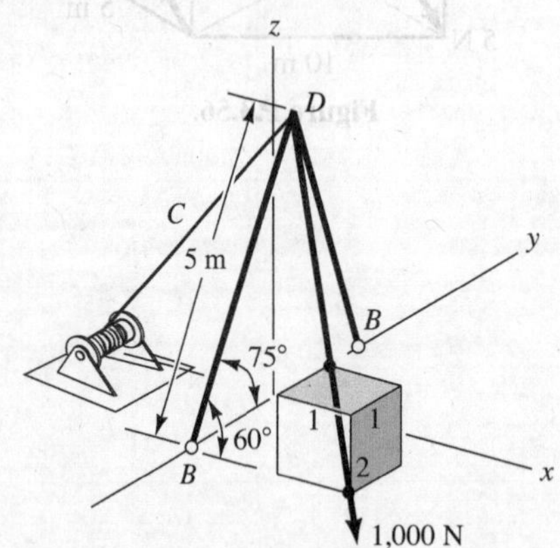

Figure P.3.57.

3.58. A plumber places his hands 18 in. apart on a pipe threader and can push (and pull) with 80 lb of force. What couple moment does he exert? How much could he exert if he moved his hands to the ends so that his hands are 24 in. apart? What force must he apply at the ends to achieve the same couple moment as when he held his hands 18 in. apart?

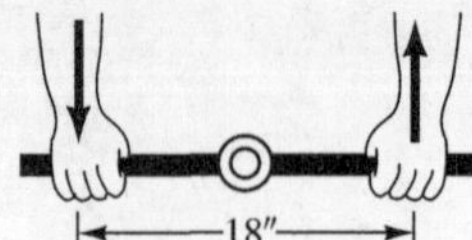

Figure P.3.58.

3.59. Find the torque about a line going from point 1 to point 2.

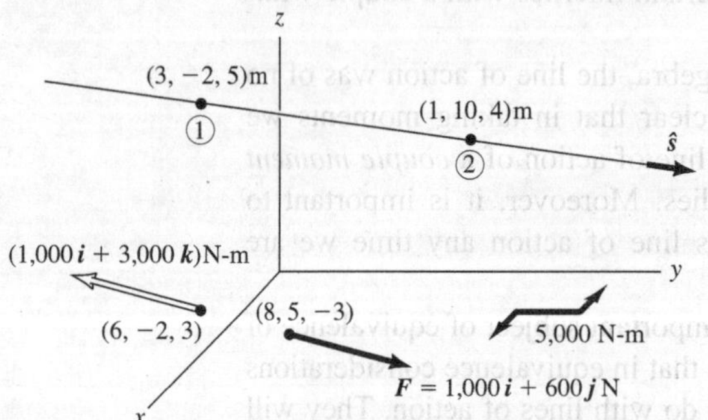

Figure P.3.59.

3.60. What is the moment about A of the 500-N force and the 3,000-N-m couple acting on the cantilever beam?

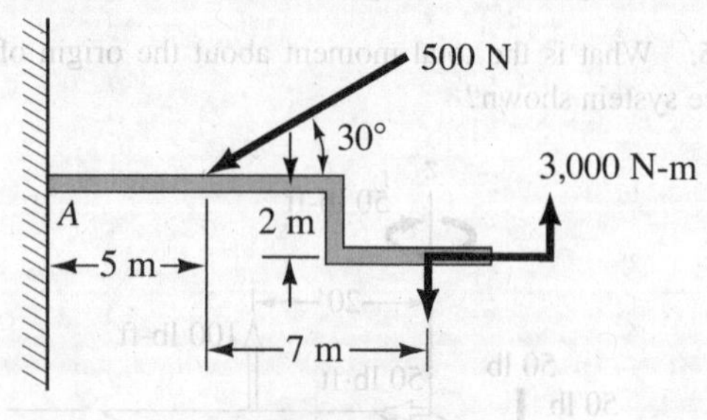

Figure P.3.60.

3.61. A force $\boldsymbol{F} = 16\boldsymbol{i} + 10\boldsymbol{j} - 3\boldsymbol{k}$ lb goes through point a having a position vector $\boldsymbol{r}_a = 16\boldsymbol{i} - 3\boldsymbol{j} + 12\boldsymbol{k}$ ft. What is the moment about an axis going through points 1 and 2 having respective position vectors given as

$$\boldsymbol{r}_1 = 6\boldsymbol{i} + 3\boldsymbol{j} - 2\boldsymbol{k} \text{ ft}$$
$$\boldsymbol{r}_2 = 3\boldsymbol{i} - 4\boldsymbol{j} + 12\boldsymbol{k} \text{ ft?}$$

3.62. Compute the torque about axis C–C from the entire force system shown. The axis goes through the origin.

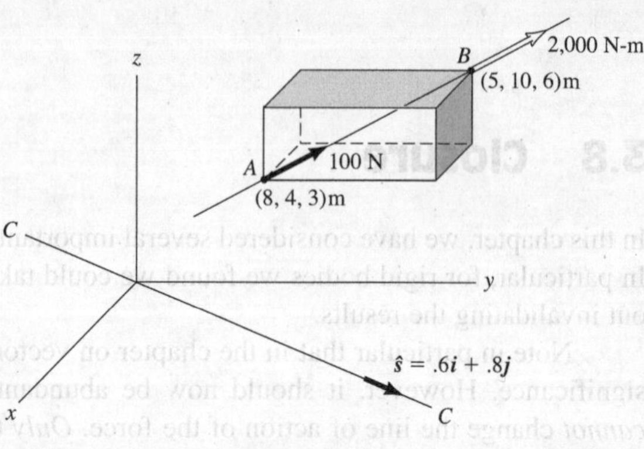

Figure P.3.62.

3.63. What is the total couple moment of the three couples shown? What is the moment of this force system about point (3, 4, 2) ft? What is the moment of this force system about the position vector $\boldsymbol{r} = 3\boldsymbol{i} + 4\boldsymbol{j} + 2\boldsymbol{k}$ ft taken as an axis? What is the total force of this system?

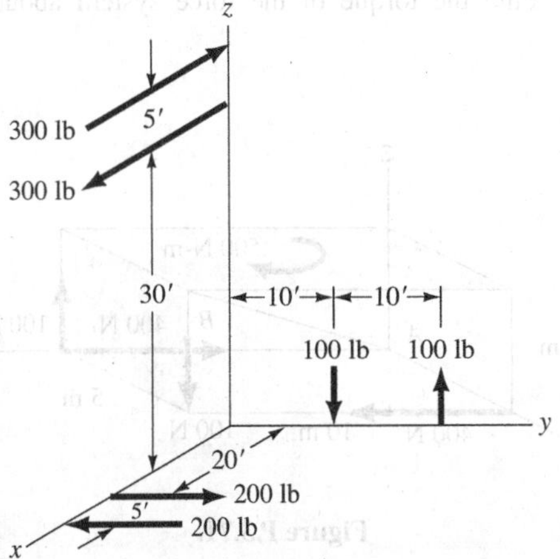

Figure P.3.63.

3.64. Find the torque about axis AB.

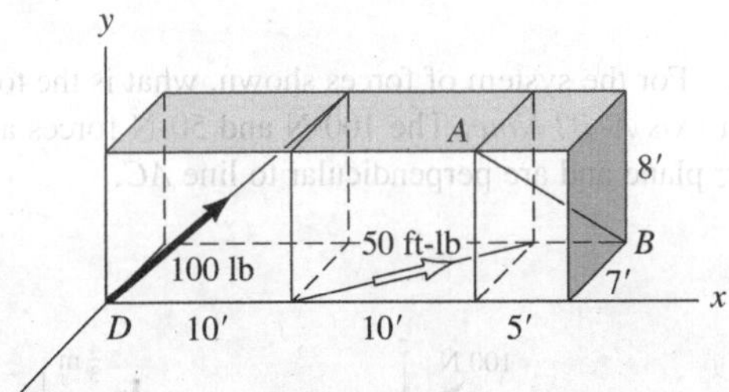

Figure P.3.64.

3.65. A force is developed by a liquid on a pipe any time the pipe changes the direction or the speed of flow as a result of an elbow or a nozzle. Such forces can be of considerable magnitude and must be taken into account in building design. We have shown three such forces. What moment stemming from these forces must be counteracted by the support at O?

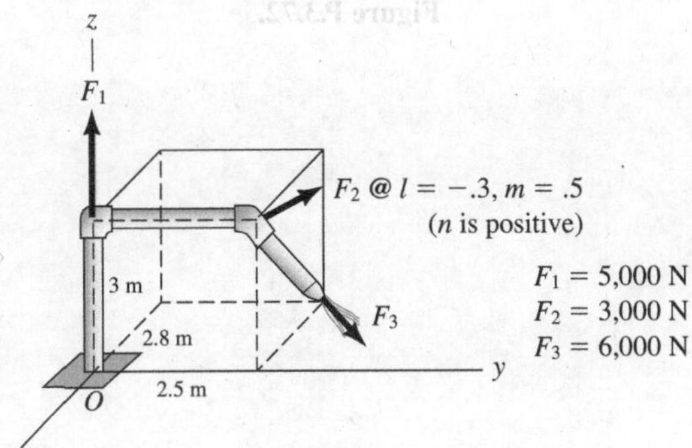

Figure P.3.65.

3.66. Compute the moment of the 300-lb force about points P_1 and P_2.

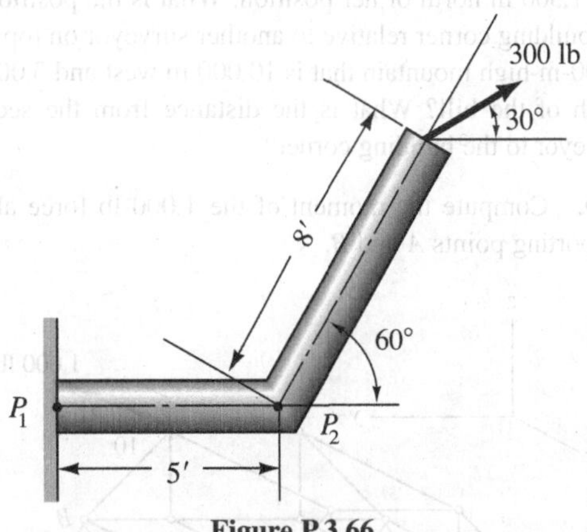

Figure P.3.66.

3.67. A tow truck is inclined at 45° to the edge A-A of a ravine with sides sloping at 45° to the vertical. The operator attaches a cable to a wrecked car in the ravine and starts the winch. The cable is oriented normal to A-A and develops a force of 15 kN. What are the moments tending to tip over the tow truck about the rear wheels (rocking backward). (*Hint:* Use the position vector from C to B in Fig. P.3.67(c).) Note that view D–D is normal to A–A and parallel to the incline.

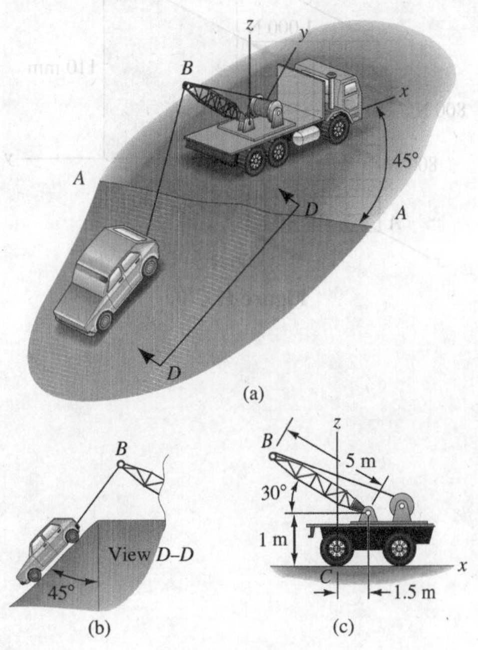

Figure P.3.67.

3.68. A surveyor on a 100-m-high hill determines that the corner of a building at the base of the hill is 600 m east and 1,500 m north of her position. What is the position of the building corner relative to another surveyor on top of a 5,000-m-high mountain that is 10,000 m west and 3,000 m south of the hill? What is the distance from the second surveyor to the building corner?

3.69. Compute the moment of the 1,000-lb force about supporting points *A* and *B*.

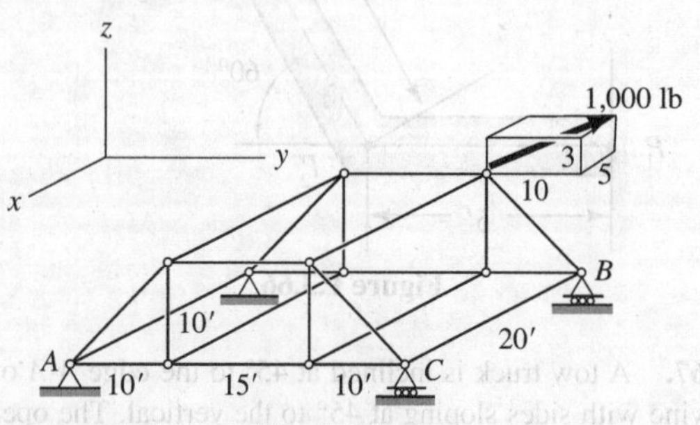

Figure P.3.69.

3.70. What is the turning action of the forces shown about the diagonal *A–D*?

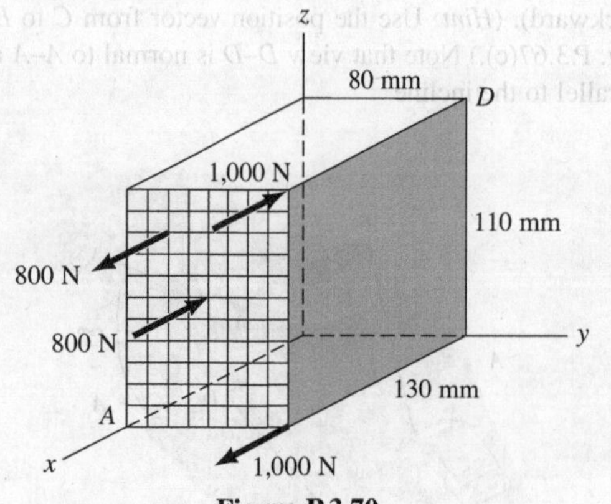

Figure P.3.70.

3.71. Find the torque of the force system about axis *A–B*.

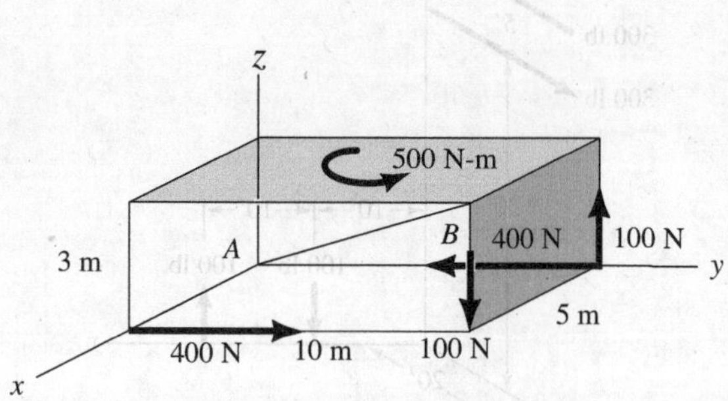

Figure P.3.71.

3.72. For the system of forces shown, what is the torque about axis *A–B*? *Note:* The 100-N and 50-N forces are in the *xz* plane and are perpendicular to line *AC*.

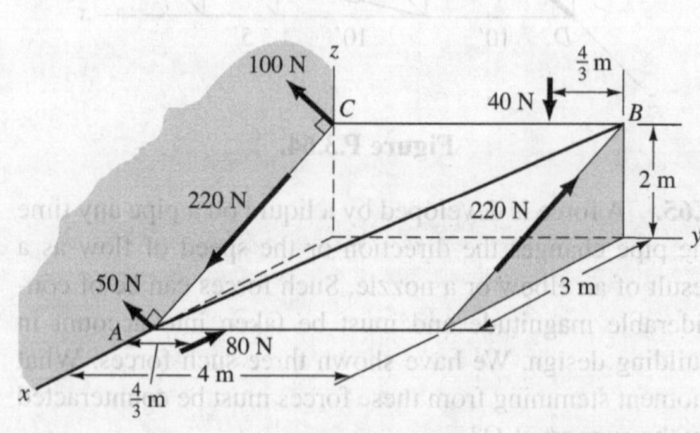

Figure P.3.72.

CHAPTER 4

Equivalent Force Systems

4.1 Introduction

In Chapter 1, we defined equivalent vectors as those that have the same capacity in some given situation. We shall now investigate an important class of situations, namely those in which a rigid-body model can be employed. Specifically, we will be concerned with equivalence requirements for force systems acting on a rigid body. Parenthetically, we will begin to see that the line of action plays a vital role in the mechanics of rigid bodies.

The effect that forces have on a rigid body is only manifested in the motion (or lack of motion) of the body induced by the forces. Two force systems, then, are equivalent if they are capable of *initiating* the same motion of the rigid body. The conditions required to give two force systems this equal capacity are:

1. Each force system must exert an equal "push" or "pull" on the body in any direction. For two systems, this requirement is satisfied if the addition of the forces in each system results in equal force vectors.
2. Each force system must exert an equal "turning" action about any point in space. This means that the moment vectors of the force systems for any chosen point must be equal.

Although these conditions will most likely be intuitively acceptable to the reader, we shall later prove them to be necessary and, for certain situations, sufficient for equivalence when we study dynamics.

As a beginning here, we shall reiterate several basic force equivalences for rigid bodies that will serve as a foundation for more complex cases. You should subject them to the tests listed above.

1. The sum of a set of concurrent forces is a single force that is equivalent to the original system. Conversely, a single force is equivalent to any complete set of its components.
2. A force may be moved along its line of action (i.e., forces are transmissible vectors).
3. The only effect that a couple develops on a rigid body is embodied in the couple moment. Since the couple moment is always a free vector, for our purposes at present the couple may be altered in any way as long as the couple moment is not changed.

Note for (1) and (2) above, we cannot change the line of action alone while maintaining equivalence.

In succeeding sections, we shall present other equivalence relations for rigid bodies and then examine perfectly general force systems with a view to replacing them with more convenient and simpler equivalent force systems. These simple replacements are often called *resultants* of the more general systems.

4.2 Translation of a Force to a Parallel Position

In Fig. 4.1, let us consider the possibility of moving a force ***F*** (solid arrow) acting on a rigid body to a parallel position at point *a* while maintaining rigid-body equivalence. If at position *a* we apply equal and opposite forces, one of which is ***F*** and the other $-\boldsymbol{F}$, a system of three forces is formed that is clearly equivalent to the original single force ***F***. Note that the original force ***F*** and the new force in the opposite sense form a couple (the pair is identified by a wavy connecting line). As usual, we represent the couple by its moment ***C***, as shown in Fig. 4.2, normal to the plane *A* of point *a* and the original force ***F***. The magnitude of the couple moment ***C*** is $|\boldsymbol{F}|d$, where *d* is the perpendicular distance between point *a* and the original line of action of the force. The couple moment may be moved to any parallel position, including the origin, as indicated in Fig. 4.2.

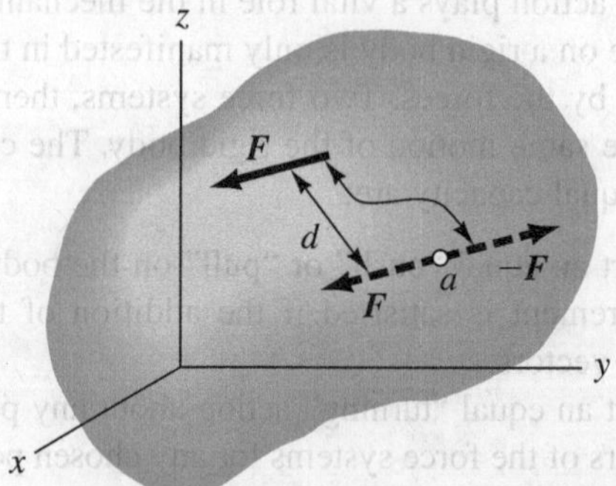

Figure 4.1. Insert equal and opposite forces at *a*.

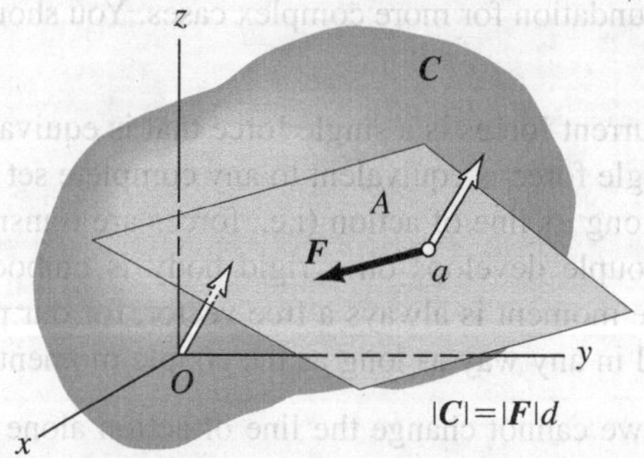

Figure 4.2. Equivalent system at *a*.

Thus, we see that *a force may be moved to any parallel position, provided that a couple moment of the correct orientation and magnitude is simultaneously provided.* There are, then, an infinite number of arrangements possible to get the equivalent effects of a single force on a rigid body.[1]

We now present a simple method for computing the couple moment developed on moving a force to a parallel position. Return to Fig. 4.1 and compute the moment $\boldsymbol{M}$ of the original force $\boldsymbol{F}$ about point a. We can express this as [see Fig. 4.3 (a)]

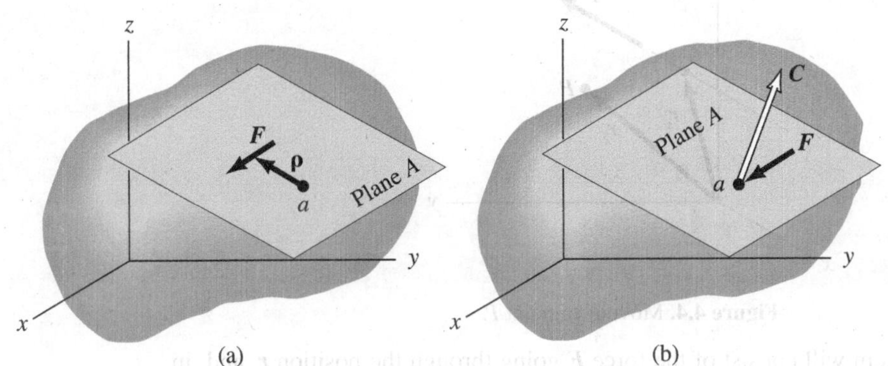

Figure 4.3. Couple moment on moving $\boldsymbol{F}$ is $\boldsymbol{\rho} \times \boldsymbol{F}$.

$$\boldsymbol{M} = \boldsymbol{\rho} \times \boldsymbol{F} \tag{4.1}$$

where $\boldsymbol{\rho}$ is a position vector from a to any point along the line of action of $\boldsymbol{F}$. Now the equivalent force system, shown in Fig. 4.3 (b), must have the *same moment,* $\boldsymbol{M}$, about point a as the original system. Clearly, the moment about point a in Fig 4.3 (b) is due only to the couple moment $\boldsymbol{C}$. That is,

$$\boldsymbol{M} = \boldsymbol{C} \tag{4.2}$$

Accordingly, we conclude, on comparing the previous two equations, that

$$\boldsymbol{C} = \boldsymbol{\rho} \times \boldsymbol{F}$$

Thus, in shifting a force to pass through some new point, *we introduce a couple whose couple moment equals the moment of the force about this new point.*

We illustrate this in the following examples.

[1]A moment of thought will give credence to the above procedure of maintaining rigid body equivalence while moving a force to a different line of action. By moving $\boldsymbol{F}$ to go through point a, you are eliminating the moment about a that existed before the move. The couple moment inserted at the time of the move restores this lost moment.

Example 4.1

A force $\boldsymbol{F} = 6\boldsymbol{i} + 3\boldsymbol{j} + 6\boldsymbol{k}$ lb goes through a point whose position vector is $\boldsymbol{r}_1 = 2\boldsymbol{i} + \boldsymbol{j} + 10\boldsymbol{k}$ ft (see Fig. 4.4). Replace this force by an equivalent force system, for purposes of rigid-body mechanics, going through position P, whose position vector is $\boldsymbol{r}_2 = 6\boldsymbol{i} + 10\boldsymbol{j} + 12\boldsymbol{k}$ ft.

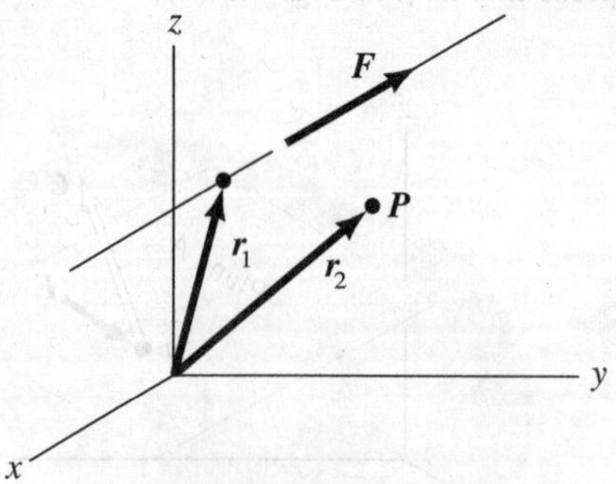

Figure 4.4. Move $\boldsymbol{F}$ to point P.

The new system will consist of the force $\boldsymbol{F}$ going through the position $\boldsymbol{r}_2$ and, in addition, there will be a couple moment $\boldsymbol{C}$ given as

$$\boldsymbol{C} = \boldsymbol{\rho} \times \boldsymbol{F} = (\boldsymbol{r}_1 - \boldsymbol{r}_2) \times \boldsymbol{F}$$

Inserting values, we have

$$\begin{aligned} \boldsymbol{C} &= [(2\boldsymbol{i} + \boldsymbol{j} + 10\boldsymbol{k}) - (6\boldsymbol{i} + 10\boldsymbol{j} + 12\boldsymbol{k})] \\ &\quad \times (6\boldsymbol{i} + 3\boldsymbol{j} + 6\boldsymbol{k}) \\ &= (-4\boldsymbol{i} - 9\boldsymbol{j} - 2\boldsymbol{k}) \times (6\boldsymbol{i} + 3\boldsymbol{j} + 6\boldsymbol{k}) \\ &= \left| \begin{array}{ccc} -4 & -9 & -2 \\ 6 & 3 & 6 \\ \boldsymbol{i} & \boldsymbol{j} & \boldsymbol{k} \\ -4 & -9 & -2 \\ 6 & 3 & 6 \end{array} \right. \end{aligned}$$

Therefore,

$$\boldsymbol{C} = -12\boldsymbol{k} - 12\boldsymbol{j} - 54\boldsymbol{i} + 6\boldsymbol{i} + 24\boldsymbol{j} + 54\boldsymbol{k}$$

$$\boldsymbol{C} = -48\boldsymbol{i} + 12\boldsymbol{j} + 42\boldsymbol{k} \text{ ft-lb}$$

Note that if the position vector $\boldsymbol{r}_1$ to a point along the line of action of $\boldsymbol{F}$ had not been given thus permitting us to get $\boldsymbol{\rho} = (\boldsymbol{r}_1 - \boldsymbol{r}_2)$, we could have used *any* positon vector $\boldsymbol{\rho}$ going from point P to *anywhere* along the line of action of $\boldsymbol{F}$.

Example 4.2

What is the equivalent force system at position A for the 100-N force shown in Fig. 4.5?

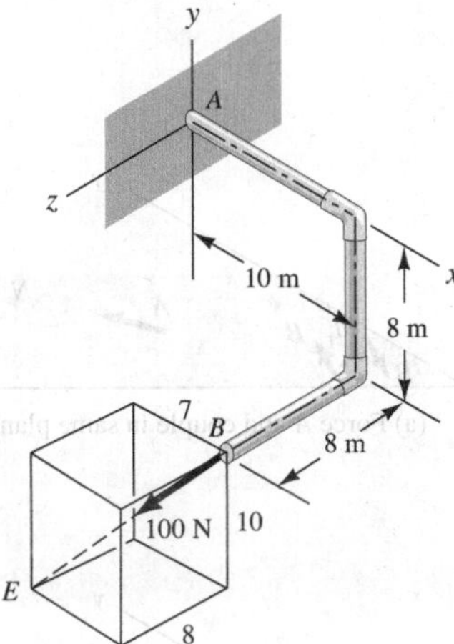

Figure 4.5. Find equivalent force system at A.

The 100-N force can be expressed vectorially as follows:

$$\boldsymbol{F} = F\frac{\overrightarrow{BE}}{|\overrightarrow{BE}|} = 100\left(\frac{-7\boldsymbol{i} - 10\boldsymbol{j} + 8\boldsymbol{k}}{\sqrt{7^2 + 10^2 + 8^2}}\right) \tag{a}$$

$$\boldsymbol{F} = -48.0\boldsymbol{i} - 68.5\boldsymbol{j} + 54.8\boldsymbol{k} \text{ N}$$

We then have the force given above at A. In addition, we have a couple moment, $\boldsymbol{C}$, found using a position vector, $\boldsymbol{r}$, from A to any point along the line of action of the 100-N force. Thus, choosing point B for $\boldsymbol{r}$, we have

$$\boldsymbol{C} = (10\boldsymbol{i} - 8\boldsymbol{j} + 8\boldsymbol{k}) \times (-48.0\boldsymbol{i} - 68.5\boldsymbol{j} + 54.8\boldsymbol{k})$$

$$= \begin{vmatrix} 10 & -8 & 8 \\ -48.0 & -68.5 & 54.8 \\ \boldsymbol{i} & \boldsymbol{j} & \boldsymbol{k} \\ 10 & -8 & 8 \\ -48.0 & -68.5 & 54.8 \end{vmatrix}$$

$$= (10)(-68.5)\boldsymbol{k} + (-48)(8)\boldsymbol{j} + (-8)(54.8)\boldsymbol{i}$$
$$- (8)(-68.5)\boldsymbol{i} - (54.8)(10)\boldsymbol{j} - (-8)(-48.0)\boldsymbol{k}$$

Therefore,

$$\boldsymbol{C} = 109.6\boldsymbol{i} - 932\boldsymbol{j} - 1{,}069\boldsymbol{k} \text{ N-m} \tag{b}$$

The reverse of the procedure just presented may be instituted in reducing a force and a couple *in the same plane* to a *single* equivalent force. This is illustrated in Fig. 4.6 (a) where a couple composed of forces ***B*** and –***B*** a distance d_1 apart and a force ***A*** are shown in plane *N*. The moment representation of the couple is shown with force ***A*** in Fig. 4.6 (b).

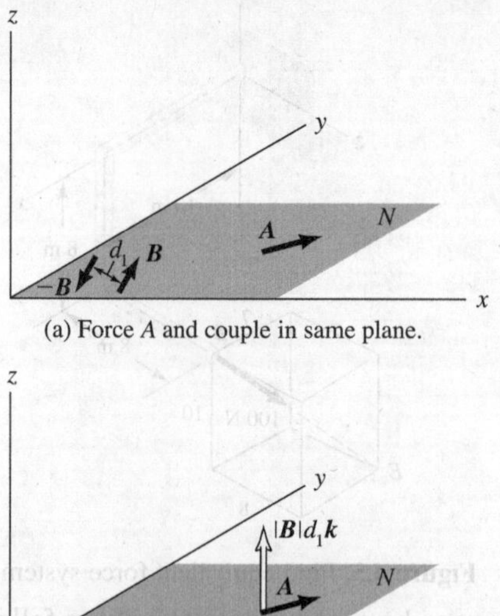

Figure 4.6. A coplanar force and couple.

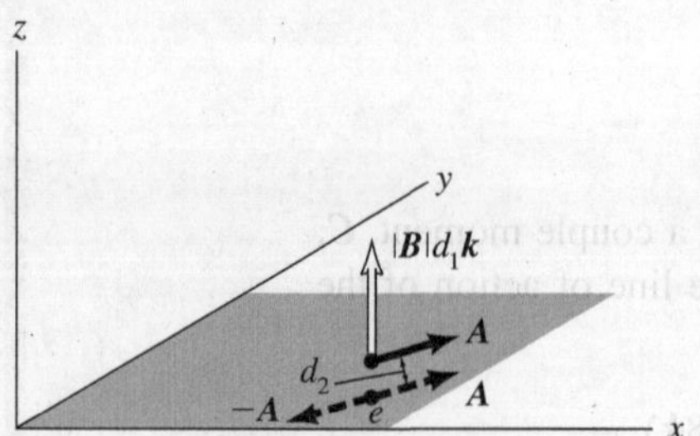

Figure 4.7. Equal and opposite forces placed at *e*.

Equal and opposite forces ***A*** and –***A*** may next be added to the system at a specific position *e* (see Fig. 4.7). The purpose of this step is to form another couple moment with a magnitude $|A|d_2$ equal to $|B|d_1$ and with a directon of turning opposite to the original couple moment (see Fig. 4.8) and this dictates the position of *e*. The couple moments then cancel each other out and we are left with only a single force ***A*** going through point *e*. Therefore, we can always reduce a force and a couple in the same plane to a single force which clearly must have a *specific line of action* for the case at hand.

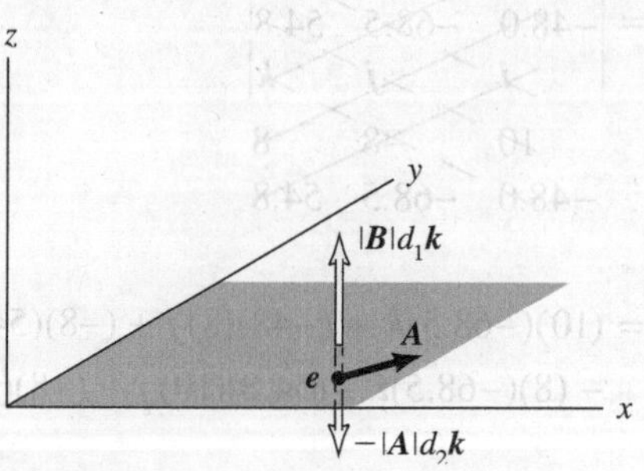

Figure 4.8. Adjust d_2 so that couple moments cancel.

We now exemplify the above procedure in the following example.

Example 4.3

In Fig. 4.9(a), we have shown a cantilever beam supporting a single force and a couple in the xy plane. We wish to reduce this system to a single force equivalent to the given system for purposes of rigid-body mechanics.

In Fig. 4.9(b), we have shown the couple moment and a point e to which we shall shift the 1,000-N force. It should be clear on inspection that the couple moment accompanying this shift will have a sign opposite to the original couple moment. Our task now is to get the correct distance d so as to effect a cancellation of the couple moments. Thus we require that

$$d\boldsymbol{i} \times (1{,}000)\,(.707\boldsymbol{i} - .707\boldsymbol{j}) + 550\boldsymbol{k} = \boldsymbol{0}$$

$$\therefore\; -(707d)\boldsymbol{k} + 550\boldsymbol{k} = \boldsymbol{0} \qquad d = .778 \text{ m}$$

Note we could have reached the above result more simply if we had resolved the force into rectangular components first. Only the vertical component has nonzero moment about e and so dispensing with vectors, we can directly say

$$-707d + 550 = 0 \qquad d = .778 \text{ m}$$

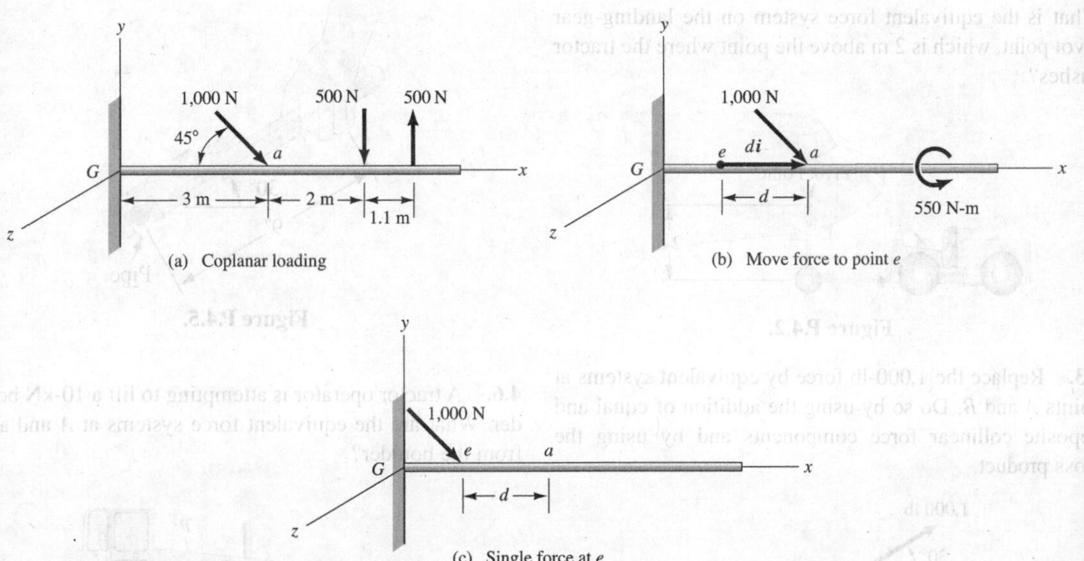

Figure 4.9. Reduction of a coplanar force and couple to a single force with a specific line of action.

PROBLEMS

In several of the problems of this set we shall concentrate the weight of a body at its center of gravity. Most likely you are used to doing this from an earlier physics course. In Section 4.5 we shall justify this procedure.

4.1. Replace the 100-lb force by an equivalent system, from a rigid-body point of view, at A. Do the same for point B. Do this problem by the technique of adding equal and opposite collinear forces and also by using the cross product.

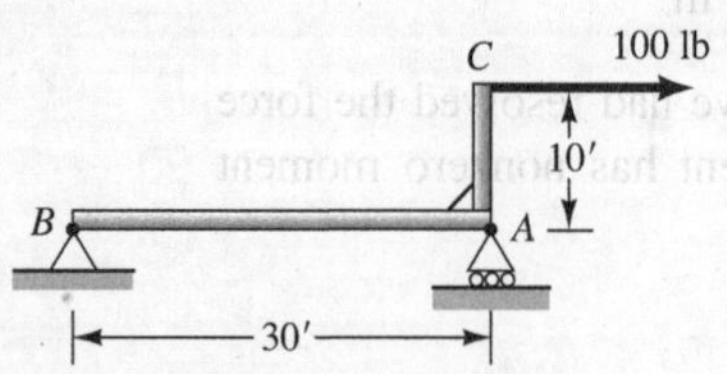

Figure P.4.1.

4.2. To back an airplane away from the boarding gate, a tractor pushes with a force of 15 kN on the nose wheels. What is the equivalent force system on the landing-gear pivot point, which is 2 m above the point where the tractor pushes?

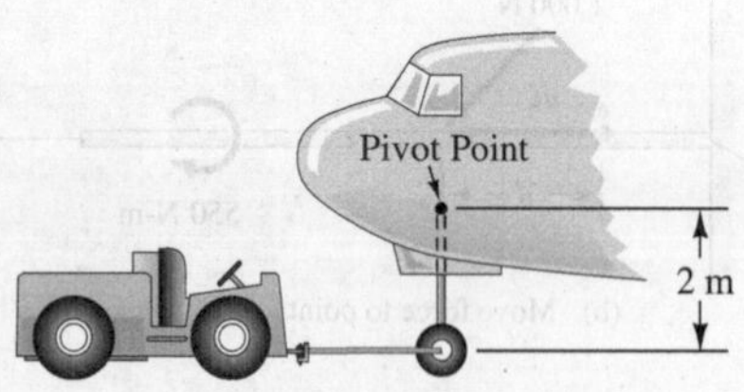

Figure P.4.2.

4.3. Replace the 1,000-lb force by equivalent systems at points A and B. Do so by using the addition of equal and opposite collinear force components and by using the cross product.

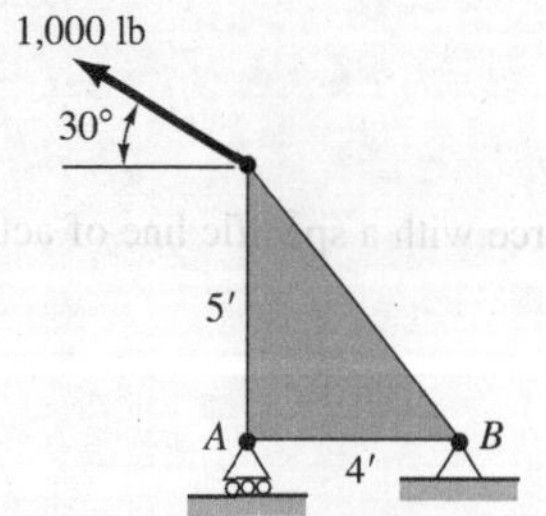

Figure P.4.3.

4.4. A parking-lot gate arms weighs 150 N. Because of the taper, the weight is concentrated at a point $1\frac{1}{4}$ m from the pivot point. What is the equivalent force system at the pivot point?

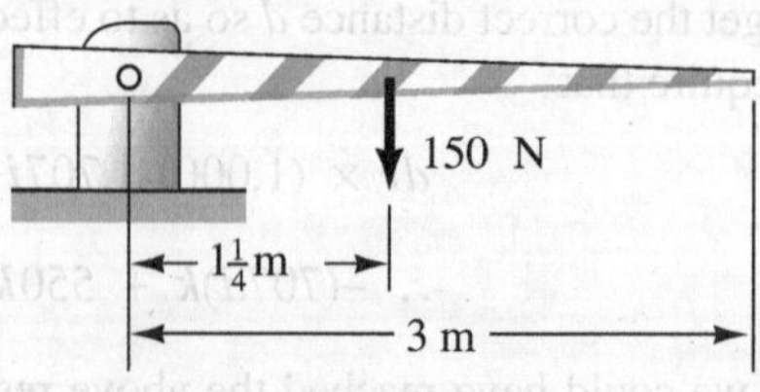

Figure P.4.4.

4.5. A plumber exerts a vertical 60-lb force on a pipe wrench inclined at 30° to the horizontal. What force and couple moment on the pipe are equivalent to the plumber's action?

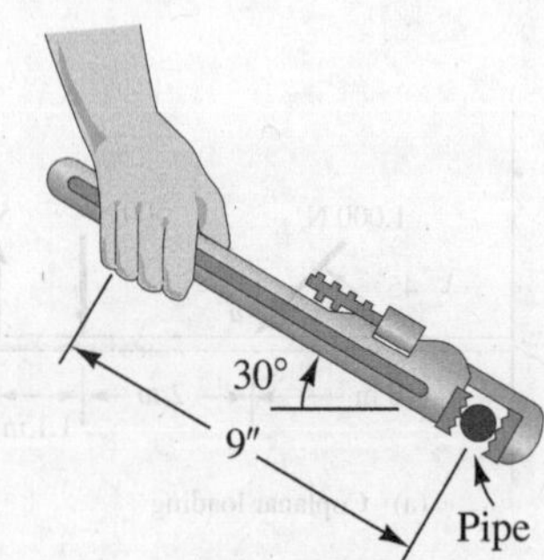

Figure P.4.5.

4.6. A tractor operator is attempting to lift a 10-kN boulder. What are the equivalent force systems at A and at B from the boulder?

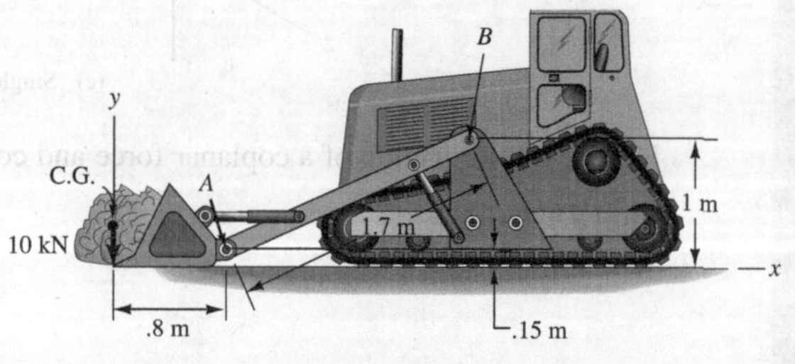

Figure P.4.6.

4.7. A small hoist has a lifting capacity of 20 kN. What are the largest and smallest equivalent force systems at A for the rated maximum capacity?

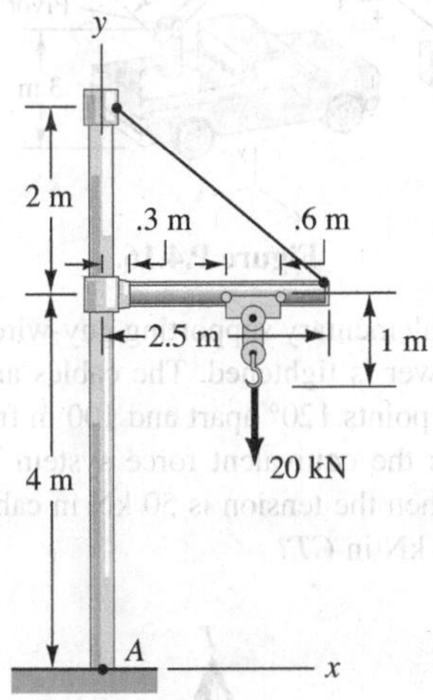

Figure P.4.7.

4.8. Replace the forces by a single equivalent force.

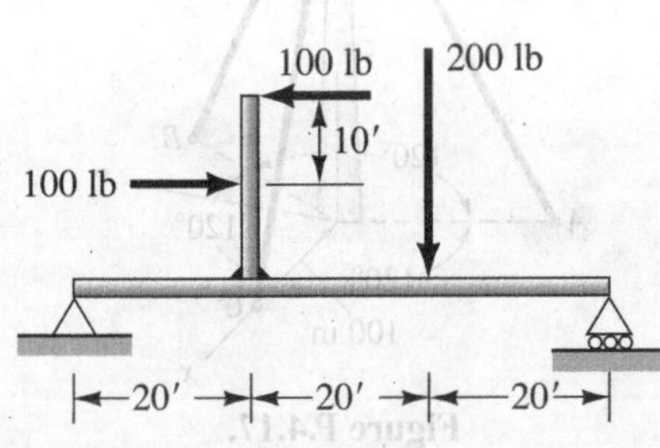

Figure P.4.8.

4.9. Replace the forces and torques shown acting on the apparatus by a single force. Carefully give the line of action of this force.

4.10. A carpenter presses down on a brace-and-bit with a 150-N force while turning the brace with a 200-N force oriented for maximum twist. What is the equivalent force system on the end of the bit at A?

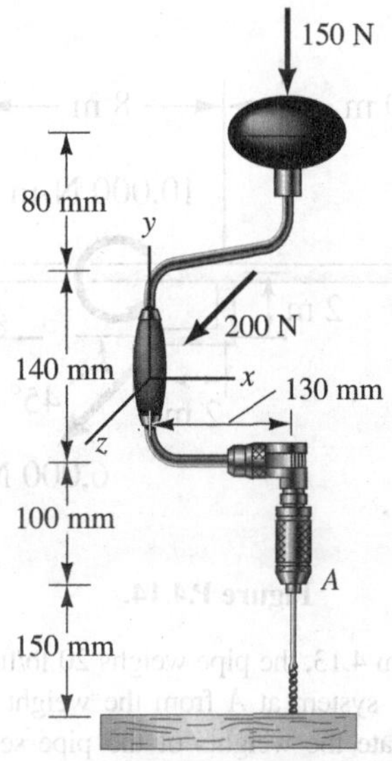

Figure P.4.10.

4.11. A force $\boldsymbol{F} = 3\boldsymbol{i} - 6\boldsymbol{j} + 4\boldsymbol{k}$ lb goes through point (6, 3, 2) ft. Replace this force by an equivalent system where the force goes through point (2, –5, 10) ft.

4.12. A force $\boldsymbol{F} = 20\boldsymbol{i} - 60\boldsymbol{j} + 30\boldsymbol{k}$ N goes through a point (10, – 5, 4) m. What is the equivalent system at point A having position vector $\boldsymbol{r}_A = 20\boldsymbol{i} + 3\boldsymbol{j} - 15\boldsymbol{k}$ m?

4.13. Find the equivalent force system at the base of the cantilever pipe system stemming from force $\boldsymbol{F}$ = 1,000 lb.

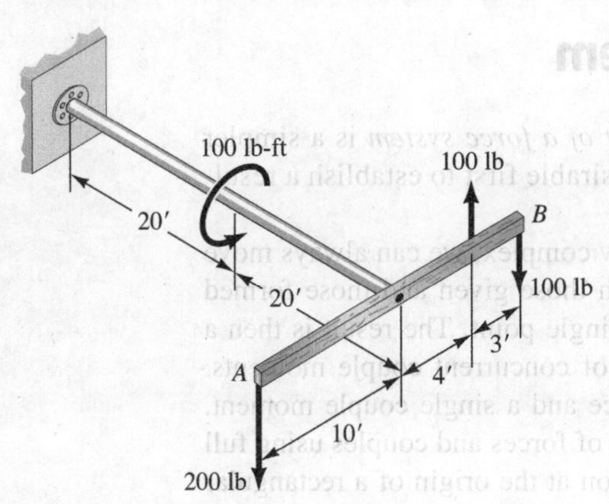

Figure P.4.9.

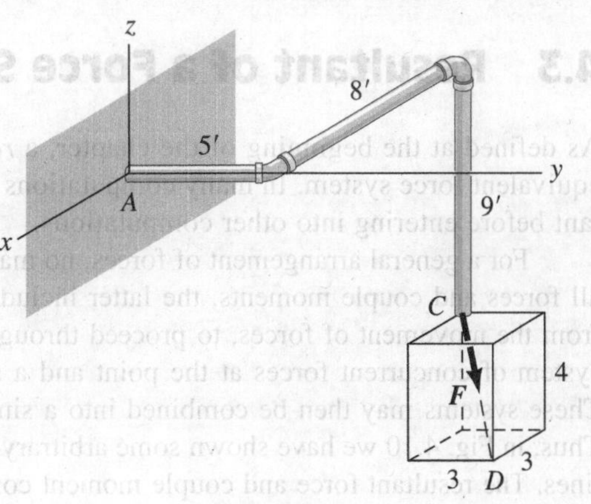

Figure P.4.13.

4.14. Replace the 6,000-N force and the 10,000-N-m couple moment by a single force. Where does this force cross the x axis?

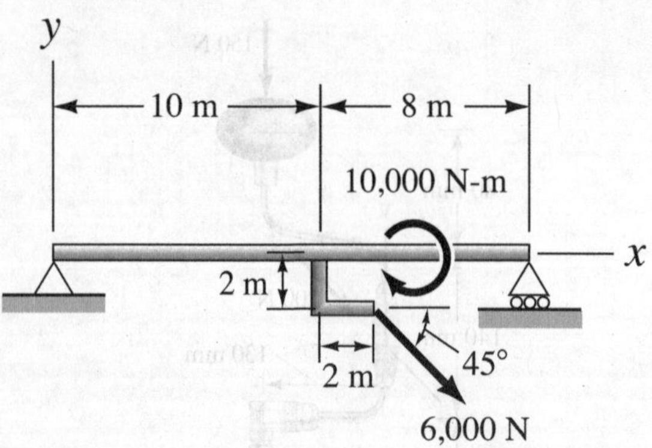

Figure P.4.14.

4.15. In Problem 4.13, the pipe weighs 20 lb/ft. What is the equivalent force system at A from the weight of the pipe? [*Hint:* Concentrate the weights of the pipe sections at the respective centers of gravity (geometric centers in this case).]

4.16. The operator of a small boom-type crane is trying to drag a chunk of concrete. The boom is 10° above the horizontal and rotated 30° clockwise as seen from above. The cable is directed as shown in the diagram and has 60 kN of tension. What is the equivalent force system at the boom pivot point?

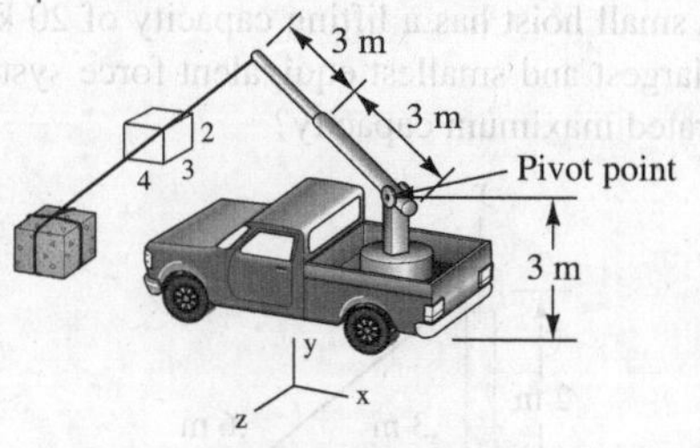

Figure P.4.16.

4.17. A supplementary supporting guy-wire system for a 200-m-tall tower is tightened. The cables are fastened to the ground at points 120° apart and 100 m from the tower base. What is the equivalent force system acting on the tower base when the tension is 50 kN in cable AT, 75 kN in BT, and 25 kN in CT?

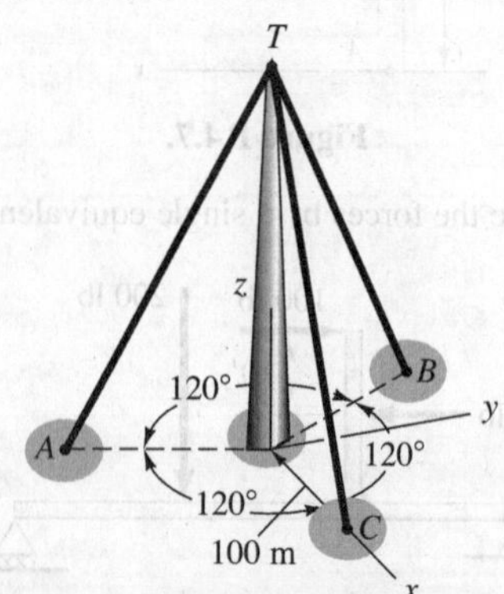

Figure P.4.17.

4.3 Resultant of a Force System

As defined at the beginning of the chapter, a *resultant of a force system* is a simpler equivalent force system. In many computations it is desirable first to establish a resultant before entering into other computations.

For a general arrangement of forces, no matter how complex, we can always move all forces and couple moments, the latter including both those given and those formed from the movement of forces, to proceed through any single point. The result is then a system of concurrent forces at the point and a system of concurrent couple moments. These systems may then be combined into a single force and a single couple moment. Thus, in Fig. 4.10 we have shown some arbitrary system of forces and couples using full lines. The resultant force and couple moment combination at the origin of a rectangular reference is shown as dashed lines.

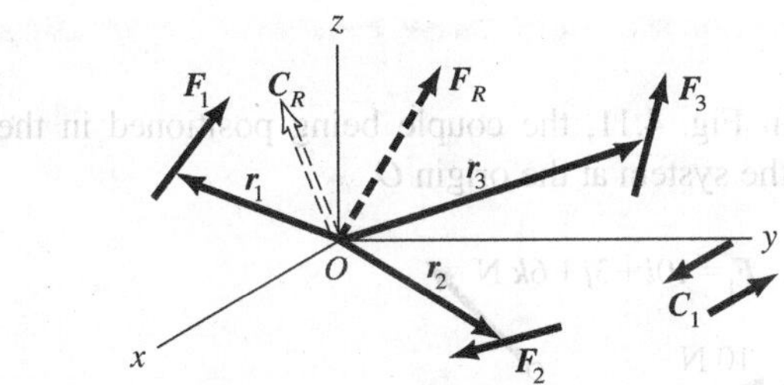

Figure 4.10. Resultant of general force system.

Thus, *any force system can be replaced at any point by equivalents no more complex than a single force and a single couple moment.* In special cases, which we shall examine shortly, we may have simpler equivalents such as a single force or a single couple moment. Finally, for *equilibrium of a body*, it is necessary that at any chosen point the simplest resultant system of force and couple moment acting on the body be zero vectors—a fact that will be discussed in dynamics.[2]

The methods of finding a resultant of forces involve nothing new. In moving to any new point, you will recall, there is no change in the force itself other than a shift of line of action; thus, any component of the *resultant* force, such as the x component, can simply be taken as the sum of the respective x components of all the forces in the system. We may then say for the resultant force

$$\boldsymbol{F}_R = \left[\sum_p (F_p)_x\right]\boldsymbol{i} + \left[\sum_p (F_p)_y\right]\boldsymbol{j} + \left[\sum_p (F_p)_z\right]\boldsymbol{k} \tag{4.3}$$

The couple moment accompanying $\boldsymbol{F}_R$ for a chosen point a may then be given as

$$\boldsymbol{C}_R = [\boldsymbol{r}_1 \times \boldsymbol{F}_1 + \boldsymbol{r}_2 \times \boldsymbol{F}_2 + \ldots] + [\boldsymbol{C}_1 + \boldsymbol{C}_2 + \ldots] \tag{4.4}$$

where the first bracketed quantities result from moving the noncouple forces to a, and the second are simply the sum of the given couple moments. The vectors $\boldsymbol{r}$ are from a to arbitrary points along the lines of action of the forces. In more compact form, the equation above becomes

$$\boldsymbol{C}_R = \sum_p \boldsymbol{r}_p \times \boldsymbol{F}_p + \sum_q \boldsymbol{C}_q \tag{4.5}$$

The following example is an illustration of the procedure.

[2]When we refer hereafter to a resultant, we shall mean the simplest resultant.

Example 4.4

Two forces and a couple are shown in Fig. 4.11, the couple being positioned in the *zy* plane. We shall find the resultant of the system at the origin *O*.

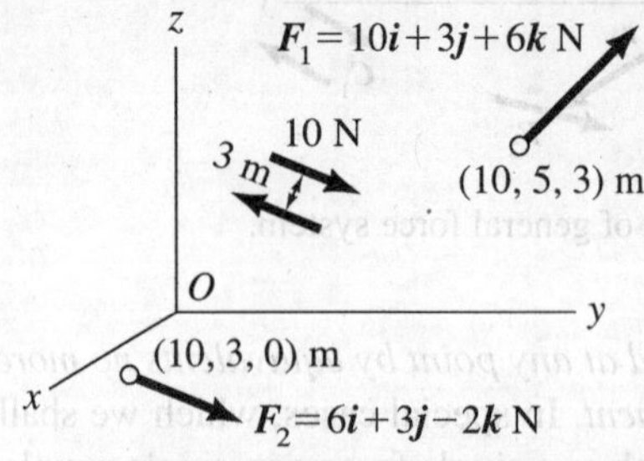

Figure 4.11. Find resultant at *O*.

At *O* we will have a set of two concurrent forces, which may be added to give $\boldsymbol{F}_R$:

$$\boldsymbol{F}_R = (10 + 6)\boldsymbol{i} + (3 + 3)\boldsymbol{j} + (6 - 2)\boldsymbol{k}$$

$$\boldsymbol{F}_R = 16\boldsymbol{i} + 6\boldsymbol{j} + 4\boldsymbol{k} \text{ N}$$

The resultant couple moment at point *O* is the vector sum of the couple-moment vectors developed by moving the two forces, plus the couple moment of the couple in the *zy* plane. Thus,

$$\boldsymbol{C}_R = \boldsymbol{r}_1 \times \boldsymbol{F}_1 + \boldsymbol{r}_2 \times \boldsymbol{F}_2 - 30\boldsymbol{i} \text{ N-m}$$

Now

$$\begin{aligned}\boldsymbol{r}_1 \times \boldsymbol{F}_1 &= (10\boldsymbol{i} + 5\boldsymbol{j} + 3\boldsymbol{k}) \times (10\boldsymbol{i} + 3\boldsymbol{j} + 6\boldsymbol{k}) \\ &= 21\boldsymbol{i} - 30\boldsymbol{j} - 20\boldsymbol{k} \text{ N-m}\end{aligned}$$

$$\begin{aligned}\boldsymbol{r}_2 \times \boldsymbol{F}_2 &= (10\boldsymbol{i} + 3\boldsymbol{j}) \times (6\boldsymbol{i} + 3\boldsymbol{j} - 2\boldsymbol{k}) \\ &= -6\boldsymbol{i} + 20\boldsymbol{j} + 12\boldsymbol{k} \text{ N-m}\end{aligned}$$

Hence,

$$\boldsymbol{C}_R = -15\boldsymbol{i} - 10\boldsymbol{j} - 8\boldsymbol{k} \text{ N-m}$$

The resultant is shown in Fig. 4.12.

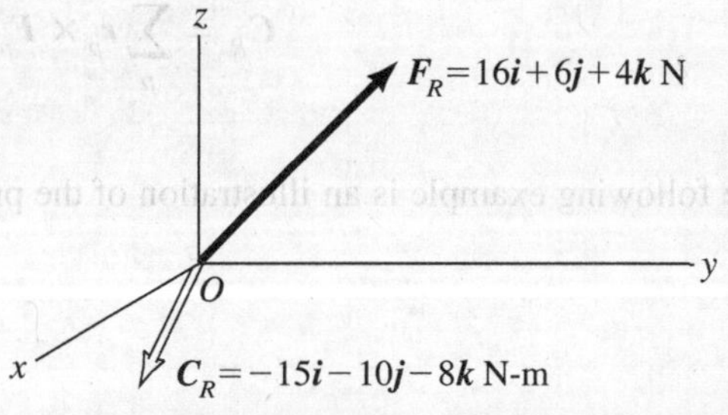

Figure 4.12. Resultant at *O*.

Example 4.5

What is the resultant at A of the applied loads acting in Fig. 4.13?

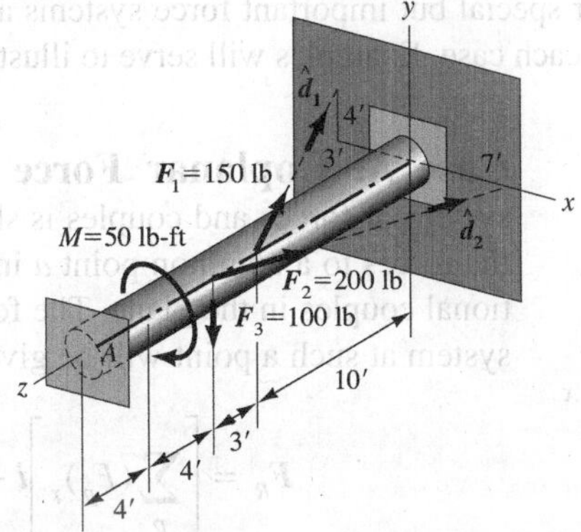

Figure 4.13. Find resultant at A; $\boldsymbol{F}_1$ and $\boldsymbol{F}_3$ are concurrent.

The forces are directed to intersect the centerline of the shaft along which we have placed the z axis. We first express the loads vectorially. Thus,

$$\boldsymbol{F}_1 = F_1\hat{\boldsymbol{d}}_1 = 150\left(\frac{-10\boldsymbol{k} - 3\boldsymbol{i} + 4\boldsymbol{j}}{\sqrt{10^2 + 3^2 + 4^2}}\right)$$

$$= -40.2\boldsymbol{i} + 53.7\boldsymbol{j} - 134.1\boldsymbol{k} \text{ lb}$$

$$\boldsymbol{F}_2 = F_2\hat{\boldsymbol{d}}_2 = 200\left(\frac{-13\boldsymbol{k} + 7\boldsymbol{i}}{\sqrt{13^2 + 7^2}}\right)$$

$$= 94.8\boldsymbol{i} - 176\boldsymbol{k} \text{ lb}$$

$$\boldsymbol{F}_3 = -100\boldsymbol{j} \text{ lb}$$

$$\boldsymbol{C} = -50\boldsymbol{k} \text{ ft-lb}$$

We can now readily find the resultant force system at A. Thus,[3]

$$\boldsymbol{F}_R = (-40.2 + 94.8)\boldsymbol{i} + (53.7 - 100)\boldsymbol{j} + (-134.1 - 176.0)\boldsymbol{k}$$

$$\boldsymbol{F}_R = 54.6\boldsymbol{i} - 46.3\boldsymbol{j} - 310\boldsymbol{k} \text{ lb}$$

$$\boldsymbol{C}_R = (-11\boldsymbol{k}) \times \boldsymbol{F}_1 + (-8\boldsymbol{k}) \times (\boldsymbol{F}_2 + \boldsymbol{F}_3) + (-50\boldsymbol{k})$$

$$= -11\boldsymbol{k} \times (-40.2\boldsymbol{i} + 53.7\boldsymbol{j} - 134.1\boldsymbol{k}) + (-8\boldsymbol{k}) \times (94.8\boldsymbol{i} - 176.0\boldsymbol{k} - 100\boldsymbol{j}) - 50\boldsymbol{k}$$

$$\boldsymbol{C}_R = -209\boldsymbol{i} - 316\boldsymbol{j} - 50\boldsymbol{k} \text{ ft-lb}$$

[3]Remember that for $\boldsymbol{C}$ resulting from a movement of a force, the position vector goes from the point (in this case point A) to the line of action of the force.

4.4 Simplest Resultants of Special Force Systems

We shall now consider special but important force systems and will establish the *simplest* resultants possible for each case. Examples will serve to illustrate the method of procedure.

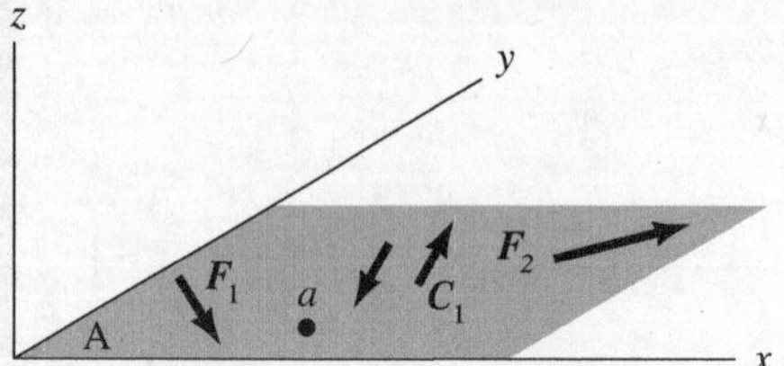

Figure 4.14. Coplanar force system.

Case A. Coplanar Force Systems. In Fig. 4.14, a system of forces and couples is shown in plane A. By moving the forces to a common point a in plane A, we will form additional couples in the plane. The force portion of the equivalent system at such a point will be given as

$$\boldsymbol{F}_R = \left[\sum_p (F_p)_x\right]\boldsymbol{i} + \left[\sum_p (F_p)_y\right]\boldsymbol{j} \tag{4.6}$$

The couple moment portion of the equivalent system can be given as (using the right hand rule for proper signs):

$$\boldsymbol{C}_R = (F_1 d_1 + F_2 d_2 + \ldots)\boldsymbol{k} + (C_1 + C_2 + \ldots)\boldsymbol{k} \tag{4.7}$$

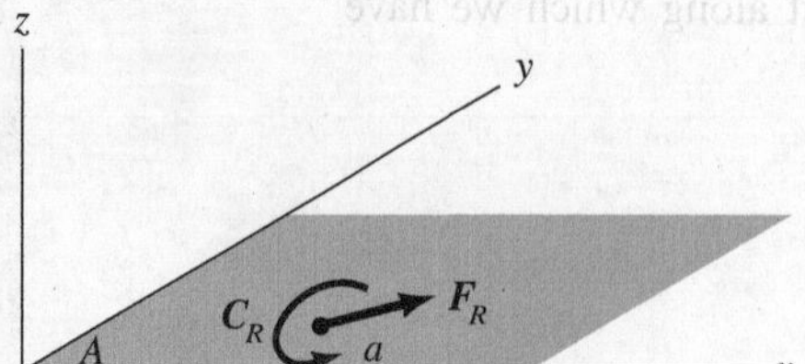

Figure 4.15. Resultant at point a.

where d_1, d_2, etc., are perpendicular distances from point a to the lines of action of the noncouple forces, and C_1, C_2, etc., are the values of the given couple moments. The resultant at a is shown in Fig. 4.15.

If $\boldsymbol{F}_R \neq \boldsymbol{0}$, (i.e., if $\sum_p F_x \neq 0$ and/or $\sum_p F_y \neq 0$) we can move the force from a to yet a new parallel position so as to introduce a second couple moment to cancel $\boldsymbol{C}_R$ of Fig. 4.15 in the manner described earlier in Section 4.2. Since the x and y directions used are arbitrary, except for the condition that they be in the plane of the forces, we can make the following conclusion. *If the force components in any direction in the plane add to other than zero, we may replace the entire coplanar system by a single force with a specific line of action.*

What happens if $\sum_p F_x = 0$ and $\sum_p F_y = 0$? Without a force at point a, we can no longer eliminate a couple in plane A. Thus, our second conclusion is that *if* $\sum_p F_x$ *and* $\sum_p F_y$ *are zero, the resultant must be a couple moment or be zero.*

In the coplanar case, therefore, the simplest equivalent force system must be a single force along a specific line of action, a single couple moment, or a null vector. The following example is used to illustrate the method of determining such a resultant directly without the intermediate steps followed in this discussion.

Example 4.6

Consider a coplanar force system shown in Fig. 4.16. The *simplest* resultant is to be found. Since $\sum_p F_x$ and $\sum_p F_y$ are not zero, we know that we can replace the system by a single force, which is

$$F_R = 6i + 13j \text{ N} \qquad \text{(a)}$$

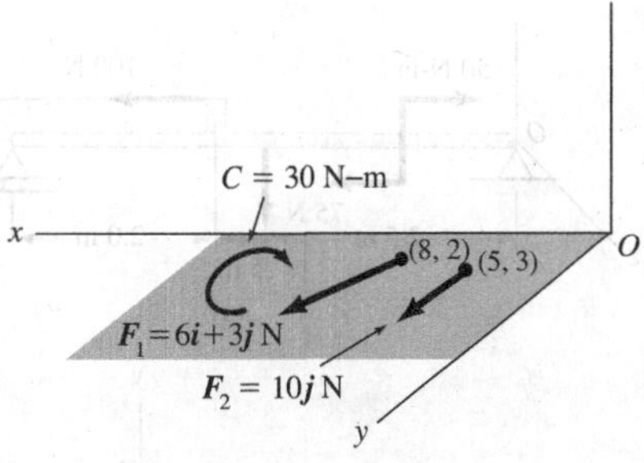

Figure 4.16. Find simplest resultant.

We now need to find the line of action in the plane that will make this single force equivalent to the given system. To be equivalent for rigid-body mechanics, this force without a couple moment must have the same turning action about any point or axis in space as that of the given system. Now the simplest resultant force must intercept the x axis at some point x.[4] We can determine x by equating the moment of the resultant force without a couple moment about the origin with that of the original system of forces and couples. Using the vector $\boldsymbol{xi}$ as a position vector from the origin to the line of action of $\boldsymbol{F_R}$ (see Fig. 4.17), we accordingly have

$$xi \times (6i + 13j) = (8i + 2j) \times (6i + 3j) + (5i + 3j) \times (10j) - 30k \qquad \text{(b)}$$

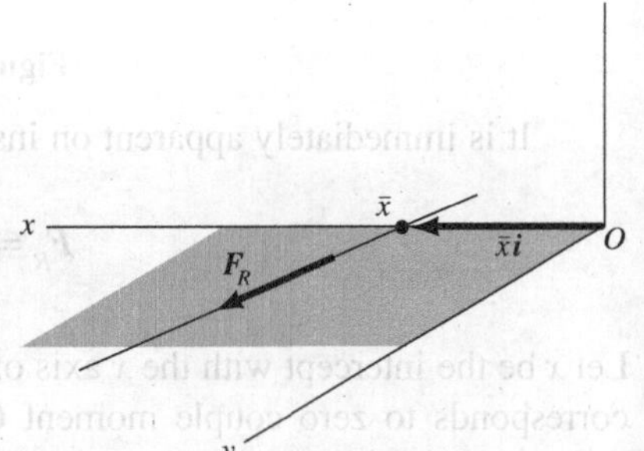

Figure 4.17. Simplest resultant.

Carrying out the cross products,

$$24k - 12k + 50k - 30k = 13xk \qquad \text{(c)}$$

Hence,

$$x = 2.46 \text{ m}$$

By specifying the x intercept, x, we fully determine the line of action of the simplest resultant force. We could have also used the intercept with the y axis, y, for this purpose. In that case, the position vector from the origin out to the line of action is yj, and we have, on equating moments about O of the resultant without a couple moment with that of the original system:

$$yj \times (6i + 13j) = (8i + 2j) \times (6i + 3j) + (5i + 3j) \times (10j) - 30k$$

$$y = -5.35 \text{ m}$$

[4]If the resultant force is parallel to the x axis, the intercept will be at infinity.

Example 4.7

Compute the *simplest* resultant for the loads shown acting on the beam in Fig. 4.18(a). Give the intercept with the x axis.

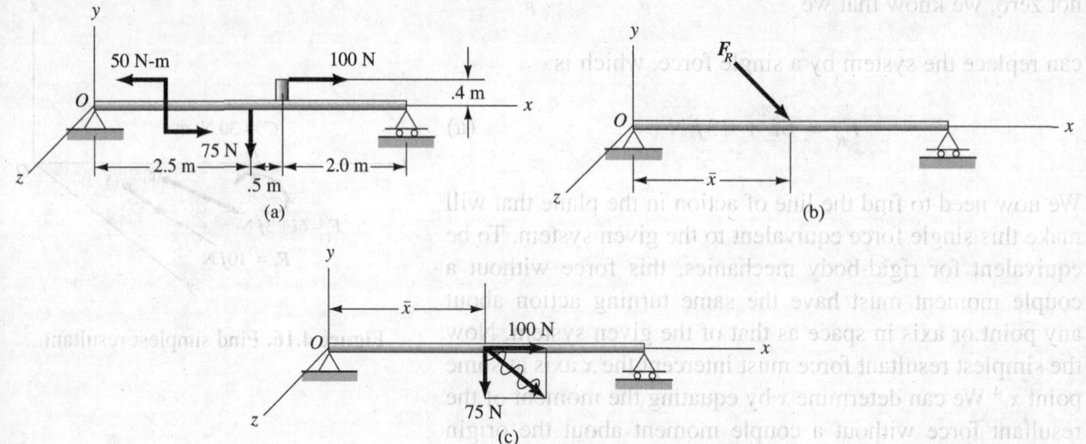

Figure 4.18. Find simplest resultant.

It is immediately apparent on inspection of the diagram that

$$F_R = 100\boldsymbol{i} - 75\boldsymbol{j} \text{ N} \quad \text{(a)}$$

Let x be the intercept with the x axis of the line of action of $\boldsymbol{F}_R$ when this line of action corresponds to zero couple moment $\boldsymbol{C}_R$ [see Fig. 4.18(b)]. In Fig. 4.18(c), we have decomposed $\boldsymbol{F}_R$ along this line of action into rectangular components so as to permit simple calculations of moments about the origin O (here we mean moments about the z axis). Accordingly, equating moments about the z axis of $\boldsymbol{F}_R$ without a couple moment, with that of the original system of loads, we get,

$$-(75)(x) = 50 - (2.5)(75) - (.4)(100)$$

$$x = 2.37 \text{ m}$$

Thus, the simplest resultant is a force $100\boldsymbol{i} - 75\boldsymbol{j}$ N intercepting the beam axis at a position $x = 2.37$ m.

As pointed out earlier, in the instance wherein $\boldsymbol{F}_R = \boldsymbol{0}$, we then possibly have as the simplest resultant a couple moment normal to the plane of the coplanar force system. There is also the possibility that there is zero couple moment, in which case the forces of the coplanar force system *completely cancel* each other's effects on a rigid body. To find the couple moment for the case where $\boldsymbol{F}_R = \boldsymbol{0}$, we simply take moments of the coplanar force system about *any point* in space. This moment, if it is not equal to zero, is clearly the couple-moment vector sought.

Example 4.8

What is the simplest resultant for the forces shown acting on beam *AB* in Fig. 4.19?

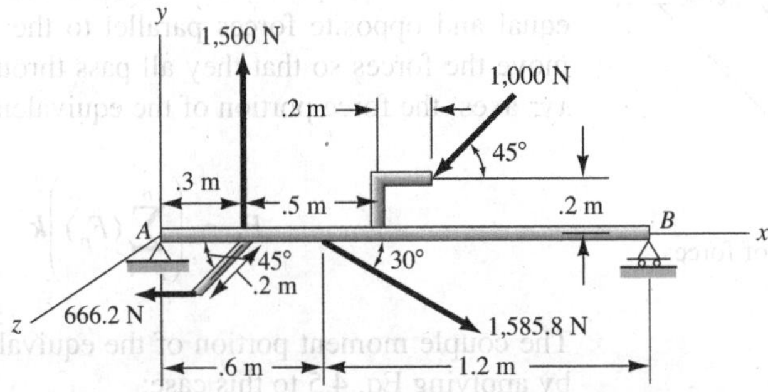

Figure 4.19. Coplanar loading on a simply supported beam.

Our first step will be to compute the resultant force by adding up the force vectors. Thus

$$\begin{aligned} \boldsymbol{F}_R = {} & 1{,}500\boldsymbol{j} - 666.2\boldsymbol{i} - (1{,}585.8)(.5)\boldsymbol{j} \\ & + (1{,}585.8)(.866)\boldsymbol{i} - 707.1\boldsymbol{i} - 707.1\boldsymbol{j} \end{aligned}$$

Collecting terms, we have

$$\boldsymbol{F}_R = (-666.2 + 1{,}373.3 - 707.1)\boldsymbol{i} + (1{,}500 - 792.9 - 701.1)\boldsymbol{j} = \boldsymbol{0}$$

The simplest resultant clearly must be either a couple moment or be a null vector. For this information, we shall take moments about point *A*.

$$\begin{aligned} \boldsymbol{C}_R = {} & \{[.3 - (.2)(.707)]\boldsymbol{i} - (.2)(.707)\boldsymbol{j}\} \times (-666.2\boldsymbol{i}) \\ & + (.3)(1{,}500)\boldsymbol{k} - (.6)(1{,}585.8)(.5)\boldsymbol{k} \\ & + (1\boldsymbol{i} + .2\boldsymbol{j}) \times (-707.1\boldsymbol{i} - 707.1\boldsymbol{j}) \end{aligned}$$

$$\begin{aligned} \boldsymbol{C}_R = {} & -94.20\boldsymbol{k} + 450\boldsymbol{k} - 475.7\boldsymbol{k} \\ & - 707.1\boldsymbol{k} + 141.4\boldsymbol{k} = \boxed{-685.6\boldsymbol{k} \text{ N-m}} \end{aligned}$$

We have a couple moment in the minus *z* direction having any arbitrary line of action as the simplest resultant.

It is important that the nature of the equivalence just instituted be clearly understood. Thus, for finding the supporting force system, we can use the undeformed geometry and hence the single force replacement. However, for finding the deflection of the beam, it should be obvious that the replacement is invalid. Note, finally, that there is only one point on the beam that will allow for a single force to be equivalent to the original system for purposes of rigid-body considerations.

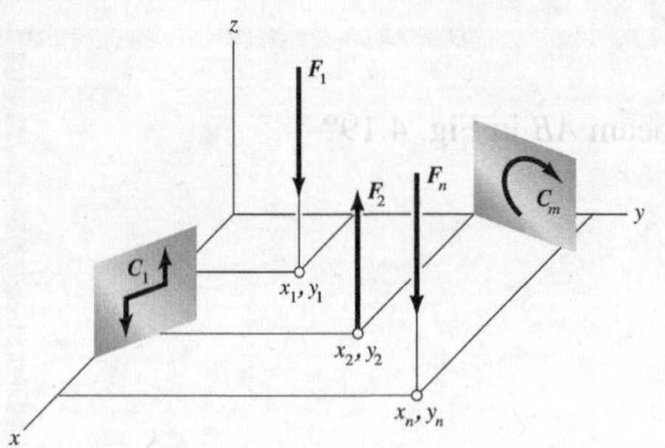

Figure 4.20. Parallel system of forces.

Case B. Parallel Force Systems in Space. Now, consider the system of n parallel forces in Fig. 4.20, where the z direction has been selected parallel to the forces. We also include m couples whose planes are parallel to the z direction because such couples can be considered to be composed of equal and opposite forces parallel to the z direction. We can move the forces so that they all pass through the origin of the xyz axes; the force portion of the equivalent system is then

$$\boldsymbol{F}_R = \left(\sum_{p=1}^{n} (F_p)\right)\boldsymbol{k} \tag{4.8}$$

The couple moment portion of the equivalent system is found by applying Eq. 4.5 to this case:

$$\boldsymbol{C}_R = \sum_{p=1}^{n}\left[\left(x_p\boldsymbol{i} + y_p\boldsymbol{j}\right) \times F_p\boldsymbol{k}\right] + \sum_{p=1}^{m}\left[\left(C_p\right)_x\boldsymbol{i} + \left(C_p\right)_y\boldsymbol{j}\right] \tag{4.9}$$

where F_p represents the noncouple forces. Carrying out the cross product, we get

$$\boldsymbol{C}_R = \sum_{p=1}^{n}\left[\left(F_py_p\right)\boldsymbol{i} - \left(F_px_p\right)\right]\boldsymbol{j} + \sum_{p=1}^{m}\left[\left(C_p\right)_x\boldsymbol{i} + \left(C_p\right)_y\boldsymbol{j}\right] \tag{4.10}$$

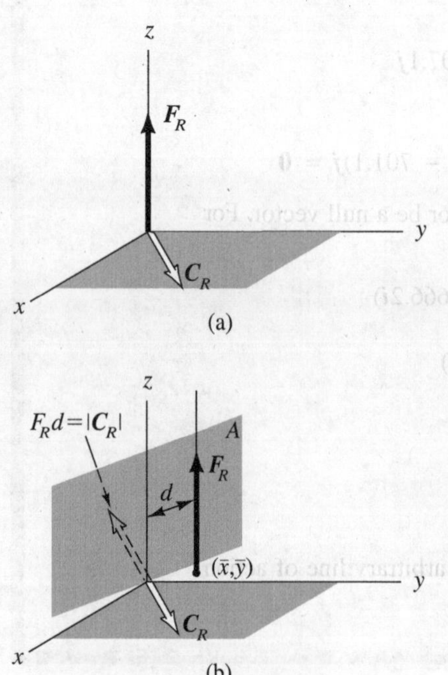

Figure 4.21. Simplest resultant for parallel force system.

From this, we see that the couple moment must always be parallel to the xy plane (i.e., perpendicular to the direction of the forces). We then have at the origin a single force and a single couple moment at right angles to each other [see Fig. 4.21(a)]. If $\boldsymbol{F}_R \neq \boldsymbol{0}$, we can move $\boldsymbol{F}_R$ again to another line of action in a plane A perpendicular to $\boldsymbol{C}_R$ [see Fig. 4.21(b)] and, choosing the proper value of d, ensure that $F_Rd = |\boldsymbol{C}_R|$ with a sense opposite to $\boldsymbol{C}_R$[5] such that we eliminate the couple moment. We thus end up with a *single* force having a particular line of action specified by the intercept x, y of the line of action of the force with the xy plane. If the summation of forces should happen to be zero, the equivalent system must then be a couple moment or a null vector.

Thus, *the simplest resultant system of a parallel force system is either a force with a specific line of action, a single couple moment, or a null vector*. The following example will illustrate how we can directly determine the simplest resultant.

[5]Here we urge some caution! The proper displacement vector to be used in computing the moment induced by moving $\boldsymbol{F}$ is found by going from the *new* position to the *original* position. The displacement vector here accordingly must go from right to left a distance d. Recall in this regard that when moving a force to a line of action through a point a, the position vector used in computing the induced couple moment always goes *from the point to the force* [see Fig. 4.3(a)].

Example 4.9

Find the simplest resultant of the parallel force system in Fig. 4.22(a).

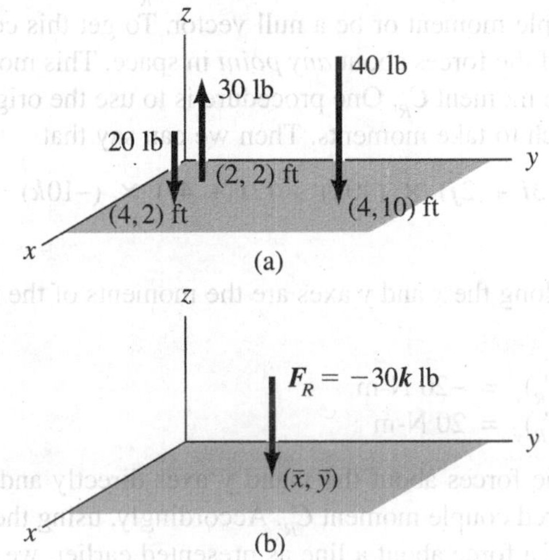

Figure 4.22. Find simplest resultant.

The sum of the forces is 30 lb in the negative z direction. Hence, a position can be found in which a single force is equivalent to the original system. Assume that this resultant force without a couple moment proceeds through the point x, y [Fig. 4.22(b)]. We can equate the moment of this resultant force about the x and y axes with the corresponding moments of the original system and thus form the scalar equations that yield the proper value of x and y. Equating moments about the x axis, we get

$$(30)(2) - (20)(2) - (40)(10) = -30y$$

Therefore,

$$y = 12.7 \text{ ft}$$

Equating moments about the y axis, we have

$$-(30)(2) + (20)(4) + (40)(4) = 30x$$

Therefore,

$$x = 6 \text{ ft}$$

You can also show, as an exercise, that the same result can be reached for x, y by equating moments of the resultant force without a couple moment about the origin with that of the original system about the origin.

Example 4.10

Consider the parallel force system in Fig. 4.23(a). What is the simplest resultant?

Here we have a case where the sum of the forces is zero and so $\boldsymbol{F}_R = \boldsymbol{0}$. Therefore, the simplest resultant must be a couple moment or be a null vector. To get this couple moment, $\boldsymbol{C}_R$, we can take moments of the forces about *any point* in space. This moment vector then equals the desired couple moment $\boldsymbol{C}_R$. One procedure is to use the origin of the reference as the point about which to take moments. Then we can say that

$$\begin{aligned} \boldsymbol{C}_R &= (4\boldsymbol{i} + 2\boldsymbol{j}) \times (-30\boldsymbol{k}) + (3\boldsymbol{i} + 2\boldsymbol{j}) \times (40\boldsymbol{k}) + (2\boldsymbol{i} + 4\boldsymbol{j}) \times (-10\boldsymbol{k}) \\ &= -20\boldsymbol{i} + 20\boldsymbol{j} \text{ N-m} \end{aligned} \quad \text{(a)}$$

The rectangular components of $\boldsymbol{C}_R$ along the x and y axes are the moments of the force system about these axes. Thus,

$$\begin{aligned} (C_R)_x &= -20 \text{ N-m} \\ (C_R)_y &= 20 \text{ N-m} \end{aligned} \quad \text{(b)}$$

We can get the moments of the forces about the x and y axes directly and thus generate the components of the desired couple moment $\boldsymbol{C}_R$. Accordingly, using the elementary definition of the moment of a force about a line as presented earlier, we have

$$\begin{aligned} (C_R)_x &= -(10)(4) + (40)(2) - (30)(2) = -20 \text{ N-m} \\ (C_R)_y &= (10)(2) - (40)(3) + (30)(4) = 20 \text{ N-m} \end{aligned}$$

Thus, the moment of the force system about the origin, and hence about any point, is then the desired couple moment (Fig. 4.23(b))

$$\boldsymbol{C}_R = -20\boldsymbol{i} + 20\boldsymbol{j} \text{ N-m}$$

Now that we have considered the concept of the simplest resultant for coplanar and parallel force systems, we wish to go back to the *general force* systems for a moment. We learned earlier that we can always replace such a system in rigid-body mechanics by a single force $\boldsymbol{F}_R$ and a single couple moment $\boldsymbol{C}_R$ at any chosen point. Is this always the very simplest system for rigid-body mechanics? No, it is not. To show this, decompose the couple moment $\boldsymbol{C}_R$ into two rectangular components $C_\perp$ and $C_\parallel$, perpendicular to the force and collinear with the force, respectively. We can now conceivably move the force to a specific parallel position and can eliminate $C_\perp$, the component of couple moment normal to the force. However, there is nothing that we can do about the $C_\parallel$ component of couple moment collinear (or parallel) to the force. The reason for this is that any movement of the force to a parallel position *always* introduces a couple moment *perpendicular* to the force. Thus the component $C_\parallel$ cannot be affected. By eliminating $C_\perp$ we end up with the force $\boldsymbol{F}_R$ and $\boldsymbol{C}_\parallel$ collinear with $\boldsymbol{F}_R$. This system is the simplest in the general case and it is called a *wrench* (see Fig. 4.24). However, we shall not generally use the wrench concept in this text and will work instead with the resultant force $\boldsymbol{F}_R$ and the couple moment $\boldsymbol{C}_R$ at any chosen point.

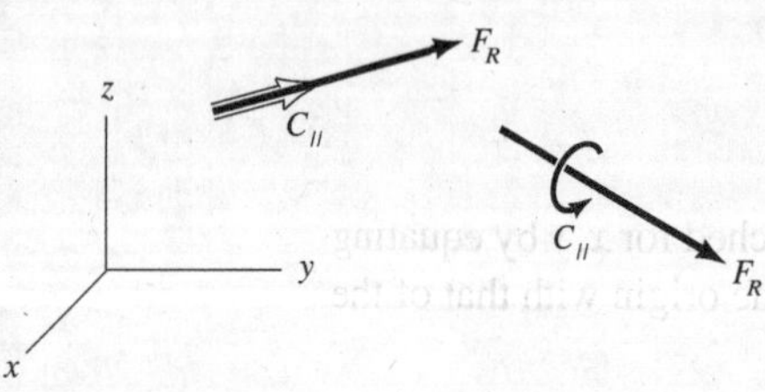

Figure 4.24. Examples of the so-called wrench. This is the simplest representation of a general force system.

PROBLEMS

In several of the problems of this set we shall concentrate the weight of a body at its center of gravity. Most likely you are used to doing this from your previous physics course. In Section 4.5 we shall justify this procedure.

4.18. Compute the resultant force system of the applied loads at positions *A* and *B*.

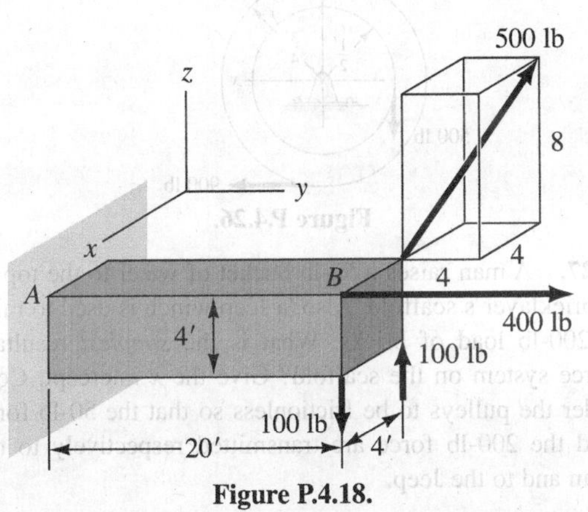

Figure P.4.18.

4.19. Compute the resultant force system at *A* stemming from the indicated 50-lb force. What is the twist developed about the axis of the shaft at *A*? The 50-lb force is normal to the wrench.

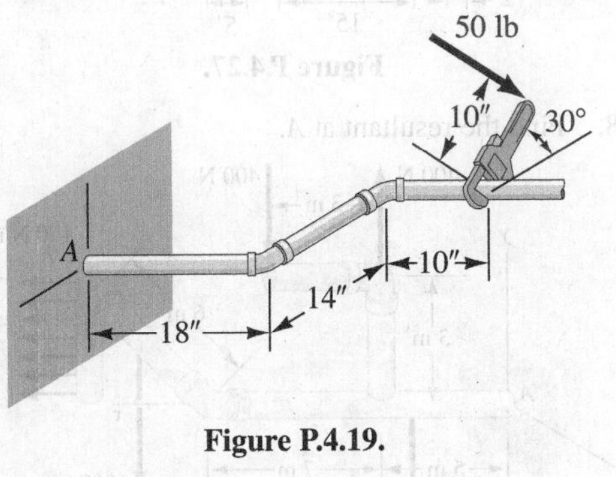

Figure P.4.19.

4.20. Find the resultant of the force system at point *A*. The 300-N, 200-N, and 900-N loads are at the centers of the pipe sections.

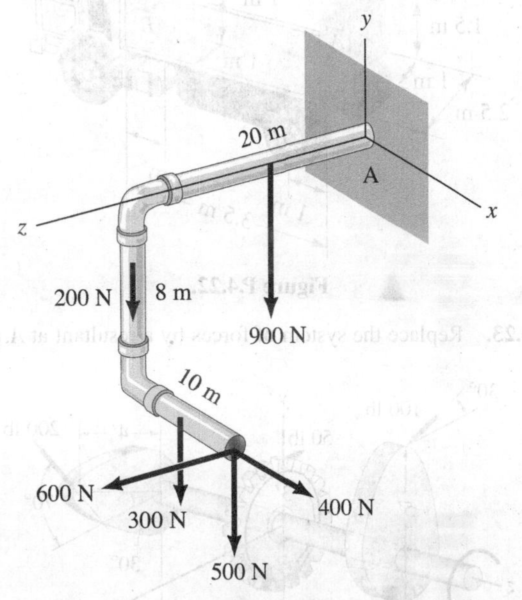

Figure P.4.20.

4.21. A 20-kN car and an 80-kN truck are stopped on a bridge. What is the resultant force system of these vehicles at the center of the bridge? At the center of the left end of the bridge? The distances given to truck and car are to respective centers of gravity where we can concentrate the weights.

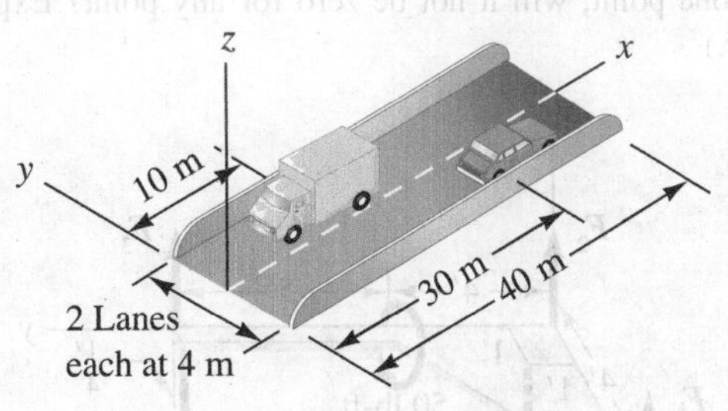

Figure P.4.21.

4.22. Two heavy machinery crates (*A* weighs 20 kN and *B* weighs 30 kN) are placed on a truck. What is the resultant force system at the center of the rear axle? The centers of gravity of the crates, where we can concentrate the weights, are at the geometric centers.

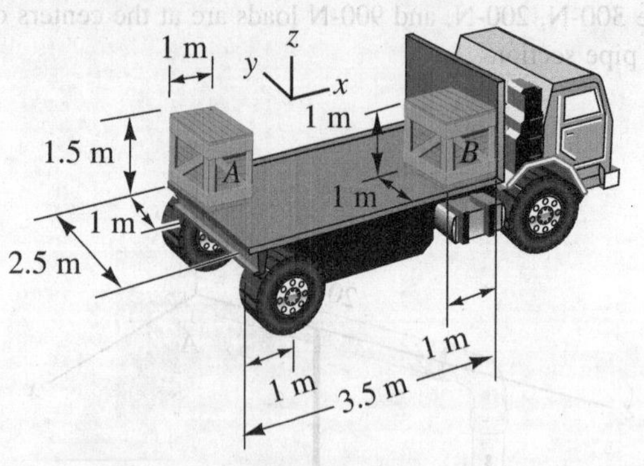

Figure P.4.22.

4.23. Replace the system of forces by a resultant at *A*.

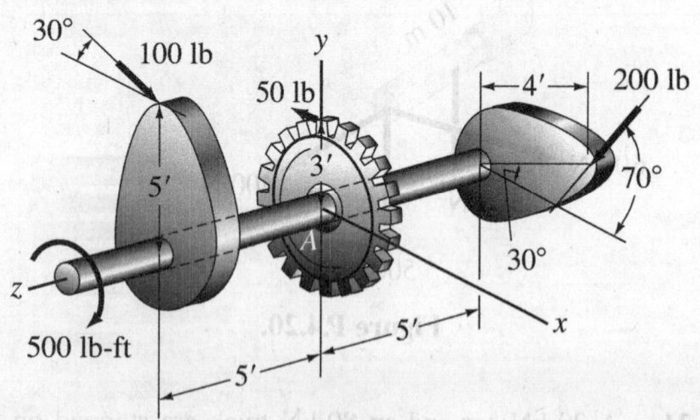

Figure P.4.23.

4.24. Evaluate Forces F_1, F_2, and F_3 so that the resultant of the forces and torque acting on the plate is zero in both force and couple moment. (*Hint:* If the resultant is zero for one point, will it not be zero for any point? Explain why.)

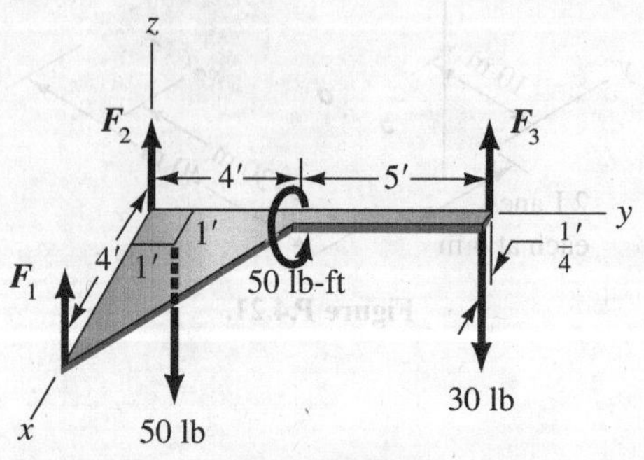

Figure P.4.24.

4.25. Find the *simplest* resultant of the forces shown acting on the beam. Give the intercept with the axis of the beam.

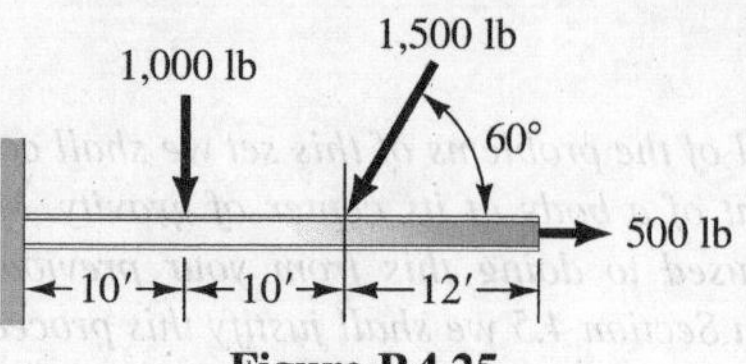

Figure P.4.25.

4.26. Find the *simplest* resultant of the forces shown acting on the pulley. Give the intercept with the *x* axis.

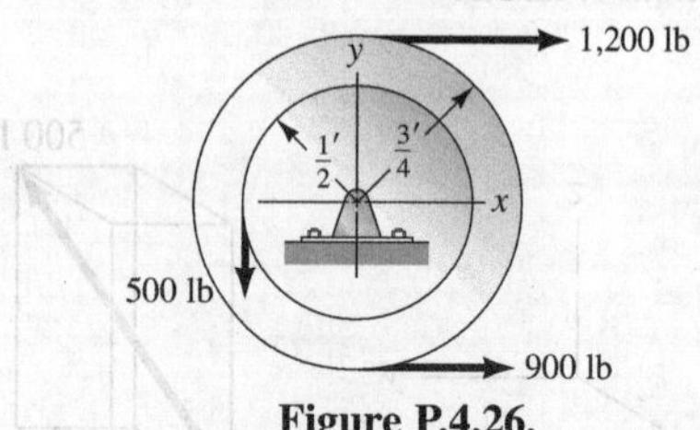

Figure P.4.26.

4.27. A man raises a 50-lb bucket of water to the top of a bricklayer's scaffold. Also, a Jeep winch is used to raise a 200-lb load of bricks. What is the *simplest* resultant force system on the scaffold? Give the *x* intercept. Consider the pulleys to be frictionless so that the 50-lb force and the 200-lb force are transmitted respectively to the man and to the Jeep.

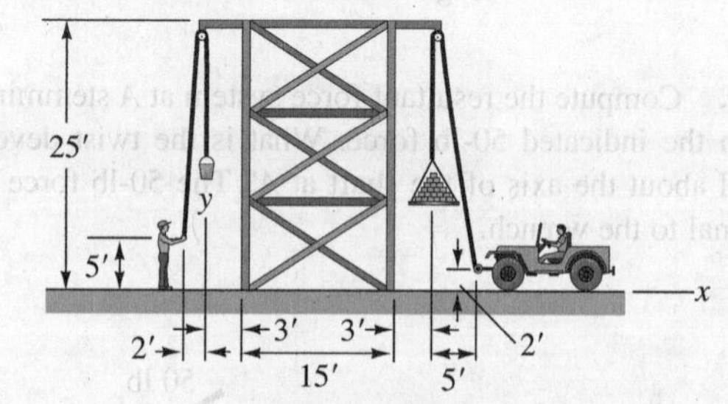

Figure P.4.27.

4.28. Find the resultant at *A*.

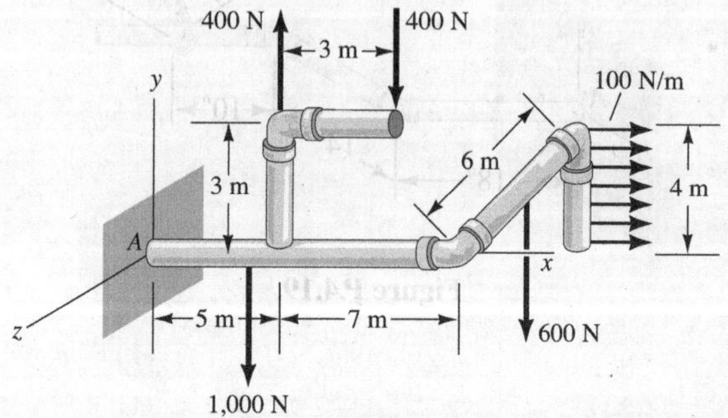

Figure P.4.28.

4.29. Compute the *simplest* resultant for the loads acting on the beam. Give the intercept with the axis of the beam.

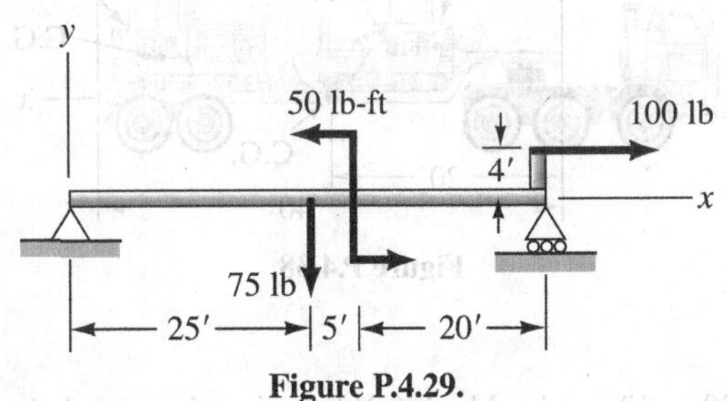

Figure P.4.29.

4.30. Find the *simplest* resultant for the forces. Give the location of this resultant clearly.

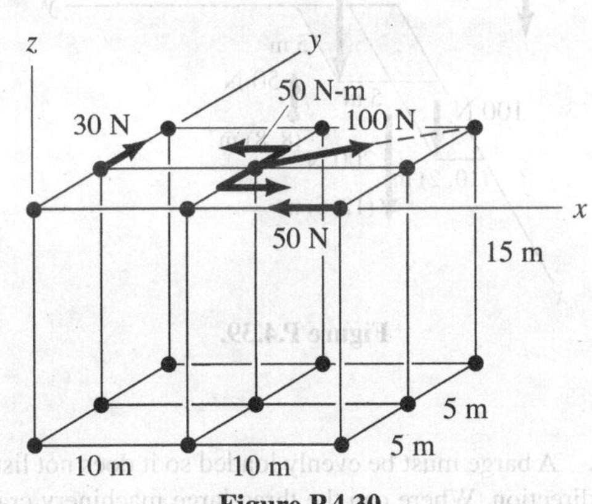

Figure P.4.30.

4.31. Replace the system of forces acting on the rivets of the plate by the *simplest* resultant. Give the intercept of this resultant with the x axis.

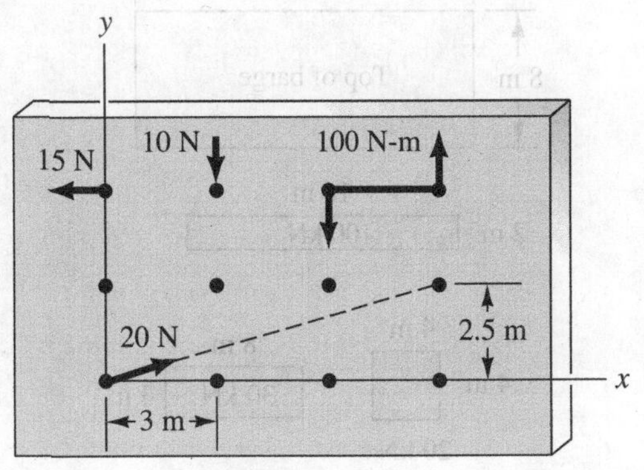

Figure P.4.31.

4.32. A parallel system of forces is such that: a 20-N force acts at position $x = 10$ m, $y = 3$ m; a 30-N force acts at position $x = 5$ m, $y = -3$m; a 50-N force acts at position $x = -2$ m, $y = 5$ m.

(a) If all forces point in the negative z direction, give the *simplest* resultant force and its line of action.

(b) If the 50-N force points in the plus z direction and the others in the negative z direction, what is the *simplest* resultant?

4.33. What is the *simplest* resultant of the three forces and couple shown acting on the shaft and disc? The disc radius is 5 ft.

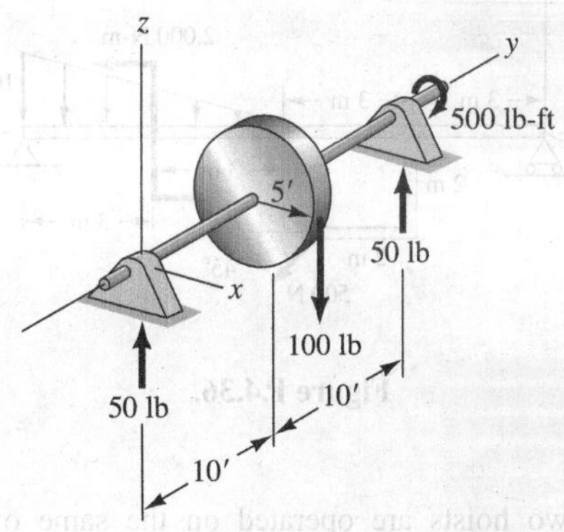

Figure P.4.33.

4.34. What is the *simplest* resultant for the system of forces? Each square is 10 mm on edge.

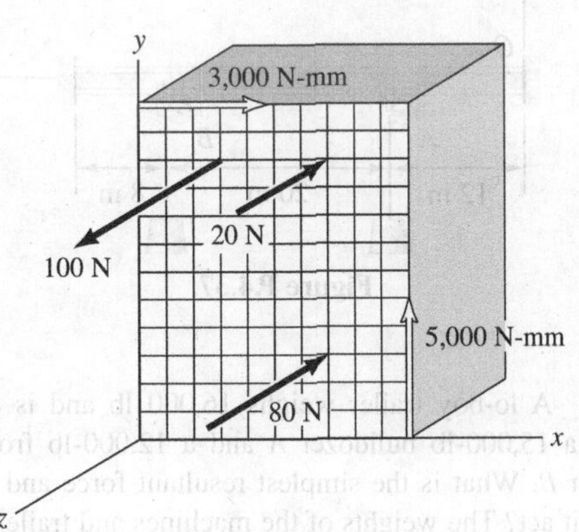

Figure P.4.34.

4.35. What is the simplest resultant? Where does its line of action cross the x axis?

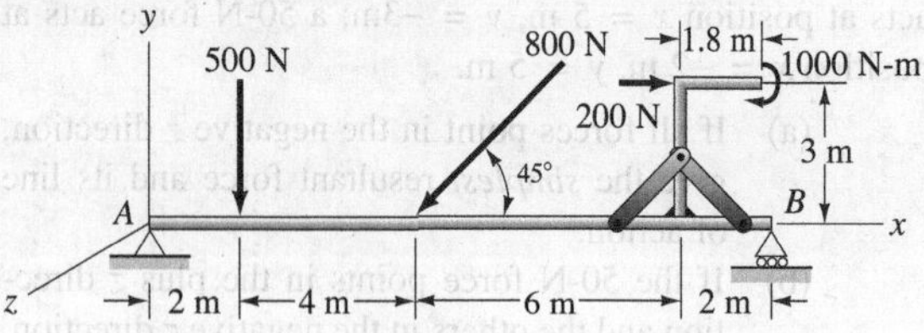

Figure P.4.35.

4.36. What is the simplest resultant of the loadings shown? Be sure to give its line of action.

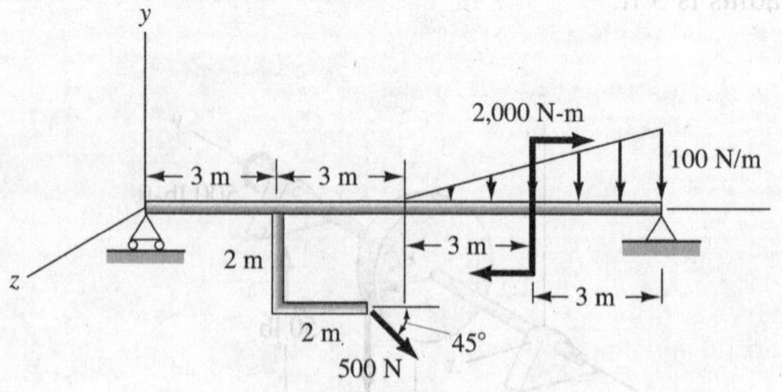

Figure P.4.36.

4.37. Two hoists are operated on the same overhead track. Hoist A has a 3,000-kN load, and hoist B has a 4,000-kN load. What is the resultant force system at the left end O of the track? Where does the *simplest* resultant force act?

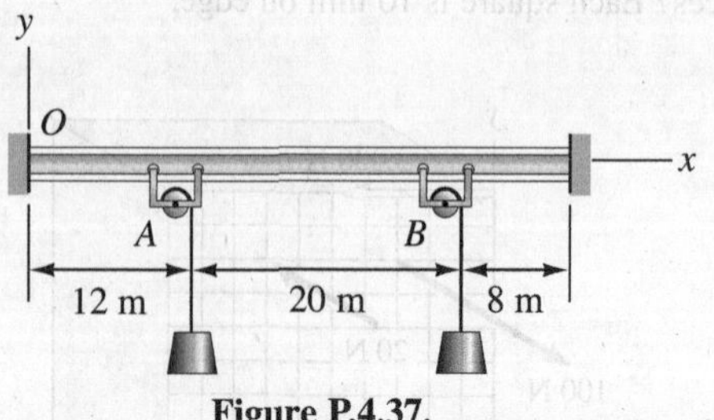

Figure P.4.37.

4.38. A lo-boy trailer weighs 16,000 lb and is loaded with a 15,000-lb bulldozer A and a 12,000-lb front-end loader B. What is the simplest resultant force and where does it act? The weights of the machines and trailer act at their respective centers of gravity (C.G.).

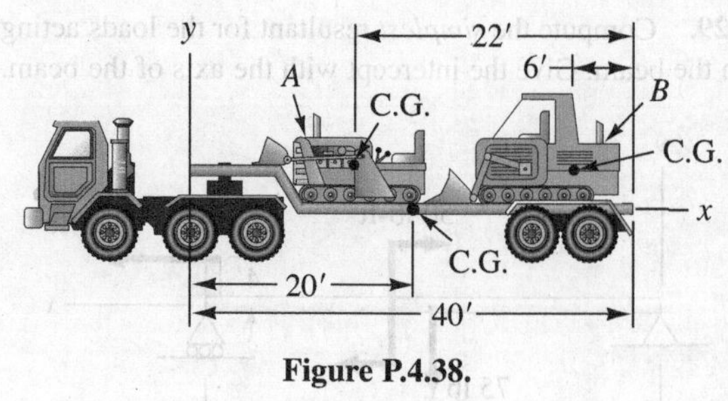

Figure P.4.38.

4.39. Where should a 100-N force in a downward direction be placed for the *simplest* resultant of all shown forces to be at position (5, 5) m?

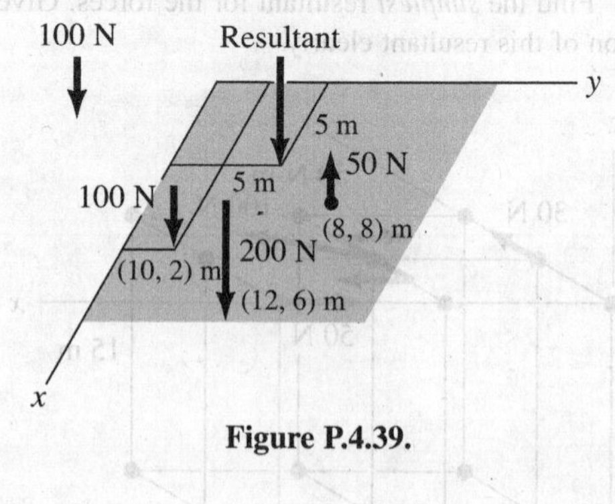

Figure P.4.39.

4.40. A barge must be evenly loaded so it does not list in any direction. Where can the three large machinery crates be placed (without either hanging over the edge, stacking, or standing on end)? Each crate is as tall as it is wide. Is there only one solution to this problem? Centers of gravity of crates correspond to geometric centers.

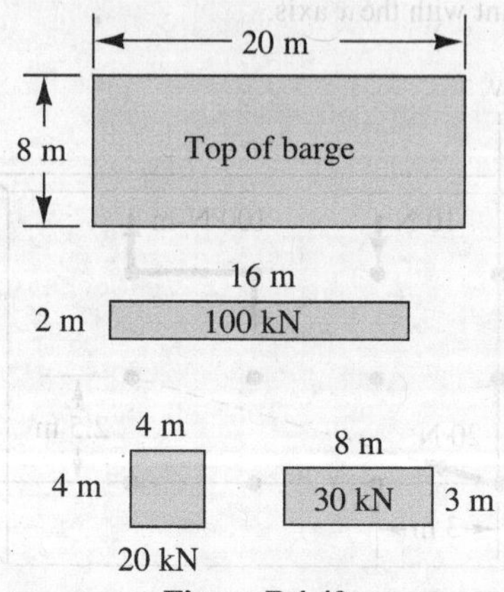

Figure P.4.40.

4.5 Distributed Force Systems

Our discussions up to now have been restricted to discrete vectors—in particular, to point forces. Vectors as well as scalars, may also be continuously distributed throughout a finite volume. Such distributions are called *vector* and *scalar fields*, respectively. A simple example of a scalar field is the temperature distribution, expressed as $T(x, y, z, t)$, where the variable t indicates that the field may be changing with time. Thus, if a position x_0, y_0, z_0 and a time t_0 are specified, we can determine the temperature at this position and time provided that we know the temperature distribution function (i.e., how T depends on the independent variables $x, y, z,$ and t). A vector field is sometimes expressed in the form $\boldsymbol{F}(x, y, z, t)$. A common example of a vector field is the gravitational force field of the earth—a field that is known to vary with elevation above sea level, among other factors. Note, however, that the gravitational field is virtually constant with time.

In place of the vector field, it is more convenient at times to employ three scalar fields that represent the orthogonal scalar components of a vector field at all points. Thus, for a force field we can say:

$$\begin{aligned}\text{force component in } x \text{ direction} &= g(x, y, z, t)\\ \text{force component in } y \text{ direction} &= h(x, y, z, t)\\ \text{force component in } z \text{ direction} &= l(x, y, z, t)\end{aligned}$$

where g, h, and l represent functions of the coordinates and time. If we substitute coordinates of a special position and the time into these functions, we get the force components F_x, F_y, and F_z for that position and time. The force field and its component scalar fields are then related in this way:

$$\boldsymbol{F}(x, y, z, t) = g(x, y, z, t)\boldsymbol{i} + h(x, y, z, t)\boldsymbol{j} + l(x, y, z, t)\boldsymbol{k}$$

More often, the notation for the equation above is written

$$\boldsymbol{F}(x, y, z, t) = F_x(x, y, z, t)\boldsymbol{i} + F_y(x, y, z, t)\boldsymbol{j} + F_z(x, y, z, t)\boldsymbol{k} \tag{4.11}$$

Vector fields are not restricted to forces but include other quantities such as velocity fields and heat-flow fields.

Force distributions, such as gravitational force, that exert influence directly on the elements of mass distributed throughout the body are termed *body force distributions* and are usually given per unit of mass that they directly influence. Thus, if $\boldsymbol{B}(x, y, z, t)$ is such a body force distribution, the force on an element dm would be $\boldsymbol{B}(x, y, z, t)dm$.

Force distributions over a *surface* are called *surface force distributions*[6] $\boldsymbol{T}(x, y, z, t)$ and are given per unit area of the surface directly influenced. A simple example is the force distribution on the surface of a body submerged in a fluid. In the case of a static fluid or of a frictionless fluid, the force from the fluid on an area element is always normal to the area element and directed in toward the body. The force per unit area stemming from such fluid action is called *pressure* and is denoted as p. Pressure is a

[6]Surface forces are often called *surface tractions* in solid mechanics.

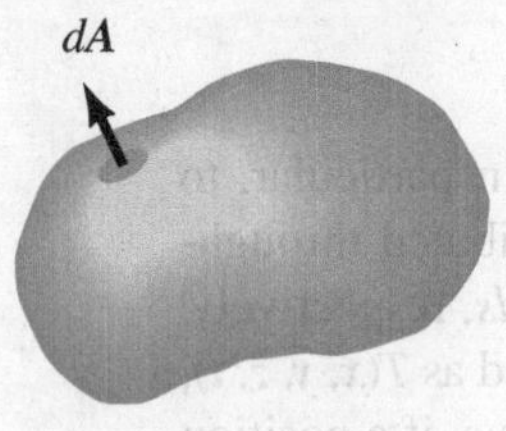

Figure 4.25. Area vector.

scalar quantity. The direction of the force resulting from a pressure on a surface is given by the orientation of the surface. [You will recall from Chapter 2 that an area element can be considered as a vector which is normal to the area element and directed outward from the enclosed body (Fig. 4.25).] The infinitesimal force on the area element is then given as

$$df = -p\, dA$$

A more specialized, but nevertheless common, force distribution is that of a continuous load on a beam. This is often a parallel loading distribution that is symmetrical about the center plane xy of a beam, as illustrated in Fig. 4.26. Various heights of bricks stacked on a beam would be an example of this kind of loading. We can replace such a loading by an equivalent coplanar distribution that acts at the center plane. The loading is given per unit length and is denoted as w, the *intensity of loading*. The force on an element dx of the beam, then, is $w\,dx$.

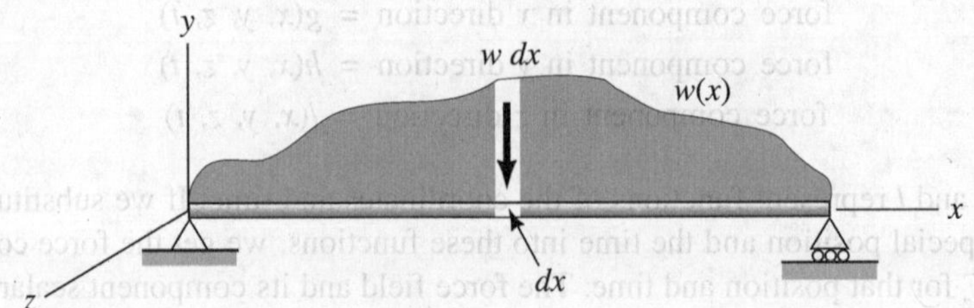

Figure 4.26. Loading on a beam.

We have thus presented force systems distributed throughout volumes (body forces), over surfaces (surface forces or tractions), and over lines. The conclusions about resultants that were reached earlier for general, parallel, and coplanar point force systems are also valid for these distributed force systems. These conclusions are true because each distributed force system can be considered as an infinite number of infinitesimal point forces of the type used heretofore. We shall illustrate the handling of force distributions in the following examples.

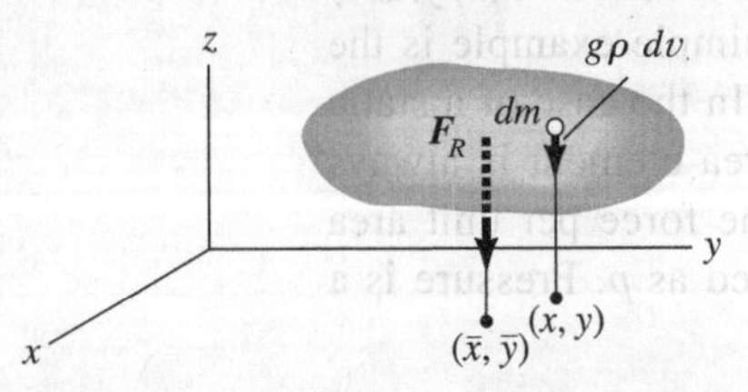

Figure 4.27. Gravity body force distribution.

Case A. Parallel Body Force System—Center of Gravity. Consider a rigid body (Fig. 4.27) whose density (mass/unit volume) is given as $\rho(x, y, z)$. It is acted on by gravity, which, for a small body, may be considered to result in a distributed parallel force field.

Since we have here a parallel system of forces in space with the same sense, we know that a single force without a couple moment along a certain line of action will be equivalent to the distribution. The gravity body force $\boldsymbol{B}(x, y, z)$ given per unit mass is $-g\boldsymbol{k}$. The infinitesimal force on a differential mass element dm, then , is $-g(\rho\, dv)\boldsymbol{k}$, where dv is the volume of the element.[7] We find the resultant force on the system by replacing the summation in Eq. 4.8 with an integration, Thus,

$$\boldsymbol{F}_R = -\int_V g(\rho\, dv)\boldsymbol{k} = -g\boldsymbol{k}\int_V \rho\, dv = -gM\boldsymbol{k}$$

where, with g as a constant, the second integral becomes simply the entire mass of the body M.

Next, we must find the line of action of this single equivalent force without a couple moment. Let us denote the intercept of this line of action with the xy plane as x, y (see Fig. 4.27). The resultant at this position must have the same moments as the distribution about the x and y axes:

$$-F_R\bar{y} = -g\int_V y\rho\, dv \qquad F_R\bar{x} = -g\int_V x\rho\, dv$$

Hence, we have

$$\bar{x} = \frac{\int x\rho\, dv}{M} \qquad \bar{y} = \frac{\int y\rho\, dv}{M}$$

Thus, we have fully established the simplest resultant. Now, the body is reoriented in space, keeping with it the line of action of the resultant as shown in Fig. 4.28. A new computation of the line of action of the simplest resultant for the second orientation yields a line that intersects the original line at a point C. It can be shown that lines of action for simplest resultants for all other orientations of the body must intersect at the same point C. We call this point the *center of gravity*. Effectively, we can say for rigid-body considerations that all the weight of the body can be assumed to be concentrated at the center of gravity.

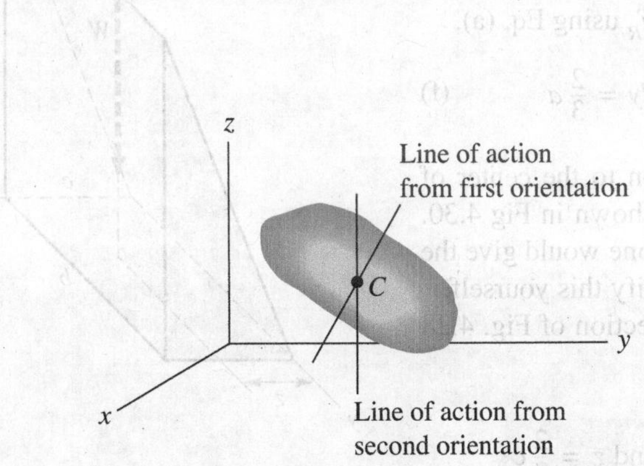

Figure 4.28. Location of center of gravity

[7]Note that $g\rho$ is the weight per unit volume, which is often given as γ, the so-called specific weight.

Example 4.11

Find the center of gravity of the triangular block having a uniform density ρ shown in Fig. 4.29.

The total weight of the body is easily evaluated as

$$F_R = g\rho\left(abc/2\right) \quad \text{(a)}$$

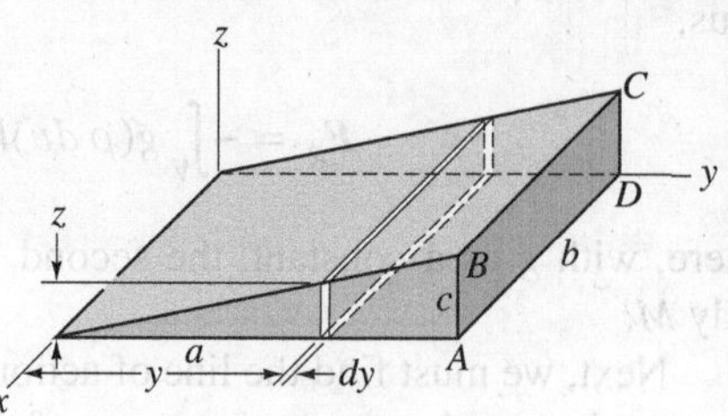

Figure 4.29. Find center of gravity.

To find $\bar{y}$, we will equate the moment of F_R about the x axis with that of the weight distribution of the block. To facilitate the latter, we shall choose within the block *infinitesimal* elements whose weights are easily computed. Also, the moment of the weight of each element about the x axis is to be likewise easily computed. Infinitesimal slices of thickness dy parallel to the xz plane fulfill our requirements nicely. The weight of such a slice is simply $(zb\ dy)\rho g$, where z is the height of the slice (see Fig. 4.29). Because all points of the slice are at the same coordinate distance y from the x axis, clearly the moment of the weight of the slice is easily computed as $-y(zb\ dy)\rho g$. By letting y run from 0 to a during an integration, we can account for all the slices in the body. Thus, we have

$$-F_R\bar{y} = -\int_0^a y(zb\ dy)\rho g \quad \text{(b)}$$

The term z can be expressed with the aid of similar triangles in terms of the integration variable y as follows:

$$\frac{z}{c} = \frac{y}{a}$$

$$z = \left(\frac{y}{a}\right)c \quad \text{(e)}$$

We then have for Eq. (b), on replacing F_R using Eq. (a),

$$\bar{y} = \frac{1}{g\rho(abc/2)}g\int_0^a \rho y^2 \frac{bc}{a}\,dy = \frac{2}{3}a \quad \text{(f)}$$

To find the coordinate in the z direction to the center of gravity, we could reorient the body as shown in Fig 4.30. A computation similar to the preceding one would give the result that $\bar{z} = \frac{2}{3}c$. You are urged to verify this yourself.

Finally, it should be clear by inspection of Fig. 4.28 that $\bar{x} = \frac{1}{2}b$.

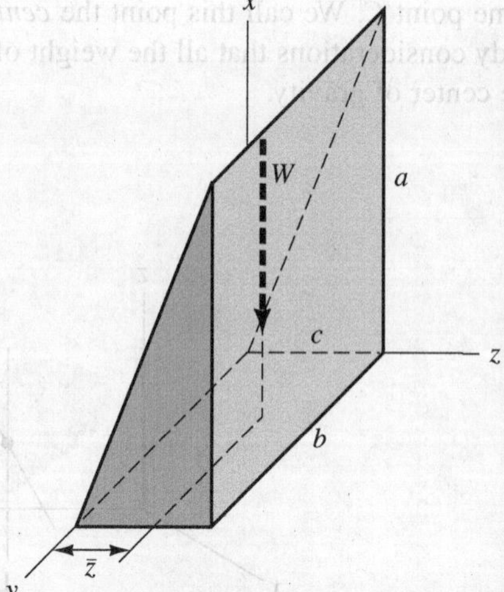

Figure 4.30. Reorientation of block.

Thus we have $\bar{x} = \dfrac{b}{2}$, $\bar{y} = \dfrac{2}{3}$a, and $\bar{z} = \dfrac{2}{3}$c

Example 4.12

Find the center of gravity for the body of revolution shown in Fig. 4.31. The radial distance of the surface from the y axis is given as

$$r = \frac{1}{20} y^2 \text{ ft} \tag{a}$$

The body has constant density ρ, is 10 ft long, and has a cylindrical hole at the right end of length 2 ft and diameter 1 ft.

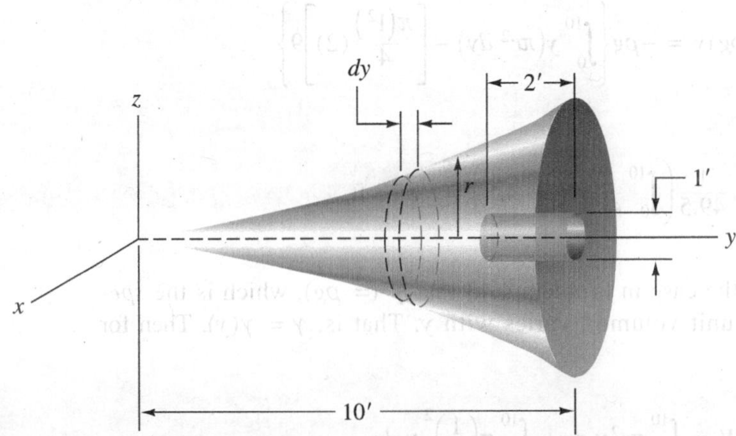

Figure 4.31. Body of revolution. Find center of gravity.

We need only compute y, since it is clear that $z = x = 0$, owing to symmetry. We first compute the weight of the body. Using slices of thickness dy, as shown in the diagram, we sum the weight of all slices in the body assuming it is whole by letting y run from 0 to 10 in an integration. We then subtract the weight of a 2-ft cylinder of diameter 1 ft to take into account the cylindrical cavity inside the body. Thus, we have, noting that the area of a circle is πr^2 or $\pi D^2/4$

$$W = \int_0^{10} \left(\pi r^2\right) dy\, \rho g - \frac{\pi\left(1^2\right)}{4}(2)(\rho g) \tag{b}$$

Using Eq. (a) to replace r^2 in terms of y, we get

$$W = \rho g\left(\pi \int_0^{10} \frac{y^4}{400}\, dy - \frac{\pi}{2}\right) = g\rho\pi\left(50 - \frac{1}{2}\right) \tag{c}$$

$$= 49.5\pi\rho g \text{ lb}$$

Example 4.12 (Continued)

To get y, we equate the moment of W about the x axis with that of the weight distribution. For the latter, we sum the moments about the x axis of the weight of all slices, assuming first no inside cavity. Then, we subtract from this the moment about the x axis of the weight of a cylinder corresponding to the cavity in the body. Because ρ is constant, the center of gravity of this latter cylinder is at its geometric center so that the moment arm from the x axis for the weight of the cylinder is clearly 9 ft. Thus, we have

$$-(49.5\pi\rho g)\bar{y} = -\rho g\left\{\int_0^{10} y\left(\pi r^2\, dy\right) - \left[\frac{\pi(1^2)}{4}(2)\right]9\right\}$$

$$\bar{y} = \frac{1}{49.5}\left(\int_0^{10} \frac{y^5}{400}\, dy - 4.5\right) = \boxed{8.33 \text{ ft}}$$

Suppose as will be the case in Problem 4.43 that γ $(= \rho g)$, which is the *specific weight* (weight per unit volume), varies with y. That is, $\gamma = \gamma(y)$. Then for this problem, note that

$$W = \int_0^{10} \pi r^2 \gamma\, dy - \int_8^{10} \pi\left(\frac{1}{2}\right)^2 \gamma\, dy$$

also

$$-W\bar{y} = -\left[\int_0^{10} y\pi r^2 \gamma\, dy - \int_8^{10} y\pi\left(\frac{1}{2}\right)^2 \gamma\, dy\right]$$

Note here we cannot take the short cuts used in the original problem where γ was a constant.

Example 4.13

A plate is shown in Fig. 4.32 lying flat on the ground. The plate is 60 mm thick and has a uniform density. The curved edge is that of a parabola with zero slope at the origin. Find the coordinates of the center of gravity.

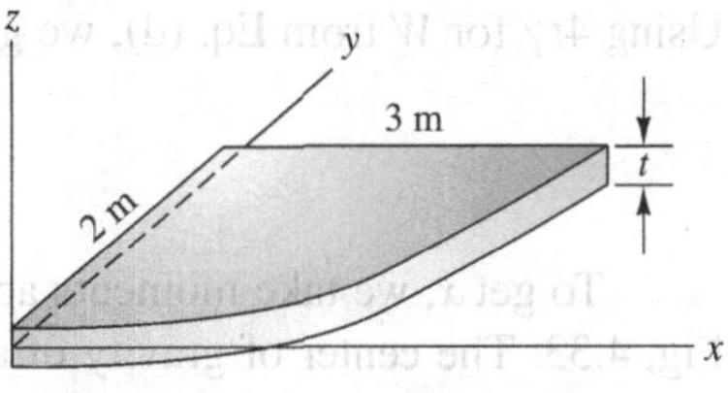

The equation of a parabola oriented like that of the curved edge of the plate is

$$y = Cx^2 \tag{a}$$

Figure 4.32. Find center of gravity of plate.

We can determine C by noting that $y = 2\ m$ when $x = 3\ m$. Hence,

$$2 = C \cdot 9 \tag{b}$$

Therefore,

$$C = \frac{2}{9}$$

The desired curve then is

$$y = \frac{2}{9}x^2$$

Therefore,

$$x = \frac{3}{\sqrt{2}}y^{1/2} \tag{c}$$

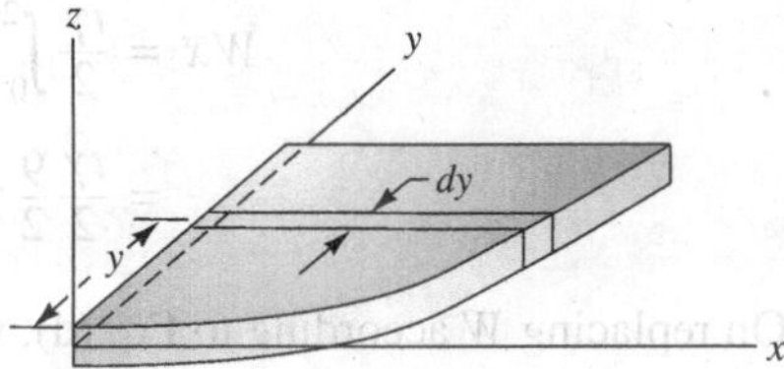

Figure 4.33. Use of horizontal strips.

We shall consider horizontal strips of the plate of width dy (see Fig. 4.33). Using the specific weight, γ, which is weight per volume and is equal to ρg, we have for the total weight W of the plate:

$$W = \int_0^2 (t\,x\,dy)\gamma$$

where t is the thickness. We replace x using Eq. (c) to get

$$W = t\gamma \int_0^2 \left(\frac{3}{\sqrt{2}}y^{1/2}\right)dy$$

Integrating, we get

$$W = t\gamma\frac{3}{\sqrt{2}}\left(y^{3/2}\right)\left(\frac{2}{3}\right)\Big|_0^2 = t\gamma\sqrt{2}(2)^{3/2} = 4t\gamma \text{ N} \tag{d}$$

We next take moments about the x axis in order to get y. Thus,

$$\begin{aligned}
-W\bar{y} &= -\int_0^2 y(t\,x\,dy)\gamma \\
&= -\gamma t\int_0^2 (y)\left(\frac{3}{\sqrt{2}}y^{1/2}\right)dy \\
&= -\gamma t\frac{3}{\sqrt{2}}\left(y^{5/2}\right)\left(\frac{2}{5}\right)\Big|_0^2 \\
&= -\gamma t\left(\frac{3}{\sqrt{2}}\right)\left(\frac{2}{5}\right)[(2^2)(2^{1/2})] \\
&= -\frac{24}{5}\gamma t
\end{aligned} \tag{e}$$

Example 4.13 (Continued)

Using $4t\gamma$ for W from Eq. (d), we get, for $\bar{y}$:

$$\bar{y} = \tfrac{6}{5} \text{ m} \tag{f}$$

To get x, we take moments about the y axis, still utilizing the horizontal strips of Fig. 4.33. The center of gravity of a strip is at its center since γ is constant and so the moment arm for a strip about the y axis is $x/2$.

$$W\bar{x} = \int_0^2 \frac{x}{2}(t\,x\,dy)\gamma \tag{g}$$

Continuing with the calculations, we have

$$W\bar{x} = \frac{t\gamma}{2}\int_0^2 x^2\,dy = \frac{t\gamma}{2}\int_0^2\left(\frac{9}{2}y\right)dy$$

$$= \frac{t\gamma}{2}\frac{9}{2}\frac{y^2}{2}\bigg|_0^2 = \frac{9t\gamma}{2}$$

On replacing W according to Eq. (d), we get for $\bar{x}$:

$$\bar{x} = \tfrac{9}{8} \text{ m} \tag{h}$$

Next, we proceed to get $\bar{x}$ using vertical strips as shown in Fig. 4.34. Equating moments of the weights of these vertical strips about the y axis with that of the total weight W at its center of gravity $\bar{x}$, we get on noting that $(2 - y)$ is the length of the strip

$$W\bar{x} = \int_0^3 x\gamma(2-y)t\,dx = \int_0^3 x\gamma\left(2 - \frac{2}{9}x^2\right)t\,dx \tag{i}$$

$$= 2\gamma t\int_0^3\left(x - \frac{1}{9}x^3\right)dx = 2\gamma t\left(\frac{3^2}{2} - \frac{1}{9}\frac{3^4}{4}\right) = \gamma t\frac{9}{2}$$

Using $W = 4\gamma t$, we can now solve for x. Thus, we again get

$$\bar{x} = \tfrac{9}{8} \text{ m}$$

Finally, it is clear that the z coordinate is zero for reference xy at the center plane of the plate.

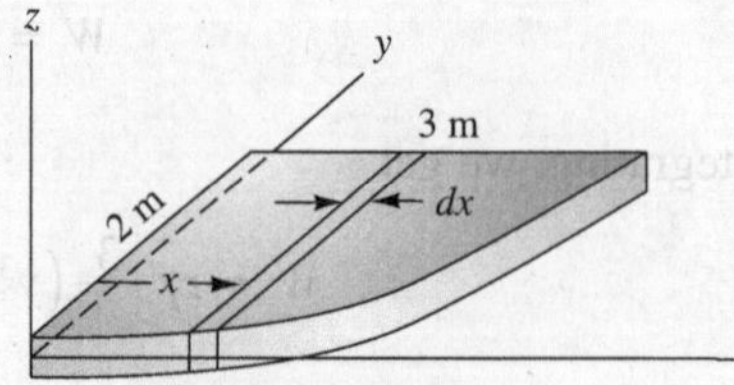

Figure 4.34. Use of vertical strips.

In the previous problems, we used slices of the body having a thicknesses dy or dx. If the specific weight were a function of position, $\gamma(x, y, z)$, we could not readily use such slices, since we cannot easily express the weight of such slices in a simple manner. The reason for this is that in the x and z directions the dimensions of the element are finite, and so γ would vary in these directions throughout the element. If, however, we choose an element that is infinitesimal in *all directions*, such as an infinitesimal rectangular parallelepiped having volume $dx\,dy\,dz$, then γ can be assumed to be constant throughout the element. The weight of the element is then easily seen to be $\gamma(dx\,dy\,dz)$, where the coordinates of γ correspond to the position of the element. We now illustrate a simple case.

Example 4.14

Consider a block (see Fig. 4.35) wherein the specific weight γ at corner A is 200 lbf/ft^3. The specific weight in the block does not change in the x direction. However, it decreases linearly by 50 lbf/ft^3 in 10 ft in the y direction, and increases linearly by 50 lbf/ft^3 in 8 ft in the z direction, as has been shown in the diagram. What are the coordinates $\bar{x}$, $\bar{y}$ of the center of gravity for this block?

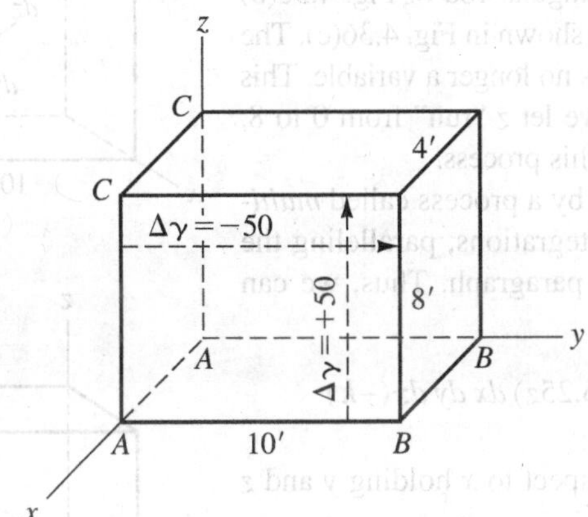

Figure 4.35. Block with varying γ.

We must first express γ at any position $P(x, y, z)$. Using simple proportions, we can say

$$\gamma = 200 - \frac{y}{10}(50) + \frac{z}{8}(50) \qquad \text{(a)}$$

$$= 200 - 5y + 6.25z \text{ lbf/ft}^3$$

We shall first compute the weight of the block (i.e., the resultant force of gravity). We do not use an infinitesimal slice or rectangular rod of the block, as we have done heretofore. With the specific weight varying with both y and z, it would not be an easy matter to compute the weight and moment of a slice or a rod. Instead, we shall use an infinitesimal rectangular parallelepiped having volume $dx\,dy\,dz$, located at a position having coordinates x, y, and z as has been shown in Fig. 4.36(a). Because of the vanishingly small size of this element, the specific weight γ can be considered constant inside the element, and so the weight dW of the element can be given as[8]

$$dW = \gamma(dx\,dy\,dz) = (200 - 5y + 6.25z)\,dx\,dy\,dz$$

[8]We are deleting higher order quantities.

Example 4.14 (Continued)

To include the weight of *all* such elements in the block, we first let x "run" from 0 to 4 ft while holding y and z fixed. The rectangular parallelepiped of Fig. 4.36(a) then becomes a rectangular rod as shown in Fig. 4.36(b). Having run its course, x is no longer a variable in this summation process. Next, let y "run" from 0 to 10 while holding z constant. The rectangular rod of Fig. 4.36(b) then becomes an infinitesimal slice, as shown in Fig. 4.36(c). The variable y has thus run its course and is no longer a variable. This leaves only the variable z, and now we let z "run" from 0 to 8. Clearly, we cover the entire block by this process.

We can do this mathematically by a process called *multiple integration*. We perform three integrations, paralleling the three steps outlined in the previous paragraph. Thus, we can formulate W as follows:

$$-W\boldsymbol{k} = \int_0^8\int_0^{10}\int_0^4 (200 - 5y + 6.25z)\,dx\,dy\,dz\,(-\boldsymbol{k})$$

We first consider integration with respect to x holding y and z constant. That is

$$\int_0^4 (200 - 5y + 6.25z)\,dx$$

As in the first step set forth in the previous paragraph, to go from a rectangular parallelepiped to a rectangular rod, we integrate with respect to x from $x = 0$ to $x = 4$ while holding y and z constant. Thus,

$$\int_0^4 (200 - 5y + 6.25z)\,dx = (200x - 5yx + 6.25zx)\Big|_0^4$$
$$= 800 - 20y + 25z$$

With x no longer a variable (since it has run its course), the equation for W becomes

$$W = \int_0^8\int_0^{10} (800 - 20y + 25z)\,dy\,dz$$

Now, we hold z constant and integrate with respect to y from 0 to 10. (This takes us from a rectangular rod to a slice.) Thus,

$$\int_0^{10} (800 - 20y + 25z)\,dy = \left(800y - 20\frac{y^2}{2} + 25zy\right)\Bigg|_0^{10}$$
$$= 8{,}000 - 1{,}000 + 250z$$

Now y has run its course, and we have

$$W = \int_0^8 (7{,}000 + 250z)\,dz$$

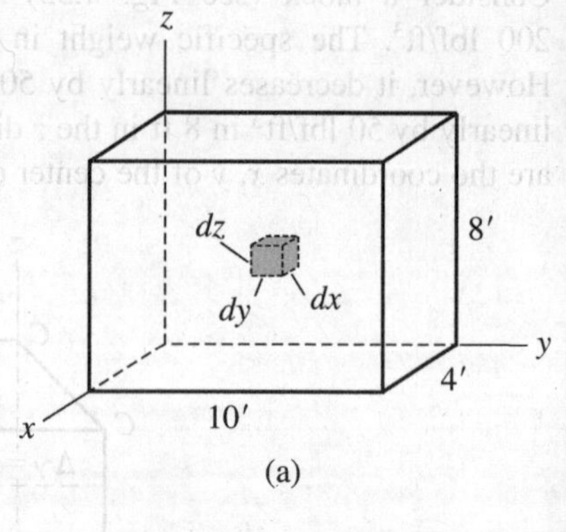

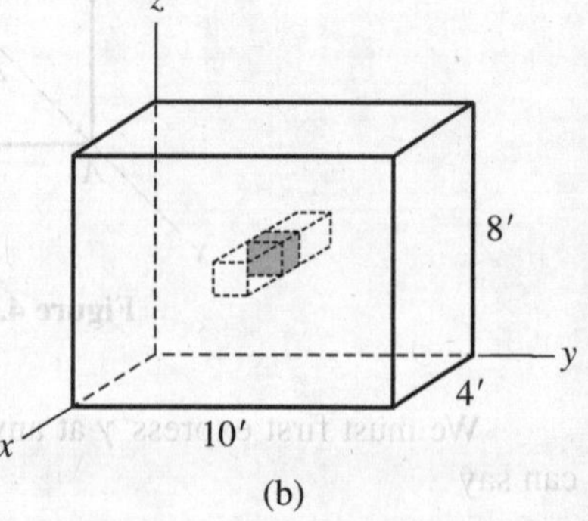

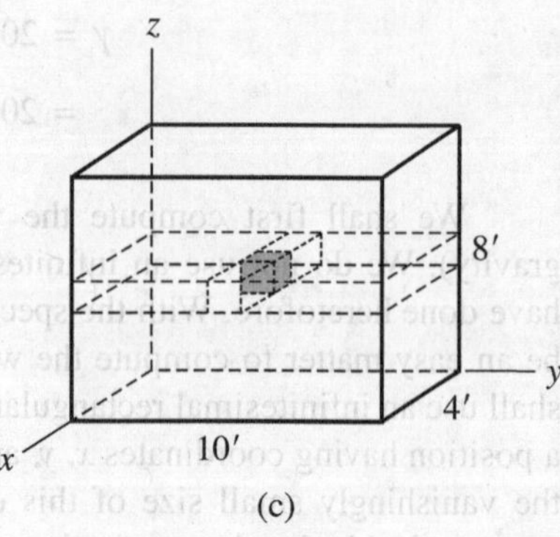

Figure 4.36. (a) Infinitesimal element at $P(x, y, z)$; (b) x runs from 0 to 4, while z and y are fixed, to form rectangular rod; (c) y runs from 0 to 10, while holding z fixed, to form slice.

Example 4.14 (Continued)

By integrating with respect to z, we sum up all the slices, and we have covered the entire block. Thus,

$$W = \left(7{,}000z + 250\frac{z^2}{2}\right)\Bigg|_0^8 = 64{,}000 \text{ lb}$$

To get $\bar{y}$, we equate the moment about the x axis of the resultant force without a couple moment with the moment of the distribution. Thus, using multiple integration as described above:

$$-(64{,}000)\bar{y} = -\int_0^8\int_0^{10}\int_0^4 y(200 - 5y + 6.25z)\,dx\,dy\,dz$$

Therefore, integrating with respect to x, then y, and then z as before, we have

$$\begin{aligned}
64{,}000\bar{y} &= \int_0^8\int_0^{10}(200yx - 5y^2x + 6.25yzx)\Big|_0^4\,dy\,dz \\
&= \int_0^8\int_0^{10}(800y - 20y^2 + 25yz)\,dy\,dz \\
&= \int_0^8\left(800\frac{y^2}{2} - \frac{20y^3}{3}x + \frac{25y^2}{2}z\right)\Bigg|_0^{10}dz \\
&= \int_0^8(40{,}000 - 6{,}667 + 1{,}250z)\,dz \\
&= \left(33{,}333z + 1{,}250\frac{z^2}{2}\right)\Bigg|_0^8 = 307{,}000
\end{aligned}$$

and

$$y = 4.79 \text{ ft}$$

Because y does not depend on x, we can directly conclude by inspection that $x = 2$ ft.

Next, in the case of a body made up of simple shapes (subbodies) such as cones, spheres, cylinders, and cubes, we can find the center of gravity of the body by using the centers of gravity of the known subbody shapes. Thus, we can say on taking moments about the y axis that

$$W_{\text{total}}(\bar{x}) = \sum_i W_i(\bar{x})_i \tag{4.12}$$

where W_i is the weight of the ith subbody and where $(x)_i$ is the x coordinate to the center of gravity of the ith subbody. Bodies made up of simple subbodies are called *composite bodies*.

Example 4.15

Find the weight and center of gravity of a large steam turbine for power generation (Fig. 4.37) needed for earthquake safety calculations. The specific weights of each turbine component are shown. Cylinder 2 has a radius of 5 m and is 14 m long. Half of this cylinder is embedded in the block 1.

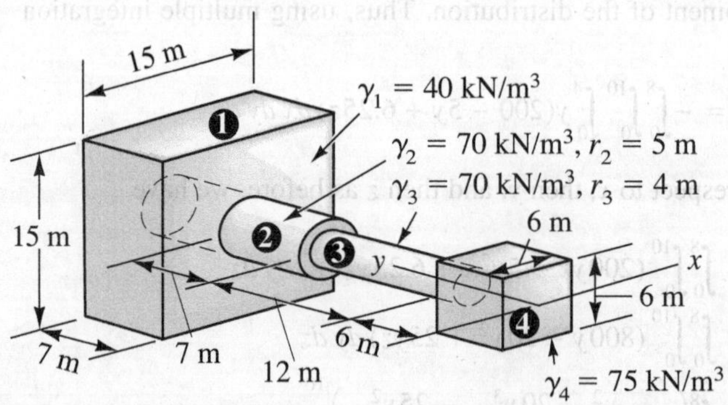

Figure 4.37. A composite body.

The weights of each component of this model are first calculated.

$$W_1 = \gamma_1 V_1 = [40][(7)(15)(15) - (7)(\pi)(5^2)] = 41 \text{ MN}$$

$$W_2 = \gamma_2 V_2 = [70][(14)(\pi)(5^2)] = 77 \text{ MN}$$

$$W_3 = \gamma_3 V_3 = [70][(12)(\pi)(4^2)] = 42.2 \text{ MN}$$

$$W_4 = \gamma_4 V_4 = [75][(6)(6)(6)] = 16.2 \text{ MN}$$

$$W_{TOTAL} = \Sigma W_i = 41 + 77 + 42.2 + 16.2 = \boxed{176.4 \text{ MN}}$$

The center of gravity lies along the axis of the shaft. Taking moments about the x axis we have

$$176.4y = (41)(28.5) + (77)(25) + (42.2)(12) + (16.2)(3) = 3{,}648.5 \text{ MN}$$

$$\boxed{y = 20.68 \text{ m}}$$

Case B. Parallel Force Distribution over a Plane Surface—Center of Pressure. Let us now consider a normal pressure distribution over a *plane* surface A in the xy plane in Fig. 4.38. The vertical ordinate is taken as a pressure ordinate, so that over the area A we have a pressure distribution $p(x, y)$ represented by the pressure surface. Since in this case there is a parallel force system with one sense of direction, we know that the simplest resultant is a single force, which is given as

$$\boldsymbol{F}_R = -\int p\,dA = -\left(\int p\,dA\right)\boldsymbol{k} \qquad (4.13)$$

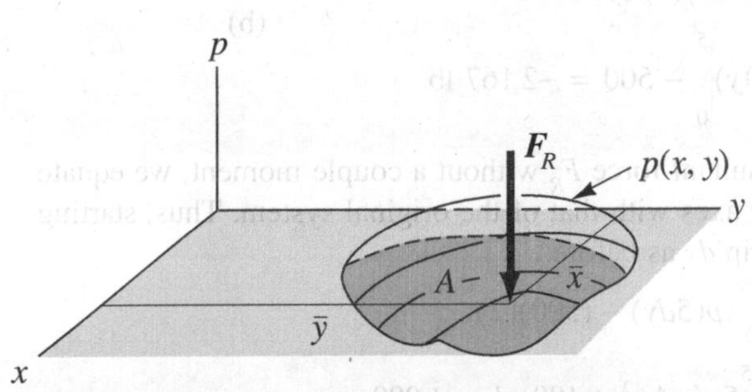

Figure 4.38. Pressure distribution.

The position x, y can be computed by equating the moments about the y and x axes of the resultant force without a couple moment with the corresponding moments of the distribution. Solving for x and y,

$$\bar{x} = \frac{\int px\,dA}{\int p\,dA}$$

$$\bar{y} = \frac{\int py\,dA}{\int p\,dA}$$

Since we know that p is a function of x and y over the surface, we can carry out the preceding integrations either analytically or numerically. The point thus determined is called the *center of pressure*.

(In later chapters, we shall consider distributed frictional forces over plane and curved surfaces. In these cases, the simplest resultant is not necessarily a single force as it was in the special case above.)

Example 4.16

A plate *ABCD* on which both distributed and point force systems act is shown in Fig. 4.39. The pressure distribution is given as

$$p = -4y^2 + 100 \text{ psf} \tag{a}$$

Find the simplest resultant for the system.

To get the resultant force, we consider a strip dy along the plate as shown in Fig. 4.39. The reason for using such a strip is that the pressure p is uniform along this strip, as can be seen from the diagram. Hence, the force from the pressure on the strip is simply $p\,dA = p(dy)(5)$. Thus, we can say that

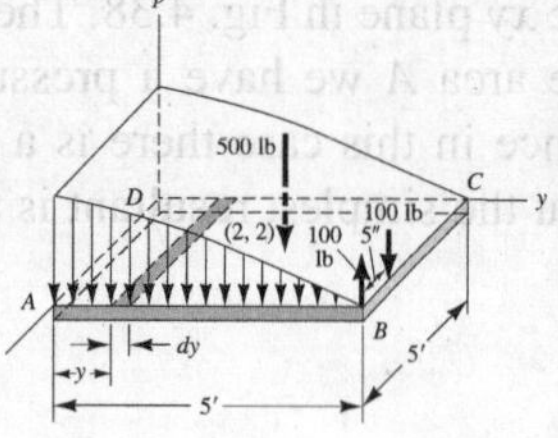

Figure 4.39. Find simplest resultant.

$$\begin{aligned} F_R &= -\int_0^5 p(5)(dy) - 500 \\ &= -\int_0^5 (-4y^2 + 100)(5)\,dy - 500 \\ F_R &= \left(20\frac{y^3}{3} - 500y\right)\Big|_0^5 - 500 = \boxed{-2{,}167 \text{ lb}} \end{aligned} \tag{b}$$

To get the position $\bar{x}$, $\bar{y}$ of the resultant force F_R without a couple moment, we equate moments of F_R about the x and y axes with that of the original system. Thus, starting with the x axis, we have using strip dy as before:

$$\begin{aligned} -2{,}167\bar{y} &= -\int_0^5 yp(5dy) - (500)(2) \\ &= -\int_0^5 5y(-4y^2 + 100)\,dy - 1{,}000 \\ &= \left(20\frac{y^4}{4} - 500\frac{y^2}{2}\right)\Big|_0^5 - 1{,}000 = -4{,}125 \end{aligned}$$

Therefore,

$$\boxed{y = 1.904 \text{ ft}}$$

Now, considering the y axis, we still use the strips dy because p is uniform along such strips. However, the force $df = p\,dA = p(5)(dy)$ may be considered acting at the center of the strip, and accordingly has a moment arm about the y axis equal to 5/2 for each strip. Hence, we can say that

$$\begin{aligned} 2{,}167\bar{x} &= \int_0^5 \frac{5}{2}p(5dy) + (500)(2) - \frac{500}{12} \\ &= \frac{25}{2}\int_0^5 (-4y^2 + 100)\,dy + 1{,}000 - 41.7 = 5{,}125 \end{aligned}$$

Therefore,

$$\boxed{x = 2.36 \text{ ft}}$$

In a stationary liquid, the pressure at the surface of the liquid is transmitted uniformly throughout the liquid. In addition, there is superposed a linearly increasing pressure with depth resulting from gravity acting on the liquid. Thus, if we have p_{atm} at the surface (often called the free surface), the pressure in the liquid is

$$p = p_{atm} + \gamma y$$

where γ is the specific weight of the liquid and y is the depth below the surface. Thus, for a given liquid the pressure is uniform at any constant distance below the free surface. With this in mind, examine the following example.

Example 4.17

In Fig. 4.40 find the force on the door AB from water whose specific weight is 9,806 N/m^3 and on whose free surface there is atmospheric pressure equal to 101,325 N/m^2 (≡ 101,325 Pa).[9] Also find the center of pressure.

The pressure on the door AB is then

$$p = p_{atm} + (\gamma)(y) = p_{atm} + (\gamma)(s)(\sin 45°)$$

where s is the distance from O along the inclined wall OR. The resultant force is then

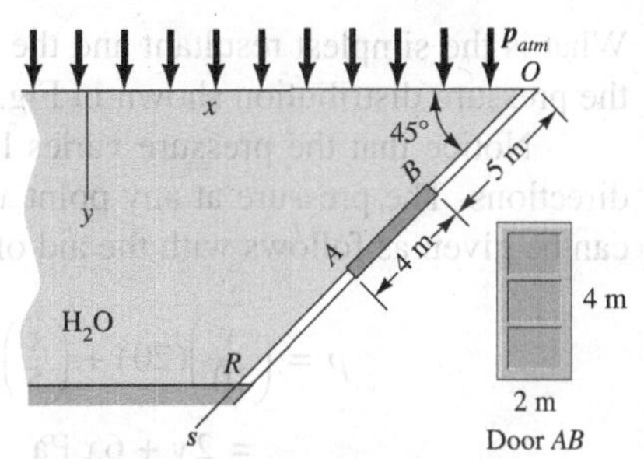

Figure 4.40. Door AB is exposed on one side to water and on the other side to air.

$$F = \int_5^9 \left[p_{atm} + (\gamma)(s)(\sin 45°)\right](2)(ds)$$

$$= \int_5^9 [101{,}325 + (9{,}806)(.707)s](2)ds$$

$$= (2)\left[101{,}325s + (9{,}806)(.707)\frac{s^2}{2}\right]_5^9$$

$$\therefore \quad F = 1.199 \times 10^6 \text{ N}$$

To get the center of pressure equate moments about O with that of the resultant. Thus using the notation $\bar{s}$ to locate the center of pressure we have

$$1.199 \times 10^6 \bar{s} = \int_5^9 s\left[p_{atm} + (\gamma)(s)(\sin 45°)\right](2)\, ds$$

$$= \left[101{,}325\frac{s^2}{2} + (9{,}806)(.707)\frac{s^3}{3}\right]_5^9 (2)$$

$$\therefore \quad s = 7.061 \text{ m}$$

Clearly the *total* force on the door from inside and outside would include the contribution of the atmospheric pressure on the outside. We can easily determine this force by deleting p_{atm} in the preceding calculations.[10]

[9]The unit of pressure in the SI system is the *pascal*, where 1 pascal = 1 Pa = 1 N/m^2.

[10]We have touched here on the subject of *hydrostatics*. For a treatment of this subject that may be similar to what you will study in your upcoming course in fluid mechanics, see Shames, I.H., *Mechanics of Fluids*, 3rd Edition, 1992, McGraw-Hill, Inc., Chapter 3

We finish this series of examples with a multiple integration problem.

*Example 4.18

What is the simplest resultant and the center of pressure for the pressure distribution shown in Fig. 4.41?

Notice that the pressure varies linearly in the x and y directions. The pressure at any point x, y in the distribution can be given as follows with the aid of similar triangles:

$$p = \left(\frac{y}{10}\right)(20) + \left(\frac{x}{5}\right)(30) \qquad \text{(a)}$$

$$= 2y + 6x \text{ Pa}$$

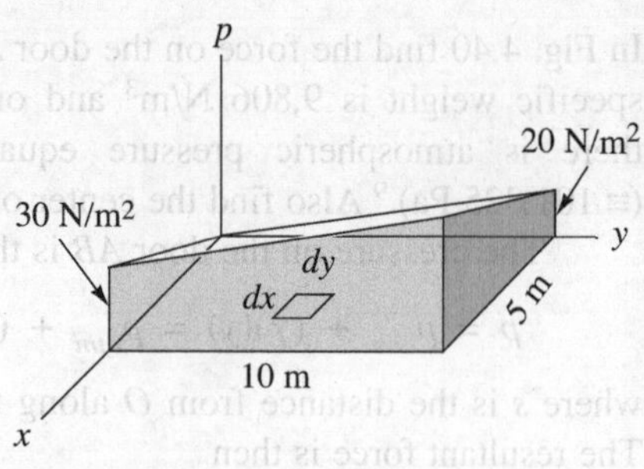

Figure 4.41. Nonuniform pressure distribution.

We cannot employ a convenient strip here along which the pressure is uniform, as in Example 4.16. For this reason we consider rectangular area element $dx\ dy$ to work with (see Fig. 4.41). For such a small area, we can assume the pressure as constant so that $p\ dx\ dy$ is the force on the element. To find the resultant force, we must integrate over the 10 × 5 rectangle. This integration involves two variables and is again a case of *multiple integration.* Thus, we can say that

$$F_R = -\int_0^{10}\int_0^5 p\,dx\,dy$$

wherein we first integrate with respect to x while holding y constant and then integrate with respect to y (in this way we cover the entire 10 × 5 rectangular area). Thus, we have

$$F_R = -\int_0^{10}\int_0^5 (2y + 6x)\,dx\,dy$$

$$= -\int_0^{10}\left(2yx + \frac{6x^2}{2}\right)\Bigg|_0^5 dy$$

$$= -\int_0^{10}(10y + 75)\,dy$$

$$= -\left[\frac{10y^2}{2} + 75y\right]_0^{10} = -1{,}250 \text{ N}$$

$$F_R = -1{,}250 \text{ N}$$

To find $\bar{y}$ for F_R without a couple moment, we equate moments of F_R about the x axis with that of the distribution. Thus,

$$-\bar{y}(1{,}250) = -\int_0^{10}\int_0^5 py\,dx\,dy$$

Example 4.18 (Continued)

Therefore,

$$\bar{y} = \frac{1}{1{,}250}\int_0^{10}\int_0^{5}(2y+6x)y\,dx\,dy$$

$$= \frac{1}{1{,}250}\int_0^{10}\left(2y^2x + \frac{6x^2}{2}y\right)\Big|_0^5 dy$$

$$= \frac{1}{1{,}250}\int_0^{10}\left(10y^2 + 75y\right)dy$$

$$= \frac{1}{1{,}250}\left(\frac{10y^3}{3} + 75\frac{y^2}{2}\right)\Big|_0^{10}$$

$$y = 5.67 \text{ m}$$

As for $\bar{x}$, we proceed as follows:

$$\bar{x}(1{,}250) = \int_5^{10}\int_0^{5} px\,dx\,dy$$

Therefore,

$$\bar{x} = \frac{1}{1{,}250}\int_0^{10}\int_0^{5}(2y+6x)x\,dx\,dy$$

$$= \frac{1}{1{,}250}\int_0^{10}\left(2y\frac{x^2}{2} + \frac{6x^3}{3}\right)\Big|_0^5 dy$$

$$= \frac{1}{1{,}250}\int_0^{10}(25y + 250)\,dy$$

$$= \frac{1}{1{,}250}\left(25\frac{y^2}{2} + 250y\right)\Big|_0^{10}$$

$$x = 3.00 \text{ m}$$

The center of pressure is thus at (3.00, 5.67) m.

Case C. Coplanar Parallel Force Distribution. As we pointed out earlier, this type of loading may be considered for beams loaded symmetrically over the longitudinal midplane of the beam. The loading is represented by an intensity function $w(x)$ as shown in Fig. 4.26. This coplanar parallel force distribution can be replaced by a single force given as

$$\boldsymbol{F}_R = -\int w(x)\,dx\boldsymbol{j}$$

We find the position of $\boldsymbol{F}_R$ without a couple moment by equating moments of $\boldsymbol{F}_R$ and the distribution w about a convenient point of the beam, usually one of the ends. Solving for $\bar{x}$, we get

$$\bar{x} = \frac{\int xw(x)\,dx}{\int w(x)\,dx}$$

Example 4.19

A simply supported beam is shown in Fig. 4.42 supporting a 1,000-lb point force, a 500 lb-ft couple, and a coplanar, parabolic, distributed load w lb/ft. Find the simplest resultant of this force system.

To express the intensity of loading for the coordinate system shown in the diagram, we begin with the general formulation

$$w^2 = ax + b \tag{a}$$

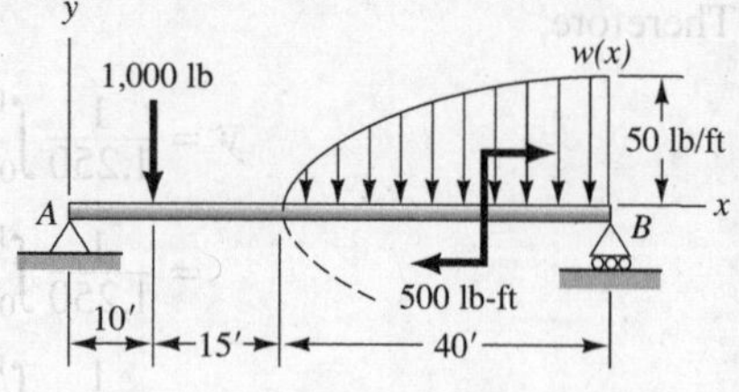

Figure 4.42. Find simplest resultant.

Note from the diagram that when $x = 25$ we have $w = 0$, and when $x = 65$, we have $w = 50$. Subjecting Eq. (a) to these conditions, we can determine a and b. Thus,

$$0 = a(25) + b \tag{b}$$

$$2{,}500 = a(65) + b \tag{c}$$

Subtracting, we can get a as follows:

$$-2{,}500 = -40a$$

Therefore,

$$a = 62.5$$

From Eq. (b), we get

$$b = -(25)(62.5) = -1{,}562.5$$

Thus, we have

$$w^2 = 62.5x - 1{,}562.5 \tag{d}$$

Summing forces, we get for F_R,

$$F_R = -1{,}000 - \int_{25}^{65} \sqrt{62.5x - 1{,}562.5}\, dx \tag{e}$$

To integrate this, we may change variables as follows:

$$\mu = 62.5x - 1{,}562.5 \tag{f}$$

Therefore,

$$d\mu = 62.5\, dx$$

Substituting into the integral in Eq. (e), we have[11]

$$F_R = -1{,}000 - \int_0^{2{,}500} \mu^{1/2} \frac{d\mu}{62.5}$$

$$= -1{,}000 - \frac{1}{62.5}\mu^{3/2}\left(\frac{2}{3}\right)\Bigg|_0^{2{,}500}$$

[11]Do not forget to change the limits for μ. Thus, from Eq. (f), the upper limit is $(62.5)(65) - 1{,}562.5 = 2{,}500$, whereas the lower limit is $(62.5)(25) - 1{,}562.5 = 0$.

Example 4.19 (Continued)

$$= -1{,}000 - \frac{1}{62.5}(2{,}500)^{3/2}\left(\frac{2}{3}\right)$$

$$F_R = -2{,}333 \text{ lb}$$

We now compute x for the resultant without a couple as follows:

$$-2{,}333\bar{x} = -(10)(1{,}000) - \int_{25}^{65} x\sqrt{62.5x - 1{,}562.5}\, dx - 500 \qquad \text{(g)}$$

We can evaluate the integral most readily by consulting the mathematical formulas in Appendix I. We find the following formula (No 6):

$$\int x\sqrt{a + bx}\, dx = -\frac{2(2a - 3bx)\sqrt{(bx + a)^3}}{15b^2}$$

In our case $b = 62.5$ and $a = -1{,}562.5$, so the indefinite integral for our case is

$$\int x\sqrt{62.5x - 1{,}562.5}\, dx = -\frac{(2)(-3{,}125 - 187.5x)\sqrt{(62.5x - 1{,}562.5)^3}}{(15)(3{,}906)}$$

Putting in limits, we have

$$\int_{25}^{65} x\sqrt{62.5x - 1{,}562.5}\, dx = -\left.\frac{(2)(-3{,}125 - 187.5x)\sqrt{(62.5x - 1{,}562.5)^3}}{(15)(3{,}906)}\right|_{25}^{65}$$

$$= 65{,}333 - 0 = 65{,}333$$

Going back to Eq. (g), we can now solve easily for $\bar{x}$. Thus,

$$\bar{x} = -\frac{1}{2{,}330}[-(10)(1{,}000) - 65{,}300 - 500]$$

$$x = 32.5 \text{ ft}$$

Before closing, it will be pointed out that, for a loading function $w(x)$, the resultant $\int_0^x w\, dx$, equals the *area* under the loading curve. This fact is particularly useful for the case of a triangular loading function such as is shown in Fig. 4.43. Hence, we can say on inspection that the resultant force has the value

$$F_R = \tfrac{1}{2}(5)(1{,}000) = 2{,}500 \text{ N}$$

Furthermore, you can readily show that the *simplest* resultant has a line of action that is (2/3) × (length of loading) from the toe of the loading.[12] Thus, F_R without a couple moment is at a position (2/3)(5) to the right of a (see Fig. 4.43). You are urged to use this information when needed.

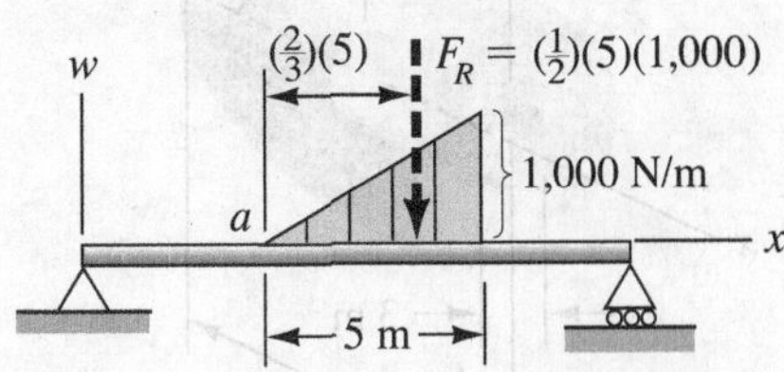

Figure 4.43. Triangular loading resultant.

[12] In Chapter 7, you will learn that the simplest resultant force for a distribution $w(x)$ goes through the *centroid* of the area under $w(x)$. The centroid will be carefully defined at that time.

PROBLEMS

4.41. A force field is given as

$$F(x, y, z, t) = (10x + 5)i + (16x^2 + 2z)j + 15k \text{ N}$$

What is the force at position (3, 6, 7) m? What is the difference between the force at this position and that at the origin?

4.42. A magnetic field is developed such that the body force on the rectangular parallelepiped of metal is given as

$$f = (.01x + \quad)k \text{ oz/lbm}$$

If the mass density of the metal is 450 lbm/ft³, what is the *simplest* resultant body force from such a field? Note that at the earth's surface the number of pounds mass equals the number of pounds force of weight.

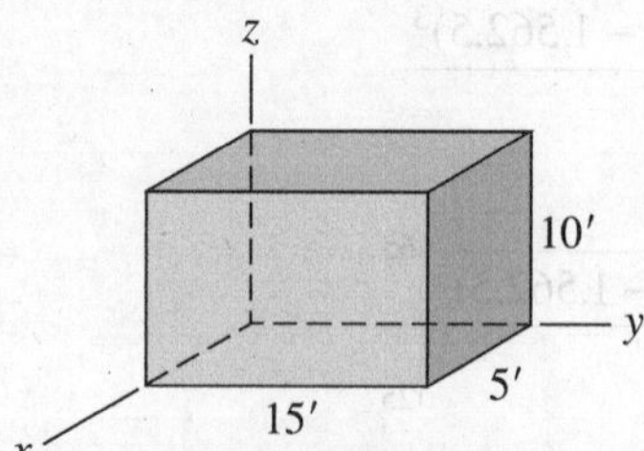

Figure P.4.42.

4.43. A body of revolution has a variable specific weight such that $\gamma = (36 + .01x^2)$ kN/m³ with x in meters. A hole of diameter 3 m and length 6 m is cut from the body as shown. Where is the center of gravity?

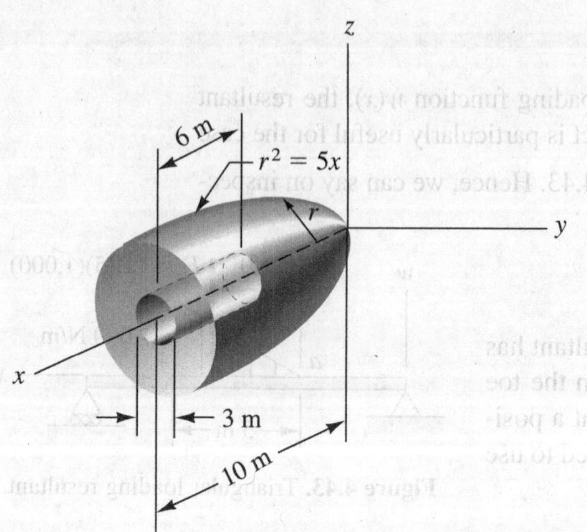

Figure P.4.43.

4.44. The specific weight γ of the material in the solid cylinder varies linearly as one goes from face A to face B. If

$$\gamma_A = 400 \text{ lbf/ft}^3, \quad \gamma_B = 500 \text{ lbf/ft}^3$$

what is the position of the center of gravity of the cylinder?

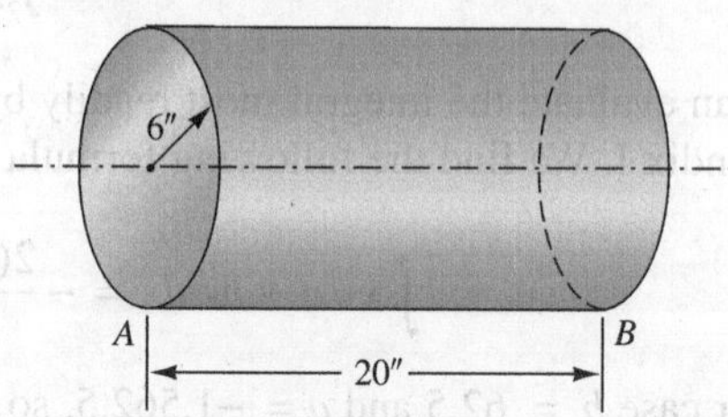

Figure P.4.44.

4.45. The specific weight of the material in a right circular cone is constant. Where is the center of gravity of the cone? *Hint:* Rotate cone 90° so that gravity is perpendicular to the z axis. Use concept of similar triangles to show that $r/R = (h - z)/h$ and solve for r needed for the integration.

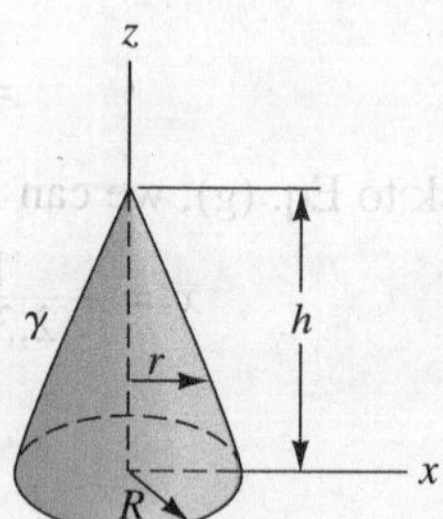

Figure P.4.45.

4.46. Show that the center of gravity of the right triangular plate of thickness t is at $x = a/3$ and $y = b/3$.

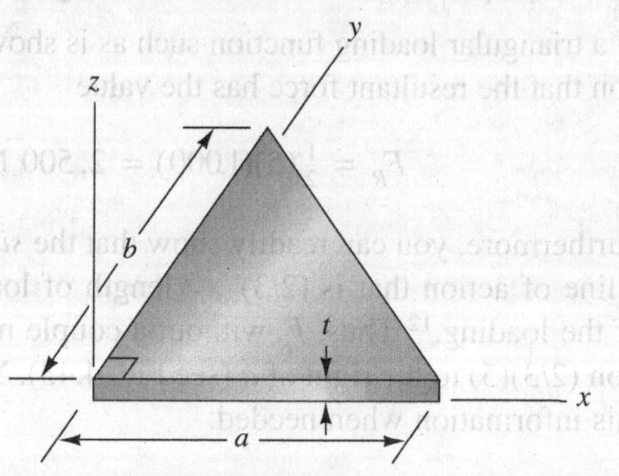

Figure P.4.46.

4.47. Show that the volume and center of gravity of the conical frustum are, respectively,

$$V = \frac{\pi h}{3}\left(r_2^2 + r_1 r_2 + r_1^2\right)$$

and

$$\bar{z} = \frac{h}{4}\,\frac{3r_2^2 + r_1^2 + 2r_1 r_2}{r_2^2 + r_1^2 + r_1 r_2}$$

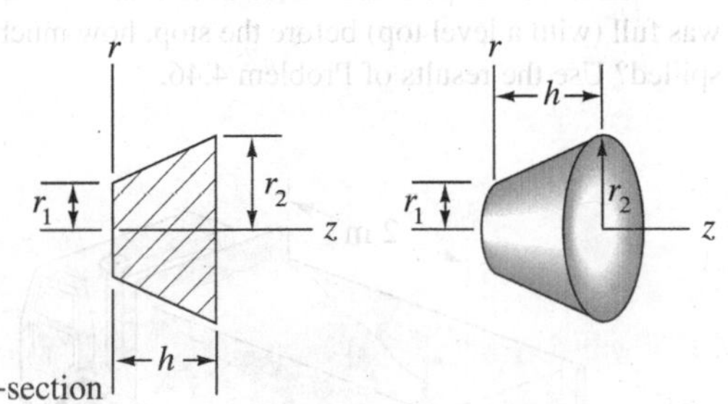

Figure P.4.47.

4.48. Find the center of gravity of the plate bounded by a straight line and a parabola.

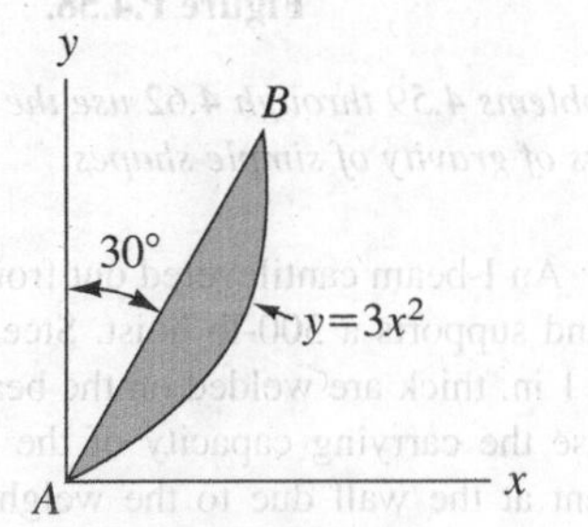

Figure P.4.48.

4.49. A massive radio-wave antenna for detection of signals from outer space is a body of revolution with a parabolic face (see the diagram). These antennas may be carved from rock in a valley away from other disturbing signals. What would the antenna weigh if made from concrete (23.6 kN/m³) for location in a remote desert area?

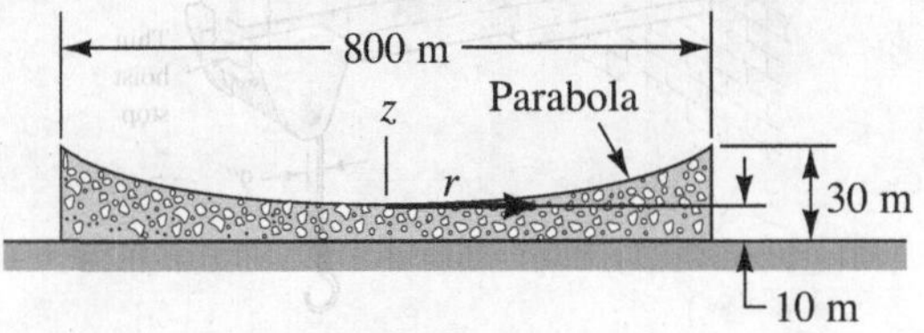

Figure P.4.49.

4.50. In Problem 4.49 find the distance from the ground to the center of gravity if the total weight is 2.37×10^8 kN.

***4.51.** A plate of thickness 30 mm has a specific weight γ that varies linearly in the x direction from 26 kN/m³ at A to 36 kN/m³ at B, and varies in the y direction as the square of y from 26 kN/m³ at A to 40 kN/m³ at C. Where is the center of gravity of the plate?

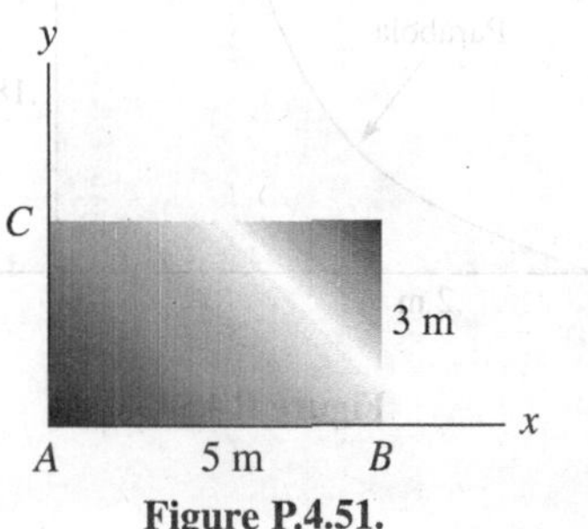

Figure P.4.51.

4.52. You are looking down on a plate with a hole in it as shown. The thickness has a constant value equal to t and the specific weight γ is constant. Find the coordinates $\bar{x}$, $\bar{y}$ of the *center of gravity*.

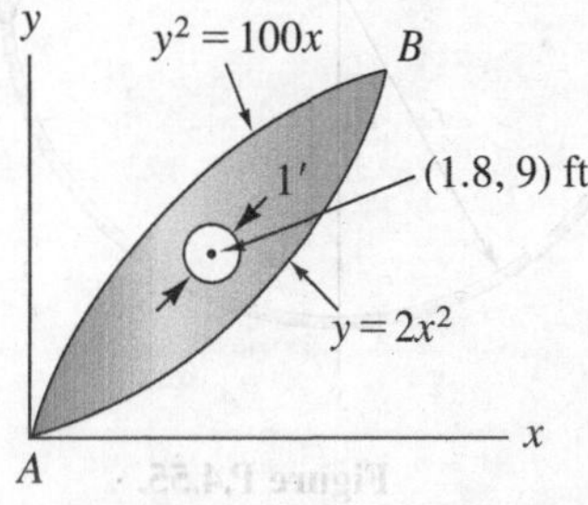

Figure P.4.52.

4.53. Find the center of gravity of the plate having uniform thickness and uniform specific weight. You are looking down onto the top of the plate.

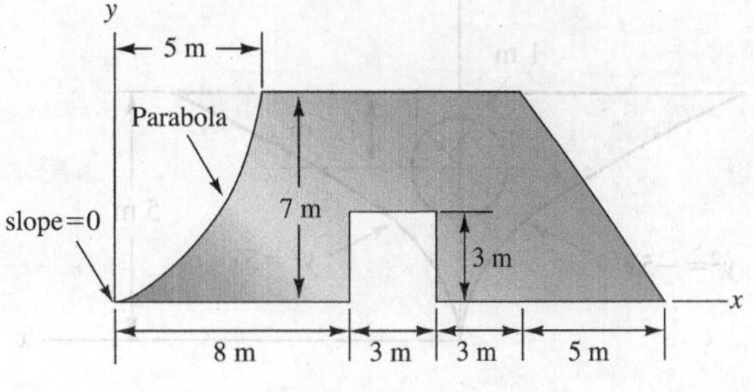

Figure P.4.53.

4.54. The top view of a plate is shown. Find center of gravity coordinates $\bar{x}$, $\bar{y}$.

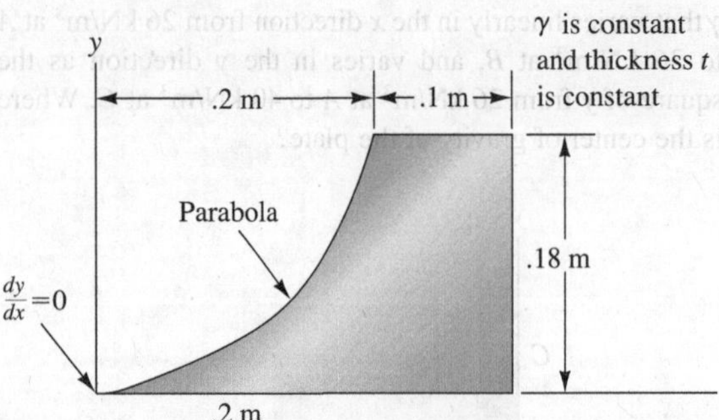

Figure P.4.54.

4.55. The thin circular rod has a weight of w N/m. What is the y coordinate of its center of gravity? The rod forms one-half of a circle.

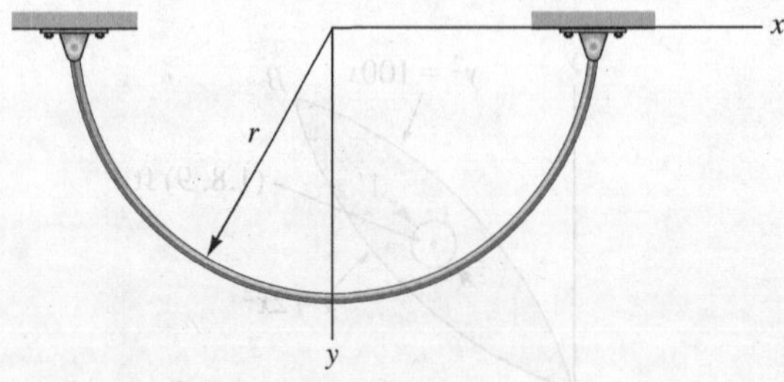

Figure P.4.55.

4.56. Find $\bar{y}$ for the center of gravity of the horizontal plate with a hole in it.

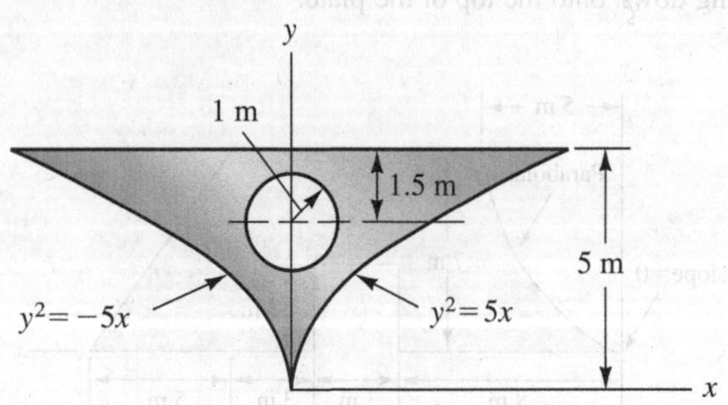

Figure P.4.56.

***4.57.** Suppose in Problem 4.42 that

$$\boldsymbol{f} = (.01x + .2y + .3z)\boldsymbol{k} \text{ oz/lbm}$$

Find the *simplest* resultant for $\boldsymbol{\rho} = 450 \text{ lbm/ft}^3$. Find the proper line of action.

4.58. After a fast stop and swerve to the left, the load of sand (specific weight = 15 kN/m^3) in a dump truck is in the position shown. What is the simplest resultant force on the truck from the sand and where does it act? If the truck was full (with a level top) before the stop, how much sand spilled? Use the results of Problem 4.46.

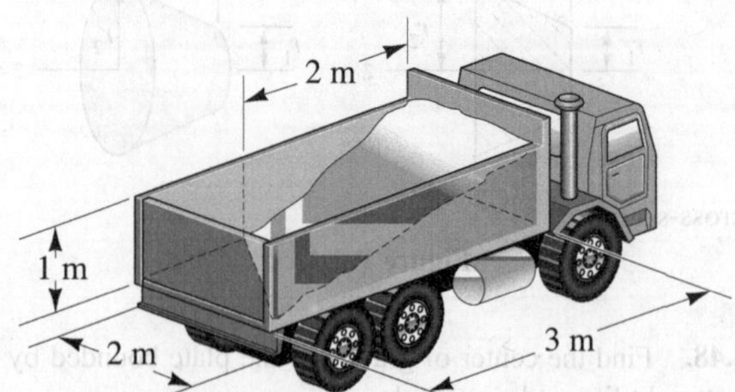

Figure P.4.58.

In Problems 4.59 through 4.62 use the known positions of centers of gravity of simple shapes.

4.59. An I-beam cantilevered out from a wall weighs 30 lb/ft and supports a 300-lb hoist. Steel (487 lb/ft^3) cover plates 1 in. thick are welded on the beam near the wall to increase the carrying capacity of the beam. What is the moment at the wall due to the weight of the reinforced beam and the hoisted load of 4,000 lb at the outermost position of the hoist? What is the *simplest* resultant force and its location?

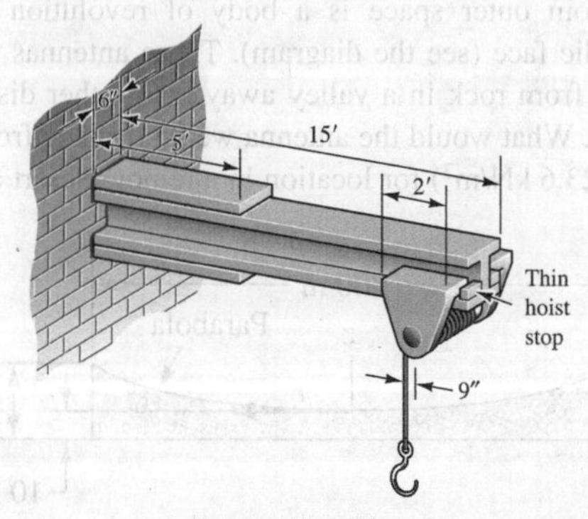

Figure P.4.59.

4.60. The bulk materials trailer weighs 10,000 lb and is filled with cement (γ = 94 lb/ft^3) in the front compartment (sections 1 and 2), and half-filled with water (γ = 62.5 lb/ft^3) in the rear compartment (sections 3 and 4). What is the *simplest* resultant force, and where does it act? What is the resultant when the water is drained? Use the center of gravity and volume results from Problem 4.47 (conical frustum).

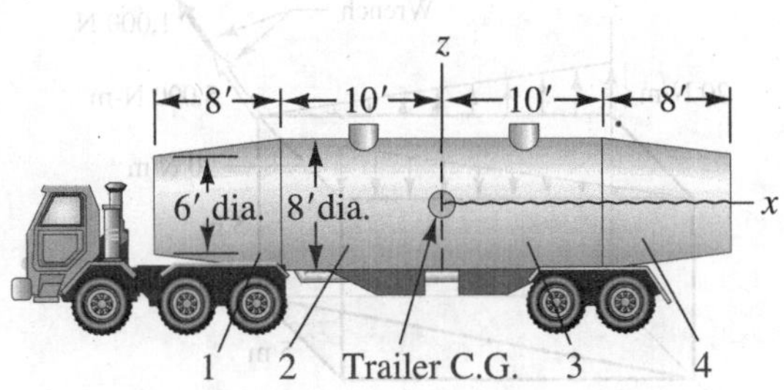

Figure P.4.60.

4.61. Find the coordinates $(x, y)_{CG}$ of the center of gravity of the loaded conveyor system. The centers of gravity of the crates C and D are at their geometric centers. W_E is the weight of the frame whose C.G. is at its geometric center.

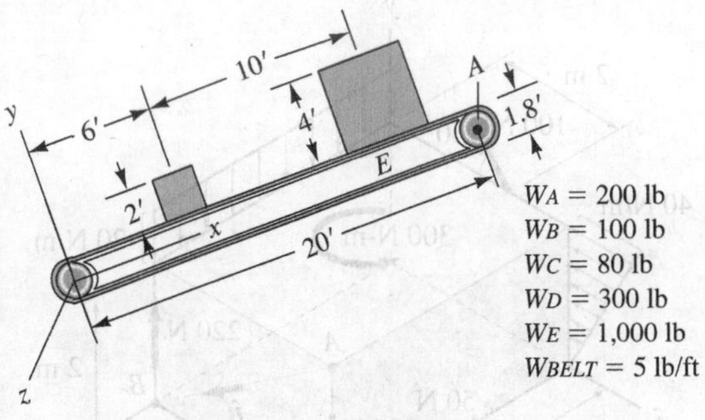

Figure P.4.61.

4.62. Find the center of gravity of the body shown. It has a constant specific weight throughout. Cone and cylinder are on block surfaces.

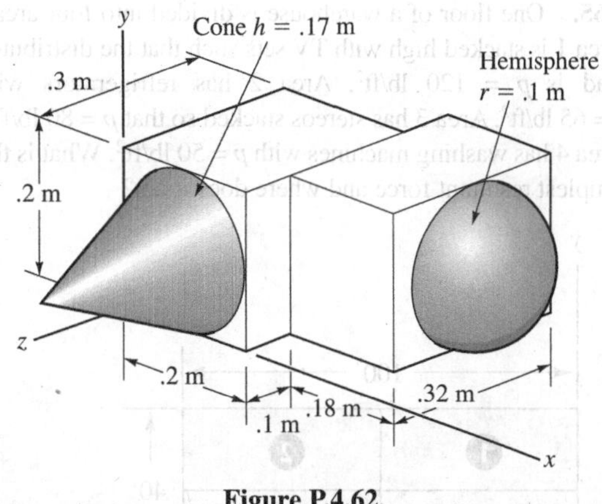

Figure P.4.62.

4.63. Find the *simplest* resultant of a normal pressure distribution over the rectangular area with sides a and b. Give the coordinates of the center of pressure.

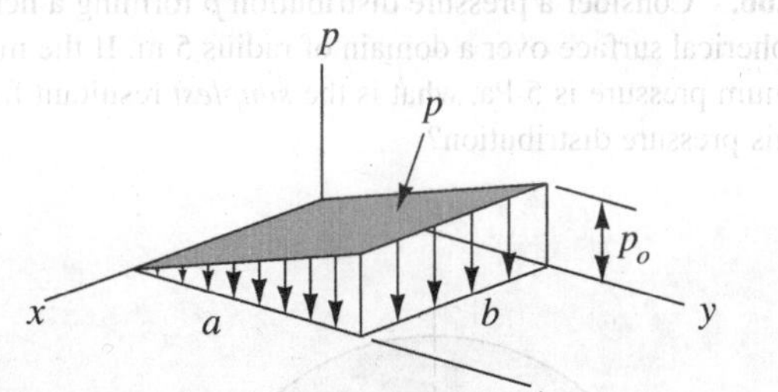

Figure P.4.63.

4.64. Find the *simplest* resultant acting on the vertical wall $ABCD$. Give the coordinates of the center of pressure. The pressure varies such that p = E/(y + psi, with y in feet, from 10 psi to 50 psi, as indicated in the diagram. E and F are constants.

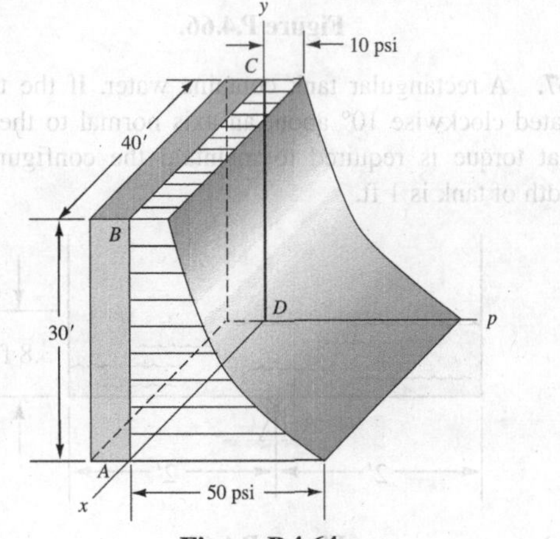

Figure P.4.64.

4.65. One floor of a warehouse is divided into four areas. Area 1 is stacked high with TV sets such that the distributed load is $p = 120$ lb/ft^2. Area 2 has refrigerators with $p = 65$ lb/ft^2. Area 3 has stereos stacked so that $p = 80$ lb/ft^2. Area 4 has washing machines with $p = 50$ lb/ft^2. What is the simplest resultant force and where does it act?

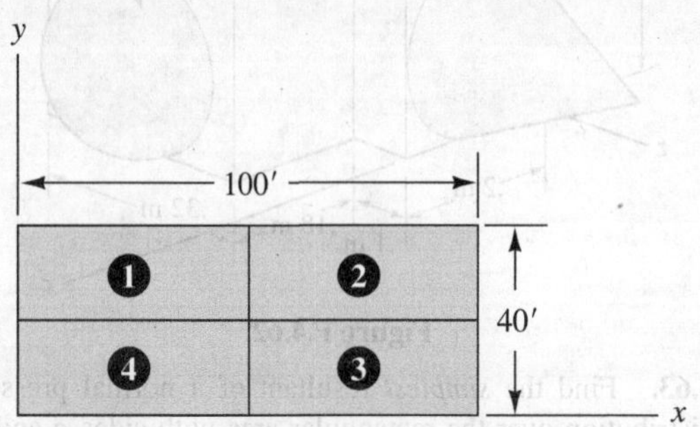

Figure P.4.65.

4.66. Consider a pressure distribution p forming a hemispherical surface over a domain of radius 5 m. If the maximum pressure is 5 Pa, what is the *simplest* resultant from this pressure distribution?

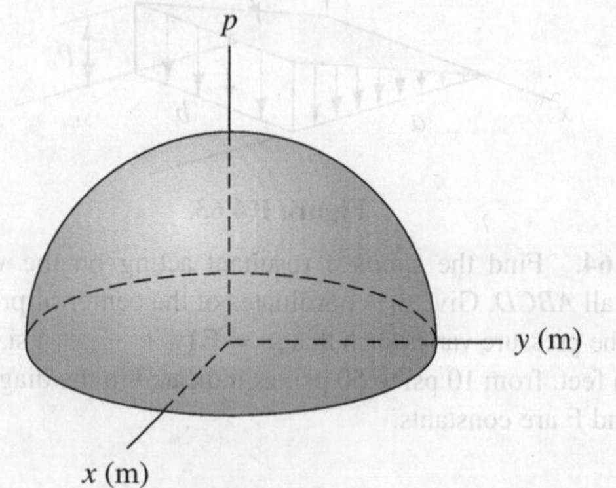

Figure P.4.66.

4.67. A rectangular tank contains water. If the tank is rotated clockwise 10° about an axis normal to the page, what torque is required to maintain the configuration? Width of tank is 1 ft.

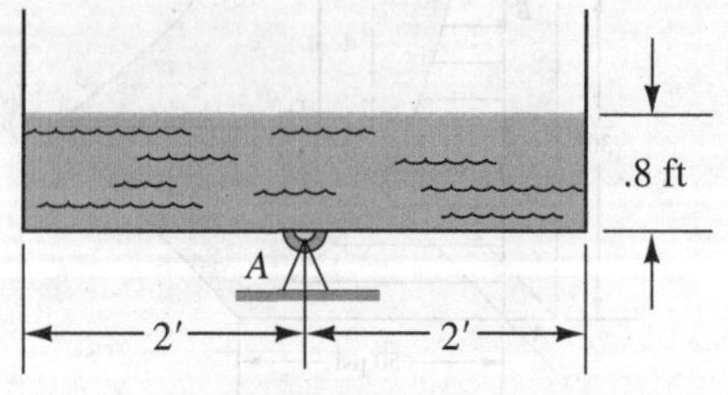

Figure P.4.67.

4.68. (a) Find the torque about axis AB from the wrench. (b) Find the torque about axis AB from the distributed loads. (*Hint:* Look down from above to help view problem.)

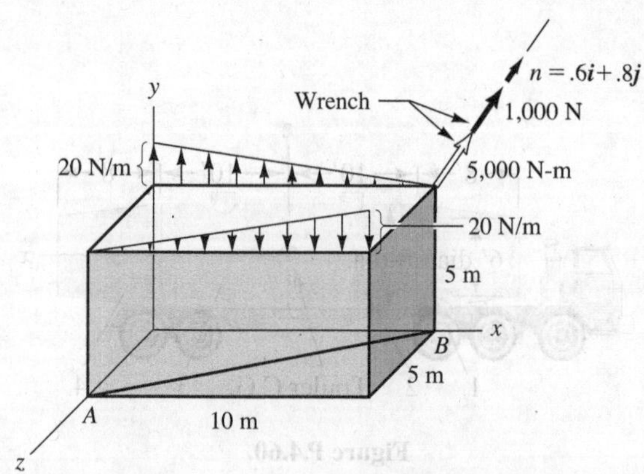

Figure P.4.68.

4.69. For the system of forces shown, determine the torque about the axis going from A to B. *Note:* The 100-N force and the triangular load distribution are in the yz plane and the 300-N-m couple is on the top face of the rectangular box.

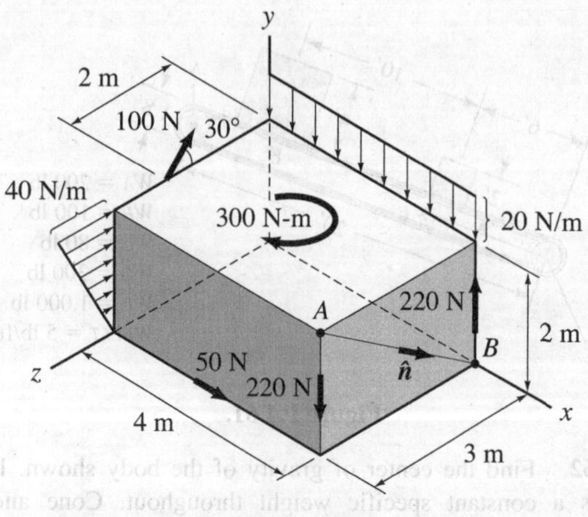

Figure P.4.69.

4.70. A *manometer* is a simple pressure measuring device. One such manometer called a *U tube* is shown in the diagram. The tank contains water including the tube to level *M* where mercury is present. *M* and *N* are at the same level. What is the gage pressure (i.e., the pressure above atmosphere) at point *a* in the tank for the following data:

$$d_1 = .2 \text{ m} \quad d_2 = .6 \text{ m} \quad \gamma_{H_2O} = 9{,}806 \text{ N/m}^3$$

$$\gamma_{Hg} = (13.7)(\gamma_{H_2O}) \text{ N/m}^3$$

[*Hint: M* and *N* are at the same level and are joined by the same fluid, namely mercury. Hence, the pressures there are equal.]

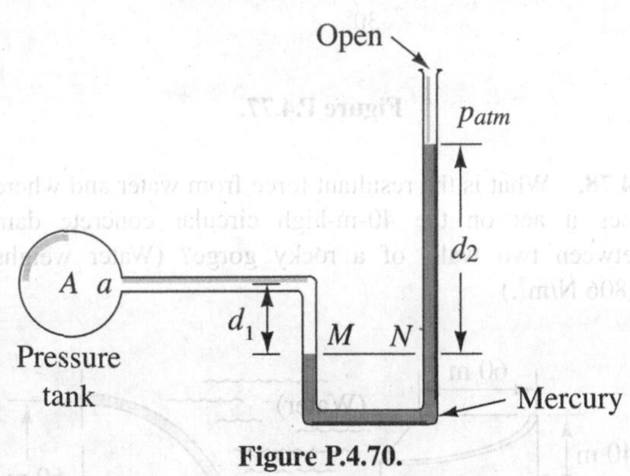

Figure P.4.70.

4.71. Imagine a liquid which when stationary stratifies in such a way that the specific weight is proportional to the square root of the pressure. At the free surface, the specific weight is known and has the value γ_0. What is the pressure as a function of depth from the free surface? What is the resultant force on one face *AB* of a rectangular plate submerged in the liquid? The width of the plate is *b*.

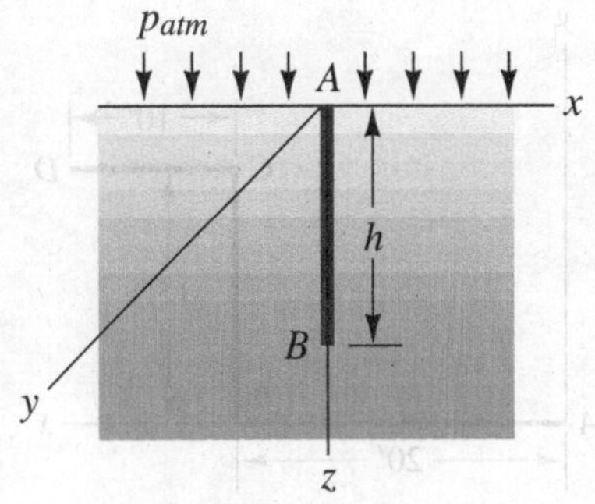

Figure P.4.71.

4.72. (a) Calculate the force on the door from all fluids inside and outside. The specific gravity of the oil is 0.8. This means that the oil has a specific weight γ which is 0.8 times that of water (γ_{water} = 62.4 lb/ft³). Note that a uniform pressure on the surface of a liquid extends undiminished throughout the depth of the liquid.

(b) Determine the distance from the surface of the oil to the *simplest* resultant force on the door from all fluids.

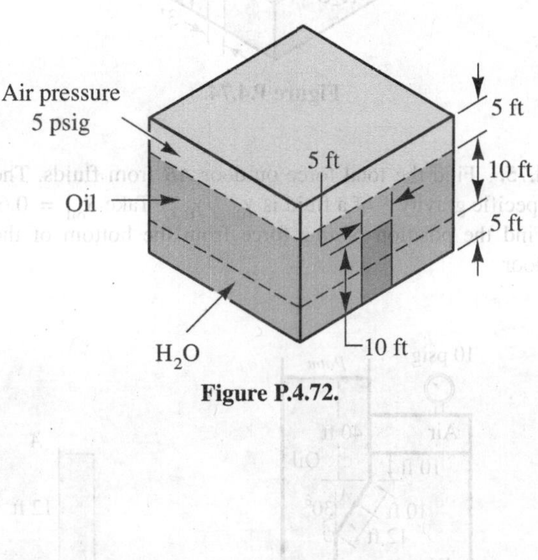

Figure P.4.72.

4.73. At what height *h* will the water cause the door to rotate clockwise? The door is 3 m wide. Neglect friction and the weight of the door.

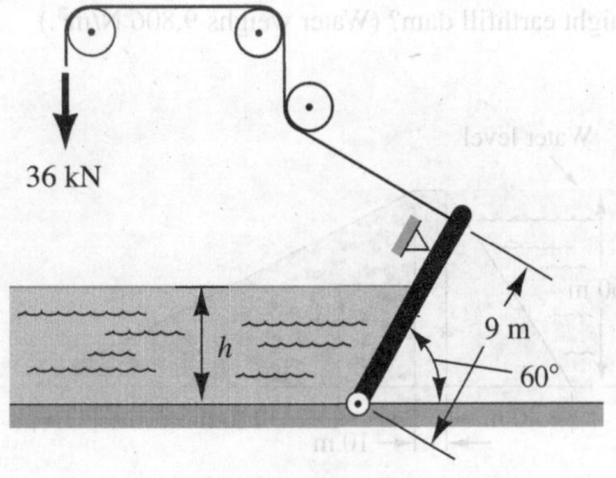

Figure P.4.73.

4.74. Find the force on the door from the inside and outside pressures. Give the position of the resultant force above the base of the door. The specific gravity S of the oil is 0.7, i.e., $\gamma_{oil} = (0.7)\gamma_{H_2O}$.

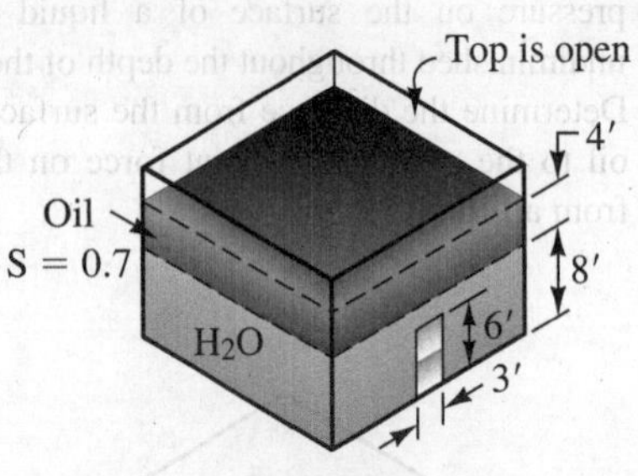

Figure P.4.74.

4.75. Find the total force on door AB from fluids. The specific gravity S of a fluid is $\gamma_{fluid}/\gamma_{H_2O}$. Take $S_{oil} = 0.6$. Find the position of this force from the bottom of the door.

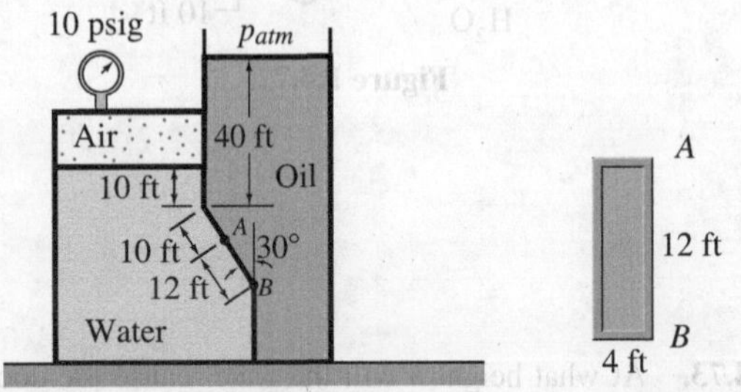

Figure P.4.75.

4.76. What is the simplest resultant force from the water and where does it act on the 60-m-high 800-m-long straight earthfill dam? (Water weighs 9,806 N/m³.)

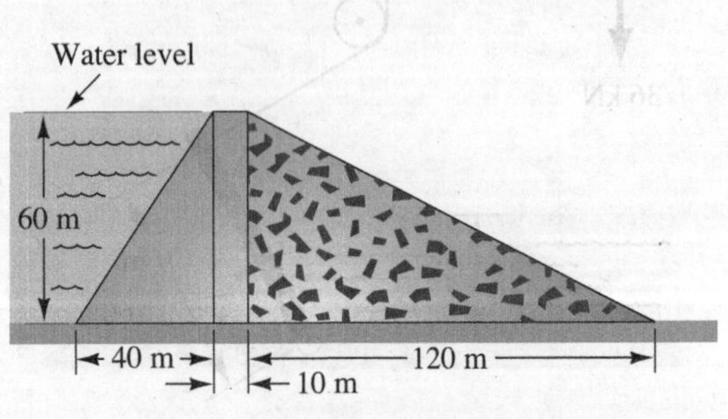

Figure P.4.76.

4.77. A block 1 ft thick is submerged in water. Compute the simplest resultant force and the center of pressure on the bottom surface. Take $\gamma = 62.4$ lb/ft³.

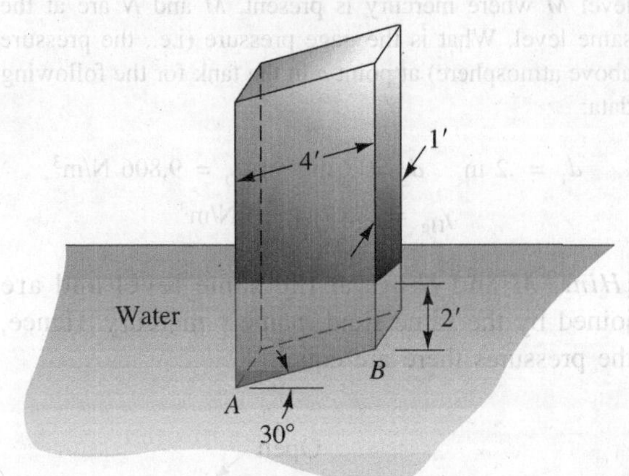

Figure P.4.77.

***4.78.** What is the resultant force from water and where does it act on the 40-m-high circular concrete dam between two walls of a rocky gorge? (Water weighs 9,806 N/m³.)

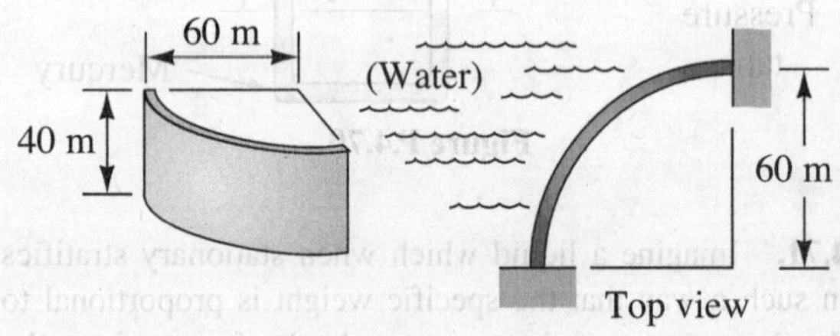

Figure P.4.78.

4.79. The weight of the wire $ABCD$ per unit length, w, increases linearly from 4 oz/ft at A to 20 oz/ft at D. Where is the center of gravity of the wire?

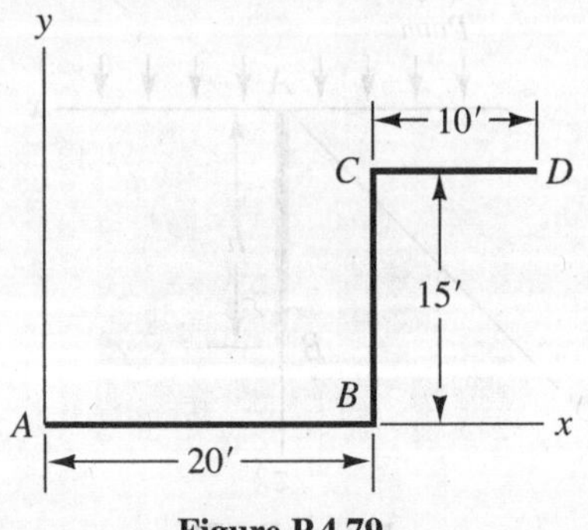

Figure P.4.79.

4.80. Find the center of gravity of the wire. The weight per unit length increases as the square of the length of wire from a value of 3 oz/ft at A until it reaches the value of 8 oz/ft at C. It then decreases 1 oz/ft for every 10 ft of length.

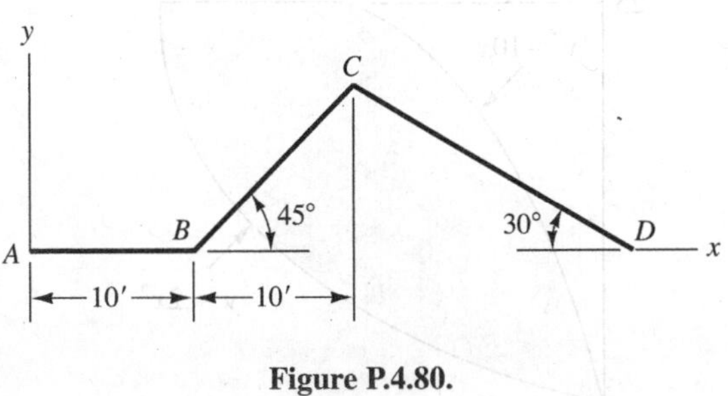

Figure P.4.80.

4.81. What is the center of gravity coordinate y_c for a thin circular rod shown in the diagram? It has a weight of w N/m. The rod is placed symmetrically about the y axis. The angle ϕ is in degrees.

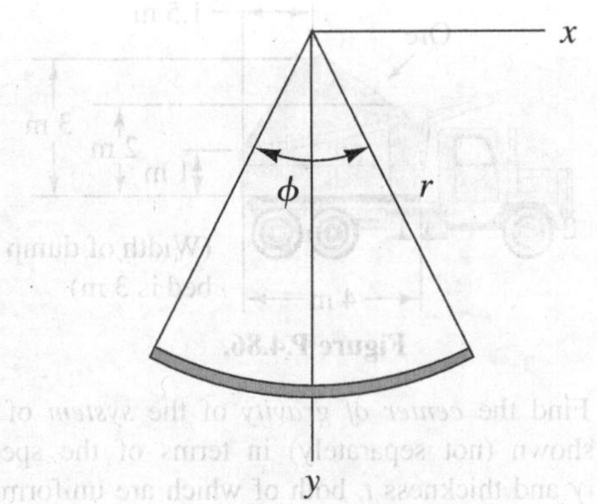

Figure P.4.81.

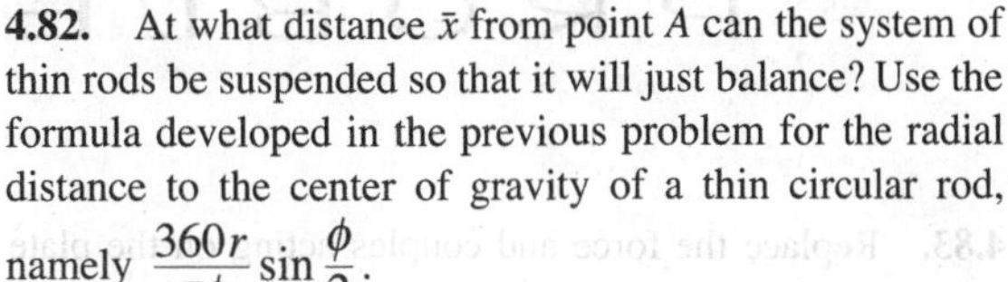

4.82. At what distance $\bar{x}$ from point A can the system of thin rods be suspended so that it will just balance? Use the formula developed in the previous problem for the radial distance to the center of gravity of a thin circular rod, namely $\frac{360r}{\pi\phi}\sin\frac{\phi}{2}$.

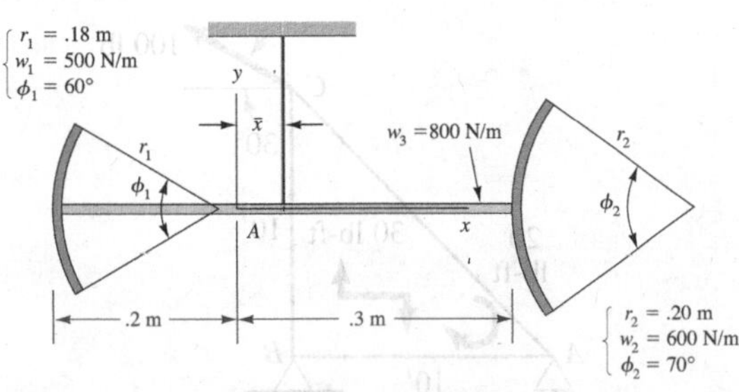

Figure P.4.82.

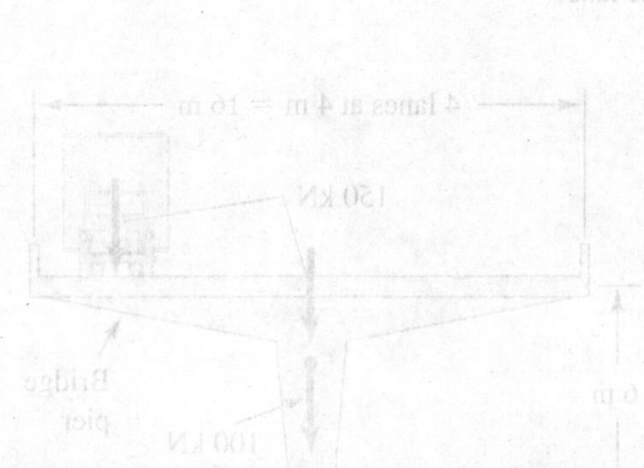

4.6 Closure

We now have the tools that enable us to replace, for purposes of rigid body mechanics, any system of forces by a resultant consisting of a force and a couple moment. These tools will prove very helpful in our computations. More important at this time, however, is the fact that in considering conditions of equilibrium for rigid bodies we need only concern ourselves with this resultant to reach conclusions valid for any force system, no matter how complex. From this viewpoint, we shall develop the fundamental equations of statics in Chapter 5 and then employ them to solve a large variety of problems.

PROBLEMS

4.83. Replace the force and couples acting on the plate by a single force. Give the intercept of the line of action of this force with the vertical edge BC of the plate.

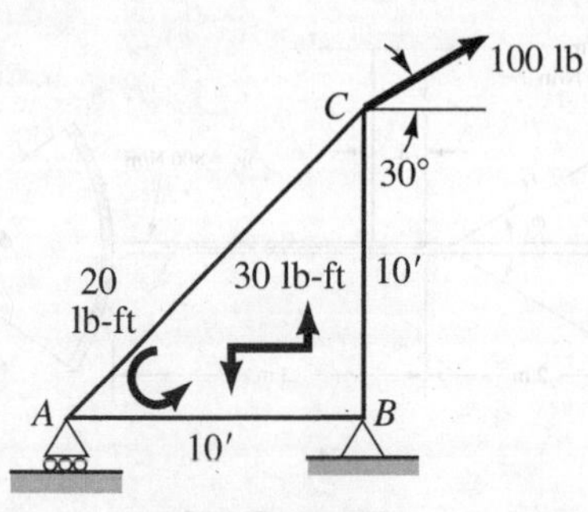

Figure P.4.83.

4.84. A 100-kN bridge pier supports a 10-m segment length of roadway weighing 150 kN and a 150-kN truck. The truck is located at the same position along the roadway as the pier. What is the equivalent force system acting on the base of the bridge pier when the truck is (a) in the center of the outside lane and (b) in the center of the inside lane?

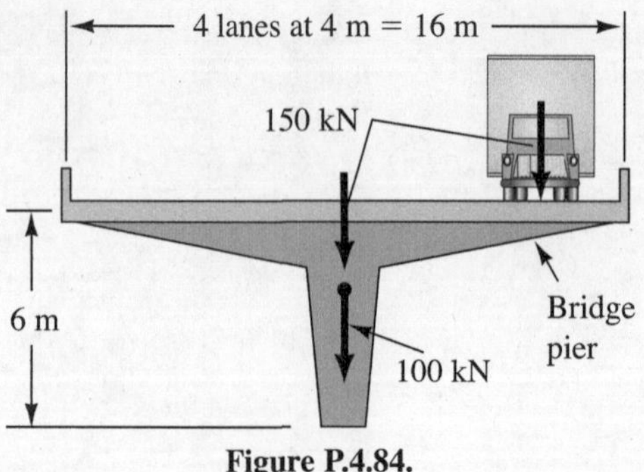

Figure P.4.84.

4.85. Find the center of gravity of a flat plate having constant specific weight γ and thickness t. You are looking down on the plate from above. Proceed as follows:

(a) Give weight of plate in terms of γt.

(b) Determine $\bar{y}$.

(c) Determine $\bar{x}$. Use vertical strips and be careful with integration limits.

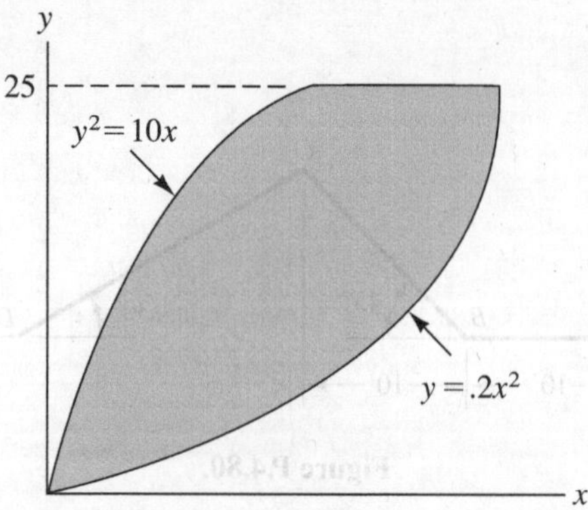

Figure P.4.85.

4.86. A heavy duty off-the-road dump truck is loaded with iron ore that weighs 51 kN/m³. What is the *simplest* resultant force on the truck and where does it act?

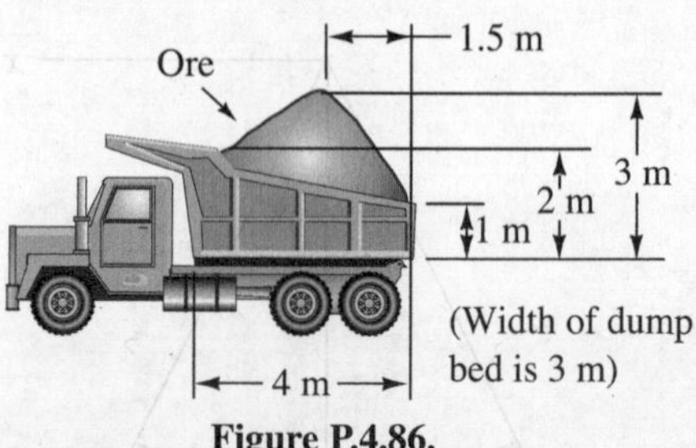

Figure P.4.86.

4.87. Find the *center of gravity* of the *system* of two plates shown (not separately) in terms of the specific weight γ and thickness t, both of which are uniform and the same for both plates. You are looking down on the plates.

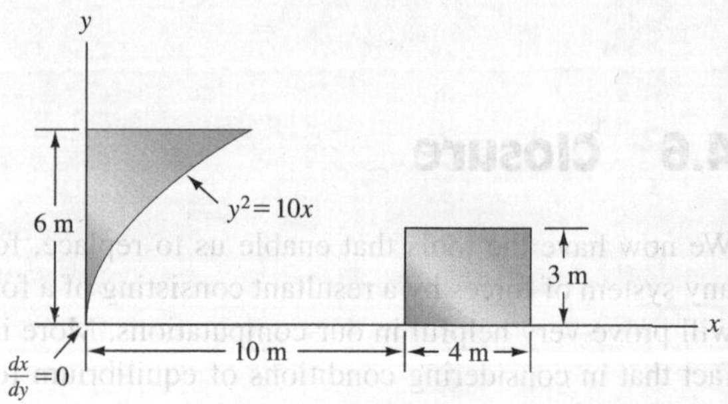

Figure P.4.87.

4.88. A Jeep weighs 11 kN and has both a front winch and a rear power take-off. The tension in the winch cable is 5 kN. The power take-off develops 300 N-m of torque T about an axis parallel to the x axis. If the driver weighs 800 N, what is the resultant force system at the indicated center of gravity of the Jeep where we can consider the weight of the Jeep to be concentrated?

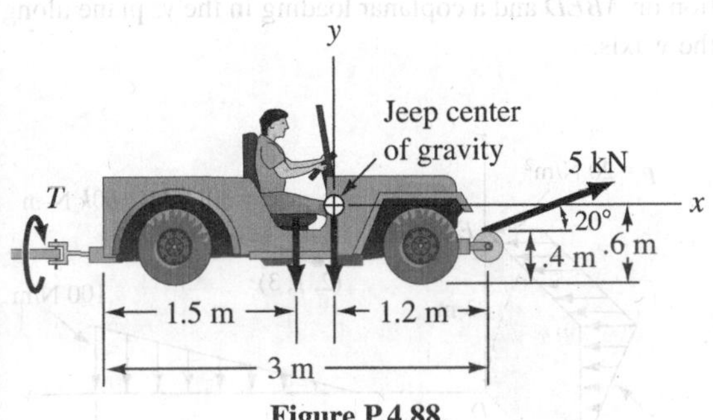

Figure P.4.88.

4.89. What is the *simplest* resultant for the forces and couple acting on the beam?

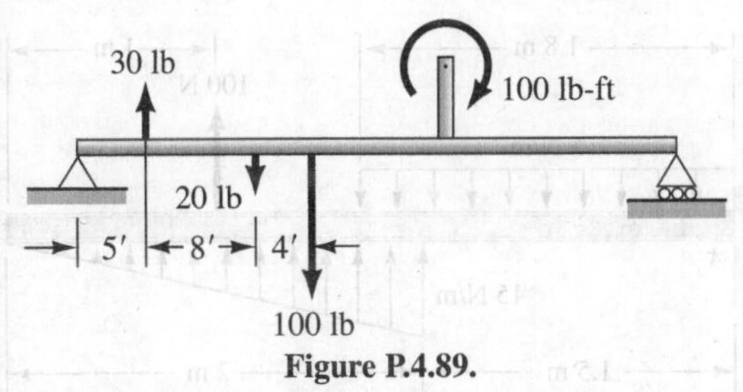

Figure P.4.89.

4.90. A parabolic body of revolution has cut out of it a second parabolic body of revolution starting at A and forming a sharp edge at B with zero slope at A.

(a) What is r' as a function of x for the cut-out body of revolution?

(b) Set up an integral for computing W (weight) and then the center of gravity coordinate $\bar{x}$.

Note: γ varies with x. Do not solve the integral.

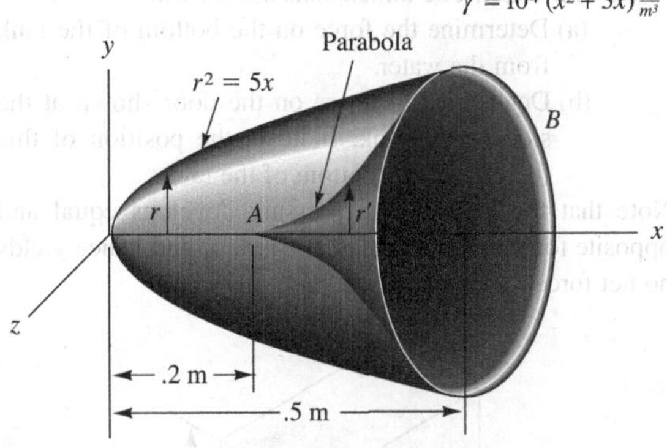

Figure P.4.90.

4.91. Find the torque about axis OB from the system of forces.

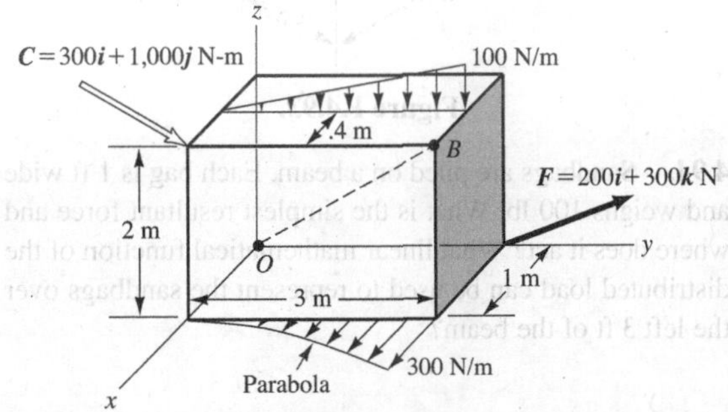

Figure P.4.91.

4.92. A rectangular plate shown as ABC can rotate about hinge B. What length l should BC be so there is zero torque about B from the water, air, and weight of the plate? Take this weight as 1,000 N/m of length. The width is 1 m. γ_{H_2O} = 9,806 N/m³.

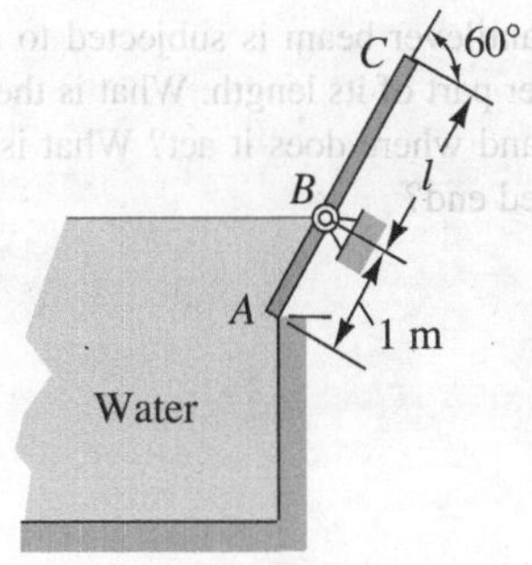

Figure P.4.92.

4.93. An open rectangular tank of water is partially filled with water. The dimensions are shown.

(a) Determine the force on the bottom of the tank from the water.

(b) Determine the force on the door shown at the side of the tank. Indicate the position of this force from the bottom of the tank.

Note that the atmospheric pressure develops equal and opposite forces on both sides of the door and hence yields no net force.

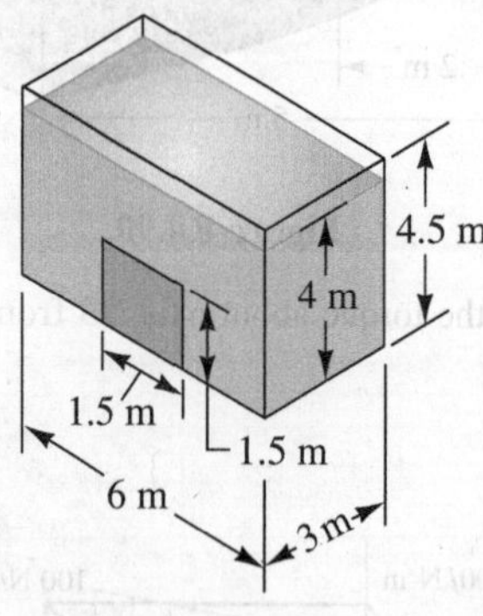

Figure P.4.93.

4.94. Sandbags are piled on a beam. Each bag is 1 ft wide and weighs 100 lb. What is the simplest resultant force and where does it act? What linear mathematical function of the distributed load can be used to represent the sandbags over the left 3 ft of the beam?

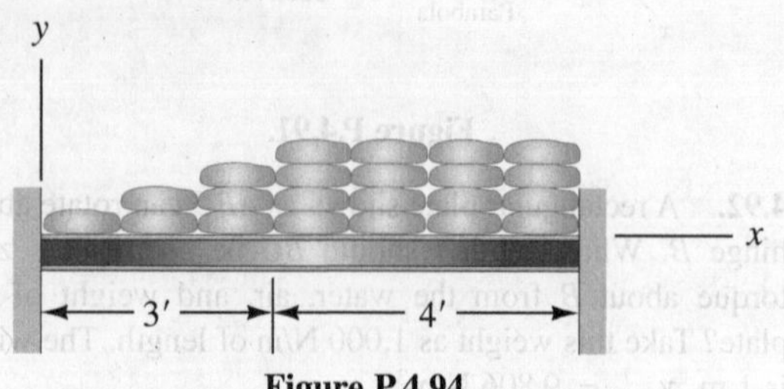

Figure P.4.94.

4.95. A cantilever beam is subjected to a linearly varying load over part of its length. What is the *simplest* resultant force, and where does it act? What is the moment at the supported end?

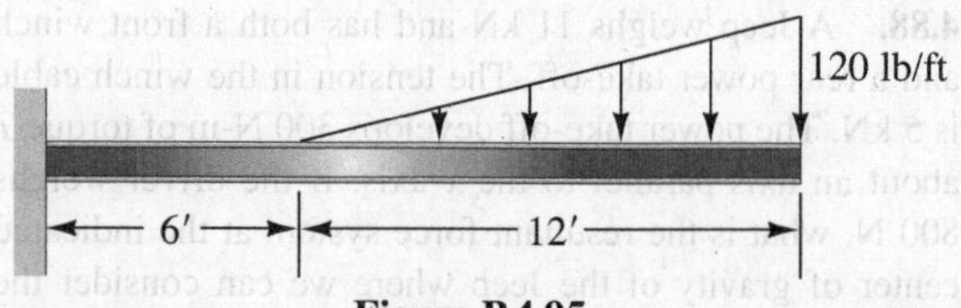

Figure P.4.95.

4.96. Find the torque about an axis along position vector $\boldsymbol{r} = 3\boldsymbol{i} + 4\boldsymbol{j} + 2\boldsymbol{k}$ m. Note we have a pressure distribution on *ABED* and a coplanar loading in the *yz* plane along the *y* axis.

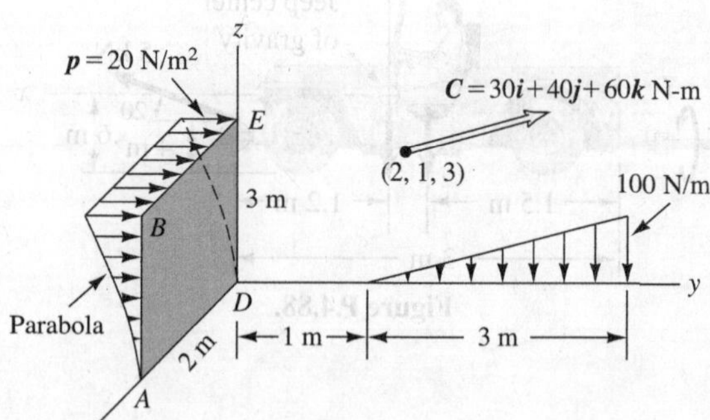

Figure P.4.96.

4.97. Compute the *simplest* resultant force for the loads acting on the cantilever beam.

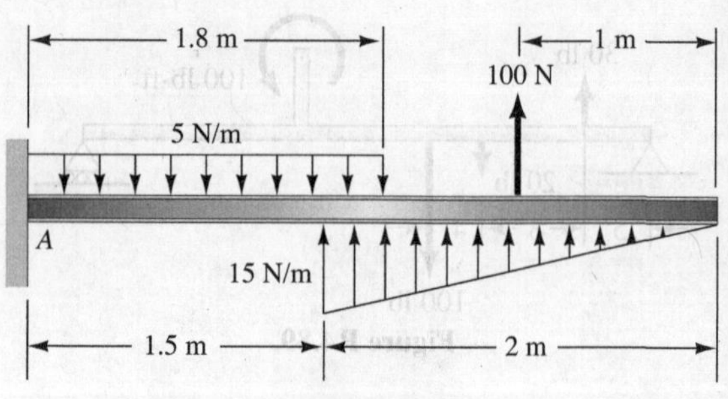

Figure P.4.97.

4.98. Find the resultant force system at *A* for the forces on the bent cantilever beam. *BC* is parallel to *z* axis.

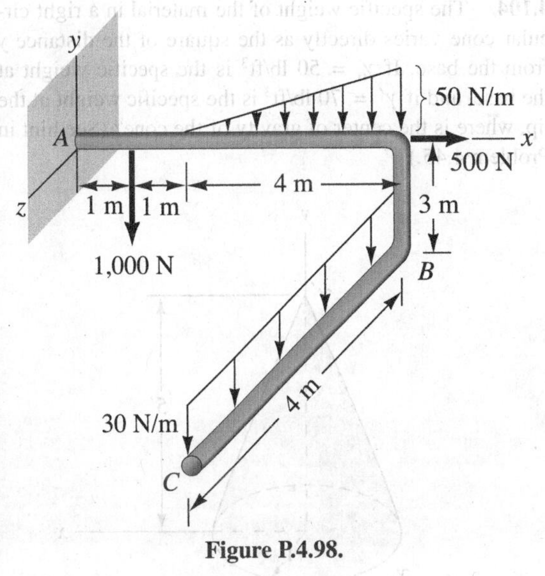

Figure P.4.98.

4.99. (a) Find equations describing both parabolas.
(b) Using *vertical* strips (and a composite body approach), find weight of plate in terms of γt.
(c) Using vertical strips, find x of C.G.

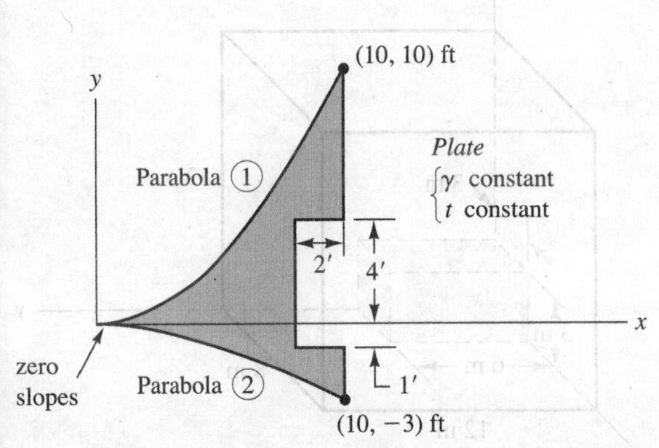

Figure P.4.99.

4.100. The L-shaped concrete post supports an elevated railroad. The concrete weighs 150 lb/ft^3. What is the simplest resultant force from the weight and the load and where does it act? Load acts at center of top surface.

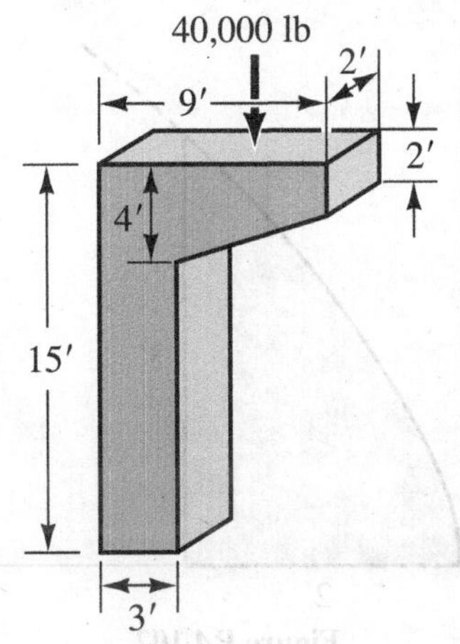

Figure P.4.100.

4.101. Explain why the system shown can be considered a system of parallel forces. Find the *simplest* resultant for this system. The grid is composed of 1-m squares.

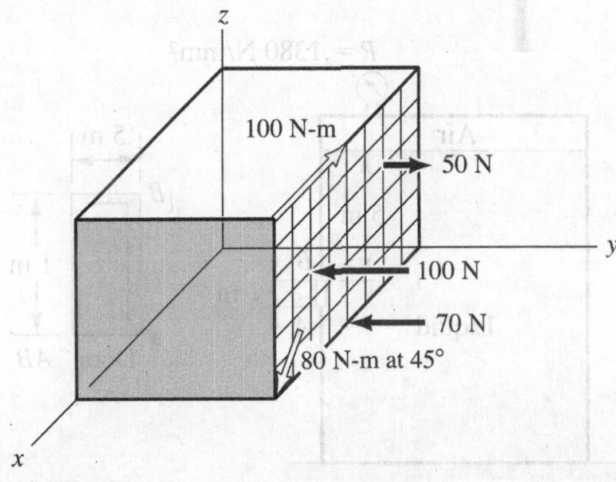

Figure P.4.101.

4.102. A plate of thickness t has as the upper edge a parabolic curve with infinite slope at the origin. Find the x, y coordinates of the center of gravity for this plate.

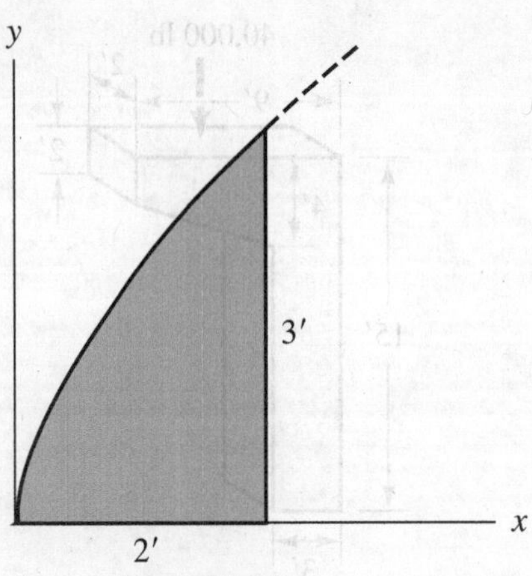

Figure P.4.102.

4.103. A rectangular tank contains a liquid. At the top of the liquid there is a pressure of .1380 N/mm² absolute. What is the simplest resultant force in the inside surface of the door AB? Where is the center of pressure relative to the bottom of the door? Take γ = 8,190 N/m³ for the liquid.

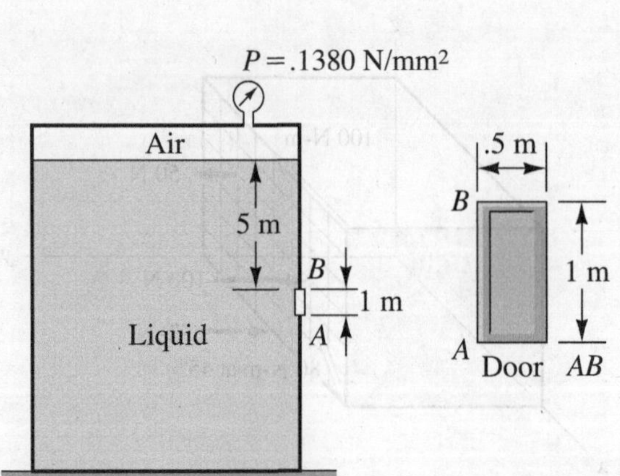

Figure P.4.103.

4.104. The specific weight of the material in a right circular cone varies directly as the square of the distance y from the base. If γ_0 = 50 lb/ft³ is the specific weight at the base, and if γ' = 70 lb/ft³ is the specific weight at the tip, where is the center of gravity of the cone? (See hint in Problem 4.45.)

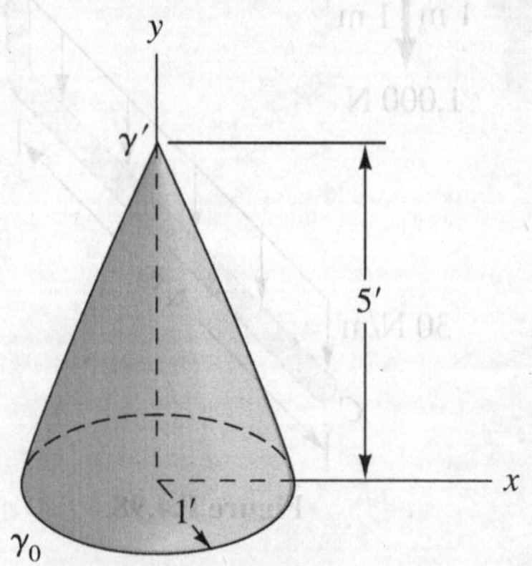

Figure P.4.104.

***4.105.** A block has a rectangular portion removed (darkened region). If the specific weight is given as

$$\gamma = (2.0x + y + 3xyz)\ \text{kN/m}^3$$

find x for the center of gravity.

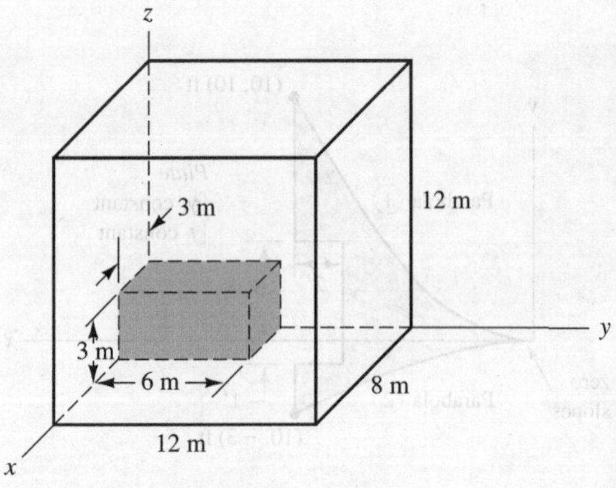

Figure P.4.105.

4.106. Compute the *simplest* resultant for the loads shown acting on the simply supported beam. Give the line of action.

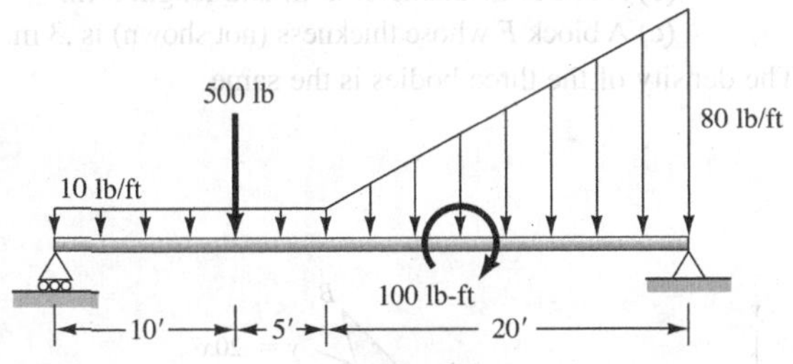

Figure P.4.106.

4.107. Find the center of gravity of the plate.

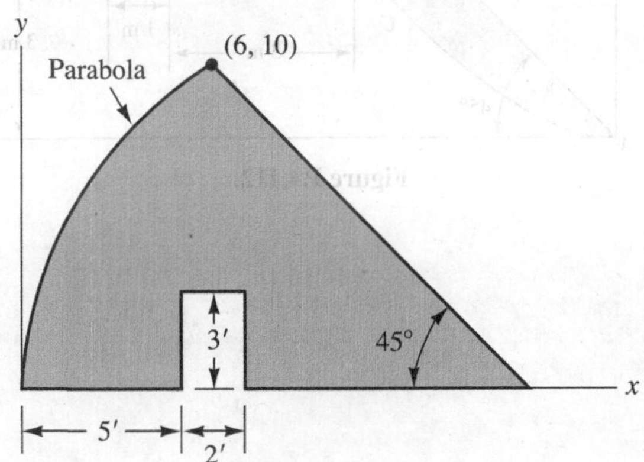

Figure P.4.107.

4.108. Find the center of gravity of the truss. All members have the same weight per unit length.

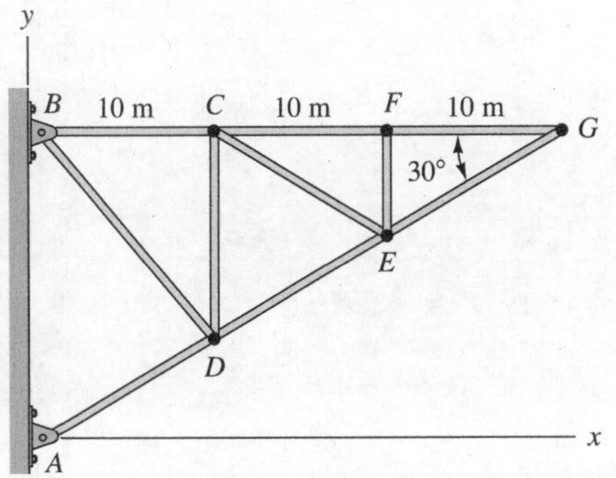

Figure P.4.108.

***4.109.** The pressure p_0 at the corner O of the plate is 50 Pa and increases linearly in the y direction by 5 Pa/m. In the x direction, it increases parabolically starting with zero slope so that in 20 m the pressure has gone from 50 Pa to 500 Pa. What is the simplest resultant for this distribution? Give the coordinates of the center of pressure.

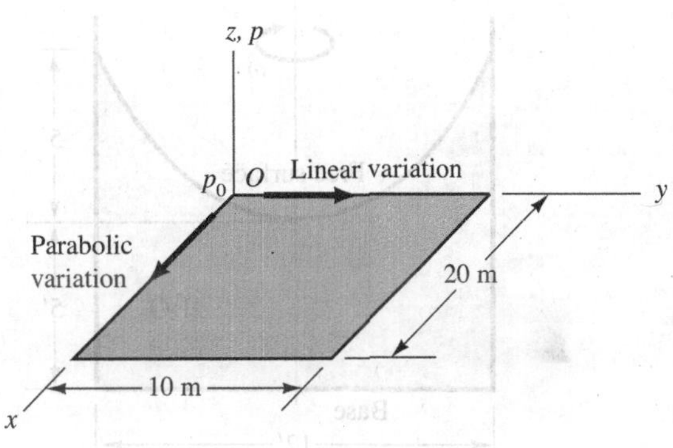

Figure P.4.109.

4.110. A sluice-gate door in a dam is 3 m wide and 3 m high. The water level in the dam is 4 m above the top of the door. The gate is opened until the water level falls 4 m. What is the simplest resultant force on the closed door at both water levels? Where do the forces act (i.e., where is the "center of pressure" in each case)? Water weighs 9,806 N/m^3.

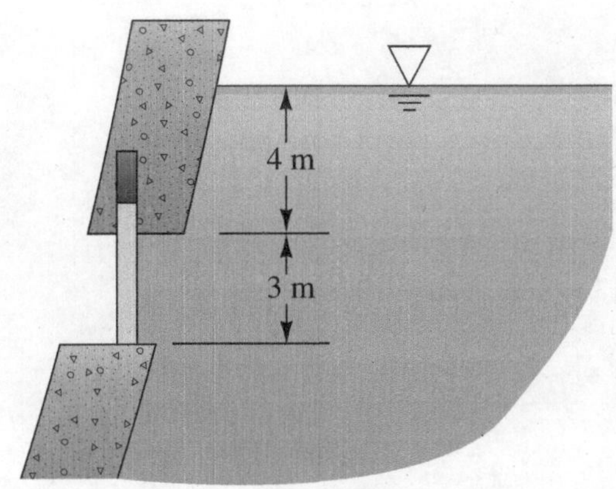

Figure P.4.110.

4.111. A cylindrical tank of water is rotated at constant angular speed ω until the water ceases to change shape. The result is a free surface which, from fluid mechanics considerations, is that of a paraboloid. If the pressure varies directly as the depth below the free surface, what is the resultant force on a quadrant of the base of the cylinder? Take $\gamma = 62.4$ lb/ft^3. [*Hint:* Use circular strip in quadrant having area $1/4(2\pi r\,dr)$.]

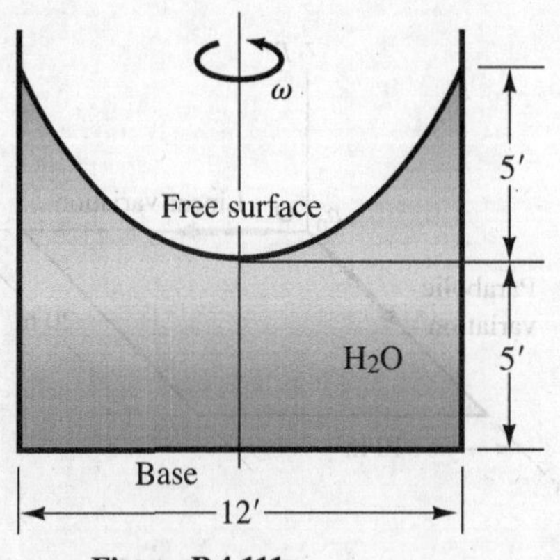

Figure P.4.111.

4.112. Find the x and y coordinates of the center of gravity of the bodies shown. These consist of:

(a) A plate ABC whose thickness $t = 50$ mm.

(b) A rod D of diameter .3 m and length 3 m.

(c) A block F whose thickness (not shown) is .3 m.

The density of the three bodies is the same.

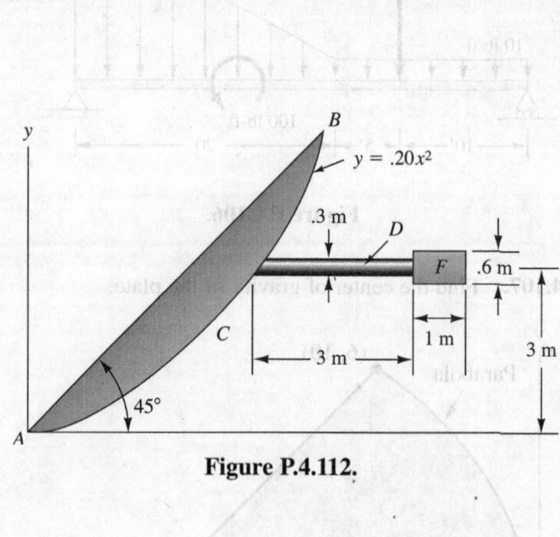

Figure P.4.112.

CHAPTER 5

Equations of Equilibrium

5.1 Introduction

You will recall from Section 1.10 that a *particle* in equilibrium is one that is stationary or that moves uniformly relative to an inertial reference. A *body* is in equilibrium if all the particles that may be considered to comprise the body are in equilibrium. It follows, then, that a rigid body in equilibrium cannot be rotating relative to an inertial reference. In this chapter, we shall consider bodies in equilibrium for which the rigid-body model is valid. For these bodies, there are certain simple equations that relate all the surface and body forces, or their equivalents, that act on the body. With these equations, we can sometimes ascertain the value of a certain number of unknown forces. For instance, in the beam shown in Fig. 5.1, we know the loads F_1 and F_2 and also the weight W of the beam, and we want to determine the forces transmitted to the earth so that we can design a foundation to support the structure properly. Knowing that the beam is in equilibrium and that the small deflection of the beam will not appreciably affect the forces transmitted to the earth, we can write rigid-body equations of equilibrium involving the unknown and known forces acting on the beam and thus arrive at the desired information.

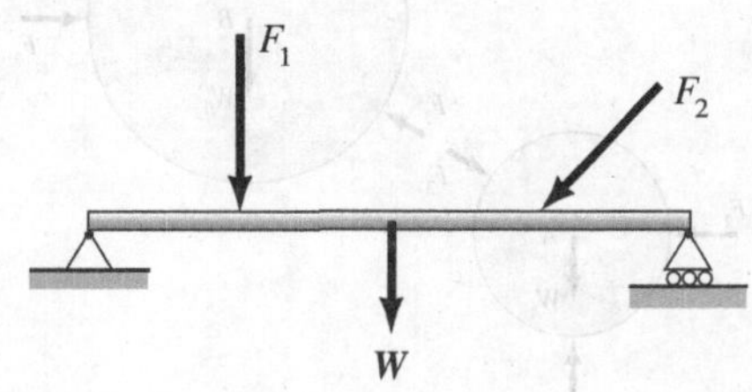

Figure 5.1. Loaded beam.

Note in the beam problem above that a number of steps are implied. First, there is the singling out of the beam itself for discussion. Then, we express certain equations of equilibrium for the beam, which we take as a rigid body. Finally, there is the evaluation of the unknowns and interpretation of the results. In this chapter, we will carefully examine each of these steps.

Of critical importance is the need to be able to isolate a body or part of a body for analysis. Such a body is called a *free body*. We will first carefully investigate the development of free-body diagrams. We urge you to pay special heed to this topic, since *it is the most important step in the solving of mechanics problems*. An incorrect free-body diagram means that all ensuing work, no matter how brilliant, will lead to wrong results. More than just a means of attacking statics problems, the free-body concept is your first

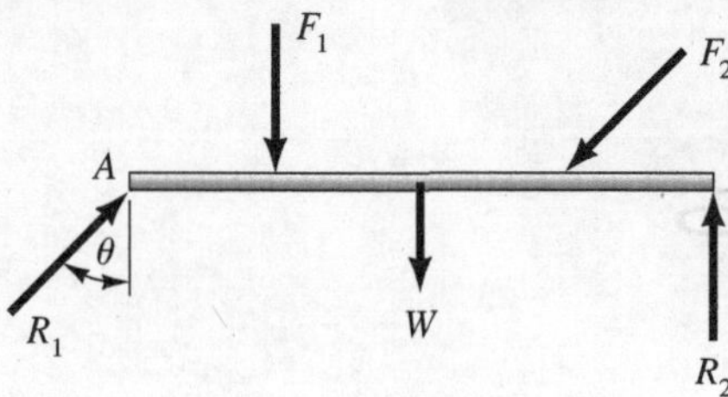

Figure 5.2. Free-body diagram of beam.

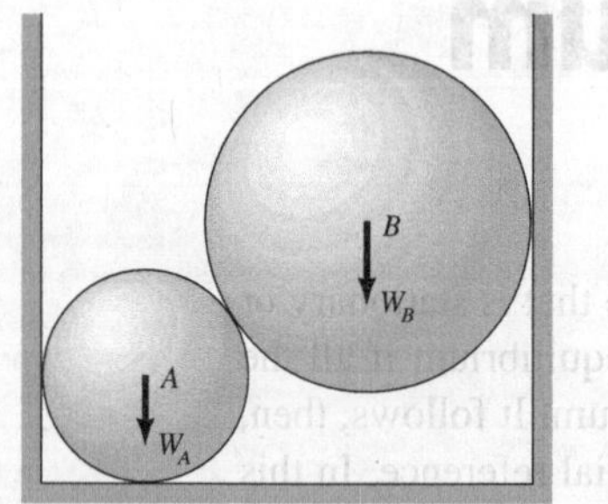

Figure 5.3. Smooth spheres in equilibrium.

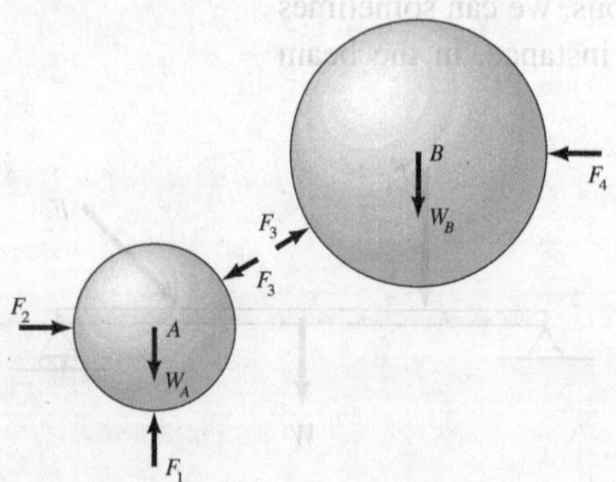

Figure 5.4. Free-body diagrams.

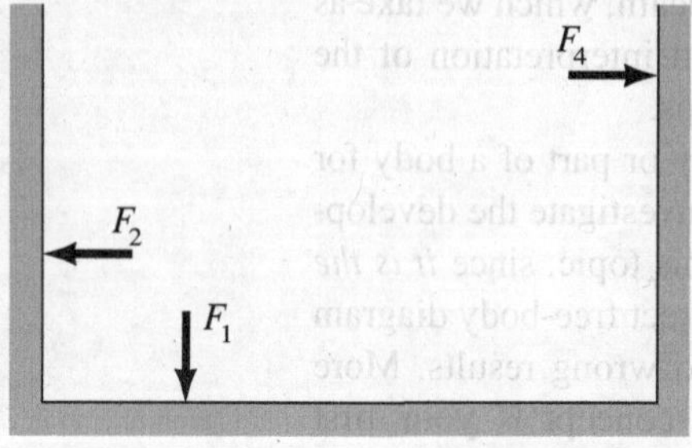

Figure 5.5. This is *not* a free-body diagram.

exposure to the overridingly important topic of *engineering analysis* in general.[1] We now examine this critical step.

5.2 The Free-Body Diagram

Since the equations of equilibrium for a particular body actually stem from the dynamic considerations of the body, we must be sure to include *all* the forces (or their equivalents) acting *on* this body, because they all affect the motion of the body and must be accounted for. To help identify all the forces and so ensure the correct use of the equations of equilibrium, we isolate the body in a simple diagram and show *all* the forces from the *surroundings* that act *on* the body. Such a diagram is called a *free-body diagram.* When we isolate the beam in our problem from its surroundings, we get Fig. 5.2. On the left end, there is an unknown force from the ground that has a magnitude denoted as R_1 and a direction denoted as θ, with a line of action going through a known point *A*. [We may also use components $(R_1)_x$ and $(R_1)_y$ as unknowns rather than R_1 and θ.] The right side involves a force in the vertical direction with an unknown magnitude denoted as R_2. The direction is vertical because the beam is on rollers to allow for thermal expansion or thermal contraction and to thus allow unhindered stretching or shrinking of the beam in the axial direction. As a result, the ground exerts a negligibly small horizontal force there. Once all the forces acting on the beam have been identified, including the three unknown quantities R_1, R_2, and θ, we can solve for these unknowns by using three equations of equilibrium.

Consider now the hard spheres shown in Fig. 5.3 in a condition of equilibrium with surfaces smooth enough to permit us to neglect friction completely. The contact forces thus must be in a direction normal to the surface of contact. The free bodies of the spheres are shown in Fig. 5.4. Notice that F_3 is the magnitude of the force from sphere *B* on sphere *A*, while the reaction, also shown as F_3 according to Newton's third law, is the magnitude of the force from sphere *A* on sphere *B*.

You might be tempted to consider a portion of the container as a free body in the manner shown in Fig. 5.5. But even if this diagram did clearly depict a body (which it does not!), it would not qualify as a free body, since all the forces acting on the body have not been shown.

In engineering problems, bodies are often in contact in a number of standard ways. In Fig. 5.6, you will find the types of

[1]The author has found through many years of experience that the absence of a free-body diagram in a student's work on a particular problem signifies that:

1. There will most likely be errors in the analysis of the problem, or
2. Even worse, the student does not have a good grasp of the problem.

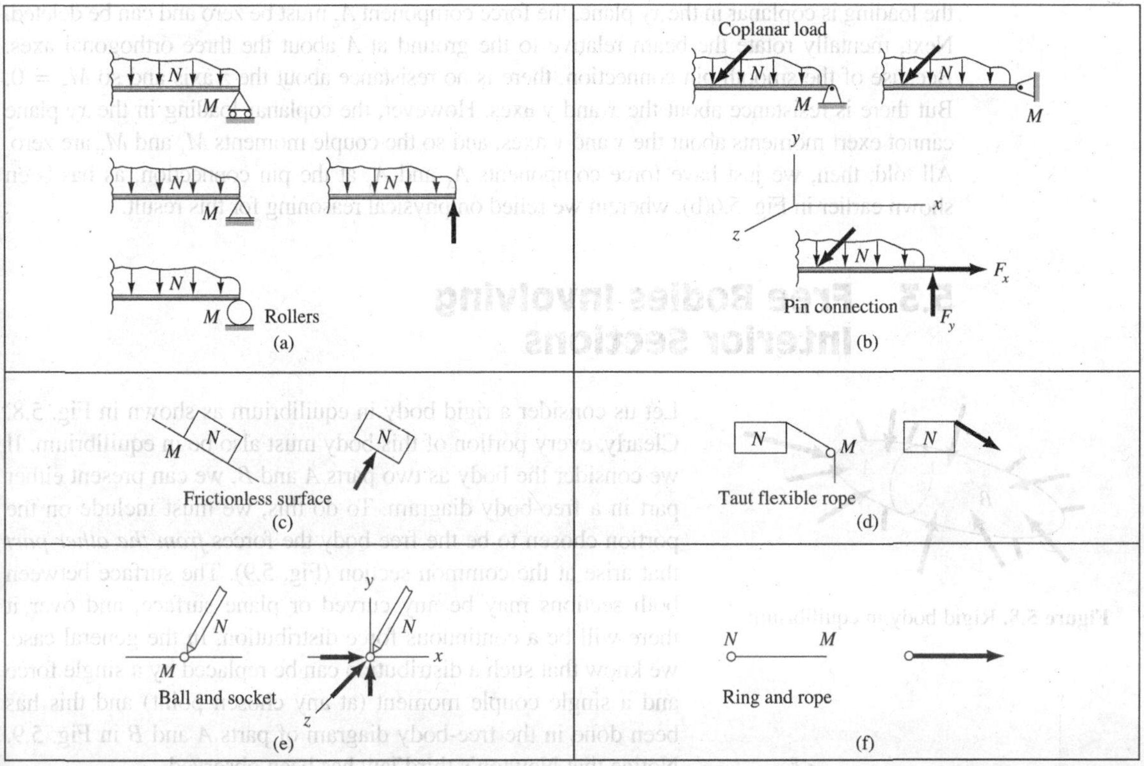

Figure 5.6. Standard connections.

forces transmitted from body M to body N for body connections that are often found in practice. (These are not free-body diagrams, since all the forces on any given body have not been shown.)

In general, to ascertain the nature of the force system that a body M is capable of transmitting to a second body N through some connector or support, we may proceed in the following manner. Mentally move the bodies relative to each other in each of three orthogonal directions. In those directions where relative motion is impeded or prevented by the connector or support, there can be a force component at this connector or support in a free-body diagram of either body M or N. Next, mentally rotate bodies M and N relative to each other about the orthogonal axes. In each direction about which relative rotation is impeded or prevented by the connector or support, there can be a couple-moment component at this connector or support in a free-body diagram of body M or N. Now as a result of equilibrium considerations of body M or N, certain force and couple-moment components that are capable of being generated at a support or connector will be zero for the particular loadings at hand. Indeed, one can often readily recognize this by inspection. For instance, consider the pin-connected beam loaded in a coplanar manner shown partially in Fig. 5.7. If we mentally move the beam relative to the ground in the x, y, and z directions, we get resistance from the pin for each direction, and so the ground at A can transmit force components A_x, A_y, and A_z. However, because

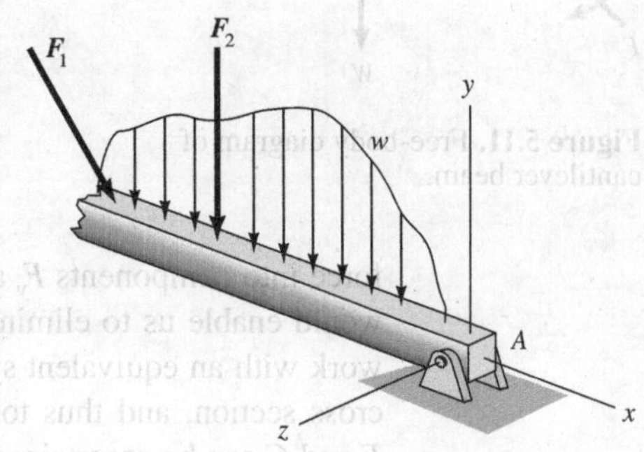

Figure 5.7. Pin connection.

the loading is coplanar in the xy plane, the force component A_z must be zero and can be deleted. Next, mentally rotate the beam relative to the ground at A about the three orthogonal axes. Because of the smooth pin connection, there is no resistance about the z axis and so $M_z = 0$. But there is resistance about the x and y axes. However, the coplanar loading in the xy plane cannot exert moments about the x and y axes, and so the couple moments M_x and M_y are zero. All told, then, we just have force components A_x and A_y at the pin connection, as has been shown earlier in Fig. 5.6(b), wherein we relied on physical reasoning for this result.

5.3 Free Bodies Involving Interior Sections

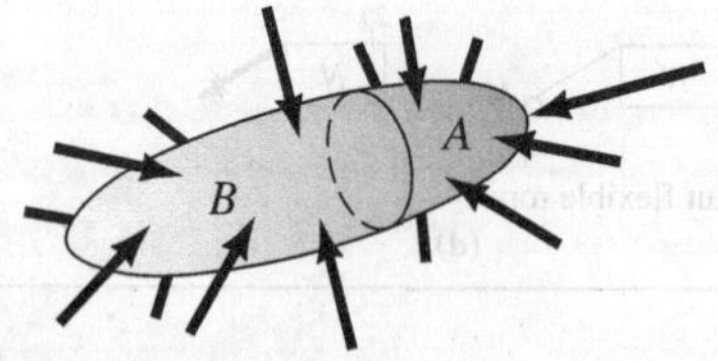

Figure 5.8. Rigid body in equilibrium.

Let us consider a rigid body in equilibrium as shown in Fig. 5.8. Clearly, every portion of this body must also be in equilibrium. If we consider the body as two parts A and B, we can present either part in a free-body diagram. To do this, we must include on the portion chosen to be the free body the forces *from the other part* that arise at the common section (Fig. 5.9). The surface between both sections may be any curved or plane surface, and over it there will be a continuous force distribution. In the general case, we know that such a distribution can be replaced by a single force and a single couple moment (at any chosen point) and this has been done in the free-body diagram of parts A and B in Fig. 5.9. Notice that Newton's third law has been observed.

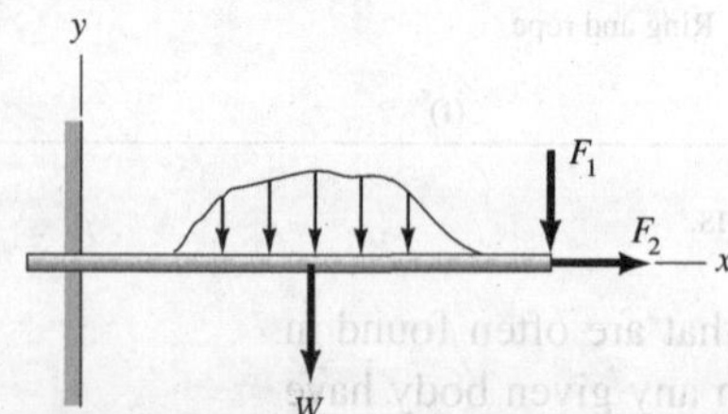

Figure 5.10. Cantilever beam.

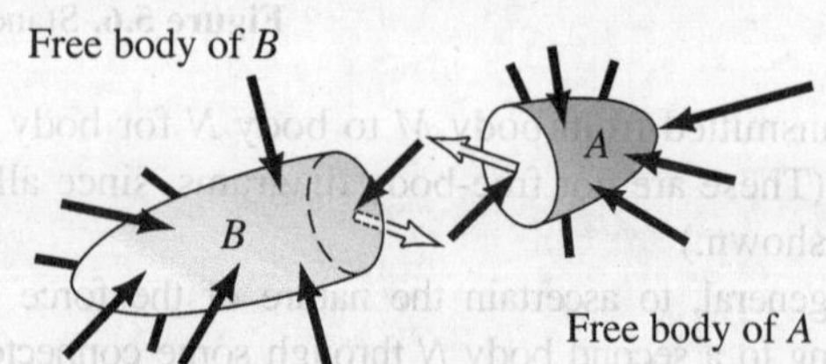

Figure 5.9. Free bodies of parts A and B.

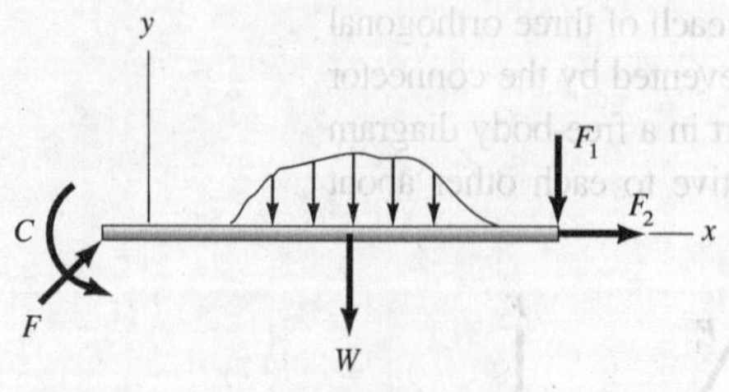

Figure 5.11. Free-body diagram of cantilever beam.

As a special case, consider a beam with one end embedded in a massive wall (cantilever beam) and loaded within the xy plane (Fig. 5.10). A free body of the portion of the beam extending from the wall is shown in Fig. 5.11. Because of the geometric symmetry of the beam about the xy plane and the fact that the loads are in this plane, the exposed forces in the cut section can be considered coplanar. Hence, these exposed forces can be replaced by a single force and a single couple moment in the center plane, and it is the usual practice to decompose the force into components F_y and F_x. Although a line of action for the force can be found that would enable us to eliminate the couple moment, it is desirable in structural problems to work with an equivalent system that has the force passing through the center of the beam cross section, and thus to have a couple moment. In the next section, we will see how F and C can be ascertained.

Example 5.1

As a further illustration of a free-body diagram, we shall now consider the frame[2] shown in Fig. 5.12, which consists of members connected by frictionless pins. The force systems acting on the assembly and its parts will be taken as coplanar. We shall now sketch free-body diagrams of the assembly and its parts.

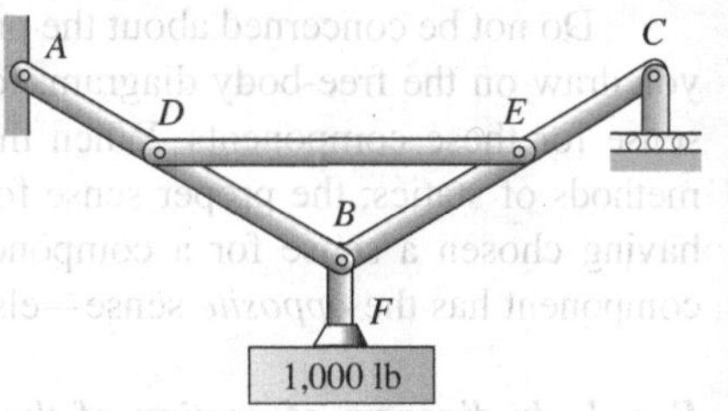

Figure 5.12. A frame.

Free-body diagram of the entire assembly. The magnitude and direction of the force at A from the wall onto the assembly is not known. However, we know that this force is in the plane of the system. Therefore, two components are shown at this point (Fig. 5.13). Since the direction of the force C is known, there are then three unknown scalar quantities, A_y, A_x, and C, for the free body.

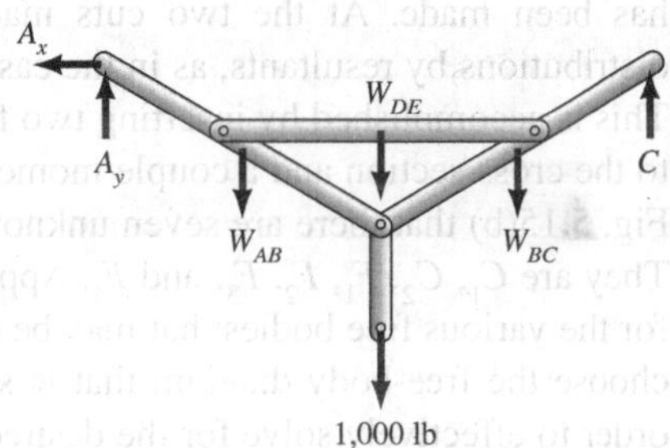

Figure 5.13. Free-body diagram of frame.

Free-body diagram of the component parts. When two members are pinned together, such as members DE and AB or DE and BC, we usually consider the pin to be part of one of the bodies. However, when more than two members are connected at a pin, such as members AB, BC, and BF at B, we often isolate the pin and consider that all members act on the pin rather than directly on each other, as illustrated in Fig. 5.14. Notice four sets of forces that form pairs of reactions have been enclosed with dashed lines. The fifth set is the 1,000-lb force on the pin and on the member BF.

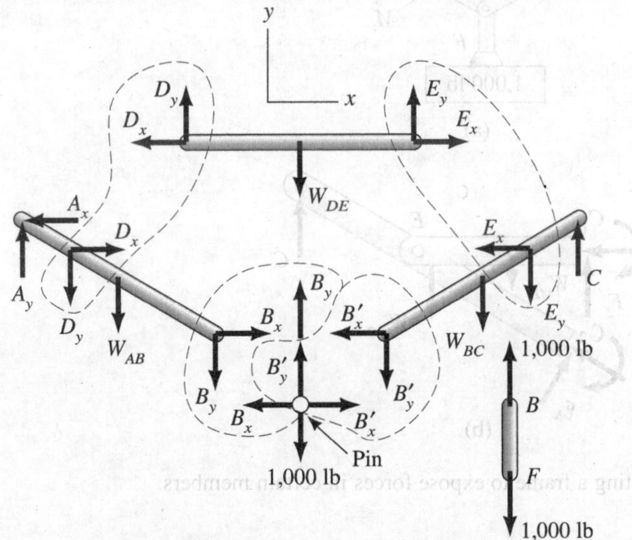

Figure 5.14. Free-body diagrams of parts.

[2]A *frame* is a system of connected straight or bent, long, slender members where some of the connecting pins are not at the ends of the members as is the case for structures that we will study later called *trusses*.

Example 5.1 (Continued)

Do not be concerned about the proper *sense* of an unknown force component that you draw on the free-body diagram, for you may choose either a positive or negative sense for these components. When the values of these quantities are ascertained by methods of statics, the proper sense for each component can then be established; but, having chosen a sense for a component, you must be sure that the *reaction* to this component has the *opposite* sense—else you will violate Newton's third law.

Free-body diagram of portion of the assembly to the right of M–M. In making a free body of the portion to the right of section M–M (see Fig. 5.15a), we must remember to put in the weight of the portions of the members remaining *after* the cut has been made. At the two cuts made by M–M we must replace coplanar force distributions by resultants, as in the case of the previously considered cantilever beam. This is accomplished by inserting two force components usually normal and tangential to the cross section and a couple moment as was done for the cantilever beam. Note in Fig. 5.15(b) that there are seven unknown scalar quantities for this free-body diagram. They are C_1, C_2, F_1, F_2, F_3, and F_4. Apparently, the number of unknowns varies widely for the various free bodies that may be drawn for the system. For this reason, you must choose the free-body diagram that is suitable for your needs with some discretion in order to effectively solve for the desired unknowns.

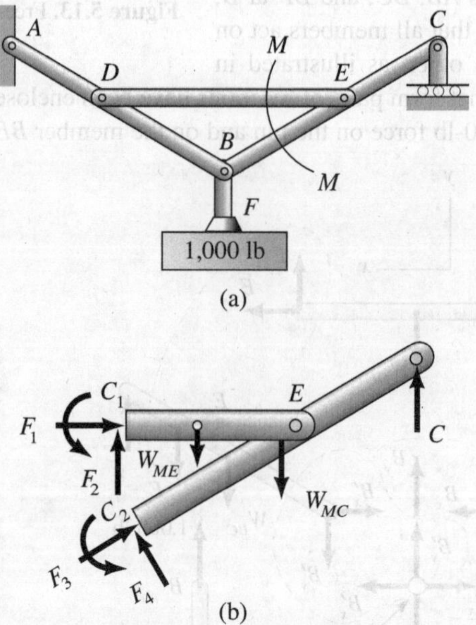

Figure 5.15. Cutting a frame to expose forces in certain members.

Example 5.2

Draw a free-body diagram of the beam *AB* and the frictionless pulley in Fig. 5.16 (a). The weight of the pulley is W_D, and the weight of the beam is W_{AB}.

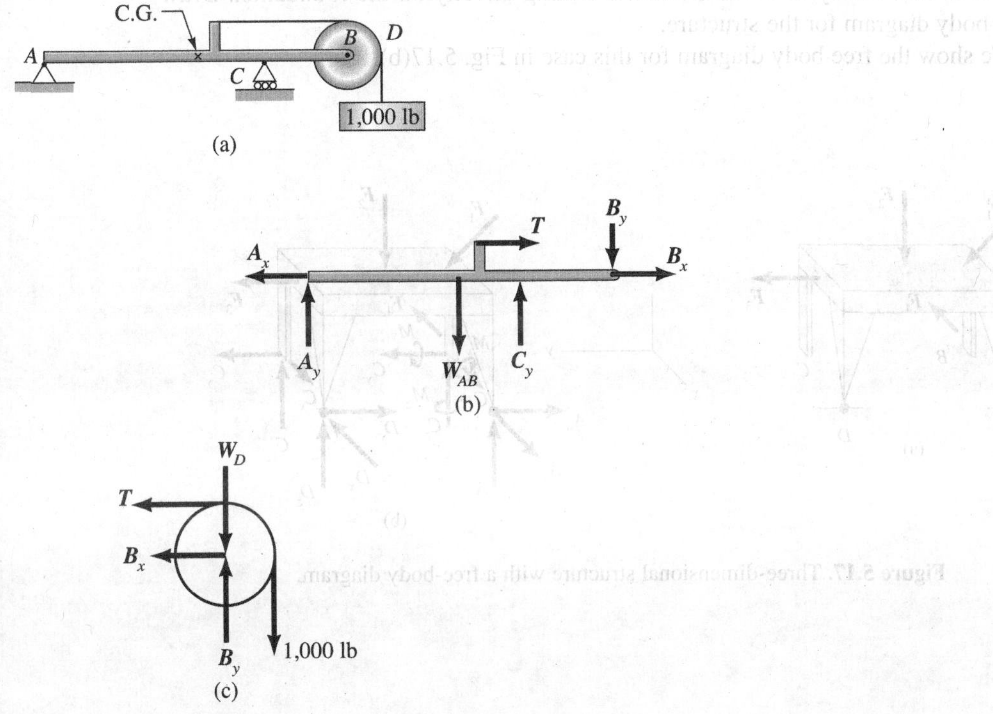

Figure 5.16. (a) Beam AB; (b) Free-body diagram of AB; (c) Free-body diagram of Pulley D

The free-body diagram of beam *AB* is shown in Fig. 5.16(b). The weight of the beam has been shown at its center of gravity. Components B_x and B_y are forces from the pulley *D* acting on the beam through the pin at *B*. The free-body diagram of the pulley is shown in Fig. 5.16(c).

Some students may be tempted to put the weight of the pulley at *B* in the free-body diagram of beam *AB*. The argument given is that this weight "goes through *B*." To put the pulley weight at *B* on free body *AB* is strictly speaking an error! The fact is that the weight of the pulley is a body force acting throughout the *pulley* and *does not* act on the *beam BD*. It so happens that the simplest resultant of this body force distribution on *D* goes through a *position* corresponding to pin *B*. This does not alter the fact that this weight acts *on the pulley* and *not on the beam*. The beam can only feel forces B_x and B_y transmitted from the pulley to the beam through pin *B*. These forces are related to the pulley weight as well as the tension in the cord around the pulley through equations of equilibrium for the free body of the pulley itself.

Example 5.3

We finish this series of free-body diagrams with a three-dimensional case. In Fig. 5.17(a) we show a structure having ball-joint connections at *A* and *D*, a fixed-end support at *B* and, finally, at *C* the column *B* resting directly on the foundation. Draw the free-body diagram for the structure.

We show the free body diagram for this case in Fig. 5.17(b).

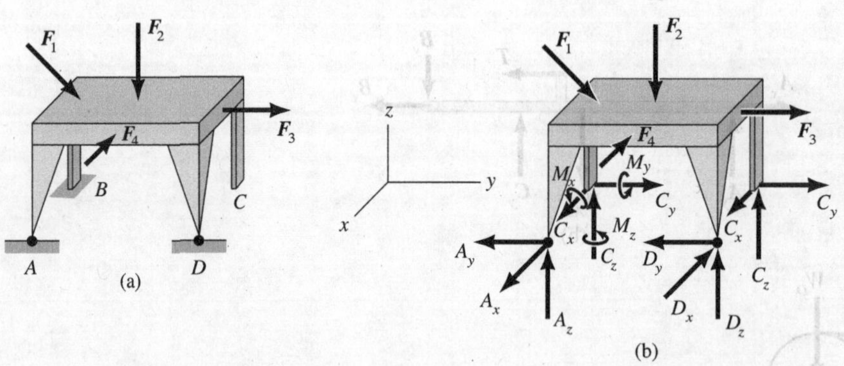

Figure 5.17. Three-dimensional structure with a free-body diagram.

PROBLEMS

5.1. Draw the free-body diagram of the gas-grill lid when it is lifted at the handle to a 45° open position.

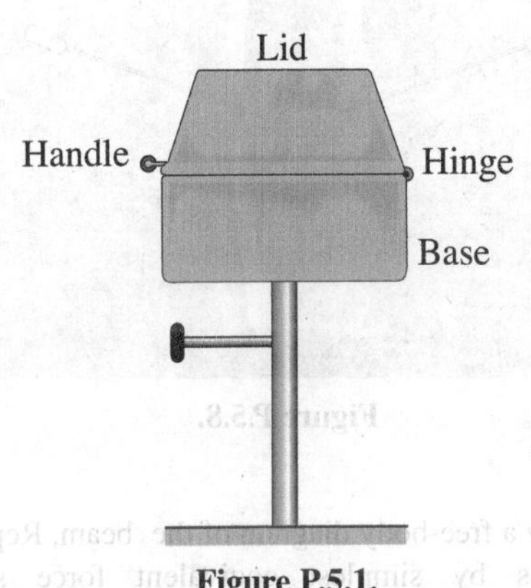

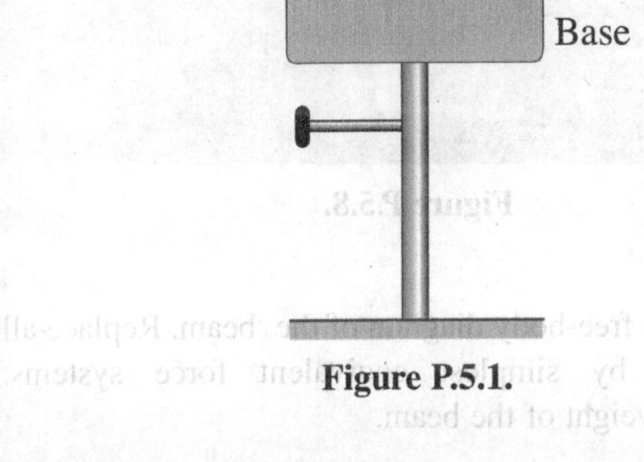

Figure P.5.1.

5.2. A large antenna is supported by three guy wires and rests on a large spherical ball joint. Draw the free-body diagram of the antenna.

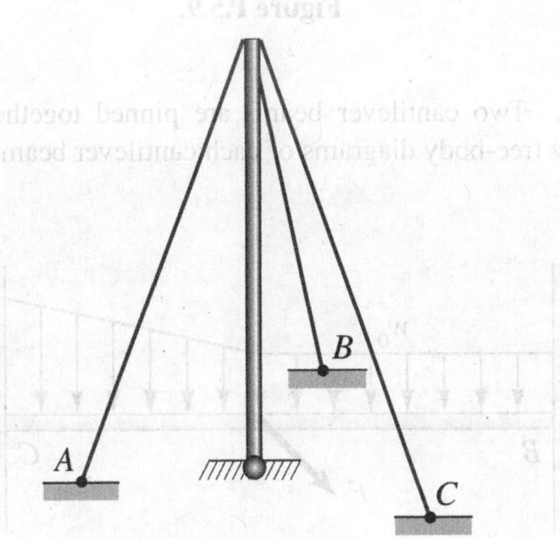

Figure P.5.2.

5.3. Draw a free-body diagram of the A-frame.

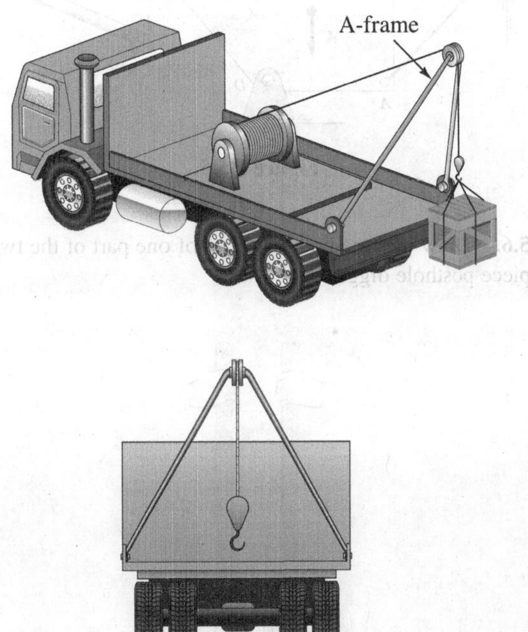

Figure P.5.3.

5.4. Draw complete free-body diagrams for the member AB and for cylinder D. Neglect friction at the contact surfaces of the cylinder. The weights of the cylinder and the member are denoted as W_D and W_{AB}, respectively.

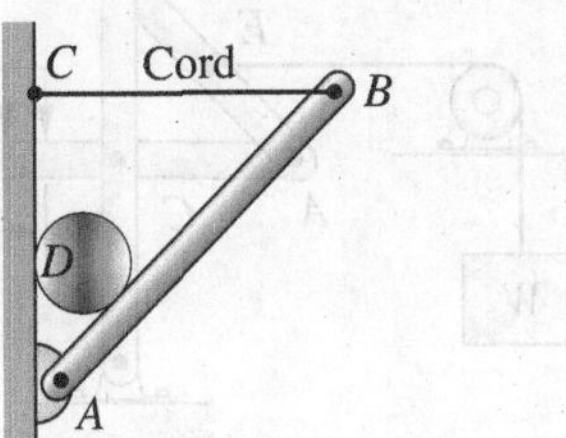

Figure P.5.4.

5.5. Draw free-body diagrams of the plate *ABCD* and the bar *EG*. Assume that there is no friction at the pulley *H* or at the contact surface *C*.

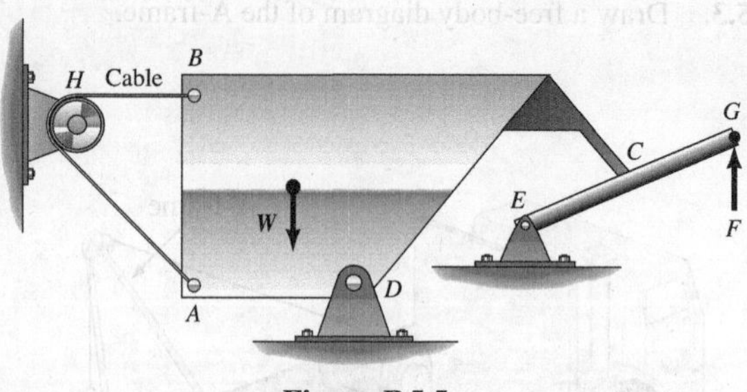

Figure P.5.5.

5.6. Draw the free-body diagram of one part of the two-piece posthole digger.

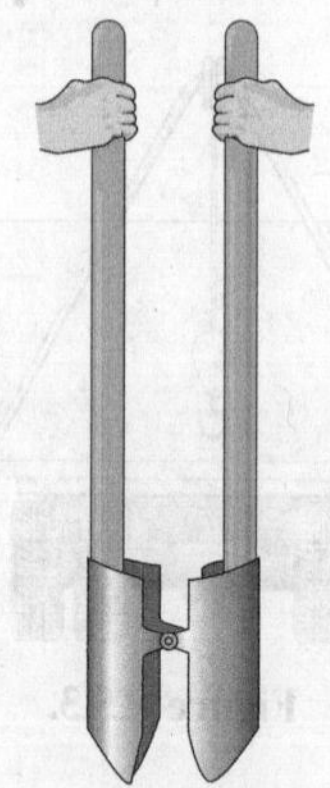

Figure P.5.6.

5.7. Draw a free-body diagram for each member of the system. Neglect the weights of the members. Replace the distributed load by a resultant.

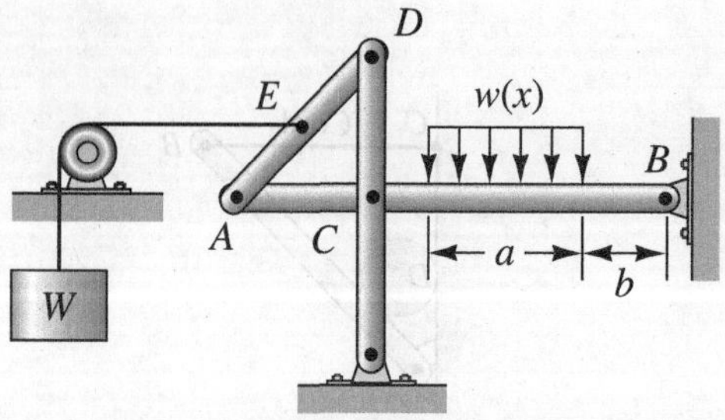

Figure P.5.7.

5.8. Draw the free-body diagrams for the oars of a rowboat when the rower pushes with one hand and pulls with the other (i.e., turns the boat).

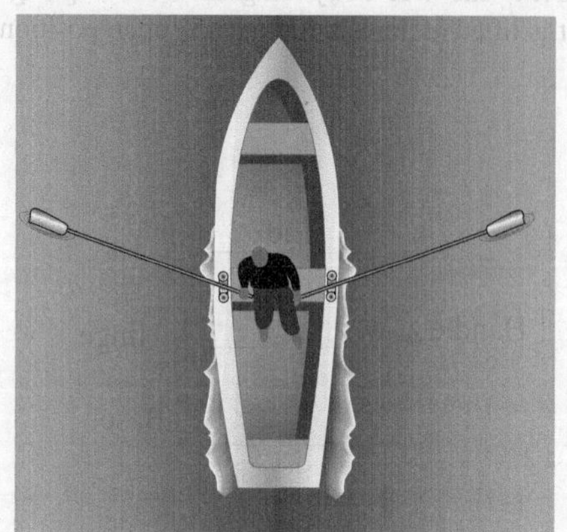

Figure P.5.8.

5.9. Draw a free-body diagram of the beam. Replace all distributions by simplest equivalent force systems. Neglect the weight of the beam.

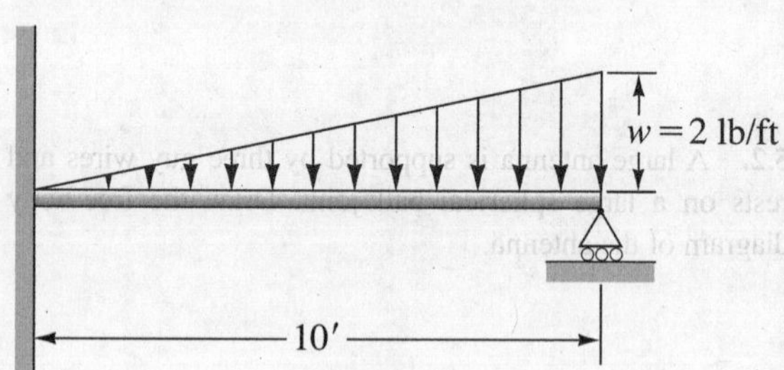

Figure P.5.9.

5.10. Two cantilever beams are pinned together at *A*. Draw free-body diagrams of each cantilever beam.

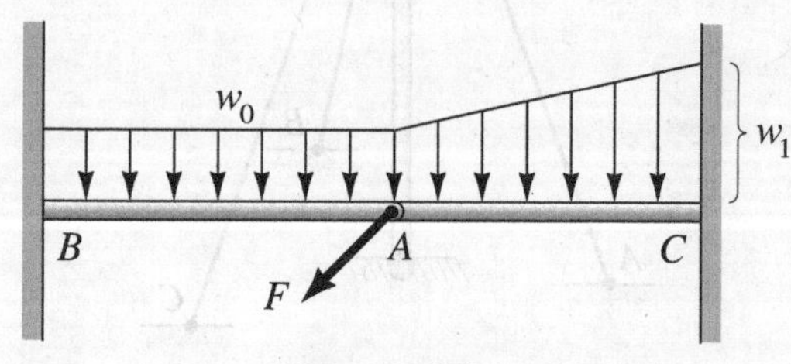

Figure P.5.10.

5.11. Draw free-body diagrams of each part of the treebranch trimmer.

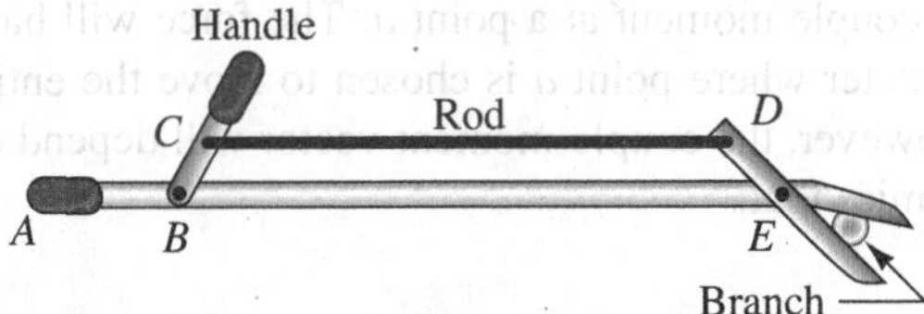

Figure P.5.11.

5.12. Draw free-body diagrams for the two booms and the body E of the power shovel. Consider the weight of each part to act at a central location. (Regard the shovel and payload as concentrated forces, W_S and W_{PL}, respectively.)

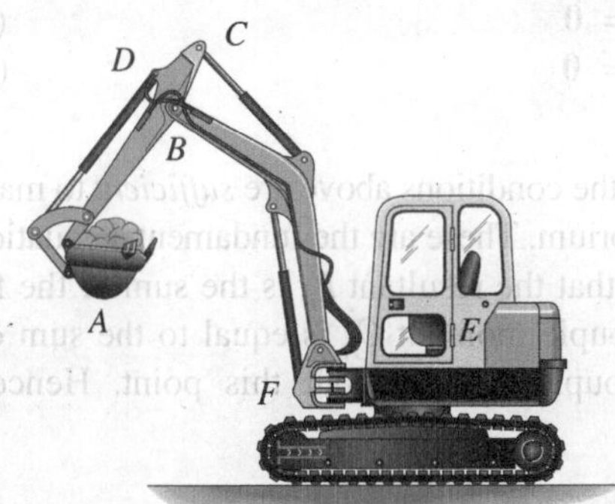

Figure P.5.12.

5.13. Draw the free-body diagram for the bulldozer, B, hydraulic ram, R, and tractor, T. Consider the weight of each part, B, R, and T.

Figure P.5.13.

5.14. Draw a free-body diagram of the whole apparatus and of each of its parts: AB, AC, BC, and D. Include the weights of all bodies. Label forces.

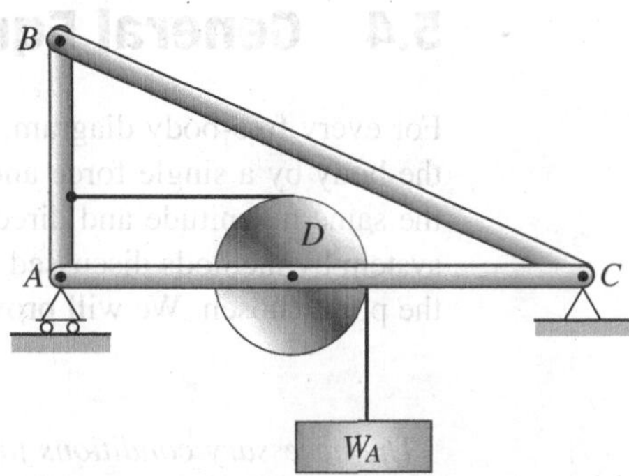

Figure P.5.14.

5.15. Draw a free-body diagram of members CG, AG, and the disc B. Include as the only weight that of disc B. Label all forces. (*Hint:* Consider the pin at G as a separate free body.)

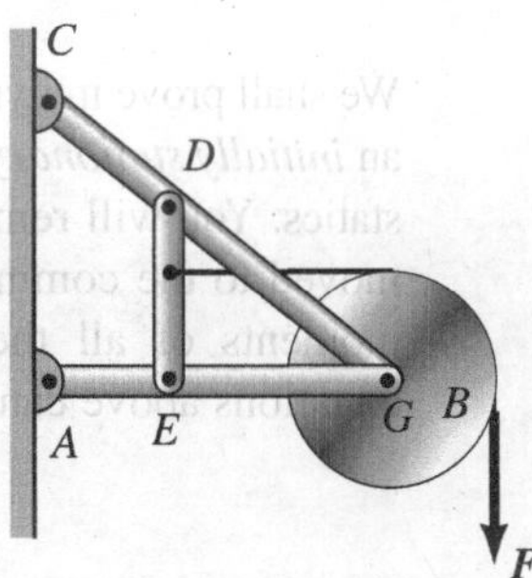

Figure P.5.15.

5.16. Draw the free-body diagram of the horizontally bent cantilevered beam. Use only *xyz* components of all vectors drawn.

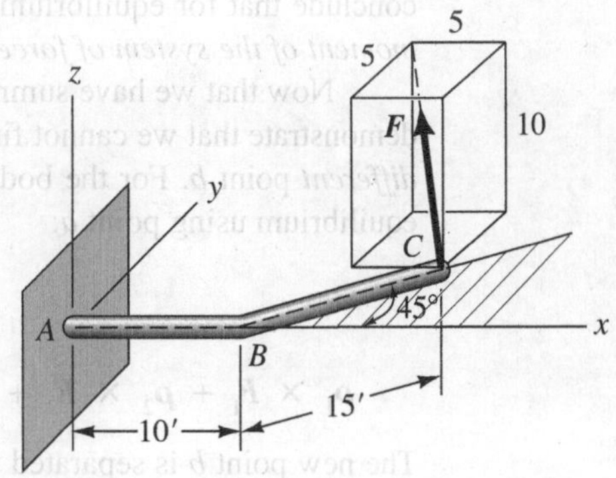

Figure P.5.16.

5.4 General Equations of Equilibrium

For every free-body diagram, we can replace the system of forces and couples acting on the body by a single force and a single couple moment at a point *a*. The force will have the same magnitude and direction, no matter where point *a* is chosen to move the entire system by methods discussed earlier. However, the couple-moment vector will depend on the point chosen. We will prove in dynamics that:

The necessary conditions for a rigid body to be in equilibrium are that the resultant force $\boldsymbol{F}_R$ and the resultant couple moment $\boldsymbol{C}_R$ for any point a be zero vectors.

That is,

$$\boldsymbol{F}_R = \boldsymbol{0} \tag{5.1a}$$
$$\boldsymbol{C}_R = \boldsymbol{0} \tag{5.1b}$$

We shall prove in dynamics, furthermore, that the conditions above are *sufficient* to maintain an *initially stationary* body in a state of equilibrium. These are the fundamental equations of statics. You will remember from Section 4.3 that the resultant $\boldsymbol{F}_R$ is the sum of the forces moved to the common point, and that the couple moment $\boldsymbol{C}_R$ is equal to the sum of the moments of all the original forces and couples taken about this point. Hence, the equations above can be written

$$\sum_i \boldsymbol{F}_i = \boldsymbol{0} \tag{5.2a}$$
$$\sum_i \boldsymbol{\rho}_i \times \boldsymbol{F}_i + \sum_i \boldsymbol{C}_i = \boldsymbol{0} \tag{5.2b}$$

where the $\boldsymbol{\rho}_i$'s are displacement vectors from the common point *a* to any point on the lines of action of the respective forces. From this form of the equations of statics, we can conclude that for equilibrium to exist, *the vector sum of the forces must be zero and the moment of the system of forces and couples about any point in space must be zero.*

Now that we have summed forces and have taken moments about a point *a*, we will demonstrate that we cannot find another *independent* equation by taking moments about a *different* point *b*. For the body in Fig. 5.18, we have initially the following equations of equilibrium using point *a*:

$$\boldsymbol{F}_1 + \boldsymbol{F}_2 + \boldsymbol{F}_3 + \boldsymbol{F}_4 = \boldsymbol{0} \tag{5.3}$$

$$\boldsymbol{\rho}_1 \times \boldsymbol{F}_1 + \boldsymbol{\rho}_2 \times \boldsymbol{F}_2 + \boldsymbol{\rho}_3 \times \boldsymbol{F}_3 + \boldsymbol{\rho}_4 \times \boldsymbol{F}_4 = \boldsymbol{0} \tag{5.4}$$

The new point *b* is separated from *a* by the position vector $\boldsymbol{d}$. The position vector (shown dashed) from *b* to the line of action of the force $\boldsymbol{F}_1$ can be given in terms of $\boldsymbol{d}$ and the displacement vector $\boldsymbol{\rho}_1$ as follows. Similarly for $(\boldsymbol{\rho}_2)_b$, which is not shown, and others.

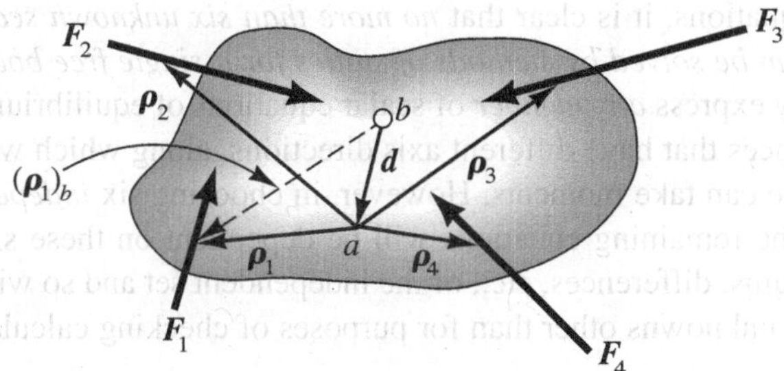

Figure 5.18. Consider moments about point *b*.

$$(\boldsymbol{\rho}_1)_b = (\boldsymbol{d} + \boldsymbol{\rho}_1)$$

$$(\boldsymbol{\rho}_2)_b = (\boldsymbol{d} + \boldsymbol{\rho}_2), \quad \text{etc.}$$

The moment equation for point *b* can then be given as

$$(\boldsymbol{\rho}_1 + \boldsymbol{d}) \times \boldsymbol{F}_1 + (\boldsymbol{\rho}_2 + \boldsymbol{d}) \times \boldsymbol{F}_2 + (\boldsymbol{\rho}_3 + \boldsymbol{d}) \times \boldsymbol{F}_3 + (\boldsymbol{\rho}_4 + \boldsymbol{d}) \times \boldsymbol{F}_4 = \boldsymbol{0}$$

Using the distributive rule for cross products, we can restate this equation as

$$\begin{aligned}(\boldsymbol{\rho}_1 \times \boldsymbol{F}_1 + \boldsymbol{\rho}_2 \times \boldsymbol{F}_2 + \boldsymbol{\rho}_3 \times \boldsymbol{F}_3 + \boldsymbol{\rho}_4 \times \boldsymbol{F}_4) \\ + \boldsymbol{d} \times (\boldsymbol{F}_1 + \boldsymbol{F}_2 + \boldsymbol{F}_3 + \boldsymbol{F}_4) = \boldsymbol{0}\end{aligned} \tag{5.5}$$

Since the expression in the second set of parentheses is zero, in accordance with Eq. 5.3, the remaining portion degenerates to Eq. 5.4, and thus we have not introduced a new equation. Therefore, *there are only two independent vector equations of equilibrium for any single free body.*

We shall now show that instead of using Eqs. 5.3 and 5.4 as the equations of equilibrium, we can instead use Eqs. 5.4 and 5.5. That is, instead of summing forces and then taking moments about a point for equilibrium, we can instead take moments about *two* points. Thus, if Eq. 5.4 is satisfied for point *a*, then for point *b* we end up in Eq. 5.5 with

$$\boldsymbol{d} \times (\boldsymbol{F}_1 + \boldsymbol{F}_2 + \boldsymbol{F}_3 + \boldsymbol{F}_4) = \boldsymbol{0} \tag{5.6}$$

If point *b* can be any point in space making $\boldsymbol{d}$ arbitrary, then the above equation indicates that the vector sum of forces is zero. If a point *b* happens to be chosen making Eq. (5.6) identically 0 = 0, (and hence useless), choose another point *b*. We then have equilibrium since $\boldsymbol{F}_R = \boldsymbol{0}$ and $\boldsymbol{C}_R = \boldsymbol{0}$.

Using the vector Eqs. 5.2, we can now express the scalar equations of equilibrium. Since, as you will recall, the rectangular components of the moment of a force about a point are the moments of the force about the orthogonal axes at the point, we may state these equations in the following manner:

$$\begin{aligned}
&\sum_i (F_x)_i = 0 \quad &\text{(a)} \qquad &\sum_i (M_x)_i = 0 \quad &\text{(d)} \\
&\sum_i (F_y)_i = 0 \quad &\text{(b)} \qquad &\sum_i (M_y)_i = 0 \quad &\text{(e)} \\
&\sum_i (F_z)_i = 0 \quad &\text{(c)} \qquad &\sum_i (M_z)_i = 0 \quad &\text{(f)}
\end{aligned} \tag{5.7}$$

From this set of equations, it is clear that *no more than six unknown scalar quantities in the general case can be solved by methods of statics for a single free body*.[3]

We can easily express *any number* of scalar equations of equilibrium for a free body by selecting references that have different axis directions, along which we can sum forces and about which we can take moments. However, in choosing six *independent* equations, we will find that the remaining equations will be dependent on these six. That is, these equations will be sums, differences, etc., of the independent set and so will be of no use in solving for desired unknowns other than for purposes of checking calculations.

5.5 Problems of Equilibrium I

We shall now examine problems of equilibrium in which the rigid-body assumption is valid. To solve such problems, we must find the value of certain unknown forces and couple moments. We first draw a free-body diagram of the entire system or portions thereof to clearly *expose* pertinent unknowns for analysis. We then write the equilibrium equations in terms of the unknowns along with the known forces and geometry. As we have seen, for any free body there is a limited number of independent scalar equations of equilibrium. Thus, at times we must employ several free-body diagrams for portions of the system to produce enough independent equations to solve all the unknowns.

For any free body, we may proceed by expressing two basic vector equations of statics. After carrying out such vector operations as cross products and additions in the equations, we form scalar equations. These scalar equations are then solved simultaneously (together with scalar equations from other free-body diagrams that may be needed) to find the unknown forces and couple moments. We can also express the scalar equations immediately by using the alternative scalar equilibrium relations that we formulated in previous sections. In the first case, we start with more compact vector equations and arrive at the expanded scalar equations by the formal procedures of vector algebra. In the latter case, we evaluate the expanded scalar equations by carrying out arithmetic operations on the free-body diagram as we write the equations. Which procedure is more desirable? It all depends on the problem and the investigator's skill in vector manipulation. It is true that many statics problems submit easily to a direct scalar approach, but the more challenging problems of statics and dynamics definitely favor an initial vector approach. In this text, we shall employ the particular procedure that the occasion warrants.

In statics problems, we must assign a sense to each component of an unknown force or couple moment in order to write the equations. If, on solving the equations, *we obtain a negative sign for a component, then we have guessed the wrong sense for that component*. Nothing need be redone should this occur. Continue with the remainder of the problem, retaining the minus sign (or signs). At the end of the problem, report the correct sense of your force components and couple-moment components.

We shall now solve and discuss a number of problems of equilibrium. These problems are divided into four classes of force systems:

[3]Keep in mind that we can also take moments about two sets of axes just as we could take moments about two points for the vector equations of equilibrium. The two sets of axes like the two points must be chosen properly so as to yield *independent* scalar equations. This can readily be done. If your second point and associated axes do not yield three additional independent equations, select another point and axes until you have your six independent equations of equilibrium.

1. Concurrent.
2. Coplanar.
3. Parallel.
4. General.

The type of *simplest* resultant for each special system of forces is most useful in determining the number of scalar equations available in a given problem. The procedure is to classify the force system, note what simplest resultant force system is associated with the classification, and then consider the number of scalar equations necessary and sufficient to guarantee this resultant to be zero. The following cases exemplify this procedure.

Case A. Concurrent System of Forces. In this case, since the simplest resultant is a single force at the point of concurrency, the only requirement for equilibrium is that this force be zero. We can ensure this condition if the orthogonal components of this force are separately equal to zero. Thus, we have *three* equations of equilibrium of the form

$$\sum_i (F_x)_i = 0, \qquad \sum_i (F_y)_i = 0, \qquad \sum_i (F_z)_i = 0 \tag{5.8}$$

As was pointed out in the general vector discussion, there are other ways of ensuring a zero resultant. Suppose that the moments of the concurrent force system are zero about three nonparallel axes: α, β, and γ. That is,

$$\sum_i (M_\alpha)_i = 0, \qquad \sum_i (M_\beta)_i = 0, \qquad \sum_i (M_\gamma)_i = 0 \tag{5.9}$$

Any one of the following three conditions must then be true:

1. The resultant force F_R is zero.
2. F_R cuts all three axes (see Fig. 5.19).
3. F_R cuts two axes and is parallel to the third (see Fig. 5.20).

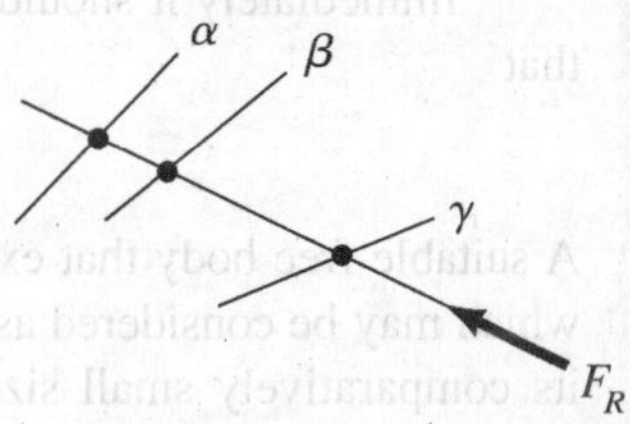

Figure 5.19. F_R cuts three axes.

We can guarantee condition 1 and thus equilibrium if we select axes α, β, and γ so that no straight line can intersect all three axes or can cut two axes and be parallel to the third. Then we can use Eqs. 5.9 as the equations of equilibrium under the aforestated conditions rather than using Eqs. 5.8. What happens if an axis used violates these conditions? The resulting equation will either be an *identity* 0 = 0 or will be dependent on a previous independent equation of equilibrium for one of the axes. No harm is done. One should use other axes until three independent equations are found.

Similarly, one can sum forces in one direction and take moments about two axes. Setting these equal to zero can yield three independent equations of equilibrium. If not, use other axes.

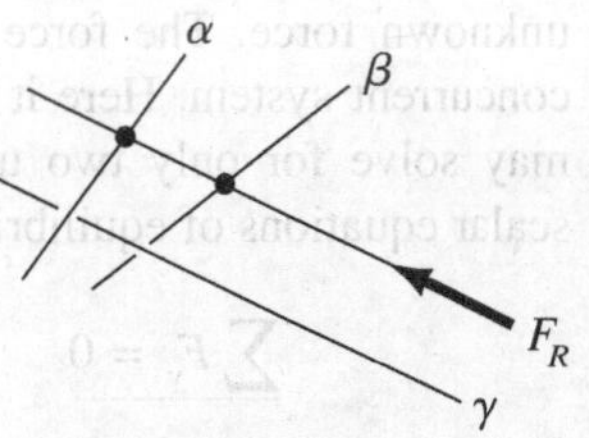

Figure 5.20. F_R cuts two axes and is parallel to third.

The essential conclusion to be drawn is that *there are three independent scalar equations of equilibrium for a concurrent force system.* For such systems it is most likely that you will always sum forces rather than take moments. However, for other force systems that we shall undertake, there will be ample opportunity to profitably employ alternate forms of equations other than those that we shall at first prescribe. The important thing to remember is that, just as in the concurrent force systems, only a definite number of equations for a given free body will be independent. Simply writing more equations beyond this number will only lead to identities and equations that will be of no use for solving for the desired unknowns.

Example 5.4

What are the tensions in cables AC and AB in Fig. 5.21? The system is in equilibrium. The following data apply:

$$W = 1{,}000 \text{ N} \qquad \beta = 50° \qquad \alpha = 37°$$

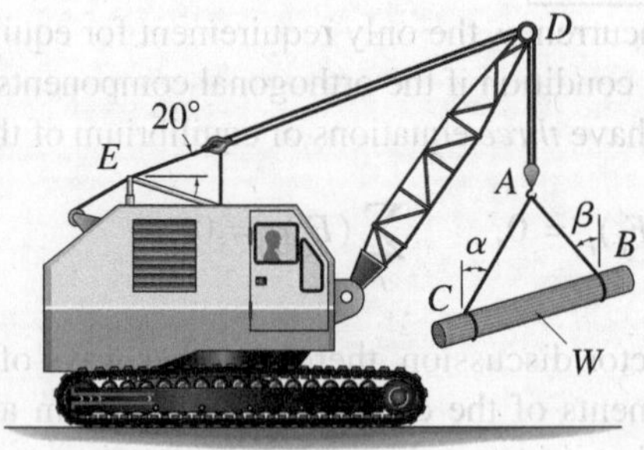

Figure 5.21. Derrick holding a beam.

Immediately it should be clear from observation of the diagram that

$$T_{AD} = 1{,}000 \text{ N (tension)}$$

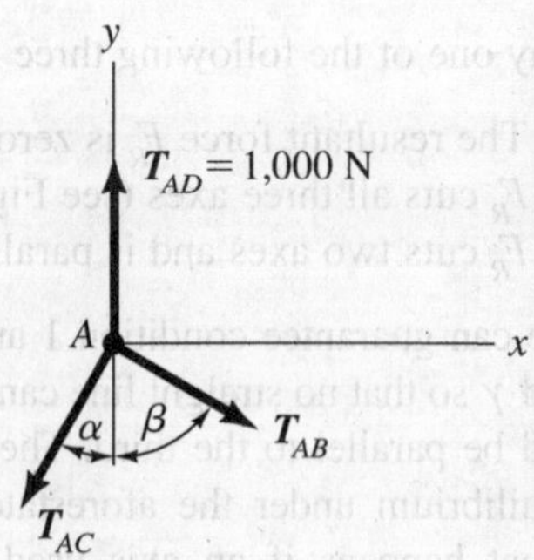

Figure 5.22. Free-body diagram of pin A.

A suitable free body that exposes the desired unknowns is the ring A, which may be considered as a particle for this computation because of its comparatively small size (Fig. 5.22). Physical intuition indicates that the cables are in tension and hence pulling away from A as we have indicated in the diagram although, as mentioned previously, it is *not* necessary to recognize at the outset the correct sense of an unknown force. The force system acting on the particle must be a concurrent system. Here it is also coplanar as well, and therefore we may solve for only two unknowns. Hence, we can proceed to the scalar equations of equilibrium. Thus,

$$\underline{\sum F_y = 0} \qquad 1{,}000 - T_{AC}\cos 37° - T_{AB}\cos 50° = 0$$

$$\therefore .7986T_{AC} + .6428T_{AB} = 1{,}000 \qquad \text{(a)}$$

$$\underline{\sum F_x = 0} \qquad -T_{AC}\sin\alpha - T_{AB}\sin\beta = 0$$

$$\therefore T_{AC} = 1.2729T_{AB} \qquad \text{(b)}$$

Example 5.4 (Continued)

Solving for T_{AC} and T_{AB} from Eqs. (a) and (b), we get

$$T_{AB} = 602.6 \text{ N} \qquad T_{AC} = 767.1 \text{ N}$$

Since the signs for T_{AC} and T_{AB} are positive, we have chosen the correct senses for the forces in the free-body diagram.

Another way of arriving at the solution, is to consider the *force polygon* that was discussed in Section 2.3. Because the forces are in equilibrium, the polygon must close; that is, the head of the final force must coincide with the tail of the initial force. In this case, we have a triangle, as shown in Fig. 5.23 approximately to scale. We can now use the law of sines.

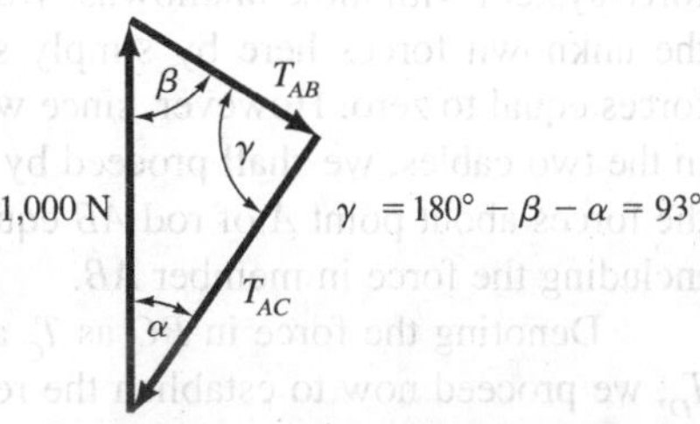

Figure 5.23. Force polygon.

$$\frac{T_{AB}}{\sin 37^\circ} = \frac{1{,}000}{\sin 93^\circ} \qquad T_{AB} = 602.6 \text{ N}$$

$$\frac{T_{AC}}{\sin 50^\circ} = \frac{1{,}000}{\sin 93^\circ} \qquad T_{AC} = 767.1 \text{ N}$$

The force polygon may thus be used to good advantage when three concurrent coplanar forces are in equilibrium.

As a final alternative, let us now initiate the computation for the unknown tensions from the basic *vector* equations of statics. First, we must express all forces in vector notation.

$$\boldsymbol{T}_{AC} = T_{AC}\,(-\sin 37^\circ\, \boldsymbol{i} - \cos 37^\circ\, \boldsymbol{j})$$

$$\boldsymbol{T}_{AB} = T_{AB}\,(\sin 50^\circ\, \boldsymbol{i} - \cos 50^\circ\, \boldsymbol{j})$$

We get the following equation when the vector sum of the forces is set equal to zero:

$$T_{AC}\,(-.6018\boldsymbol{i} - .7986\boldsymbol{j}) + T_{AB}\,(.7660\boldsymbol{i} - .6428\boldsymbol{j}) + 1{,}000\boldsymbol{j} = \boldsymbol{0}$$

Choosing point *A*, the point of concurrency, we clearly see that the sum of moments of the forces about this point is zero, so the second basic equation of equilibrium is intrinsically satisfied. We now regroup the terms of the preceding equation in the following manner:

$$(-.6018T_{AC} + .7660T_{AB})\boldsymbol{i} + (-.7986T_{AC} - .6428T_{AB} + 1{,}000)\boldsymbol{j} = \boldsymbol{0}$$

To satisfy this equation, each of the quantities in parentheses must be zero. This gives the scalar equations (a) and (b) stated earlier, from which the scalar quantities T_{AB} and T_{AC} can be solved.

The three alternative methods of solution are apparently of equal usefulness in this simple problem. However, the force polygon is only of practical use for three concurren- coplanar forces, where the trigonometric properties of a triangle can be directly used. The other methods can be readily extended to more complex concurrent problems.

Example 5.5

Find the forces in cables DB and CB in Fig. 5.24. The 500-N force is parallel to the y axis. Consider B to be a ball joint located in the xz plane. Rod AB is a compression member, with a ball joint at A.

In Fig. 5.25 we have indicated the forces acting on joint B. Clearly we have a three-dimensional concurrent force system with three unknowns. We can readily determine the unknown forces here by simply setting the sum of the forces equal to zero. However, since we only want the forces in the two cables, we shall proceed by setting the moment of the forces about point A of rod AB equal to zero, thereby not including the force in member AB.

Denoting the force in BC as T_C and the force in BD as T_D, we proceed now to establish the rectangular components of these forces.

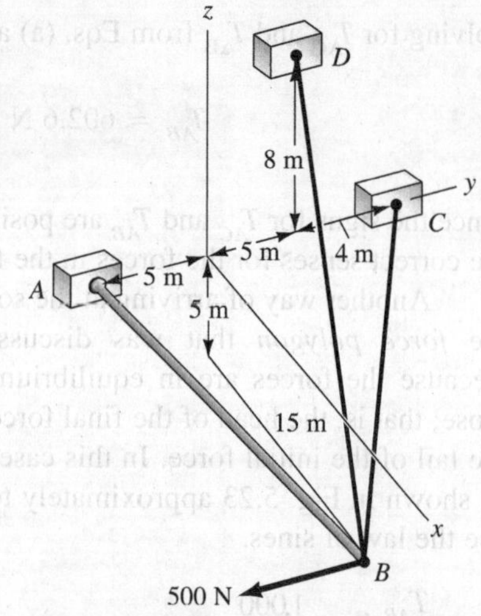

Figure 5.24. Rod AB and cables CB and DB support a 500-lb force.

$$\boldsymbol{T}_C = T_C\left(\frac{-15\boldsymbol{i} + 9\boldsymbol{j} + 5\boldsymbol{k}}{\sqrt{15^2 + 9^2 + 5^2}}\right) = T_C(-.824\boldsymbol{i} + .495\boldsymbol{j} + .275\boldsymbol{k})\text{ N}$$

$$\boldsymbol{T}_D = T_D\left(\frac{-15\boldsymbol{i} + 5\boldsymbol{j} + 13\boldsymbol{k}}{\sqrt{15^2 + 5^2 + 13^2}}\right) = T_D(-.733\boldsymbol{i} + .244\boldsymbol{j} + .635\boldsymbol{k})\text{ N}$$

The position vector that we shall use for the moment about point A is $\boldsymbol{r}_{AB}$, which is

$$\boldsymbol{r}_{AB} = 15\boldsymbol{i} + 5\boldsymbol{j} - 5\boldsymbol{k}\text{ m}$$

We now set the moments about point A equal to zero.

$$\sum \boldsymbol{M}_A = \boldsymbol{0}$$

$$(15\boldsymbol{i} + 5\boldsymbol{j} - 5\boldsymbol{k}) \times [T_C(-.824\boldsymbol{i} + .495\boldsymbol{j} + .275\boldsymbol{k}) + T_D(-.733\boldsymbol{i} + .244\boldsymbol{j} + .635\boldsymbol{k}) - 500\boldsymbol{j}] = \boldsymbol{0}$$

We simplify the calculations further by noting that cable BD has a direction inclined to the plane ACB in which the other three forces lie. This can only mean that the force T_D must have a zero value. Hence, deleting this force in the above equation and carrying out the cross products, it is easy to get the remaining nonzero force. We thus have

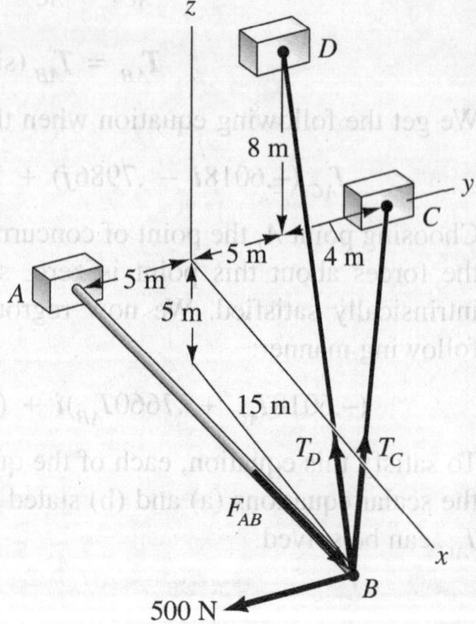

Figure 5.25. Free-body diagram of joint B.

$T_C = 649$ N $\qquad T_D = 0$ N

Example 5.5 (Continued)

Finally, as indicated at the outset, we could proceed by summing forces in the coordinate directions. The resulting scalar equations for all three unknown forces are

$$-0.824T_C - 0.733T_D + 0.905T_A = 0$$

$$0.495T_C + 0.244T_D + 0.302T_A = 500$$

$$0.275T_C + 0.635T_D - 0.302T_A = 0$$

We may now solve the simultaneous equations using *Cramer's rule*. Thus we calculate first the determinant of the coefficients of the unknowns. Thus

$$\begin{bmatrix} -0.824 & -0.733 & 0.905 \\ 0.495 & 0.244 & 0.302 \\ 0.275 & 0.635 & -0.302 \end{bmatrix} = 0.272$$

To calculate T_C we proceed as follows:

$$T_C = \frac{\begin{bmatrix} 0 & -0.733 & 0.905 \\ 500 & 0.244 & 0.302 \\ 0 & 0.635 & 0.302 \end{bmatrix}}{0.272} = 649 \text{ N}$$

Note that the first column of the determinant consists of the right side terms of the set of simultaneous equations in place of the coefficients of the desired unknown. We can solve for the other unknowns similarly. We then would have the compressive force in member *AB* which is 591 N.

Case B. Coplanar Forces System. We have shown that the simplest resultant for a coplanar force system (see Fig. 4.14) is a single force or a single couple moment normal to the plane. Thus, to ensure that the resultant force is zero, we require for a coplanar system in which all forces are in the *xy* plane:

$$\sum_i (F_x)_i = 0, \qquad \sum_i (F_y)_i = 0 \tag{5.10}$$

To ensure that the resultant couple moment is zero, we require for moments about any axis parallel to the *z* axis:

$$\sum_i (M_z)_i = 0 \tag{5.11}$$

We conclude that there are *three* scalar equations of equilibrium for a coplanar force system. Other combinations, such as two moment equations for two axes parallel to the *z* axis and a single force summation, if properly chosen, may be employed to give the three independent scalar equations of equilibrium, as was discussed in case A.

Example 5.6

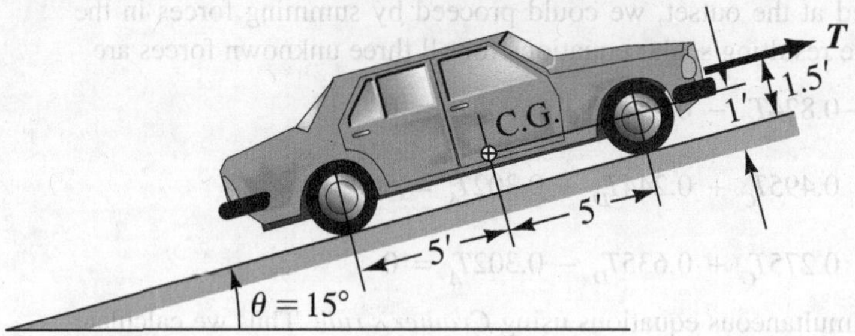

Figure 5.26. A car is being towed up an incline at a constant speed.

A car shown in Fig. 5.26 is being towed at a steady speed up an incline having an angle of 15°. The car weighs 3,600 lb. The center of gravity is located as is shown in the diagram. Calculate the supporting force on each wheel and force *T*.

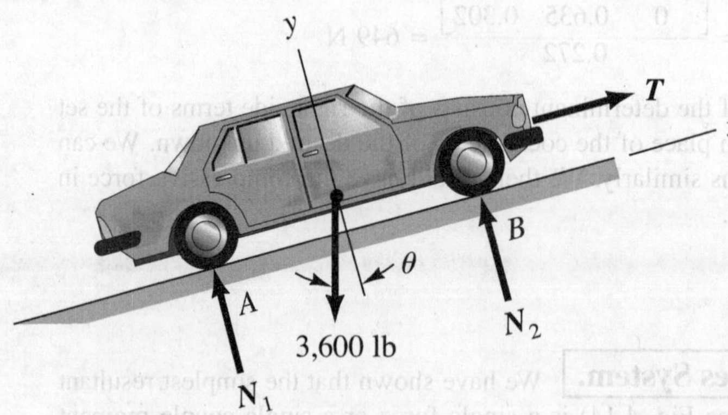

Figure 5.27. Free-body diagram of the car.

A free-body diagram of the car is shown in Fig. 5.27. The forces N_1 and N_2 are the total forces, respectively, for the rear wheels and the front wheels. Note, because the wheels are rotating at constant speed, there are no friction forces present. We have thus formed on this free body a *coplanar* force system involving three unknowns and hence the unknowns are solvable by rigid-body statics. Using axes tangent and normal to the incline we have

Example 5.6 (Continued)

$$\sum F_x = 0 \quad T - W \sin\theta = 0$$
$$\therefore T = (3{,}600)(\sin 15°) = 931.7 \text{ lb}$$

T = 931.7 lb

$$\sum F_y = 0 \quad N_1 + N_2 - W\cos\theta = 0$$
$$\therefore N_1 + N_2 = (3{,}600)(\cos 15°) = 3{,}477 \text{ lb} \qquad (1)$$

$$\sum M_A = 0 \quad (N_2)(10) + (W\sin\theta)(1) - (W\cos\theta)(5) - (T)(1.5) = 0$$
$$\therefore N_2 = \frac{1}{10}[-(3{,}600)(\sin 15°)(1) + (3{,}600)(\cos 15°)(5)$$
$$+(931.7)(1.5)] = 1{,}785 \text{ lb}$$

From Eq.(1), we can now get N_1. Thus

$$N_1 = 3{,}477 - N_2 = 3{,}477 - 1{,}785 = 1{,}692 \text{ lb}$$

Clearly each rear wheel has acting on it a normal force of $N_1/2 = 846.1$ lb and each front wheel has a normal force of $N_2/2 = 892.5$ lb.

$\therefore$ **Rear wheel support force = 846.1 lb**
Front wheel support force = 892.5 lb

We may now check this solution by using a redundant equation of equilibrium. Thus

$$\sum M_B \stackrel{?}{=} 0$$
$$-(N_1)(10) + (W\sin\theta)(1) - (W\cos\theta)(5) - (T)(1.5) = 0$$
$$\therefore -(1{,}692)(10) + (3{,}600)(\sin 15°)(1) +$$
$$(3{,}600)(\cos 15°)(5) - (931.7)(1.5)] = 0$$
$$.8634 \approx 0$$

We have here a roundoff error, which we can accept for the accuracy of the calculations taken in this problem.

Example 5.7

A frame is shown in Fig. 5.28 in which the frictionless pulley at D has a mass of 200 kg. Neglecting the weights of the bars, find the force transmitted from one bar to the other at joint C.

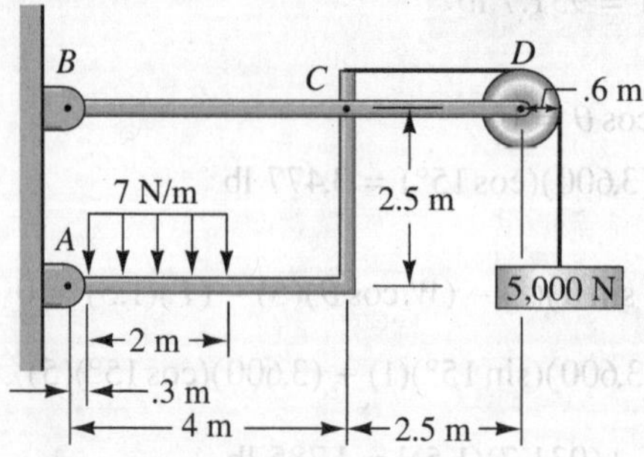

Figure 5.28. Loaded frame.

To expose force components C_x and C_y, we form the free body of bar BD. This is shown as F.B.D. I in Fig. 5.29. It is clear that for this free body we have six unknowns and only three independent equations of equilibrium.[4] The free-body diagram of the bent bar AC is then drawn (F.B.D. II in Fig. 5.29). Here, we have three more equations but we bring in three more unknowns. Finally, the free-body diagram of the pulley (F.B.D. III in Fig. 5.29) gives three more equations with no additional unknowns.

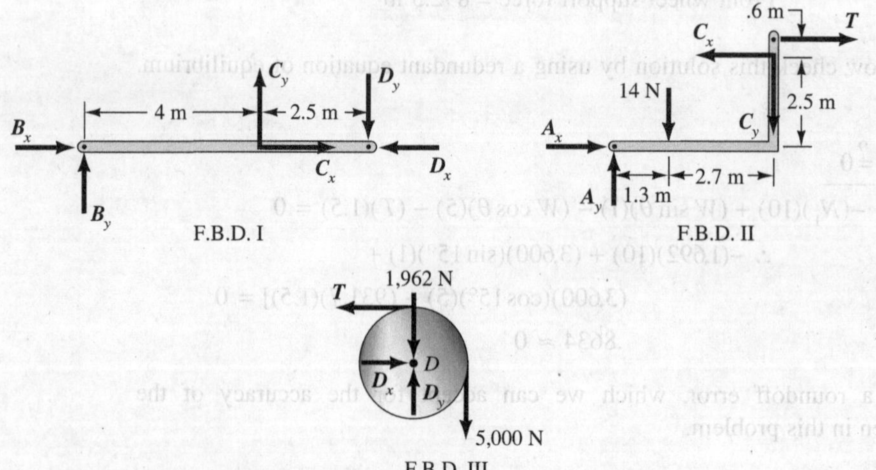

Figure 5.29. Free-body diagrams of frame parts.

[4]It should be noted that it is possible to have situations wherein there are more unknowns than independent equations of equilibrium for a given free body, but wherein some of the unknowns—perhaps the desired ones—can be still determined by the equations available. However, not all the unknowns of the free body can be solved. Accordingly, be alert for such situations, so as to minimize the work involved. In this case, we must consider other free-body diagrams.

Example 5.7 (Continued)

We now have nine equations available and nine unknowns and can proceed with confidence. Since only two of the unknowns are desired, we shall take select scalar equations from each of the free-body diagrams to arrive at the components C_x and C_y most quickly.

From F.B.D. III:

$\sum M_D = 0$:

$$(T)(.6) - (5{,}000)(.6) = 0$$

Therefore,

$$T = 5{,}000 \text{ N}$$

$\sum F_x = 0$:

$$-T + D_x = 0$$

Therefore,

$$D_x = 5{,}000 \text{ N}$$

$\sum F_y = 0$:

$$-1{,}962 - 5{,}000 + D_y = 0$$

Therefore,

$$D_y = 6{,}962 \text{ NE}$$

From F.B.D. I:

$\sum M_B = 0$:

$$(4)(C_y) - (6.5)(D_y) = 0$$

Therefore,

$$C_y = 11{,}313 \text{ N}$$

From F.B.D. II:

$\sum M_A = 0$:

$$-(1.3)(14) - (T)(3.1) - C_y(4) + C_x(2.5) = 0$$

Therefore,

$$C_x = 24{,}300 \text{ N}$$

We can give the force at *C* (transmitted from bar AC to bar BD) as

$$C = 24{,}300\boldsymbol{i} + 11.313\boldsymbol{j} \text{ N}$$

PROBLEMS

5.17. In a tug of war, when team B pulls with a 400-lb force, how much force must team C exert for a draw? With what force does team A pull?

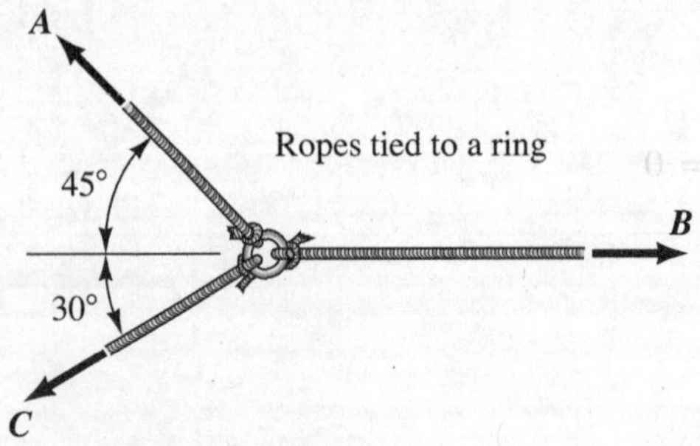

Figure P.5.17.

5.18. Find the tensile force in cables AB and CB. The remaining cables ride over frictionless pulleys E and F.

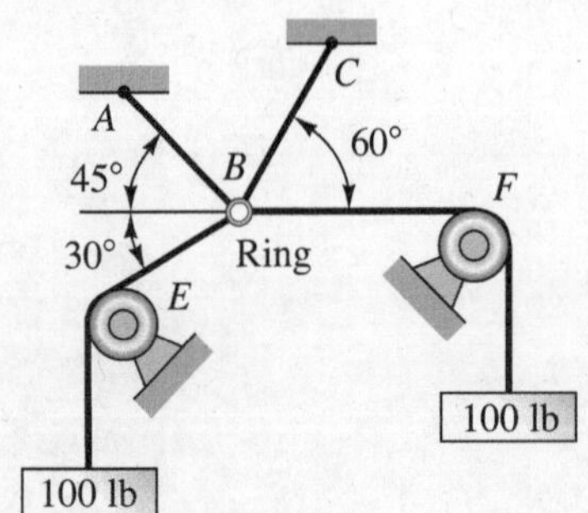

Figure P.5.18.

5.19. Find the force transmitted by wire BC. The pulley E can be assumed to be frictionless in this problem.

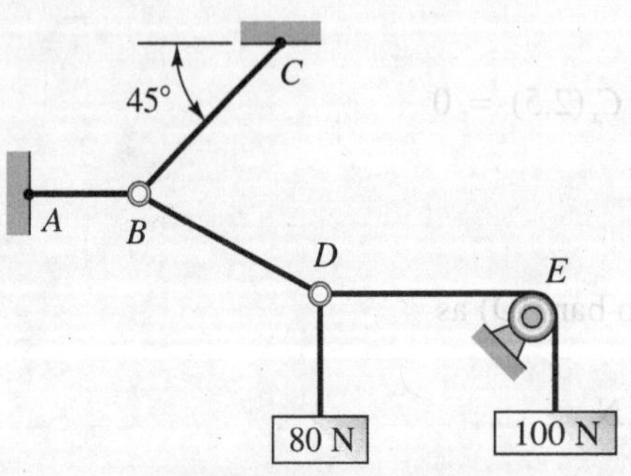

Figure P.5.19.

5.20. Find the tensions in the three cables connected to B. The entire system of cables is coplanar. The roller at E is free to turn without resistance.

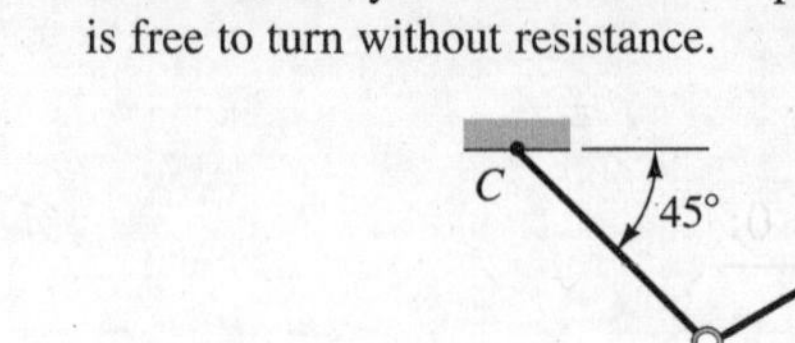

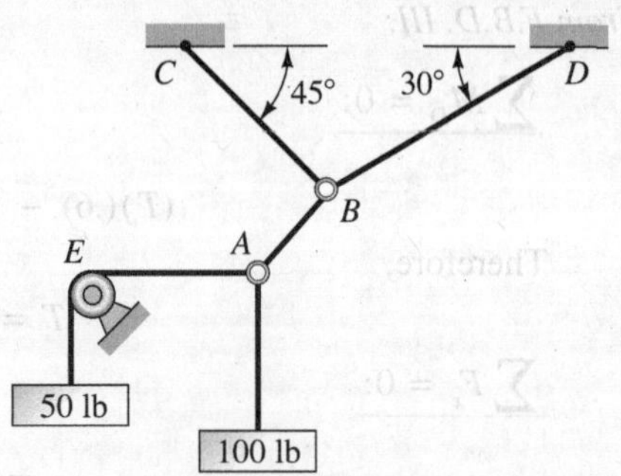

Figure P.5.20.

5.21. A 700-N circus performer causes a .15-m sag in the middle of a 12-m tightrope with a 5,000-N initial tension. What additional tension is induced in the cable? What is the cable tension when the performer is 3 m from the end and the sag is .12 m?

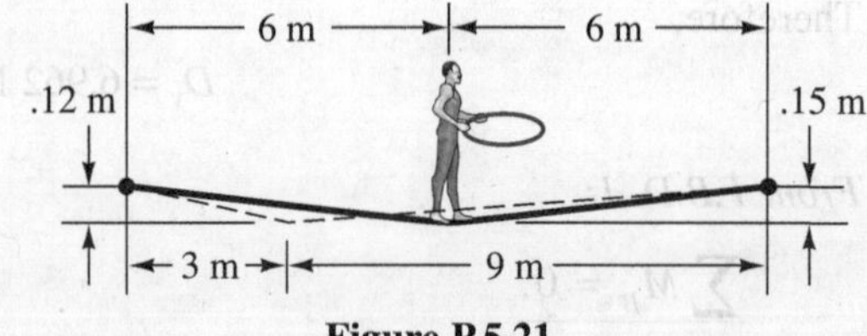

Figure P.5.21.

5.22. A 27-lb mirror is held up by a wire fastened to two hooks on the mirror frame. (a) What is the force on the wall hook and the tension in the wire? (b) If the wire will break at a tension of 32 lb, must the wall hook be moved (i.e., the wire lengthened or shortened and the 4-in. rise distance changed)? If so, to what point?

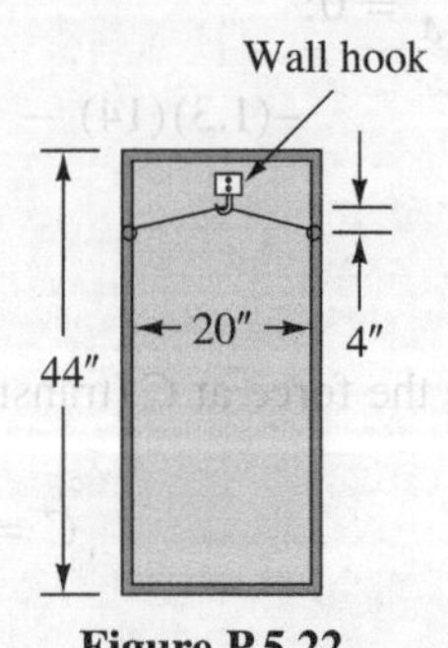

Figure P.5.22.

5.23. Explain why equilibrium of a concurrent force system is guaranteed by having $\sum_i (F_y)_i = 0$, $\sum_i (M_d)_i = 0$, and $\sum_i (M_e)_i = 0$. Axes d and e are not parallel to the xz plane. Moreover, the axes are oriented so that the line of action of the resultant force cannot intersect both axes.

5.24. Cylinders A and B weigh 500 N each and cylinder C weighs 1,000 N. Compute all contact forces.

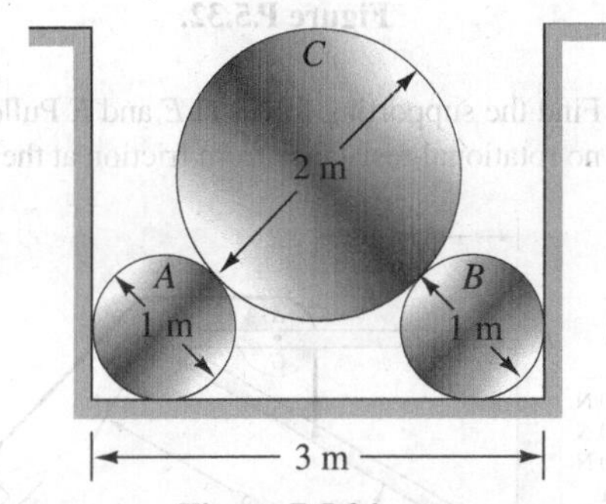

Figure P.5.24.

5.25. A block having a mass of 500 kg is held by five cables. What are the tensions in these cables? Lower cables are identical and are identically connected at ends.

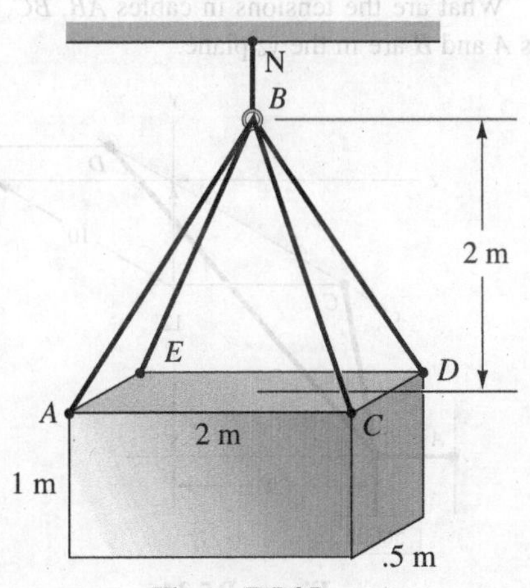

Figure P.5.25.

5.26. Gear D has a weight of 300 N while members AB and BC are light. What torque T is necessary for equilibrium?

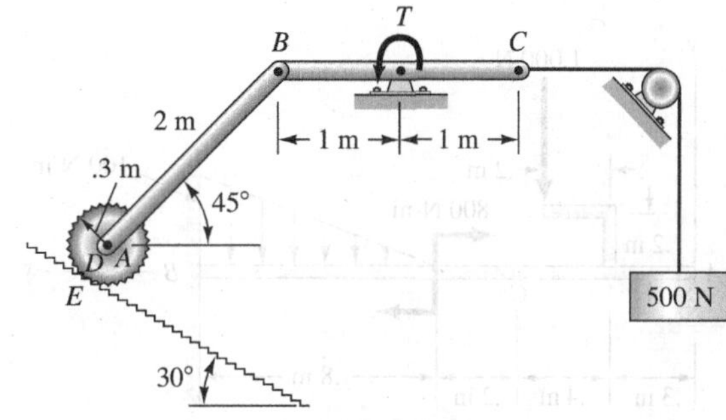

Figure P.5.26.

5.27. Find the supporting forces at A.

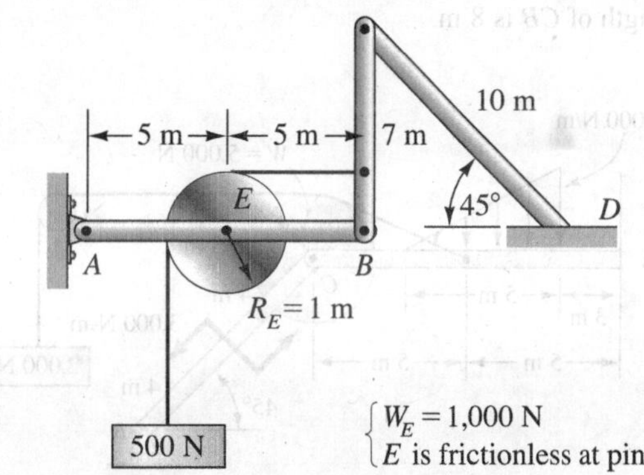

Figure P.5.27.

5.28. Find the supporting forces at A and B. At D there is a cylinder weighing 300 N.

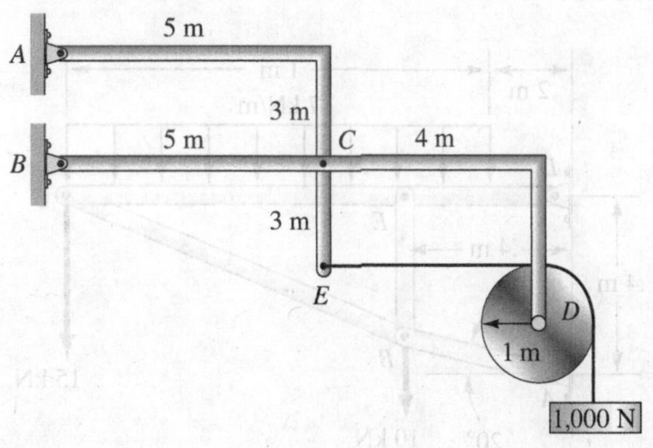

Figure P.5.28.

5.29. Find the supporting force systems for the beams shown. Note that there is a *pin connection* at C. Neglect the weights of the beams.

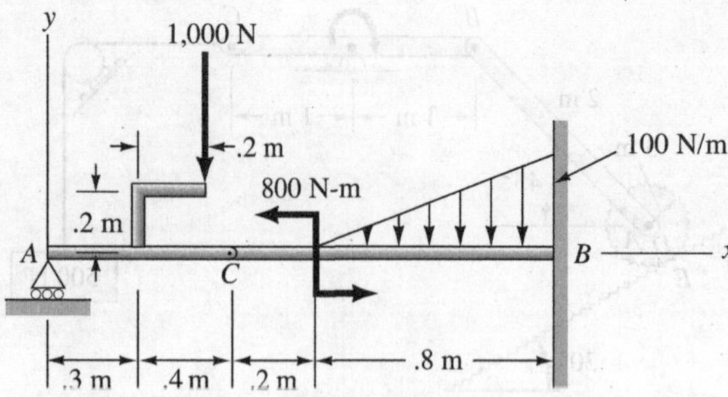

Figure P.5.29.

5.30. Find the supporting force systems at A and B. The length of CB is 8 m.

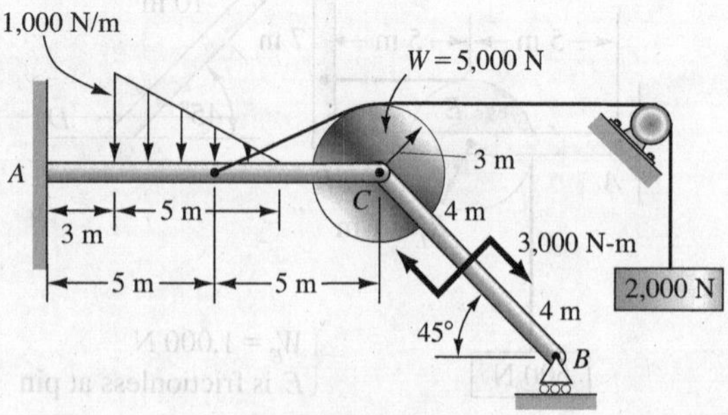

Figure P.5.30.

5.31. What are the supporting forces at A and D for the frame shown? What are the forces in members AB, BE, and BC?

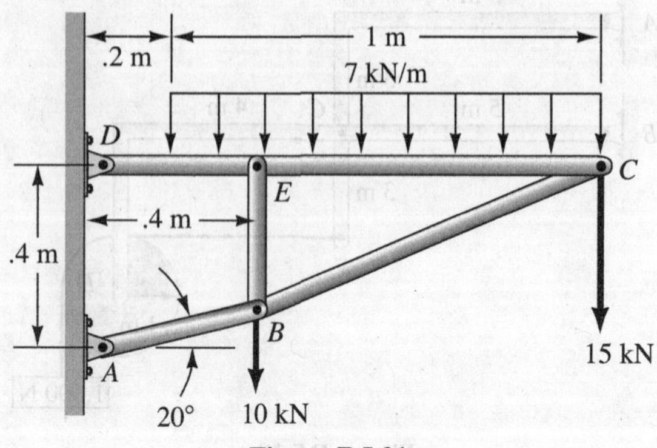

Figure P.5.31.

5.32. What are the supporting forces for the frame? Neglect all weights except the 10-kN weight. Disregard friction.

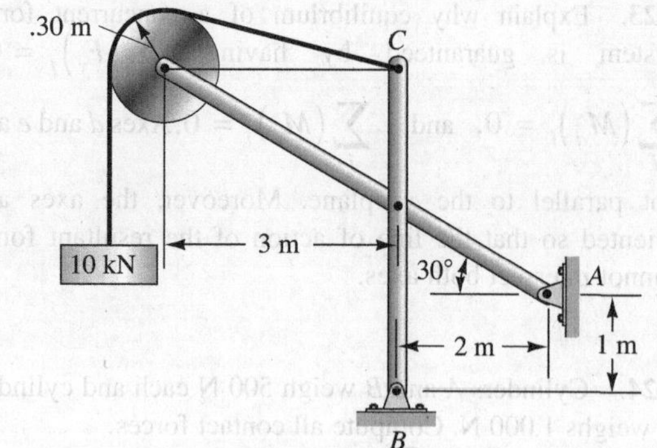

Figure P.5.32.

5.33. Find the supporting forces at E and F. Pulleys A and B offer no rotational resistance from friction at the bearings.

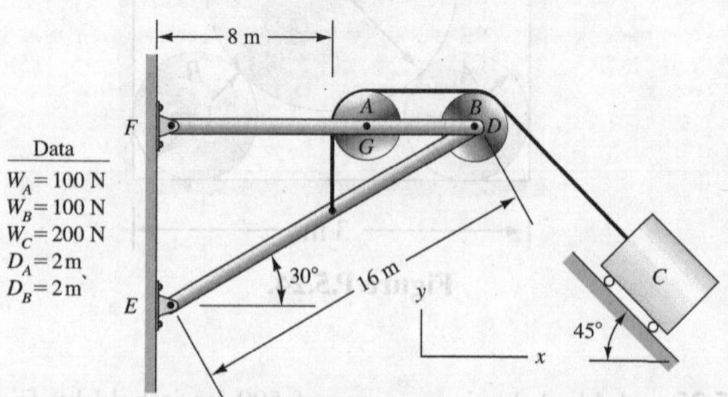

Figure P.5.33.

5.34. What are the tensions in cables AB, BC, and BD? Points A and B are in the yz plane.

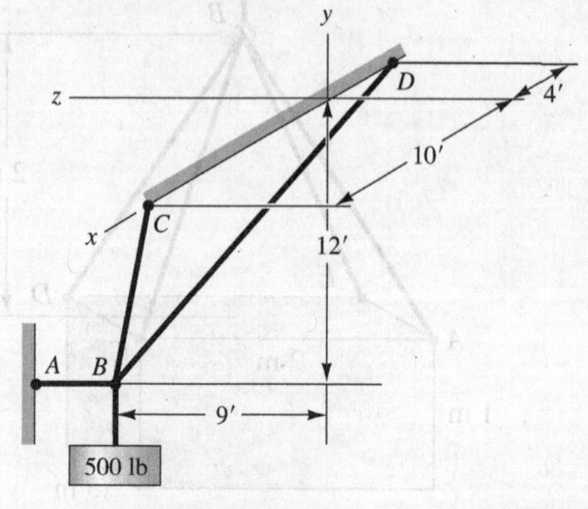

Figure P.5.34.

5.35. An elastic cord AB is just taut before the 1,000-N force is applied. If it takes 5.0 N/mm of elongation of the cord, what is the tension T in the cord after the 1,000-N force is applied? Set up the equation for T but do not solve.

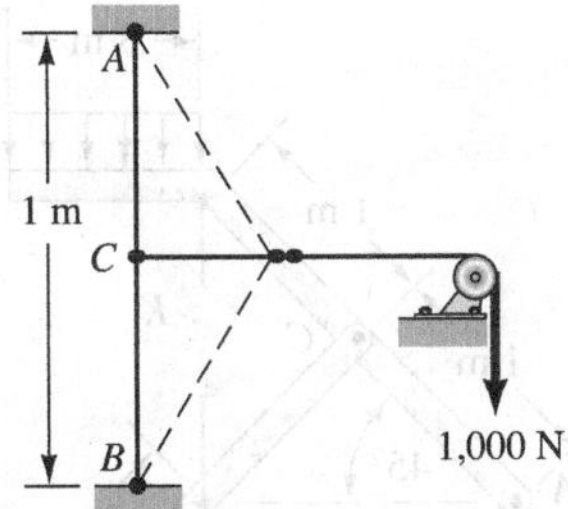

Figure P.5.35.

5.36. A thin hoop of radius 1 m and weight 500 N rests on an incline. What friction force f at A is needed for this configuration? What is the tension in wire CB?

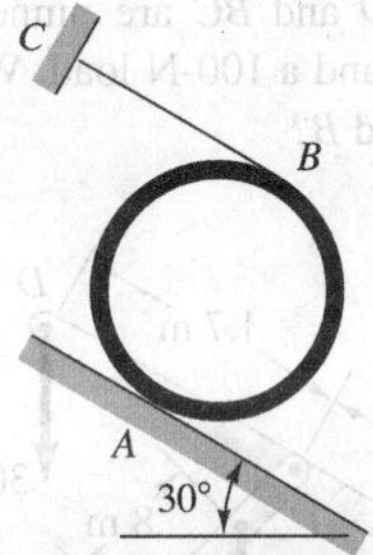

Figure P.5.36.

5.37. A stepped cylinder is pulled down an incline by a force F, which is increased from zero to 20 N very slowly while always maintaining the 30° inclination shown. If the cylinder is in equilibrium when $F = 0$, how far does O move as a result of F after equilibrium has been established with $F = 20$ N? The stepped cylinder has a mass of 10 kg. There is no slipping at the base. The force from the spring is K times the extension of the spring. For this spring, $K = 5$ N/mm. F remains parallel to the position shown.

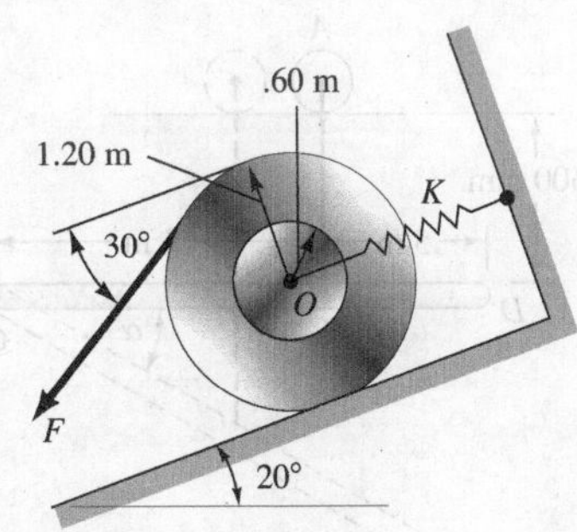

Figure P.5.37.

5.38. A 10-kg ring is supported by a smooth surface E and a wire AB. A body D having a mass of 3 kg is fixed to the ring at the orientation shown. What is the tension in the wire AB? What is its orientation α? Point A is directly above point O.

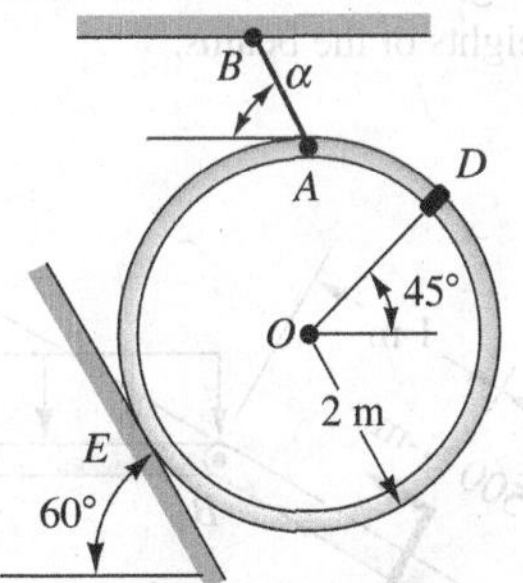

Figure P.5.38.

5.39. Find the supporting forces on the beam EF and the supporting forces at A, B, C, and D.

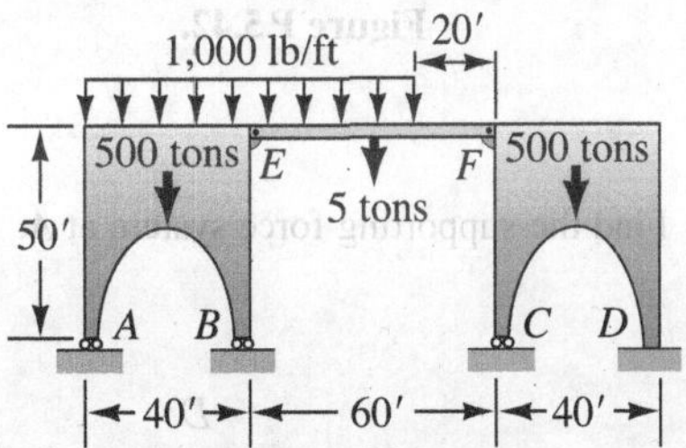

Figure P.5.39.

5.40. What is the supporting force system at A for the cantilever beam? Neglect the weight of the beam.

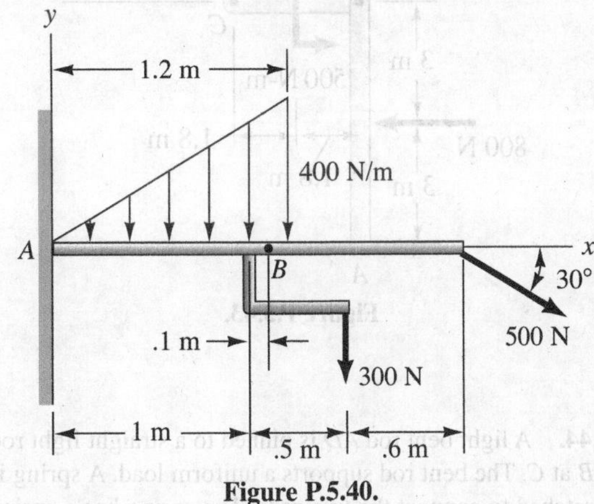

Figure P.5.40.

5.41. In Problem 5.40, find the force system transmitted through the cross section at B.

5.42. A cantilever beam AB is pinned at B to a simply supported beam BC. For the loads given, find the supporting force system at A. Determine force components that are normal and tangential to the cross-section of beam AB. Neglect the weights of the beams.

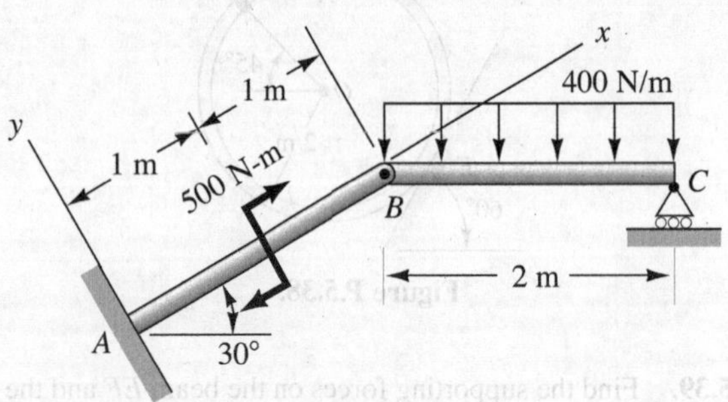

Figure P.5.42.

5.43. Find the supporting force system at A.

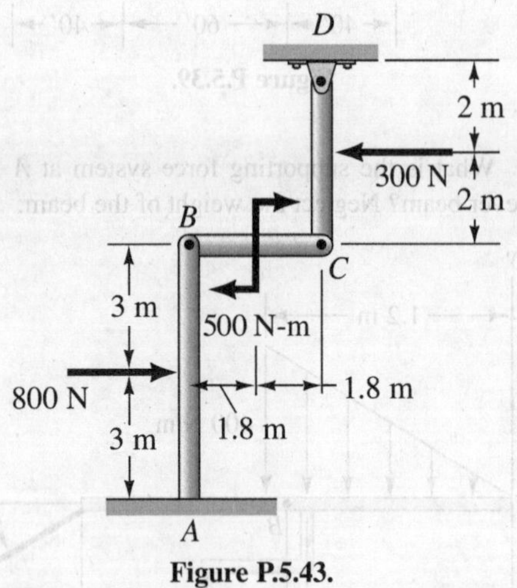

Figure P.5.43.

5.44. A light bent rod AD is pinned to a straight light rod CB at C. The bent rod supports a uniform load. A spring is stretched to connect the two rods. The spring has a spring constant of 10^4 N/m, and its unstretched length is .8 m. Find the supporting forces at A and B. The force in the spring is 10^4 times the elongation in meters.

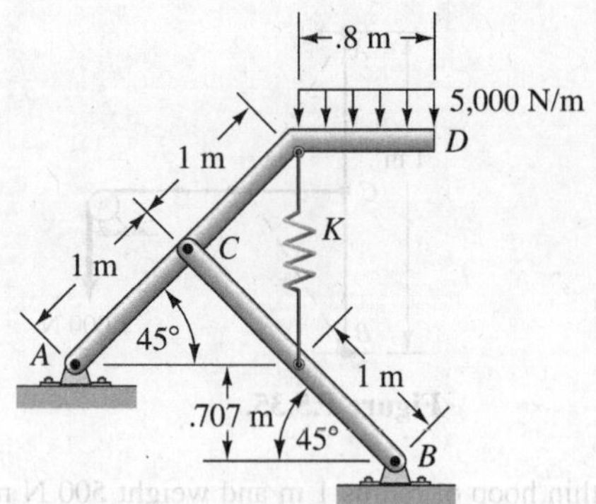

Figure P.5.44.

5.45. Light rods AD and BC are pinned together at C and support a 300-N and a 100-N load. What are the supporting forces at A and B?

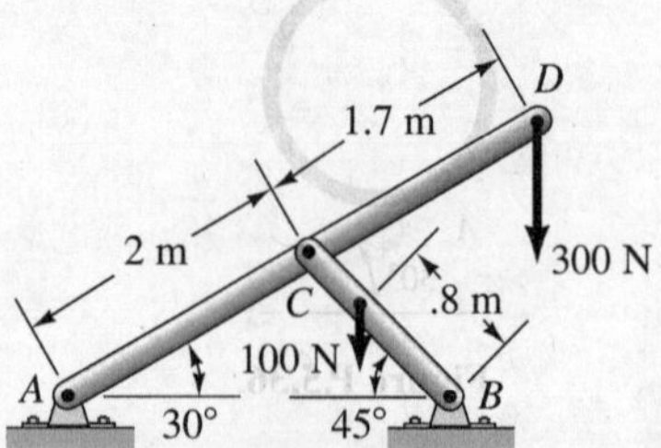

Figure P.5.45.

5.46. A light rod CD is held in a horizontal position by a strong elastic band AB (shock cord), which acts like a spring in that it takes 10^3 N per meter of elongation of the band. The upper part of the band is connected to a small wheel free to roll on a horizontal surface. What is the angle α needed to support a 200-N load as shown?

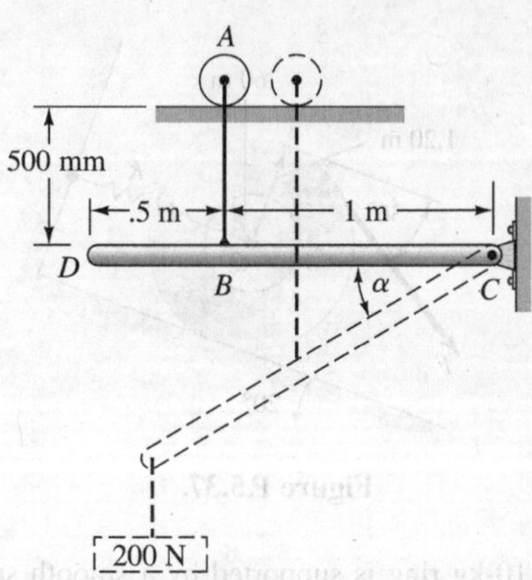

Figure P.5.46.

5.47. Solve for the supporting forces at A and C. AB weighs 100 lb, and BC weighs 150 lb.

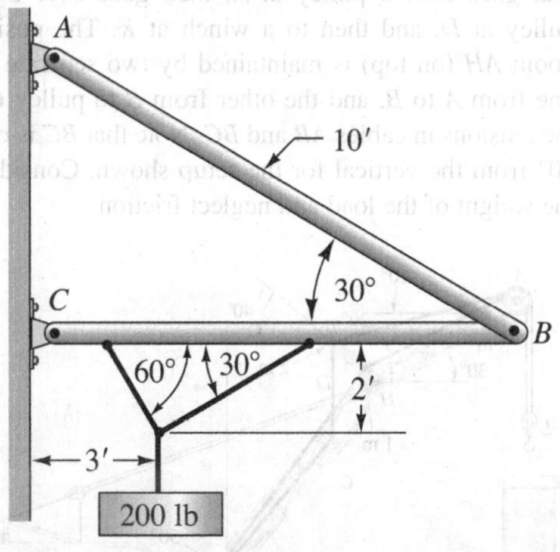

Figure P.5.47.

5.48. What torque T is needed to maintain the configuration shown for the compressor if $p_1 = 5$ psig? The system lies horizontally.

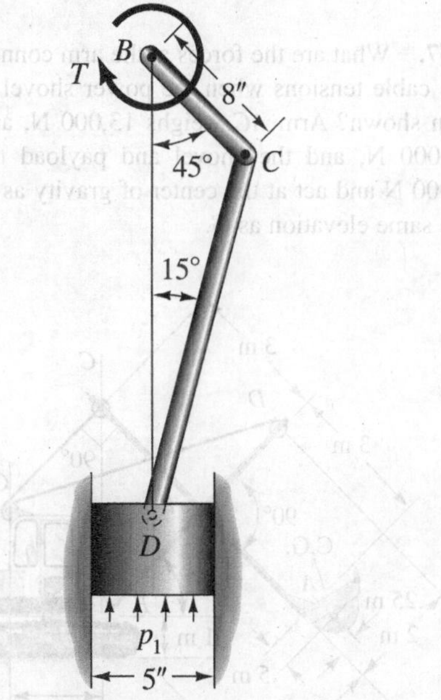

Figure P.5.48.

5.49. Work Problem 5.48 for the system oriented vertically with BC weighing 3 lb and CD weighing 5 lb.

5.50. Neglecting friction, find the angle β of line AB for equilibrium in terms of α_1, α_2 W_1, and W_2.

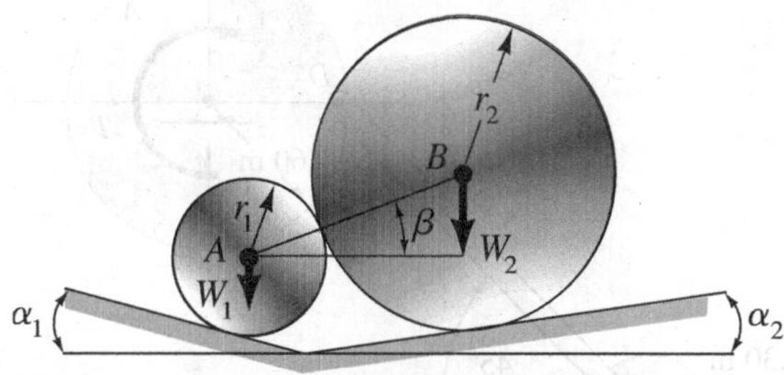

Figure P.5.50.

5.51. If the rod CD weighs 20 lb, what torque T is needed to maintain equilibrium? The system is in a vertical plane. Cylinder A weighs 10 lb and cylinder B weighs 5 lb. Disregard friction. At D there is a slot.

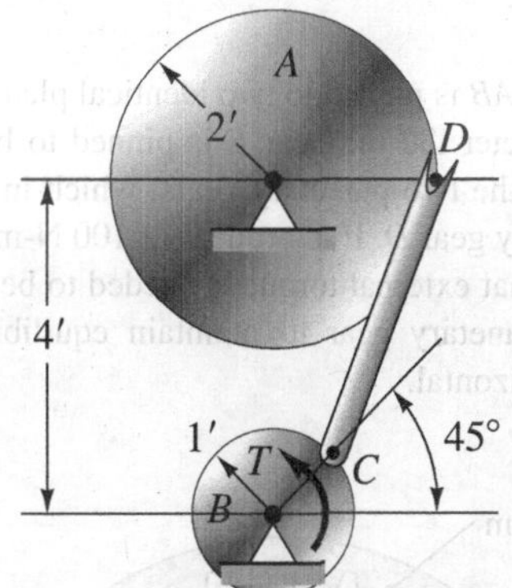

Figure P.5.51.

5.52. Find the supporting forces at A and G. The weight of W is 500 N and the weight of C is 200 N. Neglect all other weights. The cord connecting C and D is vertical.

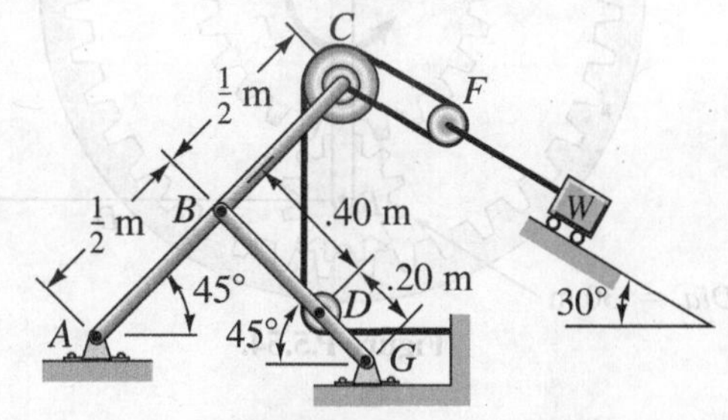

Figure P.5.52.

5.53. What torque T is needed for equilibrium if cylinder B weighs 500 N and CD weighs 300 N?

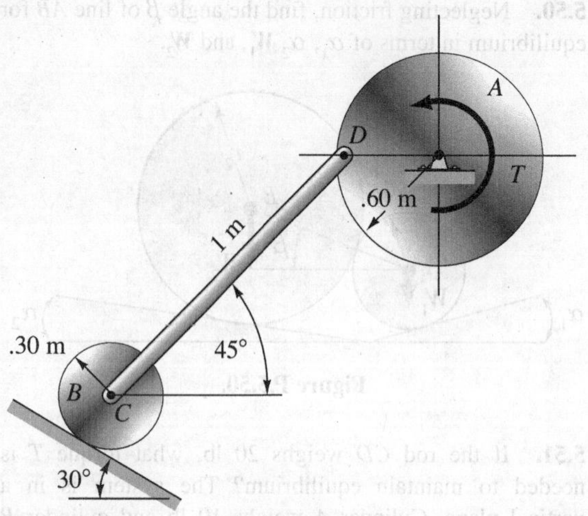

Figure P.5.53.

5.54. A bar AB is pinned to two identical planetary gears each of diameter .30 m. Gear E is pinned to bar AB and meshes with the two planetary gears, which in turn mesh with stationary gear D. If a torque T of 100 N-m is applied to bar AB, what external torque is needed to be applied to the upper planetary gear to maintain equilibrium? The system is horizontal.

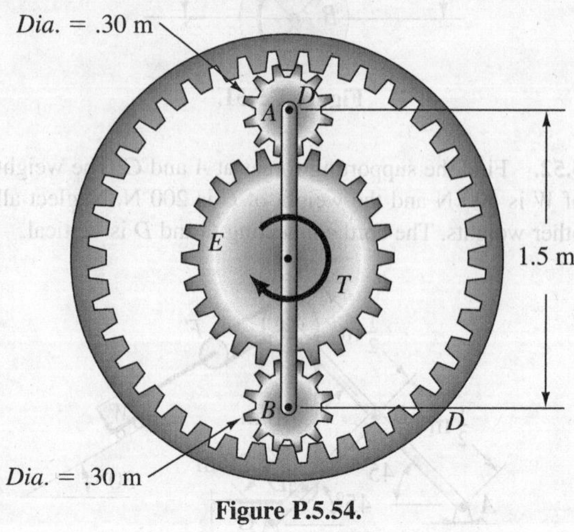

Figure P.5.54.

5.55. In Problem 5.54, equilibrium is maintained by applying a torque on gear E rather than the upper planetary gear. What is this torque?

5.56. A Bucyrus–Erie transit crane is holding a chimney having a weight of 20 kN. The chimney is held by a cable that goes over a pulley at A, then goes over a second pulley at D, and then to a winch at K. The position of boom AH (on top) is maintained by two separate cables, one from A to B, and the other from B to pulley C. Find the tensions in cables AB and BC. Note that BC is oriented 30° from the vertical for the setup shown. Consider only the weight of the load and neglect friction.

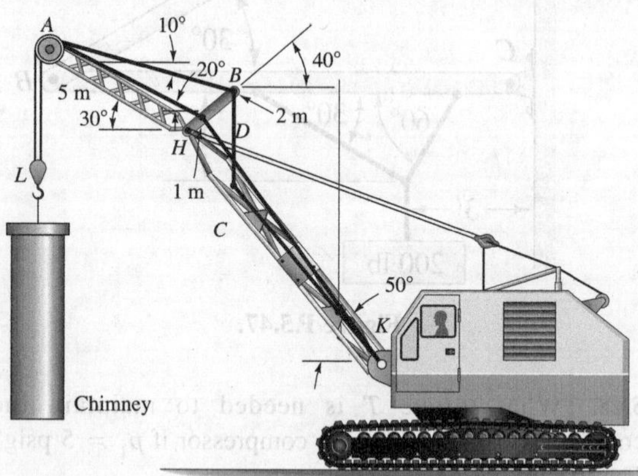

Figure P.5.56.

5.57. What are the forces at the arm connection at B and the cable tensions when the power shovel is in the position shown? Arm AC weighs 13,000 N, arm DF weighs 11,000 N, and the shovel and payload together weigh 9,000 N and act at the center of gravity as shown. B is at the same elevation as G.

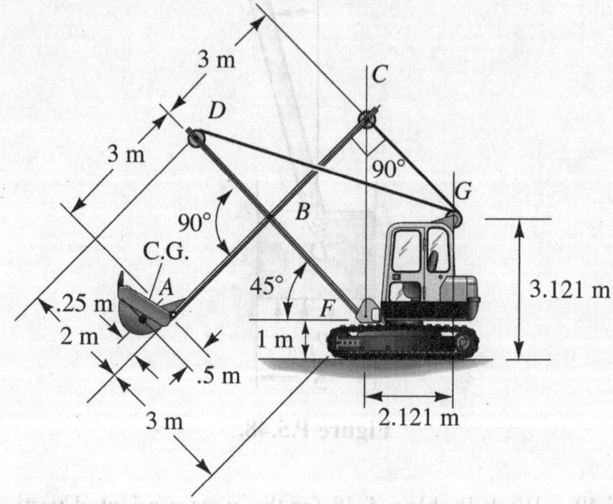

Figure P.5.57.

5.6 Problems of Equilibrium II

Case C. Parallel Forces in Space. In the case of parallel forces in space (see Fig. 4.20), we already know that the simplest resultant can be either a single force or a couple moment. If the forces are in the z direction, then

$$\sum_i (F_z)_i = 0 \tag{5.12}$$

ensures that the resultant force is zero. Also,

$$\sum_i (M_x)_i = 0, \qquad \sum_i (M_y)_i = 0 \tag{5.13}$$

guarantees that the resultant couple moment is zero, where the x and y axes may be chosen in any plane perpendicular to the direction of the forces.[5] Thus, three independent scalar equations are available for equilibrium of parallel forces in space.

A summary of the special cases discussed thus far is given below. For even simpler systems such as the concurrent-coplanar and the parallel-coplanar systems, clearly, there is one less equation of equilibrium.

Summary for Special Cases

System	*Simplest Resultant*	*Number of Equations for Equilibrium*
Concurrent (three-dimensional)	Single force	3
Coplanar	Single force or single couple moment	3
Parallel (three-dimensional)	Single force or single couple moment	3

[5]For parallel forces in the z direction, a simplest resultant consisting of a couple moment only must have this couple moment parallel to the xy plane (see diagram). Recall from Chapter 3 that the orthogonal xyz components of $\boldsymbol{C}_R$ equals the torque of the system about these axes. Hence, by setting $\sum_i (M_x)_i = \sum_i (M_y)_i = 0$, we are ensuring that $\boldsymbol{C}_R = \mathbf{0}$.

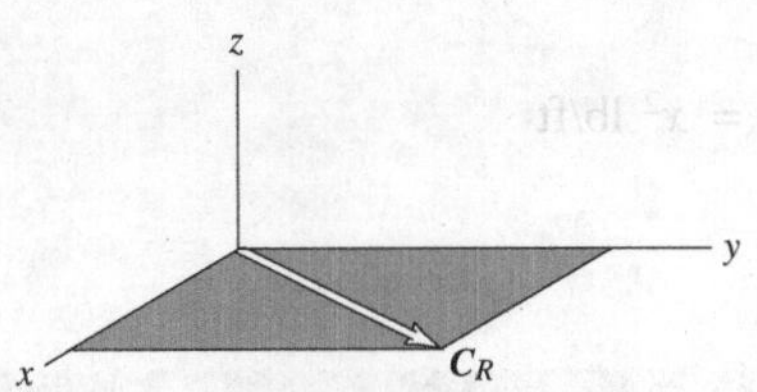

Example 5.8

Determine the forces required to support the uniform beam in Fig. 5.30 shown loaded with a couple, a point force, and a downward parabolic distribution of load having zero slope at the origin. The weight of the beam is 100 lb.

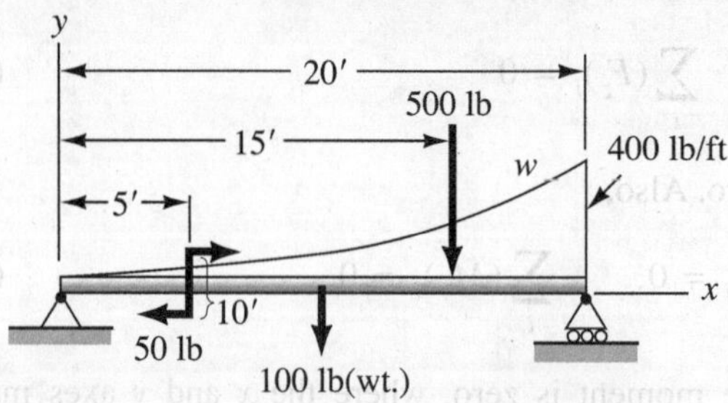

Figure 5.30. Find supporting forces.

Since a couple can be rotated without affecting the equilibrium of the body, we can orient the couple so that the forces are vertical. Accordingly, we have here a beam loaded by a system of parallel coplanar loads. Clearly, the supporting forces must be vertical, as shown in Fig. 5.31, where we have a free-body diagram of the beam. Since there are only two unknown quantities, we can handle the problem by statical consideration of this free body.

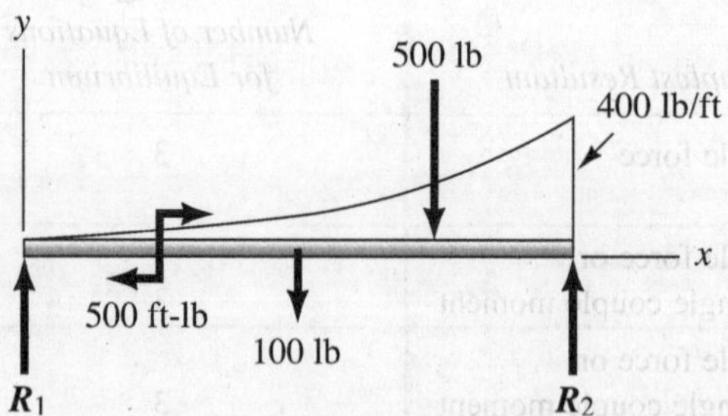

Figure 5.31. Free-body diagram.

The equation for the loading curve must be $w = ax^2 + b$, where a and b are to be determined from the loading data and the choice of reference. With an xy reference at the left end, as shown, we then have the conditions:

1. When $x = 0, w = 0$.
2. When $x = 20, w = 400$.

To satisfy these conditions, b must be zero and a must be unity; the loading function is thus given as

$$w = x^2 \text{ lb/ft}$$

Example 5.8 (Continued)

In this problem, we shall again work directly with the scalar equations. By summing moments about the left and right ends of the beam, we can then solve for the unknowns directly:

$\sum M_1 = 0$:

$$-500 - (10)(100) - (15)(500) - \int_0^{20} xy\, dx + 20R_2 = 0$$

Replacing y by x^2, then integrating and canceling terms, we get

$$-9{,}000 - \left.\frac{x^4}{4}\right|_0^{20} + 20R_2 = 0$$

By inserting limits and solving, we get one of the unknowns:

$$R_2 = 2{,}450 \text{ lb}$$

Next,

$\sum M_2 = 0$:

$$-20R_1 - 500 + (10)(100) + (5)(500) + \int_0^{20} (20 - x)y\, dx = 0$$

Replacing y by x^2, integrating and then solving for R_1, we have

$$R_1 = 817 \text{ lb}$$

As a check on these computations, we can sum forces in the vertical direction. The result must be zero (or as close to zero as the accuracy of our calculations permits):

$\sum F_y = 0$:

$$R_1 + R_2 - 100 - 500 - \int_0^{20} x^2\, dx = 0$$

$$3{,}267 - 600 - \left.\frac{x^3}{3}\right|_0^{20} = 0$$

Therefore,

$$2{,}667 - 2{,}667 = 0$$

Always take the opportunity to check a solution in this manner (i.e., by using a redundant equilibrium equation). In later problems, we shall rely heavily on calculated reactions (supporting forces); thus, we must make sure they are correct.

Example 5.9

In Fig. 5.32, find the supporting forces at *A*, *B*, and *D*. Note the pin connection at *C*. Also note that at *E* we have a welded connection.

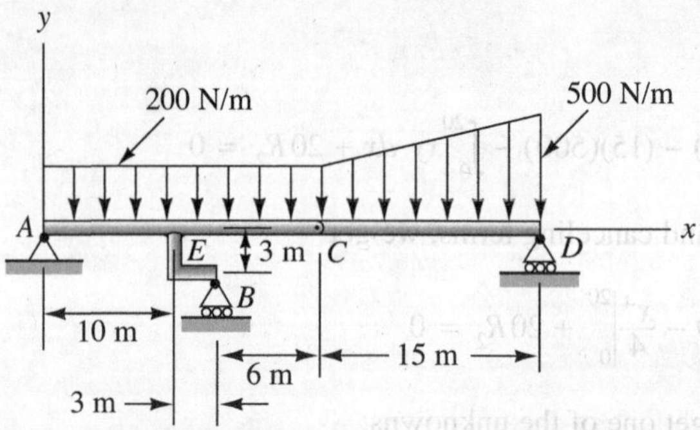

Figure 5.32. Members *AC* and *CD* are pinned at *C*.

The free-body diagram for the entire system is shown in Fig. 5.33. We have a coplanar system of forces for this free body and so we have only three independent equations of equilibrium. However we have here four unknowns. One of the unknowns, namely A_x, can be seen by inspection to be zero leaving now a coplanar parallel system with three unknowns but with only two equations of equilibrium.

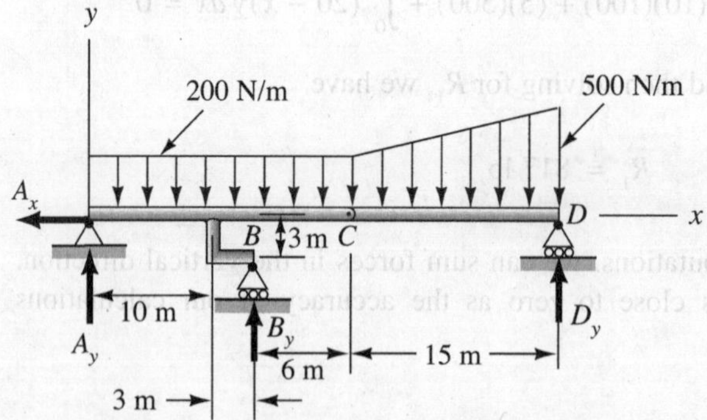

Figure 5.33. Free-body diagram I. (F.B.D. I)

We shall next consider the free body of member *CD*. This is shown in Fig. 5.34 where we have simplified the distributed loading in order to better facilitate the ensuing computations. Clearly, C_x must be zero. Thus, for F.B.D. II in Fig. 5.34, we will then have only two unknowns for which we have two equations of equilibrium. By taking moments about point *C* in Fig. 5.34, we can directly determine D_y.

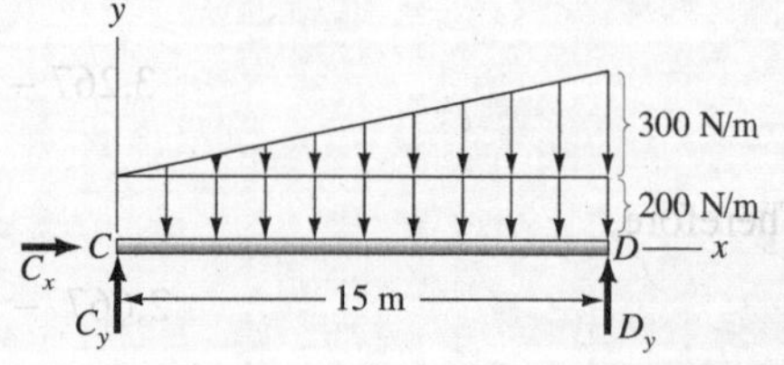

Figure 5.34. Free-body diagram II. (F.B.D. II)

Example 5.9 (Continued)

$\sum M_C = 0$:

$$(D_y)(15) - (200)(15)\left(\frac{15}{2}\right) - \left(\frac{1}{2}\right)(15)(300)\left[\left(\frac{2}{3}\right)(15)\right] = 0$$

$$D_y = 3{,}000 \text{ N}$$

We may now go back to Fig. 5.33 to solve for the remaining two unknowns. Thus, simplifying the loading between C and D as we have done in Fig. 5.34, we have

$\sum F_y = 0$:

$$A_y + B_y + 3{,}000 - (200)(34) - \frac{1}{2}(300)(15) = 0 \qquad \text{(a)}$$

$\sum M_B = 0$:

$$-A_y(13) + (3{,}000)(21) - (200)(34)\left(\frac{34}{2} - 13\right) \qquad \text{(b)}$$

$$-\frac{1}{2}(300)(15)\left[6 + \left(\frac{2}{3}\right)(15)\right] = 0$$

From Eq. (b), we get

$$A_y = -15.38 \text{ N} \qquad \text{(c)}$$

And from Eq. (a), we have

$$B_y = 6{,}065 \text{ N}$$

Note that A_y was negative indicating that we had inserted the wrong sense for this force in F.B.D. I. We did not make any changes in the ensuing calculations while going to Eq. (a) (valid for F.B.D. I) to calculate B_y (i.e., we used the result from Eq. (c) with the negative sign of A_y intact).

Also, if you were tempted in F.B.D. I to include the force C_y at pin C in this you were flirting with the prime error in the statics of rigid bodies—namely, including a force which is *internal* to the particular *free body* drawn.

Case D. General Force Systems. The simplest resultant in the general case is a force and a couple moment. Six equations of equilibrium can be given for each free-body diagram. We now examine two examples for this case.

Example 5.10

A derrick is shown in Fig. 5.35 supporting a 1,000-lb load. The vertical beam has a ball-and-socket connection into the ground at d and is held by guy wires ac and bc. Neglect the weight of the members and guy wires, and find the tensions in the guy wires ac, bc, and ce.

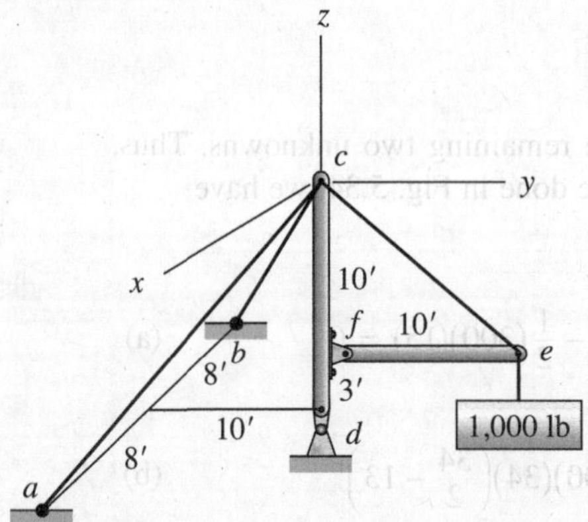

Figure 5.35. Loaded derrick.

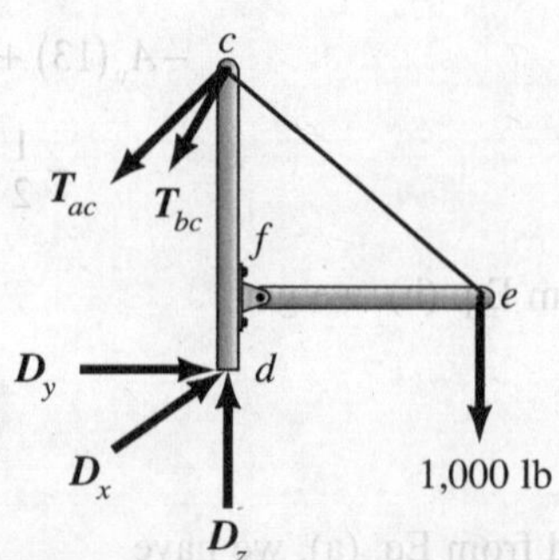

Figure 5.36. Free-body diagram 1.

If we select as a free body both members and the interconnecting guy wire ce, we shall expose two of the desired unknowns (Fig. 5.36). Note that this is a general three-dimensional force system with only five unknowns.[6] Although all these unknowns can be solved by statical considerations of this free body, you will notice that, if we take moments about point d, we will involve in a vector equation only the desired unknowns T_{bc} and T_{ac}. Accordingly, all unknown forces need not be computed for this free-body diagram. You should always look for such short cuts in situations such as this.

To determine the unknown tension T_{ce}, we must employ another free-body diagram. Either the vertical or horizontal member will expose this unknown in a manner susceptible to solution. The latter has been selected and is shown in Fig. 5.37. Note that we have here a coplanar force system with three unknowns. Again, you can see that, by taking moments about point f, we will involve only the desired unknown.

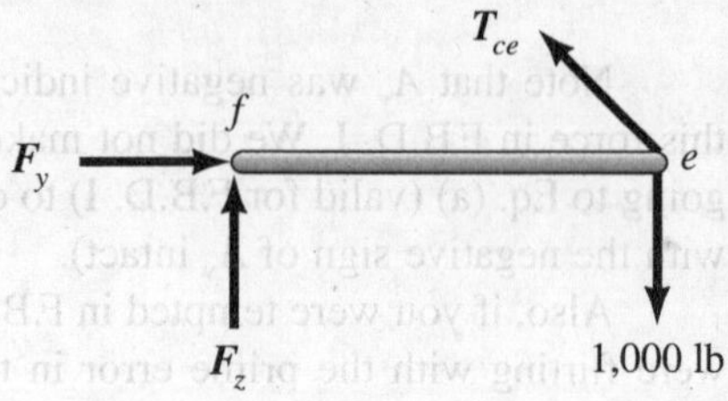

Figure 5.37. Free-body diagram 2.

[6]It should be clear on inspection of Fig. 5.35 that, due to symmetry, forces in the two supporting cables must be equal. Using this information is tantamount to using one of the equations of equilibrium. However, for practice we will not use this information and we will solve for both of these cables and demonstrate their equality. Also, note that there is a sixth equation of equilibrium that is identically satisfied. To see this, look at moments of the forces about axis cd in Fig. 5.36. Why is the total moment identically equal to zero about this axis, thus denying us an equation to help solve for unknowns? Is the derrick completely constrained? Explain.

Example 5.10 (Continued)

The vector $\boldsymbol{T}_{ac}$ may then be given as

$$\boldsymbol{T}_{ac} = T_{ac}\left[\frac{1}{\sqrt{333}}(8\boldsymbol{i} - 10\boldsymbol{j} - 13\boldsymbol{k})\right] \quad \text{(a)}$$

Similarly, we have for $\boldsymbol{T}_{bc}$,

$$\boldsymbol{T}_{bc} = T_{bc}\left[\frac{1}{\sqrt{333}}(-8\boldsymbol{i} - 10\boldsymbol{j} - 13\boldsymbol{k})\right] \quad \text{(b)}$$

Using the free-body diagram in Fig. 5.36, we now set the sum of moments about point d equal to zero. Thus, employing the relations above, we get

$$13\boldsymbol{k} \times \frac{T_{ac}}{\sqrt{333}}(8\boldsymbol{i} - 10\boldsymbol{j} - 13\boldsymbol{k}) + 13\boldsymbol{k} \times \frac{T_{bc}}{\sqrt{333}}(-8\boldsymbol{i} - 10\boldsymbol{j} - 13\boldsymbol{k}) + 10\boldsymbol{j} \times (-1{,}000\boldsymbol{k}) = \boldsymbol{0} \quad \text{(c)}$$

When we make the substitution of variable $t_1 = T_{ac}/\sqrt{333}$ and $t_2 = T_{bc}/\sqrt{333}$, the preceding equation becomes

$$[130(t_1 + t_2) - 10{,}000]\boldsymbol{i} + [104(t_1 - t_2]\boldsymbol{j} = \boldsymbol{0} \quad \text{(d)}$$

The scalar equations,

$$130(t_1 + t_2) - 10{,}000 = 0$$
$$104(t_1 - t_2) = 0$$

can now be readily solved to give $t_1 = t_2 = 38.5$. Hence, we get

$$T_{ac} = 38.5\sqrt{333} = 702 \text{ lb and } T_{bc} = 38.5\sqrt{333} = 702 \text{ lb.}$$[7]

$$\therefore \quad T_{ac} = 702 \text{ lb} \qquad T_{bc} = 702 \text{ lb}$$

Turning finally to free-body diagram 2 in Fig. 5.37, we see that, in summing moments about f, the horizontal component of the tension T_{ce} has a zero-moment arm. Thus,

$$(10)(0.707)T_{ce} - (10)(1{,}000) = 0$$

Hence,

$$T_{ce} = 1{,}414 \text{ lb}$$

[7]By taking moments in Fig. 5.36 about the line connecting points a and d (see Fig. 5.35), we could get T_{bc} directly using the scalar triple product. We suggest that you try this.

Example 5.11

A blimp is shown in Fig. 5.38 fixed at the mooring tower D by a ball-joint connection, and held by cables AB and AC. The blimp has a mass of 1,500 kg. The simplest resultant force $\boldsymbol{F}$ from air pressure (including the effects of wind) is

$$\boldsymbol{F} = 17{,}500\boldsymbol{i} + 1{,}000\boldsymbol{j} + 1{,}500\boldsymbol{k} \text{ N}$$

at a position shown in the diagram. Compute the tension in the cables as well as the force transmitted to the ball joint at the top of the tower at D. Also, what force system is transmitted to the ground at G through the mooring tower? The tower weighs 5,000 N.

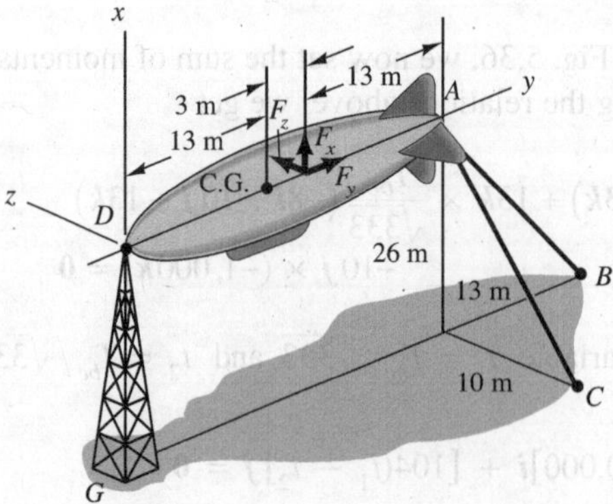

Figure 5.38. Tethered blimp.

We shall first consider a free-body diagram of the blimp, as shown in Fig. 5.39. We have five unknown forces here, and we can solve all of them by using equations of equilibrium for this free body.[8] As a first step, we express the cable tensions vectorially. That is,

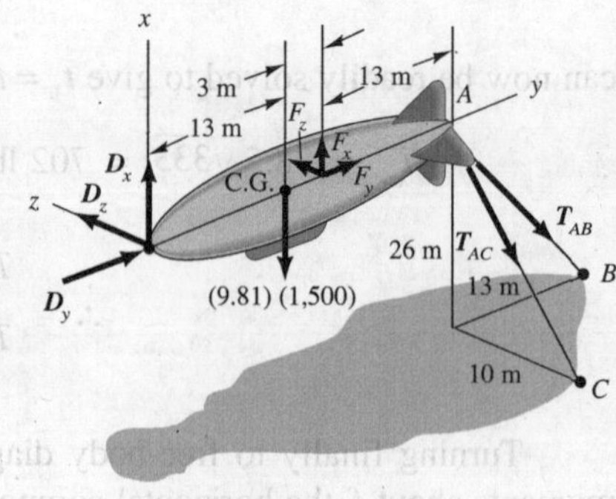

Figure 5.39. Free-body diagram of blimp.

$$\boldsymbol{T}_{AC} = T_{AC}\left(\frac{-26\boldsymbol{i} - 10\boldsymbol{k}}{\sqrt{26^2 + 10^2}}\right) = T_{AC}(-.933\boldsymbol{i} - .359\boldsymbol{k})$$

$$\boldsymbol{T}_{AB} = T_{AB}\left(\frac{-26\boldsymbol{i} + 13\boldsymbol{j}}{\sqrt{26^2 + 13^2}}\right) = T_{AB}(-.894\boldsymbol{i} + .447\boldsymbol{j})$$

We now go back to the basic vector equation of equilibrium. Thus

$$\sum \boldsymbol{F}_i = \boldsymbol{0}:$$

$$D_x\boldsymbol{i} + D_y\boldsymbol{j} + D_z\boldsymbol{k} - (1{,}500)(9.81)\boldsymbol{i} + 17{,}500\boldsymbol{i} + 1{,}000\boldsymbol{j} + 1{,}500\boldsymbol{k} + T_{AC}(-.933\boldsymbol{i} - .359\boldsymbol{k}) + T_{AB}(-.894\boldsymbol{i} + .447\boldsymbol{j}) = \boldsymbol{0}$$

[8]What equation is identically satisfied (thus forming the sixth equation of equilibrium) and thus not helpful in solving for the desired unknowns? What does this say about how the blimp can begin to move with the constraints present? See footnote 6 for the previous example.

Example 5.11 (Continued)

The scalar equations are

$$D_x + 2{,}785 - .933T_{AC} - .894T_{AB} = 0 \qquad \text{(a)}$$

$$D_y + 1{,}000 + .447T_{AB} = 0 \qquad \text{(b)}$$

$$D_z + 1{,}500 - .359T_{AC} = 0 \qquad \text{(c)}$$

Next take moments about point D.

$$\underline{\sum (\boldsymbol{M}_i)_D = \boldsymbol{0}:}$$

$$13\boldsymbol{j} \times (9.81)(1{,}500)(-\boldsymbol{i}) + 16\boldsymbol{j} \times (17{,}500\boldsymbol{i} + 1{,}000\boldsymbol{j} + 1{,}500\boldsymbol{k}) + 29\boldsymbol{j} \times T_{AC}(-.933\boldsymbol{i} - .359\boldsymbol{k}) + 29\boldsymbol{j} \times T_{AB}(-.894\boldsymbol{i} + .447\boldsymbol{j}) = \boldsymbol{0}$$

Carrying out the various cross products, we end up only with $\boldsymbol{k}$ and $\boldsymbol{i}$ components, thus generating two scalar equations.[9] They are

$$-10.41T_{AC} + 24{,}000 = 0 \qquad \text{(d)}$$

$$25.9T_{AB} + 27.1T_{AC} - 88{,}700 = 0 \qquad \text{(e)}$$

We now have five independent equations for five unknowns. We can thus solve these equations simultaneously. From Eq. (d), we have

$$T_{AC} = 2{,}305 \text{ N}$$

From Eq. (e), we have

$$T_{AB} = 1{,}012 \text{ N}$$

From Eq. (c), we have

$$D_z = -673 \text{ N}$$

From Eq. (b), we have

$$D_y = -1{,}452 \text{ N}$$

[9]The third equation is $0 = 0$. That is, there are no moments about the y axis, because all forces pass through the y axis.

Example 5.11 (Continued)

From Eq. (a), we have

$$D_x = 270 \text{ N}$$

Next, consider the mooring tower as a free body (Fig. 5.40). Notice that in showing the forces at the ball joint D, we have taken into account *both* of the negative signs shown above for D_y and D_z as well as Newton's third law.

Again, using the basic vector equations of statics, we have on summing forces

$$-270\boldsymbol{i} + 1{,}452\boldsymbol{j} + 672\boldsymbol{k} + F_x\boldsymbol{i} + F_y\boldsymbol{j} + F_z\boldsymbol{k} - 5{,}000\boldsymbol{i} = \boldsymbol{0}$$

Hence,

$$F_x = 5{,}270 \text{ N}$$
$$F_y = -1{,}452 \text{ N}$$
$$F_z = -672 \text{ N}$$

Now take moments about the base at F. We get

$$26\boldsymbol{i} \times (-270\boldsymbol{i} + 1{,}452\boldsymbol{j} + 672\boldsymbol{k}) + M_x\boldsymbol{i} + M_y\boldsymbol{j} + M_z\boldsymbol{k} = \boldsymbol{0}$$

From this, we get

$$M_x = 0$$
$$M_y = 17{,}470 \text{ N-m}$$
$$M_z = -37{,}800 \text{ N-m}$$

We can conclude that, at the center of the base, the force system from the ground is

$$\boldsymbol{F} = 5{,}270\boldsymbol{i} - 1{,}452\boldsymbol{j} - 672\boldsymbol{k} \text{ N}$$

$$\boldsymbol{C} = 17{,}470\boldsymbol{j} - 37{,}800\boldsymbol{k} \text{ N-m}$$

The force system acting on the ground at the center of the base is the reaction to the system above. Thus,

$$\boldsymbol{F}_{\text{ground}} = -5{,}270\boldsymbol{i} + 1{,}452\boldsymbol{j} + 672\boldsymbol{k} \text{ N}$$
$$\boldsymbol{C}_{\text{ground}} = -17{,}470\boldsymbol{j} + 37{,}800\boldsymbol{k} \text{ N-m}$$

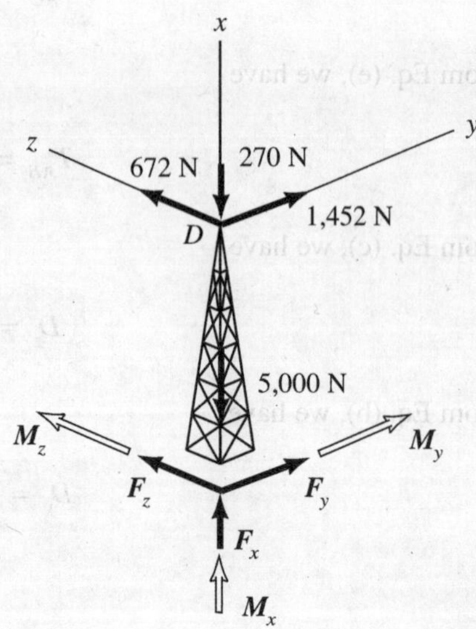

Figure 5.40. Free-body diagram of mooring tower.

PROBLEMS

5.58. The triple pulley sheave and the double pulley sheave weigh 15 lb and 10 lb, respectively. What rope force is necessary to lift a 350-lb engine? What is the force on the ceiling hook?

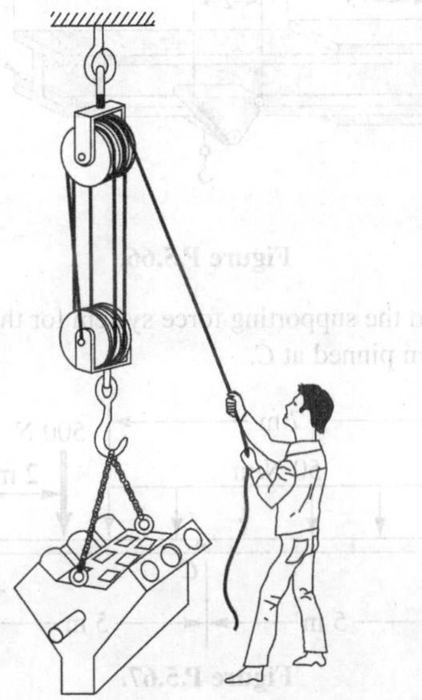

Figure P.5.58.

5.59. A multipurpose pry bar can be used to pull nails in the three positions. If a force of 400 lb is required to remove a nail and a carpenter can exert 50 lb, which position(s) must he use?

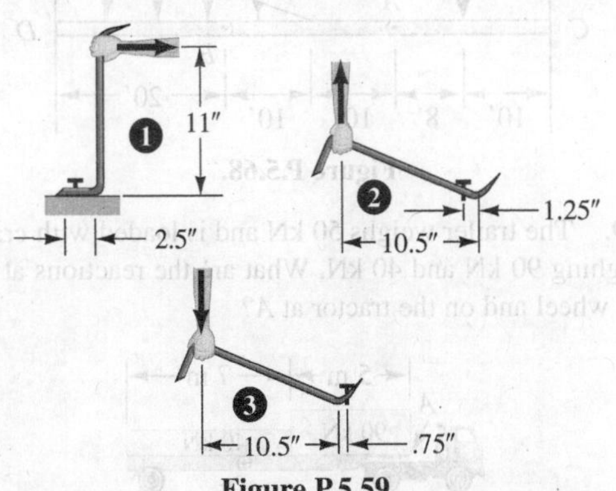

Figure P.5.59.

5.60. At what position must the operator of the counterweight crane locate the 50-kN counterweight when he lifts a 10-kN load of steel?

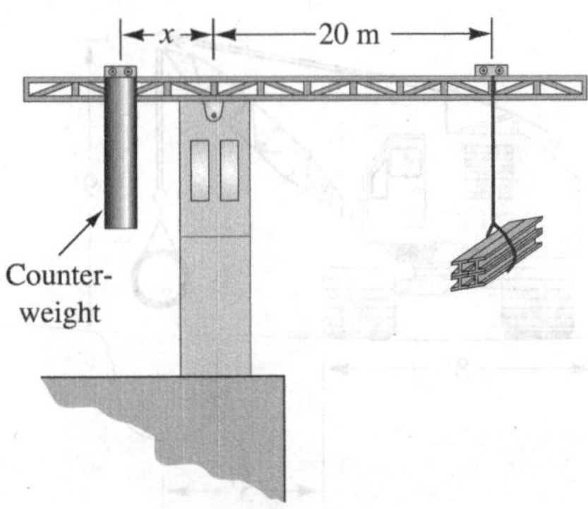

Figure P.5.60.

5.61. A Jeep winch is used to raise itself by a force of 2 kN. What are the reactions at the Jeep tires with and without the winch load? The driver weighs 800 N, and the Jeep weighs 11 kN. The center of gravity of the Jeep is shown.

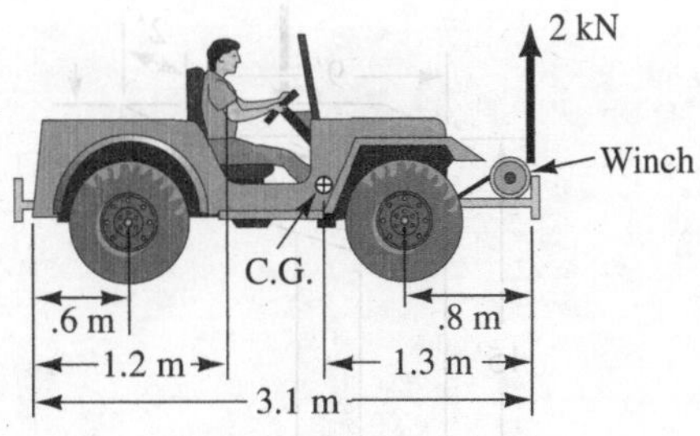

Figure P.5.61.

5.62. A *differential pulley* is shown. Compute F in terms of W, r_1, and r_2.

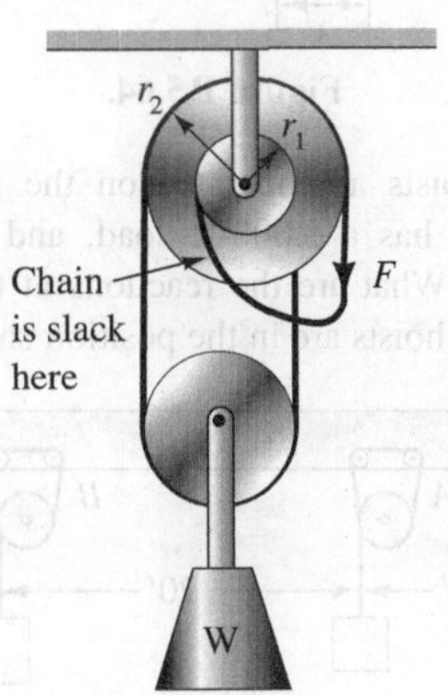

Figure P.5.62.

5.63. What is the longest portion of pipe weighing 400 lb/ft that can be lifted without tipping the 12,000-lb tractor? Take the center of gravity of the tractor at the geometric center.

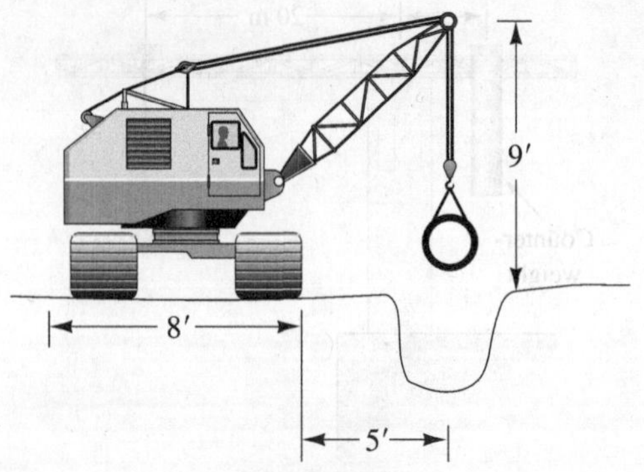

Figure P.5.63.

5.64. The L-shaped concrete post supports an elevated railroad. The concrete weighs 150 lb/ft³. What are the reactions at the base of the post?

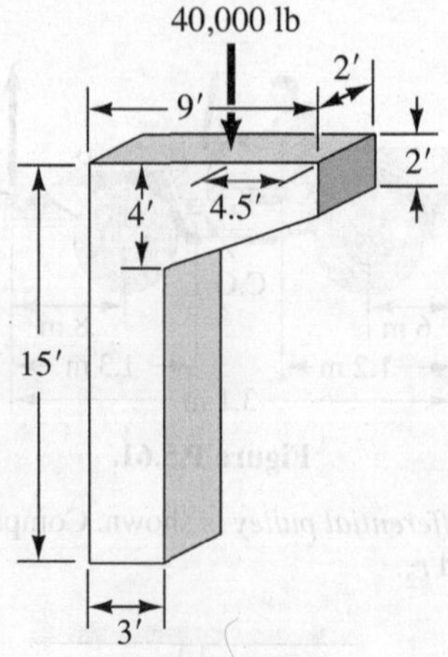

Figure P.5.64.

5.65. Two hoists are operated on the same overhead track. Hoist A has a 3,000-lb load, and hoist B has a 4,000-lb load. What are the reactions at the ends of the track when the hoists are in the position shown?

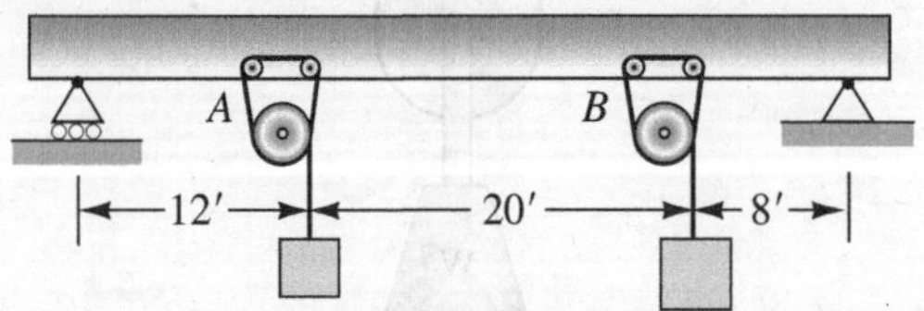

Figure P.5.65.

5.66. An I-beam cantilevered out from a wall weighs 30 lb/ft and supports a 300-lb hoist. Steel (487 lb/ft³) cover plates 1 in. thick are welded on the beam near the wall to increase the moment-carrying capacity of the beam. What are the reactions at the wall when a 400-lb load is hoisted at the outermost position of the hoist?

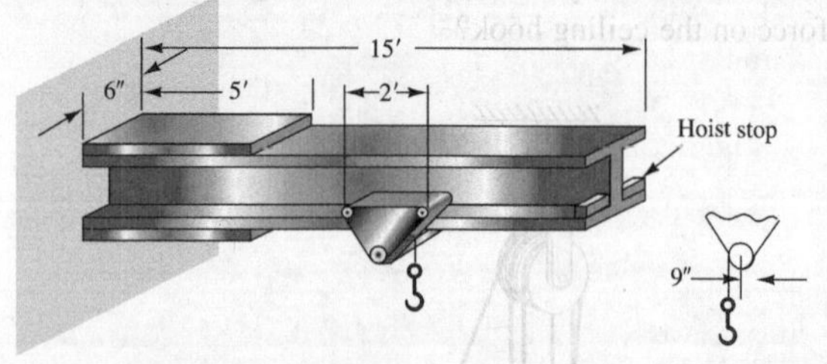

Figure P.5.66.

5.67. Find the supporting force system for the cantilever beam shown pinned at C.

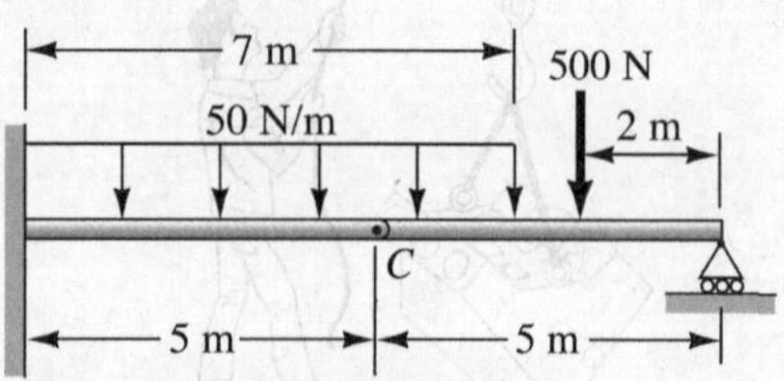

Figure P.5.67.

5.68. Find the supporting force system for the cantilever beams connected to bar AB by pins.

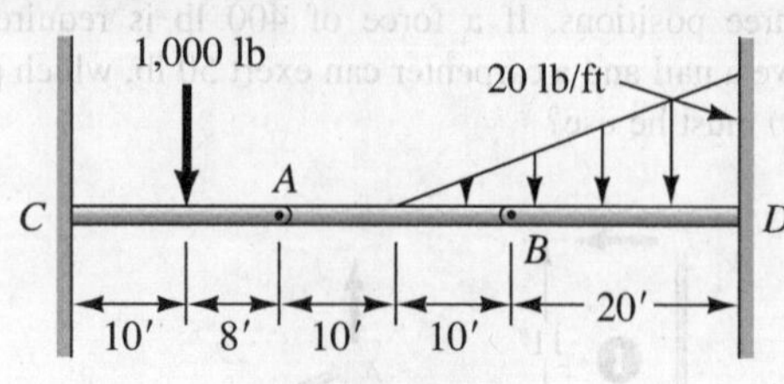

Figure P.5.68.

5.69. The trailer weighs 50 kN and is loaded with crates weighing 90 kN and 40 kN. What are the reactions at the rear wheel and on the tractor at A?

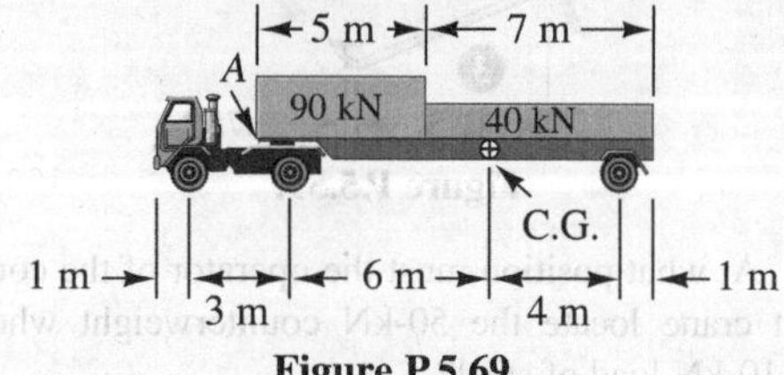

Figure P.5.69.

5.70. What load W will a pull P of 100 lb lift in the pulley system? Sheaves A, B, and C weigh 20 lb, 15 lb, and 30 lb, respectively. Assume first that the three sheaves are frictionless and find W. Then, calculate W that can be raised at constant speed for the case where the resisting torque in each of sheaves A and B is .01 times the total force at the bearing of each of sheaves A and B.

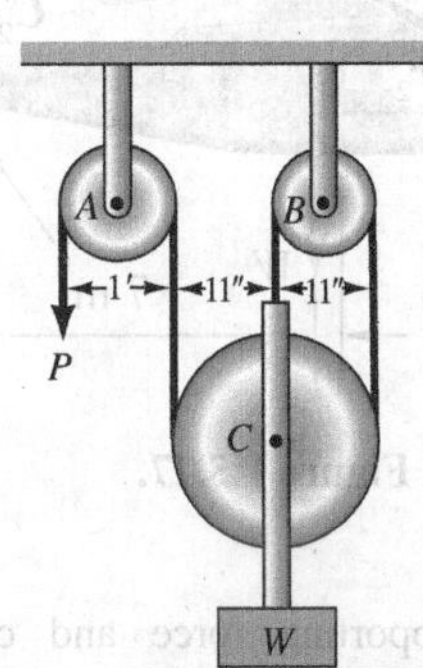

Figure P.5.70.

5.71. A piece of pop art is being developed. The weight of the body enclosed by the full lines is 2 kN. What is the smallest distance d that the artist can use for cutting a .5-m-diameter hole and still avoid tipping? The body is uniform in thickness.

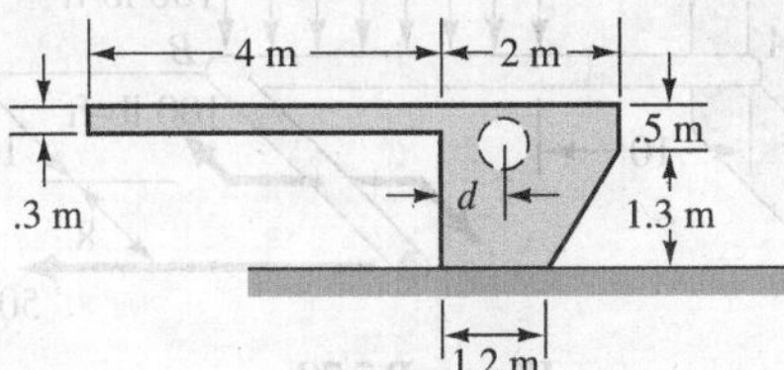

Figure P.5.71.

5.72. What is the largest weight W that the crane can lift without tipping? What are the supporting forces when the crane lifts this load? What is the force and couple-moment system transmitted through section C of the beam? Compute the force and couple-moment system transmitted through section D. The crane weighs 10 tons, having a center of gravity as shown in the diagram.

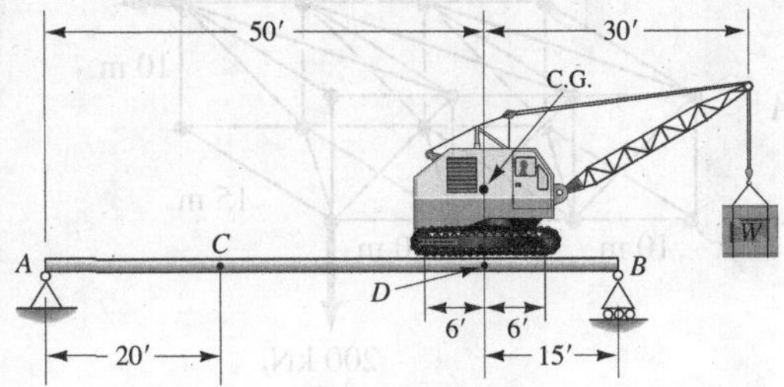

Figure P.5.72.

5.73. A 20-kN block is being raised at constant speed. If there is no friction in the three pulleys, what are forces F_1, F_2, and F_3 needed for the job? The block is not rotating in any way. The line of action of the weight vector passes through point C as shown.

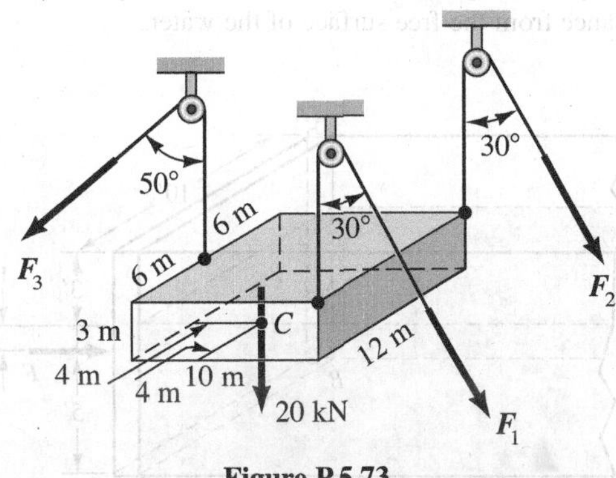

Figure P.5.73.

5.74. A 10-ton sounding rocket (used for exploring outer space) has a center of gravity shown as C.G.$_1$. It is mounted on a launcher whose weight is 50 tons with a center of gravity at C.G.$_2$. The launcher has three identical legs separated 120° from each other. Leg AB is in the same plane as the rocket and supporting arms CDE. What are the supporting forces from the ground? What torque is transmitted from the horizontal arm CD to the ramp ED by the rack and pinion at hinge D to counteract the weight of the rocket?

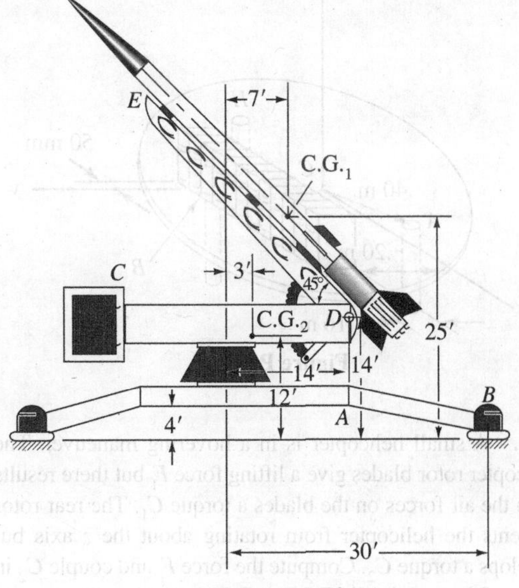

Figure P.5.74.

5.75. A door is hinged at A and B and contains water whose specific weight γ is 62.5 lb/ft^3. A force F normal to the door keeps the door closed. What are the forces on the hinges A and B and the force F to counteract the water? As noted in Chapter 4, the pressure in the water above atmosphere is given as γd, where d is the perpendicular distance from the free surface of the water.

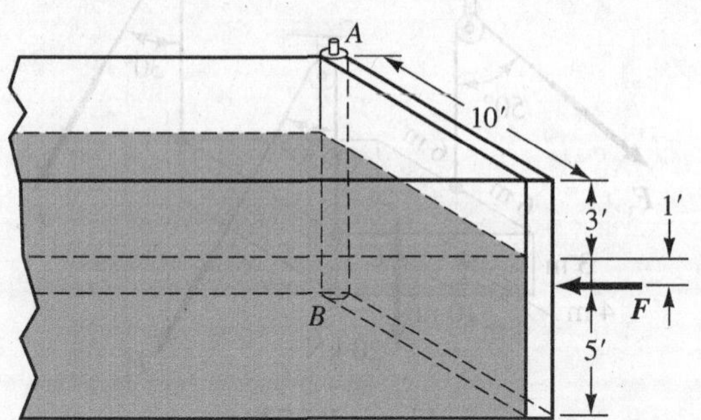

Figure P.5.75.

5.76. A row of books of length 750 mm and weighing 200 N sits on a three-legged table as shown. The legs are equidistant from each other with one leg B coinciding with the y axis. The other two legs lie along a line parallel to the x axis. If the table weighs 400 N, will it tip? If not, what are the forces on the legs?

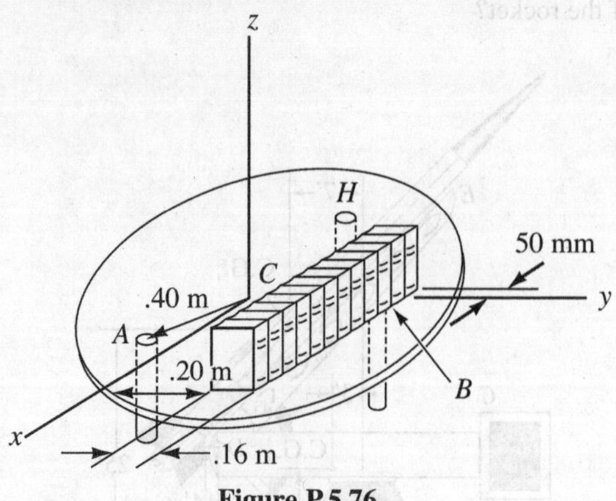

Figure P.5.76.

5.77. A small helicopter is in a hovering maneuver. The helicopter rotor blades give a lifting force F_1 but there results from the air forces on the blades a torque C_1. The rear rotor prevents the helicopter from rotating about the z axis but develops a torque C_2. Compute the force F_1 and couple C_2 in terms of the weight W. How are F_3 and C_1 related?

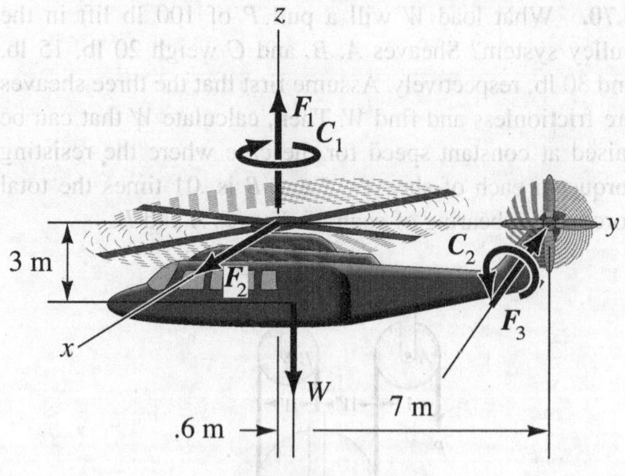

Figure P.5.77.

5.78. Find the supporting force and couple-moment system for the cantilever beam. What is the force and couple-moment system transmitted through a cross section of the beam at B?

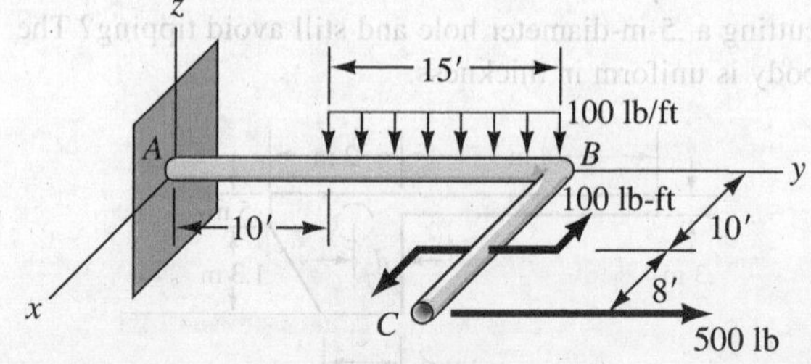

Figure P.5.78.

5.79. A structure is supported by a ball-and-socket joint at A, a pin connection at B offering no resistance in the direction AB, and a simple roller support at C. What are the supporting forces for the loads shown?

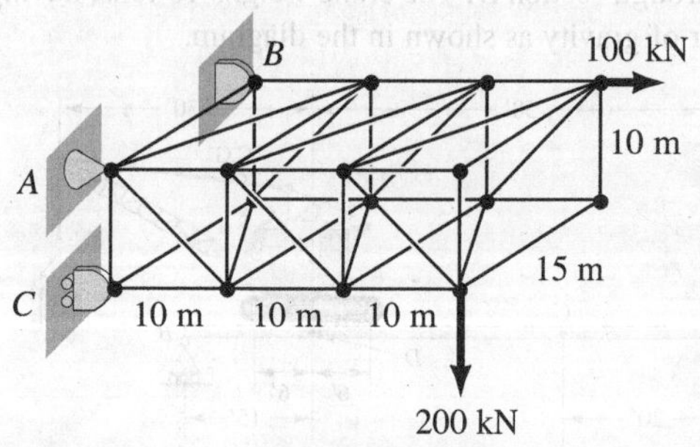

Figure P.5.79.

5.80. Compute the value of F to maintain the 200-lb weight shown. Assume that the bearings are frictionless, and determine the forces from the bearings on the shaft at A and B.

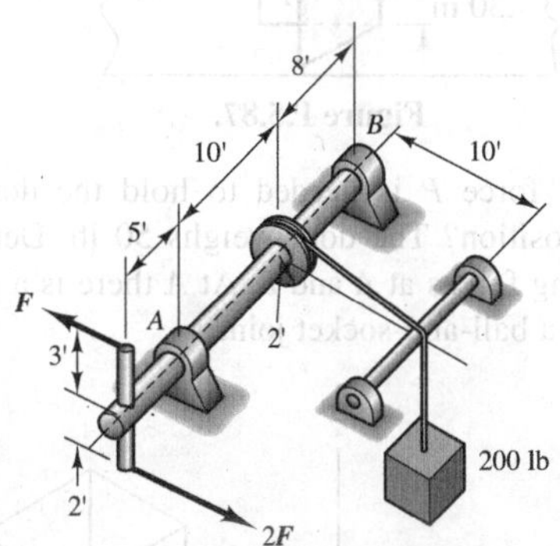

Figure P.5.80.

5.81. A bar with two right-angle bends supports a force $\boldsymbol{F}$ given as

$$\boldsymbol{F} = 10\boldsymbol{i} + 3\boldsymbol{j} + 100\boldsymbol{k} \text{ N}$$

If the bar has a weight of 10 N/m, what is the supporting force system at A?

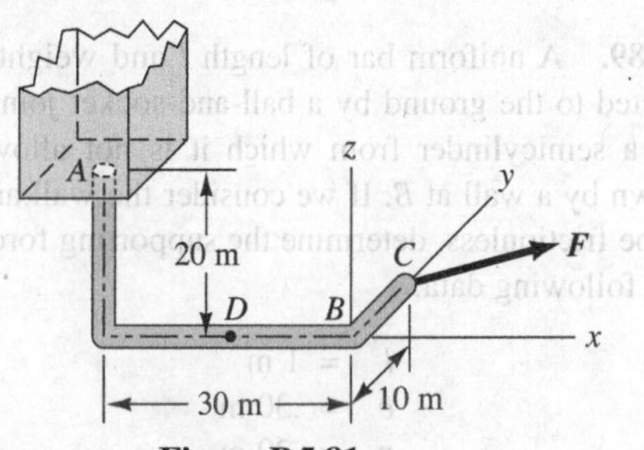

Figure P.5.81.

5.82. What is the resultant of the force system transmitted across the section at A? The couple is parallel to plane M.

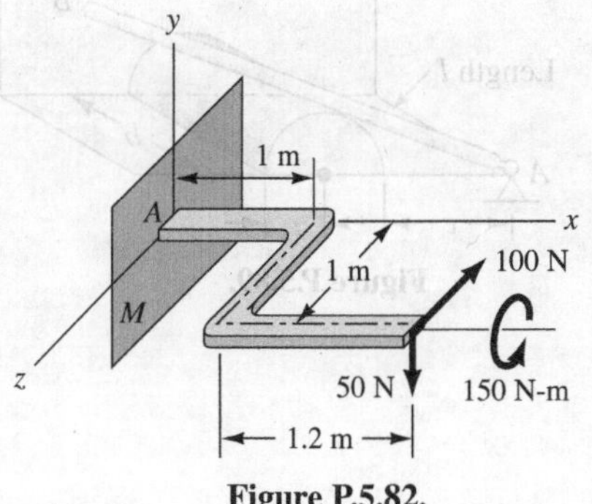

Figure P.5.82.

5.83. Determine the vertical force F that must be applied to the windlass to maintain the 100-lb weight. Also, determine the supporting forces from the bearings onto the shaft. The handle DE on which the force is applied is in the indicated xz plane.

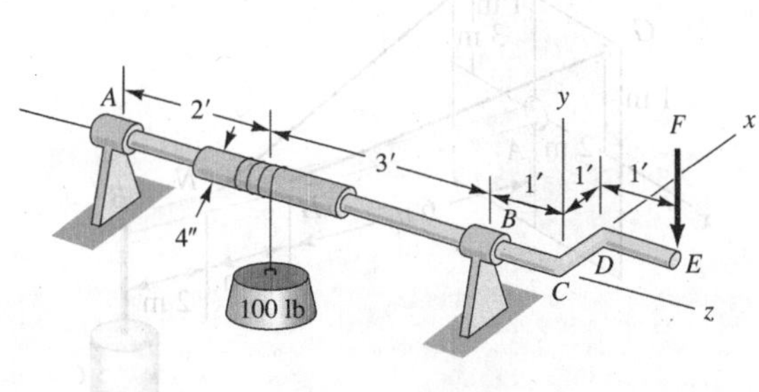

Figure P.5.83.

5.84. A transport plane has a gross weight of 70,000 lb with a center of gravity as shown. Wheels A and B are locked by the braking system while an engine is being tested under load prior to take off. A thrust T of 3,000 lb is developed by this engine. What are the supporting forces?

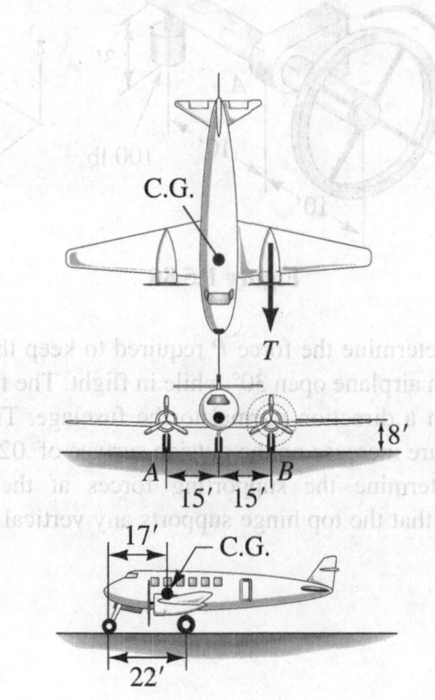

Figure P.5.84.

5.85. Two cables GH and KN support a rod AB which connects to a ball-and-socket joint support at A and supports a 500-kg body C at B. What are the tensions in the cable and the supporting forces at A?

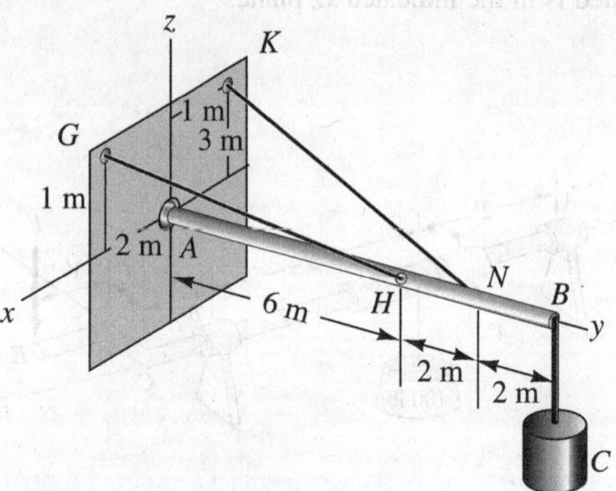

Figure P.5.85.

5.86. What change in elevation for the 100-lb weight will a couple of 300 lb-ft support if we neglect friction in the bearings at A and B? Also, determine the supporting force components at the bearings for this configuration.

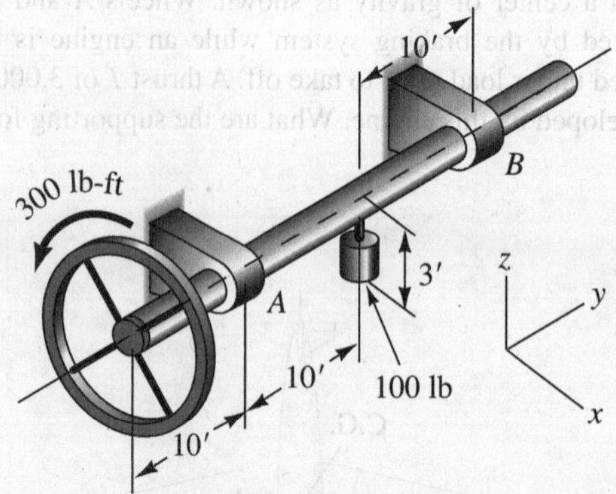

Figure P.5.86.

5.87. Determine the force P required to keep the 150-N door of an airplane open 30° while in flight. The force P is exerted in a direction normal to the fuselage. There is a net pressure increase on the outside surface of .02 N/mm^2. Also, determine the supporting forces at the hinges. Consider that the top hinge supports any vertical force on the door.

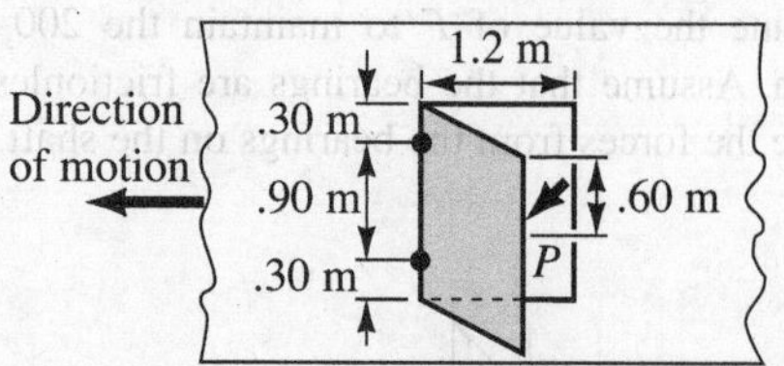

Figure P.5.87.

5.88. What force P is needed to hold the door in a horizontal position? The door weighs 50 lb. Determine the supporting forces at A and B. At A there is a pin and at B there is a ball-and-socket joint.

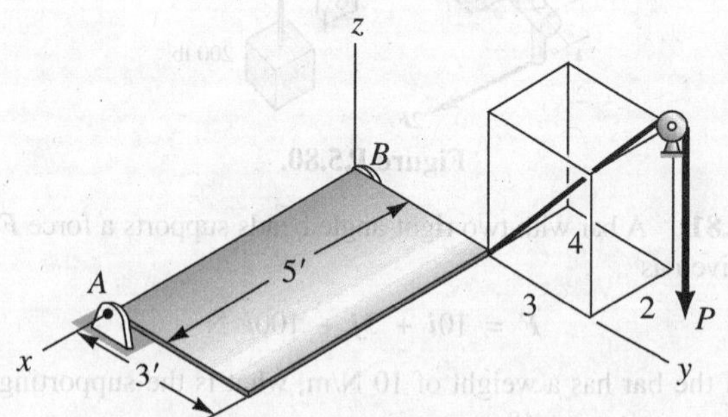

Figure P.5.88.

***5.89.** A uniform bar of length l and weight W is connected to the ground by a ball-and-socket joint, and rests on a semicylinder from which it is not allowed to slip down by a wall at B. If we consider the wall and cylinder to be frictionless, determine the supporting forces at A for the following data:

$$l = 1 \text{ m}$$
$$c = .30 \text{ m}$$
$$r = .20 \text{ m}$$
$$b = .40 \text{ m}$$
$$W = 100 \text{ N}$$

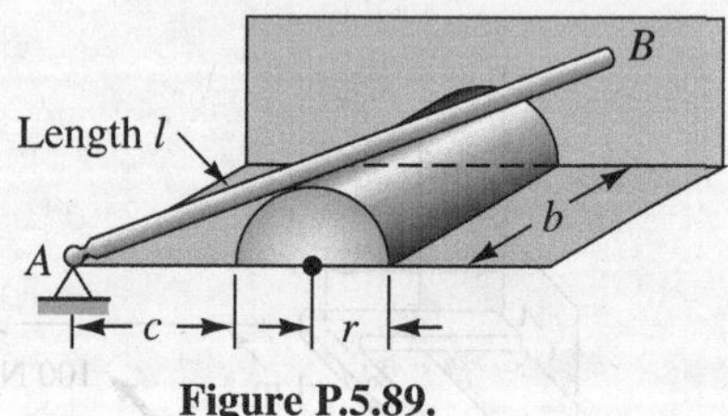

Figure P.5.89.

5.7 Two Point Equivalent Loadings

We shall now consider a simple case of equilibrium that occurs often and from which simple useful conclusions may readily be drawn.

Consider a rigid body on which the *equivalent* systems of two separate forces are respectively applied at two points *a* and *b* as shown in Fig. 5.41. If the body is in equilibrium, the first basic equation of statics, 5.1(a), stipulates that $\mathbf{F}_1 = -\mathbf{F}_2$; that is, the forces must be *equal and opposite*. The second fundamental equation of statics, 5.1(b), requires that $\mathbf{C} = \mathbf{0}$, indicating that the forces be *collinear* so as not to form a nonzero couple. With points *a* and *b* given as points of application for the two forces in Fig. 5.41, clearly the *common line of action for the forces must coincide with the line segment ab.* Such bodies, where there are only *two points of loading*, are sometimes called "two-force" members. Such members occur often in structural mechanics problems. Furthermore, there is considerable saving of time and labor if the student recognizes them at the outset of any problem.

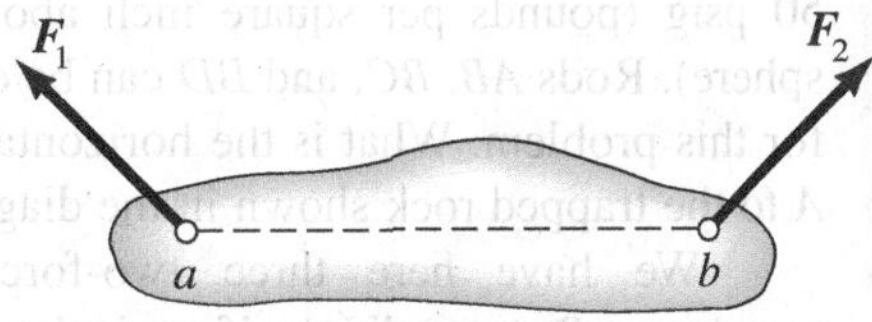

Figure 5.41. Two-force member.

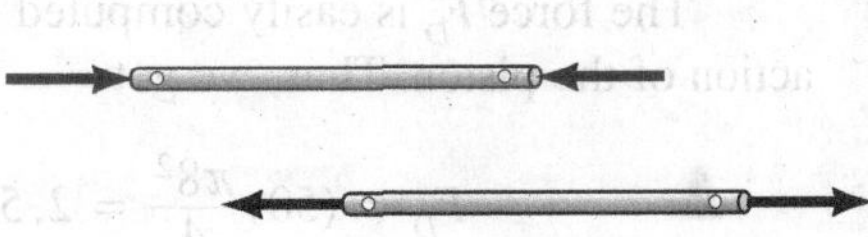
Figure 5.42. Compression and tension members.

We often have to deal with pin-connected structural members with loads applied at the pins. If we neglect friction at the pins and also the weight of the members, we can conclude that the equivalent of only two forces act on each member. These forces, then, must be equal and opposite and must have lines of action that are collinear, with the line joining the points of application of the forces. If the member is straight (see Fig. 5.42), the common line of action of the two forces coincides with the centerline of the member. The top member in Fig. 5.42 is a *compression* member, the one below a *tensile* member. Note that the bent member in Fig. 5.43, if weightless, is also a two point loaded member. The line of action of the forces must coincide with the line *ab* connecting the two points of loading. However, the beam in Fig. 5.44 is not a two point loaded member since at the left end there will be a couple moment. And, clearly, such a loading must entail two points of loading to accommodate the two equal and opposite forces comprising the couple. There are then in effect three points of loading for this member. Accordingly, use caution here when dealing with a cantilever beam.

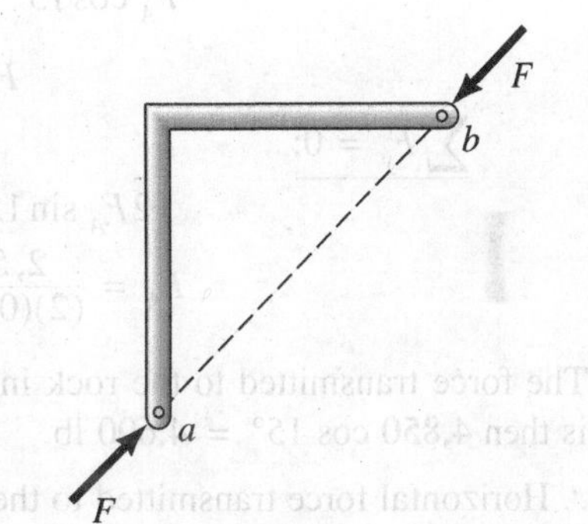

Figure 5.43. Line of action of *F* collinear with *ab*.

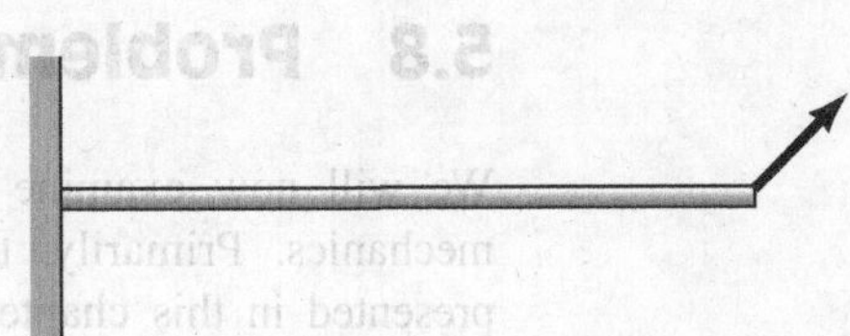
Figure 5.44. Cantilever beam is not an example of a two point loaded member.

Before considering an example, it should be emphasized that the forces F_1 and F_2 in Fig. 5.41 may be the resultants of systems of concurrent forces at *a* and *b*, respectively. Since concurrent forces are always equivalent to their resultant at the point of concurrency, the member in Fig. 5.41 is still a two-force member with the resulting restrictions on the resultants F_1 and F_2.

Example 5.12

A device for crushing rocks is shown in Fig. 5.45. A piston D having an 8-in. diameter is activated by a pressure p of 50 psig (pounds per square inch above that of the atmosphere). Rods AB, BC, and BD can be considered weightless for this problem. What is the horizontal force transmitted at A to the trapped rock shown in the diagram?

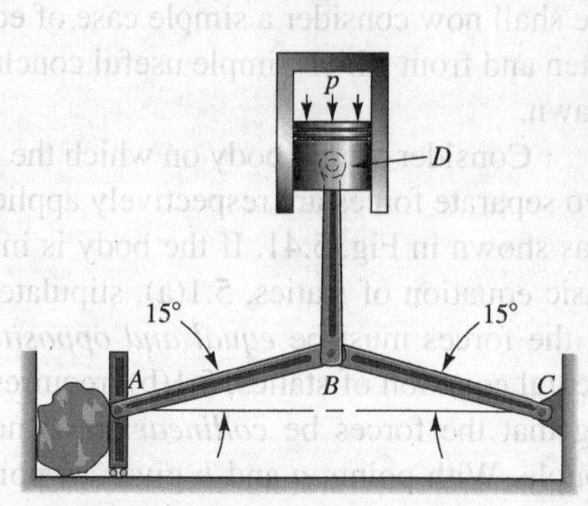

Figure 5.45. Rock crusher.

We have here three two-force members coming together at B. Accordingly, if we isolate pin B as a free body, we will have three forces acting on the pin. These forces must be collinear with the centerlines of the respective members, as explained earlier (Fig. 5.42).

The force F_D is easily computed by considering the action of the piston. Thus, we get

$$F_D = (50)\frac{\pi 8^2}{4} = 2,510 \text{ lb}$$

Summing forces at pin B:

$\sum F_x = 0$:

$$F_A \cos 15° - F_C \cos 15° = 0$$

$$F_A = F_C$$

$\sum F_y = 0$:

$$2F_A \sin 15° - 2,510 = 0$$

$$F_A = \frac{2,510}{(2)(0.259)} = 4,850 \text{ lb}$$

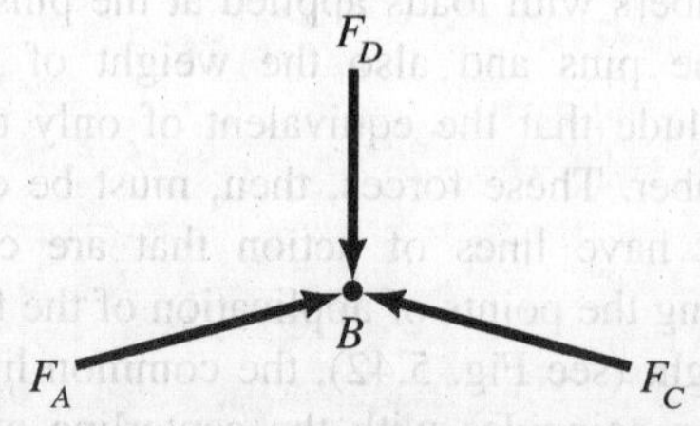

Figure 5.46. Free-body diagram of pin B.

The force transmitted to the rock in the horizontal direction is then 4,850 cos 15° = 4,690 lb.

∴ Horizontal force transmitted to the rock = **4,690 lb**

Of less direct use is the *three-force* theorem. It states that a system of three forces in equilibrium must be *coplanar* and either be *concurrent* or be *parallel*.[10]

5.8 Problems Arising From Structures

We will now examine some interesting problems that arise in studies of structural mechanics. Primarily, these problems involve procedures and theory that we have presented in this chapter with some simple additions, (which will be described in the examples to follow), and some use of freshman physics.

[10] To prove this, assume that two of the forces intersect at a point A. Show from the basic equations of equilibrium that the forces must be coplanar and concurrent. Now assume the forces do *not* intersect. Setting moments of the system equal to zero about two points along the line of action of one of the forces, show that the system must be coplanar. Now since the forces do not intersect, they must be parallel. The theorem will thus have been proved.

Example 5.13

Rod C shown in Fig. 5.47 is welded to a rigid drum A, which is rotating about its axis at a steady angular speed ω of 500 rpm. This rod has a mass per unit length w, which varies linearly from the base to the tip starting with the value of 20 kg/m at the base to 28 kg/m at the tip. If *normal stress* is defined as the normal force at a section divided by the area of the section (similar to pressure except that the force can be pulling away from the section rather than always pushing against the section), what is the normal stress at any cross section of the cylinder at a distance r from the centerline B–B of the drum due only to the motion?

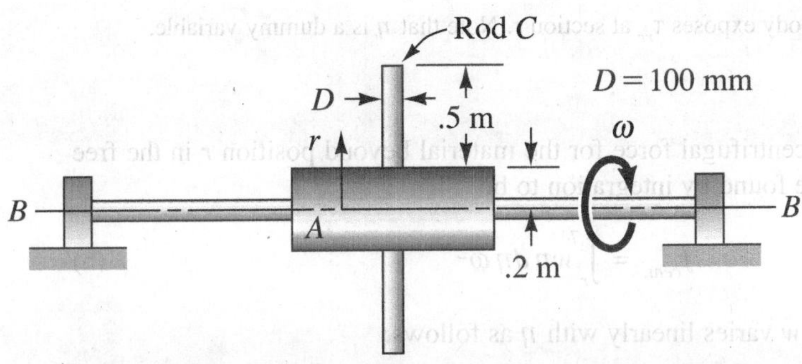

Figure 5.47. Rods attached to a rotating rigid drum.

First we expose a section of the rod at a distance r from B–B in a *free-body diagram* as shown in Fig. 5.48 in which we denote the normal stress acting on this section as τ_{rr}. The force from this section must restrain the centrifugal force stemming from the angular motion of the portion of the rod beyond position r. For this purpose, we consider an infinitesimal slice of the rod (see Fig. 5.48). We shall use the variable η to denote the distance to the slice from the centerline B–B. Furthermore, the thickness of the slice shall be $d\eta$. We shall use η to position slices between end position r of the free-body diagram and the tip of the rod. We are using this approach for bookkeeping purposes. Now you will recall from freshman physics that the centrifugal force on this slice is given by

$$df_{cent.} = (dm)(\eta)(\omega^2) = (w\ d\eta)(\eta)(\omega^2) \tag{a}$$

Example 5.13 (Continued)

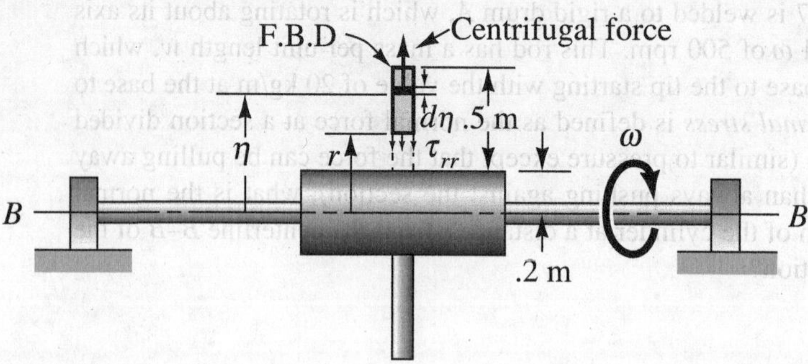

Figure 5.48. Free body exposes τ_{rr} at section r. Note that η is a dummy variable.

Consequently, the total centrifugal force for the material beyond position r in the free body of Fig. 5.48 will be found by integration to be

$$f_{cent.} = \int_r^{.7} w\eta \, d\eta \, \omega^2 \tag{b}$$

Next, note that the term w varies linearly with η as follows:

$$w = 20 + \frac{\eta - .2}{.5} 8 \text{ kg/m}$$

Clearly from this equation we see that when $\eta = .2$, we get $w = 20$ kg/m and when $\eta = .7$, we get $w = 28$ kg/m. Also, the variation is linear. Now, going back to Eq. (b), we have

$$f_{cent.} = \int_r^{.7} \left(20 + \frac{\eta - .2}{.5} 8\right) \eta \left(\frac{(500)(2\pi)}{60}\right)^2 d\eta$$

Integrating

$$\begin{aligned} f_{cent.} &= 2.742 \times 10^3 \left[(20 - 3.2) \frac{\eta^2}{2} + 16 \frac{\eta^3}{3} \right]_r^{.7} \\ &= 2.742 \times 10^3 \left[4.116 + 1.829 - 8.40r^2 - 5.333r^3 \right] \end{aligned}$$

We can give the desired stress τ_{rr} by dividing by the cross-sectional area $\pi D^2/4 = 7.854 \times 10^{-3}\ m^2$. We thus have the desired normal stress distribution for sections of the rod as follows:

$$\tau_{rr} = 3.491 \times 10^5 \, [5.945 - 8.40r^2 - 5.333r^3] \text{ N/m}^2$$

Example 5.14

Consider a *thin-walled* tank containing air at a pressure of 100 psi above that of the atmosphere [see Fig. 5.49(a)]. The outside diameter D of the tank is 2 ft and the wall thickness t is 1/4 in. We consider as a free body from the tank wall a vanishingly small element such as $ABCE$ in the diagram having the shape of a rectangular parallelepiped. What are the stresses on the cut surfaces of the element? Neglect the weight of the cylinder.

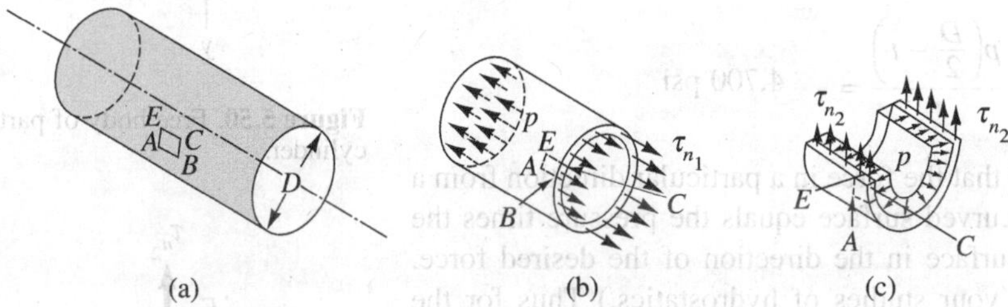

Figure 5.49. Thin-walled tank with inside gauge pressure p.

We can examine face BC by considering a free body of part of the tank, as shown in Fig. 5.49(b) exposing BC. (Note that we have not included the pressure on the inside wall). Because there is a net force from the air pressure only in the *axial* direction of the cylinder, we can expect normal stress τ_{n_1} (like pressure except that it is tensile rather than compressive) over the cut section of the cylinder, as has been indicated in the diagram. Furthermore, because the wall of the tank is *thin* compared to the diameter, we can assume that the stress τ_{n_1} is *uniform* across the thickness. Finally, for reasons of *axial symmetry* of geometry and loading, we can expect this stress to be uniform around the entire cross section. Now we may say from considerations of *equilibrium* in the axial direction that

$$\tau_{n_1}\pi\left[\frac{D^2}{4}-\frac{(D-2t)^2}{4}\right]-p\pi\frac{(D-2t)^2}{4}=0$$

$$\therefore \tau_{n_1}=\frac{p(D-2t)^2}{D^2-(D-2t)^2}=\frac{(100)(24-\frac{1}{2})^2}{24^2-23.5^2}=\boxed{2{,}325 \text{ psi}} \quad \text{(a)}$$

Hence, on face BC we have a uniform stress of 2,325 psi. Clearly, this must also be true for face AE. This is called an *axial stress*.

To expose face EC next, we consider a half-cylinder of unit length such as is shown in Fig. 5.49(c). Because it is *far* from the ends and because the wall is *thin*, we can assume as an approximation that the stress τ_{n_2} shown is uniform over the cut section.[11] A pressure p is shown acting normal to the inside wall surface. (The stress τ_{n_1} computed earlier is not shown, to avoid cluttering the diagram.) Now we consider

[11]Near the ends of the tank the stress distribution *varies* in value because of the proximity of the *complicated geometry* and the contributions toward equilibrium of the end plates.

Example 5.14 (Continued)

an end view of this body in Fig. 5.50 for equilibrium in the vertical direction. We have

$$2\left[(\tau_{n_2})(t)(1)\right] - \int_0^{\pi} p\left(\frac{D}{2} - t\right) d\theta(1) \sin\theta = 0 \quad \text{(b)}$$

$$\therefore \tau_{n_2} = \frac{1}{2t}\left[p\left(\frac{D}{2} - t\right)\right](-\cos\theta)\Big|_0^{\pi}$$

$$\tau_{n_2} = \frac{p\left(\frac{D}{2} - t\right)}{t} = \boxed{4{,}700 \text{ psi}}$$

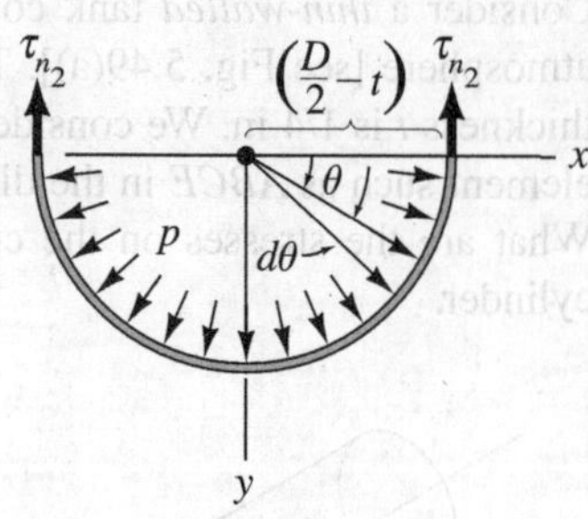

Figure 5.50. Free body of part of cylinder.

We point out now that the force in a particular direction from a uniform pressure on a curved surface equals the pressure times the projected area of this surface in the direction of the desired force. (You will learn this in your studies of hydrostatics.) Thus for the case at hand the projected area is that of a rectangle $1 \times (D - 2t)$, so that the second expression of Eq. (b) becomes $p(D - 2t)$. You may readily verify that this gives the same result as above.

The stress τ_{n_2} is called the *hoop stress*; it is about twice the *axial stress* τ_{n_1}. We show element *ABCE* with the stresses present in Fig. 5.51.

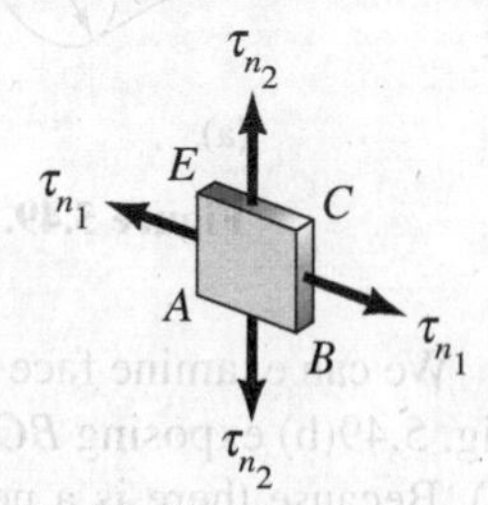

Figure 5.51. Free body of an element of the cylinder.

5.9 Static Indeterminacy

Examine the simple beam in Fig. 5.52, with known external loads and weight. If the deformation of the beam is small, and the final positions of the external loads after deformation differ only slightly from their initial positions, we can assume the beam to be rigid and, using the undeformed geometry, we can solve the supporting forces A, B_x, and B_y. This is possible since we have three equations of equilibrium available. Suppose, now, that an additional support is made available to the beam, as indicated in Fig. 5.53. The beam can still be considered a rigid body, since the applied load will shift even less because of deformation. Therefore, the resultant force coming from the ground to counteract the applied loads and weight of the beam must be the *same* as before. In the first case, in which two supports were given, however, a unique set of values for the forces A, B_x, and B_y gave us the required resultant. In other words, we were able to solve for these forces by statics alone, without further considerations. In the second case, rigid-body statics will give

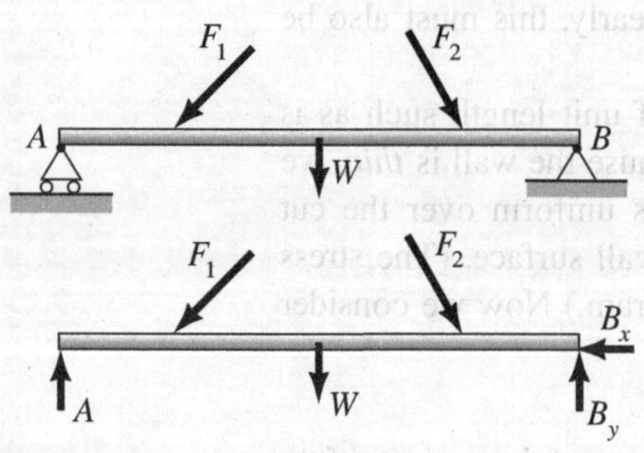

Figure 5.52. Statically determinate problem.

the required *same* resultant supporting force system, but now there are an infinite number of possible combinations of values of the supporting forces that will give us the resultant system demanded by equilibrium of rigid bodies. To decide on the proper combination of supporting forces requires additional computation. Although the deformation properties of the beam were unimportant up to this point, they now become the all-important criterion in apportioning the supporting forces. These problems are termed *statically indeterminate*, in contrast to the statically determinate type, in which statics and the rigid-body assumption suffice. For a given system of loads and masses, two models—the rigid-body model and models taught in other courses involving elastic behavior—are accordingly both employed to achieve a desired end. In summary:

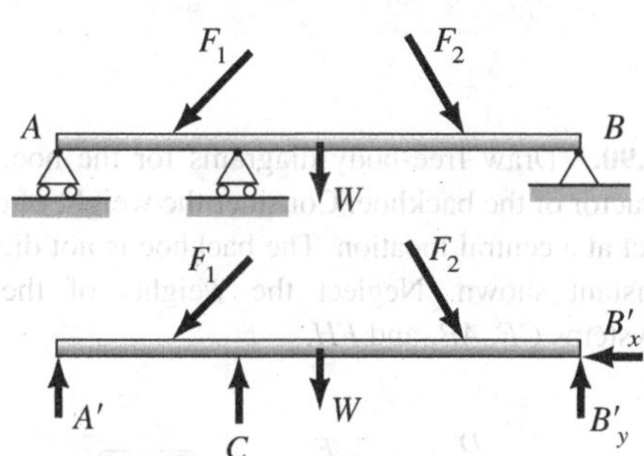

Figure 5.53. Statically indeterminate problem.

In statically indeterminate problems, we must satisfy both the equations of equilibrium for rigid bodies and the equations that stem from deformation considerations. In statically determinate problems, we need only satisfy the equations of equilibrium.

In the discussion thus far, we used a beam as the rigid body and discussed the statical determinacy of the supporting system. Clearly, the same conclusions apply to any structure that, without the aid of the external constraints, can be taken as a rigid body. If, for such a structure as a free body, there are as many unknown supporting force and couple-moment components as there are equations of equilibrium, and if these equations can be solved for these unknowns, we say that the structure is *externally statically determinate*.

On the other hand, should we desire to know the forces transmitted *between* internal members of this kind of structure (i.e., one that does not depend on the external constraints for rigidity), we then examine free bodies of these members. When all the unknown force and couple-moment components can be found by the equations of equilibrium for these free bodies, we then say that the structure is *internally statically determinate*.

There are structures that depend on the external constraints for rigidity (see the structure shown in Fig. 5.54). Mathematically speaking, we can say for such structures that the supporting force system always depends on both the internal forces and the external loads. (This is in contrast to the previous case, where the supporting forces could, for the externally statically determinate case, be related directly with the external loads without consideration of the internal forces.) In this case, we do not distinguish between internal and external statical determinacy, since the evaluation of supporting forces will involve free bodies of some or all of the internal members of the structure; hence, some or all of the internal forces and moments will be involved. For such cases, we simply state that the structure is statically determinate if, for all the unknown force and couple-moment components, we have enough equations of equilibrium that can be solved for these unknowns.

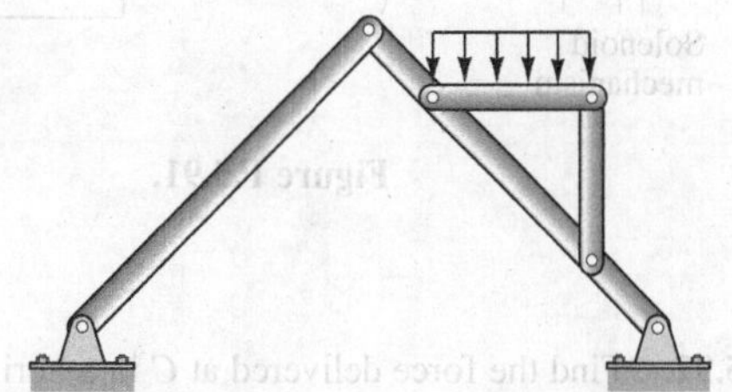
Figure 5.54. Nonrigid structure.

PROBLEMS

5.90. Draw free-body diagrams for the hoe, arms, and tractor of the backhoe. Consider the weight of each part to act at a central location. The backhoe is not digging at the instant shown. Neglect the weights of the hydraulic systems CE, AB, and FH.

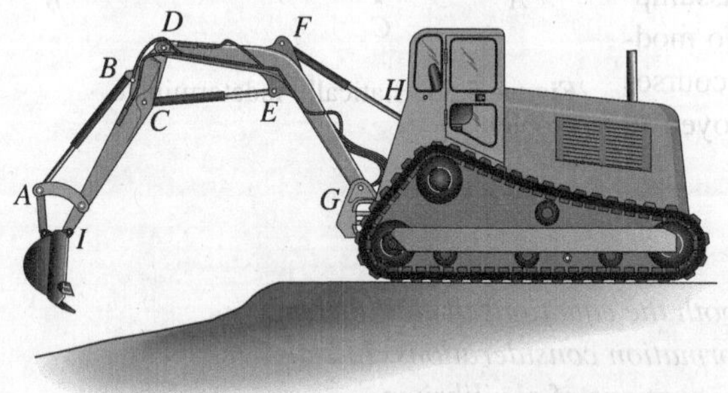

Figure P.5.90.

5.91. A parking-lot gate arm weighs 150 N. Because of the taper, the weight can be regarded as concentrated at a point 1.25 m from the pivot point. What force must be exerted by the solenoid to lift the gate? What solenoid force is necessary if a 300-N counterweight is placed .25 m to the left of the pivot point?

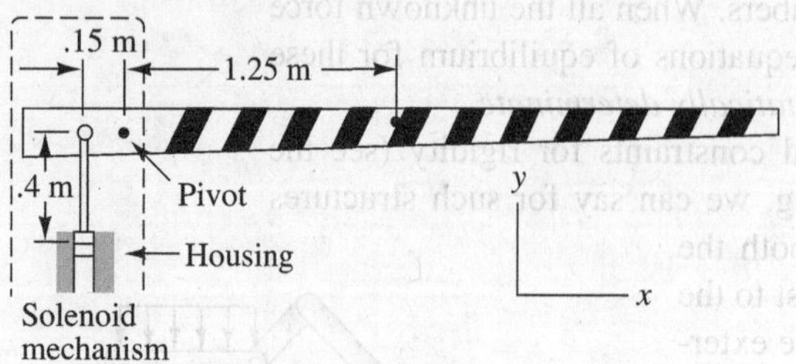

Figure P.5.91.

5.92. Find the force delivered at C in a horizontal direction to crush the rock. Pressure $p_1 = 100$ psig and $p_2 = 60$ psig (pressures measured above atmospheric pressure). The diameters of the pistons are 6 in. each. Neglect the weight of the rods.

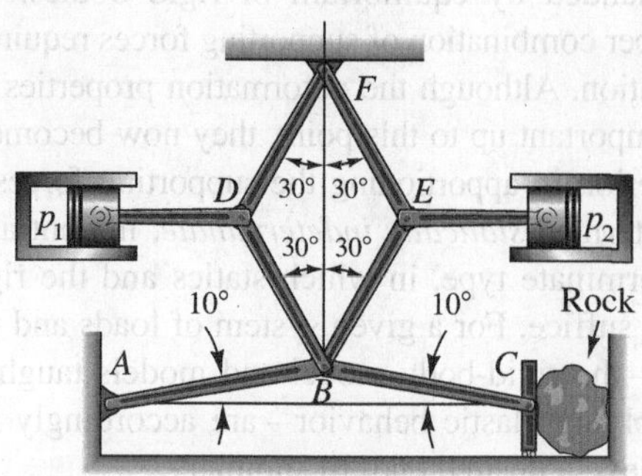

Figure P.5.92.

5.93. A Broyt X-20 digger carries a 20-kN load as shown. If hydraulic ram CB is normal to BA, where A is the axis of rotation for member E, find the force needed by ram CB. Do not consider the weights of the members.

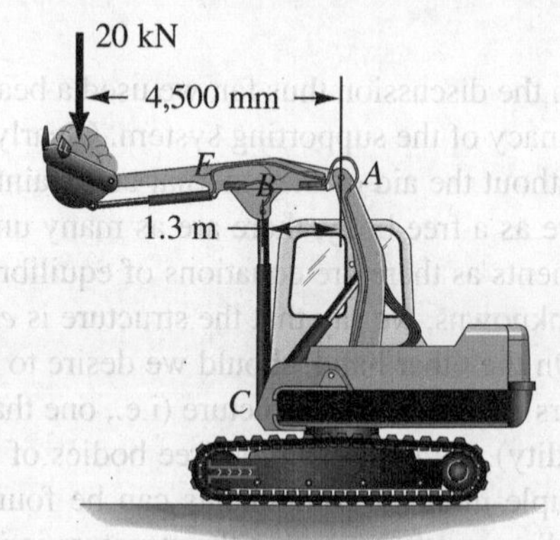

Figure P.5.93.

5.94. What force F_1 is needed for equilibrium? Neglect friction. The design is symmetrical about vertical axis. Horizontal members are pinned at E.

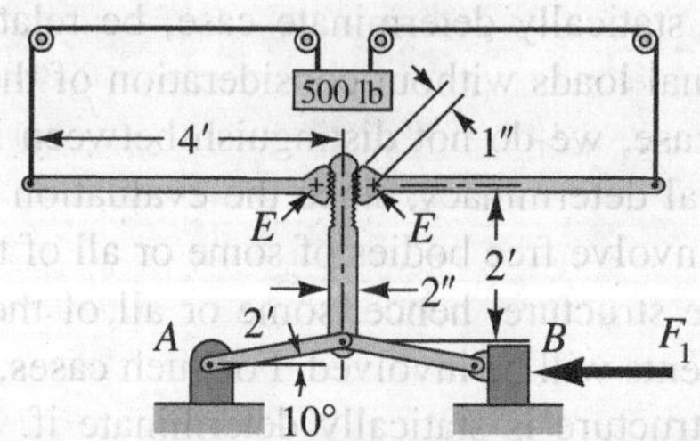

Figure P.5.94.

5.95. Find the values of F and C so that members AB and CD fail simultaneously. The maximum load for AB is 15 kN and for CD is 22 kN. Neglect the weight of the members.

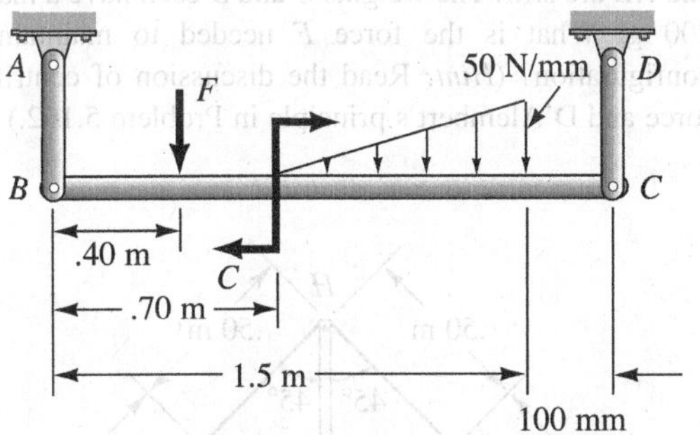

Figure P.5.95.

5.96. The landing carriage of a transport plane supports a stationary total vertical load of 200 kN. There are two wheels on each side of shock strut AB. Find the force in member EC, and the forces transmitted to the fuselage at A, if the brakes are locked and the engines are tested resulting in a thrust of 5 kN, 40% of which is resisted by this landing gear.

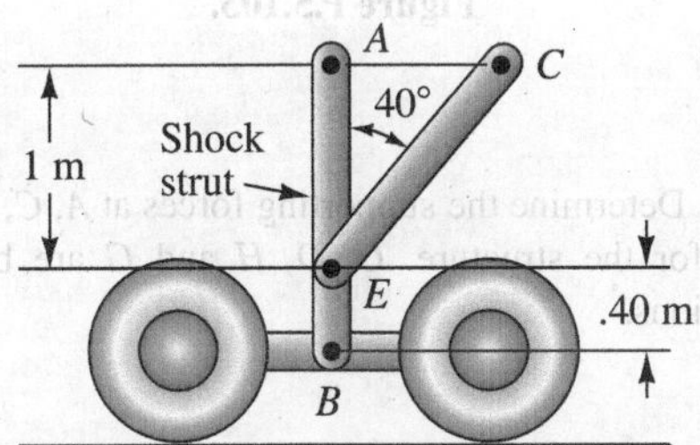

Figure P.5.96.

5.97. Find the magnitudes of the supporting forces for the frame shown. You may only use *two* free-body diagrams for this problem. Set forth a complete system of equations for solving the desired unknowns but do not carry out the algebra for actually solving these equations for the desired unknowns.

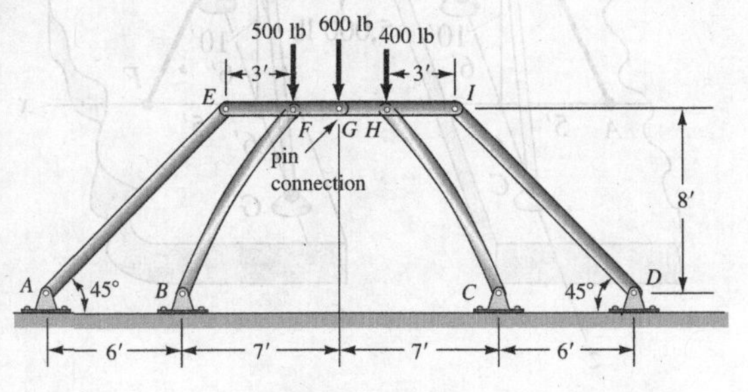

Figure P.5.97.

5.98. (a) Find the supporting forces at B. Neglect the weights of the members.
(b) What is the force in the member CB?

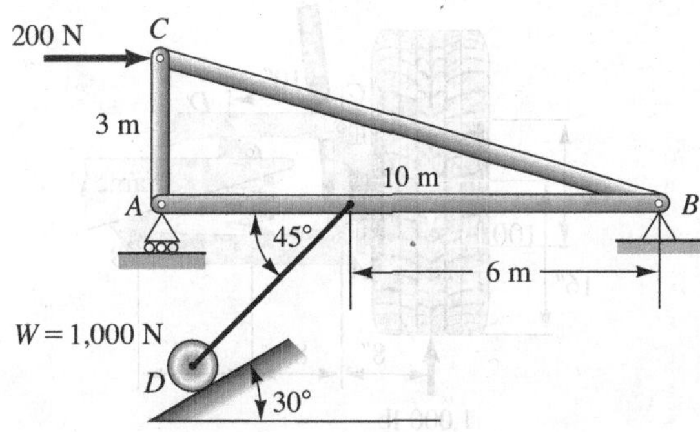

Figure P.5.98.

5.99. Find the supporting forces at B and C. Disc A weighs 200 N. Neglect the weights of the members as well as friction.

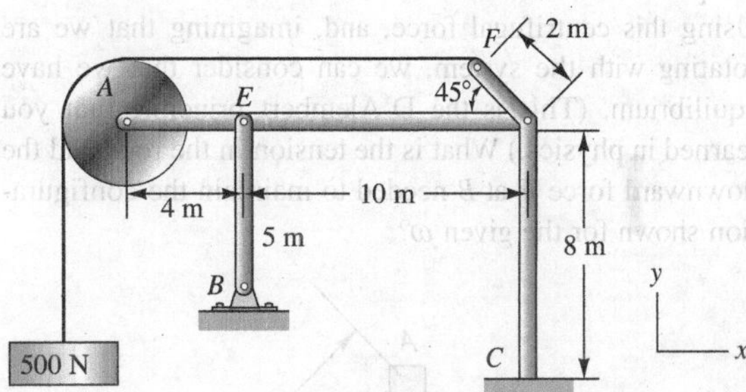

Figure P.5.99.

5.100. Find the supporting force system at C. Neglect friction.

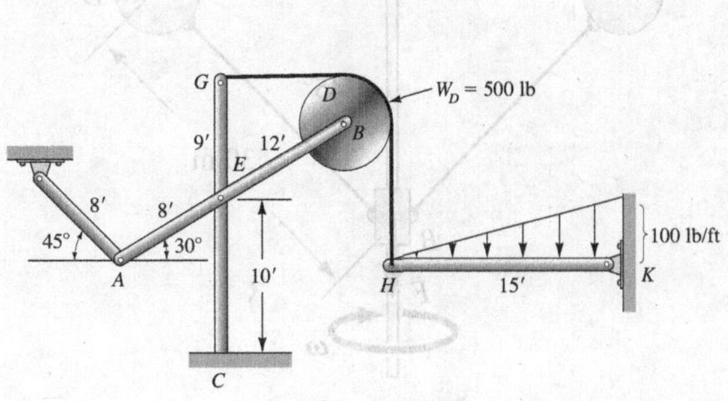

Figure P.5.100.

5.101. The pavement exerts a force of 1,000 lb on the tire. The tire, brakes, and so on, weigh 100 lb; the center of gravity is taken at the center plane of the tire. Determine the force from the spring and the compression force in *CD*.

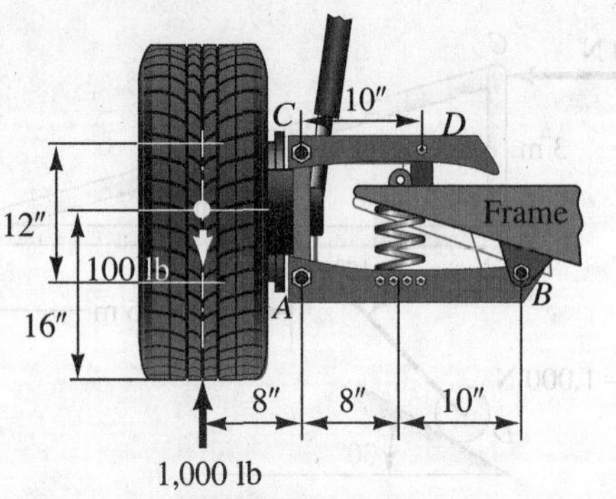

Figure P.5.101.

5.102. A *flyball governor* is shown rotating at a constant speed ω of 500 rpm. The weights *C* and *D* are each of mass 500 g and are pin-connected to light rods. The centrifugal force on the weights, you will recall from physics, is given as $mr\omega^2$, where r is the radial distance to the particle from the axis of rotation and ω is in rad/sec. Using this centrifugal force, and, imagining that we are rotating with the system, we can consider that we have equilibrium. (This is the D'Alembert principle that you learned in physics.) What is the tension in the rods and the downward force *F* at *B* needed to maintain the configuration shown for the given ω?

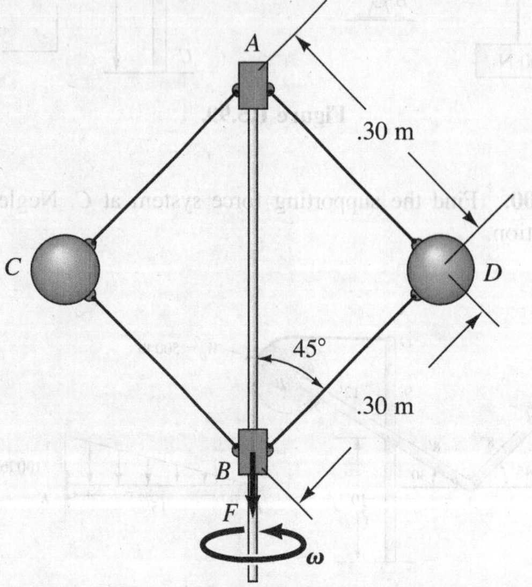

Figure P.5.102.

5.103. A flyball governor is shown below. If $\omega_1 = 3$ rad/sec, compute the tension in *AG* and *AE*. Neglect the weights of the members but assume that *HC* and *HB* are stiff. The weights *C* and *B* each have a mass of 200 g. What is the force *F* needed to maintain the configuration? (*Hint:* Read the discussion of centrifugal force and D'Alembert's principle in Problem 5.102.)

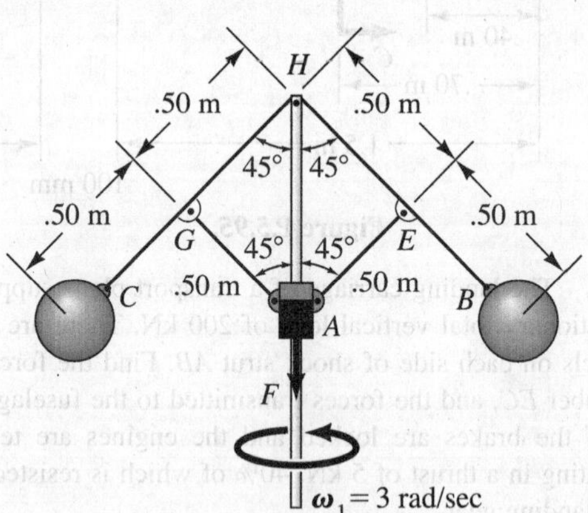

Figure P.5.103.

5.104. Determine the supporting forces at *A*, *C*, *D*, *G*, *F*, and *H* for the structure. *C*, *D*, *H* and *G* are ball joint connections.

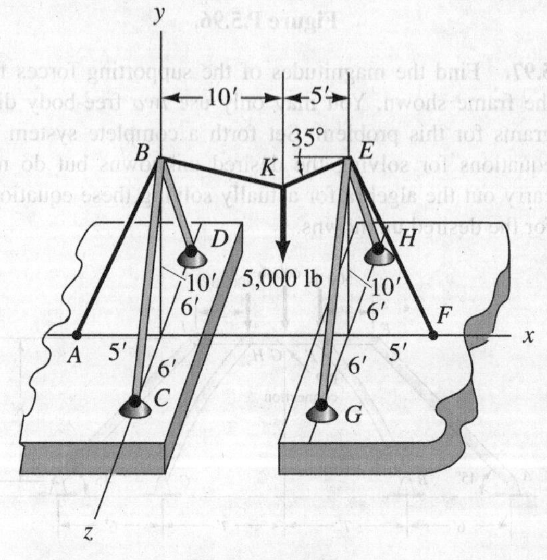

Figure P.5.104.

5.105. A trap door is kept open by a rod *CD*, whose weight we shall neglect. The door has hinges at *A* and *B* and has a weight of 200 lb. A wind blowing against the outside surface of the door creates a pressure increase of 2 lb/ft^2. Find the force in the rod, assuming that it cannot slip from the position shown. Also determine the forces transmitted to the hinges. Only hinge *B* can resist motion along direction *AB*.

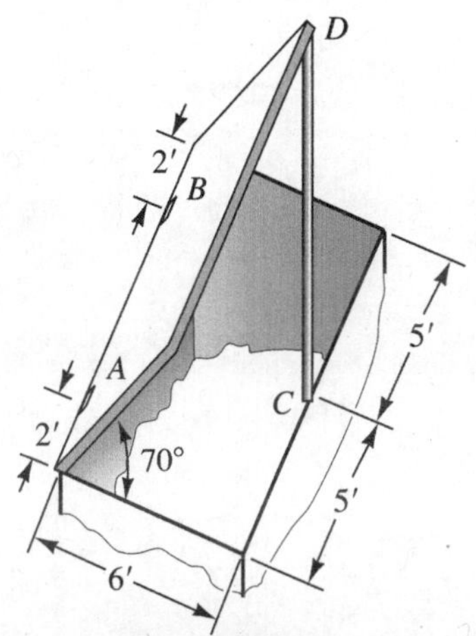

Figure P.5.105.

5.106. Find the force *BD*. All connections are ball-and-socket joints. Neglect the weights of all the members. Member *AB* has two 90° bends. Member *BD* is in the *yz* plane.

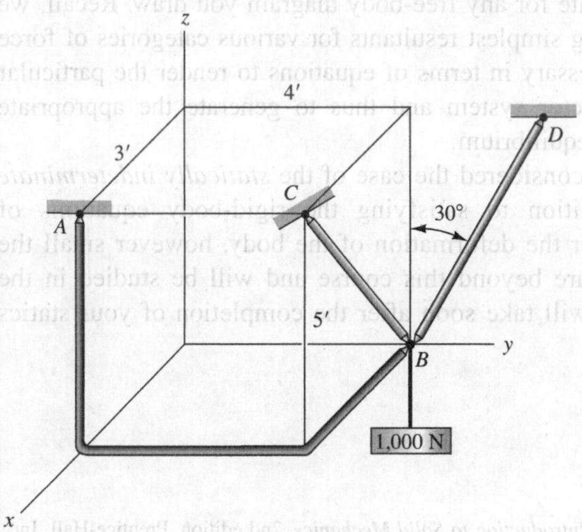

Figure P.5.106.

5.107. Find the supporting forces at *A* and *C*. You must show and use only one free-body diagram.

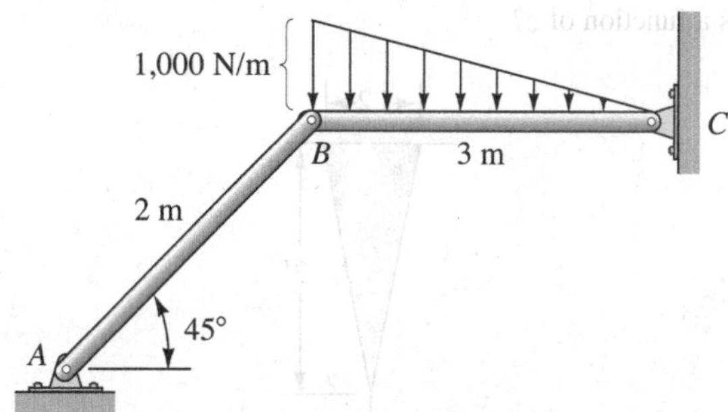

Figure P.5.107.

5.108. A coupling between two shafts transmits an axial load of 5,000 N. Four bolts having a diameter each of 13 mm, connects the two units. Before loading of the shafts, these bolts have only negligible tensile forces. Assuming that each bolt carries the same load, what is the average normal stress in each bolt stemming from the 5,000 N load?

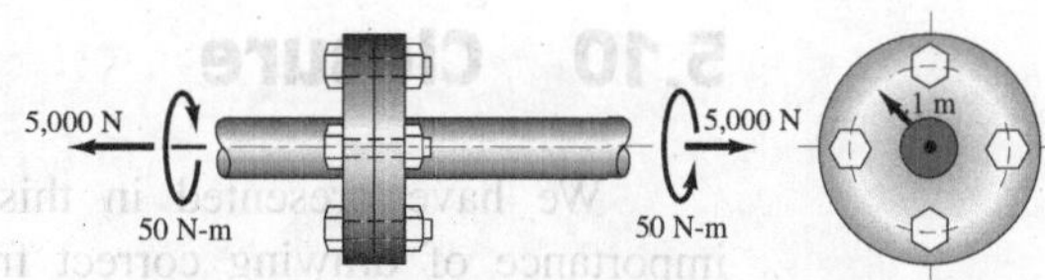

Figure P.5.108.

5.109. A circular shaft is suspended from above. The specific weight of the shaft material is 7.22×10^4 N/m^3. What is the tensile stress τ_{zz} on cross sections of the shaft as a function of *z*?

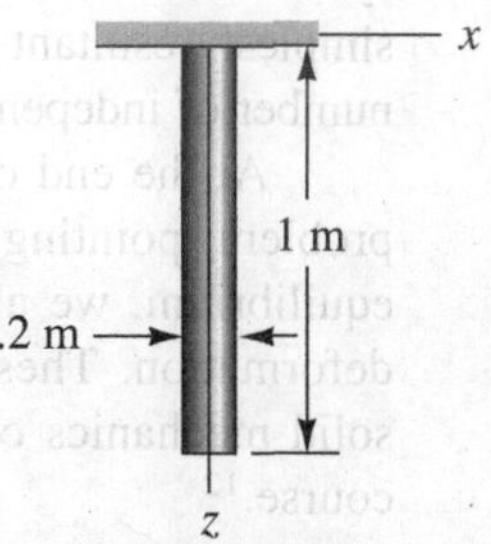

Figure P.5.109.

5.110. Do the previous problem for the case where the specific weight γ varies with the *square* of *z* starting with the value of 6.50×10^4 N/m^3 at the top and reaching a value of 7.50×10^4 N/m^3 at the bottom.

5.111. A cone is suspended from above. The specific weight γ of the material is 460 lb/ft^3. What is the average normal stress denoted as τ_{zz} for cross-sections of the cone as a function of z?

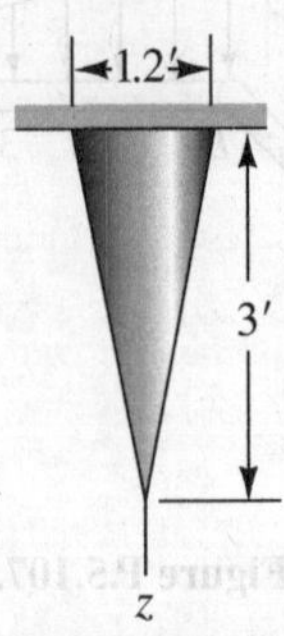

Figure P.5.111.

5.112. Solve the preceding problem for the case where the specific weight γ varies linearly from the top of the cone ($\gamma = 460$ lb/ft^3) the bottom ($\gamma = 385$ lb/ft^3). Get the normal stress τ_{zz} at cross sections of the cone as a function of z but do not bother to collect terms coming out of the integration. Finally, compute from your function the stress τ_{zz} at a position one foot from the top.

5.10 Closure

We have presented in this chapter the *free-body* concept. We hope that the importance of drawing correct free bodies has been made abundantly clear. Do not proceed in your work on a problem until correct, clearly drawn free bodies have been formed. This is the most critical step in problem solving! If you do not master this capability, you will be plagued throughout your future courses with difficulties and frustrations arising from this inadequacy. Also vital, is to know how many *independent equations of equilibrium* you can write for any free-body diagram you draw. Recall, we reached this capability by considering simplest resultants for various categories of force systems and reasoned what was necessary in terms of equations to render the particular simplest resultant system a null vector system and thus to generate the appropriate number of independent equations of equilibrium.

At the end of the chapter, we considered the case of the *statically indeterminate* problem pointing out that in addition to satisfying the rigid-body equations of equilibrium, we also had to consider the deformation of the body, however small the deformation. These considerations are beyond this course and will be studied in the solid mechanics course which you will take soon after the completion of your statics course.[12]

[12]For such problems, see I. H. Shames, *Introduction to Solid Mechanics*, 2nd edition, Prentice-Hall, Inc., Englewood Cliffs, N.J., 1989.

PROBLEMS

5.113. Determine the tensions in all the cables. Block *A* has a mass of 600 kg. Note that *GH* is in the *yz* plane.

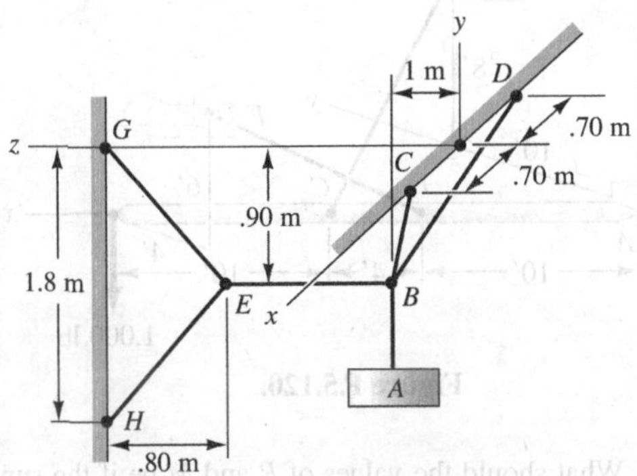

Figure P.5.113.

5.114. Determine the force components at *G*. *E* weighs 300 lb.

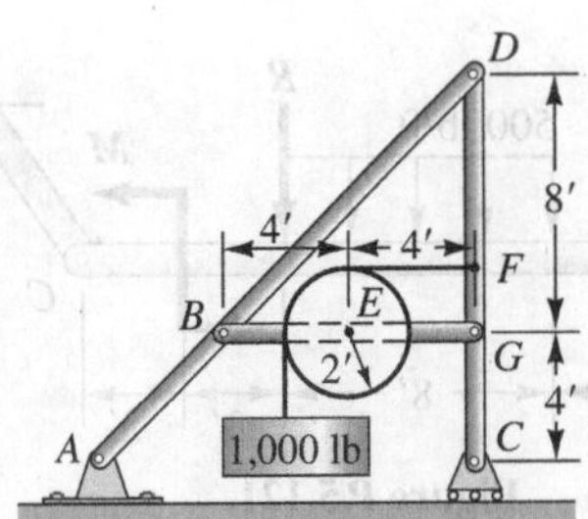

Figure P.5.114.

5.115. A scenic excursion train with cog wheels for steep inclines weighs 30 tons when fully loaded. If the cog wheels have a mean radius to the contact points of the teeth of 2 ft, what torque must be applied to the driver wheels *A* if wheels *B* run free? What force do wheels *B* transmit to the ground?

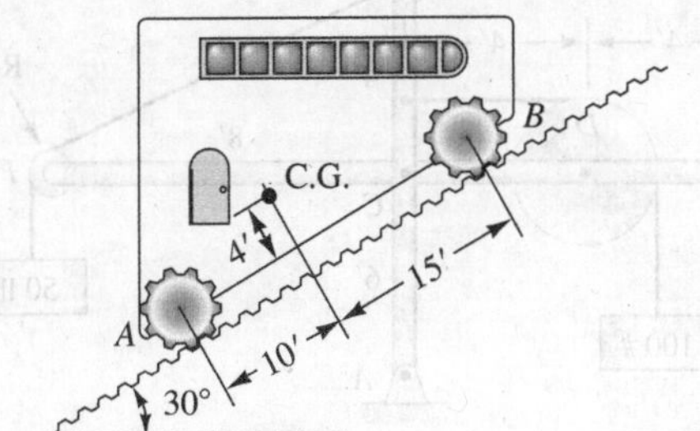

Figure P.5.115.

5.116. Find the forces on the block of ice from the hooks at *A* and *F*.

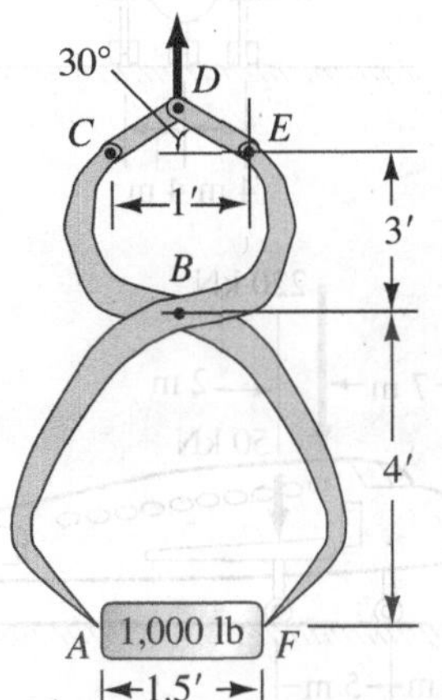

Figure P.5.116.

5.117. Members *AB* and *BC* weighing, respectively, 50 N and 200 N are connected to each other by a pin. *BC* connects to a disc *K* on which a torque T_K = 200 N-m is applied. What torque *T* is needed on *AB* to keep the system in equilibrium at the configuration shown?

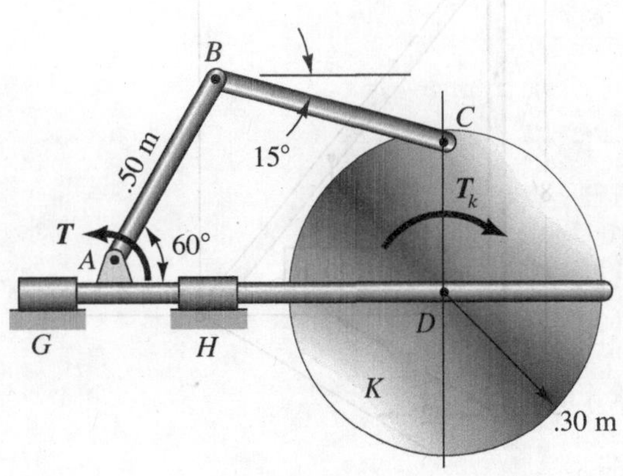

Figure P.5.117.

5.118. A transport jet plane has a weight without fuel of 220 kN. If one wing is loaded with 50 kN of fuel, what are the forces in each of the three landing gear?

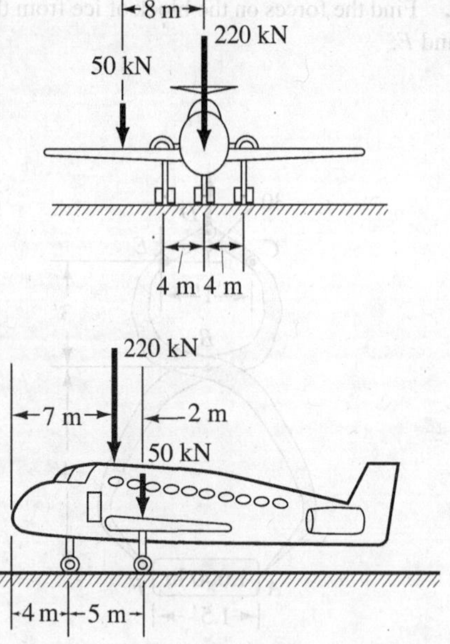

Figure P.5.118.

5.119. A rod AB is connected by a ball-and-socket joint to a frictionless sleeve at A, and by a ball-and-socket joint to a fixed position at B. What are the supporting forces at B and at A if we neglect the weight of AB? The 100-lb load is connected to the center of AB.

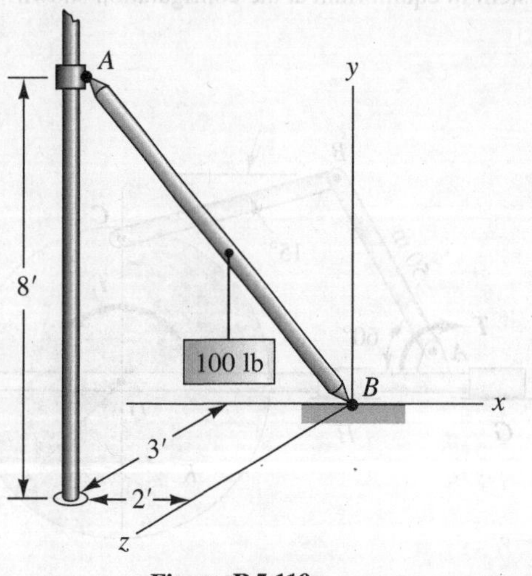

Figure P.5.119.

5.120. A beam weighing 400 lb is held by a ball-and-socket joint at A and by two cables CD and EF. Find the tension in the cables. They are attached to blocks at opposite ends of the beam as shown.

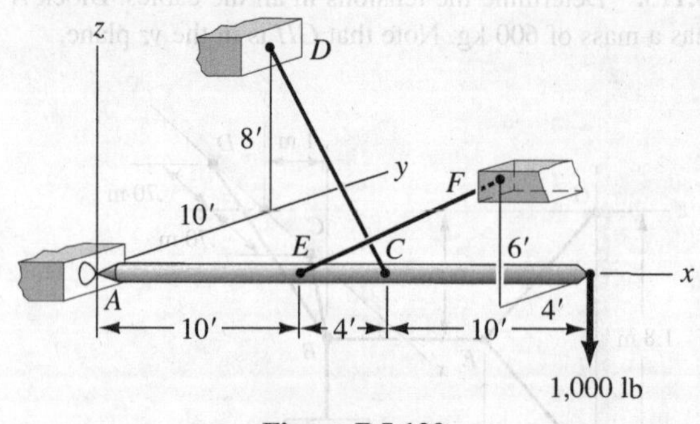

Figure P.5.120.

5.121. What should the values of R and M be if the supporting rods AB and CD are to fail simultaneously? Rod AB can withstand a 5,000-lb force, and rod CD can withstand an 8,000-lb force. Neglect the weights of the members.

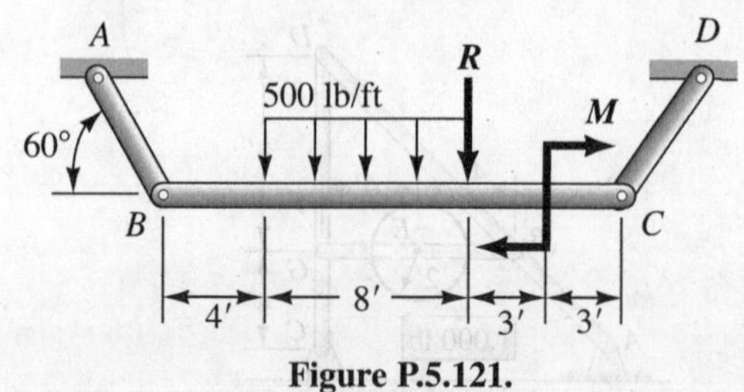

Figure P.5.121.

5.122. Find A_x and A_y at the bottom support. Do *not* determine any other unknowns. Do this using only *two* free-body diagrams and only *two* equations. Disc D weighs 30 lb and has a diameter of 2 ft. Neglect the weight of the members. Neglect friction everywhere.

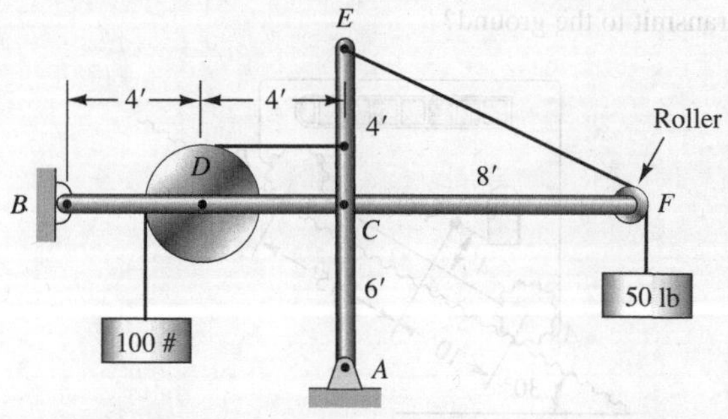

Figure P.5.122.

5.123. Light rods BC and AC are pinned together at C and support a 300-N load and a 500-N-m couple moment. What are the supporting forces at A and B?

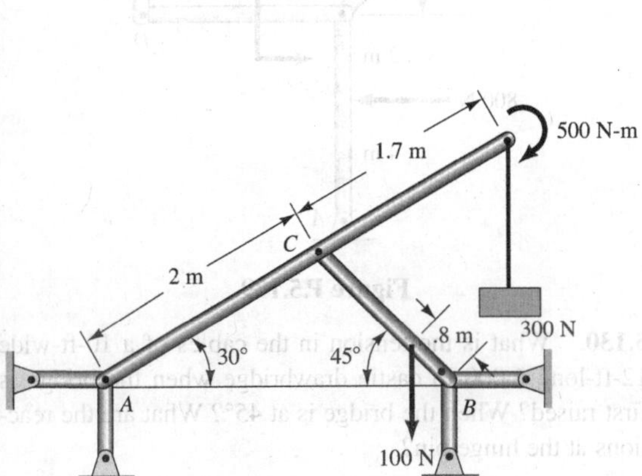

Figure P.5.123.

5.124. A rod AB is held by a ball-and-socket joint at A and supports a 100-kg mass C at B. This rod is in the zy plane and is inclined to the y axis by an angle of 15°. The rod is 16 m long and F is at its midpoint. Find the forces in cables DF and EB. Cross-hatching indicates part of the yz plane.

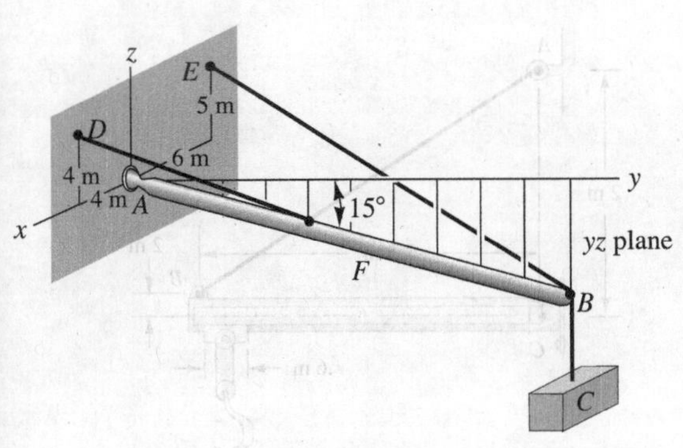

Figure P.5.124.

5.125. A bent rod $ADGB$ supports two weights—one at the center of AD and one at the center of DG. There are ball-and-socket joint supports at A and B. With one scalar equation using the triple scalar product, determine the tension in cable DC.

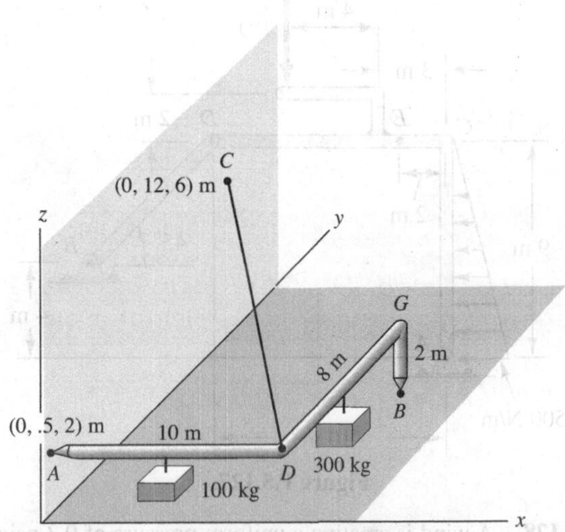

Figure P.5.125.

5.126. Find the tension in cable FH. The disc G weighs 500 lb. Use only one free body.

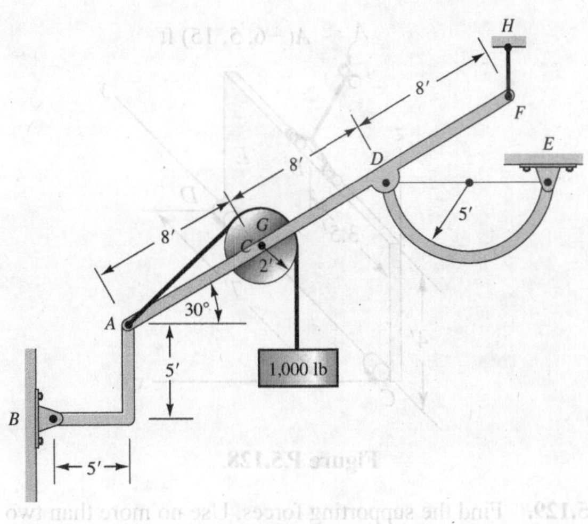

Figure P.5.126.

5.127. (a) Find the supporting force components at A using only one free-body diagram.
(b) Find the force system transmitted through the cross section at E of the beam CD. Again, use only one free-body diagram for this calculation. ACD is a bent member.

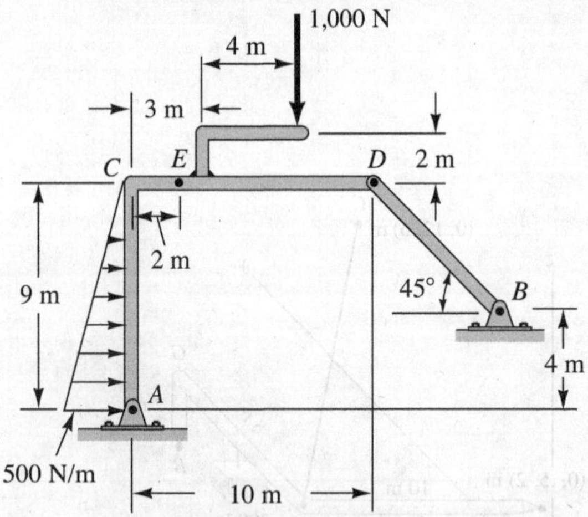

Figure P.5.127.

5.128. A wind is creating a uniform pressure of 0.3 psig on the left side of the door E and causes a uniform suction on the right side of the door of –.02 psig. A wire AB constrains the door. If hinge D allows for movement in the y direction, but hinge C does not, what are the supporting forces from the hinges? The weight of the door is 200 lb with a center of gravity at the geometric center of the door. The position of end A of the supporting wire has coordinates (–6, 5, 15) ft.

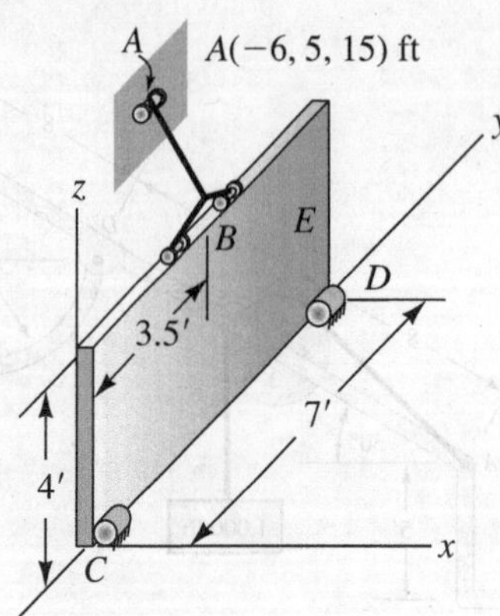

Figure P.5.128.

5.129. Find the supporting forces. Use no more than two free-body diagrams.

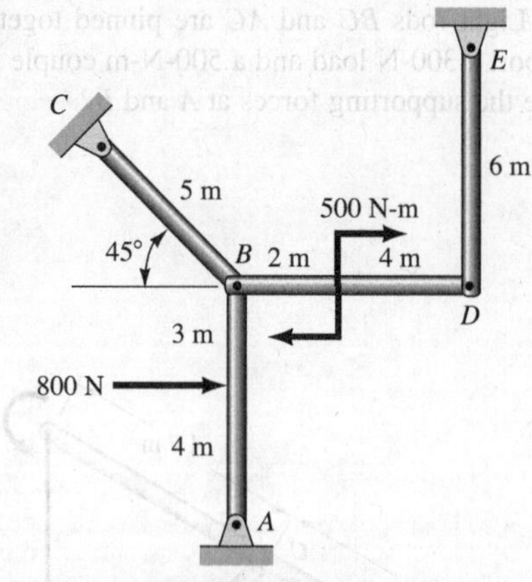

Figure P.5.129.

5.130. What is the tension in the cables of a 10-ft-wide 12-ft-long 6,000-lb castle drawbridge when the bridge is first raised? When the bridge is at 45°? What are the reactions at the hinge pin?

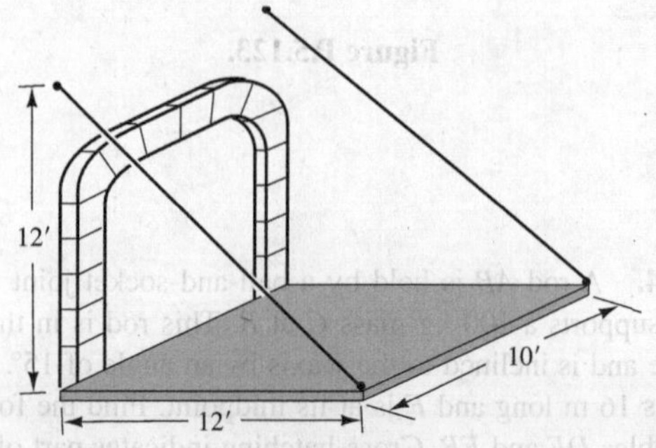

Figure P.5.130.

5.131. A small hoist has a lifting capacity of 20 kN. What is the maximum cable tension and the corresponding reactions at C? Do not consider weight of beam. At C there is a pin connection.

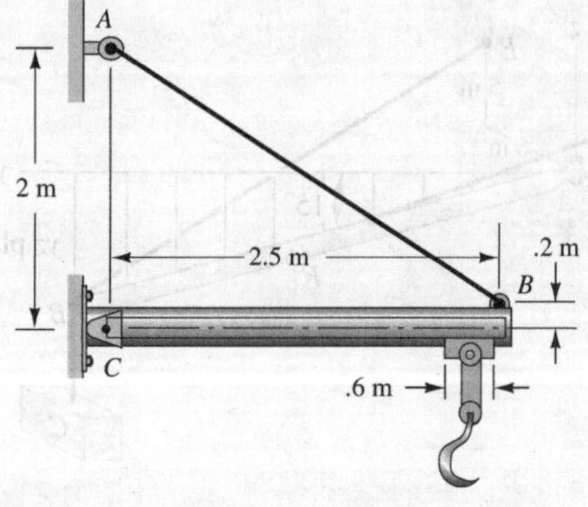

Figure P.5.131.

5.132. A uniform block weighing 500 lb is constrained by three wires. What are the tensions in these wires?

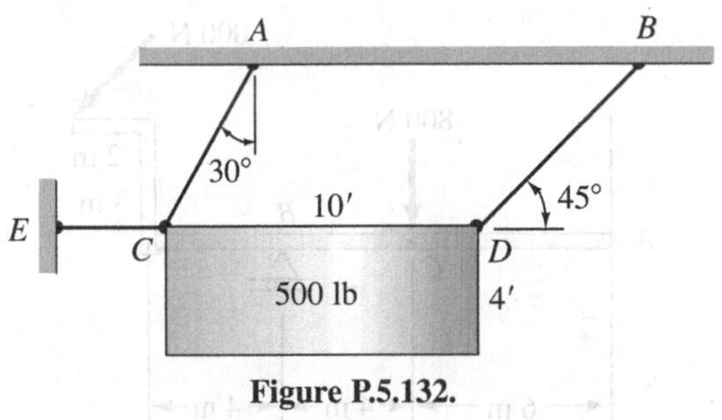

Figure P.5.132.

5.133. Find the supporting forces for the frame shown.

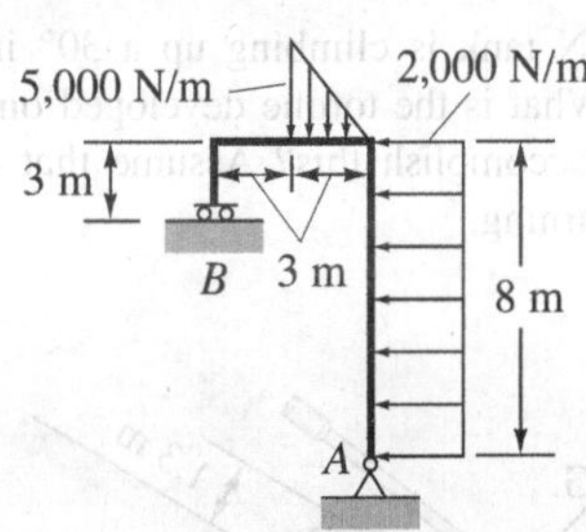

Figure P.5.133.

5.134. Four cables support a block of weight 5,000 N. The edges of the block are parallel to the coordinate axes. Point B is at (7, 7, −15). What are the forces in the cables and the direction cosines for cable CD?

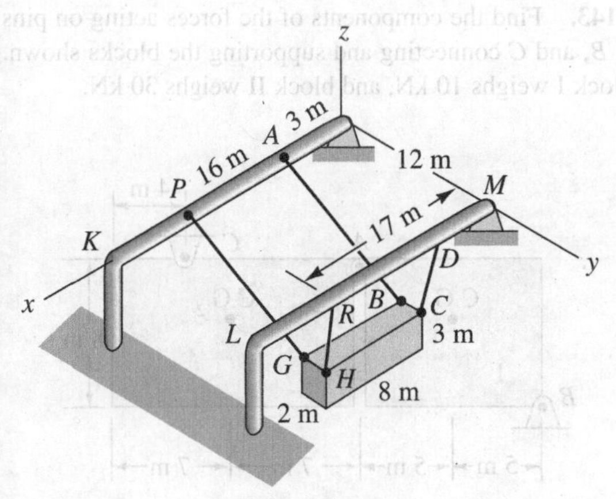

Figure P.5.134.

5.135. A mechanism consists of two weights W each of weight 50 N, four light linkage rods each of length a equal to 200 mm, and a spring K whose spring constant is 8 N/mm. The spring is unextended when $\theta = 45°$. If held vertically, what is the angle θ for equilibrium? Neglect friction. The force from the spring equals K times the compression of the spring.

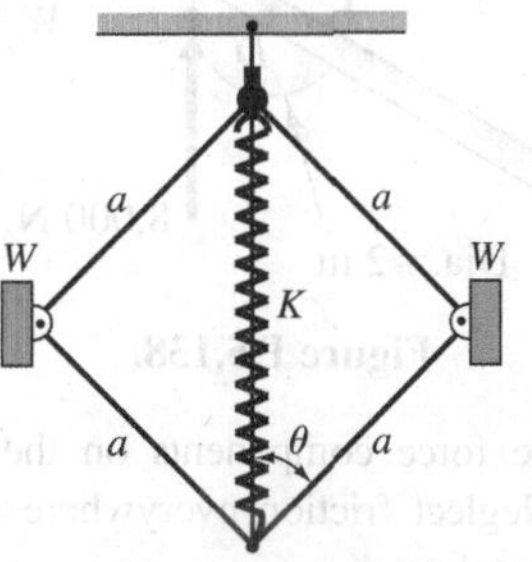

Figure P.5.135.

5.136. Find the compressive force in pawl AB. What is the resultant supporting force system at E?

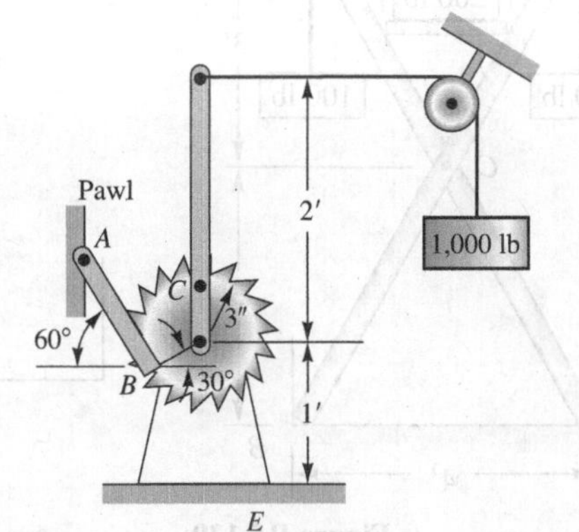

Figure P.5.136.

5.137. A 10-kN load is lifted in the front loader bucket. What are the forces at the connections to the bucket and to arm AE? Hydraulic ram DF is perpendicular to arm AE, and BC is horizontal. Points A and F are at the same height above the ground.

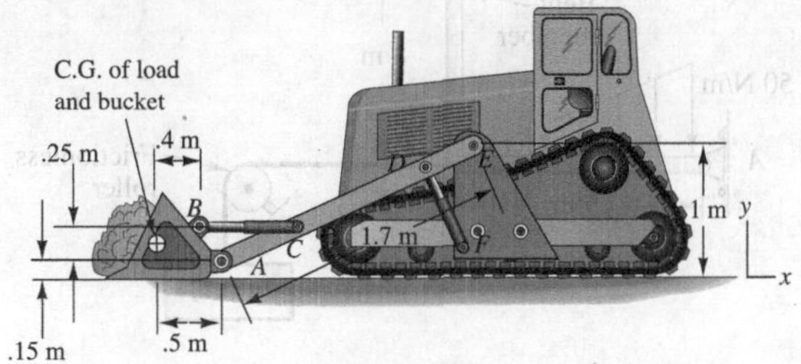

Figure P.5.137.

5.138. Determine the supporting forces at A and B.

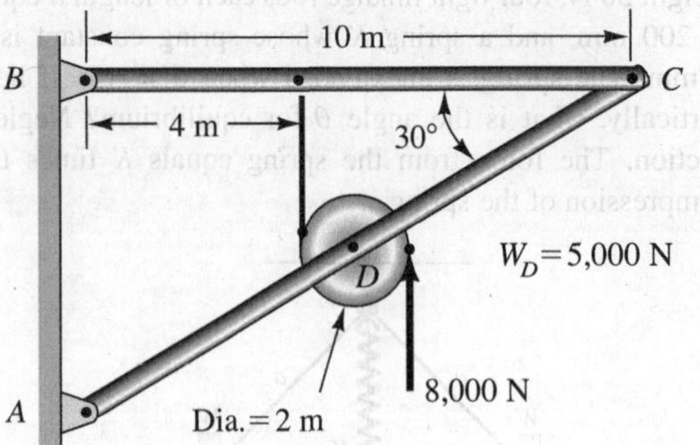

Figure P.5.138.

5.139. Find the force components on the pin C of the frame shown. Neglect friction everywhere as well as the weights of the members.

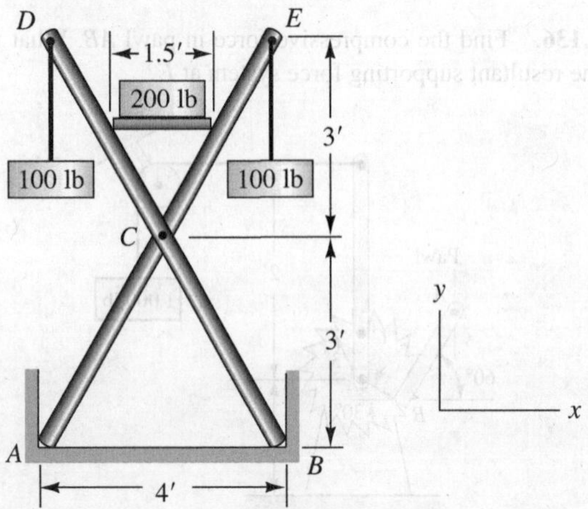

Figure P.5.139.

5.140. Find the supporting force system at A. Use only one free-body diagram.

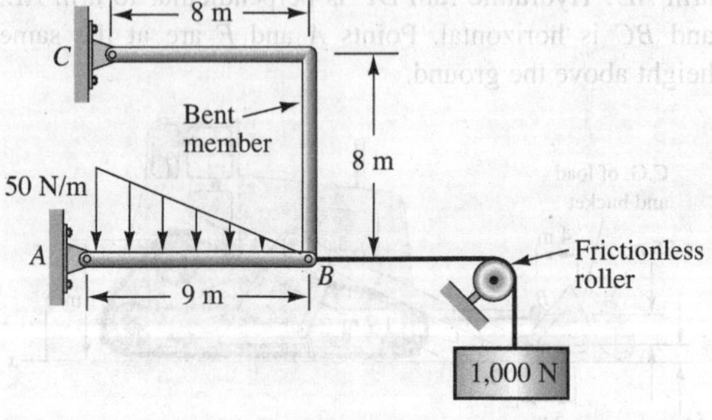

Figure P.5.140.

5.141. Find the supporting force systems at A and B. Note the pin connection at C.

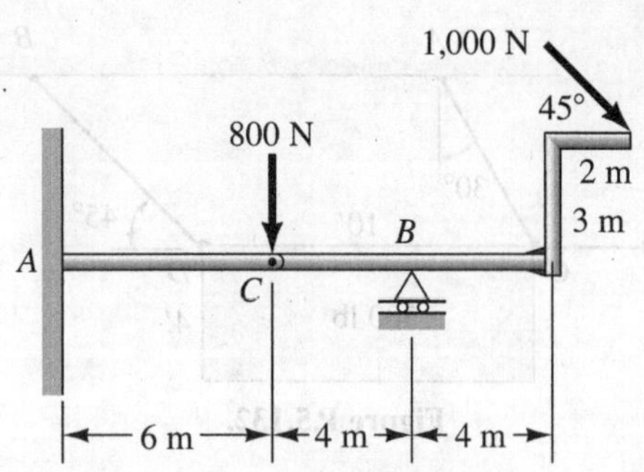

Figure P.5.141.

5.142. A 300-kN tank is climbing up a 30° incline at constant speed. What is the torque developed on the rear drive wheels to accomplish this? Assume that all other wheels are free-turning.

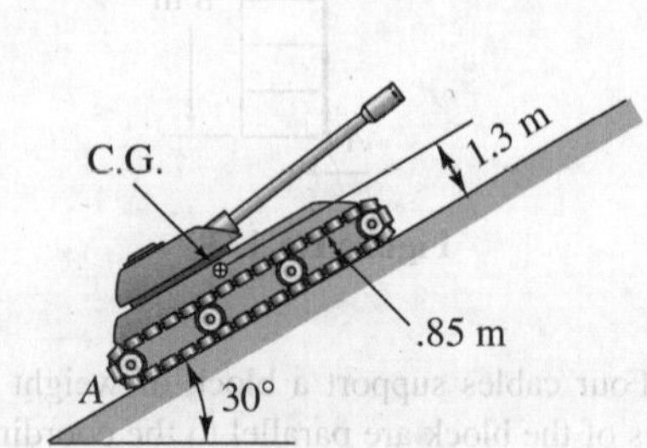

Figure P.5.142.

5.143. Find the components of the forces acting on pins A, B, and C connecting and supporting the blocks shown. Block I weighs 10 kN, and block II weighs 30 kN.

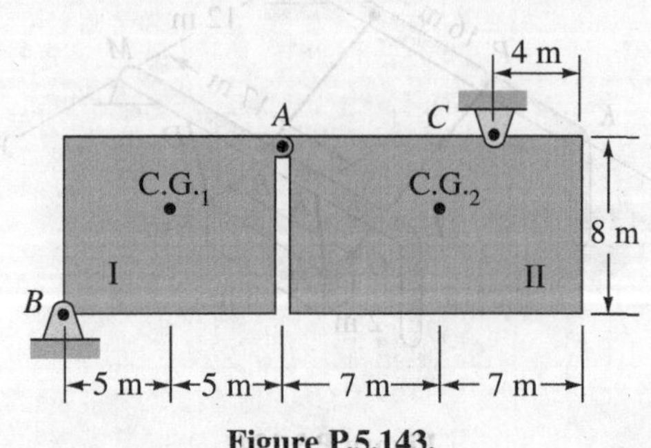

Figure P.5.143.

5.144. What force F do the pliers develop on the pipe section D? Neglect friction.

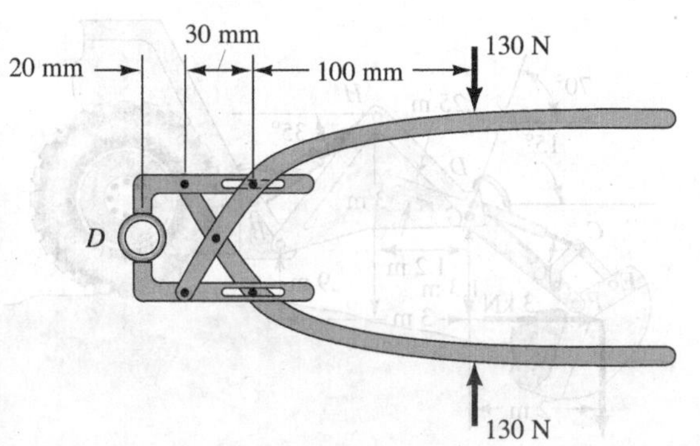

Figure P.5.144.

5.145. What are the supporting forces for the frame? Neglect all weights except the 10-kN weight.

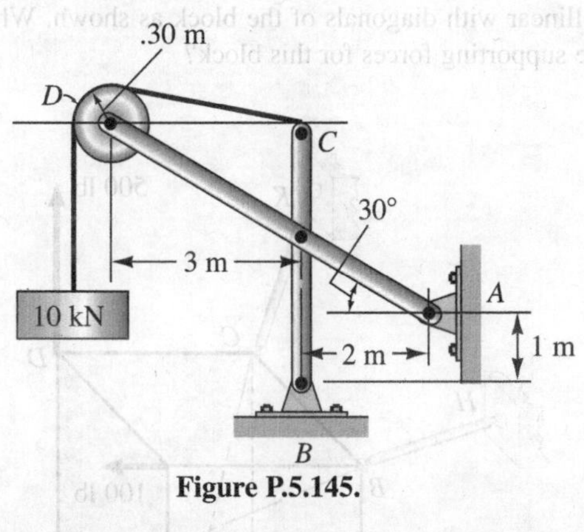

Figure P.5.145.

5.146. A 20-m circular arch must withstand a wind load given for $0 < \theta < \pi/2$ as

$$f = 5{,}000\left(1 - \frac{\theta}{\pi/2}\right) \text{ N/m}$$

where θ is measured in radians. Note that for $\theta > \pi/2$, there is no loading. What are the supporting forces? (*Hint:* What is the point for which taking moments is simplest?)

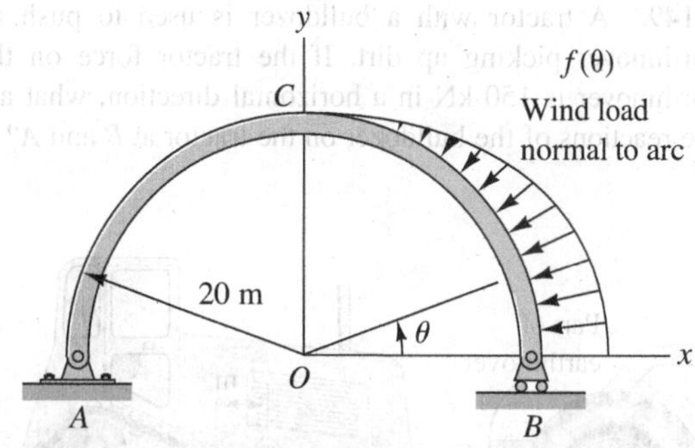

Figure P.5.146.

5.147. An arch is formed by uniform plates A and B. Plate A weighs 5 kN and plate B weighs 2 kN. What are the supporting forces at C, D, and E?

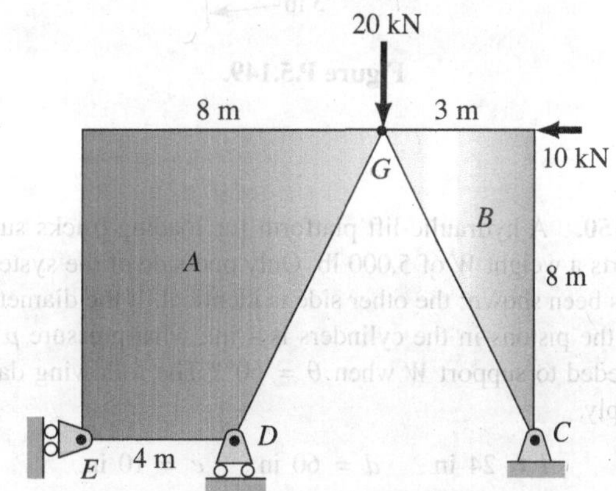

Figure P.5.147.

5.148. Find the supporting forces at the ball-and-socket connections A, D, and C. Members AB and DB are pinned together through member EC at B.

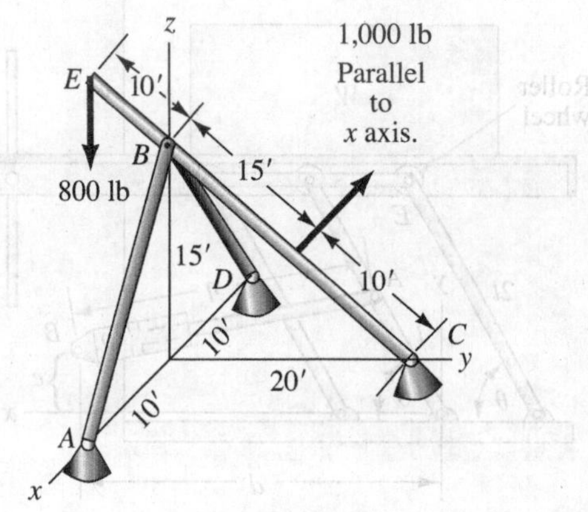

Figure P.5.148.

5.149. A tractor with a bulldozer is used to push an earthmover picking up dirt. If the tractor force on the earthmover is 150 kN in a horizontal direction, what are the reactions of the bulldozer on the tractor at B and A?

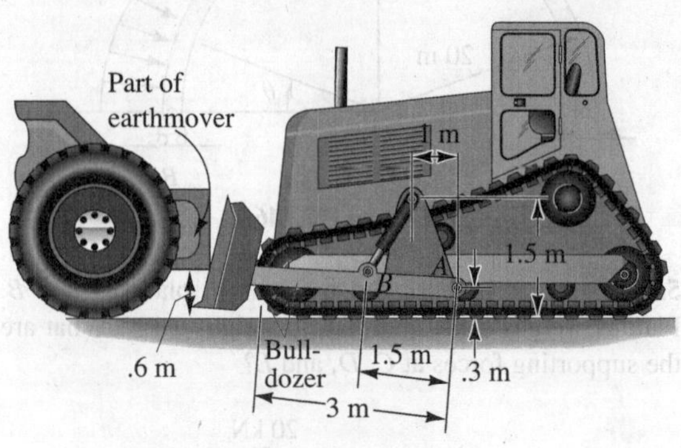

Figure P.5.149.

5.150. A hydraulic-lift platform for loading trucks supports a weight W of 5,000 lb. Only one side of the system has been shown; the other side is identical. If the diameter of the pistons in the cylinders is 4 in., what pressure p is needed to support W when $\theta = 60°$? The following data apply.

$$l = 24 \text{ in.}, \quad d = 60 \text{ in.}, \quad e = 10 \text{ in.}$$

Neglect friction everywhere. (Hint: Only two free-body diagrams need be drawn.)

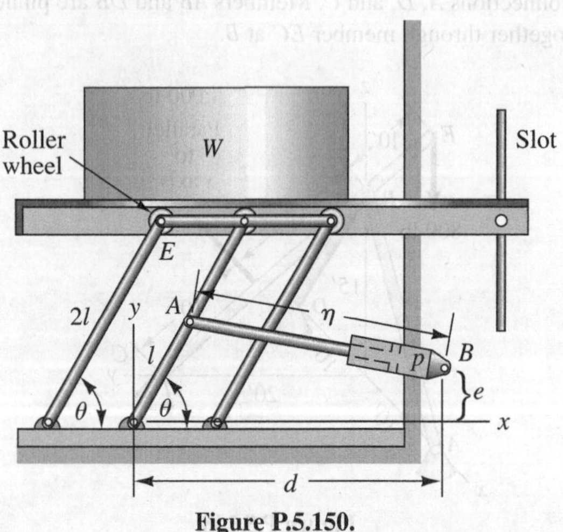

Figure P.5.150.

5.151. A Bucyrus–Erie Dynahoe digger is partially shown. To develop the indicated forces in the bucket, what forces must hydraulic cylinders HB and CD develop? Consider only the 3-kN and 5-kN loads and not the weights of the members.

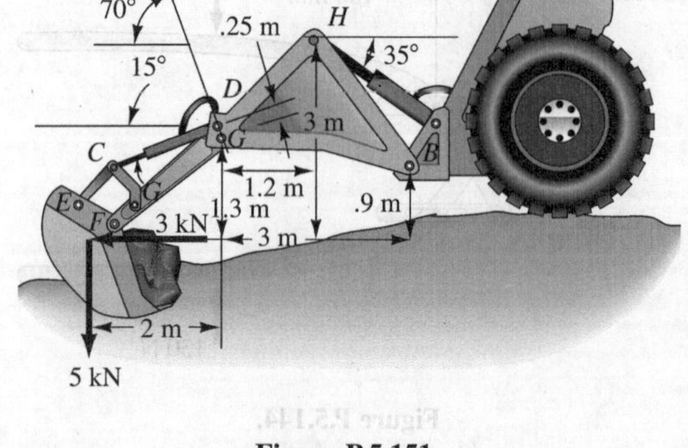

Figure P.5.151.

5.152. A block of material weighing 200-lb is supported by members KC and HB, whose weight we neglect, a ball-and-socket-joint support at A, and a smooth, frictionless support at E. Members KC and HB have directions collinear with diagonals of the block as shown. What are the supporting forces for this block?

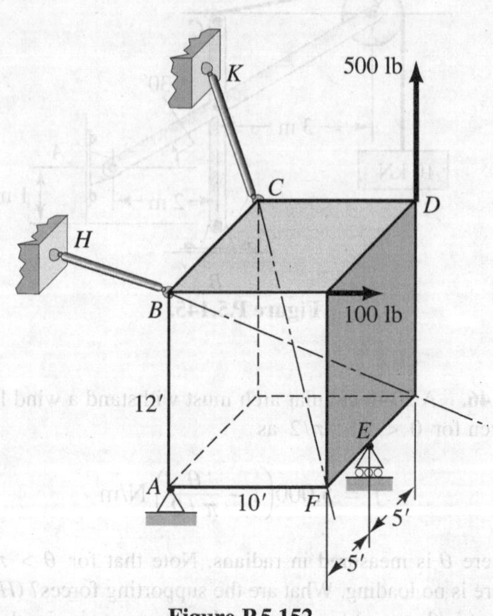

Figure P.5.152.

***5.153.** A bar can rotate parallel to plane A about an axis of rotation normal to the plane at O. A weight W is held by a cord that is attached to the bar over a small pulley that can rotate freely as the bar rotates. Find the value of C for equilibrium if $h = 300$ mm, $W = 30$ N, $\phi = 30°$, $l = 700$ mm, and $d = 500$ mm.

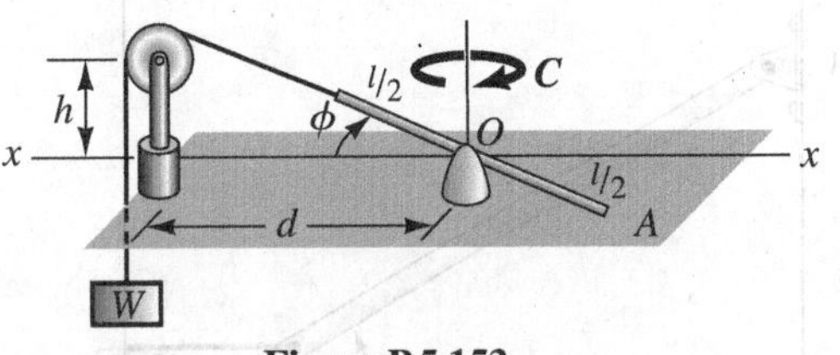

Figure P.5.153.

5.154. Find the supporting forces at A and B in the frame. Neglect weights of members.

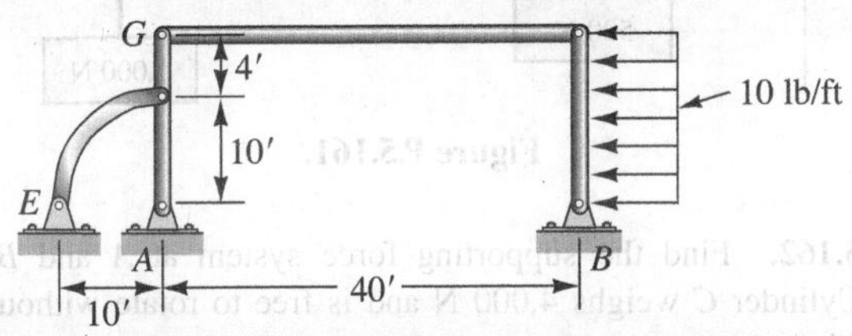

Figure P.5.154.

5.155. A bolt cutter has a force of 130 N applied at each handle. What is the force on the bolt from the cutter edge?

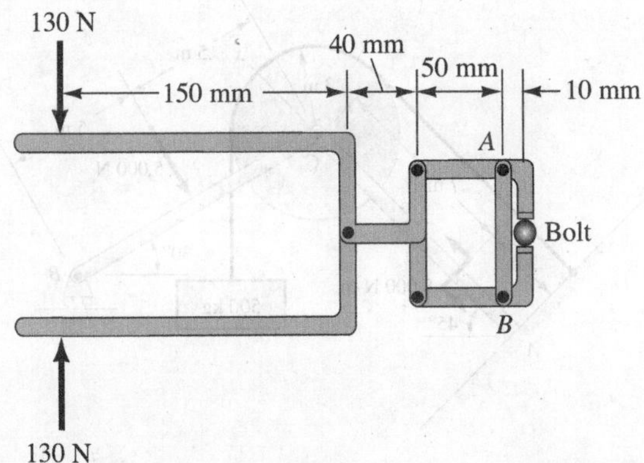

Figure P.5.155.

5.156. A steam locomotive is developing a pressure of .20 N/mm^2 gage. If the train is stationary, what is the total traction force from the two wheels shown? Neglect the weight of the various connecting rods. Neglect friction in piston system and connecting rod pins.

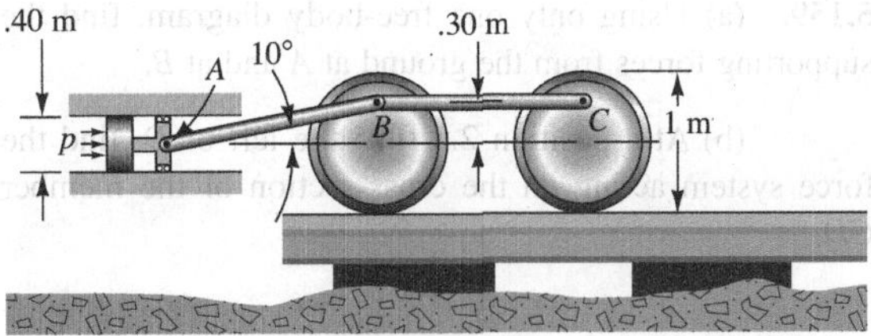

Figure P.5.156.

5.157. Find the supporting forces at A, B, and C. Neglect the weight of the rod. Use only one free-body diagram.

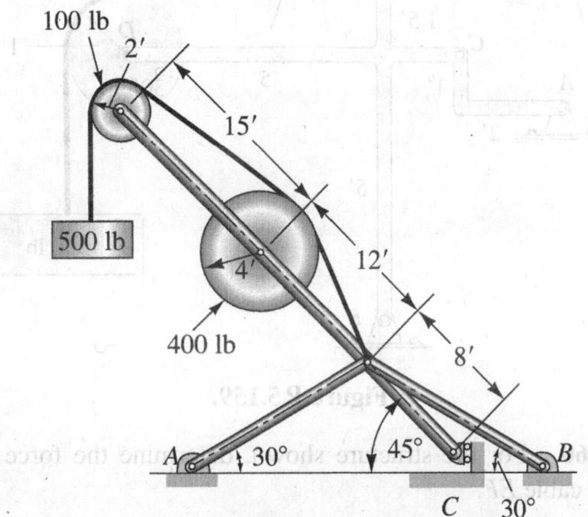

Figure P.5.157.

5.158. The 5,000 lb van A of an airline food catering truck rises straight up until its floor is level with the airplane floor. A hydraulic ram pulls on the right bottom support of the lift mechanism at which we have rollers to prevent friction. The two members of the lift mechanism are pinned at their center. The center of gravity of the van is its geometric center. What is the ram force for this position?

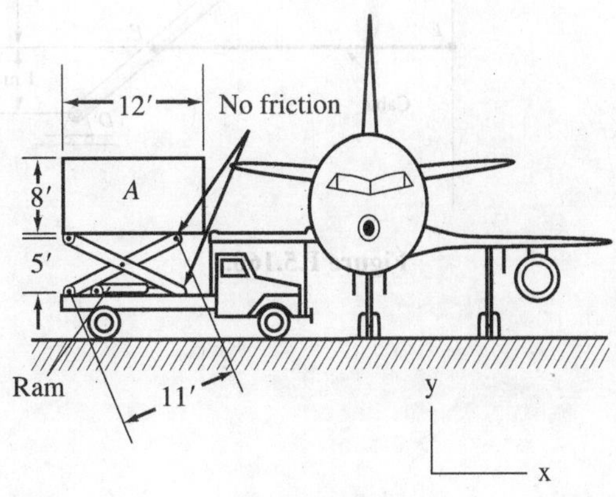

Figure P.5.158.

5.159. (a) Using only one free-body diagram, find the supporting forces from the ground at A and at B.

(b) At a position 2.5 ft to the left of D, find the force system acting on the cross-section of the member CD.

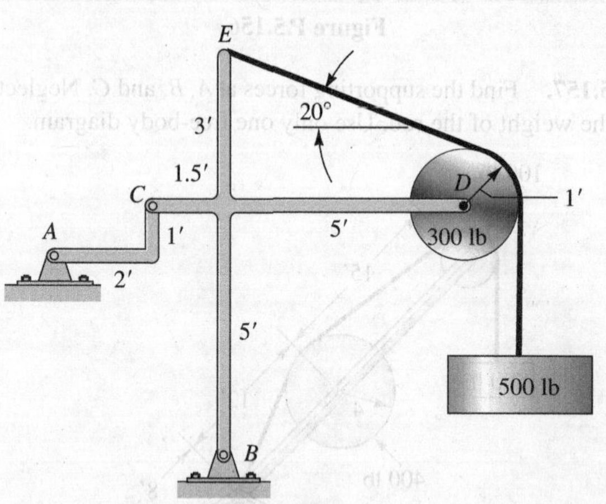

Figure P.5.159.

5.160. For the structure shown, determine the force in the cable EF.

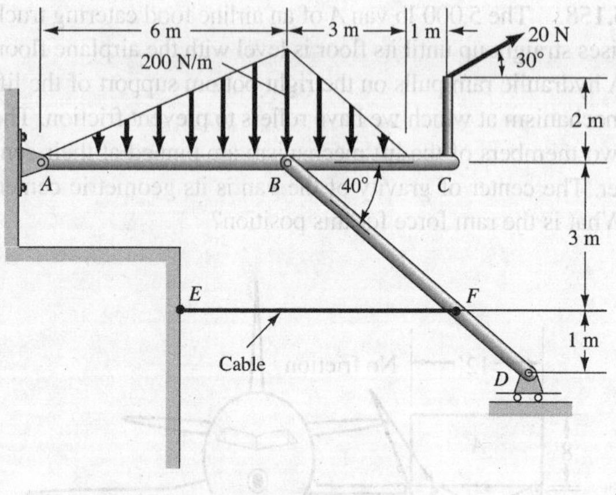

Figure P.5.160.

5.161. Find the supporting forces at A and B. The cylinder can turn freely.

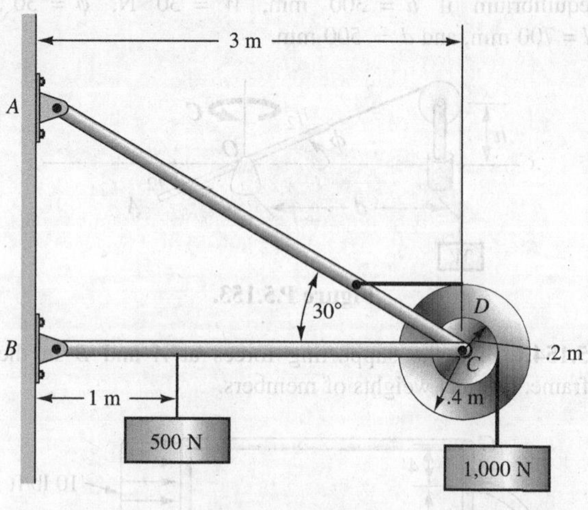

Figure P.5.161.

5.162. Find the supporting force system at A and B. Cylinder C weighs 4,000 N and is free to rotate without friction.

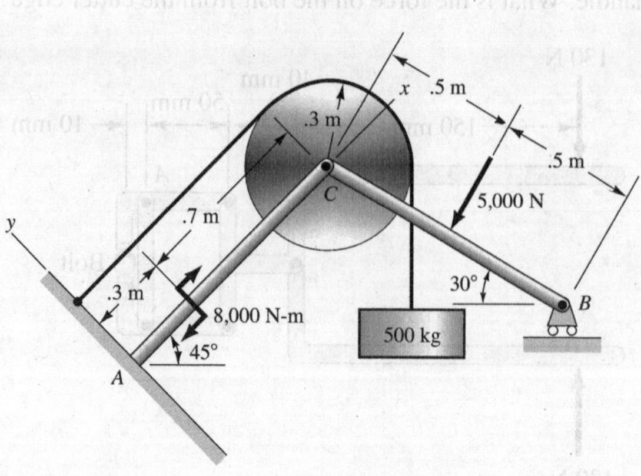

Figure P.5.162.

CHAPTER 6

Friction Forces

6.1 Introduction

Friction is the force distribution at the surface of contact between two bodies that prevents or impedes sliding motion of one body relative to the other. This force distribution is tangent to the contact surface and has, for the body under consideration, a direction at every point in the contact surface that is in opposition to the possible or existing slipping motion of the body at that point.

Frictional effects are associated with energy dissipation and are therefore sometimes considered undesirable. At other times, however, this means of changing mechanical energy to heat is a beneficial one, as for example in brakes, where the kinetic energy of a vehicle is dissipated into heat. In statics applications, frictional forces are often necessary to maintain equilibrium.

Coulomb friction is that friction which occurs between bodies having dry contact surfaces, and is not to be confused with the action of one body on another separated by a film of fluid such as oil. These latter problems are termed *lubrication problems* and are studied in the fluid mechanics courses. Coulomb, or *dry*, friction is a complicated phenomenon, and actually not much is known about its true nature.[1] The major cause of dry friction is believed to be the microscopic roughness of the surfaces of contact. Interlocking microscopic protuberances oppose the relative motion between the surfaces. When sliding is present between the surfaces, some of these protuberances either are sheared off or are melted by high local temperatures. This is the reason for the high rate of "wear" for dry-body contact and indicates why it is desirable to separate the surfaces by a film of fluid.

We have previously employed the terms "smooth" and "rough" surfaces of contact. A "smooth" surface can only support a normal force. On the other hand, a "rough" surface in addition can support a force tangent to the contact surface (i.e., a friction force). In this chapter, we shall consider certain situations whereby the friction force can be directly related to the normal force at a surface of contact. Other than including this new relationship, we use only the usual static equilibrium equations.

[1]For a more complete discussion of friction, see F. P. Bowden and D. Tabor, *The Friction and Lubrication of Solids*, Oxford University Press, New York, 1950.

6.2 Laws of Coulomb Friction

Everybody has gone through the experience of sliding furniture along a floor. We exert a continuously increasing force which is completely resisted by friction until the object begins to move—usually with a lurch. The lurch occurs because once the object begins to move, there is a decrease in frictional force from the maximum force attained under static conditions. An idealized plot of this force as a function of time is shown in Fig. 6.1 where the force P applied to the furniture, idealized as a block in Fig. 6.2, is shown to drop from the highest or limiting value to a lower value which is constant with time. This latter constant value is independent of the velocity of the object. The condition corresponding to the maximum value is termed the condition of *impending motion* or *impending slippage*.

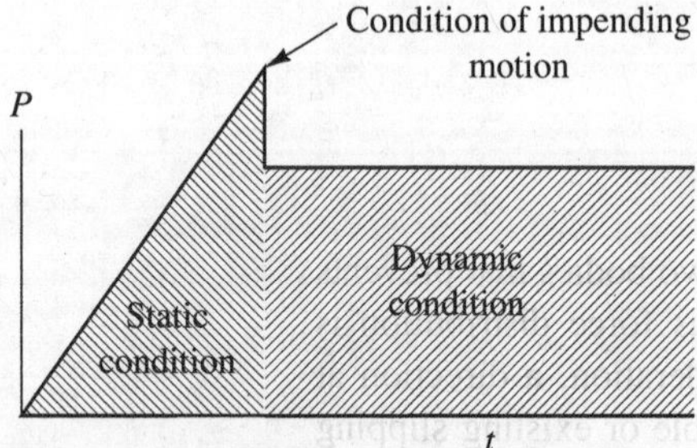

Figure 6.1. Idealized plot of applied force P.

By carrying out experiments on blocks tending to move without rotation or actually moving without rotation on flat surfaces, Coulomb in 1781 presented certain conclusions which are applicable at the condition of *impending slippage* or once *slippage has begun*. These have since become known as Coulomb's laws of friction. For block problems, he reported that:

1. The total force of friction that can be developed is independent of the magnitude of the area of contact.
2. For low relative velocities between sliding objects, the frictional force is practically independent of velocity. However, the sliding frictional force is less than the frictional force corresponding to impending slippage.
3. The total frictional force that can be developed is proportional to the normal force transmitted across the surface of contact.

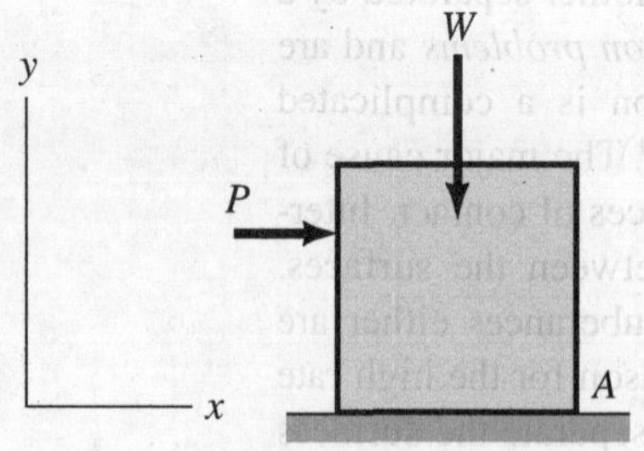

Figure 6.2. Idealization of furniture.

Conclusions 1 and 2 may come as a surprise to most of you and be contrary to your "intuition." Nevertheless, they are accurate enough statements for many engineering applications. More precise studies of friction, as was pointed out earlier, are complicated and involved. We can express conclusion 3 mathematically as:

$$f \propto N$$

Therefore,

$$f = \mu N \tag{6.1}$$

where μ is called the *coefficient of friction*.

Equation 6.1 is valid *only at conditions of impending slippage or while the body is slipping*. Since the limiting static friction force exceeds the dynamic friction force, we differentiate between coefficients of friction for those conditions. Thus, we have coefficients of *static* friction and coefficients of *dynamic* friction, μ_s and μ_d, respectively. The accompanying table is a small list of static coefficients that are commonly used. The corresponding coefficients of friction for dynamic conditions are about 25% less.

Static Coefficients of Friction[2]

Steel on cast iron	.40
Copper on steel	.36
Hard steel on hard steel	.42
Mild steel on mild steel	.57
Rope on wood	.70
Wood on wood	.20–.75

Let us consider carefully the simple block problem used to develop the laws of Coulomb. Note that we have:

1. A plane surface of contact.
2. An impending or actual motion which is in the same direction for all area elements of the contact surface. Thus, there is no impending or actual rotation between the bodies in contact.
3. The further implication that the properties of the respective bodies are uniform at the contact surface. Thus, the coefficient of friction μ is constant for all area elements of the contact surface.

What do we do if any of these conditions is violated? We can always choose an *infinitesimal* part of the area of contact between the bodies. Such an infinitesimal area can be considered plane even though the general surface of contact of which it is an infinitesimal part is not. Furthermore, the relative motion at this infinitesimal contact surface may be considered as along a straight line even though the finite surface of which it is a part may not have such a simple straight motion. Finally, for the infinitesimal area of contact, we may consider the materials to be uniform even though the properties of the material vary over the finite area of contact. In short, when conditions 1 through 3 do not prevail, we can still use Coulomb's law *in the small* (i.e., at infinitesimal contact areas) and then integrate the results. We shall call such problems *complex* surface contact problems and we shall examine a series of such problems in Section 6.4.

[2]F. P. Bowden and D. Tabor, *The Friction and Lubrication of Solids*, Oxford University Press, New York, 1950.

As a simple illustration of why curvature of the surface of contact is second order and hence negligible, consider yourself at a location on the earth where it is perfectly round as a planet. As you look around you over a small area compared to a significant portion of the earth's surface, there is no evidence to your observation or in what you normally do that indicates the presence of curvature of the earth. In the same way, an infinitesimal area can be considered flat even if is part of a finite curved surface when considering Coulomb's law.

Furthermore, a stationary, nonrotating observer in inertial space looking at a similar small area on the earth's surface at the equator sees the velocity of the area essentially as given by $R\omega$ where R is the radius of the earth and ω is its angular velocity. All parts of this small area will have this same velocity up to only second-order variation as seen by this observer. And, so from this viewpoint, all points within the area are essentially translating with the aforementioned speed. To explain further, we will now say that this small area has a maximum dimension given by the length r. The *rotational* speed of any point seen by an inertial and hence nonrotating observer at the area and translating *with* the area must be no greater than $r\omega$. Clearly then, this rotational speed is *negligibly small* when compared with $R\omega$. Hence, an infinitesimal area at a finite distance from the axis of rotation of a rotating surface can be considered as having primarily a translational velocity and so permits the use of Coulomb's law "in the small."

6.3 Simple Contact Friction Problems

We now examine simple contact problems where Coulomb's laws apply to the contact surface *as a whole* without requiring integration procedures. We shall thus consider uniform blocklike bodies akin to those used by Coulomb. Also, we shall consider bodies which however complex have very *small* contact surfaces, such as in Fig. 6.3(a). Clearly, the whole contact surface can then be considered an infinitesimal plane area and for reasons set forth earlier, we shall directly use Coulomb's laws when appropriate as has been shown in Fig. 6.3(b).

Before proceeding to the examples, we have one additional point to make. For a finite simple surface of contact, such as the block shown in Fig. 6.2, we must note that we do not generally know the line of action for the simplest resultant supporting force N, since we do not generally know the normal force distribution between the two bodies. Hence, we cannot take moments for such free-body diagrams without introducing additional unknown distances in the equation. Consequently, for such problems we limit ourselves to summing forces only. This is not true, however, when we have a *point* contact such as in Fig. 6.3(a). The line of action of the supporting force must be at the point of contact, and we can thus take moments without introducing additional unknown distances.

Two common classes of statics problems involve dry friction. In one class, we know that motion is impending, or has been established and is uniform, and we desire information about certain forces that are present. We can then express friction forces at surfaces of contact where there is impending or actual slippage as μN according to Coulomb's law and, using f_i for other friction forces, proceed by methods of statics. However, the proper direction must be given to *all* friction forces. That is, *they must oppose possible, impending, or actual relative motion at the contact surfaces*. In the second class of problems,

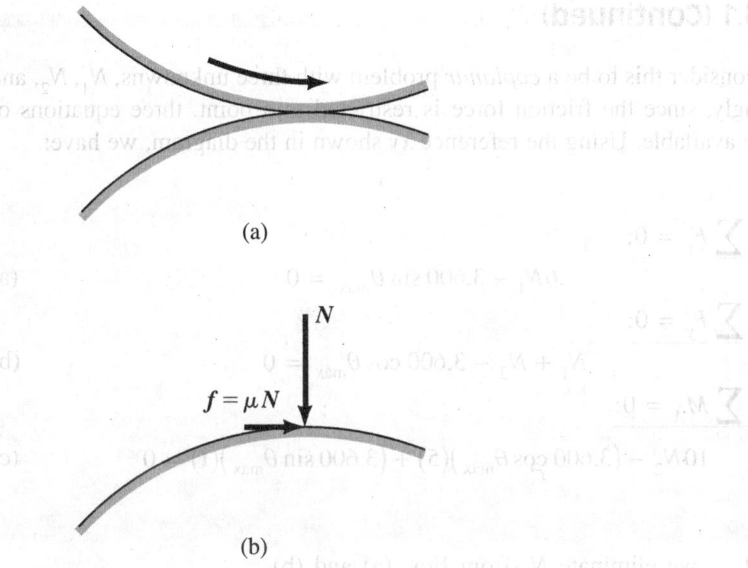

Figure 6.3. (a) Small contact surface; (b) Coulomb's laws applied.

external loads on a body are given, and we desire to determine whether the friction forces present are sufficient to maintain equilibrium. One way to attack this latter type of problem is to assume that impending motion exists in the various possible directions, and to solve for the external forces required for such conditions. By comparing the actual external forces present with those required for the various impending motions, we can then deduce whether the body can be restrained by frictional forces from sliding.

The following examples are used to illustrate the two classes of problems.

Example 6.1

An automobile is shown in Fig. 6.4(a) on a roadway inclined at an angle θ with the horizontal. If the coefficients of static and dynamic friction between the tires and the road are taken as 0.6 and 0.5, respectively, what is the maximum inclination θ_{max} that the car can climb at uniform speed? It has rear-wheel drive and has a total loaded weight of 3,600 lb. The center of gravity for this loaded condition has been shown in the diagram.

Let us assume that the drive wheels do not "spin"; that is, there is zero relative velocity between the tire surface and the road surface at the point of contact. Then, clearly, the maximum friction force possible is μ_s times the normal force at this contact surface, as has been indicated in Fig. 6.4(b).[3]

[3]You will notice that there is no friction force on the front wheels. This is so because there is no torque coming from the automobile's transmission onto these wheels, while at the same time the wheels are rotating at constant speed.

Example 6.1 (Continued)

We can consider this to be a *coplanar* problem with three unknowns, N_1, N_2, and θ_{max}. Accordingly, since the friction force is restricted to a point, three equations of equilibrium are available. Using the reference *xy* shown in the diagram, we have:

$\sum F_x = 0$:

$$.6N_1 - 3{,}600 \sin\theta_{max} = 0 \qquad \text{(a)}$$

$\sum F_y = 0$:

$$N_1 + N_2 - 3{,}600 \cos\theta_{max} = 0 \qquad \text{(b)}$$

$\sum M_A = 0$:

$$10N_2 - (3{,}600 \cos\theta_{max})(5) + (3{,}600 \sin\theta_{max})(1) = 0 \qquad \text{(c)}$$

To solve for θ_{max}, we eliminate N_1 from Eqs. (a) and (b), getting as a result the equation

$$N_2 = 3{,}600 \cos\theta_{max} - 6{,}000 \sin\theta_{max} \qquad \text{(d)}$$

Now, eliminating N_2 from Eq. (c) using Eq. (d), we get

$$18{,}000 \cos\theta_{max} - 56{,}400 \sin\theta_{max} = 0$$

Therefore,

$$\tan\theta_{max} = .320 \qquad \text{(e)}$$

Hence,

$$\theta_{max} = 17.7° \qquad \text{(f)}$$

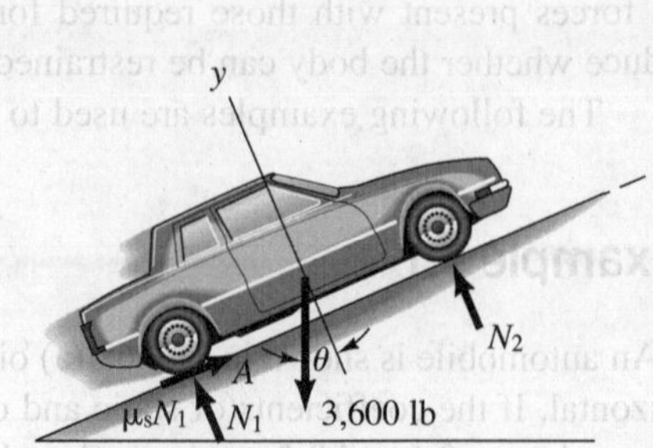

Figure 6.4. (a) Find maximum θ; (b) free-body diagram using Coulomb's law.

If the drive wheels were caused to spin, we would have to use μ_d in place of μ_s for this problem. We would then arrive at a smaller θ_{max}, which for this problem would be 14.7°.

Example 6.2

Using the data of Example 6.1, compute the torque needed by the drive wheels to move the car at a uniform speed up an incline where $\theta = 15°$. Also, assume that the brakes have "locked" while the car is in a parked position on the incline. What force is then needed to tow the car either up the incline or down the incline with the brakes in this condition? The diameter of the tire is 25 in.

A free-body diagram for the first part of the problem is shown in Fig. 6.5(a). Note that the friction force f will now be determined by Newton's law and not by Coulomb's law, since we do not have impending slippage between the wheel and the road for this case. Accordingly, we have, for f:

$$\underline{\sum F_x = 0:}$$

$$f - 3{,}600 \sin 15° = 0$$

Therefore,

$$f = 932 \text{ lb}$$

The torque needed is then computed using the rear wheels as a free body [see Fig. 6.5(a)]. Taking moments about A, we have

$$\text{torque} = (f)(r) = (932)\left(\frac{25/2}{12}\right) = \boxed{971 \text{ ft-lb}}$$

For the second part of the problem, we have shown the required free body in Fig. 6.5(b). Note that we have used Coulomb's law for the friction forces with the *dynamic* friction coefficient μ_d on all wheels. We now write the equations of equilibrium for this free body.

$$\underline{\sum F_x = 0:}$$

$$T_{\text{up}} - .5\left(N_1 + N_2\right) - 3{,}600 \sin 15° = 0 \qquad \text{(a)}$$

$$\underline{\sum F_y = 0:}$$

$$\left(N_1 + N_2\right) - 3{,}600 \cos 15° = 0 \qquad \text{(b)}$$

Solving for $N_1 + N_2$ from Eq. (b), and substituting into Eq. (a), we can now solve for T_{up}. Hence,

$$T_{\text{up}} = (.5)(3{,}600)(.966) + 932 = \boxed{2{,}670 \text{ lb}} \qquad \text{(c)}$$

Example 6.2 (Continued)

For towing the car down the incline we must reverse the direction of the friction forces as shown in Fig. 6.5(c). Solving for T_{down} as in the previous calculation, we get

$$T_{down} = (.5)(3{,}600)(.966) - 932 = \boxed{807 \text{ lb}} \qquad \text{(d)}$$

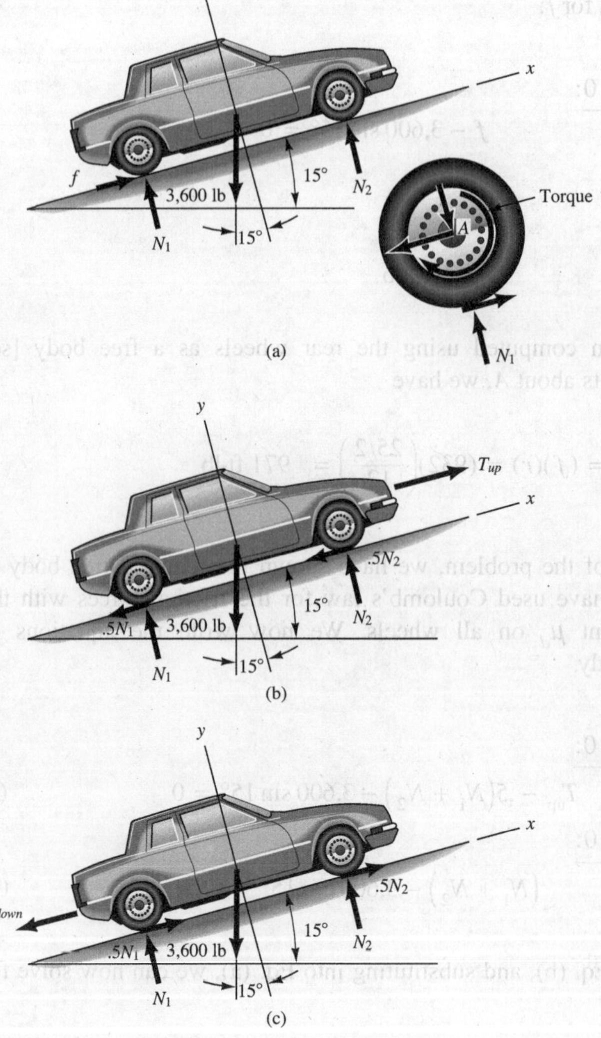

Figure 6.5. Free-body diagrams: (a) climbing at uniform speed; (b) under tow upward; (c) under tow downward.

Example 6.3

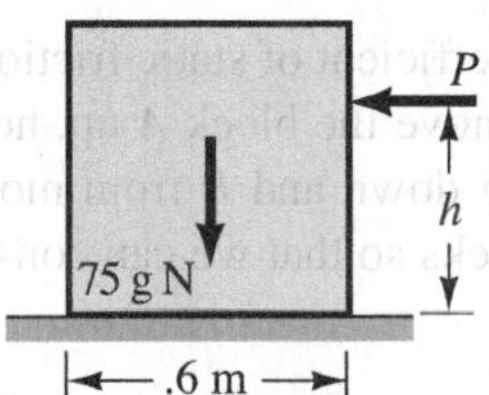

Figure 6.6. Strongbox being pushed.

In Fig. 6.6 a strongbox of mass 75 kg rests on a floor. The static coefficient of friction for the contact surface is .20. What is the largest force P and what is the highest position h for applying this force that will not allow the strongbox to either slip on the floor or to tip ?

The free-body diagram for the strongbox is shown in Fig. 6.7. The condition of *impending* motion has been recognized by the use of Coulomb's law. Furthermore, by concentrating the supporting and friction forces at the left corner, we are stipulating *impending tipping* for the problem. These two impending conditions impose the largest possible values of P and h that we are seeking.

The pertinent forces constitute a coplanar system of forces at the midplane of the strongbox. We proceed with the scalar equations of equilibrium:

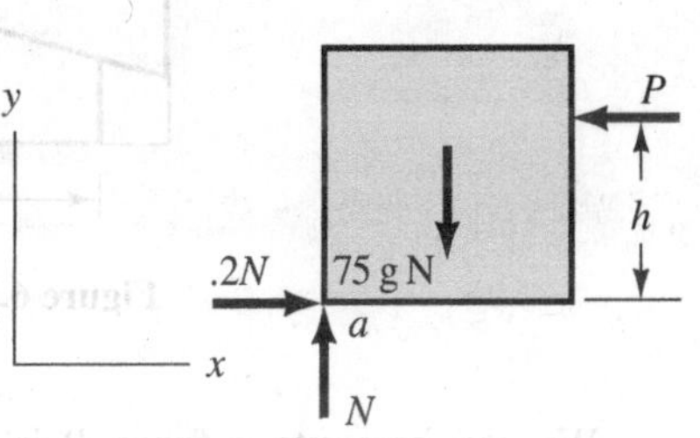

Figure 6.7. Impending tipping and slipping.

$\sum F_y = 0$:

$$N = 75g = 736 \text{ N}$$

$\sum F_x = 0$:

$$P = .2N = 147.15 \text{ N}$$

$\sum M_a = 0$:

$$-(75g)(.3) + (147.15)h = 0$$

Therefore, we get for the largest P and the largest h the following results:

$$P_{max} = 147.15 \text{ N} \qquad h_{max} = 1.50 \text{ m}$$

Thus, the height of the applied load must be less than or equal to 1.50 m in order to avoid tipping.

The three examples presented illustrated the *first* type of friction problem wherein we know the nature of the motion or impending motion present in the system and we determine certain forces or positions of certain forces. In the last example of this series, we illustrate the *second* type of friction problem set forth earlier—namely the problem of deciding whether bodies will move or not move under prescribed external forces.

Example 6.4

The coefficient of static friction for all contact surfaces in Fig. 6.8 is .2. Does the 50-lb force move the block *A* up, hold it in equilibrium, or is it too small to prevent *A* from coming down and *B* from moving out? The 50-lb force is exerted at the midplane of the blocks so that we can consider this a coplanar problem.

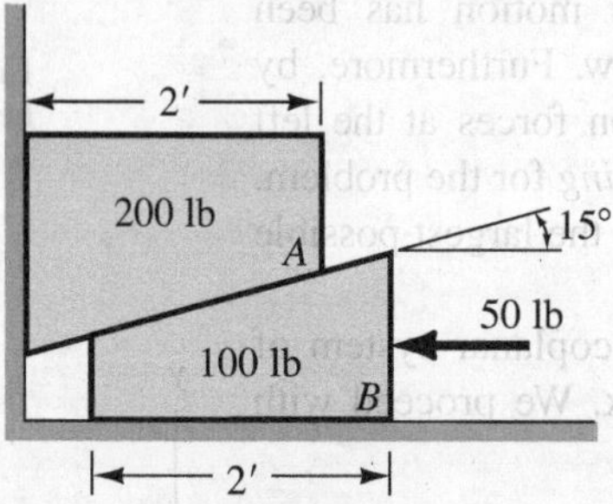

Figure 6.8. Do blocks move?

We can compute a force *P* in place of the 50-lb force to cause impending motion of block *B* to the left, and a force *P* for impending motion of block *B* to the right. In this way, we can judge by comparison the action that the 50-lb force will cause.

The free-body diagrams for impending motion of block *B* to the left have been shown in Fig. 6.9, which contains the unknown force *P* mentioned above. We need not be concerned about the correct location of the centers of gravity of the blocks, since we shall only add forces in the analysis. (We do not know the line of action of the normal forces at the contact surfaces and therefore cannot take moments.) Summing forces on block *A*, we get

$$N_2 \cos 15°\boldsymbol{j} - N_2 \sin 15°\boldsymbol{i} - .2N_1\boldsymbol{j} - 200\boldsymbol{j}$$
$$+ N_1\boldsymbol{i} - .2N_2 \cos 15°\boldsymbol{i} - .2N_2 \sin 15°\boldsymbol{j} = \boldsymbol{0}$$

The scalar equations are:

$$N_1 - .259N_2 - .1932N_2 = 0$$
$$.966N_2 - .2N_1 - 200 - .0518N_2 = 0$$

Solving simultaneously, we get

$$N_2 = 243 \text{ lb}, \qquad N_1 = 109.8 \text{ lb}$$

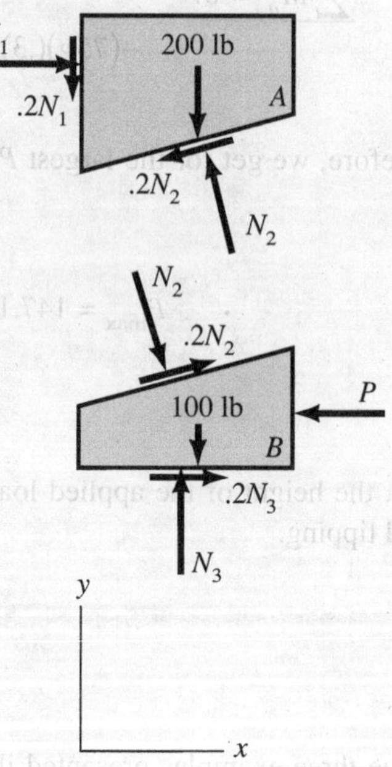

Figure 6.9. Impending motion of *B* to the left.

Example 6.4 (Continued)

For the free-body diagram of B, we have, on summing forces,

$$-N_2 \cos 15°\boldsymbol{j} + N_2 \sin 15°\boldsymbol{i} - P\boldsymbol{i} + .2N_3\boldsymbol{i} + N_3\boldsymbol{j}$$
$$- 100\boldsymbol{j} + .2N_2 \cos 15°\boldsymbol{i} + .2N_2 \sin 15°\boldsymbol{j} = \boldsymbol{0}$$

This yields the following scalar equations:

$$-P + 62.9 + .2N_3 + 46.9 = 0$$
$$-235 + N_3 - 100 + 12.6 = 0$$

Solving simultaneously, we have

$$P = 174 \text{ lb}$$

Clearly, the stipulated force of 50 lb is insufficient to induce a motion of block B to the left, so further computation is necessary.

Next, we reverse the direction of force P and compute what its value must be to move the block B to the right. The frictional forces in Fig. 6.9 are all reversed, and the vector equation of equilibrium for block A becomes

$$N_2 \cos 15°\boldsymbol{j} - N_2 \sin 15°\boldsymbol{i} + .2N_1\boldsymbol{j} - 200\boldsymbol{j}$$
$$+ N_1\boldsymbol{i} + .2N_2 \cos 15°\boldsymbol{i} + .2N_2 \sin 15°\boldsymbol{j} = \boldsymbol{0}$$

The scalar equations are:

$$N_1 - .259N_2 + .1932N_2 = 0$$
$$.966N_2 + .2N_1 - 200 + .0518N_2 = 0$$

Solving simultaneously, we get

$$N_2 = 194.1 \text{ lb}, \qquad N_1 = 12.80 \text{ lb}$$

For free body B we have, on summing forces,

$$-N_2 \cos 15°\boldsymbol{j} + N_2 \sin 15°\boldsymbol{i} + P\boldsymbol{i} - .2N_3\boldsymbol{i} + N_3\boldsymbol{j}$$
$$- 100\boldsymbol{j} - .2N_2 \cos 15°\boldsymbol{i} - .2N_2 \sin 15°\boldsymbol{j} = \boldsymbol{0}$$

The following are the scalar equations:

$$P - .2N_3 + 50.3 - 37.5 = 0$$
$$-100 + N_3 - 187.5 - 10.05 = 0$$

Solving, we get $P = 46.7$ lb. Thus, we would have to *pull to the right* to get block B to move in this direction. We can now conclude from this study that the blocks are in equilibrium.

PROBLEMS

6.1. A block has a force F applied to it. If this force has a time variation as shown in the diagram, draw a simple sketch showing the friction force variation with time. Take $\mu_s = .3$ and $\mu_d = .2$ for the problem.

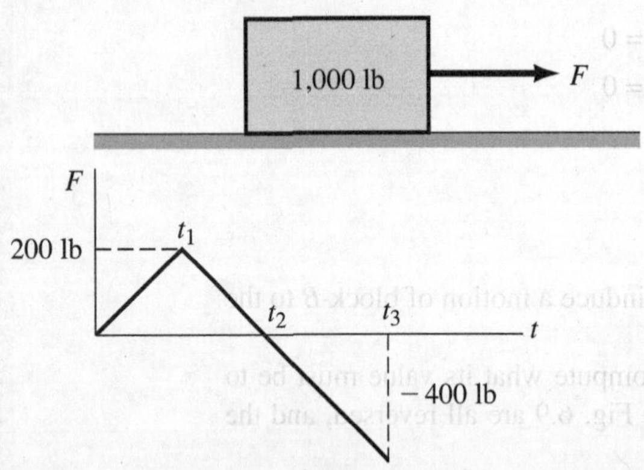

Figure P.6.1.

6.2. Show by increasing the inclination ϕ on an inclined surface until there is impending slippage of supported bodies, we reach the *angle of repose* ϕ_s so that $\tan \phi_s = \mu_s$.

6.3. To what minimum angle must the driver elevate the dump bed of the truck to cause the wooden crate of weight W to slide out? For wood on steel, $\mu_s = .6$ and $\mu_d = .4$.

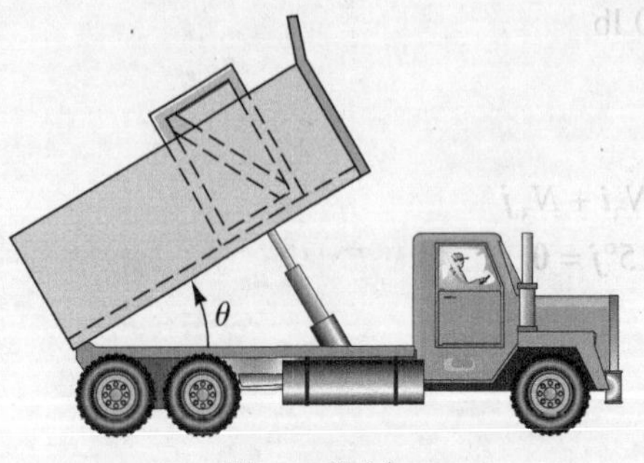

Figure P.6.3.

6.4. A platform is suspended by two ropes which are attached to blocks that can slide horizontally. At what value of W does the platform begin to descend? Will W start tipping?

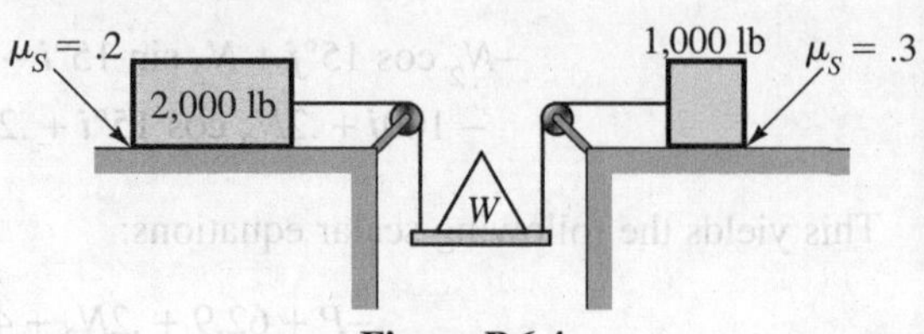

Figure P.6.4.

6.5. Explain how a violin bow, when drawn over a string, maintains the vibration of the string. Put this in terms of friction forces and the difference in static and dynamic coefficients of friction.

6.6. What is the value of the force F, inclined at 30° to the horizontal, needed to get the block just started up the incline? What is the force F needed to keep it just moving up at a constant speed? The coefficients of static and dynamic friction are .3 and .275, respectively.

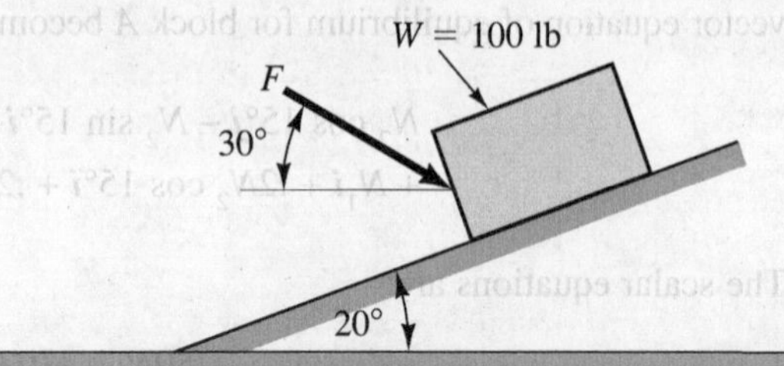

Figure P.6.6.

6.7. Bodies A and B weigh 500 N and 300 N, respectively. The platform on which they are placed is raised from the horizontal position to an angle θ. What is the *maximum* angle that can be reached before the bodies slip down the incline? Take μ_s for body B and the plane as .2 and μ_s for body A and the plane as .3.

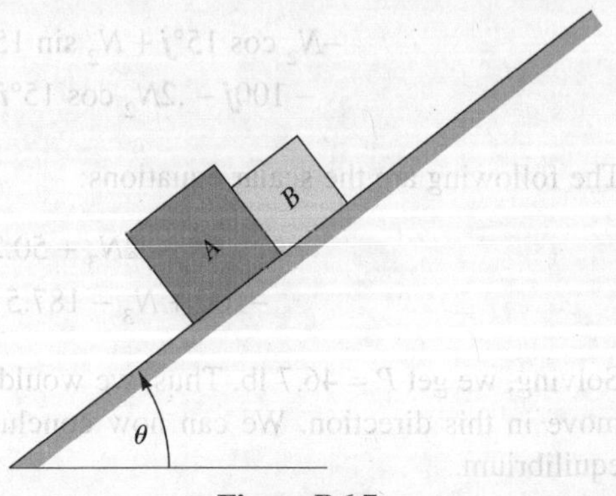

Figure P.6.7.

6.8. What is the minimum value of μ_s that will allow the rod AB to remain in place? The rod has a length of 3.3 m and it has a weight of 200 N.

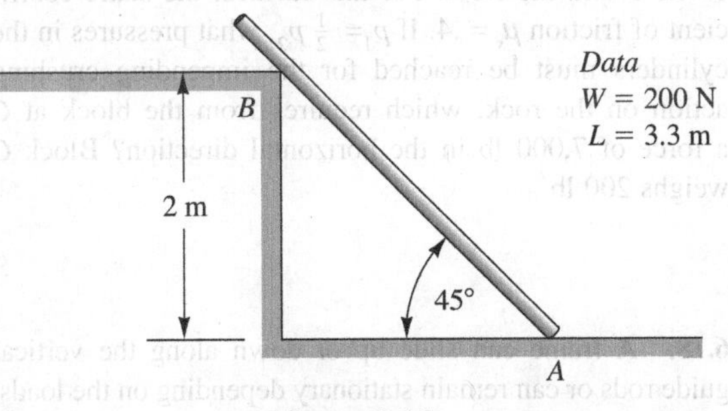

Figure P.6.8.

6.9. Find the minimum force P to get block A moving.

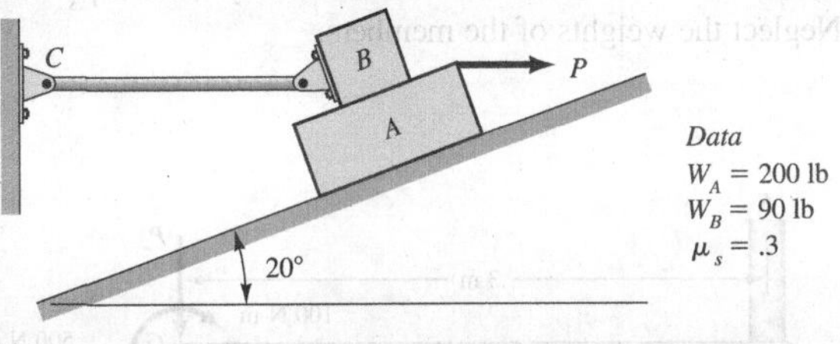

Figure P.6.9.

6.10. What minimum force F is needed to start body A moving to the right if $\mu_s = .25$ for all surfaces? The following weights are given:

$W_A = 125$ N $\quad W_B = 50$ N $\quad W_{AB} = 100$ N

The length of AB is 2.5 m.

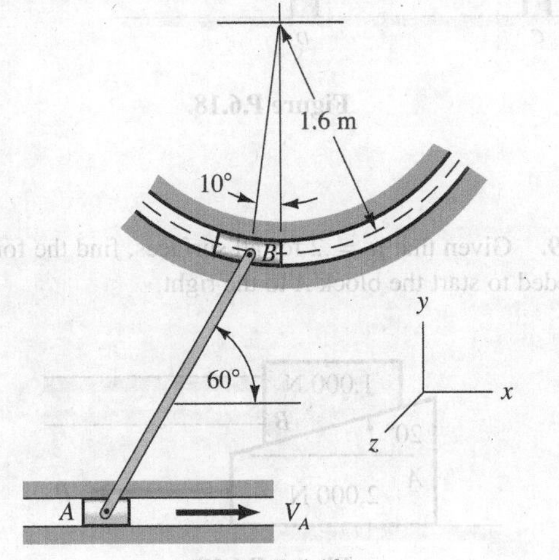

Figure P.6.10.

7.11. A 30-ton tank is moving up a 30° incline. If $\mu_s = .6$ for the contact surface between tread and ground, what *maximum* torque can be developed at the rear drive sprocket with no slipping? What maximum towing force F can the tank develop? Take the mean diameter of the rear sprocket as 2 ft.

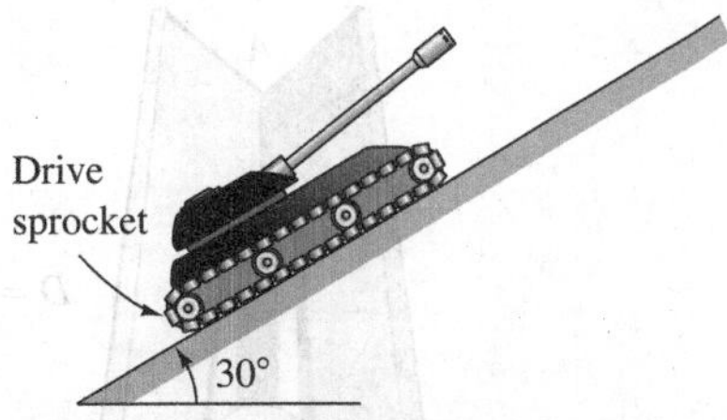

Figure P.6.11.

6.12. A 500-lb crate A rests on a 1,000-lb crate B. The centers of gravity of the crates are at the geometric centers. The coefficients of static friction between contact surfaces are shown in the diagram. The force T is increased from zero. What is the first action to occur?

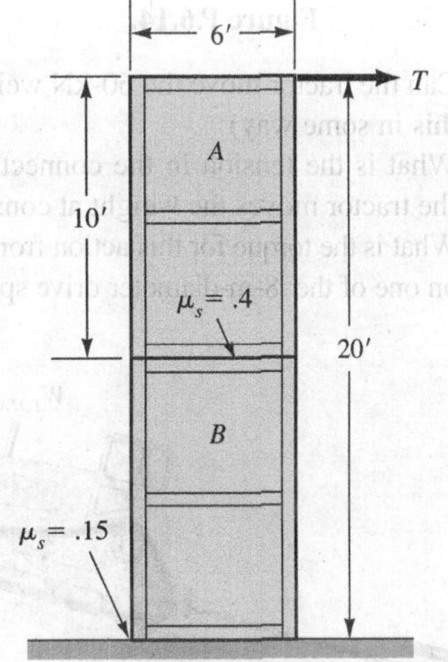

Figure P.6.12.

6.13. What force F is needed to get the 300-kg block moving to the right? The coefficient of static friction for all surfaces is .3.

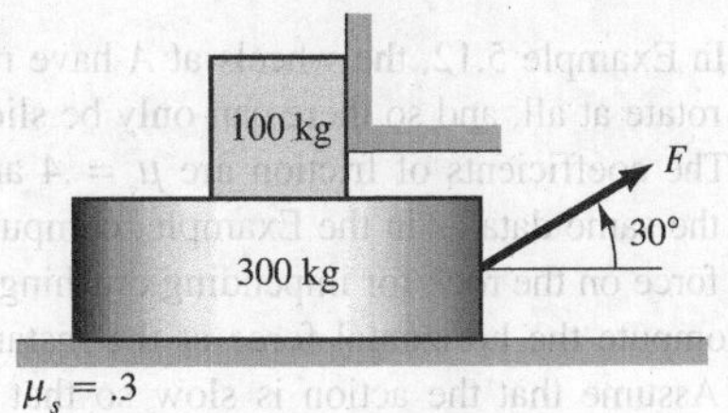

Figure P.6.13.

6.14. A chute is shown having sides that are at right angles to each other. The chute is 30 ft in length with end A 10 ft higher than end B. Cylinders weighing 200 lb are to slide down the chute. What is the *maximum* allowable static coefficient of friction so there cannot be sticking of the cylinders along the chute?

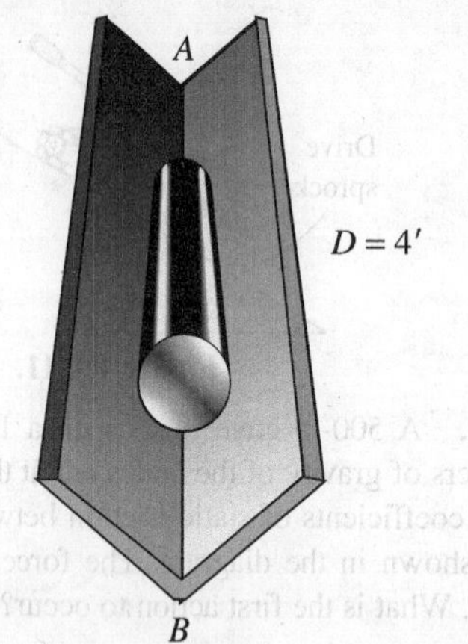

Figure P.6.14.

6.15. (a) Can the tractor move the 60-kN weight? (Prove this in some way)
(b) What is the tension in the connecting chain if the tractor moves the weight at constant speed?
(c) What is the torque for this action from the engine on one of the .8-m-diameter drive sprockets?

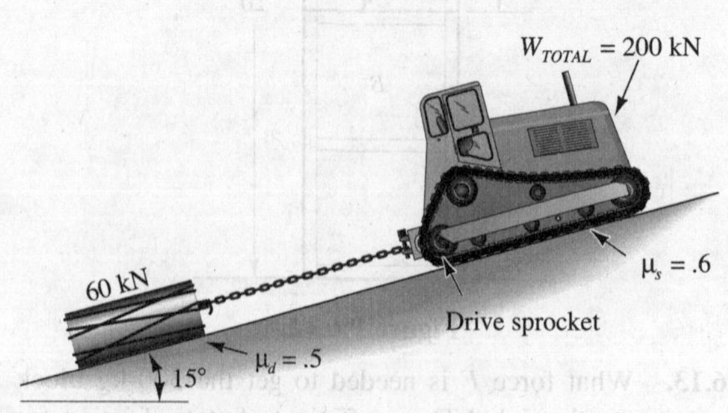

Figure P.6.15.

6.16. In Example 5.12, the wheels at A have rusted so as not to rotate at all, and so there can only be sliding on the floor. The coefficients of friction are $\mu_s = .4$ and $\mu_d = .2$. Using the same data as in the Example, compute the horizontal force on the rock for impending crushing action and then compute the horizontal force at the instant crushing starts. Assume that the action is slow so that we do not need to consider inertial effects. The plate A weighs 100 lb.

6.17. In Problem 5.92, the rollers on the block C have rusted and can no longer roll at all along the floor. There is then a dry friction at the contact surface between the block C and the floor. For this surface, the static coefficient of friction $\mu_s = .4$. If $p_1 = \frac{1}{2} p_2$, what pressures in the cylinders must be reached for the impending crushing action on the rock, which requires from the block at C a force of 7,000 lb in the horizontal direction? Block C weighs 200 lb.

6.18. A frame can slide up or down along the vertical guide rods or can remain stationary depending on the loads. What force P will cause impending slippage downward for the given applied loads? Member GA has a loose fit with circular rod EC, and member BH has a loose fit with circular rod JD. The coefficient of friction everywhere is $\mu_s = .3$. Neglect the weights of the members.

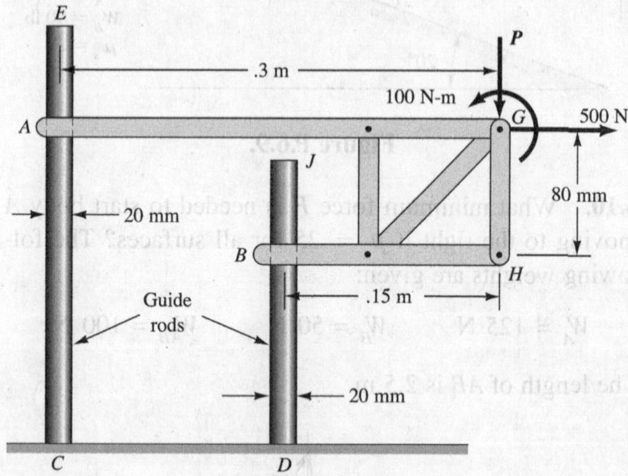

Figure P.6.18.

6.19. Given that $\mu_s = .2$ for all surfaces, find the force P needed to start the block A to the right.

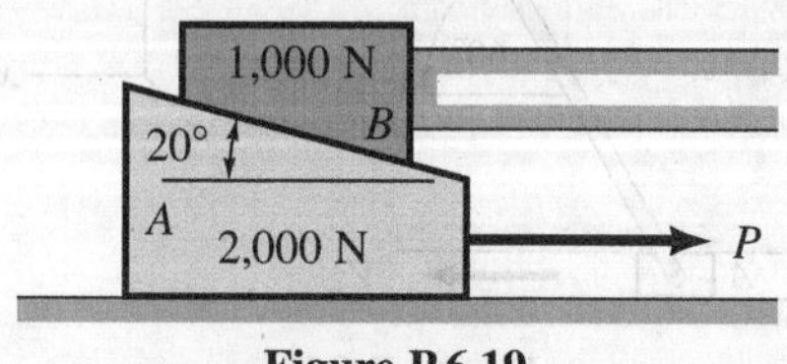

Figure P.6.19.

6.20. The cylinder shown weighs 200 N and is at rest. What is the friction force at *A*? If there is impending slippage, what is the static coefficient of friction? The supporting plane is inclined at 60° to the horizontal.

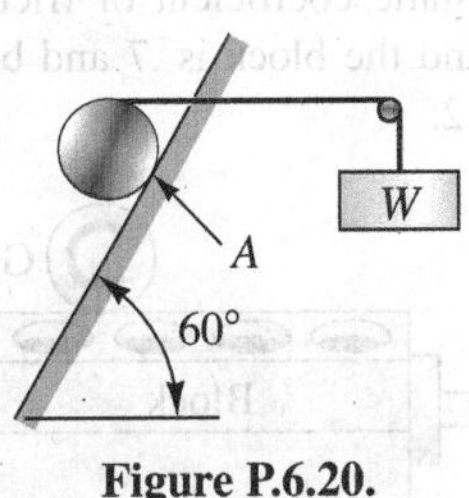

Figure P.6.20.

6.21. Armature *B* is stationary while rotor *A* rotates with angular speed ω. In armature *B* there is a braking system. If $\mu_d = .4$, what is the braking torque on *A* for a force *F* of 300 N? Note that the rod on which *F* is applied is pinned at *C* to the armature *B*. Neglect friction between *B* and the brake pads *G* and *H*.

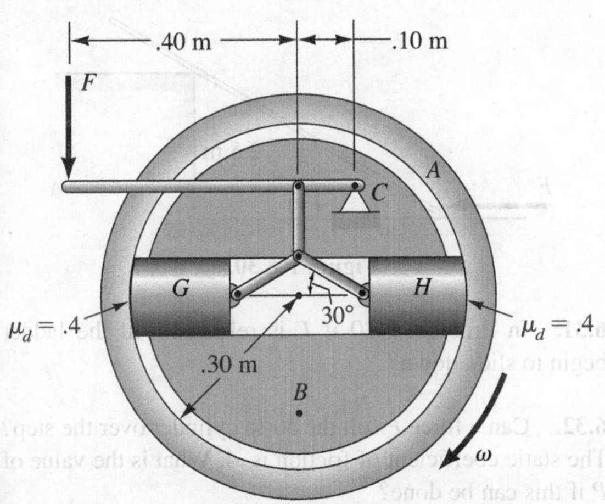

Figure P.6.21.

6.22. A 200-lb load is placed on the luggage rack of the 4,500-lb station wagon. Will the station wagon climb the hill more easily or with greater difficulty with the luggage than without? Explain. The static coefficient of friction is .55.

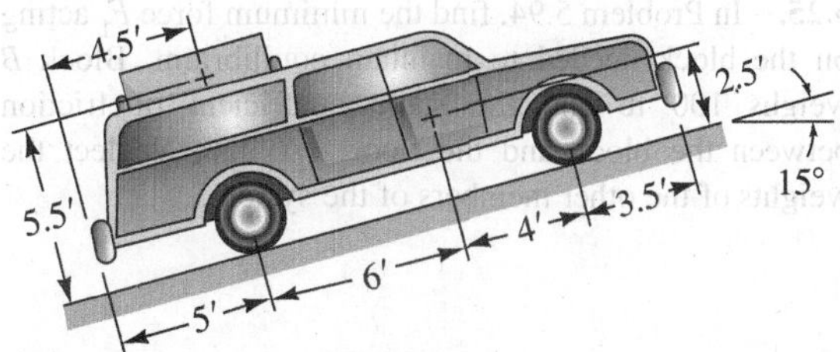

Figure P.6.22.

6.23. An insect tries to climb out of a hemispherical bowl of radius 600 mm. If the coefficient of static friction between insect and bowl is .4, how high up does the insect go? If the bowl is spun about a vertical axis, the bug gets pushed out in a radial direction by the force $mr\omega^2$, as you learned in physics. At what speed ω will the bug just be able to get out of the bowl?

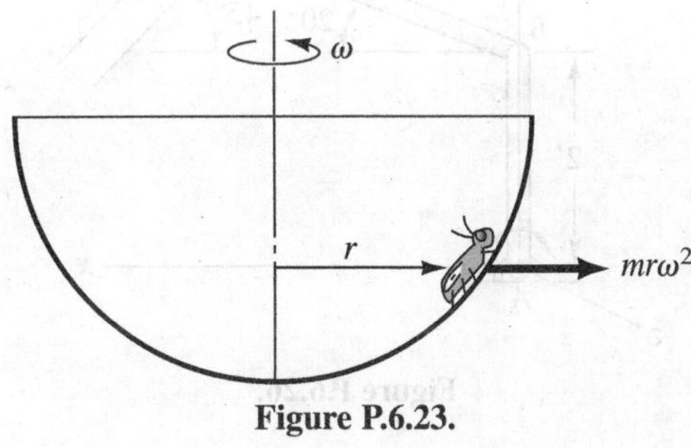

Figure P.6.23.

6.24. A block *A* of mass 500 kg rests on a stationary support *B* where the static coefficient of friction $\mu_s = .4$. On the right side, support *C* is on rollers. The dynamic coefficient of friction μ_d of the support *C* with body *A* is .2. If *C* is moved at constant speed to the left, how far does it move before body *A* begins to move?

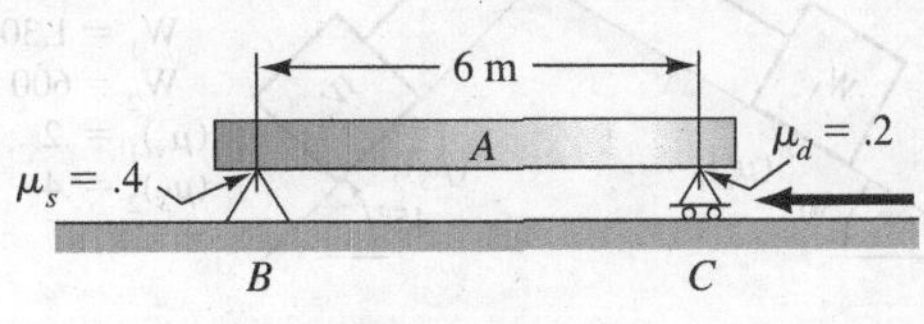

Figure P.6.24.

6.25. In Problem 5.94, find the minimum force F_1 acting on the block needed to maintain equilibrium. Block B weighs 100 lb and the static coefficient of friction between the block and the floor is 0.276. Neglect the weights of the other members of the system.

6.26. If the static coefficient of friction at C is .4, what is the minimum torque T needed to start a counterclockwise rotation of BA about hinge A? The following are the weights of the bodies involved:

$AB = 3$ lb $\quad BC = 4$ lb $\quad C = 2$ lb

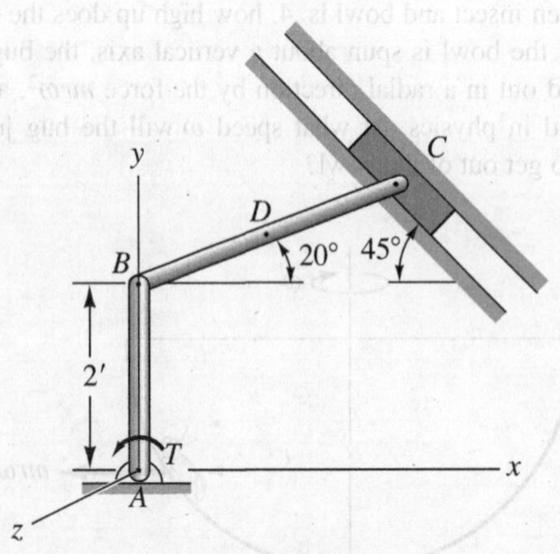

Figure P.6.26.

6.27. What will ensue if the weights are released from a state of rest? Examine all possibilities to justify your result.

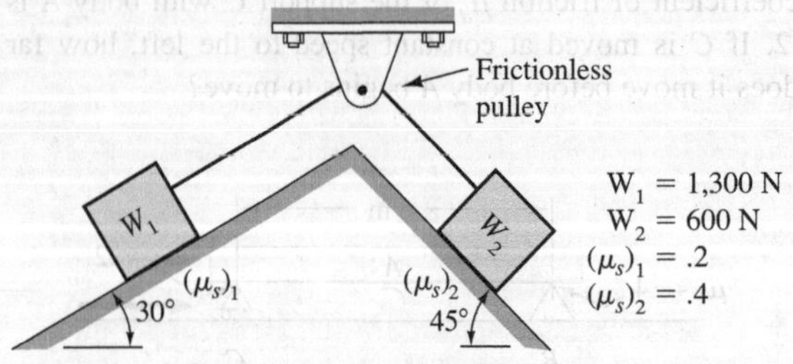

Figure P.6.27.

6.28. In a preliminary grinding operation for a 1,500-N car engine block, the grinding wheel is pushed against the block with a 500-N force. What force must be exerted by the hydraulic ram to move the block to the right if (a) the wheel rotates clockwise and (b) the wheel rotates counterclockwise? The static coefficient of friction between the grinding wheel and the block is .7 and between the table and the block is .2.

Figure P.6.28.

6.29. An 8,000-lb tow truck with four-wheel drive and 36 in. diameter tires develops a torque of 750 lb-ft at each axle. What is the *heaviest* car that can be towed up a 10° slope if $\mu_s = .3$?

6.30. A 7-m ladder weighing 250 N is being pushed by force F. What is the *minimum* force needed to get the ladder to move? The static coefficient of friction for all contact surfaces is .4.

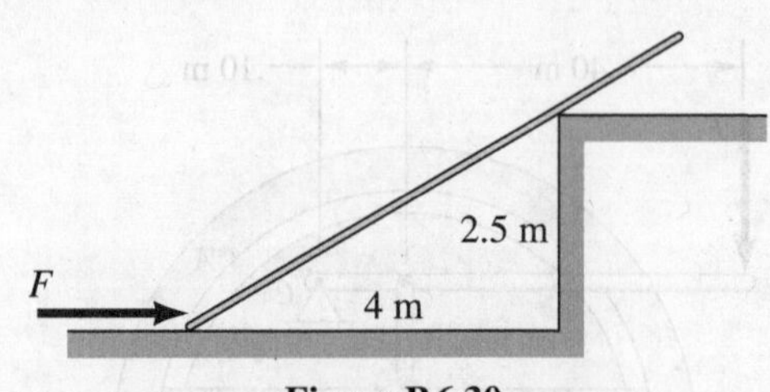

Figure P.6.30.

6.31. In Problem 6.30 if F is released will the ladder begin to slide down?

6.32. Can a force P roll the 50-lb cylinder over the step? The static coefficient of friction is .4. What is the value of P if this can be done?

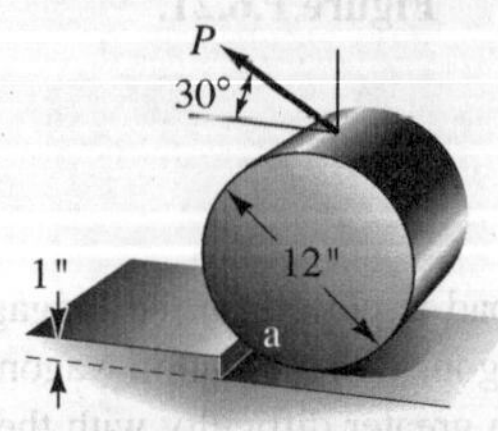

Figure P.6.32.

6.33. The block of weight W is to be moved up an inclined plane. A rod of length c with negligible weight is attached to the block and the force F is applied to the top of this rod. If the coefficient of static friction is μ_s, determine in terms of a, d, and μ_s the *maximum* length c for which the block will begin to slide rather than tip.

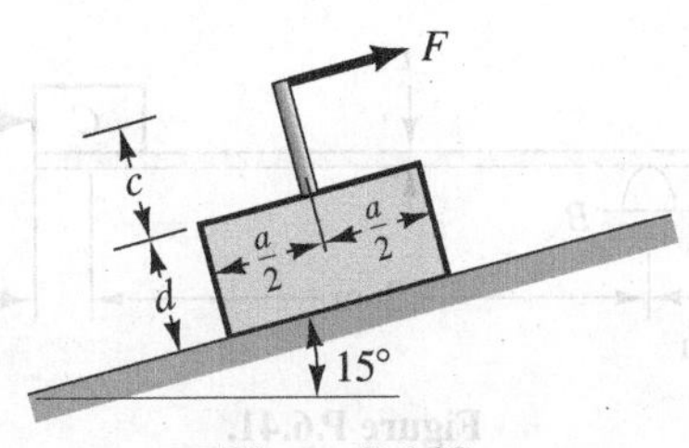

Figure P.6.33.

6.34. Determine the range of values of W_1 for which the block will either slide up the plane or slide down the plane. At what value of W_1 is the friction force zero? $W_2 = 100$ lb.

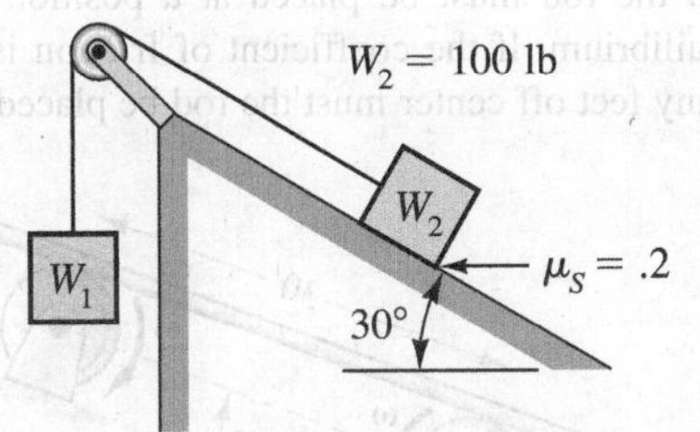

Figure P.6.34.

6.35. A 200-kN tractor is to push a 60-kN concrete beam up a 15° incline at a construction site. If $\mu_d = .5$ between beam and dirt and if μ_s is .6 between tractor tread and dirt, can the tractor do the job? If so, what torque must be developed on the tractor drive sprocket which is .8 m in diameter? What force P is then developed to push the beam?

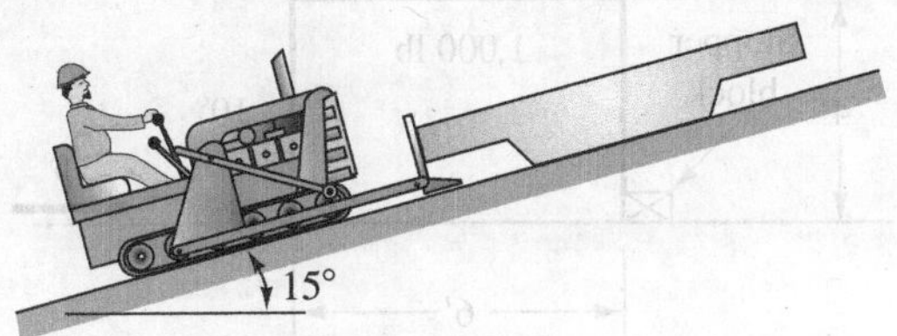

Figure P.6.35.

6.36. What is the *minimum* static coefficient of friction required to maintain the bracket and its 500-lb load in a static position? (Assume point contacts at the horizontal centerlines of the arms.) The center of gravity is 7 in. from the shaft centerline. Hint: Note that there is clearance between the vertical shaft and the horizontal arms.

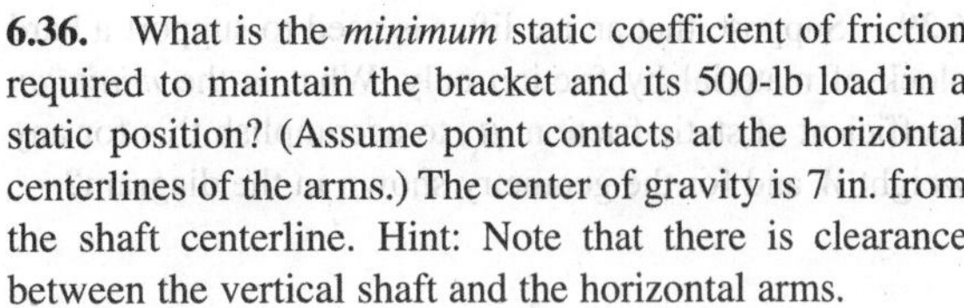

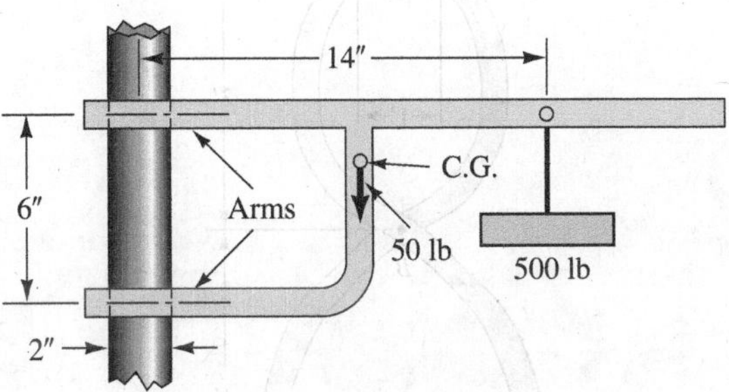

Figure P.6.36.

6.37. If the static coefficient of friction in Problem 6.36 is .2, at what *minimum* distance from the centerline of the vertical shaft can we support the 500-lb load without slipping?

6.38. A rod is held by a cord at one end. If the force $F = 200$ N, and if the rod weighs 450 N, what is the *maximum* angle α that the rod can be placed for μ_s between the rod and the floor equal to .4? The rod is 1 m in length.

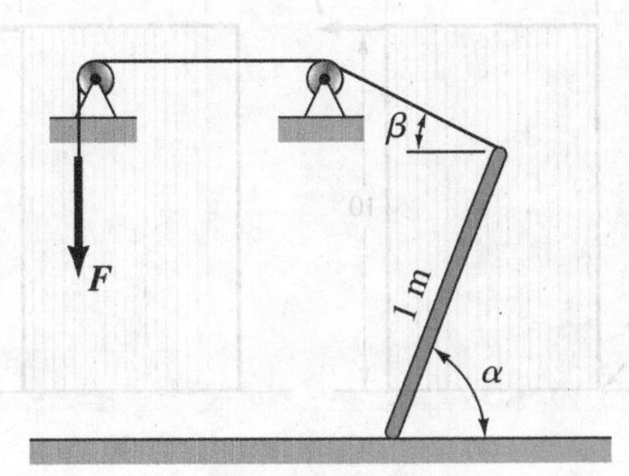

Figure P.6.38.

6.39. Suppose that an ice lifter is used to support a hard block of material by friction only. What is the *minimum* coefficient of static friction, μ_s, to accomplish this for any weight W and for the geometry shown in the diagram?

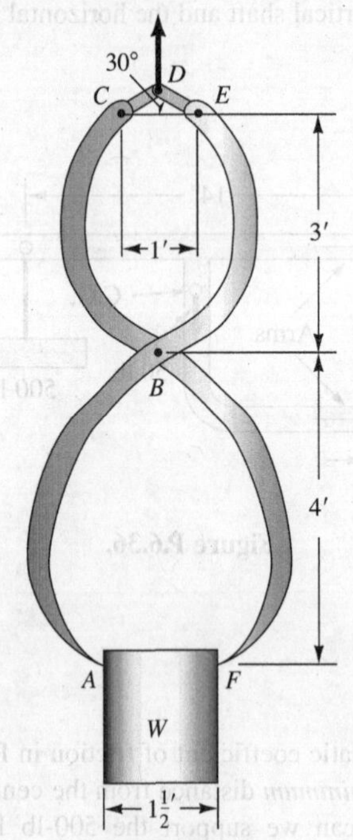

Figure P.6.39.

6.40. A rectangular case is loaded with uniform vertical thin rods such that when it is full, as shown in (a), the case has a total weight of 1,000 lb. The case weighs 100 lb when empty and has a coefficient of static friction of .3 with the floor as shown in the diagram. A force T of 200 lb is maintained on the case. If the rods are unloaded as shown in (b), what is the limiting value of x for equilibrium to be maintained?

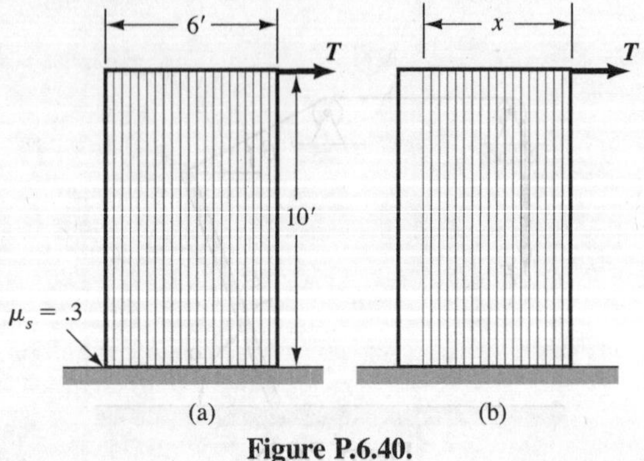

Figure P.6.40.

6.41. A beam supports load C weighing 500 N. At supports A and B, the static coefficient of friction is .2. At the contact surface between load C and the beam, the dynamic coefficient of friction is .75. If force F moves C steadily to the left, how far does it move before the beam begins to move? The beam weighs 200 N. Neglect the height t of the beam in your calculations.

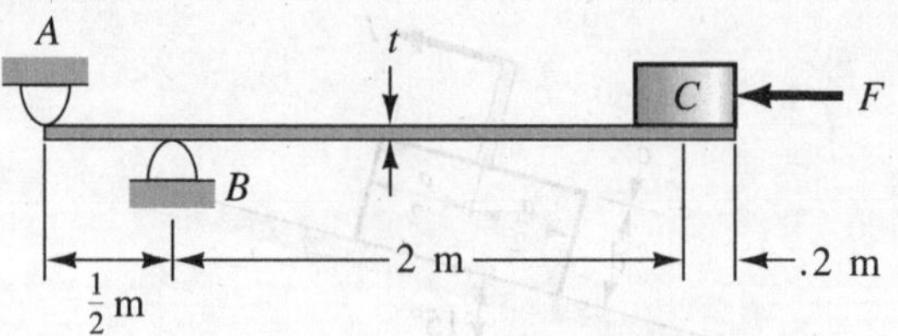

Figure P.6.41.

6.42. Do Problem 6.41 for the case where the height t is taken into account. Take $t = 120$ mm.

6.43. A rod is supported by two wheels spinning in opposite directions. If the wheels were horizontal, the rod would be placed centrally over the wheels for equilibrium. However, the wheels have an inclination of 20° as shown, and the rod must be placed at a position off center for equilibrium. If the coefficient of friction is $\mu_d = .8$, how many feet off center must the rod be placed?

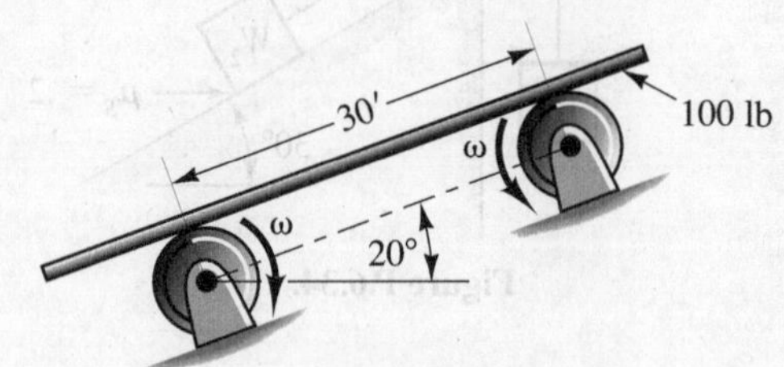

Figure P.6.43.

P.6.44. How much force F must be applied to the wedge to begin to raise the crate? Neglect changes in geometry. What force must the stopper block provide to prevent the crate from moving to the left? The static coefficient of friction between all surfaces is .3.

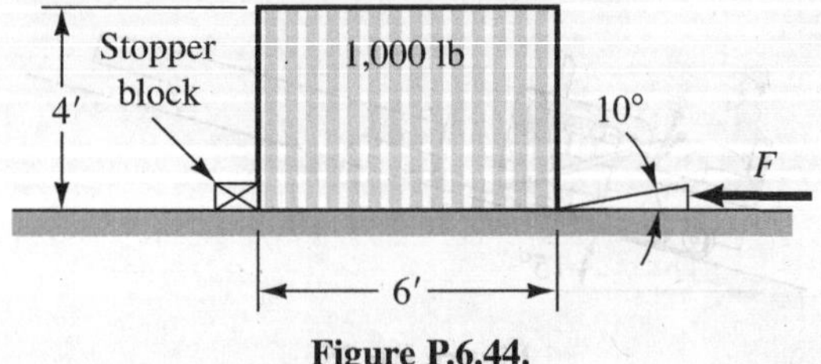

Figure P.6.44.

6.45. What is the *maximum* height x of a step so that the force P will roll the 50-lb cylinder over the step with no slipping at a? Take $\mu_s = .3$.

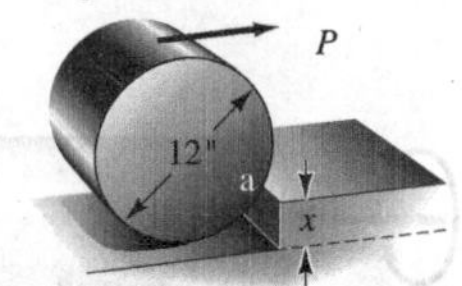

Figure P.6.45.

6.46. The rod AD is pulled at A and it moves to the left. If the coefficient of dynamic friction for the rod at A and B is .4, what must the *minimum* value of W_2 be to prevent the block from tipping when $\alpha = 20°$? With this value of W_2, determine the minimum coefficient of static friction between the block and the supporting plane needed to just prevent the block from sliding. W_1 is 100 N.

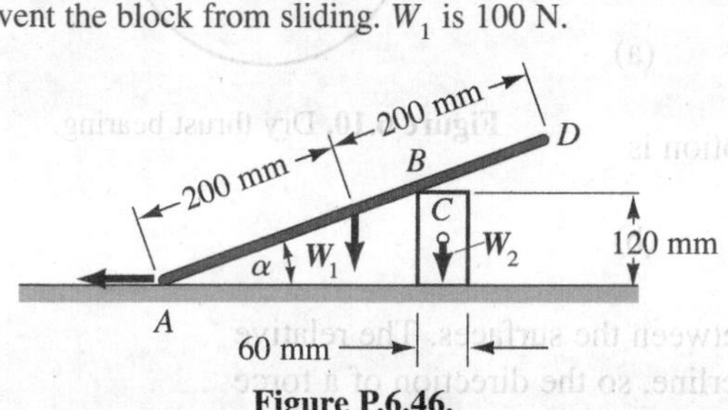

Figure P.6.46.

7.47. If we neglect friction at the rollers, and the coefficient of static friction is .2 for all surfaces, ascertain whether the 5,000-lb weight will go up, go down, or stay stationary.

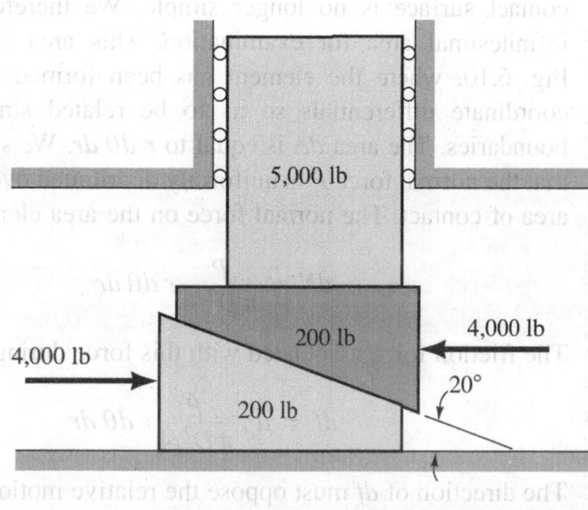

Figure P.7.47.

6.4 Complex Surface Contact Friction Problems

In the examples undertaken heretofore, the nature of the relative impending or actual motion between the plane surfaces of contact was quite simple—that of motion without rotation. We shall now examine more general types of contacts between bodies. In Example 6.5, we have a plane contact surface but with varying direction of impending or slipping motion for the area elements as a result of rotation. In such problems we shall have to apply Coulomb's laws locally to infinitesimal areas of contact and to integrate the results, for reasons explained in Section 6.2. To do this, we must ascertain the distribution of the normal force at the contact surface, an undertaking that is usually difficult and well beyond the capabilities of rigid-body statics, as explained in Chapter 5. However, we can at times *approximately* compute frictional effects by *estimating* the manner of distribution of the normal force at the surface of contact. We now illustrate this.

Example 6.5

Compute the frictional resistance to rotation of a rotating solid cylinder with an attached pad A pressing against a flat dry surface with a force P (see Fig. 6.10). The pad A and the stationary flat dry surface constitute a dry *thrust bearing*.

The direction of the frictional forces distributed over the contact surface is no longer simple. We therefore take an infinitesimal area for examination. This area is shown in Fig. 6.10, where the element has been formed from polar-coordinate differentials so as to be related simply to the boundaries. The area dA is equal to $r\,d\theta\,dr$. We shall *assume* that the normal force P is uniformly distributed over the entire area of contact. The normal force on the area element is then

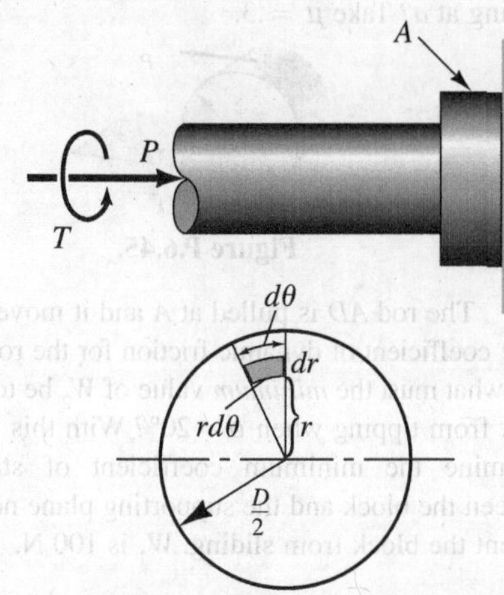

Figure 6.10. Dry thrust bearing.

$$dN = \frac{P}{\pi D^2/4} r\,d\theta\,dr \quad \text{(a)}$$

The friction force associated with this force during motion is

$$df = \mu_d \frac{P}{\pi D^2/4} r\,d\theta\,dr \quad \text{(b)}$$

The direction of df must oppose the relative motion between the surfaces. The relative motion is rotation of concentric circles about the centerline, so the direction of a force df_1 (Fig. 6.11) must lie tangent to a circle of radius r. At 180° from the position of the area element for df_1, we may carry out a similar calculation for a force df_2, which for the same r must be equal and opposite to df_1, thus forming a couple. Since the entire area may be decomposed in this way, we can conclude that there are only couples in the plane of contact. If we take moments of all infinitesimal forces about the center, we get the magnitude of the total frictional couple moment. The direction of the couple moment is along the shaft axis. First, consider area elements on the ring of radius r:

$$dM = \int_0^{2\pi} r\mu_d \frac{P}{\pi D^2/4} r\,d\theta\,dr \quad \text{(c)}$$

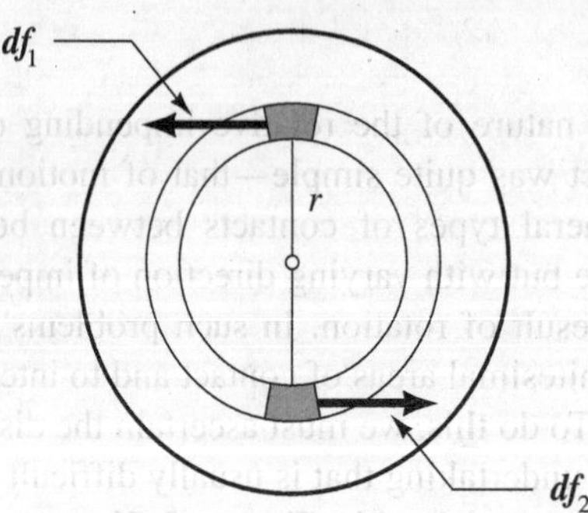

Figure 6.11. Friction forces form couples.

Example 6.5 (Continued)

Taking μ_d as constant and holding r constant, we have on integration with respect to θ:

$$dM = \mu_d \frac{P}{\pi D^2/4} 2\pi r^2 \, dr$$

We thus account for all area elements on the ring of radius r. To account for all the rings of the contact surface, we next integrate with respect to r from zero to $D/2$. Clearly, this gives us the total resisting torque M. Thus,

$$M = \mu_d \frac{8P}{D^2} \int_0^{D/2} r^2 \, dr = \frac{PD\mu_d}{3} \qquad \text{(d)}$$

What we have performed in the last three steps is *multiple integration*, which we introduced in Chapter 4 when dealing with rectangular coordinates.

6.5 Transmission of Power Through Belts

Belts are extensively used to transmit power from one shaft to another shaft. Their working is dependent on the friction acting between the pulley surface and belt surface.

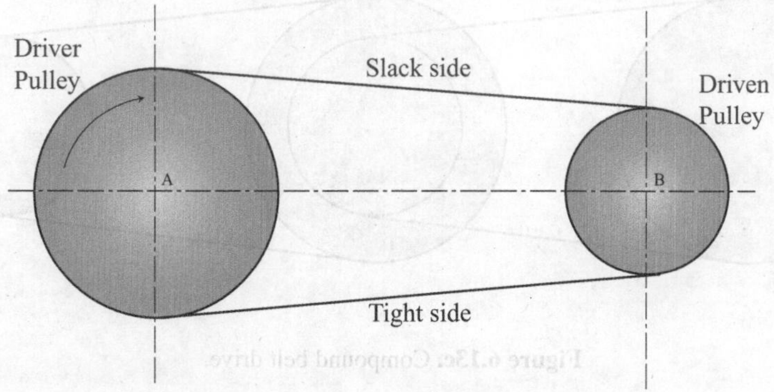

Figure 6.12. Pulley-belt system.

The belts may be flat belts, V-belts, etc. (Fig. 6.12). Ropes and chains are also used to transmit power.

Types of Flat Belts

Flat belts can be classified as (Fig. 6.13):

1. Open belt drive,
2. Cross belt drive, and
3. Compound belt drive

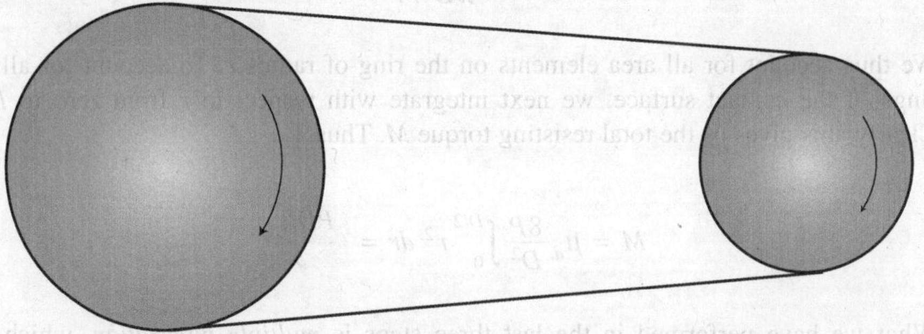

Figure 6.13a. Open belt drive.

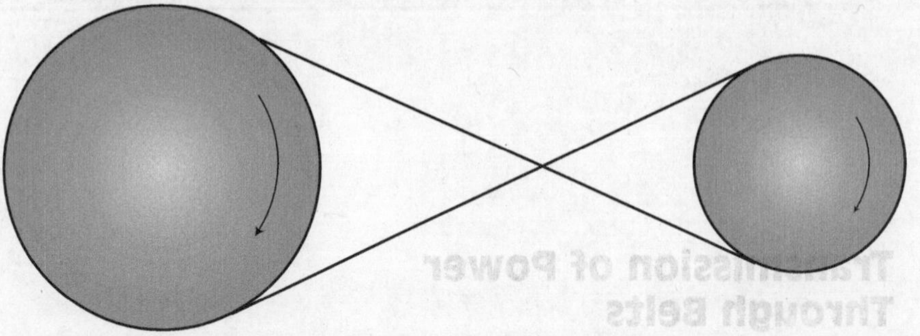

Figure 6.13b. Cross belt drive.

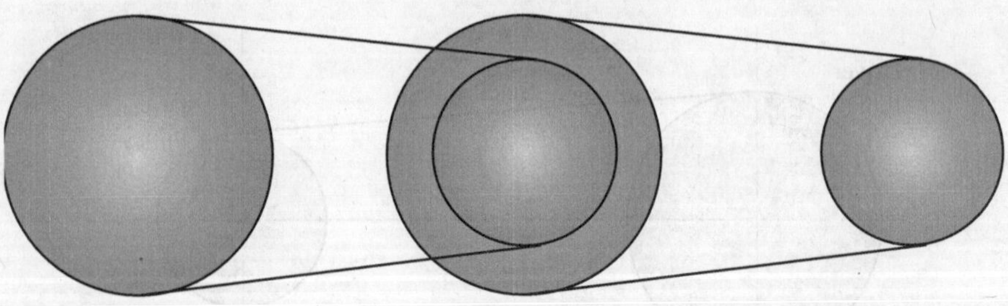

Figure 6.13c. Compound belt drive.

Open flat belt drive. It consists of two pulleys A and B. One is known as driver pulley and is keyed to the rotating shaft. The other one is known as driven pulley or follower pulley and is keyed to the shaft which is to be rotated (Fig. 6.14).

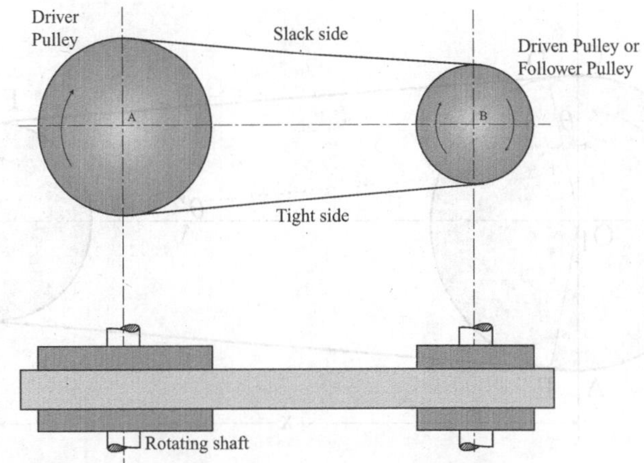

Figure 6.14. Open flat belt drive.

Shafts are placed parallel to each other and both the pulleys rotate in the same direction. The belt is pulled by the driver pulley from lower side and delivered to the upperside. This results in higher tension on the lower side than upperside.

Velocity ratio. It is defined as the ratio of the velocity of the follower (or driven) to the velocity of the driver.

Let N_1 = Speed of the driver in rpm
d_1 = Diameter of the driver
N_2, d_2 = Speed and diameter of the follower, respectively.
Length of the belt passing over the two pulley per minute will be the same

$$\pi d_1 N_1 = \pi d_2 N_2$$

$$\frac{N_2}{N_1} = \frac{d_1}{d_2}$$

Velocity ratio, $\dfrac{N_2}{N_1} = \dfrac{d_1}{d_2}$

If the thickness is of considerable amount, it may also be taken in to account.

Velocity ratio $\dfrac{N_2}{N_1} = \dfrac{d_1 + t}{d_2 + t}$

It can be inferred from the equation that the speed of the pulley is inversely proportional to the diameter of the pulley.

Length of the open belt. Total length of the belt is the sum of the length of the belt not in contact with either of the pulleys, the length of belt in contact with the larger pulley and length of the belt in contact with the smaller pulley (Fig. 6.15).

Let x = Distance between the centers of the two pulleys (i.e., $O_1 O_2$)
r_1 = Radius of the larger pulley
r_2 = Radius of the smaller pulley
L = Total length of the belt

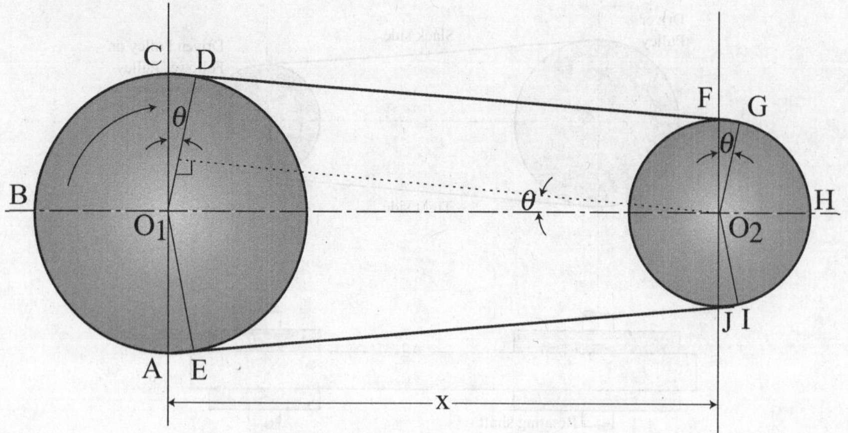

Figure 6.15. Length of the open belt.

The belt leaves the larger pulley at D and E and smaller pulley at G and I.
Now join D and E with O_1 and also G and I with O_2.
From O_2 draw O_2K parallel to GD. GD is tangent at D. Hence $\angle O_1DG = 90^\circ$. Since O_2K is parallel to GD $\angle O_1KO_2 = 90^\circ$.

Let the angle O_1O_2K be θ radians.
Now total length (L) of the belt is given by

$$L = \text{Arc EBD} + \text{DG} + \text{Arc GHI} + \text{IE}$$

$$= 2[\text{Arc BD} + \text{DG} + \text{Arc GH}]$$

$$2 = \left[r_1\left(\frac{\pi}{2}+\theta\right) + \sqrt{x^2-(r_1-r_2)^2} + r_2\left(\frac{\pi}{2}-\theta\right) \right] \tag{6.2}$$

But $\sqrt{x^2-(r_1-r_2)^2} = x\sqrt{1-\left(\frac{r_1-r_2}{x}\right)^2}$

$$= x\left[1-\left(\frac{r_1-r_2}{x}\right)^2\right]^{1/2}$$

$$= x\left[1-\frac{1}{2}\left(\frac{r_1-r_2}{x}\right)^2+\cdots\right] \quad \text{(By using Binomial expression)}$$

$$= x\left[1-\frac{(r_1-r_2)^2}{2x^2}\right] \quad \text{(By neglecting smaller terms)}$$

Substituting in Eq. (6.2) we get

$$L = 2\left[r_1\left(\frac{\pi}{2}+\theta\right)+x\left(1-\frac{(r_1-r_2)^2}{2x^2}\right)+r_2\left(\frac{\pi}{2}-\theta\right)\right]$$

$$= 2\left[\frac{\pi}{2}(r_1+r_2)+\theta(r_1-r_2)+\left(x-\frac{(r_1-r_2)^2}{2x}\right)\right]$$

$$= \pi(r_1+r_2)+2\theta(r_1-r_2)+2x-\frac{(r_1-r_2)^2}{x} \quad (6.3)$$

But $\sin\theta = \dfrac{r_1-r_2}{x}$

Since θ is very small we can assume $\sin\theta = \theta$

$$\Rightarrow \theta = \frac{r_1-r_2}{x}$$

Substituting the value of θ in Eq. (6.3)

$$L = \pi(r_1+r_2)+2\left[\frac{r_1-r_2}{x}\right](r_1-r_2)+2x-\frac{(r_1-r_2)^2}{x}$$

$$= \pi(r_1+r_2)+\frac{2}{x}(r_1-r_2)^2+2x-\frac{(r_1-r_2)^2}{x}$$

$$= \pi(r_1+r_2)+\frac{(r_1-r_2)^2}{x}+2x$$

Length of cross belt drive.

Let x = Distance between the centers of two pulley (i.e., O_1O_2) (Fig. 6.16).

r_1 = Radius of the larger pulley

r_2 = Radius of the smaller pulley

L = Total length of the cross belt

The belt leaves the larger pulley at C and D and smaller pulley at F and J.

From O_2, draw O_2K parallel to CJ.

$$\angle O_1CJ = \angle O_1KO_2 = \frac{\pi}{2}$$

Let $\angle O_1O_2K = \alpha$,

Then $\angle BO_1C = \angle FO_2G = \alpha$

Total length of cross belt (L) is given by

$$L = \text{Arc DAC} + \text{CJ} + \text{Arc JHF} + \text{DF}$$

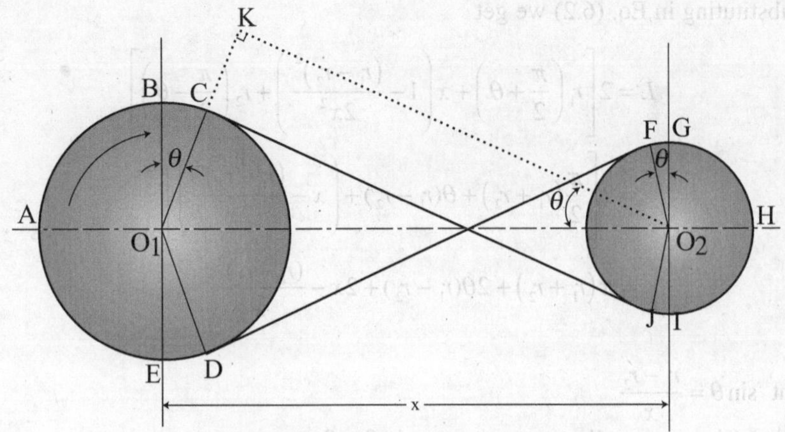

Figure 6.16. Length of cross belt drive.

$$= 2[\text{Arc ABC} + \text{CJ} + \text{Arc JIH}]$$

$$= 2\left[r_1\left(\frac{\pi}{2}+\theta\right) + \sqrt{x^2-(r_1+r_2)^2} + r_2\left(\frac{\pi}{2}+\theta\right)\right] \tag{6.4}$$

But $\sqrt{x^2-(r_1+r_2)^2} = x\sqrt{1-\left(\frac{r_1+r_2}{x}\right)^2} = x\left[1-\left(\frac{r_1+r_2}{x}\right)^2\right]^{1/2}$

$$= x\left[1-\frac{1}{2}\left(\frac{r_1+r_2}{x}\right)^2+\cdots\right] \quad \text{(By Binomial expansion)}$$

$$= x\left[1-\frac{(r_1+r_2)^2}{2x^2}\right] \quad \text{[By neglecting smaller terms]}$$

$$= x-\frac{(r_1+r_2)^2}{2x}$$

Substituting the value of x in Eq. (6.4)

$$L = 2\left[r_1\left(\frac{\pi}{2}+\theta\right) + x - \frac{(r_1+r_2)^2}{2x} + r_2\left(\frac{\pi}{2}+\theta\right)\right]$$

$$= 2\left[\frac{\pi}{2}(r_1+r_2)+\theta(r_1+r_2)+x-\frac{(r_1+r_2)^2}{2x}\right]$$

$$= \pi(r_1+r_2)+2\theta(r_1+r_2)+2x-\frac{(r_1+r_2)^2}{x} \tag{6.5}$$

But, $\sin\theta = \dfrac{r_1+r_2}{x}$

Since θ is very small we can assume $\sin\theta = \theta$

$$\Rightarrow \theta = \frac{r_1 + r_2}{x}$$

Substituting the value of θ in Eq. (6.5)

$$L = \pi(r_1 + r_2) + 2(r_1 + r_2)\frac{(r_1 + r_2)}{x} + 2x - \frac{(r_1 + r_2)^2}{x}$$

$$= \pi(r_1 + r_2) + 2x + \frac{(r_1 + r_2)^2}{x}$$

Centrifugal Tension

It is well known that when a particle of mass m is rotated in a circular path of radius r at a uniform velocity V, a centrifugal force acts radially outwards and its magnitude is equal to $\frac{mV^2}{r}$.

Since the belt material is moving with certain velocity on a pulley it develops force which acts outwards from the center of the pulley. This force is known as Centrifugal tension.

Let us consider an elemental length of the belt AB which subtends an elemental angle $\delta\alpha$ at the center of the pulley as shown in Fig 6.17.

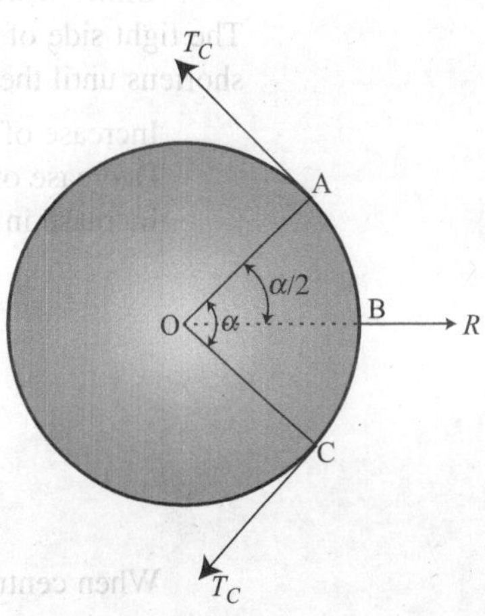

Figure 6.17. Centrifugal tension.

Let V = Velocity of the belt in m/s
r = Radius of pulley over which the belt runs
M = Mass of elemental length of belt
m = Mass of belt per meter length
T_C = Centrifugal tension acting at A and B
R = Centrifugal force acting radially outward

Elemental length of the belt, $AB = r\delta\alpha$

Mass of belt $AB = m \times r \times \delta\alpha$

Centrifugal force, $R = M\frac{V^2}{r} = m \times r \times \delta\alpha \times \frac{V^2}{r}$

Resolving the forces horizontally, we get

$$T_C \sin\frac{\delta\alpha}{2} + T_C \sin\frac{\delta\alpha}{2} = R \tag{6.6}$$

$$2T_C \sin\frac{\delta\alpha}{2} = m \times r \times \delta\alpha \times \frac{V^2}{r} \tag{6.7}$$

Since $\delta\theta$ is very small $\sin\frac{\delta\alpha}{2} = \frac{\delta\alpha}{2}$

Substituting in Eq. (6.7)

$$2T_C\frac{\delta\alpha}{2} = m \times r \times \delta\alpha \times \frac{V^2}{r}$$

$$T_C = mV^2$$

When centrifugal tension is considered
Tension on tight side $= T_1 + T_C$
Tension on stack side $= T_1 + T_C$

Initial Tension

In order to have a firm grip between the belt and pulleys under stationary conditions some amount of tension is maintained in the belt. This tension is known as initial tension.

Let T_0 = Initial tension in the belt
T_1 = Tension on the tight side of the belt
T_2 = Tension on the stack side of the belt
ε = Coefficient of increase of belt length per unit force

Under working conditions the tension in the two sides of the belt will be changed. The tight side of the belt stretches until the pull is increased from T_0 to T_1 and slack side shortens until the pull is decreased from T_0 to T_2

Increase of tension on tight side $= T_1 - T_0$
Decrease of tension on slack side $= T_0 - T_2$
Increase in length on light side = Decrease in length on slack side

$$\varepsilon(T_1 - T_0) = \varepsilon(T_0 - T_2)$$

$$(T_1 - T_0) = (T_0 - T_2)$$

$$T_0 = \frac{T_1 + T_2}{2}$$

When centrifugal tension is also considered

Initial tension, $T_0 = \dfrac{T_1 + T_2 + 2T_C}{2}$

Power Transmitted and Condition for Maximum Power Transmitted

Let T_1 = Tension on tight side
T_2 = Tension on slack side
V = Linear velocity of belt

Power transmitted by the belt is given by

$$P = (T_1 - T_2) \times V \tag{6.8}$$

But we know that $\dfrac{T_1}{T_2} = e^{\mu\theta}$ or $T_2 = \dfrac{T_1}{e^{\mu\theta}}$

Substituting the value of T_2 in Eq. (6.8)

$$P = (T_1 - \frac{T_1}{e^{\mu\theta}}) \times V \tag{6.9}$$

Let $1 - \dfrac{1}{e^{\mu\theta}} = C$

Then Eq. (6.9) becomes $P = C \times T_1 \times V$

Let the maximum tension in the belt be T_{max}

$$T_{max} = T_1 + T_C$$
$$T_1 = T_{max} - T_C$$
$$P = C\ (T_{max} - T_C) \times V$$
$$= C\ (T_{max} - mV^2) \times V$$
$$= C\ [T_{max}V - mV^3] \qquad (6.10)$$

Power transmitted will be maximum under optimum value of V.

$$\Rightarrow \frac{d(P)}{dV} = 0$$

Differentiating Eq. (6.10) with respect to V.

$$\frac{d(P)}{dV} = \frac{d}{dV}[C(T_{max}V - mV^3)] = 0$$
$$\Rightarrow \qquad (T_{max} - 3mV^2) = 0$$
$$T_{max} = 3mV^2 \qquad (6.11)$$
$$V = \sqrt{\frac{T_{max}}{3m}} \qquad (6.12)$$

Equation (6.12) gives the velocity of the belt under which the power transmitted is maximum.

From Eq. (6.11), $T_{max} = 3mV^2 = 3T_C$

Under maximum power transmission condition, $T_C = \dfrac{T_{max}}{3}$

$$T_{max} = T_1 + T_C$$
$$= T_1 + \frac{T_{max}}{3}$$
$$T_1 = \frac{2}{3}T_{max}$$

From the above analogy it can be inferred that there are two conditions for maximum power transmitted.

(i) $T_C = \dfrac{1}{3}T_{max}$

(ii) $T_1 = \dfrac{2}{3}T_{max}$

Hence, maximum power transmitted $P = T_1\left(1 - \dfrac{1}{e^{\mu\theta}}\right) \times V$

$$= \frac{2}{3}T_{max}\left(1 - \frac{1}{e^{\mu\theta}}\right)\sqrt{\frac{T_{max}}{3m}} \qquad (6.13)$$

Example 6.6

Two parallel shafts 4 m apart are to be connected by a belt running over pulleys of diameters 30 cm and 20 cm, respectively.

Determine the length of the belt considering both open and crossed belt drives.

Sol. Given:

Distance between the centers of shafts, $x = 4$ m

Radius of larger pulley, $r_1 = \dfrac{30}{2}$ cm $= 0.15$ m

Radius of smaller pulley, $r_2 = \dfrac{20}{2}$ cm $= 0.1$ m

(i) If belt is open

$$\text{Length of open belt, } L = \pi(r_1 + r_2) + \frac{(r_1 - r_2)^2}{x} + 2x$$

$$= \pi(0.15 + 0.1) + \frac{(0.15 - 0.1)^2}{4} + 2 \times 4$$

$$= 8.8479 \text{ m}$$

(ii) If belt is crossed

$$\text{Length of crossed belt, } L = \pi(r_1 + r_2) + \frac{(r_1 + r_2)^2}{x} + 2x$$

$$= \pi(0.15 + 0.1) + \frac{(0.15 + 0.1)^2}{4} + 2 \times 4$$

$$= 8.801 \text{m}$$

Example 6.7

A belt is running over a pulley or diameter 80 cm at 140 rpm. The angle of contact is 165° and coefficient of friction between the belt and pulley is 0.25. If the maximum tension in the belt is 2500 N, find the power transmitted by the belt

Sol. Given:

Diameter of pulley, $d = 80$ cm $= 0.8$ m

Speed of pulley, $N = 140$ rpm

Angle of contact, $\theta = 165° = 165 \times \dfrac{\pi}{180}$ radians

Coefficient of friction, $\mu = 0.25$

Maximum tension, $T_1 = 2500$ N

Velocity of the belt, $V = \dfrac{\pi d N}{60} = \dfrac{\pi \times 0.8 \times 140}{60} = 5.864$ m/sec

Let T_2 be the tension on the slack side of the belt.

Example 6.7 (Continued)

Then,

$$\frac{T_1}{T_2} = e^{\mu\theta}$$

$$= e^{0.25\times 165\times \frac{\pi}{180}} = 2.0543$$

$$\frac{2500}{T_2} = 2.0543$$

$$T_2 = 1217 \text{ N}$$

Power transmitted is given by

$$P = \frac{(T_1 - T_2)\times V}{1000} \text{ KW}$$

$$= \frac{(2500 - 1217)\times 5.864}{1000}$$

$$= 7.524 \text{ KW}$$

Example 6.8

A belt of density 1.5 g/cm^3 has a maximum permissible stress of 300 N/cm^2. Determine the maximum power that can be transmitted by a belt of 15 cm × 1.5 cm, if the ratio of the tension is 2.5.

Sol. Given:

Density of the belt, $\rho = 1.5 \text{ gr/cm}^3 = \frac{1.5}{1000} \text{ kg/cm}^3$

Maximum permissible stress, f = 300 N/cm^2

Width of the belt, w = 15 cm

Thickness of the belt, t = 1.5 cm

Ratio of tensions, $\frac{T_1}{T_2} = 2.5$

Mass of 1 m length of the belt,

m = density × volume of belt of one meter length

$$= \frac{1.5}{1000}\times(15\times 1.5\times 100)$$

$$= 3.375 \text{ kg}$$

Maximum tension, T_m = Maximum stress × Cross sectional area of the belt

= 300 × 15 × 1.5

= 6750 N

Example 6.8 (Continued)

Under maximum power transmission condition velocity of the belt

$$V = \sqrt{\frac{T_m}{3m}} = \sqrt{\frac{6750}{3 \times 3.375}} = 25.82 \text{ m/sec}$$

We know that $T_m = T_1 + T_C$

Under maximum power transmission condition

$$T_C = \frac{1}{3} T_m$$

Substituting we get,

$$T_m = T_1 + \frac{1}{3} T_m$$

$$T_1 = \frac{2}{3} T_m = \frac{2}{3} \times 6750 = 4500 \text{ N}$$

It is given that,

$$\frac{T_1}{T_2} = 2.5$$

$$T_2 = \frac{T_1}{2.5} = \frac{4500}{2.5} = 1800 \text{ N}$$

Now maximum power transmitted

$$P_{max} = \frac{(4500 - 1800) \times 25.82}{1000}$$

$$= 69.714 \text{ KW}$$

Example 6.9

A crossed belt connects two pulleys, the smaller pulley being of 300 mm in diameter. The angle of lap on smaller pulley is 210° and coefficient of friction between belt and pulley is 0.3. Which of the following alternatives would you prefer to increase the power transmitted.

(i) increasing the initial tension by 5%
(ii) increasing the coefficient of friction by 5%

Sol. Given:
Diameter of smaller pulley, $d_2 = 0.3$ m
Radius of smaller pulley, $r_2 = 0.15$ m

Example 6.9 (Continued)

Angle of lap, $\theta = 210° = \frac{210}{180} \times \pi$

Coefficient of friction, $\mu = 0.3$

Power transmitted, $P = (T_1 - T_2) \times V$ watts

$$\frac{T_1}{T_2} = e^{\mu\theta}$$

$$T_2 = \frac{T_1}{e^{\mu\theta}}$$

$$P = \left(T_1 - \frac{T_1}{e^{\mu\theta}}\right) \times v = \frac{T_1(e^{\mu\theta} - 1)}{e^{\mu\theta}} \times V \qquad \text{(a)}$$

We know that, initial tension, $T_0 = \frac{T_1 + T_2}{2}$

$$T_1 = 2T_0 - T_2 = 2T_0 - \frac{T_1}{e^{\mu\theta}}$$

$$T_1 + \frac{T_1}{e^{\mu\theta}} = 2T_0$$

$$T_1(e^{\mu\theta} + 1) = 2T_0 e^{\mu\theta}$$

$$T_1 = \frac{2T_0 e^{\mu\theta}}{e^{\mu\theta} + 1}$$

Substituting the value of T_1 in Eq. (a), we get

$$P = \frac{2T_0 e^{\mu\theta}(e^{\mu\theta} - 1)}{(e^{\mu\theta} + 1)e^{\mu\theta}} \times V$$

$$= 2T_0 \frac{(e^{\mu\theta} - 1)}{(e^{\mu\theta} + 1)} \times V$$

(i) When the value of T_0 is increased by 5% power will be increased by 5%.

(ii) If μ is increased by 5%.

Power transmitted when $\mu = 0.3$ and $\theta = 210°$

$$P = \frac{2T_0(e^{0.3 \times \frac{210 \times \pi}{180}} - 1)}{(e^{0.3 \times \frac{210 \times \pi}{180}} + 1)} \times V$$

$$= 1.0007T_0 V$$

Power transmitted when $\mu = 0.315$ and $\theta = 210°$

Example 6.9 (Continued)

$$P' = \frac{2T_0(e^{0.315\times\frac{210\times\pi}{180}} - 1)}{(e^{0.315\times\frac{210\times\pi}{180}} + 1)} \times V$$

$$= 1.0414T_0V$$

Percentage of increase in power,

$$= \frac{P' - P}{P} \times 100$$

$$= \frac{(1.0414T_0V - 1.0007T_0V)}{1.0007T_0V} \times 100$$

$$= 4.067\%$$

The power increase in the first case is 5% whereas it is 4.067% in the second case.

6.6. Screw Jack

A screw jack is used for lifting heavy weights or loads with the help of little effort.

Figure 6.18 shows a simple screw jack. It consists of a nut, a screw with square threads, and a handle fitted to the head of the screw.

The load to be lifted is placed directly on the head of the screw or on a specially designed platform which is attached to the screw. An effort P is applied at the end of the handle which is attached to the screw head. The working principle of screw jack is same as that of an inclined plane.

Let W = Load placed on the screw head
P = Effort applied at the end of the handle
L = Length of handle
p = Pitch of the screw
d = Mean diameter of the screw
α = Angle of the screw
φ = Angle of friction
μ = Coefficient of friction between screw and nut = $\tan\varphi$.

When the handle is rotated through one complete turn, the screw is also rotated through one turn and hence the screw will have vertical displacement by a height p.

The development of one complete turn of the screw is shown in Fig. 6.19. The developed surface is similar to an inclined plane. The distance AB will be equal to the circumference and distance BC will be equal to the pitch (p) of the screw.

$$\tan\alpha = \frac{BC}{AC} = \frac{p}{\Pi d}$$

Let P' = Effort applied horizontally at the mean radius of the screw jack to left the load W,

r = Mean radius of the screw jack = $\frac{d}{2}$

The present case is similar to that of lifting a load W up an inclined plane by a horizontal force P' (Fig. 6.20)

Let N = Normal reaction

F = Force of friction = μN

Resolving the forces along the inclined plane

$$F + W\sin\alpha = P'\cos\alpha \tag{6.14}$$

Resolving the forces normal to the inclined plane

$$N = W\cos\alpha + P'\sin\alpha$$

Substituting the value of N in Eq. (6.14) we get

$$\mu[W\cos\alpha + P'\sin\alpha] + W\sin\alpha = P'\cos\alpha$$

$$\Rightarrow \frac{\sin\phi}{\cos\phi}[W\cos\alpha + P'\sin\alpha] + W\sin\alpha = P'\cos\alpha$$

Multiplying by $\cos\varphi$

$$\Rightarrow W\sin\phi\cos\alpha + P'\sin\phi\sin\alpha + W\sin\alpha\cos\phi - P'\cos\alpha\cos\phi$$

$$\Rightarrow W[\sin\alpha\cos\phi + \cos\alpha\sin\phi] = P'[\cos\alpha\cos\phi - \sin\alpha\sin\phi]$$

$$P' = W\frac{\sin(\alpha+\phi)}{\cos(\alpha+\phi)} = W\tan(\alpha+\phi) \tag{6.15}$$

Moment of P' about the axis of the screw = Moment of P about the axis of the screw (Fig. 6.21)

$$P' \times \frac{d}{2} = P \times L$$

$$P = P' \times \frac{d}{2L} \tag{6.16}$$

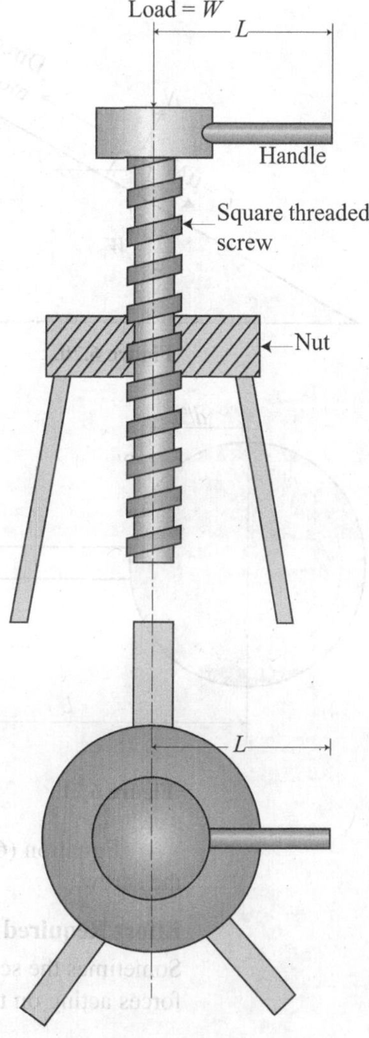

Figure 6.18. Screw jack.

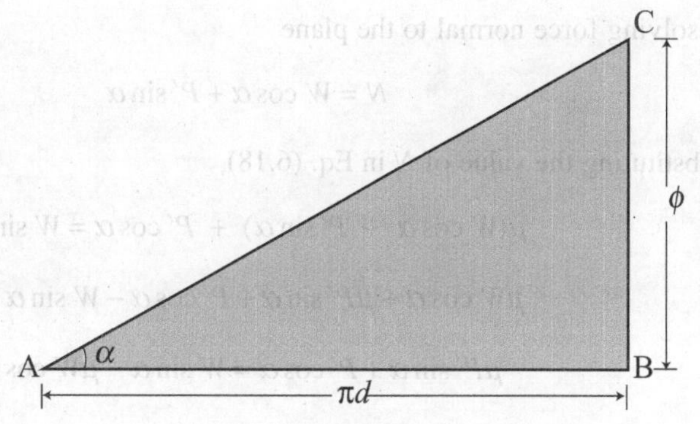

Figure 6.19.

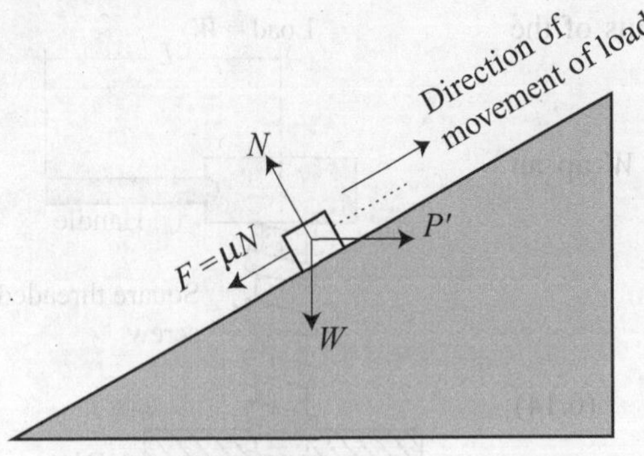

Figure 6.20.

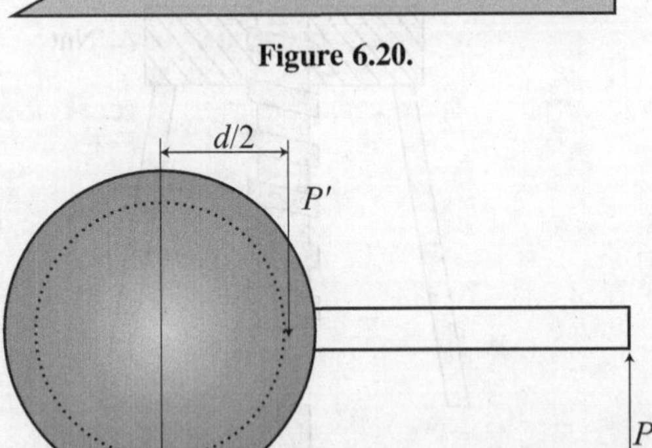

Figure 6.21.

Substituting the value of P' from Eq. (6.15) in Eq. (6.16)

$$P = \frac{d}{2L} \times W \tan(\alpha + \phi)$$

Torque required,

$$T = P \times L$$

$$= \frac{d}{2} W \tan(\alpha + \phi)$$

$$P = \frac{Wd}{2L} \tan(\alpha + \phi)$$

$$= \frac{Wd}{2L} \frac{\tan\alpha + \tan\phi}{1 - \tan\alpha \tan\phi}$$

$$= \frac{Wd}{2L} \frac{\frac{p}{\Pi d} + \mu}{1 - \frac{p}{\Pi d}\mu}$$

$$= \frac{Wd}{2L}\left[\frac{p + \mu \Pi d}{\Pi d - p\mu}\right] \tag{6.17}$$

Equation (6.17) gives the value of P in terms of coefficient of friction and pitch of the screw.

Effort Required at the End of the Handle of the Screw Jack to Lower the Load W

Sometimes the screw jack can be used to lower heavy loads slowly. Figure 6.22 shows the forces acting on the body. Resolving the forces along the inclined plane

$$F + P' \cos\alpha = W \sin\alpha$$

$$\mu N + P' \cos\alpha = W \sin\alpha \tag{6.18}$$

Resolving force normal to the plane

$$N = W \cos\alpha + P' \sin\alpha$$

Substituting the value of N in Eq. (6.18)

$$\mu(W \cos\alpha + P' \sin\alpha) + P' \cos\alpha = W \sin\alpha$$

$$\mu W \cos\alpha + \mu P' \sin\alpha + P' \cos\alpha - W \sin\alpha = 0$$

$$\mu P' \sin\alpha + P' \cos\alpha = W \sin\alpha - \mu W \cos\alpha$$

$$P'[\mu \sin\alpha + \cos\alpha] = W[\sin\alpha - \mu \cos\alpha]$$

$$\mu = \tan\phi = \frac{\sin\phi}{\cos\phi}$$

$$P'\left[\frac{\sin\phi}{\cos\phi}\sin\alpha + \cos\alpha\right] = W\left[\sin\alpha - \frac{\sin\phi}{\cos\phi}\cos\alpha\right]$$

Multiplying by $\cos\varphi$, we get

$$P'[\sin\phi\sin\alpha + \cos\alpha\cos\phi] = W[\sin\alpha\cos\phi - \sin\phi\cos\alpha]$$

$$P'[\cos(\phi - \alpha)] = W[\sin(\phi - \alpha)]$$

$$P' = W\frac{\sin(\phi - \alpha)}{\cos(\phi - \alpha)} = W\tan(\phi - \alpha)$$

$$\text{If } \alpha > \phi \text{ then } P' = W\tan(\alpha - \phi)$$

Equating the moment of P and P' about the axis of the screw

$$P' \times \frac{d}{2} = P \times L$$

$$P = P'\frac{d}{2L} = \frac{d}{2L}W\tan(\alpha - \phi)$$

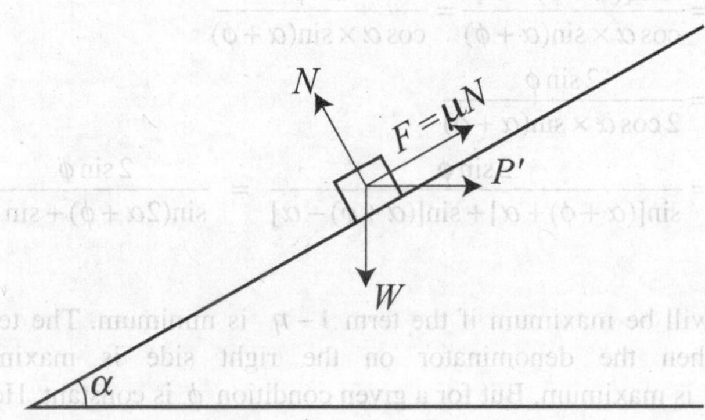

Figure 6.22. Forces acting on the body.

Efficiency of a Screw Jack for Raising a Load

Let P_{ideal} = Ideal effort required at the end of the handle of the screw jack to lift the load W.

P_{ideal} will be obtained by making friction zero.

$$P_{ideal} = \frac{d}{2L}W\tan(\alpha + \phi) \text{ where } \tan\phi = 0$$

$$= \frac{d}{2L}W\frac{\tan\alpha + \tan\phi}{1 - \tan\alpha\tan\phi} = \frac{d}{2L}W\tan\alpha \qquad (6.19)$$

Efficiency of the screw jack is given by

$$\eta = \frac{\text{Ideal effort}}{\text{Actual effort}} = \frac{P_{\text{ideal}}}{P} = \frac{\frac{d}{2L} W \tan\alpha}{\frac{d}{2L} W \tan(\alpha+\phi)} = \frac{\tan\alpha}{\tan(\alpha+\phi)} \tag{6.20}$$

It can be inferred from the equation that efficiency of the screw jack is independent of the weight lifted or effort applied but dependent on helix angle and friction angle.

Equation (6.20) can be written as

$$\eta = \frac{\frac{\sin\alpha}{\cos\alpha}}{\frac{\sin(\alpha+\phi)}{\cos(\alpha+\phi)}} = \frac{\sin\alpha}{\cos\alpha} \times \frac{\cos(\alpha+\phi)}{\sin(\alpha+\phi)}$$

Subtracting each side of the above equation from 1.

$$1-\eta = 1-\frac{\sin\alpha \times \cos(\alpha+\phi)}{\cos\alpha \times \sin(\alpha+\phi)}$$

$$= \frac{\cos\alpha\sin(\alpha+\phi) - \sin\alpha\cos(\alpha+\phi)}{\cos\alpha \times \sin(\alpha+\phi)}$$

$$= \frac{\sin[(\alpha+\phi)-\alpha]}{\cos\alpha \times \sin(\alpha+\phi)} = \frac{\sin\phi}{\cos\alpha \times \sin(\alpha+\phi)}$$

$$= \frac{2\sin\phi}{2\cos\alpha \times \sin(\alpha+\phi)}$$

$$= \frac{2\sin\phi}{\sin[(\alpha+\phi)+\alpha] + \sin[(\alpha+\phi)-\alpha]} = \frac{2\sin\phi}{\sin(2\alpha+\phi)+\sin\phi}$$

Efficiency will be maximum if the term $1-\eta$ is minimum. The term $1-\eta$ will be minimum when the denominator on the right side is maximum, that is, $[\sin(2\alpha+\phi)+\sin$ is maximum. But for a given condition ϕ is constant. Hence, the value of $[\sin(2\alpha+\phi)+\sin$ is maximum when $\sin(2\alpha+\phi$ is maximum. This is possible when

$$2\alpha+\phi = 90° \text{ or } 2\alpha = 90° - \phi$$

$$\alpha = 45 - \frac{\phi}{2} \tag{6.21}$$

By substituting the above value of α in Eq. (6.20)

$$\eta_{\max} = \frac{\tan\left(45-\frac{\phi}{2}\right)}{\tan\left(45-\frac{\phi}{2}+\phi\right)} = \frac{\tan\left(45-\frac{\phi}{2}\right)}{\tan\left(45+\frac{\phi}{2}\right)}$$

$$= \frac{\left[\dfrac{\left(\tan 45 - \tan\dfrac{\phi}{2}\right)}{1+\tan 45 \tan\dfrac{\phi}{2}}\right]}{\left[\dfrac{\tan 45 + \tan\dfrac{\phi}{2}}{1-\tan 45 \tan\dfrac{\phi}{2}}\right]}$$

$$= \frac{1-\tan\dfrac{\phi}{2}}{1+\tan\dfrac{\phi}{2}} \times \frac{1-\tan\dfrac{\phi}{2}}{1+\tan\dfrac{\phi}{2}} = \frac{\left(1-\tan\dfrac{\phi}{2}\right)^2}{\left(1+\tan\dfrac{\phi}{2}\right)^2}$$

$$= \frac{\left[1-\dfrac{\sin\dfrac{\phi}{2}}{\cos\dfrac{\phi}{2}}\right]^2}{\left[1+\dfrac{\sin\dfrac{\phi}{2}}{\cos\dfrac{\phi}{2}}\right]^2} = \frac{\left[\cos\dfrac{\phi}{2}-\sin\dfrac{\phi}{2}\right]^2}{\left[\cos\dfrac{\phi}{2}+\sin\dfrac{\phi}{2}\right]^2}$$

$$= \frac{\cos^2\dfrac{\phi}{2}+\sin^2\dfrac{\phi}{2}-2\sin\dfrac{\phi}{2}\cos\dfrac{\phi}{2}}{\cos^2\dfrac{\phi}{2}+\sin^2\dfrac{\phi}{2}+2\sin\dfrac{\phi}{2}\cos\dfrac{\phi}{2}}$$

$$= \frac{1-2\sin\dfrac{\phi}{2}\cos\dfrac{\phi}{2}}{1+2\sin\dfrac{\phi}{2}\cos\dfrac{\phi}{2}} = \frac{1-\sin\phi}{1+\sin\phi}$$

$$\eta_{max} = \frac{1-\sin\phi}{1+\sin\phi}$$

Note 1: Velocity ratio: It is defined as the ratio of the distance moved by the effort to the distance moved by the load. It is denoted by V.R.

$$\text{V.R.} = \frac{\text{Distance moved by the effort}}{\text{Distance moved by the load}} = \frac{y}{x}$$

Mechanical advantage: It is defined as the ratio of the load lifted to the effort applied.

$$\text{M.A.} = \frac{\text{Load lifted}}{\text{Effort applied}} = \frac{W}{P}$$

$$\text{Efficiency} = \frac{\text{Output}}{\text{Input}} = \frac{W \times X}{P \times Y} = \frac{W / P}{y / x} = \frac{\text{M.A.}}{\text{V.R.}}$$

Note 2: Sometimes a screw jack is not capable of rotating in the opposite direction (will not come down) when the external effort is removed. Such a screw jack is known as self locking screw jack. Efficiency of a self locking screw jack will not be more than 50%.

Differential Screw Jack

The differential screw jack possesses two screws, S_1 and S_2. S_1 is threaded on the outside only, whereas S_2 is threaded both on inside and outside.

The external threads of S_2 mesh with the threads of nut. The internal threads of S_2 gear with the external threads of S_1.

The screw S_1 does not rotate but free to move in vertical direction. It carries the load when the effort is applied at the end of the lever S_2. S_2 raises up and S_1 goes down making the net lift of the load as the algebric sun of the motions of the screws S_1 and S_2.

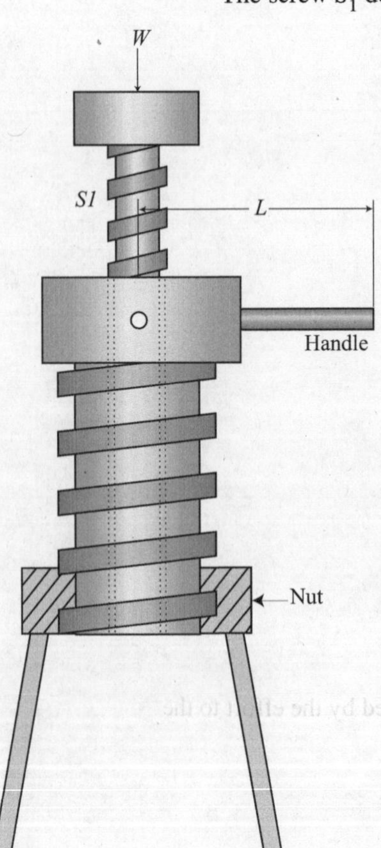

Figure 6.23. Differential screw jack.

Let p_1 = Pitch of S_1

p_2 = Pitch of S_2

L = Length of the lever arm

W = Load lifted

P = Effort applied to lift the load

For one revolution of the lever arm

Distance moved by the effort = $2\Pi L$

Upward distance moved by $S_2 = p_2$

Downward distance move by $S_1 = p_1$

Lift of the load $= p_2 - p_1$

$$\text{Velocity ratio} = \frac{\text{Distance moved by the effort}}{\text{Distance moved by the load}} = \frac{2\Pi L}{p_2 - p_1}$$

$$\text{Mechanical Advantage, M.A.} = \frac{W}{P}$$

$$\text{Mechanical Advantage, M.A.} = \frac{W}{P}$$

$$\eta = \frac{\text{M.A.}}{\text{V.R.}}$$

Example 6.10

Find the effort required to apply at the end of a handle fitted to a screw head of a screw jack to lift a load of 1000 N. The length of the handle is 50 cm. The mean diameter and the pitch of the screw jack are 5 cm and 1 cm, respectively. The coefficient of friction is given as 0.085. Also, determine the effort required to apply at the end of the handle to lower the load.

Sol. Given:

Load $W = 1000$ N

Length of handle, $L = 0.5$ m

Mean diameter of screw jack, $d = 5$ cm $= 0.05$ m

Pitch of the screw jack, $P = 1$ cm$=0.01$ m

Coefficient of friction, $\mu = 0.085$

Effort required at the end of the handle to raise the load

$$P = \frac{Wd}{2L}\left(\frac{p + \mu \Pi d}{\Pi d - p\mu}\right) = \frac{1000 \times 0.05}{2 \times 0.5}\left(\frac{0.01 + 0.085 \times \Pi \times 0.05}{\Pi \times 0.05 - 0.01 \times 0.085}\right)$$

Effort required at the end of the handle to lower the load

$$P = \frac{Wd}{2L}\left(\frac{\mu \Pi d - p}{\Pi d + \mu p}\right) = \left(\frac{1000 \times 0.05}{2 \times 0.5}\right)\left(\frac{0.085 \times \Pi \times 0.05 - 0.01}{\Pi \times 0.05 + 0.085 \times 0.01}\right)$$

Example 6.11

A screw jack has a square thread with 6.8 cm mean diameter and 1.3 cm pitch. The coefficient of friction at the screw thread is 0.035. Find the tangential force to be applied to the jack at 25 cm radius so as to lift a load of 750 N. State whether the jack is self-locking. If it is, find the torque required to lower the load. Otherwise find the torque required to keep the load from descending.

Sol. Given:

Mean diameter, d = 6.8 cm = 0.068 m

Pitch of threads, p = 1.3 cm = 0.013 m

Coefficient of friction, μ = 0.035

Let P = Tangential force required at 25 cm radius to lift the load

W = Load lifted = 750 N

L = Length of handle = 25 cm = 0.25 cm

$$P = \frac{Wd}{2L}\left(\frac{p+\mu\Pi d}{\Pi d - p\mu}\right)$$

$$= \frac{750\times0.068}{2\times0.25}\left(\frac{0.013+0.035\times\Pi\times0.068}{\Pi\times0.068-0.013\times0.035}\right)$$

$$= 9.8\,\text{N}$$

For a self-locking screw jack the efficiency will be less than 50%

$$\eta = \frac{\tan\alpha}{\tan(\alpha+\phi)}$$

$$\tan\alpha = \frac{p}{\Pi d} = \frac{0.013}{\Pi\times0.068}$$

$$\tan\phi = \mu = 0.035$$

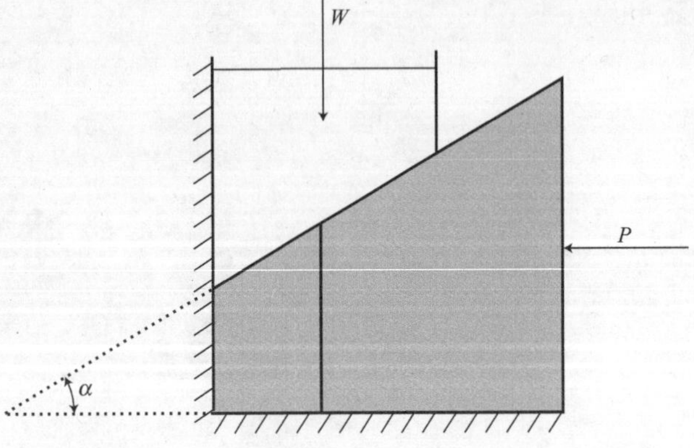

Figure 6.24. Wedge

6.7 Wedge

A wedge can be defined as piece of wood or metal with triangular or trapazoidal cross section. It is used for lifting loads through small distances or for slight adjustments in the position of a body. Effort is applied in direction which is perpendicular to the direction moment of the load.

To lift a heavy load the wedge is placed below the load and a horizontal force P is applied as shown in the Fig. 6.24. When enough force is applied on the wedge it moves to its left thus lifting the load. During this process sliding of one surface over the other takes place.

Equilibrium of Wedge

Resolving the forces horizontally

$$R_1 \sin\phi_1 + R_2 \sin(\phi_2 + \alpha) = P$$

Resolving the forces vertically

$$R_1 \cos\phi_1 = R_2 \cos(\phi_2 + \alpha)$$

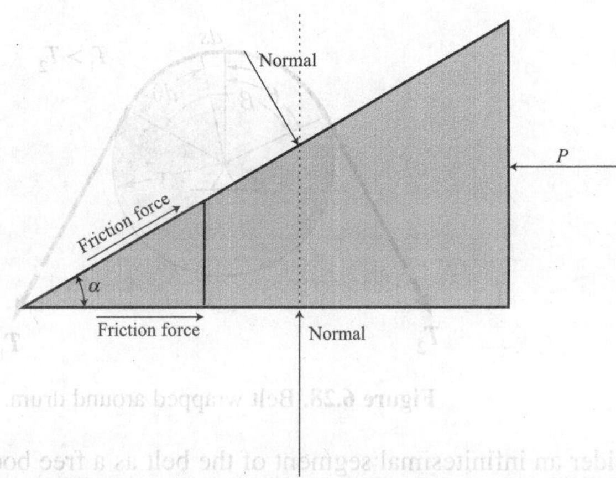

Figure 6.25.

Equilibrium of Body

Resolving the forces horizontally

$$R_3 \cos\phi_3 = R_2 \sin(\alpha + \phi_2)$$

Resolving forces vertically

$$W + R_3 \sin\phi_3 = R_2 \cos(\alpha + \phi_2)$$

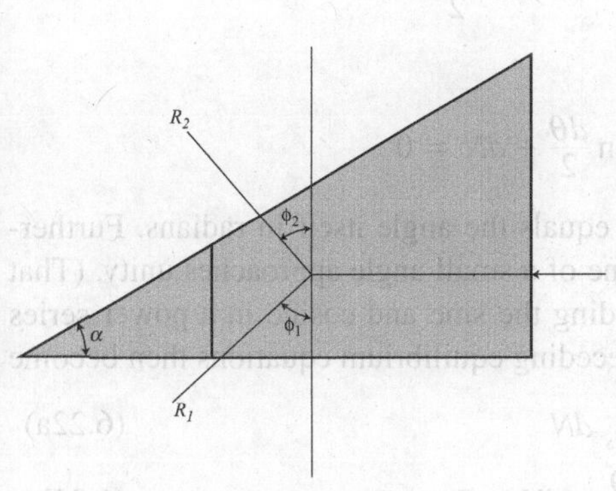

Figure 6.26.

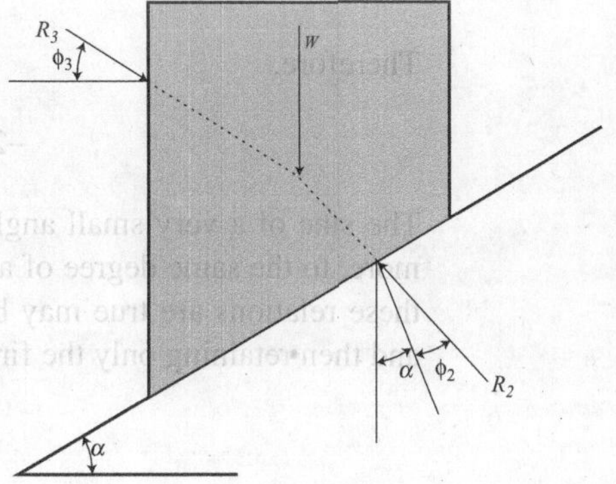

Figure 6.27.

6.8 Belt Friction

A flexible belt is shown in Fig. 6.28 wrapped around a portion of a drum, with the amount of wrap indicated by angle β. The angle β is called the *angle of wrap*. Assume that the drum is stationary and tensions T_1 and T_2 are such that motion is impending between the belt and the drum. We shall take the impending motion of the belt to be clockwise relative to the drum, and therefore the tension T_1 exceeds tension T_2.

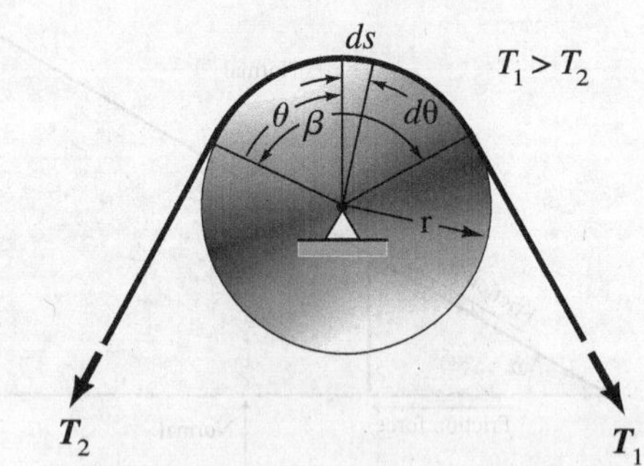

Figure 6.28. Belt wrapped around drum.

Consider an infinitesimal segment of the belt as a free body. This segment subtends an angle $d\theta$ at the drum center as shown in Fig. 6.29. Summing force components in the radial and transverse directions and equating them to zero as per *equilibrium*, we get the following scalar equations:

$\sum F_t = 0$:

$$-T\cos\frac{d\theta}{2} + (T + dT)\cos\frac{d\theta}{2} - \mu_s dN = 0$$

Therefore,

$$dT\cos\frac{d\theta}{2} = \mu_s\, dN$$

$\sum F_n = 0$:

$$-T\sin\frac{d\theta}{2} - (T + dT)\sin\frac{d\theta}{2} + dN = 0$$

Therefore,

$$-2T\sin\frac{d\theta}{2} - dT\sin\frac{d\theta}{2} + dN = 0$$

The sine of a very small angle approximately equals the angle itself in radians. Furthermore, to the same degree of accuracy, the cosine of a small angle approaches unity. (That these relations are true may be seen by expanding the sine and cosine in a power series and then retaining only the first terms.) The preceding equilibrium equations then become

$$dT = \mu_s\, dN \tag{6.22a}$$

$$-Td\theta - dT\frac{d\theta}{2} + dN = 0 \tag{6.22b}$$

In the last equation, we have an expression involving the product of two infinitesimals. This quantity may be considered negligible compared to the other terms of the equation involving only one differential. Thus, we have for this equation:

$$T\,d\theta = dN \tag{6.23}$$

From Eqs. 6.22a and 6.23, we may form an equation involving T and θ. Thus, by eliminating dN from the equations, we have

$$dT = \mu_s T\,d\theta$$

Hence,

$$\frac{dT}{T} = \mu_s\,d\theta$$

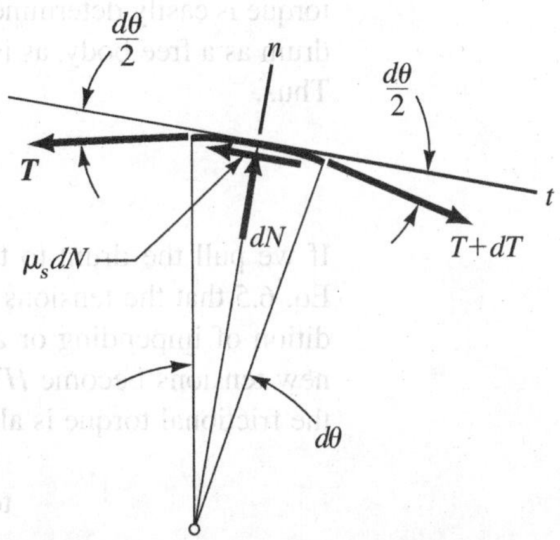

Figure 6.29. Free-body diagram of segment of belt; impending slippage.

Integrating both sides around the portion of the belt in contact with the drum,

$$\int_{T_2}^{T_1} \frac{dT}{T} = \int_0^{\beta} \mu_s\,d\theta$$

we get

$$\ln\frac{T_1}{T_2} = \mu_s\beta$$

or

$$\frac{T_1}{T_2} = e^{\mu_s\beta} \tag{6.24}$$

We therefore have established a relation between the tensions on each part of the belt at a condition of impending motion between the belt and the drum. The same relation can be reached for a *rotating* drum with impending slippage between the belt and the drum *if we neglect centrifugal effects on the belt*. Furthermore, by using the *dynamic* coefficient of friction in the formula above, we have the case of the belt slipping at constant speed over either a rotating or stationary drum (again neglecting centrifugal effects on the belt). Thus, for all such cases, we have

$$\frac{T_1}{T_2} = e^{\mu\beta} \tag{6.25}$$

where the proper coefficient of friction must be used to suit the problem, and the angle β must be expressed in *radians*. Note that *the ratio of tensions depends only on the angle of wrap* β and the coefficient of friction μ. Thus, if the drum A is forced to the right, as shown in Fig. 6.30, the tensions will increase, but if β is not affected by the action, the ratio of T_1/T_2 for impending or actual constant speed slippage is *not* affected by this action. However, the *torque* developed by the belt on the drum as a result of friction *is* affected by the force F. The

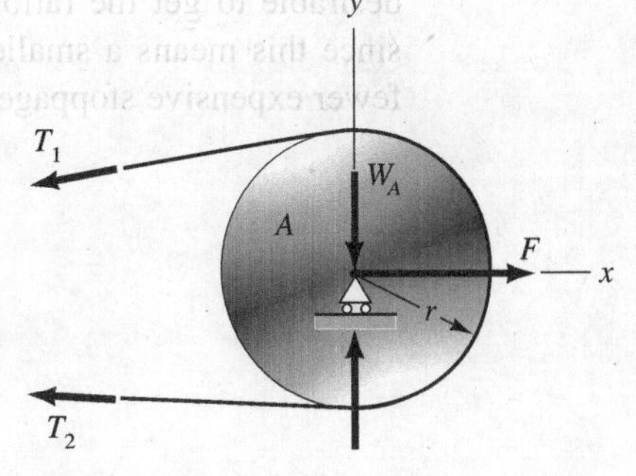

Figure 6.30. Force F affects T_1 and T_2 but not T_1/T_2.

torque is easily determined by using the drum and the portion of the belt in contact with the drum as a free body, as is shown in Fig. 6.30.
Thus,

$$\text{torque} = T_1 r - T_2 r = (T_1 - T_2)r \tag{6.26}$$

If we pull the drum to the right without disturbing the angle of wrap, we can see from Eq. 6.5 that the tensions T_1 and T_2 must increase by the same factor to maintain the condition of impending or actual constant speed slippage. And if we call this factor H, the new tensions become HT_1 and HT_2, respectively. Substituting into Eq. 6.26, we see that the frictional torque is also increased by the same factor:

$$\text{torque} = H(T_1 - T_2)r = H(\text{torque})_{\text{original}}$$

If we sum forces on the free body diagram in Fig. 6.31, we can determine the force F needed to maintain the tensions T_1 and T_2 as follows:

$$(T_1)_x + (T_2)_x = F \tag{6.27}$$

At impending slippage or at constant speed slippage, Eq. 6.25 is valid, and with Eq. 6.27 and knowing F, we can solve for T_1 and T_2. With Eq. 6.26, the torque that the belt is capable of developing on the drum now becomes a simple computation.

Accordingly, we have three equations at our disposal. Thus, there is the geometry determining for us the angle of wrap and there is the coefficient of friction. These can be combined and used in the *first* equation [Eq. (6.25)] to get the correct *ratio* of tensions for the conditions of impending slippage or for actual slippage. Note, because of the use of *equilibrium* in the derivation of this equation, both of these conditions (impending slippage and actual slippage) require that the belt be stationary or be moving at constant speed in inertial space. These results serve as benchmarks for the design engineer, particularly the impending slippage condition. If the ratio of the belt tensions is less than $e^{\mu_s \beta}$ and the belt has not started to slip, we can say that there will be no slipping until this ratio of the belt tensions is exceeded,[4] at which time for steady speed of the belt, the ratio of the belt tensions will have to become equal to the value $e^{\mu_d \beta}$. In the *second* of the three equations [Eq. (6.26)], the torque on or from the drum is related to the belt tensions via equilibrium. Finally, the *third* equation [Eq. (6.27)], again from equilibrium, gives us the required force F on the drum needed for the aforestated torque. Note that it is desirable to get the ratio of the tensions close to the condition of impending slippage since this means a smaller F and hence a longer life for the belt. This also will mean fewer expensive stoppages for assembly lines.

[4]This is like the example given at the outset of moving a piece of furniture where, as the driving force is increased from zero, no movement occurs until the maximum static friction force has been exceeded.

Example 6.12

A drum (see Fig. 6.31) requires a torque of 200 N-m to get it to start rotating. If the static coefficient of friction μ_s between the belt and the drum is .35, what is the *minimum* axial force F on the drum required to create enough tension in the belt to start the rotation of the drum?

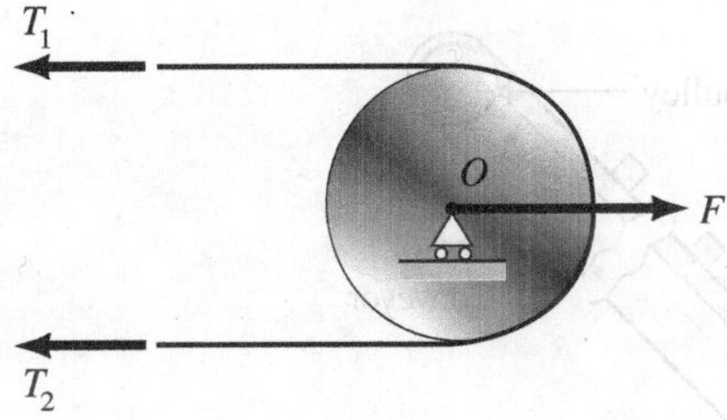

Figure 6.31. Drum is driven by a belt.

The angle of wrap β clearly is π radians. To get the minimum force F, we use the condition of impending slippage between belt and drum so that we can say

$$\frac{T_1}{T_2} = e^{.35\pi} = 3.00 \qquad \text{(a)}$$

For the condition of starting rotation we have

$$\sum M_0 = \text{torque required} = 200 \text{ N-m}$$

$$\therefore (T_1 - T_2)(.25) = 200 \qquad \text{(b)}$$

It is now a simple matter to solve the previous two equations simultaneously to get

$$T_1 = 1{,}200 \text{ N} \qquad T_2 = 400 \text{ N}$$

Finally we can determine the minimum value of F for this problem by summing forces in the x direction as follows:

$$\sum F_x = 0:$$

$$F_{min} = 1{,}200 + 400 = \boxed{1{,}600 \text{ N}}$$

Example 6.13

A conveyor is moving ten 50-lb boxes at a constant speed at a 45° setting (Fig. 6.32). The dynamic coefficient of friction between the belt and the bed of the conveyor is .05. Furthermore, the static coefficient of friction between the driving pulley and the belt is .4. The idler pulley is moved along the direction of the conveyor by a crank mechanism so that the idler pulley is subject to a force F of 500 lb. Compute the maximum tension found in the belt and ascertain if there will be slipping on the driving pulley. Neglect the weight of the belt.

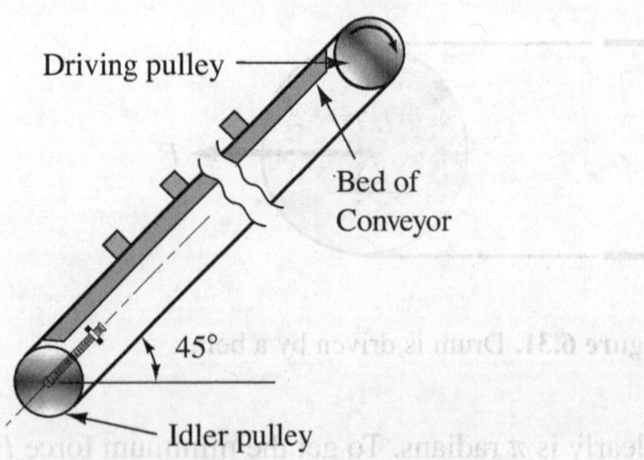

Figure 6.32. Conveyor.

In Fig. 6.33, we have shown free-body diagrams of various parts of the conveyor.[5] Consider the portion of the belt on the conveyor frame. From *equilibrium* we set the sums of forces normal and tangent to the belt equal to zero. Thus

$\sum F_n = 0$:

$$N - (10)(50)(.707) = 0$$

Therefore,

$$N = 354 \text{ lb} \quad \text{(a)}$$

$\sum F_t = 0$:

$$T_1 - T_3 - (10)(50)(.707) - (.05)(354) = 0$$

Therefore,

$$T_1 - T_3 = 371 \text{ lb} \quad \text{(b)}$$

[5]The weights of the pulleys have been counteracted by supporting forces at the axles and have not been shown.

Example 6.13 (Continued)

For the idler pulley (no resisting or applied torques), we have

$\sum M_0 = 0$:

$$T_3 = T_2 \qquad \text{(c)}$$

$\sum F_t = 0$:

$$T_3 + T_2 = 500 \qquad \text{(d)}$$

From Eqs. (c) and (d), we conclude that

$$T_2 = T_3 = 250 \text{ lb} \qquad \text{(e)}$$

From Eq. (b) we now get for the maximum tension T_1:

$$T_1 = T_3 + 371 = \boxed{621 \text{ lb}} \qquad \text{(f)}$$

We must next check the driving pulley to ensure that there is no slippage occurring. For the condition of impending slippage, we have, using as T_2 the value of 250 lb and solving for T_1

$$T_1 = T_2 e^{.4\pi} = (250)(3.51) = 878 \text{ lb}$$

Clearly, since the T_1 needed is only 621 lb (see Eq. (f)), we do not have slippage at the driving pulley, and we conclude that the maximum tension is indeed 621 lb.

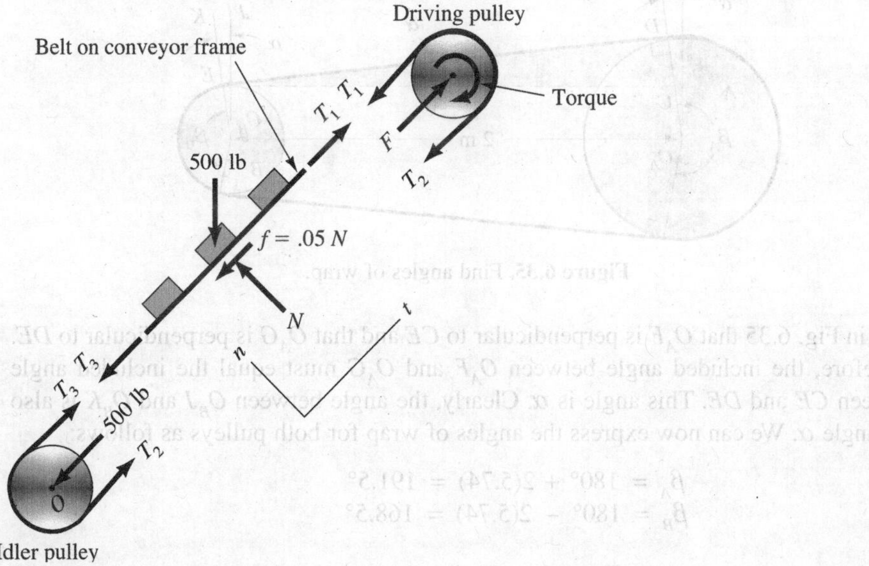

Figure 6.33. Various free-body diagrams of parts of conveyor.

Example 6.14

An electric motor (not shown) in Fig. 6.34 drives at constant speed the pulley B, which connects to pulley A by a belt. Pulley A is connected to a compressor (not shown) which requires 700 N-m torque to drive it at constant speed ω_A. If μ_s for the belt and either pulley is .4, what minimum value of the indicated force F is required to have no slipping anywhere?

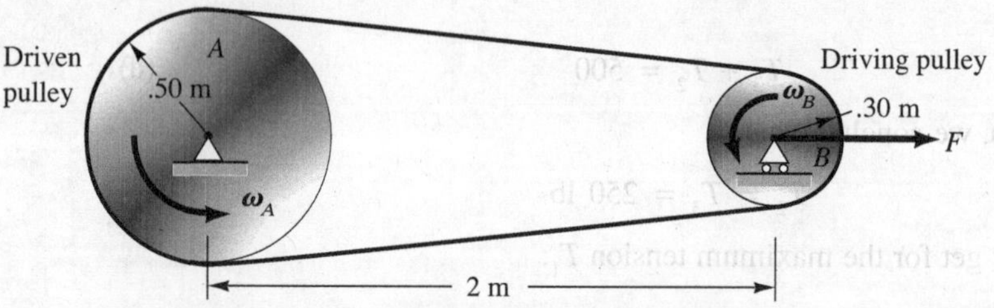

Figure 6.34. Belt-driven compressor.

As a first step, we determine the angles of wrap β for the respective pulleys. For this purpose, we first compute α (Fig. 6.35). Note that the radii O_AD and O_BE, being perpendicular to the same line DE, are therefore parallel to each other. Drawing EC parallel to O_AO_B, we then form α in the shaded triangle. Hence, we can say:

$$\alpha = \sin^{-1}\frac{CD}{CE} = \sin^{-1}\frac{r_A - r_B}{O_AO_B} = \sin^{-1}\frac{.50 - .30}{2} = 5.74°$$

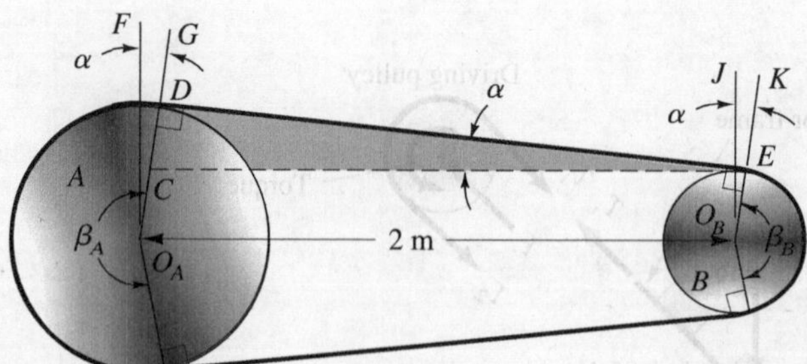

Figure 6.35. Find angles of wrap.

Note in Fig. 6.35 that O_AF is perpendicular to CE and that O_AG is perpendicular to DE. Therefore, the included angle between O_AF and O_AG must equal the included angle between CE and DE. This angle is α. Clearly, the angle between O_BJ and O_BK is also this angle α. We can now express the angles of wrap for both pulleys as follows:

$$\beta_A = 180° + 2(5.74) = 191.5°$$
$$\beta_B = 180° - 2(5.74) = 168.5°$$

Example 6.14 (Continued)

Now consider pulley A as a free body in Fig. 6.36. Note that the minimum force F corresponds to the condition of impending slippage. Accordingly, for this condition at A, we have

$$\frac{(T_1)_A}{(T_2)_A} = e^{\mu_s \beta_A} = e^{(.4)[(191.5/360)2\pi]} = 3.81 \qquad \text{(a)}$$

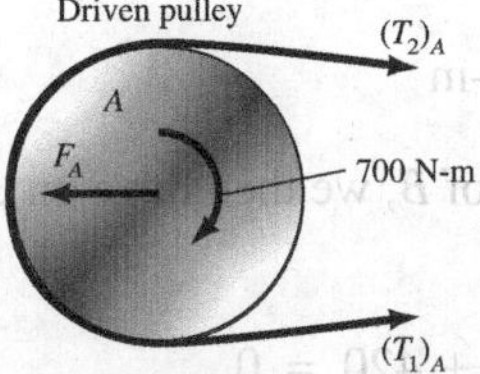

Figure 6.36. Free-body diagram of A.

Also, summing moments about the center of the pulley, we have

$$[(T_1)_A - (T_2)_A](.50) - 700 = 0 \qquad \text{(b)}$$

Therefore,

$$(T_1)_A - (T_2)_A = 1{,}400$$

Solving Eqs. (a) and (b) simultaneously, we get

$$(T_1)_A = 1{,}898 \text{ N}; \qquad (T_2)_A = 498 \text{ N}$$

From *equilibrium*, we can compute force F_A as follows:

$$(1{,}898 + 498) \cos 5.74° - F_A = 0 \qquad \text{(c)}$$

Therefore,

$$F_A = 2{,}384 \text{ N}$$

Now go to pulley B to see what minimum force F_B is needed so that the belt does not slip on it during operations. Consider in *Fig. 6.37* the free-body diagram of pulley B. For impending slipping on pulley B, we have

$$\frac{(T_1)_B}{(T_2)_B} = e^{\mu_s \beta_B} = e^{(.4)[(168.5/360)2\pi]} = 3.24 \qquad \text{(d)}$$

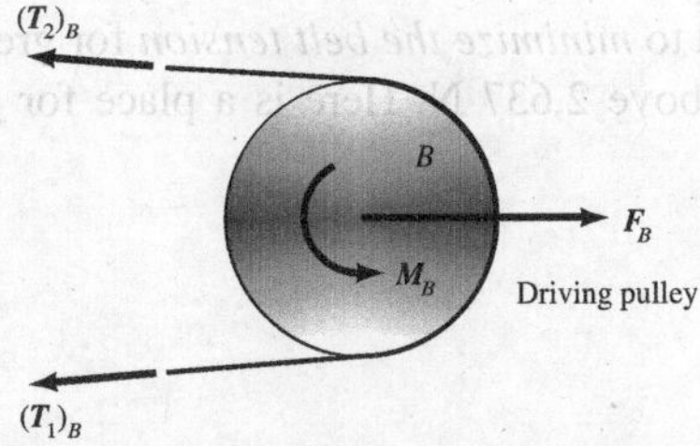

Figure 6.37. Free-body diagram of B.

Example 6.14 (Continued)

The torque for pulley B needed to develop 700 N-m on pulley A is next computed. Thus,[6]

$$M_B = \frac{r_B}{r_A} M_A = \frac{.30}{.50} M_A = \frac{.30}{.50}(700)$$

Therefore,

$$M_B = 420 \text{ N-m}$$

Summing moments in Fig. 7.21 about the center of B, we then have on using the above result

$$-[(T_1)_B - (T_2)_B](.30) + 420 = 0$$

Therefore,

$$(T_1)_B - (T_2)_B = 1{,}400 \qquad \text{(e)}$$

Solving Eqs. (d) and (e) simultaneously, we get

$$(T_1)_B = 2{,}025 \text{ N}; \qquad (T_2)_B = 625 \text{ N}$$

Hence, the minimum F_B needed for pulley B is

$$F_B = (2{,}025 + 625) \cos 5.74° = \boxed{2{,}637 \text{ N}}$$

Note that the ratios of the tensions for *impending slippage* for the two cylinders are

Driven cylinder A $\quad \frac{T_1}{T_2} = 3.81 \quad F_A = 2{,}384$ N

Driving cylinder B $\quad \frac{T_1}{T_2} = 3.24 \quad F_B = 2{,}637$ N

If we use impending slippage for driven cylinder A, we will have slipping for driving cylinder B, which we cannot tolerate. Thus, if we use the larger force of 2,637 N, we will be well under the impending slippage condition of the driven cylinder A. Clearly, the optimum result to *avoid slippage* and to *minimize the belt tension* for greater life of the belt is to have a force somewhere above 2,637 N. Here is a place for good engineering judgement and experience.

[6]Note that the ratio of transmitted torques M_2/M_1 between directly connected pulleys and gears will equal r_2/r_1 or D_2/D_1 of the pulleys or gears. Can you verify this yourself?

PROBLEMS

6.48. Two parallel shafts 100 cm apart are to have a velocity ratio of 2 and are to be connected by a flat belt. The pulley on the high speed shaft is 20 cm in diameter. Determine the length of the belt required: (i) if the belt is open, and (ii) if the belt is crossed.

6.49. An open belt drive connects two pulleys 100 cm and 40 cm diameter on parallel shafts 3.5 m apart. The belt material has a mass 0.7 kg/m length and maximum tension in the belt is not to exceed 1500 N. The coefficient of friction is 0.25. The 100 cm pulley, which is the driver runs at 180 rpm. Calculate the torque on each of the two shafts, the power transmitted.

6.50. An open belt drive connects two parallel shafts 1m apart. The driving and driven shafts rotate at 300 rpm and 120 rpm, respectively and the driven pulley is 50 cm in diameter. The belt is 6 mm thick and 10 cm wide coefficient of friction between belt and pulley is 0.25 and maximum permissible tension in the belt is 120 N/cm^2. Determine (i) diameter of driving pulley, (ii) maximum power that may be transmitted by the belt, and (iii) required initial tension in the belt neglect centrifugal tension.

6.51. Compute the frictional resisting torque for the concentric dry thrust bearing. The coefficient of friction is taken as μ_d.

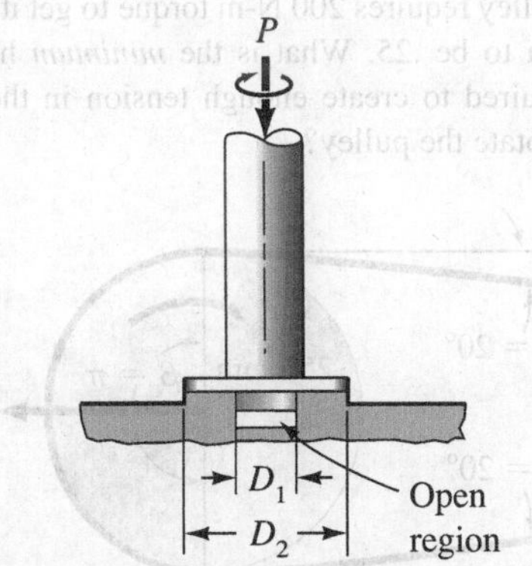

Figure P.6.51.

6.52. The support end of a dry thrust bearing is shown. Four pads form the contact surface. If a shaft creates a 100-N thrust uniformly distributed over the pads, what is the resisting torque for a dynamic coefficient of friction of .1?

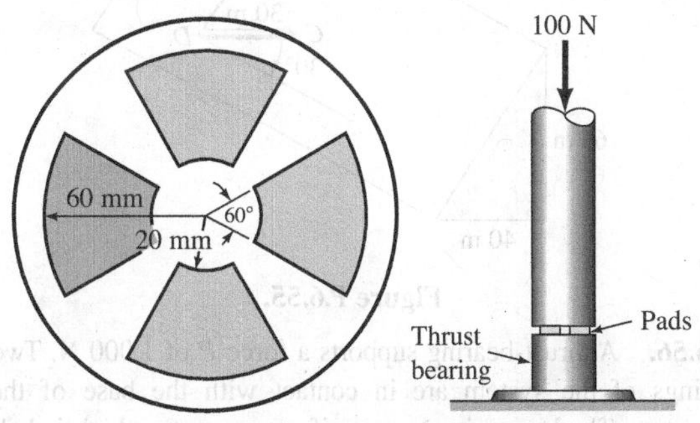

Figure P.6.52.

6.53. In Example 6.5, the normal force distribution at the contact surface is not uniform but, as a result of wear, is inversely proportional to the radius r. What, then, is the resisting torque M?

6.54. Compute the frictional torque needed to rotate the truncated cone relative to the fixed member. The cone has a 20-mm-diameter base and a 60° cone angle and is cut off 3 mm from the cone tip. The dynamic coefficient of friction is .2.

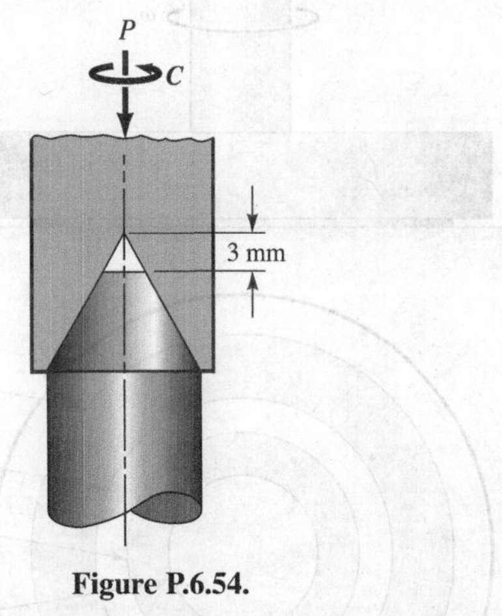

Figure P.6.54.

6.55. A 1,000-N block is being lowered down an inclined surface. The block is pinned to the incline at C, and at B a cord is played out so as to cause the body to rotate at uniform speed about C. Taking μ_d to be .3 and assuming the contact pressure is uniform along the base of the block, compute T for the configuration shown in the diagram.

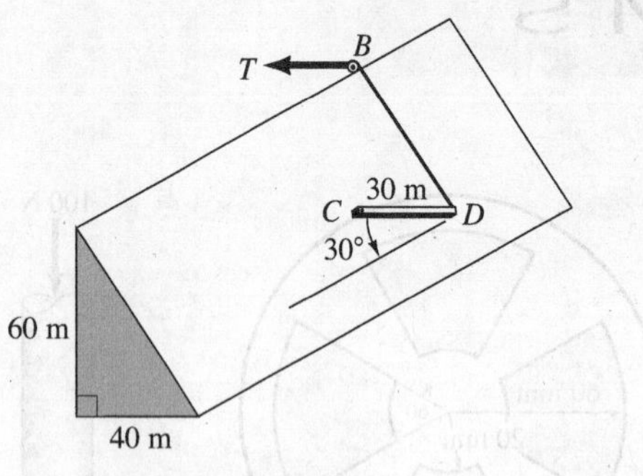

Figure P.6.55.

6.56. A thrust-bearing supports a force P of 1,000 N. Two rings of the system are in contact with the base of the system. The larger ring has a uniform pressure which is half the value of the uniform pressure on the inner ring. The shaft is rotating with an angular speed ω. If the dynamic coefficient of friction μ_d is 0.3, what is the resisting torque generated by this thrust bearing? Consider P to include the weights of the shaft and the pads attached to it.

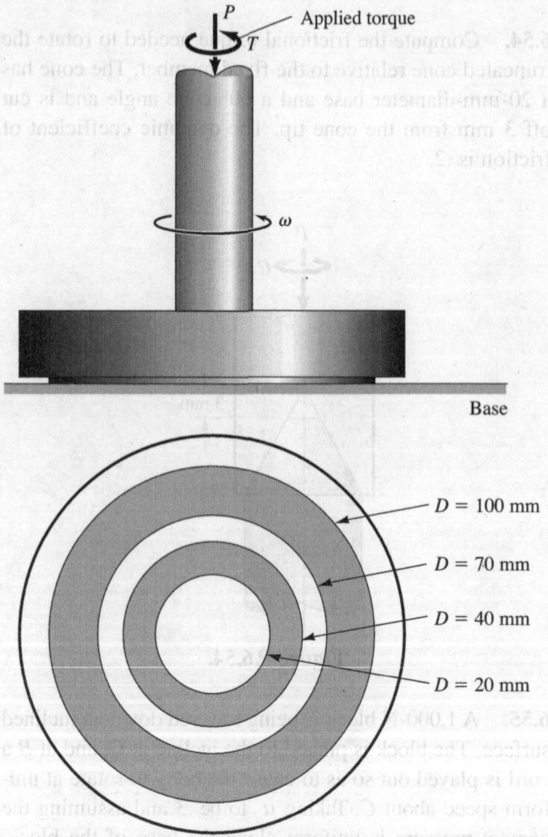

Figure P.6.56.

6.57. A shaft has three dry thrust-bearing surfaces which push, respectively, on three stationary rigid surfaces, namely rigid plates A and B and the rigid base C. The total load P coming onto the shaft including weights of all parts is 6,000 N. What is the minimal torque T needed to start rotation if the pressure on each disc contact surface is uniform for each disc but is proportional to the outside diameter squared of the disc? The static coefficient of friction between the bearing surfaces and the plates A and B as well as the base C is 0.2. The proportionality constant for pressure is the *same value* for all 3 discs. The shaft passes through A and B but is welded to the top surface of C.

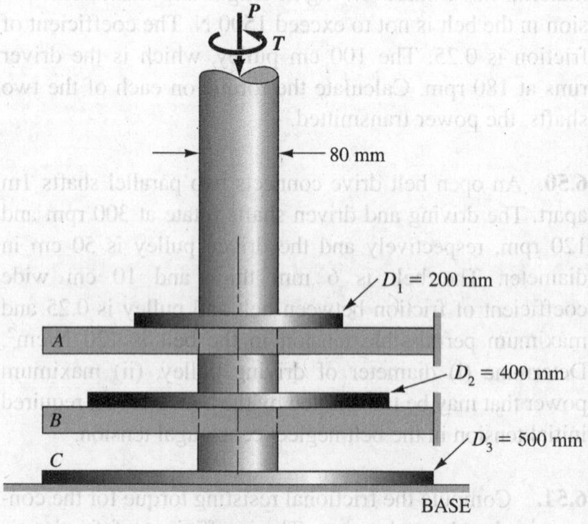

Figure P.6.57.

6.58. A pulley requires 200 N-m torque to get it rotating. μ_s is known to be .25. What is the *minimum* horizontal force F required to create enough tension in the belt so that it can rotate the pulley?

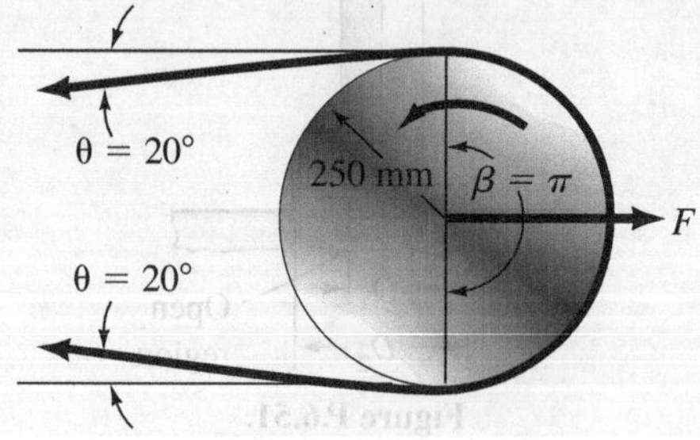

Figure P.6.58.

6.59. If in Problem 7.55, $\theta = 0$ and the belt is wrapped $2\frac{1}{2}$ times around the pulley, what is the *minimum* horizontal force F needed to rotate the pulley?

6.60. The seaman pulls with 100-N force and wants to stop the motorboat from moving away from the dock under power. How few wraps n of the rope must he make around the post if the motorboat develops 3,500 N of thrust and the static coefficient of friction between the rope and the post is .2?

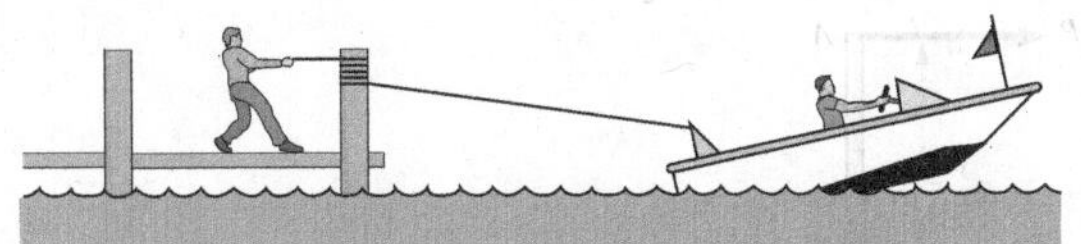

Figure P.6.60.

6.61. A length of belt rests on a flat surface and runs over a quarter of the drum. A load W rests on the horizontal portion of the belt, which in turn is supported by a table. If the static coefficient of friction for all surfaces is .3, compute the *maximum* weight W that can be moved by rotating the drum.

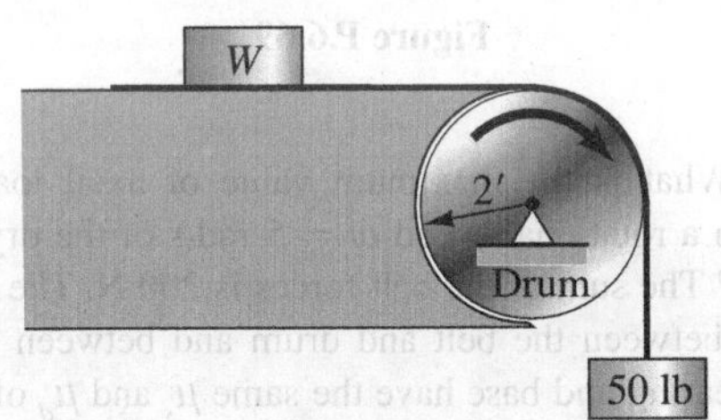

Figure P.6.61

6.62. The rope holding the 50-lb weight E passes over the drum and is attached at A. The weight of C is 60 lb. What is the *minimum* static coefficient of friction between the rope and the drum to maintain equilibrium of the drum?

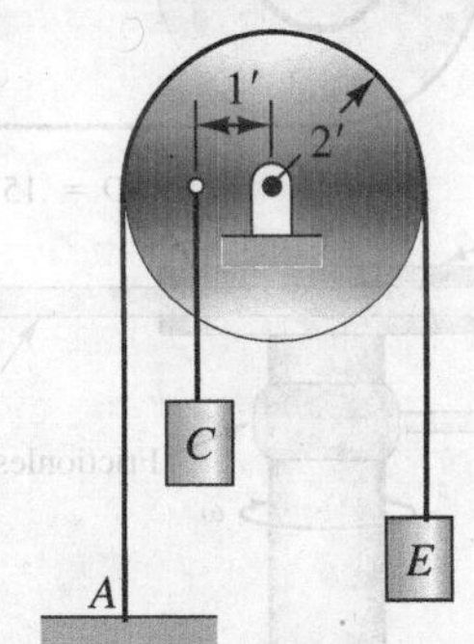

Figure P.6.62.

6.63. What is the *maximum* weight that can be supported by the system in the position shown? Pulley B *cannot* turn. Bar AC is fixed to cylinder A, which weighs 500 N. The coefficient of static friction for all contact surfaces is .3.

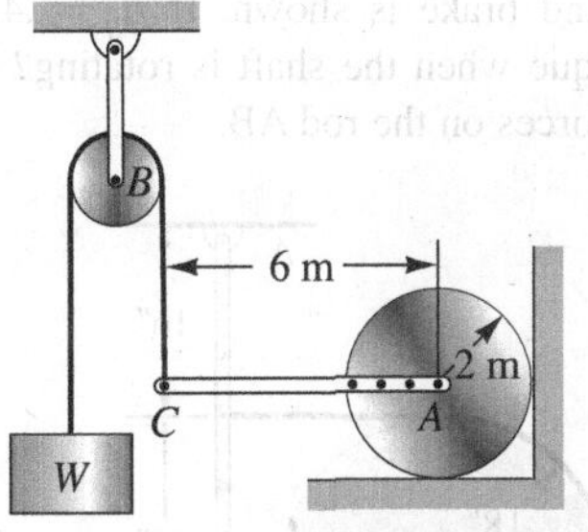

Figure P.6.63.

6.64. A mountain climber of weight W hangs freely suspended by one rope that is fastened at one end to his waist, wrapped one-half turn about a rock with μ_s = .2, and held at the other end in his hand. What *minimum* force in terms of W must he pull with to maintain his position? What minimum force must he pull the rope into himself with to gain altitude?

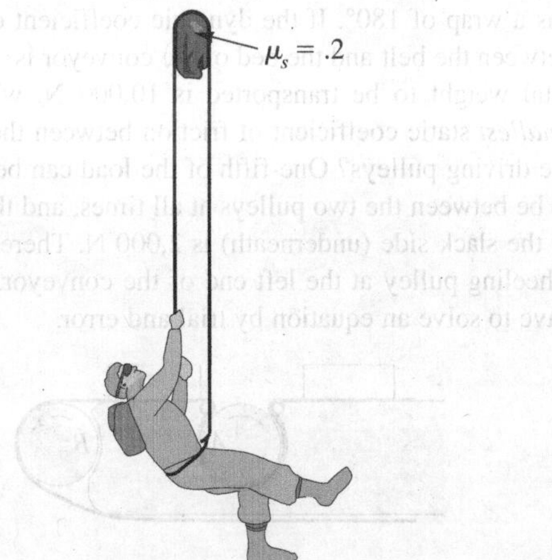

Figure P.6.64.

6.65. Pulley B is turned by a diesel engine and drives pulley A connected to a generator. If the torque that A must transmit to the generator is 500 N-m, what is the *minimum* static coefficient of friction between the belt and pulleys for the case where the force F is 2,000 N?

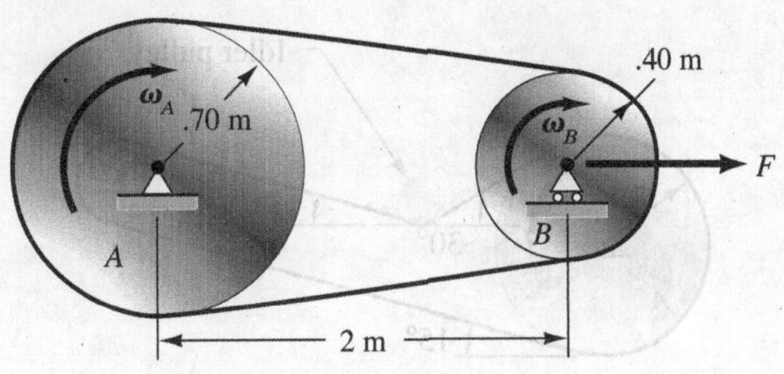

Figure P.6.65.

6.66. A hand brake is shown. If $\mu_d = .4$, what is the resisting torque when the shaft is rotating? What are the supporting forces on the rod AB.

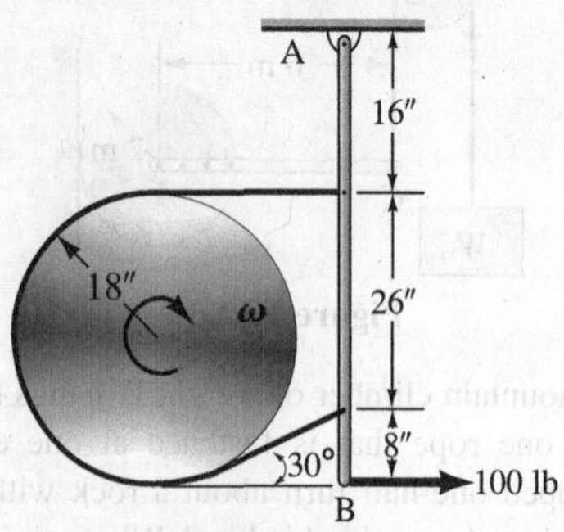

Figure P.6.66.

6.67. A conveyor is shown with two driving pulleys *A* and *B*. Driver *A* has an angle of wrap of 330°, whereas *B* has a wrap of 180°. If the dynamic coefficient of friction between the belt and the bed of the conveyor is .1, and the total weight to be transported is 10,000 N, what is the *smallest* static coefficient of friction between the belt and the driving pulleys? One-fifth of the load can be assumed to be between the two pulleys at all times, and the tension in the slack side (underneath) is 2,000 N. There is a free-wheeling pulley at the left end of the conveyor. You will have to solve an equation by trial and error.

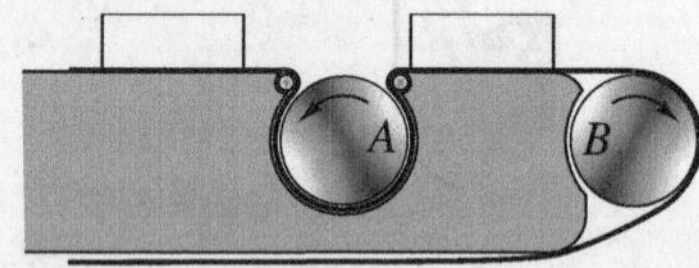

Figure P.6.67.

6.68. A freely turning idler pulley is used to increase the angle of wrap for the pulleys shown. If the tension in the slack side above is 200 lb, find the *maximum* torque that can be transmitted by the pulleys for a static coefficient of friction of .3.

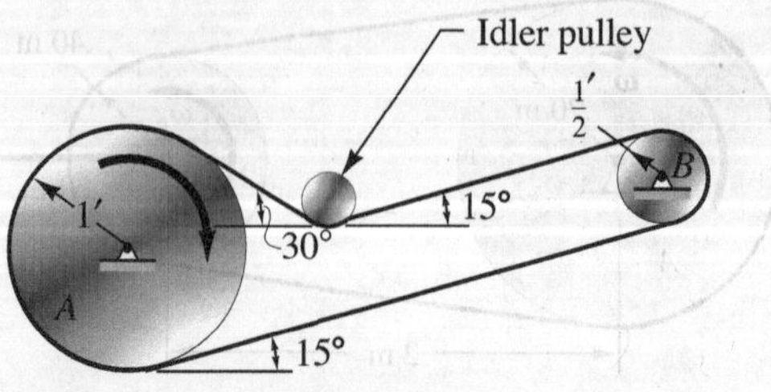

Figure P.6.68.

6.69. (a) What force *P* is needed to develop a resisting torque of 65 N-m on the rotating drum? The dynamic coefficient of friction μ_d is 0.4.
(b) With the same force *P* from part (a), what must the value of μ_d be in order to increase the resisting torque by 10 N-m?

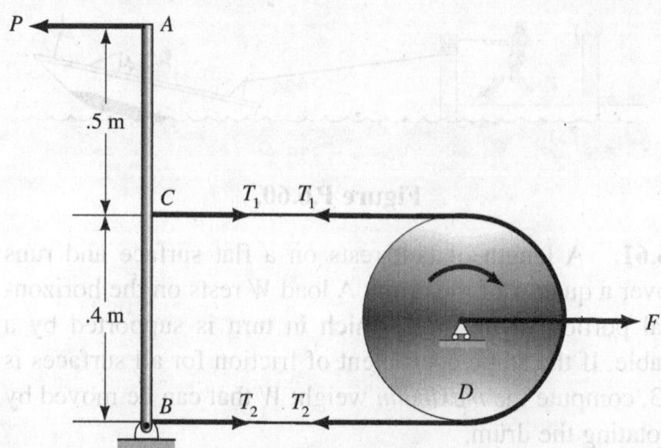

Figure P.6.69.

6.70. What is the maximum value of axial load *P* to maintain a rotational speed ω = 5 rad/s of the dry thrust bearing? The sum of the belt forces is 200 N. The contact surface between the belt and drum and between the dry thrust bearing and base have the same μ_s and μ_d of .4 and .3, respectively.

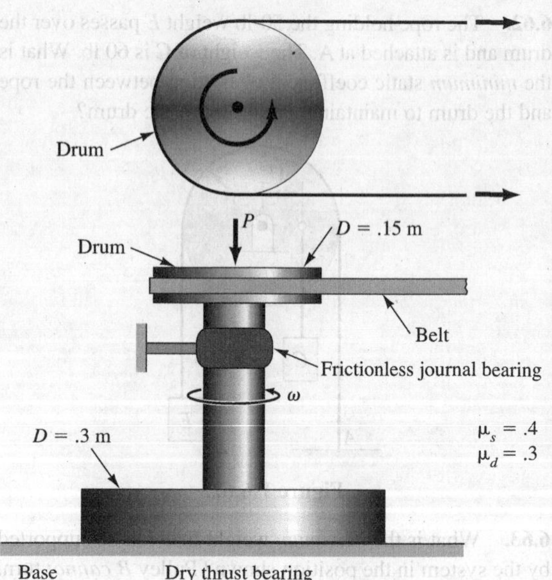

Figure P.6.70.

6.71. Rod *AB* weighing 200 N is supported by a cable wrapped around a semicylinder having a coefficient of friction μ_s equal to .2. A weight *C* having a mass of 10 kg can slide on rod *AB*. What is the maximum range *x* from the centerline that the center of *C* can be placed without causing slippage?

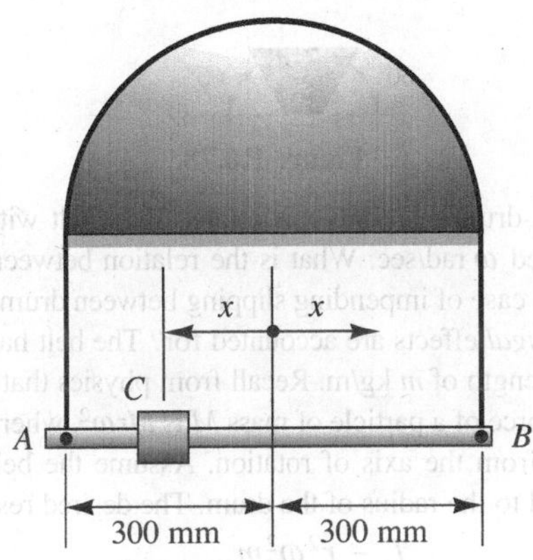

Figure P.6.71.

6.72. The cable mechanism shown is similar to that used to move the station indicator on a radio. If the indicator jams, what force is developed at the indicator base to free the jam when the required torque applied to the turning nob is 10 lb-in? Also, what are the forces in the various regions of the cable? The static coefficient of friction is .15.

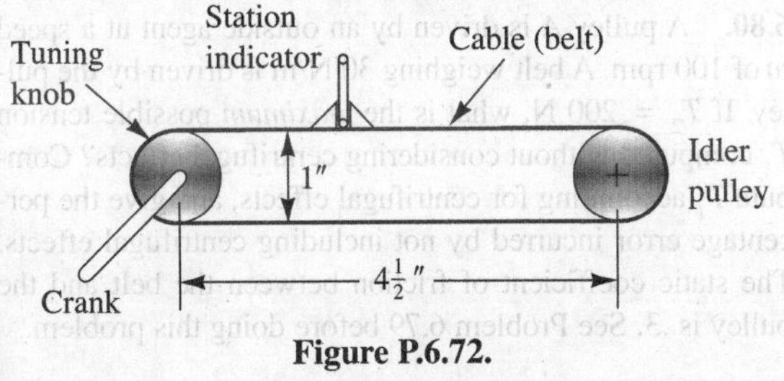

Figure P.6.72.

6.73. What are the *minimum* possible supporting force components needed for pulley *B* as a result of the action of the belt? The static coefficient of friction between the belt and pulley *B* is .3 and between the belt and pulley *A* is .4. The torque that the belt delivers to *A* is 200 N-m.

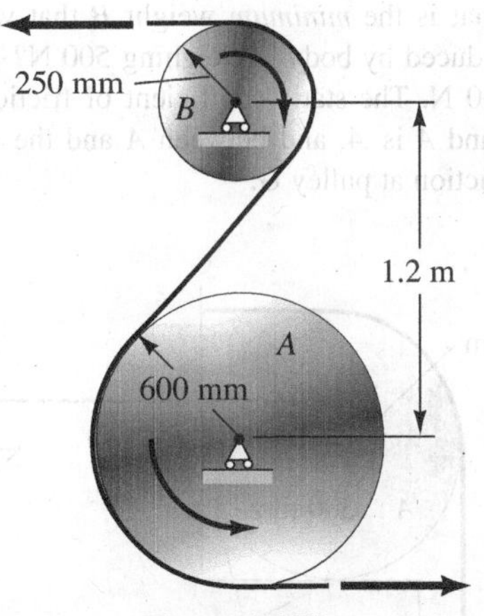

Figure P.6.73.

6.74. From first principles, show that the normal force per unit length, *w*, acting on a drum from a belt is given as

$$w = \frac{T_2}{r} e^{\mu\theta}$$

Use the indicated diagram as an aid. [*Hint:* Start with Eq. 6.2(a) and use Eq. 6.4 for any point *a*.]

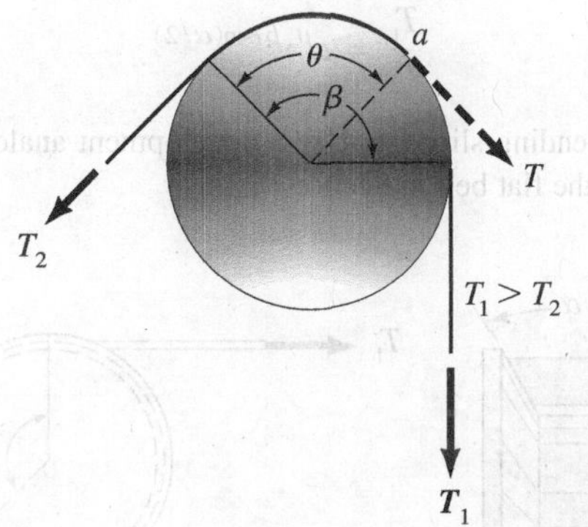

Figure P.6.74.

6.75. What *minimum* force *F* is needed so that drum *A* can transmit a clockwise torque of 500 N-m without slipping? The coefficient of friction, μ_s, between *A* and the belt is .4. What *minimum* coefficient of static friction is needed between drum *B* and belt for no slipping?

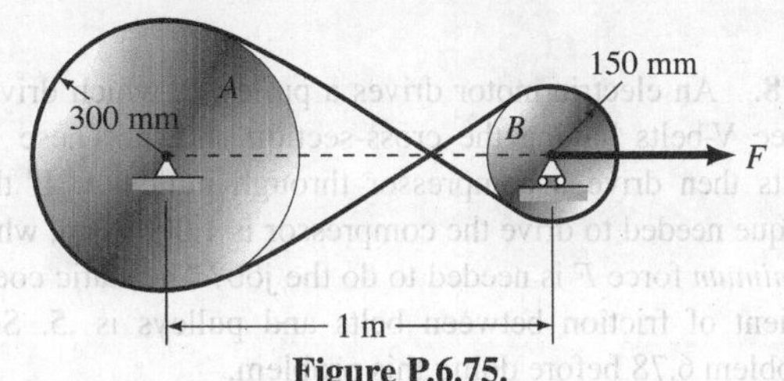

Figure P.6.75.

6.76. What is the *minimum* weight B that will prevent rotation induced by body C weighing 500 N? The weight of A is 100 N. The static coefficient of friction between the belts and A is .4, and between A and the walls is .1. Neglect friction at pulley G.

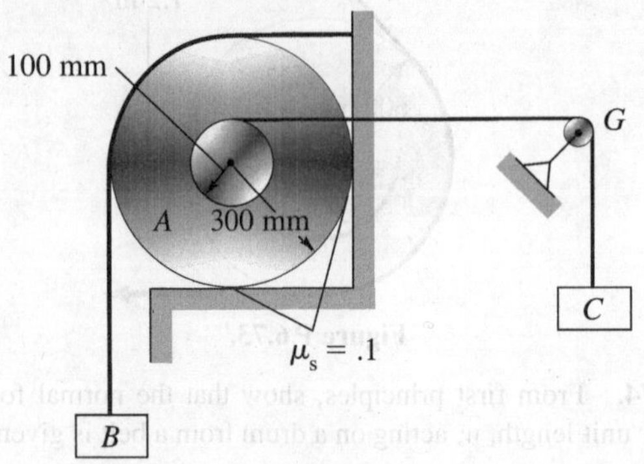

Figure P.6.76.

6.77. A V-belt is shown. Show that

$$\frac{T_1}{T_2} = e^{\mu_s \beta / \sin(\alpha/2)}$$

for impending slippage. Use a development analogous to that of the flat belt in Section 7.8.

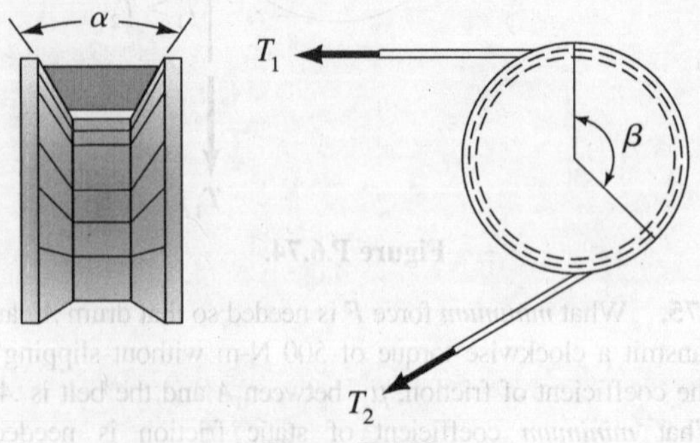

Figure P.6.78.

6.78. An electric motor drives a pulley B, which drives three V-belts having the cross-section shown. These V-belts then drive a compressor through pulley A. If the torque needed to drive the compressor is 1,000 N-m, what *minimum* force F is needed to do the job? The static coefficient of friction between belts and pulleys is .5. See Problem 6.78 before doing this problem.

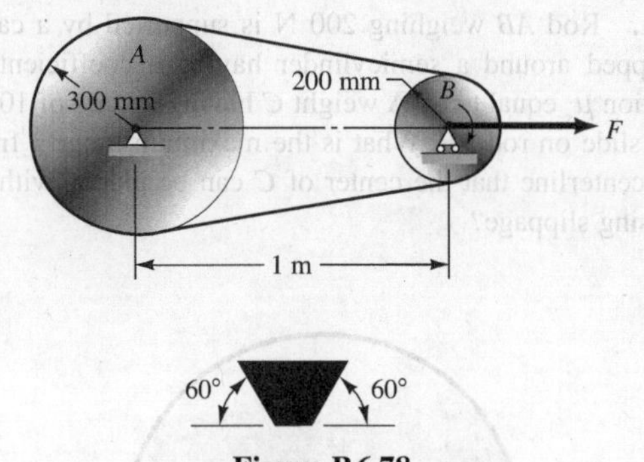

60° 60°

Figure P.6.78.

6.79. A drum of radius r is rotated by a belt with a constant speed ω rad/sec. What is the relation between T_1 and T_2 for the case of impending slipping between drum and belt if *centrifugal* effects are accounted for? The belt has a mass per unit length of m kg/m. Recall from physics that the centrifugal force of a particle of mass M is $Mr\omega^2$, where r is the distance from the axis of rotation. Assume the belt is thin compared to the radius of the drum. The desired result is

$$\frac{T_1 - r^2\omega^2 m}{T_2 - r^2\omega^2 m} = e^{\mu_s \beta}$$

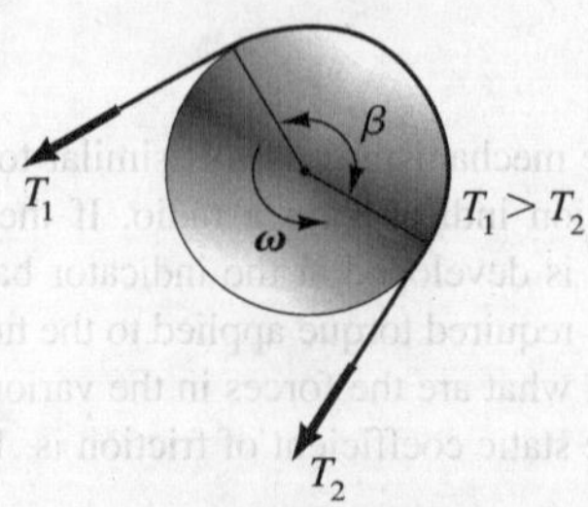

Figure P.6.79.

6.80. A pulley A is driven by an outside agent at a speed ω of 100 rpm. A belt weighing 30 N/m is driven by the pulley. If $T_2 = 200$ N, what is the *maximum* possible tension T_1 computed without considering centrifugal effects? Compute T_1 accounting for centrifugal effects, and give the percentage error incurred by not including centrifugal effects. The static coefficient of friction between the belt and the pulley is .3. See Problem 6.79 before doing this problem.

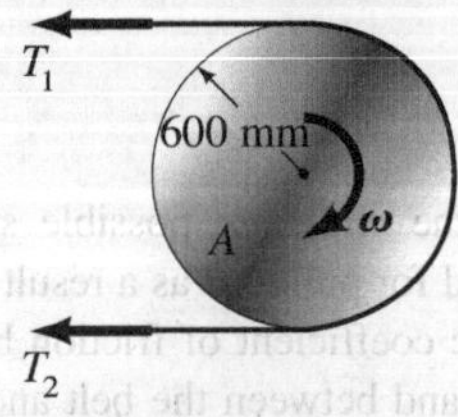

Figure P.6.80.

6.9 The Square Screw Thread

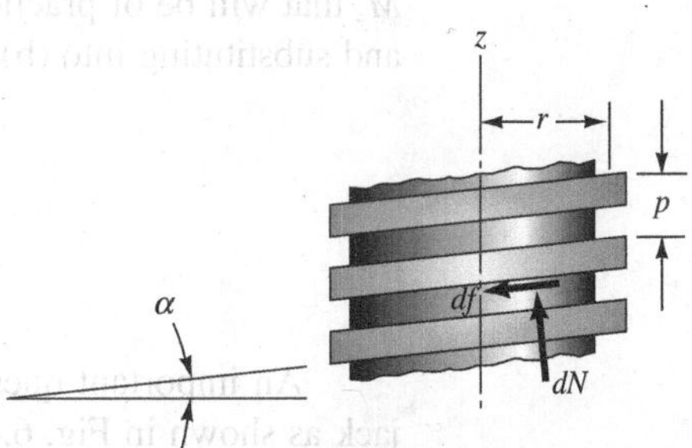

Figure 6.38. Square screw thread.

We shall now consider the action of a nut on a screw that has square threads (Fig. 6.38). Let us take r as the mean radius from the centerline of the screw to the thread. The *pitch*, p, is the distance along the screw between adjacent threads, and the *lead*, L, is the distance that a nut will advance in the direction of the axis of the screw in one revolution. For screw threads that are single-threaded, L equals p. For an n-threaded screw, the lead L is np.

Forces are transmitted from screw to nut over several revolutions of thread, and hence we have a distribution of normal and friction forces. However, because of the narrow width of the thread, we may consider the distribution to be confined at a distance r from the centerline, thus forming a "loading" strip winding around the centerline of the screw. Figure 6.38 illustrates infinitesimal normal and frictional forces on an infinitesimal part of the strip. The local slope $\tan \alpha$ as one looks in radially is determined by considering the definition of L, the lead. Thus,

$$\text{slope} = \tan \alpha = \frac{L}{2\pi r} = \frac{np}{2\pi r}$$

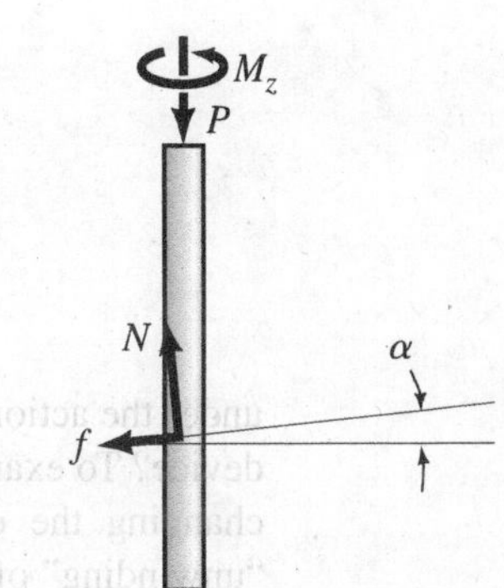

Figure 6.39. Free-body diagram.

All elements of the proposed distribution have the same inclination (direction cosine) relative to the z direction. In the summation of forces in this direction, therefore, we can consider the distribution to be replaced by a single normal force N and a single friction force f at the inclinations shown in Fig. 6.39 at a position anywhere along the thread. And, since the elements of the distribution have the same moment arm about the centerline in addition to the common inclination, we may use the concentrated forces mentioned above in taking moments about the centerline. There is thus a "limited equivalence" between N and f and the force distribution from the nut onto the screw. The other forces on the screw will be considered as an axial load P and a torque M_z collinear with P (Fig. 6.40). For equilibrium at a condition of *impending motion* to raise the screw, we then have the following scalar equations:[7]

$\underline{\sum F_z = 0:}$

$$-P + N\cos\alpha - \mu_s N \sin\alpha = 0 \qquad \text{(a)}$$

$\underline{\sum M_z = 0:}$

$$-\mu_s N \cos\alpha\, r - N\sin\alpha\, r + M_z = 0 \qquad \text{(b)}$$

[7]The equations also apply to *steady rotation* of the nut on the screw, in which case one uses the dynamic coefficient of friction μ_d in the equations.

These equations may be used to eliminate the force N and so get a relation between P and M_z that will be of practical significance. This may readily be done by solving for N in (a) and substituting into (b). The result is

$$M_z = \frac{Pr(\mu_s \cos\alpha + \sin\alpha)}{\cos\alpha - \mu_s \sin\alpha} \tag{6.28}$$

An important question arises when we employ the screw and nut in the form of a jack as shown in Fig. 6.40. Once having raised a load P by applying the torque M_z to the jackscrew, does the device maintain the load at the raised position when the applied torque is released, or does the screw unwind

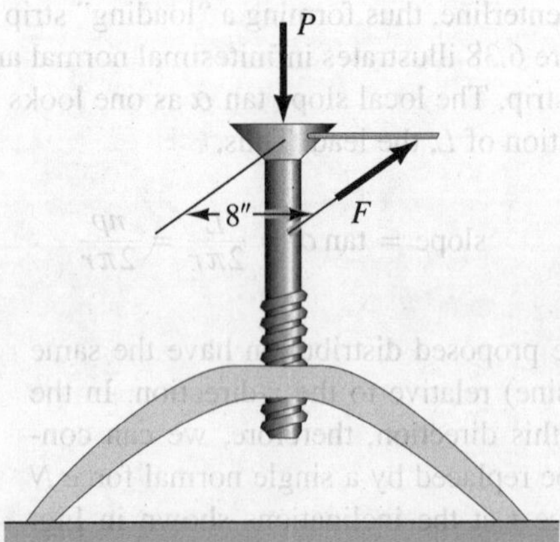

Figure 6.40. Jackscrew.

under the action of the load and thus lower the load? In other words, is this a *self-locking* device? To examine this, we go back to the equations of equilibrium. Setting $M_z = 0$ and changing the direction of the friction forces, we have the condition for impending "unwinding" of the screw. Eliminating N from the equations, we get

$$\frac{Pr(-\mu_s \cos\alpha + \sin\alpha)}{\cos\alpha + \mu_s \sin\alpha} = 0$$

This requires that

$$-\mu_s \cos\alpha + \sin\alpha = 0$$

Therefore,

$$\mu_s = \tan\alpha \tag{6.29}$$

We can conclude that, if the coefficient of friction μ_s equals or exceeds $\tan\alpha$, we will have a self-locking condition. If μ_s is less than $\tan\alpha$, the screw will unwind and will not support a load P without the proper external torque.

Example 6.15

A jackscrew with a double thread of mean diameter 2 in. is shown in Fig. 7.24. The pitch is .2 in. If a force F of 40 lb is applied to the device, what load W can be raised? With this load on the device, what will happen if the applied force F is released? Take $\mu_s = .3$ for the surfaces of contact.

The applied torque M_z is clearly:

$$M_z = \tfrac{8}{12}(40) = 26.7 \text{ lb-ft} \tag{a}$$

The angle α for this screw is given as

$$\tan\alpha = \frac{(2)(.2)}{(2\pi)(1)} = .0636 \tag{b}$$

Therefore,

$$\alpha = 3.64°$$

Using Eq. 6.8 we can solve for P. Thus,

$$P = \frac{M_z(\cos\alpha - \mu_s \sin\alpha)}{r(\mu_s \cos\alpha + \sin\alpha)} \tag{c}$$

$$= \frac{(26.7)[.998 - (.3)(.0635)]}{\frac{1}{12}[(.3)(.998) + .0635]} = \boxed{864 \text{ lb}}$$

The load W is 864 lb. The device is self-locking since μ_s exceeds $\tan\alpha = .0636$.

$$\boxed{\mu_s > \tan\alpha \quad \therefore \text{ self-locking}}$$

To lower the load requires a reverse torque. We may readily compute this torque by using Eq. 7.8 with the friction forces reversed. Thus,

$$(M_z)_{\text{down}} = \frac{864(\frac{1}{12})[-(.3)(.998) + .0635]}{.998 + (.3)(.0635)} = -16.71 \text{ lb-ft} \tag{d}$$

PROBLEMS

6.81. A simple C-clamp is used to hold two pieces of metal together. The clamp has a single square thread with a pitch of .12 in. and a mean diameter of .75 in. The static coefficient of friction is .30. Find the torque required if a 1,000-lb compressive load is required on the blocks. If the thread is a double thread, what is the required torque?

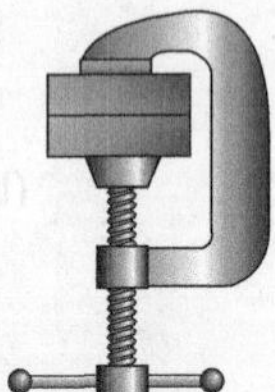

Figure P.6.81.

6.82. The mast of a sailboat is held by wires called shrouds, as shown in the diagram. Racing sailors are careful to get the proper tension in the shrouds by adjusting the turnbuckle at the bottom of the shrouds. When we do this we say we are "tuning" the boat. If a tension of 150 N exists in the shroud, what torque is needed to start tightening further by turning the turnbuckle? The pitch of the single threaded screw is 1.5 mm and the mean diameter is 8.0 mm. The static coefficient of friction is .2.

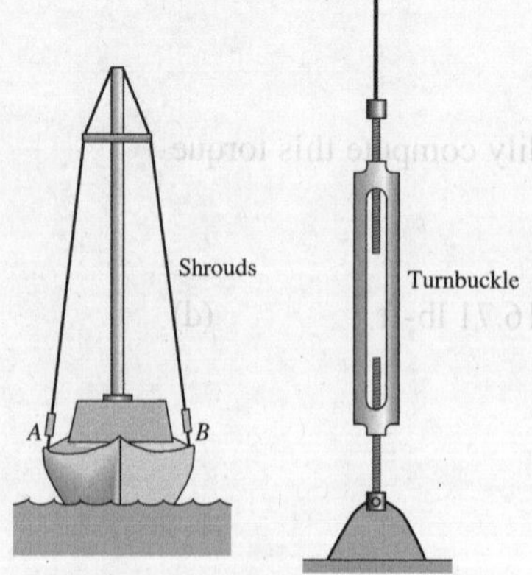

Figure P.6.82.

6.83. Forces F of 50 lb are applied to the jackscrew shown. The thread diameter is 2 in. and the pitch is $\frac{1}{2}$ in. The static coefficient of friction for the thread is .05. The weight W and collar are not permitted to rotate and so the collar must rotate on the shaft of the screw. If the static coefficient of friction between the collar and shaft is .1, determine the weight W that can be lifted by this system.

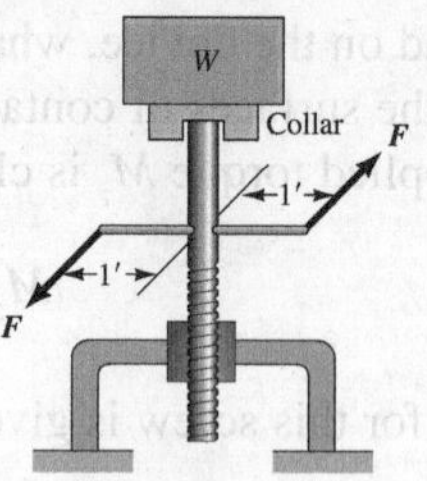

Figure P.6.83.

6.84. A brake is shown. Force is developed at the brake shoes by turning A, which has a single right-handed square screw thread at B and a single left-handed screw thread at C. The diameter of the screw thread is $1\frac{1}{2}$ in. and the pitch is .3 in. If the static coefficient of friction is .1 for the thread and the dynamic coefficient of friction is .4 for the brake shoes, what resisting torque is developed on the wheel by a 100 in.-lb torque at A?

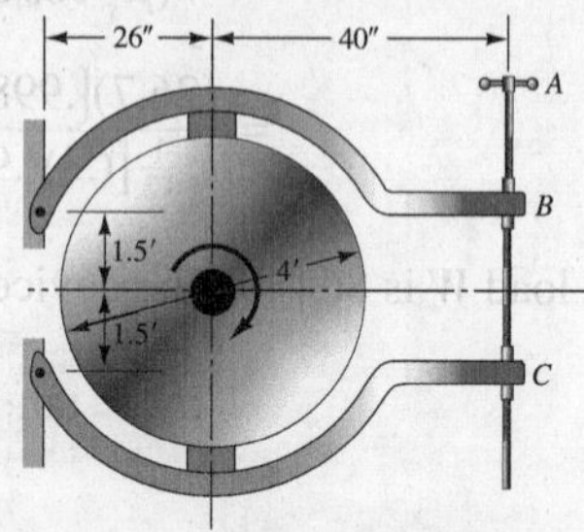

Figure P.6.84.

6.85. A triangular-threaded screw is shown. In a manner paralleling the development of the square-thread formulation in Section 6.9, show that

$$M_z = \frac{rP(\mu_s \cos\alpha + \cos\theta\tan\alpha)}{\cos\theta - \mu_s \sin\alpha}$$

where

$$\cos\theta = \frac{1}{\sqrt{\tan^2\alpha + \tan^2\gamma + 1}}$$

and

$$\gamma = \beta - \alpha$$

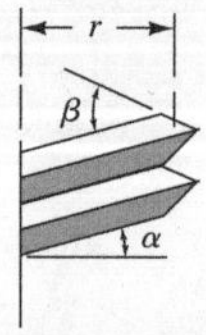

Figure P.6.85.

6.86. Consider a single-threaded screw where the pitch p = 4.5 mm and the mean radius is 20 mm. For a coefficient of friction μ_s = .3, what torque is needed on the nut for it to turn under a load of 1,000 N? Compute this for a square thread and then do it for a triangular thread where the angle β is 30°. See Problem 7.82 before doing this problem.

6.87. If the coefficient of rolling resistance of a cylinder on a flat surface is .05 in., at what inclination of the surface will the cylinder of radius r = 1 ft roll with uniform velocity?

6.88. A 65-kN vehicle designed for polar expeditions is on a very slippery ice surface for which the static coefficient of friction between tires and ice is .005. Also, the coefficient of rolling resistance is known to be .8 mm. Will the vehicle be able to move? The vehicle has four-wheel drive.

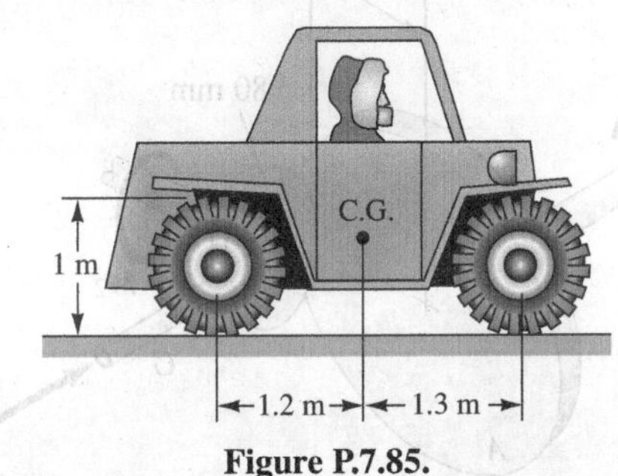

Figure P.7.85.

6.89. In Problem 6.88, suppose there is only rear-wheel drive available. What is the minimum static coefficient of friction needed between tires and ground for the vehicle to move?

6.90. A roller thrust bearing is shown supporting a force P of 2.5 kN. What torque T is need to turn the shaft A at constant speed if the only resistance is that from the ball bearings? The coefficient of rolling resistance for the balls and the bearing surfaces is .01270 mm. The mean radius from the centerline of the shaft to the balls is 30 mm.

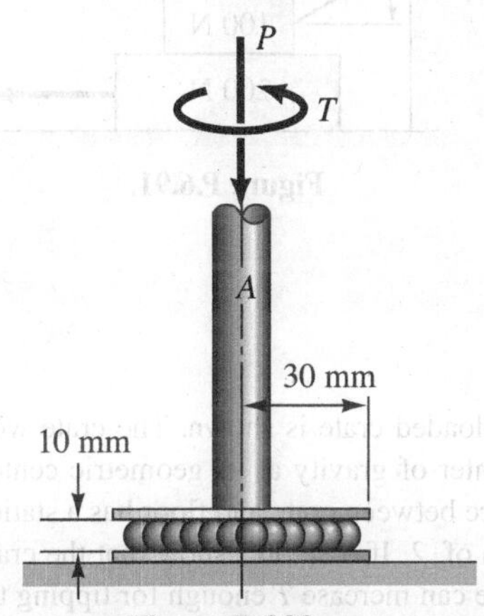

Figure P.6.90.

6.10 Closure

In this chapter, we have examined the results of two independent experiments: that of impending or actual sliding of one body over another and that of a cylinder or sphere rolling at constant speed over a flat surface. Without any theoretical basis, the results of such experiments must be used in situations that closely parallel the experiments themselves.

In the case of a rolling cylinder, both rolling resistance and sliding resistance are present. However, for a cylinder accelerating with any appreciable magnitude, only sliding friction need be accounted for. With no acceleration on a horizontal surface, only rolling resistance need be considered. Most situations fall into these categories. For very small accelerations, both effects are present and must be taken into account. We can then expect only a crude result for such computations. Transmission of power through belts and screw jack have been explained in detail.

Before going further, we must carefully define certain properties of plane surfaces in order to facilitate later computations in mechanics where such properties are most useful. These plane surface properties and other related topics will be studied in Chapter 7.

PROBLEMS

6.91. If the static coefficient of friction for all surfaces is .35, find the force F needed to start the 200-N weight moving to the right.

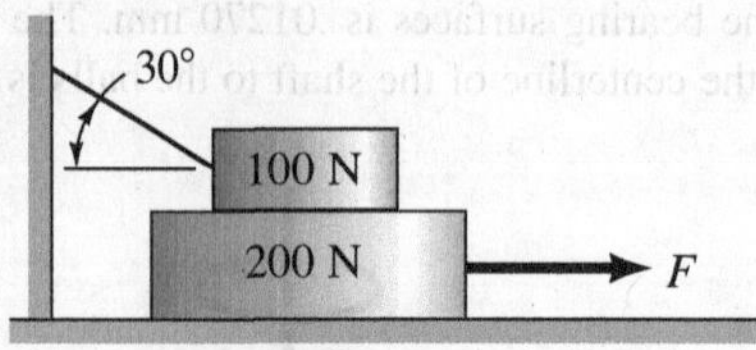

Figure P.6.91.

6.92. A loaded crate is shown. The crate weighs 500 lb with a center of gravity at its geometric center. The contact surface between crate and floor has a static coefficient of friction of .2. If $\theta = 90°$, show that the crate will slide before one can increase T enough for tipping to occur. If a stop is to be inserted in the floor at A to prevent slipping so that the crate could be tipped, what *minimum* horizontal force will be exerted on the stop?

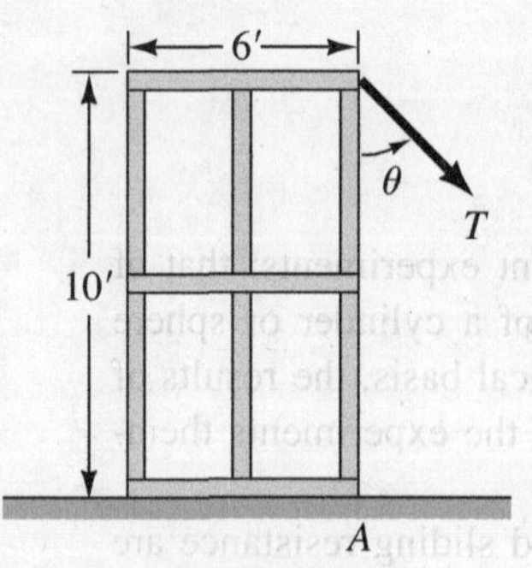

Figure P.6.92.

6.93. In Problem 6.92, compute a value of θ and T where slipping and tipping will occur simultaneously. If the actual angle θ is smaller than this value of θ, is there any further need of the stop at A to prevent slipping?

6.94. A friction drive is shown with A the driver disc and B the driven disc. If force F pressing B onto A is 150 N, what is the *maximum* torque M_2 that can be developed? For this torque, what is the torque M_1 needed for the drive disc A? The static coefficient of friction between A and B is .7. What vertical force must rod G withstand for the action described above?

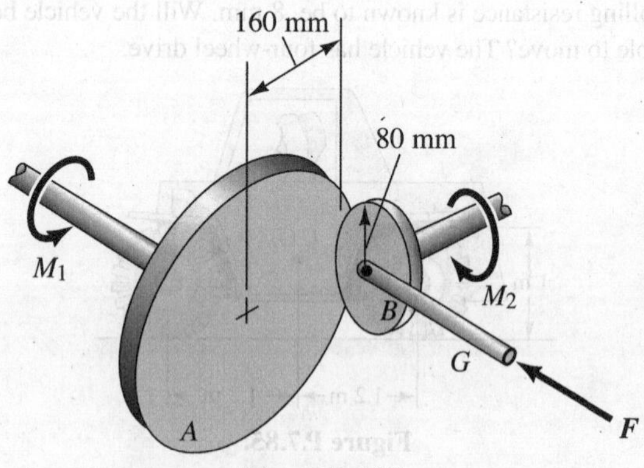

Figure P.6.94.

6.95. Determine the weight of block A for impending motion to the right. The static coefficient of friction between the cable and the surfaces which it contacts is 0.2. The static coefficient of friction between block A and the surface upon which it rests is 0.4. The two posts are circular with a diameter of 0.25 m.

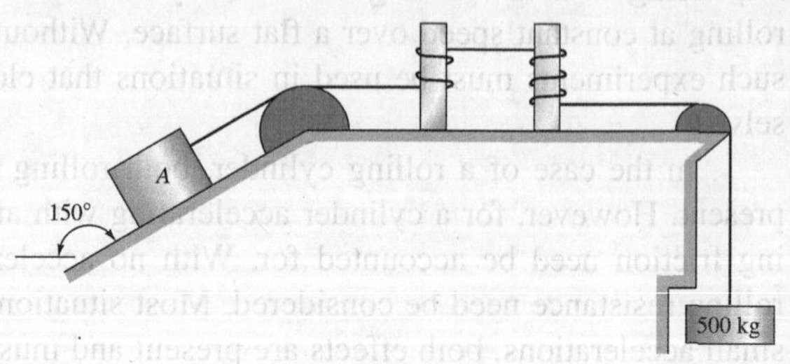

Figure P.6.95.

6.96. Do Problem 6.95 for impending slippage of body A to the left.

6.97. A tug is pushing a barge into a berth. After the barge turns clockwise and touches the sides of the pilings, what thrust must the tug develop to move it at uniform speed of 2 knots farther into the berth? The dynamic coefficient of friction between the barge and the sides of the berth is .4. The drag from the water is 3,000 N along the centerline of the barge.

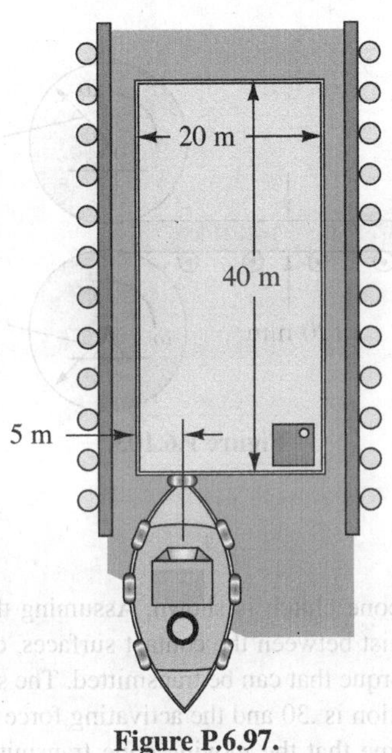

Figure P.6.97.

6.98. The static and dynamic coefficients of friction for the upper surface of contact A of the cylinder are $\mu_s = .4$, $\mu_d = .3$, and for the lower surface of contact B are $\mu_s = .1$ and $\mu_d = .08$. What is the *minimum* force P needed to just get the cylinder moving?

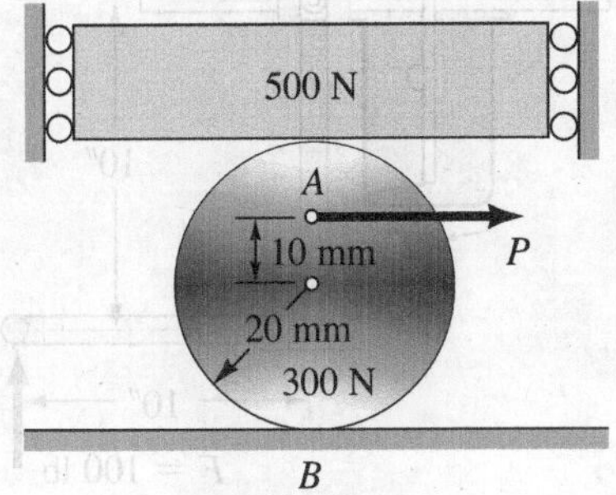

Figure P.6.98.

6.99. The drum is driven by a motor with a maximum torque capability of 500 lb-ft. The static coefficient of friction between the drum and the braking strap (belt) is .4. How much force P must an operator exert to stop the drum if it rotates (1) clockwise and (2) counterclockwise? What are the belt forces in each case?

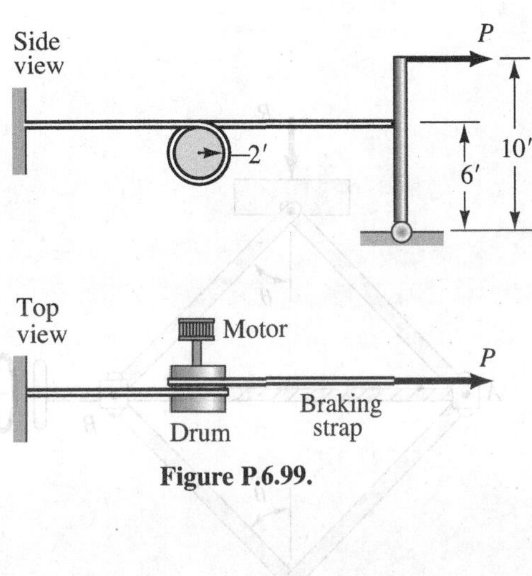

Figure P.6.99.

6.100. The four drive pulleys shown are used to transmit a torque from pulley A to pulley D on an electric typewriter. If the static coefficient of friction between the belts and the pulleys is .3, what is the torque available at pulley D if 10 lb-in. of torque is input to the shaft of pulley A? What are the belt forces?

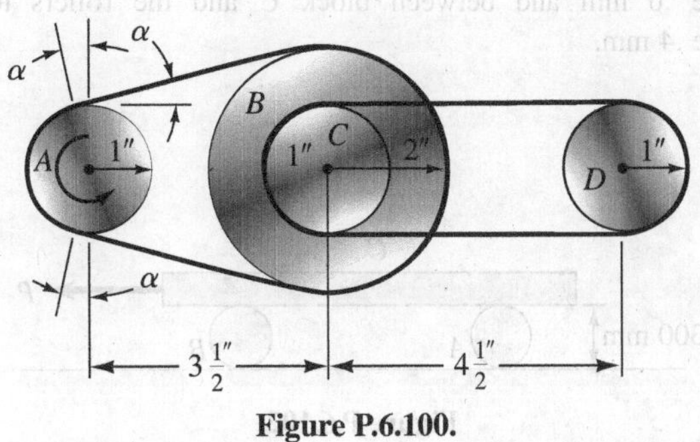

Figure P.6.100.

6.101. A scissors jack is shown lifting the end of a car so that $R = 6.67$ kN. What torque T is needed for this operation? Note that A is merely a bearing and at B we have a nut. The screw is single-threaded with a pitch of 3 mm and a mean diameter of 20 mm. The static coefficient of friction between the screw and nut at B is .3. Neglect the weight of the members and evaluate T for $\theta = 45°$ and for $\theta = 60°$.

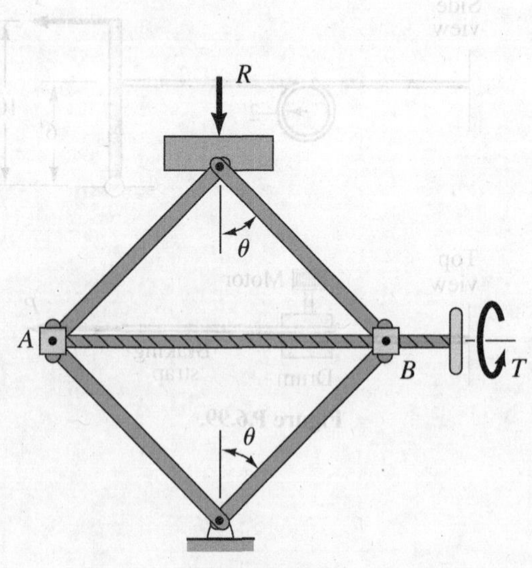

Figure P.6.101.

6.102. A block C weighing 10 kN is being moved on rollers A and B each weighing 1 kN. What force P is needed to maintain steady motion? Take the coefficient of rolling resistance between the rollers and the ground to be .6 mm and between block C and the rollers to be .4 mm.

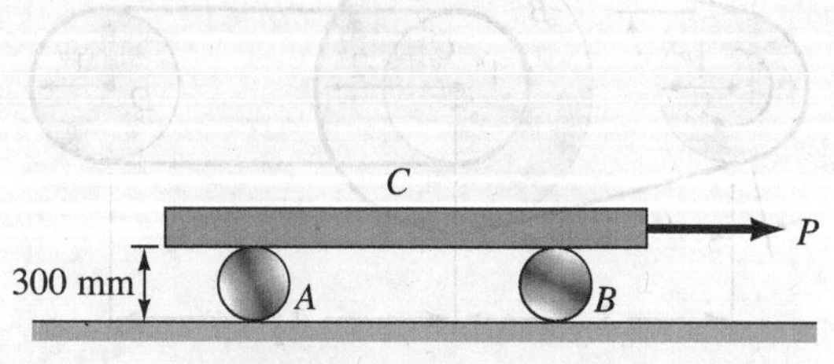

Figure P.6.102.

6.103. A hot rectangular metal ingot is to be flattened by passing through cylindrical rollers. If the ingot is to be drawn into the rollers by friction once it touches the rollers, what is the *minimum* thickness t of the ingot that can be achieved by this process on one pass? The static coefficient of friction for the contact between ingot and cylinder is .3. The cylinders rotate as shown with angular speed ω.

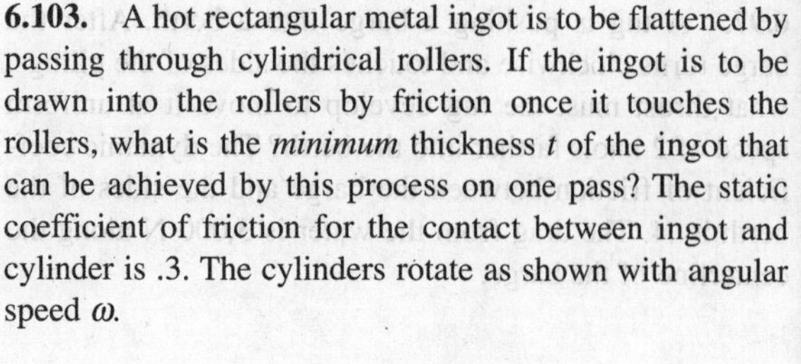

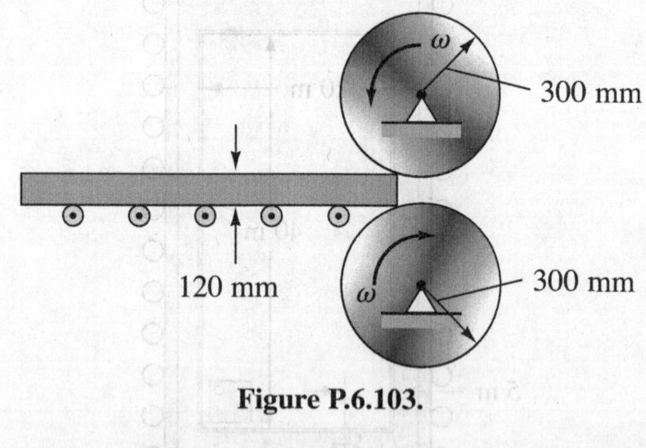

Figure P.6.103.

6.104. A cone clutch is shown. Assuming that uniform pressures exist between the contact surfaces, compute the *maximum* torque that can be transmitted. The static coefficient of friction is .30 and the activating force F is 100 lb. [*Hint:* Assume that the moving cone transmits its 100-lb axial force to the stationary cone by pressure primarily. That is, we will neglect the friction-force component on the cone surface normal to the transverse direction.]

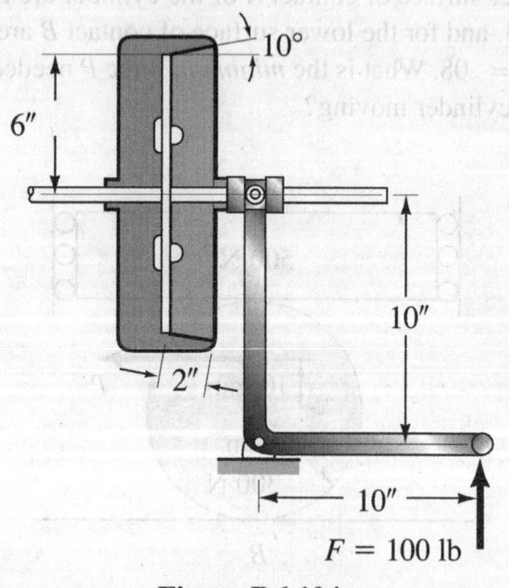

Figure P.6.104.

6.105. In Fig. P.6.18, delete the external forces and couple at point G, and replace with a force of 6,000 N at point G at an angle α with the horizontal going from left to right. For the condition of impending slippage in the downward direction, what should the angle α be?

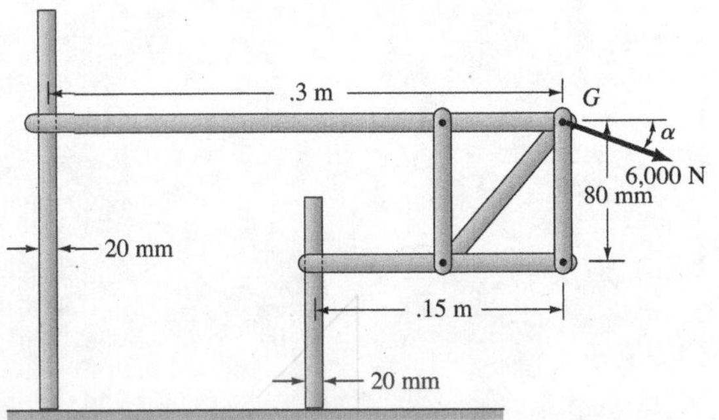

Figure P.6.105.

6.106. In Problem 6.39, what is the minimum angle between the supporting links CD and ED to support any weight W if the static coefficient of friction is $\mu_s = .4$?

6.107. A shaft AB rotates at constant angular speed in a pair of *dry* journal bearings. The total weight of the shaft and the cylinder it is supporting is W. A torque T is needed to maintain the steady angular motion. There is a small clearance between the shaft and the journal, resulting in a *point contact* between these bodies at some position E as shown. Form a two-dimensional force system acting on the shaft by moving W, the total friction force, the total normal force on the shaft surface, and the torque to a plane normal to the centerline of the shaft and at a location at the center of the system. View this system along the centerline.

(a) From considerations of equilibrium, what and where is the resultant force vector from the friction force and the normal force onto the shaft?

(b) If the angle ϕ in the diagram is very small and if the coefficient of dynamic friction is μ_d, explain how we can give the following approximate equation

$$T = W\mu_d r$$

where r is the radius of the shaft.

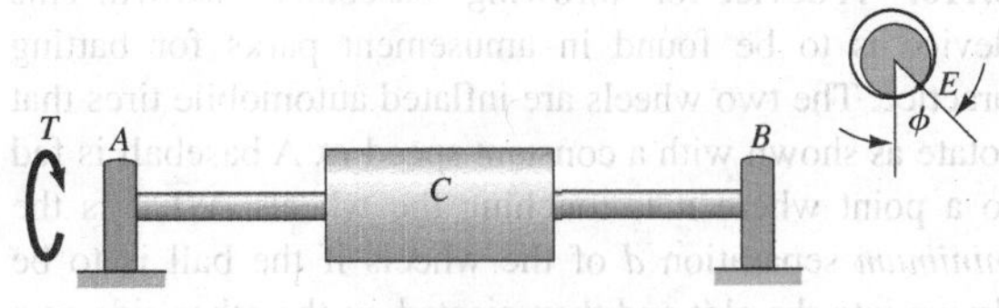

Figure P.6.107.

6.108. Shaft CD rotates at constant angular speed in a set of dry journal bearings (as a result of poor maintenance). Approximately what torque is needed to maintain the angular motion for the following data?

$M_A = 50$ kg $M_B = 80$ kg Shaft $CD = 40$ kg $\mu_d = .2$

See Problem 6.107 before doing this problem.

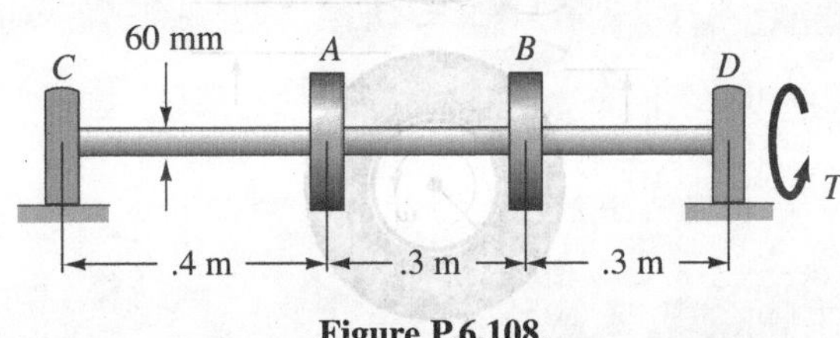

Figure P.6.108.

6.109. Two identical light rods are pinned together at B. End C of rod BC is pinned while end A of rod AB rests on a rough floor having a coefficient of friction with the rod of $\mu_d = .5$. The spring requires a force of 5 N/mm of stretch. A load is applied slowly at B and then maintained constant at $F = 300$ N. What is the angle θ when the system ceases to move? The spring is unstretched when $\theta = 45°$. [*Hint:* You will have to solve an equation by trial and error.]

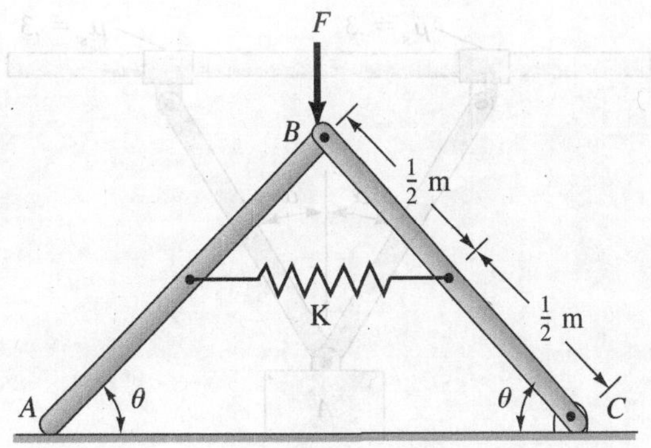

Figure P.6.109

6.110. A device for "throwing" baseballs is shown. This device is to be found in amusement parks for batting practice. The two wheels are inflated automobile tires that rotate as shown with a constant speed ω. A baseball is fed to a point where it is touching the wheels. What is the *minimum* separation d of the wheels if the ball is to be drawn into the slot and then ejected on the other side as a pitched ball? The coefficient of static friction at the contact surfaces is .4.

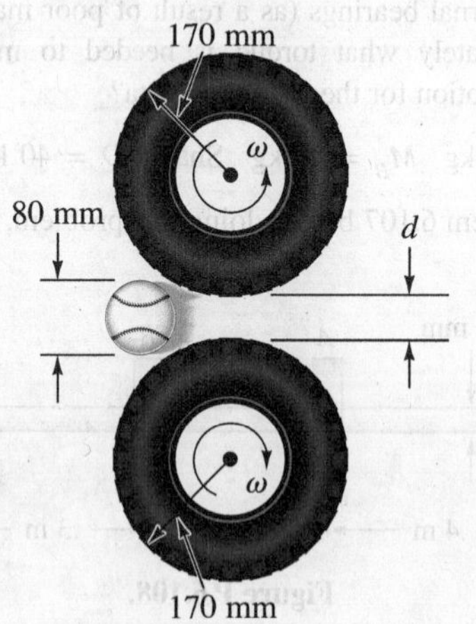

Figure P.6.110.

6.111. What is the *maximum* angle α for which there will be equilibrium if A has a mass of 50 kg. The static coefficient of friction between the supports and the horizontal rod is equal to .3. What is the force in each of the supporting members?

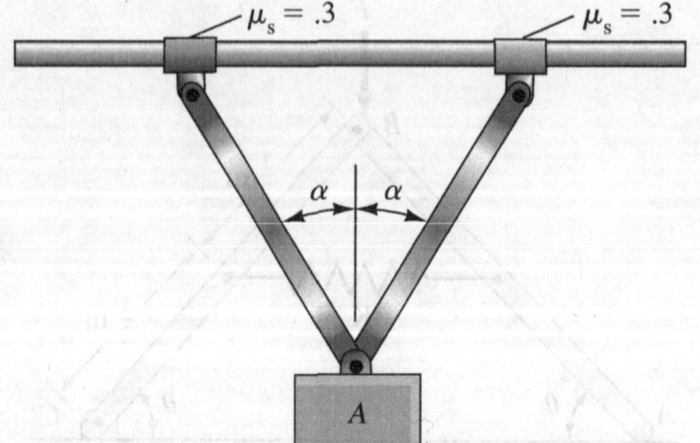

Figure P.6.111.

6.112. What is the *maximum* angle α for which there will be equilibrium if A weighs 1,000 N and if μ_s between the supports and the curved rod is .3? The rods are each 1.3 m long. You will have to solve a transcendental equation by trial and error.

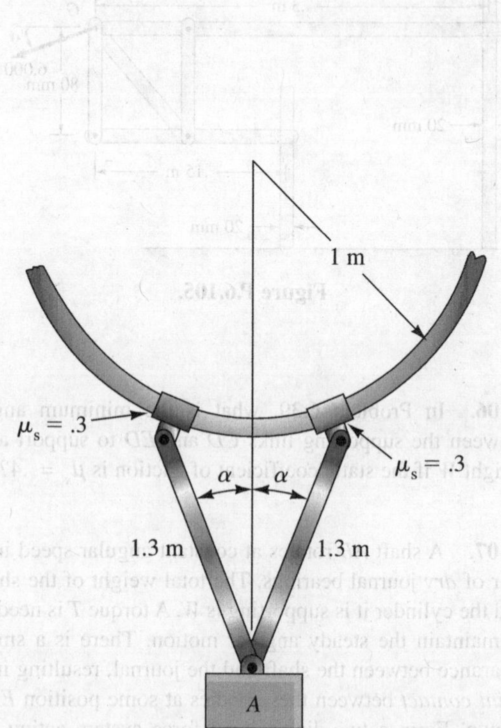

Figure P.6.112.

6.113. A 500-kN crate is being lowered slowly down an elevator shaft, which as shown is slightly wider than the crate. The cord wraps around a freely turning pulley and then has 2 rotations around a capstan. Find the tension T needed for the operation. The center of gravity of the crate is at its geometric center. [*Hint:* The crate will rotate so as to rub against the elevator shaft at points A and B.]

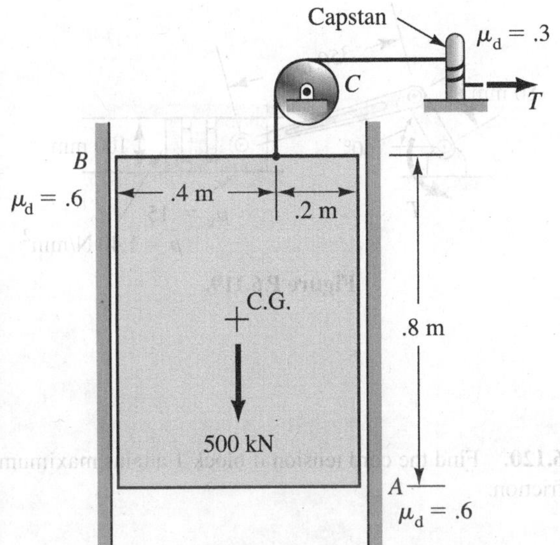

Figure P.6.113.

6.114. A triangular pile of width 1 ft is being driven into the ground slowly by a force P of 50,000 lb. There is pressure on the lateral surfaces of the pile. This pressure varies linearly from 0 at A to p_0 at B, as has been shown in the diagram. If the coefficient of dynamic friction between the pile and the soil is 0.6, what is the maximum pressure p_0?

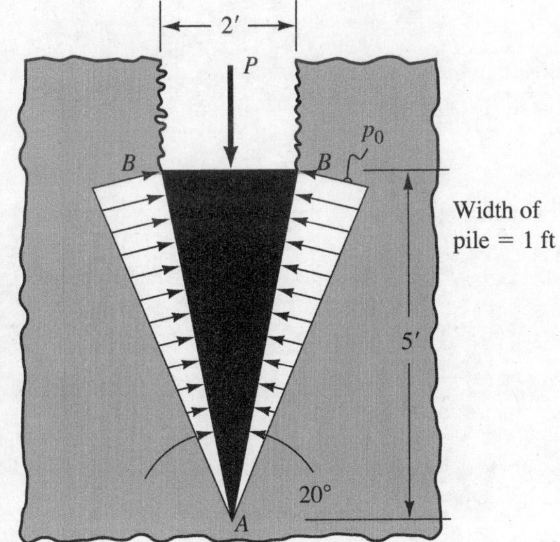

Figure P.6.114

6.115. In Problem 6.27, what maximum value should W_2 be in order to start moving the system to the left? All other data are unchanged.

6.116. A block rests on a surface for which there is a coefficient of friction $\mu_s = .2$. Over what range of angle β will there be no movement of the block for the 150-N force? (You will have to solve an equation by trial and error.)

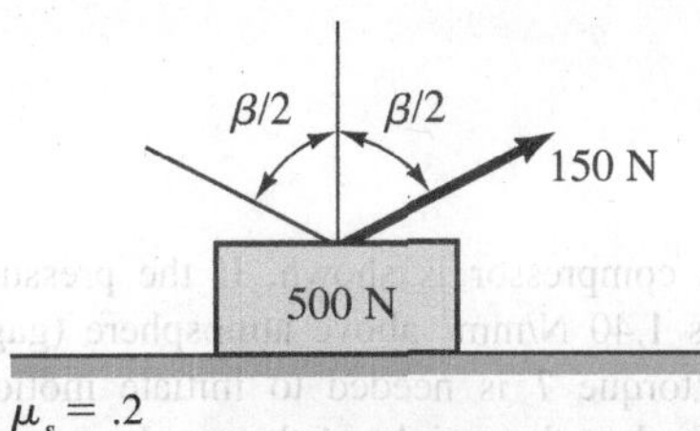

Figure P.6.116.

6.117. What is the *largest* load that can be suspended without moving blocks A and B? The static coefficient of friction for all plane surfaces of contact is .3. Block A weighs 500 N and block B weighs 700 N. Neglect friction in the pulley system.

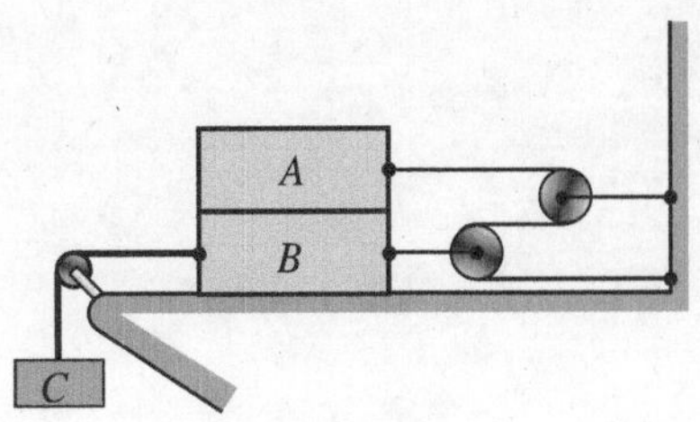

Figure P.6.117.

6.118. What is the *minimum* force F to hold the cylinders, each weighing 100 lb? Take $\mu_s = .2$ for all surfaces of contact.

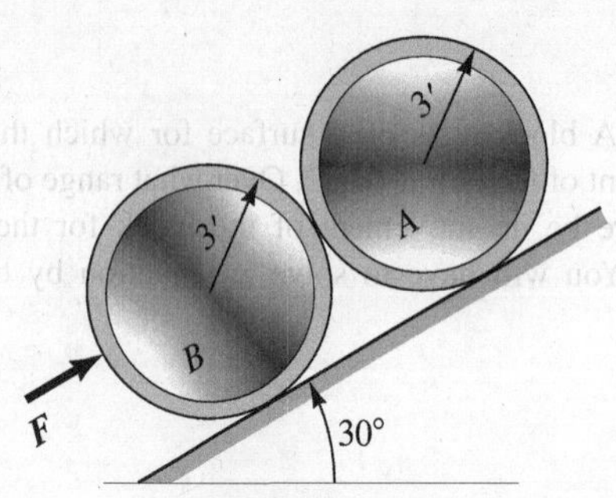

Figure P.6.118.

6.119. A compressor is shown. If the pressure in the cylinder is 1.40 N/mm^2 above atmosphere (gage), what *minimum* torque T is needed to initiate motion in the system? Neglect the weight of the crank and connecting rod as far as their contribution toward moving the system. Consider friction only between the piston and cylinder walls where the coefficient of friction $\mu_s = .15$.

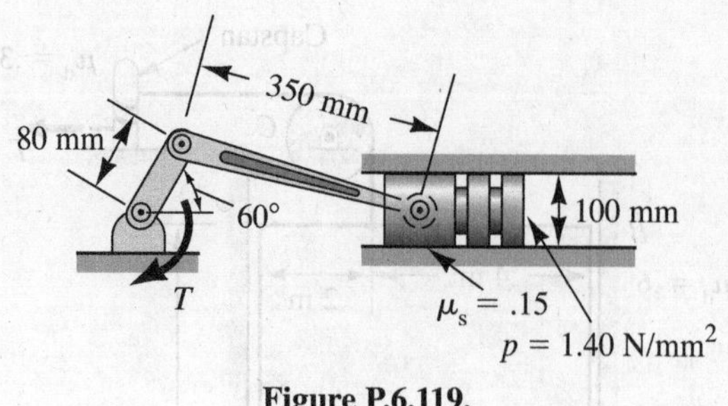

Figure P.6.119.

6.120. Find the cord tension if block 1 attains maximum friction.

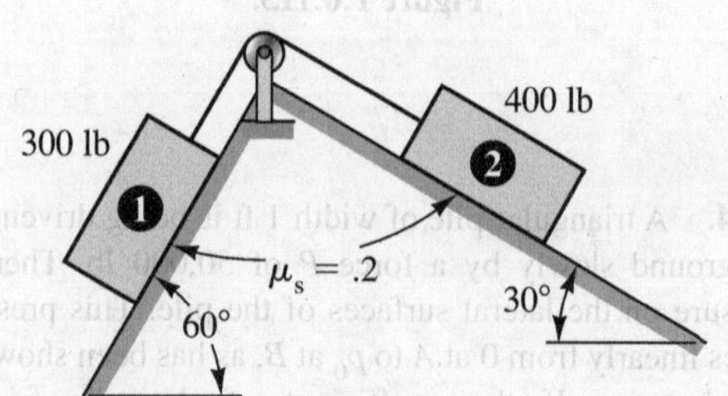

Figure P.6.120.

CHAPTER 7

Properties of Surfaces

7.1 Introduction

If we are buying a tract of land, we certainly want to consider the size and, with equal interest, the shape and orientation of the earth's surface, and possibly its agricultural, geological, or aesthetic potentials. The size of a surface (i.e., the area) is a familiar concept and has been used in the previous section. Certain aspects of the shape and orientation of a surface will be examined in this chapter. There are a number of formulations that convey meaning about the shape and disposition of a surface relative to some reference. To be sure, these formulations are not used by real estate people, but in engineering work, where a variety of quantitative descriptions are necessary, these formulations will prove most useful. In general, we shall restrict our attention to coplanar surfaces.

7.2 First Moment of an Area and the Centroid

A coplanar surface of area A and a reference xy in the plane of the surface are shown in Fig. 7.1. We define the *first moment of area A* about the x axis as

$$M_x = \int_A y\,dA \quad (7.1)$$

and the first moment about the y axis as

$$M_y = \int_A x\,dA \quad (7.2)$$

These two quantities convey a certain knowledge of the shape, size, and orientation of the area, which we can use in many analyses of mechanics.

You will no doubt notice the similarity of the preceding integrals to those which would occur for computing moments about the x and y axes from a parallel force distribution oriented

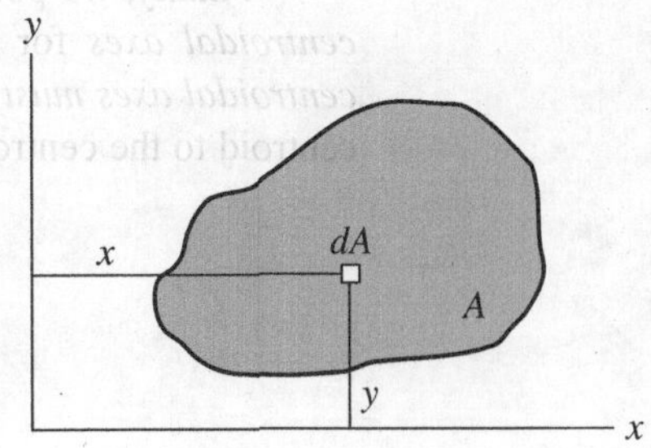

Figure 7.1. Plane area.

normal to the area A in Fig. 7.1. The moment of such a force distribution has been shown for the purposes of rigid-body calculations to be equivalent to that of a single resultant force located at a particular point x, y. Similarly, we can concentrate the entire area A at a position x_c, y_c, called the *centroid,*[1] where, for computations of first moments, this new arrangement is equivalent to the original distribution (Fig. 7.2). The coordinates x_c and y_c are usually called the *centroidal coordinates*. To compute these coordinates, we simply equate moments of the distributed area with that of the concentrated area about both axes:

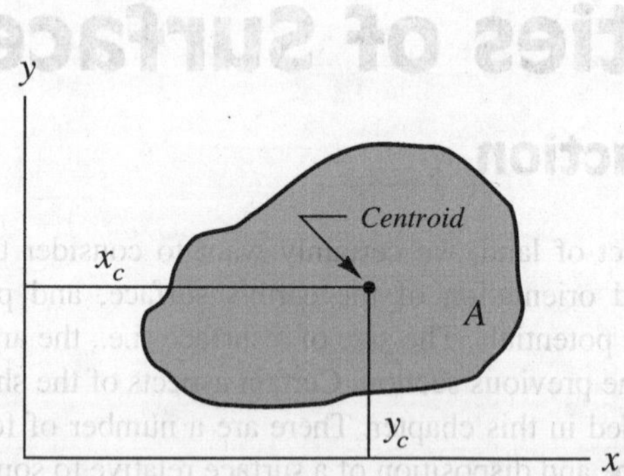

Figure 7.2. Centroidal coordinates.

$$Ay_c = \int_A y\,dA; \qquad \text{therefore, } y_c = \frac{\int_A y\,dA}{A} = \frac{M_x}{A} \tag{7.3a}$$

$$Ax_c = \int_A x\,dA; \qquad \text{therefore, } x_c = \frac{\int_A x\,dA}{A} = \frac{M_y}{A} \tag{7.3b}$$

The location of the centroid of an area can readily be shown to be *independent* of the *reference axes* employed. That is, the centroid is a *property* only of the *area* itself. We have asked the reader to prove this in Problem 7.1.

If the axes xy have their origin at the centroid, then these axes are called *centroidal axes* and clearly the first moments about these axes must be zero.

Finally, we point out that *all* axes going through the centroid of an area are called *centroidal axes* for that area. Clearly, the *first moments of an area about any of its centroidal axes must be zero*. This must be true since the perpendicular distance from the centroid to the centroidal axis must be zero.

[1]The concept of the centroid can be used for any geometric quantity. In the next section, we shall consider centroids of volumes and arcs.

Example 7.1

A plane surface is shown in Fig. 7.3 bounded by the x axis, the curve $y^2 = 25x$, and a line parallel to the y axis. What are the first moments of the area about the x and y axes and what are the centroidal coordinates?

We shall first compute M_x and M_y for this area. Using vertical infinitesimal area elements of width dx and height y, we have noting that $y = 5\sqrt{x}$

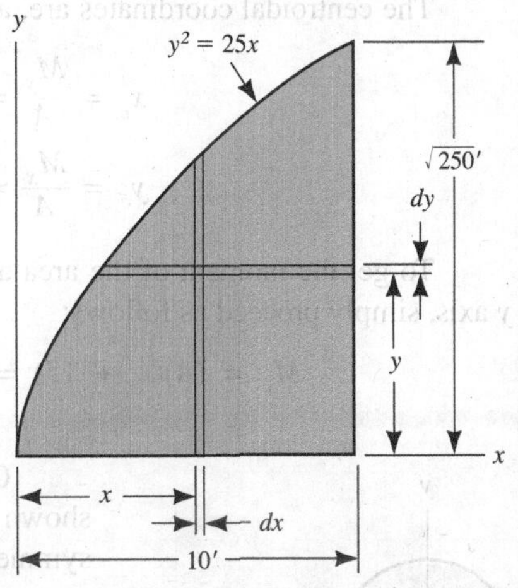

Figure 7.3. Find centroid.

$$M_y = \int_0^{10} x(y\,dx) = \int_0^{10} x(5\sqrt{x})\,dx$$

$$= \left.\frac{5x^{5/2}}{\frac{5}{2}}\right|_0^{10} = \boxed{632 \text{ ft}^3}$$

To compute M_x, we use horizontal area elements of width dy and length $(10-x)$ as shown in the diagram. Thus

$$M_x = \int_0^{\sqrt{250}} y[(10-x)dy]$$

$$= \int_0^{\sqrt{250}} \left(10y - \frac{y^3}{25}\right)dy$$

$$= \left.\left(5y^2 - \frac{y^4}{100}\right)\right|_0^{\sqrt{250}} = \boxed{625 \text{ ft}^3}$$

We could also have used vertical strips for computing M_x as follows using centroids of the vertical strips:

$$M_x = \int_0^{10} \frac{y}{2}(y\,dx) = \int_0^{10} \frac{25x}{2}\,dx$$

$$= (12.5)\left.\left(\frac{x^2}{2}\right)\right|_0^{10} = 625 \text{ ft}^3$$

To compute the position of the centroid (x_c, y_c), we will need the area A of the surface. Thus, using vertical strips:

$$A = \int_0^{10} y\,dx = \int_0^{10} 5\sqrt{x}\,dx = \left.\frac{5x^{3/2}}{\frac{3}{2}}\right|_0^{10}$$

$$= 105.4 \text{ ft}^2$$

Example 7.1 (Continued)

The centroidal coordinates are, accordingly,

$$x_c = \frac{M_y}{A} = \frac{632}{105.4} = 6.00 \text{ ft}$$

$$y_c = \frac{M_x}{A} = \frac{625}{105.4} = 5.93 \text{ ft}$$

To get the moment of the area about an axis y', which is 15 ft to the left of the y axis, simply proceed as follows:

$$M_{y'} = (A)(x_c + 15) = 105.4(6.00 + 15) = 2{,}213 \text{ ft}^3$$

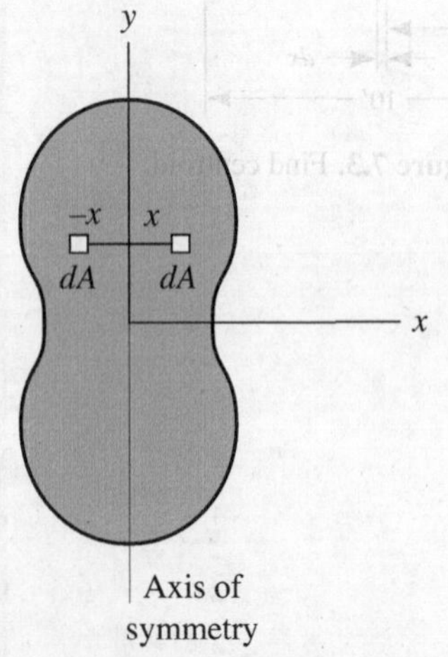

Figure 7.4. Area with one axis of symmetry.

Consider now a plane area with an *axis of symmetry* such as is shown in Fig. 7.4, where the y axis is collinear with the axis of symmetry. In computing x_c for this area, we have

$$x_c = \frac{1}{A}\int_A x\,dA$$

In evaluating the integral above, we can consider area elements in symmetric pairs such as shown in Fig. 7.4, where we have shown a pair of area elements which are mirror images of each other about the axis of symmetry. Clearly, the first moment of such a pair about the axis of symmetry is zero. And, since the entire area can be considered as composed of such pairs, we can conclude that $x_c = 0$. Thus, the centroid of an area with one axis of symmetry must therefore lie somewhere along this axis of symmetry. The axis of symmetry then is a centroidal axis, which is another indication that the first moment of area must be zero about such an axis. With two orthogonal axes of symmetry, the centroid must lie at the intersection of these axes. Thus, for such areas as circles and rectangles, the centroid is easily determined by inspection.

In many problems, the area of interest can be considered formed by the addition or subtraction of simple familiar areas whose centroids are known by inspection as well as by other familiar areas, such as triangles and sectors of circles whose centroids and areas are given in handbooks. We call areas made up of such simple areas *composite* areas. (A listing of familiar areas is given for your convenience on the inside back cover of this text.) For such problems, we can say that

$$x_c = \frac{\sum_i A_i \bar{x}_i}{A}$$

$$y_c = \frac{\sum_i A_i \bar{y}_i}{A}$$

where $\bar{x}_i$ and $\bar{y}_i$, (with proper signs) are the centroidal coordinates to simple area A_i, and where A is the total area.

Example 7.2

Find the centroid of the shaded section shown in Fig. 7.5.

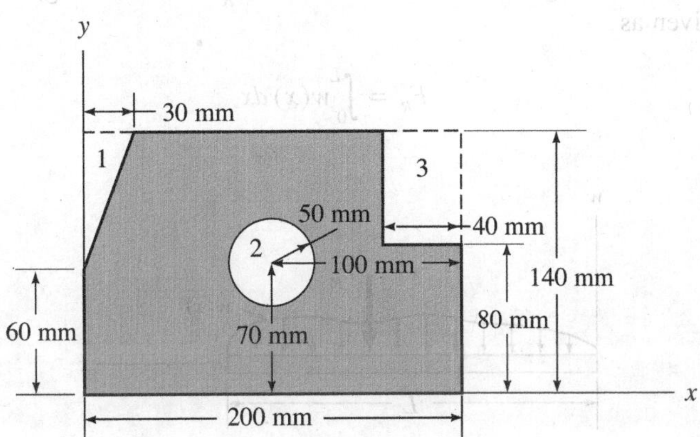

Figure 7.5. Composite area.

We may consider four separate areas. These are the triangle (1), the circle (2), and the rectangle (3) all cut from an original rectangular 200 × 140 mm² area which we denote as area (4). In composite-area problems, we urge you to set up a format of the kind we shall now illustrate. Using the positions of the centroid of a right triangle as given in the inside covers of this text, we have:

	A_i	x_i	$A_i\bar{x}_i$	y_i	$A_i y_i$
$A_1 = -\frac{1}{2}(30)(80)$	$= -1{,}200$	10	−12,000	113.3	−136,000
$A_2 = -\pi 50^2$	$= -7{,}850$	100	−785,000	70	−549,780
$A_3 = -(40)(60)$	$= -2{,}400$	180	−432,000	110	−264,000
$A_4 = (200)(140)$	$= 28{,}000$	100	2,800,000	70	1,960,000
	$A = 16{,}550 \text{ mm}^2$		$\sum_i A_i\bar{x}_i = 1.571 \times 10^6 \text{ mm}^3$		$\sum_i A_i\bar{y}_i = 1.011 \times 10^6 \text{ mm}^3$

Therefore,

$$x_c = \frac{\sum A_i\bar{x}_i}{A} = \frac{1.571 \times 10^6}{16{,}550} = 94.9 \text{ mm}$$

$$y_c = \frac{\sum A_i\bar{y}_i}{A} = \frac{1.011 \times 10^6}{16{,}550} = 61.1 \text{ mm}$$

We now illustrate how we can use the composite-area approach for finding the centroid of an area composed of familiar parts as just described.

In closing, we would like to point out that the centroid concept can be of use in finding the simplest resultant of a distributed loading. Thus, consider the distributed loading *w(x)* shown in Fig. 7.6. The resultant force F_R of this loading, also shown in the diagram, is given as

$$F_R = \int_0^L w(x)\,dx \tag{7.4}$$

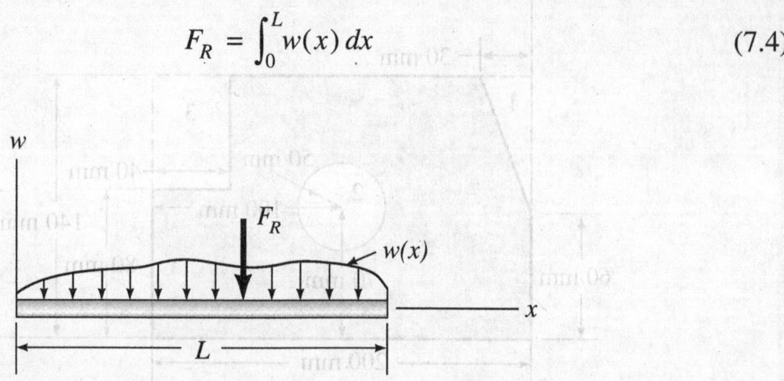

Figure 7.6. Loading curve $w(x)$ and its resultant F_R.

From the equation above, we can readily see that the *resultant force equals the area under the loading curve*. To get the position of the *simplest* resultant for the loading, we then say that

$$F_R\bar{x} = \int_0^L xw(x)\,dx$$

Therefore,

$$\bar{x} = \frac{\int_0^L xw(x)\,dx}{F_R} \tag{7.5}$$

The preceding result shows that $\bar{x}$ is actually the centroidal coordinate of the loading curve area from reference *xy*. Thus, the *simplest resultant force of a distributed load acts at the centroid of the area under the loading curve*. Accordingly, for a triangular load such as is shown in Fig. 7.7, we can replace the loading for free bodies on which the entire loading acts by a force *F* equal to $(\frac{1}{2})(w_0)(b-a)$ at a position $\frac{2}{3}(b-a)$ from the left end of the loading.

You will recall that we pointed this out in Chapter 4.

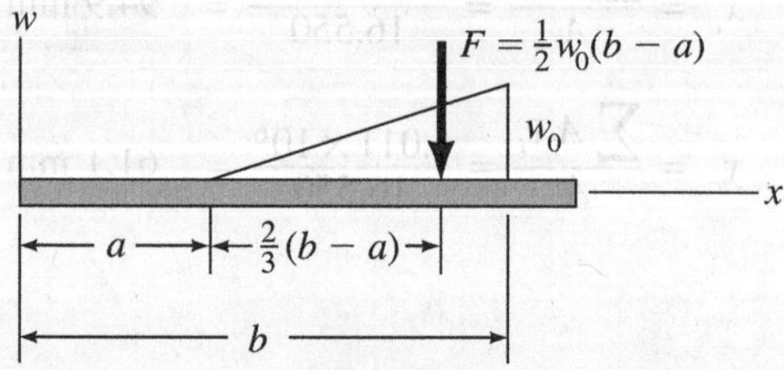

Figure 7.7. Triangular loading with simplest resultant.

PROBLEMS

7.1. Show that the centroid of area A is the same point for axes xy and $x'y'$. Thus, the position of the centroid of an area is a property only of the area.

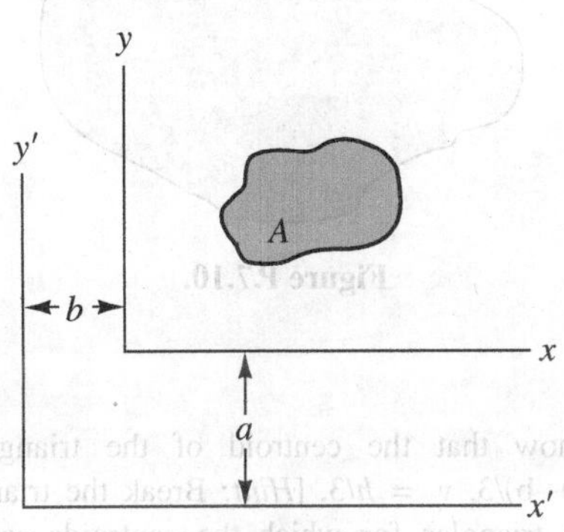

Figure P.7.1.

7.2. Show that the centroid of the right triangle is $x_c = 2a/3$, $y_c = b/3$.

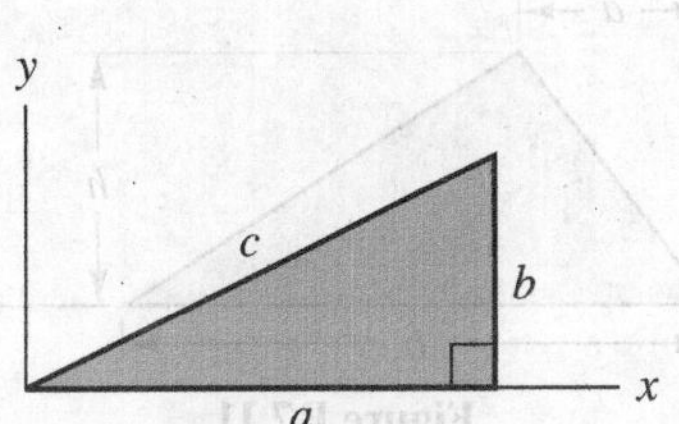

Figure P.7.2.

7.3. Find the centroid of the area under the half-sine wave. What is the first moment of this area about axis A–A?

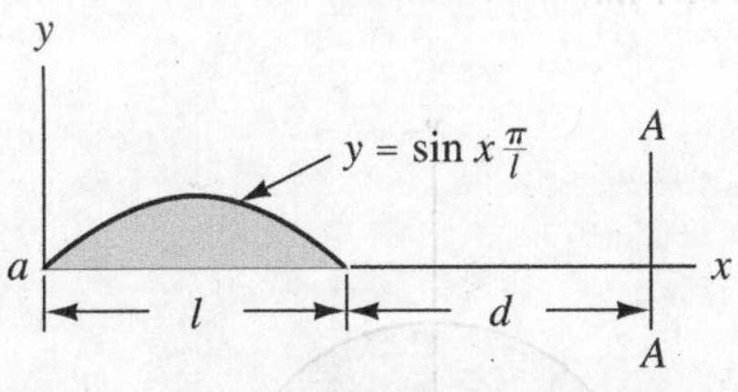

Figure P.7.3.

7.4. What are the first moments of the area about the x and y axes? The curved boundary is that of a parabola. [*Hint:* The general equation for parabolas of the shape shown is $y^2 = ax + b$.]

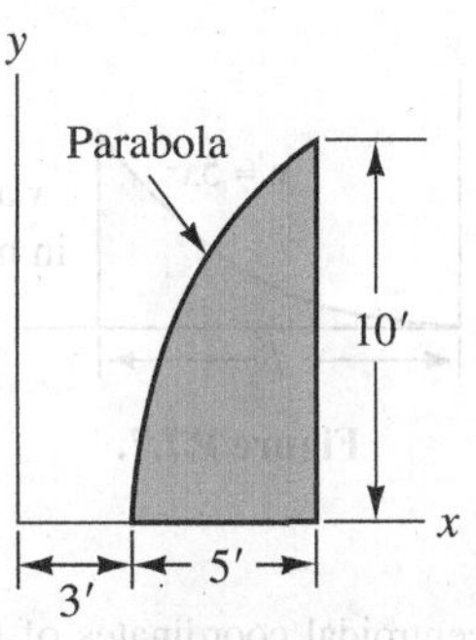

Figure P.7.4.

7.5. What are the centroidal coordinates for the shaded area? The curved boundary is that of a parabola. [*Hint:* The general equation for parabolas of the shape shown is $y = ax^2 + b$.]

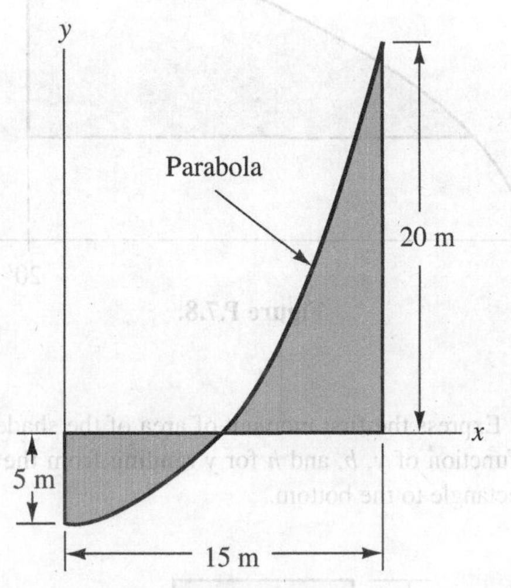

Figure P.7.5.

7.6. Show that the centroid of the area under a semicircle is as shown in the diagram.

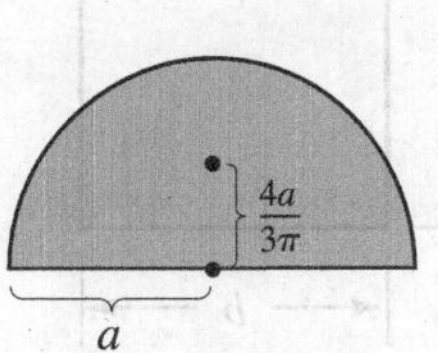

Figure P.7.6.

7.7. What is the first moment of the area under the parabola about an axis through the origin and going through point $\boldsymbol{r} = 6\boldsymbol{i} + 7\boldsymbol{j}$ m. Take $l = 10$ m.

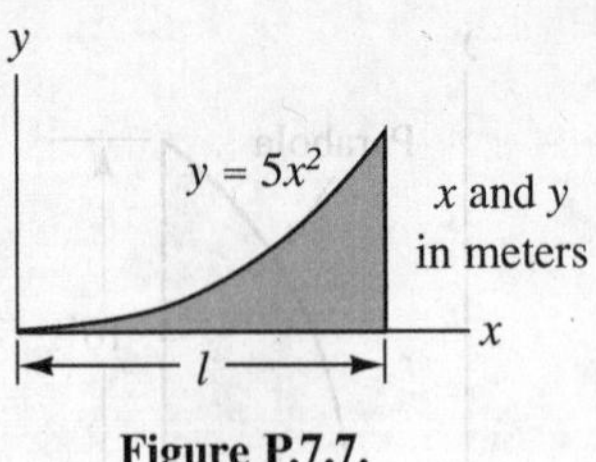

Figure P.7.7.

7.8. Find the centroidal coordinates of the shaded area.

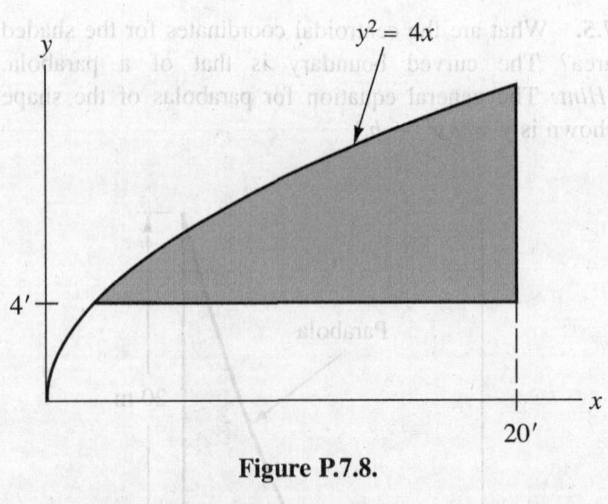

Figure P.7.8.

7.9. Express the first moment of area of the shaded area as a function of y, b, and h for y ranging from the top of the rectangle to the bottom.

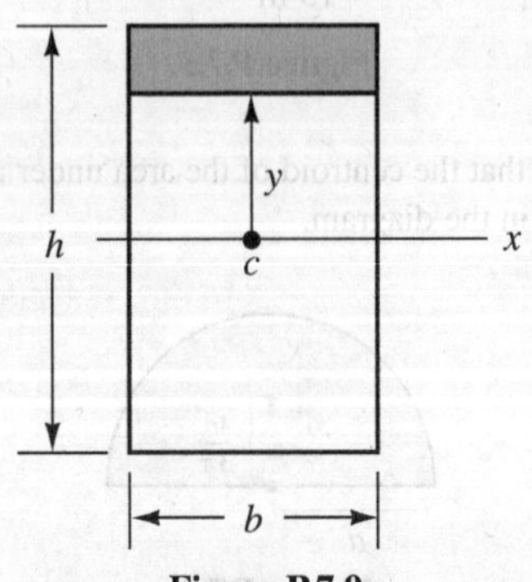

Figure P.7.9.

7.10. Suppose you had an irregular area and you had to *approximate* the location of the centroid. How could you accomplish this using only a straight edge, a ruler, and a pencil?

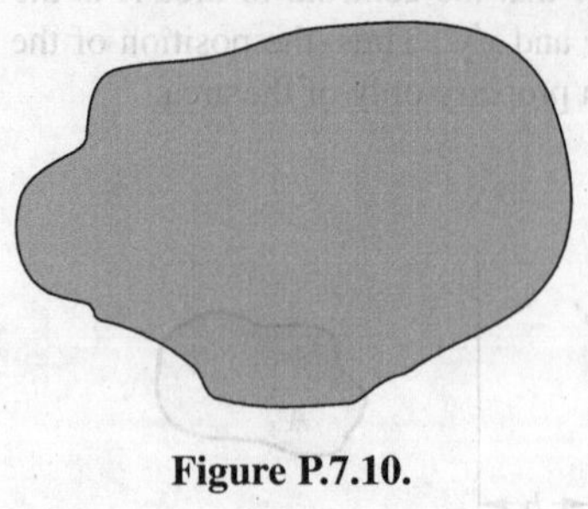

Figure P.7.10.

7.11. Show that the centroid of the triangle is at $x_c = (a + b)/3$, $y_c = h/3$. [*Hint:* Break the triangle into two right triangles for which the centroids are known from Problem 7.2.]

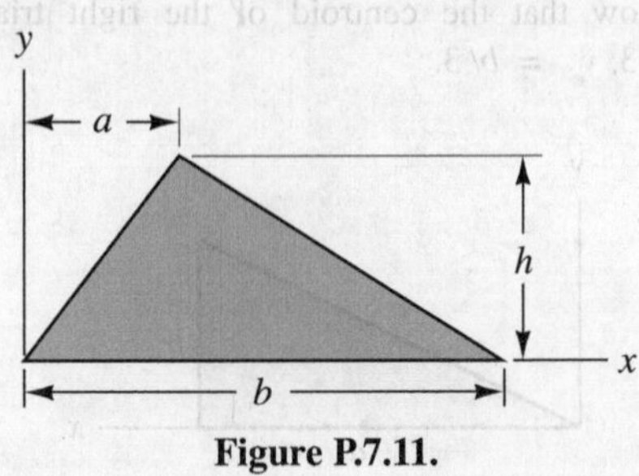

Figure P.7.11.

7.12. What are the centroidal coordinates for the shaded area? The outer boundary is that of a circle having radius of 1 m.

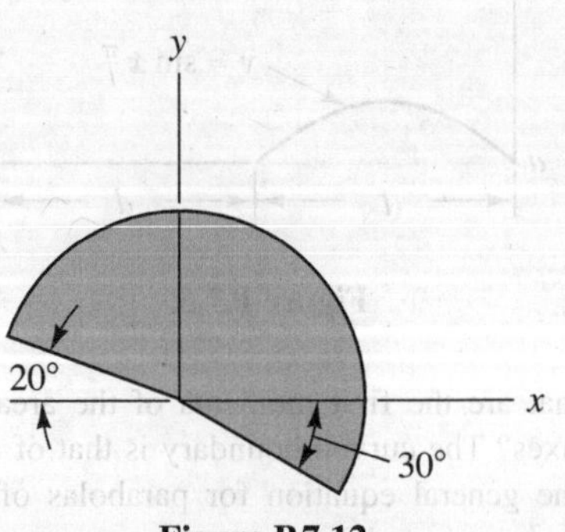

Figure P.7.12.

7.13. What are the coordinates of the centroid of the shaded area? The parabola is given as $y^2 = 2x$ with y and x in millimeters.

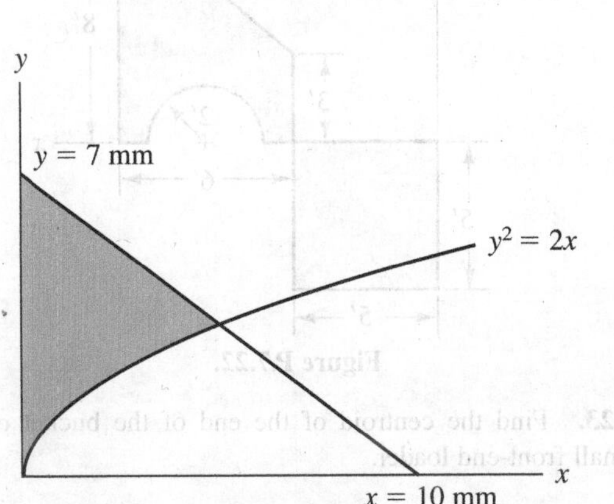

Figure P.7.13.

7.14. Find the centroid of the shaded area. The equation of the curve is $y = 5x^2$ with x and y in millimeters. What is the first moment of the area about line AB?

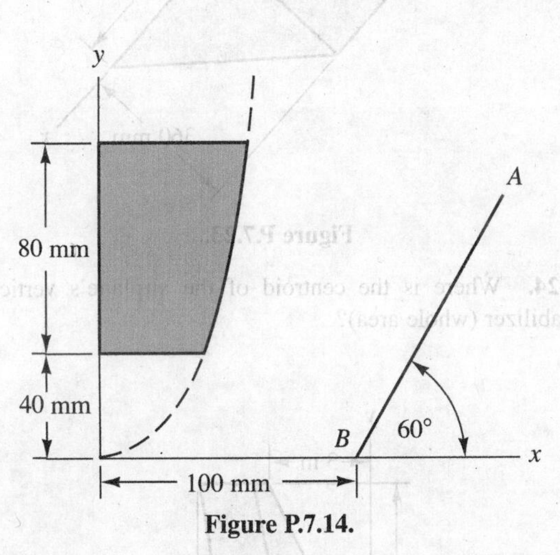

Figure P.7.14.

7.15. Find the centroid of the shaded area. What is the first moment of this area about line A–A. The upper boundary is a parabola $y^2 = 50x$ with x and y in millimeters.

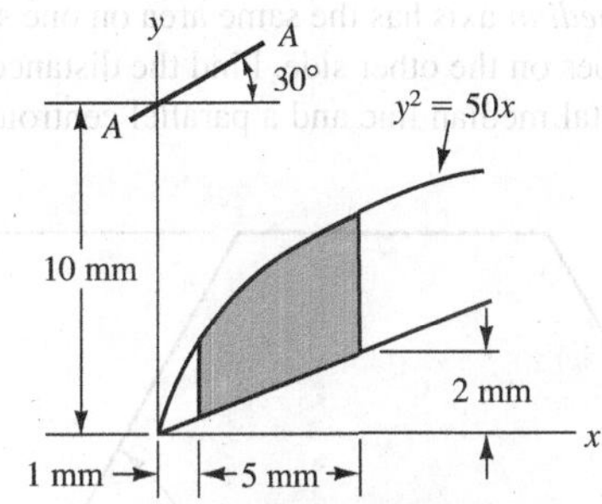

Figure P.7.15.

In the remaining problems of this section, use centroidal positions of simple areas as found in the inside covers.

7.16. Find the centroid of the end shield of a bulldozer blade.

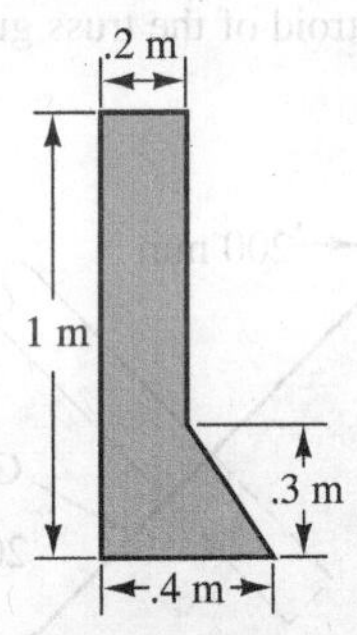

Figure P.7.16.

7.17. In Example 2.6, determine z_C for the centroid of the triangular faces of the pyramid plus the base area $ABCE$. The height of the pyramid is 300 ft.

7.18. A parallelogram and an ellipse have been cut from a rectangular plate, What are the centroidal coordinates of what is left of the plate? What is $M_{x'}$ for this area? Use formulas given on the inside of the back covers.

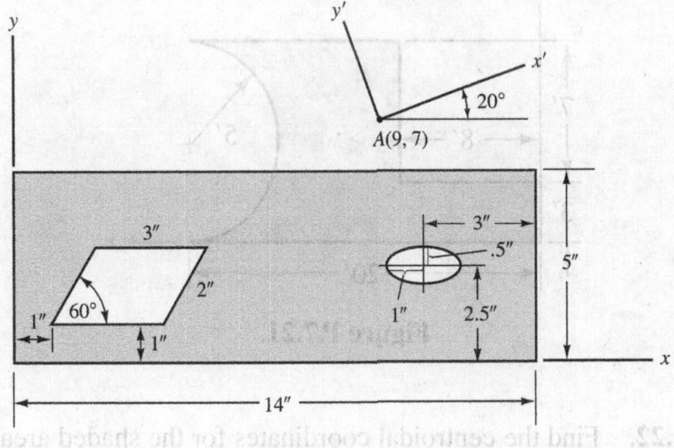

Figure P.7.18.

7.19. A *median* axis has the same area on one side of the axis as it does on the other side. Find the distance between the horizontal median line and a parallel centroidal axis.

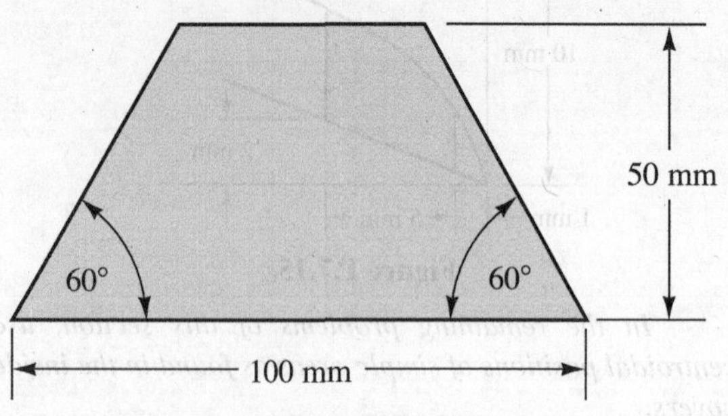

Figure P.7.19.

7.20. Find the centroid of the truss gusset plate.

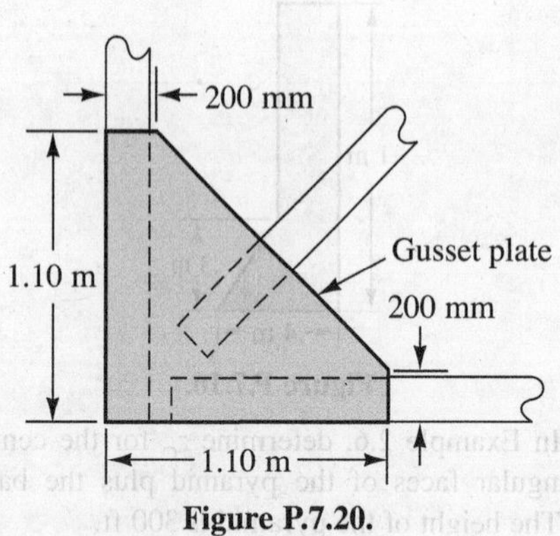

Figure P.7.20.

7.21. Find the centroid of the indicated area.

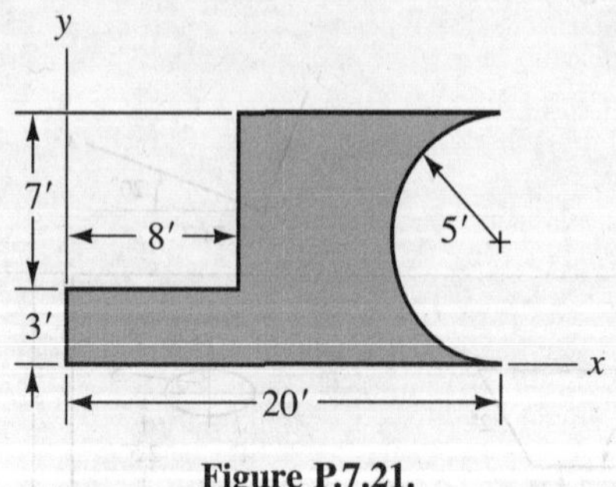

Figure P.7.21.

7.22. Find the centroidal coordinates for the shaded area shown. Give the results in meters. [*Hint:* See Fig. P.7.6.]

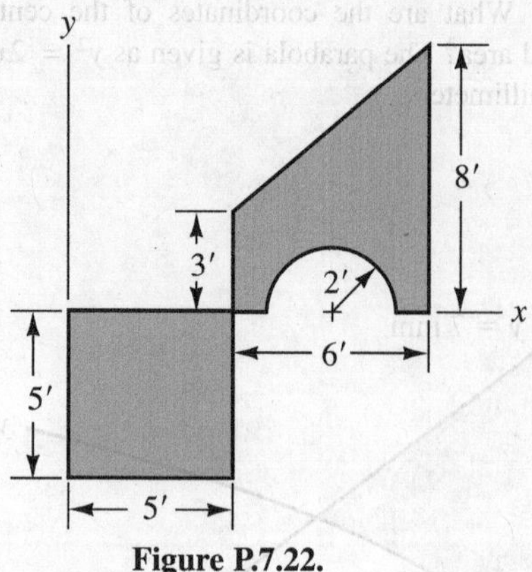

Figure P.7.22.

7.23. Find the centroid of the end of the bucket of a small front-end loader.

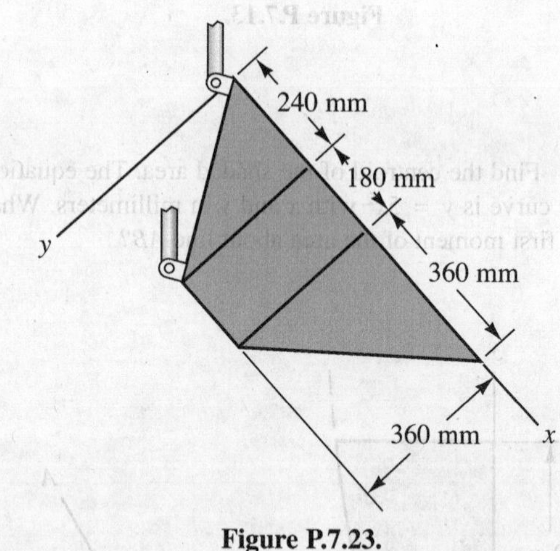

Figure P.7.23.

7.24. Where is the centroid of the airplane's vertical stabilizer (whole area)?

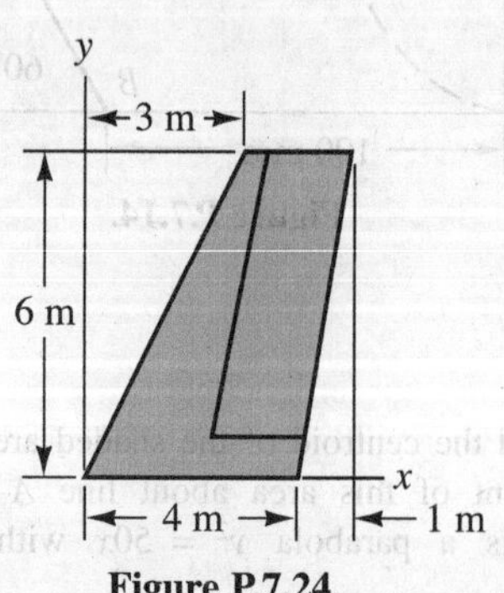

Figure P.7.24.

7.25. What is the first moment of the shaded area about the diagonal A–A? [*Hint:* Consider symmetry.]

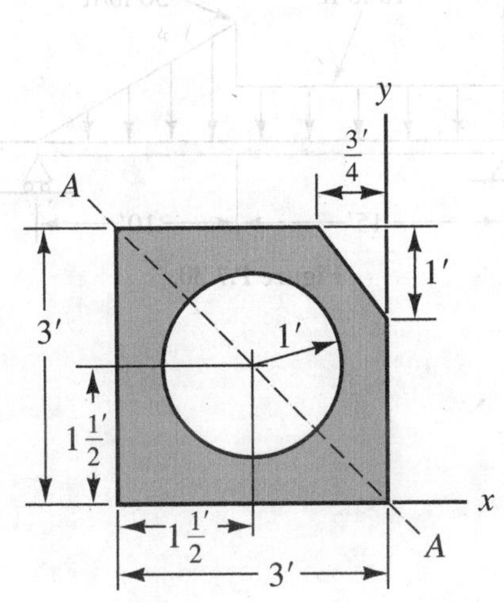

Figure P.7.25.

7.26. A built-up beam is shown with four 120-mm by 120-mm by 20-mm angles. Find the vertical distance above the base for the centroid of the cross-section.

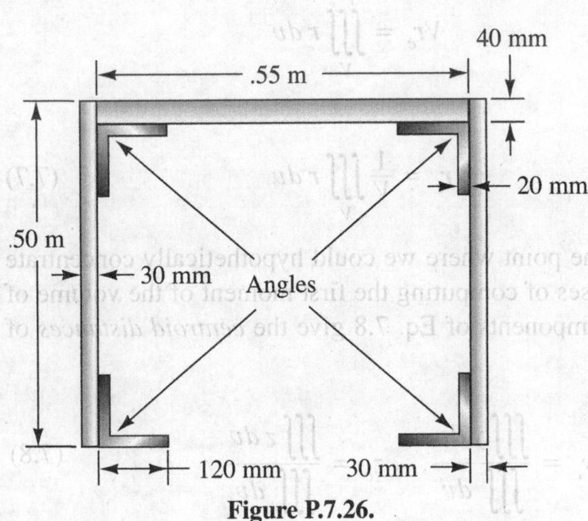

Figure P.7.26.

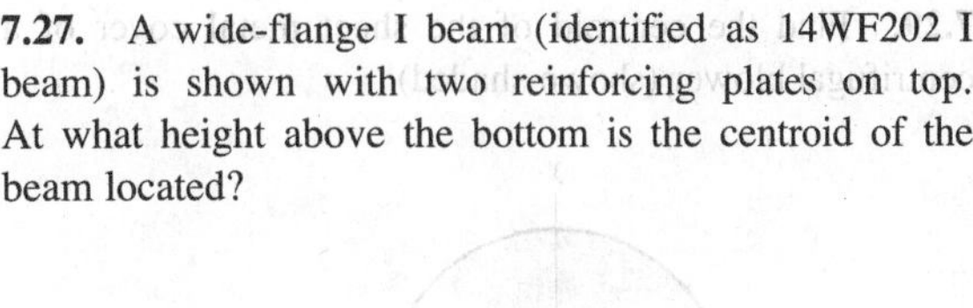

7.27. A wide-flange I beam (identified as 14WF202 I beam) is shown with two reinforcing plates on top. At what height above the bottom is the centroid of the beam located?

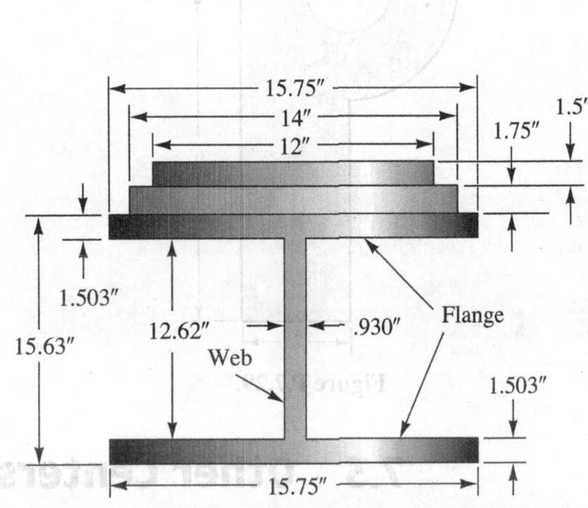

Figure P.7.27.

7.28. Compute the position of the centroid of the shaded area. [*Hint:* See Fig. P.7.6.]

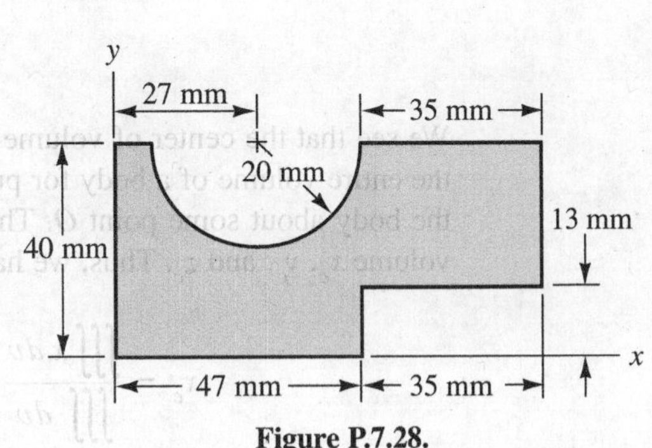

Figure P.7.28.

7.29. Find the centroid of the sheet metal cover of a centrifugal blower (shown shaded).

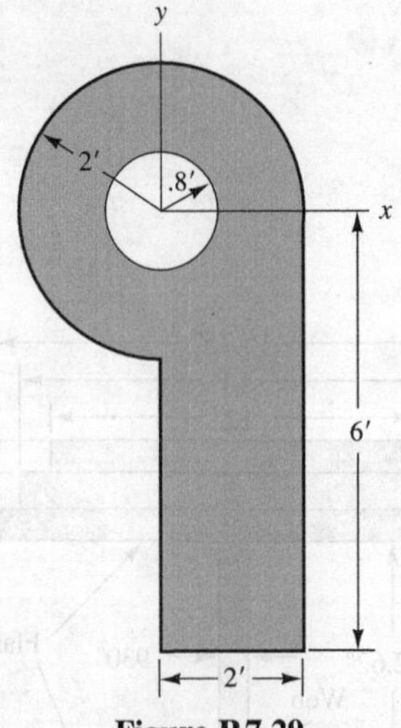

Figure P.7.29.

7.30. What is the position from the left end of the simplest resultant force of the distribution shown?

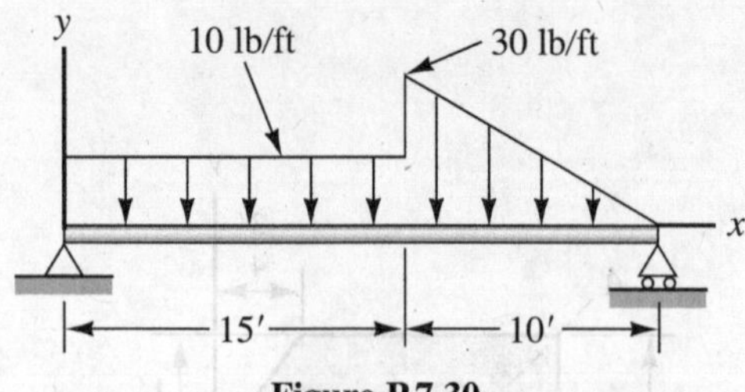

Figure P.7.30.

7.3 Other Centers

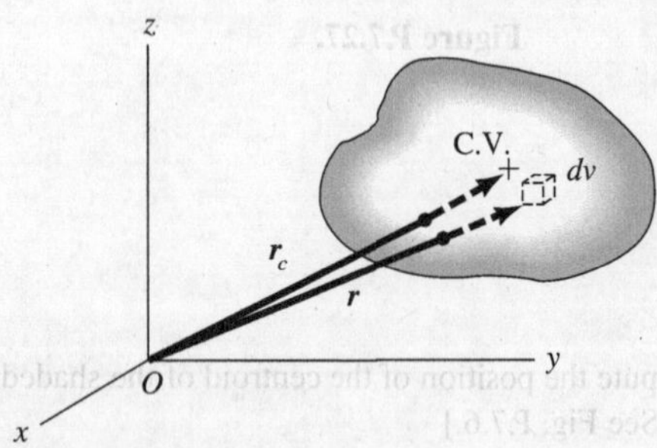

Figure 7.8. Center of volume, C.V., of a body.

We employ the concepts of moments and centroids in mechanics for three-dimensional bodies as well as for plane areas. Thus, we introduce now the first moment of a volume, V, of a body (see Fig. 7.8) about a point O where we have shown a reference xyz. We say that the first *moment of volume* V about O is

$$\text{moment vector of volume} \equiv \iiint_V \mathbf{r}\, dv \tag{7.6}$$

The *center of volume*, $\mathbf{r}_c$, is then defined as follows:

$$V\mathbf{r}_c = \iiint_V \mathbf{r}\, dv$$

Therefore,

$$\mathbf{r}_c = \frac{1}{V}\iiint_V \mathbf{r}\, dv \tag{7.7}$$

We see that the center of volume is the point where we could hypothetically concentrate the entire volume of a body for purposes of computing the first moment of the volume of the body about some point O. The components of Eq. 7.8 give the *centroid distances* of volume x_c, y_c, and z_c. Thus, we have

$$x_c = \frac{\iiint x\, dv}{\iiint dv}, \quad y_c = \frac{\iiint y\, dv}{\iiint dv}, \quad z_c = \frac{\iiint z\, dv}{\iiint dv} \tag{7.8}$$

The integral $\iiint x\, dv$, it should be noted, gives the first moment of volume about the yz plane, etc.

If we replace dv by $dm = \rho\, dv$ in Eq. 7.6, where ρ is the mass *density*, we get the *first moment of mass* about O. That is,

$$\text{moment vector of mass} \equiv \iiint_V \boldsymbol{r}\, \rho\, dv \tag{7.9}$$

The *center of mass* $\boldsymbol{r}_c$ is then given as

$$\boldsymbol{r}_c = \frac{1}{M} \iiint_V \boldsymbol{r}\, \rho\, dv \tag{7.10}$$

where M is the total mass of the body. The center of mass is the point in space where hypothetically we could concentrate the entire mass for purposes of computing the first moment of mass about a point O. Using the components of Eq. 7.10, we can say that

$$x_c = \frac{\iiint x\rho\, dv}{\iiint \rho\, dv}, \quad y_c = \frac{\iiint y\rho\, dv}{\iiint \rho\, dv}, \quad z_c = \frac{\iiint z\rho\, dv}{\iiint \rho\, dv}$$

In our work in dynamics, we shall consider the center of mass of a system of n particles (see Fig. 7.9). We will then say:

$$\left(\sum_{i=1}^{n} m_i\right)\boldsymbol{r}_c = \sum_{i=1}^{n} m_i \boldsymbol{r}_i$$

Therefore,

$$\boldsymbol{r}_c = \frac{\sum_{i=1}^{n} m_i \boldsymbol{r}_i}{M} \tag{7.11}$$

where M is the total mass of the system. Clearly, if the particles are of infinitesimal mass and constitute a continuous body, we get back Eq. 7.10.

Finally, if we replace dv by $\gamma\, dv$, where $\gamma\,(= \rho g)$ is the *specific weight*, we arrive at the concept of *center of gravity* discussed in Chapter 4. We have used the center of gravity of a body in many calculations thus far as a point to concentrate the entire weight of a body.

You should have no trouble in concluding from Eq. 7.10 that if ρ is constant throughout a body, the center of mass coincides with the center of volume. Furthermore, if $\gamma (= \rho g)$ is constant throughout a body, the center of gravity of the body corresponds to the center of volume of the body. If, finally, ρ and g are each constant for a body, all three points coincide for the body.

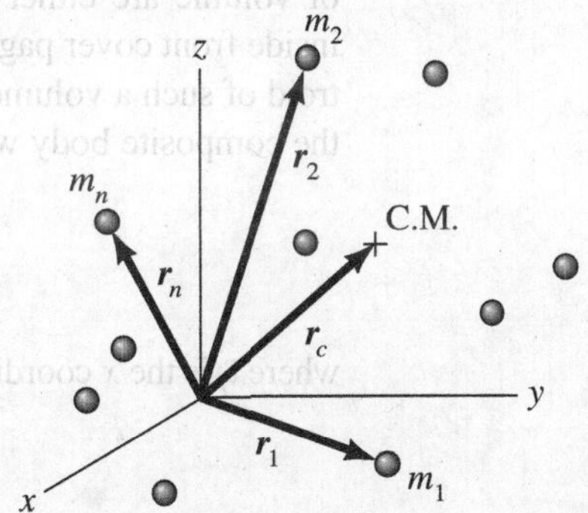

Figure 7.9. System of n particles showing center of mass, C.M.

We now illustrate the computation of the center of volume. Computation for the center of mass follows similar lines, and we have already computed centers of gravity in Chapter 4.

Example 7.3

Consider a volume of revolution formed by revolving the area shown in Fig. 7.3 about the x axis. This volume has been shown in Fig. 7.10. Clearly, the centroid of this volume must lie somewhere along the x axis. Determine the centroidal distance x_c.

Using r, θ, and x as coordinates (cylindrical coordinates), we then have, using slices of thickness dx as volume elements:

$$V = \int_0^{10} (\pi r^2)\, dx = \int_0^{10} (\pi)(25x)\, dx$$

where we have replaced r^2 with $25x$ according to the equation for the boundary of the generating area. Integrating, we get

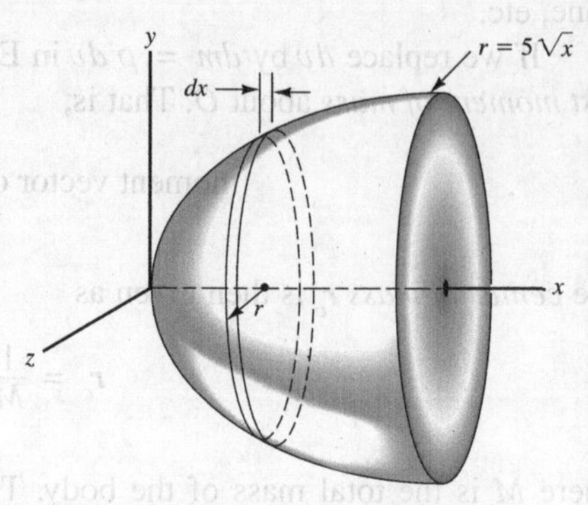

Figure 7.10. Body of revolution.

$$V = 25\pi \frac{x^2}{2}\bigg|_0^{10} = 3{,}927 \text{ ft}^3$$

Now we compute x_c by using infinitesimal slices of the body of the kind employed for the computation of V. The centroid of each such slice is at the intercept of the slice with the x axis. Thus, we have

$$x_c = \frac{1}{V}\int_0^{10} x(\pi r^2\, dx) = \frac{1}{3{,}927}\int_0^{10} x(\pi)(25x)\, dx$$

$$= \frac{25\pi}{3{,}927}\frac{x^3}{3}\bigg|_0^{10} = \boxed{6.67 \text{ ft}}$$

Many volumes are composed of a number of simple familiar shapes whose centers of volume are either known by inspection or can be found in handbooks (also see the inside front cover page). Such volumes may be called *composite volumes*. To find the centroid of such a volume, we use the known centroids of the composite parts. Thus, for x_c of the composite body whose total volume is V, we have

$$x_c = \frac{\sum_i \bar{x}_i V_i}{V}$$

where $\bar{x}_i$ is the x coordinate to the centroid of the ith composite body of volume V_i. Similarly,

$$y_c = \frac{\sum_i \bar{y}_i V_i}{V}$$

$$z_c = \frac{\sum_i \bar{z}_i V_i}{V}$$

We now illustrate the use of these formulas.

Example 7.4

What is the coordinate x_c for the center of volume of the body of revolution shown in Fig. 7.11? Note that a cone has been cut away from the left end while, at the right end, we have a hemispherical region.

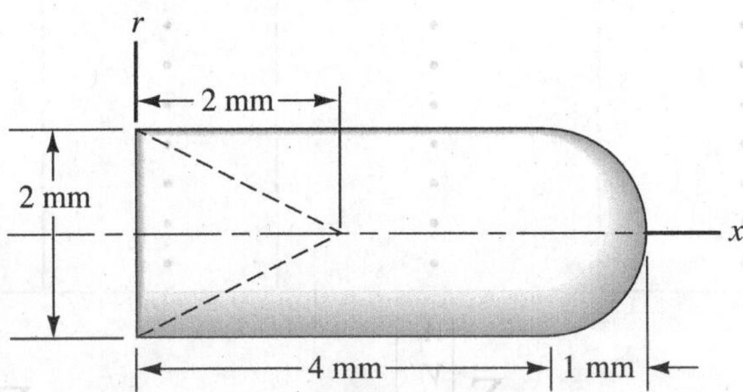

Figure 7.11. Composite volume.

We have a composite body consisting of three simple domains—a cone (body 1), a cylinder (body 2), and a hemisphere (body 3). Using formulas from the inside covers, we have:

V_i (mm³)		$\bar{x}_i$ (mm)	$V_i\bar{x}_i$ (mm⁴)
1. $-(\frac{1}{3})(\pi)(1^2)(2)$	$= -2.09$	$\frac{2}{4}$	-1.047
2. $(\pi)(1^2)(4)$	$= 12.57$	2	25.14
3. $\frac{2}{3}(\pi)(1^3)$	$= 2.09$	$4 + \frac{3}{8}(1) = 4.38$	9.15
	$V = 12.57$		$\sum_i V_i\bar{x}_i = 33.24$

Therefore,

$$x_c = \frac{\sum_i V_i\bar{x}_i}{V} = \frac{33.24}{12.57} = \boxed{2.64 \text{ mm}}$$

We have presented a number of three-dimensional problems for determining the center of volume, center of mass, and the center of gravity of composite bodies. We will leave it to the student to work his/her way through these problems, working from first principles, without the need of examples. However, we ask that you follow the following format, which clearly is an extension of what we have been doing up to this point.

Area, Volume, Mass, or Weight No.	*Area, Volume, Mass, or Weight* ***Value***	$(r_c)_i$	$(r_c)_i \begin{Bmatrix} A_i \\ V_i \\ M_i \\ W_i \end{Bmatrix}$
1	•	•	•
2	•	•	•
•	•	•	•
•	•	•	•
•	•	•	•
•	•	•	•
•	•	•	•
•	•	•	•
•	•	•	•

$$\sum_i \begin{Bmatrix} A_i \\ V_i \\ M_i \\ W_i \end{Bmatrix} = \qquad \sum_i (r_c)_i \begin{Bmatrix} A_i \\ V_i \\ M_i \\ W_i \end{Bmatrix} =$$

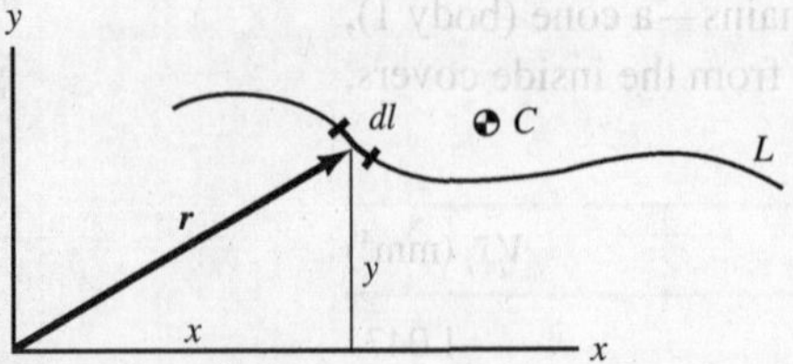

Figure 7.12. Centroid for curved line.

In closing, we wish to point out further that curved surfaces and lines have centroids. Since we shall have occasion in the next section to consider the centroid of a line, we simply point out now (see Fig. 7.12) that

$$x_c = \frac{\int x\,dl}{L} \tag{7.12a}$$

$$y_c = \frac{\int y\,dl}{L} \tag{7.12b}$$

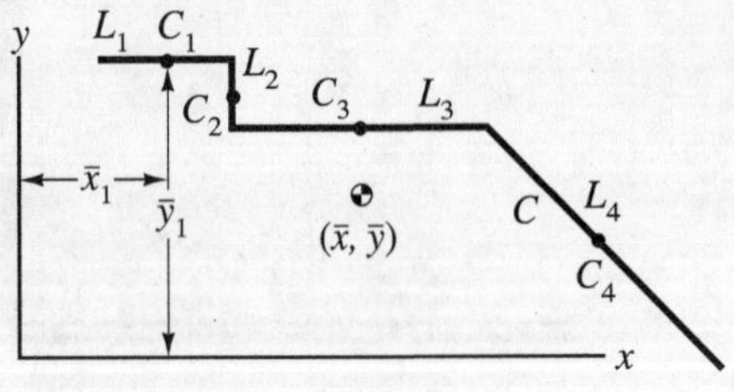

Figure 7.13. Centroid for composite line.

where L is the length of the line. Note that the centroid C will not generally lie along the line.

Consider next a curve made up of simple curves each of whose centroids is known. Such is the case shown in Fig. 7.13, made up of straight lines. The line segment L_1, has for instance centroid C_1 with coordinates x_1, y_1, as has been shown in the diagram. We can then say for the entire curve that

$$x_c = \frac{\sum_i \bar{x}_i L_i}{L}$$

$$y_c = \frac{\sum_i \bar{y}_i L_i}{L} \tag{7.13}$$

*7.4 Theorems of Pappus–Guldinus

The theorems of Pappus–Guldinus were first set forth by Pappus about 300 A.D. and then restated by the Swiss mathematician Paul Guldinus about 1640. These theorems are concerned with the relation of a surface of revolution to its generating curve, and the relation of a volume of revolution to its generating area.

The first of the theorems may be stated as follows:

> *Consider a coplanar generating curve and an axis of revolution in the plane of this curve (see Fig. 7.14). The generating curve can touch but must not cross the axis of revolution. The surface of revolution developed by revolving the generating curve about the axis of revolution has an* area *equal to the product of the* length *of the* generating curve *times the* circumference *of the* circle *formed by the* centroid *of the* generating curve *in the process of generating a surface of revolution.*

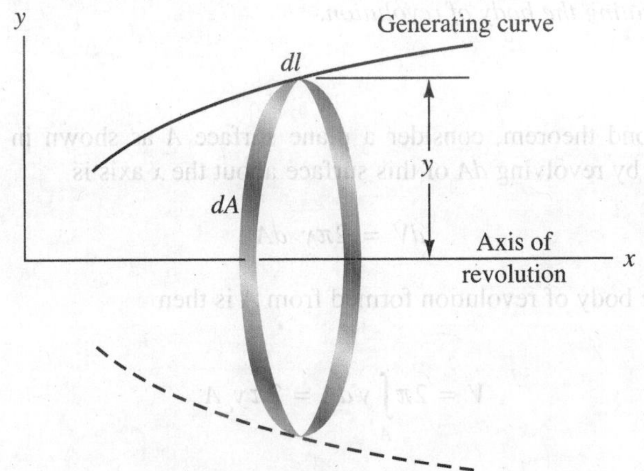

Figure 7.14. Coplanar generating curve.

To prove this theorem, consider first an element dl of the generating curve shown in Fig. 7.14. For a single revolution of the generating curve about the x axis, the line segment dl traces an area

$$dA = 2\pi y\ dl$$

For the entire curve this area becomes the surface of revolution given as

$$A = 2\pi \int y\,dl = 2\pi y_c L \tag{7.14}$$

where L is the length of the curve and y_c is the centroidal coordinate of the curve. But $2\pi y_c$ is the circumferential length of the circle formed by having the centroid of the curve rotate about the x axis. The first theorem is thus proved.

Another way of interpreting Eq. 7.14 is to note that the area of the body of revolution is equal to 2π times the *first moment* of the generating curve about the axis of revolution. If the generating curve is composed of simple curves, L_i, whose centroids are known, such as the case shown in Fig. 7.13, then we can express A as follows:

$$A = 2\pi\left(\sum_i L_i \bar{y}_i\right) \tag{7.15}$$

where y_i is the centroidal coordinate to the ith line segment L_i.

The second theorem may be stated as follows:

Consider a plane surface and an axis of revolution coplanar with the surface but oriented such that the axis can intersect the surface only as a tangent at the boundary or have no intersection at all. The volume of the body of revolution developed by rotating the plane surface about the axis of revolution equals the product of the area *of the surface times the* circumference *of the circle formed by the* centroid *of the surface in the process of generating the body of revolution.*

To prove the second theorem, consider a plane surface A as shown in Fig. 7.15. The volume generated by revolving dA of this surface about the x axis is

$$dV = 2\pi y \ dA$$

The volume of the body of revolution formed from A is then

$$V = 2\pi \int_A y \, dA = 2\pi y_c A \tag{7.16}$$

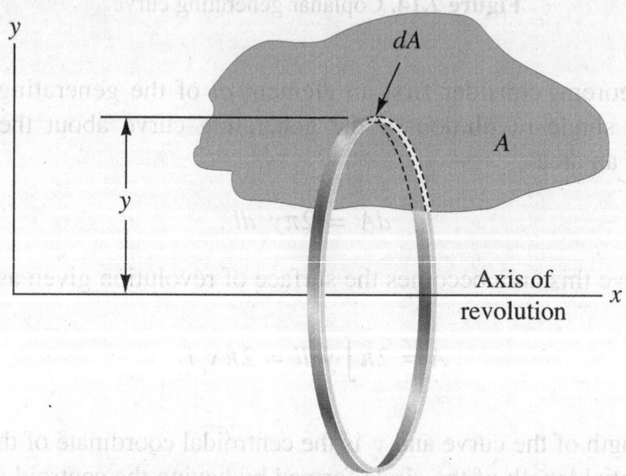

Figure 7.15. Plane surface A coplanarwith xy plane.

Thus, the volume V equals the area of the generating surface A times the circumferential length of the circle of radius y_c. The second theorem is thus also proved.[2]

Another way to interpret Eq. 7.16 is to note that V equals 2π times the *first moment* of the generating area A about the axis of revolution. If this area A is made up of simple areas A_i, we can say that

$$V = 2\pi\left(\sum_i A_i \bar{y}_i\right) \tag{7.17}$$

where y_i, is the centroidal coordinate to the ith area A_i.

We now illustrate the use of the theorems of Pappus and Guldinus. As we proceed, it will be helpful to remember the theorems by noting that you multiply a length (or area) of the generator by the distance moved by the centroid of the generator.

[2]It is to be pointed out that the centroid of a volume of revolution will not be coincident with the centroid of a longitudinal cross-section taken along the axis of the volume. Example: a cone and its triangular, longitudinal cross-section.

Example 7.5

Determine the surface area and volume of the bulk materials trailer shown in Fig. 7.16.

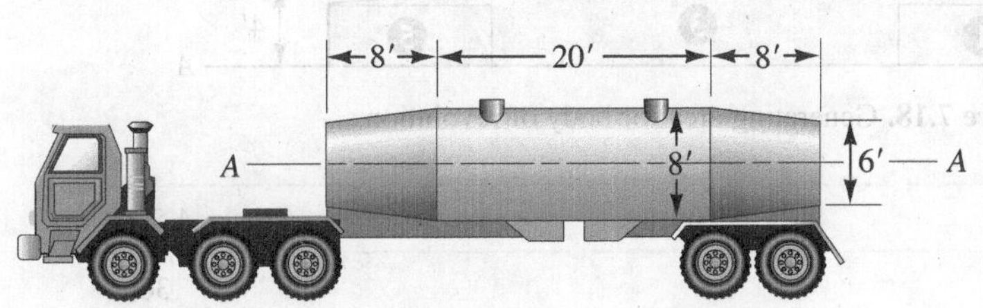

Figure 7.16. Bulk materials trailer.

We shall first determine the surface area by considering the first moment about the centerline A–A (see Fig. 7.17) of the generating curve of the surface of revolution.

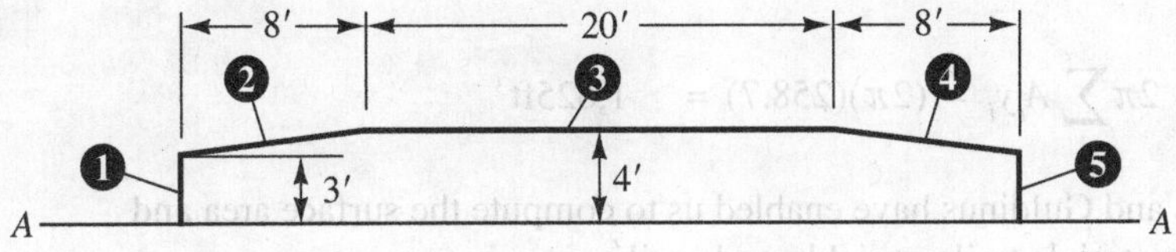

Figure 7.17. Generating curve for surface of revolution.

Example 7.5 (Continued)

This curve is a set of 5 straight lines each of whose centroids is easily known by inspection. Accordingly we may use Eq. 7.15. For clarity, we use a column format for the data as follows:

L_i (ft)	$\bar{y}_i$ (ft)	$L_i\bar{y}_i$ (ft^2)
1. 3	1.5	4.5
2. $\sqrt{8^2+1^2} = 8.06$	3.5	28.21
3. 20	4	80
4. 8.06	3.5	28.21
5. 3	1.5	4.5
		$\sum_i L_i\bar{y}_i = 145.43$

Therefore,

$$A = (2\pi)(145.43) = 914 \text{ ft}^2$$

To get the volume, we next show in Fig. 7.18 the generating area for the body of revolution. Notice it has been decomposed into simple composite areas. We shall employ Eq. 7.17 and hence we shall need the first moment of area about the axis A–A of the composite areas. Again, we shall employ a column format for the data.

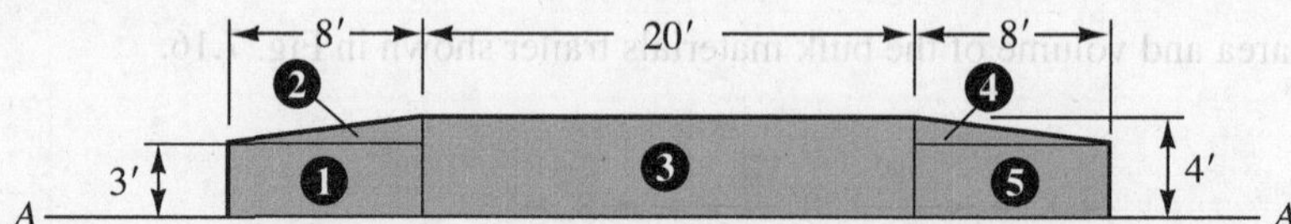

Figure 7.18. Generating area for body of revolution.

A_i (ft^2)	$\bar{y}_i$ (ft)	$A_i\bar{y}_i$ (ft^3)
1. 24	1.5	36
2. $\frac{1}{2}(8)(1) = 4$	$3 + \frac{1}{3} = 3.33$	13.33
3. 80	2	160
4. 4	3.333	13.33
5. 24	1.5	36
		$\sum_i A_i\bar{y}_i = 258.7$

Therefore,

$$V = 2\pi\sum_i A_i\bar{y}_i = (2\pi)(258.7) = 1{,}625\text{ft}^3$$

The theorems of Pappus and Guldinus have enabled us to compute the surface area and the volume of the bulk materials trailer quickly and easily.

PROBLEMS

7.31. If $r^2 = ax$ in the body of revolution shown, compute the centroidal distance x_c of the body.

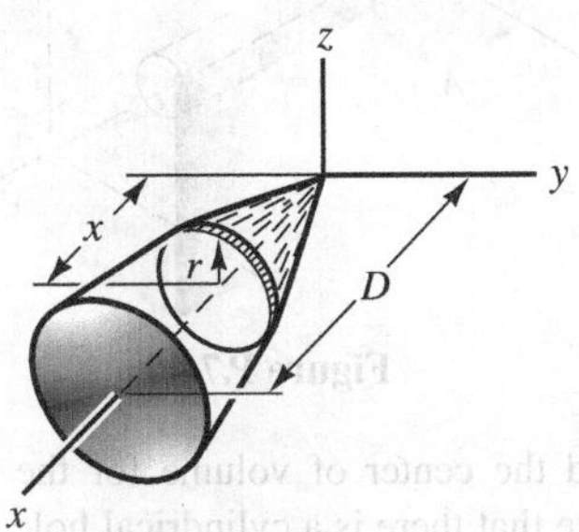

Figure P.7.31.

7.32. Using vertical elements of volume as shown, compute the centroidal coordinates x_c, y_c of the body. Then, using horizontal elements, compute z_c.

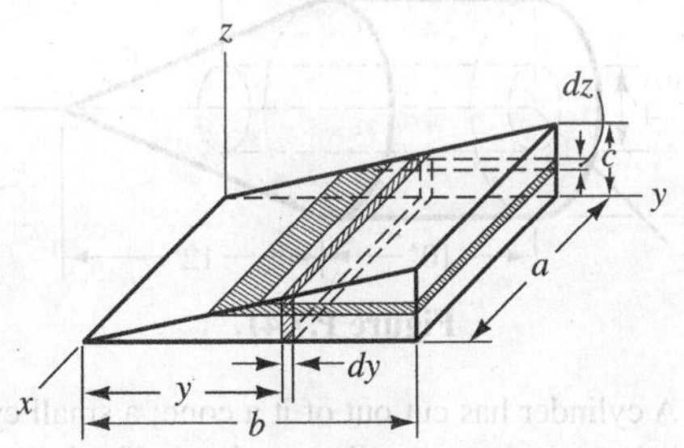

Figure P.7.32.

7.33. Compute the center of volume of a right circular cylinder of height h and radius at the base r.

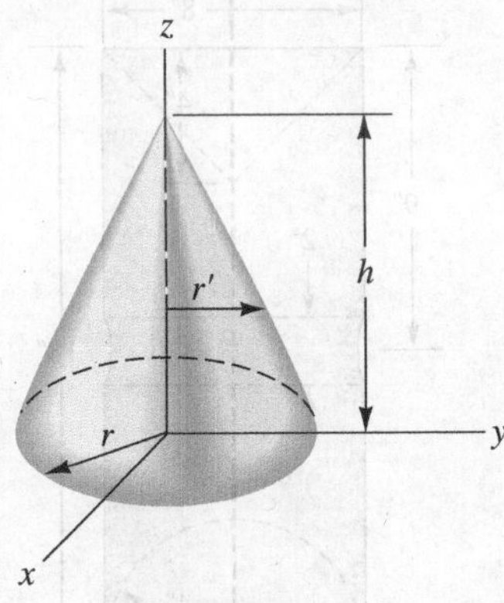

Figure P.7.33.

7.34. Determine the position of the center of mass of the solid hemisphere having a uniform mass density ρ and with a radius a.

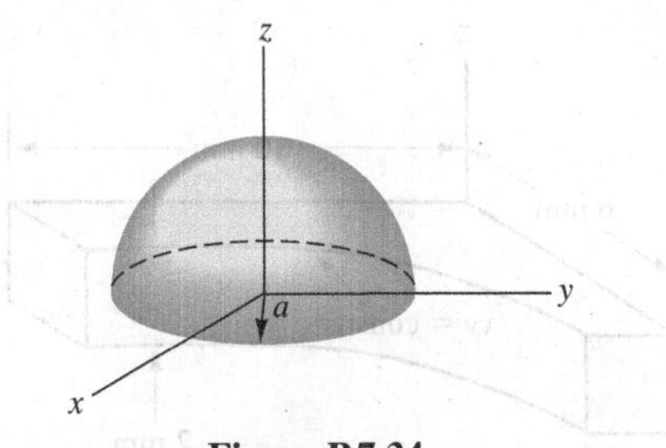

Figure P.7.34.

7.35. Find the center of mass for the paraboloid of revolution having a uniform density ρ.

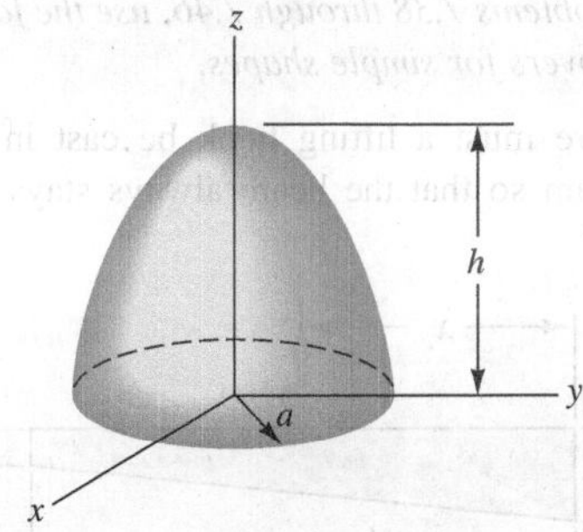

Figure P.7.35.

7.36. A small bomb has exploded at position O. Four pieces of the bomb move off at high speed. At $t = 3$ sec, the following data apply:

m	(kg)	r (m)
1.	.2	$2i + 3j + 4k$
2.	.1	$4i + 4j - 6k$
3.	.15	$-3i + 2j - 3k$
4.	.22	$2i - 3j + 2k$

What is the position of the center of mass?

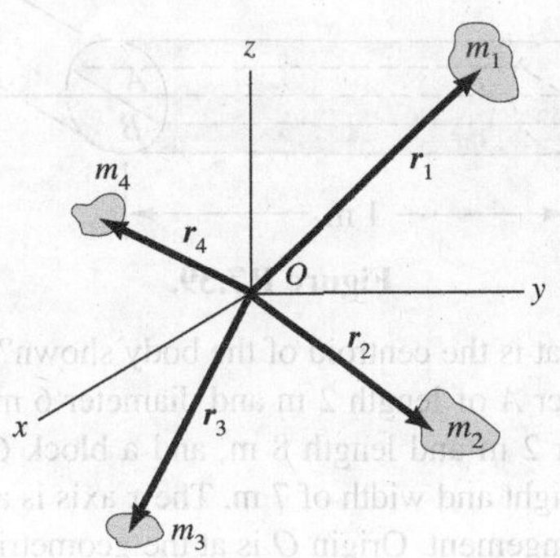

Figure P.7.36.

7.37. A plate of uniform thickness and density has for its curved edge a rectangular hyperbola (xy = constant). Find the centroid of the upper surface. Find the centers of mass, volume, and gravity for the plate.

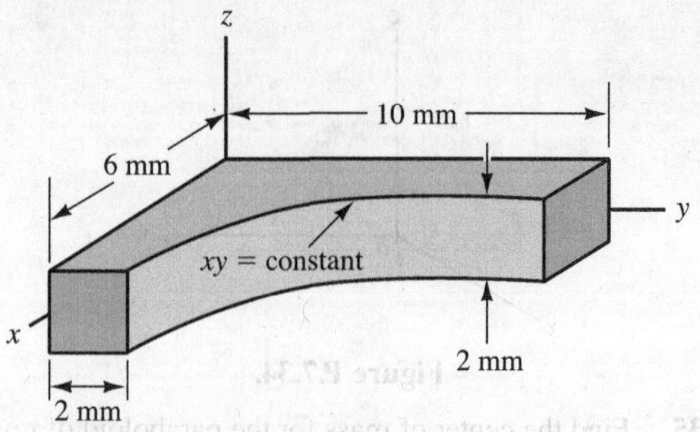

Figure P.7.37.

In Problems 7.38 through 7.46, use the formulas on the inside covers for simple shapes.

7.38. Where must a lifting hook be cast in a tapered concrete beam so that the beam always stays horizontal when lifted?

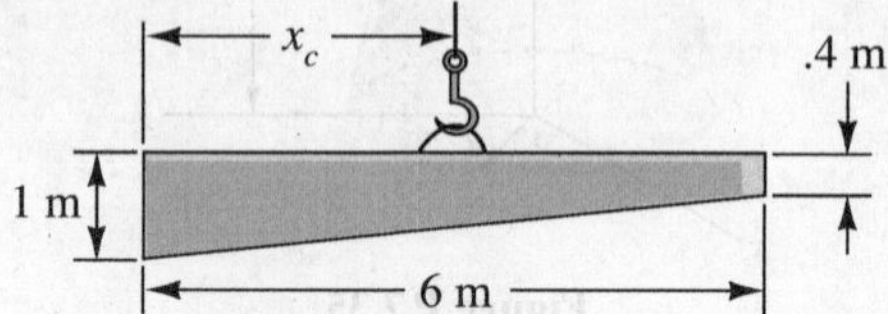

Figure P.7.38.

7.39. Two solid semicylinders are glued together. Body A has a uniform mass density of 6.54 kN/m^3, while body B has a uniform mass density of 10 kN/m^3. Determine:

(a) Center of volume
(b) Center of mass
(c) Center of gravity

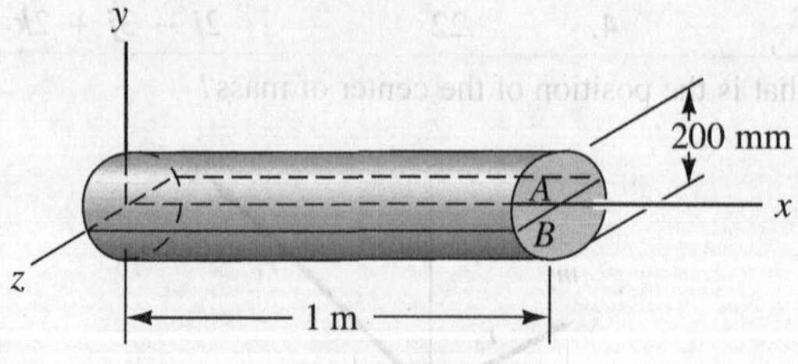

Figure P.7.39.

7.40. What is the centroid of the body shown? It consists of a cylinder A of length 2 m and diameter 6 m, a shaft B of diameter 2 m and length 8 m, and a block C of length 4 m and height and width of 7 m. The x axis is a centerline for the arrangement. Origin O is at the geometric center of cylinder A.

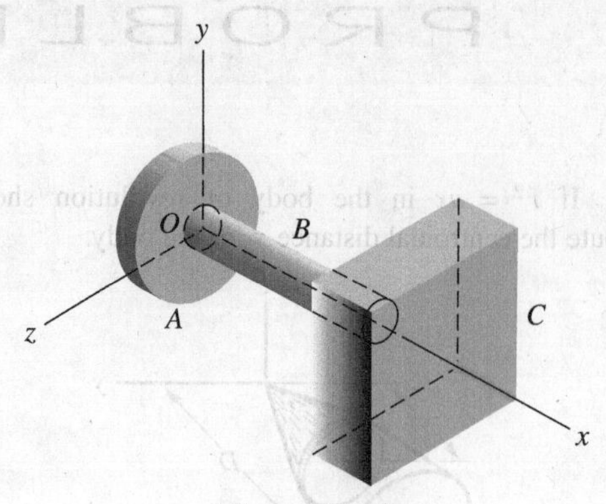

Figure P.7.40.

7.41. Find the center of volume for the cone–cylinder shown. Note that there is a cylindrical hole of length 16 ft and diameter 4 ft cut into the body.

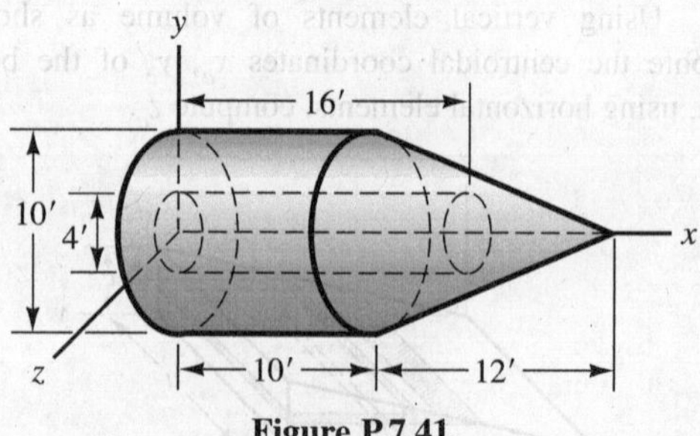

Figure P.7.41.

7.42. A cylinder has cut out of it a cone; a small cylinder directed through the centerline; and a half-sphere as has been shown in the diagram. Determine the center of gravity.

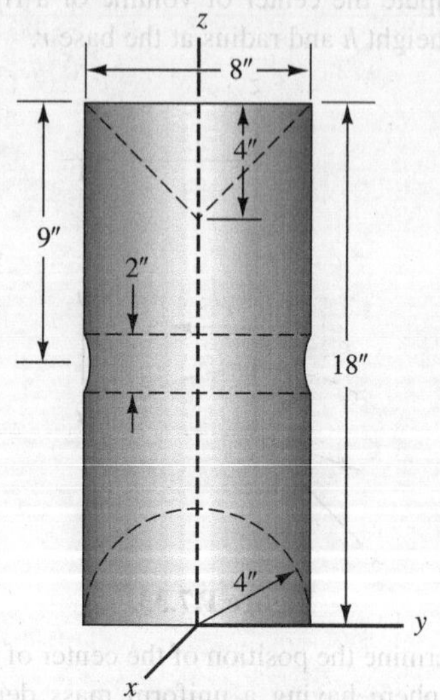

Figure P.7.42.

7.43. Find the center of gravity of the bent plate. The rectangular cutout occurs at the geometric center of the surface in the xz plane.

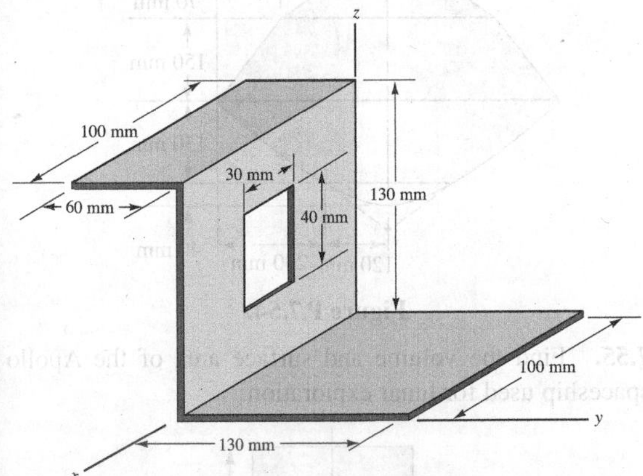

Figure P.7.43.

7.44. A bent aluminum rod weighing 30 N/m is fitted into a plastic cylinder weighing 200 N, as shown. What are the centers of volume, mass, and gravity?

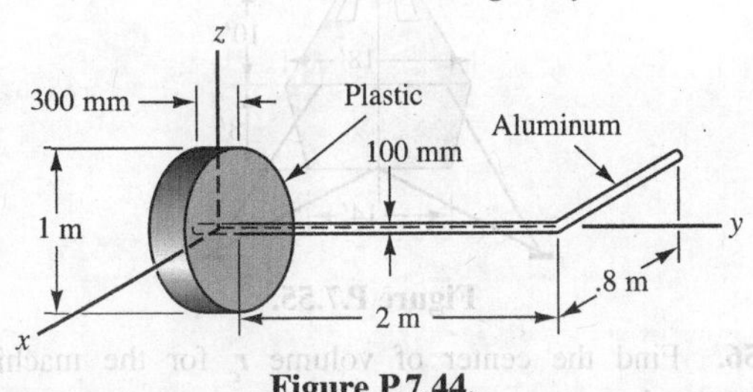

Figure P.7.44.

7.45. An aluminum cylinder fits snugly into a brass block. The brass weighs 43.2 kN/m^3 and the aluminum weighs 30 kN/m^3. Find the center of volume, the center of mass, and the center of gravity.

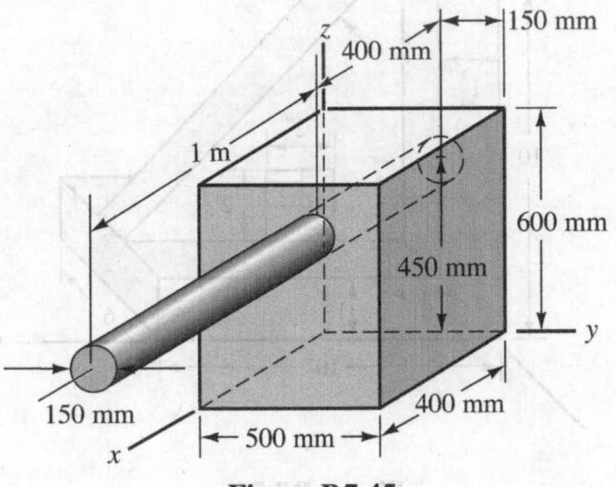

Figure P.7.45.

7.46. Two thin plates are welded together. One has a circle of radius 200 mm cut out as shown. If each plate weighs 450 N/m^2, what is the position of the center of mass?

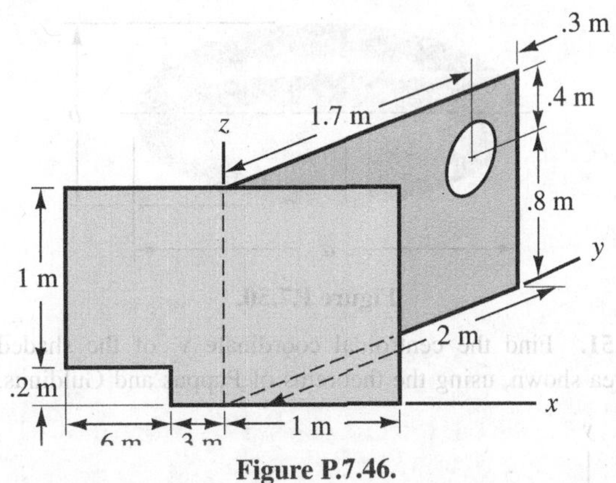

Figure P.7.46.

7.47. Where is the center of mass of the bent wire if it weighs 10 N/m?

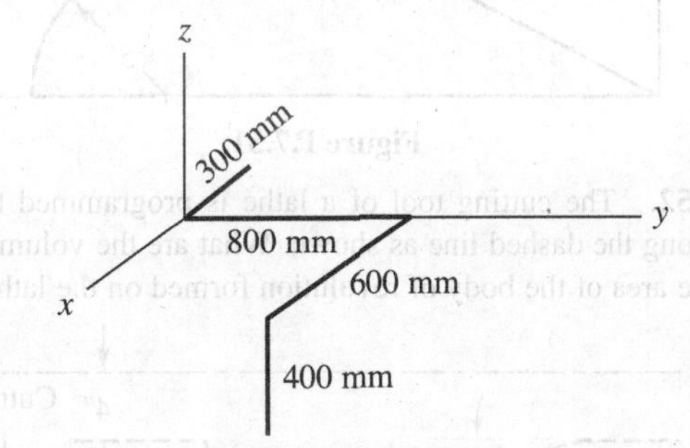

Figure P.7.47.

7.48. Find the center of mass of the bent wire shown in the zy plane. The wire weighs 15 N/m.

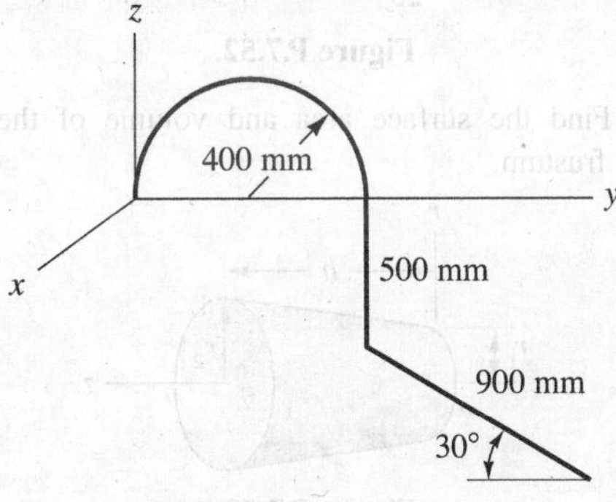

Figure P.7.48.

7.49. In Problem 7.41, involving a wooden cone–cylinder with a cylindrical hole, find the center of mass for the case where the cylinder has a density of 46.0 lbm/ft^3 and the cone has a density of 30.0 lbm/ft^3.

7.50. The volume of an ellipsoidal body of revolution is known from calculus to be $\frac{1}{6}\pi ab^2$. If the area of an ellipse is $\pi ab/4$, find the centroid of the area for a semiellipse.

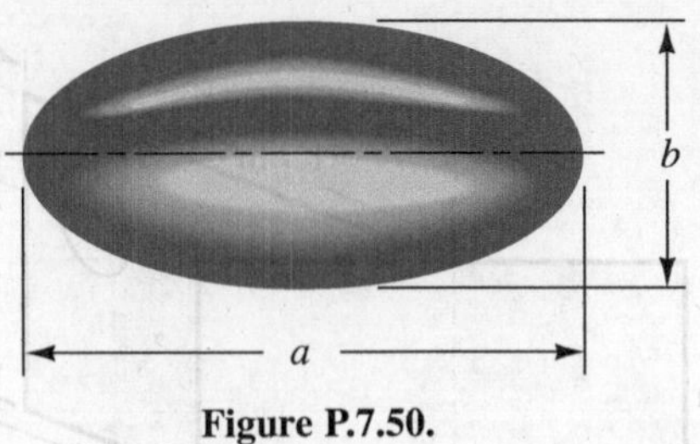

Figure P.7.50.

7.51. Find the centroidal coordinate y_c of the shaded area shown, using the theorems of Pappus and Guldinus.

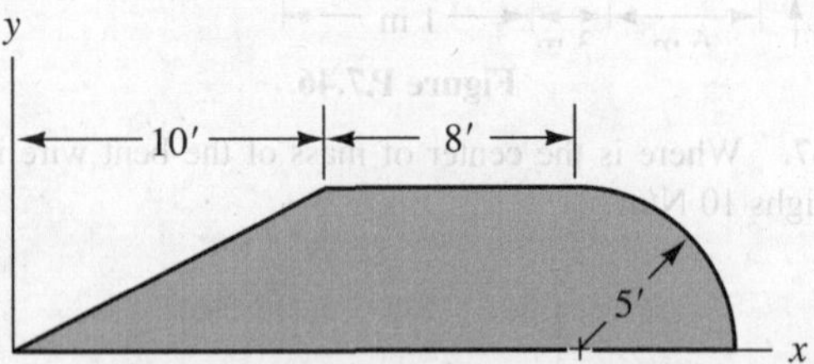

Figure P.7.51.

7.52. The cutting tool of a lathe is programmed to cut along the dashed line as shown. What are the volume and the area of the body of revolution formed on the lathe?

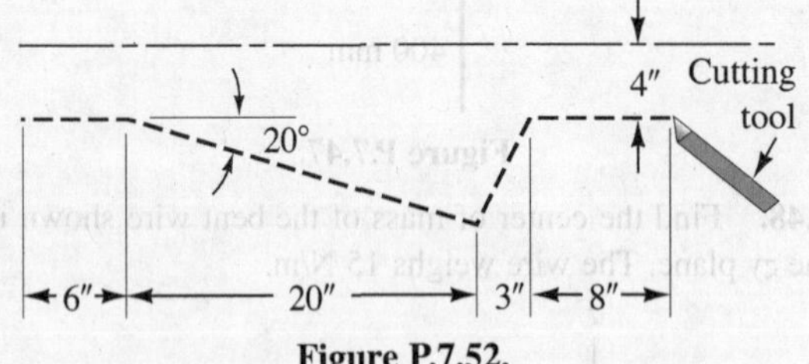

Figure P.7.52.

7.53. Find the surface area and volume of the right conical frustum.

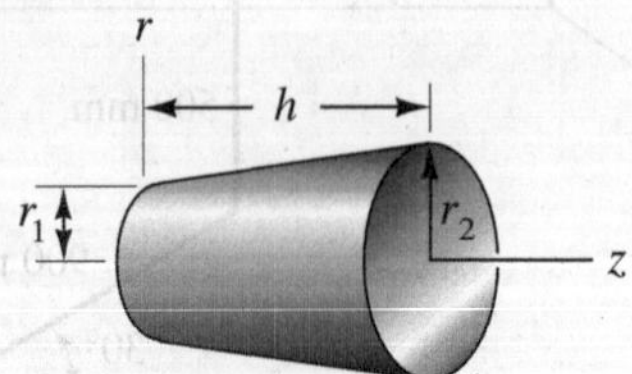

Figure P.7.53.

7.54. Find the surface area and volume of the Earth entry capsule for an unmanned Mars sampling mission. Approximate the rounded nose with a pointed nose as shown with the dashed lines.

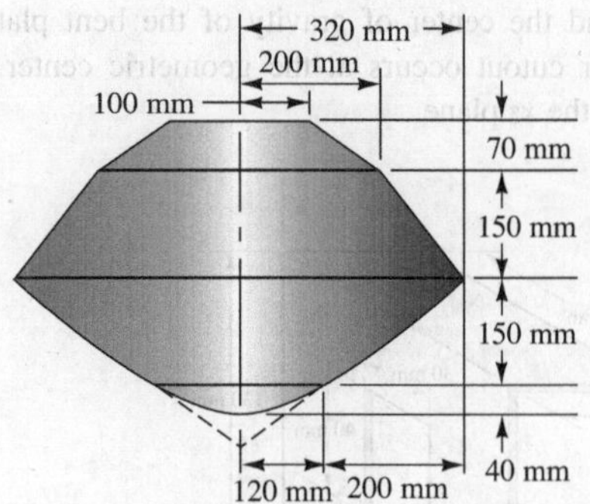

Figure P.7.54.

7.55. Find the volume and surface area of the Apollo spaceship used for lunar exploration.

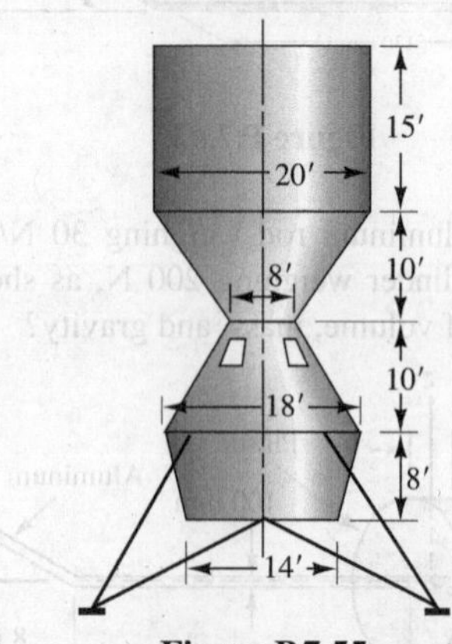

Figure P.7.55.

7.56. Find the center of volume r_c for the machine element shown.

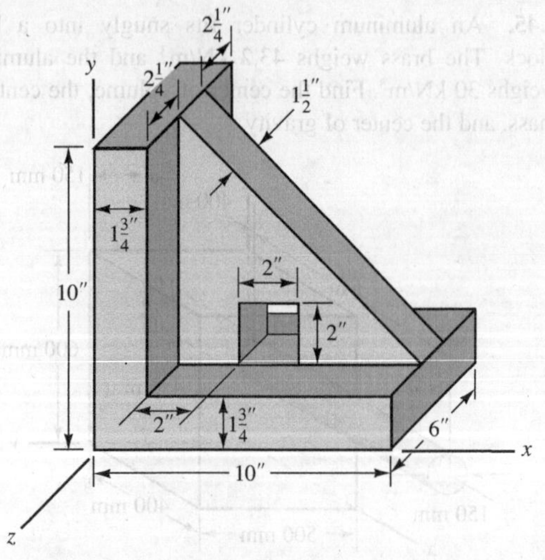

Figure P.7.56.

7.5 Second Moments and the Product of Area[3] of a Plane Area

We shall now consider other properties of a plane area relative to a given reference. The *second moments* of the area A about the x and y axes (Fig. 7.19), denoted as I_{xx} and I_{yy}, respectively, are defined as

$$I_{xx} = \int_A y^2 \, dA \tag{7.18a}$$

$$I_{yy} = \int_A x^2 \, dA \tag{7.18b}$$

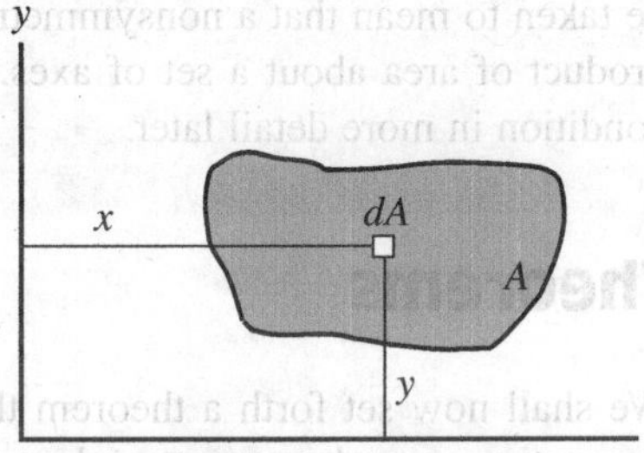

Figure 7.19. Plane surface.

The second moment of area cannot be negative, in contrast to the first moment. Furthermore, because the square of the distance from the axis is used, elements of area that are farthest from the axis contribute most to the second moment of area.

In an analogy to the centroid, the entire area may be concentrated at a single point (k_x, k_y) to give the same second moment of area for a given reference. Thus,

$$\begin{aligned} Ak_x^2 &= I_{xx} = \int_A y^2 \, dA; \quad \text{therefore, } k_x^2 = \frac{\int_A y^2 \, dA}{A} \\ Ak_y^2 &= I_{yy} = \int_A x^2 \, dA; \quad \text{therefore, } k_y^2 = \frac{\int_A x^2 \, dA}{A} \end{aligned} \tag{7.19}$$

The distances k_x and k_y are called the *radii of gyration. This point will have a position that depends not only on the shape of the area but also on the position of the reference.* This situation is unlike the centroid, whose location is independent of the reference position.

The *product of area* relates an area directly to a set of axes and is defined as

$$I_{xy} = \int_A xy \, dA \tag{7.20}$$

This quantity may be negative.

[3] We often use the expressions *moment* and *product of inertia* for second moment and product of area, respectively. However, we shall also use the former expressions in Chapter 8 in connection with mass distributions.

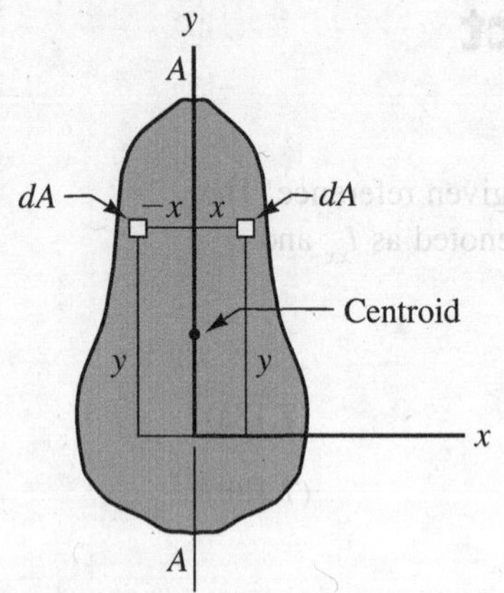

Figure 7.20. Area symmetric about y axis.

If the area under consideration has an axis of symmetry, the product of area for this axis and any axis orthogonal to this axis must be zero. You can readily reach this conclusion by considering the area in Fig. 7.20, which is symmetrical about the axis A–A. Notice that the centroid is somewhere along this axis. (Why ?) The axis of symmetry has been indicated as the y axis, and an arbitrary x axis coplanar with the area has been shown. Also indicated are two elemental areas that are positioned as mirror images about the y axis. The contribution to the product of area of each element is $xy\,dA$, but with opposite signs, and so the net result is zero. Since the entire area can be considered to be composed of such pairs, it becomes evident that the product of area for such cases is zero. This *should not* be taken to mean that a nonsymmetric area cannot have a zero product of area about a set of axes. We shall discuss this last condition in more detail later.

7.6 Transfer Theorems

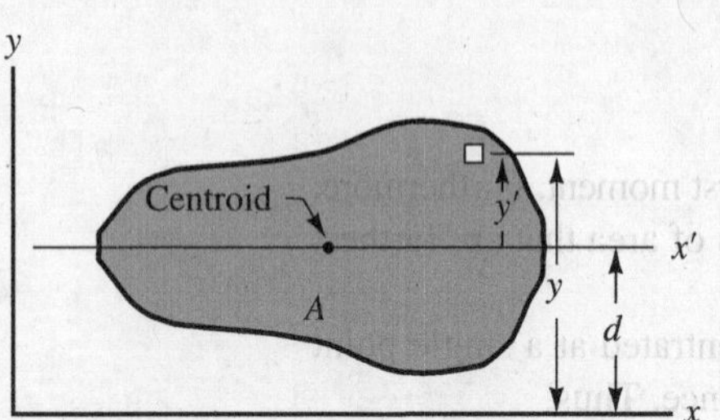

Figure 7.21. x and x' are parallel axes.

We shall now set forth a theorem that will be of great use in computing second moments and products of area for areas that can be decomposed into simple parts (composite areas). With this theorem, we can find second moments or products of area about any axis in terms of second moments or products of area about a *parallel* set of axes going through the *centroid* of the area in question.

An x axis is shown in Fig. 7.21 parallel to and at a distance d from an axis x' going through the centroid of the area. The latter axis you will recall is a *centroidal axis*. The second moment of area about the x axis is

$$I_{xx} = \int_A y^2\,dA = \int_A (y' + d)^2\,dA$$

where the distance y has been replaced by $(y' + d)$. Carrying out the squaring operation and integrating, leads to the result

$$I_{xx} = \int_A y'^2\,dA + 2d\int_A y'\,dA + Ad^2$$

The first term on the right-hand side is by definition $I_{x'x'}$. The second term involves the first moment of area about the x' axis. But the x' axis here is a centroidal axis, and so the second term is zero. We can now state the transfer theorem (frequently called the parallel-axis theorem):

$$I_{\text{about any axis}} = I_{\substack{\text{about a parallel}\\ \text{axis at centroid}}} + Ad^2 \qquad (7.21)$$

where d is the perpendicular distance between the axis for which I is being computed and the parallel centroidal axis.

In strength of materials, a course generally following statics, second moments of area about noncentroidal axes are commonly used. The areas involved are complicated and not subject to simple integration. Accordingly, in structural handbooks, the areas and second moments about various centroidal axes are listed for many of the practical configurations with the understanding that designers will use the parallel-axis theorem for axes not at the centroid.

Let us now examine the product of area in order to establish a parallel-axis theorem for this quantity. Accordingly, two references are shown in Fig. 7.22, one (x', y') at the centroid and the other (x, y) positioned arbitrarily but *parallel* relative to $x'y'$. Note that c and d are the x and y *coordinates*, respectively, of the centroid of A as measured from reference xy. These coordinates accordingly must have the proper signs, dependent on what quadrant the centroid of A is in relative to xy. The product of area about the noncentroidal axes xy can then be given as

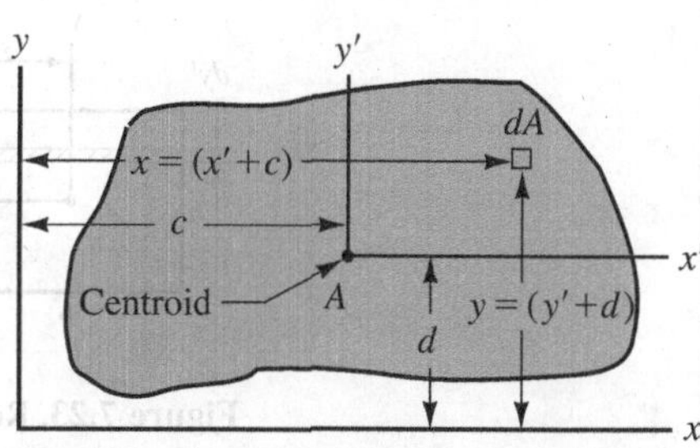

Figure 7.22. c and d measured from xy.

$$I_{xy} = \int_A xy\,dA = \int_A (x' + c)(y' + d)\,dA$$

Carrying out the multiplication, we get

$$I_{xy} = \int_A x'y'\,dA + c\int_A y'\,dA + d\int_A x'\,dA + Adc$$

Clearly, the first term on the right side by definition is $I_{x'y'}$, whereas the next two terms are zero since x' and y' are centroidal axes. Thus, we arrive at a parallel-axis theorem for products of area of the form:

$$I_{xy \text{ for any set of axes}} = I_{x'y' \text{ for a parallel set of axes at centroid}} + Adc \qquad (7.22)$$

It is important to remember that c and d are measured *from the xy axes to the centroid* and must have the appropriate sign. This will be carefully pointed out again in the examples of Section 7.7.

7.7 Computations Involving Second Moments and Products of Area

We shall examine examples for the computation of second moments and products of an area.

Example 7.6

A rectangle is shown in Fig. 7.23. Compute the second moments and products of area about the centroidal $x'y'$ axes as well as about the xy axes.

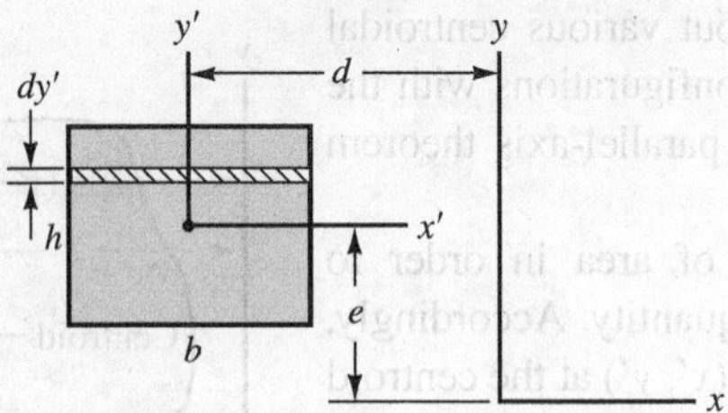

Figure 7.23. Rectangle: base b, height h.

$I_{x'x'}, I_{y'y'}, I_{x'y'}$. For computing $I_{x'x'}$, we can use a strip of width dy' at a distance y' from the x' axis. The area dA then becomes $b\,dy'$. Hence, we have

$$I_{x'x'} = \int_{-h/2}^{+h/2} y'^2 b\,dy' = b\left.\frac{y'^3}{3}\right|_{-h/2}^{+h/2} = \frac{b}{3}\left(\frac{h^3}{8}+\frac{h^3}{8}\right) = \frac{1}{12}bh^3 \qquad \text{(a)}$$

This is a common result and should well be remembered since it occurs so often. Verbally, for such an axis, the second moment of area is equal to 1⁄12 the base b times the height h cubed. The second moment of area for the y' axis can immediately be written as

$$I_{y'y'} = \tfrac{1}{12}hb^3 \qquad \text{(b)}$$

where the base and height have simply been interchanged.

As a result of the previous statements on symmetry, we immediately note that

$$I_{x'y'} = 0 \qquad \text{(c)}$$

I_{xx}, I_{yy}, I_{xy}. Employing the transfer theorems, we get

$$I_{xx} = \tfrac{1}{12}bh^3 + bhe^2 \qquad \text{(d)}$$

$$I_{yy} = \tfrac{1}{12}hb^3 + bhd^2 \qquad \text{(e)}$$

In computing the product of area, we must be careful to employ the proper signs for the transfer distances. In checking the derivation of the transfer theorem, we see that these distances are measured from the noncentroidal axes to the centroid C. Therefore, in this problem the transfer distances are $(+e)$ and $(-d)$. Hence, the computation of I_{xy} becomes

$$I_{xy} = 0 + (bh)(+e)(-d) = -bhed \qquad \text{(f)}$$

and is thus a negative quantity.

Example 7.7

What are I_{xx}, I_{yy}, and I_{xy} for the area under the parabolic curve shown in Fig. 7.24?

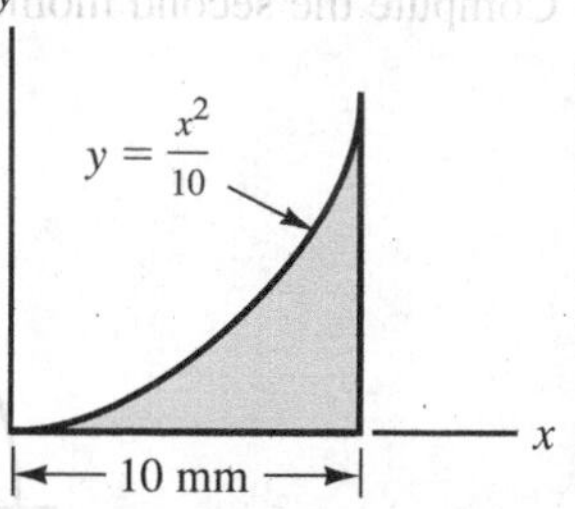

Figure 7.24. Plane area.

To find I_{xx}, we may use horizontal strips of width dy as shown in Fig. 7.25. We can then say for I_{xx}:

$$I_{xx} = \int_0^{10} y^2[dy(10 - x)]$$

But

$$x = \sqrt{10}y^{1/2}$$

Therefore,

$$I_{xx} = \int_0^{10} y^2(10 - \sqrt{10}y^{1/2})\,dy$$

$$= \left[10\frac{y^3}{3} - \sqrt{10}y^{7/2}\left(\frac{2}{7}\right)\right]_0^{10}$$

$$= \frac{10(10^3)}{3} - \sqrt{10}(10^{7/2})\left(\frac{2}{7}\right) = 476.2 \text{ mm}^4$$

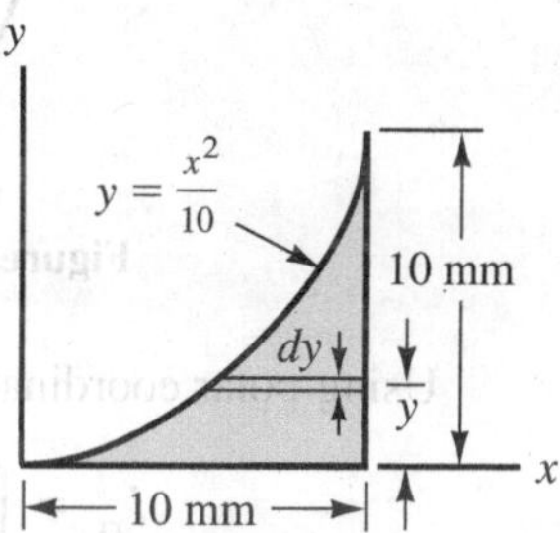

Figure 7.25. Horizontal strip.

As for I_{yy}, we use vertical infinitesimal strips as shown in Fig. 7.26. We can, accordingly, say:

$$I_{yy} = \int_0^{10} x^2(y\,dx) = \int_0^{10} \frac{x^4}{10}\,dx$$

$$= \left.\frac{x^5}{50}\right|_0^{10} = 2{,}000 \text{ mm}^4$$

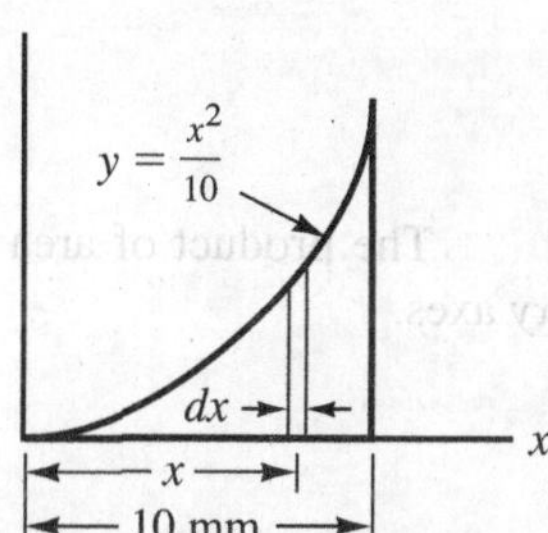

Figure 7.26. Vertical strip.

Finally, for I_{xy} we use an infinitesimal area element $dx\,dy$ shown in Fig. 7.27. We must now perform multiple integration.[4] Thus, we have

$$I_{xy} = \int_0^{10}\int_{y=0}^{y=x^2/10} xy\,dy\,dx$$

Notice by holding x constant and letting y first run from $y = 0$ to the curve $y = x^2/10$ we cover the vertical strip of thickness dx at position x such as is shown in Fig. 7.26. Then by letting x run from zero to 10, we cover the entire area. Accordingly, we first integrate with respect to y holding x constant. Thus,

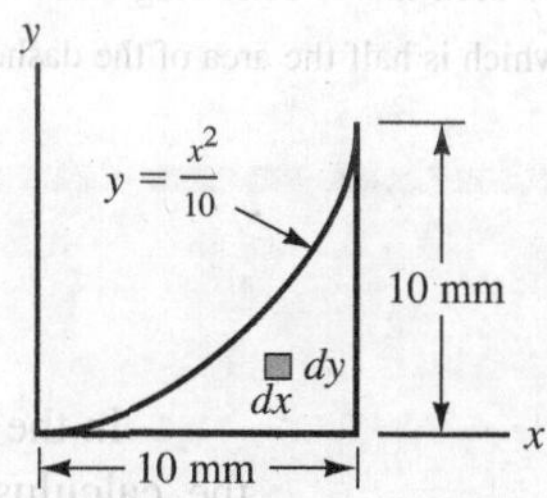

Figure 7.27. Element for multiple integration.

$$I_{xy} = \int_0^{10} x\left(\frac{y^2}{2}\right)\Bigg|_0^{x^2/10} dx = \int_0^{10} \frac{x^5}{200}\,dx$$

Next, integrating with respect to x, we have

$$I_{xy} = \left.\frac{x^6}{1{,}200}\right|_0^{10} = 833 \text{ mm}^4$$

[4]This multiple integration involves boundaries requiring some *variable* limits, in contrast to previous multiple integrations.

Example 7.8

Compute the second moment of area of a circular area about a diameter (Fig. 7.28).

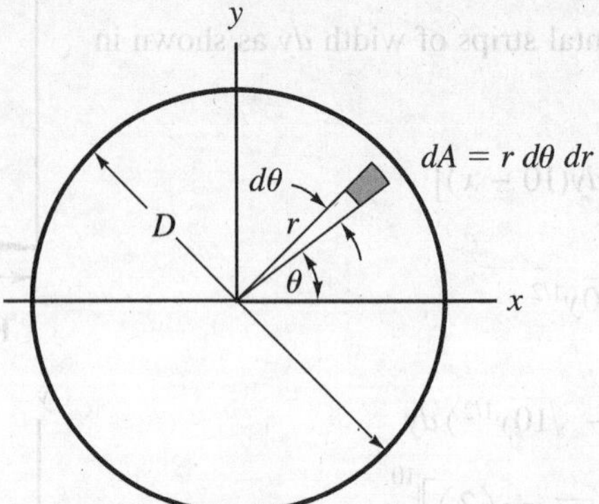

Figure 7.28. Circular area with polar coordinates.

Using polar coordinates, we have[5] for I_{xx}:

$$I_{xx} = \int_0^{D/2} \int_0^{2\pi} (r \sin\theta)^2 r \, d\theta \, dr = \int_0^{D/2} \pi r^3 \, dr$$

Completing the integration, we have

$$I_{xx} = \frac{r^4}{4}\pi \Big|_0^{D/2} = \pi \frac{D^4}{64}$$

The product of area I_{xy} must be zero, owing to symmetry of the area about the xy axes.

[5]The integral $\int_0^{2\pi} \sin^2\theta \, d\theta$ may be evaluated by methods of substitution or may readily be seen in the following manner. $\int_0^{2\pi} \sin^2\theta \, d\theta$ equals the area under the curve shown, which is half the area of the dashed rectangle. Hence, this integral equals π.

In the previous examples, we computed second moments and products of area using the calculus. Many problems of interest involve an area that may be subdivided into simpler *component* areas. Such an area has been referred to in earlier discussions as a composite area. The second moments and products of area for certain centroidal axes of many simple areas may be found in engineering handbooks (also see the inside covers). Using these formulas combined with the parallel-axis theorems, we can easily compute desired second moments and products of area for composite areas as we have done earlier for first moments of area. The following example illustrates this procedure.

Example 7.9

Find the centroid of the area of the unequal-leg Z section shown in Fig. 7.29. Next, determine the second moment of area about the centroidal axes parallel to the sides of the Z section. Finally, determine the product of area for the aforementioned centroidal axes.

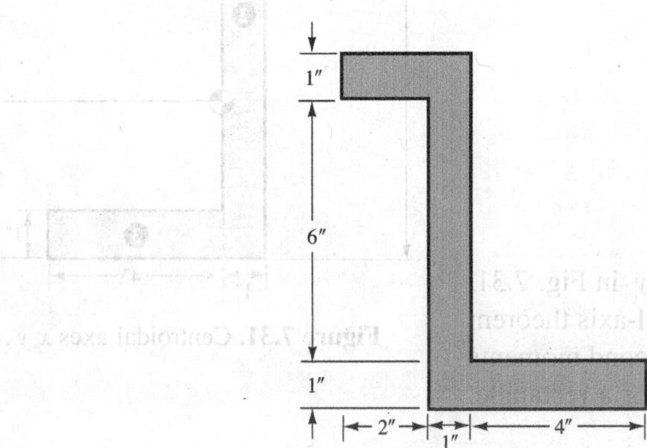

Figure 7.29. Unequal-leg Z section.

We shall subdivide the Z section into three rectangular areas, as shown in Fig. 7.30. Also, we shall insert a convenient reference *xy*, as shown in the diagram. To find the centroid, we proceed in the following manner:

A_i (in.^{2})	$\bar{x}_i$ (in.)	$\bar{y}_i$ (in.)	$A_i\bar{x}_i$ (in.^{3})	$A_i\bar{y}_i$ (in.^{3})
1. (2)(1) = 2	1	7.50	2	15
2. (8)(1) = 8	2.50	4	20	32
3. (4)(1) = 4	5	.50	20	2
$\sum_i A_i = 14$			$\sum_i A_i\bar{x}_i = 42$	$\sum_i A_i\bar{y}_i = 49$

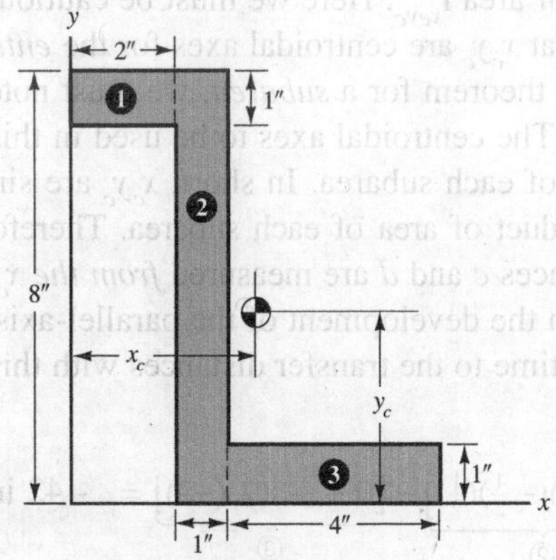

Figure 7.30. Composite area.

Example 7.9 (Continued)

Therefore,

$$x_c = \frac{\sum_i A_i \bar{x}_i}{\sum_i A_i} = \frac{42}{14} = \boxed{3 \text{ in.}}$$

$$y_c = \frac{\sum_i A_i \bar{y}_i}{\sum_i A_i} = \frac{49}{14} = \boxed{3.5 \text{ in.}}$$

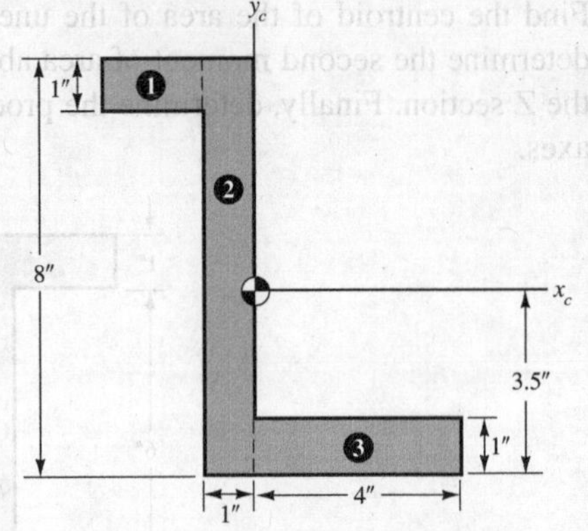

Figure 7.31. Centroidal axes $x_c y_c$.

We have shown the centroidal axes $x_c y_c$ in Fig. 7.31. We now find $I_{x_c x_c}$ and $I_{y_c y_c}$ using the parallel-axis theorem and the formulas $\frac{1}{12}bh^3$ and $\frac{1}{12}hb^3$ for the second moments of area about centroidal axes of symmetry of a rectangle.

$$I_{x_c x_c} = \underbrace{\left[(\tfrac{1}{12})(2)(1^3) + (2)(4^2)\right]}_{①} + \underbrace{\left[(\tfrac{1}{12})(1)(8^3) + (8)(\tfrac{1}{12})^2\right]}_{②} + \underbrace{\left[(\tfrac{1}{12})(4)(1^3) + (4)(3^2)\right]}_{③} = \boxed{113.2 \text{ in.}^4}$$

$$I_{y_c y_c} = \underbrace{\left[(\tfrac{1}{12})(1)(2^3) + (2)(2^2)\right]}_{①} + \underbrace{\left[(\tfrac{1}{12})(8)(1^3) + (8)(\tfrac{1}{2})^2\right]}_{②} + \underbrace{\left[(\tfrac{1}{12})(1)(4^3) + (4)(2^2)\right]}_{③} = \boxed{32.67 \text{ in.}^4}$$

Finally, we consider the product of area $I_{x_c y_c}$. Here we must be cautious in using the parallel-axis theorem. Remember that $x_c y_c$ are centroidal axes for the *entire* area of the Z section. In using the parallel-axis theorem for a *subarea*, we must note that $x_c y_c$ are not centroidal axes for the subarea. The centroidal axes to be used in this problem for subareas are the axes of symmetry of each subarea. In short, $x_c y_c$ are simply axes about which we are computing the product of area of each subarea. Therefore, in the parallel-axis theorem, the transfer distances c and d are measured *from the $x_c y_c$ axes to the centroid* in each subarea, as noted in the development of the parallel-axis theorem. The proper sign must be assigned each time to the transfer distances with this in mind. We have for $I_{x_c y_c}$:

$$I_{x_c y_c} = \underbrace{\left[0 + (2)(-2)(4)\right]}_{①} + \underbrace{\left[0 + (8)(-\tfrac{1}{2})(\tfrac{1}{2})\right]}_{②} + \underbrace{\left[0 + (4)(2)(-3)\right]}_{③} = \boxed{-42 \text{ in.}^4}$$

PROBLEMS

7.57. Find I_{xx}, I_{yy}, and I_{xy} for the triangle shown. Give the results in feet.

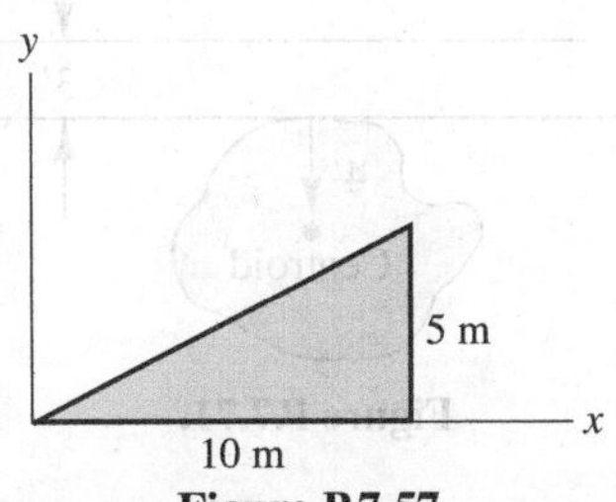

Figure P.7.57.

7.58. What are the second moments and products of area of the ellipse for reference xy? [*Hint:* Can you work with one quadrant and then multiply by 4 for the second moments?]

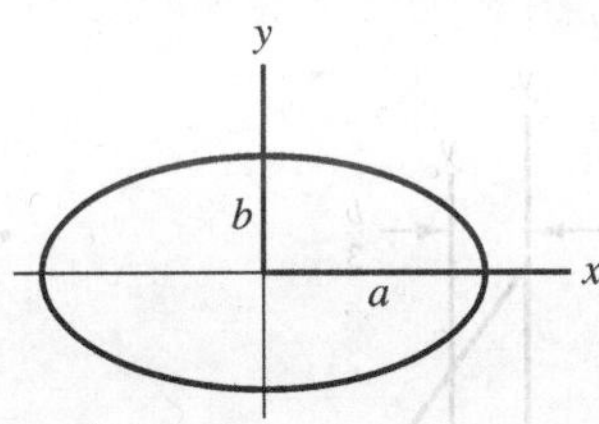

Figure P.7.58.

7.59. Find I_{xx} and I_{yy} for the quarter circle of radius 5 m.

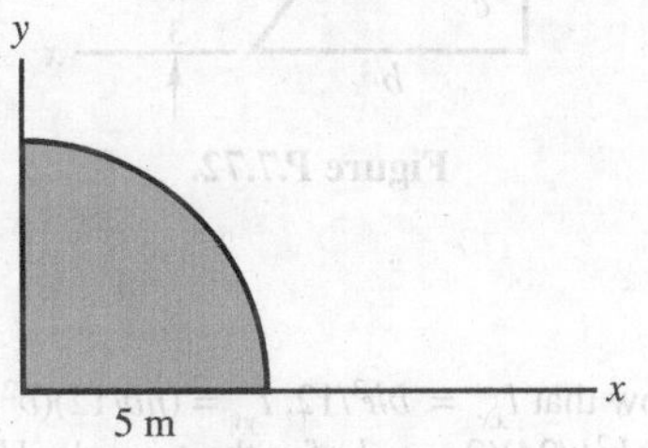

Figure P.7.59.

7.60. Find I_{xx}, I_{yy}, and I_{xy} for the shaded area.

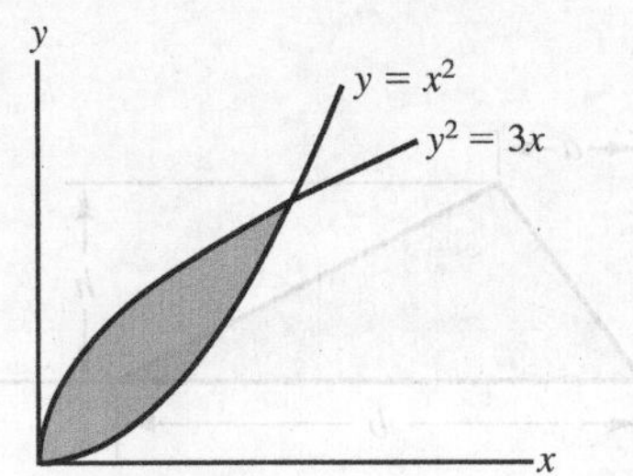

Figure P.7.60.

7.61. Find I_{yy} for the shaded area. You must first determine the constant c.

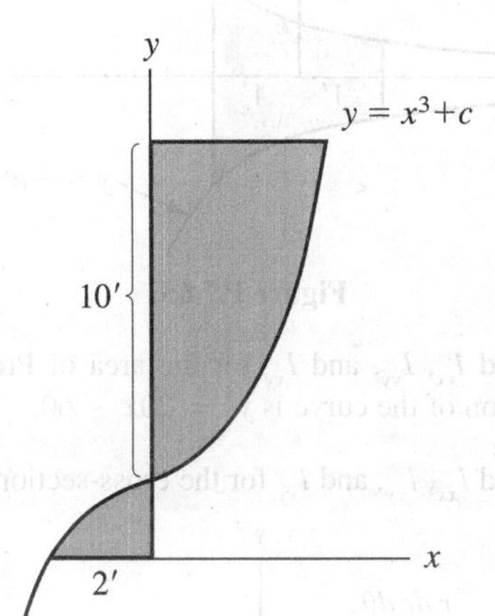

Figure P.7.61.

7.62. Find I_{yy} for the area between the curves

$$y = 2 \sin x \text{ ft}$$
$$y = \sin 2x \text{ ft}$$

from $x = 0$ to $x = \pi$ ft.

7.63. Find I_{yy} for the areas enclosed between curves $y = \cos x$ and $y = \sin x$ and the lines $x = 0$ and $x = \pi/2$.

7.64. Show that $I_{xx} = bh^3/12$, $I_{yy} = b^3h/12$, and $I_{xy} = b^2h^2/24$ for the right triangle.

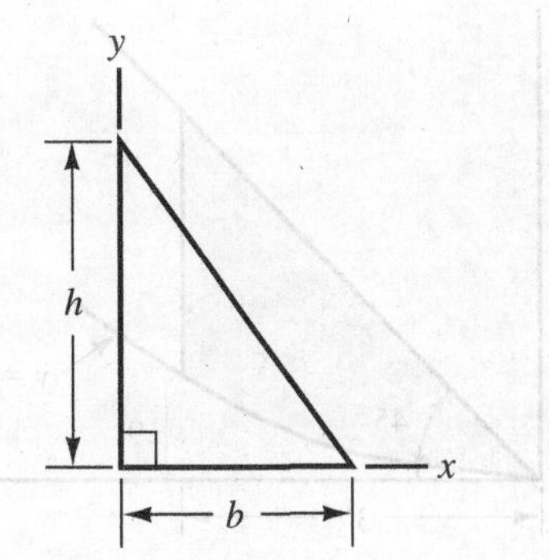

Figure P.7.64.

7.65. Find I_{xx}, I_{yy}, and I_{xy} for the shaded area.

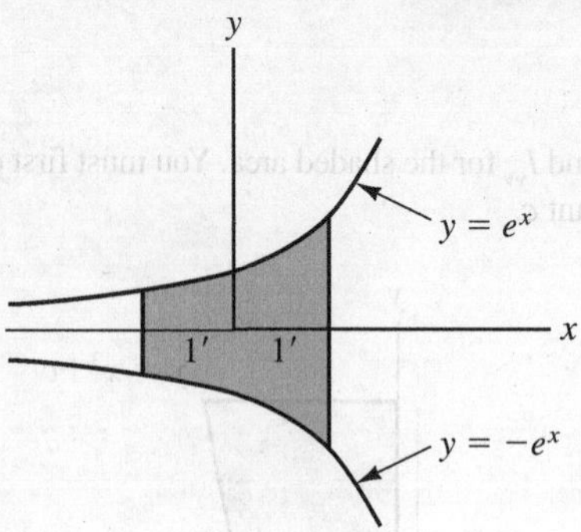

Figure P.7.65.

7.66. Find I_{xx}, I_{yy}, and I_{xy} for the area of Problem 7.4. The equation of the curve is $y^2 = 20x - 60$.

7.67. Find I_{xx}, I_{yy}, and I_{xy} for the cross-section shown.

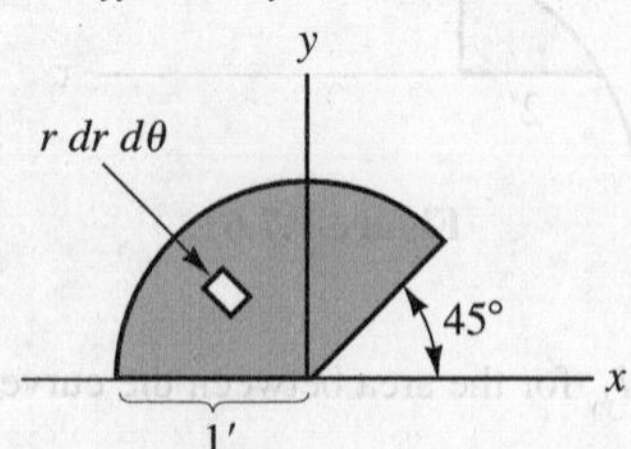

Figure P.7.67.

7.68. Find I_{xx} and I_{yy} for the area of Problem 7.5. The equation of the parabola is $y = (x^2/9) - 5$. [*Hint:* The area of a vertical element in the region below the x axis is $(0 - y)\,dx$.]

7.69. In Problem 7.68 determine I_{xy} using multiple integration.

7.70. Find the second moments of area about axes xy for the shaded area shown.

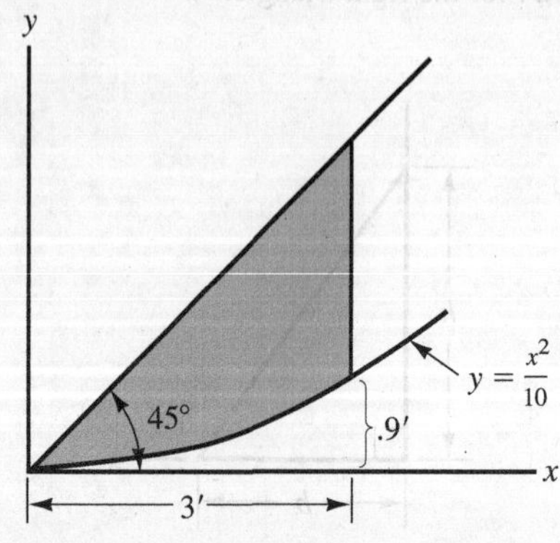

Figure P.7.70.

7.71. If the second moment of area about axis A–A is known to be 600 ft^4, what is the second moment of area about a parallel axis B–B a distance 3 ft from A–A, for an area of 10 ft^2? The centroid of this area is 4 ft from B–B.

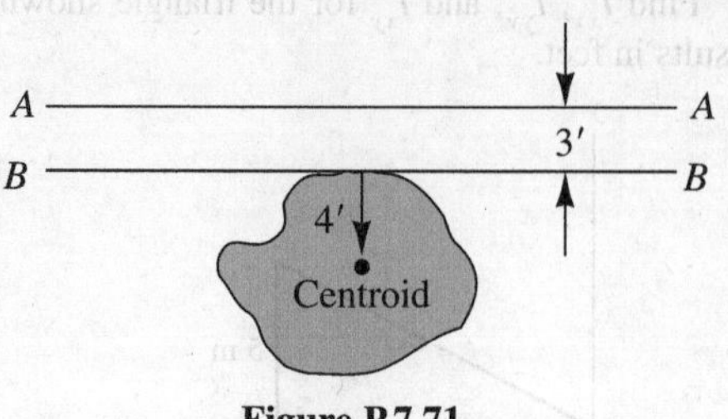

Figure P.7.71.

7.72. Using the results of Problem 7.64, show that $I_{x_cx_c} = bh^3/36$, $I_{y_cy_c} = hb^3/36$, and $I_{x_cy_c} = -b^2h^2/72$ for the right triangle shown.

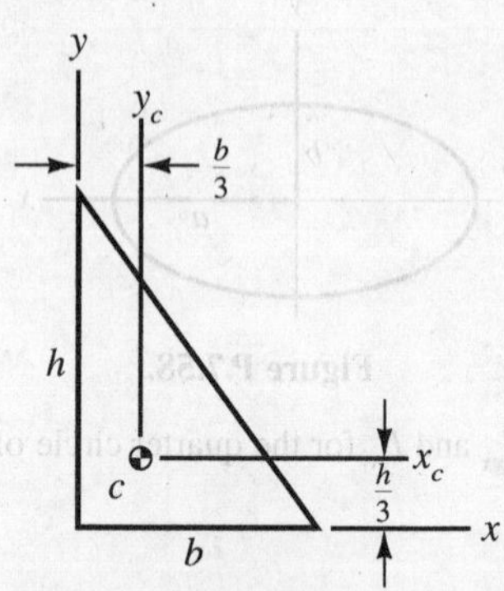

Figure P.7.72.

7.73. Show that $I_{xx} = bh^3/12$, $I_{yy} = (hb/12)(b^2 + ab + a^2)$, and $I_{xy} = (h^2b/24)(2a + b)$ for the triangle. [*Hint:* Break the triangle into two right triangles for which the various moments are known. (See Problem 7.72.)]

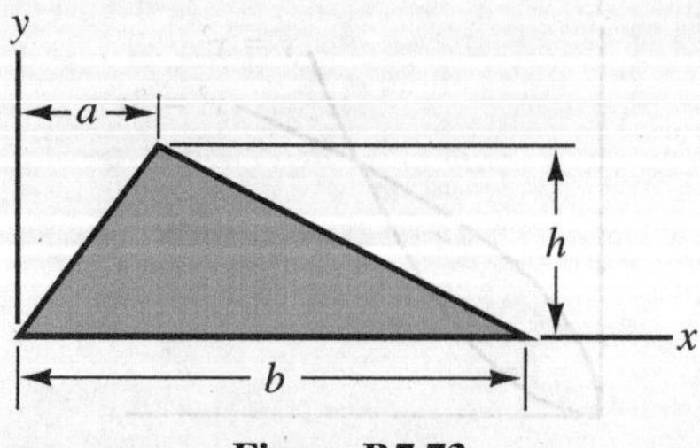

Figure P.7.73.

7.74. In Problem 7.73, show that $I_{x_cx_c} = bh^3/36$, $I_{y_cy_c} = (bh/36)(b2 - ab + a^2)$, and $I_{x_cy_c} = (h^2b/72)(2a - b)$ for the triangle. [*Hint:* Use the results of Problems 7.11 and 7.73 and the parallel-axis theorem.]

7.75. Find I_{xx}, I_{yy}, and I_{xy} of the extruded section. Disregard all rounded edges. Do this problem using 4 areas. Check using 2 areas.

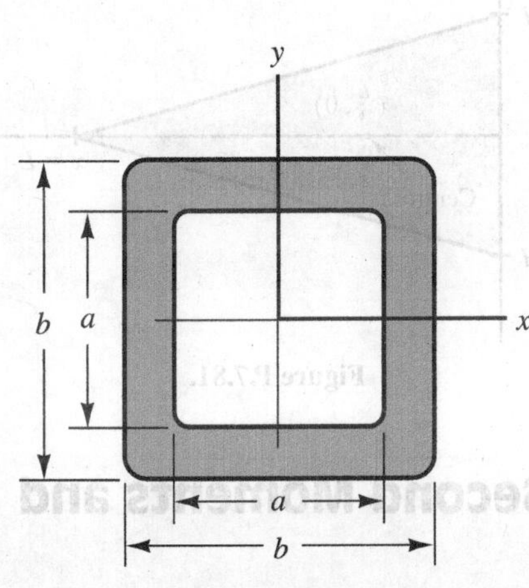

Figure P.7.75.

7.76. Find the second moment of area of the rectangle (with a hole) about the base of the rectangle. Also, determine the product of area about the base and left side.

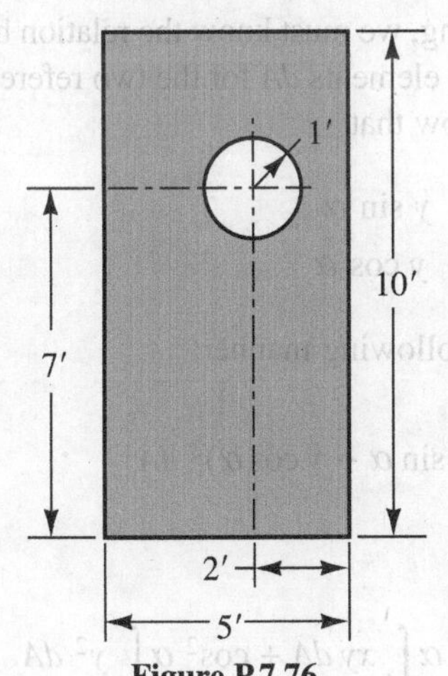

Figure P.7.76.

7.77. Find I_{xx}, I_{yy}, $I_{x_cx_c}$, and $I_{y_cy_c}$ for the structural "hat" section. Disregard all rounded edges.

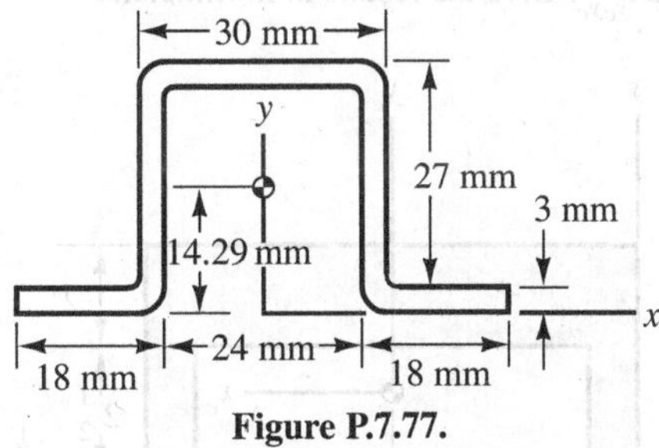

Figure P.7.77.

7.78. Find I_{xx}, I_{yy}, and I_{xy} of the hexagon.

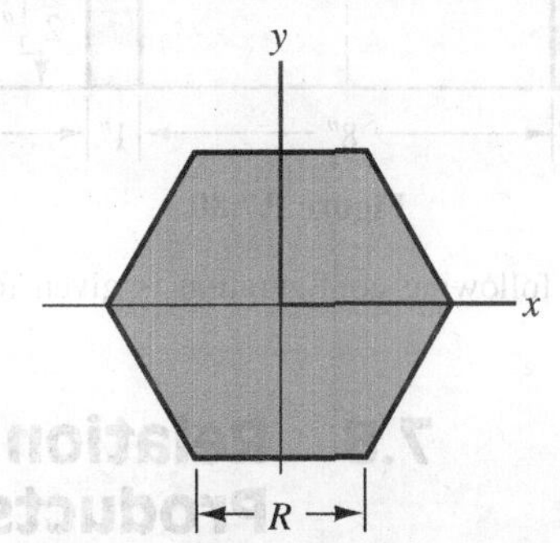

Figure P.7.78.

7.79. A beam cross-section is made up of an I-shaped section with an additional thick plate welded on. Find the second moments of area for the centroidal axes x_cy_c of the beam cross-section. What is $I_{x_cy_c}$? Give the results in millimeters.

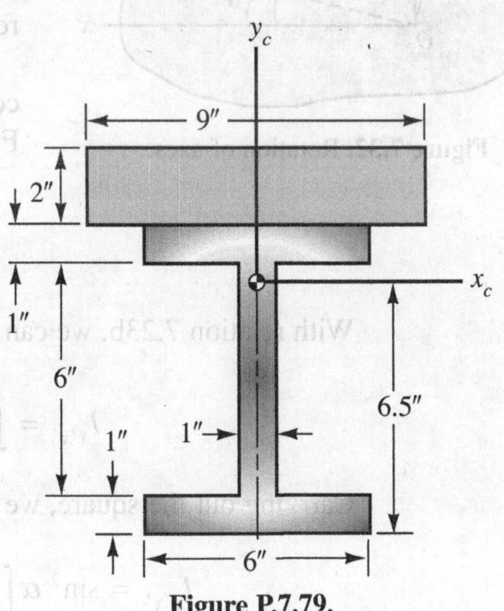

Figure P.7.79.

7.80. Find the second moments of the area shown about centroidal axes parallel to the x and y axes. That is, find $I_{x_c x_c}$ and $I_{y_c y_c}$. Give the results in millimeters.

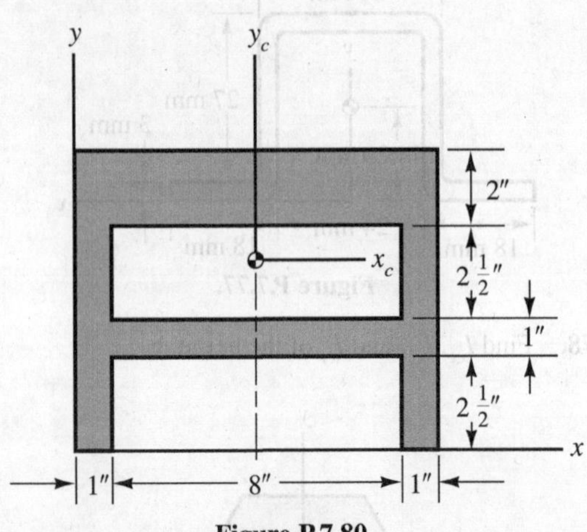

Figure P.7.80.

7.81. The following configuration is given for a plane area.

(a) Find the first moments of area about the x and y axes.

(b) Find the second moments of area for these axes.

(c) Find the product of area for these axes.

(d) What are the radii of gyration, k_x and k_y, for this area?

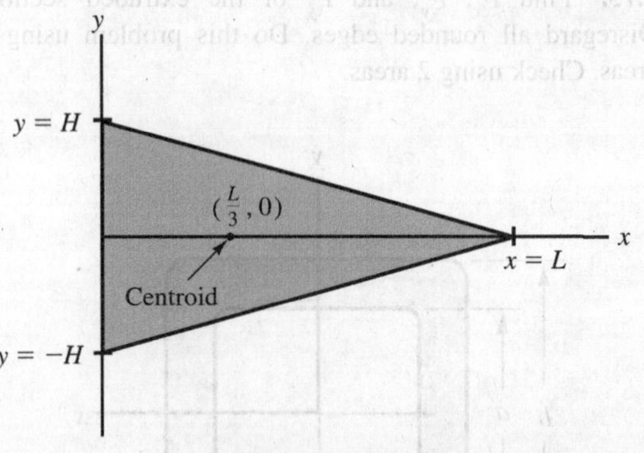

Figure P.7.81.

7.8 Relation Between Second Moments and Products of Area

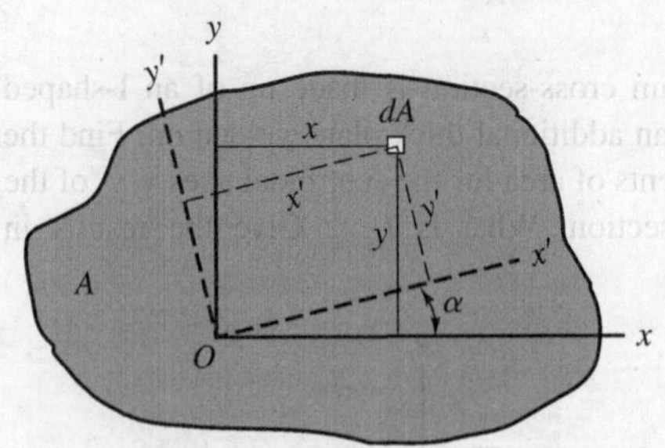

Figure 7.32. Rotation of axes.

We shall now show that we can ascertain second moments and product of area relative to a rotated reference $x'y'$ if we know these quantities for reference xy that has the *same origin.* Such a reference $x'y'$ rotated an angle α from xy (counterclockwise as positive) is shown in Fig. 7.32. We shall assume that the second moments and product of area for the unprimed reference are known.

Before proceeding, we must know the relation between the coordinates of the area elements dA for the two references. From Fig. 7.32, you may show that

$$x' = x \cos \alpha + y \sin \alpha \tag{7.23a}$$

$$y' = -x \sin \alpha + y \cos \alpha \tag{7.23b}$$

With relation 7.23b, we can express $I_{x'x'}$ in the following manner:

$$I_{x'x'} = \int_A (y')^2 \, dA = \int_A (-x \sin \alpha + y \cos \alpha)^2 \, dA \tag{7.24}$$

Carrying out the square, we have

$$I_{x'x'} = \sin^2 \alpha \int_A x^2 \, dA - 2 \sin \alpha \cos \alpha \int_A xy \, dA + \cos^2 \alpha \int_A y^2 \, dA$$

Therefore,

$$I_{x'x'} = I_{yy} \sin^2 \alpha + I_{xx} \cos^2 \alpha - 2I_{xy} \sin \alpha \cos \alpha \tag{7.25}$$

A more common form of the desired relation can be formed by using the following trigonometric identities:

$$\cos^2 \alpha = \tfrac{1}{2}(1 + \cos 2\alpha) \tag{a}$$

$$\sin^2 \alpha = \tfrac{1}{2}(1 - \cos 2\alpha) \tag{b}$$

$$2 \sin \alpha \cos \alpha = \sin 2\alpha \tag{c}$$

We then have[6]

$$I_{x'x'} = \frac{I_{xx} + I_{yy}}{2} + \frac{I_{xx} - I_{yy}}{2} \cos 2\alpha - I_{xy} \sin 2\alpha \tag{7.26}$$

To determine $I_{y'y'}$, we need only replace the α in the preceding result by $(\alpha + \pi/2)$. Thus,

$$I_{y'y'} = \frac{I_{xx} + I_{yy}}{2} + \frac{I_{xx} - I_{yy}}{2} \cos(2\alpha + \pi) - I_{xy} \sin(2\alpha + \pi)$$

Note that $\cos(2\alpha + \pi) = -\cos 2\alpha$ and $\sin(2\alpha + \pi) = -\sin 2\alpha$. Hence, the equation above becomes

$$I_{y'y'} = \frac{I_{xx} + I_{yy}}{2} - \frac{I_{xx} - I_{yy}}{2} \cos 2\alpha + I_{xy} \sin 2\alpha \tag{7.27}$$

Next, the product of area $I_{x'y'}$ can be computed in a similar manner:

$$I_{x'y'} = \int_A x'y' \, dA = \int_A (x \cos \alpha + y \sin \alpha)(-x \sin \alpha + y \cos \alpha) \, dA$$

This becomes

$$I_{x'y'} = \sin \alpha \cos \alpha \, (I_{xx} - I_{yy}) + (\cos^2 \alpha - \sin^2 \alpha) I_{xy}$$

Utilizing the previously defined trigonometric identities, we get

$$I_{x'y'} = \frac{I_{xx} - I_{yy}}{2} \sin 2\alpha + I_{xy} \cos 2\alpha \tag{7.28}$$

Thus, we see that, if we know the quantities I_{xx}, I_{yy}, and I_{xy} for some reference xy at point O, the second moments and products of area for every set of axes at point O can be computed. And if, in addition, we employ the transfer theorems, we can compute second moments and products of area for *any* reference in the plane of the area.

[6]Equations 7.26, 7.27, and 7.28 are called *transformation* equations. They will appear in the next chapter and in your upcoming solid mechanics course for variables other than second moments and products of area. In the remaining portions of this chapter, you will see that a number of important properties of second moments and products of area are deducible *directly* from these transformation equations. This primarily accounts for the importance of the transformation equations. Chapter 8 will give you additional insight into this topic.

Example 7.10

Find $I_{x'x'}$, $I_{y'y'}$ and $I_{x'y'}$ for the cross section of the beam shown in Fig. 7.33. The origin of $x'y'$ is at the centroid of the cross-section.

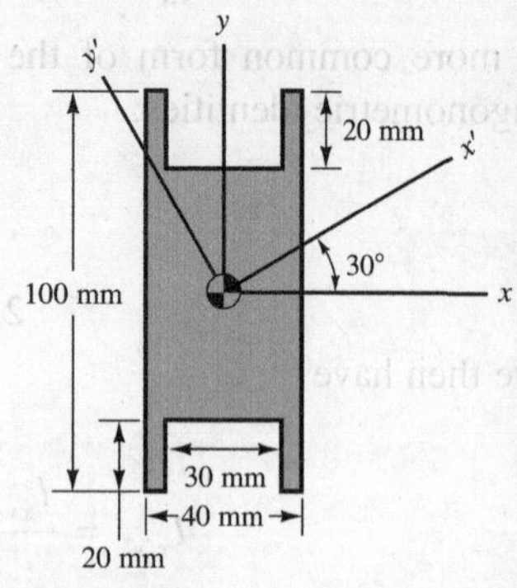

Figure 7.33. A composite area.

We shall consider three rectangles labeled in Fig. 7.34. The dimensions of these rectangles are listed below.

Rect. (1) 100 mm × 40 mm (outside outline)
Rect. (2) 20 mm × 30 mm
Rect. (3) 20 mm × 30 mm

We shall find I_{xx}, I_{yy}, and I_{xy} as a first step. Thus

$$I_{xx} = \left(\tfrac{1}{12}\right)(40)(100)^3 - 2\left[\tfrac{1}{12}(30)(20)^3 + (20)(30)(40)^2\right] = 1.373 \times 10^6 \text{ mm}^4$$

$$I_{yy} = \left(\tfrac{1}{12}\right)(100)(40)^3 - 2\left[\tfrac{1}{12}(20)(30)^3\right] = 4.43 \times 10^5 \text{ mm}^4$$

$$I_{xy} = 0 \text{ (symmetry)}$$

Hence, we can now go to the transformation equations for the desired information.[7]

$$I_{x'x'} = \frac{1.373 \times 10^6 + 4.43 \times 10^5}{2} + \frac{1.373 \times 10^6 - 4.43 \times 10^5}{2}\cos 60° + 0$$

$$= 1.141 \times 10^6 \text{ mm}^4$$

$$I_{y'y'} = \frac{1.373 \times 10^6 + 4.43 \times 10^5}{2} - \frac{1.373 \times 10^6 - 4.43 \times 10^5}{2}\cos 60° + 0$$

$$= 6.76 \times 10^5 \text{ mm}^4$$

$$I_{x'y'} = \frac{1.373 \times 10^6 - 4.43 \times 10^5}{2}\sin 60° = 4.03 \times 10^5 \text{ mm}^4$$

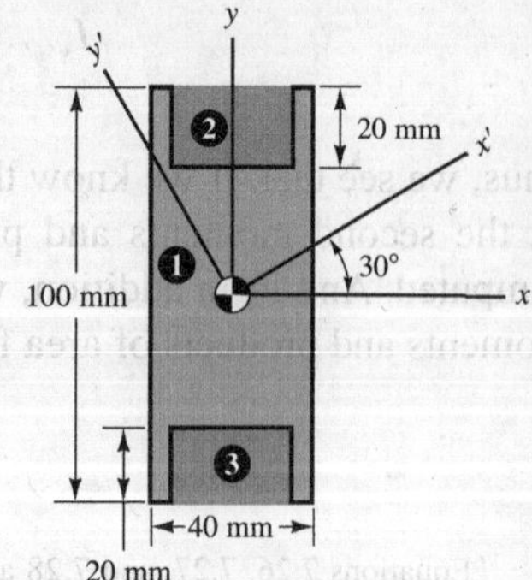

Figure 7.34. Composite area with three subareas.

[7]It will be helpful to note that in going from the computation of $I_{x'x'}$ to the computation of $I_{y'y'}$, one needs only to change the sign associated with the second and third expressions for $I_{x'x'}$.

7.9 Polar Moment of Area

In the previous section, we saw that the second moments and product of area for an orthogonal reference determined all such quantities for *any* orthogonal reference having the same origin. We shall now show that the sum of the pairs of second moments of area is a constant for all such references at a point. Thus, in Fig. 7.35 we have a reference *xy* associated with point *a*. Summing I_{xx} and I_{yy} we have

$$\begin{aligned} I_{xx} + I_{yy} &= \int_A y^2\, dA + \int_A x^2\, dA \\ &= \int_A (x^2 + y^2)\, dA = \int_A r^2\, dA \end{aligned}$$

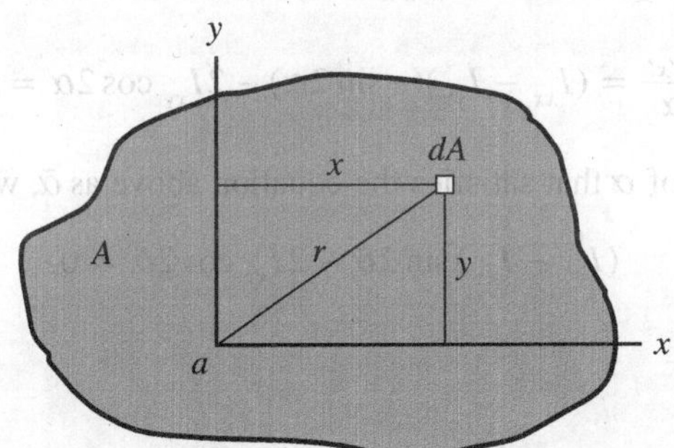

Figure 7.35. $J = I_{xx} + I_{yy}$.

Since r^2 is independent of the orientation of the coordinate system, the sum $I_{xx} + I_{yy}$ is independent of the orientation of the reference. Therefore, the sum of second moments of area about orthogonal axes is a function only of the position of the origin *a* for the axes. This sum is termed the *polar moment of area*, *J*.[8] We can then consider *J* to be a scalar field. Mathematically, this statement is expressed as

$$J = J(x', y') \tag{7.29}$$

where x' and y' are the coordinates as measured from some convenient reference $x'y'$ for the point of interest.

That the quantity $(I_{xx} + I_{yy})$ does not change on rotation of axes can also be deduced by summing transformation equations 7.26 and 7.27 as we suggest you do. This group of terms is accordingly termed an *invariant*. Parenthetically, we can similarly show that $(I_{xx}I_{yy} - I_{xy}^2)$ is also invariant under a rotation of axes.

[8]Quite often I_p is used for the polar moment of area.

7.10 Principal Axes

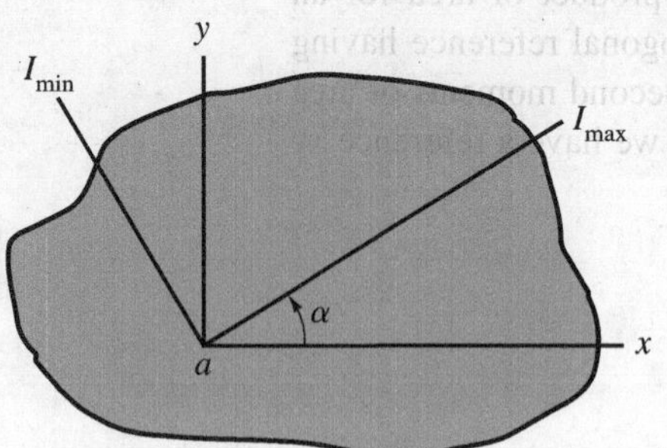

Figure 7.36. Principal axes.

Still other conclusions may be drawn about second moments and products of area associated with a point in an area. In Fig. 7.36 an area is shown with a reference *xy* having its origin at point *a*. We shall assume that I_{xx}, I_{yy}, and I_{xy} are known for this reference, and shall ask at what angle α we shall find an axis having the *maximum* second moment of area. Since the sum of the second moments of area is constant for any reference with origin at *a*, the *minimum* second moment of area must then correspond to an axis at *right angles* to the axis having the maximum second moment. Since second moments of area have been expressed in Eqs. 7.26 and 7.27 as functions of the variable α at a point, these extremes may readily be determined by setting the partial derivative of $I_{x'x'}$ with respect to α equal to zero. Thus,

$$\frac{\partial I_{x'x'}}{\partial \alpha} = (I_{xx} - I_{yy})(-\sin 2\alpha) - 2I_{xy}\cos 2\alpha = 0$$

If we denote the value of α that satisfies the equation above as $\tilde{\alpha}$, we have

$$(I_{yy} - I_{xx})\sin 2\tilde{\alpha} - 2I_{xy}\cos 2\tilde{\alpha} = 0$$

Hence,

$$\tan 2\tilde{\alpha} = \frac{2I_{xy}}{I_{yy} - I_{xx}} \tag{7.30}$$

This formulation gives us the angle $\tilde{\alpha}$, which corresponds to an extreme value of $I_{x'x'}$ (i.e., to a maximum or minimum value). Actually, there are two possible values of $2\tilde{\alpha}$ which are π radians apart that will satisfy the equation above. Thus,

$$2\tilde{\alpha} = \beta \quad \text{where } \beta = \tan^{-1}\frac{2I_{xy}}{I_{yy} - I_{xx}}$$

or

$$2\tilde{\alpha} = \beta + \pi$$

This means that we have two values of $\tilde{\alpha}$, given as

$$\tilde{\alpha}_1 = \frac{\beta}{2}, \qquad \tilde{\alpha}_2 = \frac{\beta}{2} + \frac{\pi}{2}$$

Thus, there are two axes orthogonal to each other having extreme values for the second moment of area at *a*. On one of these axes is the maximum second moment of area and, as pointed out earlier, the minimum second moment of area must appear on the other axis. These axes are called the *principal axes*.

Let us now substitute the angle $\tilde{\alpha}$ into Eq. 7.28 for $I_{x'y'}$:

$$I_{x'y'} = \frac{I_{xx} - I_{yy}}{2} \sin 2\tilde{\alpha} + I_{xy} \cos 2\tilde{\alpha} \tag{7.31}$$

If we now form a right triangle with legs $2I_{xy}$ and $(I_{yy} - I_{xx})$ and angle $2\tilde{\alpha}$ such that Eq. (7.30) is satisfied we can readily express the sine and cosine expressions needed in the preceding equation. Thus

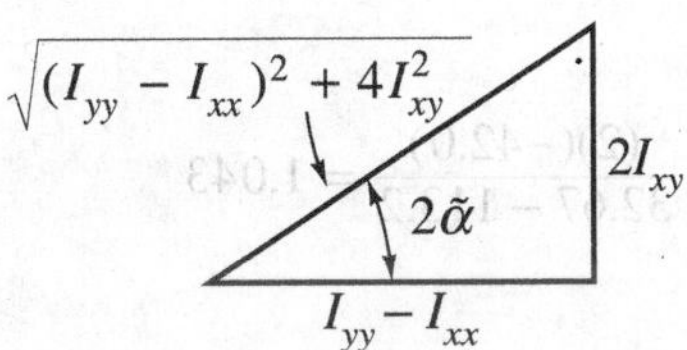

$$\sin 2\tilde{\alpha} = \frac{2I_{xy}}{\sqrt{\left(I_{yy} - I_{xx}\right)^2 + 4I_{xy}^2}}$$

$$\cos 2\tilde{\alpha} = \frac{I_{yy} - I_{xx}}{\sqrt{\left(I_{yy} - I_{xx}\right)^2 + 4I_{xy}^2}}$$

By substituting these results into Eq. 7.31, we get

$$I_{x'y'} = -\left(I_{yy} - I_{xx}\right)\frac{I_{xy}}{\left[\left(I_{yy} - I_{xx}\right)^2 + 4I_{xy}^2\right]^{1/2}} + I_{xy}\frac{I_{yy} - I_{xx}}{\left[\left(I_{yy} - I_{xx}\right)^2 + 4I_{xy}^2\right]^{1/2}}$$

Hence,

$$I_{x'y'} = 0$$

Thus, we see that the *product of area corresponding to the principal axes is zero.* If we set $I_{x'y'}$ equal to zero in Eq. 7.28, you can demonstrate the converse of the preceding statement by solving for α and comparing the result with Eq. 7.30. That is, if the product of area is zero for a set of axes at a point, these axes must be the principal axes at that point. Consequently, if one axis of a set of axes at a point is symmetrical for the area, the axes are principal axes at that point.

The concept of principal axes will appear again in the following chapter in connection with the inertia tensor. Thus, the concept is not an isolated occurrence but is characteristic of a whole family of quantities. We shall, then, have further occasion to examine some of the topics introduced in this chapter from a more general viewpoint.

Example 7.11

Find the principal second moments of area at the centroid of the Z section of Example 7.9.

We have from this example the following results that will be of use to us:

$$I_{x_c x_c} = 113.2 \text{ in.}^4$$
$$I_{y_c y_c} = 32.67 \text{ in.}^4$$
$$I_{x_c y_c} = -42.0 \text{ in.}^4$$

Hence, we have

$$\tan 2\tilde{\alpha} = \frac{2I_{x_c y_c}}{I_{y_c y_c} - I_{x_c x_c}} = \frac{(2)(-42.0)}{32.67 - 113.2} = 1.043$$

$$2\tilde{\alpha} = 46.21°;\ 226.2°$$

For $2\tilde{\alpha} = 46.21°$:

$$I_1 = \frac{113.2 + 32.67}{2} + \frac{113.2 - 32.67}{2}\cos(46.21°) - (-42)\sin 46.21°$$
$$= 72.9 + 27.9 + 30.3 = \boxed{131.1 \text{ in.}^4}$$

For $2\tilde{\alpha} = 226.2°$:

$$I_2 = 72.9 - 27.9 - 30.3 = \boxed{14.75 \text{ in.}^4}$$

As a check on our work, we note that the sum of the second moments of area are invariant at a point for a rotation of axes. This means that

$$I_{x_c x_c} + I_{y_c y_c} = I_1 + I_2$$
$$113.2 + 32.7 = 131.1 + 14.75$$

Therefore,

$$145.9 = 145.9$$

We thus have a check on our work.

Before closing, we wish to point out that there is a graphical construction called *Mohr's circle* relating second moments and products of area for all possible axes at a point. However, in this text we shall use the analytical relations thus far presented rather than Mohr's circle. You will see Mohr circle construction in your strength of materials course where its use in conjunction with the important topics of plane stress and plane strain is very helpful.[9]

[9]See I. H. Shames, *Introduction to Solid Mechanics*, Second Edition, Prentice-Hall, Inc., Englewood Cliffs, N.J., 1989.

PROBLEMS

7.82. It is known that area A is 10 ft^2 and has the following moments and products of area for the centroidal axes shown:

$$I_{xx} = 40 \text{ ft}^4, \quad I_{yy} = 20 \text{ ft}^4, \quad I_{xy} = -4 \text{ ft}^4$$

Find the moments and products of area for the $x'y'$ reference at point a.

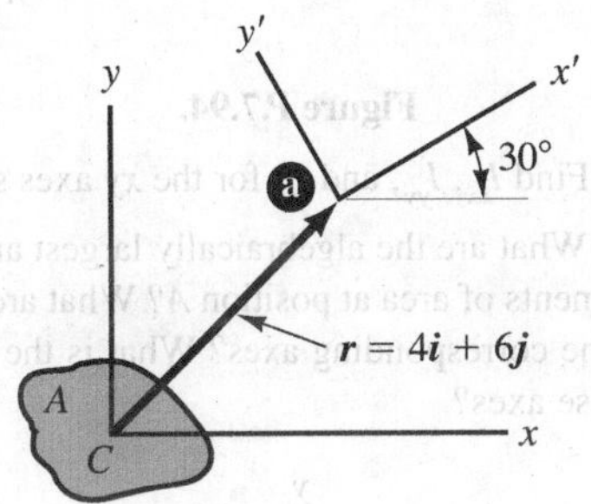

Figure P.7.82.

7.83. The cross-section of a beam is shown. Compute $I_{x'x'}$, $I_{y'y'}$, and $I_{x'y'}$ in the simplest way without using formulas for second moments and products of area for a triangle.

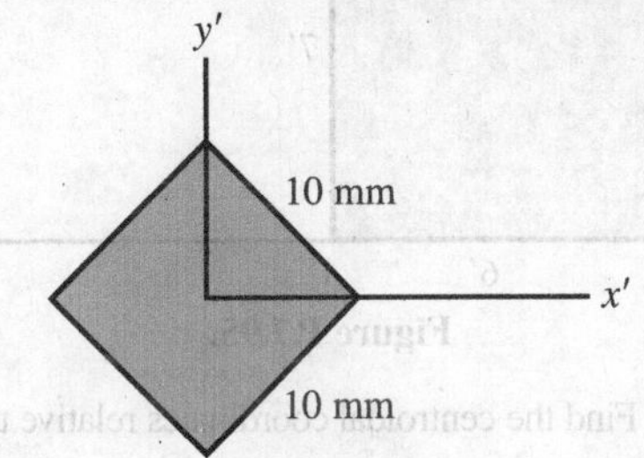

Figure P.7.83.

7.84. Find I_{xx}, I_{yy}, and I_{xy} for the rectangle. Also, compute the polar moment of area at points a and b.

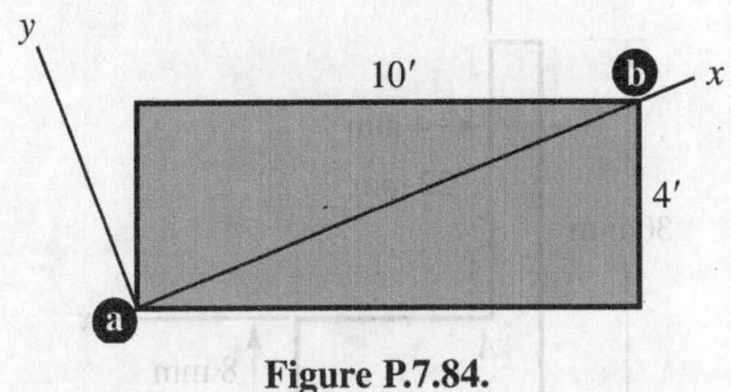

Figure P.7.84.

7.85. Express the polar moment of area of the square as a function of x, y, the coordinates of points about which the polar moment is taken.

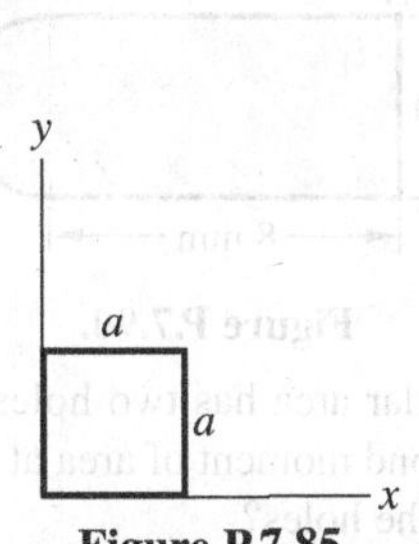

Figure P.7.85.

7.86. Use the calculus to show that the polar moment of area of a circular area of radius r is $\pi r^4/2$ at the center.

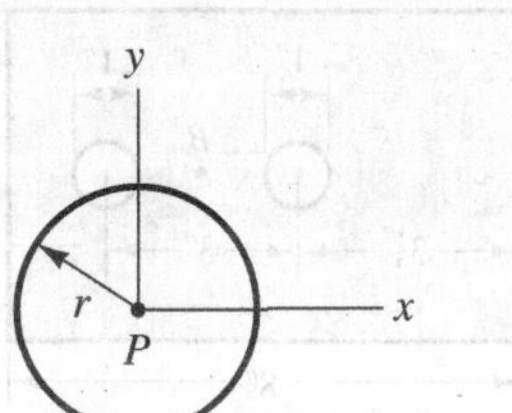

Figure P.7.86.

7.87. Find the direction of the principal axes for the angle section at point A.

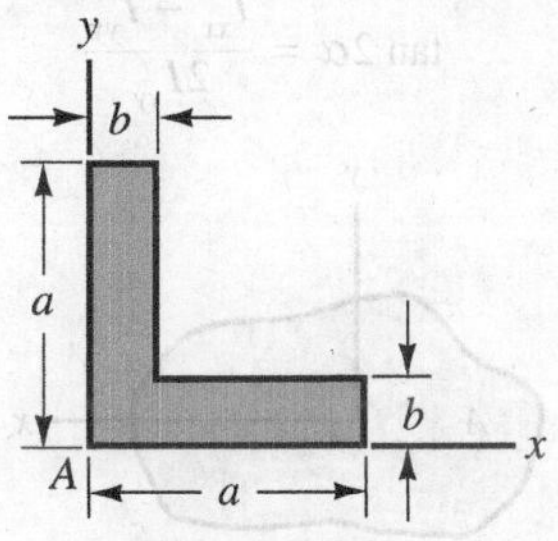

Figure P.7.87.

7.88. What are the principal second moments of area at the origin for the area of Example 7.7?

7.89. Find the principal second moments of area at the centroid for the area shown.

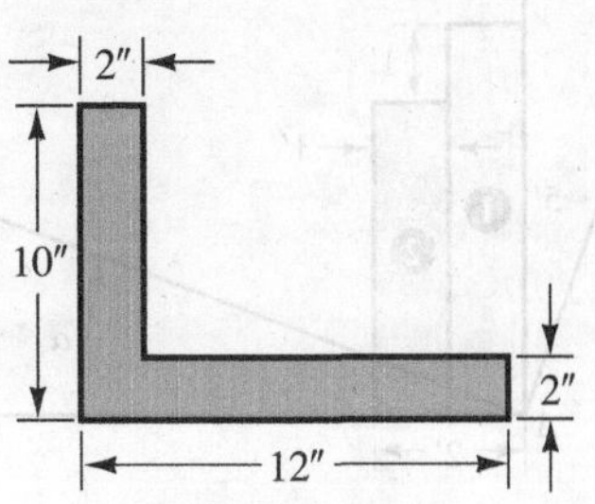

Figure P.7.89.

7.90. Determine the principal second moments of area at point A.

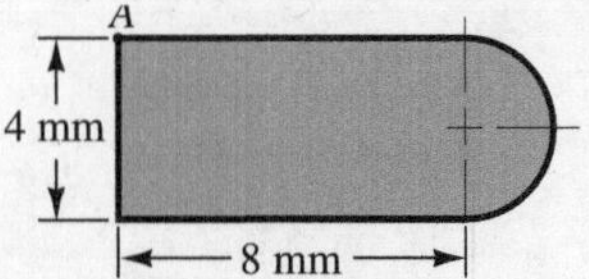

Figure P.7.90.

7.91. A rectangular area has two holes cut out. What is the maximum second moment of area at A? What is it at B halfway between the holes?

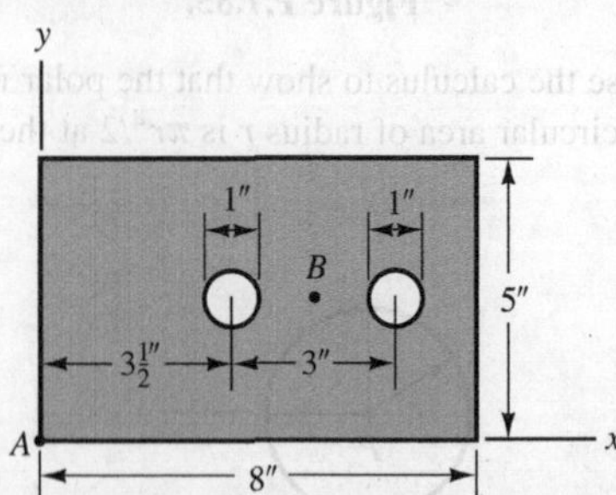

Figure P.7.91.

7.92. Show that the axes for which the product of area is a maximum are rotated from xy by an angle α so that

$$\tan 2\alpha = \frac{I_{xx} - I_{yy}}{2I_{xy}}$$

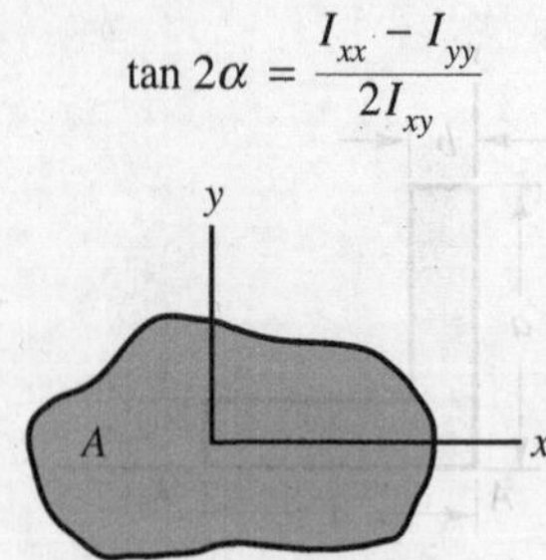

Figure P.7.92.

7.93. What is the value of the angle α for the principal axes at A.

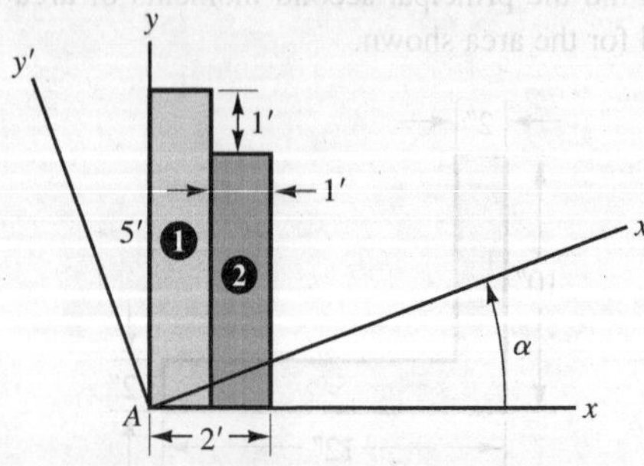

Figure P.7.93.

7.94. Find the principal second moments of area at A.

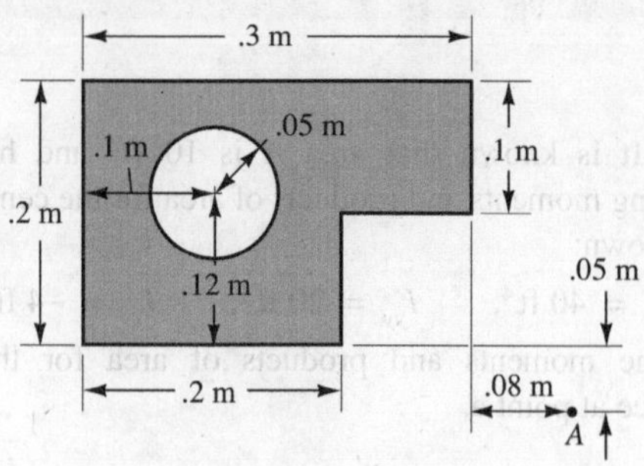

Figure P.7.94.

7.95. (a) Find I_{xx}, I_{yy}, and I_{xy} for the xy axes shown.

(b) What are the algebraically largest and smallest second moments of area at position A? What are the orientations of the corresponding axes? What is the product of area for these axes?

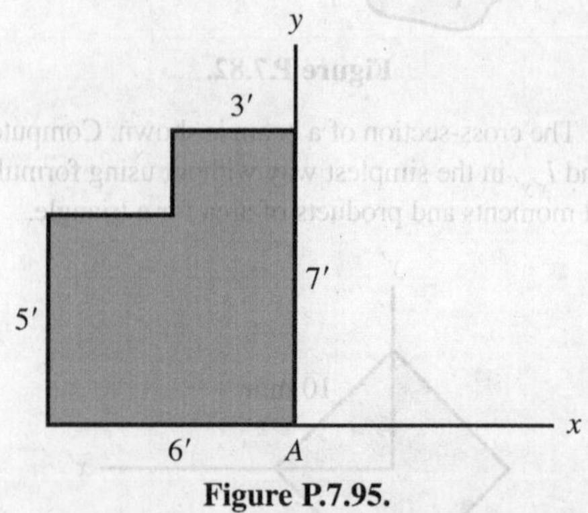

Figure P.7.95.

7.96. (a) Find the centroidal coordinates relative to axes xy.

(b) Find the principal axes and the corresponding maximum and minimum second moments of area at A.

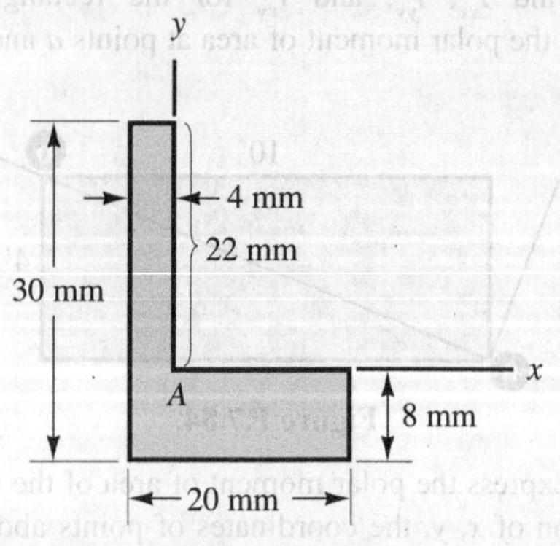

Figure P.7.96.

7.11 Closure

In this chapter, we discussed primarily the first and second moments of plane areas as well as the product of plane areas. These formulations give certain kinds of evaluations of the distribution of area relative to a plane reference *xy*. You will most certainly make much use of these quantities in your later courses in strength of materials.

In this chapter, we have touched on subject matter that you will encounter in the next chapter and also, most assuredly, in later courses. Specifically, in Chapter 8 you will be introduced to the so-called second-order inertia tensor having nine terms which change (or transform) in a certain particular way when we rotate coordinate axes at a point. These particular transformation equations define the inertia tensor. Any other set of nine symmetric terms that transform via the same form of equations, are symmetric second-order tensors. You will also learn that the second moments and products of *area* form a two-dimensional simplification of the inertia tensor. The transformation equations 7.26 through 7.28 are thus special cases of the three- dimensional defining equations of the inertia tensor. Take note that certain vital results emerged from these simplified two-dimensional transformation equations. They included

1. Invariant property at a point for the sum $(I_{xx} + I_{yy})$ on rotation of axes.
2. Principal axes and principal moments of area at a point.
3. A graphical construction called Mohr's circle that depicts the transformation equations. This topic has been omitted in this chapter but will be described in your solids course where you will study the two-dimensional simplifications of the stress and strain tensors. At that time one can make much use of the Mohr's circle construction and it is a simple matter then to describe Mohr's circle for second moments and products of area.

We will emphasize in the next chapter that there are many symmetric sets of nine terms that transform exactly like the terms of the inertia tensor and we classify all of them as second-order tensors. Identifying these sets as second-order tensors immediately yields vital properties common to all of them such as those presented above for the two-dimensional moments and products of area.

PROBLEMS

7.97. Find the position of the centroid of the shaded area under the curve $y = \sin^2 x$ m. Find $M_{x'}$ and $M_{y'}$ of this area.

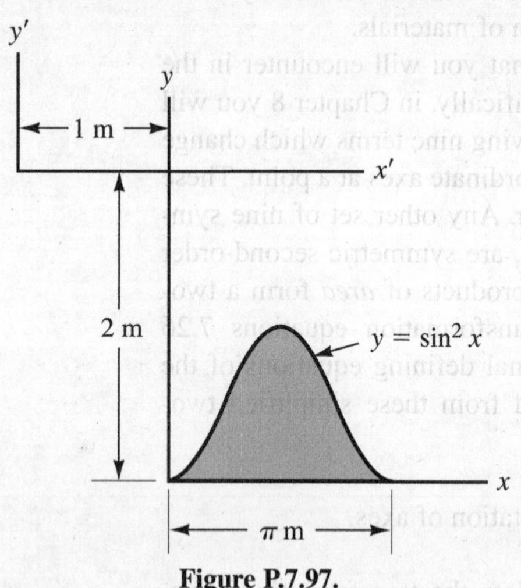

Figure P.7.97.

7.98. Find the center of volume of the body of revolution with a cylindrical cavity.

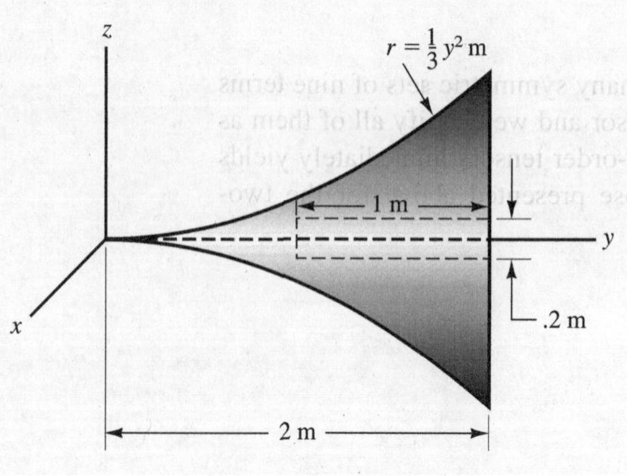

Figure P.7.98.

7.99. Locate the center of volume, center of mass, and center of gravity of the wooden rectangular block and the plastic semicylinder. The wood weighs .0003 N/mm³ and the plastic weighs .0005 N/mm³.

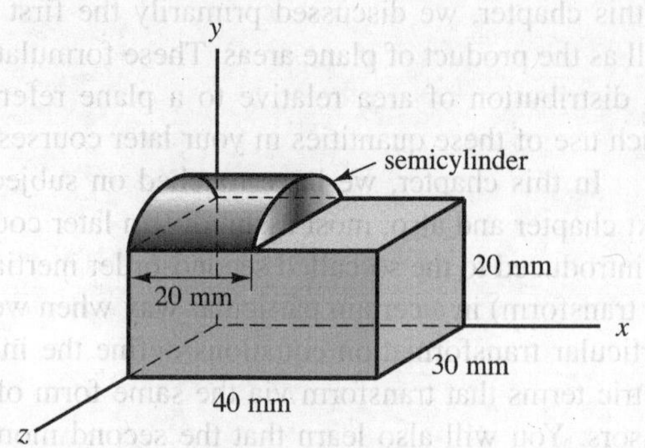

Figure P.7.99.

7.100. Find I_{xx}, I_{yy}, I_{xy}, $I_{x_cx_c}$, $I_{y_cy_c}$, and $I_{x_cy_c}$ of the unequal-leg rolled channel section.

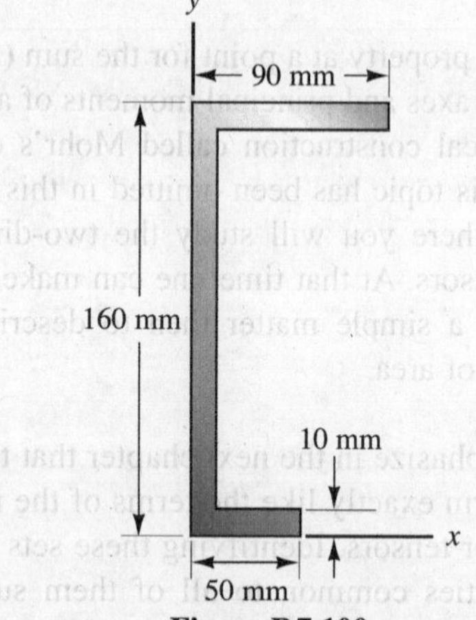

Figure P.7.100.

7.101. Find the centroid of the half cone. Use the known formula for the volume of a half cone, $V = (1/6)\,\pi r^2 h$.

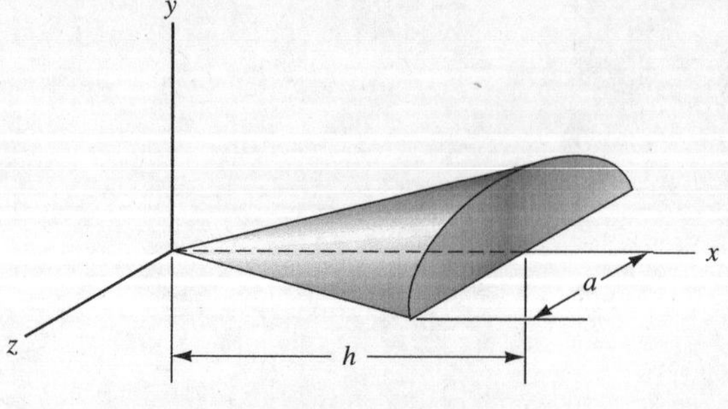

Figure P.7.101.

7.102. A half body of revolution is shown with the xz plane as a plane of symmetry. Determine the centroidal coordinates. The radius at any section x varies as the square of x.

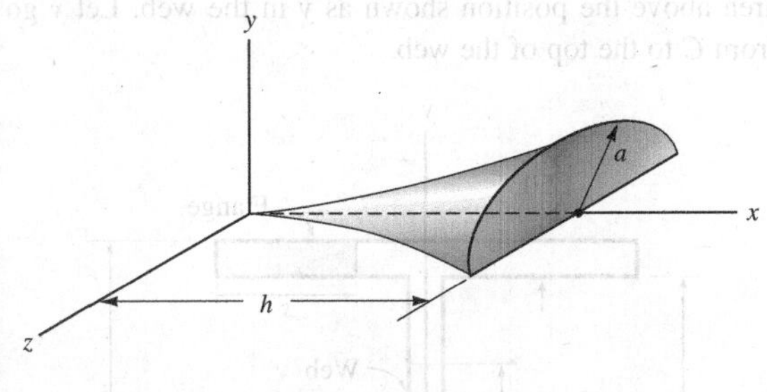

Figure P.7.102.

7.103. (a) Find I_{xx}, I_{yy}, and I_{xy} for the xy axes at position A.

(b) Find the principal second moments of area at point A.

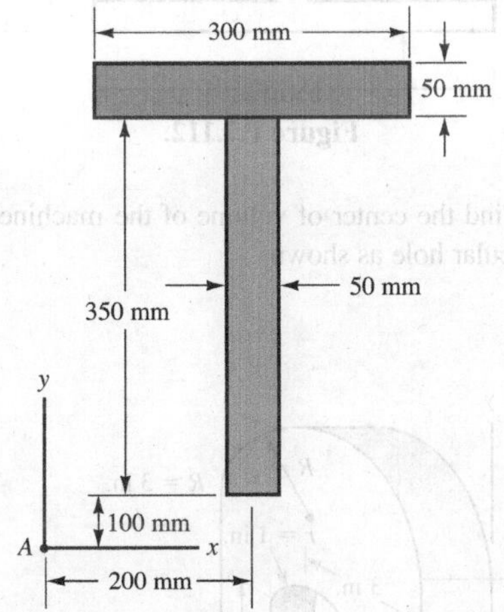

Figure P.7.103.

7.104. What are the directions of the principal axes at point A?

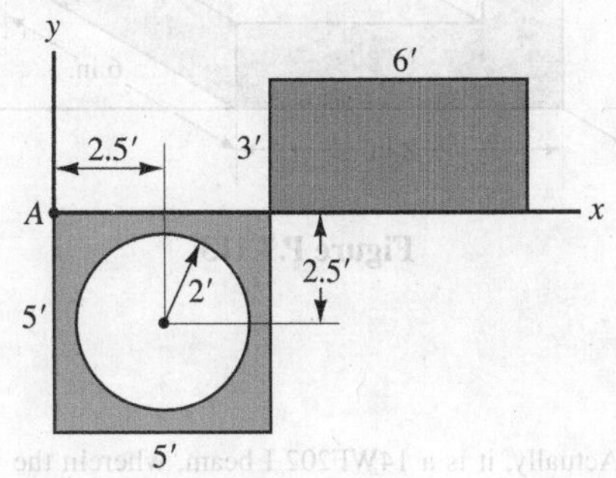

Figure P.7.104.

7.105. Using the theorems of Pappus and Guldinus, find the centroid of the area of a quarter-circle.

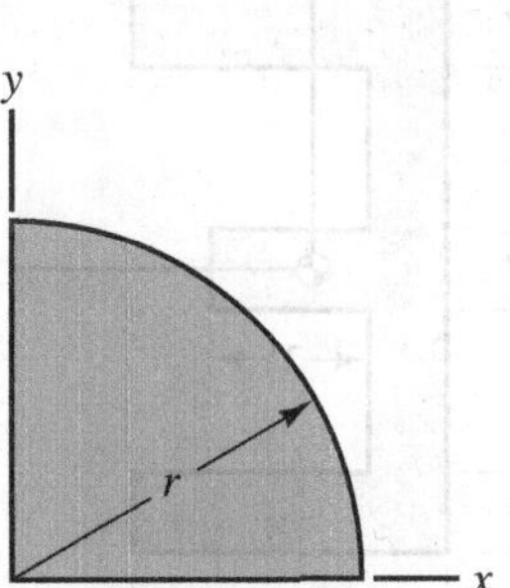

Figure P.7.105.

7.106. A tank has a semispherical dome at the left end. Using the theorems of Pappus and Guldinus, compute the surface and volume of the tank. Give the results in meters.

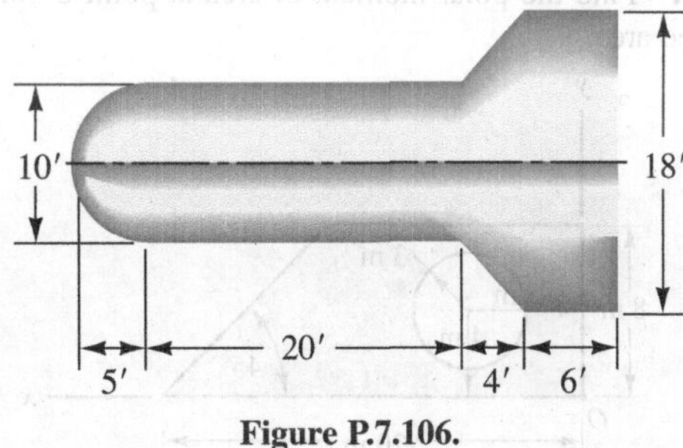

Figure P.7.106.

7.107. Find $I_{x'x'}$, $I_{y'y'}$, and $I_{x'y'}$ for the set of axes at point A for the rectangular area.

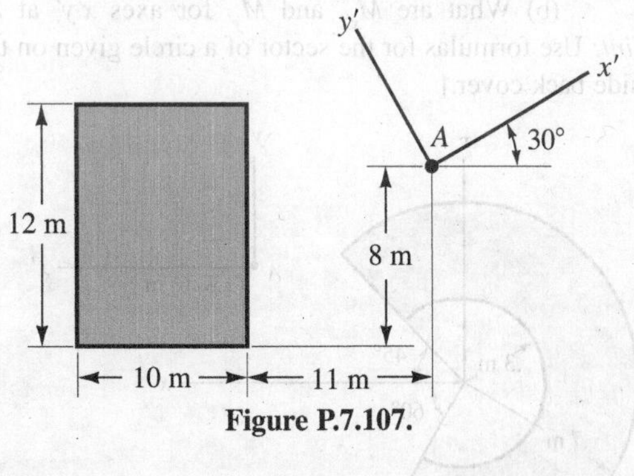

Figure P.7.107.

7.108. Find the centroid of the area, and then find the second moments of the area about centroidal axes parallel to the sides of the area.

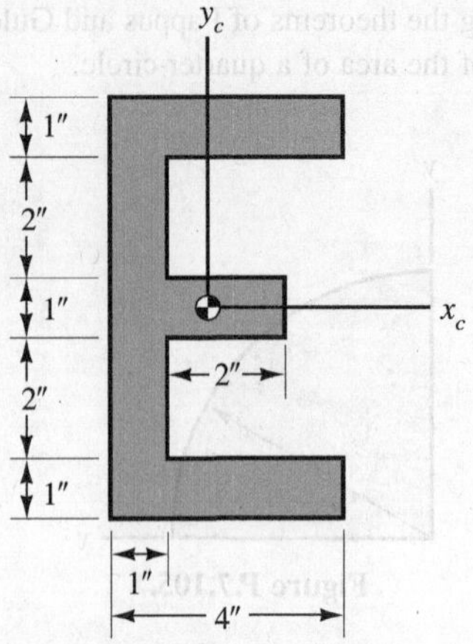

Figure P.7.108.

7.109. Find the principal second moments of area at a point where $I_{xy} = 321 \text{ in}^4$, $I_{xx} = 118.4 \text{ in}^4$, and $I_{yy} = 1{,}028 \text{ in}^4$.

7.110. Find the polar moment of area at point O for the shaded area.

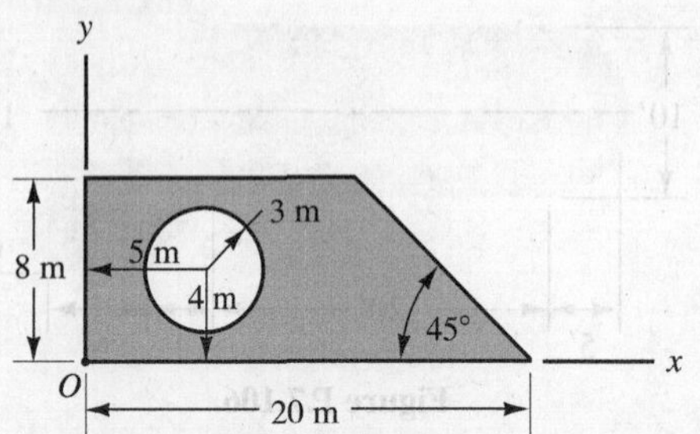

Figure P.7.110.

7.111. (a) What are the centroidal coordinates of the shaded area?

(b) What are $M_{x'}$, and $M_{y'}$ for axes $x'y'$ at A? [*Hint:* Use formulas for the sector of a circle given on the inside back cover.]

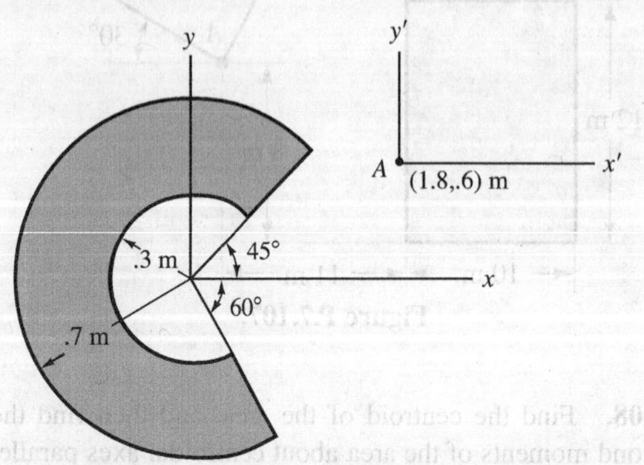

Figure P.7.111.

7.112. A wide-flanged I beam is shown.[10] For the upper flange, what is the first moment of the shaded area as a function of s measured from $s = 0$ at the right end to where s approaches the web? Next get M_x for the entire area above the position shown as y in the web. Let y go from C to the top of the web.

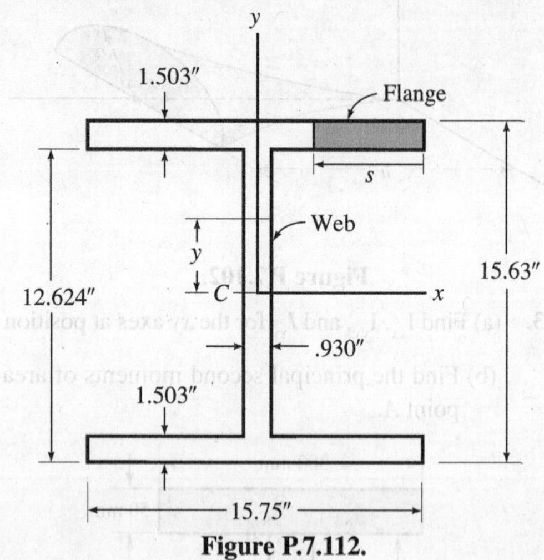

Figure P.7.112.

7.113. Find the center of volume of the machine element with a circular hole as shown.

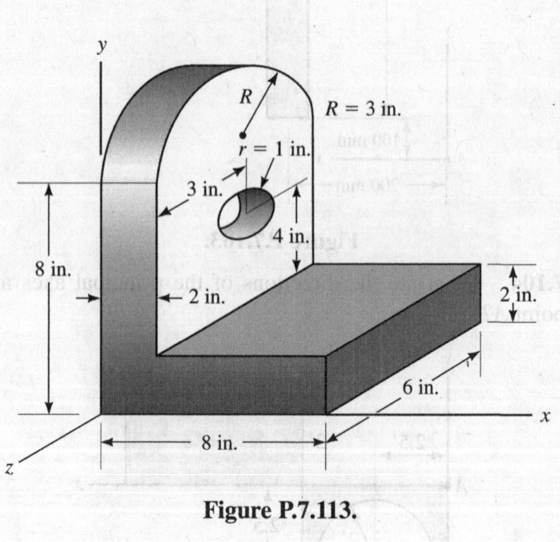

Figure P.7.113.

[10]Actually, it is a 14WF202 I beam, wherein the weight is 202 lb/ft.

CHAPTER 8

Moments and Products of Inertia[1]

8.1 Introduction

In this chapter, we shall consider certain measures of mass distribution relative to a reference. These quantities are vital for the study of the dynamics of rigid bodies. Because these quantities are so closely related to second moments and products of area, we shall consider them at this early stage rather than wait for dynamics. We shall also discuss the fact that these measures of mass distribution—the second moments of inertia of mass and the products of inertia of mass—are components of what we call a second-order tensor. Recognizing this fact early will make more simple and understandable your future studies of stress and strain, since these quantities also happen to be components of second-order tensors.

8.2 Formal Definition of Inertia Quantities

We shall now formally define a set of quantities that give information about the distribution of mass of a body relative to a Cartesian reference. For this purpose, a body of mass M and a reference xyz are presented in Fig. 8.1. This reference and the body may have any motion whatever relative to each other. The ensuing discussion then holds for the instantaneous orientation shown at time t. We shall consider that the body is composed of a continuum of particles, each of which has a mass given by $\rho\, dv$. We now present the following definitions:

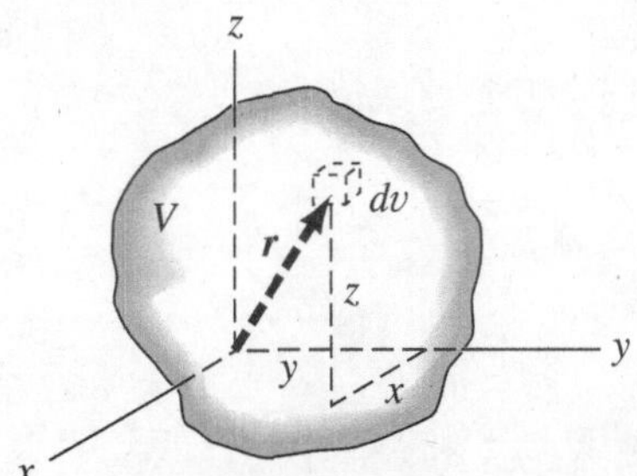

Figure 8.1. Body and reference at time t.

[1]This chapter may be covered at a later stage when studying dynamics. In that case, it should be covered directly after Chapter 13.

$$I_{xx} = \iiint\limits_V (y^2 + z^2)\rho\, dv \tag{8.1a}$$

$$I_{yy} = \iiint\limits_V (x^2 + z^2)\rho\, dv \tag{8.1b}$$

$$I_{zz} = \iiint\limits_V (x^2 + y^2)\rho\, dv \tag{8.1c}$$

$$I_{xy} = \iiint\limits_V xy\, \rho\, dv \tag{8.1d}$$

$$I_{xz} = \iiint\limits_V xz\, \rho\, dv \tag{8.1e}$$

$$I_{yz} = \iiint\limits_V yz\, \rho\, dv \tag{8.1f}$$

The terms I_{xx}, I_{yy}, and I_{zz} in the set above are called the *mass moments of inertia* of the body about the *x,y,* and *z* axes, respectively.[2] Note that in each such case we are integrating the mass elements $\rho\, dv$, times the *perpendicular distance squared* from the mass elements to the coordinate axis about which we are computing the moment of inertia. Thus, if we look along the *x* axis toward the origin in Fig. 8.1, we would have the view shown in Fig. 8.2. The quantity $y^2 + z^2$ used in Eq. 8.1a for I_{xx} is clearly d^2, the perpendicular distance squared from dv to the *x* axis (now seen as a dot). Each of the terms with mixed indices is called the *mass product of inertia* about the pair of axes given by the indices. Clearly, from the definition of the product of inertia, we could reverse indices and thereby form three additional products of inertia for a reference. The additional three quantities formed in this way, however, are equal to the corresponding quantities of the original set. That is,

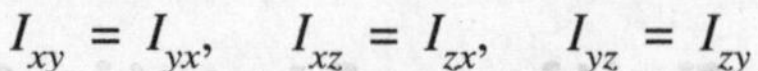

$$I_{xy} = I_{yx}, \quad I_{xz} = I_{zx}, \quad I_{yz} = I_{zy}$$

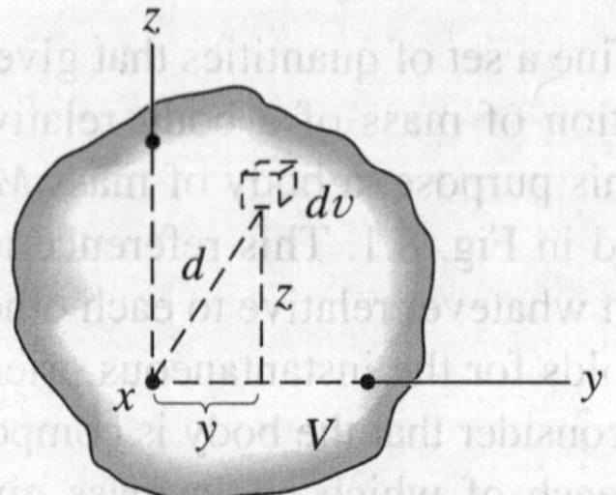

Figure 8.2. View of body along *x* axis.

We now have nine inertia terms at a point for a given reference at this point. The values of the set of six independent quantities will, for a given body, depend on the *position* and *inclination* of the reference relative to the body. You should also understand that the reference may be established anywhere in space and *need not* be situated in the rigid body

[2]We use the same notation as was used for second moments and products of area, which are also sometimes called moments and products of inertia. This is standard practice in mechanics. There need be no confusion in using these quantities if we keep the context of discussions clearly in mind.

of interest. Thus there will be nine inertia terms for reference xyz at point O outside the body (Fig. 8.3) computed using Eqs. 8.1, where the domain of integration is the volume V of the body. As will be explained later, the nine moments and products of inertia are components of the inertia tensor.

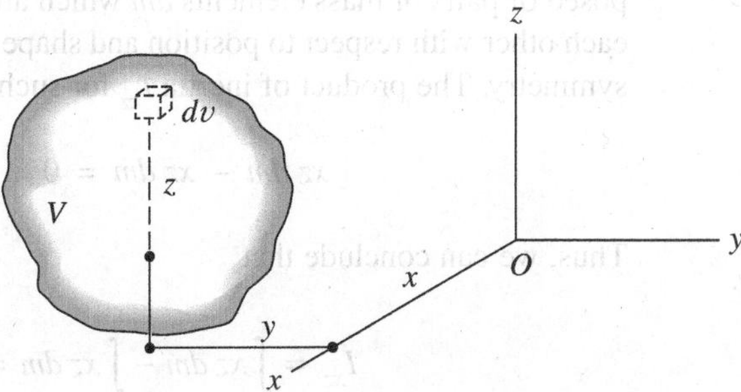

Figure 8.3. Origin of xyz outside body.

It will be convenient, when referring to the nine moments and products of inertia for reference xyz at a point, to list them in a matrix array, as follows:

$$I_{ij} = \begin{pmatrix} I_{xx} & I_{xy} & I_{xz} \\ I_{yx} & I_{yy} & I_{yz} \\ I_{zx} & I_{zy} & I_{zz} \end{pmatrix}$$

Notice that the first subscript gives the row and the second subscript gives the column in the array. Furthermore, the left-to-right downward diagonal in the array is composed of mass moment of inertia terms while the products of inertia, oriented at mirror-image positions about this diagonal, are equal. For this reason we say that the array is *symmetric*.

We shall now show that the sum of the mass moments of inertia for a set of orthogonal axes is independent of the orientation of the axes and depends only on the position of the origin. Examine the sum of such a set of terms:

$$I_{xx} + I_{yy} + I_{zz} = \iiint_V (y^2 + z^2)\rho\, dv + \iiint_V (x^2 + z^2)\rho\, dv + \iiint_V (x^2 + y^2)\rho\, dv$$

Combining the integrals and rearranging, we get

$$I_{xx} + I_{yy} + I_{zz} = \iiint_V 2(x^2 + y^2 + z^2)\rho\, dv = \iiint_V 2|r|^2 \rho\, dv \qquad (8.2)$$

But the magnitude of the position vector from the origin to a particle is *independent* of the inclination of the reference at the origin. Thus, *the sum of the moments of inertia at a point in space for a given body clearly is an invariant with respect to rotation of axes.*

Clearly, on inspection of the equations 8.1, it is clear that the moments of inertia must always be positive, while the products of inertia may be positive or negative. Of interest is the case where one of the coordinate planes is a *plane of symmetry* for the mass distribution of the body. Such a plane is the zy plane shown in Fig. 8.4 cutting a body into two parts, which, by definition of symmetry, are mirror images of each other. For the

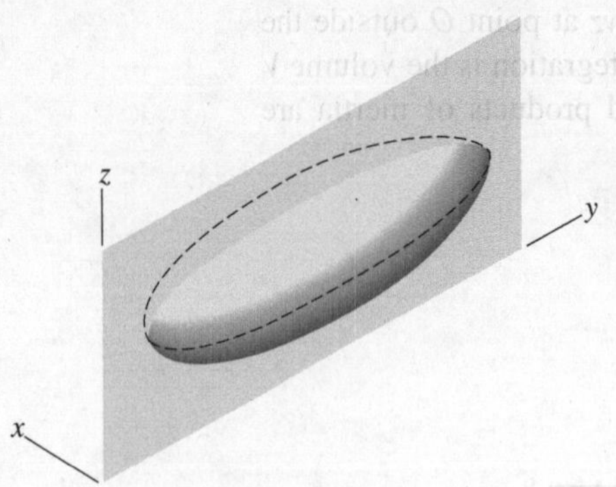

Figure 8.4. zy is plane of symmetry.

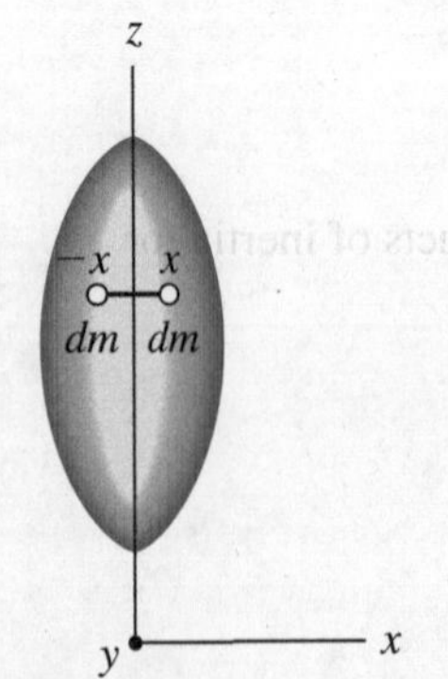

Figure 8.5. View along y axis.

computation of I_{xz}, each half will give a contribution of the same magnitude but of opposite sign. We can most readily see that this is so by looking along the y axis toward the origin. The plane of symmetry then appears as a line coinciding with the z axis (see Fig. 8.5). We can consider the body to be composed of pairs of mass elements dm which are mirror images of each other with respect to position and shape about the plane of symmetry. The product of inertia I_{xz} for such a pair is then

$$xz\,dm - xz\,dm = 0$$

Thus, we can conclude that

$$I_{xz} = \underbrace{\int xz\,dm}_{\text{right domain}} - \underbrace{\int xz\,dm}_{\text{left domain}} = 0$$

This conclusion is also true for I_{xy}. We can say that $I_{xy} = I_{xz} = 0$. But on consulting Fig. 8.4, you should be able to readily decide that the term I_{zy} will have a positive value. Note that those products of inertia having x as an index are zero and that the x coordinate axis is normal to the plane of symmetry. Thus, we can conclude that *if two axes form a plane of symmetry for the mass distribution of a body, the products of inertia having as an index the coordinate that is normal to the plane of symmetry will be zero.*

Consider next a body of *revolution*. Take the z axis to coincide with the axis of symmetry. It is easy to conclude for the origin O of xyz anywhere along the axis of symmetry that

$$I_{xz} = I_{yz} = I_{xy} = 0$$
$$I_{xx} = I_{yy} = \text{constant}$$

for all possible xy axes formed by rotating about the z axis at O. Can you justify these conclusions?

Finally, we define *radii of gyration* in a manner analogous to that used for second moments of area in Chapter 7. Thus:

$$I_{xx} = k_x^2 M$$

$$I_{yy} = k_y^2 M$$

$$I_{zz} = k_z^2 M$$

where k_x, k_y, and k_z are the radii of gyration and M is the total mass.

Example 8.1

Find the nine components of the inertia tensor of a rectangular body of uniform density ρ about point O for a reference xyz coincident with the edges of the block as shown in Fig. 8.6.

We first compute I_{xx}. Using volume elements $dv = dx\,dy\,dz$, we get on using simple multiple integration:

$$\begin{aligned} I_{xx} &= \int_0^a \int_0^b \int_0^c (y^2 + z^2)\rho\,dx\,dy\,dz \\ &= \int_0^a \int_0^b (y^2 + z^2)c\rho\,dy\,dz = \int_0^a \left(\frac{b^3}{3} + z^2 b\right)c\rho\,dz \\ &= \left(\frac{ab^3c}{3} + \frac{a^3bc}{3}\right)\rho = \frac{\rho V}{3}(b^2 + a^2) \end{aligned} \quad \text{(a)}$$

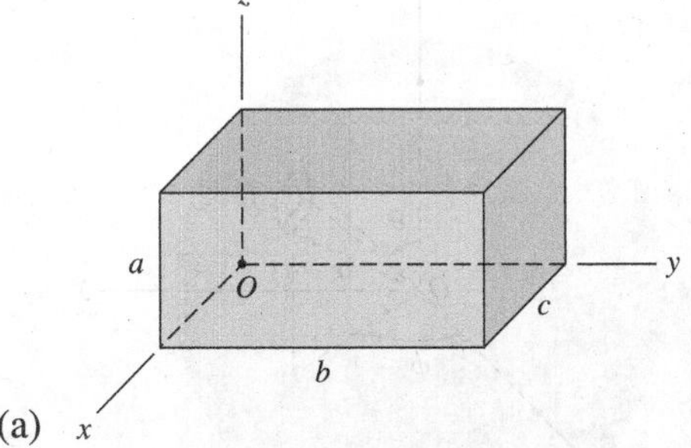

Figure 8.6. Find I_{ij} at O.

where V is the volume of the body. Note that the x axis about which we are computing the moment of inertia I_{xx} is *normal* to the plane having sides of length a and b, i.e., along the z and y axes. Similarly:

$$I_{yy} = \frac{\rho V}{3}(c^2 + a^2) \quad \text{(b)}$$

$$I_{zz} = \frac{\rho V}{3}(b^2 + c^2) \quad \text{(c)}$$

We next compute I_{xy}.

$$\begin{aligned} I_{xy} &= \int_0^a \int_0^b \int_0^c xy\,\rho\,dx\,dy\,dz = \int_0^a \int_0^b \frac{c^2}{2}y\rho\,dy\,dz \\ &= \int_0^a \frac{c^2b^2}{4}\rho\,dz = \frac{ac^2b^2}{4}\rho = \frac{\rho V}{4}cb \end{aligned} \quad \text{(d)}$$

Note for I_{xy}, we use the lengths of the sides along the x and y axes.

$$I_{xz} = \frac{\rho V}{4}ac \quad \text{(e)}$$

$$I_{yz} = \frac{\rho V}{4}ab \quad \text{(f)}$$

We accordingly have, for the inertia tensor:

$$I_{ij} = \begin{pmatrix} \frac{\rho V}{3}(b^2 + a^2) & \frac{\rho V}{4}cb & \frac{\rho V}{4}ac \\ \frac{\rho V}{4}cb & \frac{\rho V}{3}(c^2 + a^2) & \frac{\rho V}{4}ab \\ \frac{\rho V}{4}ac & \frac{\rho V}{4}ab & \frac{\rho V}{3}(b^2 + c^2) \end{pmatrix} \quad \text{(g)}$$

Example 8.2

Compute the components of the inertia tensor at the center of a solid sphere of uniform density ρ as shown in Fig. 8.7.

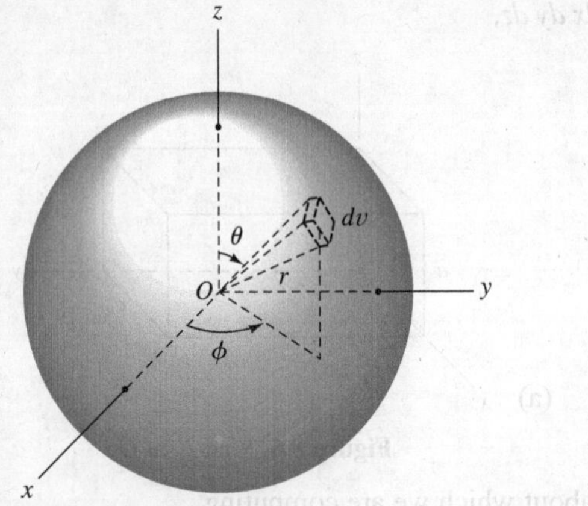

Figure 8.7. Find I_{ij} at O

Figure 8.8. $dv = (r \sin\theta\, d\phi)(dr)(r\, d\theta) = r^2 \sin\theta\, d\theta\, d\phi\, dr$.

We shall first compute I_{yy}. Using spherical coordinates, we have[3]

$$
\begin{aligned}
I_{yy} &= \iiint_V (x^2 + z^2)\rho\, dv \\
&= \int_0^R \int_0^{2\pi} \int_0^{\pi} \left[(r \sin\theta \cos\phi)^2 + (r\cos\theta)^2 \right] \rho \left(r^2 \sin\theta\, d\theta\, d\phi\, dr \right) \\
&= \int_0^R \int_0^{2\pi} \int_0^{\pi} \left(r^4 \sin^3\theta \cos^2\phi \right) \rho\, d\theta\, d\phi\, dr \\
&\qquad + \int_0^R \int_0^{2\pi} \int_0^{\pi} \left(r^4 \cos^2\theta \sin\theta \right) \rho\, d\theta\, d\phi\, dr \\
&= \rho \int_0^R \int_0^{2\pi} \left(r^4 \cos^2\phi \right) \left(\int_0^{\pi} \sin^3\theta\, d\theta \right) d\phi\, dr \\
&\qquad + \rho \int_0^R \int_0^{2\pi} r^4 \left(\int_0^{\pi} \cos^2\theta \sin\theta\, d\theta \right) d\phi\, dr
\end{aligned}
$$

With the aid of integration formulas from Appendix I, we have

$$
\begin{aligned}
I_{yy} &= \rho \int_0^R \int_0^{2\pi} r^4 \cos^2\phi \left[-\tfrac{1}{3} \cos\theta (\sin^2\theta + 2) \right]_0^{\pi} d\phi\, dr \\
&\qquad + \rho \int_0^R \int_0^{2\pi} r^4 \left(-\frac{\cos^3\theta}{3} \right) \Bigg|_0^{\pi} d\phi\, dr \\
&= \rho \int_0^R \int_0^{2\pi} r^4 \cos^2\phi\, \tfrac{4}{3}\, d\phi\, dr + \rho \int_0^R \int_0^{2\pi} (r^4)\, \tfrac{2}{3}\, d\phi\, dr
\end{aligned}
$$

Integrating next with respect to ϕ, we get

[3]For those unfamiliar with spherical coordinates, we have shown in Fig. 8.8a more detailed study of the volume element used. The volume dv is simply the product of the three edges of the element shown in the diagram.

Example 8.2 (Continued)

$$I_{yy} = \rho\int_0^R (r^4)(\tfrac{4}{3})(\pi)\,dr + \rho\int_0^R (r^4)(\tfrac{2}{3})(2\pi)\,dr$$

Finally, we get

$$I_{yy} = \rho\frac{R^5}{5}\frac{4}{3}\pi + \rho\frac{R^5}{5}\frac{4}{3}\pi$$

$$\therefore I_{yy} = \frac{8}{15}\rho\pi R^5$$

But

$$M = \rho\tfrac{4}{3}\pi R^3$$

Hence,

$$I_{yy} = \tfrac{2}{5}MR^2$$

Because of the point symmetry about point O, we can also say that

$$I_{xx} = I_{zz} = \tfrac{2}{5}MR^2$$

Because the coordinate planes are all planes of symmetry for the mass distribution, the products of inertia are zero. Thus, the inertia tensor can be given as

$$I_{ij} = \begin{pmatrix} \frac{2}{5}MR^2 & 0 & 0 \\ 0 & \frac{2}{5}MR^2 & 0 \\ 0 & 0 & \frac{2}{5}MR^2 \end{pmatrix}$$

8.3 Relation Between Mass-Inertia Terms and Area-Inertia Terms

We now relate the second moment and product of area studied in Chapter 7 with the inertia tensor. To do this, consider a plate of constant thickness t and uniform density ρ (Fig. 8.9). A reference is selected so that the xy plane is in the midplane of this plate. The components of the inertia tensor are rewritten for convenience as

$$I_{xx} = \rho\iiint_V (y^2 + z^2)\,dv, \qquad I_{xy} = \rho\iiint_V xy\,dv$$

$$I_{yy} = \rho\iiint_V (x^2 + z^2)\,dv, \qquad I_{xz} = \rho\iiint_V xz\,dv$$

$$I_{zz} = \rho\iiint_V (x^2 + y^2)\,dv, \qquad I_{yz} = \rho\iiint_V yz\,dv \tag{8.3}$$

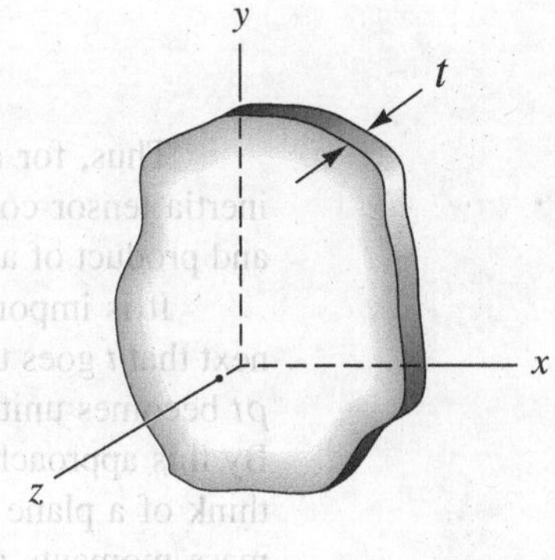

Figure 8.9. Plate of thickness t

Now consider that the thickness t is *small* compared to the lateral dimensions of the plate. This means that z is restricted to a range of values having a small magnitude. As a result, we can make two simplifications in the equations above. First, we shall set z equal to zero whenever it appears on the right side of the equations above. Second, we shall express dv as

$$dv = t\,dA$$

where dA is an area element on the *surface* of the plate, as shown in Fig. 8.10. Equations 8.3 then become

$$I_{xx} = \rho t \iint_A y^2\, dA, \qquad I_{xy} = \rho t \iint_A xy\, dA$$

$$I_{yy} = \rho t \iint_A x^2\, dA, \qquad I_{xz} = 0$$

$$I_{zz} = \rho t \iint_A (x^2 + y^2)\, dA, \qquad I_{yz} = 0$$

Notice, now, that the integrals on the right sides of the equations above are moments and products of *area* as presented in Chapter 7. Denoting mass-moment and product of inertia

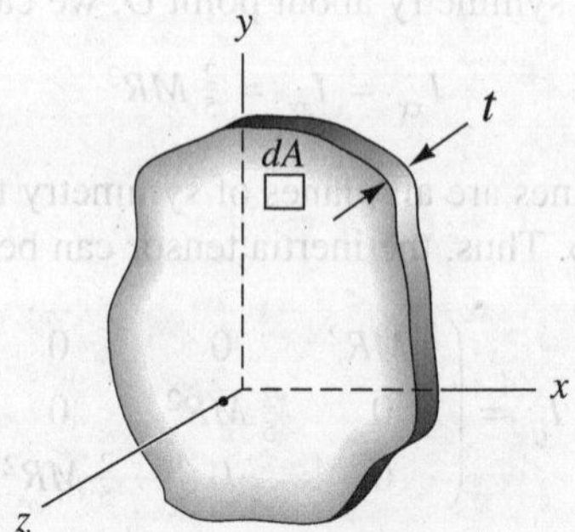

Figure 8.10. Use volume elements $t\ dA$.

terms with a subscript M and second moment and product of area terms with a subscript A, we can then say for the nonzero expressions:

$$(I_{xx})_M = \rho t (I_{xx})_A$$
$$(I_{yy})_M = \rho t (I_{yy})_A$$
$$(I_{zz})_M = \rho t (J)_A$$
$$(I_{xy})_M = \rho t (I_{xy})_A$$

Thus, for a thin plate with a constant product ρt throughout, we can compute the inertia tensor components for reference xyz (see Fig. 8.9) by using the second moments and product of area of the surface of the plate relative to axes xy.

It is important to point out that ρt is the *mass per unit area* of the plate. Imagine next that t goes to zero and simultaneously ρ goes to infinity at rates such that the product ρt becomes unity in the limit. One might think of the resulting body to be a *plane area.* By this approach, we have thus formed a plane area from a plate and in this way we can think of a plane area as a special mass. This explains why we use the same notation for mass moments and products of inertia as we use for second moments and products of area. However the units clearly will be different.We now examine a plate problem.

Example 8.3

Determine the inertia tensor components for the thin plate (Fig. 8.11) relative to the indicated axes xyz. The weight of the plate is .002 N/mm.[2] For the top edge, $y = 2\sqrt{x}$ with x and y in millimeters.

It is clear that for ρt we have, remembering that this product represents mass per unit area:

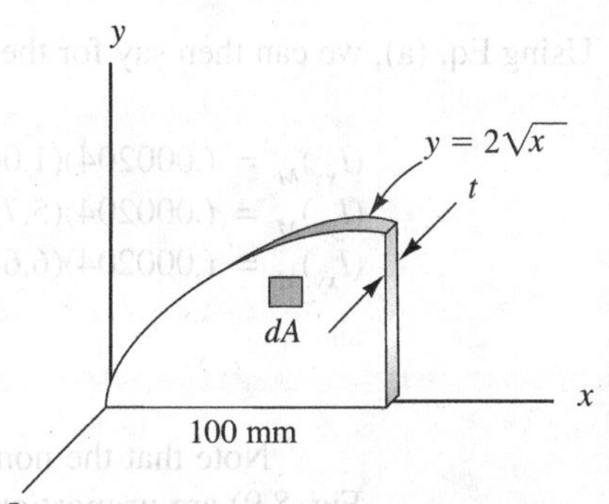

Figure 8.11. Plate of thickness t.

$$\rho t = \frac{.002}{9.81} = .000204 \text{ kg/mm}^2 \qquad \text{(a)}$$

We now examine the second moments and product of area for the surface of the plate about axes xy. Thus,[4]

$$(I_{xx})_A = \int_0^{100}\int_{y=0}^{y=2\sqrt{x}} y^2\, dy\, dx$$

$$= \int_0^{100} \left.\frac{y^3}{3}\right|_0^{2\sqrt{x}} dx = \int_0^{100} \frac{8}{3} x^{3/2} dx$$

$$= \left.\frac{8}{3}\frac{x^{5/2}}{\frac{5}{2}}\right|_0^{100} = \left(\tfrac{8}{3}\right)\left(\tfrac{2}{5}\right)\left(100^{5/2}\right)$$

$$= 1.067 \times 10^5 \text{ mm}^4$$

$$(I_{yy})_A = \int_0^{100}\int_{y=0}^{y=2\sqrt{x}} x^2\, dy\, dx$$

$$= \int_0^{100} x^2 y\Big|_0^{2\sqrt{x}} dx = \int_0^{100} x^2\left(2\sqrt{x}\right) dx$$

$$= 2\left.\frac{x^{7/2}}{\frac{7}{2}}\right|_0^{100} = 2\left(\tfrac{2}{7}\right)\left(100^{7/2}\right)$$

$$= 5.71 \times 10^6 \text{ mm}^4$$

$$(I_{xy})_A = \int_0^{100}\int_{y=0}^{y=2\sqrt{x}} xy\, dy\, dx$$

$$= \int_0^{100} x\left.\frac{y^2}{2}\right|_0^{2\sqrt{x}} dx = \int_0^{100} 2x^2\, dx$$

$$= 2\left(\frac{100^3}{3}\right) = 6.67 \times 10^5 \text{ mm}^4$$

[4]Note we have multiple integration where one of the boundaries is variable. The procedure to follow should be evident from the example.

Example 8.3

Using Eq. (a), we can then say for the nonzero inertia tensor components:

$$(I_{xx})_M = (.000204)(1.067 \times 10^5) = 21.76 \text{ kg-mm}^2$$
$$(I_{yy})_M = (.000204)(5.71 \times 10^6) = 1{,}165 \text{ kg-mm}^2$$
$$(I_{xy})_M = (.000204)(6.67 \times 10^5) = 136.1 \text{ kg-mm}^2$$

Note that the nonzero inertia tensor components for a reference xyz on a plate (see Fig. 8.9) are *proportional* through ρt to the corresponding area-inertia terms for the plate surface. This means that all the formulations of Chapter 7 apply to the aforementioned nonzero inertia tensor components. Thus, on rotating the axes about the z axis we may use the transformation equations of Chapter 7. Consequently, the concept of *principal axes* in the midplane of the plate at a point applies. For such axes, the product of inertia is zero. One such axis then gives the maximum moment of inertia for all axes in the midplane at the point, the other the minimum moment of inertia. We have presented such problems at the end of this section.

What about principal axes for the inertia tensor at a point in a general three-dimensional body? Further will learn that there are *three principal axes* at a point in the general case. These axes are *mutually orthogonal* and the *products of inertia are all zero* for such a set of axes at a point.[5] Furthermore, one of the axes will have a maximum moment of inertia, another axis will have a minimum moment of inertia, while the third axis will have an intermediate value. The sum of these three inertia terms must have a value that is common for all sets of axes at the point.

If, perchance, a set of axes xyz at a point is such that xy and xz form *two planes of symmetry* for the mass distribution of the body, then, as we learned earlier, since the z axis and the y axis are normal to the planes of symmetry, $I_{xy} = I_{xz} = I_{yz} = 0$. Thus, all products of inertia are zero. This would also be true for *any* two sets of axes of xyz forming two planes of symmetry. Clearly, axes forming two planes of symmetry must be *principal axes*. This information will suffice in most instances when we have to identify principal axes. On the other hand, consider the case where there is only *one plane of symmetry* for the mass distribution of a body at some point A. Let the xy plane at A form this plane of symmetry. Then, clearly, the products of inertia between the z axis that is normal to the plane of symmetry xy and *any axis* in the xy plane at A must be zero, as pointed out earlier. Obviously, the z axis must be a principal axis. The other two principal axes must be in the plane of symmetry, but generally cannot be located by inspection.

[5]The third principal axis for a plate at a point in the midplane is the z axis normal to the plate. Note that $(I_{zz})_M$ must always equal $(I_{xx})_M + (I_{yy})_M$. Why?

PROBLEMS

8.1. A uniform homogeneous slender rod of mass M is shown. Compute I_{xx} and $I_{x'x'}$.

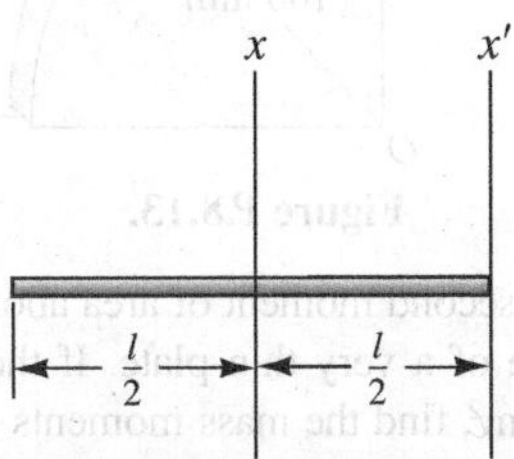

Figure P.8.1.

8.2. Find I_{xx} and $I_{x'x'}$ for the thin rod of Problem 8.1 for the case where the mass per unit length at the left end is 5 lbm/ft and increases linearly so that at the right end it is 8 lbm/ft. The rod is 20 ft in length.

8.3. Compute I_{xy} for the thin homogeneous hoop of mass M.

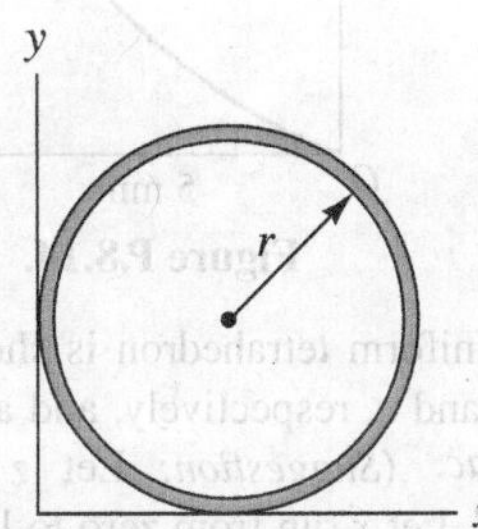

Figure P.8.3.

8.4. Compute I_{xx}, I_{yy}, I_{zz}, and I_{xy} for the homogeneous rectangular parallelepiped.

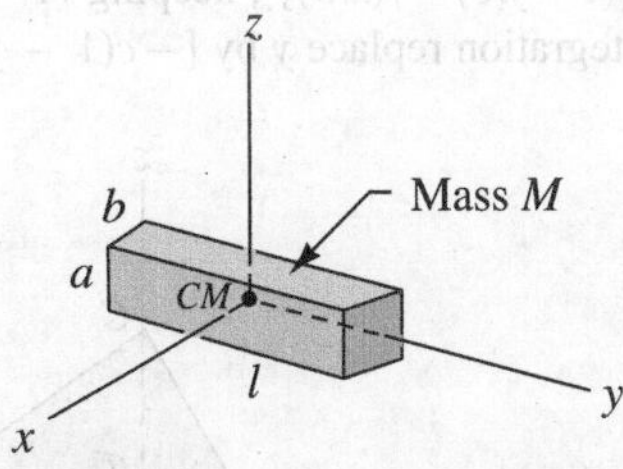

Figure P.8.4.

8.5. A wire having the shape of a parabola is shown. The curve is in the yz plane. If the mass of the wire is .3 N/m, what are I_{yy} and I_{xz}? [*Hint:* Replace ds along the wire by $\sqrt{(dy/dz)^2 + 1}\, dz$.]

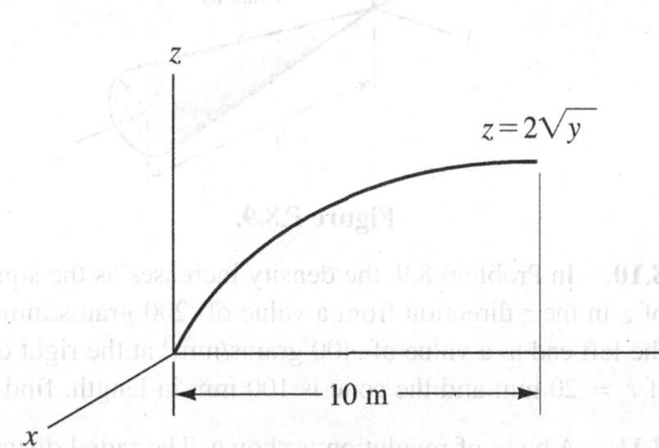

Figure P.8.5.

8.6. Compute the moment of inertia, I_{BB}, for the half-cylinder shown. The body is homogeneous and has a mass M.

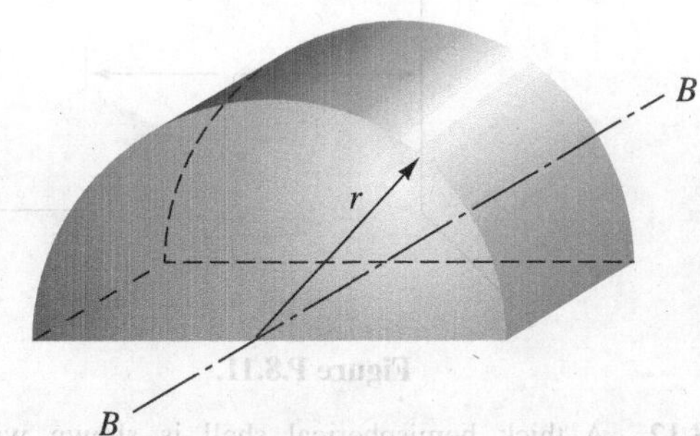

Figure P.8.6.

8.7. Find I_{zz} and I_{xx} for the homogeneous right circular cylinder of mass M.

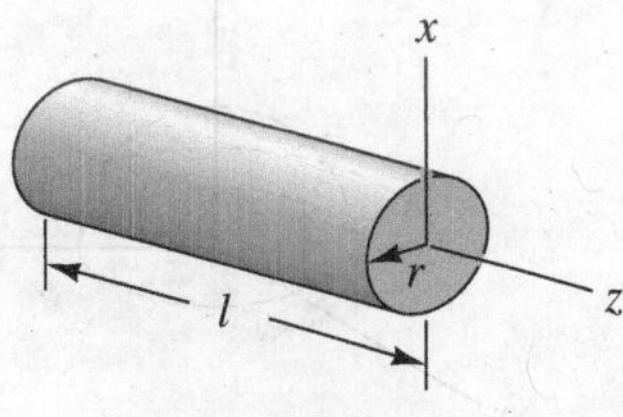

Figure P.8.7.

8.8. For the cylinder in Problem 8.7, the density increases linearly in the z direction from a value of .100 grams/mm^3 at the left end to a value of .180 grams/mm^3 at the right end. Take $r = 30$ mm and $l = 150$ mm. Find I_{xx} and I_{zz}.

8.9. Show that I_{zz} for the homogeneous right circular cone is $\frac{3}{10}MR^2$.

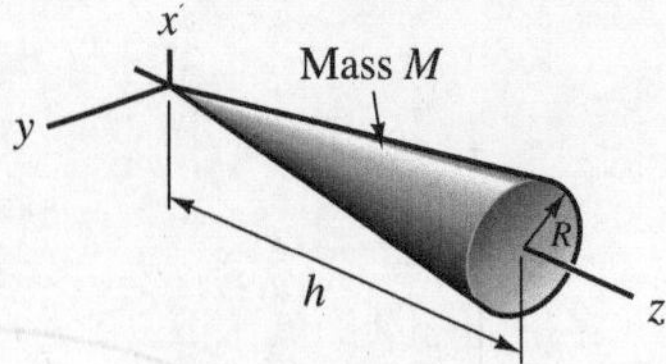

Figure P.8.9.

8.10. In Problem 8.9, the density increases as the square of z in the z direction from a value of .200 grams/mm^3 at the left end to a value of .400 grams/mm^3 at the right end. If $r = 20$ mm and the cone is 100 mm in length, find I_{zz}.

8.11. A body of revolution is shown. The radial distance r of the boundary from the x axis is given as $r = .2x^2$ m. What is I_{xx} for a uniform density of 1,600 kg/m^3?

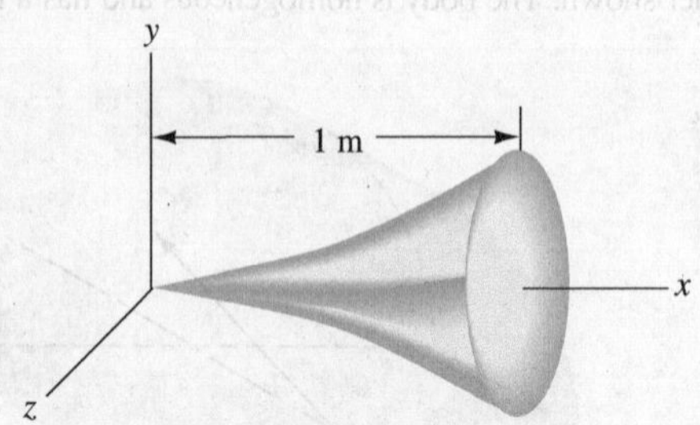

Figure P.8.11.

8.12. A thick hemispherical shell is shown with an inside radius of 40 mm and an outside radius of 60 mm. If the density ρ is 7,000 kg/m,^{3} what is I_{yy}?

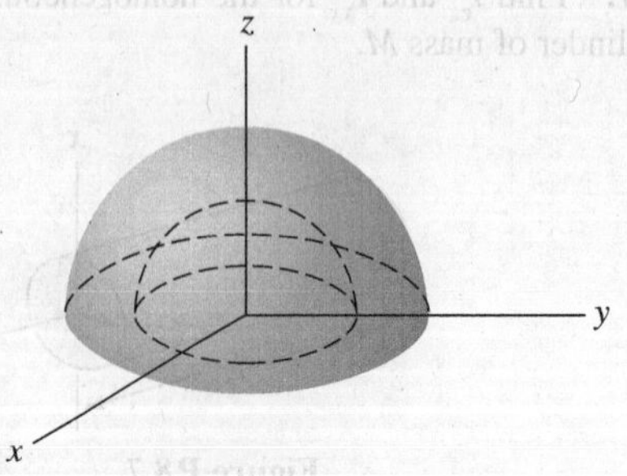

Figure P.8.12.

8.13. Find the mass moment of inertia I_{xx} for a very thin plate forming a quarter-sector of a circle. The plate weighs .4 N. What is the second moment of area about the x axis? What is the product of inertia? Axes are in the midplane of the plate.

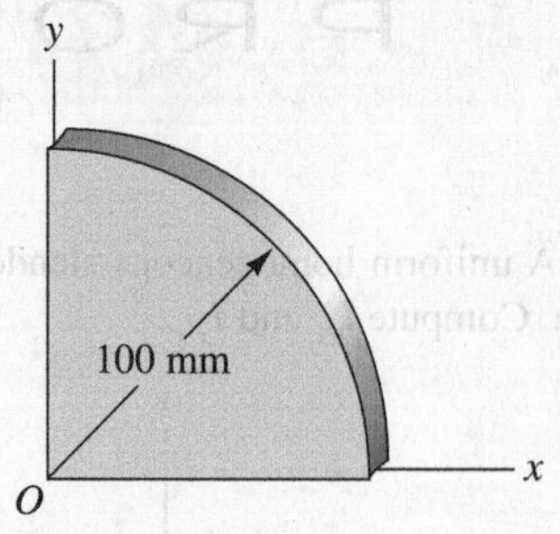

Figure P.8.13.

8.14. Find the second moment of area about the x axis for the front surface of a very thin plate. If the weight of the plate is .02 N/mm^2, find the mass moments of inertia about the x and y axes. What is the mass product of inertia I_{xy}?

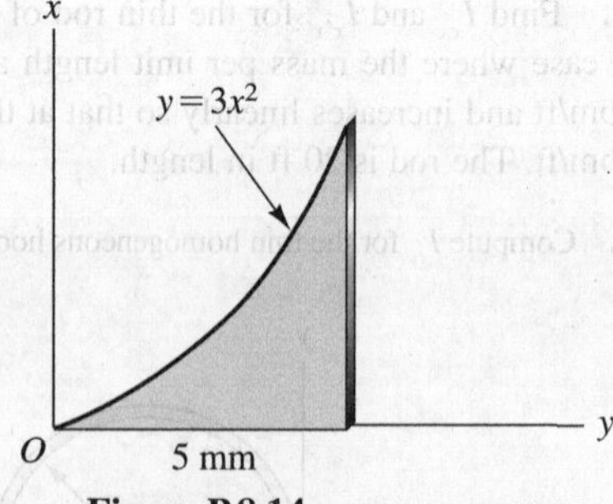

Figure P.8.14.

***8.15.** A uniform tetrahedron is shown having sides of length a, b, and c, respectively, and a mass M. Show that $I_{yz} = \frac{1}{20}Mac$. (*Suggestion:* Let z run from zero to surface ABC. Let x run from zero to line AB. Finally, let y run from zero to B. Note that the equation of a plane surface is $z = \alpha x + \beta y + \gamma$, where α, β, and γ are constants. The mass of the tetrahedron is $\rho abc/6$. It will be simplest in expanding $(1 - x/b - y/c)^2$ to proceed in the form $[(1 - y/c) - (x/b)]^2$, keeping $(1 - y/c)$ intact. In the last integration replace y by $[-c(1 - y/c) + c]$, etc.)

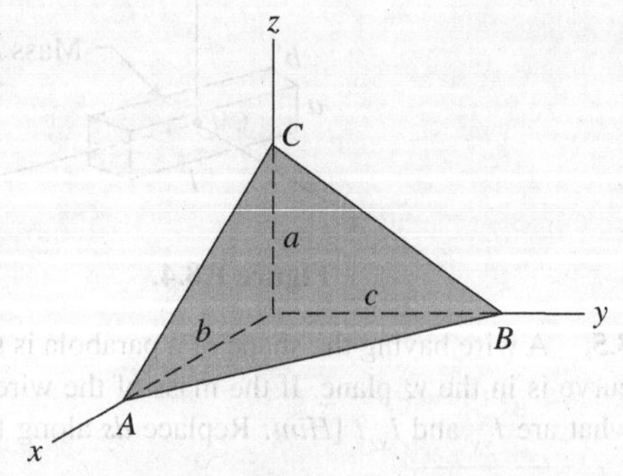

Figure P.8.15.

8.16. In Problem 8.13, find the three principal mass moments of inertia at O. Use the following results from problem 8.13

$$(I_{xx})_M = 101.9 \text{ kg-mm}^2$$
$$(I_{xy})_M = 64.9 \text{ kg-mm}^2$$

8.17. In Problem 8.14, compute the values of the three principal mass moments of inertia at O. From Problem 8.14 we have the results

$$(I_{xx})_M = 205 \text{ kg-mm}^2$$
$$(I_{yy})_M = 3.82 \text{ kg-mm}^2$$
$$(I_{xy})_M = 23.9 \text{ kg-mm}^2$$

8.18. Can you identify by inspection any of the principal axes of inertia at A? At B? Explain. The density of the material is uniform.

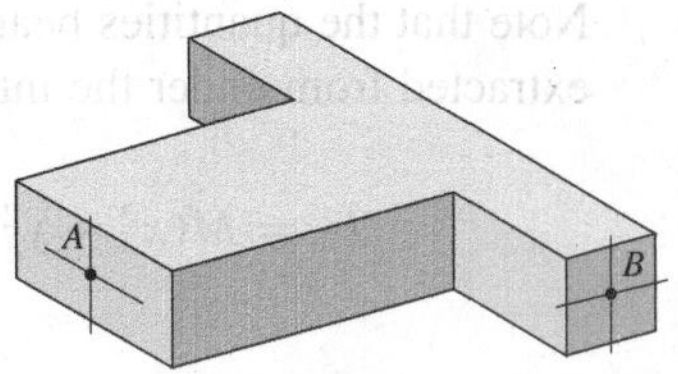

Figure P.8.18.

8.19. By inspection, identify as many principal axes as you can for mass moments of inertia at positions A, B, and C. Explain your choices. The mass density of the material is uniform throughout.

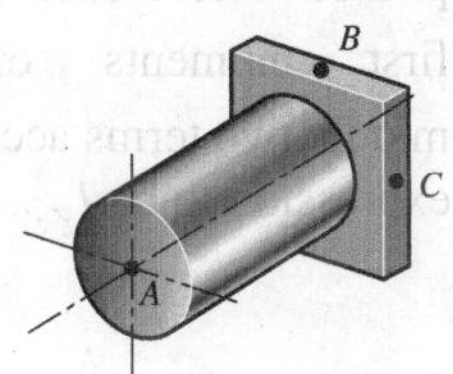

Figure P.8.19.

8.4 Translation of Coordinate Axes

In this section, we will compute mass moment and product of inertia quantities for a reference xyz that is displaced under a translation (no rotation) from a reference $x'y'z'$ at the center of mass (Fig. 8.12) for which the inertia terms are presumed known. Let us first compute the mass moment of inertia I_{zz}.

Observing Fig. 8.12, we see that

$$\boldsymbol{r} = \boldsymbol{r}_c + \boldsymbol{r}'$$

Hence,

$$x = x_c + x'$$
$$y = y_c + y'$$
$$z = z_c + z'$$

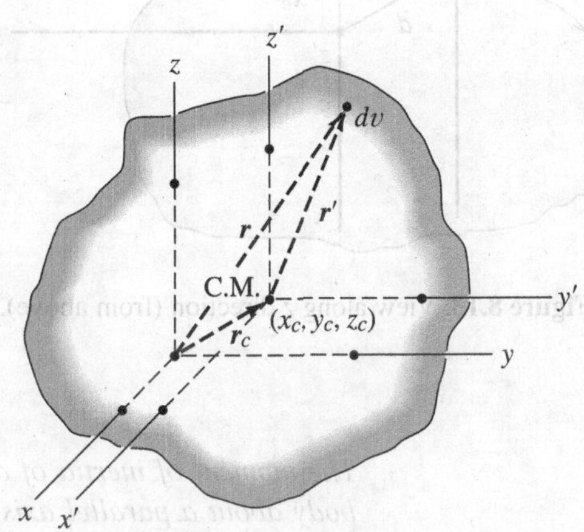

Figure 9.12. xyz translated from $x'y'z'$ at C.M.

We can now formulate I_{zz} in the following way:

$$I_{zz} = \iiint_V (x^2 + y^2)\rho\, dv = \iiint_V \left[(x_c + x')^2 + (y_c + y')^2\right]\rho\, dv \qquad (8.4)$$

Carrying out the squares and rearranging, we have

$$I_{zz} = \iiint_V (x_c^2 + y_c^2)\rho\, dv + 2\iiint_V x_c x' \rho\, dv$$
$$+2\iiint_V y_c y' \rho\, dv + \iiint_V (x'^2 + y'^2)\rho\, dv \qquad (8.5)$$

Note that the quantities bearing the subscript c are constant for the integration and can be extracted from under the integral sign. Thus,

$$I_{zz} = M(x_c^2 + y_c^2) + 2x_c \iiint_V x'\,dm + 2y_c \iiint_V y'\,dm + \iiint_V (x'^2 + y'^2)\rho\,dv \qquad (8.6)$$

where $\rho\,dv$ has been replaced in some terms by dm, and the integration $\iiint_V \rho\,dv$ in the first integral has been evaluated as M, the total mass of the body. The origin of the primed reference being at the center of mass requires of the first moments of mass that $\iiint x'\,dm = \iiint y'\,dm = \iiint z'\,dm = 0$. The middle two terms accordingly drop out of the expression above, and we recognize the last expression to be $I_{z'z'}$. Thus, the desired relation is

$$I_{zz} = I_{z'z'} + M(x_c^2 + y_c^2) \qquad (8.7)$$

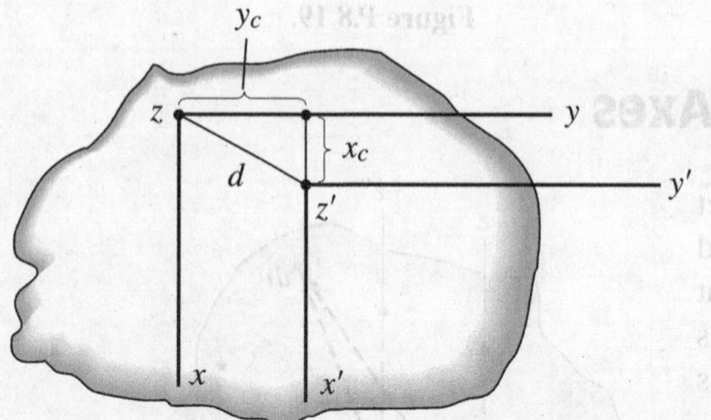

Figure 8.13. View along z direction (from above).

By observing the body in Fig. 8.12 along the z and z' axes (i.e., from directly above), we get a view as is shown in Fig 8.13. From this diagram, we can see that $y_c^2 + x_c^2 = d^2$, where d is the perpendicular distance between the z' axis through the center of mass and the z axis about which we are taking moments of inertia. We may then give the result above as

$$I_{zz} = I_{z'z'} + Md^2 \qquad (8.8)$$

Let us generalize from the previous statement.

The moment of inertia of a body about any axis equals the moment of inertia of the body about a parallel axis that goes through the center of mass, plus the total mass times the perpendicular distance between the axes squared.

We leave it to you to show that for products of inertia a similar relation can be reached. For I_{xy}, for example, we have

$$I_{xy} = I_{x'y'} + Mx_c y_c \qquad (8.9)$$

Here, we must take care to put in the proper signs of x_c and y_c as measured *from* the xyz reference. Equations 8.8 and 8.9 comprise the well-known *parallel-axis theorems* analogous to those formed in Chapter 7 for areas. You can use them to advantage for bodies composed of simple familiar shapes, as we now illustrate.

Example 8.4

Find I_{xx} and I_{xy} for the body shown in Fig. 8.14. Take ρ as constant for the body. Use the formulations for moments and products of inertia at the center of mass as given on the inside front cover page.

We shall consider first a solid rectangular prism having the outer dimensions given in Fig. 8.14, and we shall then subtract the contribution of the cylinder and the rectangular block that have been cut away. Thus, we have, for the overall rectangular block which we consider as body 1,

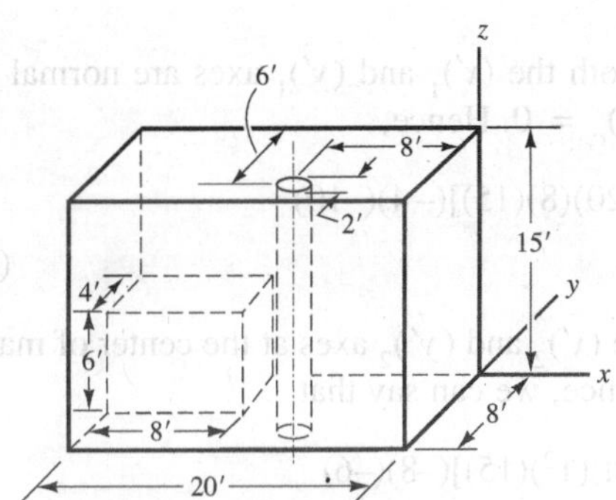

Figure 8.14. Find I_{xx} and I_{xy}.

$$\begin{aligned}(I_{xx})_1 &= (I_{xx})_c + Md^2 = \tfrac{1}{12}M(a^2+b^2) + Md^2 \\ &= \tfrac{1}{12}[(\rho)(20)(8)(15)](8^2+15^2) + [(\rho)(20)(8)(15)](4^2+7.5^2) \qquad \text{(a)} \\ &= 231{,}200\rho\end{aligned}$$

From this, we shall take away the contribution of the cylinder, which we denote as body 2. Using the formulas from the inside front cover page,

$$\begin{aligned}(I_{xx})_2 &= \tfrac{1}{12}M(3r^2+h^2) + Md^2 \\ &= \tfrac{1}{12}[\rho\pi(1)^2(15)][3(1^2)+15^2] + [\rho\pi(1)^2(15)][6^2+7.5^2] \qquad \text{(b)} \\ &= 5{,}243\rho\end{aligned}$$

Also, we shall take away the contribution of the rectangular cutout (body 3):

$$\begin{aligned}(I_{xx})_3 &= \tfrac{1}{12}M(a^2+b^2) + Md^2 \\ &= \tfrac{1}{12}[\rho(8)(6)(4)](4^2+6^2) + [\rho(8)(6)(4)](2^2+3^2) \qquad \text{(c)} \\ &= 3{,}328\rho\end{aligned}$$

The quantity I_{xx} for the body with the rectangular and cylindrical cavities is then

Example 8.4

$$I_{xx} = (231{,}200 - 5{,}243 - 3{,}328)\rho$$

$$I_{xx} = \boxed{223{,}000\rho} \qquad \text{(d)}$$

We follow the same procedure to obtain I_{xy}. Thus, for the block as a whole, we have

$$(I_{xy})_1 = (I_{xy})_c + Mx_c y_c$$

At the center of mass of the block, both the $(x')_1$ and $(y')_1$ axes are normal to planes of symmetry. Accordingly, $(I_{xy})_c = 0$. Hence,

$$(I_{xy})_1 = 0 + [\rho(20)(8)(15)](-4)(-10)$$
$$= 96{,}000\rho \qquad \text{(e)}$$

For the cylinder, we note that both the $(x')_2$ and $(y')_2$ axes at the center of mass are normal to planes of symmetry. Hence, we can say that

$$(I_{xy})_2 = 0 + [\rho(\pi)(1^2)(15)](-8)(-6)$$
$$= 2{,}262\rho \qquad \text{(f)}$$

Finally, for the small cutout rectangular parallelepiped, we note that the $(x')_3$ and $(y')_3$ axes at the center of mass are perpendicular to planes of symmetry. Hence, we have

$$(I_{xy})_3 = 0 + [\rho(8)(6)(4)](-2)(-16)$$
$$= 6{,}144\rho \qquad \text{(g)}$$

The quantity I_{xy} for the body with the rectangular and cylindrical cavities is then

$$I_{xy} = (96{,}000 - 2{,}262 - 6{,}144)\rho = \boxed{87{,}600\rho} \qquad \text{(h)}$$

If ρ is given in units of lbm/ft,^{3} the inertia terms have units of lbm-ft.^{2}

PROBLEMS

In the following problems, use the formulas for moments and products of inertia at the mass center to be found in the inside front cover page.

8.20. What are the moments and products of inertia for the *xyz* and *x'y'z'* axes for the cylinder?

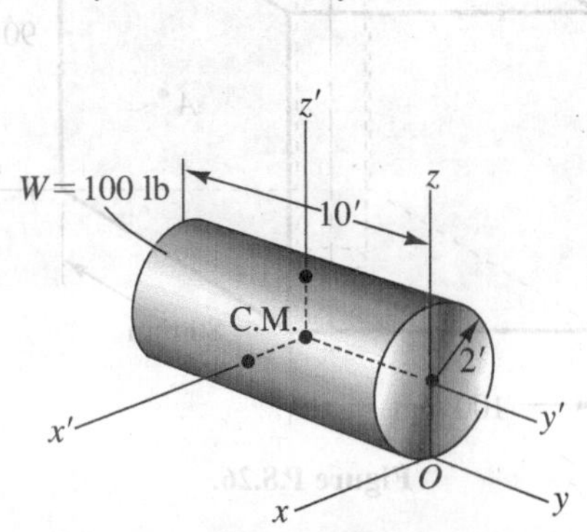

Figure P.8.20.

8.21. For the uniform block, compute the inertia tensor at the center of mass, at point *a*, and at point *b* for axes parallel to the *xyz* reference. Take the mass of the body as *M* kg.

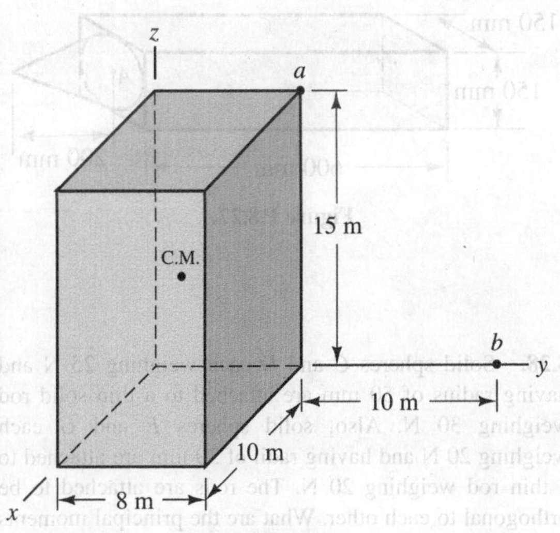

Figure P.8.21.

8.22. Determine $I_{xx} + I_{yy} + I_{zz}$ as a function of *x*, *y*, and *z* for all points in space for the uniform rectangular parallelepiped. Note that *xyz* has its origin at the center of mass and is parallel to the sides.

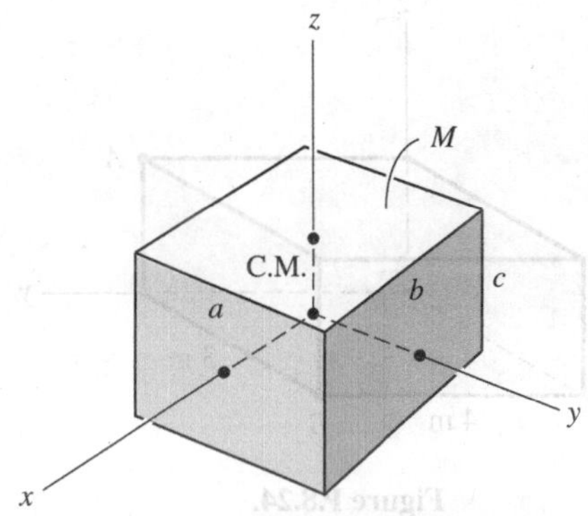

Figure P.8.22.

8.23. A thin plate weighing 100 N has the following mass moments of inertia at mass center *O*:

$$I_{xx} = 15 \text{ kg-m}^2$$
$$I_{yy} = 13 \text{ kg-m}^2$$
$$I_{xy} = -10 \text{ kg-m}^2$$

What are the moments of inertia $I_{x'x'}$, $I_{y'y'}$, and $I_{z'z'}$ at point *P* having the position vector:

$$\boldsymbol{r} = .5\boldsymbol{i} + .2\boldsymbol{j} + .6\boldsymbol{k} \text{ m}$$

Also determine $I_{x'z'}$ at *P*.

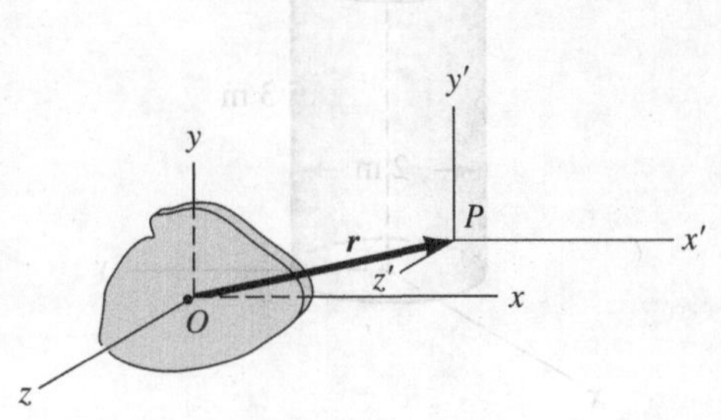

Figure P.8.23.

8.24. A crate with its contents weighs 20 kN and has its center of mass at

$$r_c = 1.3\boldsymbol{i} + 3\boldsymbol{j} + .8\boldsymbol{k}\text{ m}$$

It is known that at corner A,

$$I_{x'x'} = 5{,}500\text{ kg-m}^2$$
$$I_{x'y'} = -1{,}500\text{ kg-m}^2$$

for primed axes parallel to xyz. At point B, find $I_{x''x''}$ and $I_{x''y''}$ for double-primed axes parallel to xyz.

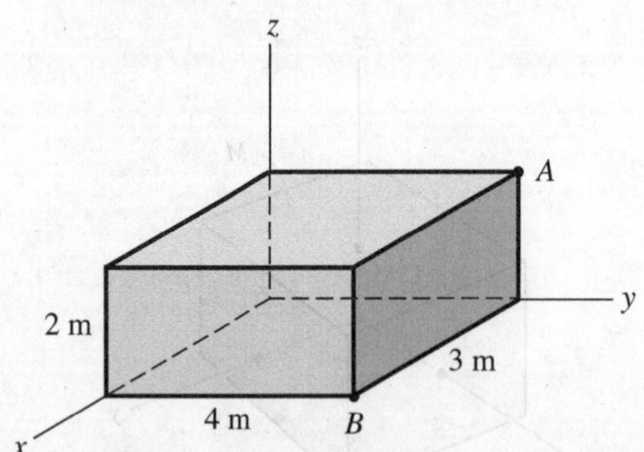

Figure P.8.24.

8.25 A cylindrical crate and its contents weigh 500 N. The center of mass is at

$$r_c = .6\boldsymbol{i} + .7\boldsymbol{j} + 2\boldsymbol{k}\text{ m}$$

It is known that at A,

$$(I_{yy})_A = 85\text{ kg-m}^2$$
$$(I_{yz})_A = -22\text{ kg-m}^2$$

Find I_{yy} and I_{zy} at B.

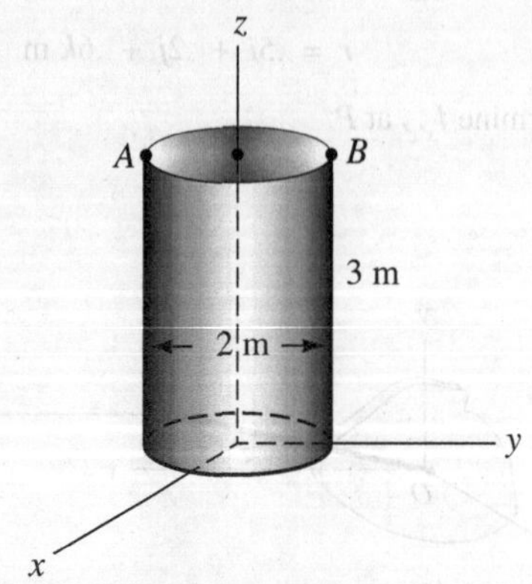

Figure P.8.25.

8.26. A block having a uniform density of 5 grams/cm³ has a hole of diameter 40 mm cut out. What are the principal moments of inertia at point A at the centroid of the right face of the block?

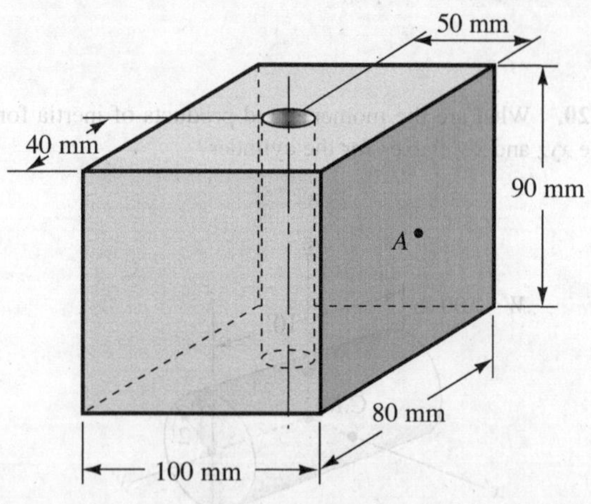

Figure P.8.26.

8.27. Find maximum and minimum moments of inertia at point A. The block weighs 20 N and the cone weighs 14 N.

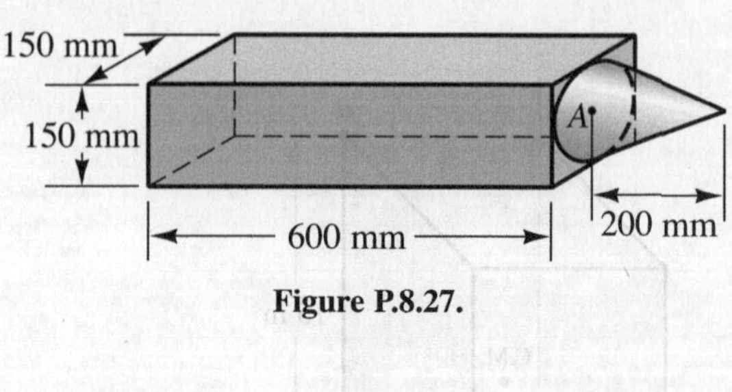

Figure P.8.27.

8.28. Solid spheres C and D each weighing 25 N and having radius of 50 mm are attached to a thin solid rod weighing 30 N. Also, solid spheres E and G each weighing 20 N and having radii of 30 mm are attached to a thin rod weighing 20 N. The rods are attached to be orthogonal to each other. What are the principal moments of inertia at point A?

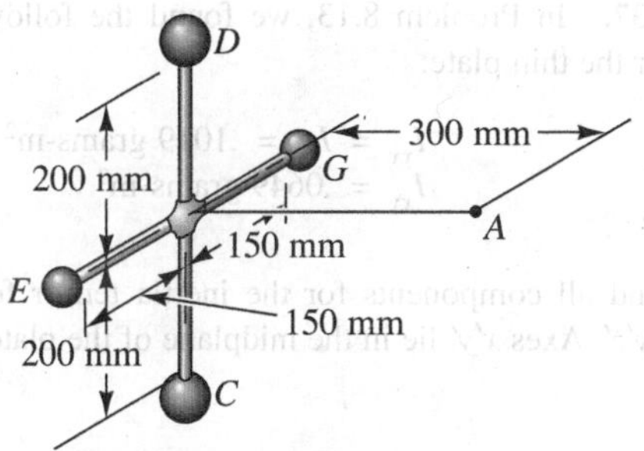

Figure P.8.28.

8.29. A cylinder is shown having a conical cavity oriented along the axis A–A and a cylindrical cavity oriented normal to A–A. If the density of the material is 7,200 kg/m^3, what is I_{AA}?

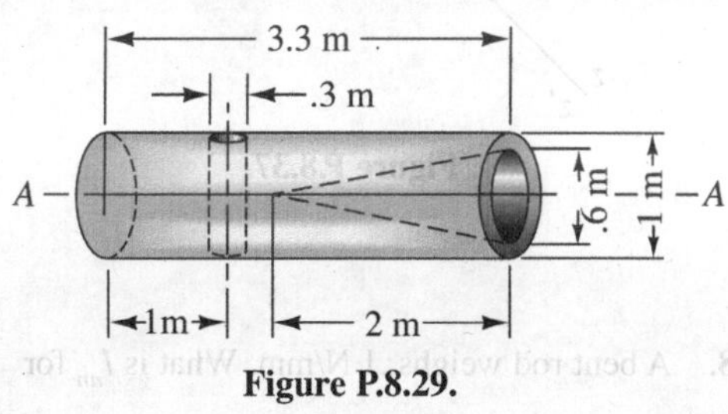

Figure P.8.29.

8.30. A flywheel is made of steel having a specific weight of 490 lb/ft.^{3} What is the moment of inertia about its geometric axis? What is the radius of gyration?

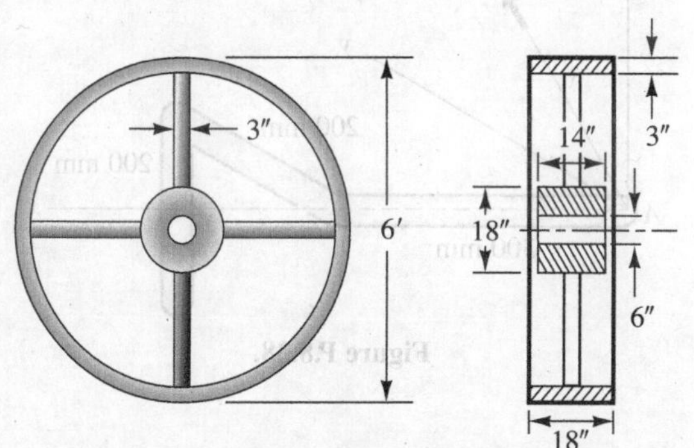

Figure P.8.30.

8.31. Compute I_{yy} and I_{xy} for the right circular cylinder, which has a mass of 50 kg, and the square rod, which has a mass of 10 kg, when the two are joined together so that the rod is radial to the cylinder. The x axis lies along the bottom of the square rod.

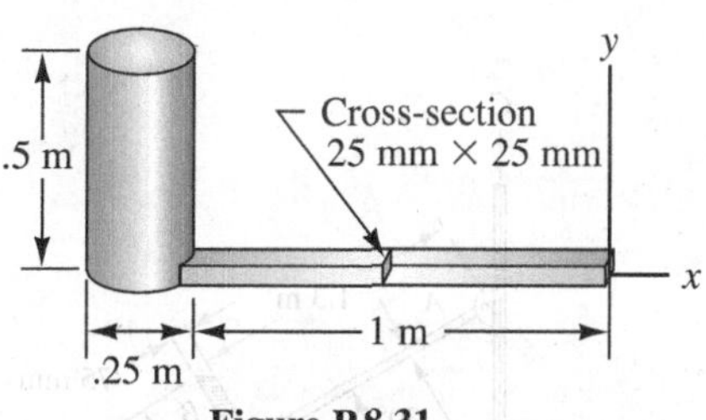

Figure P.8.31.

8.32. Compute the moments and products of inertia for the xy axes. The specific weight is 490 lb/ft^3 throughout.

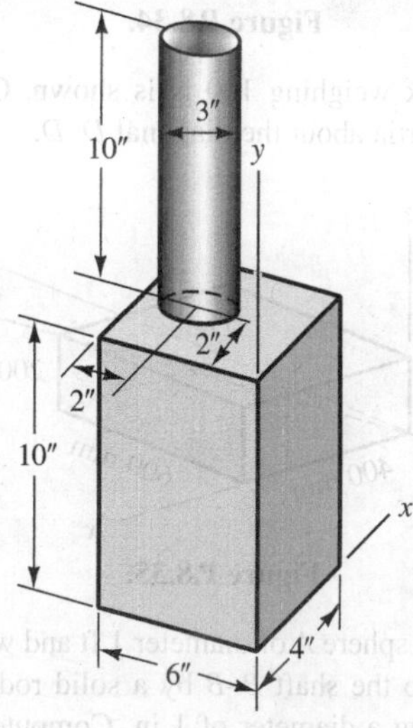

Figure P.8.32.

8.33. A disc A is mounted on a shaft such that its normal is oriented 10° from the centerline of the shaft. The disc has a diameter of 2 ft, is 1 in. in thickness, and weighs 100 lb. Compute the moment of inertia of the disc about the centerline of the shaft.

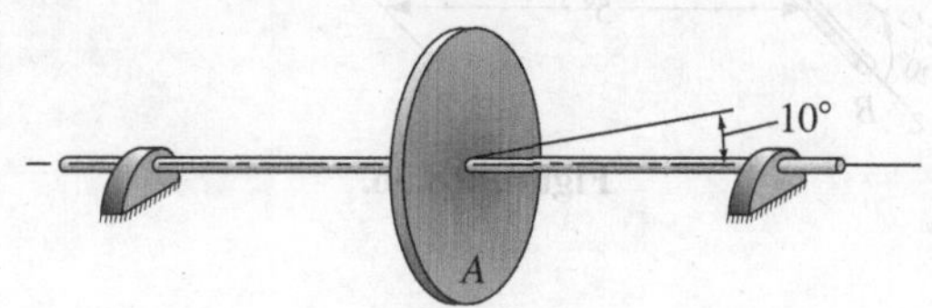

Figure P.8.33.

8.34. A gear B having a mass of 25 kg rotates about axis C–C. If the rod A has a mass distribution of 7.5 kg/m, compute the moment of inertia of A and B about the axis C–C.

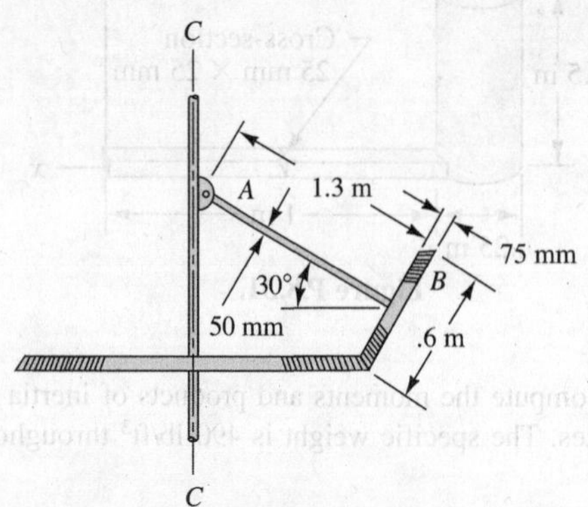

Figure P.8.34.

8.35. A block weighing 100 N is shown. Compute the moment of inertia about the diagonal D–D.

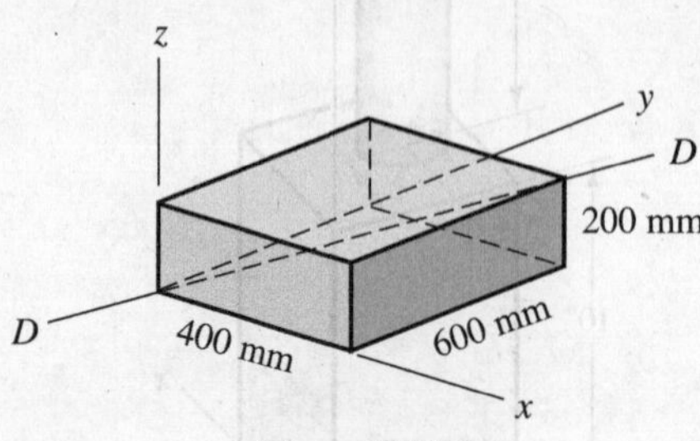

Figure P.8.35.

8.36. A solid sphere A of diameter 1 ft and weight 100 lb is connected to the shaft B–B by a solid rod weighing 2 lb/ft and having a diameter of 1 in. Compute $I_{z'z'}$ for the rod and ball.

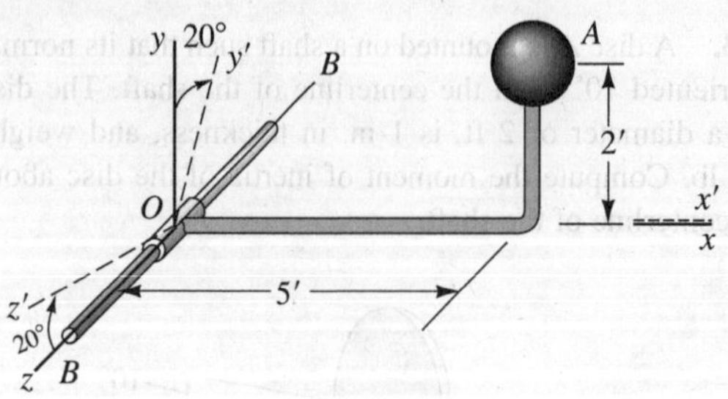

Figure P.8.36.

8.37. In Problem 8.13, we found the following results for the thin plate:

$$I_{xx} = I_{yy} = .1019 \text{ grams-m}^2$$
$$I_{xy} = .0649 \text{ grams-m}^2$$

Find all components for the inertia tensor for reference $x'y'z'$. Axes $x'y'$ lie in the midplane of the plate.

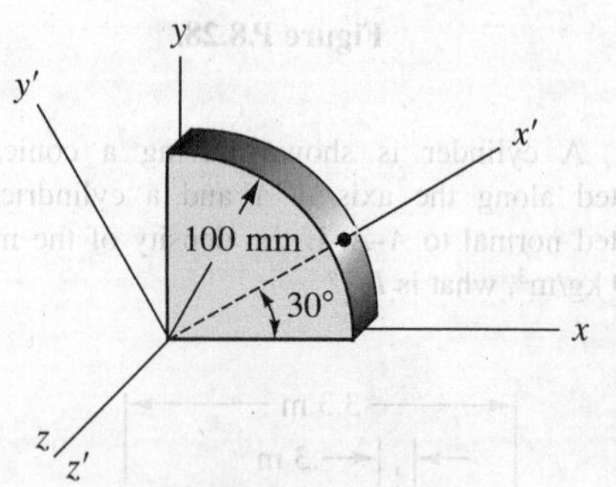

Figure P.8.37.

8.38. A bent rod weighs .1 N/mm. What is I_{nn} for

$$\epsilon_n = .30i + .45j + .841k?$$

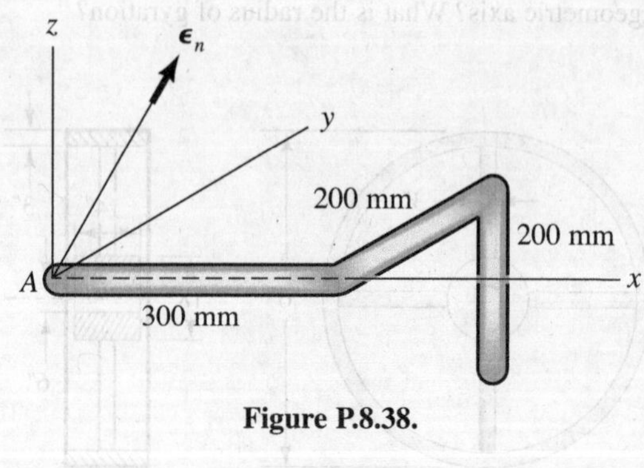

Figure P.8.38.

8.39. Evaluate the matrix of direction cosines for the primed axes relative to the unprimed axes.

$$a_{ij} = \begin{pmatrix} a_{x'x} & a_{x'y} & a_{x'z} \\ a_{y'x} & a_{y'y} & a_{y'z} \\ a_{z'x} & a_{z'y} & a_{z'z} \end{pmatrix}$$

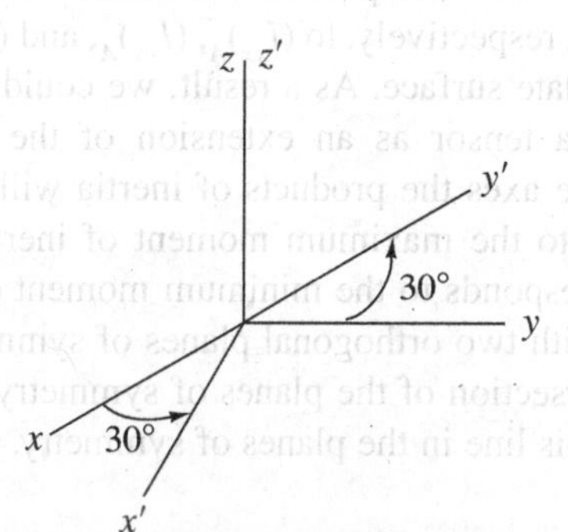

Figure P.8.39.

8.40. The block is uniform in density and weighs 10 N. Find $I_{y'z'}$.

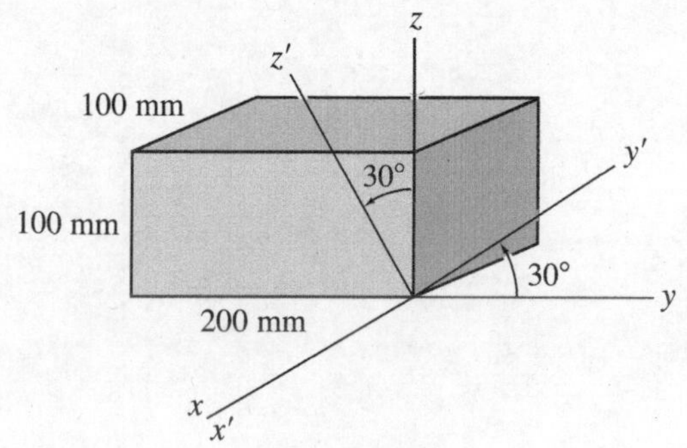

Figure P.8.40.

***8.41.** A thin rod of length 300 mm and weight 12 N is oriented relative to $x'y'z'$ such that

$$\epsilon_n = .4i' + .3j' + .866k'$$

What is $I_{x'y'}$?

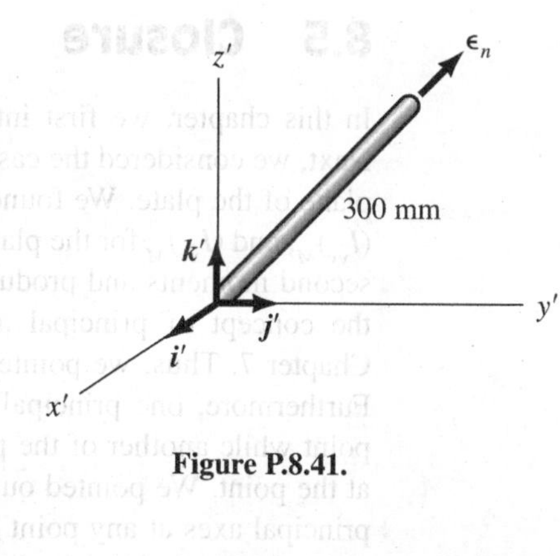

Figure P.8.41.

***8.42.** Show that the transformation equation for the inertia tensor components at a point when there is a rotation of axes (i.e., Eqs. 8.14 and 8.20) can be given as follows:

$$I_{kq} = \sum_j \sum_i a_{ki} a_{qj} I_{ij}$$

where k can be x', y', or z' and q can be x', y', or z', and where i and j go from x to y to z. The equation above is a compact definition of *second-order tensors*. Remember that in the inertia tensor you must have a minus sign in front of each product of inertia term (i.e., $-I_{xy}$, $-I_{yz}$, etc.). [*Hint:* Let $i = x$; then sum over j; then let $i = y$ and sum again over j; etc.]

***8.43.** In Problem 8.42, express the transformation equation to get $I_{y'z'}$ in terms of the inertia tensor components for reference xyz having the same origin as $x'y'z'$.

8.5 Closure

In this chapter, we first introduced the nine components comprising the inertia tensor. Next, we considered the case of the very thin flat plate in which the xy axes form the mid-plane of the plate. We found that the mass moments and products of inertia terms $(I_{xx})_M$, $(I_{yy})_M$, and $(I_{xy})_M$ for the plate are proportional, respectively, to $(I_{xx})_A$, $(I_{yy})_A$, and $(I_{xy})_A$, the second moments and product of area of the plate surface. As a result, we could set forth the concept of principal axes for the inertia tensor as an extension of the work in Chapter 7. Thus, we pointed out that for these axes the products of inertia will be zero. Furthermore, one principal axis corresponds to the maximum moment of inertia at the point while another of the principal axes corresponds to the minimum moment of inertia at the point. We pointed out that for bodies with two orthogonal planes of symmetry, the principal axes at any point on the line of intersection of the planes of symmetry must be along this line of intersection and normal to this line in the planes of symmetry.

PROBLEMS

8.44. Find I_{zz} for the body of revolution having uniform density of .2 kg/mm³. The radial distance out from the z axis to the surface is given as

$$r^2 = -4\,z \text{ mm}^2$$

where z is in millimeters. [*Hint:* Make use of the formula for the moment of inertia about the axis of a disc, $\frac{1}{2}Mr^2$.]

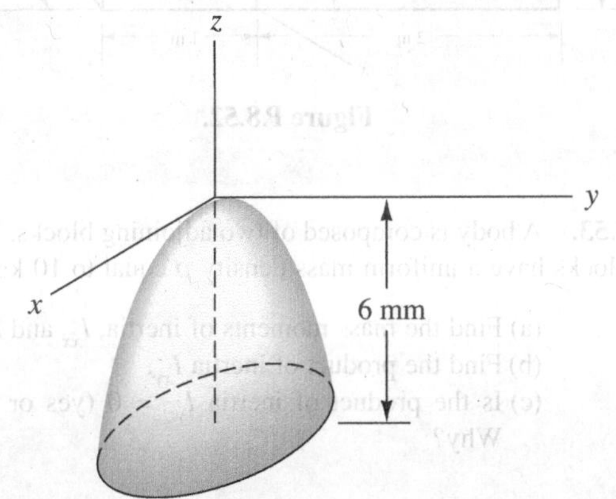

Figure P.8.44.

8.45. In Problem 8.44, determine I_{zz} without using the disc formula but using multiple integration instead.

8.46. What are the inertia tensor components for the thin plate about axes xyz? The plate weighs 2 N.

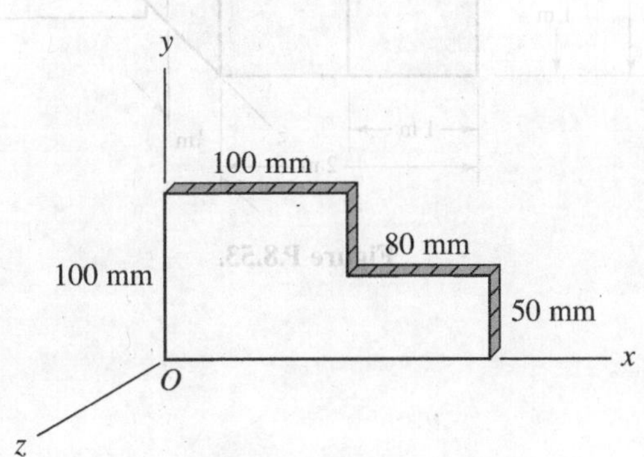

Figure P.8.46.

8.47. In Problem 8.46, what are the principal axes and the principal moments of inertia for the inertia tensor at O?

8.48. What are the principal mass moments of inertia at point O? Block A weighs 15 N. Rod B weighs 6 N and solid sphere C weighs 10 N. The density in each body is uniform. The diameter of the sphere is 50 mm.

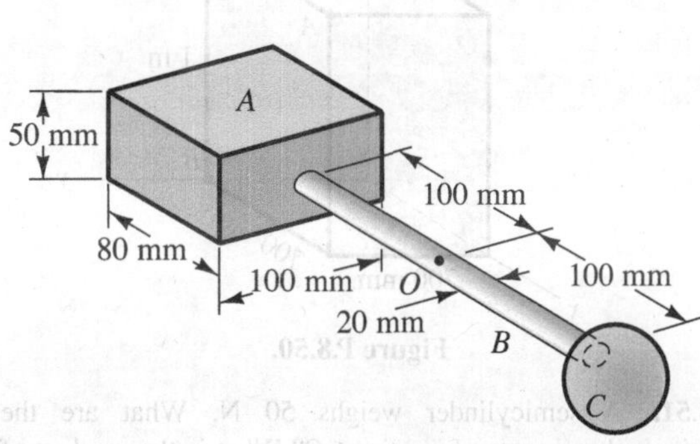

Figure P.8.48.

8.49. The block has a density of 15 kg/m.³ Find the moment of inertia about axis AB.

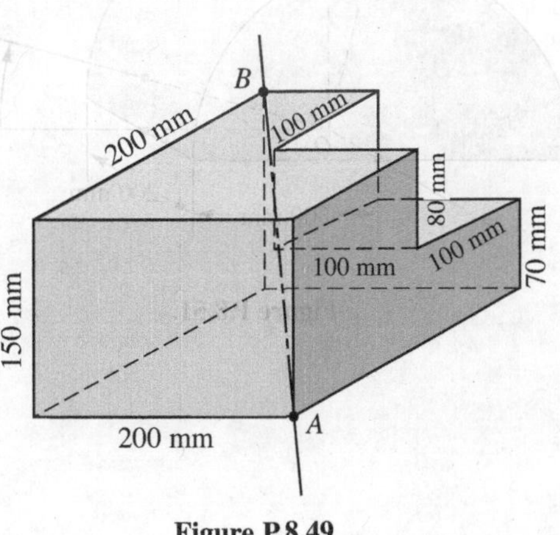

Figure P.8.49

8.50. A crate and its contents weighs 10 kN. The center of mass of the crate and its contents is at

$$\boldsymbol{r}_c = .40\boldsymbol{i} + .30\boldsymbol{j} + .60\boldsymbol{k} \text{ m}$$

If at A we know that

$$I_{yy} = 800 \text{ kg-m}^2$$
$$I_{yz} = 500 \text{ kg-m}^2$$

find I_{yy} and I_{yz} at B.

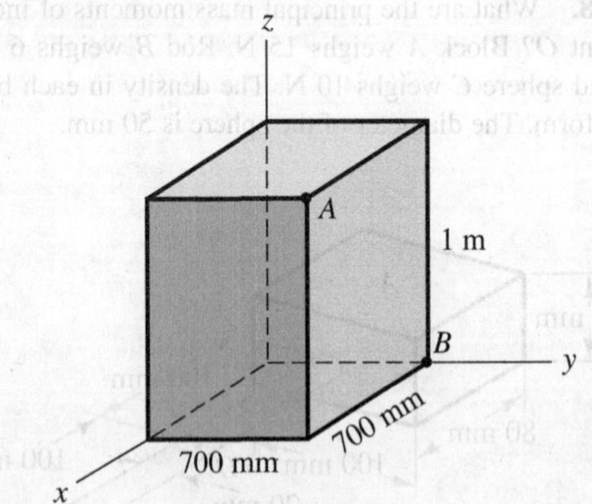

Figure P.8.50.

8.51. A semicylinder weighs 50 N. What are the principal moments of inertia at O? What is the product of inertia $I_{y'z'}$? What conclusion can you draw about the direction of principal axes at O?

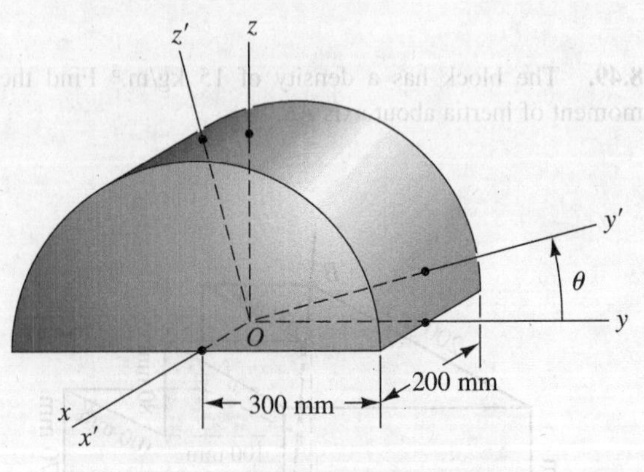

Figure P.8.51.

8.52. Find I_{yy} and I_{yz}. The diameter of A is 0.3 m. B is the center of the right face of the block. Take $\rho = \rho_0$ kg/m.[3]

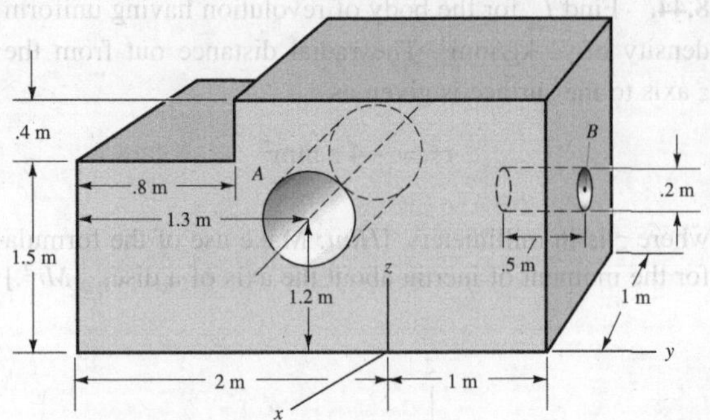

Figure P.8.52.

8.53. A body is composed of two adjoining blocks. Both blocks have a uniform mass density ρ equal to 10 kg/m.[3]

(a) Find the mass moments of inertia, I_{xx} and I_{zz}

(b) Find the product of inertia I_{xy}.

(c) Is the product of inertia $I_{yz} = 0$ (yes or no)? Why?

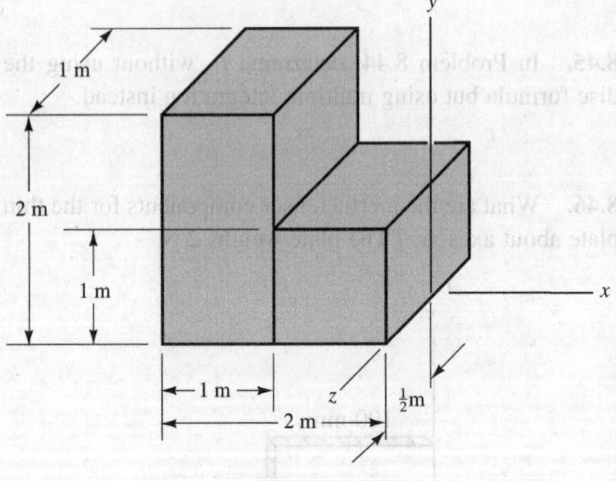

Figure P.8.53.

Dynamics

CHAPTER 9

Kinematics of a Particle—Simple Relative Motion

9.1 Introduction

Kinematics is that phase of mechanics concerned with the study of the motion of particles and rigid bodies without consideration of what has caused the motion. We can consider kinematics as the geometry of motion. Once kinematics is mastered, we can smoothly proceed to the relations between the factors causing the motion and the motion itself. The latter area of study is called dynamics. Dynamics can be conveniently separated into the following divisions, most of which we shall study in this text:

1. Dynamics of a single particle. (You will remember from our chapters on statics that a particle is an idealization having no volume but having mass.)
2. Dynamics of a system of particles. This follows division 1 logically and forms the basis for the motion of continuous media such as fluid flow and rigid-body motion.
3. Dynamics of a rigid body. A large portion of this text is concerned with this important part of mechanics.
4. Dynamics of a system of rigid bodies.
5. Dynamics of a continuous deformable medium.

Clearly, from our opening statements, the particle plays a vital role in the study of dynamics. What is the connection between the particle, which is a completely hypothetical concept, and the finite bodies encountered in physical problems? Briefly the relation is this: In many problems, the size and shape of a body are not relevant in the discussion of certain aspects of its motion; only the mass of the object is significant for such computations. For example, in towing a truck up a hill, as shown in Fig. 9.1, we would only be concerned with the mass of the truck and not with its shape or size (if we neglect forces from the wind, etc., and the rotational effects of the wheels). The truck can just as well be considered a particle in computing the necessary towing force.

Figure 9.1. Truck considered as a particle.

We can present this relationship more precisely in the following manner. The equation of motion of the center of mass of any body can be formed by:

1. Concentrating the entire mass at the mass center of the body.
2. Applying the total resultant force acting on the body to this hypothetical particle.

When the motion of the mass center characterizes all we need to know about the motion of the body, we employ the particle concept (i.e., we find the motion of the mass center). Thus, if all points of a body have the same velocity at any time t (this is called *translatory motion*), we need only know the motion of the mass center to fully characterize the motion. (This was the case for the truck, where the rotational inertia of the wheels was neglected.) If, additionally, the size of a body is small compared to its trajectory (as in planetary motion, for example), the motion of the center of mass is all that might be needed, and so again we can use the particle concept for such bodies.

Part A: General Notions

9.2 Differentiation of a Vector with Respect to Time

In the study of statics, we dealt with vector quantities. We found it convenient to incorporate the directional nature of these quantities in a certain notation and set of operations. We called the totality of these very useful formulations "vector algebra." We shall again expand our thinking from scalars to vectors—this time for the operations of differentiation and integration with respect to any scalar variable t (such as time).

For scalars, we are concerned only with the variation in magnitude of some quantity that is changing with time. The scalar definition of the time derivative, then, is given as

$$\frac{df(t)}{dt} = \lim_{\Delta t \to 0} \left[\frac{f(t + \Delta t) - f(t)}{\Delta t} \right] \tag{9.1}$$

This operation leads to another function of time, which can once more be differentiated in this manner. The process can be repeated again and again, for suitable functions, to give higher derivatives.

In the case of a vector, the variation in time may be a change in magnitude, a change in direction, or both. The formal definition of the derivative of a vector $\boldsymbol{F}$ with respect to time has the same form as Eq. 9.1:

$$\frac{d\boldsymbol{F}}{dt} = \lim_{\Delta t \to 0} \left[\frac{\boldsymbol{F}(t + \Delta t) - \boldsymbol{F}(t)}{\Delta t} \right] \tag{9.2}$$

If $\boldsymbol{F}$ has no change in direction during the time interval, this operation differs little from the scalar case. However, when $\boldsymbol{F}$ changes in direction, we find for the derivative of $\boldsymbol{F}$ a new vector, having a magnitude as well as a direction, that is different from $\boldsymbol{F}$ itself. This directional consideration can be somewhat troublesome.

Let us consider the rate of change of the position vector for a reference *xyz* of a particle with respect to time; this rate is defined as the *velocity vector*, $\boldsymbol{V}$, of the particle relative to *xyz*. Following the definition given by Eq. 9.2, we have

$$\frac{d\boldsymbol{r}}{dt} = \lim_{\Delta t \to 0} \left[\frac{\boldsymbol{r}(t + \Delta t) - \boldsymbol{r}(t)}{\Delta t} \right]$$

The position vectors given in brackets are shown in Fig. 9.2. The subtraction between the two vectors gives rise to the displacement vector $\Delta\boldsymbol{r}$, which is shown as a chord connecting two points Δs apart along the trajectory of the particle. Hence, we can say (using the chain rule) that

$$\frac{d\boldsymbol{r}}{dt} = \lim_{\Delta t \to 0} \left(\frac{\Delta \boldsymbol{r}}{\Delta t} \right) = \lim_{\Delta t \to 0} \left(\frac{\Delta \boldsymbol{r}}{\Delta s} \frac{\Delta s}{\Delta t} \right)$$

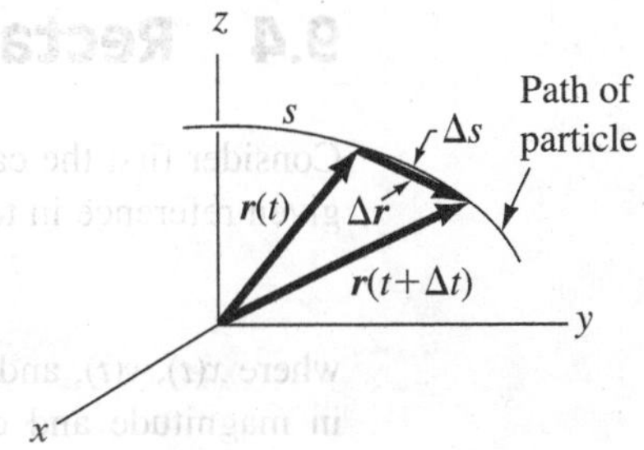

Figure 9.2. Particle at times t and $t + \Delta t$.

where we have multiplied and divided by Δs in the last expression. As Δt goes to zero, the direction of $\Delta\boldsymbol{r}$ approaches tangency to the trajectory at position $\boldsymbol{r}(t)$ and approaches Δs in magnitude. Consequently, in the limit, $\Delta\boldsymbol{r}/\Delta s$ becomes a unit vector $\boldsymbol{\epsilon}_t$, tangent to the trajectory. That is

$$\frac{\Delta \boldsymbol{r}}{\Delta s} \to \frac{\Delta s \boldsymbol{\epsilon}_t}{\Delta s} = \boldsymbol{\epsilon}_t \qquad \therefore \frac{d\boldsymbol{r}}{ds} = \boldsymbol{\epsilon}_t \tag{9.3}$$

We can then say

$$\frac{d\boldsymbol{r}}{dt} = \boldsymbol{V} = \lim_{\Delta t \to 0} \left[\left(\frac{\Delta s}{\Delta t} \right) \left(\frac{\Delta \boldsymbol{r}}{\Delta s} \right) \right] = \frac{ds}{dt} \boldsymbol{\epsilon}_t \tag{9.4}$$

Therefore, $d\boldsymbol{r}/dt$ leads to a vector having a magnitude equal to the speed of the particle and a direction tangent to the trajectory. Keep in mind that there can be any angle between the position vector and the velocity vector. Students seem to want to limit this angle to 90°, which actually restricts you to a circular path. The acceleration vector of a particle can then be given as

$$\boldsymbol{a} = \frac{d\boldsymbol{V}}{dt} = \frac{d^2\boldsymbol{r}}{dt^2} \tag{9.5}$$

The differentiation and integration of vectors $\boldsymbol{r}$, $\boldsymbol{V}$, and $\boldsymbol{a}$ will concern us throughout the text.

Part B: Velocity and Acceleration Calculations

9.3 Introductory Remark

As you know from statics, we can express a vector in many ways. For instance, we can use rectangular components, or, as we will shortly explain, we can use cylindrical components. In evaluating derivatives of vectors with respect to time, we must proceed in accordance with the manner in which the vector has been expressed. In Part B of this chapter, we will therefore examine certain differentiation processes that are used extensively in mechanics. Other differentiation processes will be examined later at appropriate times.

We have already carried out a derivative operation in Section 9.2 directly on the vector $\boldsymbol{r}$. You will see in Section 9.5 that the approach used gives the derivative in terms of *path variables*. This approach will be one of several that we shall now examine with some care.

9.4 Rectangular Components

Consider first the case where the position vector $\boldsymbol{r}$ of a moving particle is expressed for a given reference in terms of rectangular components in the following manner:

$$\boldsymbol{r}(t) = x(t)\boldsymbol{i} + y(t)\boldsymbol{j} + z(t)\boldsymbol{k} \tag{9.6}$$

where $x(t)$, $y(t)$, and $z(t)$ are scalar functions of time. The unit vectors $\boldsymbol{i}, \boldsymbol{j}$, and $\boldsymbol{k}$ are fixed in magnitude and direction at all times, and so we can obtain $d\boldsymbol{r}/dt$ in the following straightforward manner:

$$\frac{d\boldsymbol{r}}{dt} = \boldsymbol{V}(t) = \frac{dx(t)}{dt}\boldsymbol{i} + \frac{dy(t)}{dt}\boldsymbol{j} + \frac{dz(t)}{dt}\boldsymbol{k} = \dot{x}(t)\boldsymbol{i} + \dot{y}(t)\boldsymbol{j} + \dot{z}(t)\boldsymbol{k} \tag{9.7}$$

A second differentiation with respect to time leads to the acceleration vector:

$$\frac{d^2\boldsymbol{r}}{dt^2} = \boldsymbol{a} = \ddot{x}(t)\boldsymbol{i} + \ddot{y}(t)\boldsymbol{j} + \ddot{z}(t)\boldsymbol{k} \tag{9.8}$$

By such a procedure, we have formulated velocity and acceleration vectors in terms of components parallel to the coordinate axes.

Up to this point, we have formulated the rectangular velocity components and the rectangular acceleration components, respectively, by differentiating the position vector once and twice with respect to time. Quite often, we know the acceleration vector of a particle as a function of time in the form

$$\boldsymbol{a}(t) = \ddot{x}(t)\boldsymbol{i} + \ddot{y}(t)\boldsymbol{j} + \ddot{z}(t)\boldsymbol{k} \tag{9.9}$$

and wish to have for this particle the velocity vector or the position vector or any of their components at any time. We then integrate the time function $\ddot{x}(t)$, $\ddot{y}(t)$, and $\ddot{z}(t)$, remembering to include a constant of integration for each integration. For example, consider $\ddot{x}(t)$. Integrating once, we obtain the velocity component $V_x(t)$ as follows:

$$V_x(t) = \int \ddot{x}(t)\,dt + C_1 \tag{9.10}$$

where C_1 is the constant of integration. Knowing V_x at some time t_0, we can determine C_1 by substituting t_0 and $(V_x)_0$ into the equation above and determining C_1. Similarly, for $x(t)$ we obtain from the above:

$$x(t) = \int\left[\int \ddot{x}(t)\,dt\right]dt + C_1 t + C_2 \tag{9.11}$$

where C_2 is the second constant of integration. Knowing x at some time t, we can determine C_2 from Eq. 9.11. The same procedure involving additional constants applies to the other acceleration components.

We now illustrate the procedures described above in the following series of examples.

Example 9.1

Pins A and B must always remain in the vertical slot of yoke C, which moves to the right at a constant speed of 6 ft/sec in Fig. 9.3. Furthermore, the pins cannot leave the elliptic slot. (a) What is the speed at which the pins approach each other when the yoke slot is at $x = 5$ ft? (b) What is the rate of change of speed toward each other when the yoke slot is at $x = 5$ ft?

The equation of the elliptic path in which the pins must move is seen by inspection to be

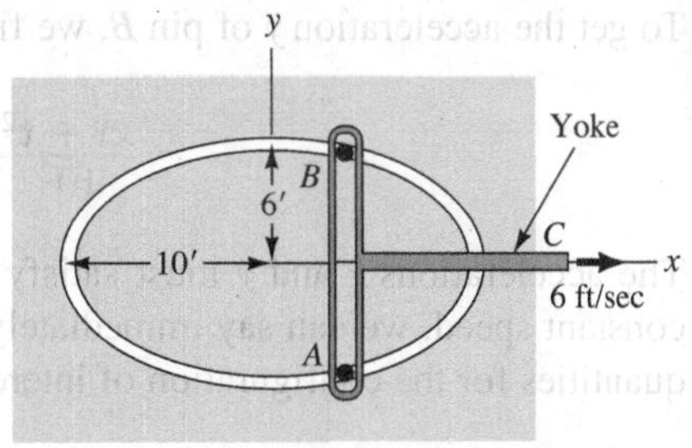

Figure 9.3. Pin slides in slot and yoke.

$$\frac{x^2}{10^2} + \frac{y^2}{6^2} = 1 \quad \text{(a)}$$

Clearly, if coordinates (x, y) are to represent the coordinates of pin B, they must be time functions such that for any time t the values $x(t)$ and $y(t)$ satisfy Eq. (a). Also, $\dot{x}(t)$ and $\dot{y}(t)$ must be such that pin B moves at all times in the elliptic path. We can satisfy these requirements by first differentiating Eq. (a) with respect to time. Canceling the factor 2, we obtain

$$\frac{x\dot{x}}{10^2} + \frac{y\dot{y}}{6^2} = 0 \quad \text{(b)}$$

Now $x(t)$, $y(t)$, $\dot{x}(t)$, and $\dot{y}(t)$ must satisfy Eq. (b) for all values of t to ensure that B remains in the elliptic path.

We can now proceed to solve part (a) of this problem. We know that pin B must have a velocity $\dot{x} = 6$ ft/sec because of the yoke. Furthermore, when $x = 5$ ft, we know from Eq. (a) that

$$\frac{5^2}{10^2} + \frac{y^2}{6^2} = 1 \quad \text{(c)}$$

$$\therefore y = 5.20 \text{ ft}$$

Now going to Eq. (b), we can solve for $\dot{y}$ at the instant of interest.

$$\frac{(5)(6)}{10^2} + \frac{(5.20)(\dot{y})}{6^2} = 0$$

Therefore,

$$\dot{y} = -2.08 \text{ ft/sec}$$

Thus, pin B moves downward with a speed of 2.08 ft/sec. Clearly, pin A must move upward with the same speed of 2.08 ft/sec. The pins approach each other at the instant of interest at a speed of 4.16 ft/sec.

Example 9.1 (Continued)

To get the acceleration $\ddot{y}$ of pin B, we first differentiate Eq. (b) with respect to time.

$$\frac{x\ddot{x} + \dot{x}^2}{10^2} + \frac{y\ddot{y} + \dot{y}^2}{6^2} = 0 \qquad \text{(d)}$$

The accelerations $\ddot{x}$ and $\ddot{y}$ must satisfy the equation above. Since the yoke moves at constant speed, we can say immediately that $\ddot{x} = 0$. And using for x, y, $\dot{x}$, and $\dot{y}$ known quantities for the configuration of interest, we can solve for $\ddot{y}$ from Eq. (d). Thus,

$$\frac{0 + 6^2}{10^2} + \frac{5.20\ddot{y} + 2.08^2}{6^2} = 0$$

Therefore,

$$\ddot{y} = -3.32 \text{ ft/sec}^2$$

Pin B must be accelerating downward at a rate of 3.32 ft/sec^2 while pin A accelerates upward at the same rate. The pins accelerate toward each other, then, at a rate of 6.64 ft/sec^2 at the configuration of interest.

In the motion of particles near the earth's surface, such as the motion of shells or ballistic missiles, we can often simplify the problem by neglecting air resistance and taking the acceleration of gravity g as constant (32.2 ft/sec^2 or 9.81 m/sec^2). For such a case (see Fig. 9.4), we know immediately that $\ddot{y}(t) = -g$ and $\ddot{x}(t) = \ddot{z}(t) = 0$. On integrating these accelerations, we can often determine for the particle useful information as to velocities or positions at certain times of interest in the problem. We illustrate this procedure in the following examples.

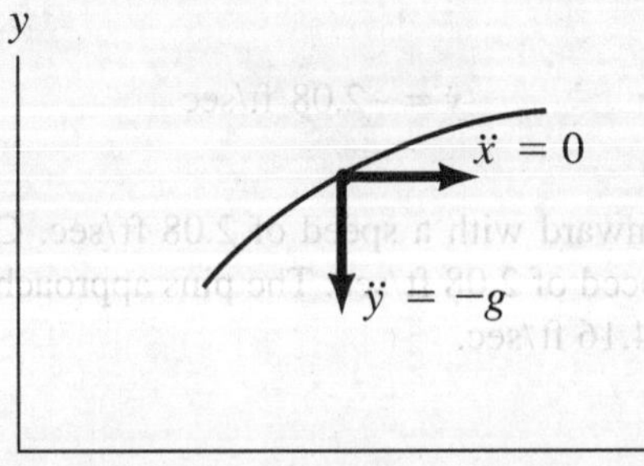

Figure 9.4. Simple ballistic motion of a shell.

Example 9.2

Ballistics Problem 1. A shell is fired from a hill 500 ft above a plain. The angle α of firing (see Fig. 9.5) is 15° above the horizontal, and the muzzle velocity V_0 is 3,000 ft/sec. At what horizontal distance, d, will the shell hit the plain if we neglect friction of the air? What is the maximum height of the shell above the plain? Finally, determine the trajectory of the shell [i.e., find $y = f(x)$].

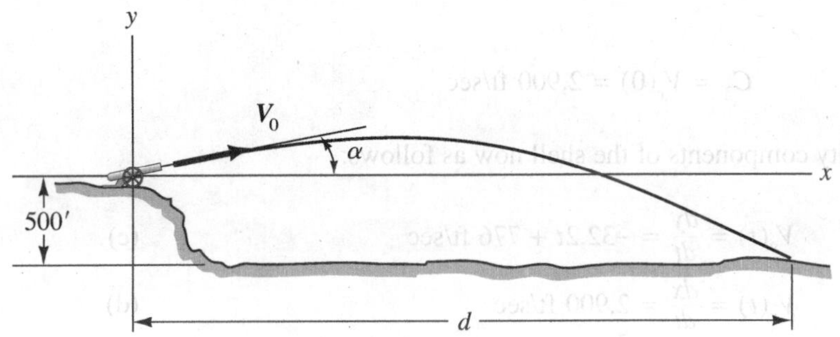

Figure 9.5 Ballistics problem: find d.

We know immediately that

$$\ddot{y}(t) = \frac{dV_y}{dt} = -32.2 \text{ ft/sec}^2 \tag{a}$$

$$\ddot{x}(t) = \frac{dV_x}{dt} = 0 \tag{b}$$

We need not bother with $\ddot{z}(t)$, since the motion is coplanar with $\dot{z}(t) = z = 0$ at all times. We next separate the velocity variables from the time variables by bringing dt to the right sides of the previous equations. Thus

$$dV_y = -32.2\, dt$$
$$dV_x = 0\, dt$$

Integrating the above equations, we get

$$V_y(t) = -32.2t + C_1$$
$$V_x(t) = C_2$$

We shall take $t = 0$ at the instant the cannon is fired. At this instant, we know V_y and V_x and can determine C_1 and C_2. Thus,

$$V_y(0) = 3{,}000 \sin 15° = (-32.2)(0) + C_1$$

Example 9.2 (Continued)

Therefore,

$$C_1 = V_y(0) = 776 \text{ ft/sec}$$

Also

$$V_x(0) = 3{,}000 \cos 15° = C_2$$

Therefore,

$$C_2 = V_x(0) = 2{,}900 \text{ ft/sec}$$

We can give the velocity components of the shell now as follows:

$$V_y(t) = \frac{dy}{dt} = -32.2t + 776 \text{ ft/sec} \qquad \text{(c)}$$

$$V_x(t) = \frac{dx}{dt} = 2{,}900 \text{ ft/sec} \qquad \text{(d)}$$

Thus, the horizontal velocity is constant. Separating the position and time variables and then integrating, we get the x and y coordinates of the shell.

$$y(t) = -32.2\frac{t^2}{2} + 776t + C_3 \qquad \text{(e)}$$

$$x(t) = 2{,}900t + C_4 \qquad \text{(f)}$$

When $t = 0$, $y = x = 0$. Thus, from Eqs. (e) and (f), we clearly see that $C_3 = C_4 = 0$. The coordinates of the shell are then

$$y(t) = -16.1t^2 + 776t \qquad \text{(g)}$$

$$x(t) = 2{,}900t \qquad \text{(h)}$$

To determine *distance d*, first find the time t for the impact of the shell on the plain. That is, set $y = -500$ in Eq. (g) and solve for the time t. Thus,

$$-500 = -16.1t^2 + 776t$$

Therefore,

$$16.1t^2 - 776t - 500 = 0$$

Using the quadratic formula, we get for t:

$$t = 48.8 \text{ sec}$$

Example 9.2 (Continued)

Substituting this value of t into Eq. (h), we get

$$d = (2{,}900)(48.8) = \boxed{141{,}500 \text{ ft}}$$

To get the *maximum height* y_{max}, first find the time t when $V_y = 0$. Thus, from Eq. (c) we get

$$0 = -32.2t + 776$$

Therefore,

$$t = 24.1 \text{ sec}$$

Now substitute $t = 24.1$ sec into Eq. (g). This gives us y_{max}.

$$y_{max} = -(16.1)(24.1)^2 + (776)(24.1)$$

$$\therefore \quad \boxed{y_{max} = 9{,}350 \text{ ft}}$$

Finally, to get the *trajectory* of the shell (i.e., y as a function of x), solve for t in Eq. (h) and substitute this into Eq. (g). We then have

$$y = -(16.1)\left(\frac{x}{2{,}900}\right)^2 + (776)\left(\frac{x}{2{,}900}\right)$$

Therefore,

$$\boxed{y = -1.917 \times 10^{-6} x^2 + .268x}$$

Clearly, the trajectory is that of a *parabola*.

Example 9.3

Ballistics Problem 2. A gun emplacement is shown on a cliff in Fig. 9.6. The muzzle velocity of the gun is 1,000 m/sec. At what angle α must the gun point in order to hit target A shown in the diagram? Neglect friction.

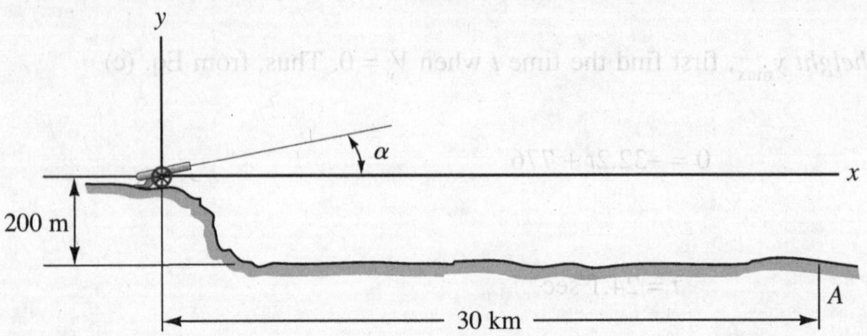

Figure 9.6. Find α to hit A.

Newton's law for the shell is given as follows for a reference xy having its origin at the gun.

$$\ddot{y}(t) = -9.81$$
$$\ddot{x}(t) = 0$$

Integrating, we get

$$\dot{y}(t) = V_y(t) = -9.81t + C_1 \quad \text{(a)}$$
$$\dot{x}(t) = V_x(t) = C_2 \quad \text{(b)}$$

When $t = 0$, we have $\dot{y} = 1{,}000 \sin \alpha$ and $\dot{x} = 1{,}000 \cos \alpha$. Applying these conditions to Eqs. (a) and (b), we solve for C_1 and C_2. Thus,

$$1{,}000 \sin \alpha = 0 + C_1$$

Therefore,

$$C_1 = 1{,}000 \sin \alpha$$

Also,

$$1{,}000 \cos \alpha = C_2$$

Therefore,

$$C_2 = 1{,}000 \cos \alpha$$

Hence, we have

$$\dot{y}(t) = -9.81t + 1{,}000 \sin \alpha$$
$$\dot{x}(t) = 1{,}000 \cos \alpha$$

Example 9.3 (Continued)

Integrating again, we get

$$y(t) = -9.81\frac{t^2}{2} + 1{,}000 \sin \alpha\, t + C_3$$

$$x(t) = 1{,}000 \cos \alpha\, t + C_4$$

When $t = 0$, $x = y = 0$. Hence, it is clear that $C_3 = C_4 = 0$. Thus, we have

$$y = -4.095t^2 + 1{,}000 \sin \alpha\, t \quad \text{(c)}$$

$$x = 1{,}000 \cos \alpha\, t \quad \text{(d)}$$

To get the *trajectory*, we solve for t in Eq. (d) and substitute into Eq. (c).

$$y = -4.095 \frac{x^2}{(1{,}000 \cos \alpha)^2} + 1{,}000 \sin \alpha \frac{x}{(1{,}000 \cos \alpha)}$$

$$= -4.095 \times 10^{-6} \frac{x^2}{\cos^2 \alpha} + x \tan \alpha \quad \text{(e)}$$

where we have replaced $\sin \alpha/\cos \alpha$ by $\tan \alpha$. When $x = 30$ km (i.e., 30,000 m), $y = -200$ m. Hence, we have on substituting these data into Eq. (e):

$$-200 = -4.095 \times 10^{-6} \frac{(30{,}000)^2}{\cos^2 \alpha} + 30{,}000 \tan \alpha$$

Replace $1/\cos^2 \alpha$ by $\sec^2 \alpha = \left(1 + \tan^2 \alpha\right)$:

$$-200 = -4.095 \times 10^{-6}(30{,}000)^2(1 + \tan^2 \alpha) + 30{,}000 \tan \alpha$$

Therefore,

$$\tan^2 \alpha - 6.796 \tan \alpha + .955 = 0 \quad \text{(f)}$$

Using the quadratic formula, we find the following angles:

$$\alpha_1 = 8.17°$$
$$\alpha_2 = 81.44°$$

There are thus *two* possible firing angles that will permit the shell to hit the target, as shown in Fig. 9.7.

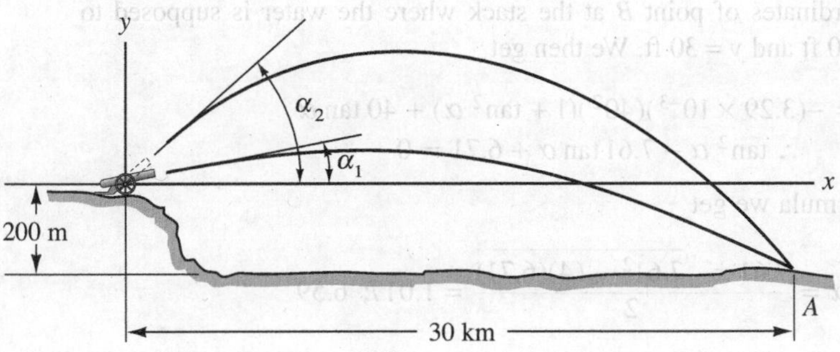

Figure 9.7. Two firing angles are possible.

Example 9.4

The engine room of a freighter is on fire. A fire-fighting tugboat has drawn alongside and is directing a stream of water to enter the stack of the freighter as shown in Fig. 9.8. If the initial speed of the jet of water is 70 ft/sec, is there a value of α of the issuing jet of water that will do the job? If so, what should α be?

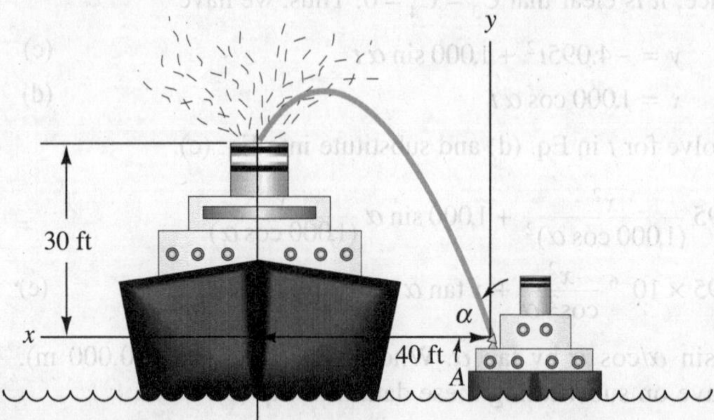

Figure 9.8. Fire-fighting tugboat directing a jet of water into the stack of a freighter.

Consider a particle within the stream of water. Neglecting friction, **Newton's law** for the particle is given as follows:

$$\ddot{y} = -32.2 \text{ ft/sec}^2 \qquad \ddot{x} = 0 \text{ ft/sec}^2$$

Integrating twice, and using initial conditions at A, we get

$$\dot{y} = -32.2t + 70 \sin \alpha \text{ ft/sec} \quad \text{(a)} \qquad \dot{x} = 70 \cos \alpha \text{ ft/sec} \quad \text{(c)}$$

$$y = -16.1t^2 + 70 \sin \alpha\, t \text{ ft} \quad \text{(b)} \qquad x = 70 \cos \alpha\, t \text{ ft} \quad \text{(d)}$$

Solve for t from Eq. (d) and substitute into Eq. (b) to get

$$y = -16.1\left[\frac{x}{70 \cos \alpha}\right]^2 + 70 \sin \alpha \left[\frac{x}{70 \cos \alpha}\right]$$

Replace $\cos^2 \alpha$ by $1/(1 + \tan^2 \alpha)$ and $(\sin \alpha/\cos \alpha)$ by $\tan \alpha$ in the previous equation and then substitute the coordinates of point B at the stack where the water is supposed to reach. That is, set $x = 40$ ft and $y = 30$ ft. We then get

$$30 = -(3.29 \times 10^{-3})(40^2)(1 + \tan^2 \alpha) + 40 \tan \alpha$$

$$\therefore \tan^2 \alpha - 7.61 \tan \alpha + 6.71 = 0$$

Using the quadratic formula we get

$$\tan \alpha = \frac{7.61 \pm \sqrt{7.61^2 - (4)(6.71)}}{2} = 1.017;\ 6.59$$

Example 9.4 (Continued)

We thus have two angles for α, each of which will theoretically cause the stream to go to point B of the stack. These angles are

$$\alpha_1 = 45.50^\circ \qquad \alpha_2 = 81.37^\circ$$

Does one, none, or both angles above yield a stream of water that will come down at B so as to enter the stack? We can determine this by finding the maximum value of y and locating the position x for this maximum value. To do this, we set $\dot{y} = 0$ and solve for t using each α. Thus we have

$$0 = -32.2t + 70 \sin\begin{Bmatrix} 45.50^\circ \\ 81.37^\circ \end{Bmatrix}$$

$$\therefore\ t = \begin{Bmatrix} 1.551 \\ 2.149 \end{Bmatrix} \text{ sec}$$

Now get the position x for maximum elevation for each case as well as the elevation maximum, $y_{\max}$.

For $\alpha = 45.50^\circ$:

$$x = (70)(\cos 45.50^\circ)(1.551) = 76.1 \text{ ft}$$

$$y_{\max} = -(16.1)(1.551)^2 + (70)(\sin 45.50^\circ)(1.551) = 38.7 \text{ ft}$$

For $\alpha = 81.37^\circ$:

$$x = (70)(\cos 81.37^\circ)(2.149) = 22.6 \text{ ft}$$

$$y_{\max} = -(16.1)(2.149)^2 + (70)(\sin 81.37^\circ)(2.149) = 74.4 \text{ ft}$$

A sketch of the two possible trajectories is shown in Fig. 9.9. Clearly the shallow trajectory will hit the side of the stack and is unacceptable, while the high trajectory will deposit water inside the stack and is thus the desired trajectory. Thus,

$$\boxed{\alpha = 81.37^\circ}$$

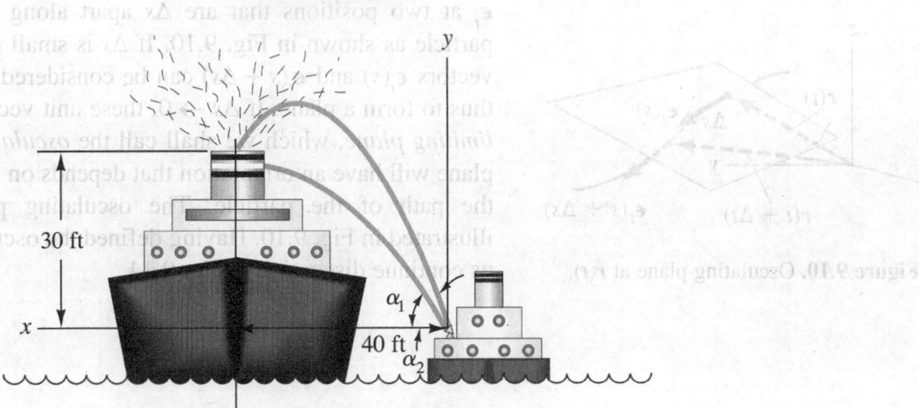

Figure 9.9. Two possible trajectories of the jet.

We do not always know the variation of the position vector with time in the form of Eq. 9.6. Furthermore, it may be that the components of velocity and acceleration that we desire are not those parallel to a fixed Cartesian reference. The evaluation of $\boldsymbol{V}$ and $\boldsymbol{a}$ for certain other circumstances will be considered in the following sections.

9.5 Velocity and Acceleration in Terms of Path Variables

We have formulated velocity and acceleration for the case where the rectangular coordinates of a particle are known as functions of time. We now explore another approach in which the formulations are carried out in terms of the path variables of the particle, that is, in terms of geometrical parameters of the path and the speed and the rate of change of speed of the particle along the path. These results are particularly useful when a particle moves along a path that we know apriori (such as the case of a roller coaster).

As a matter of fact, in Section 9.2 (Eq. 9.4) we expressed the velocity vector in terms of path variables in the following form:

$$\boldsymbol{V} = \frac{ds}{dt}\boldsymbol{\epsilon}_t \tag{9.12}$$

where ds/dt represents the speed along the path and $\boldsymbol{\epsilon}_t = d\boldsymbol{r}/ds$ (see Eq. 9.3) is the unit vector tangent to the path (and hence collinear with the velocity vector). The acceleration becomes

$$\frac{d\boldsymbol{V}}{dt} = \boldsymbol{a} = \frac{d^2 s}{dt^2}\boldsymbol{\epsilon}_t + \frac{ds}{dt}\frac{d\boldsymbol{\epsilon}_t}{dt} \tag{9.13}$$

Replace $d\boldsymbol{\epsilon}_t/dt$ in this expression by $(d\boldsymbol{\epsilon}_t/ds)(ds/dt)$, the validity of which is assured by the chain rule of differentiation. We then have

$$\boldsymbol{a} = \frac{d^2 s}{dt^2}\boldsymbol{\epsilon}_t + \left(\frac{ds}{dt}\right)^2 \frac{d\boldsymbol{\epsilon}_t}{ds} \tag{9.14}$$

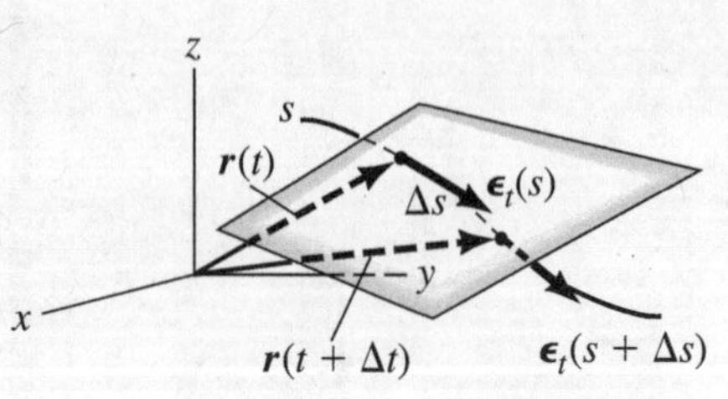

Figure 9.10. Osculating plane at $\boldsymbol{r}(t)$.

Before proceeding further, let us consider the unit vector $\boldsymbol{\epsilon}_t$ at two positions that are Δs apart along the path of the particle as shown in Fig. 9.10. If Δs is small enough, the unit vectors $\boldsymbol{\epsilon}_t(s)$ and $\boldsymbol{\epsilon}_t(s + \Delta s)$ can be considered to intersect and thus to form a plane. If $\Delta s \to 0$, these unit vectors then form a *limiting plane*, which we shall call the *osculating plane*.[1] The plane will have an orientation that depends on the position s on the path of the particle. The osculating plane at $\boldsymbol{r}(t)$ is illustrated in Fig. 9.10. Having defined the osculating plane, let us continue discussion of Eq. 9.14.

[1]From the definition, it should be apparent that the osculating plane at position s along a curve is actually *tangent* to the curve at position s. Since osculate means to kiss, the plane "kisses" the curve, as it were, at s.

Since we have not formally carried out the differentiation of a vector with respect to a spatial coordinate, we shall carry out the derivative $d\boldsymbol{\epsilon}_t/ds$ needed in Eq. 9.14 from the basic definition. Thus,

$$\frac{d\boldsymbol{\epsilon}_t}{ds} = \lim_{\Delta s \to 0}\left[\frac{\boldsymbol{\epsilon}_t(s+\Delta s) - \boldsymbol{\epsilon}_t(s)}{\Delta s}\right] = \lim_{\Delta s \to 0}\left(\frac{\Delta \boldsymbol{\epsilon}_t}{\Delta s}\right) \tag{9.15}$$

The vectors $\boldsymbol{\epsilon}_t(s)$ and $\boldsymbol{\epsilon}_t(s + \Delta s)$ are shown in Fig. 9.11(a) along the path and are also shown (enlarged) with $\Delta\boldsymbol{\epsilon}_t$ as a vector triangle in Fig. 9.11(b). As pointed out earlier, for small enough Δs the lines of action of the unit vectors $\boldsymbol{\epsilon}_t(s)$ and $\boldsymbol{\epsilon}_t(s + \Delta s)$ will intersect to form a plane as shown in Fig. 9.11(a). Now in this plane, draw normal lines to the aforementioned vectors at the respective positions s and $s + \Delta s$. These lines will intersect at some point O, as shown in the diagram. Next, consider what happens to the plane and to point O as $\Delta s \to 0$. Clearly, the limiting plane is our osculating plane at s [see Fig. 9.11(c)]. Furthermore, the limiting position arrived at for point O is *in the osculating plane* and is called the *center of curvature* for the path at s. The distance between O and s is denoted as R and is called the *radius of curvature*. Finally, the vector $\Delta\boldsymbol{\epsilon}_t$ (see Fig. 9.11 (b)), in the limit as $\Delta s \to 0$, ends up in the osculating plane normal to the path at s and directed toward the center of curvature. The unit vector collinear with the limiting vector for $\Delta\boldsymbol{\epsilon}_t$ is denoted as $\boldsymbol{\epsilon}_n$ and is called the *principal normal vector*.

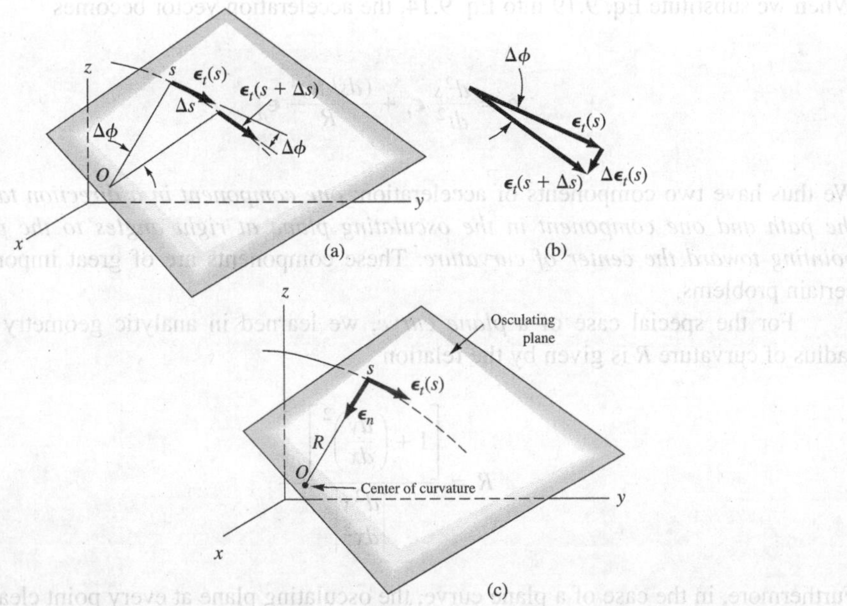

Figure 9.11. Development of the osculating plane and the center of curvature.

With the limiting *direction* of $\Delta\boldsymbol{\epsilon}_t$ established, we next evaluate the *magnitude* of $\Delta\boldsymbol{\epsilon}_t$ as an approximate value that becomes correct as $\Delta s \to 0$. Observing the vector triangle in Fig. 9.11(b), we can accordingly say:

$$|\Delta\boldsymbol{\epsilon}_t| \approx |\boldsymbol{\epsilon}_t|\Delta\phi = \Delta\phi \tag{9.16}$$

Next, we note in Fig. 9.11(a) that the lines from point O to the points s and $s + \Delta s$ along the trajectory form the same angle $\Delta\phi$ as is between the vectors $\boldsymbol{\epsilon}_t(s)$ and $\boldsymbol{\epsilon}_t(s + \Delta s)$ in the vector triangle, and so we can say:

$$\Delta\phi = \frac{\Delta s}{Os} \approx \frac{\Delta s}{R} \tag{9.17}$$

Hence, we have for Eq. 9.16:

$$|\Delta\boldsymbol{\epsilon}_t| \approx \frac{\Delta s}{R} \tag{9.18}$$

We thus have the magnitude of $\Delta\boldsymbol{\epsilon}_t$ established in an approximate manner. Using $\boldsymbol{\epsilon}_n$, the principal normal at s, to approximate the direction of $\Delta\boldsymbol{\epsilon}_t$ we can write

$$\Delta\boldsymbol{\epsilon}_t \approx \frac{\Delta s}{R}\boldsymbol{\epsilon}_n$$

If we use this result in the limiting process of Eq. 9.15 (where it becomes exact), the evaluation of $d\boldsymbol{\epsilon}_t/ds$ becomes

$$\frac{d\boldsymbol{\epsilon}_t}{ds} = \lim_{\Delta s \to 0}\left(\frac{\Delta\boldsymbol{\epsilon}_t}{\Delta s}\right) = \lim_{\Delta s \to 0}\left[\frac{(\Delta s/R)\boldsymbol{\epsilon}_n}{\Delta s}\right] = \frac{\boldsymbol{\epsilon}_n}{R} \tag{9.19}$$

When we substitute Eq. 9.19 into Eq. 9.14, the acceleration vector becomes

$$\boldsymbol{a} = \frac{d^2 s}{dt^2}\boldsymbol{\epsilon}_t + \frac{(ds/dt)^2}{R}\boldsymbol{\epsilon}_n \tag{9.20}$$

We thus have two components of acceleration: *one component in a direction tangent to the path and one component in the osculating plane at right angles to the path and pointing toward the center of curvature*. These components are of great importance in certain problems.

For the special case of a *plane curve*, we learned in analytic geometry that the radius of curvature R is given by the relation

$$R = \frac{\left[1 + \left(\dfrac{dy}{dx}\right)^2\right]^{3/2}}{\left|\dfrac{d^2y}{dx^2}\right|} \tag{9.21}$$

Furthermore, in the case of a plane curve, the osculating plane at every point clearly must correspond to the plane of the curve, and the computation of unit vectors $\boldsymbol{\epsilon}_n$ and $\boldsymbol{\epsilon}_t$ is quite simple, as will be illustrated in Example 9.5.

How do we get the principal normal vector $\boldsymbol{\epsilon}_n$, the radius of curvature R, and the direction of the osculating plane for a three-dimensional curve? One procedure is to evaluate $\boldsymbol{\epsilon}_t$ as a function of s and then differentiate this vector with respect to s. Accordingly, from Eq. 9.19 we can then determine $\boldsymbol{\epsilon}_n$ as well as R. We establish the direction of the osculating plane by taking the cross product $\boldsymbol{\epsilon}_n \times \boldsymbol{\epsilon}_t$, to get a unit vector normal to the osculating plane. This vector is called the *binormal vector*. This is illustrated in starred problem 9.7.

Example 9.5

A particle is moving along a circular path in the xy plane (Fig. 9.12). When the particle crosses the x axis, it has an acceleration along the path of 5 ft/sec^2 and is moving with the speed of 20 ft/sec in the negative y direction. What is the total acceleration of the particle?

Clearly, the osculating plane must be the plane of the path. Hence, R is 2 ft, as is shown in the diagram. We need simply to employ Eq. 9.20 for the desired result. Thus,

$$\boldsymbol{a} = 5\boldsymbol{\epsilon}_t + \frac{20^2}{2}\boldsymbol{\epsilon}_n \text{ ft/sec}^2$$

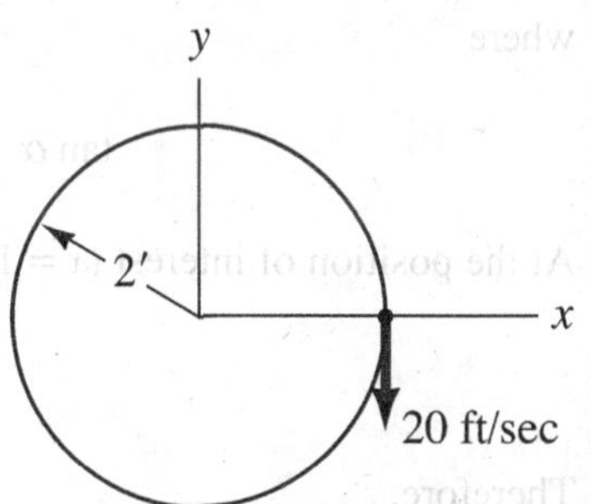

Figure 9.12. Particle on circular path.

For the xy reference, the acceleration is

$$\boldsymbol{a} = -5\boldsymbol{j} - 200\boldsymbol{i} \text{ ft/sec}^2$$

Example 9.6

A particle is moving in the xy plane along a parabolic path given as $y = 1.22\sqrt{x}$ (see Fig. 9.13) with x and y in meters. At position A, the particle has a speed of 3 m/sec and has a rate of change of speed of 3 m/sec^2 along the path. What is the acceleration vector of the particle at this position?

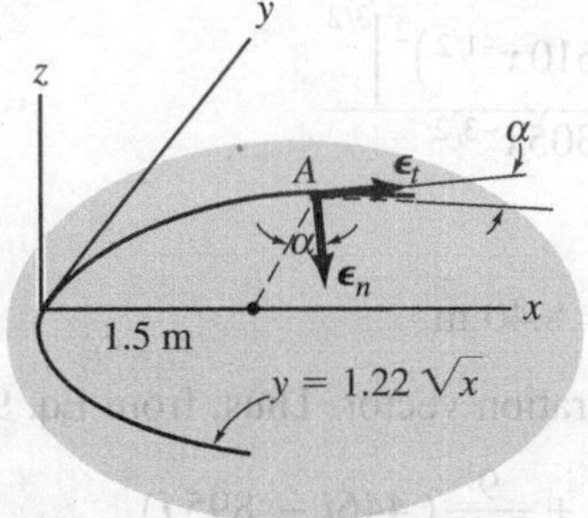

Figure 9.13. Particle on a parabolic path.

We first find $\boldsymbol{\epsilon}_t$ by noting from the diagram that

$$\boldsymbol{\epsilon}_t = \cos\alpha\,\boldsymbol{i} + \sin\alpha\,\boldsymbol{j} \quad \text{(a)}$$

Example 9.6 (Continued)

where

$$\tan\alpha = \frac{dy}{dx} = \frac{d}{dx}\left(1.22\sqrt{x}\right) = \frac{.610}{\sqrt{x}} \tag{b}$$

At the position of interest ($x = 1.5$ m) we have

$$\tan\alpha = \frac{.610}{\sqrt{1.5}} = \frac{1}{2}$$

Therefore,

$$\alpha = 26.5°$$

Hence,

$$\boldsymbol{\epsilon}_t = .895\boldsymbol{i} + .446\boldsymbol{j} \tag{c}$$

As for $\boldsymbol{\epsilon}_n$, we see from the diagram that

$$\boldsymbol{\epsilon}_n = \sin\alpha\,\boldsymbol{i} - \cos\alpha\,\boldsymbol{j}$$

Therefore,

$$\boldsymbol{\epsilon}_n = .446\boldsymbol{i} - .895\boldsymbol{j} \tag{d}$$

Next, employing Eq. 9.21, we can find R. We shall need the following results for this step:

$$\frac{dy}{dx} = .610x^{-1/2} \tag{e}$$

$$\frac{d^2y}{dx^2} = -.305x^{-3/2} \tag{f}$$

Substituting Eqs. (e) and (f) into Eq. 9.21, we have for R:

$$R = \frac{\left[1 + \left(.610x^{-1/2}\right)^2\right]^{3/2}}{.305x^{-3/2}} \tag{g}$$

At the position of interest, $x = 1.5$, we get

$$R = 8.40 \text{ m} \tag{h}$$

We can now give the desired acceleration vector. Thus, from Eq. 9.20, we have

$$\boldsymbol{a} = 3(.895\boldsymbol{i} + .446\boldsymbol{j}) + \frac{9}{8.40}(.446\boldsymbol{i} - .895\boldsymbol{j}) \tag{i}$$

$$\therefore \quad \boldsymbol{a} = 3.16\boldsymbol{i} + .379\boldsymbol{j} \text{ m/sec}^2$$

*Example 9.7

A particle is made to move along a spiral path, as is shown in Fig. 9.14.

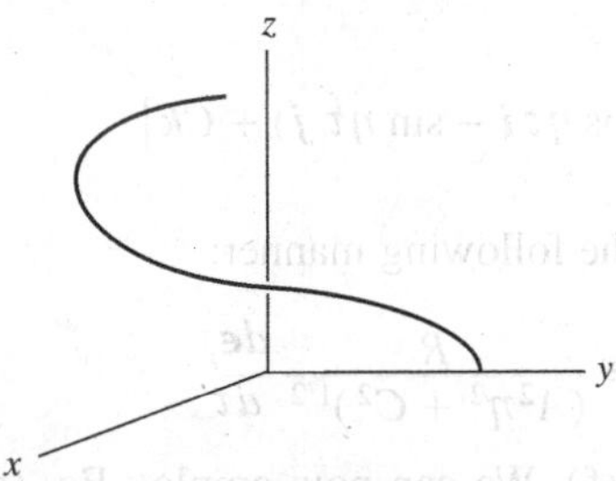

Figure 9.14.

The equations representing the *path* are given parametrically in terms of the variable τ in the following manner:

$$x_p = A \sin \eta\tau$$

$$y_p = A \cos \eta\tau \quad (A,\ \eta,\ C \text{ are known constants}) \tag{a}$$

$$z_p = C\tau$$

where the subscript p is to remind the reader that these relations refer to a fixed path. When the particle is at the xy plane ($z = 0$), it has a speed of V_0 ft/sec and a rate of change of speed of N ft/sec^2. What is the acceleration of the particle at this position?

To answer this, we must ascertain $\boldsymbol{\epsilon}_t$, $\boldsymbol{\epsilon}_n$, and R. To get $\boldsymbol{\epsilon}_t$ we write:

$$\boldsymbol{\epsilon}_t = \frac{d\boldsymbol{r}_p}{ds} = \frac{dx_p}{ds}\boldsymbol{i} + \frac{dy_p}{ds}\boldsymbol{j} + \frac{dz_p}{ds}\boldsymbol{k} \tag{b}$$

But:

$$\frac{dx_p}{ds} = \frac{dx_p}{d\tau}\frac{d\tau}{ds} \text{ and } \frac{dy_p}{ds} = \frac{dy_p}{d\tau}\frac{d\tau}{ds}, \text{ etc.}$$

Solving for $dx_p/d\tau$, $dy_p/d\tau$, and $dz_p/d\tau$ from Eq. (a), we can express Eq. (b) as:

$$\boldsymbol{\epsilon}_t = (A\eta \cos \eta\tau \boldsymbol{i} - A\eta \sin \eta\tau \boldsymbol{j} + C\boldsymbol{k})\frac{d\tau}{ds} \tag{c}$$

But:

$$ds = \sqrt{(dx_p)^2 + (dy_p)^2 + (dx_p)^2} \tag{d}$$

Solving for the differentials dx_p, dy_p, and dz_p from Eq. (a) and substituting into Eq. (d), we get:

$$ds = \left[(A\eta \cos \eta\tau)^2 + (A\eta \sin \eta\tau)^2 + C^2\right]^{1/2} d\tau \tag{e}$$

Solving for $d\tau/ds$ from the above equation, we have:

$$\frac{d\tau}{ds} = \frac{1}{\left[(A\eta)^2(\cos^2 \eta\tau + \sin^2 \eta\tau) + C^2\right]^{1/2}} = \frac{1}{(A^2\eta^2 + C^2)^{1/2}} \tag{f}$$

Example 9.7 (Continued)

in which we replaced ($\cos^2 \eta\tau + \sin^2 \eta\tau$) by unity. Returning to Eq. (c), we can thus say:

$$\boldsymbol{\epsilon}_t = \frac{1}{(A^2\eta^2 + C^2)^{1/2}}\left[A\eta(\cos\eta\tau\, \boldsymbol{i} - \sin\eta\tau\, \boldsymbol{j}) + C\boldsymbol{k}\right] \qquad \text{(g)}$$

To get $\boldsymbol{\epsilon}_n$ and R we employ Eq. 9.19, but in the following manner:

$$\boldsymbol{\epsilon}_n = R\frac{d\boldsymbol{\epsilon}_t}{ds} = R\frac{d\boldsymbol{\epsilon}_t/d\tau}{ds/d\tau} = \frac{R}{(A^2\eta^2 + C^2)^{1/2}}\frac{d\boldsymbol{\epsilon}_t}{d\tau} \qquad \text{(h)}$$

in which we have replaced $ds/d\tau$ using Eq. (f). We can now employ Eq. (g) to find $d\boldsymbol{\epsilon}_t/d\tau$:

$$\frac{d\boldsymbol{\epsilon}_t}{d\tau} = -\frac{A\eta^2}{(A^2\eta^2 + C^2)^{1/2}}(\sin\eta\tau\, \boldsymbol{i} + \cos\eta\tau\, \boldsymbol{j}) \qquad \text{(i)}$$

When we substitute this relation for $d\boldsymbol{\epsilon}_t/d\tau$ in Eq. (h), the principal normal vector $\boldsymbol{\epsilon}_n$ becomes:

$$\boldsymbol{\epsilon}_n = -\frac{RA\eta^2}{A^2\eta^2 + C^2}(\sin\eta\tau\, \boldsymbol{i} + \cos\eta\tau\, \boldsymbol{j}) \qquad \text{(j)}$$

If we take the magnitude of each side, we can solve for R:

$$R = \frac{A^2\eta^2 + C^2}{A\eta^2} \qquad \text{(k)}$$

We now have $\boldsymbol{\epsilon}_t$ and $\boldsymbol{\epsilon}_n$ at any point of the curve in terms of the parameter τ. As the particle goes through the xy plane, this means that the z coordinate of the position of the particle is zero and z_p of the path corresponding to the position of the particle is zero. When we note the last of Eqs. (a), it is clear, that τ must be zero for this position. Thus $\boldsymbol{\epsilon}_n$ and $\boldsymbol{\epsilon}_t$ for the point of interest are:

$$\boldsymbol{\epsilon}_t = \frac{1}{(A^2\eta^2 + C^2)^{1/2}}(A\eta\, \boldsymbol{i} + C\boldsymbol{k}) \qquad \text{(l)}$$

$$\boldsymbol{\epsilon}_n = -\frac{RA\eta^2}{A^2\eta^2 + C^2}\boldsymbol{j} = -\boldsymbol{j} \qquad \text{(m)}$$

where we have used Eq. (k) to replace R in Eq. (m). We can now express the acceleration vector using Eq. 9.20. Thus:

$$\boldsymbol{a} = \frac{N}{(A^2\eta^2 + C^2)^{1/2}}(A\eta\, \boldsymbol{i} + C\boldsymbol{k}) - \frac{V_0^2 A\eta^2}{A^2\eta^2 + C^2}\boldsymbol{j} \qquad \text{(n)}$$

The direction of the osculating plane can be found by taking the cross product of $\boldsymbol{\epsilon}_t$ and $\boldsymbol{\epsilon}_n$.

PROBLEMS

9.1. A mass is supported by four springs. The mass is given a vibratory movement in the horizontal (x) direction and simultaneously a vibratory movement in the vertical (y) direction. These motions are given as follows:

$$x = 2 \sin 2t \text{ mm}$$
$$y = 2 \cos (2t + .3) \text{ mm}$$

What is the value of the acceleration vector at $t = 4$ sec? How many g's of acceleration does this correspond to?

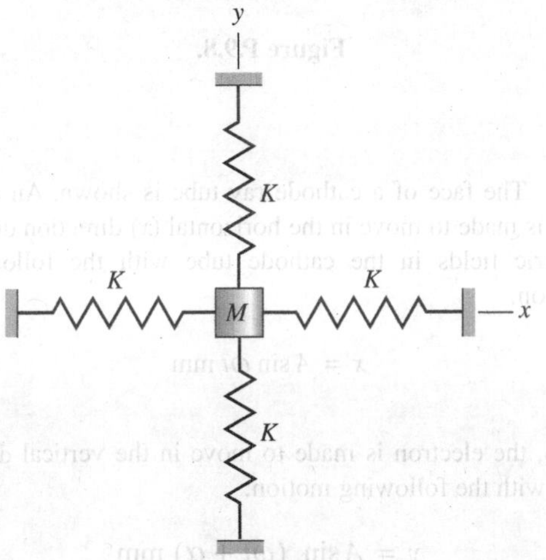

Figure P.9.1.

9.2. A particle moves along a plane circular path of radius r equal to 1 ft. The position OA is given as a function of time as follows:

$$\theta = 6 \sin 5t \text{ rad}$$

where t is in seconds. What are the rectangular components of velocity for the particle at time $t = \frac{1}{5}$ sec?

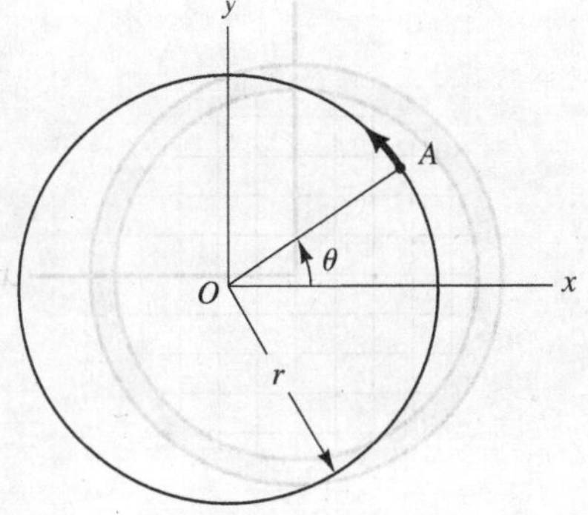

Figure P.9.2.

9.3. A particle with an initial position vector $\boldsymbol{r} = 5\boldsymbol{i} + 6\boldsymbol{j} + \boldsymbol{k}$ m has an acceleration imposed on it, given as

$$\boldsymbol{a} = 6t\boldsymbol{i} + 5t^2\boldsymbol{j} + 10\boldsymbol{k} \text{ m/sec}^2$$

If the particle has zero velocity initially, what are the acceleration, velocity, and position of the particle when $t = 10$ sec?

9.4. The position of a particle at times $t = 10$ sec, $t = 5$ sec, and $t = 2$ sec is known to be, respectively:

$$\boldsymbol{r}(10) = 10\boldsymbol{i} + 5\boldsymbol{j} - 10\boldsymbol{k} \text{ ft}$$
$$\boldsymbol{r}(5) = 3\boldsymbol{i} + 2\boldsymbol{j} + 5\boldsymbol{k} \text{ ft}$$
$$\boldsymbol{r}(2) = 8\boldsymbol{i} - 20\boldsymbol{j} + 10\boldsymbol{k} \text{ ft}$$

What is the acceleration of the particle at time $t = 5$ sec if the acceleration vector has the form

$$\boldsymbol{a} = C_1 t\boldsymbol{i} + C_2 t^2\boldsymbol{j} + C_3 \ln t\boldsymbol{k} \text{ ft/sec}^2$$

where C_1, C_2, and C_3 are constants and t is in seconds?

9.5. A highly idealizeĬam is shown of an *accelerometer*, a device for measuring the acceleration component of motion along a certain direction—in this case the indicated x direction. A mass B is constrained in the accelerometer case so that it can only move against linear springs in the x direction. When the accelerometer case accelerates in this direction, the mass assumes a displaced position, shown dashed, at a distance δ from its original position. This configuration is such that the force in the springs gives the mass B the acceleration corresponding to that of the accelerometer case. The shift δ of the mass in the case is picked up by an electrical sensor device and is plotted as a function of time. The damping fluid present eliminates extraneous oscillations of the mass. If a plot of a_x versus time has the form shown, what is the speed of the body after 10 sec, 30 sec, and 45 sec? The acceleration a_x is measured in g's—i.e., in units of 32.2 ft/sec^2 or 9.81 m/sec^2. Assume that the body starts from rest at $x = 0$.

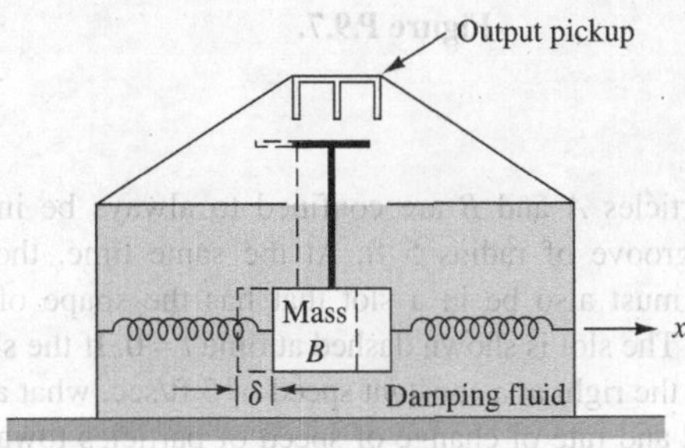

Figure P.9.5 *(Continued)*

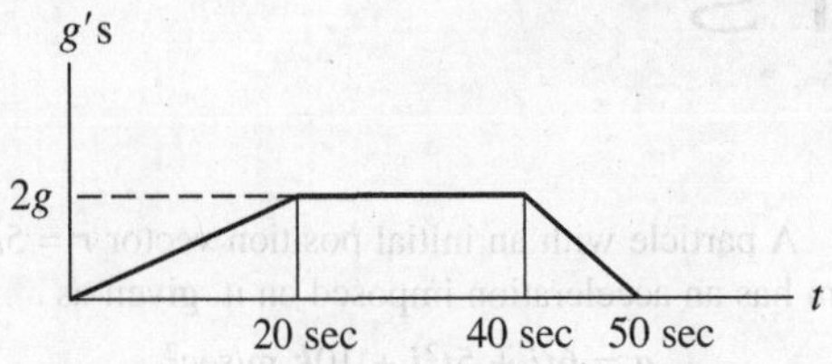

Figure P.9.5.

9.6. The position vector of a particle is given as

$$\boldsymbol{r} = 6t\boldsymbol{i} + (5t + 10)\boldsymbol{j} + 6t^2\boldsymbol{k} \text{ m}$$

What is the acceleration of the particle at $t = 3$ sec? What distance has been traveled by the particle during this time? [*Hint:* Let $dr = \sqrt{dx^2 + dy^2 + dz^2}$ and divide and multiply by dt in second half of problem. Look up integration form $\int \sqrt{a^2 + t^2}\, dt$ in Appendix I.]

9.7. In Example 9.1, what is the acceleration vector for pin B if the yoke C is accelerating at the rate of 10 ft/sec² at the instant of interest?

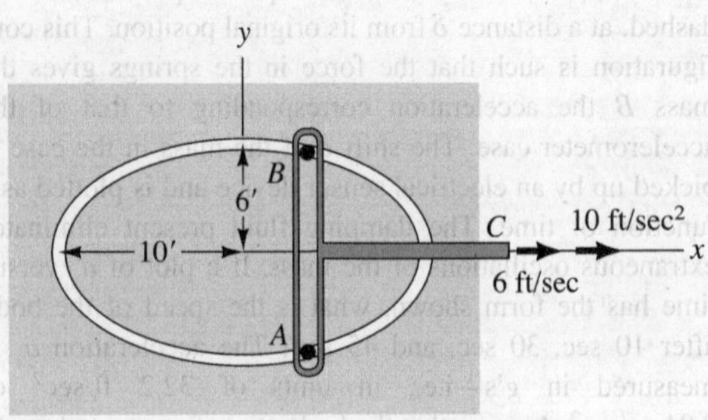

Figure P.9.7.

9.8. Particles A and B are confined to always be in a circular groove of radius 5 ft. At the same time, these particles must also be in a slot that has the shape of a parabola. The slot is shown dashed at time $t = 0$. If the slot moves to the right at a constant speed of 3 ft/sec, what are the speed and rate of change of speed of particles toward each other at $t = 1$ sec?

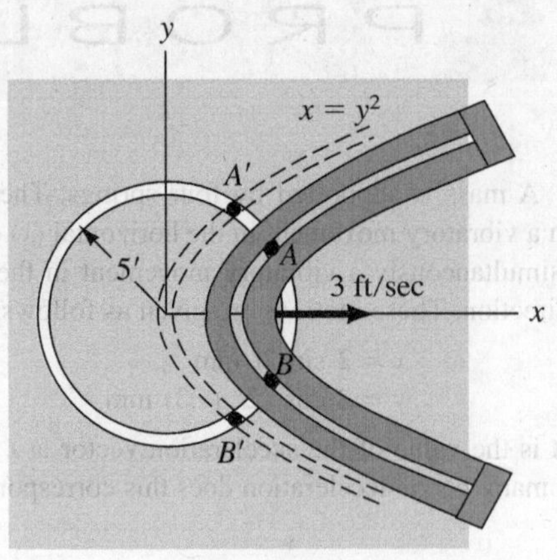

Figure P.9.8.

9.9. The face of a cathode ray tube is shown. An electron is made to move in the horizontal (x) direction due to electric fields in the cathode tube with the following motion:

$$x = A \sin \omega t \text{ mm}$$

Also, the electron is made to move in the vertical direction with the following motion:

$$y = A \sin\ (\omega t + \alpha) \text{ mm}$$

Show that for $\alpha = \pi/2$, the trajectory on the screen is that of a *circle* of radius A mm. If $\alpha = \pi$, show that the trajectory is that of a *straight line* inclined at $-45°$ to the xy axes. Finally, give the formulations for the directions of velocity and acceleration of the electron in the xy plane.

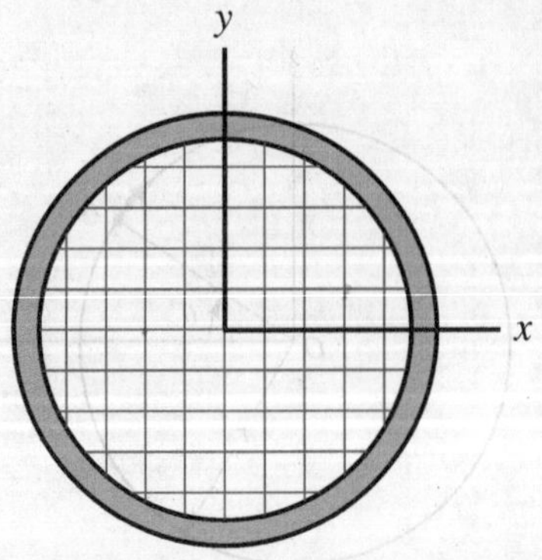

Figure P.9.9.

9.10. A yoke A moves to the right at a speed $V = 2$ m/s and a rate of change of speed $\dot{V} = .6\ \text{m/s}^2$ when the yoke is at a position $d = .27$ m from the y axis. A pin is constrained to move inside a slot in the yoke and is forced by a spring in the slot to slide on a parabolic surface. What are the velocity and acceleration vectors for the pin at the instant of interest? What is the acceleration normal to the parabolic surface at the position shown?

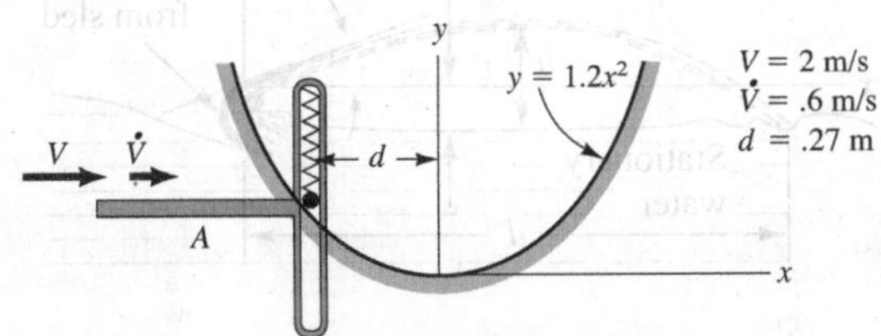

Figure P.9.10.

9.11. A flexible inextensible cord restrains mass M. Both pins A acting on top of the cord move downward at a constant speed V_2 while pin B acting on the bottom of the cord moves upward at a constant speed V_1. The cord is free to slide along the pins without friction. Starting from the horizontal orientation of the cord, what is the velocity of the mass M as a function of time?

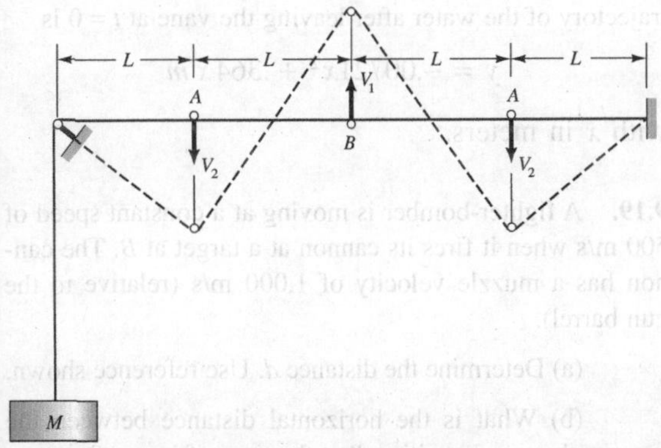

Figure P.9.11.

9.12. Mass M is held by an inextensible cord. What is the velocity of M as a function of α; the time t; the constant velocities V_A and V_C; and the distance h? Disc G is free to turn.

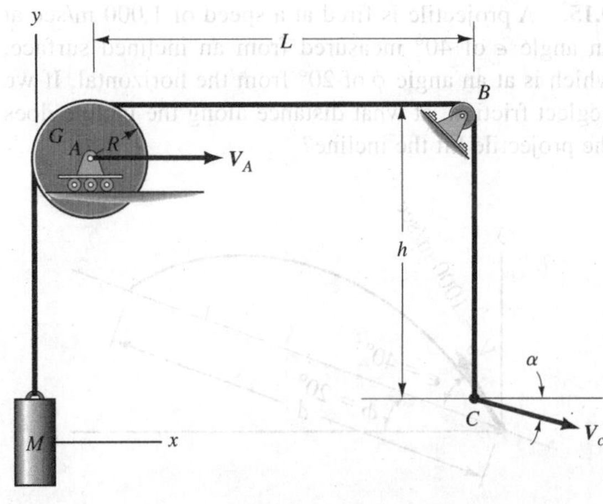

Figure P.9.12.

9.13. A stunt motorcyclist is to attempt a "jump" over a deep chasm. The distance between jump-off point and landing is 100 m. As technical advisor to this stunt man, what minimum speed do you tell him to exceed at the jump-off point A? The cycle is highly streamlined to minimize wind resistance. Give the result in km/hr.

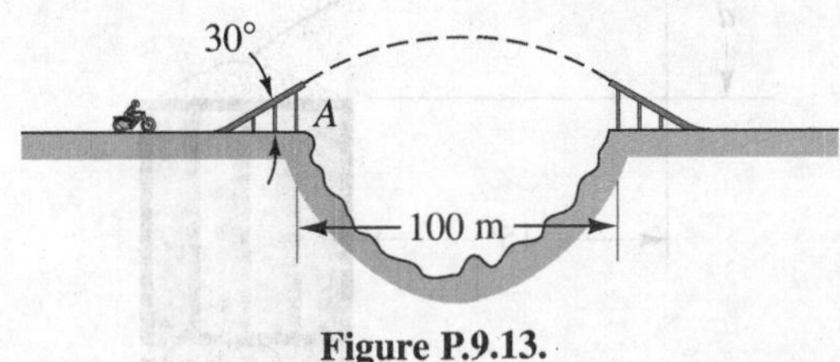

Figure P.9.13.

9.14. A charged particle is shot at time $t = 0$ at an angle of 45° with a speed 10 ft/sec. If an electric field is such that the body has an acceleration $-200t^2\mathbf{j}$ ft/sec², what is the equation for the trajectory? What is the value of d for impact?

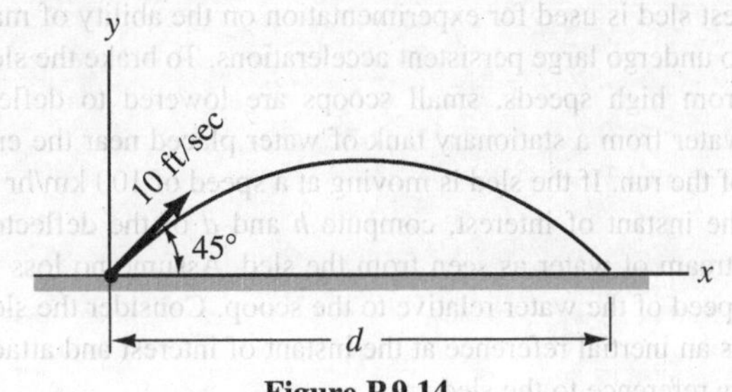

Figure P.9.14.

9.15. A projectile is fired at a speed of 1,000 m/sec at an angle ϵ of 40° measured from an inclined surface, which is at an angle ϕ of 20° from the horizontal. If we neglect friction, at what distance along the incline does the projectile hit the incline?

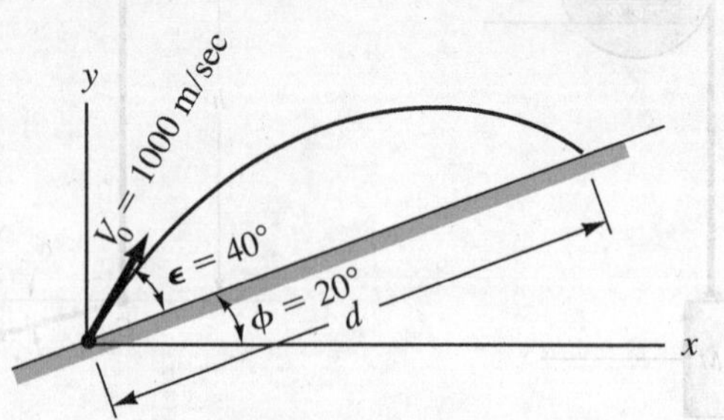

Figure P.9.15.

9.16. Grain is being blown into an open train container at a speed V_0 of 20 ft/sec. What should the minimum and maximum elevations d be to ensure that all the grain gets into the train? Neglect friction and winds.

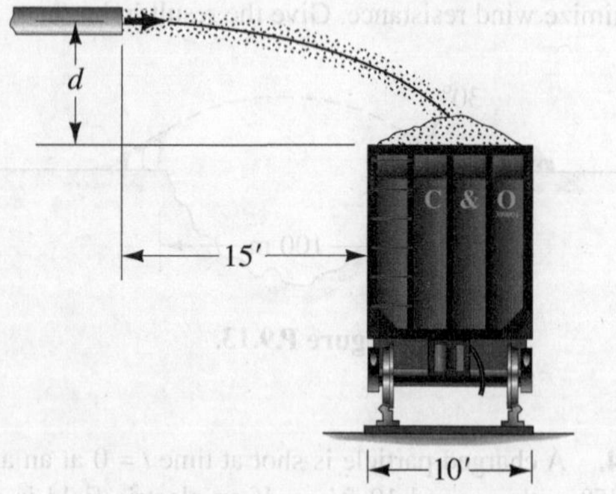

Figure P.9.16.

9.17. A rocket-powered test sled slides over rails. This test sled is used for experimentation on the ability of man to undergo large persistent accelerations. To brake the sled from high speeds, small scoops are lowered to deflect water from a stationary tank of water placed near the end of the run. If the sled is moving at a speed of 100 km/hr at the instant of interest, compute h and d of the deflected stream of water as seen from the sled. Assume no loss in speed of the water relative to the scoop. Consider the sled as an inertial reference at the instant of interest and attach xy reference to the sled.

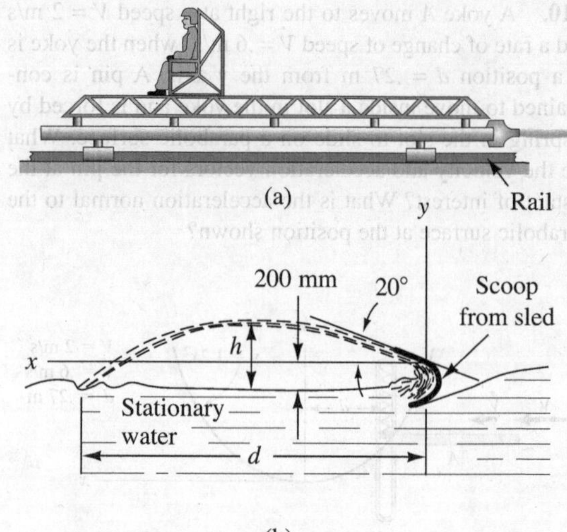

Figure P.9.17.

9.18. In the previous problem, the vane has a velocity given relative to the ground reference XY as

$$V = -5t^2 + 27.8 \text{ m/s}$$

What is the distance δ between the vane and the position of impact of the water that left the vane at time $t = 0$. Use the trajectory of the preceding problem, which relates x and y for a reference xy attached to the vane and moving to the left at $t = 0$ at a speed of 100 km/hr = 27.8 m/s. The trajectory of the water after leaving the vane at $t = 0$ is

$$y = -.00721x^2 + .364x \ m$$

with x in meters.

9.19. A fighter-bomber is moving at a constant speed of 500 m/s when it fires its cannon at a target at B. The cannon has a muzzle velocity of 1,000 m/s (relative to the gun barrel).

(a) Determine the distance d. Use reference shown.

(b) What is the horizontal distance between the plane and position B at the time of impact?

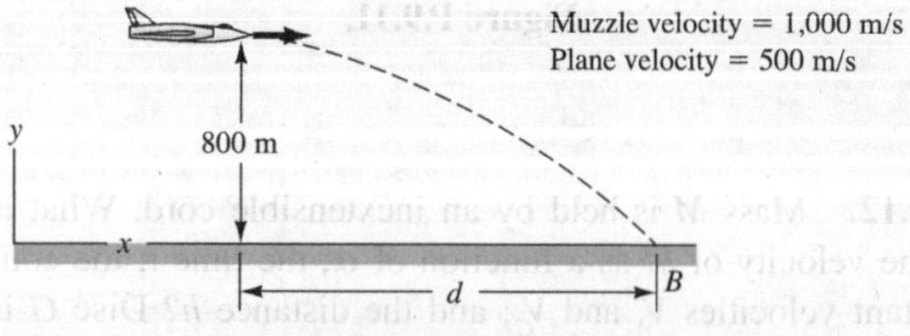

Figure P.9.19.

9.20. A golfer has the bad luck of having his golf ball strike a nearby tree while having a shallow trajectory. The ball bounces off at a speed that is 60 percent of the preimpact speed. If it moves in the same plane as the initial trajectory, compute the distance d at which the ball hits the ground with respect to the tee at A.

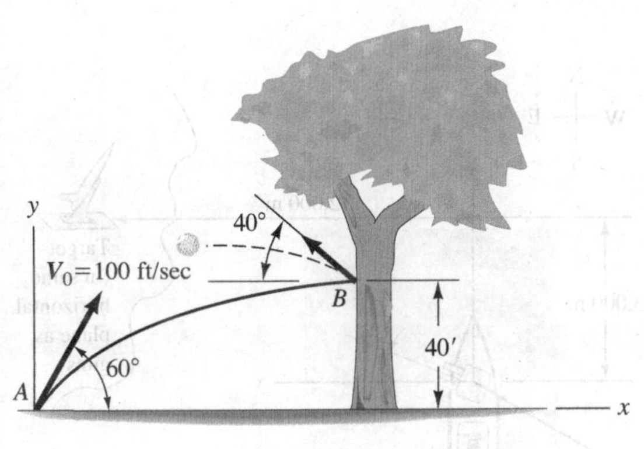

Figure P.9.20.

9.21. What angle α will result in the longest distance d at impact? The muzzle velocity is V_0. The surface is flat.

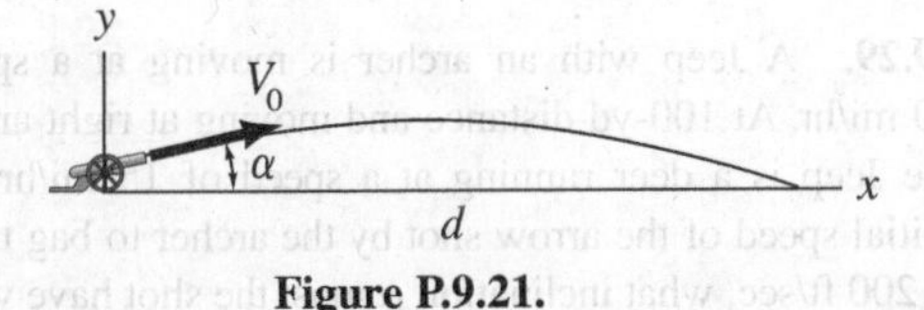

Figure P.9.21.

9.22. A sportsman in a valley is trying to shoot a deer on a hill. He quickly estimates the distance of the deer along his line of sight as 500 yd and the height of the hill as 100 yd. His gun has a muzzle velocity of 3,000 ft/sec. If he has no graduated sight, how many feet above the deer should he aim his rifle in order to hit it? (Neglect friction.)

9.23. A fireman is directing water from a hose into the broken window of a burning house. The velocity of the water is 15 m/sec as it leaves the hose. What are the angles α needed to do the job?

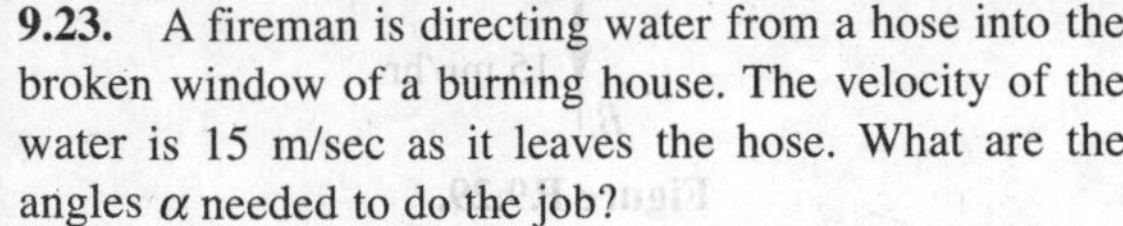

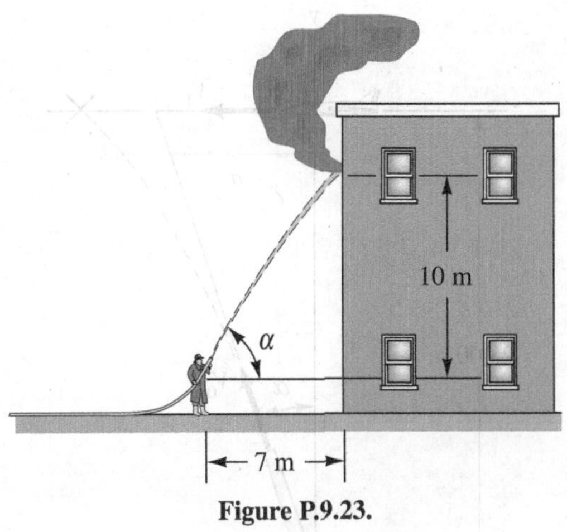

Figure P.9.23.

9.24. A long range gun is shown for which the muzzle velocity is 1,000 m/s. If we neglect friction, at what position x, y does the shell hit the ground?

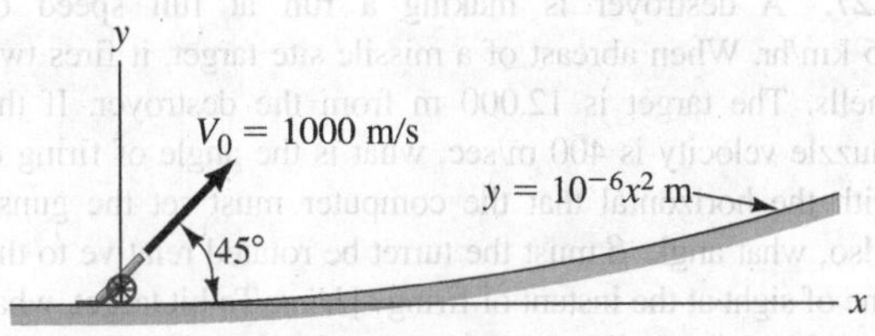

Figure P.9.24.

9.25. An archer in a Jeep is chasing a deer. The Jeep moves at 30 mi/hr and the deer moves at 15 mi/hr along the same direction. At what inclination must the arrow be shot if the deer is 100 yd ahead of the Jeep and if the initial speed of the arrow is 200 ft/sec relative to archer? (Neglect friction.)

9.26. A fighter plane is directly over an antiaircraft gun at time $t = 0$. The plane has a speed V_1 of 500 km/hour. A shell is fired at $t = 0$ in an attempt to hit the plane. If the muzzle velocity V_0 is 1,000 m/sec, how many meters d should the gun be aimed ahead of the plane to hit it? What is the time of impact?

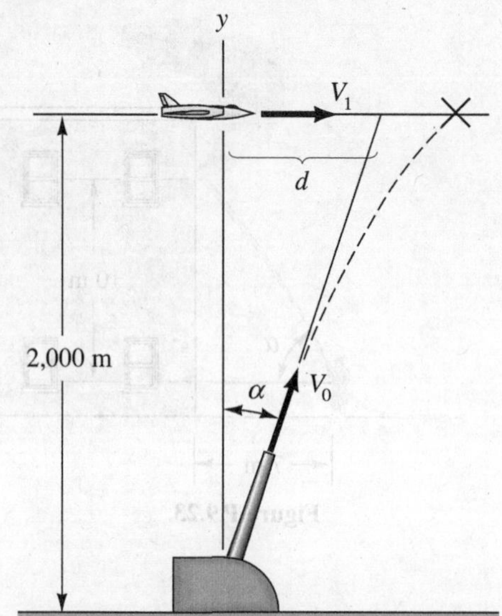

Figure P.9.26.

9.27. A destroyer is making a run at full speed of 75 km/hr. When abreast of a missile site target, it fires two shells. The target is 12,000 m from the destroyer. If the muzzle velocity is 400 m/sec, what is the angle of firing α with the horizontal that the computer must set the guns? Also, what angle β must the turret be rotated relative to the line of sight at the instant of firing? [*Hint:* To hit target, what must V_y of the shell be? Result: $\alpha = 23.7°$ and $\beta = 3.26°$.]

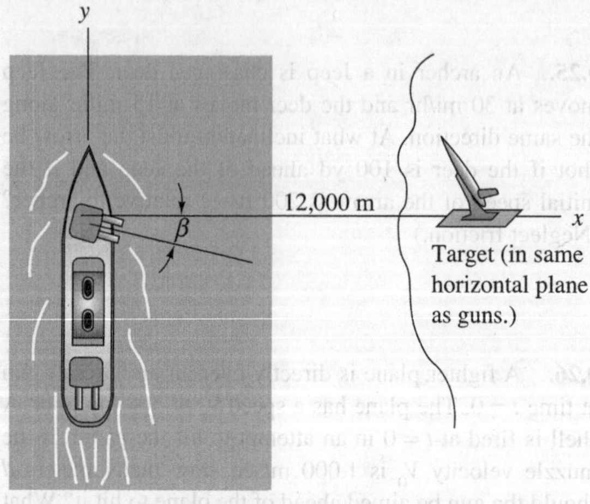

Figure P.9.27.

9.28. In the preceding problem, a second smaller destroyer is firing at the target as shown in the diagram. The data of the preceding problem applies with the following additional data. Due to strong wind and current, the destroyer has a drift velocity of 6 km/hr in a northeast direction in addition to its full speed of 75 km/hr. Form two simultaneous transcendental equations for α and β and verify that $\alpha = 21.39°$ and that $\beta = 10.727°$.

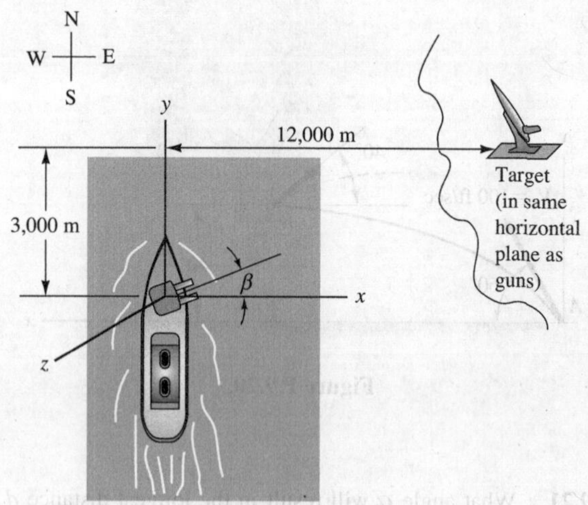

Figure P.9.28.

***9.29.** A Jeep with an archer is moving at a speed of 30 mi/hr. At 100-yd distance and moving at right angles to the Jeep is a deer running at a speed of 15 mi/hr. If the initial speed of the arrow shot by the archer to bag the deer is 200 ft/sec, what inclination α must the shot have with the horizontal and what angle β must the vertical projection of the shot onto the ground have relative to the line *AB*?

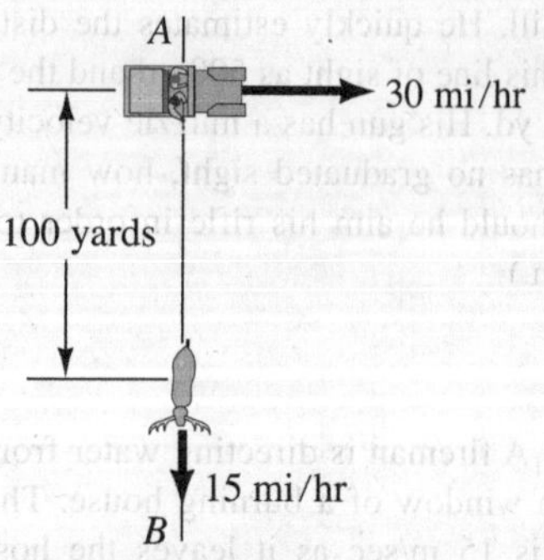

Figure P.9.29.

9.30. A particle moves with constant speed of 5 ft/sec along the path. Compute the acceleration at points 1, 2, and 3.

9.30. A particle moves with a constant speed of 5 ft/sec along the path. Compute the acceleration at points 1, 2, and 3.

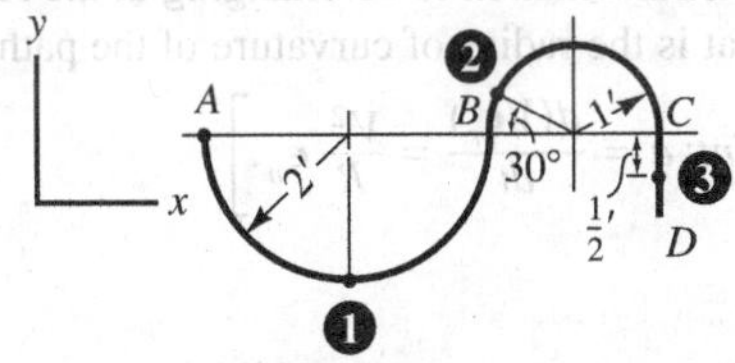

Figure P.9.30.

9.31. If, in Problem 9.30 the speed is 5 ft/sec only at point A, and it increases 5 ft/sec for each foot traveled, compute the acceleration at points 1, 2, and 3.

9.32. A car is moving at a speed of 88 km/hr along a highway. At a curve in the highway, the radius of curvature is 1,300 m. What is the acceleration of the car? To decrease this acceleration by 30%, what must its speed be?

9.33. A high-speed train is running at 100 km/hr. It goes into a curve having a minimum radius of curvature of 2,000 m. What is the acceleration that sitting passengers are subjected to? If the radius of curvature were to be doubled, at what constant speed could the train then go with the same acceleration?

9.34. An amusement park ride consists of a cockpit in which a passenger is strapped in a seated position. The cockpit rotates about A with angular speed ω. The average person's head is 10 ft from the axis of rotation at A. We know that if a person's head is subjected to an acceleration of 3 g's or more in a direction from shoulders to head for any length of time, he/she will be uncomfortable and perhaps black out. What, then, is the maximum value of ω in rpm to prevent these effects, using a safety factor of 3? [*Hint:* You will soon learn that the speed in a circular path is $R\omega$.]

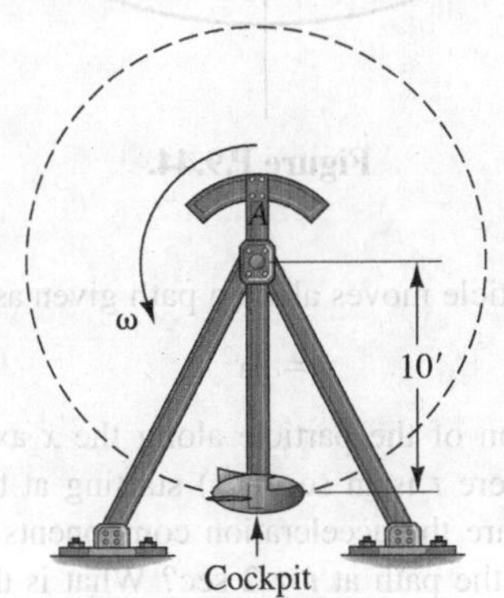

Figure P.9.34.

9.35. A motorcyclist is moving along a circular path having a radius of curvature of 400 m. He is increasing his speed along the path at the rate of 5 km/hr/sec. If he enters the curve at a speed of 48 km/hr, what is his total acceleration after traveling 10 sec along this path?

9.36. What is the direction of the normal vector and the value of the radius of curvature at a position a of the curve?

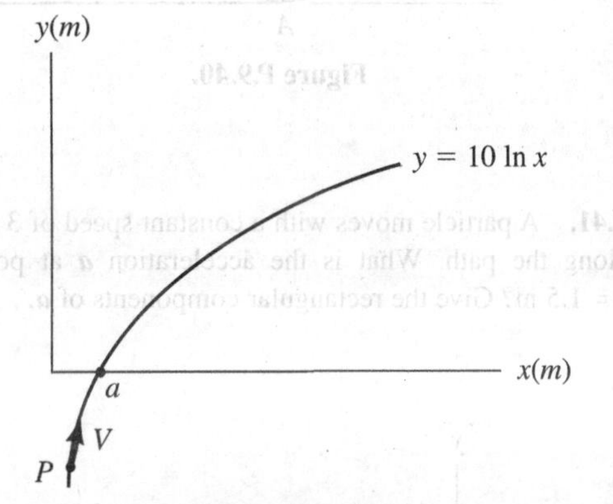

Figure P.9.36.

9.37. A particle P moves with constant speed V along the curve $y = 10 \ln x$ m. At what position x does the particle have the maximum acceleration? What is the value of this acceleration if $V = 1$ m/sec?

9.38. A car is moving along a circularly curved road of radius 2,000 ft so as to merge with traffic on a highway. If the car accelerates at a constant rate of 7 mi/hr/sec, what will be the total acceleration of the car when it is going 50 mph?

9.39. A motorcycle is moving along a circular flat road and is accelerating at a uniform rate of 5 m/s^2. At what speed will the total acceleration be 6 m/s^2? The radius of the path of the motorcycle is 220 m.

9.40. A fighter plane is in a diving maneuver along a trajectory that is approximately a parabola. If the maximum number of g's that the pilot can withstand is 5 g's from shoulder to head from the dynamics of the plane, what is the maximum allowable speed for this maneuver when the plane reaches A?

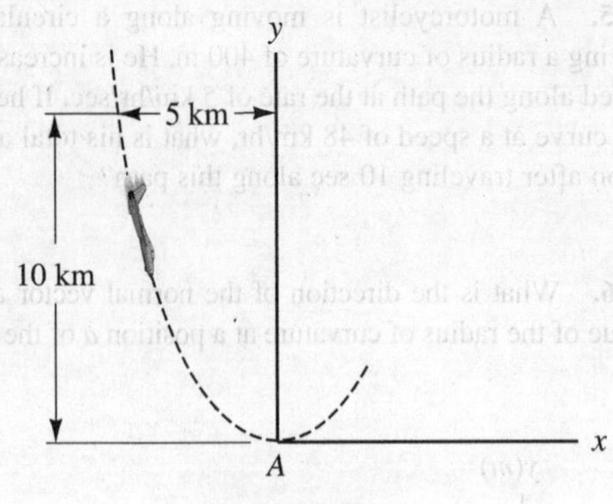

Figure P.9.40.

9.41. A particle moves with a constant speed of 3 m/sec along the path. What is the acceleration $\boldsymbol{a}$ at position $x = 1.5$ m? Give the rectangular components of $\boldsymbol{a}$.

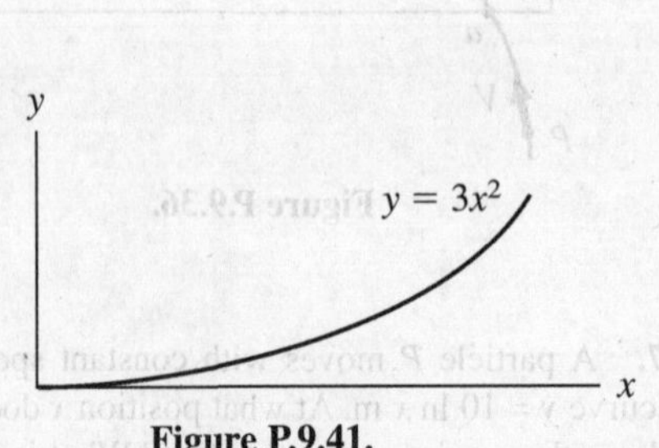

Figure P.9.41.

9.42. A particle moves along a sinusoidal path. If the particle has a speed of 10 ft/sec and a rate of change of speed of 5 ft/sec^2 at A, what is the *magnitude* of the acceleration? What is the magnitude and direction of the acceleration of the particle at B, if it has a speed of 20 ft/sec and a rate of change of speed of 3 ft/sec^2 at this point?

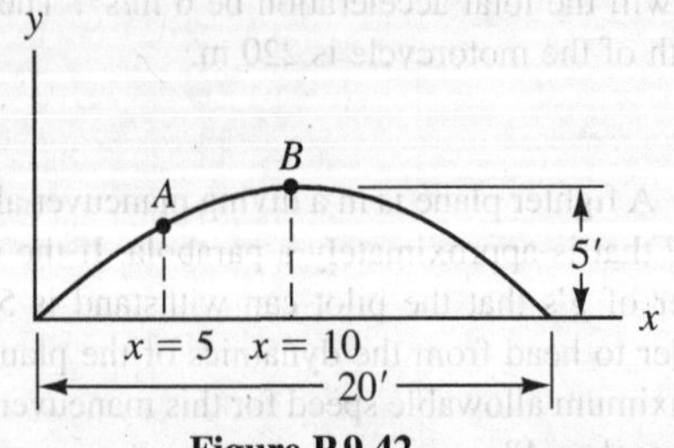

Figure P.9.42.

9.43. A passenger plane is moving at a constant speed of 200 km/hr in a holding pattern at a constant elevation. At the instant of interest, the angle β between the velocity vector and the x axis is 30°. The vector is known through on-board gyroscopic instrumentation to be changing at the rate $\dot{\beta}$ of $-5°$/sec. What is the radius of curvature of the path at this instant? $\left[\textit{Hint}: \boldsymbol{a} = \frac{d(V\boldsymbol{\epsilon}_t)}{dt} = \frac{V^2}{R}\boldsymbol{\epsilon}_n.\right]$

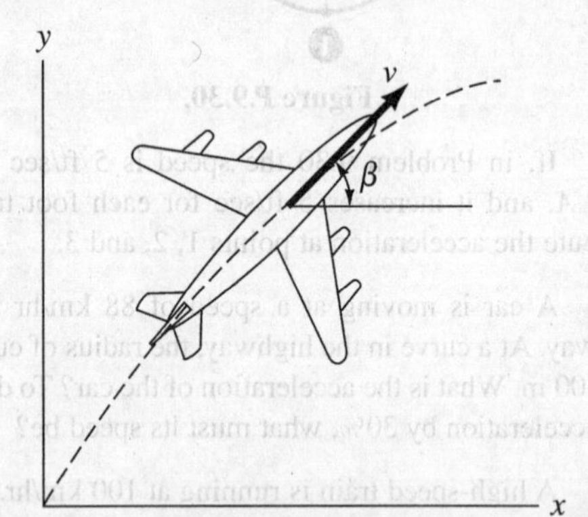

Figure P.9.43.

9.44. At what position along the ellipse shown does the normal vector have a set of direction cosines (.707, .707, 0)? Recall that the equation for an ellipse in the position shown is $x^2/a^2 + y^2/b^2 = 1$.

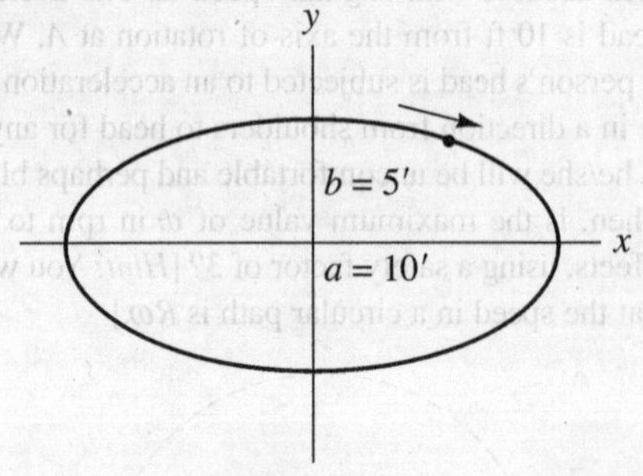

Figure P.9.44.

9.45. A particle moves along a path given as

$$y = 3x^2 \text{ ft}$$

The projection of the particle along the x axis varies as $\sqrt{.2}t^2$ ft (where t is in seconds) starting at the origin at $t = 0$. What are the acceleration components normal and tangential to the path at $t = 2$ sec? What is the radius of curvature at this point?

9.46. A particle moves along a path $y^2 = 10x$ with x and y in meters. The distance traversed along this path starting from the origin is given by S such that

$$S = \frac{t}{2} + \frac{t^2}{100} \text{ m}$$

where t is measured in units of seconds. What are the normal and tangential acceleration components of the particle when $y = 10$ m?
[*Hint:* $\int \sqrt{y^2 + a^2}\, dy$ is presented in Appendix I.] Also, note that

$$ds = \sqrt{dx^2 + dy^2} = \sqrt{\left(\frac{dx}{dy}\right)^2 + 1}\, dy.$$

9.47. Show by arguments similar to those used in the text for deriving the relation $d\boldsymbol{\epsilon}_t/ds = (1/R)\boldsymbol{\epsilon}_n$ that $d\boldsymbol{\epsilon}_n/ds = -(1/R)\boldsymbol{\epsilon}_t$.

9.48. (a) For coplanar paths in the xy plane, find the formula for $\dot{\mathbf{a}}$, that is, the "jerk."

(b) If a particle moves on a plane circular path of radius 5 m at a speed of 5 m/sec, and if the rate of change of speed is 2 m/sec², what is $\dot{\mathbf{a}}$ for the particle if the second derivative of its speed along the path is 10 m/sec³? [*Hint:* Use the result of Problem 9.47.]

9.49. A beebee gun shoots a pellet as shown. Determine the radius of curvature of the trajectory as a function of x.

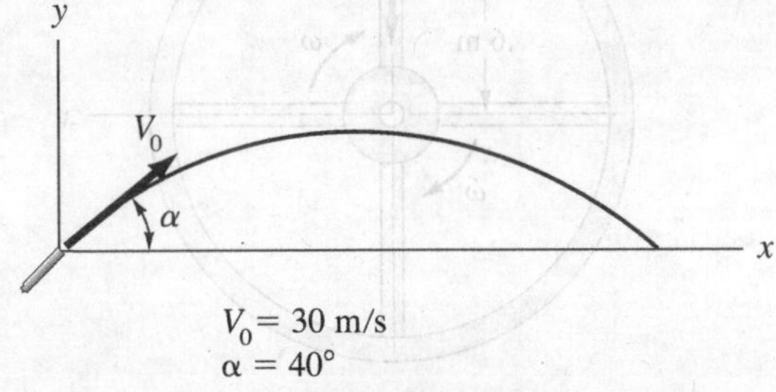

Figure P.9.49.

9.50. A car is moving along a circular track of radius 40 ft. The position S along the path is given as

$$S = 3t^2 + \frac{t^3}{6} \text{ ft}$$

The time t is given in seconds. What are the angular velocity and angular acceleration of the car at $t = 5$ sec?

9.51. A point P fixed on a rotating plate has an acceleration in the x direction of −10 m/sec². If r for the point is 1 m, what is the angular acceleration of the plate? The angular speed at the instant of interest is 2 rad/sec counterclockwise.

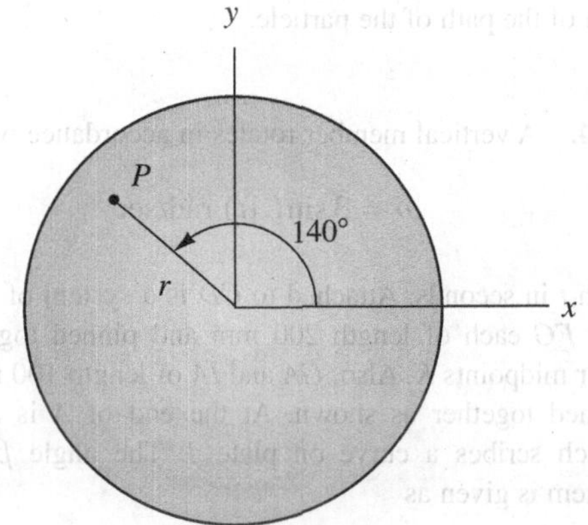

Figure P.9.51.

9.52. A flat disc A with a rubber surface is driven by bevel gears having diameters $D_1 = 8$ in. and $D_2 = 3$ in. A second rubber disc B of diameter $D_3 = 2$ in. is turned by the friction contact with A. We thus have a *friction drive* system. At the instant of interest, $\omega = 5$ rad/sec and $\dot{\omega} = 3$ rad/sec². If wheel B is moved downward at a speed $V_B = 3$ in./sec at the instant of interest, what is the rotational speed Ω and the rate of change of rotational speed $\dot{\Omega}$ of the small disc B? Slipping between B and A occurs only in the radial direction of disc A. The distance r is 4 in. at the instant of interest.

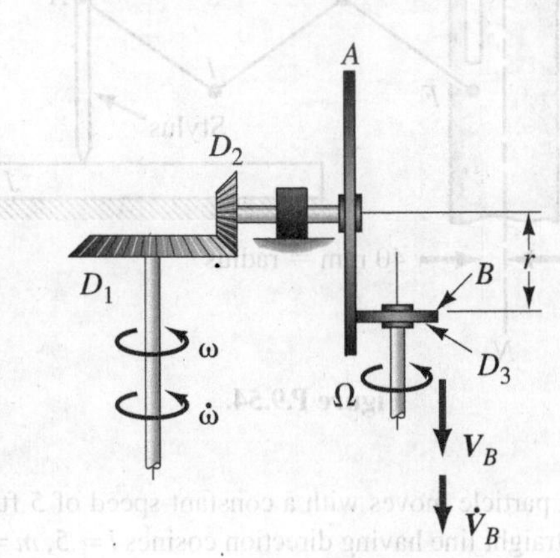

Figure P.9.52.

9.53. What are the velocity and acceleration components in the axial, transverse, and radial directions for a particle moving relative to xyz in the following way:

$$\bar{r} = 10e^{-2t} \text{ m} \qquad \theta = .2t \text{ rad} \qquad z = .6t \text{ m}$$

with t in seconds. Make a rough sketch of the early portion of the path of the particle.

9.54. A vertical member rotates in accordance with:

$$\omega = 3\sin(.1t) \text{ rad/sec}$$

with t in seconds. Attached to CD is a system of rods HI and FG each of length 200 mm and pinned together at their midpoints K. Also, GA and IA of length 100 mm, are pinned together as shown. At the end of A is a stylus which scribes a curve on plate J. The angle β of the system is given as

$$\beta = 1.3 - \frac{t}{10} \text{ rad}$$

with t in seconds. What are the radial and transverse velocity and acceleration components of the stylus at time $t = 5$ sec about axis N–N? (*Note:* Pin F is fixed but pin H moves vertically in a slot as shown.)

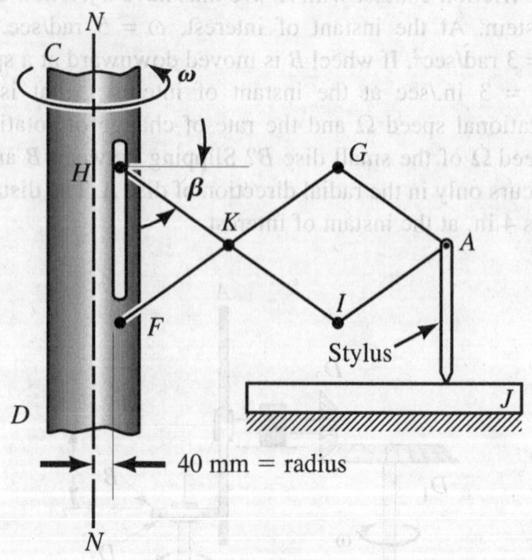

Figure P.9.54.

9.55. A particle moves with a constant speed of 5 ft/sec along a straight line having direction cosines $l = .5$, $m = .3$. What are the cylindrical coordinates when $|\boldsymbol{r}| = 20$ ft? What are the axial and transverse velocities of the particle at this position?

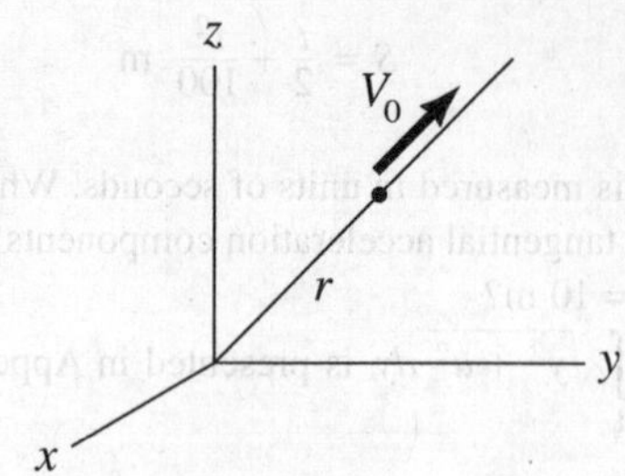

Figure P.9.55.

9.56. A wheel is rotating at time t with an angular speed ω of 5 rad/sec. At this instant, the wheel also has a rate of change of angular speed of 2 rad/sec^2. A body B is moving along a spoke at this instant with a speed of 3 m/sec relative to the spoke and is increasing in its speed at the rate of 1.6 m/sec^2. These data are given when the spoke, on which B is moving, is vertical and when B is .6 m from the center of the wheel, as shown in the diagram. What are the velocity and acceleration of B at this instant relative to the fixed reference xyz?

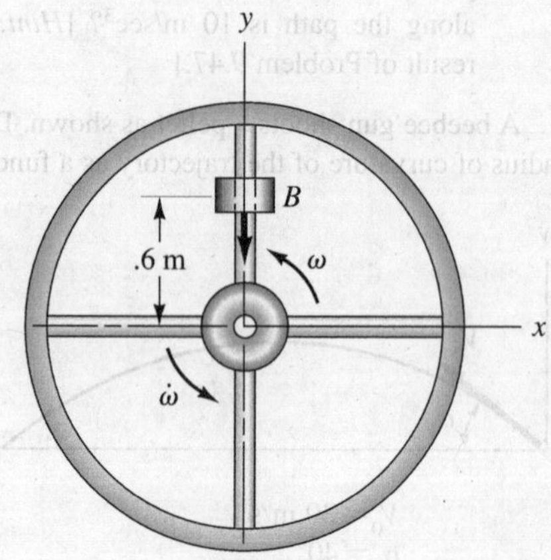

Figure P.9.56.

9.57. A plane is shown in a dive-bombing mission. It has at the instant of interest a speed of 485 km/hr and is increasing its speed downward at a rate of 81 km/hr/sec. The propeller is rotating at 150 rpm and has a diameter of 4 m. What is the velocity of the tip of the propeller shown at A and its acceleration at the instant of interest? Use cylindrical velocity components.

Figure P.9.57.

9.58. The motion of a particle relative to a reference xyz is given as follows:

$$\bar{r} = .2 \sinh t \text{ m} \qquad \theta = .5 \sin \pi t \text{ rad} \qquad z = 6t^2 \text{ m}$$

with t in seconds. What are the magnitudes of the velocity and acceleration vectors at time $t = 2$ sec? Note that $\sinh 2 = 3.6269$ and $\cosh 2 = 3.7622$.

9.59. Given the following cylindrical coordinates for the motion of a particle:

$$\bar{r} = 20 \text{ m} \qquad \theta = 2\pi t \text{ rad} \qquad z = 5t \text{ m}$$

with t in seconds. Sketch the path. What is this curve? Determine the velocity and acceleration vectors.

9.60. A grain of plutonium is being tracked in a turbulent atmosphere. Relative to reference xyz, the displacement components are

$$x = 6t \text{ ft} \qquad y = 10t \text{ ft} \qquad z = t^3 + 10 \text{ ft}$$

Express the position, velocity, and acceleration vectors of the particle using cylindrical coordinates with components in the axial, transverse, and radial directions.

9.61. The motion of a particle in cylindrical coordinates is given by the following parametric equations:

$$\bar{r} = 3 \sin \pi t \text{ m}$$
$$\theta = 6t + 3t^2 \text{ rad}$$
$$z = 5 \cos \pi t + 3 \text{ m} \quad (t \text{ in seconds})$$

Determine the velocity and acceleration of the particle at $t = .35$ sec.

9.62. A flyball governor has the following data at the instant of interest:

$$\omega = .2 \text{ rad/s} \qquad \dot{\omega} = .04 \text{ rad/s}^2 \qquad \alpha = 45°$$
$$\dot{\alpha} = 5 \text{ rad/s} \qquad \ddot{\alpha} = .2 \text{ rad/s}^2$$

If at this instant, the arms are in the xz plane, give the velocity and acceleration vectors of the spheres using cylindrical coordinates for the axial, transverse, and radial directions.

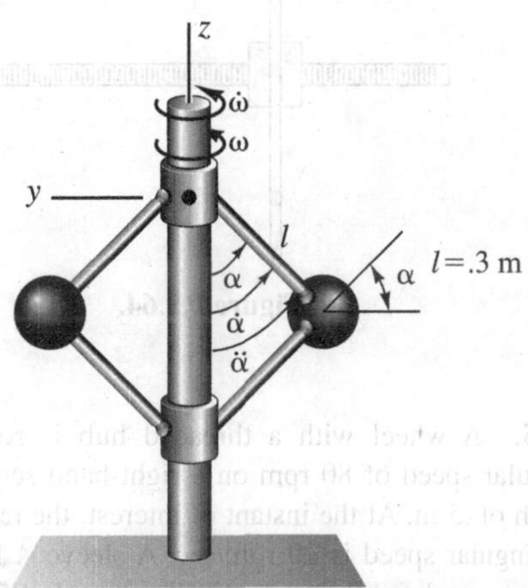

Figure P.9.62.

9.63. A stray tomahawk missile is being tracked. It is moving at a constant speed of 500 mph along a straight path having direction cosines $l = .23$ and $m = .64$. When

$$\mathbf{r} = 10\mathbf{i} + 6\mathbf{j} + 8\mathbf{k} \text{ mi}$$

express the position in cylindrical coordinates. Also, give the velocity vector and the acceleration vector using axial, transverse, and radial components.

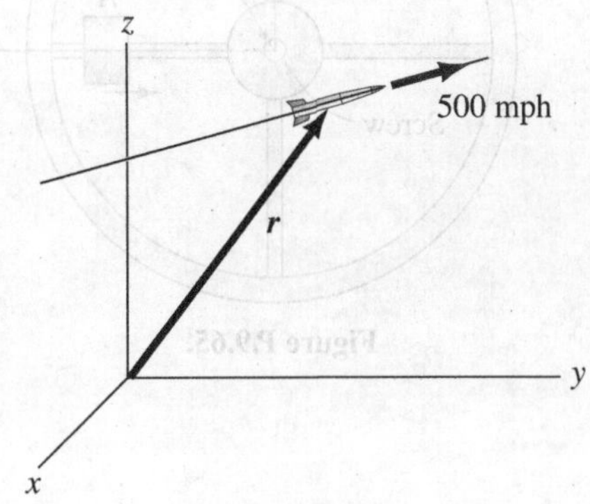

Figure P.9.63.

9.64. A wheel of diameter 2 ft is rotated at a speed of 2 rad/sec and is increasing its rotational speed at the rate of 3 rad/sec^2. It advances along a screw having a pitch of .5 in.[2] What is the acceleration of elements on the rim in terms of cylindrical components?

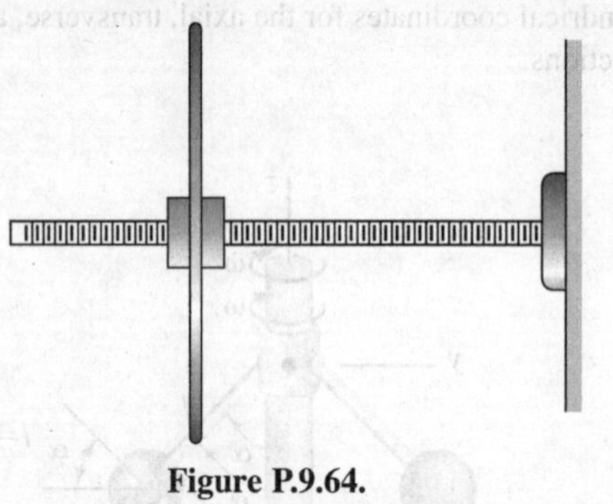

Figure P.9.64.

9.65. A wheel with a threaded hub is rotating at an angular speed of 80 rpm on a right-hand screw having a pitch of .5 in. At the instant of interest, the rate of change of angular speed is 20 rpm/sec. A sleeve A is advancing along a spoke at this instant with a speed of 5 ft/sec and a rate of increase of speed of 5 ft/sec^2. The sleeve is 2 ft from the centerline at O at the instant of interest. What are the velocity vector and the acceleration vector of the sleeve at this instant? Use cylindrical coordinates. (See footnote for definition of pitch.)

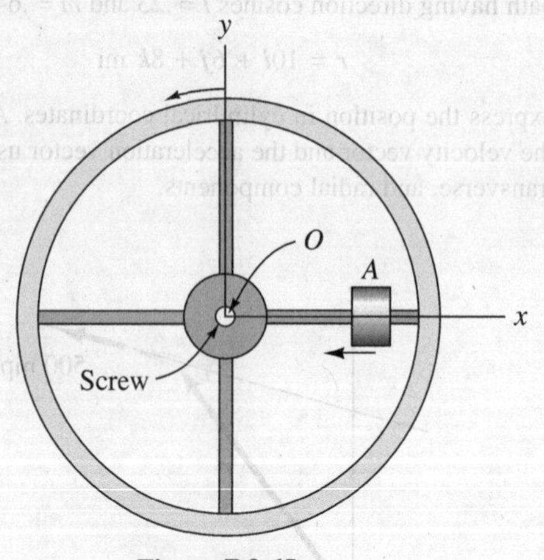

Figure P.9.65.

9.66. A simple garden sprinkler is shown. Water enters at the base and leaves at the end at a speed of 3 m/sec as seen from the rotor of the sprinkler. Furthermore, it leaves upward relative to the rotor at an angle of 60° as shown in the diagram. The rotor has an angular speed ω of 2 rad/sec. As seen from the ground, what are the axial, transverse, and radial velocity and acceleration components of the water just as it leaves the rotor?

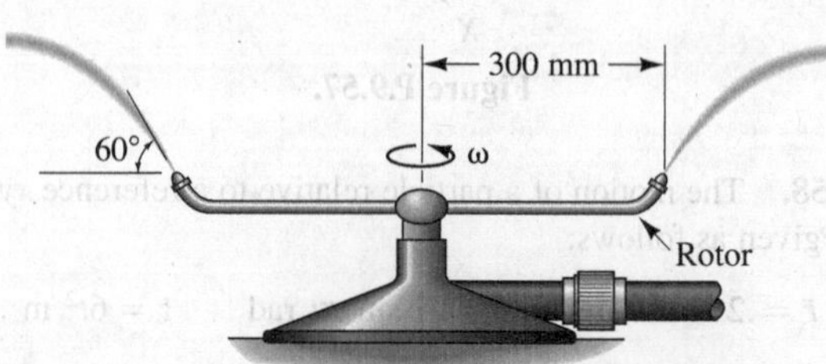

Figure P.9.66.

9.67. The acceleration of gravity on the surface of Mars is .385 times the acceleration of gravity on earth. The radius R of Mars is about .532 times that of the earth. What is the time of flight of one cycle for a satellite in a circular parking orbit 803 miles from the surface of Mars? [*Note:* $GM = gR^2$.]

9.68. A threaded rod rotates with angular position $\theta = 315t^2$ rad. On the rod is a nut which rotates relative to the rod at the rate $\omega = .4t$ rad/sec. When $t = 0$, the nut is at a distance 2 ft from A. What is the velocity and acceleration of the nut at $t = 10$ sec? The thread has a pitch of .2 in (see footnote 2). Give results in radial and transverse directions.

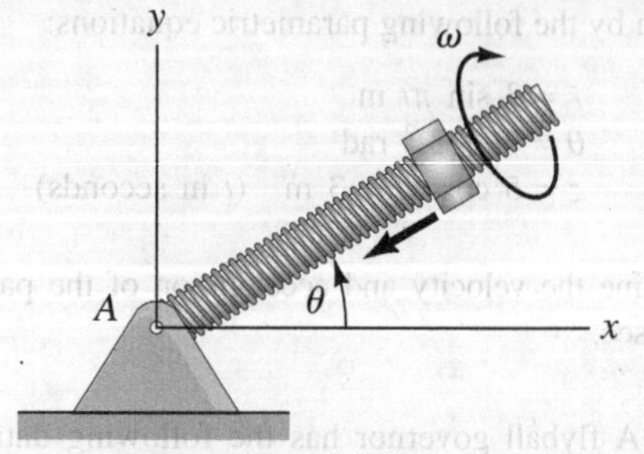

Figure P.9.68.

[2]Pitch is the distance advanced along a screw for one revolution.

9.69. Underwater cable is being laid from an ocean-going ship. The cable is unwound from a large spool A at the rear of the ship. The cable must be laid so that is *not dragged* on the ocean bottom. If the ship is moving at a speed of 3 knots, what is the necessary angular speed ω of the spool A when the cable is coming off at a radius of 3.2 m? What is the average rate of change of ω for the spool required for proper operation? The cable has a diameter of 150 mm.

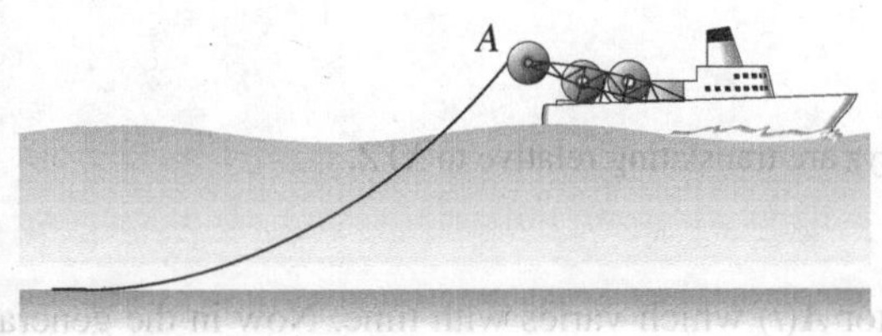

Figure P.9.69.

9.70. A variable diameter drum is rotated by a motor at a constant speed ω of 10 rpm. A rope of diameter d of .5 in. wraps around this drum and pulls up a weight W. It is desired that the velocity of the weight's *upward* movement be given as

$$\dot{X} = .4 + \frac{t^2}{8{,}000} \text{ ft/sec}$$

where for $t = 0$ the rope is just about to start wrapping around the drum at $Z = 0$. What should the radius $\bar{r}$ of the drum be as a function of Z to accomplish this? What are the velocity components $\dot{Y}$ and $\dot{Z}$ of the weight W when $t = 100$ sec?

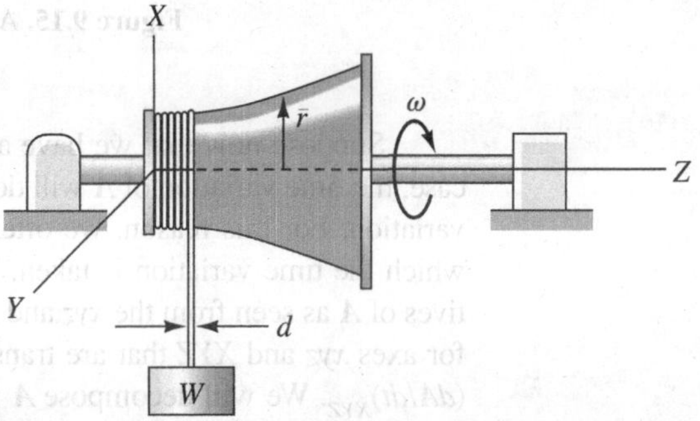

Figure P.9.70.

Part C: Simple Kinematical Relations and Applications

9.6 Simple Relative Motion

Up to now, we have considered only a single reference in our kinematical considerations. There are times when two or more references may be profitably employed in describing the motion of a particle. We shall consider in this section a very simple case that will fulfill our needs in the early portion of the text.

As a first step, consider two references xyz and XYZ (Fig. 9.15) moving in such a way that the direction of the axes of xyz always retain the same orientation relative to XYZ such as has been suggested by the dashed references giving successive positions of xyz. Such a motion of xyz relative to XYZ is called *translation*.

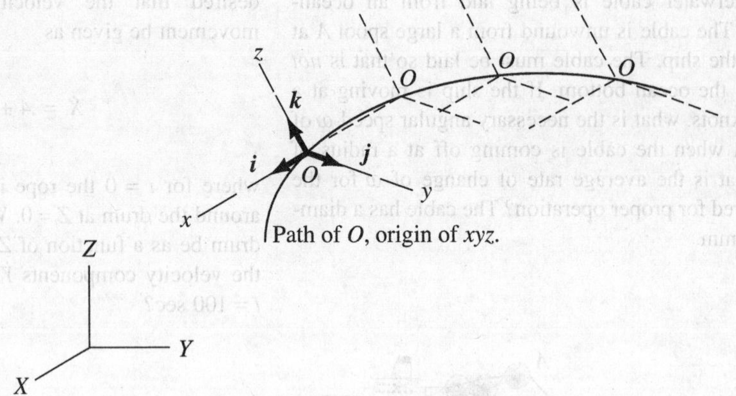

Figure 9.15. Axes *xyz* are translating relative to *XYZ*.

Suppose now that we have a vector $\boldsymbol{A}(t)$ which varies with time. Now in the general case, the time variation of $\boldsymbol{A}$ will depend on from which reference we are observing the time variation. For this reason, we often include subscripts to identify the reference relative to which the time variation is taken. Thus, we have $(d\boldsymbol{A}/dt)_{xyz}$ and $(d\boldsymbol{A}/dt)_{XYZ}$ as time derivatives of $\boldsymbol{A}$ as seen from the *xyz* and *XYZ* axes, respectively. How are these derivatives related for axes *xyz* and *XYZ* that are translating relative to each other? For this purpose, consider $(d\boldsymbol{A}/dt)_{XYZ}$. We will decompose $\boldsymbol{A}$ into components parallel to the *xyz* axes and so we have

$$\left(\frac{d\boldsymbol{A}}{dt}\right)_{XYZ} = \left[\frac{d}{dt}(A_x\boldsymbol{i} + A_y\boldsymbol{j} + A_z\boldsymbol{k})\right]_{XYZ} \tag{9.22}$$

where A_x, A_y, and A_z are the scalar components of $\boldsymbol{A}$ along the *xyz* axes. Because *xyz* translates relative to *XYZ* (see Fig. 9.21), the unit vectors of *xyz*, which we have denoted as $\boldsymbol{i}$, $\boldsymbol{j}$, and $\boldsymbol{k}$, are *constant vectors* as seen from *XYZ*. That is, whereas these vectors may change their lines of action, they *do not* change *magnitude* and *direction* as seen from *XYZ* and are thus constant vectors as seen from *XYZ*. We then have, for the equation above:

$$\left(\frac{d\boldsymbol{A}}{dt}\right)_{XYZ} = \left(\frac{dA_x}{dt}\right)_{XYZ}\boldsymbol{i} + \left(\frac{dA_y}{dt}\right)_{XYZ}\boldsymbol{j} + \left(\frac{dA_z}{dt}\right)_{XYZ}\boldsymbol{k} \tag{9.23}$$

But A_x, A_y, and A_z are *scalars* and a time derivative of a scalar, as you may remember from the calculus, is not dependent on a reference of observation.[3] We could readily replace $(dA_x/dt)_{XYZ}$ by $(dA_x/dt)_{xyz}$, etc., with no change in meaning—or we could leave off the subscripts entirely for these terms. Thus, we can say now:

$$\left(\frac{d\boldsymbol{A}}{dt}\right)_{XYZ} = \left(\frac{dA_x}{dt}\right)\boldsymbol{i} + \left(\frac{dA_y}{dt}\right)\boldsymbol{j} + \left(\frac{dA_z}{dt}\right)\boldsymbol{k} \tag{9.24}$$

[3]Clearly, the time variation of the temperature $T(x,y,z)$, a scalar, at any position in the classroom does not depend on the motion of an observer in the classroom who might be interested in the temperature at a particular position at a particular time.

Now consider $(dA/dt)_{xyz}$. Again, decomposing $\boldsymbol{A}$ into components along the xyz axes and noting that $\boldsymbol{i}, \boldsymbol{j}$, and $\boldsymbol{k}$ are constant vectors as seen from xyz, we can conclude that

$$\left(\frac{dA}{dt}\right)_{xyz} = \left(\frac{dA_x}{dt}\right)_{xyz} \boldsymbol{i} + \left(\frac{dA_y}{dt}\right)_{xyz} \boldsymbol{j} + \left(\frac{dA_z}{dt}\right)_{xyz} \boldsymbol{k}$$

$$= \left(\frac{dA_x}{dt}\right)\boldsymbol{i} + \left(\frac{dA_y}{dt}\right)\boldsymbol{j} + \left(\frac{dA_z}{dt}\right)\boldsymbol{k} \tag{9.25}$$

where as discussed earlier we have dropped the xyz subscripts. Observing Eqs. 9.24 and 9.25, we conclude that

$$\left(\frac{dA}{dt}\right)_{XYZ} = \left(\frac{dA}{dt}\right)_{xyz} \tag{9.26}$$

We can conclude that

$$\left(\frac{d}{dt}\right)_{XYZ} = \left(\frac{d}{dt}\right)_{xyz} \tag{9.27}$$

That is, the *time derivative of a vector is the same for all reference axes that are translating relative to each other.*

Note in the discussion that the fact that the unit vectors of xyz were *constant* relative to XYZ resulted in the simple relation 9.27. If xyz were *rotating* relative to XYZ, the unit vectors of xyz would not be constant as seen from XYZ and a more complex relationship would exist between $(dA/dt)_{XYZ}$ and $(dA/dt)_{xyz}$. We shall develop this relationship later in the text.

9.7 Motion of a Particle Relative to a Pair of Translating Axes

A pair of references xyz and XYZ are shown now in Fig. 9.16 moving in translation relative to each other. The *velocity vector* of any particle P depends on the reference from which the motion is observed. More precisely, we say that the velocity of particle P relative to reference XYZ is the time rate of change of the position vector $\boldsymbol{r}$ for this reference, where this rate of change is viewed from the XYZ reference. This can be stated mathematically as

$$\boldsymbol{V}_{XYZ} = \left(\frac{d\boldsymbol{r}}{dt}\right)_{XYZ} \tag{9.28}$$

Similarly, for the velocity of particle P as seen from reference xyz, we have

$$\boldsymbol{V}_{xyz} = \left(\frac{d\boldsymbol{\rho}}{dt}\right)_{xyz} \tag{9.29}$$

where we now use position vector $\boldsymbol{\rho}$ for reference xyz and view the change from the xyz reference (see Fig. 9.16). By the same token, $(d\boldsymbol{R}/dt)_{XYZ}$ is the velocity of the origin of the xyz reference as seen from XYZ. Since all points of the xyz reference have the same velocity relative to XYZ at any time t for this case (translation of xyz), we can say that $(d\boldsymbol{R}/dt)_{XYZ}$ is the velocity of reference xyz as seen from XYZ.

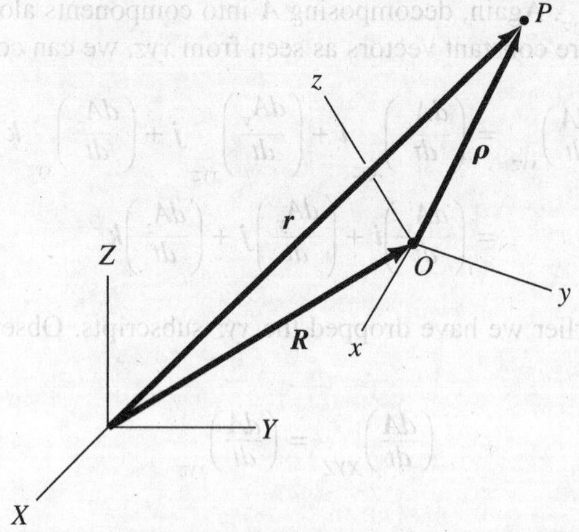

Figure 9.16. Axes xyz are translating relative to XYZ.

From Fig. 9.16 we can relate position vectors $\boldsymbol{\rho}$ and $\boldsymbol{r}$ by the equation

$$\boldsymbol{r} = \boldsymbol{R} + \boldsymbol{\rho} \tag{9.30}$$

Now take the time rate of change of these vectors as seen from XYZ. We get

$$\left(\frac{d\boldsymbol{r}}{dt}\right)_{XYZ} = \left(\frac{d\boldsymbol{R}}{dt}\right)_{XYZ} + \left(\frac{d\boldsymbol{\rho}}{dt}\right)_{XYZ} \tag{9.31}$$

The term on the left side of this equation is $\boldsymbol{V}_{XYZ}$, as indicated earlier, and we shall use the notation $\dot{\boldsymbol{R}}$ for $(d\boldsymbol{R}/dt)_{XYZ}$. We can replace the last term by the derivative $(d\boldsymbol{\rho}/dt)_{xyz}$ in accordance with Eq. 9.27 since the axes are in translation relative to each other. But $(d\boldsymbol{\rho}/dt)_{xyz}$ is simply $\boldsymbol{V}_{xyz}$, the velocity of P relative to xyz. Thus, we have

$$\boldsymbol{V}_{XYZ} = \boldsymbol{V}_{xyz} + \dot{\boldsymbol{R}} \tag{9.32}$$

By the same reasoning, we can show that the acceleration of particle P is related to references XYZ and xyz as follows[4]:

$$\boldsymbol{a}_{XYZ} = \boldsymbol{a}_{xyz} + \ddot{\boldsymbol{R}} \tag{9.33}$$

[4]As you no doubt will anticipate, the acceleration of a particle as seen from reference XYZ is

$$\boldsymbol{a}_{XYZ} = \left(\frac{d\boldsymbol{V}_{XYZ}}{dt}\right)_{XYZ}$$

Similarly, we have for $\boldsymbol{a}_{xyz}$,

$$\boldsymbol{a}_{xyz} = \left(\frac{d\boldsymbol{V}_{xyz}}{dt}\right)_{xyz}$$

Equations 9.32 and 9.33 convey the physically simple picture that the motion of a particle relative to *XYZ* is the sum of the motion of the particle relative to *xyz* plus the motion of *xyz* relative to *XYZ*.

It must be kept clearly in mind that the equations which we have developed apply only to references which have a *translatory* motion relative to each other. In Chapter 13 we shall consider references which have arbitrary motion relative to each other. (Since a reference is a rigid system, we shall need to examine at that time the kinematics of rigid bodies in order to develop these general considerations of relative motion.) The equations presented here will then be special cases.

How can we make use of multiple references? In many problems the motion of a particle is known relative to a given rigid body, and the motion of this body is known relative to the ground or other convenient reference. We can fix a reference *xyz* to the body, and if the body is in translation relative to the ground, we can then employ the given relations presented in this section to express the motion of the particle relative to the ground.

If, in ensuing chapters, we talk about the "motion of particles relative to a point," such as, for example, the center of mass of the system, then it will be understood that this motion is relative to a *hypothetical reference* moving with the center of mass in a *translatory manner* or, in other words, relative to a nonrotating observer moving with the center of mass.[5]

We illustrate these remarks in the following examples.

[5]Using a point to convey information about relative motion of a particle only allows you to convey information as to how far or how near the particle is to the point and also as to the speed and rate of change of speed of the particle toward or away from the point. The important information regarding *direction* is entirely left out, requiring a reference frame in order to give this kind of information.

Example 9.8

A jet airliner is shown in Fig. 9.17 flying at a speed of 600 mi/hr in a translatory manner relative to the ground reference *XYZ*. At the instant of interest, a downdraft causes the plane to accelerate downward at a rate of 50 mi/hr/sec. While this is happening, the pilot cuts back on the throttle so that the plane is decelerating in the *Y* direction at the rate of 30 mi/hr/sec. Thus, the plane has an acceleration given as

$$\boldsymbol{a} = -50\boldsymbol{k} - 30\boldsymbol{j} \text{ mi/hr/sec} \tag{a}$$

while maintaining a translatory attitude. While this is happening, a solenoid is operated to close a valve gate that weighs $\frac{1}{2}$ lb. What is the force on the valve gate from the plane at the instant when the valve gate is moving downward

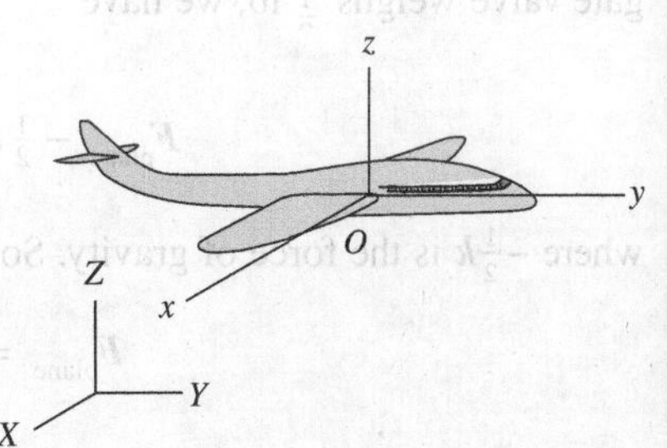

Figure 9.17. Plane translates relative to *XYZ*.

Example 9.8 (Continued)

relative to the airplane at a speed of 10 ft/sec and accelerating downward relative to the plane at a rate of 16.1 ft/sec^2?

We must find the acceleration of the valve relative to the ground reference XYZ, which may be taken in the problem to be an *inertial reference*. This information will permit us to use the familiar form of Newton's law. It will be convenient in this undertaking to *fix* a reference xyz, having the same unit vectors as reference XYZ, to the airplane at any convenient location (see Fig. 9.17). Thus

Fix xyz to the plane
Fix XYZ to the ground

We can then say for the motion of the valve gate relative to xyz:

$$\boldsymbol{a}_{xyz} = -16.1\boldsymbol{k} \text{ ft/sec}^2 \tag{b}$$

The acceleration of O, the origin of xyz relative to XYZ, is

$$\ddot{\boldsymbol{R}} = -50\boldsymbol{k} - 30\boldsymbol{j} \text{ mi/hr/sec} \tag{c}$$

Since the references are translating relative to each other, we can employ Eq. 9.33 to get $\boldsymbol{a}_{XYZ}$, the acceleration of the valve gate relative to inertial space. Thus,

$$\begin{aligned}\boldsymbol{a}_{XYZ} &= (-50\boldsymbol{k} - 30\boldsymbol{j})\left(\frac{5{,}280}{3{,}600}\right) + (-16.1\boldsymbol{k})\\ &= -44\boldsymbol{j} - 89.5\boldsymbol{k} \text{ ft/sec}^2\end{aligned} \tag{d}$$

We can now employ **Newton's law** in the form

$$\boldsymbol{F} = m\boldsymbol{a}_{XYZ} \tag{e}$$

Thus, denoting the total force from the airplane as $\boldsymbol{F}_{\text{plane}}$, and remembering that the gate valve weighs $\frac{1}{2}$ lb, we have

$$\boldsymbol{F}_{\text{plane}} - \tfrac{1}{2}\boldsymbol{k} = \frac{\frac{1}{2}}{g}(-44\boldsymbol{j} - 89.5\boldsymbol{k}) \tag{e}$$

where $-\frac{1}{2}\boldsymbol{k}$ is the force of gravity. Solving for $\boldsymbol{F}_{\text{plane}}$ we get

$$\boldsymbol{F}_{\text{plane}} = -.684\boldsymbol{j} - .890\boldsymbol{k} \text{ lb} \tag{f}$$

Example 9.9

The freighter in Fig. 9.18 is moving at a steady speed V_1 of 15 km/hr relative to the water. The freighter is 200 m long at the waterline with point A at midship. A stalking submerged submarine fires a torpedo when the submarine and freighter are at the positions shown in the diagram. The torpedo maintains a steady speed V_2 of 40 km/hr relative to the water. Will the torpedo hit the freighter?

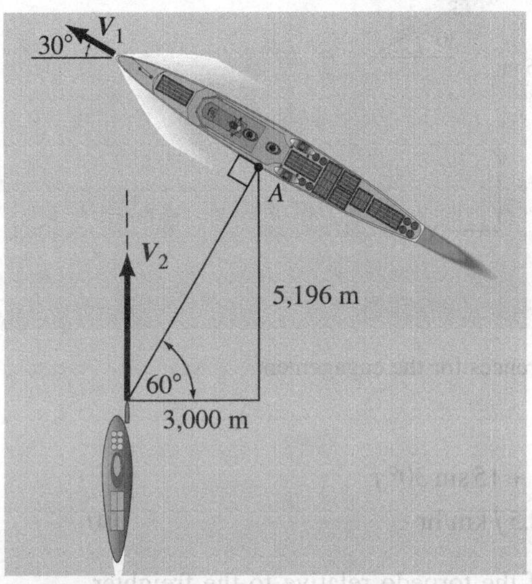

V_1 = 15 km/hr
V_2 = 40 km/hr
Freighter length = 200 m

Figure 9.18. A torpedo is fired toward a freighter. Does it hit or miss?

A key feature in solving this problem (and others like it) is that we can readily tell whether there is a hit or a miss and, if there is a hit, exactly where this takes place. This is done by simply observing the torpedo from a vantage point of the freighter. The torpedo velocity *relative to the freighter* (i.e., the motion seen by an on-board observer) will point directly to the position of potential contact with the freighter or will indicate a miss.

We accordingly make the following reference fixes:

Fix *xyz* to the freighter
Fix *XYZ* to the water

This is shown in Fig. 9.19. The velocity of *xyz*, and, hence the freighter, relative to *XYZ* (i.e., $\dot{\boldsymbol{R}}$) is $(-15 \cos 30°\boldsymbol{i} + 15 \sin 30°\boldsymbol{j})$ km/hr. The velocity of the torpedo relative to *XYZ* is $40\boldsymbol{j}$ km/hr. We can then say

$$\boldsymbol{V}_{XYZ} = \boldsymbol{V}_{xyz} + \dot{\boldsymbol{R}}$$

Example 9.9 (Continued)

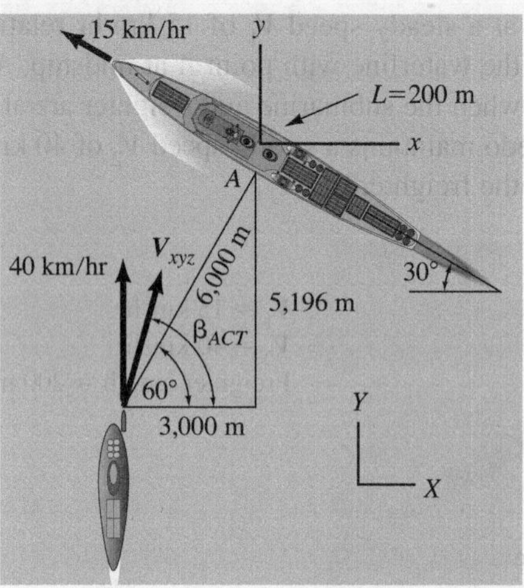

Figure 9.19. Velocity vectors and references for the engagement.

Hence,

$$40\boldsymbol{j} = \boldsymbol{V}_{xyz} - 15\cos 30°\boldsymbol{i} + 15\sin 30°\boldsymbol{j}$$

$$\therefore \boldsymbol{V}_{xyz} = 12.99\boldsymbol{i} + 32.5\boldsymbol{j} \text{ km/hr} \qquad \text{(a)}$$

To just miss the freighter, the velocity vector of the torpedo relative to the freighter, $\boldsymbol{V}_{xyz}$, must have a course such that this vector forms an angle β_0 with the horizontal axis given as (see Fig. 9.20)

$$\beta_0 = \alpha + 60° = \tan^{-1}\frac{100}{6{,}000} + 60° = 60.95° \qquad \text{(b)}$$

Now go back to Eq. (a) to obtain the actual angle, β_{ACT} (see Fig. 9.19), for the actual relative velocity vector $\boldsymbol{V}_{xyz}$.

$$\beta_{ACT} = \tan^{-1}\frac{\left(V_{xyz}\right)_y}{\left(V_{xyz}\right)_x} = \tan^{-1}\frac{32.5}{12.99} = 68.21°$$

Thus we may all relax; the torpedo just misses the freighter since $\beta_{ACT} > \beta_0$.

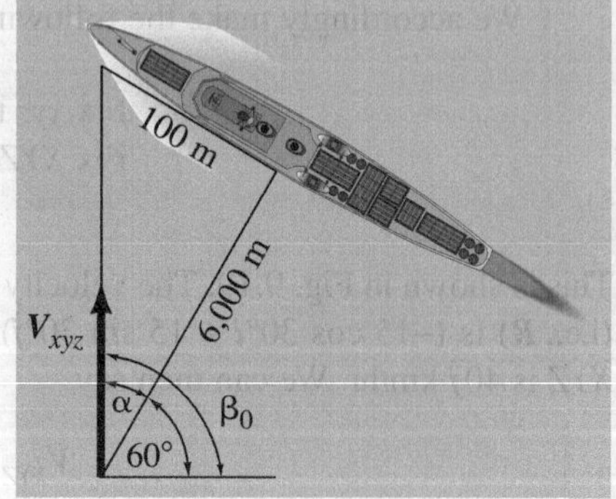

Figure 9.20. Relative velocity vector $\boldsymbol{V}_{xyz}$ for just missing the freighter.

PROBLEMS

9.71. Two wheels rotate about stationary axes each at the same angular velocity, $\dot{\theta} = 5$ rad/sec. A particle A moves along the spoke of the larger wheel at the speed V_1 of 5 ft/sec relative to the spoke and at the instant shown is decelerating at the rate of 3 ft/sec^2 relative to the spoke. What are the velocity and acceleration of particle A as seen by an observer on the hub of the smaller wheel? What are the velocity and acceleration of particle A as seen by an observer on the hub of the smaller wheel if the axis of the larger wheel moves at the instant of interest to the left with a speed of 10 ft/sec while decelerating at the rate of 2 ft/sec^2? Both wheels maintain equal angular speeds.

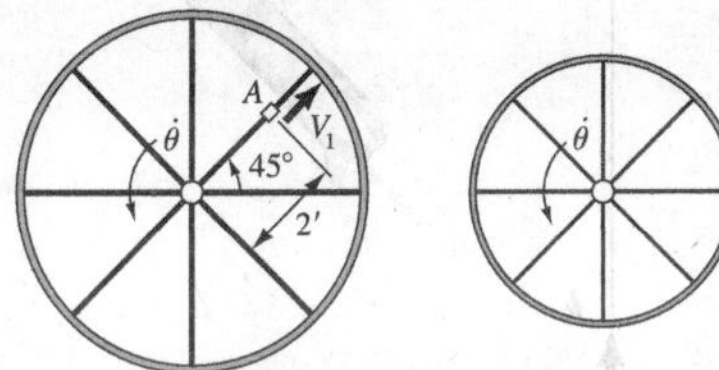

Figure P.9.71.

9.72. Four particles of equal mass undergo coplanar motion in the xy plane with the following velocities:

$$V_1 = 2 \text{ m/sec}$$
$$V_2 = 3 \text{ m/sec}$$
$$V_3 = 2 \text{ m/sec}$$
$$V_4 = 5 \text{ m/sec}$$

We showed in Section 7.3 that the velocity of the center of mass can be found as follows:

$$\left(\sum_i m_i\right) V_c = \sum_i m_i V_i$$

where V_c is the velocity of the center of mass. What are the velocities of the particles relative to the center of mass?

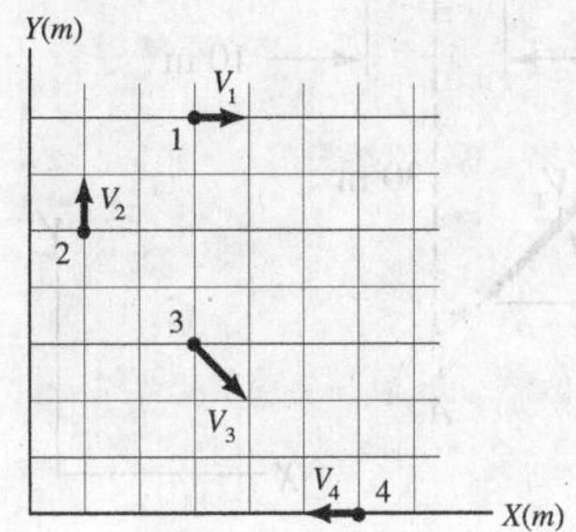

Figure P.9.72.

9.73. A sled, used by researchers to test man's ability to perform during large accelerations over extended periods of time, is powered by a small rocket engine in the rear and slides on lubricated tracks. If the sled is accelerating at $6g$, what force does the man need to exert on a 3-ounce body to give it an acceleration relative to the sled of

$$30\boldsymbol{i} + 20\boldsymbol{j} \text{ ft/sec}^2$$

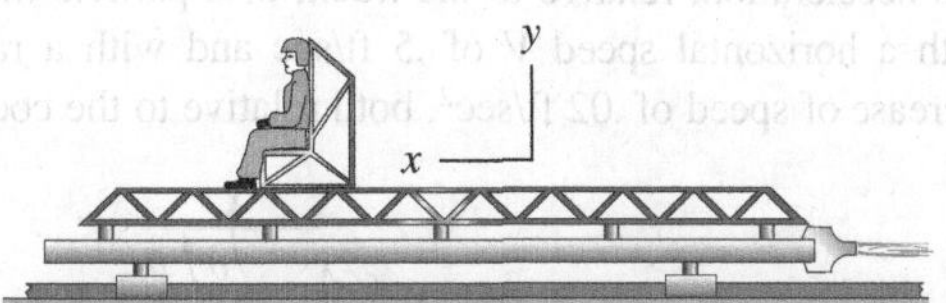

Figure P.9.73.

9.74. On the sled of Problem 9.73 is a device (see the diagram) on which mass M rotates about a horizontal axis at an angular speed ω of 5,000 rpm. If the inclination θ of the arm BM is maintained at 30° with the vertical plane C–C, what is the total force on the mass M at the instant it is in its uppermost position? The sled is undergoing an acceleration of $5g$. Take M as having a mass of .15 kg.

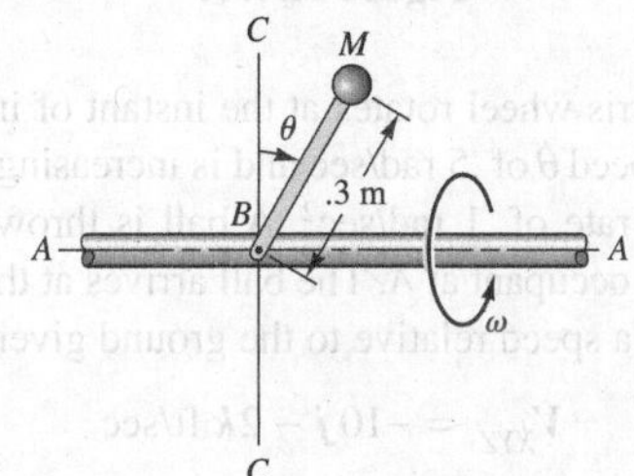

Figure P.9.74.

9.75. A vehicle, wherein a mass M of 1 lbm rotates with an angular speed ω equal to 5 rad/sec, moves with a speed V given as $V = 5 \sin \Omega t$ ft/sec relative to the ground with t in seconds. When $t = 1$ sec, the rod AM is in the position shown. At this instant, what is the dynamic force exerted by the mass M along the axis of rod AM if $\Omega = 3$ rad/sec?

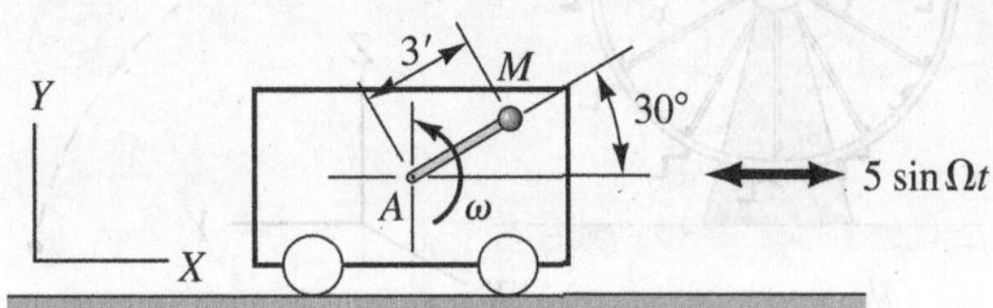

Figure P.9.75.

9.76. In Problem 9.75, what is the frequency of oscillation, Ω, of the vehicle and the value of ω if, at the instant shown, there is a force on the mass M given as

$$F = 25\boldsymbol{i} - 35\boldsymbol{j} \text{ lb}$$

9.77. A cockpit C is used to carry a worker for service work on road lighting systems. The cockpit is moved always in a translatory manner relative to the ground. If the angular speed ω of arm AB is 1 rad/min when $\theta = 30°$, what are the velocity and acceleration of any point in the cockpit body relative to the truck? At this instant, what are the velocity and acceleration, relative to the truck, of a particle moving with a horizontal speed V of .5 ft/sec and with a rate of increase of speed of .02 ft/sec^2, both relative to the cockpit?

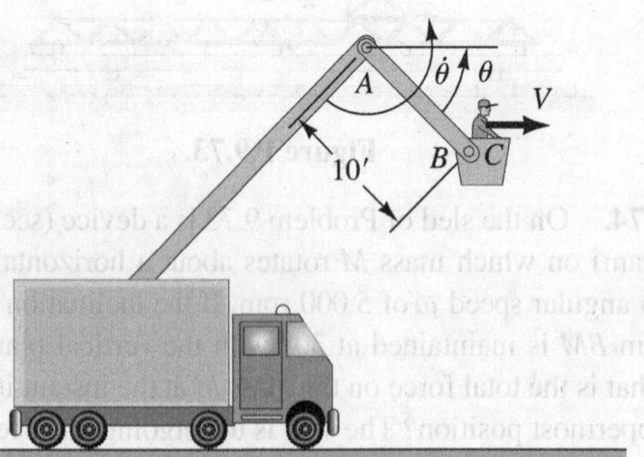

Figure P.9.77.

9.78. A Ferris wheel rotates at the instant of interest with an angular speed $\dot{\theta}$ of .5 rad/sec and is increasing its angular speed at the rate of .1 rad/sec^2. A ball is thrown from the ground to an occupant at A. The ball arrives at the instant of interest with a speed relative to the ground given as

$$\boldsymbol{V}_{XYZ} = -10\boldsymbol{j} - 2\boldsymbol{k} \text{ ft/sec}$$

What are the velocity and the acceleration of the ball relative to the occupant at seat A provided that this seat is not "swinging?" The radius of the wheel is 20 ft.

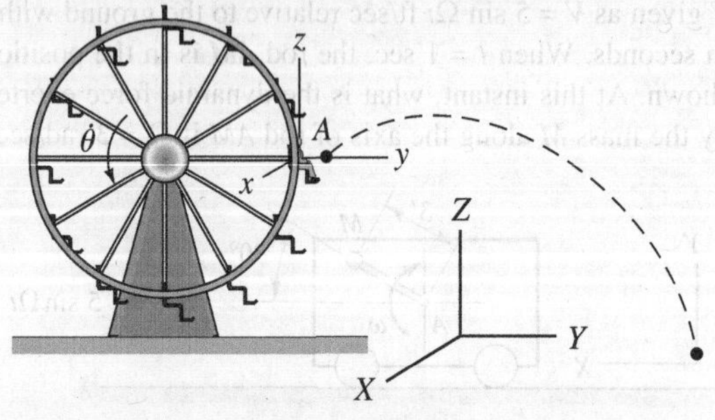

Figure P.9.78.

9.79. A rocket moves at a speed of 700 m/sec and accelerates at a rate of $5g$ relative to the ground reference XYZ. The products of combustion at A leave the rocket at a speed of 1,700 m/sec relative to the rocket and are accelerating at the rate of 30 m/sec^2 relative to the rocket. What are the speed and acceleration of an element of the combustion products as seen from the ground? The rocket moves along a straight-line path whose direction cosines for the XYZ reference are $l = .6$ and $m = .6$.

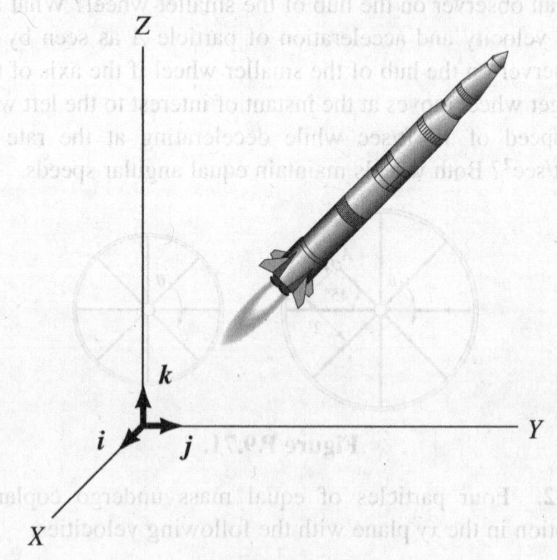

Figure P.9.79.

9.80. Two 10-m international-class sailboats are racing. They are on different tacks. If there is no change in course, will A hit B? If so, where from the center of B does this happen?

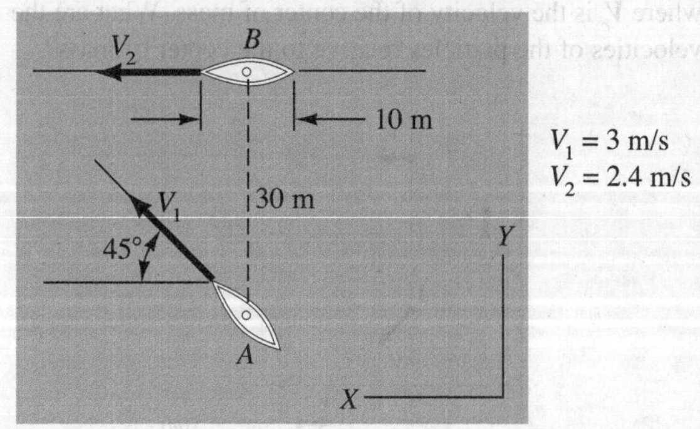

Figure P.9.80.

9.81. A train is moving at a speed of 10 km/hr. What speed should car A have to just barely miss the front of the train? How long does it take to reach this position? Use a multireference approach only.

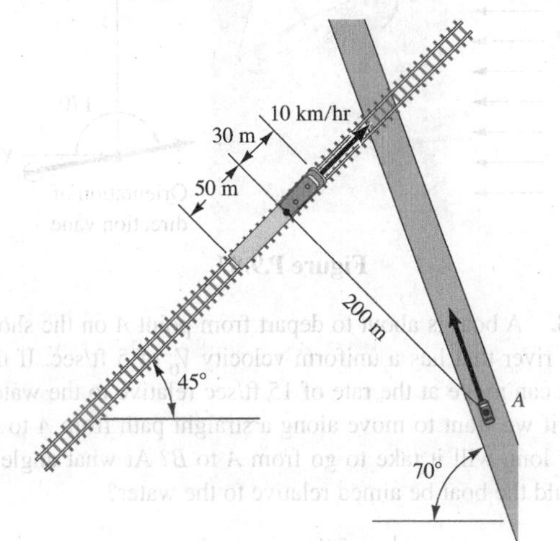

Figure P.9.81.

9.82. A Tomahawk missile is being tested for its effect on a naval vessel. A destroyer is towing an old expendable naval frigate at a speed of 15 knots. The missile is shown at time t moving along a straight line at a constant speed of 500 mi/hr, the guidance system having been shut off to avoid an accident involving the towing destroyer. Does the missile hit the target and if so where does the impact occur? The missile moves at a constant elevation of 10 ft above the surface of the water.

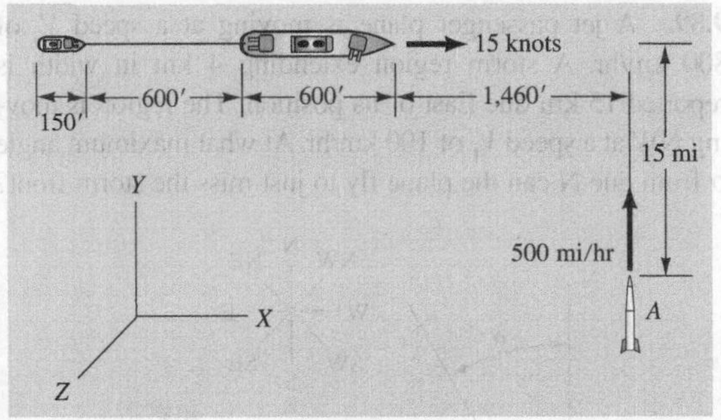

Figure P.9.82.

9.83. On a windy day, a hot air balloon is moving in a translatory manner relative to the ground with the following acceleration:

$$\boldsymbol{a} = 2\boldsymbol{i} - 5\boldsymbol{j} + 3\boldsymbol{k} \text{ m/s}^2$$

Simultaneously, a man in the balloon basket is swinging a small device for measuring the dew point. The device of mass 5 kg is connected to a massless rod. At the instant shown, $\omega = 2$ rad/sec and $\dot{\omega} = 3$ rad/sec^2 both relative to the balloon. What force does the rod exert on the device at this instant? Give the result in vector and scalar form. The rod is in a horizontal position (see elevation view) at the instant shown and has a length of .5 m.

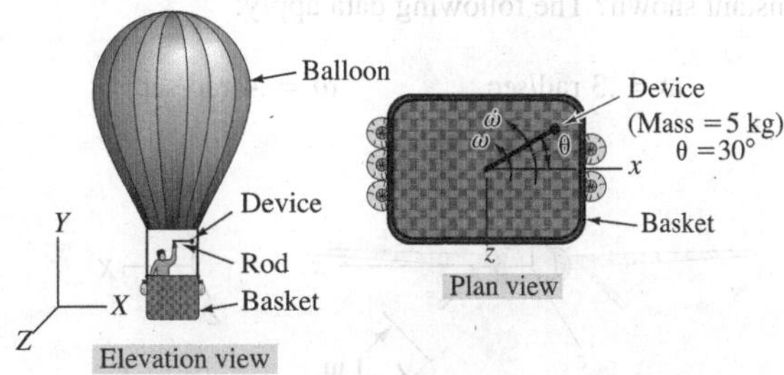

Figure P.9.83.

9.84. A submarine is moving at a constant horizontal speed of 15 knots below the surface of the ocean. At the same time, the sub is descending downward by discharging air with an acceleration of $.023g$'s while remaining horizontal. In the submarine, a flyball governor operates with weights having a mass of 500 g each. The governor is rotating with speed ω of 5 rad/sec. If at time t, $\theta = 30°$, $\dot{\theta} = .2$ rad/sec, and $\ddot{\theta} = 1$ rad/sec^2, what is the force developed on the support of the governor system as a result solely of the motion of the weights at this instant? [*Hint:* What is the *acceleration* of the *center of mass* of the spheres relative to *inertial* space?]

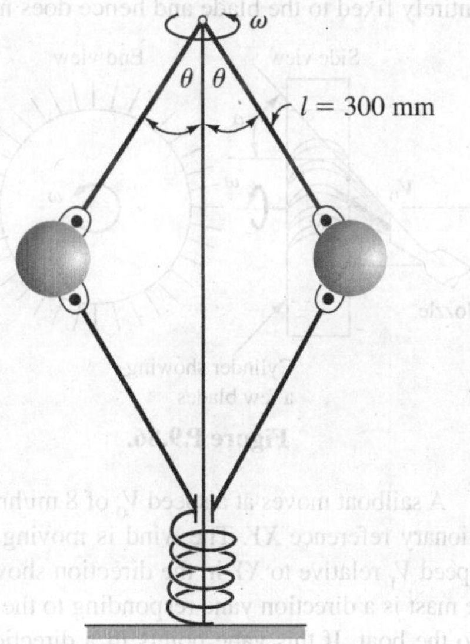

Figure P.9.84.

9.85. A fighter plane is landing and has the following acceleration relative to the ground while moving in a translatory manner:

$$\boldsymbol{a} = -.2g\boldsymbol{k} - .1g\boldsymbol{j} \text{ m/s}^2$$

The wheels are being let down as shown. What is the dynamic force acting at the center of the wheel at the instant shown? The following data apply:

$$\omega = .3 \text{ rad/sec} \qquad \dot{\omega} = .4 \text{ rad/sec}^2$$

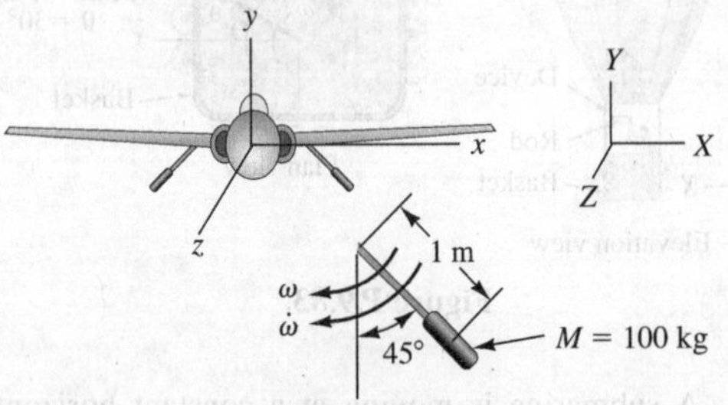

Figure P.9.85.

9.86. In a steam turbine, steam is expended through a stationary nozzle at a speed V_0 of 3,000 m/sec at an angle of 30°. The steam impinges on a series of blades mounted all around the periphery of a cylinder, which is rotating at a speed Ω of 5,000 rpm. The steam impinges on the blades at a radial distance of 1.20 m from the axis of rotation of the cylinder. What angle α should the left side of the blades have for the steam to enter the region between blades most smoothly? [*Hint:* Let *xyz* move with a blade in a translatory manner relative to the ground (*xyz* is thus not entirely fixed to the blade and hence does not rotate).]

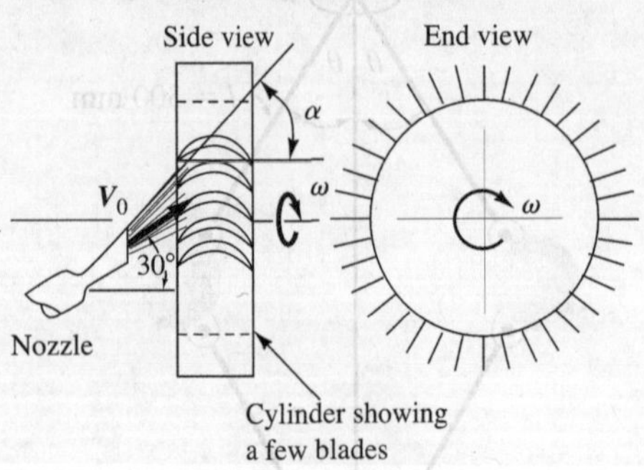

Figure P.9.86.

9.87. A sailboat moves at a speed V_0 of 8 mi/hr relative to a stationary reference *XY*. The wind is moving uniformly at a speed V_1 relative to *XY* in the direction shown. On top of the mast is a direction vane responding to the wind relative to the boat. If this vane points in a direction of 170° from the *X* axis, what is the velocity V_1 of the wind?

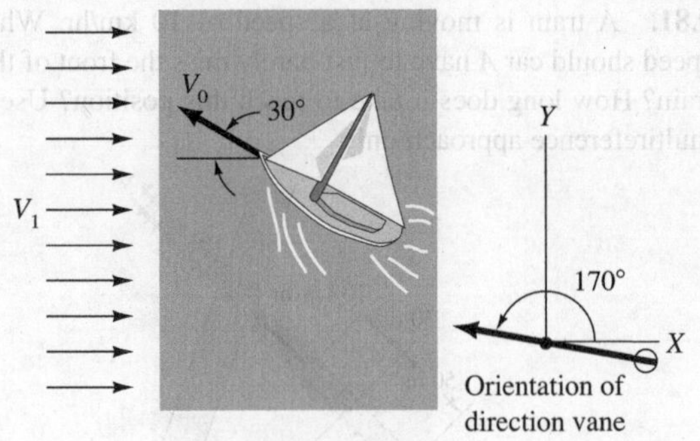

Figure P.9.87.

9.88. A boat is about to depart from point *A* on the shore of a river that has a uniform velocity V_0 of 5 ft/sec. If the boat can move at the rate of 15 ft/sec relative to the water, and if we want to move along a straight path from *A* to *B*, how long will it take to go from *A* to *B*? At what angle β should the boat be aimed relative to the water?

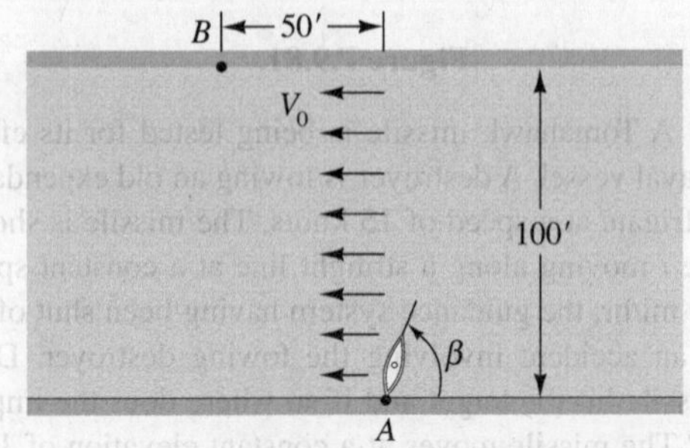

Figure P.9.88.

9.89. A jet passenger plane is moving at a speed V_0 of 800 km/hr. A storm region extending 4 km in width is reported 15 km due East of its position. The region is moving NW at a speed V_1 of 100 km/hr. At what maximum angle α from due N can the plane fly to just miss the storm front?

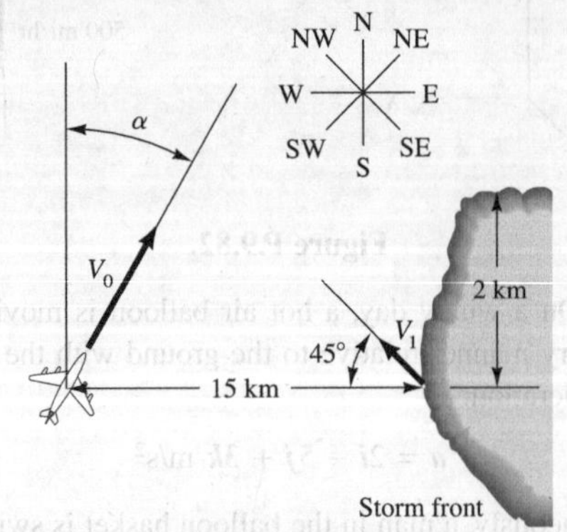

Figure P.9.89.

9.90. Mass M of 3 kg rotates about point O in an accelerating rocket in the xy plane. At the instant shown, what is the force from the rod onto the mass? Include the effects of gravity if $g = 7.00 \text{ m/s}^2$ at the elevation of the rocket.

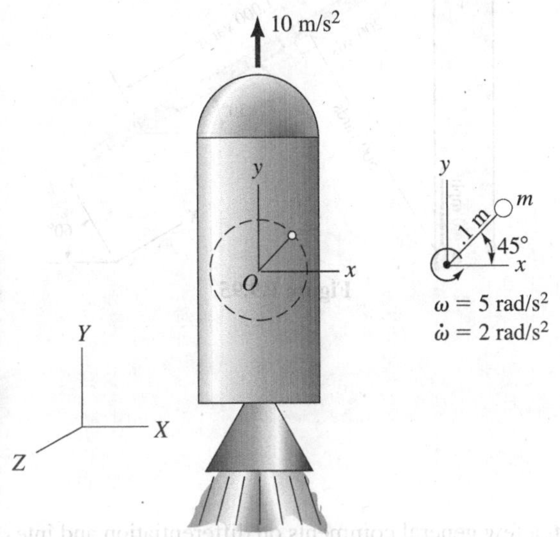

Figure P.9.90.

9.91. A light plane is approaching a runway in a cross-wind. This cross-wind has a uniform speed V_0 of 33 mi/hr. The plane has a velocity component V_1 parallel to the ground of 70 mi/hr relative to the wind at an angle β of 30°. The rate of descent is such that the plane will touch down somewhere along A–A. Will this touchdown occur on the runway or off the runway for the data given?

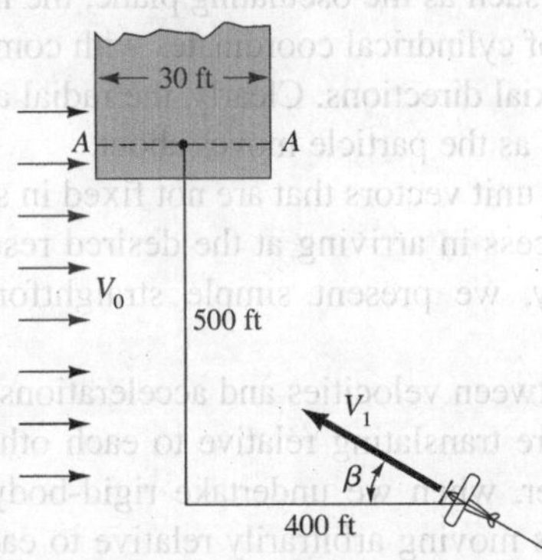

Figure P.9.91.

9.92. A helicopter is shown moving relative to the ground with the following motion:

$$\boldsymbol{V} = 130\boldsymbol{i} + 70\boldsymbol{j} + 20\boldsymbol{k} \text{ km/hr}$$
$$\boldsymbol{a} = 10\boldsymbol{i} + 16\boldsymbol{j} + 7\boldsymbol{k} \text{ km/hr/s}$$

The helicopter blade is rotating relative to the helicopter in the following manner at the instant of interest:

$$\omega_1 = 100 \text{ rpm} \qquad \dot{\omega}_1 = 10.3 \text{ rpm/sec}$$

The blade is 10 m long. What is the velocity and the acceleration of the tip B relative to the ground reference XYZ? Give your results in meters and seconds. The blade is parallel to the X axis at the instant of interest.

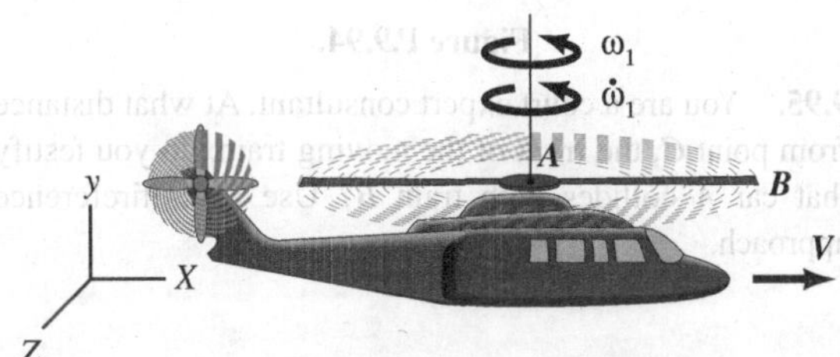

Figure P.9.92.

9.93. A destroyer in rough seas has the following translational acceleration as seen from inertial reference XYZ when it is firing its main battery in the YZ plane:

$$\boldsymbol{a} = 5\boldsymbol{j} + 2\boldsymbol{k} \text{ m/s}^2$$

What must ω_1 and $\dot{\omega}_1$ of the gun barrel be relative to the ship at this instant so that tip A of the barrel has zero acceleration relative to XYZ?

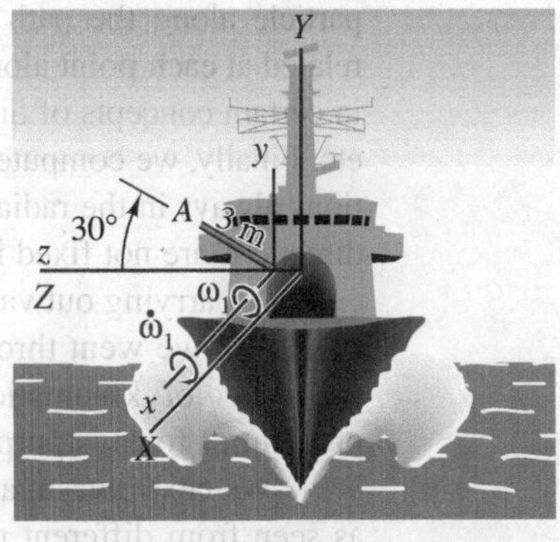

Figure P.9.93.

9.94. A small elevator E in an ocean-going vessel has the following motion relative to the ship:

$$\boldsymbol{a}_{Elev.} = .2g\boldsymbol{k} \text{ m/s}^2$$

The ship has the following motion relative to nearby land:

$$\boldsymbol{a}_{Ship} = .2\boldsymbol{i} + 3\boldsymbol{j} + .6\boldsymbol{k} \text{ m/s}^2$$

If the weight of the elevator including passengers is 8,000 N, what is the force on the ship from the elevator?

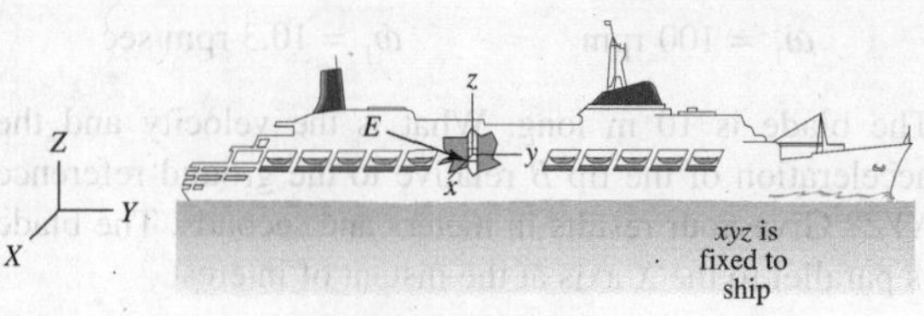

Figure P.9.94.

9.95. You are a court expert consultant. At what distance from point C, the front of the moving train, do you testify that car A collides with train B? Use a multireference approach.

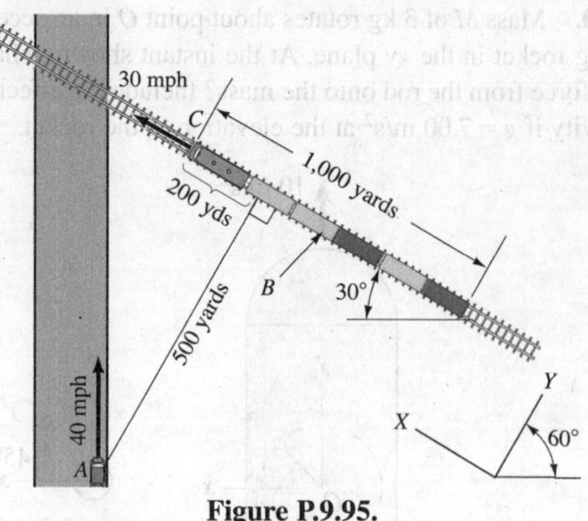

Figure P.9.95.

9.9 Closure

In this chapter, we have presented, first, a few general comments on differentiation and integration of vectors. We then carried out differentiations in a variety of ways. In the first case, the vector $\boldsymbol{r}$ was expressed in terms of rectangular scalar components and the fixed unit vectors $\boldsymbol{i}$, $\boldsymbol{j}$, and $\boldsymbol{k}$. The procedure for finding $\dot{\boldsymbol{r}}$ and $\ddot{\boldsymbol{r}}$ in terms of rectangular scalar components is straightforward and involves only the familiar differentiation operations of scalar calculus. We next considered the kinematics of a particle moving along some given path. Here, we obtained $\dot{\boldsymbol{r}}$ and $\ddot{\boldsymbol{r}}$ in terms of speeds and rates of changes of speeds of the particle along the path with component directions no longer fixed in space but instead related at each point along the path to the geometry of the path. For this reason, we brought in certain concepts of differential geometry such as the osculating plane, the normal vector, etc. Finally, we computed $\dot{\boldsymbol{r}}$ and $\ddot{\boldsymbol{r}}$ in terms of cylindrical coordinates with component directions always in the radial, transverse, and axial directions. Clearly, the radial and transverse directions are not fixed in space and change as the particle moves about.

In carrying out various derivatives of unit vectors that are not fixed in space, such as $\boldsymbol{\epsilon}_r$ and $\boldsymbol{\epsilon}_\theta$, we went through a limiting process in arriving at the desired results. Later, in the study of kinematics of a rigid body, we present simple straightforward formal procedures for this purpose.

We next investigated the relations between velocities and accelerations of a particle, as seen from different references, which are translating relative to each other. We called such motions simple relative motion. Later, when we undertake rigid-body motion, we shall consider the case involving references moving arbitrarily relative to each other. It is vital to remember that we must measure $\boldsymbol{a}$ relative to an *inertial reference* when we employ **Newton's law** in the form $\boldsymbol{F} = m\boldsymbol{a}$. We may at times find it convenient to employ two references in this connection where one reference is the inertial reference needed for the desired acceleration vector. This situation is illustrated in Example 9.10.

In Chapter 10, we shall consider the *dynamics* of motion of a particle. We shall then have ample opportunity to employ the kinematics of Chapter 9.

PROBLEMS

9.96. A particle at position (3, 4, 6) ft at time $t_0 = 1$ sec is given a constant acceleration having the value $6\boldsymbol{i} + 3\boldsymbol{j}$ ft/sec^2. If the velocity at the time t_0 is $16\boldsymbol{i} + 20\boldsymbol{j} + 5\boldsymbol{k}$ ft/sec, what is the velocity of the particle 20 sec later? Also give the position of the particle.

9.97. A pin is confined to slide in a circular slot of radius 6 m. The pin must also slide in a straight slot that moves to the right at a constant speed, V, of 3 m/sec while maintaining a constant angle of 30° with the horizontal. What are the velocity and acceleration of the pin A at the instant shown?

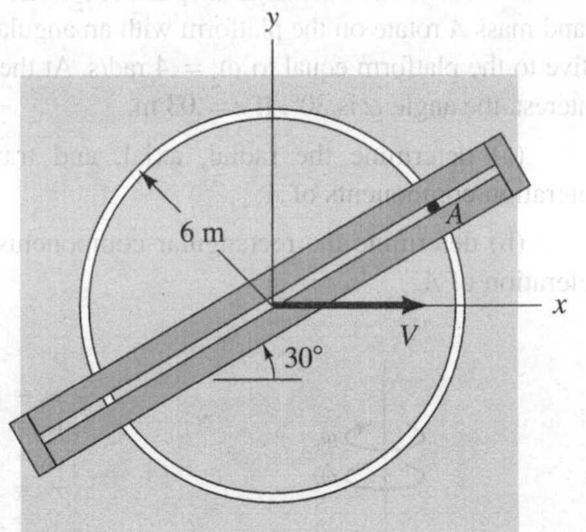

Figure P.9.97.

9.98. A freighter is moving in a river at a speed of 5 knots relative to the water. A small boat A is moving relative to the water at a speed of 3 knots in a direction as shown in the diagram. The river is moving at a uniform speed of .6 knots relative to the ground. Will the boat hit the freighter and, if so, where will the impact occur?

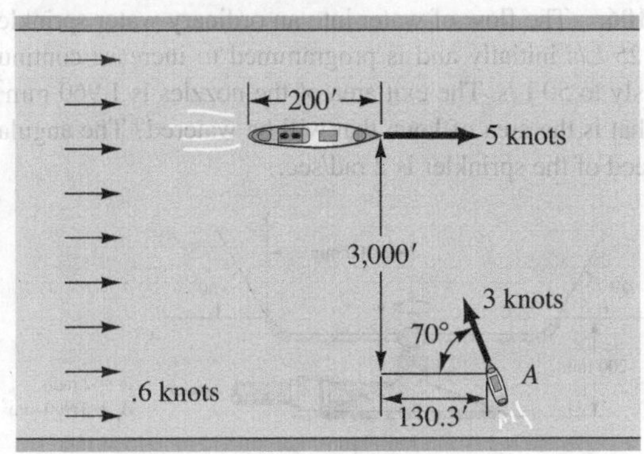

Figure P.9.98.

9.99. A light line attached to a streamlined weight A is "shot" by a line rifle from a small boat C to a large boat D in heavy seas. The weight must travel a distance of 20 yd horizontally and reach the larger boat's deck, which is 20 ft higher than the deck of boat C. If the angle α of firing is 40°, what minimum velocity V_0 is needed? At the instant of firing, boat C is dipping down into the water at a speed of 5 ft/sec. Assume that the larger boat remains essentially fixed at constant level.

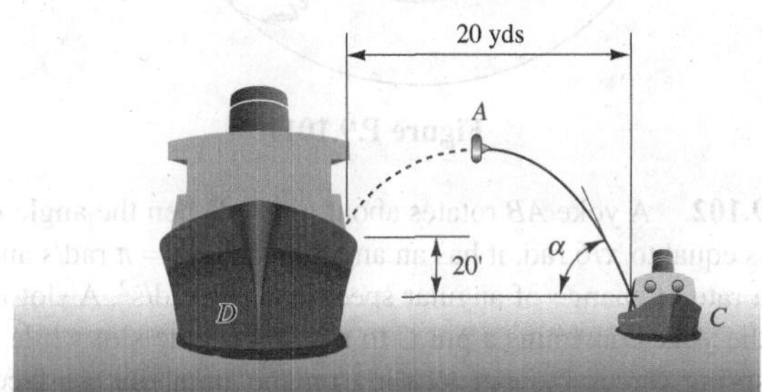

Figure P.9.99.

9.100. A projectile is fired at an angle of 60° as shown. At what elevation y does it strike the hill whose equation has been estimated as $y = 10^{-5}x^2$ m? Neglect air friction and take the muzzle velocity as 1,000 m/sec.

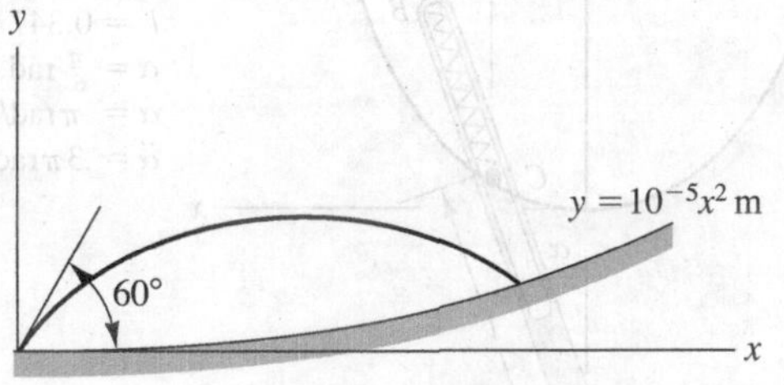

Figure P.9.100.

9.101. A proposed space laboratory, in order to simulate gravity, rotates relative to an inertial reference XYZ at a rate ω_1. For occupant A in the living quarters to be comfortable, what should the approximate value of ω_1 be? Clearly, at the center, there is close to zero gravity for zero-g experiments. A conveyor connects the living quarters with the zero-g laboratory. At the instant of interest, a package D has a speed of 5 m/s and a rate of change of speed of 3 m/s^2 relative to the space station, both toward the laboratory. What are the axial, transverse, and radial velocity and acceleration components at the instant of interest relative to the inertial reference? What are the rectangular components of the acceleration vector?

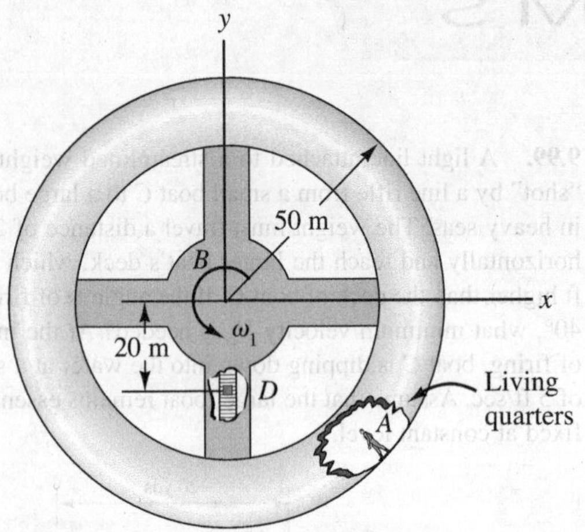

Figure P.9.101.

9.102. A yoke AB rotates about pin A. When the angle α is equal to $\pi/6$ rad, it has an angular speed $\dot{\alpha} = \pi$ rad/s and a rate of change of angular speed $\ddot{\alpha} = .3\pi$ rad/s^2. A slot in the yoke constrains a pin C to move with the slot while a spring forces the pin to slide on the parabolic surface. What are the velocity and acceleration vectors at the instant of interest?

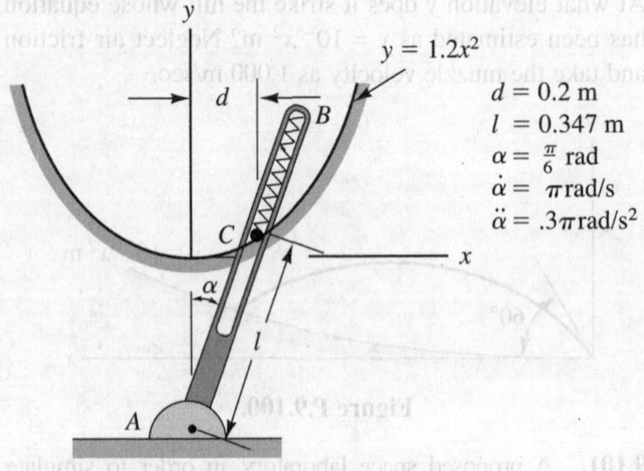

Figure P.9.102.

9.103. In Problem 9.17, what is the change of radius of curvature of a stream of water as it goes out from a circular vane of radius .30 m into free flight? What and where is the minimum radius of curvature?

9.104. A batter has hit a home run ball that just clears the fence. It has a line drive trajectory. A fan attempts to catch the ball. Neglecting friction, what is the speed of the ball as the fan attempts to catch it?

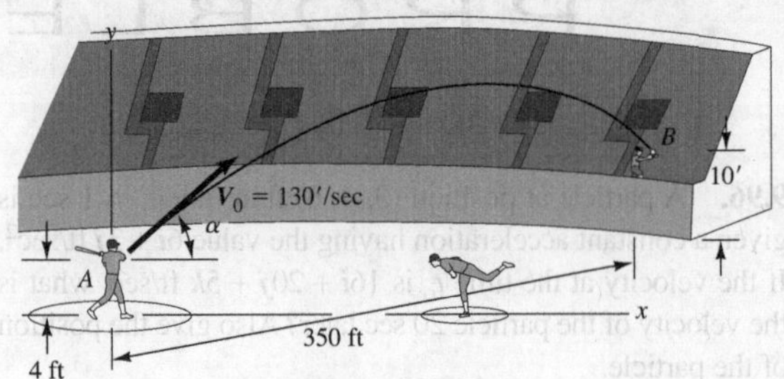

Figure P.9.104.

9.105. A platform is rotating relative to a stationary reference xyz with angular speed $\omega_1 = 1$ rad/s and a rate of change of angular speed $\dot{\omega}_1 = .2$ rad/s^2 at the instant of interest. Also at this instant, the platform is being raised at a speed $V_z = .47$ m/s with a rate of change of speed of $\dot{V}_z = .26$ m/s^2. A rod and mass A rotate on the platform with an angular speed relative to the platform equal to $\omega_2 = .4$ rad/s. At the instant of interest, the angle α is 30°. If $r = .03$ m,

(a) determine the radial, axial, and transverse acceleration components of A;

(b) determine the rectangular components of the acceleration of A.

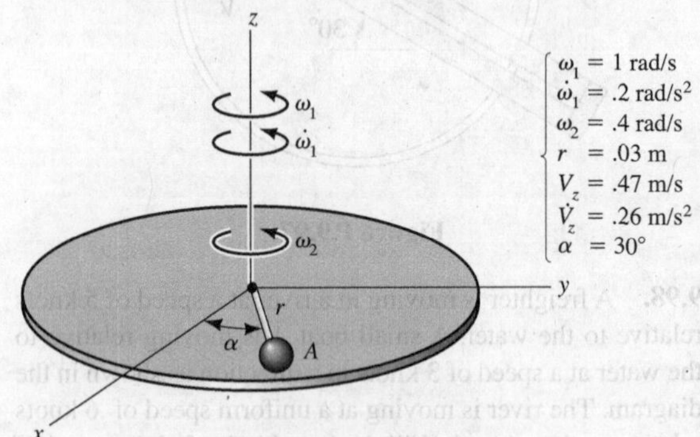

Figure P.9.105.

9.106. The flow of water into an ordinary water sprinkler is 25 L/s initially and is programmed to increase continuously to 50 L/s. The exit area of the nozzles is 1,960 mm^2. What is the area of lawn that will be watered? The angular speed of the sprinkler is 2 rad/sec.

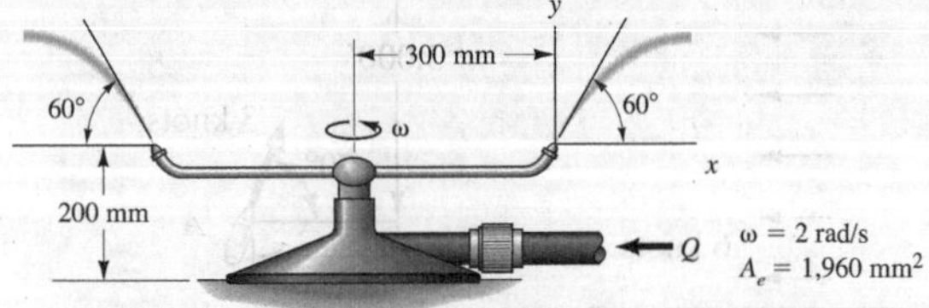

Figure P.9.106.

9.107. Pilots of fighter planes wear special suits designed to prevent blackouts during a severe maneuver. These suits tend to keep the blood from draining out of the head when the head is accelerated in a direction from shoulders to head. With this suit, a flier can take $5g$'s of acceleration in the aforementioned direction. If a flier is diving at a speed of 1,000 km/hr, what is the minimum radius of curvature that he can manage at pullout without suffering bad physiological effects?

9.108. A particle moves with constant speed of 1.5 m/sec along a path given as $x = y^2 - \ln y$ m. Give the acceleration vector of the particle in terms of rectangular components when the particle is at position $y = 3$ m. Do the problem by using path coordinate techniques and then by Cartesian-component techniques. How many g's of acceleration is the particle subject to?

9.109. A submarine is moving in a translatory manner with the following velocity and acceleration relative to an inertial reference:

$$\boldsymbol{V} = 6\boldsymbol{i} + 7.5\boldsymbol{j} + 2\boldsymbol{k} \text{ knots} \qquad \boldsymbol{a} = .2\boldsymbol{i} - .24\boldsymbol{j} + .52\boldsymbol{k} \text{ knots/s}$$

A device inside the submarine consists of an arm and a mass at the end of the arm. At the instant of interest, the arm is rotating in a vertical plane with the following angular speed and angular acceleration:

$$\omega = 10 \text{ rad/s} \qquad \dot{\omega} = 3 \text{ rad/s}^2$$

The arm is vertical at this instant. The mass at the end of the rod may be considered to be a particle having a mass of 5 kg. What are the velocity and acceleration vectors for the motion of the particle at this instant relative to the inertial reference? Use units of meters and seconds. What must be the force vector from the arm onto the particle at this instant?

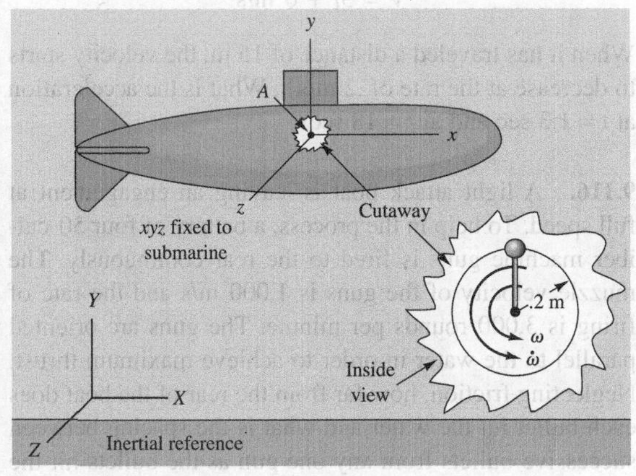

Figure P.9.109.

9.110. A mechanical "arm" for handling radioactive materials is shown. The distance r can be varied by telescoping action of the arm. The arm can be rotated about the vertical axis A–A. Finally, the arm can be raised or lowered by a worm gear drive (not shown). What is the velocity and acceleration of the object C if the end of the arm moves out radially at a rate of 1 ft/sec while the arm turns at a speed ω of 2 rad/sec? Finally, the arm is raised at a rate of 2 ft/sec. The distance r at the instant of interest is 5 ft. What is the acceleration in the direction $\boldsymbol{\epsilon} = .8\boldsymbol{i} + .6\boldsymbol{j}$?

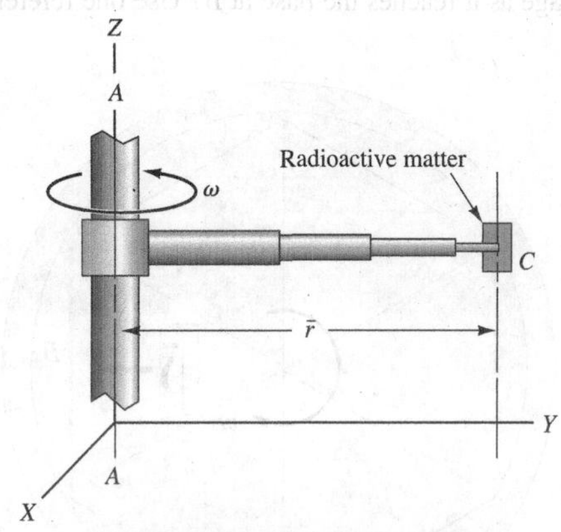

Figure P.9.110.

9.111. A top-section view of a water sprinkler is shown. Water enters at the center from below and then goes through four passageways in an impeller. The impeller is rotating at constant speed ω of 8 rpm. As seen from the impeller, the water leaves at a speed of 10 ft/sec at an angle of 30° relative to r. What is the velocity and acceleration as seen from the ground of the water as it leaves the impeller and becomes free of the impeller? Give results in the radial, axial, and transverse directions. Use one reference only.

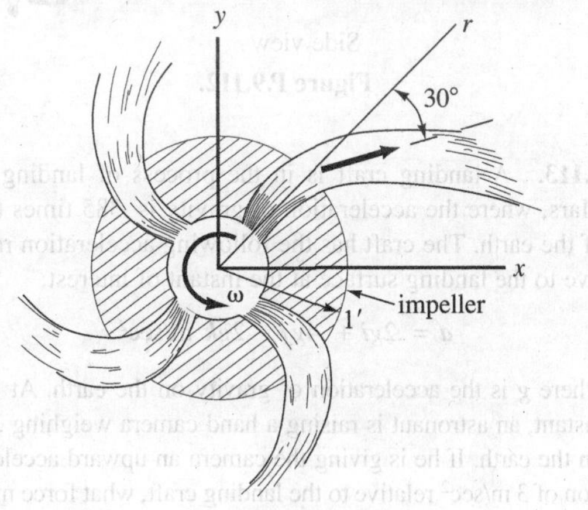

Figure P.9.111.

9.112. A luggage dispenser at an airport resembles a pyramid with six flat segments as sides as shown in the diagram. The system rotates with an angular speed ω of 2 rpm. Luggage is dropped from above and slides down the faces to be picked up by travelers at the base.

A piece of luggage is shown on a face. It has just been dropped at the position indicated. It has at this instant zero velocity as seen from the rotating face but has at this instant and thereafter an acceleration of $.2g$ along the face. What is the total acceleration, as seen from the ground, of the luggage as it reaches the base at B? Use one reference only.

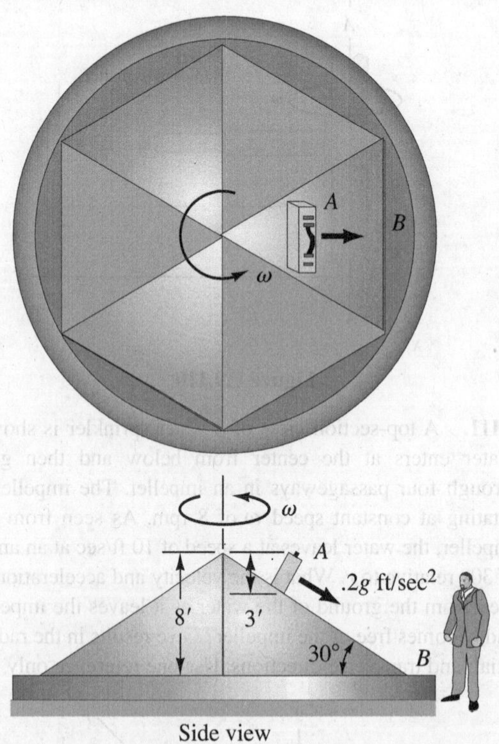

Figure P.9.112.

9.113. A landing craft is in the process of landing on Mars, where the acceleration of gravity is .385 times that of the earth. The craft has the following acceleration relative to the landing surface at the instant of interest:

$$\boldsymbol{a} = .2g\boldsymbol{i} + .4g\boldsymbol{j} - 2g\boldsymbol{k} \text{ m/sec}^2$$

where g is the acceleration of gravity on the earth. At this instant, an astronaut is raising a hand camera weighing 3 N on the earth. If he is giving the camera an upward acceleration of 3 m/sec^2 relative to the landing craft, what force must the astronaut exert on the camera at the instant of interest?

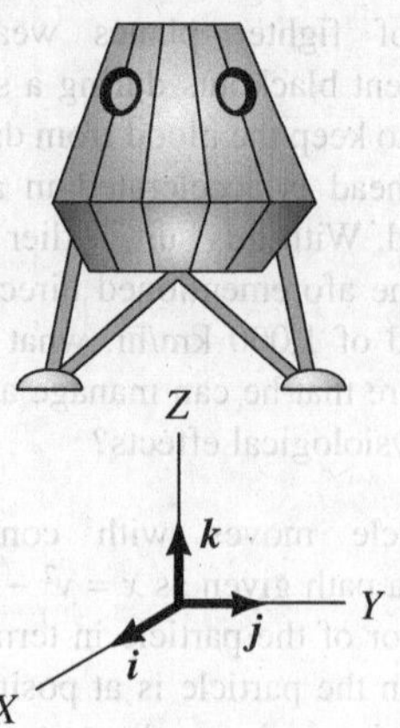

Figure P.9.113.

9.114. A jet of water has a speed at the nozzle of 20 m/s. At what position does it hit the parabolic hill? What is its speed at that point? Do not include friction.

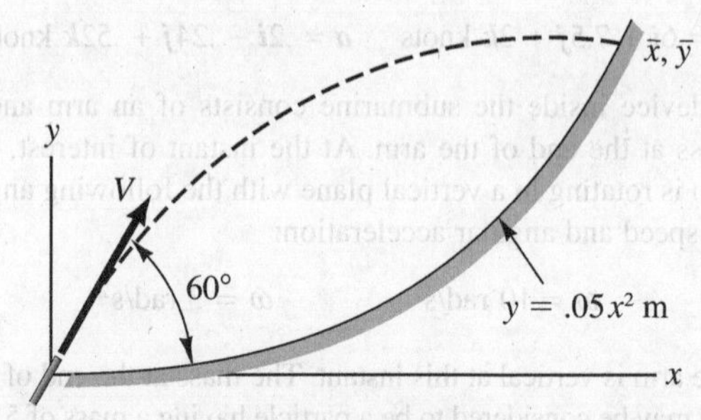

Figure P.9.114.

9.115. A particle moves along a circular path of diameter 10 m such that

$$V = 3t + 6 \text{ m/s}$$

When it has traveled a distance of 15 m, the velocity starts to decrease at the rate of .2 m/s/s. What is the acceleration at $t = 1.3$ sec and at $t = 18$ sec?

9.116. A light attack boat is leaving an engagement at full speed. To help in the process, a battery of four 50-caliber machine guns is fired to the rear continuously. The muzzle velocity of the guns is 1,000 m/s and the rate of firing is 3,000 rounds per minute. The guns are oriented parallel to the water in order to achieve maximum thrust. Neglecting friction, how far from the rear of the boat does each bullet hit the water and what is the spacing between successive bullets from any one gun as the bullets hit the water? The boat is moving at a constant speed of 45 knots.

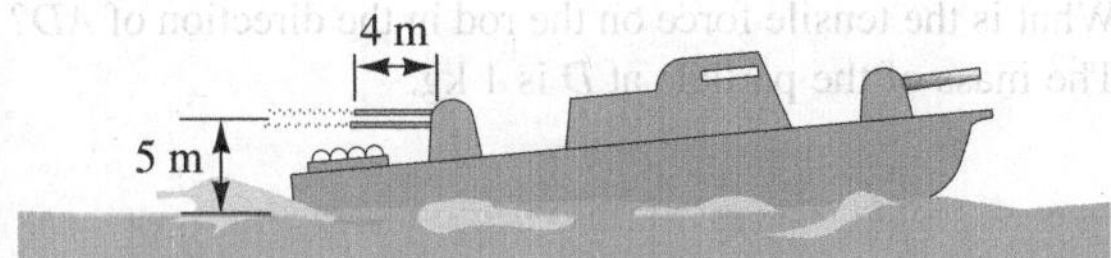

Figure P.9.116.

***9.117.** A particle has a variable velocity $V(t)$ along a helix wrapped around a cylinder of radius e. The helix makes a constant angle α with plane A perpendicular to the z axis. Express the acceleration $\boldsymbol{a}$ of the particle using cylindrical coordinates. Next, express $\boldsymbol{\epsilon}_t$ using cylindrical unit vectors and note that the sum of the transverse and axial components of $\boldsymbol{a}$ (just computed) can be given simply as $\dot{V}\boldsymbol{\epsilon}_t$. Next, express the acceleration of the particle using path coordinates. Finally, noting that $\boldsymbol{\epsilon}_n = -\boldsymbol{\epsilon}_{\bar{r}}$, show that the radius of curvature is given as $R = \dfrac{e}{\cos^2 \alpha}$.

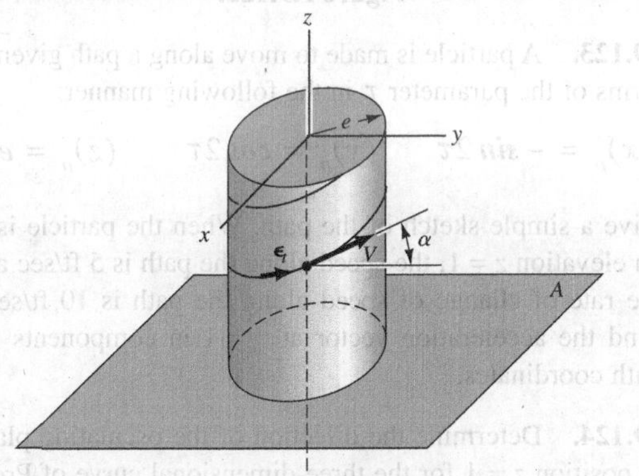

Figure P.9.117.

9.118. An eagle is diving at a constant speed of 40 ft/sec to catch a 10-ft snake that is moving at a constant speed of 15 ft/sec. What should α be so that the eagle hits the small head of the snake? The eagle and the snake are moving in a vertical plane.

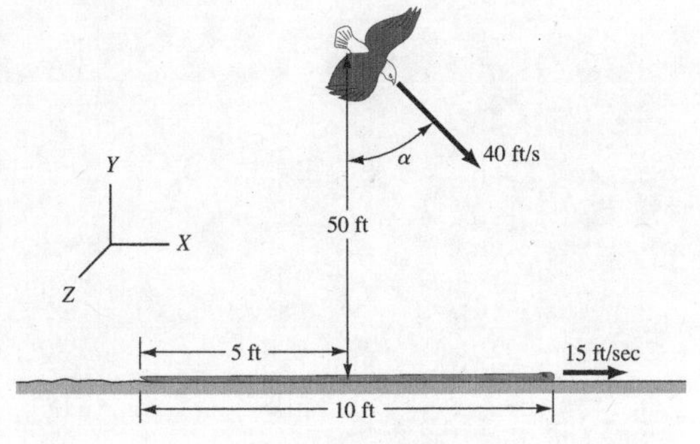

Figure P.9.118.

9.119. A tube, most of whose centerline is that of an ellipse given as

$$\frac{z^2}{1.8^2} + \frac{\bar{r}^2}{.32^2} = 1$$

has a cross-sectional diameter $D = 100$ mm. The tube has the following rotational motion at the instant of interest:

$$\omega = .15 \text{ rad/s} \qquad \dot{\omega} = .036 \text{ rad/s}^2$$

Water is flowing through the tube at the following rate at the instant of interest:

$$Q = .18 \text{ L/s} \qquad \dot{Q} = .025 \text{ L/s}^2$$

The tube is in the vertical plane at the instant of interest. What is the acceleration of the water particles at the centerline of the tube at point C using cylindrical coordinates and cylindrical components? Assume over the cross-section of the tube that the water velocity and acceleration are uniform.

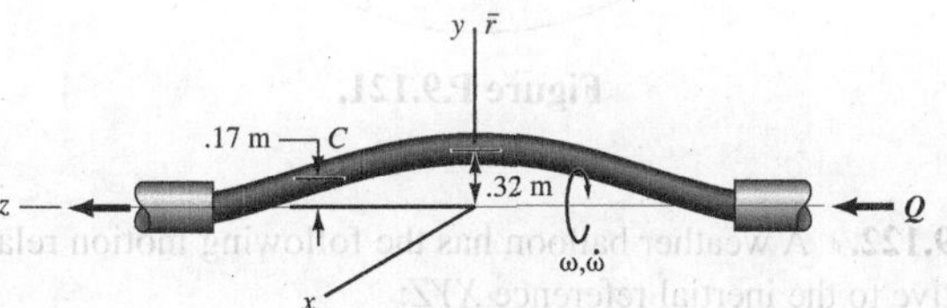

Figure P.9.119.

9.120. A World War I fighter plane is in level flight moving at a speed of 60 km/hr. At time t_0 it has an acceleration given as:

$$\boldsymbol{a} = .2g\boldsymbol{i} - .3g\boldsymbol{j} + 2g\boldsymbol{k} \text{ m/s}^2$$

Also at this time, the co-pilot is raising a camera upward with an acceleration of $0.1g$ relative to the plane. If the camera has a mass of .01 kg, what force must the co-pilot exert on the camera to give it the desired motion at time t_0? Note that the plane never rotates during this action. Take $g = 9.81$ m/s^2.

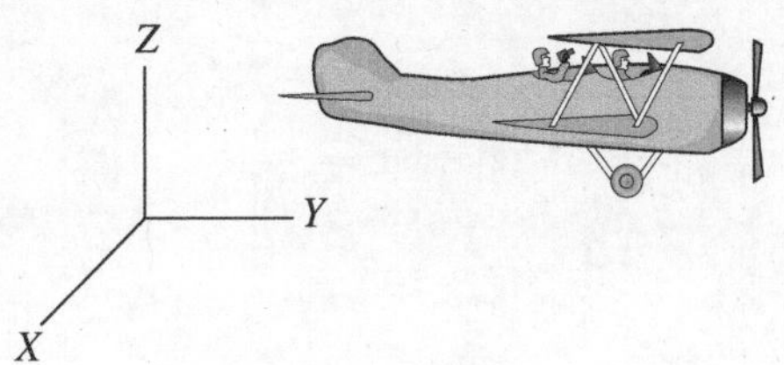

Figure P.9.120.

9.121. A space device has a velocity of $.500\boldsymbol{i} + .200\boldsymbol{j}$ m/s and an acceleration of $.200\boldsymbol{i} + .300\boldsymbol{k}$ m/s^2, both relative to the ground reference XYZ. Rod CD has an angular motion relative to the space device equal to $\omega = 2$ rad/s and $\dot{\omega} = 3$ rad/s^2. What are the velocity and the acceleration of mass M relative to the ground? CD is in the vertical zy plane at the instant of interest. Reference xyz shown is fixed to rod AB. CD is at 60° from AB.

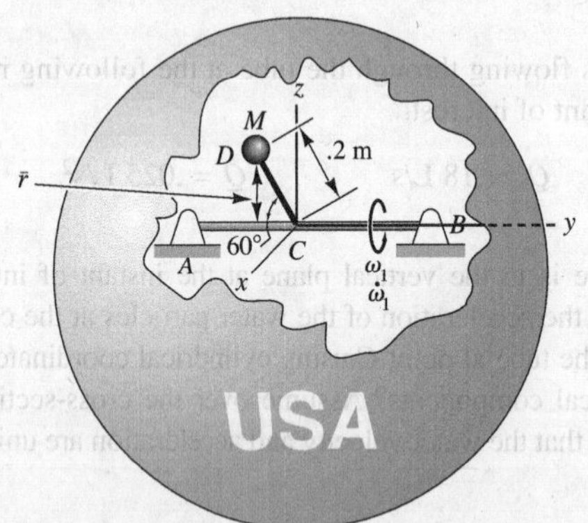

Figure P.9.121.

9.122. A weather balloon has the following motion relative to the inertial reference XYZ:

$$\boldsymbol{V} = 150\boldsymbol{i} + 200\boldsymbol{j} + 60\boldsymbol{k} \text{ m/s}$$
$$\boldsymbol{a} = 20\boldsymbol{i} - 40\boldsymbol{j} + 38\boldsymbol{k} \text{ m/s}^2$$

A light rod at A is connected to particle D and is rotating relative to the balloon at the instant of interest with angular speed $\omega_1 = 5$ rad/s and rate of change of speed $\dot{\omega}_1 = 2$ rad/s^2. What is the velocity of the particle at D relative to XYZ at the instant of interest? The distance $AD = .2$ m. What is the tensile force on the rod in the direction of AD? The mass of the particle at D is 1 kg.

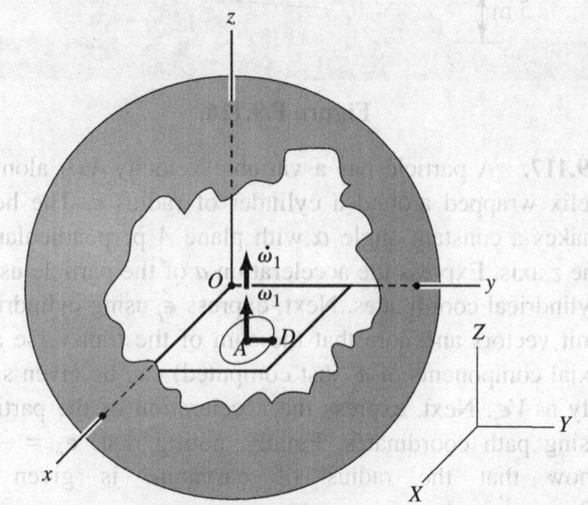

Figure P.9.122.

***9.123.** A particle is made to move along a path given in terms of the parameter τ in the following manner:

$$(x)_p = -\sin 2\tau \qquad (y)_p = \cos 2\tau \qquad (z)_p = e^{-\tau}$$

Give a simple sketch of the path. When the particle is at an elevation $z = 1$, the speed along the path is 5 ft/sec and the rate of change of speed along the path is 10 ft/sec^2. Find the acceleration vector at $z = 1$ in components for path coordinates.

***9.124.** Determine the direction of the osculating plane at position $z = 1$ for the three-dimensional curve of Problem 9.123. We got the following results from the previous problem:

$$\boldsymbol{\epsilon}_n = R(.08\boldsymbol{i} - .80\boldsymbol{j} + .16\boldsymbol{k})$$
$$\boldsymbol{\epsilon}_t = .894\boldsymbol{i} - .447\boldsymbol{k}$$
$$R = .8198 \text{ ft}$$

CHAPTER 10

Particle Dynamics

10.1 Introduction

In Chapter 9, we examined the geometry of motion—the kinematics of motion. In particular, we considered various kinds of coordinate systems: rectangular coordinates and path coordinates. In this chapter, we shall consider Newton's law for the three coordinate systems mentioned above, as applied to the motion of a particle.

Before embarking on this study, we shall review notions concerning units of mass presented earlier in Chapter 1. Recall that a pound mass (lbm) is the amount of matter attracted by gravity at a specified location on the earth's surface by a force of 1 pound (lbf). A slug, on the other hand, is the amount of matter that will accelerate relative to an inertial reference at the rate of 1 ft/sec^2 when acted on by a force of 1 lbf. Note that the slug is defined via Newton's law, and therefore the slug is the proper unit to be used in Newton's law. The relation between the pound mass (lbm) and the slug is

$$M\,(\text{slugs}) = \frac{M\,(\text{lbm})}{32.2} \tag{10.1}$$

Note also that the weight of a body in pounds force near the earth's surface will numerically equal the mass of the body in pounds mass. It is vital in using Newton's law that the mass of the body in pounds mass be properly converted into slugs via Eq. 10.1.

In SI units, recall that a kilogram is the mass that accelerates relative to an inertial reference at the rate of 1 meter/sec^2 when acted on by a force of 1 newton (which is about one-fifth of a pound). If the weight W of a body is given in terms of newtons, we must divide by 9.81 to get the mass in kilograms needed for Newton's law. That is,

$$M\,(\text{kg}) = \frac{W\,(\text{N})}{9.81} \tag{10.2}$$

We are now ready to consider Newton's law in rectangular coordinates.

Part A: Rectangular Coordinates; Rectilinear Translation

10.2 Newton's Law for Rectangular Coordinates

In rectangular coordinates, we can express Newton's law as follows:

$$
\begin{aligned}
F_x &= ma_x = m\frac{dV_x}{dt} = m\frac{d^2x}{dt^2} \\
F_y &= ma_y = m\frac{dV_y}{dt} = m\frac{d^2y}{dt^2} \\
F_z &= ma_z = m\frac{dV_z}{dt} = m\frac{d^2z}{dt^2}
\end{aligned}
\tag{10.3}
$$

If the motion is known relative to an inertial reference, we can easily solve for the rectangular components of the resultant force on the particle. The equations to be solved are just algebraic equations. The *inverse* of this problem, wherein the forces are known over a time interval and the motion is desired during this interval, is not so simple. For the inverse case, we must get involved generally with integration procedures.

In the next section, we shall consider situations in which the resultant force on a particle has the same direction and line of action at all times. The resulting motion is then confined to a straight line and is usually called *rectilinear translation.*

10.3 Rectilinear Translation

For rectilinear translation, we may consider the line of action of the motion to be collinear with one axis of a rectilinear coordinate system. Newton's law is then one of the equations of the set 10.3. We shall use the x axis to coincide with the line of action of the motion. The resultant force F (we shall not bother with the x subscript here) can be a constant, a function of time, a function of speed, a function of position, or any combination of these. At this time, we shall examine some of these cases, leaving others to Chapter 16, where, with the aid of the students' knowledge of differential equations,[1] we shall be more prepared to consider them.

Case 1. Force Is a Function of Time or a Constant. A particle of mass m acted on by a time-varying force $F(t)$ is shown in Fig. 10.1. The plane on which the body moves is frictionless. The force of gravity is equal and opposite to the normal force from the plane so

[1]Most students studying dynamics will concurrently be taking a course in differential equations.

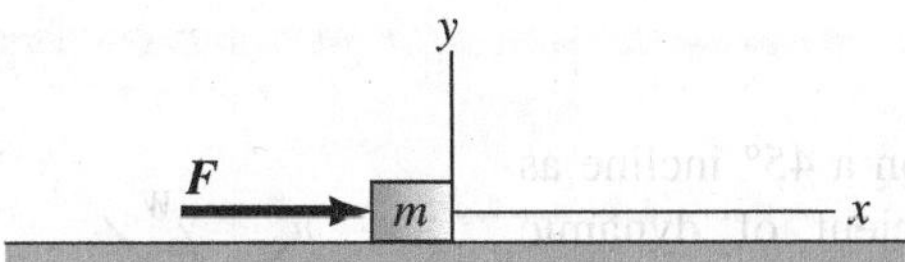

Figure 10.1. Rectilinear translation.

that $F(t)$ is the resultant force acting on the mass. Newton's law can then be given as follows:

$$F(t) = m\frac{d^2x}{dt^2}$$

Therefore,

$$\frac{d^2x}{dt^2} = \frac{F(t)}{m} \tag{10.4}$$

Knowing the acceleration in the x direction, we can readily solve for $F(t)$.

The inverse problem, where we know $F(t)$ and wish to determine the motion, requires integration. For this operation, the function $F(t)$ must be piecewise continuous.[2] To integrate, we rewrite Eq. 10.4 as follows:

$$\frac{d}{dt}\left(\frac{dx}{dt}\right) = \frac{F(t)}{m}$$

$$d\left(\frac{dx}{dt}\right) = \frac{F(t)}{m}\,dt$$

Now integrating both sides we get

$$\frac{dx}{dt} = V = \int \frac{F(t)}{m}\,dt + C_1 \tag{10.5}$$

where C_1 is a constant of integration. Integrating once again after bringing dt from the left side of the equation to the right side, we get

$$x = \int\left[\int \frac{F(t)}{m}\,dt\right]dt + C_1t + C_2 \tag{10.6}$$

We have thus found the velocity of the particle and its position as functions of time to within two arbitrary constants. These constants can be readily determined by having the solutions yield a certain velocity and position at given times. Usually, these conditions are specified at time $t = 0$ and are then termed initial conditions. That is, when $t = 0$,

$$V = V_0 \quad \text{and} \quad x = x_0 \tag{10.7}$$

These equations can be satisfied by substituting the initial conditions into Eqs. 10.5 and 10.6 and solving for the constants C_1 and C_2.

Although the preceding discussion centered about a force that is a function of time, the procedures apply directly to a force that is a constant. The following examples illustrate the procedures set forth.

[2]That is, the function has only a finite number of finite discontinuities.

Example 10.1

A 100-lb body is initially stationary on a 45° incline as shown in Fig. 10.2(a). The coefficient of dynamic friction μ_d between the block and incline is .5. What distance along the incline must the weight slide before it reaches a speed of 40 ft/sec?

A free-body diagram is shown in Fig. 10.2(b). Since the acceleration is zero in the direction normal to the incline, we have from **equilibrium** that

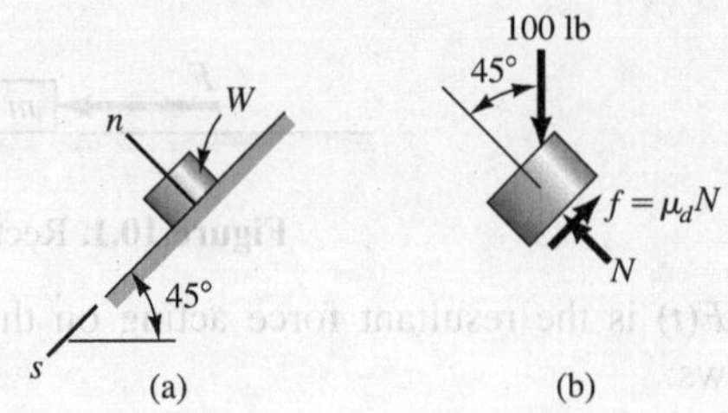

Figure 10.2. Body slides on an incline.

$$100 \cos 45° = N = 70.7 \text{ lb} \quad \text{(a)}$$

Now applying **Newton's law** in a direction along the incline, we have

$$\frac{100}{g}\frac{d^2s}{dt^2} = 100 \sin 45° - \mu_d N$$

Therefore,

$$\frac{d^2s}{dt^2} = 11.38 \quad \text{(b)}$$

Rewriting Eq. (b) we have

$$d\left(\frac{ds}{dt}\right) = 11.38\, dt$$

Integrating, we get

$$\frac{ds}{dt} = 11.38t + C_1 \quad \text{(c)}$$

$$s = 11.38\frac{t^2}{2} + C_1 t + C_2 \quad \text{(d)}$$

When $t = 0$, $s = ds/dt = 0$, and thus $C_1 = C_2 = 0$. When $ds/dt = 40$ ft/sec, we have for t from Eq. (c) the result

$$40 = 11.38t$$

Therefore,

$$t = 3.51 \text{ sec}$$

Substituting this value of t in Eq. (d), we can get the distance traveled to reach the speed of 40 ft/sec as follows:

$$s = 11.38\frac{(3.51)^2}{2} = 70.4 \text{ ft}$$

Example 10.2

A charged particle is shown in Fig. 10.3 at time $t = 0$ between large parallel condenser plates separated by a distance d in a vacuum. A time-varying voltage V (notation not to be confused with velocity) given as

$$V = 6 \sin \omega t \qquad \text{(a)}$$

is applied to the plates. What is the motion of the particle if it has a charge q coulombs and if we do not consider gravity?

As we learned in physics, the electric field E becomes for this case

$$E = \frac{V}{d} \qquad \text{(b)}$$

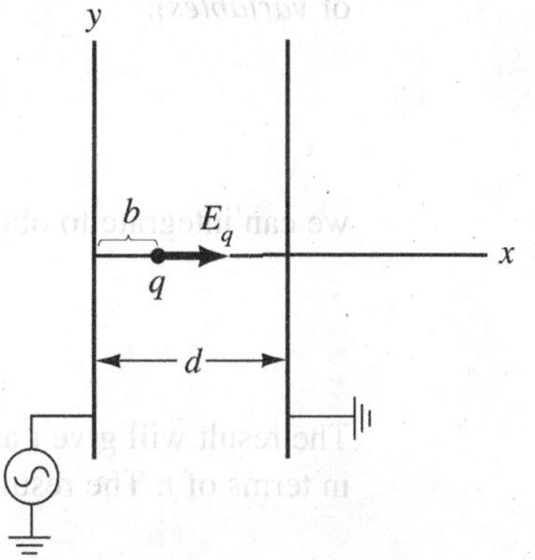

Figure 10.3. Charged particle between condenser plates.

The force on the particle is qE and the resulting motion is that of rectilinear translation. Using **Newton's law** we accordingly have

$$\frac{d^2x}{dt^2} = q\frac{6 \sin \omega t}{md} \qquad \text{(c)}$$

Rewriting Eq. (c), we have

$$d\left(\frac{dx}{dt}\right) = q\frac{6 \sin \omega t}{md}\, dt$$

Integrating, we get

$$\frac{dx}{dt} = -\frac{6q}{\omega md} \cos \omega t + C_1 \qquad \text{(d)}$$

$$x = -\frac{6q}{\omega^2 md} \sin \omega t + C_1 t + C_2 \qquad \text{(e)}$$

Applying the initial conditions $x = b$ and $dx/dt = 0$ when $t = 0$, we see that $C_1 = 6q/m\omega d$ and $C_2 = b$. Thus, we get

$$x = -\frac{6q}{\omega^2 md} \sin \omega t + \frac{6q}{m\omega d} t + b$$

The motion of the charged particle will be that of sinusoidal oscillation in which the center of the oscillation drifts from left to right.

Case 2. Force Is a Function of Speed. We next consider the case where the resultant force on the particle depends only on the value of the speed of the particle. An example of such a force is the aerodynamic drag force on an airplane or missile.

We can express *Newton's law* in the following form:

$$\frac{dV}{dt} = \frac{F(V)}{m} \qquad (10.8)$$

where $F(V)$ is a piecewise continuous function representing the force in the positive x direction. If we rearrange the equation in the following manner (this is called *separation* of *variables*):

$$\frac{dV}{F(V)} = \frac{1}{m}\,dt$$

we can integrate to obtain

$$\int \frac{dV}{F(V)} = \frac{1}{m}t + C_1 \tag{10.9}$$

The result will give t as a function of V. However, we will generally prefer to solve for V in terms of t. The result will then have the form

$$V = H(t, C_1)$$

where H is a function of t and the constant of integration C_1. A second integration may now be performed by first replacing V by dx/dt and bringing dt over to the right side of the equation. We then get on integration

$$x = \int H(t, C_1)\,dt + C_2 \tag{10.10}$$

The constants of integration are determined from the initial conditions of the problem.

Example 10.3

A high-speed land racer (Fig. 10.4) is moving at a speed of 100 m/sec. The resistance to motion of the vehicle is primarily due to aerodynamic drag, which for this speed can be approximated as $.2V^2$ N with V in m/sec. If the vehicle has a mass of 4,000 kg, what distance will it coast before its speed is reduced to 70 m/sec?

We have, using **Newton's law** for this case,

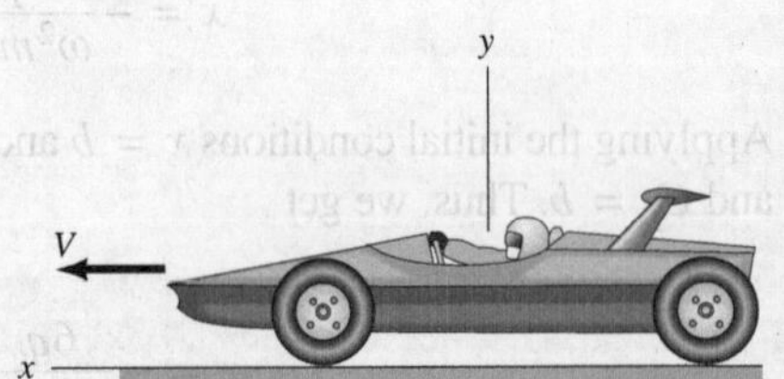

Figure 10.4. High-speed racer.

$$\frac{dV}{dt} = -\frac{.2V^2}{4{,}000} = -5 \times 10^{-5} V^2 \tag{a}$$

Separating the variables, we get

$$\frac{dV}{V^2} = -5 \times 10^{-5}\,dt \tag{b}$$

Example 10.3 (Continued)

Integrating, we have

$$-\frac{1}{V} = -5\times 10^{-5}t + C_1 \tag{c}$$

Taking $t = 0$ when $V = 100$, we get $C_1 = -1/100$. Replacing V by dx/dt, we have next

$$\frac{1}{V} = \frac{dt}{dx} = 5\times 10^{-5}t + \frac{1}{100} \tag{d}$$

Separating variables once again, we get

$$\frac{dt}{5\times 10^{-5}t + (1/100)} = dx$$

To integrate, we perform a change of variable. Thus

$$\eta = 5\times 10^{-5}t + (1/100)$$
$$\therefore\ d\eta = 5\times 10^{-5}\,dt$$

We then have as a replacement for our equation

$$\frac{d\eta}{\eta} = 5\times 10^{-5}\,dx$$

Now integrating and replacing η, we get

$$\ln\left(5\times 10^{-5}t + \frac{1}{100}\right) = 5\times 10^{-5}x + C_2$$

When $t = 0$, we take $x = 0$ and so $C_2 = \ln(1/100)$. We then have on combining the logarithmic terms:

$$\ln\left(5\times 10^{-3}t + 1\right) = 5\times 10^{-5}x \tag{e}$$

Substitute $V = 70$ in Eq. (d); solve for t. We get $t = 85.7$ sec. Finally, find x for this time from Eq. (e). Thus,

$$\ln\left[(5\times 10^{-3})(85.7) + 1\right] = 5\times 10^{-5}x$$

Therefore,

$$x = 7.13 \text{ km}$$

The distance traveled is then 7.13 km.

Example 10.4

A conveyor is inclined 20° from the horizontal as shown in Fig. 10.5. As a result of spillage of oil on the belt, there is a viscous friction force between body D and the belt. This force equals 0.1 lbf per unit relative velocity between body D and the belt. The belt moves at a constant speed V_B up the conveyor while initially body D has a speed $(V_D)_0 = 2$ ft/sec relative to the ground in a direction down the conveyor. What speed V_B^* should the belt have in order for body D to be able to eventually approach a zero velocity relative to the ground? For belt speed V_B^*, and for the given initial speed of body D, namely $(V_D)_0 = 2$ ft/sec, determine the time when body D attains a speed of 1 ft/sec relative to the ground. The mass of D is 5 lbm.

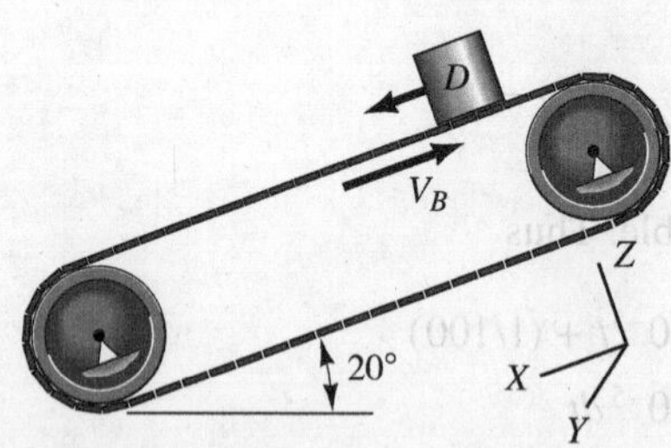

Figure 10.5. A body slides down a conveyor belt wet with oil.

We begin by assigning axes for the problem as follows (see Fig. 10.6):

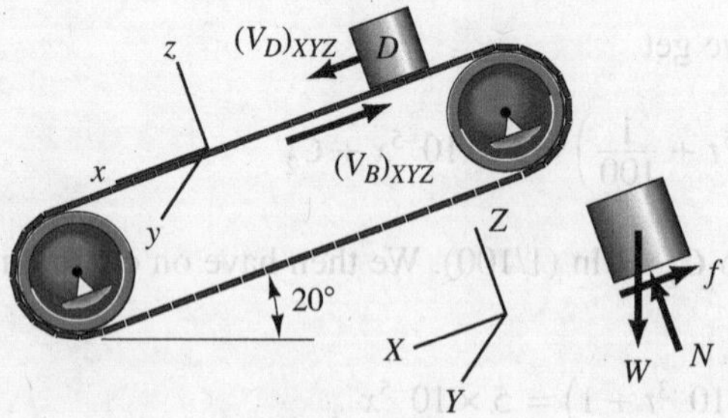

Figure 10.6. Friction force f is 0.1 times the relative velocity between body D and the belt.

> Fix xyz to the belt
> Fix XYZ to the ground

From **kinematics** we can say

$$(\boldsymbol{V}_D)_{XYZ} = (\boldsymbol{V}_D)_{xyz} + \dot{\boldsymbol{R}} \qquad \text{Note that } \dot{\boldsymbol{R}} = -(V_B)_{XYZ}\boldsymbol{i}$$

$$\therefore (\boldsymbol{V}_D)_{xyz} = (\boldsymbol{V}_D)_{XYZ} - \dot{\boldsymbol{R}} = \left[V_D - (-V_B)\right]_{XYZ}\boldsymbol{i} = (V_D + V_B)_{XYZ}\boldsymbol{i}$$

Example 10.4 (Continued)

For the friction force f we have

$$f = -(.1)(V_D)_{xyz} = -(.1)(V_D + V_B)_{XYZ}\boldsymbol{i}$$

We may now use **Newton's law** for body D in the x direction. Since all velocities from here on will be relative to the ground, we can dispense with the reference subscripts. Thus

$$\frac{5}{g}\frac{dV_D}{dt} = -(.1)(V_D + V_B) + 5\sin 20° \qquad \text{(a)}$$

When body D attains a theoretical permanent zero velocity relative to the ground, V_D and $\frac{dV_D}{dt}$ are equal to zero. This gives us (noting that V_B now becomes V_B^*)

$$0 = -(.1)(0 + V_B^*) + 5\sin 20° \qquad \therefore \quad \boxed{V_B^* = 17.10 \text{ ft./sec}}$$

Now determine the time for body D to attain a velocity of 1 ft/sec relative to the ground for a belt speed of 17.10 ft/sec. For this we go back to Eq. (a).

$$\frac{5}{g}\frac{dV_D}{dt} = (-.1)(V_D + 17.10) + 5\sin 20° = -.1V_D$$

$$\frac{dV_D}{V_D} = -\left(\frac{g}{5}\right)(.1)\,dt = -\frac{3.22}{5}\,dt$$

$$\therefore \ln V_D = -\frac{3.22}{5}t + C_1$$

When

$$t = 0, \qquad V_D = 2 \text{ ft/sec}, \qquad \therefore C_1 = \ln 2$$

Hence, on combining log terms[3]

$$\ln\left(\frac{V_D}{2}\right) = -.644t$$

Set $V_D = 1$ ft/sec. Solve for t.

$$t = -\frac{1}{.644}\ln(.500) = \boxed{1.076 \text{ sec}}$$

[3]Note from this equation that $V_D = 2e^{-0.644t}$ and that $\dot{V}_D = -1.288e^{-0.644t}$ and so we see that as t approaches infinity both of these quantities approach zero. Thus, theoretically body D could approach a permanent zero velocity relative to the ground.

Case 3. Force Is a Function of Position. As the final case of this series, we now consider the rectilinear motion of a body under the action of a force that is expressible as a function of position. Perhaps the simplest example of such a case is the frictionless mass–spring system shown in Fig. 10.7. The body is shown at a position where the spring is unstrained. The horizontal force from the spring at all positions of the body clearly will be a function of position x.

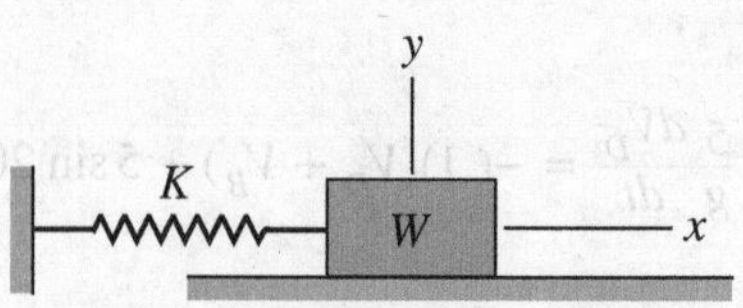

Figure 10.7. Mass–spring system.

Newton's law for position-dependent forces can be given as

$$m\frac{dV}{dt} = F(x) \tag{10.11}$$

We cannot separate the variables for this form of the equation as in previous cases since there are three variables (V, t, and x). However, by using the chain rule of differentiation, we can change the left side of the equation to a more desirable form in the following manner:

$$m\frac{dV}{dt} = m\frac{dV}{dx}\frac{dx}{dt} = mV\frac{dV}{dx}$$

We can now separate the variables in Eq. 10.11 as follows:

$$mV\,dV = F(x)\,dx$$

Integrating, we get

$$\frac{mV^2}{2} = \int F(x)\,dx + C_1 \tag{10.12}$$

Solving for V and using dx/dt in its place, we get

$$\frac{dx}{dt} = \left[\frac{2}{m}\int F(x)\,dx + C_1\right]^{1/2}$$

Separating variables and integrating again, we get

$$t = \int \frac{dx}{\left[\frac{2}{m}\int F(x)\,dx + C_1\right]^{1/2}} + C_2 \tag{10.13}$$

For a given $F(x)$, V and x can accordingly be evaluated as functions of time from Eqs. 10.12 and 10.13. The constants of integration C_1 and C_2 are determined from the initial conditions.

A very common force that occurs in many problems is the *linear restoring force*. Such a force occurs when a body W is constrained by a linear spring (see Fig. 10.7). The force from such a spring will be proportional to x measured from a position of W corresponding to the *undeformed configuration* of the system. Consequently, the force will have a magnitude of $|Kx|$, where K, called the *spring constant*, is the force needed on the spring per unit elongation or compression of the spring. Furthermore, when x has a positive value, the spring force points in the negative direction, and when x is negative, the spring force points in the positive direction. That is, it always points toward the position $x = 0$ for which the spring is undeformed. The spring force is for this reason called a *restoring* force and must be expressed as $-Kx$ to give the proper direction for all values of x.

For a *nonlinear* spring, K will not be constant but will be a function of the elongation or shortening of the spring. The spring force is then given as

$$F_{\text{spring}} = -\int_0^x K(x)\,dx \tag{10.14}$$

In the following example and in the homework problems, we examine certain limited aspects of spring–mass systems to illustrate the formulations of case 3 and to familiarize us with springs in dynamic systems. A more complete study of spring–mass systems will be made in Chapter 16. The motion of such systems, we shall later learn, centers about some stationary point. That is, the motion is *vibratory* in nature. We shall study vibrations in Chapter 16, wherein time-dependent and velocity-dependent forces are present simultaneously with the linear restoring force. We are deferring this topic so as to make maximal use of your course in differential equations that you are most likely studying concurrently with dynamics. It is important to understand, however, that even though we defer vibration studies until later, such studies are not something apart from the general particle dynamics undertaken in this chapter.

Example 10.5

A cart A (see Fig. 10.8) having a mass of 200 kg is held on an incline so as to just touch an undeformed spring whose spring constant K is 50 N/mm. If body A is released very slowly, what distance down the incline must A move to reach an equilibrium configuration? If body A is released suddenly, what is its speed when it reaches the aforementioned equilibrium configuration for a slow release?

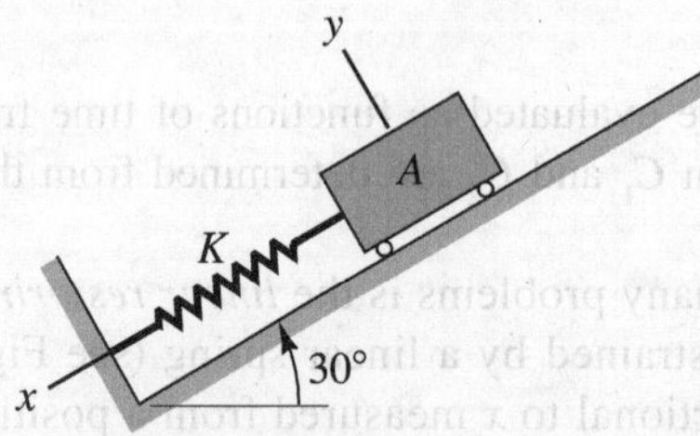

Figure 10.8. Cart–spring system.

As a first step, we have shown a free body of the vehicle in Fig. 10.9. To do the first part of the problem, all we need do is utilize the definition

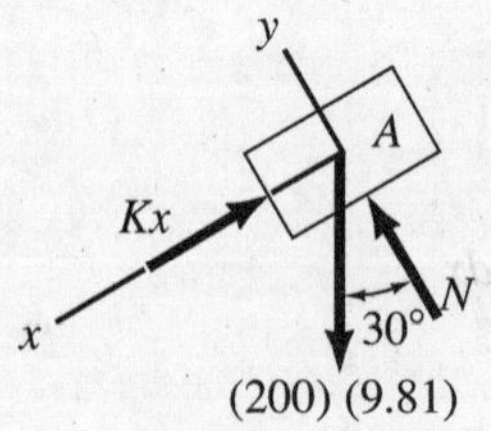

Figure 10.9. Free-body diagram of cart.

of the spring constant. Thus, if δ represents the compression of the spring, we can say:

$$K = \frac{F}{\delta}$$

Example 10.5 (Continued)

Therefore,

$$\delta = \frac{F}{K} = \frac{(200)(9.81)\sin 30^\circ}{50}$$

$$\delta = 19.62 \text{ mm}$$

Thus, the spring will be compressed .01962 m by the cart if it is allowed to move down the incline very slowly.

For the case of the quick release, we use **Newton's law**. Thus, using x in meters so that K is (50)(1,000) N/m:

$$200\ddot{x} = (200)(9.81)\sin 30^\circ - (50)(1{,}000)(x)$$

Therefore,

$$\ddot{x} = 4.905 - 250x$$

Rewriting $\ddot{x}$, we have

$$V\frac{dV}{dx} = 4.905 - 250x$$

Separating variables and integrating,

$$\frac{V^2}{2} = 4.905x - 125x^2 + C_1$$

To determine the constant of integration C_1, we set $x = 0$ when $V = 0$. Clearly, $C_1 = 0$. As a final step, we set $x = .01962$ m and solve for V.

$$V = \left\{2\left[(4.905)(.01962) - (125)(.01962)^2\right]\right\}^{1/2}$$

$$V = .310 \text{ m/sec}$$

The following example illustrates an interesting device used by the U.S. Navy to test small devices for high, prolonged acceleration. Hopefully, the length of the problem will not intimidate you. Use is made of the gas laws presented in your elementary chemistry courses.

Example 10.6

An *air gun* is used to test the ability of small devices to withstand high prolonged accelerations. A "floating piston" A (Fig. 10.10), on which the device to be tested is mounted, is held at position C while region D is filled with highly compressed air. Region E is initially at atmospheric pressure but is entirely sealed from the outside. When "fired," a quick-release mechanism releases the piston and it accelerates rapidly toward the other end of the gun, where the trapped air in E "cushions" the motion so that the piston will begin eventually to return. However, as it starts back, the high pressure developed in E is released through valve F and the piston only returns a short distance.

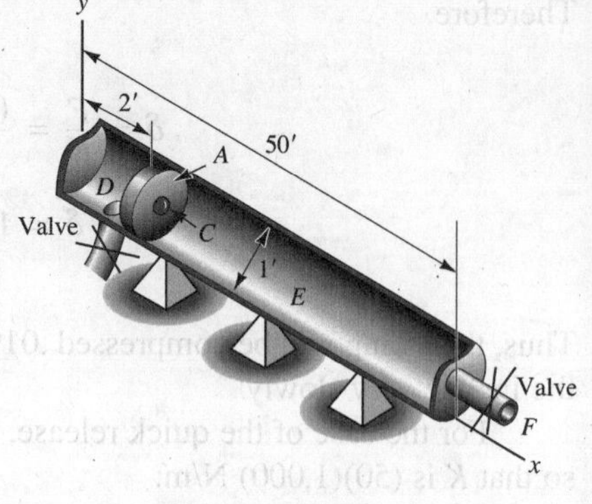

Figure 10.10. Air gun.

Suppose that the piston and its test specimen have a combined mass of 2 lbm and the pressure initially in the chamber D is 1,000 psig (above atmosphere). Compute the speed of the piston at the halfway point of the air gun if we make the simple assumption that the air in D expands according to pv = constant and the air in E is compressed also according to pv = constant.[4] Note that v is the specific volume (i.e., the volume per unit mass). Take v of this fluid at D to be initially .207 ft^3/lbm and v in E to be initially 13.10 ft^3/lbm. Neglect the inertia of the air.

The force on the piston results from the pressures on each face, and we can show that this force is a function of x (see Fig. 10.10 for reference axes). Thus, examining the pressure p_D first for region D, we have, from initial conditions,

$$(p_D v_D)_0 = [(1{,}000 + 14.7)(144)](.207) = 30{,}300 \qquad \text{(a)}$$

Furthermore, the mass of air D given as M_D is determined from initial data as

$$M_D = \frac{(V_D)_0}{(v_D)_0} = \frac{(2)\left(\frac{\pi}{4}\right)(1^2)}{.207} = 7.58 \text{ lbm} \qquad \text{(b)}$$

where $(V_D)_0$ is the volume of the air in D initially. Noting that pv = const. and then using the right side of Eq. (a) for $p_D v_D$ as well as the first part of Eq. (b) for v_D, we can determine p_D at any position x of the piston:

$$p_D = \frac{30{,}300}{v_D} = \frac{30{,}300}{V_D/M_D} = \frac{30{,}300}{(\pi/4)(1^2)(x)/7.58}$$

Therefore,

$$p_D = \frac{293{,}000}{x} \qquad \text{(c)}$$

[4]You should recall from your earlier work in physics and chemistry that we are using here the isothermal form of the equation of state for a perfect gas. Two factors of caution should be pointed out relative to the use of this expression. First, at the high pressures involved in part of the expansion, the perfect gas model is only an approximation for the gas, and so the equation of state of a perfect gas that gives us pv = constant is only approximate. Furthermore, the assumption of isothermal expansion gives only an approximation of the actual process. Perhaps a better approximation is to assume an adiabatic expansion (i.e, no heat transfer). This is done in Problem 10.130.

Example 10.6 (Continued)

We can similarly get p_E as a function of x for region E. Thus,

$$(p_E v_E)_0 = (14.7)(144)(13.10) = 27{,}700$$

and

$$M_E = \frac{(V_E)_0}{(v_E)_0} = \frac{(48)\left(\frac{\pi}{4}\right)(1^2)}{13.10} = 2.88 \text{ lbm}$$

Hence, at position x of the piston

$$p_E = \frac{27{,}700}{v_E} = \frac{27{,}700}{V_E/M_E} = \frac{27{,}700}{(\pi/4)(1^2)(50-x)/2.88}$$

Therefore,

$$p_E = \frac{101{,}600}{50-x}$$

Now we can write **Newton's law** for this case. Noting that V without subscripts is velocity and not volume,

$$MV\frac{dV}{dx} = \frac{\pi 1^2}{4}\left(p_D - p_E\right) = \frac{\pi}{4}\left(\frac{293{,}000}{x} - \frac{101{,}600}{50-x}\right) \qquad \text{(d)}$$

where M is the mass of piston and load. Separating variables and integrating, we get

$$\frac{MV^2}{2} = \frac{\pi}{4}[293{,}000 \ln x + 101{,}600 \ln(50-x)] + C_1 \qquad \text{(e)}$$

To get the constant C_1, set $V = 0$ when $x = 2$ ft. Hence,

$$C_1 = -\frac{\pi}{4}(293{,}000 \ln 2 + 101{,}600 \ln 48)$$

Therefore,

$$C_1 = -468{,}000$$

Substituting C_1 in Eq. (e), we get

$$V = \left(\frac{2}{M}\right)^{1/2}\left\{\frac{\pi}{4}[293{,}000 \ln x + 101{,}600 \ln(50-x)] - 468{,}000\right\}^{1/2}$$

We may rewrite this as follows noting that $M = 2$ lbm/g:

$$V = 566[23 \ln x + 7.98 \ln(50-x) - 46.8]^{1/2}$$

At $x = 25$ ft, we then have for V the desired result:

$$V = 566(23 \ln 25 + 7.98 \ln 25 - 46.8)^{1/2}$$

$$V = 4{,}120 \text{ ft/sec}$$

*Example 10.7

A light stiff rod is pinned at A and is constrained by two linear springs, $K_1 = 1{,}000$ N/m and $K_2 = 1{,}200$ N/m. The springs are unstretched when the rod is horizontal. At the right end of the rod, a mass $M = 5$ kg is attached. If the rod is rotated 12° *clockwise* from a horizontal configuration and then released, what is the speed of the mass when the rod returns to a position corresponding to the *static equilibrium* position with mass M attached?

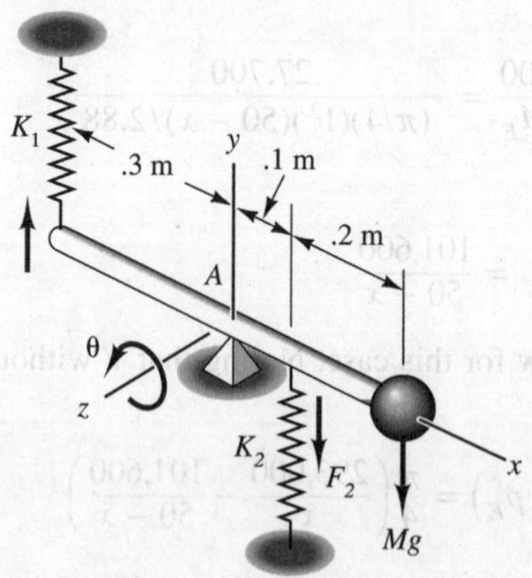

Figure 10.11. Two linear springs and a particle on a weightless rigid rod. Spring forces shown for positive θ.

A free-body diagram of the system for *positive* θ is shown in Fig. 10.12(a) and a free-body diagram of the particle M is shown in Fig. 10.12(b). The spring forces F_1 and F_2 on the rod are given as follows for small positive rotations θ:

$$F_1 = (.3)(\theta)(K_1) = 300\,\theta\,\text{N}$$

$$F_2 = -(.1)(\theta)(K_2) = -120\,\theta\,\text{N} \qquad \text{(a)}$$

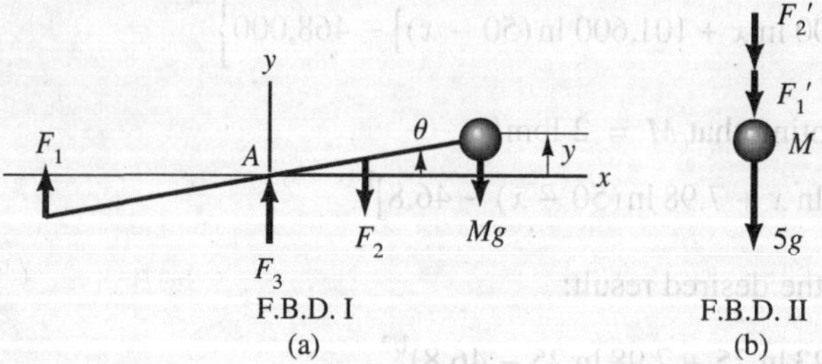

Figure 10.12. Free-body diagrams of the system and the particle for positive θ.

Example 10.7 (Continued)

where θ is in radians. In the first free body, we will think of the rod as a massless perfectly rigid lever as studied in high school or perhaps even earlier. Then we can say for the forces on the second of our free bodies stemming from the springs[5]

$$F_1' \quad (\text{from } F_1) = -F_1 = -300\,\theta\,\text{N}$$

$$F_2' \quad (\text{from } F_2) = \frac{.1}{.3}(F_2) = -40\,\theta\,\text{N}$$

We can now give **Newton's law** for M as follows using y for the vertical coordinate of the particle:

$$5\ddot{y} = -5g - 300\,\theta - 40\,\theta$$

$$\therefore\ \ddot{y} = -g - \frac{340}{5}\,\theta \qquad \text{(b)}$$

Next, from **kinematics**, we can say for small rotation

$$\theta = \frac{y}{.3}$$

Now going back to Eq.(b) we replace $\ddot{y}$ by

$$\frac{d\dot{y}}{dt} = \left(\frac{d\dot{y}}{dy}\right)\left(\frac{dy}{dt}\right) = \dot{y}\left(\frac{d\dot{y}}{dy}\right)$$

in order to be able to separate variables. Also replace θ by $y/.3$. We then may say

$$\dot{y}\,d\dot{y} = (-227y - g)\,dy$$

Integrating

$$\frac{\dot{y}^2}{2} = (-227)\left(\frac{y^2}{2}\right) - gy + C_1 \qquad \text{(c)}$$

When $\theta = -12° = -\left(\frac{12}{360}\right)(2\pi)$ rad $= -\ .2094$ rad, $\dot{\theta} = \dot{y} = 0$.

We can then solve for the constant of integration using $y = .3\ \theta$.

$$C_1 = (227)\left[\frac{[(.3)(-.2094)]^2}{2}\right] + (9.81)(.3)(-.2094) = -.1683$$

Hence,

$$\dot{y}^2 = 2\left[(-227)\left(\frac{y^2}{2}\right) - gy - .1683\right] \qquad \text{(d)}$$

[5]Note that a positive θ gives negative values for F_1' and F_2' on M and vice versa. It is for this reason that we require the minus signs.

Example 10.7 (Continued)

For the static **equilibrium** configuration of the rod, we require from Fig. 10.12(a)

$$\underline{\sum M_A = 0:}$$

$$-(Mg)(.3) - F_1(.3) - F_2(.1) = 0$$

Substituting values from Eqs. (a) and noting that we are only using the magnitudes of the forces above for the required negative moments we get

$$-(5)(9.81)(.3) - (300\,\theta_{Eq})(.3) - (120\,\theta_{Eq})(.1) = 0$$

Solving for θ_{Eq}

$$\theta_{Eq} = -.1443 \text{ rad}$$

Hence

$$y_{Eq} = (.3)(-.1443) = -.04328 \text{ m}$$

Now go to Eq. (d) and substitute y_{Eq}. We get

$$\dot{y}_{Eq} = \sqrt{2}\left[(-113.5)(-.04328)^2 - (9.81)(-.04328) - .1683\right]^{1/2}$$

The desired result is then

$$\dot{y}_{Eq} = 0.968 \text{ m/s}$$

10.4 A Comment

In Part A, we have considered only rectilinear motions of particles. Actually in Chapter 9, we considered the coplanar motion of particles having a constant acceleration of gravity in the minus y direction and zero acceleration in the x direction. These were the *ballistic* problems. We treated them earlier in Chapter 9 because the considerations were primarily kinematic in nature. In the present chapter, they correspond to the coplanar motion of a particle having a constant force in the minus y direction along with an initial velocity component in this direction, plus a zero force in the x direction, with a possible initial velocity component in this direction.

PROBLEMS

10.1. A particle of mass 1 slug is moving in a constant force field given as

$$F = 3i + 10j - 5k \text{ lb}$$

The particle starts from rest at position (3, 5, –4). What is the position and velocity of the particle at time $t = 8$ sec? What is the position when the particle is moving at a speed of 20 ft/sec?

10.2. A particle of mass m is moving in a constant force field given as

$$F = 2mi - 12mj \text{ N}$$

Give the vector equation for $r(t)$ of the particle if, at time $t = 0$, it has a velocity V_0 given as

$$V_0 = 6i + 12j + 3k \text{ m/sec}$$

Also, at time $t = 0$, it has a position given as

$$r_0 = 3i + 2j + 4k \text{ m}$$

What are coordinates of the body at the instant that the body reaches its maximum height, y_{max}?

10.3. A block is permitted to slide down an inclined surface. The coefficient of friction is .05. If the velocity of the block is 30 ft/sec on reaching the bottom of the incline, how far up was it released and how many seconds has it traveled?

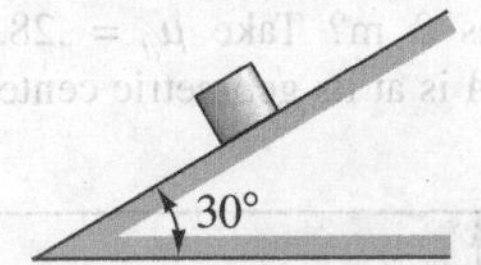

Figure P.10.3.

10.4. An arrow is shot upward with an initial speed of 80 ft/sec. How high up does it go and how long does it take to reach the maximum elevation if we neglect friction?

10.5. A mass D at $t = 0$ is moving to the left at a speed of .6 m/sec relative to the ground on a belt that is moving at constant speed to the right at 1.6 m/sec. If there is coulombic friction present with $\mu_d = .3$, how long does it take before the speed of D relative to the belt is .3 m/sec to the left?

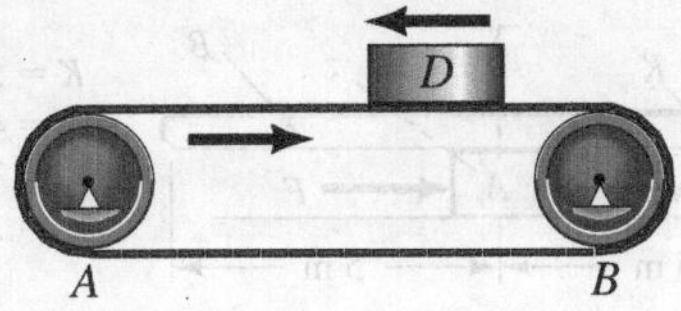

Figure P.10.5.

10.6. Do Problem 10.5 with the belt system inclined 15° with the horizontal so that end B is above end A.

10.7. A drag racer can develop a torque of 200 ft-lb on each of the rear wheels. If we assume that this maximum torque is maintained and that there is no wind friction, what is the time to travel a quarter mile from a standing start? What is the speed of the vehicle at the quarter-mile mark? The weight of the racer and the driver combined is 1,600 lb. For simplicity, neglect the rotational effects of the wheels.

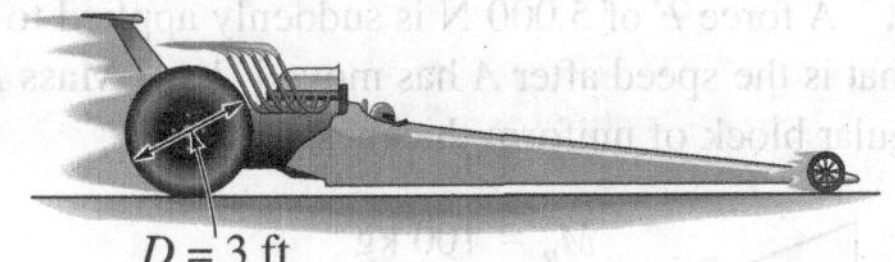

Figure P.10.7.

10.8. A truck is moving down a 10° incline. The driver strongly applies his brakes to avoid a collision and the truck decelerates at the steady rate of 1 m/sec^2. If the static coefficient of friction μ_s between the load W and the truck trailer is .3, will the load slide or remain stationary relative to the truck trailer? The weight of W is 4,500 N and it is not held to the truck by cables.

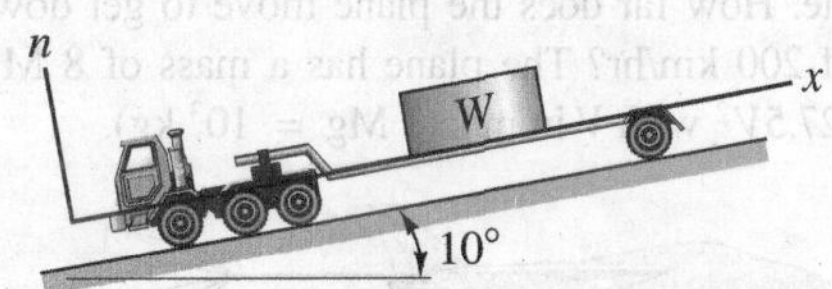

Figure P.10.8.

10.9. A simple device for measuring reasonably uniform accelerations is the pendulum. Calibrate θ of the pendulum for vehicle accelerations of 5 ft/sec^2, 10 ft/sec^2, and 20 ft/sec^2. The bob weighs 1 lb. The bob is connected to a post with a flexible string.

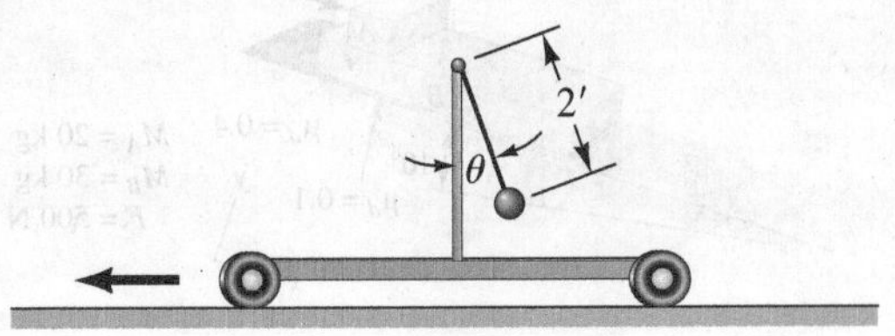

Figure P.10.9.

10.10. A piston is being moved through a cylinder. The piston is moved at a constant speed V_p of .6 m/sec relative to the ground by a force F. The cylinder is free to move along the ground on small wheels. There is a coulombic friction force between the piston and the cylinder such that $\mu_d = .3$. What distance d must the piston move relative to the ground to advance .01 m along the cylinder if the cylinder is stationary at the outset? The piston has a mass of 2.5 kg and the cylinder has a mass 5 kg.

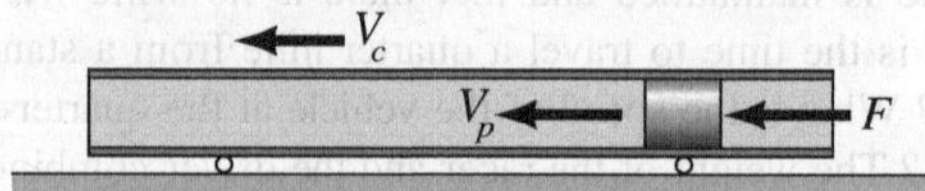

Figure P.10.10.

10.11. A force F of 5,000 N is suddenly applied to mass A. What is the speed after A has moved .1 m? Mass B is a triangular block of uniform thickness.

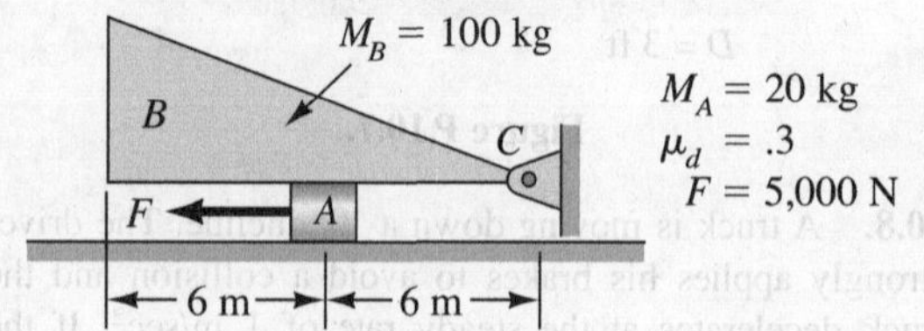

Figure P.10.11.

10.12. A fighter plane is moving on the ground at a speed of 350 km/hr when the pilot deploys the braking parachute. How far does the plane move to get down to a speed of 200 km/hr? The plane has a mass of 8 Mg. The drag is $27.5V^2$ with V in m/s (1 Mg = 10^3 kg).

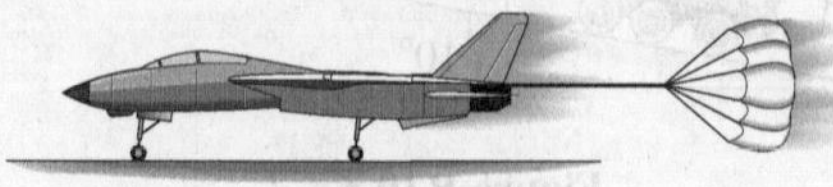

Figure P.10.12.

10.13. Blocks A and B are initially stationary. How far does A move along B if A moves .2 m relative to the ground?

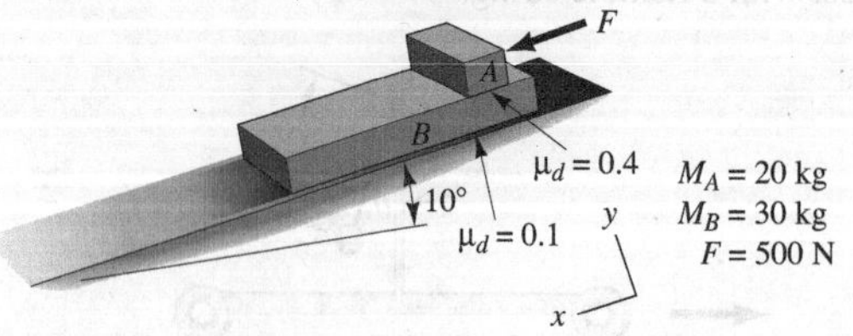

Figure P.10.13.

10.14. A 30-N block at the position shown has a force $F = 100$ N applied suddenly. What is its velocity after moving 1 m? Also, how far does the block move before stopping? Member AB weighs 200M.

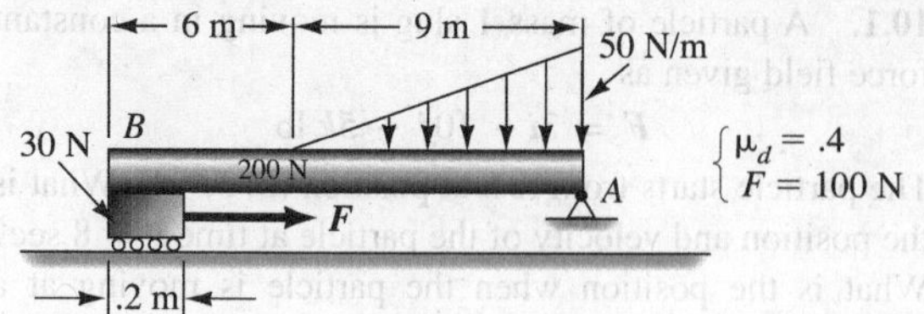

Figure P.10.14.

10.15. A block B of mass M is being pulled up an incline by a force F. If μ_d is .3, at what angle α will the force F cause the maximum steady acceleration?

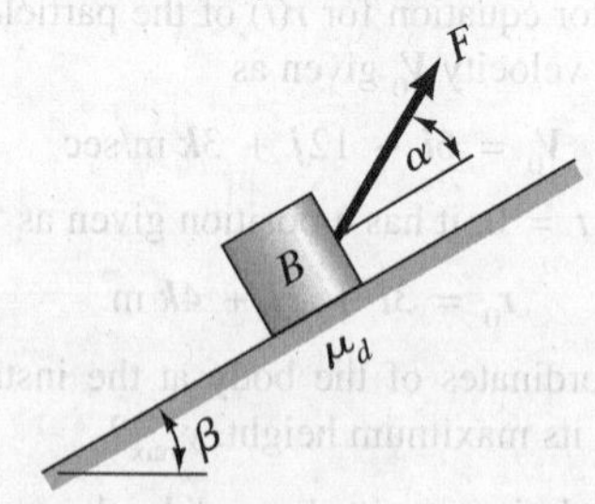

Figure P.10.15.

10.16. A 10-kN force is applied to body B whose mass is 15 kg. Body A has a mass of 20 kg. What is the speed of B after it moves 3 m? Take $\mu_d = .28$. The center of gravity of body A is at its geometric center.

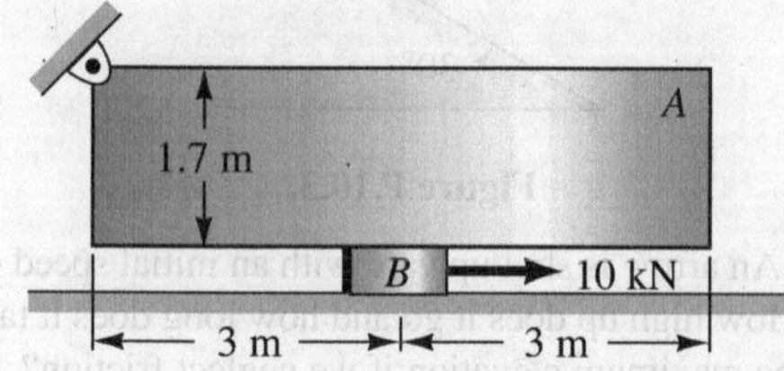

Figure P.10.16.

10.17. A constant force F is applied to the body A when it is in the position shown. What should F be if A is to attain a velocity of 2 m/s after moving 1 m? The spring is unstretched at the position shown.

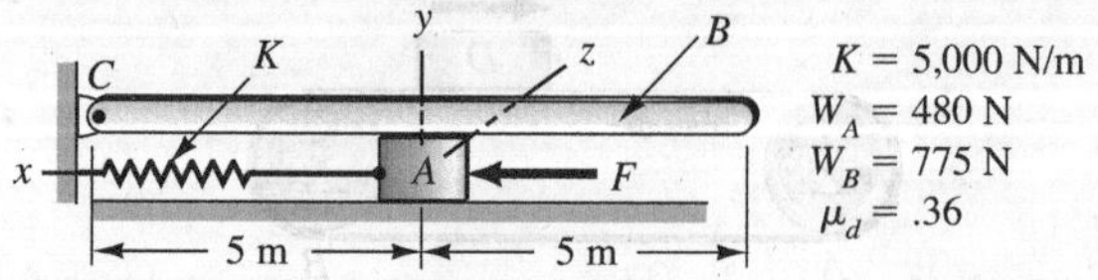

Figure P.10.17.

10.18. Two slow moving steam roller vehicles are moving in opposite directions on a straight path. They start at A and B at the time $t = 0$. How far from point A do they pass each other? What are their speeds when this happens? [*Hint:* Show that the time for this is 1.5 hours.] Note t is in hours.

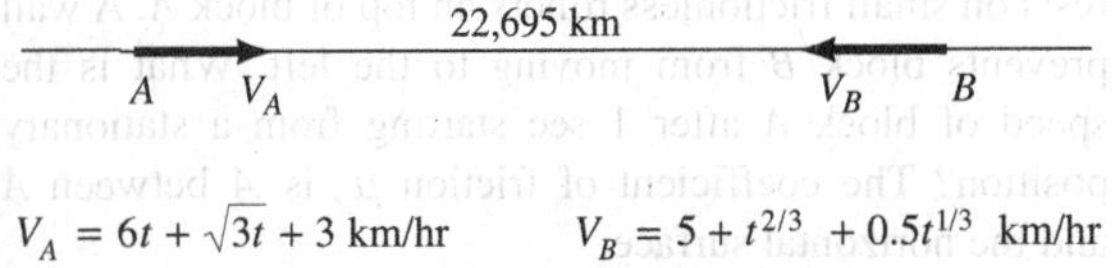

22,695 km

A V_A V_B B

$V_A = 6t + \sqrt{3t} + 3$ km/hr $\qquad V_B = 5 + t^{2/3} + 0.5t^{1/3}$ km/hr

Figure P.10.18.

10.19. As you learned in chemistry, the *coefficient of viscosity* μ is a measure, roughly speaking, of the "stickiness" of a fluid. To measure this property for a highly viscous liquid-like oil, we let a small sphere of metal of radius R descend in a container of the liquid. From fluid mechanics, we know that a drag force will be developed from the oil given by the formula

$$F = 6\pi\mu VR$$

This relation is called *Stoke's law*. The other forces acting on the sphere are its weight (take the density of the sphere as ρ_{Sphere}) and the buoyant force, which is the weight of the oil displaced (take the density of the oil as ρ_{Oil}). The sphere will reach a constant velocity called the *terminal velocity* denoted as $V_{\text{Term.}}$. Show that

$$\mu = \frac{2}{9}\frac{gR^2}{V_{\text{Term.}}}\left(\rho_{\text{Sphere}} - \rho_{\text{Oil}}\right)$$

10.20. A force F is applied to a system of light pulleys to pull body A. If F is 10 kN and A has a mass of 5,000 kg, what is the speed of A after 1 sec starting from rest?

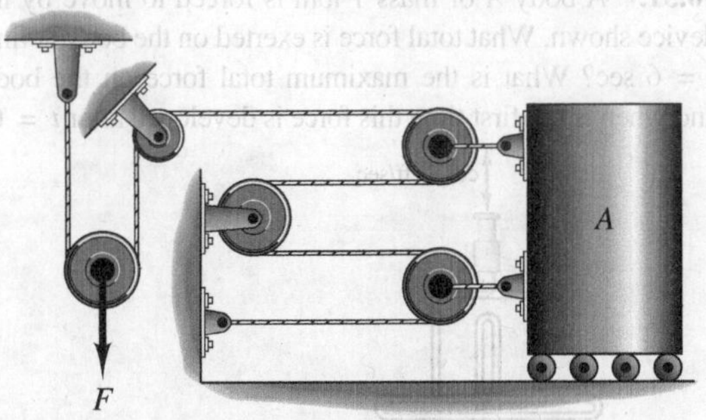

Figure P.10.20.

10.21. A force represented as shown acts on a body having a mass of 1 slug. What is the position and velocity at $t = 30$ sec if the body starts from rest at $t = 0$?

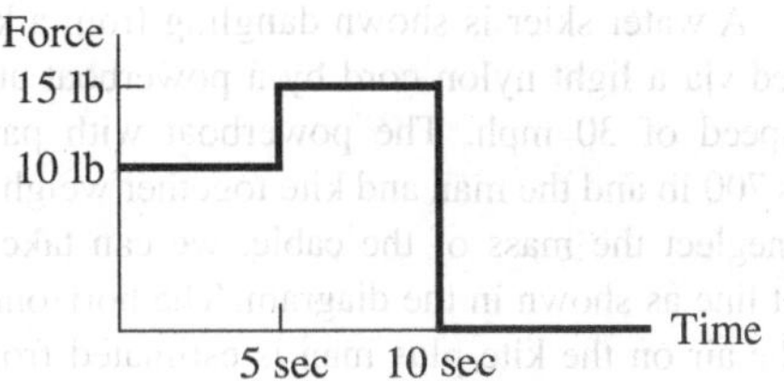

Figure P.10.21.

10.22. A body of mass 1 kg is acted on by a force as shown in the diagram. If the velocity of the body is zero at $t = 0$, what is the velocity and distance traversed when $t = 1$ min? The force acts for only 45 sec.

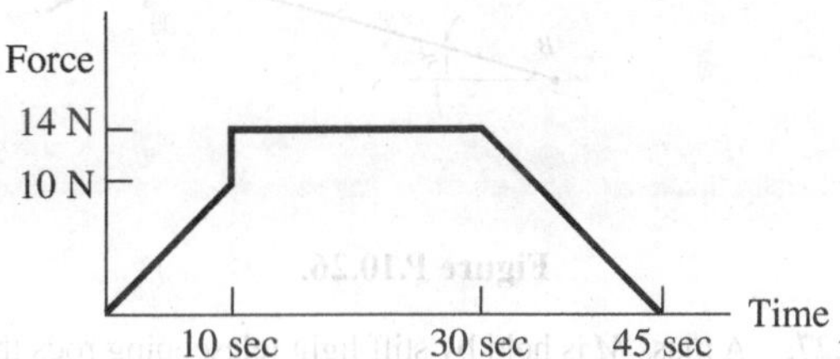

Figure P.10.22.

10.23. Three coupled streetcars are moving down an incline at a speed of 20 km/hr when the brakes are applied for a panic stop. All the wheels lock except for car B, where due to a malfunction all the brakes on the front end of the car do not operate. How far does the system move and what are the forces in the couplings between the cars? Each streetcar weighs 220 kN and the coefficient of dynamic friction μ_d between wheel and rail is .30. Weight is equally distributed on the wheels.

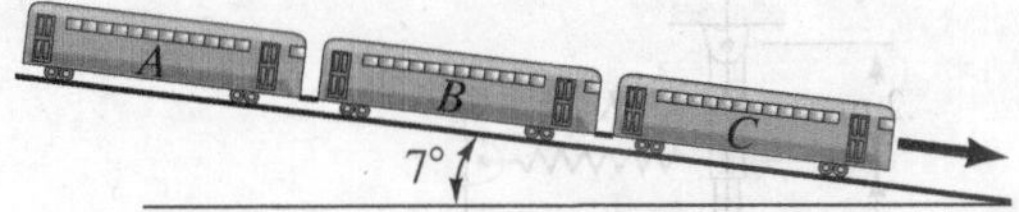

Figure P.10.23.

10.24. A body having a mass of 30 lbm is acted on by a force given by

$$F = 30t^2 + e^{-t} \text{ lb}$$

If the velocity is 10 ft/sec at $t = 0$, what is the body's velocity and the distance traveled when $t = 2$ sec?

10.25. A body of mass 10 kg is acted on by a force in the x direction, given by the relation $F = 10 \sin 6t$ N. If the body has a velocity of 3 m/sec when $t = 0$ and is at position $x = 0$ at that instant, what is the position reached by the body from the origin at $t = 4$ sec? Sketch the displacement-versus-time curve.

10.26. A water skier is shown dangling from a kite that is towed via a light nylon cord by a powerboat at a constant speed of 30 mph. The powerboat with passenger weighs 700 lb and the man and kite together weigh 270 lb. If we neglect the mass of the cable, we can take it as a straight line as shown in the diagram. The horizontal drag from the air on the kite plus man is estimated from fluid mechanics to be 80 lb. What is the tension in the cable? If the cable suddenly snaps, what is the instantaneous horizontal relative acceleration between the kite system and the powerboat?

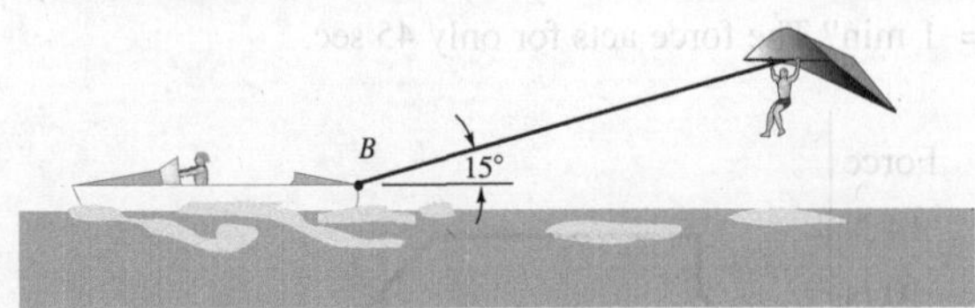

Figure P.10.26.

10.27. A mass M is held by stiff light telescoping rods that can elongate or shorten freely but cannot bend. Each rod is pin connected at the ends A, B, C, and D. The system is on a horizontal, frictionless surface. Two linear springs having spring constants K_1 = 880 N/m and K_2 = 1,400 N/m are connected to the rods as shown in the diagram. If mass M = 3 kg is moved .003 m to the right and is released from rest, what is the equation for the velocity in the x direction as a function of x? What is the speed of the mass when it returns to the vertical position of the rods?

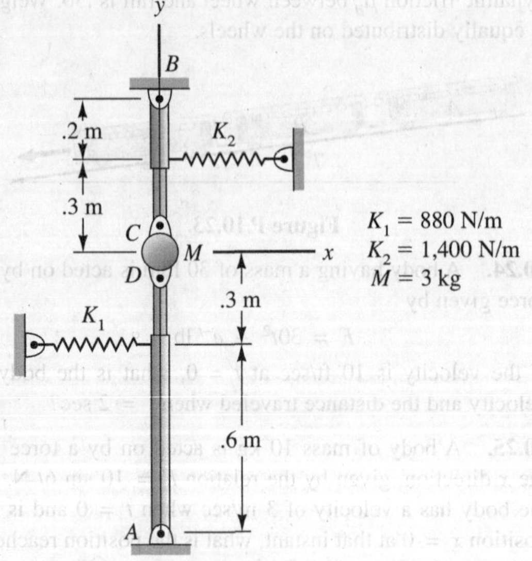

Figure P.10.27.

10.28. A force given as 5 sin 3t lb acts on a mass of 1 slug. What is the position of the mass at t = 10 sec? Determine the total distance traveled. Assume the motion started from rest.

10.29. A block A of mass 500 kg is pulled by a force of 10,000 N as shown. A second block B of mass 200 kg rests on small frictionless rollers on top of block A. A wall prevents block B from moving to the left. What is the speed of block A after 1 sec starting from a stationary position? The coefficient of friction μ_d is .4 between A and the horizontal surface.

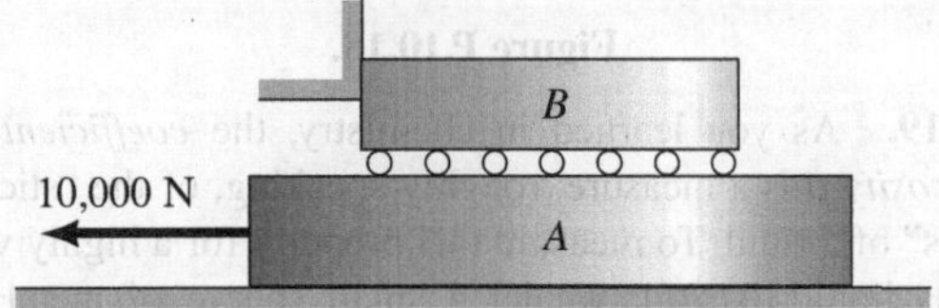

Figure P.10.29.

10.30. Block B weighing 500 N rests on block A, which weighs 300 N. The dynamic coefficient of friction between contact surfaces is .4. At wall C there are rollers whose friction we can neglect. What is the acceleration of body A when a force F of 5,000 N is applied?

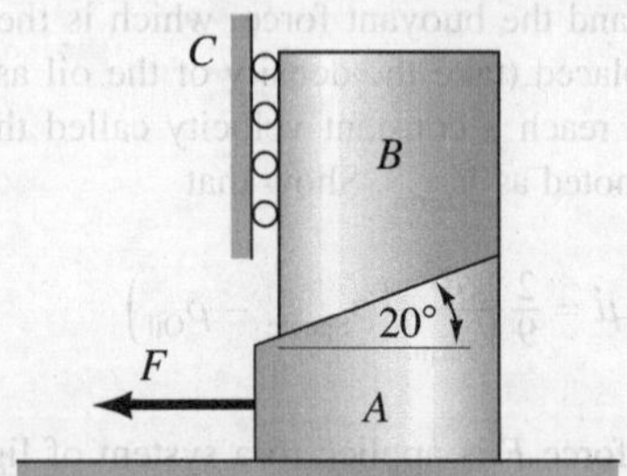

Figure P.10.30.

10.31. A body A of mass 1 lbm is forced to move by the device shown. What total force is exerted on the body at time t = 6 sec? What is the maximum total force on the body, and when is the first time this force is developed after t = 0?

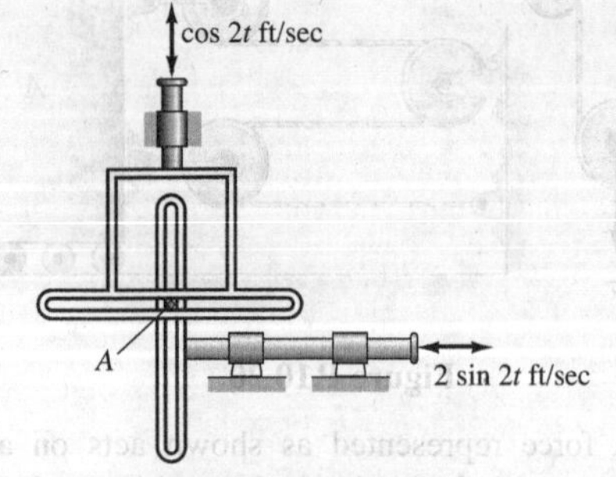

Figure P.10.31.

10.32. Do Problem 10.10 for the case where there is viscous friction between piston and cylinder given as 150 N/m/sec of relative speed. Also, what is the maximum distance l the piston can advance relative to the cylinder?

10.33. The high-speed aerodynamic drag on a car is $.02V^2$ lb with V in ft/sec. If the initial speed is 100 mi/hr, how far will the car move before its speed is reduced to 60 mi/hr? The mass of the car is 2,000 lbm.

10.34. A block slides on a film of oil. The resistance to motion of the block is proportional to the speed of the block relative to the incline at the rate of 7.5 N/m/sec. If the block is released from rest, what is the *terminal speed*? What is the distance moved after 10 sec?

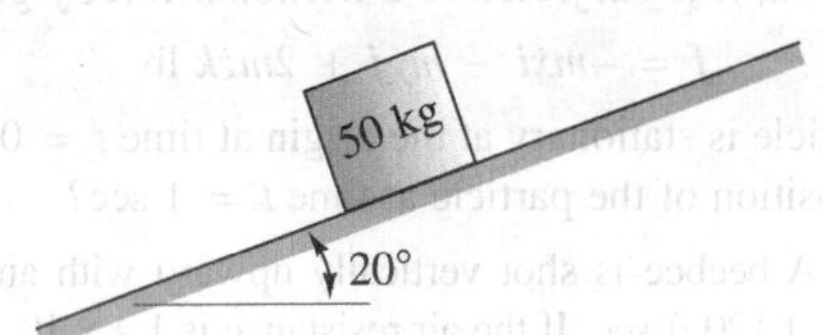

Figure P.10.34.

10.35. When you study fluid mechanics, you will learn that the drag D on a body when moving through a fluid with mass density ρ is given as $\frac{1}{2}C_D\rho V^2 A$ where V is the velocity of the body relative to the fluid; A is the frontal area of the object; and C_D is the so-called *coefficient of drag* usually determined by experiment.

A racing plane on landing is moving at a speed of 350 km/hr when a braking parachute is deployed. This parachute has a frontal area of 30 m² and a $C_D = 1.2$. The plane has a frontal area of 20 m² and a $C_D = 0.4$. If the plane and parachute have a combined mass of 8 Mg, how long does it take to go from 350 km/hr to 200 km/hr by just coasting? Take $\rho = 1.2475$ kg/m³ and neglect rolling resistance from the tires. There is no wind.

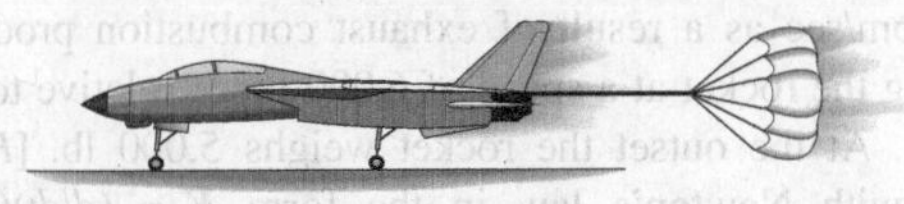

Figure P.10.35.

10.36. In the previous problem, what is the largest frontal area of the braking parachute if the maximum deceleration of the plane is to be 5g's when at a speed of 350 km/s the parachute is first deployed?

10.37. Mass B is on small rollers and moves down the incline. It is connected to a linear spring, which at the position shown is stretched from its undeformed length of 2 m to a length of 5 m. What is the speed of B after it moves 1 m? Use Newton's law as well as the x coordinate shown in the diagram.

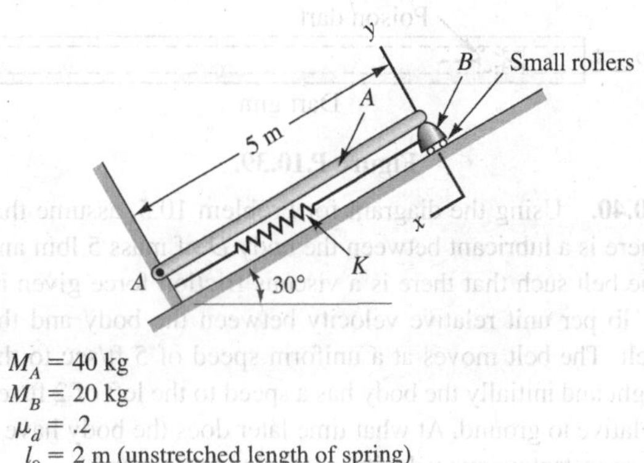

Figure P.10.37.

10.38. A wedge of wood having a specific gravity of 0.6 is forced into the water by a 150-lb force. The wedge is 2 ft in width.

(a) What is the depth d?

(b) What is the speed of the wedge when it has moved upward 0.48 ft after releasing the 150-lb force assuming the wedge does not turn as it rises? Recall, a buoyant force equals the weight of the volume displaced (Archimedes).

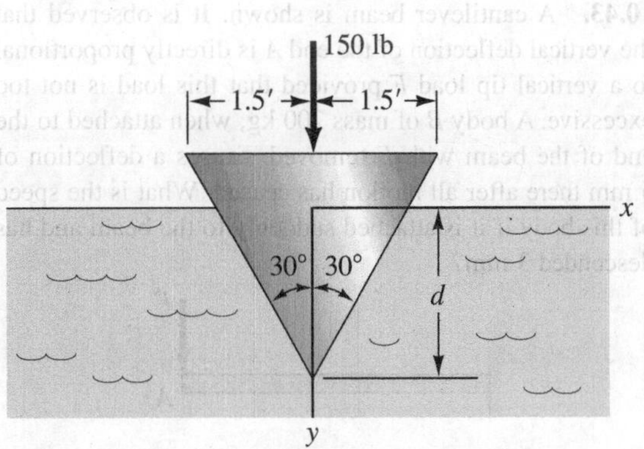

Figure P.10.38.

10.39. A poison dart gun is shown. The cross-sectional area inside the tube is 1 in^2. The dart being blown weighs 3 oz. The dart gun bore has a viscous resistance given as .3 oz per unit velocity in ft/sec. The hunter applies a constant pressure p at the mouth of the gun. Express the relation between p, V (velocity), and t. What constant pressure p is needed to cause the dart to reach a speed of 60 ft/sec in 2 sec? Assume the dart gun is long enough.

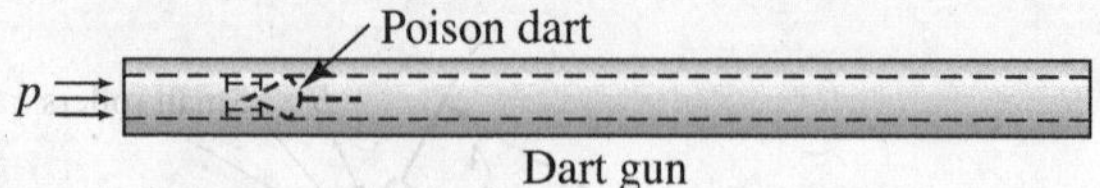

Figure P.10.39.

10.40. Using the diagram for Problem 10.5, assume that there is a lubricant between the body D of mass 5 lbm and the belt such that there is a viscous friction force given as .1 lb per unit relative velocity between the body and the belt. The belt moves at a uniform speed of 5 ft/sec to the right and initially the body has a speed to the left of 2 ft/sec relative to ground. At what time later does the body have a zero instantaneous velocity relative to the ground?

10.41. In Problem 10.40 assume that the belt system is inclined 20° from the horizontal with end B above end A. What minimum belt speed is required so that a body of mass M moving downward will come to a permanent halt relative to the ground? For this belt speed, how long does it take for the body to slow down to half of its initial speed of 2 ft/sec relative to the ground?

10.42. One of the largest of the supertankers in the world today is the *S.S. Globtik London*, having a weight when fully loaded of 476,292 tons. The thrust needed to keep this ship moving at 10 knots is 50 kN. If the drag on the ship from the water is proportional to the speed, how long will it take for this ship to slow down from 10 knots to 5 knots after the engines are shut down? (The answer may make you wonder about the safety of such ships.)

10.43. A cantilever beam is shown. It is observed that the vertical deflection of the end A is directly proportional to a vertical tip load F provided that this load is not too excessive. A body B of mass 200 kg, when attached to the end of the beam with F removed, causes a deflection of 5 mm there after all motion has ceased. What is the speed of this body if it is attached suddenly to the beam and has descended 3 mm?

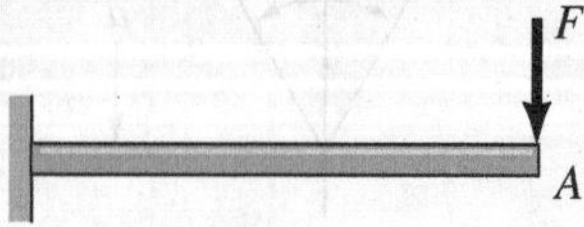

Figure P.10.43.

10.44. The spring shown is nonlinear. That is, K is not a constant, but is a function of the extension of the spring. If $K = 2x + 3$ lb/in. with x measured in inches, what is the speed of the mass when $x = 0$ after it is released from a state of rest at a position 3 in. from the equilibrium position? The mass of the body is 1 slug.

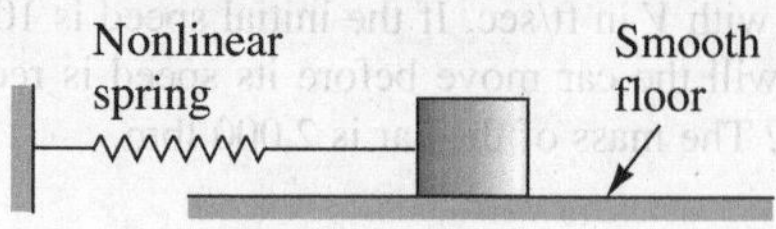

Figure P.10.44.

10.45. A particle of mass m is subject to the following force field:

$$\boldsymbol{F} = m\boldsymbol{i} + 4m\boldsymbol{j} + 16m\boldsymbol{k} \text{ lb}$$

In addition, it is subjected to a frictional force $\boldsymbol{f}$ given as

$$\boldsymbol{f} = -m\dot{x}\boldsymbol{i} - m\dot{y}\boldsymbol{j} + 2m\dot{z}\boldsymbol{k} \text{ lb}$$

The particle is stationary at the origin at time $t = 0$. What is the position of the particle at time $t = 1$ sec?

10.46. A beebee is shot vertically upward with an initial velocity of 120 ft/sec. If the air resistance is $1.4 \times 10^{-5}V^2$ lb, how much time elapses for the projectile to reach its maximum elevation? How high does it go? The beebee weighs .85 oz.

10.47. If in the previous problem, the beebee has reached a maximum height of 92.75 ft, what is the speed when it returns to the ground, assuming it does not reach its terminal velocity? If it has reached the terminal velocity, what is your answer?

10.48. A rocket weighing 5,000 lb is fired vertically from a test stand on the ground. A constant thrust of 20,000 lb is developed for 20 seconds. If just as an exercise, we do not take into account the amount of fuel burned, and if we neglect air resistance, how high up does this hypothetical rocket go? Note that neglecting fuel consumption is a serious error! In the next problem we will investigate the case of the variable mass problem.

***10.49.** Calculate the velocity after 20 seconds for the case where there is a *decrease* of mass of a rocket of 100 lbm/sec as a result of exhaust combustion products leaving the rocket at a speed of 6,000 ft/sec relative to the rocket. At the outset the rocket weighs 5,000 lb. [*Hint:* Start with Newton's law in the form $\boldsymbol{F} = (d/dt)(m\boldsymbol{V})$ where $\boldsymbol{F}$ is the weight, a variable that decreases as fuel is burned. The first term on the right side of this equation is $m(dV/dt)$ where m is the instantaneous mass of the rocket and unburned fuel. Now there is a force on the 100 lbm/sec of combustion products being expelled from the rocket at a speed relative to the rocket of 6,000 ft/sec. The rate of change of linear momentum associated with this force clearly must be $(dm/dt)(6{,}000)$. The *reaction* to this force for this momentum change is on the rocket in the direction of flight of the rocket and must be added to $m(dV/dt)$. The force exerted by the exhaust gases on the rocket is a propulsive force and is called the *thrust* of the rocket. Again, neglect drag of the atmosphere since it will be small at the outset because of low velocity and small later because of the thinness of the atmosphere.]

10.50. We start with a cylindrical tank with diameter 50 ft containing water up to a depth of 10 ft. Initially the solid movable cylindrical piston A having a diameter of 20 ft and a centerline colinear with the centerline of the tank is positioned so that its top is flush with the bottom of the tank. Now the cylinder is moved upward so that the following data apply at the instant of interest assuming the free surface of the water remains flat;

$h_2 = 2$ ft $\qquad \dot{h}_2 = 5$ ft/sec $\qquad \ddot{h}_2 = 3$ ft/sec^2

What is the external force from the *ground* support on the water needed for this condition not including the force required to support the dead weight of the water?

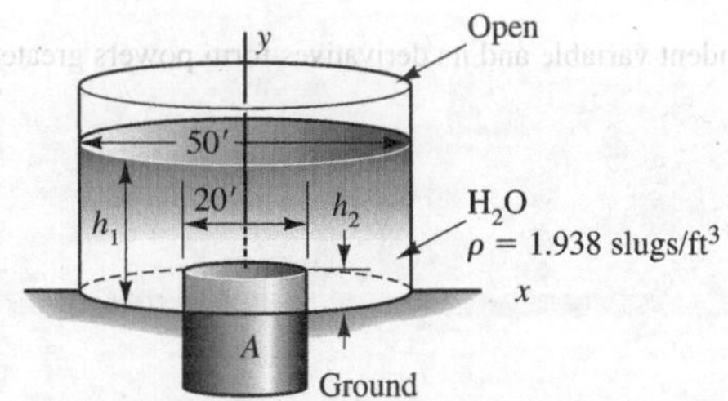

Figure P.10.50.

10.51. A sleeve slides downward along a pipe on which there is dry friction with $\mu_d = .35$. A wire having a constant tension of 80 N is attached to the sleeve and moves with it always retaining the same angle α with the horizontal. If the sleeve weighs 60 N, what should α be for the sleeve to move for 10 seconds before stopping after starting downward with an initial speed of 5 m/s?

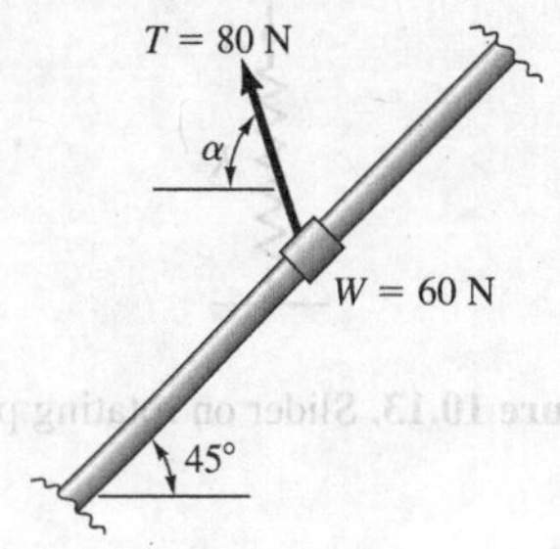

Figure P.10.51.

10.52. An electron having a charge of $-e$ coulombs is moving between two parallel plates in a vacuum with an impressed voltage E. If at $t = 0$, the electron has a velocity V_0 at an angle α_0 with the horizontal in the xy plane, what will be the trajectory equation taking the initial conditions to be at the origin of xy? Show that

$$y = \frac{eE}{2m} + \frac{x^2}{(V_0 \cos \alpha_0)^2} + x \tan \alpha_0$$

where m is the mass of the electron. Note we have neglected gravity here since it is very small compared with the electrostatic force.

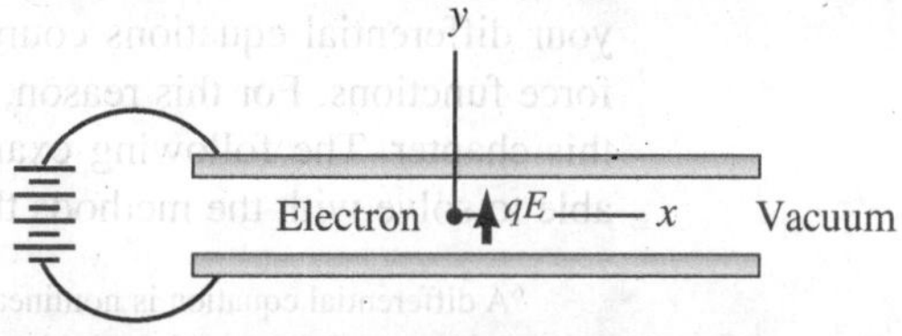

Figure P.10.52.

10.53. A system of light pulleys and inextensible wire connects bodies A, B, and C as shown. If the coefficient of friction between C and the support is .4, what is the acceleration of each body? Take M_A as 100 kg, M_B as 300 kg, and M_C as 80 kg.

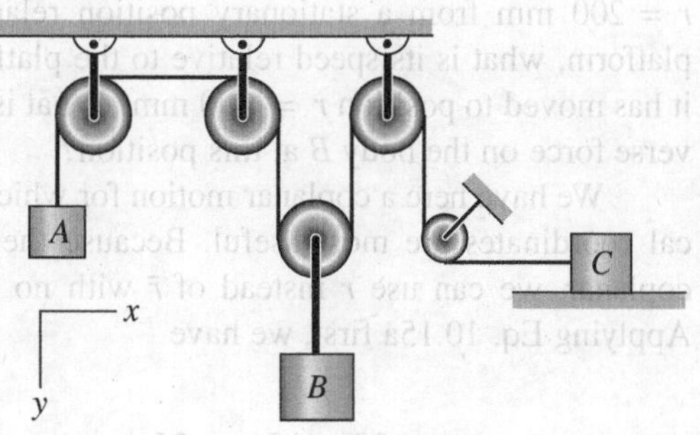

Figure P.10.53.

Part B: Cylindrical Coordinates; Central Force Motion

10.5 Newton's Law for Cylindrical Coordinates

In cylindrical coordinates we can express Newton's law as follows:

$$F_{\bar{r}} = m(\ddot{\bar{r}} - \bar{r}\dot{\theta}^2) \tag{10.15a}$$

$$F_\theta = m(\bar{r}\ddot{\theta} - 2\dot{\bar{r}}\dot{\theta}) \tag{10.15b}$$

$$F_z = m\ddot{z} \tag{10.15c}$$

If the motion is known, it is a simple matter to ascertain the force components using Eqs. 10.15. The inverse problem of determining the motion given the forces is particularly difficult in this case. The reason for this difficulty, as you may have already learned in your differential equations course, is that Eqs. 10.15a and 10.15b are *nonlinear*[6] for all force functions. For this reason, we cannot present integration procedures as in Part A of this chapter. The following example will serve to illustrate the kind of problem we are able to solve with the methods thus far presented in this chapter.

[6]A differential equation is nonlinear if the dependent variable and its derivatives form powers greater than unity or form products anywhere in the equation.

Example 10.8

A platform shown in Fig. 10.13 has a constant angular velocity ω equal to 5 rad/sec. A mass B of 2 kg slides in a frictionless chute attached to the platform. The mass is connected via a light inextensible cable to a linear spring having a spring constant K of 20 N/m. A swivel connector at A allows the cable to turn freely relative to the spring. The spring is unstretched when the mass B is at the center C of the platform. If the mass B is released at r = 200 mm from a stationary position relative to the platform, what is its speed relative to the platform when it has moved to position r = 400 mm? What is the transverse force on the body B at this position?

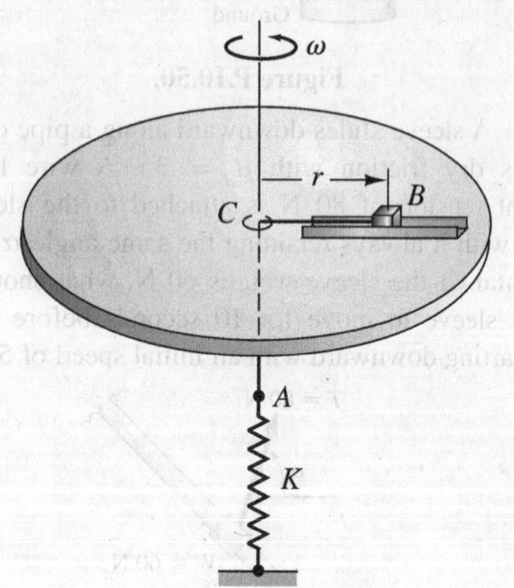

Figure 10.13. Slider on rotating platform.

We have here a coplanar motion for which cylindrical coordinates are most useful. Because the motion is coplanar, we can use r instead of $\bar{r}$ with no ambiguity. Applying Eq. 10.15a first, we have

$$-20r = 2(\ddot{r} - 25r)$$

Example 10.8 (Continued)

Therefore,

$$\ddot{r} = 15r \tag{a}$$

As in Example 10.5, we can replace $\ddot{r}$ so as to allow for a separation of variables.

$$\ddot{r} \equiv \frac{dV_r}{dt} \equiv \frac{dV_r}{dr}\frac{dr}{dt} \equiv V_r \frac{dV_r}{dr} = 15r$$

Therefore,

$$V_r\, dV_r = 15r\, dr$$

Integrating, we get

$$\frac{V_r^2}{2} = \frac{15r^2}{2} + C_1 \tag{b}$$

To determine C_1, note that, when $r = .20m$, $V_r = 0$. Hence,

$$C_1 = -\frac{.600}{2}$$

Equation (b) then becomes

$$V_r^2 = 15r^2 - .600 \tag{c}$$

When $r = .40$ m, we get for V_r from Eq. (c):

$$V_r = 1.342 \text{ m/sec} \tag{d}$$

This is the desired velocity relative to the platform.

To get the transverse force F_θ, go to Eq. 10.15b. Substituting the known data into the equation, we have

$$F_\theta = 2[(.40)(0) + (2)(1.342)(5)]$$

$$F_\theta = 26.84 \text{ N}$$

This is the transverse force on the mass B.

Although you will be asked to solve problems similar to the preceding example, the main use of cylindrical coordinates in Part B of this chapter will be for gravitational central force motion. We shall first present the basic physics underlying this motion expressing certain salient characteristics of the motion, and then we shall arrive at a point where we can effectively employ cylindrical coordinates to describe the motion.

10.6 Central Force Motion—An Introduction

At this time, we shall consider the motion of a particle on which the resultant force is always *directed toward some point fixed in inertial space.* Such forces are termed *central forces* and the resulting motion of this particle is called *central force motion.* A simple example of this is the case of a space vehicle moving with its engine off in the vicinity of a large planet (see Fig. 10.14).

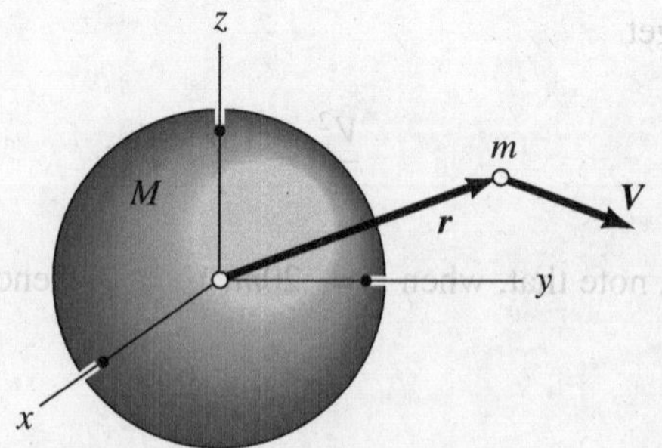

Figure 10.14. Body *m* moving about a planet.

The space vehicle is very small compared to the planet and may be considered to be a particle. Away from the planet's atmosphere, this vehicle will experience no frictional forces, and, if no other astronautical bodies are reasonably close, the only force acting on the vehicle will be the gravitational attraction of the fixed planet.[7] This force is directed toward the center of the planet and, from the gravitational law, is given as

$$\boldsymbol{F} = -G\frac{M_{\text{planet}}m_{\text{body}}}{r^2}\hat{\boldsymbol{r}} \tag{10.16}$$

In the ensuing problems for this chapter and also for Chapter 12, we shall need to compute the quantity *GM* in the equation above. For this purpose, note that, for any particle of mass *m* at the surface of any planet of mass *M* and radius *R*, by the law of gravitation:

$$W = mg = \frac{GMm}{R^2}$$

[7]We are neglecting drag developed from collisions of the space vehicle with solar dust particles.

where g is the acceleration of gravity at the surface of the planet. Solving for GM, we get

$$GM = gR^2 \tag{10.17}$$

Thus, knowing g and R for a planet, it is a simple matter to find GM needed for orbit calculations around this planet.

As pointed out earlier, the motion of a space vehicle with power off is an important example of a central force motion—more precisely a *gravitational* central force motion. The vehicle is usually launched from a planet and accelerated to a high speed outside the planet's atmosphere by multistage rockets (see Fig. 10.15). The velocity at the final instant of powered flight is called the *burnout* velocity. After burnout, the vehicle undergoes gravitational central force motion. Depending on the position and velocity at burnout, the vehicle can go into an orbit around the earth (elliptic and circular orbits are possible), or it can depart from the earth's influence on a parabolic or a hyperbolic trajectory. In all cases, the motion must be coplanar.

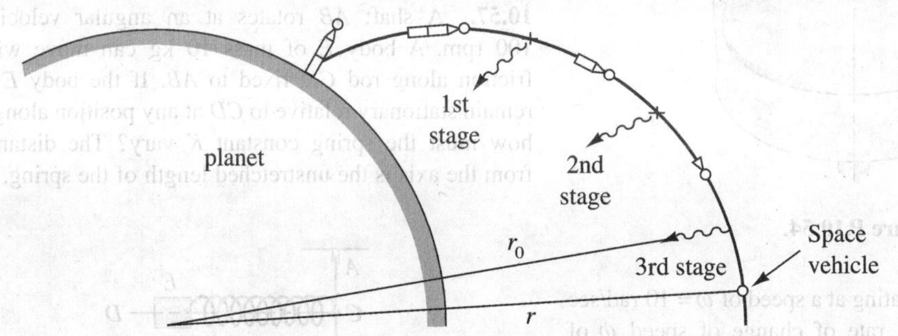

Figure 10.15. Launching a space vehicle.

In the following sections, we shall make a careful detailed study of gravitational central force trajectories. Those who do not have the time for such a detailed study of the trajectories can still make many useful and interesting calculations in Chapter 10 using energy and momentum methods that we shall soon undertake.

PROBLEMS

10.54. A device used at amusement parks consists of a circular room that is made to revolve about its axis of symmetry. People stand up against the wall, as shown in the diagram. After the whole room has been brought up to speed, the floor is lowered. What minimum angular speed is required to ensure that a person will not slip down the wall when the floor is lowered? Take $\mu_s = .3$.

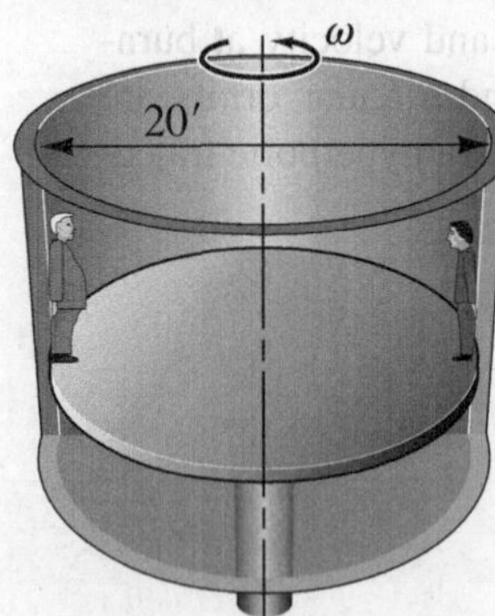

Figure P.10.54.

10.55. A flywheel is rotating at a speed of $\omega = 10$ rad/sec and has at this instant a rate of change of speed $\dot{\omega}$ of 5 rad/sec^2. A solenoid at this instant moves a valve toward the centerline of the flywheel at a speed of 1.5 m/sec and is decelerating at the rate of .6 m/sec^2. The valve has a mass of 1 kg and is .3 m from the axis of rotation at the time of interest. What is the total force on the valve?

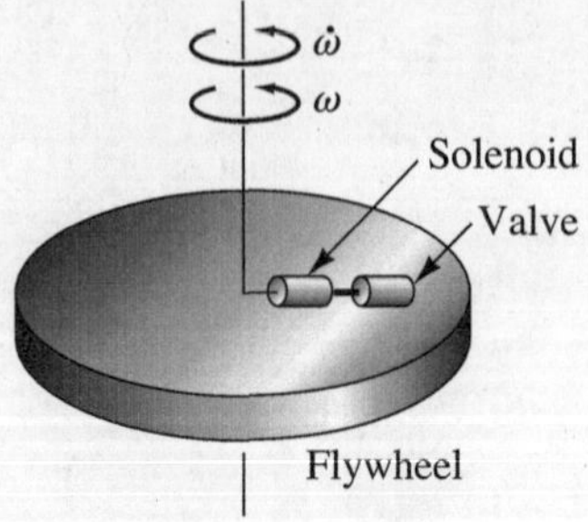

Figure P.10.55.

10.56. A conical pendulum of length l is shown. The pendulum is made to rotate at a constant angular speed of ω about the vertical axis. Compute the tension in the cord if the pendulum bob has weight W. What is the distance of the plane of the trajectory of the bob from the support at O?

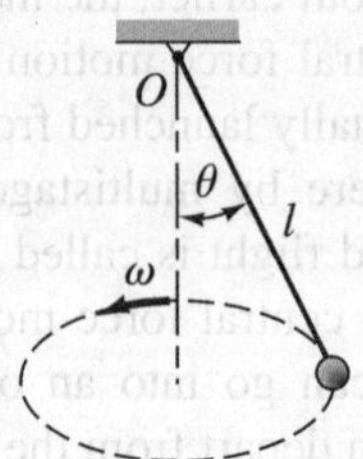

Figure P.10.56.

10.57. A shaft AB rotates at an angular velocity of 100 rpm. A body E of mass 10 kg can move without friction along rod CD fixed to AB. If the body E is to remain stationary relative to CD at any position along CD, how must the spring constant K vary? The distance r_0 from the axis is the unstretched length of the spring.

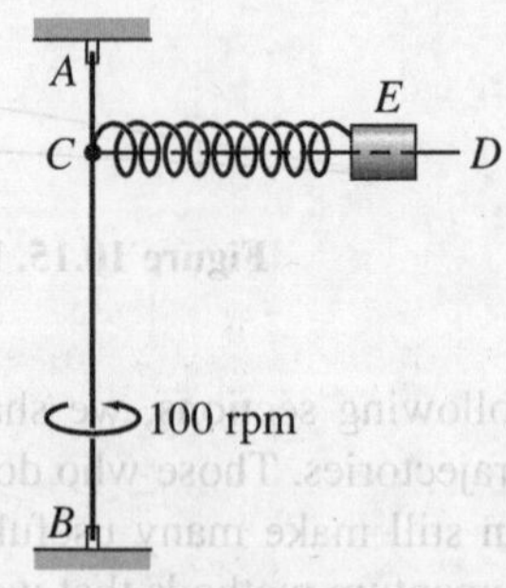

Figure P.10.57.

10.58. A device consists of three small masses, three weightless rods, and a linear spring with $K = 200$ N/m. The system is rotating in the given fixed configuration at a constant speed $\omega = 10$ rad/s in a horizontal plane. The following data apply:

$$M_A = 2 \text{ kg} \qquad M_B = 3 \text{ kg} \qquad M_C = 2 \text{ kg}$$

If the spring is stretched by an amount .025 m, determine the total force components acting on the mass at C and the tensile force in member DA. [*Hint:* Consider a single particle, then a system of particles; DA and DB are pin connected.]

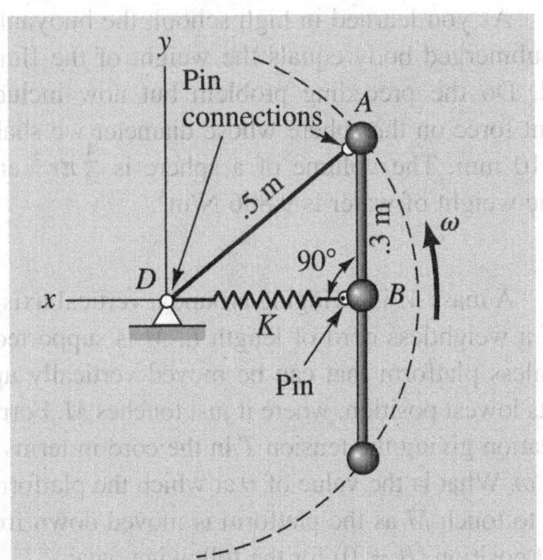

Figure P.10.58.

10.59. In the preceding problem consider that member *DA* is welded to the sphere at *A*. Now, at the instant of interest, there is also an angular acceleration of the system having the value of .28 rad/sec^2 counterclockwise. What are the force components acting on particle *C*, and what are the force components from rod *AD* acting on particle *A*? See hint given in the preceding problem.

10.60. A device called a *flyball governor* is used to regulate the speed of such devices as steam engines and turbines. As the governor is made to rotate through a system of gears by the device to be controlled, the balls will attain a configuration given by the angle θ, which is dependent on both the angular speed ω of the governor and the force *P* acting on the collar bearing at *A*. The up-and-down motion of the bearing at *A* in response to a change in ω is then used to open or close a valve to regulate the speed of the device. Find the angular velocity required to maintain the configuration of the flyball governor for $\theta = 30°$. Neglect friction.

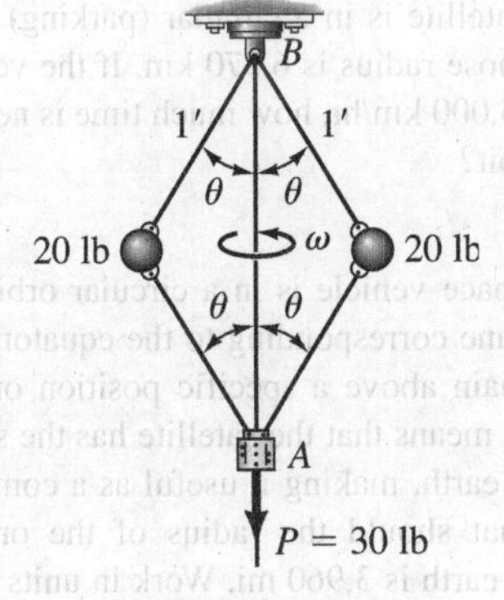

Figure P.10.60.

10.61. A platform rotates at 2 rad/sec. A body *C* weighing 450 N rests on the platform and is connected by a flexible weightless cord to a mass weighing 225 N, which is prevented from swinging out by part of the platform. For what range of values of *x* will bodies *C* and *B* remain stationary relative to the platform? The static coefficient of friction for all surfaces is .4.

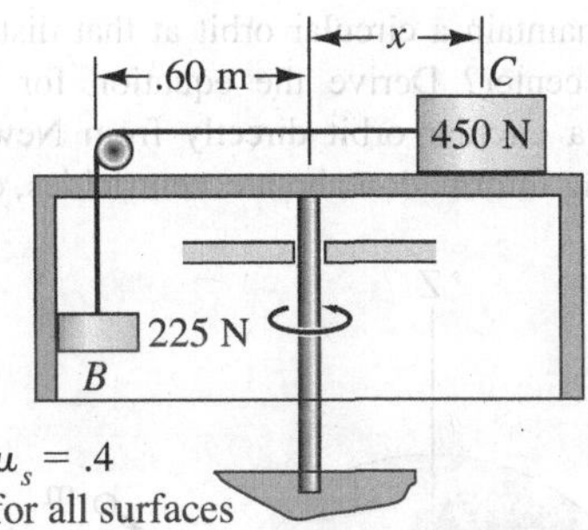

Figure P.10.61.

10.62. A particle moves under gravitational influence about a body *M*, the center of which can be taken as the origin of an inertial reference. The mass of the particle is 50 slugs. At time *t*, the particle is at a position 4,500 mi from the center of *M* with direction cosines $l = .5$, $m = -.5$, $n = .707$. The particle is moving at a speed of 17,000 mi/hr along the direction $\boldsymbol{\epsilon}_t = .8\boldsymbol{i} + .2\boldsymbol{j} + .566\boldsymbol{k}$. What is the direction of the normal to the plane of the trajectory?

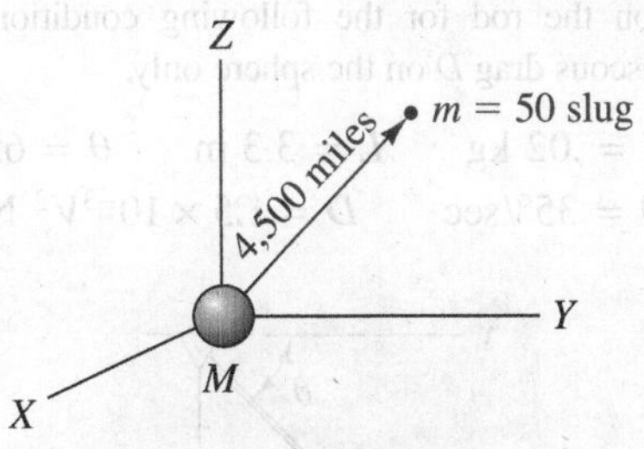

Figure P.10.62.

10.63. If the position of the particle in Problem 10.62 were to reach a distance of 4,300 mi from the center of body *M*, what would the transverse velocity V_θ of the particle be?

10.64. Use Eqs. 10.38b and 10.40 to show that if the eccentricity is zero, the trajectory must be that of a circle.

10.65. A satellite has at one time during its flight around the earth a radial component of velocity 3,200 km/hr and a transverse component of 26,500 km/hr. If the satellite is at a distance of 7,040 km from the center of the earth, what is its areal velocity?

10.66. Compute the escape velocity at a position 8,000 km from the center of the earth. What speed is needed to maintain a circular orbit at that distance from the earth's center? Derive the equation for the speed needed for a circular orbit directly from Newton's law without using information about eccentricities, etc.

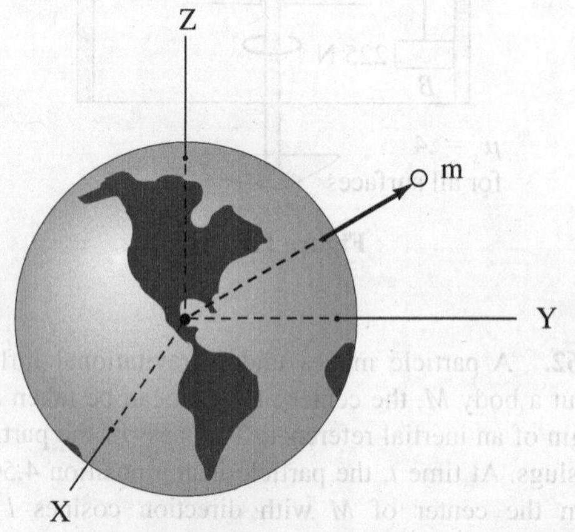

Figure P.10.66.

10.67. A small sphere is swinging in the xy plane at the end of a thin light rod while in a tank of water. If we neglect the buoyant force (see next problem) on the sphere and on the rod, determine the angular acceleration $\ddot{\theta}$ and the tensile force T on the rod for the following conditions, which include viscous drag D on the sphere only.

$$M = .02 \text{ kg} \qquad L = 3.3 \text{ m} \qquad \theta = 65^\circ$$
$$\dot{\theta} = 35^\circ/\text{sec} \qquad D = 1.5 \times 10^{-3} V^2 \text{ N}$$

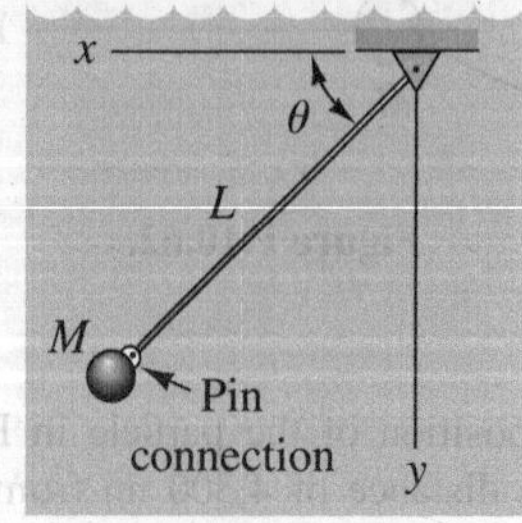

Figure P.10.67.

10.68. As you learned in high school, the buoyant force on a submerged body equals the weight of the fluid displaced. Do the preceding problem but now include the buoyant force on the sphere whose diameter we shall take to be 10 mm. The volume of a sphere is $\frac{4}{3}\pi r^3$ and the specific weight of water is 9,806 N/m³.

10.69. A mass M is swinging around a vertical axis at the end of a weightless cord of length L. M is supported by a frictionless platform that can be moved vertically upward from its lowest position, where it just touches M. Formulate an equation giving the tension T in the cord in terms of M, L, and ω. What is the value of θ at which the platform first ceases to touch M as the platform is moved down from its highest position ($\theta = 0$) for the following data:

$$\omega = 8 \text{ rad/s} \qquad M = 3.1 \text{ kg} \qquad L = .42 \text{ m}$$

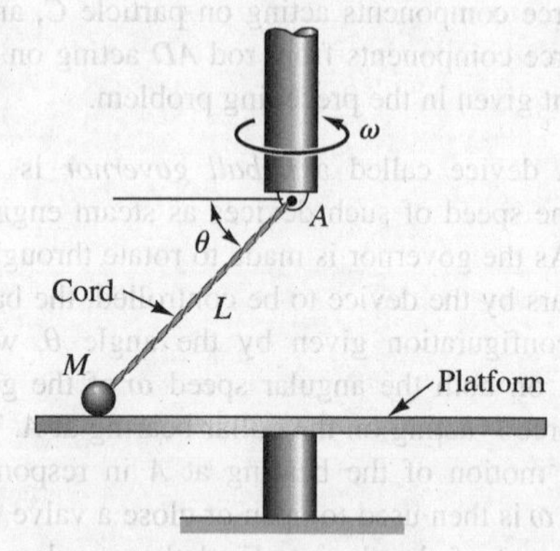

Figure P.10.69.

10.70. A satellite is in a circular (parking) orbit around the earth, whose radius is 6,370 km. If the velocity of the satellite is 25,000 km/hr, how much time is needed for one complete orbit?

10.71. A space vehicle is in a circular orbit around the earth in a plane corresponding to the equator and moving so as to remain above a specific position on the earth's surface. This means that the satellite has the same angular speed as the earth, making it useful as a communications satellite. What should the radius of the orbit be? The radius of the earth is 3,960 mi. Work in units of miles and hours.

10.72. Consider a satellite of mass m in a circular orbit around the earth at a radius R_0 from the center of the earth. Using the universal law of gravitation (Eq. 1.11) with M as the mass of the earth and using *Newton's law* in a direction normal to the path, show that

$$V_{Circ.} = \sqrt{\frac{GM}{R_0}}$$

for a circular orbit. Now at the earth's surface use the universal gravitational law again and the weight of the body, to show that $GM = gR^2_{Earth}$, where g is the acceleration of gravity.

10.73. The acceleration of gravity on the planet Mars is about .385 times the acceleration of gravity on earth, and the radius of Mars is about .532 times that of the earth. What is the escape velocity from Mars at a position 100 mi from the surface of the planet?

10.74. In 1971 Mariner 9 was placed in orbit around Mars with an eccentricity of .5. At the lowest point in the orbit, Mariner 9 is 320 km from the surface of Mars.

(a) Compute the maximum velocity of the space vehicle relative to the center of Mars.

(b) Compute the time of one cycle.

Use the data in Problem 10.73 for Mars.

10.75. A man is in orbit around the earth in a space-shuttle vehicle. At his lowest possible position, he is moving with a speed of 18,500 mi/hr at an altitude of 200 mi. When he wants to come back to earth, he fires a retro-rocket straight ahead when he is at the aforementioned lowest position and slows himself down. If he wishes subsequently to get within 50 mi from the earth's surface during the first cycle after firing his retro-rocket, what must his decrease in velocity be? (Neglect air resistance.)

10.76. The Pioneer 10 space vehicle approaches the planet Jupiter with a trajectory having an eccentricity of 3. The vehicle comes to within 1,000 mi of the surface of Jupiter. What is the speed of the vehicle at this instant? The acceleration of gravity of Jupiter is 90.79 ft/sec^2 at the surface and the radius is 43,400 mi.

10.77. If the moon has a motion about the earth that has an eccentricity of .0549 and a period of 27.3 days, what is the closest distance of the moon to the earth in its trajectory?

10.78. The satellite Hyperion about the planet Saturn has a motion with an eccentricity known to be .1043. At its closest distance from Saturn, Hyperion is 1.485×10^6 km away (measured from center to center). What is the period of Hyperion about Saturn? The acceleration of gravity of Saturn is 13.93 m/sec^2 at its surface. The radius of Saturn is 57,600 km.

10.79. Two satellite stations, each in a circular orbit around the earth, are shown. A small vehicle is shot out of the station at A tangential to the trajectory in order to "hit" station B when it is at a position E 120° from the x axis as shown in the diagram. What is the velocity of the vehicle relative to station A when it leaves? The circular orbits are 200 miles and 400 miles, respectively, from the earth's surface.

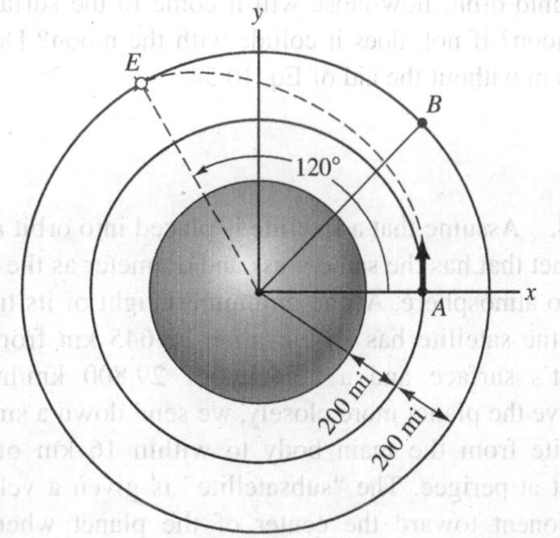

Figure P.10.79.

10.80. In Problem 10.79, determine the total velocity of the vehicle as it arrives at E as seen by an observer in the satellite B. The values of C and D for the vehicle from Problem 10.79 are 7.292×10^7 mi^2/hr and 7.373×10^{-6} mi^{-1}, respectively.

10.81. The Viking I space probe is approaching Mars. When it is 80,650 km from the center of Mars, it has a speed of 16,130 km/hr with a component (V_r) toward the center of Mars of 15,800 km/hr. Does Viking I crash into Mars, go into orbit, or have one pass in the vicinity of Mars? If there is no crash, how close to Mars does it come? The acceleration of gravity on the surface of Mars is 4.13 m/sec^2, and its radius is 3,400 km.

10.82. A meteor is moving at a speed of 20,000 mi/hr relative to the center of the earth when it is 350 mi from the surface of the earth. At that time, the meteor has a radial velocity component of 4,000 mi/hr toward the center of the earth. How close does it come to the earth's surface?

10.83. The moon's radius is about .272 times that of the earth, and its acceleration of gravity at the surface is .165 times that of the earth at the earth's surface. A space vehicle approaches the moon with a velocity component toward the center of the moon of 3,200 km/hr and a transverse component of 8,000 km/hr relative to the center of the moon. The vehicle is 3,200 km from the center of the moon when it has these velocity components. Will the vehicle go into orbit around the moon if we consider only the gravitational effect of the moon on the vehicle? If it goes into orbit, how close will it come to the surface of the moon? If not, does it collide with the moon? Do this problem without the aid of Eq. 10.54.

10.84. Assume that a satellite is placed into orbit about a planet that has the same mass and diameter as the earth but no atmosphere. At the minimum height of its trajectory, the satellite has an elevation of 645 km from the planet's surface and a velocity of 29,800 km/hr. To observe the planet more closely, we send down a smaller satellite from the main body to within 16 km of this planet at perigee. The "subsatellite" is given a velocity component toward the center of the planet when the main satellite is at its lowest position. What is this radial velocity, and what is the eccentricity of the trajectory of the subsatellite? What is a better way to get closer to the planet?

10.85. Suppose that you are on a planet having no atmosphere. This planet rotates once every 6 hr about its axis relative to an inertial reference XYZ at its center. The planet has a radius of 1,600 km, and the acceleration of gravity at the surface is 7 m/sec^2. A bullet is fired by a man at the equator in a direction normal to the surface of the planet as seen by this man. The muzzle velocity of the gun is 1,500 m/sec. What is the eccentricity of the trajectory and the maximum height h of the bullet above the surface of the planet?

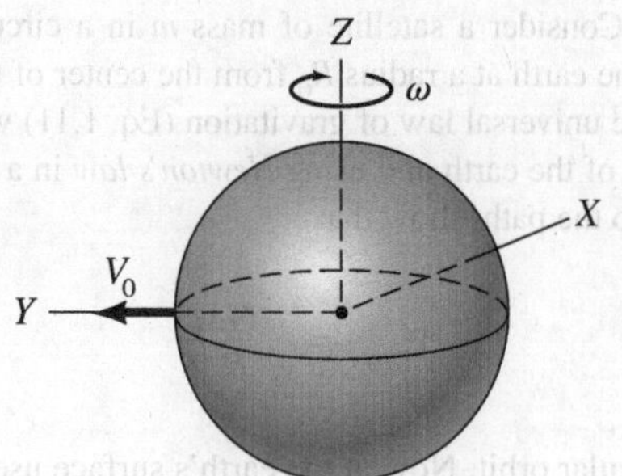

Figure P.10.85.

***10.86.** A satellite is launched at A. We wish to determine the time required, Δt, to get to position B. Show that for this calculation we can employ the formulation

$$\Delta t = \frac{1}{C}\int_{\theta_0}^{\theta_B} r^2\, d\theta$$

For integration purposes, show that the formulation above becomes

$$\Delta t = \frac{1}{C}\int_{\theta_0}^{\theta_B} \frac{d\theta}{[(GM/C^2) + D\cos\theta]^2}$$

Carry out the integration.

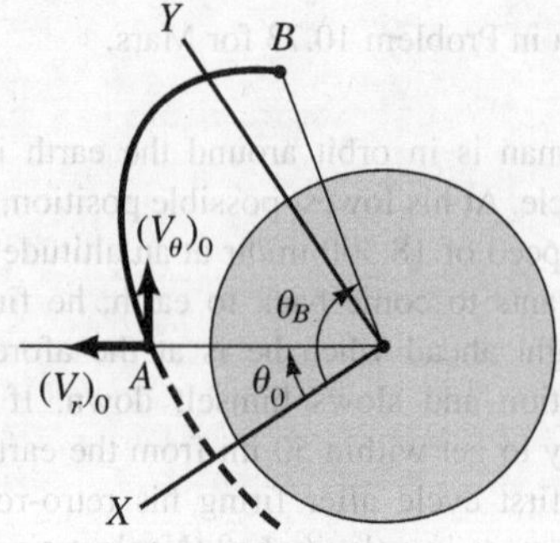

Figure P.10.86.

***10.87.** A satellite is launched at a speed of 20,000 mi/hr relative to the earth's center at an altitude of 340 mi above the earth's surface. The guidance system has malfunctioned, and the satellite has a direction 20° up from the tangent plane to the earth's surface. Will the satellite go into orbit? Give the time required for one cycle if it goes into orbit or the time it takes before it strikes the earth after firing. Neglect friction in both cases. (See Problem 10.86 before doing this problem.)

Part C: A System of Particles

10.7 The General Motion of a System of Particles

Let us examine a system of n particles (Fig. 10.16) that has interactions between the particles for which *Newton's third law* of motion (action equals reaction) applies. *Newton's second law* for any particle (let us say the ith particle) is then

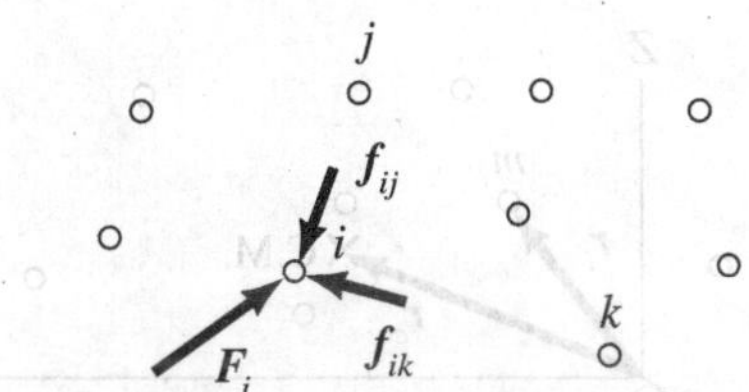

Figure 10.16. Forces on ith particle of the system.

$$m_i \frac{d^2 \boldsymbol{r}_i}{dt^2} = \boldsymbol{F}_i + \sum_{\substack{j=1 \\ i \neq j}}^{n} \boldsymbol{f}_{ij} \tag{10.18}$$

where $\boldsymbol{f}_{ij}$ is the force on particle i from particle j and is thus considered an *internal* force for the system of particles. Clearly, the $j = i$ term of the summation must be deleted since the ith particle cannot exert force on itself. The force $\boldsymbol{F}_i$ represents the resultant force on the ith particle from the forces *external* to the system of particles.

If these equations are added for all n particles, we have

$$\sum_{i=1}^{n} m_i \frac{d^2 \boldsymbol{r}_i}{dt^2} = \sum_{i=1}^{n} \boldsymbol{F}_i + \sum_{i=1}^{n} \sum_{j=1}^{n} \boldsymbol{f}_{ij} \tag{10.19}$$

Carrying out the double summation and excluding terms with repeated indexes, such as $\boldsymbol{f}_{11}$, $\boldsymbol{f}_{22}$, etc., we find that for each term with any one set of indexes there will be a term with the reverse of these indexes present. For example, for the force $\boldsymbol{f}_{12}$, a force $\boldsymbol{f}_{21}$ will exist. Considering the meaning of the indexes, we see that $\boldsymbol{f}_{ij}$ and $\boldsymbol{f}_{ji}$ represent action and reaction forces between a pair of particles. Thus, as a result of *Newton's third law*, the double summation in Eq. 10.19 should add up to zero. *Newton's second law* for a system of particles then becomes:

$$\boldsymbol{F} = \sum_{i=1}^{n} m_i \frac{d^2 \boldsymbol{r}_i}{dt^2} = \frac{d^2}{dt^2} \sum_{i=1}^{n} m_i \boldsymbol{r}_i \tag{10.20}$$

where $\boldsymbol{F}$ now represents the vector sum of all the *external* forces acting on *all* the particles of the system.

To make further useful simplifications, we use the first moment of mass of a system of n particles about a fixed point A in inertial space given as

$$\text{first moment vector} \equiv \sum_{i=1}^{n} m_i \boldsymbol{r}_i$$

where $\boldsymbol{r}_i$ represents the position vector from the point A to the ith particle (Fig. 10.19). As explained in Chapter 8, we can find a position, called the

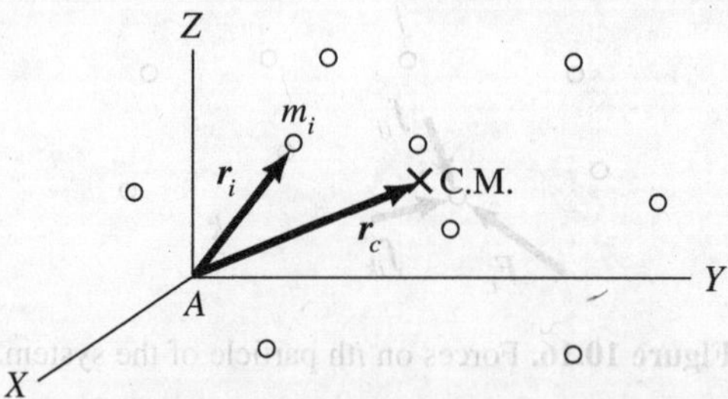

Figure 10.17. Center of mass of system.

center of mass of the system, with position vector $\boldsymbol{r}_c$, where the entire mass of the system of particles can be concentrated to give the correct first moment. Thus,

$$\boldsymbol{r}_c \sum_{i=1}^{n} m_i = \sum_{i=1}^{n} m_i \boldsymbol{r}_i$$

Therefore,

$$\boldsymbol{r}_c = \frac{\sum m_i \boldsymbol{r}_i}{\sum m_i} = \frac{\sum m_i \boldsymbol{r}_i}{M} \tag{10.21}$$

Let us reconsider Newton's law using the center-of-mass concept. To do this, replace $\sum m_i \boldsymbol{r}_i$ by $M\boldsymbol{r}_c$ in Eq. 10.20. Thus,

$$\boldsymbol{F} = \frac{d^2}{dt^2}(M\boldsymbol{r}_c) = M\frac{d^2\boldsymbol{r}_c}{dt^2} \tag{10.22}$$

We see that *the center of mass of any aggregate of particles has a motion that can be computed by methods already set forth, since this is a problem involving a single hypothetical particle of mass M*. You will recall that we have alluded to this important relationship several times earlier to justify the use of the particle concept in the analysis of many dynamics problems. We must realize for such an undertaking that $\boldsymbol{F}$ is the total *external* force acting on *all* the particles.

Example 10.9

Three charged particles in a vacuum are shown in Fig. 10.18. Particle 1 has a mass of 10^{-5} kg and a charge of 4×10^{-3} C (coulombs) and is at the origin at the instant of interest. Particles 2 and 3 each have a mass of 2×10^{-5} kg and a charge of 5×10^{-5} C

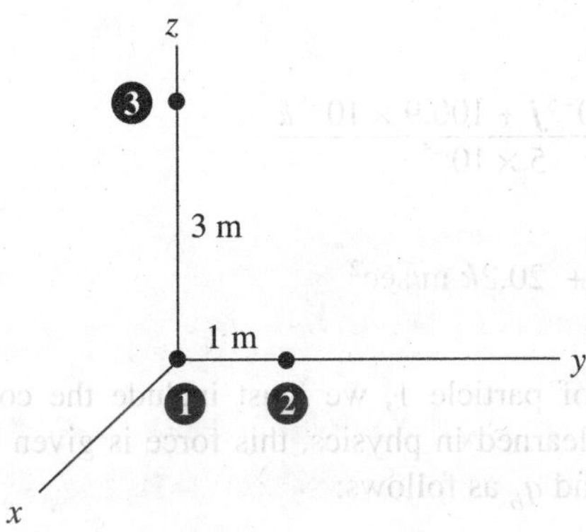

Figure 10.18. Charged particles in field $\boldsymbol{E}$.

and are located, respectively, at the instant of interest 1 m along the y axis and 3 m along the z axis. An electric field $\boldsymbol{E}$ given as

$$\boldsymbol{E} = 2x\boldsymbol{i} + 3z\boldsymbol{j} + 3(y + z^2)\boldsymbol{k} \text{ N/C} \tag{a}$$

is imposed from the outside. Compute: (a) the position of the center of mass for the system, (b) the acceleration of the center of mass, and (c) the acceleration of particle 1.

To get the position of the center of mass, we merely equate moments of the masses about the origin with that of a particle having a mass equal to the sum of masses of the system. Thus,

$$(1 + 2 + 2) \times 10^{-5}\boldsymbol{r}_c = (2 \times 10^{-5})\boldsymbol{j} + (2 \times 10^{-5})3\boldsymbol{k}$$

Therefore,

$$\boldsymbol{r}_c = .4\boldsymbol{j} + 1.2\boldsymbol{k} \text{ m} \tag{b}$$

To get the acceleration of the mass center, we must find the sum of the *external* forces acting on the particles. Two external forces act on each particle: the force of gravity and the electrostatic force from the external field. Recall from physics that this electrostatic force is given as $q\boldsymbol{E}$, where q is the charge on the particle. Hence, the total external force for each particle is given as follows:

$$\boldsymbol{F}_1 = -(9.81)(10^{-5})\boldsymbol{k} + \boldsymbol{0} \text{ N} \tag{c}$$
$$\boldsymbol{F}_2 = -(9.81)(2 \times 10^{-5})\boldsymbol{k} + (5 \times 10^{-5})(3\boldsymbol{k}) \text{ N} \tag{d}$$
$$\boldsymbol{F}_3 = -(9.81)(2 \times 10^{-5})\boldsymbol{k} + (5 \times 10^{-5})(9\boldsymbol{j} + 27\boldsymbol{k}) \text{ N} \tag{e}$$

Example 10.9 (Continued)

The sum of these forces F_T is

$$F_T = 45 \times 10^{-5}\boldsymbol{j} + 100.9 \times 10^{-5}\boldsymbol{k}\ \text{N} \tag{f}$$

Accordingly, we have for $\ddot{r}_c$:

$$\ddot{r}_c = \frac{45 \times 10^{-5}\boldsymbol{j} + 100.9 \times 10^{-5}\boldsymbol{k}}{5 \times 10^{-5}}$$

$$\ddot{r}_c = 9\boldsymbol{j} + 20.2\boldsymbol{k}\ \text{m/sec}^2 \tag{g}$$

Finally, to get the acceleration of particle 1, we must include the coulombic forces from particles 2 and 3. As you learned in physics, this force is given between two particles a and b with charges q_a and q_b as follows:

$$f_{\text{coul}} = -\frac{q_a q_b}{4\pi\epsilon_0 r^2}\hat{\boldsymbol{r}}$$

where $\hat{\boldsymbol{r}}$ is the unit vector between the particles, and ϵ_0 is the dielectric constant equal to 8.854 × 10^{-12} F/m (farads per meter) for a vacuum. Note that the coulombic force is repulsive between like charges. The total coulombic force F_C from particles 2 and 3 is

$$F_C = -\frac{(4 \times 10^{-5})(5 \times 10^{-5})}{(4\pi\epsilon_0)(1^2)}\boldsymbol{j} - \frac{(4 \times 10^{-5})(5 \times 10^{-5})}{(4\pi\epsilon_0)(3^2)}\boldsymbol{k} \tag{h}$$

$$= -18\boldsymbol{j} - 2\boldsymbol{k}\ \text{N}$$

The total force acting on particle 1 is then

$$(F_1)_T = \underbrace{-(9.81)(10^{-5})\boldsymbol{k}}_{\text{from weight}} + \underbrace{\boldsymbol{0}}_{\text{from external field}} + \underbrace{(-18\boldsymbol{j} - 2\boldsymbol{k})}_{\text{from internal field}}\ \text{N} \tag{i}$$

Clearly, the internal field dominates here. **Newton's law** then gives us

$$\ddot{r}_1 = \frac{-18\boldsymbol{j} - 2\boldsymbol{k}}{10^{-5}}$$

$$\ddot{r}_1 = -18 \times 10^5\boldsymbol{j} - 2 \times 10^5\boldsymbol{k}\ \text{m/sec}^2 \tag{j}$$

We see here from Eqs. (g) and (j) that although the particles tend to "scramble" away from each other due to very strong internal coulombic forces, the center of mass accelerates slowly by comparison.

Example 10.10

A young man is standing in a canoe awaiting a young lady (Fig. 10.19). The man weighs 150 lb. and, as shown, is positioned near the end of the canoe, which weighs 200 lb. When the young lady appears, he quickly scrambles forward to greet her, but when he has moved 20 ft to the forward end of the canoe, he finds (not having studied mechanics) that he cannot reach her. How far is the tip of the canoe from the dock after our hero has made the 20-ft dash? The canoe is in no way tied to the dock and there are no water currents. Neglect friction from the water on the canoe.

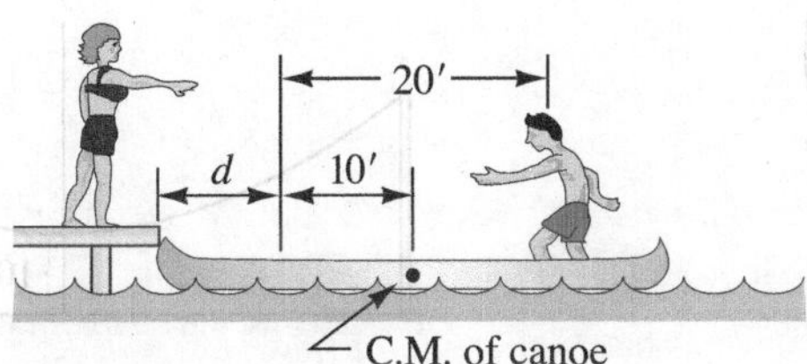

Figure 10.19. Man in canoe awaits his date.

The center of mass of the man plus the canoe cannot change position during this action since there is no net external force acting on this system during this action. Hence the first moment of mass about any fixed position must remain constant during this action. In Fig. 10.20 we have shown the man in the forward position and we choose the position at the tip of the dock to equate moment of mass at the beginning of the action and just when the man has moved the 20 ft. We then can say, noting that we are denoting the unspecified distance between the tip of the canoe and the forward position of the man as d as shown in Figs. 10.19 and 10.20,

$$\frac{200}{g}(d+10)+\frac{150}{g}(d+20)=\frac{150}{g}(x+d)+\frac{200}{g}(x+d+10)$$

Figure 10.20. Man rushes forward to greet his date.

Canceling terms where possible we then have

$$350x = 3{,}000 \qquad \therefore \qquad x = 8.571 \text{ ft.}$$

PROBLEMS

10.88. A warrior of old is turning a sling in a vertical plane. A rock of mass .3 kg is held in the sling prior to releasing it against an enemy. What is the minimum speed ω to hold the rock in the sling?

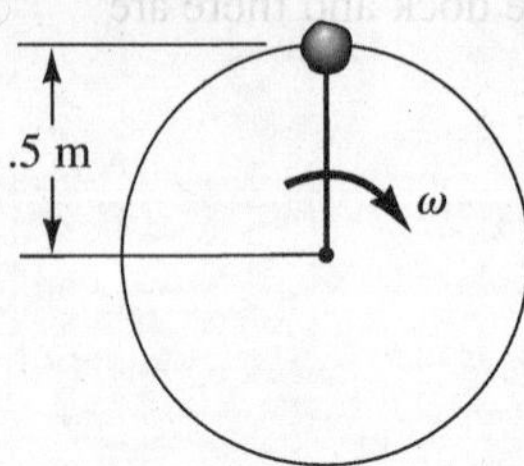

Figure P.10.88.

10.89. A car is traveling at a speed of 55 mi/hr along a banked highway having a radius of curvature of 500 ft. At what angle should the road be banked in order that a zero friction force is needed for the car to go around this curve?

10.90. A car weighing 20 kN is moving at a speed V of 60 km/hr on a road having a vertical radius of curvature of 200 m as shown. At the instant shown, what is the maximum deceleration possible from the brakes along the road for the vehicle if the coefficient of dynamic friction between tires and the road is .55?

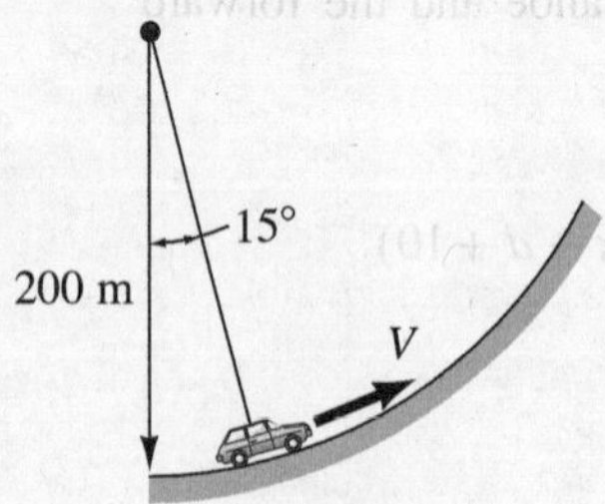

Figure P.10.90.

10.91. A particle moves at uniform speed of 1 m/sec along a plane sinusoidal path given as

$$y = 5 \sin \pi x \text{ m}$$

What is the position between $x = 0$ and $x = 1$ m for the maximum force normal to the curve? What is this force if the mass of the particle is 1 kg?

10.92. A catenary curve is formed by the cable of a suspension bridge. The equation of this curve relative to the axes shown can be given as

$$y = \frac{a}{2}(e^{ax} + e^{-ax}) = a \cosh ax$$

with x and y in feet. A small one-passenger vehicle is designed to move along the catenary to facilitate repair and painting of the bridge. Consider that the vehicle moves at uniform speed of 10 ft/sec along the curve. If the vehicle and passenger have a combined mass of 250 lbm, what is the force normal to the curve as a function of position x?

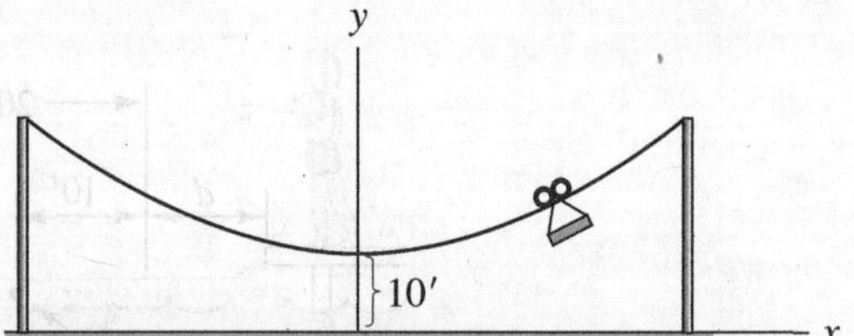

Figure P.10.92.

10.93. A rod CD rotates with shaft G–G at an angular speed ω of 300 rpm. A sleeve A of mass 500 g slides on CD. If no friction is present between A and CD, what is the distance S for no relative motion between A and CD?

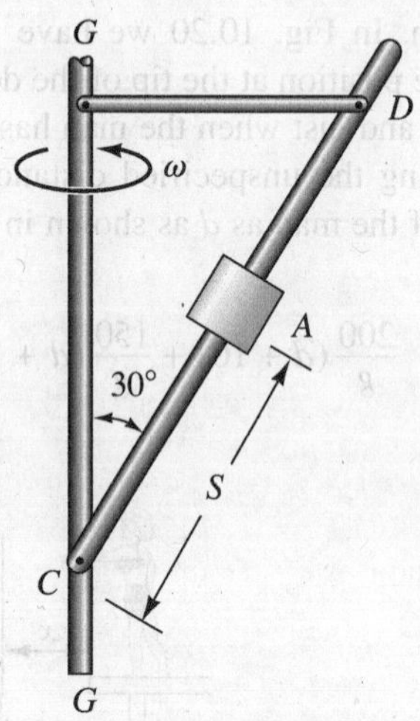

Figure P.10.95.

10.94. In Problem 10.93, what is the range of values for S for which A will remain stationary relative to CD if there is coulombic friction between A and CD such that $\mu_s = .4$?

10.97. A circular rod EB rotates at constant angular speed ω of 50 rpm. A sleeve A of mass 2 lbm slides on the circular rod. At what position θ will sleeve A remain stationary relative to the rod EB if there is no friction?

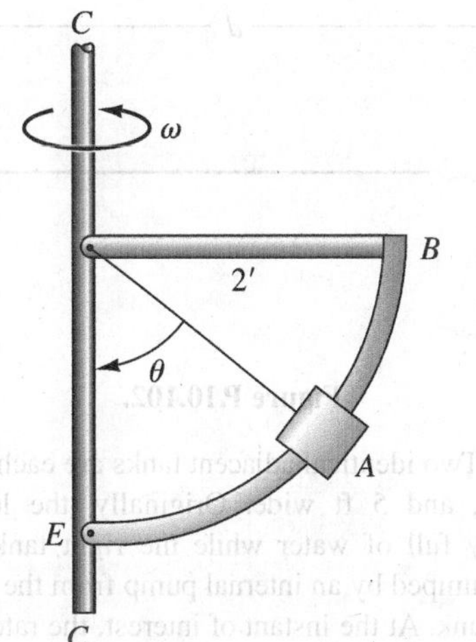

Figure P.10.95.

10.96. In Problem 10.95 assume that there is coulombic friction between A and EB with $\mu_s = .3$. Show that the minimum value of θ for which the sleeve will remain stationary relative to the rod is 75.45°.

10.97. The following data for a system of particles are given at time $t = 0$:

$$M_1 = 50 \text{ kg at position } (1, 1.3, -3) \text{ m}$$

$$M_2 = 25 \text{ kg at position } (-.6, 1.3, -2.6) \text{ m}$$

$$M_3 = 5 \text{ kg at position } (-2.6, 5.3, 1) \text{ m}$$

The particles are acted on by the following respective external forces:

$$F_1 = 50\boldsymbol{j} + 10t\boldsymbol{k} \text{ N (particle 1)}$$

$$F_2 = 50\boldsymbol{k} \text{ N} \quad \text{(particle 2)}$$

$$F_3 = 5t^2\boldsymbol{i} \text{ N} \quad \text{(particle 3)}$$

What is the velocity of M_1 relative to the mass center after 5 sec, assuming that at $t = 0$, the particles are at rest?

***10.98.** Given the following force field:

$$\boldsymbol{F} = -2x\boldsymbol{i} + 3\boldsymbol{j} - z\boldsymbol{k} \text{ lb/slug}$$

what is the force on any particle in the field per unit mass of the particle. If we have two particles initially stationary in the field with position vectors

$$\boldsymbol{r}_1 = 3\boldsymbol{i} + 2\boldsymbol{j} \text{ ft}$$

$$\boldsymbol{r}_2 = 4\boldsymbol{i} - 2\boldsymbol{j} + 4\boldsymbol{k} \text{ ft}$$

what is the velocity of each particle relative to the center of mass of the system after 2 sec have elapsed? Each particle has a weight of .1 oz.

10.99. A stationary uniform block of ice is acted on by forces that maintain constant magnitude and direction at all times. If

$$F_1 = (25g) \text{ N}$$

$$F_2 = (10g) \text{ N}$$

$$F_3 = (15g) \text{ N}$$

what is the velocity of the center of mass of the block after 10 sec? Neglect friction. The density of ice is 56 lbm/ft³.

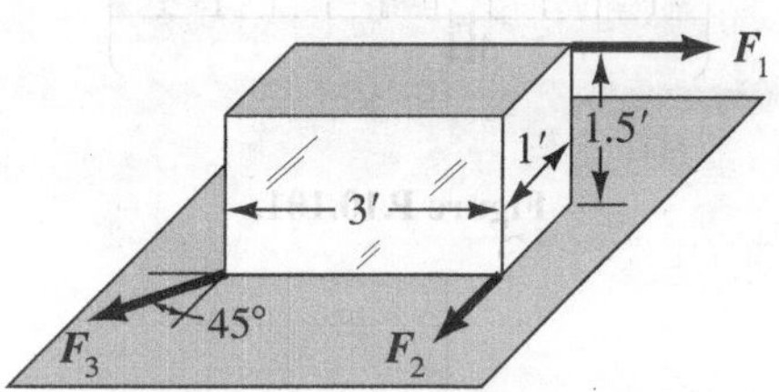

Figure P.10.99.

10.100. A space vehicle decelerates downward (Z-direction) at 1,613 km/hr/sec while moving in a translatory manner relative to inertial space. Inside the vehicle is a rod BC rotating in the plane of the paper at a rate of 50 rad/sec relative to the vehicle. Two masses rotate at the rate of 20 rad/sec around BC on rod EF. The masses are each 300 mm from C. Determine the force transmitted at C between BC and EF if the mass of each of the rotating bodies is 5 kg and the mass of rod EF is 1 kg. BC is in the vertical position at the time of interest. Neglect gravity.

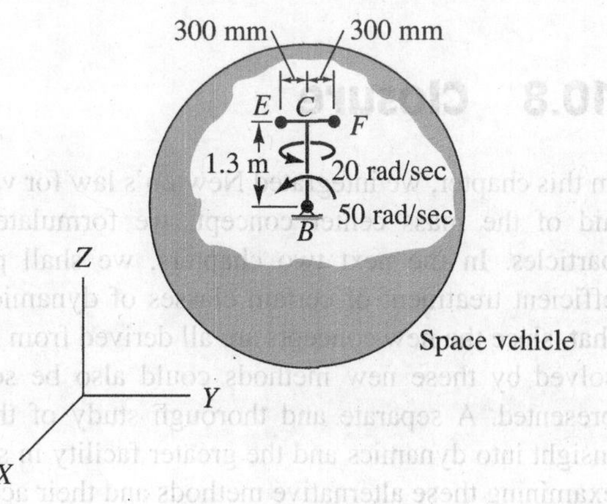

Figure P.10.100.

10.101. Two men climb aboard a barge at A to shift a load with the aid of a fork lift. The barge has a mass of 20,000 kg and is 10 m long. The load consists of four containers each with a mass of 1,300 kg and each having a length of 1 m. The men shift the containers to the opposite end of the barge, put the fork lift where they found it, and prepare to step off the barge at A, where they came on. If the barge has not been constrained and if we neglect water friction, currents, wind, and so on, how far has the barge shifted its position? The fork lift has a mass of 1,000 kg.

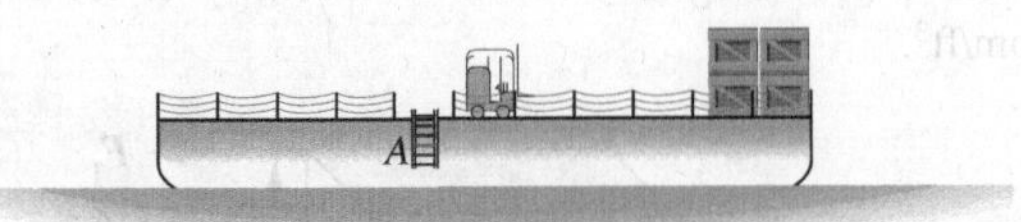

Figure P.10.101.

10.102. An astronaut on a space walk pulls a mass A of 100 kg toward him and shortens the distance d by 5 m. If the astronaut weighs 660 N on earth, how far does the mass A move from its original position? Neglect the mass of the cord.

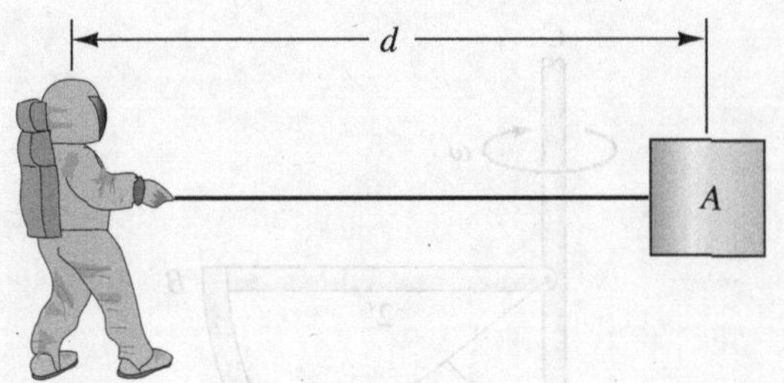

Figure P.10.102.

10.103. Two identical adjacent tanks are each 10 ft long, 5 ft high, and 5 ft wide. Originally, the left tank is completely full of water while the right tank is empty. Water is pumped by an internal pump from the left tank to the right tank. At the instant of interest, the rate of flow Q is 20 ft³/sec, while $\dot{Q}$ is 5 ft³/sec². What horizontal force on the tanks is needed at this instant from the foundation? Assume that the water surface in the tanks remains horizontal. The specific weight of water is 62.4 lb/ft³.

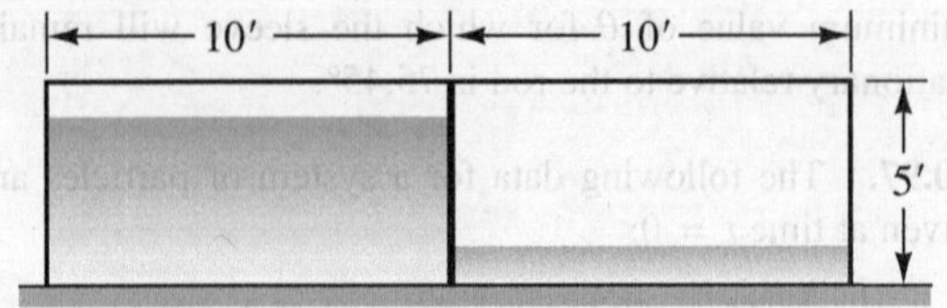

Figure P.10.103.

10.8 Closure

In this chapter, we integrated Newton's law for various coordinate systems. Also, with the aid of the mass center concept, we formulated Newton's law for any aggregate of particles. In the next two chapters, we shall present alternative procedures for more efficient treatment of certain classes of dynamics problems for particles. You will note that, since the new concepts are all derived from Newton's law, whatever problems can be solved by these new methods could also be solved by the methods we have already presented. A separate and thorough study of these topics is warranted by the gain in insight into dynamics and the greater facility in solving problems that can be achieved by examining these alternative methods and their accompanying concepts. As in this chapter, we will make certain generalizations applicable to any aggregate of particles.

PROBLEMS

10.104. A block A of mass 10 kg rests on a second block B of mass 8 kg. A force F equal to 100 N pulls block A. The coefficient of friction between A and B is .5; between B and the ground, .1. What is the speed of block A relative to block B in 0.1 sec if the system starts from rest?

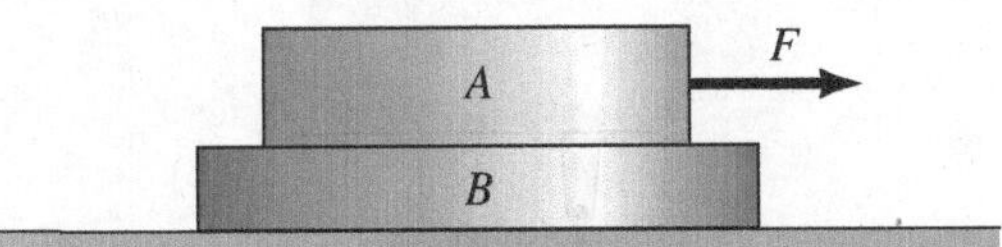

Figure P.10.104.

***10.105.** A block B slides from A to F along a rectangular chute where there is coulombic friction on the faces of the chute. The coefficient of dynamic friction is .4. The bottom face of the chute is parallel to face $EACF$ (a plane surface) and the other two faces are perpendicular to $EACF$. The body weighs 5 lb. How long does it take B to go from A to F starting from rest?

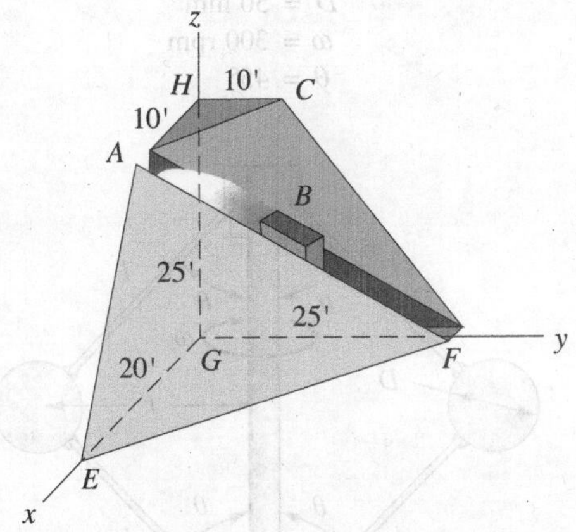

Figure P.10.105.

10.106. A tugboat is pushing a barge at a steady speed of 8 knots. The thrust from the tugboat needed for this motion is 800 lb. The barge with load weighs 100 tons. If the water resistance to the barge is proportional to the speed of the barge, how long will it take the barge to slow to 5 knots after the tugboat ceases to push? (*Note:* 1 knot equals 1.152 mi/hr.)

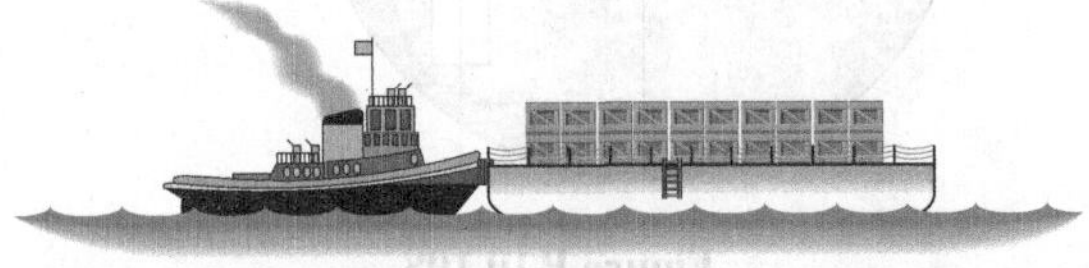

Figure P.10.106.

10.107. A spring requires a force x^2 N for a deflection of x mm, where x is the deflection of the spring from the undeformed geometry. Because the deflection is not proportional to x to the first power, the spring is called a *nonlinear* spring. If a 100-kg block is suddenly released on the undeformed spring, what is the speed of the block after it has descended 10 mm?

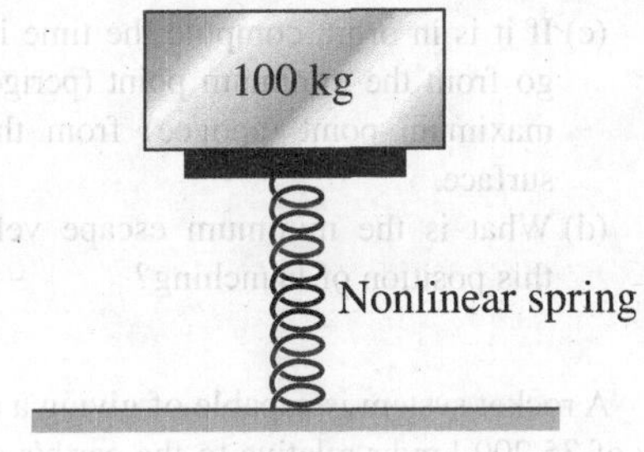

Figure P.10.107.

10.108. A horizontal platform is rotating at a constant angular speed ω of 5 rad/s. Fixed to the platform is a frictionless chute in which two identical masses each of 2 kg are constrained by a pair of linear springs each of spring constant $K = 250$ N/m. If the unstretched length l_0 of each of the springs is .18 m, show that at steady state the angle θ must have the value 36.87°. Springs are fixed to the platform at A.

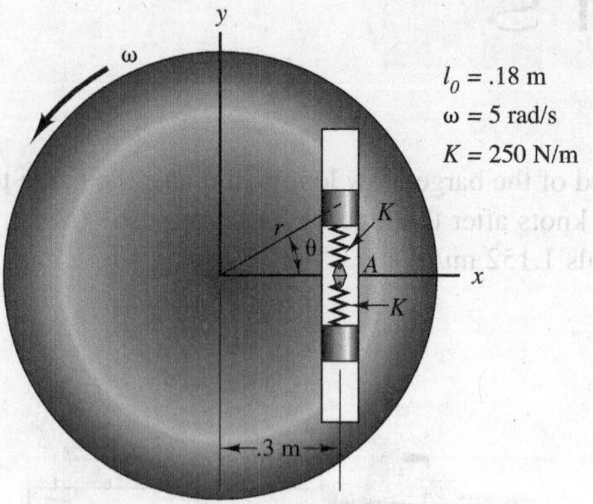

Figure P.10.108.

10.109. What is the velocity and altitude of a communications satellite that remains in the same position above the equator relative to the earth's surface?

10.110. A satellite is launched and attains a velocity of 19,000 mi/hr relative to the center of the earth at a distance of 240 mi from the earth's surface. The satellite has been guided into a path that is parallel to the earth's surface at burnout.

(a) What kind of trajectory will it have?

(b) What is its farthest position from the earth's surface?

(c) If it is in orbit, compute the time it takes to go from the minimum point (perigee) to the maximum point (apogee) from the earth's surface.

(d) What is the minimum escape velocity for this position of launching?

10.111. A rocket system is capable of giving a satellite a velocity of 35,200 km/hr relative to the earth's surface at an elevation of 320 km above the earth's surface. What would be its maximum distance h from the surface of the earth if it were launched (1) from the North Pole region or (2) from the equator, utilizing the spin of the earth as an aid?

10.112. A space vehicle is to change from a circular parking orbit 320 km above the surface of Venus to one that is 1,620 km above this surface. This motion will be accomplished by two firings of the rocket system of the vehicle. The first firing causes the vehicle to attain an apogee that is 1,620 km above the surface of Venus. At this apogee, a second firing is accomplished so as to achieve the desired circular orbit. What is the change in speed demanded for each firing if the thrust is maintained in each instance over a small portion of the trajectory of the vehicle? Neglect friction. The radius of Venus is 6,160 km, and the escape velocity at the surface is 1.026×10^4 m/sec^2.

10.113. Weights A and B are held by light pulleys. If released from rest, what is the speed of each weight after 1 sec? Weight A is 10 lb and weight B is 40 lb.

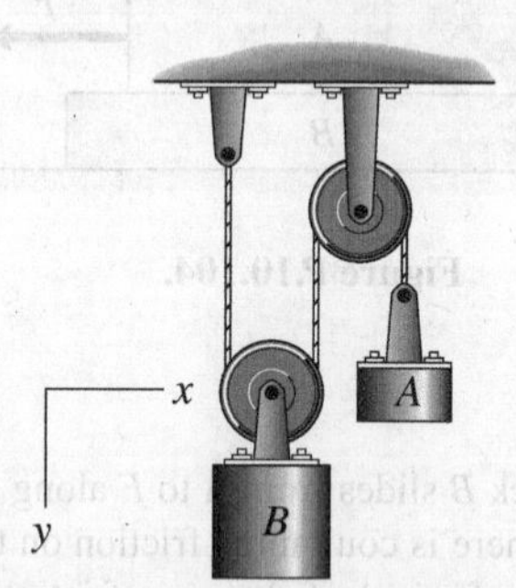

Figure P.10.115.

10.114. The following data are given for the flyball governor (read Problem 10.60 for details on how the governor works):

$$l = .275 \text{ m}$$
$$D = 50 \text{ mm}$$
$$\omega = 300 \text{ rpm}$$
$$\theta = 45°$$

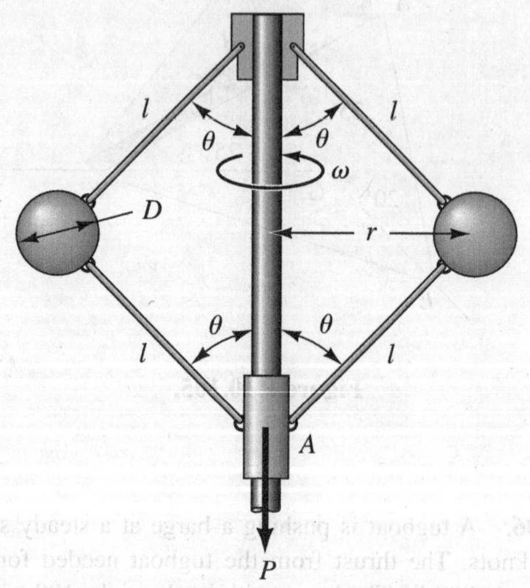

Figure P.10.114.

What is the force P acting on frictionless collar A if each ball has a mass of 1 kg and we neglect the weight of all other moving members of the system?

10.115. A spy satellite to observe the United States is put into a circular orbit about the North and South Poles. The satellite is to make 10 cycles/day (24 hr). What must be the distance from the surface of the earth for this satellite?

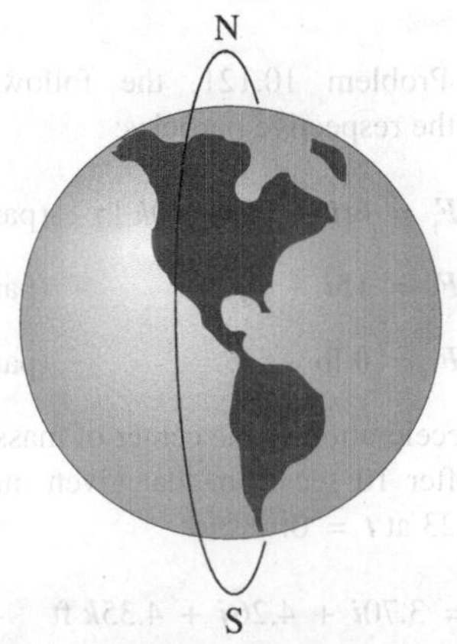

Figure P.10.115.

10.116. A skylab is in a circular orbit about the earth at a distance of 500 km above the earth's surface. A space shuttle has rendezvoused with the skylab and now, wishing to depart, decouples and fires its rockets to move more slowly than the skylab. If the rockets are fired over a short time interval, what should the relative speed between the space shuttle and skylab be at the end of rocket fire if the space shuttle is to come as close as 100 km to the earth's surface in subsequent ballistic (rocket motors off) flight?

10.117. A space vehicle is launched at a speed of 19,000 mi/hr relative to the earth's center at a position 250 mi above the earth's surface. If the vehicle has a radial velocity component of 3,000 mi/hr toward the earth's center, what is the eccentricity of the trajectory? What is the maximum elevation above the earth's surface reached by the vehicle? Do not use Eq. 10.54.

10.118. A skier is moving down a hill at a speed of 30 mi/hr when he is at the position shown. If the skier weighs 180 lb, what total force do his skis exert on the snow surface? Assume that the coefficient of friction is .1. The hill can be taken as a parabolic surface.

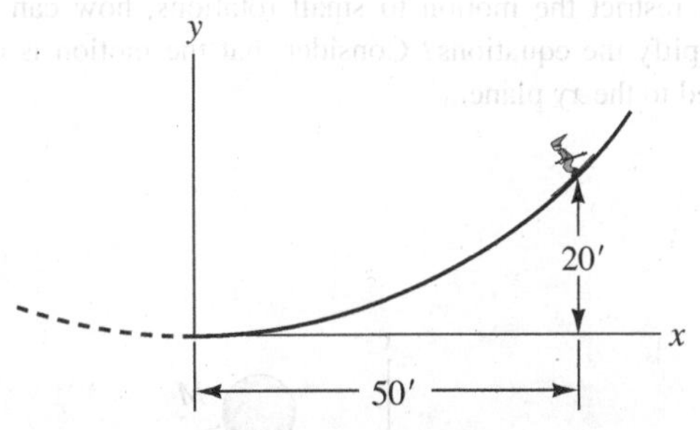

Figure P.10.118.

10.119. A submarine is moving at constant speed of 15 knots below the surface of the ocean. The sub is at the same time descending downward while remaining horizontal with an acceleration of $.023g$. In the submarine a flyball governor operates with weights having a mass each of 500g. The governor is rotating with speed ω of 5 rad/sec. If at time t, $\theta = 30°$, $\dot{\theta} = .2$ rad/sec, and $\ddot{\theta} = 1$ rad/sec^2, what is the force developed on the support of governor system as a result solely of the motion of the weights at this instant?

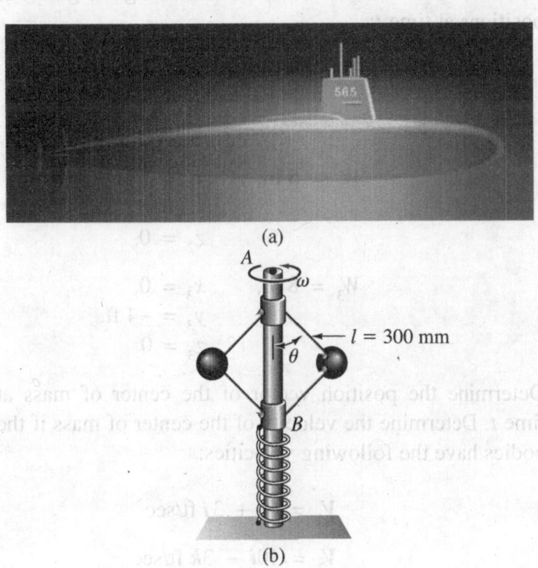

Figure P.10.119.

10.120. A mass spring system is shown. Write two simultaneous differential equations describing the motion of the mass. The spring has an unstretched length r_0. Consider that the spring does not bend and only changes length. Neglect all masses except that of the particle. If you restrict the motion to small rotations, how can you simplify the equations? Consider that the motion is confined to the xy plane.

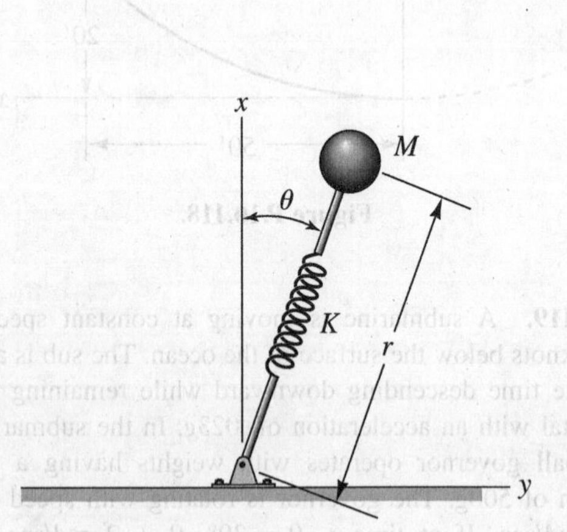

Figure P.10.120.

10.121. Three bodies have the following weights and positions at time t:

$$W_1 = 10\text{ lb}, \quad x_1 = 6\text{ft}, \quad y_1 = 10\text{ ft}, \quad z_1 = 10\text{ ft}$$

$$W_2 = 5\text{ lb}, \quad x_2 = 5\text{ ft}, \quad y_2 = 6\text{ ft}, \quad z_2 = 0$$

$$W_3 = 8\text{ lb}, \quad x_3 = 0, \quad y_3 = -4\text{ ft}, \quad z_3 = 0$$

Determine the position vector of the center of mass at time t. Determine the velocity of the center of mass if the bodies have the following velocities:

$$V_1 = 6\boldsymbol{i} + 3\boldsymbol{j}\text{ ft/sec}$$

$$V_2 = 10\boldsymbol{i} - 3\boldsymbol{k}\text{ ft/sec}$$

$$V_3 = 6\boldsymbol{k}\text{ ft/sec}$$

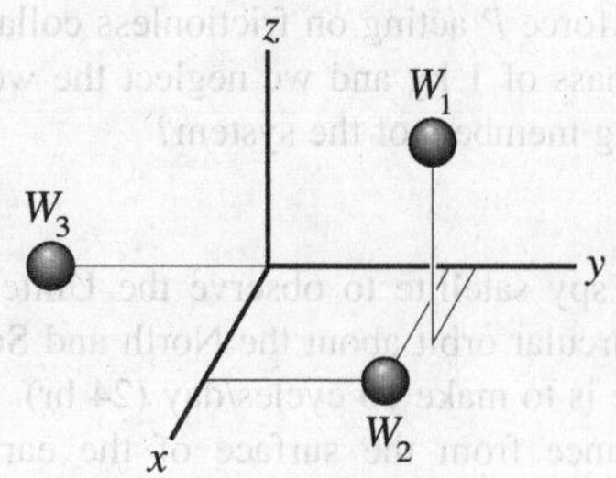

Figure P.10.121.

10.122. In Problem 10.121, the following external forces act on the respective particles:

$$F_1 = 6t\boldsymbol{i} + 3\boldsymbol{j} - 10\boldsymbol{k}\text{ lb} \quad \text{(particle 1)}$$

$$F_2 = 15\boldsymbol{i} - 3\boldsymbol{j}\text{ lb} \quad \text{(particle 2)}$$

$$F_3 = \boldsymbol{0}\text{ lb} \quad \text{(particle 3)}$$

What is the acceleration of the center of mass, and what is its position after 10 sec from that given initially? From Problem 10.123 at $t = 0$:

$$r_C = 3.70\boldsymbol{i} + 4.26\boldsymbol{j} + 4.35\boldsymbol{k}\text{ ft}$$

$$V_C = 4.78\boldsymbol{i} + 1.304\boldsymbol{j} + 1.435\boldsymbol{k}\text{ ft/sec}$$

10.123. A small body M of mass 1 kg slides along a wire from A to B. There is coulombic friction between the mass M and the wire. The dynamic coefficient of friction is .4. How long does it take to go from A to B?

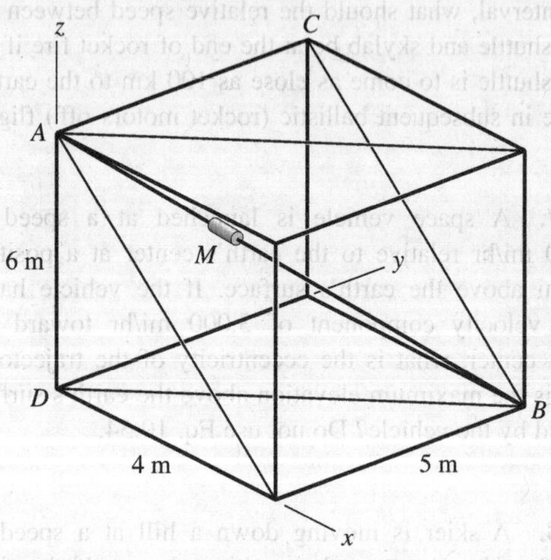

Figure P.10.123.

10.124. For $M = 1$ slug and $K = 10$ lb/in., what is the speed at $x = 1$ in. if a force of 5 lb in the x direction is applied suddenly to the mass–spring system and then maintained constant? Neglect the mass of the spring and friction.

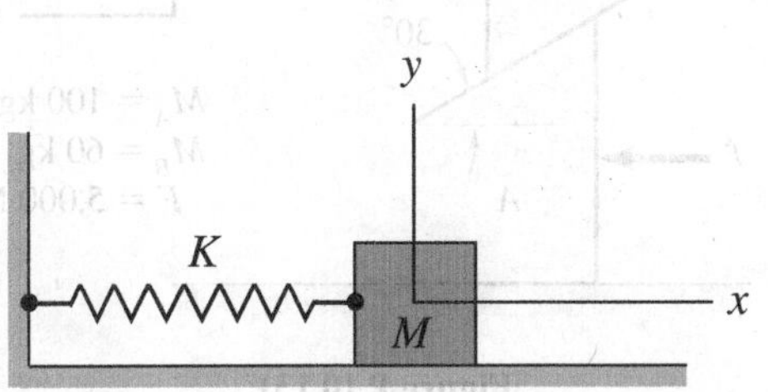

Figure P.10.124.

10.125. A rod B of mass 500 kg rests on a block A of mass 50 kg. A force F of 10,000 N is applied suddenly to block A at the position shown. If the coefficient of friction μ_d is .4 for all contact surfaces, what is the speed of A when it has moved 3 m to the end of the rod?

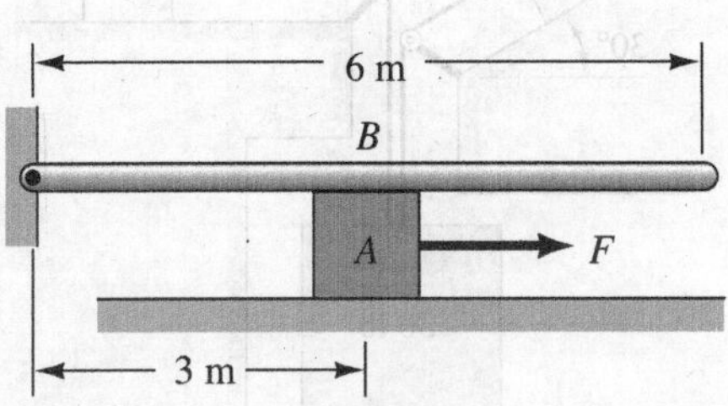

Figure P.10.125.

10.126. A simply supported beam is shown. You will learn in your course on strength of materials that a vertical force F applied at the center causes a deflection δ at the center given as

$$\delta = \frac{1}{48}\frac{FL^3}{EI}$$

If a mass of 200 lbm, fastened to the beam at its midpoint, is suddenly released, what will its speed be when the deflection is $\frac{1}{8}$ in.? Neglect the mass of the beam. The length of the beam, L, is 20 ft. Young's modulus E is 30×10^6 psi, and the moment of inertia of the cross section I is 20 in.^{4}.

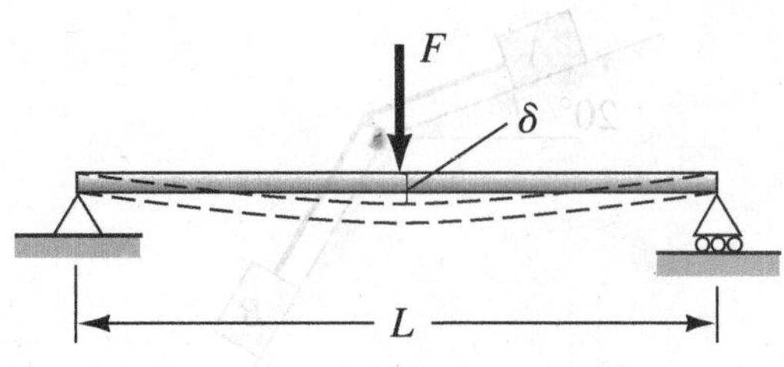

Figure P.10.126.

10.127. A piston is shown maintaining air at a pressure of 8 psi above that of the atmosphere. If the piston is allowed to accelerate to the left, what is the speed of the piston after it moves 3 in.? The piston assembly has a mass of 3 lbm. Assume that the air expands *adiabatically* (i.e., with no heat transfer). This means that at all times pV^k = constant, where V is the volume of the gas and k is a constant which for air equals 1.4. Neglect the inertial effects of the air.

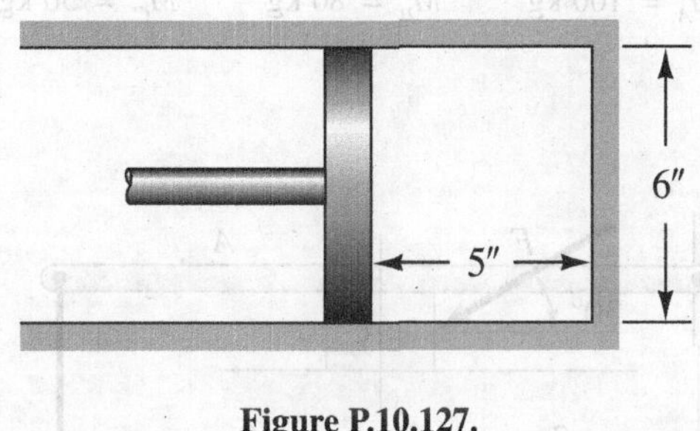

Figure P.10.127.

***10.128.** In Example 10.6 assume that there are adiabatic expansions and compressions of the gases (i.e., that pv^k = constant with $k = 1.4$). Compare the results for the speed of the piston. Explain why your result should be higher or lower than for the isothermal case.

10.129. Body A and body B are connected by an inextensible cord as shown. If both bodies are released simultaneously, what distance do they move in $\frac{1}{2}$ sec? Take $M_A = 25$ kg and $M_B = 35$ kg. The coefficient of friction μ_d is .3.

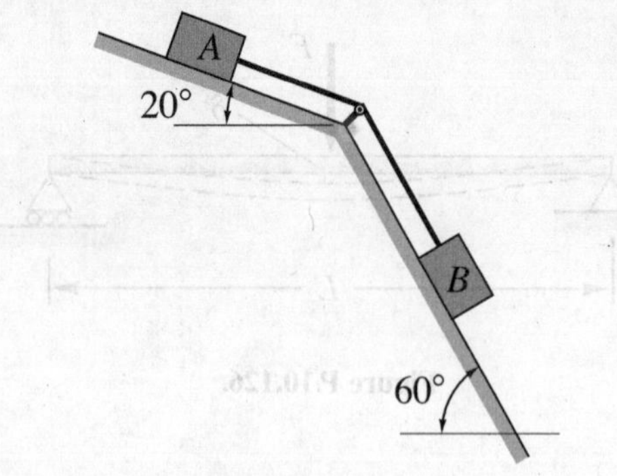

Figure P.10.129.

10.130. A force F of 2 kN is exerted on body C. If μ_d for all surface contacts is .2, what is the speed of C after it moves 1 m? The body C is initially stationary at the position shown, when the force F is applied. Solve using Newton's law. The following are the masses of the three bodies involved.

$$M_A = 100 \text{ kg} \qquad M_B = 80 \text{ kg} \qquad M_C = 50 \text{ kg}$$

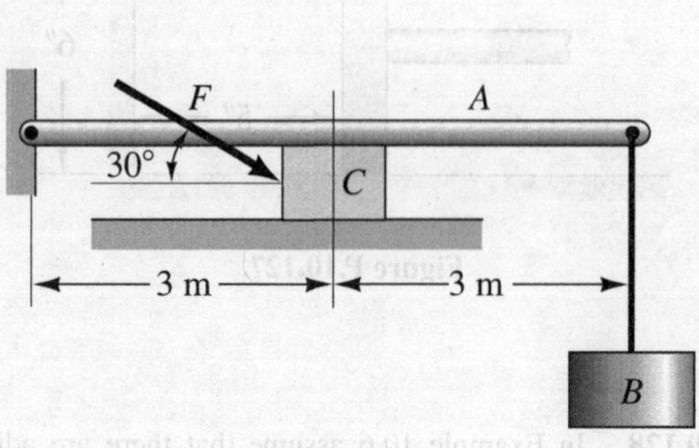

Figure P.10.130.

10.131. A constant force F of 5,000 N acts on block A. If we do not have friction anywhere, what is the acceleration of block A?

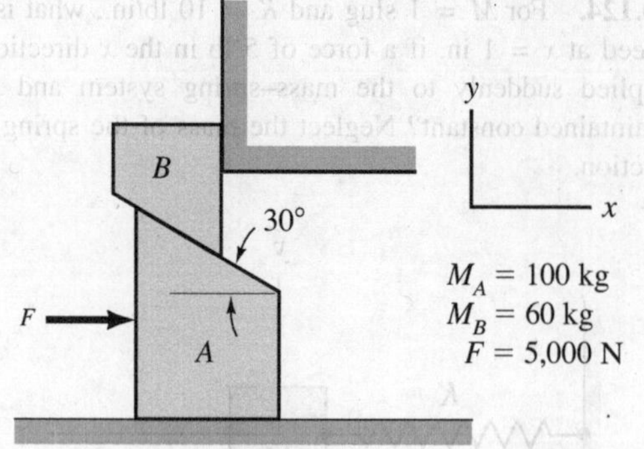

Figure P.10.131.

10.132. The system shown is released from rest. What distance does the body C drop in 2 sec? The cable is inextensible. The coefficient of dynamic friction μ_d is .4 for contact surfaces of bodies A and B.

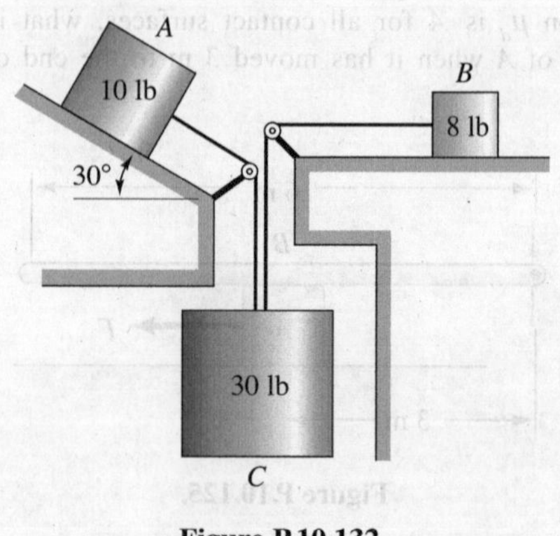

Figure P.10.132.

10.133. Do Problem 10.134 for the case where there is viscous damping for the contact surfaces of bodies A and B given as $.5V$ lb, with V in ft/sec.

10.134. Two bodies A and B are shown having masses of 40 kg and 30 kg, respectively. The cables are inextensible. Neglecting the inertia of the cable and pulleys at C and D, what is the speed of the block B 1 sec after the system has been released from rest? The dynamic coefficient of friction μ_d for the contact surface of body A is .3. [*Hint:* From your earlier work in physics, recall that pulley D is instantaneously rotating about point a and hence point c moves at a speed that is twice that of point b.]

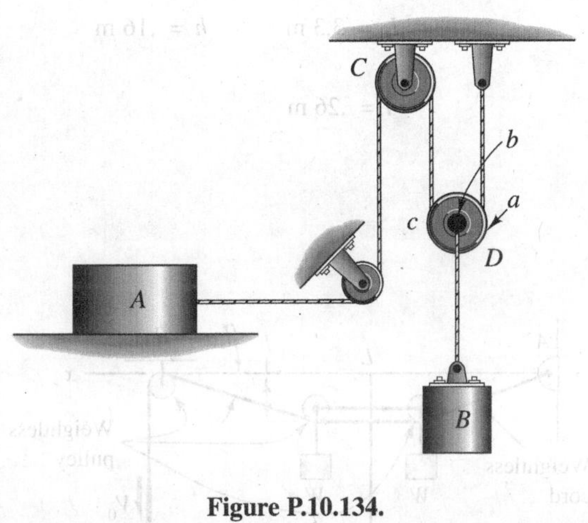

Figure P.10.134.

10.135. Bodies A, B, and C have weights, of 100 lb, 200 lb, and 150 lb, respectively. If released from rest, what are the respective speeds of the bodies after 1 sec? Neglect the weight of pulleys.

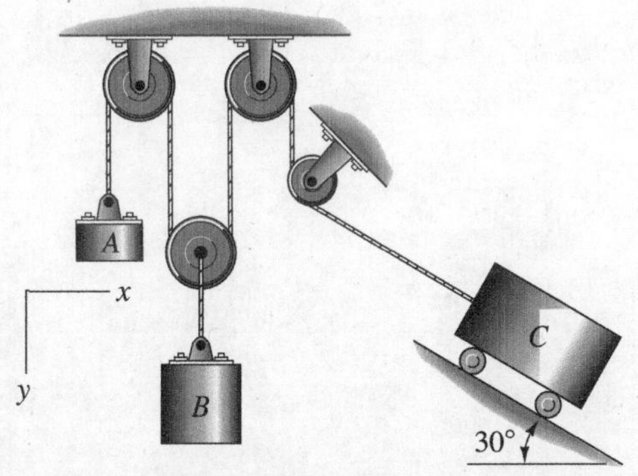

Figure P.10.135.

10.136. A car is moving at a constant speed of 65 km/hr on a road part of which ($A \rightarrow B$) is parabolic and part of which ($C \rightarrow D$) is circular with a radius of 3 km. If the car has an anti-lock braking system and the static coefficient of friction μ_s between the road and the tires is 0.6, what is the maximum deceleration possible at the $x = 2$ km position and at the $x = 10$ km position? The total vehicle weight is 12,000 N.

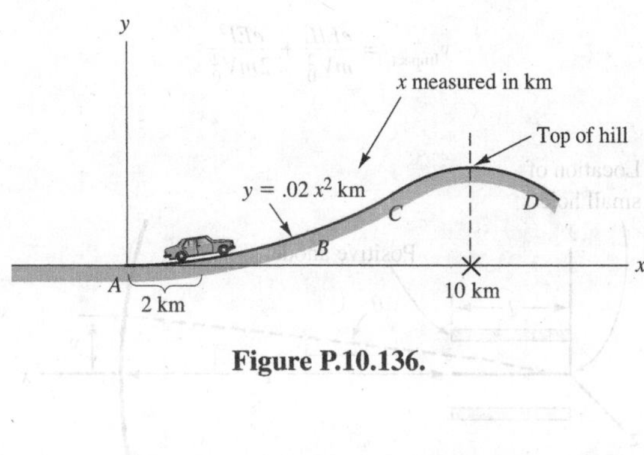

Figure P.10.136.

10.137. A mass of 3 kg is moving along a vertically oriented parabolic rod whose equation is $y = 3.4x^2$. A linear spring with $K = 550$ N/m connects to the mass and is unstretched when the mass is at the bottom of the rod having an unstretched length $l_0 = 1$ m. When the spring centerline is 30° from the vertical, as shown in the diagram, the mass is moving at 2.8 m/s. At this instant, what is the force component on the rod directed normal to the rod?

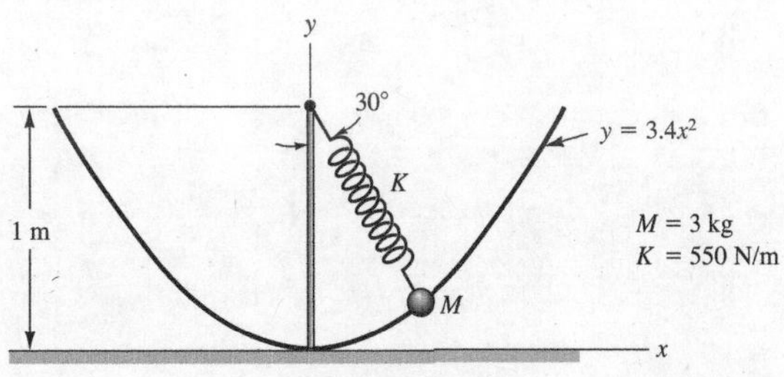

Figure P.10.137.

10.138. A heated cathode gives off electrons which are attracted to the positive anode. Some go through a small hole and enter the parallel plates at an angle with the horizontal of $\alpha_0 = 0$ and a velocity of V_0. Determine the horizontal and vertical motions of the electron inside the plates as a function of time. Letting $x = l$, find the time that the electron is in the parallel plate region and then obtain the exit vertical velocity. Assuming straight-line motion until the electron hits the screen, show that the vertical position of impact, assuming the screen is flat, is

$$y_{\text{Impact}} = \frac{eElL}{mV_0^2} + \frac{eEl^2}{2mV_0^2}$$

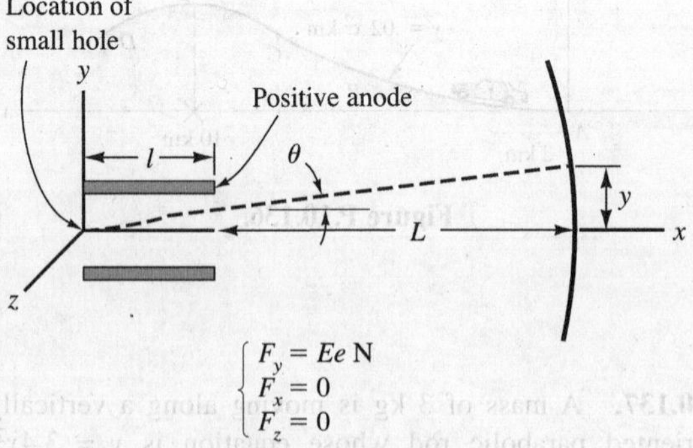

$$\begin{cases} F_y = Ee \text{ N} \\ F_x = 0 \\ F_z = 0 \end{cases}$$

Figure P.10.138.

***10.139.** A weightless cord supports two identical masses each of weight W. The cord is being pulled at a constant speed V_0 by a force F. Formulate an equation for F in terms of V_0, L, l, W, and h. Determine F for the following conditions:

$V_0 = 2.2$ m/s $\quad W = 40$ N

$L = 3.3$ m $\quad h = .16$ m

$l = .26$ m

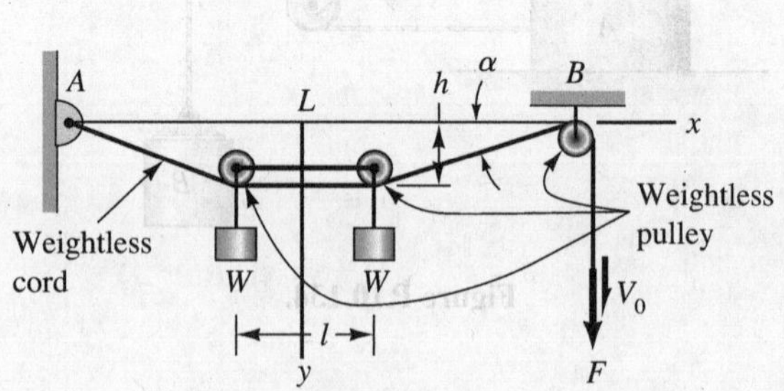

Figure P.10.139.

CHAPTER 11

Energy Methods for Particles

Part A: Analysis for a Single Particle

11.1 Introduction

In Chapter 10, we integrated the differential equation derived from Newton's law to yield velocity and position as functions of time. At this time, we shall present an alternative procedure, that of the method of energy, and we shall see that certain classes of problems can be more easily handled by this method in that we shall not need to integrate a differential equation.

To set forth the basic equation underlying this approach, we start with *Newton's law* for a particle moving relative to an inertial reference, as shown in Fig. 11.1. Thus,

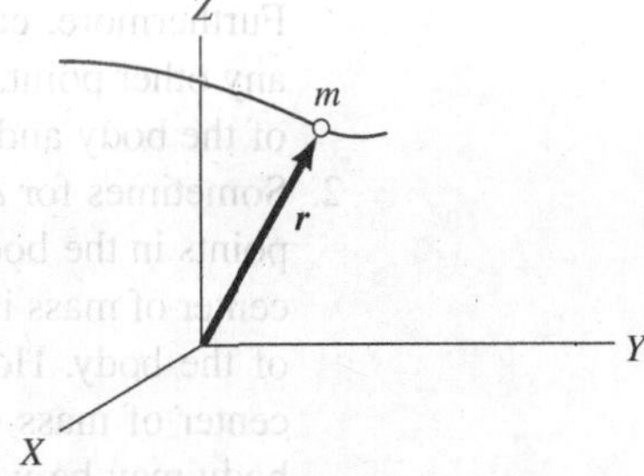

Figure 11.1. Particle moving relative to an inertial reference.

$$F = m\frac{d^2r}{dt^2} = m\frac{dV}{dt} \tag{11.1}$$

Multiply each side of this equation by dr as a dot product and integrate from r_1 to r_2 along the path of motion:

$$\int_{r_1}^{r_2} F \bullet dr = m\int_{r_1}^{r_2} \frac{dV}{dt} \bullet dr = m\int_{t_1}^{t_2} \frac{dV}{dt} \bullet \frac{dr}{dt}\,dt$$

In the last integral, we multiplied and divided by dt, thus changing the variable of integration to t. Since $dr/dt = V$, we then have

$$\begin{aligned}\int_{r_1}^{r_2} \mathbf{F} \bullet d\mathbf{r} &= m\int_{t_1}^{t_2}\left(\frac{d\mathbf{V}}{dt} \bullet \mathbf{V}\right)dt = \frac{1}{2}\,m\int_{t_1}^{t_2} \frac{d}{dt}(\mathbf{V} \bullet \mathbf{V})\,dt \\ &= \frac{1}{2}\,m\int_{t_1}^{t_2} \frac{d}{dt}V^2\,dt = \frac{1}{2}\,m\int_{V_1}^{V_2} d(V^2)\end{aligned}$$

On carrying out the integration, we arrive at the familiar equation

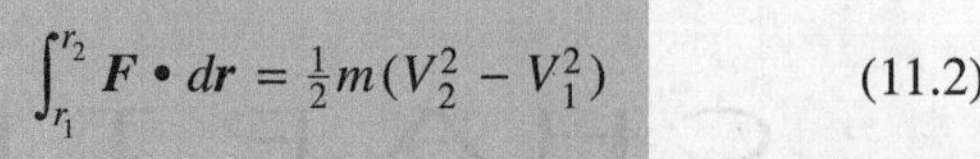

$$\int_{r_1}^{r_2} \boldsymbol{F} \bullet d\boldsymbol{r} = \frac{1}{2}m(V_2^2 - V_1^2) \tag{11.2}$$

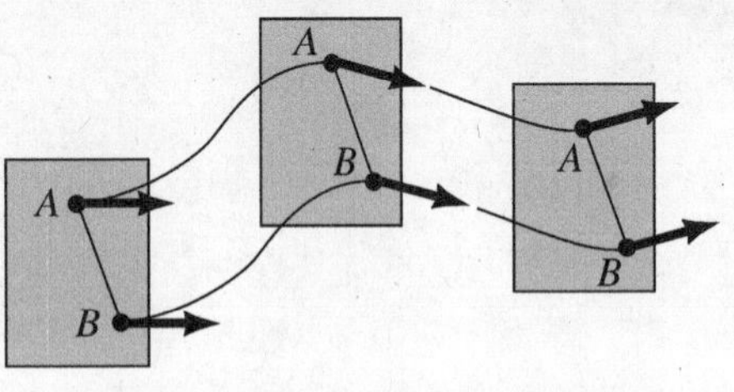

Figure 11.2. Translating body.

where the left side is the well-known expression for *work* (to be denoted at times as $\mathcal{W}_{1-2}$)[1] and the right side is clearly the change in *kinetic energy* as the mass moves from position $\boldsymbol{r}_1$ to position $\boldsymbol{r}_2$.

We shall see in Section 11.6 that for any system of particles, including, of course, rigid bodies, we get a work–energy equation of the form 11.2, where the velocity is that of the mass center, the force is the resultant external force on the system, and the path of integration is that of the mass center. Clearly, then, we can use a single particle model (and consequently Eq. 11.2) for:

1. *A rigid body moving without rotation*. Such a motion was discussed in Chapter 9 and is called translation. Note that lines in a translating body remain parallel to their original directions, and points in the body move over a path which has identically the same form for all points. This condition is illustrated in Fig. 11.2 for two points *A* and *B*. Furthermore, each point in the body has at any instant of time *t* the same velocity as any other point. Clearly the motion of the center of mass fully characterizes the motion of the body and Eq. 11.2 will be used often for this situation.
2. Sometimes for *a body whose size is small compared to its trajectory*. Here the paths of points in the body differ very little from that of the mass center and knowing where the center of mass is tells us with sufficient accuracy all we need to know about the position of the body. However, keep in mind that the *velocity* and *acceleration* relative to the center of mass of a part of the body may be *very large*, irrespective of how small the body may be when compared to the trajectory of its center of mass. Then, information about the velocity and acceleration of this part of the body relative to the center of mass would require a more detailed consideration beyond a simple one-particle model centered around the center of mass.

Thus, as in our considerations of Newton's law in Chapter 10, when the motion of the mass center characterizes with sufficient accuracy what we want to know about the motion of a body, we use a particle at the mass center for energy considerations.

Next, suppose that we have a component of Newton's law in one direction, say the *x* direction:

$$F_x \boldsymbol{i} = m\frac{dV_x}{dt}\boldsymbol{i} \tag{11.3}$$

[1]It is important to note that the work done by a force system depends on the path over which the forces move, except in the case of conservative forces to be considered in Section 11.3. Thus, W_{1-2} is called a *path function* in thermodynamics. However, kinetic energy depends only on the instantaneous state of motion of the particle and is independent of the path. Kinetic energy is called, accordingly, a *point function* in thermodynamics.

Taking the dot product of each side of this equation with $d x\mathbf{i} + d y\mathbf{j} + d z\mathbf{k}$ (= $d\mathbf{r}$), we get, after integrating in the manner set forth at the outset:

$$\int_{x_1}^{x_2} F_x\, dx = \frac{m}{2}\left[(V_x)_2^2 - V_x)_1^2)\right] \tag{11.3a}$$

Similarly,

$$\int_{y_1}^{y_2} F_y\, dy = \frac{m}{2}\left[(V_y)_2^2 - V_y)_1^2)\right] \tag{11.3b}$$

$$\int_{z_1}^{z_2} F_z\, dz = \frac{m}{2}\left[(V_z)_2^2 - V_z)_1^2)\right] \tag{11.3c}$$

Thus, the foregoing equations demonstrate that the work done on a particle in any direction equals the change in kinetic energy associated with the component of velocity in that direction.

Instead of employing Newton's law, we can now use the energy equations developed in this section for solving certain classes of problems. This energy approach is particularly handy when velocities are desired and forces are functions of position. However, please understand that any problem solvable with the energy equation can be solved from Newton's law; the choice between the two is mainly a question of convenience and the manner in which the information is given.

Example 11.1

An automobile is moving at 60 mi/hr (see Fig. 11.3) when the driver jams on his brakes and goes into a skid in the direction of motion. The car weighs 4,000 lb, and the dynamic coefficient of friction between the rubber tires and the concrete road is .60. How far, l, will the car move before stopping?

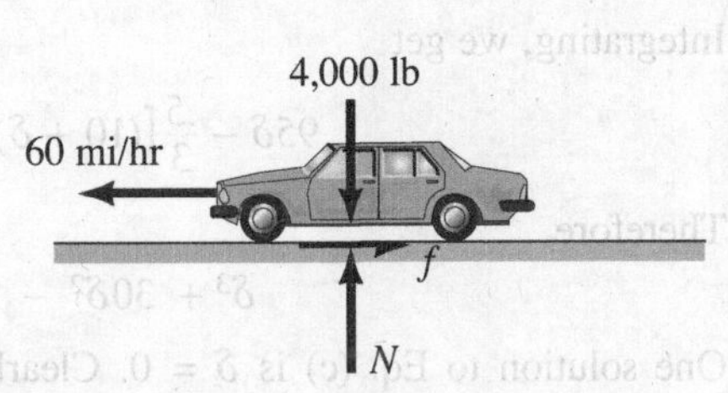

Figure 11.3. Car moving with brakes locked.

A constant friction force acts, which from Coulomb's law is $\mu_d N$ = (.60)(4,000) = 2,400 lb. This force is the only force performing work, and clearly it is changing the kinetic energy of the vehicle from that corresponding to the speed of 60 mi/hr (or 88 ft/sec) to zero. (You will learn in thermodynamics that this work facilitates a transfer of kinetic energy of the vehicle to an increase of internal energy of the vehicle, the road, and the air, as well as the wear of brake parts) From the **work–energy equation** 11.2, we get[2]

$$-2{,}400l = \frac{1}{2}\frac{4{,}000}{g}(0 - 88^2)$$

Hence,

$$l = 200 \text{ ft}$$

(Perhaps every driver should solve this problem periodically.)

[2]Note that the sign of the work done is negative since the friction force is opposite in sense to the motion.

Example 11.2

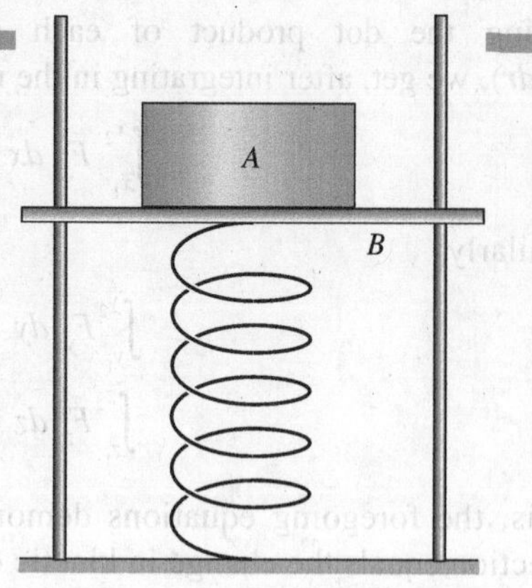

Figure 11.4. Preloaded nonlinear spring.

Shown in Fig. 11.4 is a light platform B guided by vertical rods. The platform is positioned so that the spring has been compressed 10 mm. In this configuration a body A weighing 100 N is placed on the platform and released suddenly. If the guide rods give a total constant resistance force f to downward movement of the platform of 5 N, what is the largest distance that the weight falls? The spring used here is a *nonlinear* spring requiring $.5x^2$ N of force for a deflection of x mm.

We take as the position of interest for the body the location δ below the initial configuration at which location the body A reaches zero velocity for the first time after having been released. The change in kinetic energy over the interval is accordingly zero. Thus, zero net work is done by the forces acting on the body A during displacement δ. These forces comprise the force of gravity, the friction force from the guides, and finally the force from the spring. Using as the origin for our measurements the *undeformed* top end position of the spring,[3] we can say:

$$\int_{10}^{(10+\delta)} \boldsymbol{F} \bullet d\boldsymbol{r} = \int_{10}^{(10+\delta)} \left(W_A - f - .5x^2\right) dx$$

$$= \int_{10}^{(10+\delta)} \left(100 - 5 - .5x^2\right) dx = 0 \qquad \text{(a)}$$

Integrating, we get

$$95\delta - \frac{.5}{3}\left[(10+\delta)^3 - 10^3\right] = 0 \qquad \text{(b)}$$

Therefore,

$$\delta^3 + 30\delta^2 - 270\delta = 0 \qquad \text{(c)}$$

One solution to Eq. (c) is $\delta = 0$. Clearly, no work is done if there is no deflection. But this solution has no meaning for this problem since the force in the spring is only $.5x^2 = .5(10)^2 = 50$ N, when the weight of 100 N is released. Therefore, there must be a nonzero positive value of δ that satisfies the equation and has physical meaning. Factoring out one δ from the equation, we then set the resulting quadratic expression equal to zero. Two roots result and the positive root $\delta = 7.25$ mm is the one with physical meaning.

$$\delta = 7.25 \text{ mm}$$

[3]Since the force in the spring is a function of the elongation of the spring from its *undeformed* geometry, we must put the origin of our reference at a position corresponding to the undeformed geometry. At this position, both x and the spring force are zero simultaneously.

In the following example we deal with two bodies which can be considered as particles, rather than with one body as has been the case in the previous examples. We shall deal with these bodies separately in this example. Later in the chapter, we shall consider *systems* of particles, and in that context we will be able to consider this problem as a system of particles with less work needed to reach a solution.

Example 11.3

In Fig. 11.5, we have shown bodies A and B interconnected through a block and pulley system. Body B has a mass of 100 kg, whereas body A has a mass of 900 kg. Initially the system is stationary with B held at rest. What speed will B have when it reaches the ground at a distance h = 3 m below after being released? What will be the corresponding speed of A? Neglect the masses of the pulleys and the rope. Consider the rope to be inextensible.

You will note from Fig. 11.5 that, as the bodies move, only the distances l_B and l_A change; the other distances involving the ropes do not change. And because the rope is taken as inextensible, we conclude that at all times

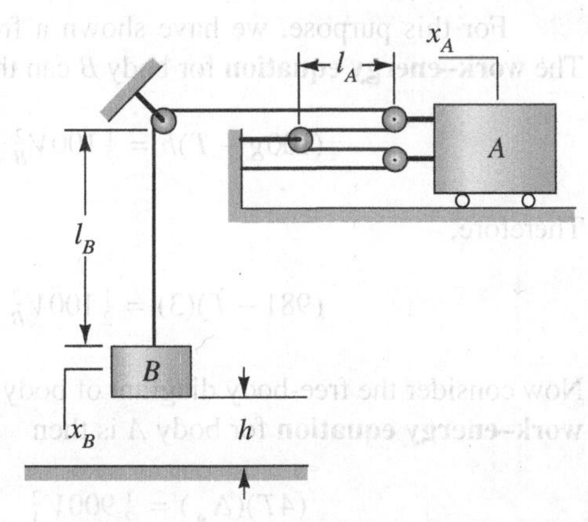

Figure 11.5. System of blocks and pulleys.

$$l_B + 4l_A = \text{constant} \tag{a}$$

Differentiating with respect to time, we can find that

$$\dot{l}_B + 4\dot{l}_A = 0$$

Therefore,

$$\dot{l}_B = -4\dot{l}_A \tag{b}$$

On inspecting Fig. 11.5, you should have no trouble in concluding that $\dot{l}_A = -V_A$ and that $\dot{l}_B = V_B$. Hence, from Eq. (b), we can conclude that

$$V_B = 4V_A \tag{c}$$

Next take the differential of Eq. (a):

$$dl_B + 4dl_A = 0$$

Therefore,

$$dl_B = -4dl_A \tag{d}$$

Note that $dx_B = dl_B$ and that $dx_A = -dl_A$. Hence we see from Eq. (d) that a movement magnitude Δ_A of body A results in a movement magnitude, $4\Delta_A$, of body B:

$$\Delta_B = 4\Delta_A \tag{e}$$

With these kinematical conclusions as a background, we are now ready to proceed with the work–energy considerations.

Example 11.3 (Continued)

For this purpose, we have shown a free-body diagram of body B in Fig. 11.6. The **work–energy equation** for body B can then be given as follows:

$$(100g - T)h = \tfrac{1}{2}100V_B^2$$

Therefore,

$$(981 - T)(3) = \tfrac{1}{2}100V_B^2 \qquad \text{(f)}$$

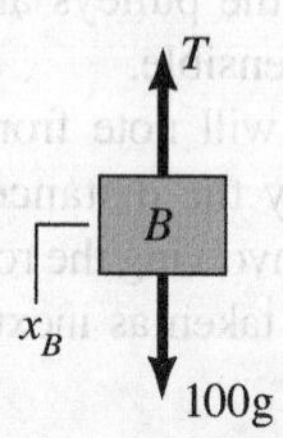

Figure 11.6. Free-body diagram of B.

Now consider the free-body diagram of body A in Fig. 11.7. The **work–energy equation** for body A is then

$$(4T)(\Delta_A) = \tfrac{1}{2}900V_A^2 \qquad \text{(g)}$$

But according to Eq. (e),

$$\Delta_A = \frac{\Delta_B}{4} = \frac{h}{4} = \frac{3}{4}\text{ m} \qquad \text{(h)}$$

And according to Eq. (c),

$$V_A = \tfrac{1}{4}V_B \qquad \text{(i)}$$

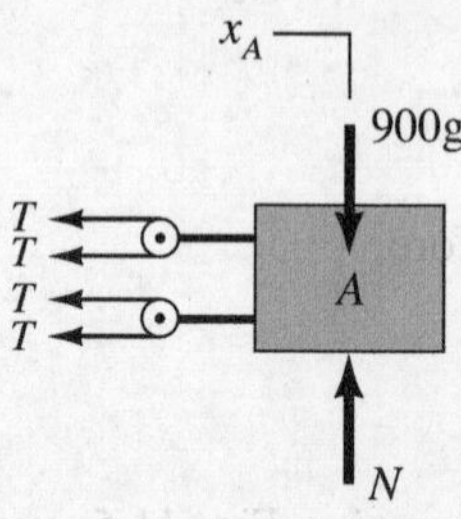

Figure 11.7. Free-body diagram of A.

Substituting the results from Eqs. (h) and (i) into (g), we get

$$(4T)\left(\frac{3}{4}\right) = \frac{1}{2}900\left(\frac{V_B^2}{16}\right) \qquad \text{(j)}$$

Adding Eqs.(f) and (j), we can eliminate T to form the following equation with V_B as the only unknown:

$$(981)(3) = \tfrac{1}{2}(V_B^2)(100 + \tfrac{900}{16})$$

Therefore,

$$V_B = 6.14\text{ m/sec downward}$$

Hence,

$$V_A = 1.534\text{ m/sec to the left}$$

PROBLEMS

11.1. What value of constant force P is required to bring the 100-lb body, which starts from rest, to a velocity of 30 ft/sec in 20 ft? Neglect friction.

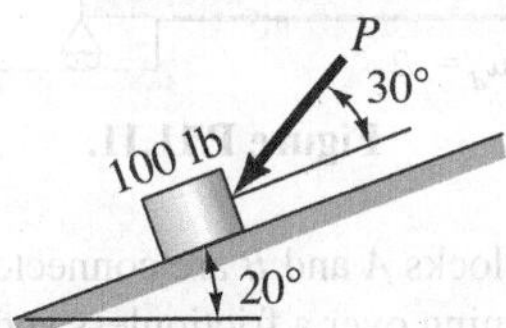

Figure P.11.1.

11.2. A light cable passes over a frictionless pulley. Determine the velocity of the 100-lb block after it has moved 30 ft from rest. Neglect the inertia of the pulley. The dynamic coefficient of friction between block and incline is 0.2.

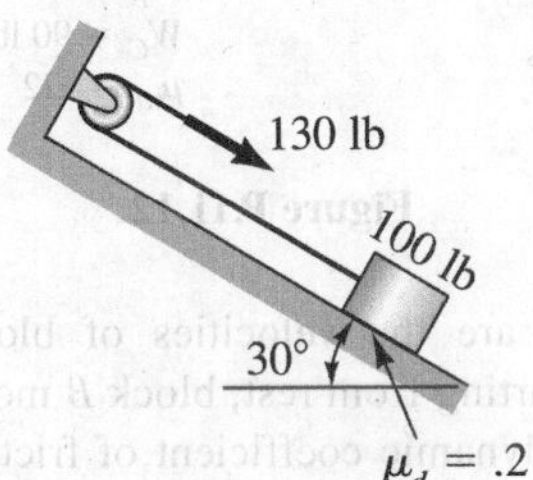

Figure P.11.2.

11.3. In Problem 11.2, the pulley has a radius of 1 ft and has a resisting torque at the bearing of 10 lb-ft. Neglect the inertia of the pulley and the mass of the cable. Compute the kinetic energy of the 100-lb block after it has moved 30 ft from rest.

11.4. A light cable is wrapped around two drums fixed between a pair of blocks. The system has a mass of 50 kg. If a 250-N tension is exerted on the free end of the cable, what is the velocity change of the system after 3 m of travel down the incline? The body starts from rest. Take μ_d for all surfaces as .05.

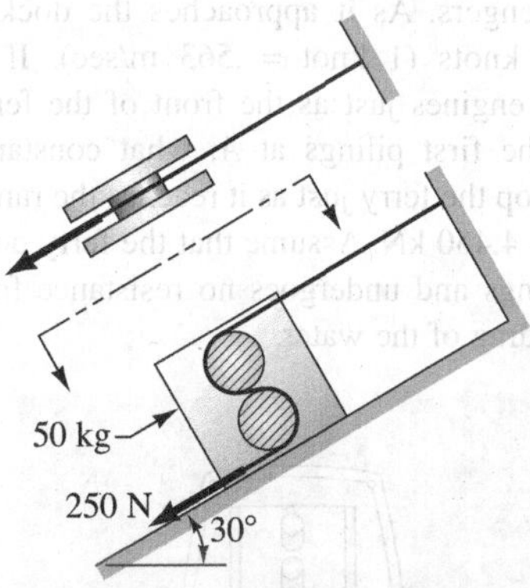

Figure P.11.4.

11.5. A 50-kg mass on a spring is moved so that it extends the spring 50 mm from its unextended position. If the dynamic coefficient of friction between the mass and the supporting surface is .3,

(a) What is the velocity of the mass as it returns to the undeformed configuration of the spring?

(b) How far will the spring be compressed when the mass stops instantaneously before starting to the left?

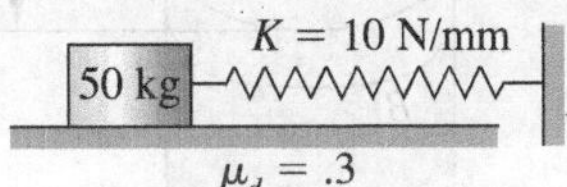

Figure P.11.5.

11.6. A truck–trailer is shown carrying three crushed junk automobile cubes each weighing 2,500 lb. An electromagnet is used to pick up the cubes as the truck moves by. Suppose the truck starts at position 1 by applying a constant 600 in.-lb total torque on the drive wheels. The magnet picks up only one cube C during the process. What will the velocity of the truck be when it has moved a total of 100 ft? The truck unloaded weighs 5,000 lb and has a tire diameter of 18 in. Neglect the rotational effects of the tires and wind friction.

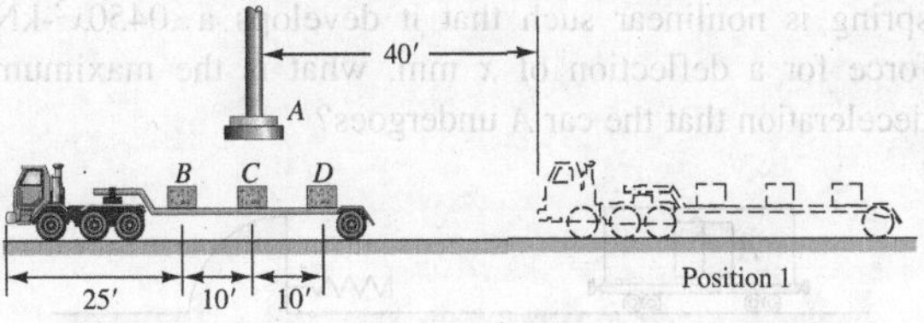

Figure P.11.6.

11.7. Do Problem 11.6 if the first cube B and the last cube D are removed as they go by the magnet.

11.8. A passenger ferry is shown moving into its dock to unload passengers. As it approaches the dock, it has a speed of 3 knots (1 knot = .563 m/sec). If the pilot reverses his engines just as the front of the ferry comes abreast of the first pilings at A, what constant reverse thrust will stop the ferry just as it reaches the ramp B? The ferry weighs 4,450 kN. Assume that the ferry does not hit the side pilings and undergoes no resistance from them. Neglect the drag of the water.

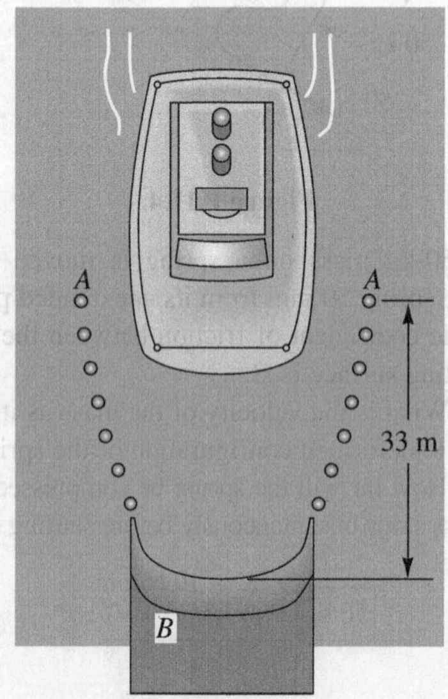

Figure P.11.8.

11.9. Do Problem 11.8 assuming that the ferry rubs against the pilings as a result of a poor entrance and undergoes a resistance against its forward motion given as

$$f = 9(x + 50) \text{ N}$$

where x is measured in meters from the first pilings at A to the front of the ferry.

11.10. A freight car weighing 90 kN is rolling at a speed of 1.7 m/sec toward a spring-stop system. If the spring is nonlinear such that it develops a $.0450x^2$-kN force for a deflection of x mm, what is the maximum deceleration that the car A undergoes?

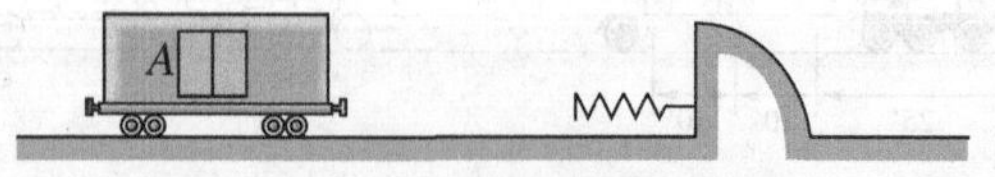

Figure P.11.10.

11.11. A 1,000 N force is applied to a 3,000-N block at the position shown. What is the speed of the block after it moves 2 m? There is Coulomb friction present. Assume at all times that the pressure at the bottom of the block is uniform. Neglect the height of the block in your calculations. Roller at right end moves with the block.

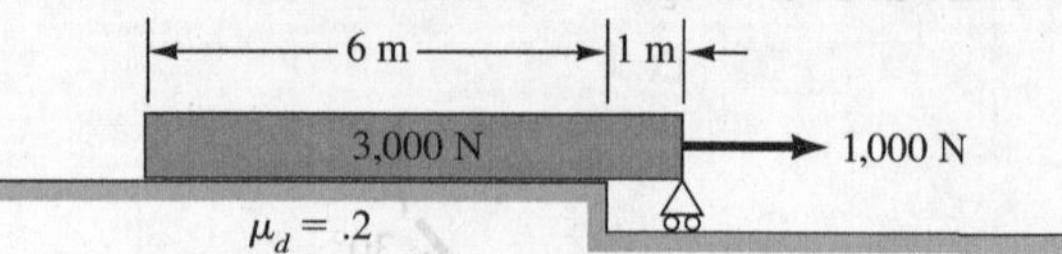

Figure P.11.11.

11.12. Two blocks A and B are connected by an inextensible chord running over a frictionless and massless pulley at E. The system starts from rest. What is the velocity of the system after it has moved 3 ft? The coefficient of dynamic friction μ_d equals .22 for bodies A and B.

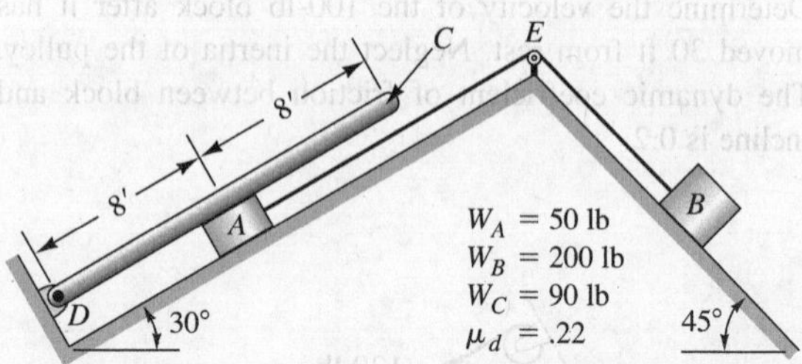

Figure P.11.12.

11.13. What are the velocities of blocks A and B when, after starting from rest, block B moves a distance of .3 ft? The dynamic coefficient of friction is .2 at all surfaces.

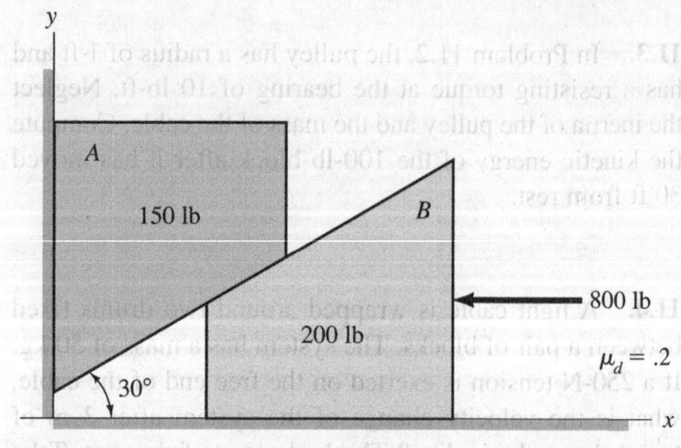

Figure P.11.13.

11.14. A particle of mass 10 lbm is acted on by the following force field:

$$F = 5x\boldsymbol{i} + (16 + 2y)\boldsymbol{j} + 20\boldsymbol{k} \text{ lb}$$

When it is at the origin, the particle has a velocity V_0 given as

$$V_0 = 5\boldsymbol{i} + 10\boldsymbol{j} + 8\boldsymbol{k} \text{ ft/sec}$$

What is its kinetic energy when it reaches position (20, 5, 10) while moving along a frictionless path? Does the shape of the path between the origin and (20, 5, 10) affect the result?

11.15. A plate AA is held down by screws C and D so that a force of 245 N is developed in each spring. Mass M of 100 kg is placed on plate AA and released suddenly. What is the maximum distance that plate AA descends if the plate can slide freely down the vertical guide rods? Take K = 3,600 N/m.

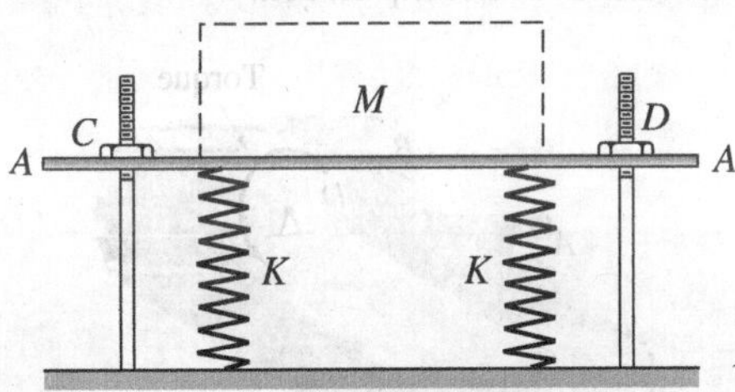

Figure P.11.15.

11.16. A 200-lb block is dropped on the system of springs. If K_1 = 600 lb/ft and K_2 = 200 lb/ft, what is the maximum force developed on the body?

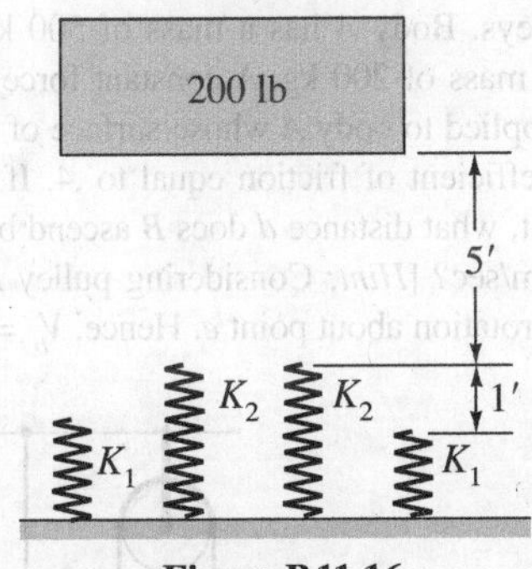

Figure P.11.16.

11.17. A block weighing 50 lb is shown on an inclined surface. The block is released at the position shown at a rest condition. What is the maximum compression of the spring? The spring has a spring constant K of 10 lb/in., and the dynamic coefficient of friction between the block and the incline is .3.

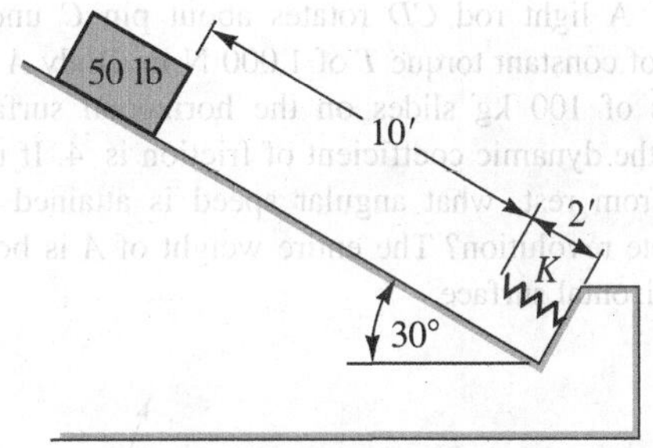

Figure P.11.17.

11.18. A classroom demonstration unit is used to illustrate vibrations and interactions of bodies. Body A has a mass of .5 kg and is moving to the left at a speed of 1.6 m/sec at the position indicated. The body rides on a cushion of air supplied from the tube B through small openings in the tube. If there is a constant friction force of .1 N, what speed will A have when it returns to the position shown in the diagram? There are two springs at C, each having a spring constant of 15 N/m.

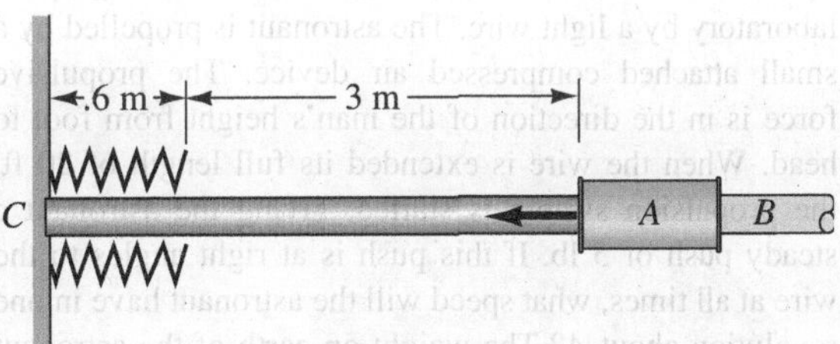

Figure P.11.18.

11.19. An electron moves in a circular orbit in a plane at right angles to the direction of a uniform magnetic field B. If the strength of B is slowly changed so that the radius of the orbit is halved, what is the ratio of the final to the initial angular speed of the electron? Explain the steps you take. The force $\boldsymbol{F}$ on a charged particle is $q\boldsymbol{V} \times \boldsymbol{B}$, where q is the charge and $\boldsymbol{V}$ is the velocity of the particle.

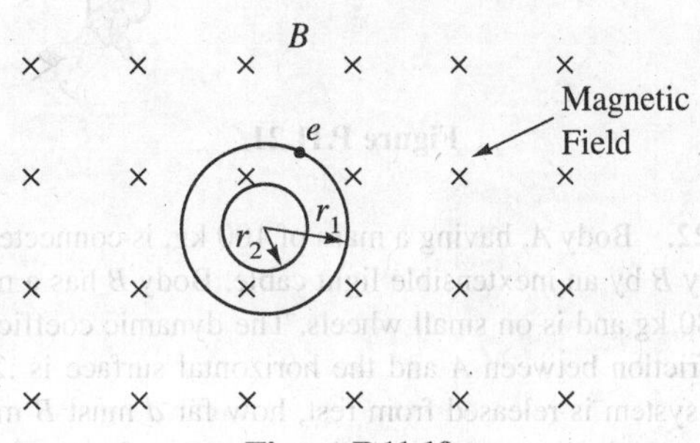

Figure P.11.19.

11.20. A light rod CD rotates about pin C under the action of constant torque T of 1,000 N-m. Body A having a mass of 100 kg slides on the horizontal surface for which the dynamic coefficient of friction is .4. If rod CD starts from rest, what angular speed is attained in one complete revolution? The entire weight of A is borne by the horizontal surface.

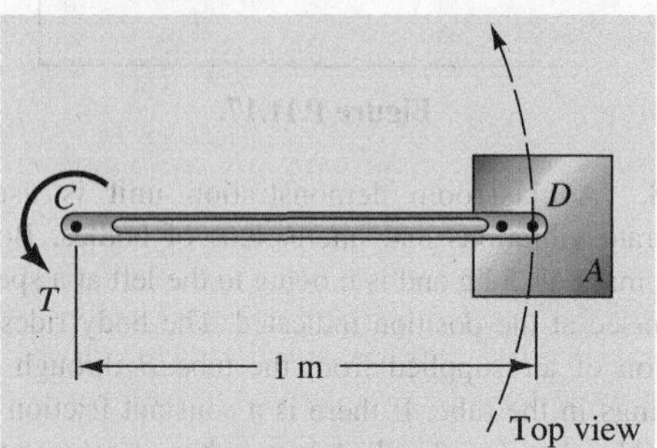

Figure P.11.20.

11.21. An astronaut is attached to his orbiting space laboratory by a light wire. The astronaut is propelled by a small attached compressed air device. The propulsive force is in the direction of the man's height from foot to head. When the wire is extended its full length of 20 ft, the propulsion system is started, giving the astronaut a steady push of 5 lb. If this push is at right angles to the wire at all times, what speed will the astronaut have in one revolution about A? The weight on earth of the astronaut plus equipment is 250 lb. The mass of the laboratory is large compared to that of the man and his equipment.

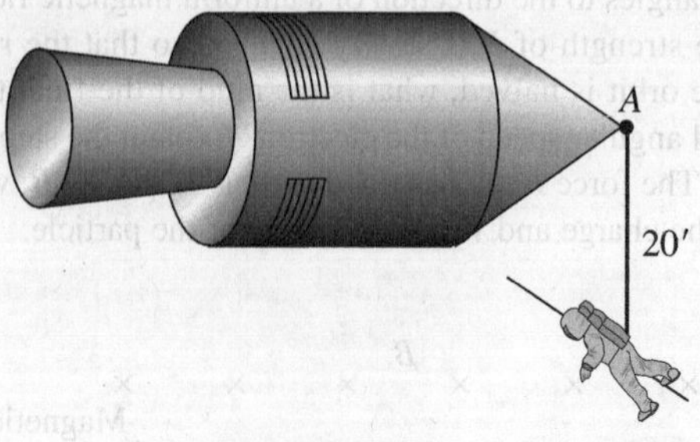

Figure P.11.21.

11.22. Body A, having a mass of 100 kg, is connected to body B by an inextensible light cable. Body B has a mass of 80 kg and is on small wheels. The dynamic coefficient of friction between A and the horizontal surface is .2. If the system is released from rest, how far d must B move along the incline before reaching a speed of 2 m/sec?

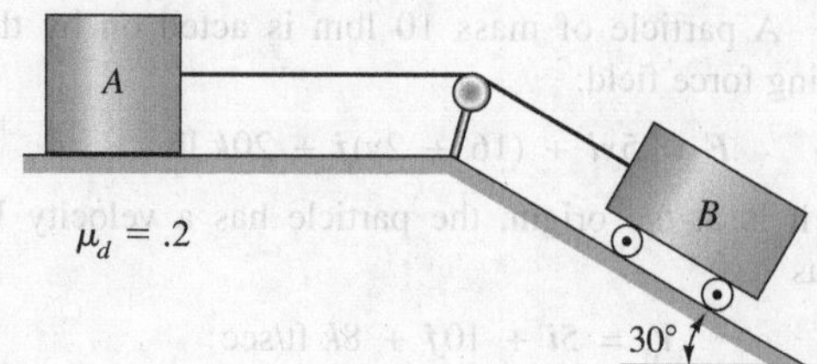

Figure P.11.22.

11.23. A conveyor has drum D driven by a torque of 50 ft-lb. Bodies A and B on the conveyor each weigh 30 lb. The dynamic coefficient of friction between the conveyor belt and the conveyor bed is .2. If the conveyor starts from rest, how fast along the conveyor do A and B move after traveling 2 ft? Drum C rotates freely, and the tension in the belt on the underside of the conveyor is 20 lb. The diameter of both drums is 1 ft. Neglect the mass of drums and belt. A and B do not slip on belt.

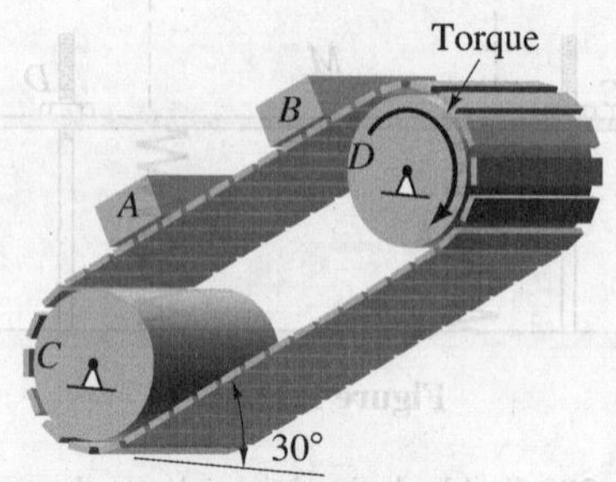

Figure P.11.23.

11.24. Bodies A and B are connected to each other through two light pulleys. Body A has a mass of 500 kg, whereas body B has a mass of 200 kg. A constant force F of value 10,000 N is applied to body A whose surface of contact has a dynamic coefficient of friction equal to .4. If the system starts from rest, what distance d does B ascend before it has a speed of 2 m/sec? [*Hint:* Considering pulley E, we have instantaneous rotation about point e. Hence, $V_b = 1V_c$.]

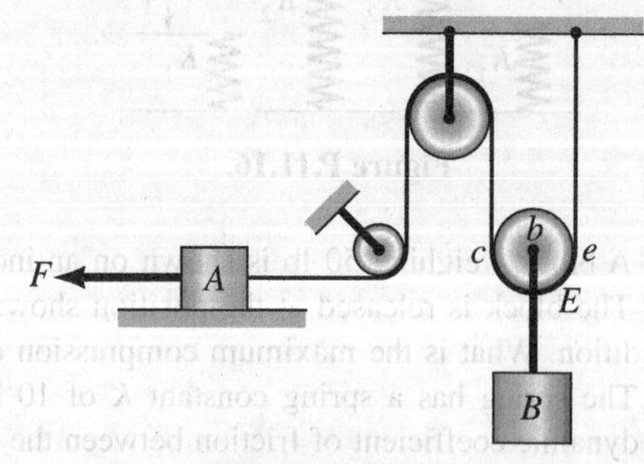

Figure P.11.24.

11.25. A rope tow for skiers is shown pulling 20 skiers up a 20° incline. The driving pulley A has a diameter of 5 ft. The idler pulley B rotates freely. The system has been stopped to allow a fallen skier to untangle himself. The driving pulley starts from rest and is given a torque of 5,000 ft-lb. With this torque, what distance d do skiers move before their speed is 15 ft/sec? The tension on the slack side of the tow can be taken as zero. The coefficient of friction between skis and slope is .15 and the average weight of the skiers is 150 lb. Neglect the mass of the rope and the pulleys.

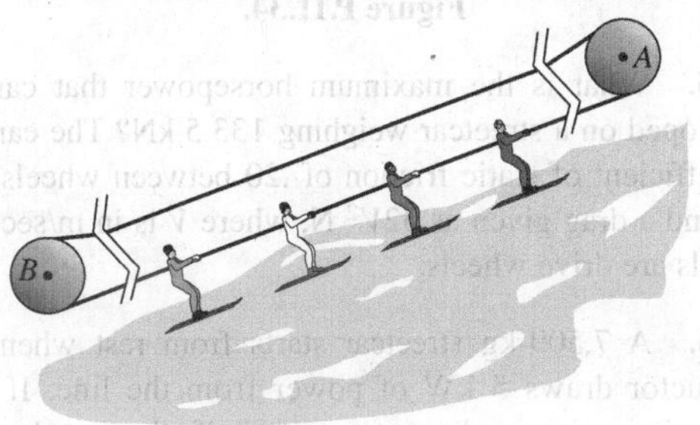

Figure P.11.25.

11.26. A uniform block A has a mass of 25 kg. The block is hinged at C and is supported by a small block B as shown in the diagram. A constant force F of 400 N is applied to block B. What is the speed of B after it moves 1.6 m? The mass of block B is 2.5 kg and the dynamic coefficient of friction for all contact surfaces is .3.

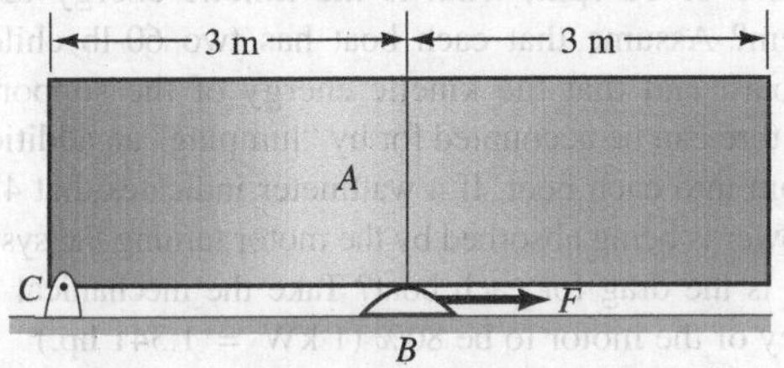

Figure P.11.26.

11.27. Block A weighs 200 lb and block B weighs 150 lb. If the system starts from rest, what is the speed of block B after it moves 1 ft? Neglect the weight of the pulleys.

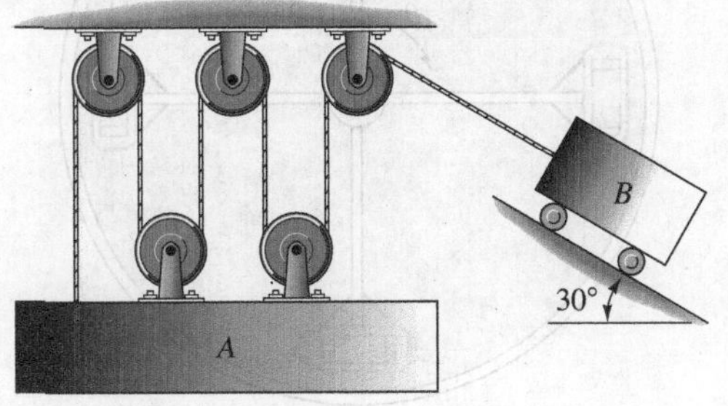

Figure P.11.27.

11.28. A weight W is to be lowered by a man. He lets the rope slip through his hands while maintaining a tension of 130 N on the rope. What is the maximum weight W that he can handle if the weight is not to exceed a speed of 5 m/sec starting from rest and dropping 3 m? Use the coefficients of friction shown in the diagram. Neglect the mass of the rope. There are three wraps of rope around the post.

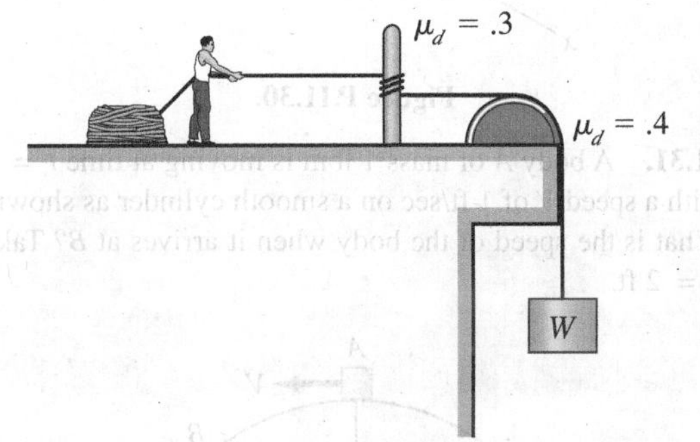

Figure P.11.28.

11.29. A vehicle B is being let down a 30° incline. The vehicle is attached to a weight A that restrains the motion. Vehicle B weighs 2,000 lb. What should the minimum restraining weight A be if, after starting from rest, the system does not exceed 8 ft/sec after moving 10 ft? There are two wraps around the post.

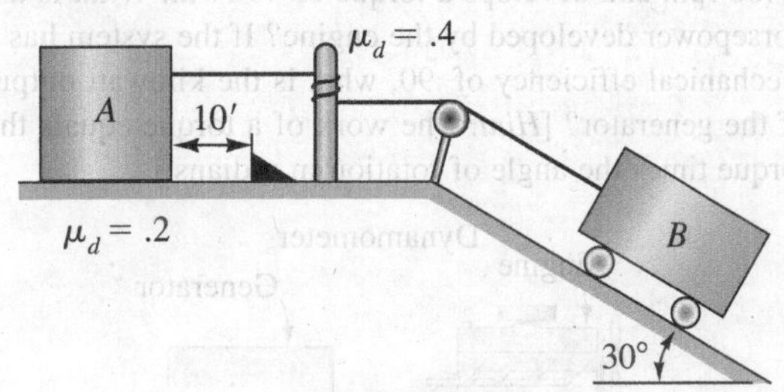

Figure P.11.29.

***11.30.** A spiral path is given parametrically in terms of the parameter τ as follows:

$$x_p = A \sin \eta\tau \text{ ft}$$
$$y_p = A \cos \eta\tau \text{ ft}$$
$$z_p = C\tau \text{ ft}$$

where A, η, and C are known constants. A particle P of mass 1 lbm is released from a position of rest 1 ft above the xy plane. The particle is constrained by a spring (K = 2 lb/ft) coiled around the path. The spring is unstretched when P is released. Neglect friction and find how far P drops. Take $\eta = \pi/2$, $A = C = 1$.

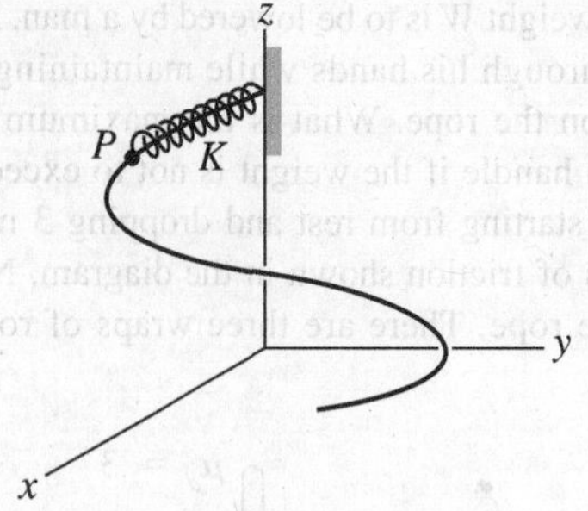

Figure P.11.30.

11.31. A body A of mass 1 lbm is moving at time $t = 0$ with a speed V of 1 ft/sec on a smooth cylinder as shown. What is the speed of the body when it arrives at B? Take $r = 2$ ft.

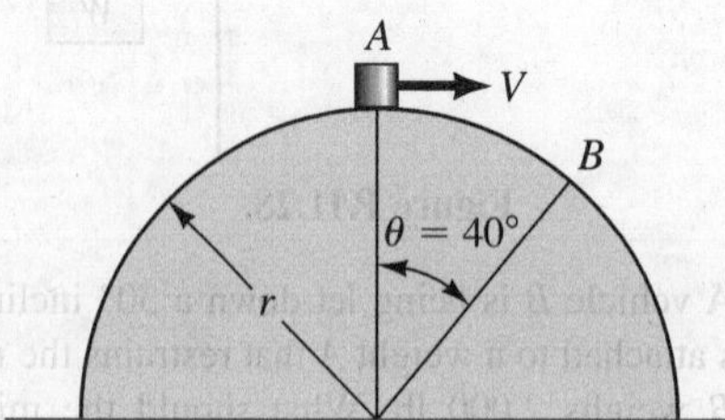

Figure P.11.31.

11.32. An automobile engine under test is rotating at 4,400 rpm and develops a torque of 40 N-m. What is the horsepower developed by the engine? If the system has a mechanical efficiency of .90, what is the kilowatt output of the generator? [*Hint:* The work of a torque equals the torque times the angle of rotation in radians.]

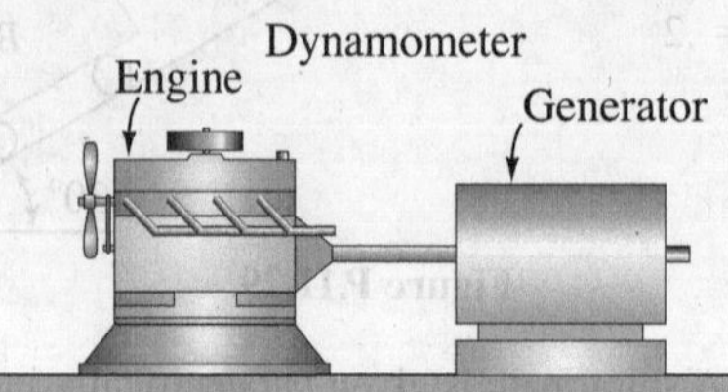

Figure P.11.32.

11.33. A rocket is undergoing static thrust tests in a test stand. A thrust of 300,000 lb is developed while 300 gal of fuel (specific gravity .8) is burned per second. The exhaust products of combustion have a speed of 5,000 ft/sec relative to the rocket. What power is being developed on the rocket? What is the power developed on the exhaust gases? (1 gal = .1337 ft^3.)

11.34. A 15-ton streetcar accelerates from rest at a constant rate a_0 until it reaches a speed V_1, at which time there is zero acceleration. The wind resistance is given as κV^2. Formulate expressions for power developed for the stated ranges of operation.

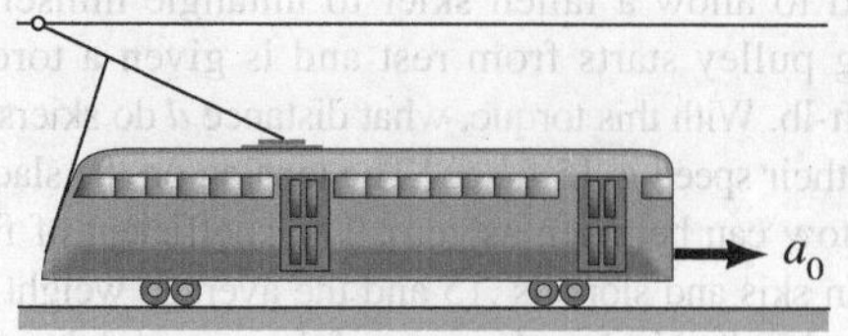

Figure P.11.34.

11.35. What is the maximum horsepower that can be developed on a streetcar weighing 133.5 kN? The car has a coefficient of static friction of .20 between wheels and rail and a drag given as $32V^2$ N, where V is in m/sec. All wheels are drive wheels.

11.36. A 7,500-kg streetcar starts from rest when the conductor draws 5 kW of power from the line. If this input is maintained constant and if the mechanical efficiency of the motors is 90%, how long does the streetcar take to reach a speed of 10 km/hr? Neglect wind resistance. (1 kW = 1.341 hp.)

11.37. A children's boat ride can be found in many amusement parks. Small boats each weighing 100 lb are rotated in a tank of water. If the system is rotating with a speed $\dot{\theta}$ of 10 rpm, what is the kinetic energy of the system? Assume that each boat has two 60-lb children on board and that the kinetic energy of the supporting structure can be accounted for by "lumping" an additional 30 lbm into each boat. If a wattmeter indicates that 4 kW of power is being absorbed by the motor turning the system, what is the drag for each boat? Take the mechanical efficiency of the motor to be 80% (1 kW = 1.341 hp.)

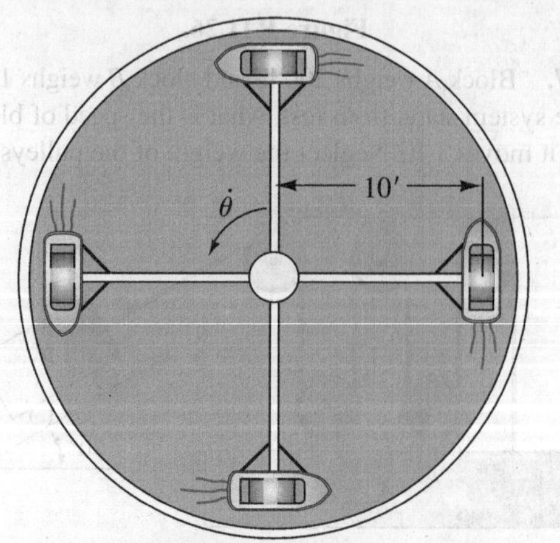

Figure P.11.37.

11.2 Conservative Force Fields

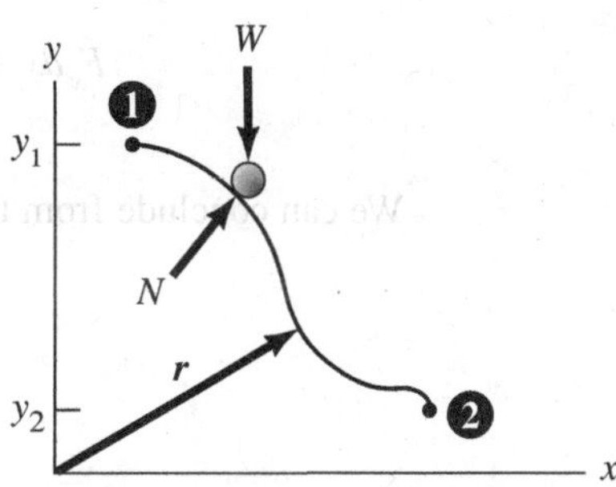

Figure 11.8. Particle moving along frictionless path.

An important class of forces called conservative forces are discussed here.

Consider first a body acted on only by gravity W as an active force (i.e., a force that can do work) and moving along a frictionless path from position 1 to position 2, as shown in Fig. 11.8. The work done by gravity $\mathcal{W}_{1-2}$ is then

$$\mathcal{W}_{1-2} = \int_1^2 \boldsymbol{F} \bullet d\boldsymbol{r} = \int_1^2 (-W\boldsymbol{j}) \bullet d\boldsymbol{r} = -W\int_1^2 dy$$
$$= -W(y_2 - y_1) = W(y_1 - y_2) \qquad (11.4)$$

Note that the work done *does not depend* on the path, but depends only on the positions of the end points of the path. *Force fields whose work like gravity is independent of the path are called conservative force fields.* In general, we can say for conservative force field $\boldsymbol{F}(x, y, z)$ that, along any path between positions 1 and 2, the work is

$$\mathcal{W}_{1-2} = \int_1^2 \mathbf{F} \bullet d\mathbf{r} = \mathcal{V}_1(x,y,z) - \mathcal{V}_2(x,y,z) \qquad (11.5)$$

where $\mathcal{V}$ is a function of position of the end points and is called the *potential energy function.* We may rewrite Eq. 11.5 as follows:

$$-\int_1^2 \boldsymbol{F} \bullet d\boldsymbol{r} = \mathcal{V}_2(x, y, z) - \mathcal{V}_1(x, y, z) = \Delta\mathcal{V} \qquad (11.6)$$

Note that the potential energy, $\mathcal{V}(x, y, z)$, depends on the reference xyz used or, as we shall often say, the *datum* used. However, the *change* in potential energy, $\Delta\mathcal{V}$, is *independent* of the datum used. Since we shall be using the change in potential energy, the datum is arbitrary and is chosen for convenience. From Eq. 11.6, we can say that *the change in potential energy*, $\Delta\mathcal{V}\,(= \mathcal{V}_2 - \mathcal{V}_1)$, of a conservative force field is *the negative of the work done by this conservative force field on a particle in going from position 1 to position 2 along any path.* For any *closed* path, clearly the work done by a conservative force field $\boldsymbol{F}$ is then

$$\oint \boldsymbol{F} \bullet d\boldsymbol{r} = 0 \qquad (11.7)$$

Hence, this is a second way to define a conservative force field. How is the potential energy function $\mathcal{V}$ related to $\boldsymbol{F}$? To answer this query, consider that an infinitesimal path $d\boldsymbol{r}$ starts from point 1. We can then give Eq. 11.6 as

$$\boldsymbol{F} \bullet d\boldsymbol{r} = -d\mathcal{V} \qquad (11.8)$$

Expressing the dot product on the left side in terms of components, and expressing $d\mathcal{V}$ as a total differential, we get

[4]We shall also use the notation P.E. or simply PE for $\mathcal{V}$. Note that we used V for potential energy in Statics as is common practice. Here in Dynamics we have switched to $\mathcal{V}$ to avoid confusion with V the velocity.

[5]Thus, considering Eq. 11.4, the value of y itself for a particle at any time depends on the position of the origin O of the xyz reference. However, changing the position of O but keeping the same direction of the xyz axes (i.e., *changing the datum*) does not affect the value of $y_2 - y_1$.

$$F_x\,dx + F_y\,dy + F_z\,dz = -\left(\frac{\partial \mathcal{V}}{\partial x}dx + \frac{\partial \mathcal{V}}{\partial y}dy + \frac{\partial \mathcal{V}}{\partial z}dz\right) \tag{11.9}$$

We can conclude from this equation that

$$\begin{aligned} F_x &= -\frac{\partial \mathcal{V}}{\partial x} \\ F_y &= -\frac{\partial \mathcal{V}}{\partial y} \\ F_z &= -\frac{\partial \mathcal{V}}{\partial z} \end{aligned} \tag{11.10}$$

In other words,

$$\begin{aligned} \boldsymbol{F} &= -\left(\frac{\partial \mathcal{V}}{\partial x}\boldsymbol{i} + \frac{\partial \mathcal{V}}{\partial y}\boldsymbol{j} + \frac{\partial \mathcal{V}}{\partial z}\boldsymbol{k}\right) \\ &= -\left(\frac{\partial}{\partial x}\boldsymbol{i} + \frac{\partial}{\partial y}\boldsymbol{j} + \frac{\partial}{\partial z}\boldsymbol{k}\right)\mathcal{V} \\ &= -\mathbf{grad}\,\mathcal{V} = -\boldsymbol{\nabla}\mathcal{V} \end{aligned} \tag{11.11}$$

The operator **grad** or **∇** that we have introduced is called the *gradient* operator[6] and is given as follows for rectangular coordinates:

$$\mathbf{grad} \equiv \boldsymbol{\nabla} \equiv \left(\frac{\partial}{\partial x}\boldsymbol{i} + \frac{\partial}{\partial y}\boldsymbol{j} + \frac{\partial}{\partial z}\boldsymbol{k}\right) \tag{11.12}$$

We can now say as a third definition that a *conservative force field must be a function of position and expressible as the gradient of a scalar function.* The *inverse* to this statement is also valid. That is, *if a force field is a function of position and the gradient of a scalar field, it must then be a conservative force field.*

Two examples of conservative force fields will now be presented and discussed.

Constant Force Field. If the force field is constant at all positions, it can always be expressed as the gradient of a scalar function of the form $\mathcal{V} = -(ax + by + cz)$, where a, b, and c are constants. The constant force field, then, is $\boldsymbol{F} = a\boldsymbol{i} + b\boldsymbol{j} + c\boldsymbol{k}$.

In limited changes of position near the earth's surface (a common situation), we can consider the gravitational force on a particle of mass, m, as a constant force field given by $-mg\boldsymbol{k}$ (or $-W\boldsymbol{k}$). Thus, the constants for the general force field given above are $a = b = 0$ and $c = -mg$. Clearly, PE $= mgz$ for this case.

Force Proportional to Linear Displacements. Consider a body limited by constraints to move along a straight line. Along this line is developed a force directly proportional to the

[6]The gradient operator comes up in many situations in engineering and physics. In short, the gradient represents a *driving action.* Thus, in the present case, the gradient is a driving action to cause mass to move. And, the gradient of temperature causes heat to flow. Finally, the gradient of electric potential causes electric charge to flow.

displacement of the body from some position O at $x = 0$ along the line. Furthermore, this force is always directed toward point O; it is then termed a *restoring* force. We can give this force as

$$\boldsymbol{F} = -Kx\boldsymbol{i} \tag{11.13}$$

where x is the displacement from point O. An example of this force is that of the linear spring (Fig.11.9) discussed in Section 10.3. The potential energy of this force field is given as follows wherein x is measured from the **undeformed** geometry (don't forget this important factor) of the spring:

$$\text{PE} = \frac{Kx^2}{2} \tag{11.14}$$

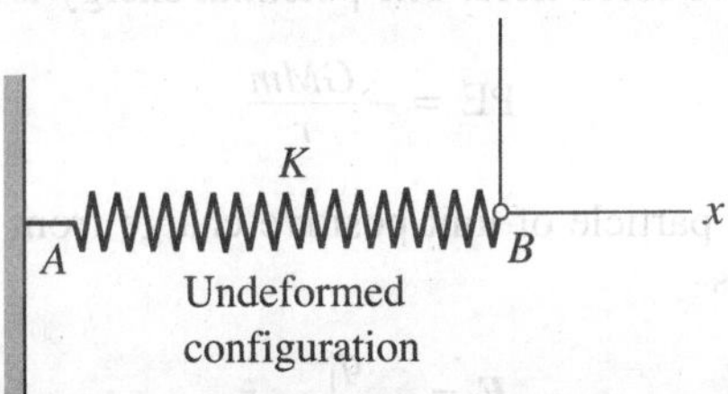

Figure 11.9. Linear spring.

What is the physical meaning of the term PE? Note that the change in potential energy has been defined (see Eq. 11.6) as the *negative* of the work done by a conservative force as the particle on which it acts goes from one position to another. Clearly, the change in the potential energy is then *directly equal* to the work done by the *reaction* to the conservative force during this displacement. In the case of the *spring*, the reaction force would be the force *from* the surroundings acting *on* the spring at point B (Fig. 11.9). During extension or compression of the spring from the undeformed position, this force (from the surroundings) does a *positive* amount of work. This work can be considered as a measure of the energy *stored* in the spring. Why? Because when allowed to return to its original position, the spring will do this amount of positive work on the surroundings at B, provided that the return motion is slow enough to prevent oscillations; and so on. Clearly then, since PE equals work of the surroundings on the spring, then PE is in effect the stored energy in the spring. In a general case, PE is the energy stored in the force field as measured from a given datum.

In previous chapters, several additional force fields were introduced: the gravitational central force field, the electrostatic field, and the magnetic field. Let us see which we can add to our list of conservative force fields.

Consider first the central gravitational force field where particle m, shown in Fig. 11.10, experiences a force given by the equation

$$\boldsymbol{F} = -G\frac{Mm}{r^2}\hat{\boldsymbol{r}} \tag{11.15}$$

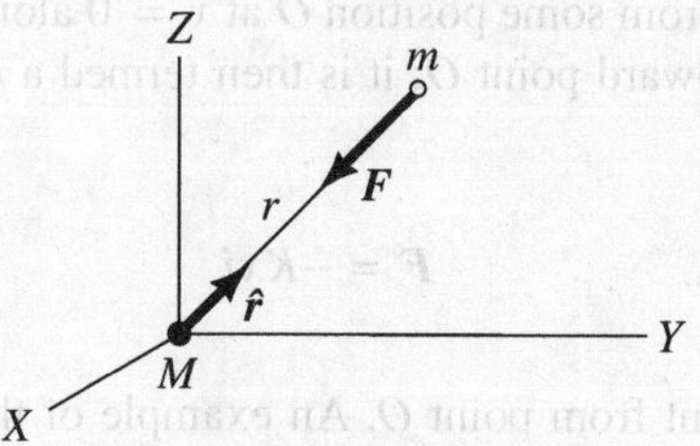

Figure 11.10. Central force on *m*.

Clearly, this force field is a function of spatial coordinates and can easily be expressed as the gradient of a scalar function in the following manner:

$$\boldsymbol{F} = -\mathbf{grad}\left(-\frac{GMm}{r}\right) \tag{11.16}$$

Hence, this is a conservative force field. The potential energy is then

$$\text{PE} = -\frac{GMm}{r} \tag{11.17}$$

Next, the force on a particle of unit positive charge from a particle of charge q_1 is given by Coulomb's law as

$$\boldsymbol{E} = \frac{q_1}{4\pi\epsilon_0 r^2}\hat{\boldsymbol{r}} \tag{11.18}$$

Since this equation has the same form as Eq. 11.15 (i.e., is also a function of $1/r^2$), we see immediately that the force field from q_1 is conservative. The potential energy per unit charge is then

$$\text{PE} = \frac{q_1}{4\pi\epsilon_0 r} \tag{11.19}$$

The remaining field introduced was the magnetic field where $\boldsymbol{F} = q\boldsymbol{V} \times \boldsymbol{B}$. For this field, the force on a charged particle depends on the velocity of the particle. The condition that the force be a function of position is not satisfied, therefore, and the magnetic field does *not* form a conservative force field.

11.3 Conservation of Mechanical Energy

Let us now consider the motion of a particle upon which only a conservative force field does work. We start with Eq. 11.2:

$$\int_{r_1}^{r_2} \boldsymbol{F} \bullet d\boldsymbol{r} = \tfrac{1}{2}mV_2^2 - \tfrac{1}{2}mV_1^2 \tag{11.20}$$

Using the definition of potential energy, we replace the left side of the equation in the following manner:

$$(\text{PE})_1 - (\text{PE})_2 = \tfrac{1}{2}mV_2^2 - \tfrac{1}{2}mV_1^2 \tag{11.21}$$

Rearranging terms, we reach the following useful relation:

$$(PE)_1 + \tfrac{1}{2}mV_1^2 = (PE)_2 + \tfrac{1}{2}mV_2^2 \qquad (11.22)$$

Since positions 1 and 2 are arbitrary, obviously *the sum of the potential energy and the kinetic energy for a particle remains constant at all times during the motion of the particle.* This statement is sometimes called the *law of conservation of mechanical energy for conservative systems.* The usefulness of this relation can be demonstrated by the following examples.

Example 11.4

A particle is dropped with zero initial velocity down a frictionless chute (Fig. 11.11). What is the magnitude of its velocity if the vertical drop during the motion is h ft?

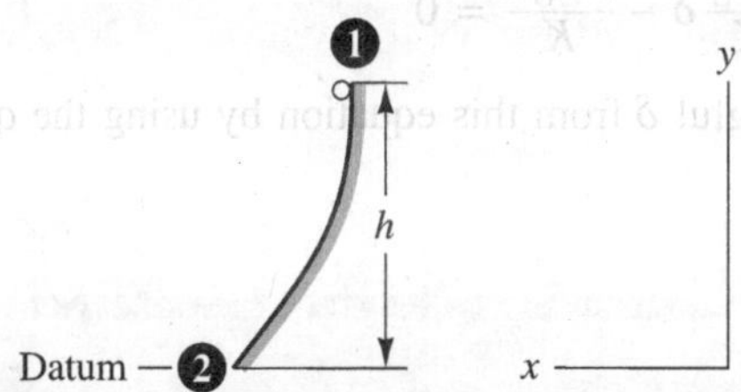

Figure 11.11. Particle on frictionless chute.

For small trajectories, we can assume a uniform force field $-mg\mathbf{j}$. Since this is the only force that can perform work on the particle (the normal force from the chute does no work), we can employ the **conservation-of-mechanical-energy** equation. If we take position 2 as a datum, we then have from Eq. 11.22:

$$mgh + 0 = 0 + \frac{1}{2}mV_2^2$$

Solving for V_2, we get

$$V_2 = \sqrt{2gh}$$

The advantages of the energy approach for conservative fields become apparent from this problem. That is, not all the forces need be considered in computing velocities, and the path, however complicated, is of no concern. If friction were present, a nonconservative force would perform work, and we would have to go back to the general relation given by Eq. 11.2 for the analysis.

Example 11.5

A mass is dropped onto a spring that has a spring constant K and a negligible mass (see Fig. 11.12). What is the maximum deflection δ? Neglect the effects of permanent deformation of the mass and any vibration that may occur.

In this problem, only conservative forces act on the body as it falls. Using the lowest position of the body as a datum, we see that the body falls a distance $h + \delta$. We shall equate the *mechanical energies* at the uppermost and lowest positions of the body. Thus,

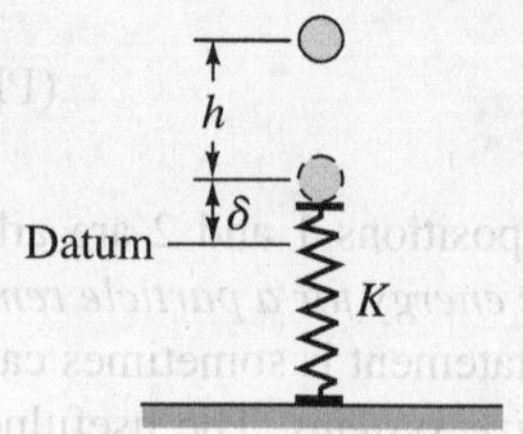

Figure 11.12. Mass dropped on spring

$$\underbrace{mg(h+\delta)}_{\text{PE gravity}} + \underbrace{0}_{\text{PE spring}} + \underbrace{0}_{\text{KE}} = \underbrace{0}_{\text{PE gravity}} + \underbrace{\tfrac{1}{2}K\delta^2}_{\text{PE spring}} + \underbrace{0}_{\text{KE}} \qquad \text{(a)}$$

Rearranging the terms,

$$\delta^2 - \frac{2mg}{K}\delta - \frac{2mgh}{K} = 0 \qquad \text{(b)}$$

We may solve for a physically meaningful δ from this equation by using the quadratic formula.

Example 11.6

A ski jumper moves down the ramp aided only by gravity (Fig. 11.13). If the skier moves 33 m in the horizontal direction and is to land very smoothly at B, what must be the angle θ for the landing incline? Neglect friction. Also determine h.

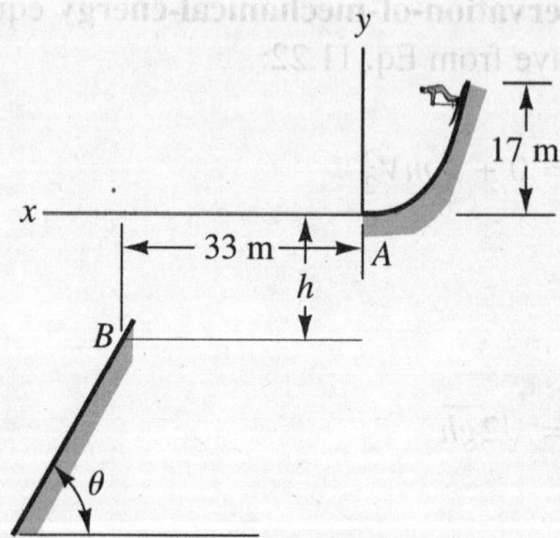

Figure 11.13. A ski jump with a landing ramp at an angle θ to be determined.

We first use **conservation of mechanical energy** along the ramp. Thus

$$(mg)(17) = \tfrac{1}{2}mV^2$$

$$\therefore V = \sqrt{(2g)(17)} = 18.26 \text{ m/s}$$

Example 11.6 (Continued)

Using a reference *xy* at *A* as shown in Fig. 11.13 and measuring time from the instant that the skier is at the origin, we now use **Newton's law** for the free flight. Thus

$$\ddot{y} = -9.81$$
$$\dot{y} = -9.81t + C_1$$
$$y = -9.81\frac{t^2}{2} + C_1 t + C_2$$

When $t = 0, \dot{y} = 0$, and we take $y = 0$. Hence,

$$C_1 = C_2 = 0$$

Also,

$$\ddot{x} = 0$$
$$\dot{x} = C_3$$
$$x = C_3 t + C_4$$

When $t = 0, \dot{x} = 18.26$, and $x = 0$,

$$\therefore C_3 = 18.26 \qquad C_4 = 0$$

Thus we have

$$\dot{y} = -9.81t \quad \text{(a)} \qquad \dot{x} = 18.26 \quad \text{(c)}$$
$$y = -9.81\frac{t^2}{2} \quad \text{(b)} \qquad x = 18.26t \quad \text{(d)}$$

To get *h*, set $x = 33$ in Eq. (d) and solve for the time *t*.

$$\therefore 33 = 18.26t \qquad t = 1.807 \text{ sec}$$

Hence, going to Eq. (b) we get

$$h = \left| -9.81\left(\frac{1.807^2}{2}\right) \right| = 16.01 \text{ m}$$

Now get $\dot{y}$ at landing. Using Eq. (a) we have

$$\dot{y} = -(9.81)(1.807) = -17.73 \text{ m/s}$$

Also, we have at all times

$$\dot{x} = 18.26 \text{ m/s}$$

For best landing, **V** is parallel to incline

$$\therefore \tan\theta = \frac{-\dot{y}}{\dot{x}} = \frac{17.73}{18.26}$$

$$\theta = 44.15°$$

Example 11.7

A block A of mass .200 kg slides on a frictionless surface as shown in Fig. 11.14. The spring constant K_1 is 25 N/m and initially, at the position shown, it is stretched .40 m. An elastic cord connects the top support to point C on A. It has a spring constant K_2 of 10.26 N/m. Furthermore, the cord disconnects from C at the instant that C reaches point G at the end of the straight portion of the incline. If A is released from rest at the indicated position, what value of θ corresponds to the end position B where A just loses contact with the surface? The elastic cord (at the top) is initially unstretched.

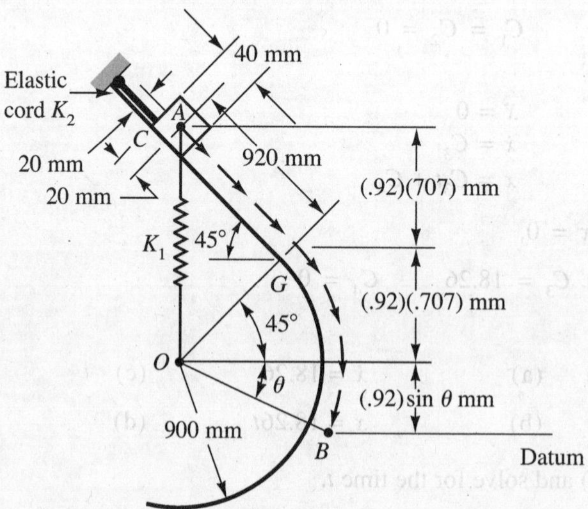

Figure 11.14. Mass A slides along frictionless surface.

We have conservative forces performing work on A so we have **conservation of mechanical energy.** Using the datum at B and using l_0 as the unstretched length of the spring with δ as the elongation of the spring, we then say that

$$mgz_1 + \frac{mV_1^2}{2} + \frac{1}{2}K_1\delta_1^2 = mgz_2 + \frac{mV_2^2}{2} + \frac{1}{2}K_1\delta_2^2 + \frac{1}{2}K_2(CG)^2$$

where the last term is the energy in the elastic cord when it disconnects at G. Therefore, noting that CG = .94 m and that OB = .92 m, we have on observing vertical distances in Fig. 11.14:

$$(.200)(9.81)[(.92)(.707) + (.92)(.707) + (.92)\sin\theta] + 0 + \tfrac{1}{2}(25)(.40)^2$$
$$= 0 + \tfrac{1}{2}(.20)V_2^2 + \tfrac{1}{2}(25)(.92 - l_0)^2 + \tfrac{1}{2}(10.26)(.94)^2 \qquad \text{(a)}$$

Example 11.7 (Continued)

To get l_0, examine the initial configuration of the system. With an initial stretch of .40 m for the spring, we can say observing again vertical distances in Fig. 11.14:

$$l_0 = [(.92)(.707) + (.92)(.707)] - .40$$
$$= .901 \text{ m}$$

Equation (a) can then be written as

$$V_2^2 = .1490 + 18.05 \sin\theta \tag{b}$$

We now use **Newton's law** at the point of interest B where A just loses contact. This condition is shown in detail in Fig. 11.15, where you will notice that the contact force N has been taken as zero and thus deleted from the free-body diagram. In the radial direction, we have

$$-F_{sp} + (.200)(g)\sin\theta = -.200\left(\frac{V_2^2}{.92}\right)$$

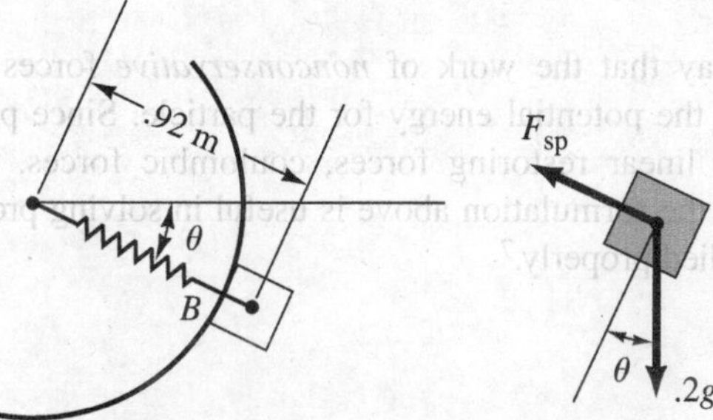

Figure 11.15. Contact is first lost at θ.

Therefore,

$$-(25)(.92 - .901) + (.200)(9.81)\sin\theta = -\left(\frac{V_2^2}{4.60}\right)$$

This equation can be written as

$$V_2^2 = 2.20 - 9.03 \sin\theta \tag{c}$$

Solving Eqs. (b) and (c) simultaneously for θ, we get

$$\theta = 4.34°$$

11.4 Alternative Form of Work-Energy Equation

With the aid of the material in Section 11.3, we shall now set forth an alternative energy equation which has much physical appeal and which resembles the *first law of thermodynamics* as used in other courses. Let us take the case where certain of the forces acting on a particle are conservative while others are not. Remember that for conservative forces the negative of the change in potential energy between positions 1 and 2 equals the work done by these forces as the particle goes from position 1 to position 2 along any path. Thus, we can restate Eq. 11.2 in the following way:

$$\int_1^2 \boldsymbol{F} \bullet d\boldsymbol{r} - \Delta(\text{PE})_{1,2} = \Delta(\text{KE})_{1,2} \qquad (11.23)$$

where the integral represents the work of *nonconservative* forces and the Δ represents the final state minus the initial state. Calling the integral $\mathcal{W}_{1-2}$, we than have, on rearranging the equation:

$$\Delta(\text{KE} + \text{PE}) = \mathcal{W}_{1-2} \qquad (11.24)$$

In this form, we say that the work of *nonconservative* forces goes into changing the kinetic energy plus the potential energy for the particle. Since potential energies of such common forces as linear restoring forces, coulombic forces, and gravitational forces are so well known, the formulation above is useful in solving problems if it is understood thoroughly and applied properly.[7]

[7]Equation 11.24, you may notice, is actually a form of the first law of thermodynamics for the case of no heat transfer.

Example 11.8

Three coupled streetcars (Fig. 11.16) are moving at a speed of 32 km/hr down a 7° incline. Each car has a weight of 198 kN. Specifications from the buyer requires that

Figure 11.16 Coupled streetcars.

Example 11.8 (Continued)

the cars must stop within 50 m beyond the position where the brakes are fully applied so as to cause the wheels to lock. What is the maximum number of brake failures that can be tolerated and still satisfy this specification? We will assume for simplicity that the weight is loaded equally among all the wheels of the system. There are 24 brake systems, one for each wheel. Take $\mu_d = .45$.

The friction force f on any one wheel where the brake has operated is ascertained from **Coulomb's law** as

$$f = \frac{198{,}000 \cos 7°}{8}(.45) = 11{,}050 \text{ N}$$

We now consider the **work–energy relation** 11.24 for the case where a minimum number of good brakes, n, just causes the trains to stop in 50 m. We shall neglect the kinetic energy due to rotation of the rather small wheels. This assumption permits us to use a single particle to represent the three cars, wherein this particle moves a distance of 50 m. Using the end configuration of the train as the datum for potential energy of gravity, we have for Eq. 11.24:

$$\Delta KE + \Delta PE = \mathcal{W}_{1-2}$$

$$\left(0 - 3\left\{\frac{1}{2}\frac{198{,}000}{g}\left[\frac{(32)(1{,}000)}{3{,}600}\right]^2\right\}\right) + [0 - (3)(198{,}000)(50)\sin 7°]$$
$$= -(n)(11{,}050)(50)$$

$$n = 10.89$$

The number of brake failures that can accordingly be tolerated is 24 − 11 = 13.

Number of brake failures to be tolerated = 13

Another example of conservation of mechanical energy will be in the next section (Example 11.10) for the case of a system of particles.

PROBLEMS

11.38. A railroad car traveling 5 km/hr runs into a stop at a railroad terminal. A vehicle having a mass of 1,800 kg is held by a linear restoring force system that has an equivalent spring constant of 20,000 N/m. If the railroad car is assumed to stop suddenly and if the wheels in the vehicle are free to turn, what is the maximum force developed by the spring system? Neglect rotational inertia of the wheels of the vehicle.

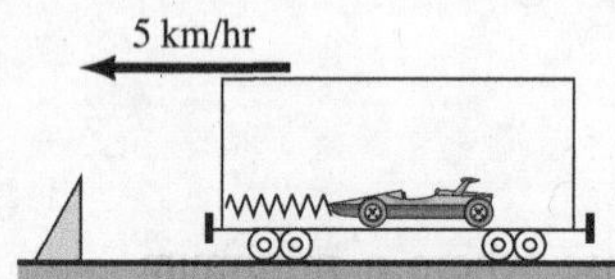

Figure P.11.38.

11.39. A mass of one slug is moving at a speed of 50 ft/sec along a horizontal frictionless surface, which later inclines upward at an angle 45°. A spring of constant $K = 5$ lb/in. is present along the incline. How high does the mass move?

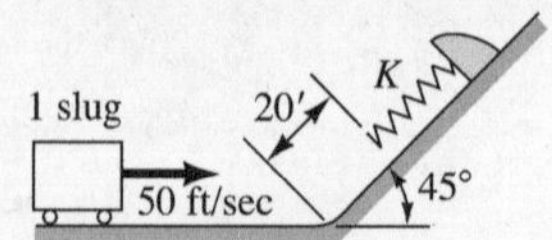

Figure P.11.39.

11.40. A block weighing 10 lb is released from rest where the springs acting on the body are horizontal and have a tension of 10 lb each. What is the velocity of the block after it has descended 4 in. if each spring has a spring constant $K = 5$ lb/in.?

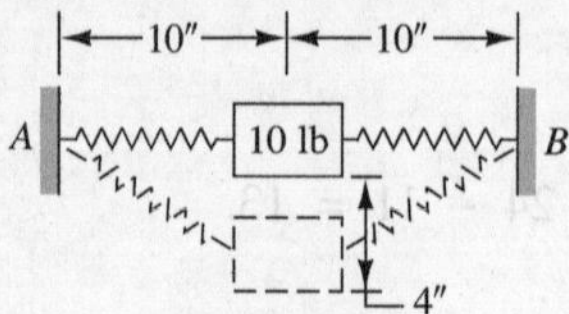

Figure P.11.40.

11.41. A nonlinear spring develops a force given as $.06x^2$ N, where x is the amount of compression of the spring in millimeters. Does such a spring develop a conservative force? If so, what is the potential energy stored in the spring for a deflection of 60 mm?

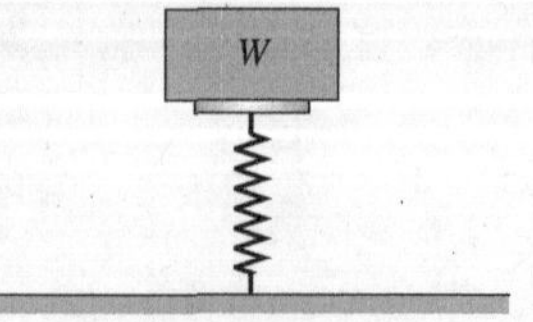

Figure P.11.41.

11.42. In Problem 11.41, a weight W of 225 N is released suddenly from rest on the nonlinear spring. What is the maximum deflection of the spring?

11.43. A vector operator that you will learn more about in fluid mechanics and electromagnetic theory is the *curl* vector operator. This operator is defined for rectangular coordinates in terms of its action on $\boldsymbol{V}$ as follows:

$$\textbf{curl}\,\boldsymbol{V}(x, y, z) = \left(\frac{\partial V_z}{\partial y} - \frac{\partial V_y}{\partial z}\right)\boldsymbol{i} + \left(\frac{\partial V_x}{\partial z} - \frac{\partial V_z}{\partial x}\right)\boldsymbol{j} + \left(\frac{\partial V_y}{\partial x} - \frac{\partial V_x}{\partial y}\right)\boldsymbol{k}$$

(When the curl is applied to a fluid velocity field $\boldsymbol{V}$ as above, the resulting vector field is twice the angular velocity field of infinitesimal elements in the flow.) Show that if $\boldsymbol{F}$ is expressible as $\boldsymbol{\nabla}\phi(x, y, z)$, then it must follow that $\textbf{curl}\,\boldsymbol{F} = \boldsymbol{0}$. The converse is also true, namely that *if* $\textbf{curl}\,\boldsymbol{F} = \boldsymbol{0}$, *then* $\boldsymbol{F} = \boldsymbol{\nabla}\phi\,(x, y, z)$ *and is thus a conservative force field.*

11.44. Determine whether the following force fields are conservative or not.

(a) $\boldsymbol{F} = (10z + y)\boldsymbol{i} + (15yz + x)\boldsymbol{j} + \left(10x + \dfrac{15y^2}{2}\right)\boldsymbol{k}$

(b) $\boldsymbol{F} = (z \sin x + y)\boldsymbol{i} + (4yz + x)\boldsymbol{j} + \left(2y^2 - 5 \cos x\right)\boldsymbol{k}$

See Problem 11.43 before doing this problem.

11.45. Given the following conservative force field:

$$\boldsymbol{F} = (10z + y)\boldsymbol{i} + (15yz + x)\boldsymbol{j} + \left(10x + \frac{15y^2}{2}\right)\boldsymbol{k}\ N$$

find the force potential to within an arbitrary constant. What work is done by the force field on a particle going from $\boldsymbol{r}_1 = 10\boldsymbol{i} + 2\boldsymbol{j} + 3\boldsymbol{k}$ m to $\boldsymbol{r}_2 = -2\boldsymbol{i} + 4\boldsymbol{j} - 3\boldsymbol{k}$ m? [*Hint:* Note that if $\partial\phi/\partial x$ equals some function $(xy^2 + z)$, then we can say on integrating that

$$\phi = \frac{x^2y^2}{2} + zx + g(y, z)$$

where $g(y, z)$ is an arbitrary function of y and z. Note we have held y and z constant during the integration.]

11.46. If the following force field is conservative,

$$\boldsymbol{F} = (5z \sin x + y)\boldsymbol{i} + (4yz + x)\boldsymbol{j} + (2y^2 - 5 \cos x)\boldsymbol{k}\ \text{lb}$$

(where x, y, and z are in ft), find the force potential up to an arbitrary constant. What is the work done on a particle starting at the origin and moving in a circular path of radius 2 ft to form a semicircle along the positive x axis? (See the hint in Problem 11.45.)

11.47. A body A can slide in a frictionless manner along a stiff rod CD. At the position shown, the spring along CD has been compressed 6 in. and A is at a distance of 4 ft from D. The spring connecting A to E has been elongated 1 in. What is the speed of A after it moves 1 ft? The spring constants are K_1 = 1.0 lb/in. and K_2 = .5 lb/in. The mass of A is 30 lbm.

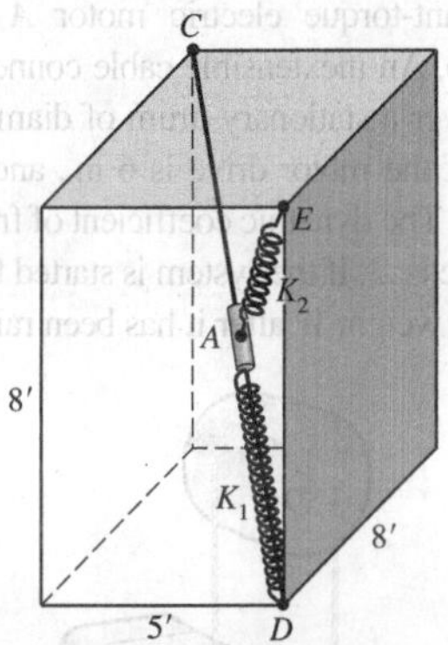

Figure P.11.47.

11.48. A collar A of mass 10 lbm slides on a frictionless tube. The collar is connected to a linear spring whose spring constant K is 5.0 lb/in. If the collar is released from rest at the position shown, what is its speed when the spring is at elevation EF? The spring is stretched 3 in. at the initial position of the collar.

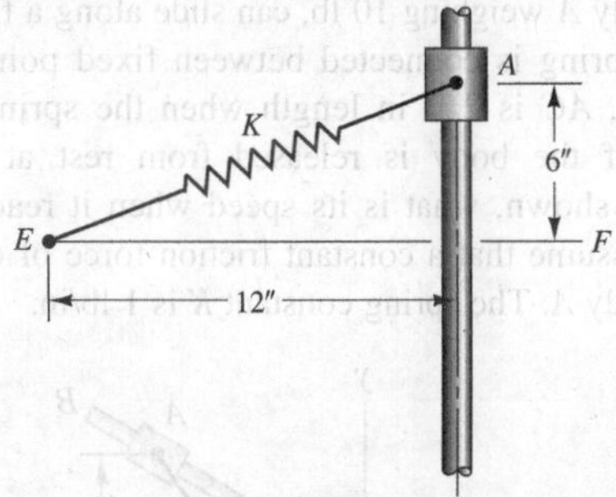

Figure P.11.48.

11.49. A mass M of 20 kg slides with no friction along a vertical rod. Two springs each of spring constant K_1 = 2 N/mm and a third spring having a spring constant K_2 = 3 N/mm are attached to the mass M. At the starting position when $\theta = 30°$, the springs are unstretched. What is the velocity of M after it descends a distance d of .02 m?

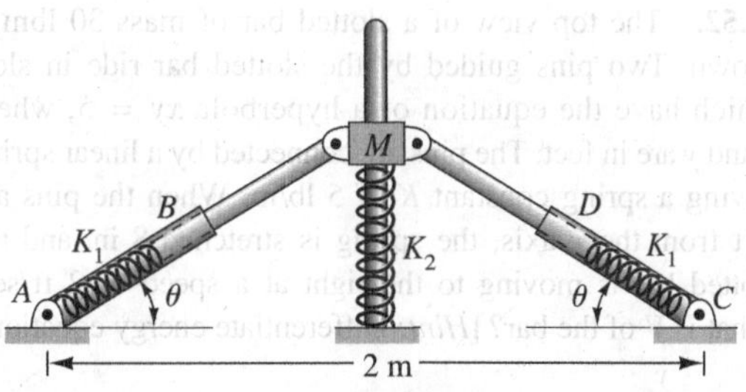

Figure P.11.49.

11.50. A collar A having a mass of 5 kg can slide without friction on a pipe. If released from rest at the position shown, where the spring is unstretched, what speed will the collar have after moving 50 mm? The spring constant is 2,000 N/m.

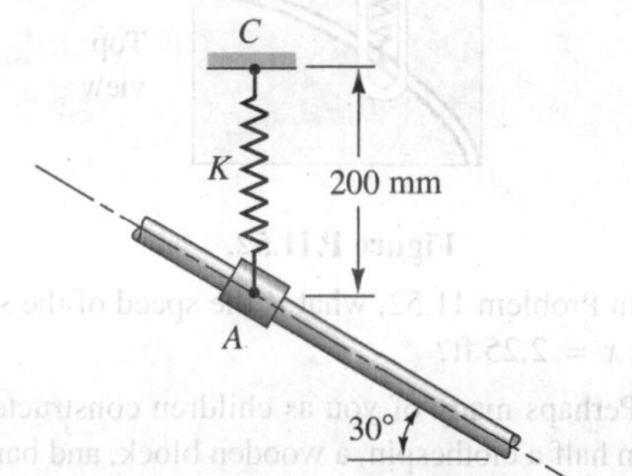

Figure P.11.50.

11.51. A slotted rod A is moving to the left at a speed of 2 m/sec. Pins are moved to the left by this rod. These pins must slide in a slot under the rod as shown in the diagram. The pins are connected by a spring having a spring constant K of 1,500 N/m. The spring is unstretched in the configuration shown. What distance d do the pins reach before stopping instantaneously? The mass of the slotted rod is 10 kg. The spring is held in the slotted rod so as not to buckle outward. Neglect the mass of the pins.

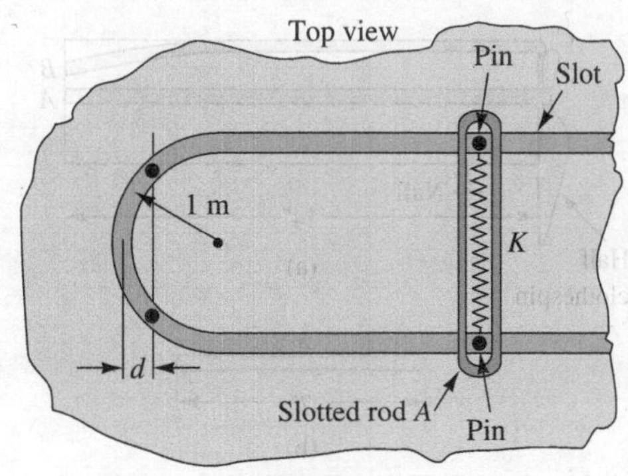

Figure P.11.51.

11.52. The top view of a slotted bar of mass 30 lbm is shown. Two pins guided by the slotted bar ride in slots which have the equation of a hyperbola $xy = 5$, where x and y are in feet. The pins are connected by a linear spring having a spring constant K of 5 lb/in. When the pins are 2 ft from the y axis, the spring is stretched 8 in. and the slotted bar is moving to the right at a speed of 2 ft/sec. What is $\dot{V}$ of the bar? [*Hint:* Differentiate energy equation.]

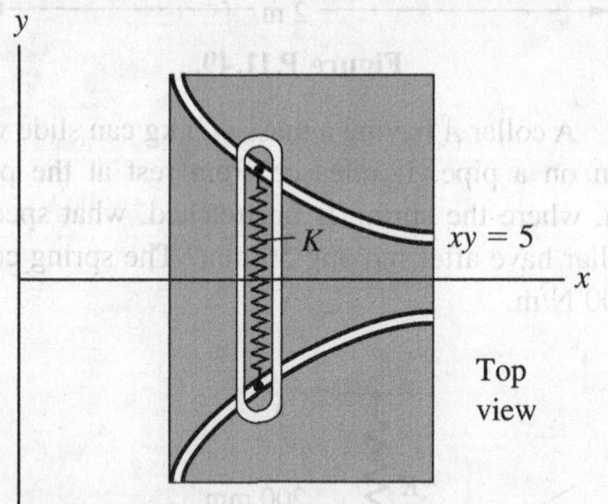

Figure P.11.52.

11.53. In Problem 11.52, what is the speed of the slotted bar when $x = 2.25$ ft?

11.54. Perhaps many of you as children constructed toy guns from half a clothespin, a wooden block, and bands of rubber cut from the inner tube of an automobile tire [see diagram (a)]. Rubber band A holds the half-clothespin to the wooden "gun stock." The "ammunition" is a rubber band B held by the clothespin at C by friction and stretched to go around the block at the other end. The rubber band B when laid flat as in (b) has a length of 7 in. To "load the ammunition" takes a force of 20 lb at C. If the gun is pointed upward, estimate how high the fired rubber band will go when "fired" if it weighs 4 oz. To "fire" the gun you push lowest part of clothespin toward the nail (see diagram) to release at C.

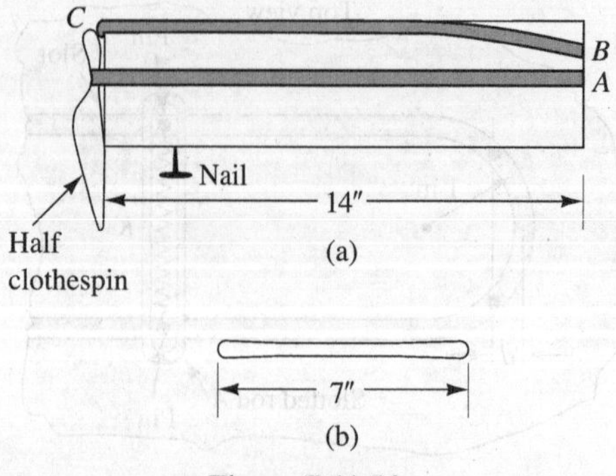

Figure P.11.54.

11.55. A meteor has a speed of 56,000 km/hr when it is 320,000 km from the center of the earth. What will be its speed when it is 160 km from the earth's surface?

11.56 Do Problem 11.2 using the energy equation in the usual form of the first law of thermodynamics.

11.57. Do Problem 11.5 using the energy equation in the usual form of the first law of thermodynamics.

11.58. Do Problem 11.17 using the energy equation in the usual form of the first law of thermodynamics.

11.59. Do Problem 11.18 using the energy equation in the usual form of the first law of thermodynamics.

11.60. A constant-torque electric motor A is hoisting a weight W of 30 lb. An inextensible cable connects the weight W to the motor over a stationary drum of diameter $D = 1$ ft. The diameter d of the motor drive is 6 in., and the delivered torque is 150 lb-ft. The dynamic coefficient of friction between the drum and cable is .2. If the system is started from rest, what is the speed of the weight W after it has been raised 5 ft?

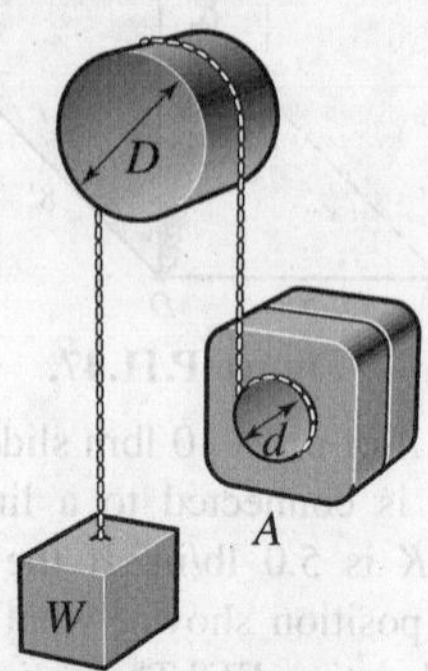

Figure P.11.60.

11.61. A body A weighing 10 lb, can slide along a fixed rod B–B. A spring is connected between fixed point C and the mass. AC is 2 ft in length when the spring is unextended. If the body is released from rest at the configuration shown, what is its speed when it reaches the y axis? Assume that a constant friction force of 6 oz acts on the body A. The spring constant K is 1 lb/in.

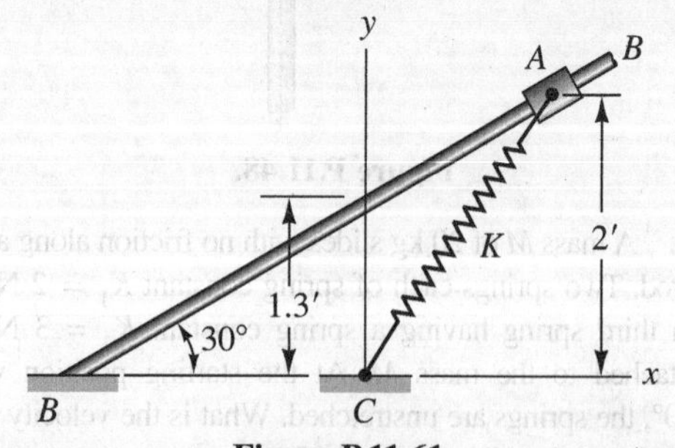

Figure P.11.61.

11.62. A body A is released from rest on a vertical circular path as shown. If a constant resistance force of 1 N acts along the path, what is the speed of the body when it reaches B? The mass of the body is .5 kg and the radius r of the path is 1.6 m.

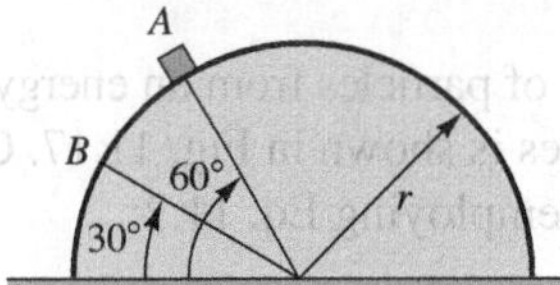

Figure P.11.62.

11.63. A cylinder slides down a rod. What is the distance δ that the spring is deflected at the instant that the disk stops instantaneously? Take $\mu_d = .3$.

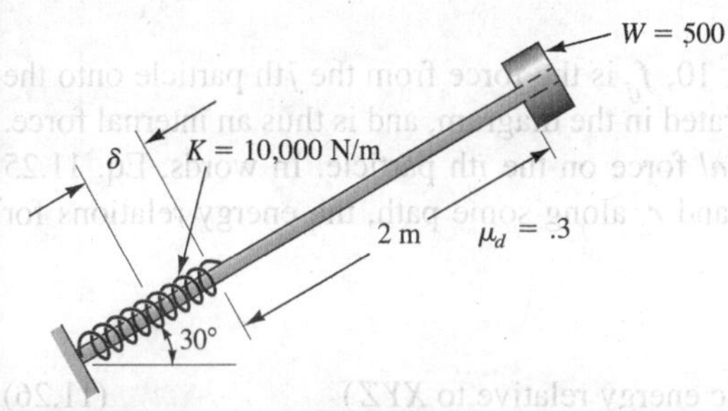

Figure P.11.63.

11.64. In ordnance work a very vital test for equipment is the *shock* test, in which a piece of equipment is subjected to a certain level of acceleration of short duration. A common technique for this test is the *drop test.* The specimen is mounted on a rigid carriage, which upon release is dropped along guide rods onto a set of lead pads resting on a heavy rigid anvil. The pads deform and absorb the energy of the carriage and specimen. We estimate through other tests that the energy E absorbed by a pad versus compression distance δ is given as shown, where the curve can be taken as a parabola. For four such pads, each placed directly on the anvil, and a height h of 3 m, what is the compression of the pads? The carriage and specimen together weigh $50g$ N. Neglect the friction of the guides. (*Note:* 1 J = 1 N-m.)

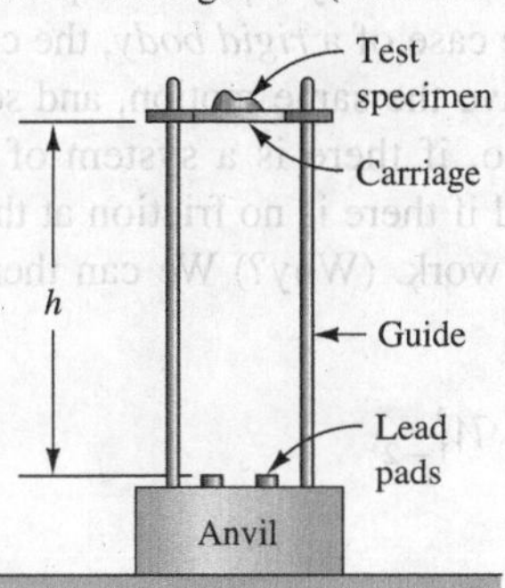

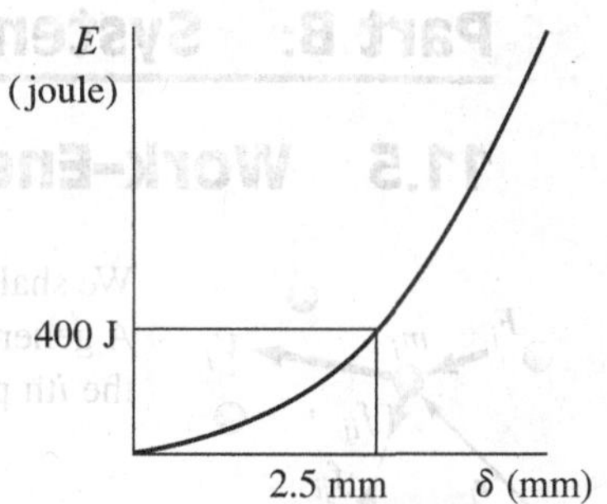

Figure P.11.64.

11.65. Two bodies are connected by an inextensible cord over a frictionless pulley. If released from rest, what velocity will they reach when the 500-lb body has dropped 5 ft?

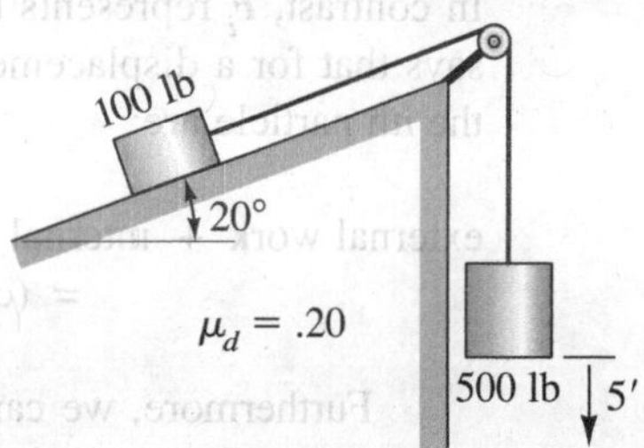

Figure P.11.65.

11.66. Suppose in Example 11.9 that only the brakes on train A operate and lock. What is the distance d before stopping? Also, determine the force in each coupling of the system.

11.67. A large constant force F is applied to a body of weight W resting on an inclined surface for which the coefficient of dynamic friction is μ_d. The body is acted on by a spring having a spring constant K. If initially the spring is compressed a distance δ, compute the velocity of the body in terms of F and the other parameters that are given, when the body has moved from rest a distance up the incline of $\frac{3}{2}\delta$.

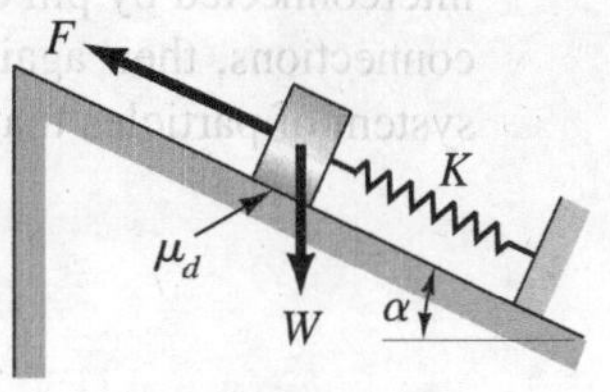

Figure P.11.67.

Part B: Systems of Particles

11.5 Work–Energy Equations

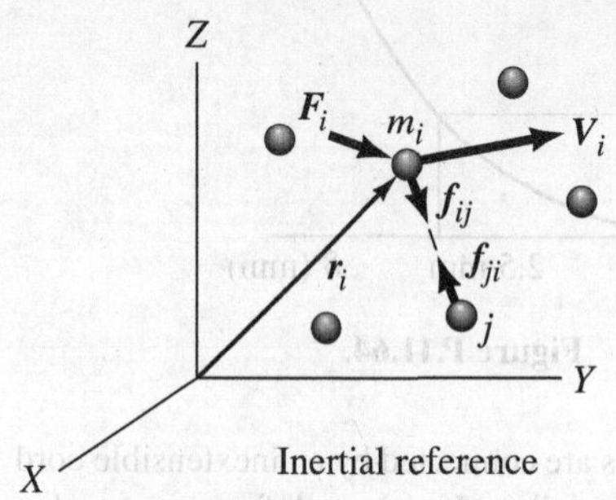

Figure 11.17. System of particles.

We shall now examine a system of particles from an energy viewpoint. A general aggregate of n particles is shown in Fig. 11.17. Considering the ith particle, we can say, by employing Eq. 11.2:

$$\int_1^2 \boldsymbol{F}_i \cdot d\boldsymbol{r}_i + \int_1^2 \left(\sum_{\substack{j=1 \\ j \neq i}}^{n} \boldsymbol{f}_{ij} \right) \cdot d\boldsymbol{r}_i = \left(\tfrac{1}{2} m_i V_i^2\right)_2 - \left(\tfrac{1}{2} m_i V_i^2\right)_1 \qquad (11.25)$$

where, as in Chapter 10, $\boldsymbol{f}_{ij}$ is the force from the jth particle onto the ith particle, as illustrated in the diagram, and is thus an internal force. In contrast, $\boldsymbol{F}_i$ represents the total *external* force on the ith particle. In words, Eq. 11.25 says that for a displacement between $\boldsymbol{r}_1$ and $\boldsymbol{r}_2$ along some path, the energy relations for the ith particle are:

external work + internal work
= (change in kinetic energy relative to XYZ) (11.26)

Furthermore, we can adopt the point of view set forth in Section 11.4 and identify conservative forces, both external and internal, so as to utilize potential energies for these forces in the energy equation. To qualify as a conservative force, an internal force would have to be a function of only the spatial configuration of the system and expressible as the gradient of a scalar function. Clearly, forces arising from the gravitational attraction between the particles, electrostatic forces from electric charges on the particles, and forces from elastic connectors between the particles (such as springs) are all conservative internal forces.

We now sum Eqs. 11.25 for all the particles in the system to get the energy equation for a *system of particles*. We do *not* necessarily get a cancellation of contributions of the internal forces as we did for Newton's law in Chapter 10 because we are now adding the *work* done by each internal force on each particle. And even though we have pairs of internal forces that are equal and opposite, the *movements* of the corresponding particles in general are *not* equal. The result is that the work done by a pair of equal and opposite internal forces is not always zero. However, in the case of a *rigid body*, the contact forces between pairs of particles making up the body have the same motion, and so in this case the internal work is *zero* from such forces.[8] Also, if there is a system of rigid bodies interconnected by pin or ball joint connections, and if there is no friction at these movable connections, then again there will be no internal work. (Why?) We can then say for the system of particles that

$$\Delta(\text{KE} + \text{PE}) = \mathcal{W}_{1-2} \qquad (11.27)$$

[8]We shall show this more directly in Chapter 15.

where $\mathcal{W}_{1-2}$ represents the net work done by *internal and external* nonconservative forces, and PE represents the total potential energy of the conservative *internal and external* forces. Clearly, if there are no nonconservative forces present then Eq. (11.27) degenerates to the conservation-of-mechanical-energy principle. As pointed out earlier, since we are employing the *change* in potential energy, the datums chosen for measuring PE are of little significance here.[9] For instance, any convenient datum for measuring the potential energy due to gravity of the earth yields the same result for the term ΔPE.

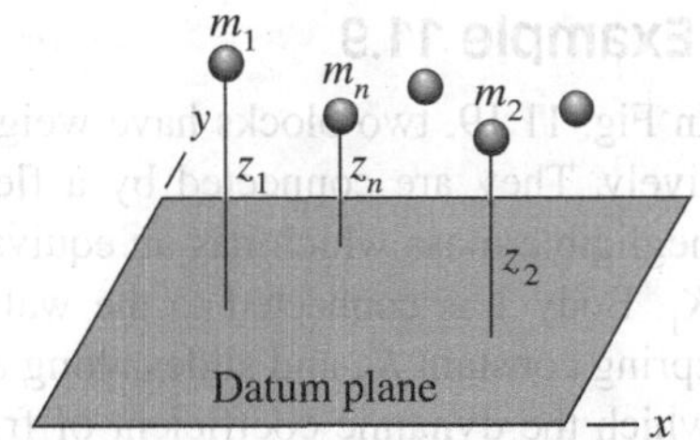

Figure 11.18. Particles above reference plane.

Looking back on Eq. (11.25), which on summation over all the particles gave rise to Eq. (11.27), namely the equation to be used for a system of particles, we wish to make the following point. It is the fact that the work contribution of each force stems from the *movement of each force with its specific point of application.* This should be clear from the use of $d\mathbf{r}_i$, with i identifying each particle. This will be an important consideration later.

Let us now consider the action of gravity on a system of particles. The potential energy relative to a datum plane, xy, for such a system (see Fig. 11.18) is simply

$$\text{PE} = \sum_i m_i g z_i$$

Note that the right side of this equation represents the first moment of the weight of the system about the xy plane. This quantity can be given in terms of the center of gravity and the entire weight of W as follows:

$$\text{PE} = W z_c$$

where z_c is the vertical distance from the datum plane to the center of gravity. Note that if g is constant, the center of gravity corresponds to the center of mass. And so for any system of particles, the change in potential energy is readily found by concentrating the entire weight at the center of gravity or, as is almost always the case, at the center of mass.

Before proceeding with the problems we wish to emphasize certain salient features governing the work–energy principle for a system of particles.

1. In computing work, we must remember to have the forces move **with their points of application** (see Eq. 11.25).
2. Both **internal** and **external** forces may be present as conservative and as nonconservative forces and must be accounted for.
3. The kinetic energy must be the **total** kinetic energy and not just that of the mass center.

[9] One precaution in this regard must again be brought to your attention. You will remember that in the spring-force formula, $-Kx$, the term x represents the elongation or contraction of the spring from the *undeformed* condition. This condition must not be violated in the potential-energy expression $\frac{1}{2}Kx^2$.

Example 11.9

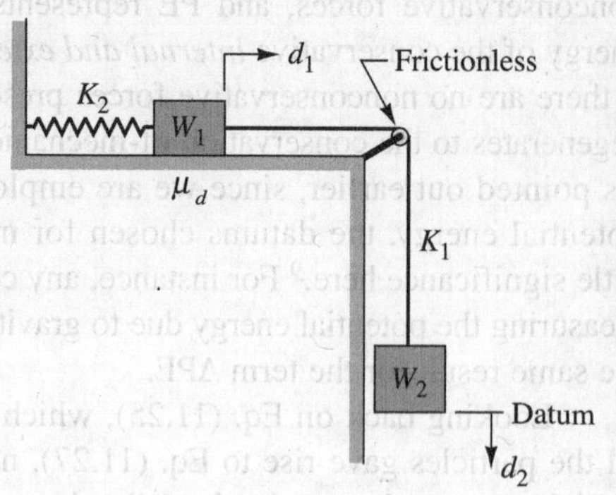

Figure 11.19. Elastically connected bodies.

In Fig. 11.19, two blocks have weights W_1 and W_2, respectively. They are connected by a flexible, *elastic* cable of negligible mass which has an equivalent spring constant of K_1. Body 1 is connected to the wall by a spring having a spring constant K_2 and slides along a horizontal surface for which the dynamic coefficient of friction with the body is μ_d. Body 2 is supported initially by some external agent so that, at the outset of the problem, the spring and cable are unstretched. What is the total kinetic energy of the system when, after release, body 2 has moved a distance d_2 and body 1 has moved a smaller distance d_1?

Use Eq. 11.27. Only one nonconservative force exists in the system, the external friction force on body 1. Therefore, the work term of the equation becomes

$$\mathcal{W}_{1-2} = -W_1 \mu_d d_1 \tag{a}$$

Three conservative forces are present; the spring force and the gravitational force are *external* and the force from the elastic cable is *internal*. (We neglect mutual gravitational forces between the bodies.) Using the initial position of W_2 as the datum for gravitational potential energy, we have, for the total change in potential energy:

$$\Delta\text{PE} = \left[\tfrac{1}{2}K_2 d_1^2 - 0\right] + \left[\tfrac{1}{2}K_1\left(d_2 - d_1\right)^2 - 0\right] + \left[0 - W_2 d_2\right]$$

We can compute the desired change in kinetic energy from Eq. 11.27 as

$$\Delta\text{KE} = -W_1 \mu_d d_1 - \tfrac{1}{2}K_2 d_1^2 - \tfrac{1}{2}K_1\left(d_2 - d_1\right)^2 + W_2 d_2 \tag{b}$$

As an additional exercise, you should arrive at this result by using the basic Eq. 11.26, where you cannot rely on familiar formulas for potential energies.

In the following example, we will see how using a system of particles approach can make for great simplification in a problem over the procedure of dealing with particles individually. Also we will have a case where there is no nonconservative work, which then results in a conservation of mechanical energy.

Example 11.10

Masses A and B, each having a mass of 75 kg, are constrained to move in frictionless slots (see Fig. 11.20). They are connected by a light rod of length l = 300 mm. Mass B is connected to two massless linear springs, each having a spring constant K = 900 N/m. The springs are unstretched when the connecting rod to masses A and B is vertical. What are the velocities of B and A when A descends a distance of 25 mm? There is no friction in the end connections of the rod.[10]

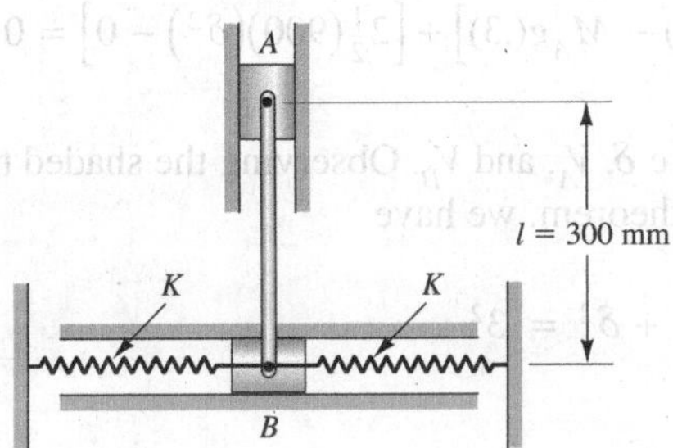

Figure 11.20. Two interconnected masses constrained by linear springs.

We shall use a **system of particles** approach for this case. This will eliminate the need to calculate the work of the rod on each mass that we would need had we elected to deal with each mass separately. For a system of particles approach, this work is internal between rigid bodies, having zero value as a result of **Newton's third law** and having ideal pin-connected joints.

We next note that only conservative forces are present (gravitational force and spring forces) so the first law form of our energy equation degenerates to conservation of mechanical energy. In Fig. 11.21, we show the system in a configuration where mass A

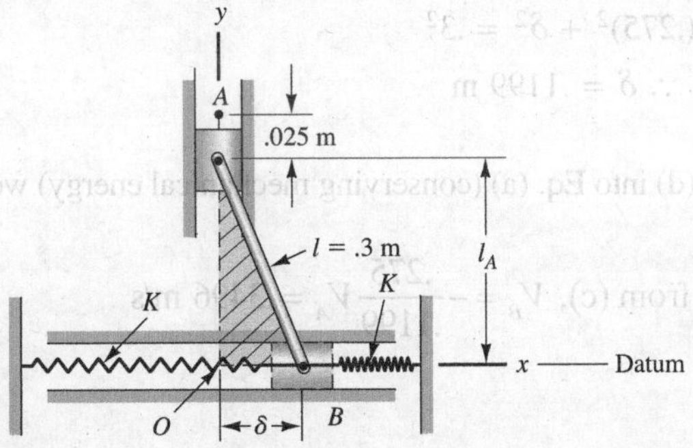

Figure 11.21. System in a configuration wherein mass A has dropped 25 mm.

[10]This is an example of an **unstable equilibrium** (a topic considered in Chapter 8) in that any slight movement of block B from its central position to the right or to the left causes A to start to accelerate downward.

Example 11.10 (Continued)

has dropped a distance of .025 m. We can then say for the beginning and end configurations using the datum shown in Fig. 11.21,

$$\Delta[\text{KE} + \text{PE}] = \mathcal{W}_{1-2} = 0$$

$$\left(\tfrac{1}{2}M_A V_A^2 - 0\right) + \left(\tfrac{1}{2}M_B V_B^2 - 0\right) + \left[M_A g(.3 - .025) - M_A g(.3)\right] + \left[2\tfrac{1}{2}(900)(\delta^2) - 0\right] = 0 \qquad \text{(a)}$$

We have three unknowns here. They are δ, V_A, and V_B. Observing the shaded triangle in Fig. 11.21 and using the Pythagorean theorem, we have

$$l_A^2 + \delta^2 = .3^2 \qquad \text{(b)}$$

Taking the time derivative we get

$$2l_A\dot{l}_A + 2\delta\dot{\delta} = 0$$

We note that $\dot{l}_A = V_A$ and that $\dot{\delta} = V_B$. We then see from the preceding equation that

$$V_B = -\frac{l_A}{\delta}V_A \qquad \text{(c)}$$

Now, returning to Fig. 11.21, we can compute δ for the case at hand. Noting that A has descended a distance of .025 m, thus making $l_A = .3 - .025 = .275$ m, we next go to Eq. (b) to get δ. Thus

$$(.275)^2 + \delta^2 = .3^2$$
$$\therefore \delta = .1199 \text{ m} \qquad \text{(d)}$$

Substituting data from Eqs. (c) and (d) into Eq. (a) (conserving mechanical energy) we get

$$V_A = -.1524 \text{ m/s} \quad \therefore \text{from (c), } V_B = -\frac{.275}{.1199}V_A = .3496 \text{ m/s}$$

Thus,

$V_A = .1524$ m/s downward $\qquad V_B = .3496$ m/s to the right or the left

11.6 Kinetic Energy Expression Based on Center of Mass

In this and the next section, we shall introduce the center of mass into our discussion in order to develop useful expressions for the kinetic energy of an aggregate of particles. Also, we shall develop the work–energy equation for the center of mass set forth at the outset of this chapter.

Consider a system of n particles, shown in Fig. 11.22. The total kinetic energy relative to xyz of the system of particles can be given as

$$\text{KE} = \sum_{i=1}^{n} \tfrac{1}{2} m_i V_i^2 \tag{11.28}$$

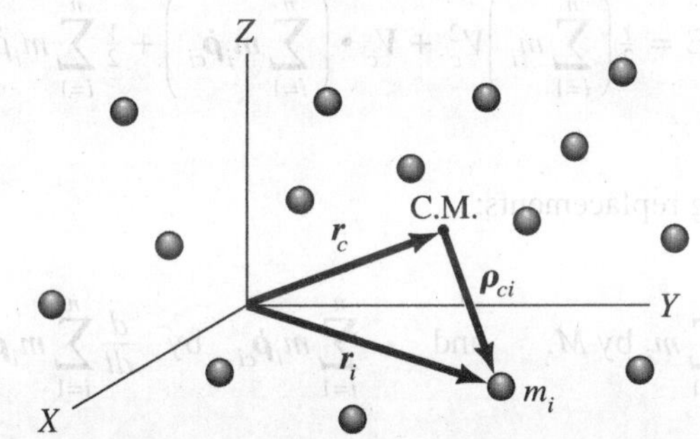

Figure 11.22. System of particles with center of mass.

We shall now express Eq. 11.28 in another way by introducing the mass center. Note in the diagram we have employed the vector $\boldsymbol{\rho}_{ci}$ as the displacement vector from the center of mass to the ith particle. We can accordingly say:

$$\boldsymbol{r}_i = \boldsymbol{r}_c + (\boldsymbol{r}_i - \boldsymbol{r}_c) = \boldsymbol{r}_c + \boldsymbol{\rho}_{ci} \tag{11.29}$$

Differentiating with respect to time, we get

$$\dot{\boldsymbol{r}}_i = \dot{\boldsymbol{r}}_c + (\dot{\boldsymbol{r}}_i - \dot{\boldsymbol{r}}_c) = \dot{\boldsymbol{r}}_c + \dot{\boldsymbol{\rho}}_{ci}$$

Therefore,

$$\boldsymbol{V}_i = \boldsymbol{V}_c + \dot{\boldsymbol{\rho}}_{ci} \tag{11.30}$$

From our earlier discussions on simple relative motion we can say that $\dot{\boldsymbol{\rho}}_{ci}$ is the motion of the ith particle *relative to the mass center*.[11] Substituting the relation above into the expression for kinetic energy, Eq. 11.28, we get

[11]Note that

$$\dot{\boldsymbol{\rho}}_{ci} = \dot{\boldsymbol{r}}_i - \dot{\boldsymbol{r}}_c$$

That is, $\dot{\boldsymbol{\rho}}_{ci}$ is the *difference* between the velocity of the ith particle and that of the mass center. This is then the velocity of the particle *relative* to the center of the mass (i.e., relative to a reference translating with c or to a nonrotating observer moving with c.)

$$\text{KE} = \sum_{i=1}^{n} \tfrac{1}{2} m_i (\boldsymbol{V}_c + \dot{\boldsymbol{\rho}}_{ci})^2 = \sum_{i=1}^{n} \tfrac{1}{2} m_i (\boldsymbol{V}_c + \dot{\boldsymbol{\rho}}_{ci}) \bullet (\boldsymbol{V}_c + \dot{\boldsymbol{\rho}}_{ci})$$

Carrying out the dot product, we have

$$\text{KE} = \tfrac{1}{2} \sum_{i=1}^{n} m_i V_c^2 + \sum_{i=1}^{n} m_i \boldsymbol{V}_c \bullet \dot{\boldsymbol{\rho}}_{ci} + \tfrac{1}{2} \sum_{i=1}^{n} m_i \dot{\rho}_{ci}^2 \tag{11.31}$$

Since $\boldsymbol{V}_c$ is common for all values of the summation index, we can extract it from the summation operation, and this leaves

$$\text{KE} = \tfrac{1}{2} \left(\sum_{i=1}^{n} m_i \right) V_c^2 + \boldsymbol{V}_c \bullet \left(\sum_{i=1}^{n} m_i \dot{\boldsymbol{\rho}}_{ci} \right) + \tfrac{1}{2} \sum_{i=1}^{n} m_i \dot{\rho}_{ci}^2 \tag{11.32}$$

Perform the following replacements:

$$\sum_{i=1}^{n} m_i \text{ by } M, \quad \text{and} \quad \sum_{i=1}^{n} m_i \dot{\boldsymbol{\rho}}_{ci} \text{ by } \frac{d}{dt} \sum_{i=1}^{n} m_i \boldsymbol{\rho}_{ci}$$

We then have

$$\text{KE} = \tfrac{1}{2} M V_c^2 + \boldsymbol{V}_c \bullet \frac{d}{dt} \sum_{i=1}^{n} m_i \boldsymbol{\rho}_{ci} + \tfrac{1}{2} \sum_{i=1}^{n} m_i \dot{\rho}_{ci}^2 \tag{11.33}$$

But the expression

$$\sum_{i=1}^{n} m_i \boldsymbol{\rho}_{ci}$$

represents the first moment of mass of the system about the center of mass for the system. Clearly by definition, this quantity must always be zero. The expression for kinetic energy becomes

$$\text{KE} = \tfrac{1}{2} M V_c^2 + \tfrac{1}{2} \sum_{i=1}^{n} m_i \dot{\rho}_{ci}^2 \tag{11.34}$$

Thus, we see that the *kinetic energy* for some reference *can be considered to be composed of two parts: (1) the kinetic energy of the total mass moving relative to that reference with the velocity of the mass center, plus (2) the kinetic energy of the motion of the particles relative to the mass center.*

Example 11.11

A hypothetical vehicle is moving at speed V_0 in Fig. 11.23. On this vehicle are two bodies each of mass m sliding along a horizontal rod at a speed v relative to the rod. This rod is rotating at an angular speed ω rad/sec relative to the vehicle. What is the kinetic energy of the two bodies relative to the ground (XYZ) when they are at a distance r from point A?

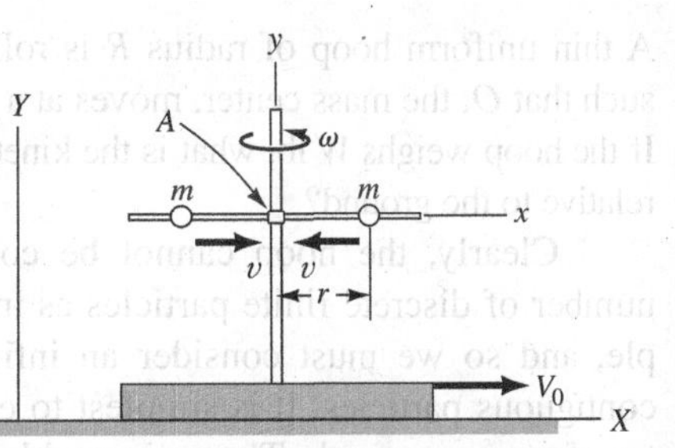

Figure 11.23. Moving device.

Clearly, the center of mass corresponds to point A and is thus moving at a speed V_0 relative to the ground. Hence, we have as part of the kinetic energy the term

$$\tfrac{1}{2}MV_c^2 = mV_0^2 \qquad \text{(a)}$$

The velocity of each ball relative to the center of mass is easily formed using cylindrical components. Thus, imagining a reference xyz at A translating with the vehicle relative to XYZ, we have for the velocity of each ball relative to xyz:

$$\dot{\rho}^2 = \dot{r}^2 + (\omega r)^2 = v^2 + (\omega r)^2 \qquad \text{(b)}$$

The total kinetic energy of the two masses relative to the ground is then

$$\text{KE} = mV_0^2 + m\left[v^2 + (\omega r)^2\right] \qquad \text{(c)}$$

In Example 11.11, we considered a case where the bodies involved constituted a finite number of *discrete* particles. In the next example, we consider a case where we have a *continuum* of particles forming a rigid body. The formulation given by Eq. 11.34 can still be used but now, instead of summing for a finite number of discrete particles, we must integrate to account for the infinite number of infinitesimal particles comprising the system. We are thus taking a glimpse, for simple cases, of rigid-body dynamics to be studied later in the text. Those that do not have time for studying such energy problems in detail will be able to solve simple but useful rigid-body dynamics problems on the basis of these examples as well as later examples in this chapter.

Example 11.12

A thin uniform hoop of radius R is rolling without slipping such that O, the mass center, moves at a speed V (Fig. 11.24). If the hoop weighs W lb, what is the kinetic energy of the hoop relative to the ground?

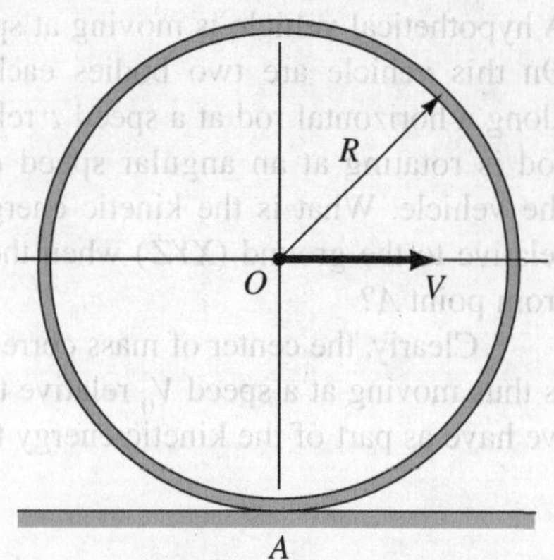

Figure 11.24. Rolling hoop.

Clearly, the hoop cannot be considered as a finite number of discrete finite particles as in the previous example, and so we must consider an infinity of infinitesimal contiguous particles. It is simplest to employ here the center-of-mass approach. The main problem then is to find the kinetic energy of the hoop relative to the mass center O, that is, relative to a reference xy translating with the mass center as seen from the ground reference XY (see Fig. 11.25). The motion relative to xy is clearly simple rotation; accordingly, we must find the angular velocity of the hoop for this reference. The no slipping condition means that the point of contact of the hoop with the ground has instantaneously a zero velocity. Observe the motion from a stationary reference XY. As you may have learned in physics, and as will later be shown (Chapter 13), the body has a *pure instantaneous rotational* motion about the point of contact. The angular velocity ω for this motion is then easily evaluated by considering point O rotating about the instantaneous center of rotation A. Thus,

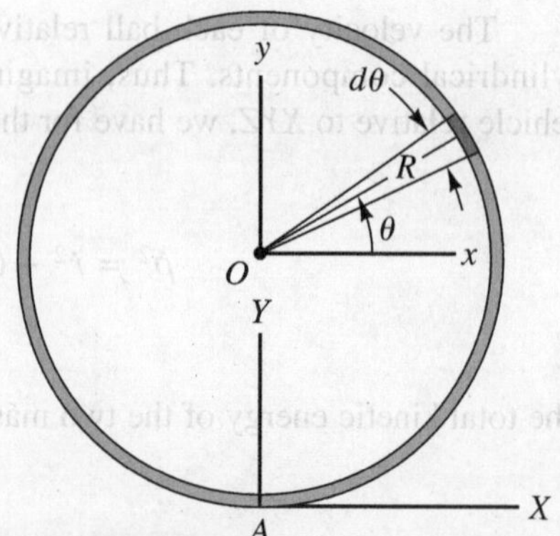

Figure 11.25. xy translates with O relative to XY.

$$\omega = \frac{V}{R} \tag{a}$$

Since reference xy *translates* relative to reference XY, an observer on xy sees the *same angular velocity* ω for the hoop as the observer on XY. Accordingly, we can now readily evaluate the second term on the right side of Eq. 11.34. As particles, use elements of the hoop which are $R\,d\theta$ in length, as shown in Fig. 11.25, and which have a mass per unit length of $W/(2\pi Rg)$. We then have, on replacing summation by integration, the result

$$\begin{aligned}\frac{1}{2}\sum_{i=1}^{n} m_i \dot{\rho}_{ci}^2 &= \frac{1}{2}\int_0^{2\pi}\left[\left(\frac{W}{2\pi Rg}\right)(R\,d\theta)\right](\omega R)^2 \\ &= \frac{1}{2}\int_0^{2\pi}\left[\frac{W}{2\pi Rg}(R\,d\theta)\right]\left(\frac{V}{R}R\right)^2 \\ &= \frac{1}{2}\frac{W}{g}V^2\end{aligned} \tag{b}$$

The kinetic energy of the hoop is then in accordance with Eq. 11.34:

Example 11.12 (Continued)

$$\text{KE} = \underbrace{\frac{1}{2}\frac{W}{g}V^2}_{(\text{KE})_{\text{C.M.}}} + \underbrace{\frac{1}{2}\frac{W}{g}V^2}_{(\text{KE})_{\text{Rel. to C.M.}}} = \frac{W}{g}V^2 \quad \text{(c)}$$

Suppose that the body were a generalized cylinder of mass M (see Fig. 11.26) such as a tire of radius R having O as the center of mass with axisymmetrical distribution of mass about the axis at O. Then, we would express Eq. (b) as follows:

$$\frac{1}{2}\sum_i^n m_i\dot{\rho}_{ci}^2 = \frac{1}{2}\iiint_M (dm)(r\omega)^2 \quad \text{(d)}$$

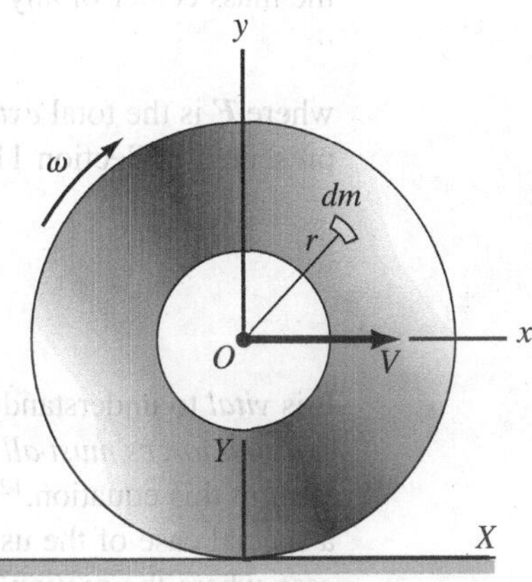

Figure 11.26. Rolling generalized cylinder of mass M.

You will recall from Chapter 8 that

$$\iiint_M r^2\, dm$$

is the *second moment of inertia* of the body taken about the z axis at O. That is,

$$I_{zz} = \iiint_M r^2\, dm$$

Thus, we have for the kinetic energy of such a body:

$$\text{KE} = \frac{1}{2}MV^2 + \frac{1}{2}I_{zz}\omega^2 \quad \text{(e)}$$

You may also recall from Chapter 8 that we could employ the *radius of gyration k* to express I_{zz} as follows:

$$I_{zz} = k^2 M \quad \text{(f)}$$

Hence, Eq. (e) can be given as

$$\text{KE} = \frac{1}{2}MV^2 + \frac{1}{2}k^2 M\omega^2$$

We shall examine the kinetic energy formulations of rigid bodies carefully in Chapter 15. Here, we have used certain familiar results from physics pertaining to kinematics of plane motion of a nonslipping rolling rigid body. For a more general undertaking, we shall have to carefully consider more general aspects of kinematics of rigid-body motion. This will be done in Chapter 13. Also, in the last example we see one term of the inertia tensor I_{ij} showing up. The vital role of the inertia tensor in the dynamics of rigid bodies will soon be seen.

11.7 Work–Kinetic Energy Expressions Based on Center of Mass

The work–kinetic energy expressions of Section 11.5 were developed for a system of particles without regard to the mass center. We shall now introduce this point into the work–kinetic energy formulations. You will recall from Chapter 10 that *Newton's law* for the mass center of any system of particles is

$$\boldsymbol{F} = M\ddot{\boldsymbol{r}}_c \tag{11.35}$$

where $\boldsymbol{F}$ is the total *external* force on the system of particles. By the same development as presented in Section 11.1, we can readily arrive at the following equation:

$$\int_1^2 \boldsymbol{F} \bullet d\boldsymbol{r}_c = \left(\tfrac{1}{2}MV_c^2\right)_2 - \left(\tfrac{1}{2}MV_c^2\right)_1 \tag{11.36}$$

It is *vital* to understand from the left side of Eq. 11.36, where we note the term $d\boldsymbol{r}_c$, that the *external forces must all move with the center of mass* for the computation of the proper work term in this equation.[12] We wish next to point out that the single particle model represents a special case of the use of Eq. 11.36. Specifically, the single particle model represents the case where the motion of the center of mass of a body sufficiently describes the motion of the body and where the external forces on the body essentially move with the center of mass of the body. Such cases were set forth in Section 11.1.

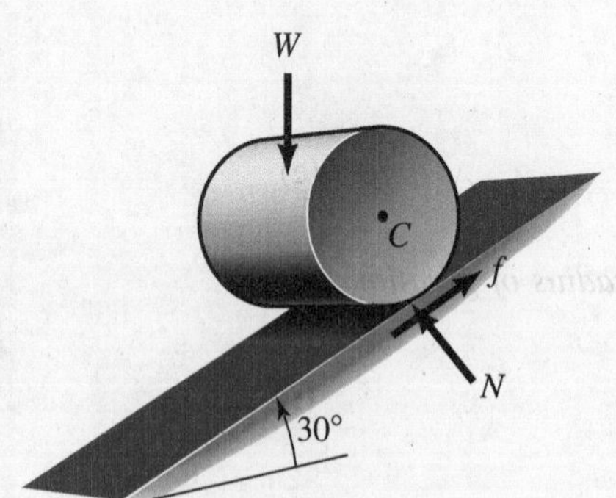

Figure 11.27. Cylinder on incline.

Before proceeding to the examples, let us consider for a moment the case of the cylinder rolling without slipping down an incline (see Fig. 11.27). We shall consider the cylinder as an *aggregate of particles* which form a rigid body—namely a cylinder. When using such an approach, we require that *all the forces both external and internal must move with their respective points of application.* Let us then consider the external work done on the particles making up the cylinder other than the work done by gravity. Clearly, only particles on the *rim* of the cylinder are acted on by external forces other than gravity. Consider one such particle during one rotation of the cylinder. This particle will have acting on it a friction force f and a normal force N at the *instant* when the particle is in *contact* with the inclined surface. The particle will have *zero* external force (except for gravity) at all other positions during the cycle. Now, at the instant of this contact, the normal force N has zero velocity in its direction because of the rigidity of the bodies. Therefore, N transmits no power and does no work on the particle during the cycle under consideration. Also, the friction force f acts on a particle having zero velocity at the instant of contact because of the *no slipping* condition. Accordingly, f transmits no

[12]This is in direct contrast to the work–energy equation for a system of particles wherein as was pointed out emphatically earlier, each external force moves with its *actual* point of application. Also, only external forces are involved for the center-of-mass approach, in contrast to the system of particles where internal forces may also be involved. Note that in Examples 11.1, 11.2, and 11.18 we were using a particle approach and thus were really considering the motion of the center of mass. The friction forces then moved with the center of mass.

power and does no work on the particle during this cycle.[13] This result must be true for each and every particle on the rim of the cylinder. Thus, clearly, f and N do no work when the cylinder rolls down the incline. Also, because of the rigidity of the body the internal forces do no work as pointed out earlier. Thus, only gravity does work.

However, in considering the motion of the *center of mass* C of the cylinder in Fig. 11.27, we note that force f now *moves* with C and hence *does* work.

At the risk of being repetitive, we now summarize the key features for properly using the center-of-mass approach.

1. Only **external** forces are involved.
2. Forces all move with the **center of mass** when computing work (see Eq. 13.36).
3. Only the kinetic energy of the center of mass is used.

[13]We multiply and divide the usual expression for work by dt as follows:

$$\int_1^2 \boldsymbol{F} \bullet d\boldsymbol{r} = \int_1^2 \boldsymbol{F} \bullet \frac{d\boldsymbol{r}}{dt}\, dt = \int_1^2 \boldsymbol{F} \bullet \boldsymbol{V}\, dt$$

We can now clearly see with $\boldsymbol{V} = \boldsymbol{0}$ at the point of contact that there will be zero work from the friction force f there.

Example 11.13

A cylinder with a mass of 25 kg is released from rest on an incline, as shown in Fig. 11.28. The diameter of the cylinder is .60 m. If the cylinder rolls without slipping, compute the speed of the centerline C after it has moved 1.6 m along the incline . Also, ascertain the friction force acting on the cylinder. Use the result from Problem 11.76 that the kinetic energy of a cylinder rotating about its own stationary axis is $\frac{1}{4}MR^2\omega^2$, where ω is the angular speed in rad/sec.

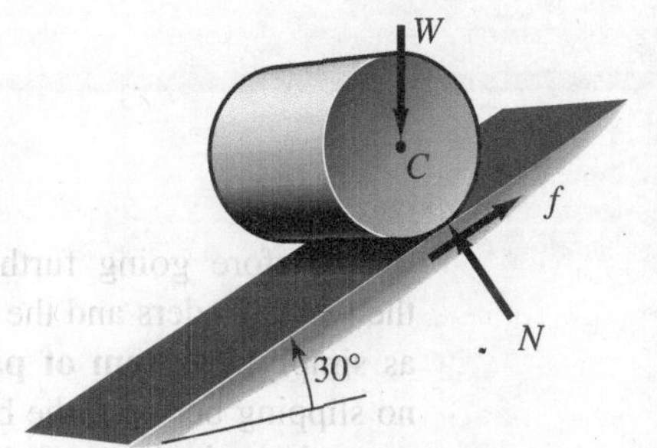

Figure 11.28. Free-body diagram of cylinder.

In Fig. 11.28 we have shown the free body of the cylinder. We proceed to use the **work–energy equation** for a **system of particles.** Recall that we can concentrate the weight at the center of gravity (Section 11.5). Accordingly, using the lowest position as a datum and noting from our earlier discussion that the friction force f does no work we have

$$\Delta(\text{PE} + \text{KE}) = \mathcal{W}_{1-2} \tag{a}$$

$$[0 - (25)(9.81)(1.6)\sin 30°] + \left\{\left[\tfrac{1}{2}(25)V_c^2 + \tfrac{1}{4}(25)(.30)^2(\omega^2)\right] - 0\right\} = 0$$

Example 11.13 (Continued)

where the kinetic energy of the cylinder is given as the kinetic energy of the mass taken at the mass center (straight line motion of C) plus kinetic energy of the cylinder relative to the center of mass (pure rotation about C). Noting from Example 11.12 that

$$\omega = \frac{V}{R} = \frac{V}{.30}$$

we substitute into Eq. (a) and solve for V_c. We get

$$V_c = 3.23 \text{ m/sec} \qquad \text{(b)}$$

Now to find f, we *consider the motion of the mass center of the cylinder*. This means that we use Eq. 11.36 for the **center of mass**. Now *all external forces must move with the center of mass*; thus, f does work. Since the center of mass moves along a path always at right angles to N, this force still does no work. Accordingly, we can say:

$$-f(1.6) + W(1.6 \sin 30°) = \tfrac{1}{2}MV_c^2$$
$$-f(1.6) + (25)(9.81)(1.6)\sin 30° = \tfrac{1}{2}(25)(3.23^2)$$

$$f = 41.1 \text{ N}$$

Before going further, let us consider the two cylinders and the block in Fig. 11.29 as simply a **system of particles**. If there is no slipping between the block and the cylinders, the velocities of the particles on the block and the cylinders at the points of contact between these bodies have the *same* velocity at any time t. Furthermore, the friction force on the cylinder from the block is *equal* and *opposite* to the friction force on the block from the cylinder at the point of contact. We can then conclude that there is zero net work done by the friction forces between block and cylinders when considering them as an aggregate of particles.

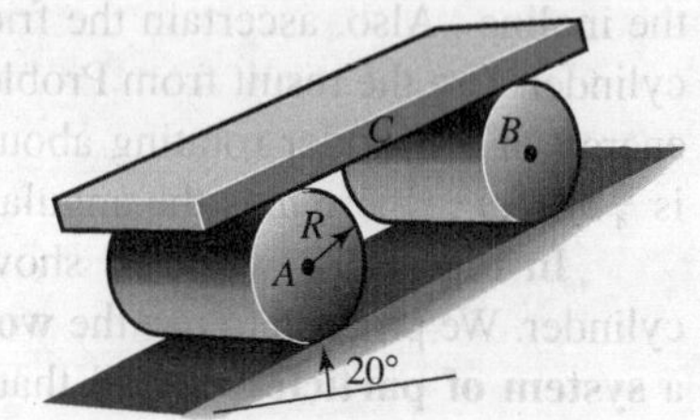

Figure 11.29. Three rigid bodies moving without slipping at any of the contact points.

Also, in the next problem, we will consider as a system of particles, rigid bodies which are joined by rigid connectors with frictionless interconnections.

In the next example, we shall consider a case where *internal forces* do work.

Example 11.14

An external torque T of 50 N-m is applied to a solid cylinder B (see Fig. 11.30), which has a mass of 30 kg and a radius of .2 m. The cylinder rolls without slipping. Block A, having a mass of 20 kg, is dragged up the 15° incline. The dynamic coefficient of friction μ_d between block A and the incline is .25. The connections at C and D are frictionless.

(a) What is the velocity of the system after moving a distance d of 2 m?
(b) What is the friction force on the cylinder?

Neglect the mass of the connecting rod.

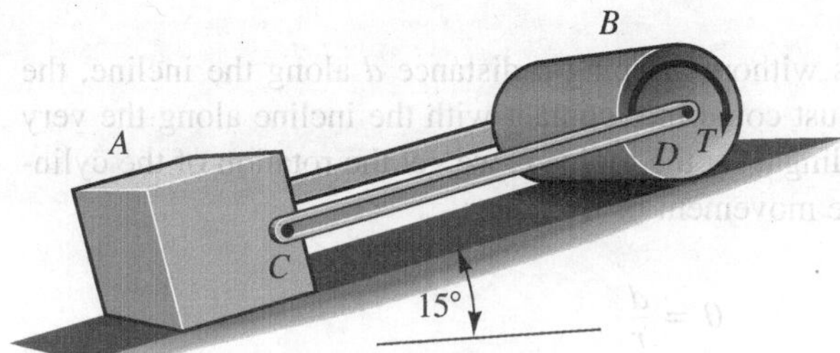

Figure 11.30. Torque-driven cylinder moves without slipping.

We show a free-body diagram of the system in Fig. 11.31. We begin by employing a **system of particles** point of view. Note there are pairs of internal forces present between the rod CD and body A at the contact point

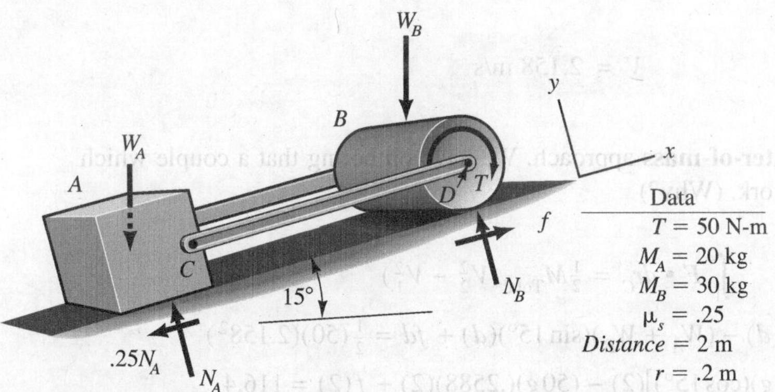

Figure 11.31. Free-body diagram of the system.

Example 11.14 (Continued)

and similarly between CD and cylinder B. These force pairs are equal and opposite because of Newton's third law. And because the forces in each pair move exactly the same distance at the respective points of contact, there will be zero internal work from these force pairs. Hence, using the uppermost configuration as the datum,

$$\mathcal{W}_{1\to 2} = \Delta\text{PE} + \Delta\text{KE}$$

$$T\theta - (\mu_d N_A)(d) = \left[(W_A + W_B)(d)(\sin 15°) - 0\right] \qquad \text{(a)}$$
$$+\left[(M_A + M_B)\frac{V^2}{2} + \frac{1}{4}M_B r_B^2 \omega_B^2 - 0\right]$$

Note that as the cylinder moves without slipping a distance d along the incline, the circumference of the cylinder must come into contact with the incline along the very same distance d. Hence, by dividing d by the radius r, we get the rotation of the cylinder in radians associated with the movement of its center.

$$\theta = \frac{d}{r}$$

We then get for Eq. (a), on substituting data for the problem

$$(50)\left(\frac{2}{.2}\right) - (.25)[(20g)(\cos 15°)](2) = [(50g)(2)(.2588)]$$
$$+\left[(50)\frac{V^2}{2} + \frac{1}{4}(30)(.2^2)\frac{V^2}{(.2)^2}\right]$$

$$V = 2.158 \text{ m/s}$$

Next, use the **center-of-mass** approach. We have on noting that a couple which is translating does no work. (Why?)

$$\int_1^2 \boldsymbol{F} \bullet d\boldsymbol{r}_C = \tfrac{1}{2}M_{\text{Total}}(V_2^2 - V_1^2)$$

$$-(.25)(N_A)(d) - (W_A + W_B)(\sin 15°)(d) + fd = \tfrac{1}{2}(50)(2.158^2)$$

$$-(.25)[(20g)(\cos 15°)](2) - (50g)(.2588)(2) + f(2) = 116.4$$

$$f = 232.5 \text{ N}$$

Example 11.15

In Example 11.14, suppose that cylinder B is *slipping*. What is the dynamic coefficient of friction $(\mu_d)_B$ between the cylinder and the incline so that the system reaches a speed of 1.5 m/s after moving a distance $d = 2$ m starting from rest?

Using the **center-of-mass** approach, we have

$$\int_1^2 \boldsymbol{F} \bullet d\boldsymbol{r}_C = \left(\tfrac{1}{2}MV_C^2\right)_2 - \left(\tfrac{1}{2}MV_C^2\right)_1$$

$$-(W_A \cos 15°)(.25)(d) + (W_B \cos 15°)(\mu_d)_B(d) - (W_A + W_B)(\sin 15°)(d) = \frac{1}{2}\frac{W_A + W_B}{g}V^2 - 0$$

$$-(20g)(\cos 15°)(.25)(2) + (\mu_d)_B(30g)(\cos 15°)(2) - (50g)(\sin 15°)(2) = \frac{1}{2}(50)(1.5^2) - 0$$

$$(\mu_d)_B = .7122$$

In the next example, we shall consider a case where *internal forces* do work.

Example 11.16

A diesel-powered electric train moves up a 7° grade in Fig. 13.32. If a torque of 750 N-m is developed at each of its six pairs of drive wheels, what is the increase of speed of the train after it moves 100 m? Initially, the train has a speed of 16 km/hr. The train weighs 90 kN. The drive wheels have a diameter of 600 mm. Neglect the rotational energy of the drive wheels.

Figure 11.32. Diesel–electric train.

We shall consider the train as a **system of particles** including the 6 pairs of wheels and the body. We have shown the train in Fig. 11.33 with the external forces, W, N, and f. In addition, we have shown certain internal torques T.[15] The torques shown act on the *rotors* of the motors, and, as the train moves, these torques rotate and accordingly do work. The *reactions* to these torques are equal and opposite to T according to Newton's third law and act on the *stators* of the motors (i.e., the field coils). The stators are stationary, and so

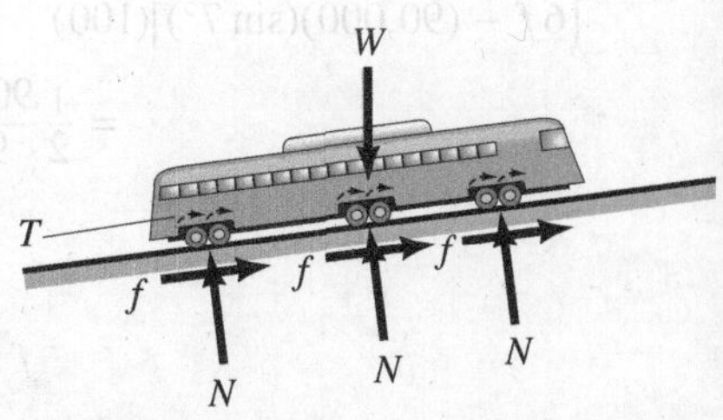

Figure 11.33. External and internal forces and torques.

[14]Figure 11.33. accordingly, is *not* a free-body diagram.

Example 11.16 (Continued)

the reactions to T do *no* work as the train moves. Thus we have an example wherein, using a **system of particles approach**, internal forces perform a nonzero amount of work. We now employ Eq. 11.28. Thus

$$\Delta \text{PE} + \text{KE} = \mathcal{W}_{1\to 2} \qquad \text{(a)}$$

For the rolling without slipping condition, the friction forces f do no work. We then have

$$\left[(90{,}000)(100\sin 7^\circ - 0)\right] + \left\{\frac{1}{2}\frac{90{,}000}{g}V^2 - \frac{1}{2}\frac{90{,}000}{g}\left[\frac{(16)(1{,}000)}{3{,}600}\right]^2\right\}$$
$$= (6)(750)(\theta) \qquad \text{(b)}$$

where θ is the clockwise rotation of the rotor in radians. Assuming direct drive from rotor to wheel, we can compute θ as follows for the 100-m distance d over which the train moves:

$$\theta = \frac{d}{r} = \frac{100}{.3} = 333.3 \text{ rad} \qquad \text{(c)}$$

Substituting into Eq. (b) and solving for V, we get

$$V = 10.38 \text{ m/sec}$$

Hence, the increase of speed of the train is

$$\Delta V = \frac{(10.38)(3{,}600)}{1{,}000} - 16 = \boxed{21.4 \text{ km/hr}} \qquad \text{(d)}$$

To determine the friction forces f, we now adopt a **center-of-mass** approach. Thus all forces now move with the center of mass. And they must be *external forces*. Accordingly we have

$$[6f - (90{,}000)(\sin 7^\circ)](100)$$
$$= \frac{1}{2}\frac{90{,}000}{9.81}(10.38)^2 - \frac{1}{2}\frac{90{,}000}{9.81}\left(\frac{(16)(1{,}000)}{3{,}600}\right)^2$$

$$\boxed{f = 2{,}500 \text{ N}}$$

PROBLEMS

11.68. A chain of total length L is released from rest on a smooth support as shown. Determine the velocity of the chain when the last link moves off the horizontal surface. In this problem, neglect the friction. Also, do not attempt to account for centrifugal effects stemming from the chain links rounding the corner.

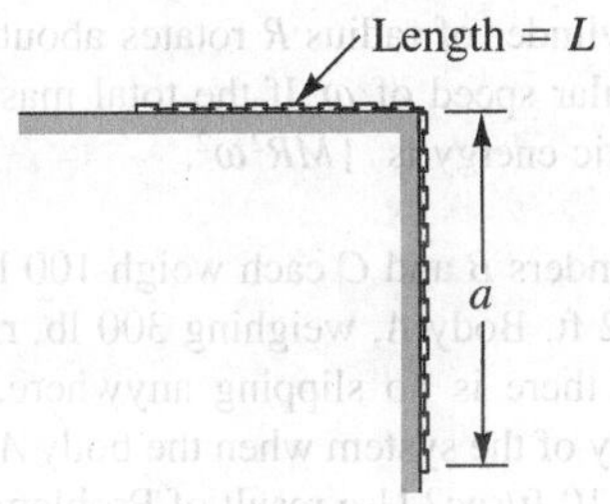

Figure P.11.68.

11.69. A chain is 50 ft long and weighs 100 lb. A force P of 80 lb has been applied at the configuration shown. What is the speed of the chain after force P has moved 10 ft? The dynamic coefficient of friction between the chain and the supporting surface is .3. Utilize an approximate analysis.

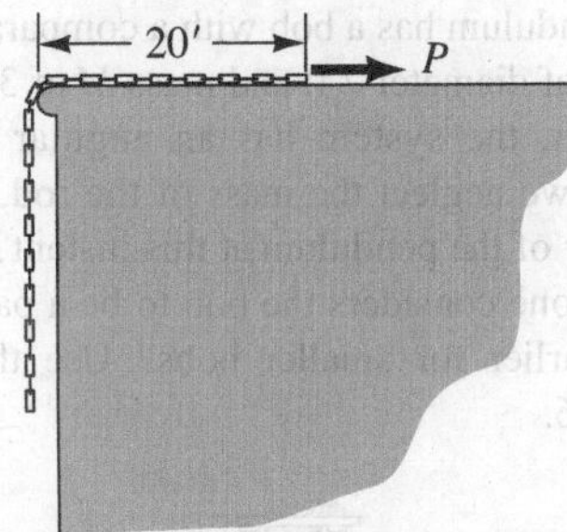

Figure P.11.69.

11.70. A bullet of weight W_1 is fired into a block of wood weighing W_2 lb. The bullet lodges in the wood, and both bodies then move to the dashed position indicated in the diagram before falling back. Compute the amount of internal work done during the action. Discuss the effects of this work. The bullet has a speed V_0 before hitting the block. Neglect the mass of the supporting rod and friction at A.

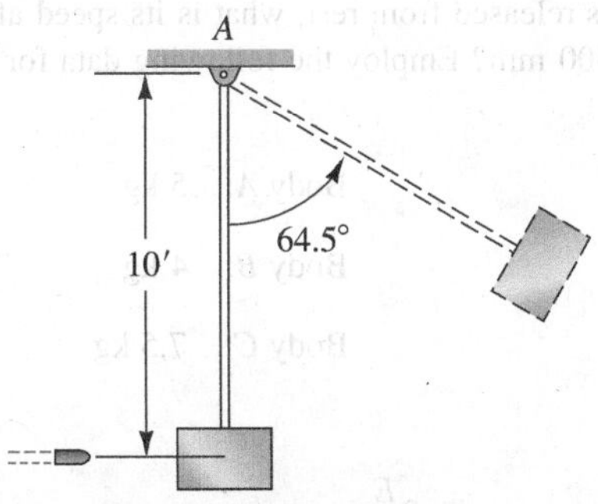

Figure P.11.70.

11.71. A device is mounted on a platform that is rotating with an angular speed of 10 rad/sec. The device consists of two masses (each is .1 slug) rotating on a spindle with an angular speed of 5 rad-sec relative to the platform. The masses are moving radially outward with a speed of 10 ft/sec, and the entire platform is being raised at a speed of 5 ft/sec. Compute the kinetic energy of the system of two particles when they are 1 ft from the spindle.

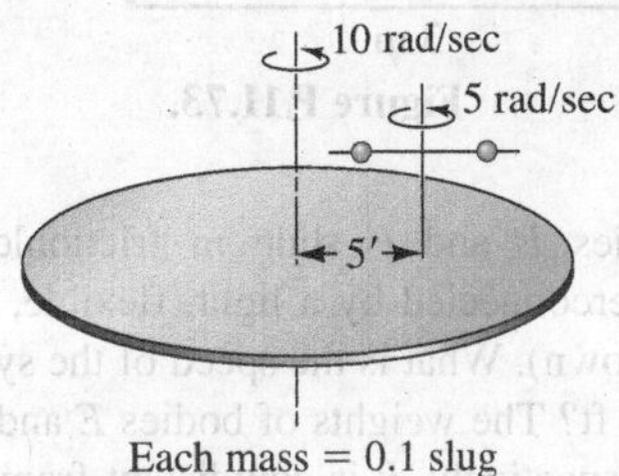

Figure P.11.71.

11.72. A hoop, with four spokes, rolls without slipping such that the center C moves at a speed V of 1.7 m/sec. The diameter of the hoop is 3.3 m and the weight per unit length of the rim is 14 N/m. The spokes are uniform rods also having a weight per unit length of 14 N/m. Assume that rim and spokes are thin. What is the kinetic energy of the body?

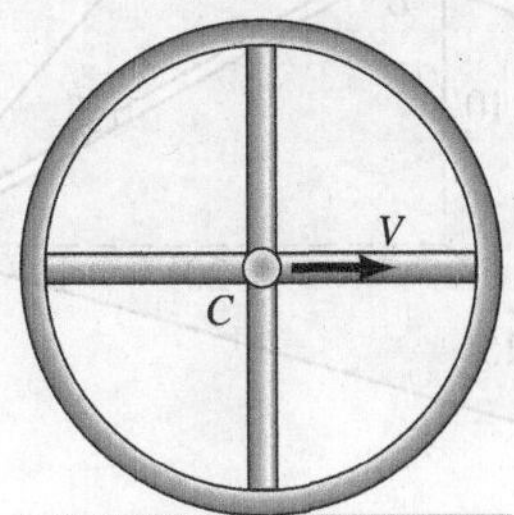

Figure P.11.72.

11.73. Three weights A, B, and C slide frictionlessly along the system of connected rods. The bodies are connected by a light, flexible, inextensible wire that is directed by frictionless small pulleys at E and F. If the system is released from rest, what is its speed after it has moved 300 mm? Employ the following data for the body masses:

Body A: 5 kg

Body B: 4 kg

Body C: 7.5 kg

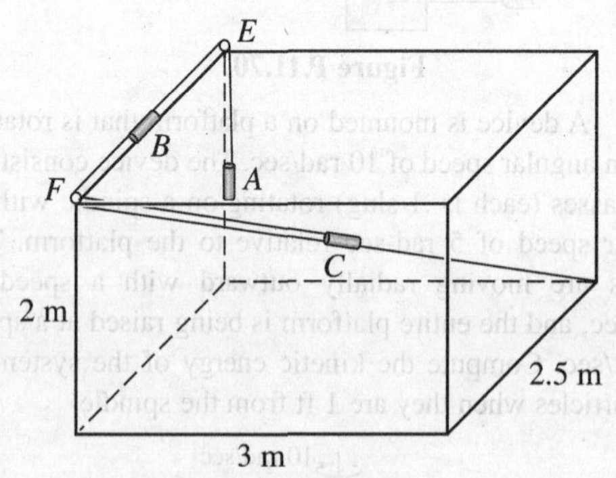

Figure P.11.73.

11.74. Bodies E and F slide in frictionless grooves. They are interconnected by a light, flexible, inextensible cable (not shown). What is the speed of the system after it has moved 2 ft? The weights of bodies E and F are 10 lb and 20 lb, respectively. B is equidistant from A and C. E remains in top groove.

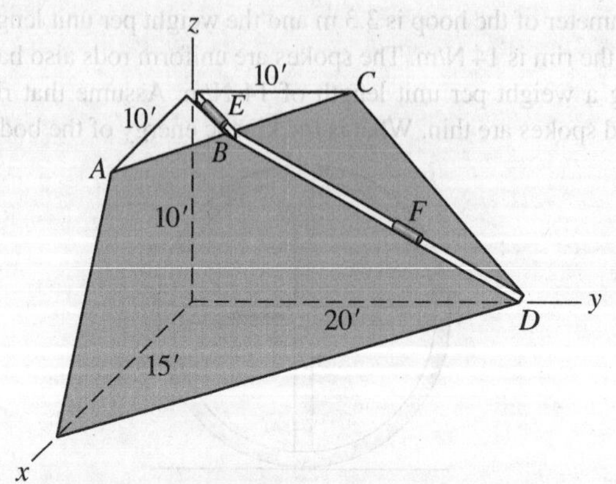

Figure P.11.74.

11.75. A tank is moving at the speed V of 16 km/hr. What is the kinetic energy of each of the treads for this tank if they each have a mass per unit length of 300 kg/m?

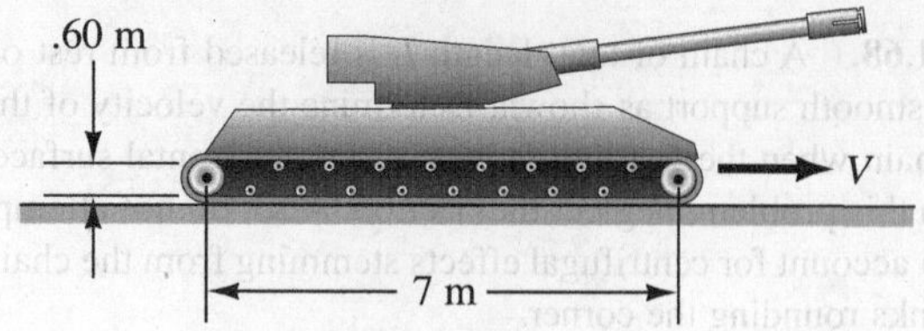

Figure P.11.75.

11.76. A cylinder of radius R rotates about its own axis with an angular speed of ω. If the total mass is M, show that the kinetic energy is $1MR^2\omega^2$.

11.77. Cylinders B and C each weigh 100 lb and have a diameter of 2 ft. Body A, weighing 300 lb, rides on these cylinders. If there is no slipping anywhere, what is the kinetic energy of the system when the body A is moving at a speed V of 10 ft/sec? Use result of Problem 11.76.

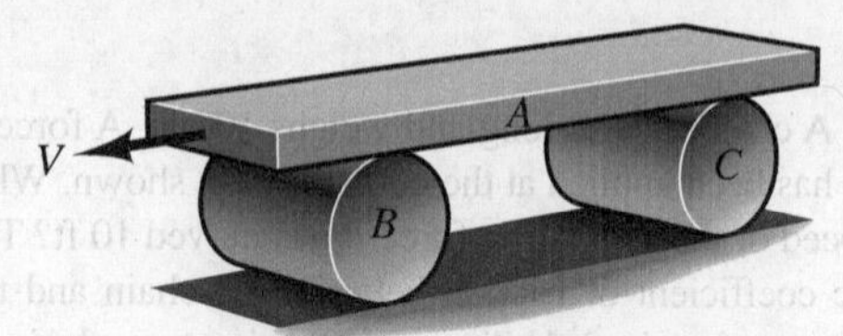

Figure P.13.77.

11.78. A pendulum has a bob with a comparatively large uniform disc of diameter 2 ft and mass M of 3 lbm. At the instant shown, the system has an angular speed θ of .3 rad/sec. If we neglect the mass of the rod, what is the kinetic energy of the pendulum at this instant? What error is incurred if one considers the bob to be a particle as we have done earlier for smaller bobs? Use the result of Problem 11.76.

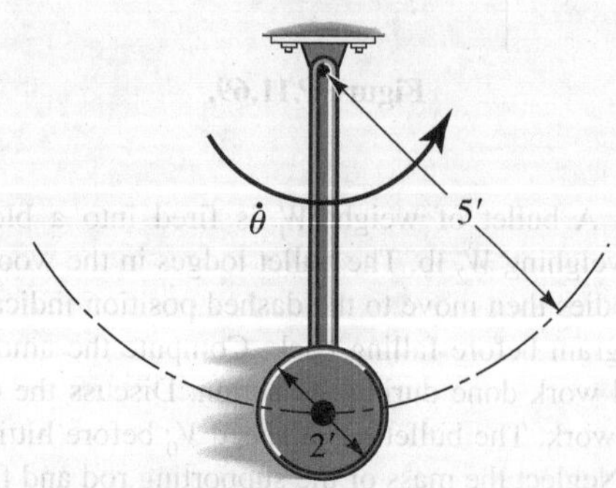

Figure P.11.78.

11.79. In Problem 11.78 compute the maximum angle that the pendulum rises.

11.80. Do Example 11.3 by treating as an aggregate of particles.

11.81. Do Problem 11.27 by treating as an aggregate of particles.

11.82. Do Problem 11.22 by treating as an aggregate of particles.

11.83. Do Problem 11.24 by treating as an aggregate of particles.

11.84. A constant force F is applied to the axis of a cylinder, as shown, causing the axis to increase its speed from 1 ft/sec to 3 ft/sec in 10 ft without slipping. What is the friction force acting on the cylinder? The cylinder weighs 100 lb.

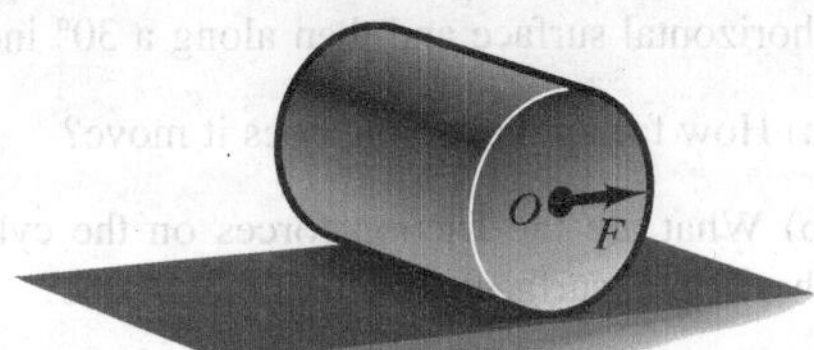

Figure P.11.84.

11.85. A cylinder with a mass of 25 kg is released from rest on an incline, as shown. The inner diameter D of the cylinder is 300 mm. If the cylinder rolls without slipping, compute the speed of the centerline O after the cylinder has moved 1.6 m along the incline. Ascertain the friction force acting on the cylinder. The radius of gyration k at O is $.30/\sqrt{2}$ m.

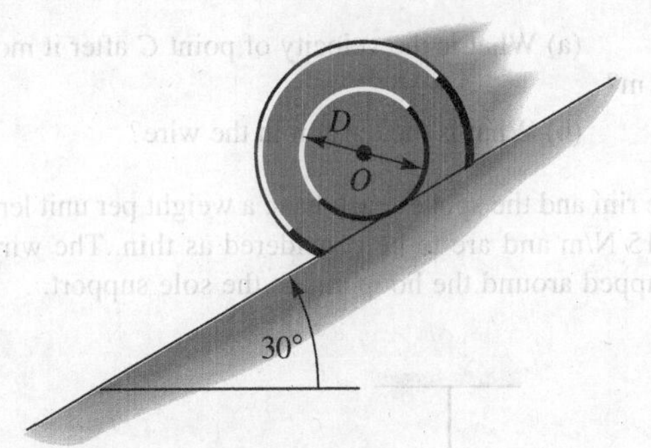

Figure P.11.85.

11.86. A uniform cylinder having a diameter of 2 ft and a weight of 100 lb rolls down a 30° incline without slipping, as shown. What is the speed of the center after it has moved 20 ft? Compare this result with that for the case when there is no friction present. [*Hint:* Use the result of Problem 11.76.]

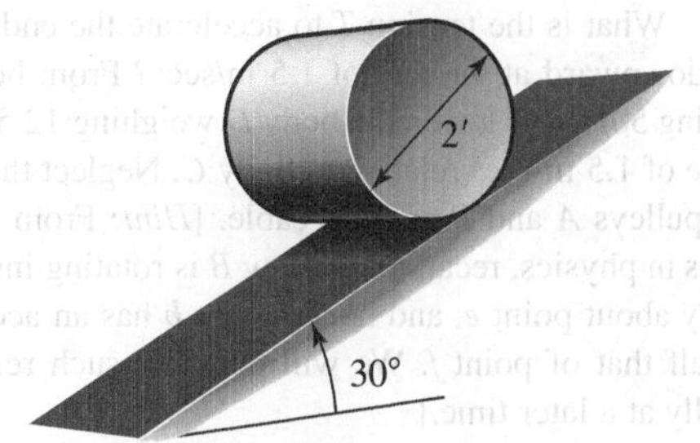

Figure P.11.86.

11.87. Cylinders A and B each have a mass of 25 kg and a diameter of 300 mm. Block C, riding on A and B, has a mass of 100 kg. If the system is released from rest at the configuration shown, what is the speed of C after the cylinders have made half a revolution? Use the result of Problem 11.76.

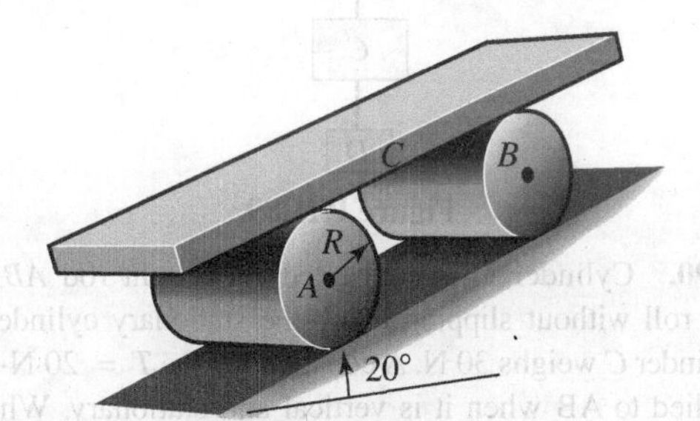

Figure P.11.87.

11.88. Shown are two identical blocks A and B, each weighing 50 lb. A force F of 100 lb is applied to the lower block, causing it to move to the right. Block A, however, is restrained by the wall C. If block B reaches a speed of 10 ft/sec in 2 ft starting from rest at the position shown in the diagram, what is the restraining force from the wall? The dynamic coefficient of friction between B and the ground surface is .3. Do this problem first by using Eq. 11.28. Then check the result by using separate free-body diagrams, and so on.

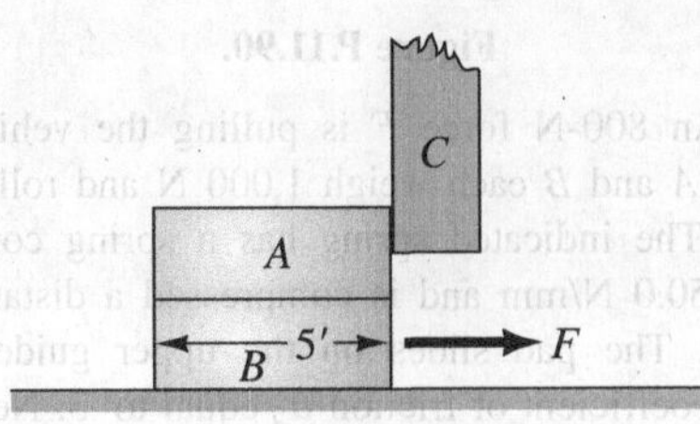

Figure P.11.88.

11.89. What is the tension T to accelerate the end of the cable downward at the rate of 1.5 m/sec²? From body C, weighing $50g$ N, is lowered a body D weighing $12.5g$ N at the rate of 1.5 m/sec² relative to body C. Neglect the inertia of pulleys A and B and the cable. [*Hint:* From earlier courses in physics, recall that pulley B is rotating instantaneously about point e, and hence point b has an acceleration half that of point f. We will consider such relations carefully at a later time.]

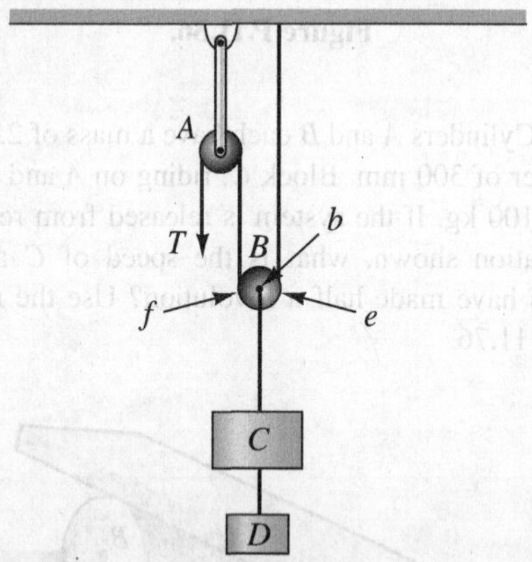

Figure P.11.89.

11.90. Cylinder C is connected by a light rod AB and can roll without slipping along the stationary cylinder D. Cylinder C weighs 30 N. A constant torque T = 20 N-m is applied to AB when it is vertical and stationary. What is the angular speed of AB when it has rotated 90°? The system of bodies is in the vertical plane. Recall from physics that a body which is rolling without slipping has instantaneous rotation about the point of contact.

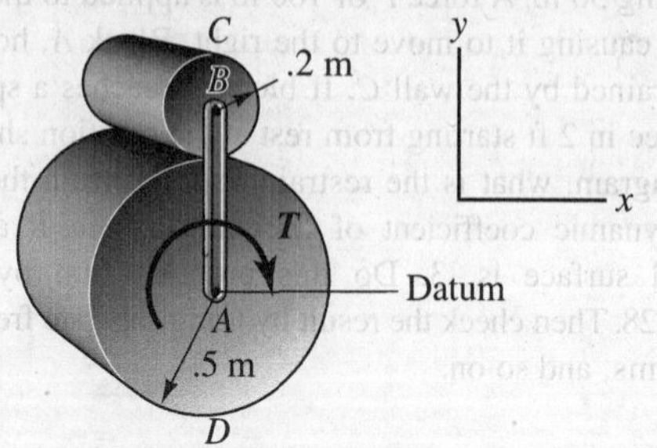

Figure P.11.90.

11.91. An 800-N force F is pulling the vehicle. The cylinders A and B each weigh 1,000 N and roll without slipping. The indicated spring has a spring constant K equal to 50.0 N/mm and is compressed a distance δ of 20.0 mm. The pad slides on the upper guide with a dynamic coefficient of friction μ_d equal to .3. Neglect all masses except the cylinders, whose diameter D is .2 m.

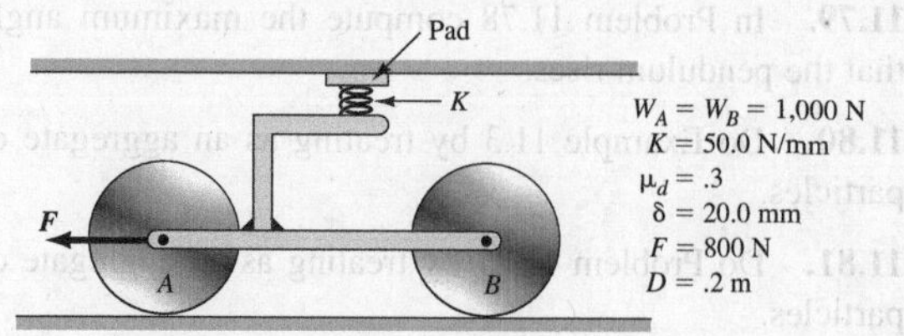

Figure P.11.91.

(a) What is the velocity of the vehicle after it moves a distance of 1.7 m starting from rest?

(b) What is the total friction force f on the cylinders from the ground?

11.92. A cylinder weighing 500 N rolls without slipping, first on a horizontal surface and then along a 30° incline.

(a) How far up the incline does it move?

(b) What are the friction forces on the cylinder along the horizontal surface and along the incline?

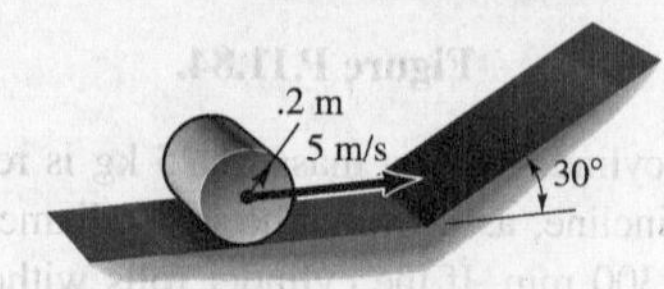

Figure P.11.92.

11.93. A hoop with four spokes is released from rest from a vertical position.

(a) What is the velocity of point C after it moves 1.3 m?

(b) What is the tension in the wire?

The rim and the spokes each have a weight per unit length of 15 N/m and are to be considered as thin. The wire is wrapped around the hoop and is the sole support.

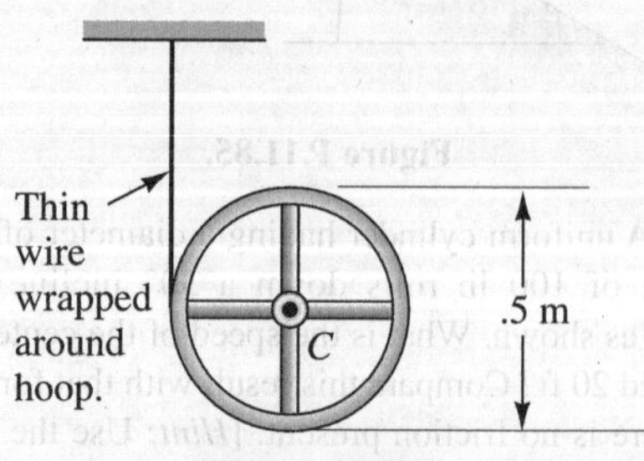

Figure P.11.93.

11.94. Three cylinders roll without slipping starting from rest. What is the speed of the system after moving .3 m? What is the *total* friction from the two walls on the system?

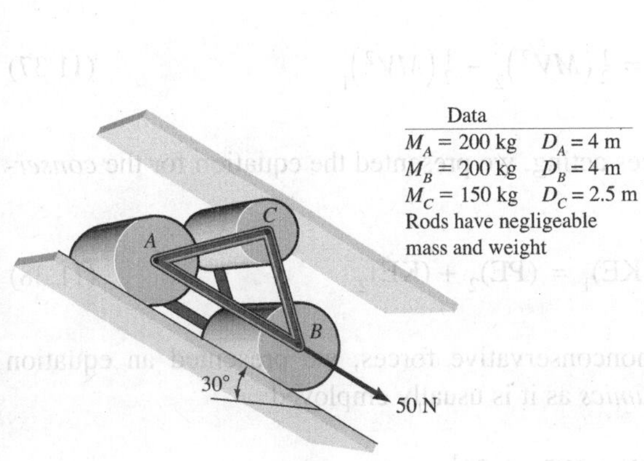

Figure P.11.94.

11.95. Find the velocity V_A after starting from rest and moving 5 m. What is the friction force from the incline on cylinder *A*? There is only rolling without slipping for the cylinders.

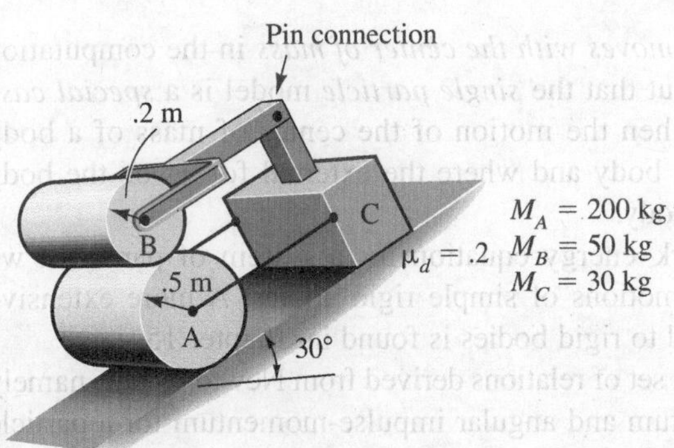

Figure P.11.95.

11.96. A cylinder is about to roll down an incline without slipping. It is connected to a linear spring. What is the angular speed of the cylinder after it rotates 20° starting from rest? The spring is originally unstretched.

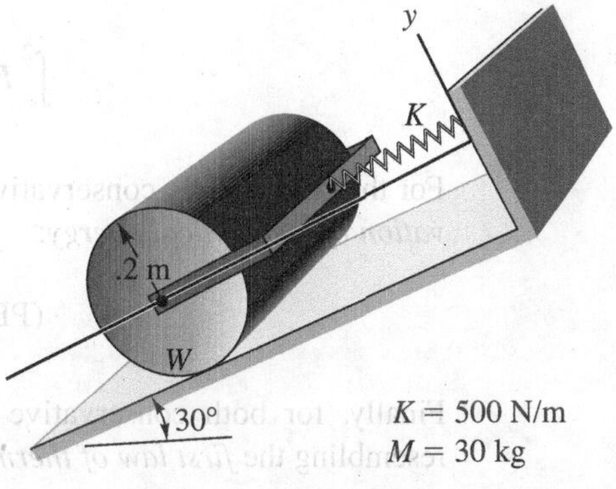

K = 500 N/m
M = 30 kg

Figure P.11.96.

11.97. Three cylinders are connected together by light rods. Cylinders *A* have a mass of 5 kg each and cylinder *B* has a mass of 3 kg. If there is no slipping anywhere,

(a) What is the speed of the system after moving .8 m? The system starts from rest.

(b) What are the friction forces from the ground on each cylinder *A*?

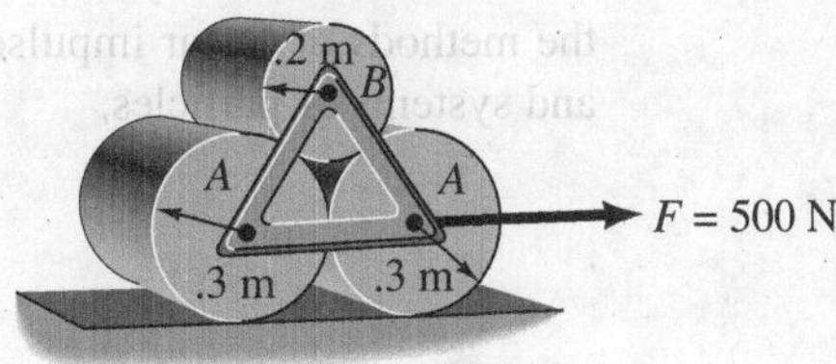

Figure P.11.97.

11.8 Closure

In this chapter, we presented the energy method as applied to particles. In Part A, we presented three forms of the energy equation applied to a *single* particle. The basic equation was

$$\int_1^2 \boldsymbol{F} \bullet d\boldsymbol{r} = \tfrac{1}{2}\left(MV^2\right)_2 - \tfrac{1}{2}\left(MV^2\right)_1 \tag{11.37}$$

For the case of only conservative forces acting, we presented the equation for the *conservation of mechanical energy*:

$$(\text{PE})_1 + (\text{KE})_1 = (\text{PE})_2 + (\text{KE})_2 \tag{11.38}$$

Finally, for both conservative and nonconservative forces, we presented an equation resembling the *first law of thermodynamics* as it is usually employed:

$$\Delta(\text{PE} + \text{KE}) = \mathcal{W}_{1-2} \tag{11.39}$$

In Part B, we considered a *system of particles* and presented the above equation again, but this time the work and potential-energy terms are from both *internal* and *external* force systems.[15] Furthermore, all work and potential-energy terms are evaluated by using the *actual movement* of the points of application of internal and external forces.

Next, we presented the work–energy equation for the *center of mass* of any system of particles:

$$\int_1^2 \boldsymbol{F} \bullet d\boldsymbol{r}_c = \tfrac{1}{2}\left(MV_c^2\right)_2 - \tfrac{1}{2}\left(MV_c^2\right)_1 \tag{11.40}$$

where $\boldsymbol{F}$, the resultant *external force*, *moves with the center of mass* in the computation of the work expression. We pointed out that the *single particle* model is a *special case* of the use of Eq. 11.40 applicable when the motion of the center of mass of a body sufficiently describes the motion of a body and where the external forces on the body move with the center of mass of the body.

To illustrate the use of the work–energy equation for a system of particles, we considered various elementary plane motions of simple rigid bodies. A more extensive treatment of the energy method applied to rigid bodies is found in Chapter 15.

We now turn to yet another useful set of relations derived from Newton's law, namely the methods of linear impulse-momentum and angular impulse-momentum for a particle and systems of particles.

[15]As will be seen in Chapter 12, this equation for a system of particles is the *only one* that involves internal forces. Note, however, that for a *rigid body* the internal forces *do no work*.

PROBLEMS

11.98. A tractor exerts a force of 800 lb on a block A, which has a dynamic coefficient of friction with block B of .7. Block B has a dynamic coefficient of friction of .2 with the ground. If block A weighs 400 lb and block B weighs 600 lb, what is the speed of the block A when, after starting from rest, the tractor has moved 2 ft? What is the acceleration of block B?

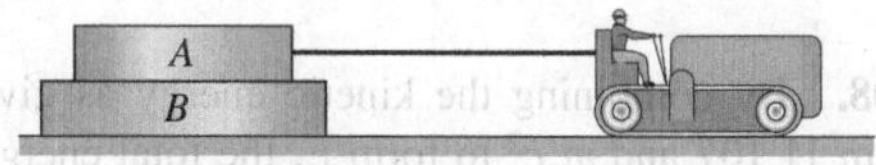

Figure P.11.98.

11.99. A body A is released from a condition of rest on a frictionless circular surface. The body then moves on a horizontal surface CD whose dynamic coefficient of friction with the body is .2. A spring having a spring constant K = 900 N/m is positioned at C as shown in the diagram. How much will the spring be compressed? The body has a mass of 5 kg.

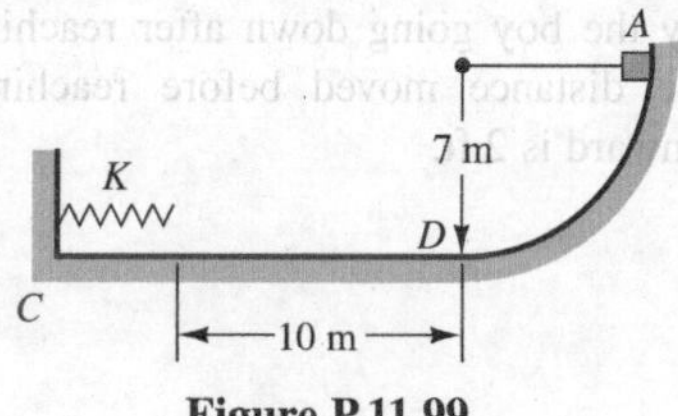

Figure P.11.99.

11.100. A cylinder is about to roll down an incline dragging block B. After starting from rest, what is the angular speed of the cylinder when it has moved .5 m? Use the following data:

$$(R_A)_{OUTSIDE} = 2.5 \text{ m} \qquad (R_A)_{INSIDE} = 1 \text{ m}$$
$$M_A = 100 \text{ kg} \qquad M_B = 30 \text{ kg}$$

The wire is thin and wraps around the inner cylinder of A. The kinetic energy of the compound cylinder due to rotation about its centerline is given as 0.8 times that of a solid cylinder of outside radius r = 2.5 m.

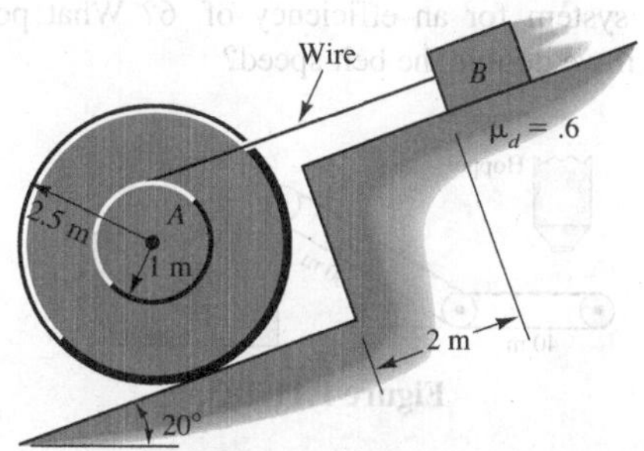

Figure P.11.100.

11.101. The cylinders in the system roll without slipping.

(a) What is the velocity of the system after it moves 1 m starting from rest?

(b) What is the total friction force f_{TOT} for the two cylinders?

(c) What is the acceleration of the system?

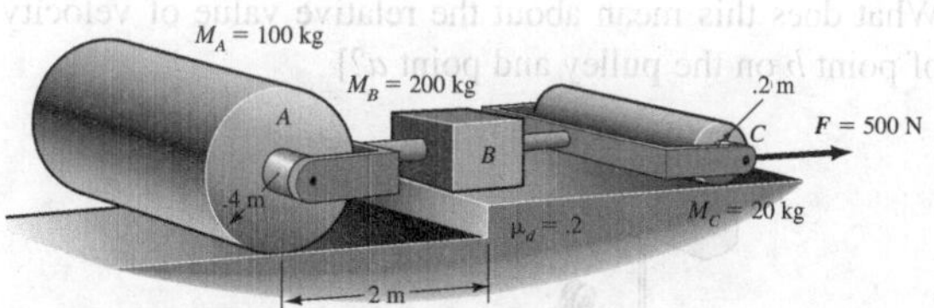

Figure P.11.101.

11.102. A 180-lb man runs up an escalator while it is not in operation in 10 sec. What is the horsepower developed by the man? If the escalator is moving at a speed of 2 ft/sec and carrying, on the average, 2,000 people per hour, what is the power requirement on the driving motor assuming that the average weight of a passenger is 150 lb? Take the mechanical efficiency of the drive system to be 80%. Assume that passengers enter and leave at the same speed of 2 ft/sec and that there are equal numbers of passengers on the escalator at any one time.

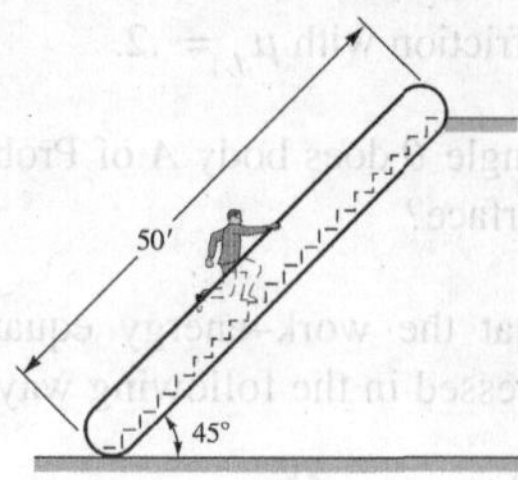

Figure P.11.102.

11.103. Grain is coming out of a hopper at the rate of 7,200 kg/hr and falls onto a conveyor system that takes the grain into a bin. The conveyor belt moves at a steady speed of 2 m/sec. What power in watts is needed to operate the system for an efficiency of .6? What power is needed if we double the belt speed?

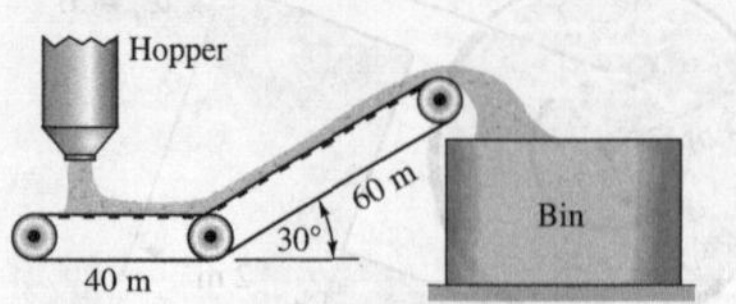

Figure P.11.103.

11.104. A self-propelled vehicle A has a weight of $\frac{1}{2}$ ton. A gasoline engine develops torque on the drive wheels to help move A up the incline. A counterweight B of 300 lb is also shown in the diagram. What horsepower is needed when A is moving up at a speed of 2 ft/sec and has an acceleration of 3 ft/sec^2? Neglect the weight of the pulley. [*Hint:* The pulley rolls along cord dg without slipping. It therefore has an instantaneous center of rotation at d. What does this mean about the relative value of velocity of point b on the pulley and point a?]

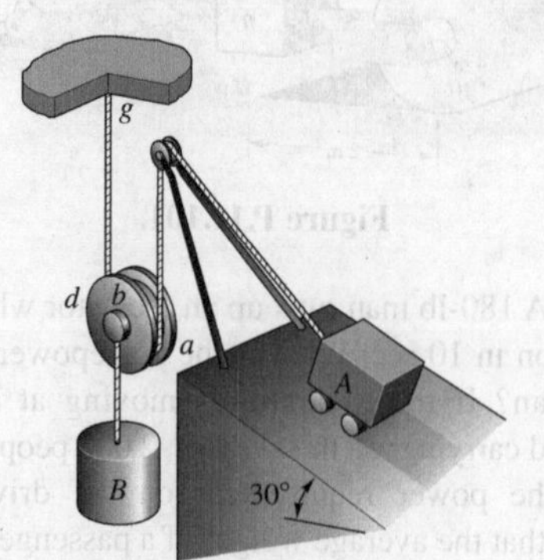

Figure P.11.104.

***11.105.** Set up an integro-differential equation (involving derivatives and integrals) for θ in Problem 11.31 if there is Coulombic friction with $\mu_d = .2$.

11.106. At what angle θ does body A of Problem 11.31 leave the circular surface?

***11.107.** Show that the work–energy equation for a particle can be expressed in the following way:

$$\int_0^x F\,dx = \int_0^V V\,d(mV)$$

Integrating the right side by parts,[17] and using relativistic mass $m_0/\sqrt{1 - V^2/c^2}$, where m_0 is the *rest mass* and c is the speed of light, show that a relativistic form of this equation can be given as

$$\int_0^x F\,dx = \frac{m_0c^2}{\sqrt{1 - V^2/c^2}} - m_0c^2 = mc^2 - m_0c^2$$

so that the *relativistic kinetic energy* is

$$\text{KE} = mc^2 - m_0c^2$$

***11.108.** By combining the kinetic energy as given in Problem 11.107 and m_0c^2 to form E, the total energy, we get the famous formula of Einstein:

$$E = mc^2$$

in which energy is equated with mass. How much energy is equivalent to 6×10^{-8} lbm of matter? How high could a weight of 100 lb be lifted with such energy?

11.109. A 100-lb boy climbs up a rope in gym in 10 sec and slides down in 4 sec after he reaches uniform speed downward. What is the horsepower developed by the boy going up? What is the average horsepower dissipated on the rope by the boy going down after reaching uniform speed? The distance moved before reaching uniform speed downward is 2 ft.

---

[17] To integrate by parts, note that

$$d(uv) = u\,dv + v\,du$$

Now integrate these terms:

$$\int_1^2 d(uv) = \int_1^2 u\,dv + \int_1^2 v\,du$$

Therefore,

$$\int_1^2 u\,dv = (uv)\Big|_1^2 - \int_1^2 v\,du$$

The last formulation is called integration by parts.

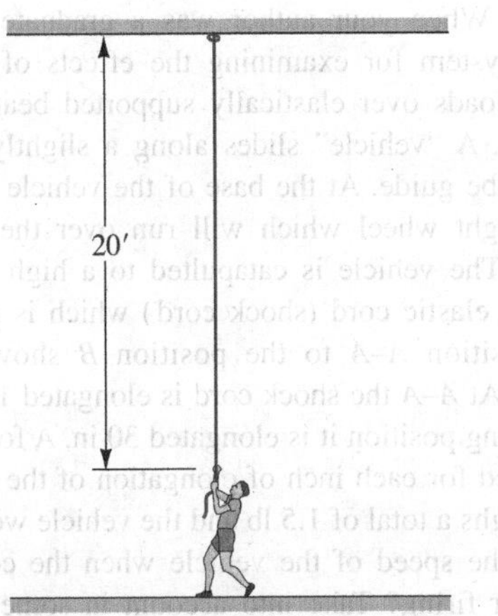

Figure P.11.109.

11.110. An aircraft carrier is shown in the process of launching an airplane via a catapult mechanism. Before leaving the catapult, the plane has a speed of 192 km/hr relative to the ship. If the plane is accelerating at the rate of 1g and if it has a mass of 18,000 kg, what horsepower is being developed by the catapult system at the end of launch on the plane if we neglect drag? The thrust from the jet engines of the plane is 100,000 N.

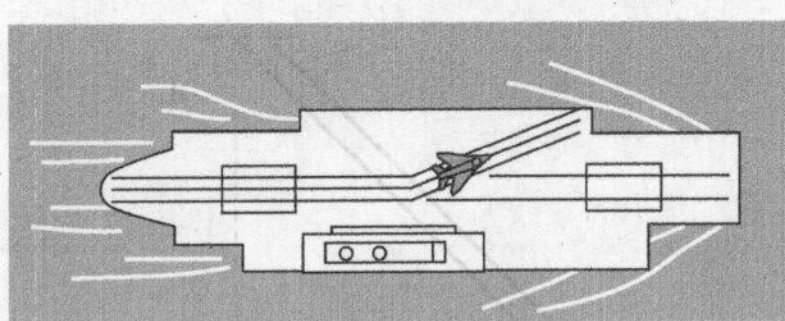

Figure P.11.110.

11.111. Vehicle B, weighing 25 kN, is to go down a 30° incline. The vehicle is connected to body A through light pulleys and a capstan. What should body A weigh if starting from rest it restricts body B to a speed of 5 m/sec when B moves 3 m? There are two wraps of rope around the capstan.

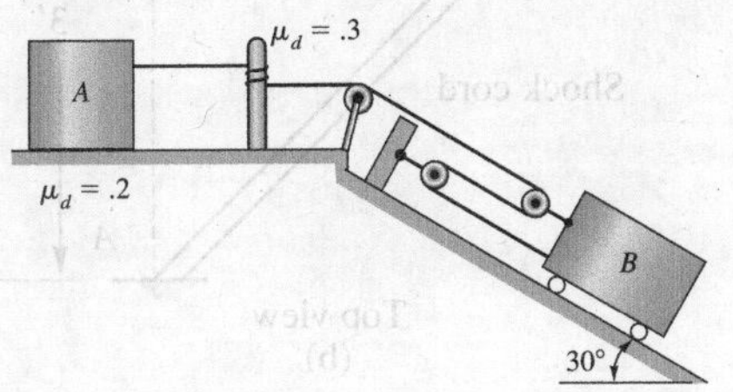

Figure P.11.111.

11.112. A jet passenger plane is moving along the runway for a takeoff. If each of its four engines is developing 44.5 kN of thrust, what is the horsepower developed when the plane is moving at a speed of 240 km/hr?

11.113. Block B, with a mass of 200 kg, is being pulled up an incline. A motor C pulls on one cable, developing 4 hp. The other cable is connected to a counterweight A having a mass of 150 kg. If B is moving at a speed of 2 m/sec, what is its acceleration? [*Hint:* Start with Newton's law for A and B.]

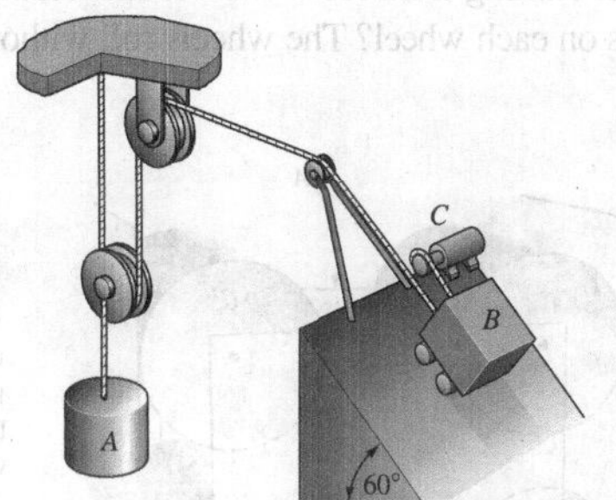

Figure P.11.113.

11.114. A block G slides along a frictionless path as shown. What is the minimum initial speed that G should have along the path if it is to remain in contact when it gets to A, the uppermost position of the path? The block weighs 9 N. What is the normal force on the path when for the condition described the block is at position B?

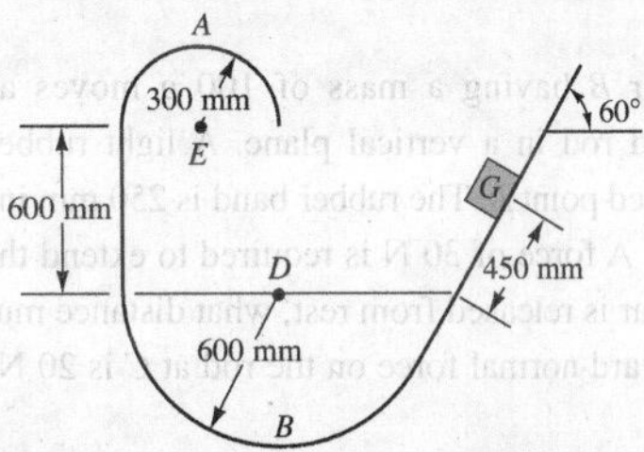

Figure P.11.114.

11.115. Cylinders A and B have masses of 50 kg each. Cylinder A can only rotate about a stationary axis while cylinder B rolls without slipping. Block C has a mass of 100 kg. Starting from rest, what is the speed of C after moving .1 m? Force P is 500 N and the diameter of the cylinders is .2 m.

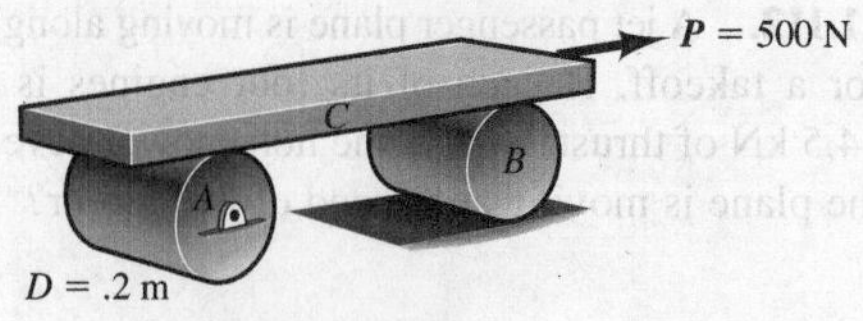

Figure P.11.115.

11.116. A system of 4 solid cylinders and a heavy block move vertically downward aided by a 1,000-N force F. What is the angular speed of the wheels after the system descends .5 m after starting from rest? What is the friction force from the walls on each wheel? The wheels roll without slipping.

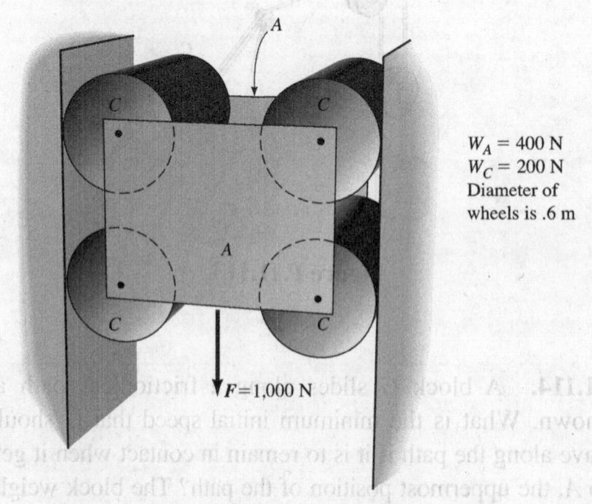

Figure P.11.116.

11.117. A collar B having a mass of 100 g moves along a frictionless curved rod in a vertical plane. A light rubber band connects B to a fixed point A. The rubber band is 250 mm in length when unstretched. A force of 30 N is required to extend the band 50 mm. If the collar is released from rest, what distance must d be so that the downward normal force on the rod at C is 20 N?

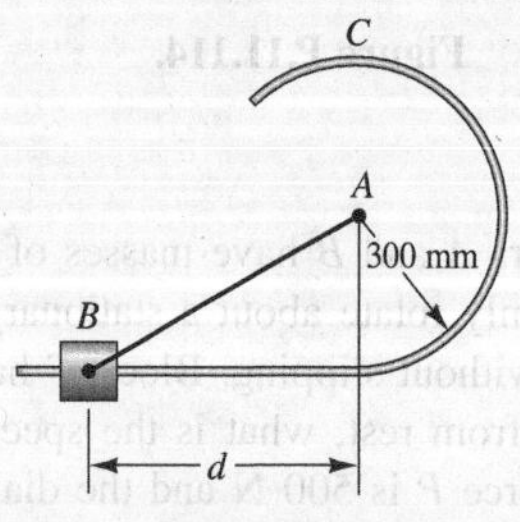

Figure P.11.117.

11.118. When your author was a graduate student he built a system for examining the effects of high-speed moving loads over elastically supported beams (see the diagram). A "vehicle" slides along a slightly lubricated square tube guide. At the base of the vehicle is a spring-loaded light wheel which will run over the beam (not shown). The vehicle is catapulted to a high speed by a stretched elastic cord (shock cord) which is pulled back from position A–A to the position B shown prior to "firing." At A–A the shock cord is elongated 10 in., while at the firing position it is elongated 30 in. A force of 10 lb is required for each inch of elongation of the cord. If the cord weighs a total of 1.5 lb and the vehicle weighs 10 oz, what is the speed of the vehicle when the cord reaches A–A after firing? Take into account in some reasonable way the kinetic energy of the cord, but neglect friction.

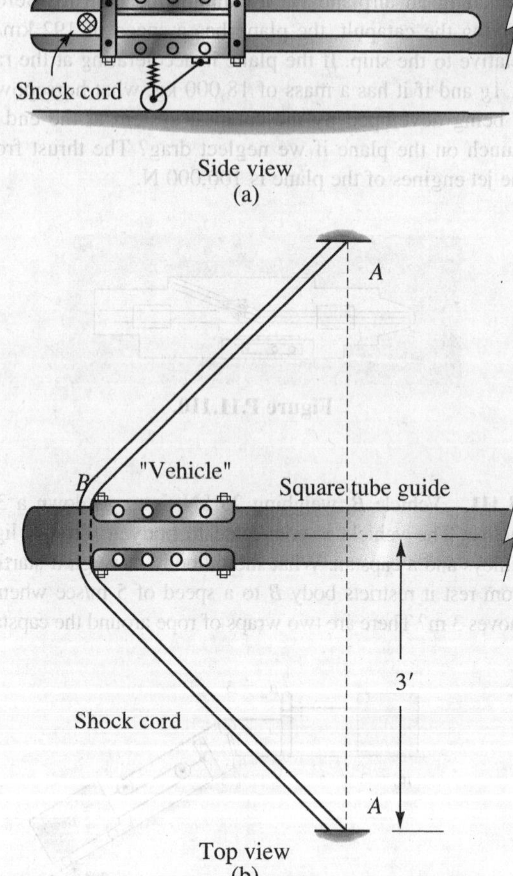

Figure P.11.118.

11.119. A body B of mass 60 kg slides in a frictionless slot on an inclined surface as shown. An elastic cord connects B to A. The cord has a "spring constant" of 360 N/m. If the body B is released from rest from a position where the elastic cord is unstretched, what is body B's speed after it moves .3 m?

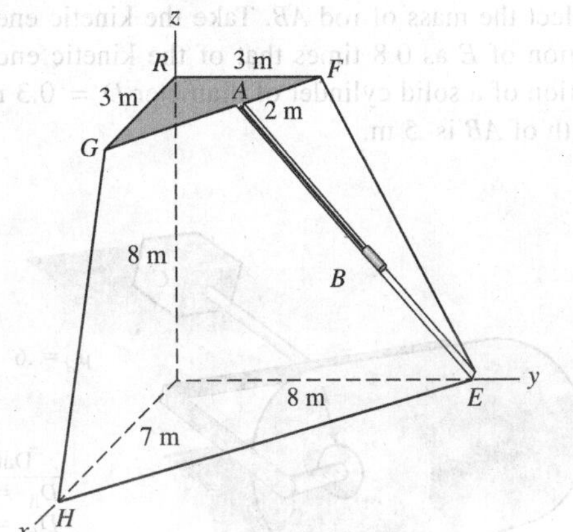

Figure P.11.119.

11.120. A collar slides on a frictionless tube as shown. The spring is unstretched when in the horizontal position and has a spring constant of 1.0 lb/in. What is the minimum weight of A to just reach A' when released from rest from the position shown in the diagram? What is the force on the tube when A has traveled half the distance to A'?

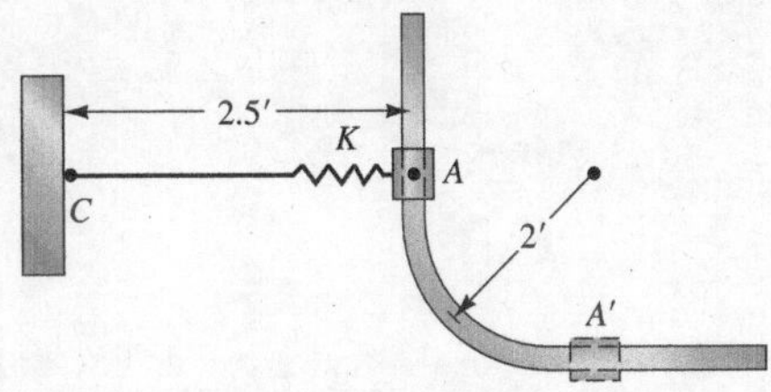

Figure P.11.120.

11.121. A 15-kg vehicle has two bodies (each with mass 1 kg) mounted on it, and these bodies rotate at an angular speed of 50 rad/sec relative to the vehicle. If a 500-N force acts on the vehicle for a distance of 17 m, what is the kinetic energy of the system, assuming that the vehicle starts from rest and the bodies in the vehicle have constant rotational speed? Neglect friction and the inertia of the wheels.

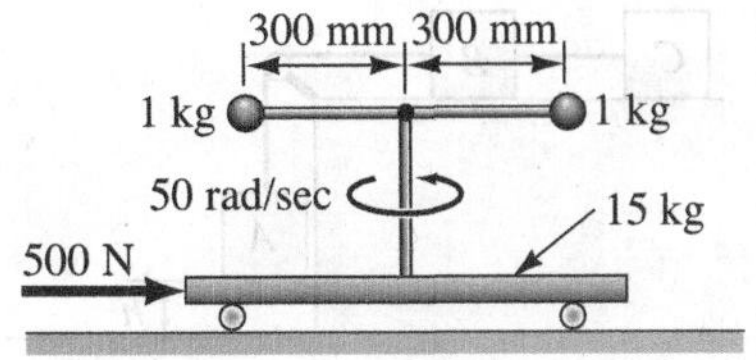

Figure P.11.121.

11.122. Two identical solid cylinders each weighing 100 N support a load A weighing 50 N. If a force F of 300 N acts as shown, what is the speed of the vehicle after moving 5 m? Also, what is the total friction force on each wheel? Neglect the mass of the supporting system connecting the cylinders. Note that the kinetic energy of the angular motion of a cylinder about its own axis is $\frac{1}{4}MR^2\omega^2$. The system starts from rest.

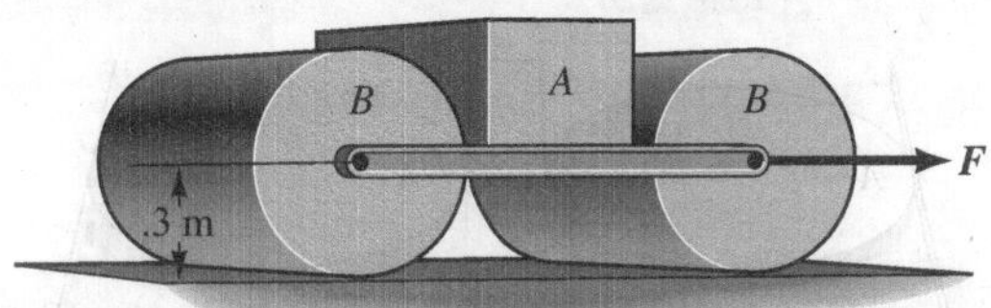

Figure P.11.122.

11.123. A triangular block of uniform density and total weight 100 lb rests on a hinge and on a movable block B. If a constant force F of 150 lb is exerted on the block B, what will be its speed after it moves 10 ft? The mass of block B is 10 lbm, and the dynamic coefficient of friction for all contact surfaces is .3.

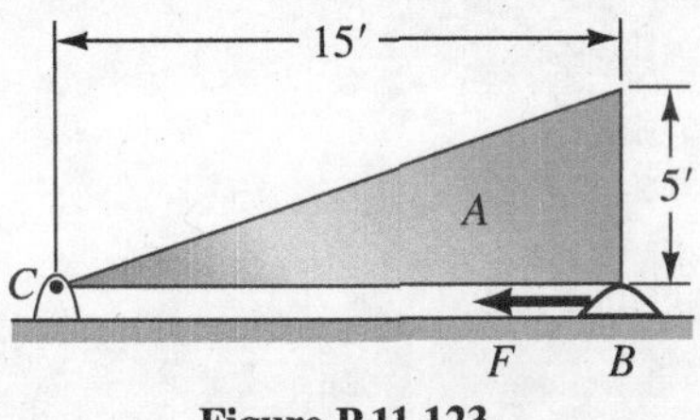

Figure P.11.123.

11.124. Three blocks are connected by an inextensible flexible cable. The blocks are released from a rest configuration with the cable taut. If A can only fall a distance h equal to 2 ft, what is the velocity of bodies C and B after each has moved a distance of 3 ft? Each body weighs 100 lb. The coefficient of dynamic friction for body C is .3 and for body B is .2.

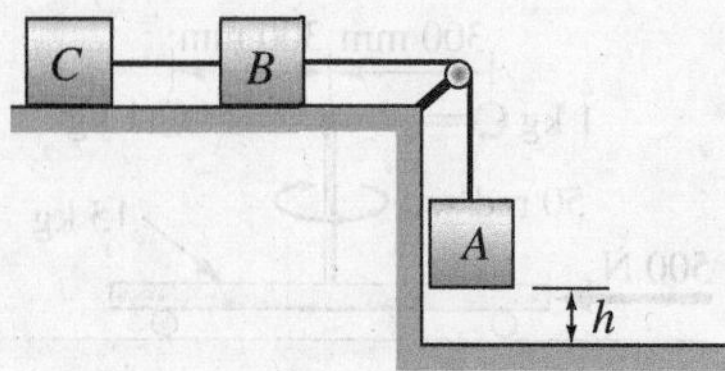

Figure P.11.124.

11.125. Two discs move on a horizontal frictionless surface shown looking down from above. Each disc weighs 20 N. A rectangular member B weighing 50 N is pulled by a force F having a value of 200 N. If there is no slipping anywhere except on the horizontal support surface, what is the speed of B after it moves 18 cm? Determine the friction forces from the walls onto the cylinders.

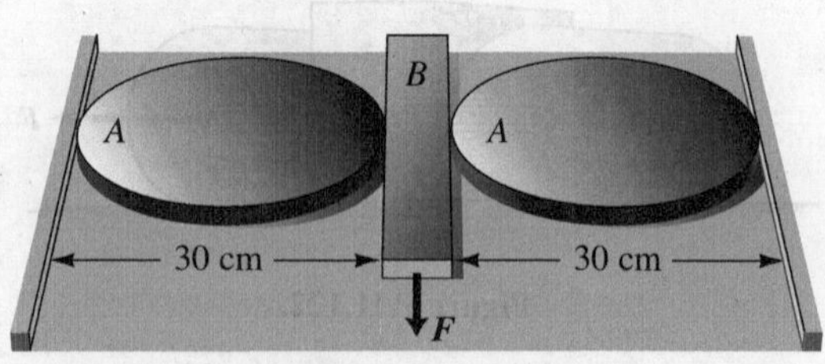

Figure P.13.125.

11.126. Rod AB is pinned to block C and is welded to cylinder D. Cylinder E rolls without slipping along the incline and rotates around cylinder D, which does not rotate at all. There is a constant friction torque between D and E of 25 N-m. Starting from rest, what is the speed of the system after moving .1 m along the incline? What is the frictional force between the ground and cylinder E? Neglect the mass of rod AB. Take the kinetic energy of rotation of E as 0.8 times that of the kinetic energy of rotation of a solid cylinder of diameter D = 0.3 m. The length of AB is .5 m.

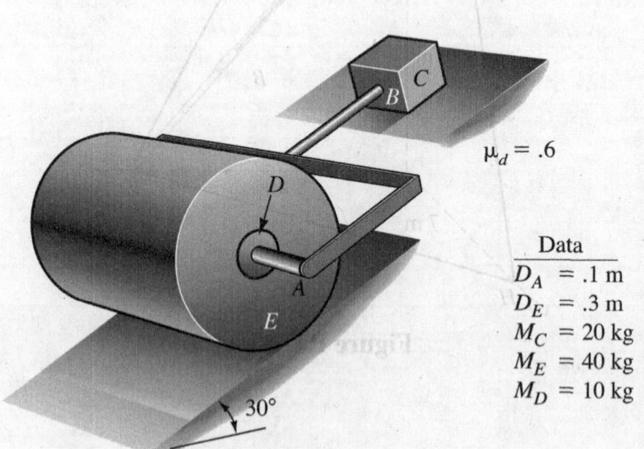

Figure P.11.126.

CHAPTER 12

Methods of Momentum for Particles

Part A: Linear Momentum

12.1 Impulse and Momentum Relations for a Particle

In Section 10.3, we integrated differential equations of motion for particles that are acted upon by forces that are functions of time. In this chapter, we shall again consider such problems and shall present alternative formulations, called *methods of momentum*, for handling certain of these problems in a convenient and straightforward manner. We start by considering Newton's law for a particle:

$$\boldsymbol{F} = m\frac{d\boldsymbol{V}}{dt} \tag{12.1}$$

Multiply both sides by dt and integrate from some initial time t_i to some final time t_f:

$$\int_{t_i}^{t_f} \boldsymbol{F}\,dt = \int_{t_i}^{t_f} m\frac{d\boldsymbol{V}}{dt}\,dt = m\boldsymbol{V}_f - m\boldsymbol{V}_i \tag{12.2}$$

Note first that this is a vector equation, in contrast to the work–kinetic energy equation 11.2. The integral

$$\int_{t_i}^{t_f} \boldsymbol{F}\,dt$$

which we shall denote as $\boldsymbol{I}$, is called the *impulse* of the force $\boldsymbol{F}$ during the time interval $t_f - t_i$, whereas $m\boldsymbol{V}$ is the *linear-momentum vector* of the particle. Equation 12.2, then, states that *the impulse* $\boldsymbol{I}$ *over a time interval equals the change in linear momentum of a particle during that time interval.* As we shall demonstrate later, the impulse of a force may be known even though the force itself is not known.

Finally, you must remember that to produce an impulse, a force need only exist for a time interval. Sometimes we use the work integral so much that we tend to think—erroneously—that a force acting on a stationary body does not produce an impulse.

We now illustrate the use of the impulse-momentum equation.

Example 12.1

A particle initially at rest is acted on by a force whose variation with time is shown graphically in Fig. 12.1. If the particle has a mass of 1 slug and is constrained to move rectilinearly in the direction of the force, what is the speed after 15 sec?

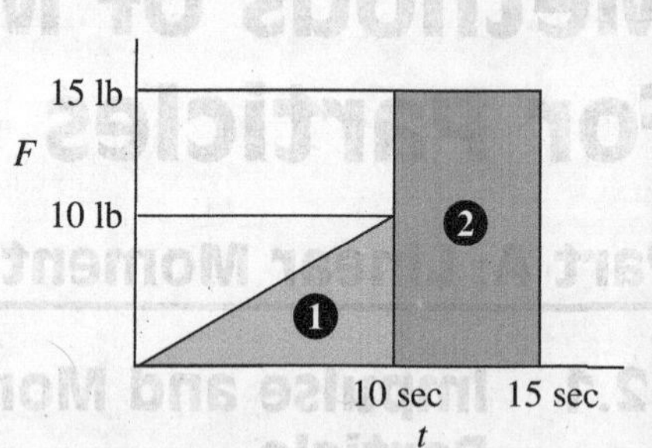

Figure 12.1. Force-versus-time plot.

From the definition of the impulse, the area under the force–time curve will, in the one-dimensional example, equal the impulse magnitude. Thus, we simply compute this area between the times $t = 0$ and $t = 15$ sec:

$$\text{impulse} = \underbrace{\tfrac{1}{2}(10)(10)}_{\text{area 1}} + \underbrace{(5)(15)}_{\text{area 2}} = 125 \text{ lb-sec}$$

The final velocity, then, is given as

$$125 = (1)(V_f) - 0$$

Therefore,

$$V_f = 125 \text{ ft/sec}$$

Note that the impulse-momentum equation is useful when the force variation during a time interval is a curve that cannot be conveniently expressed mathematically. The impulse, which is the area under an F versus t curve, can then be found with the help of a *planimeter*, thus permitting a quick solution of the velocity change during the time interval.[1]

[1]A planimeter is a mechanical device for measuring the area of a plane region bounded by an arbitrary curve.

Example 12.2

A particle *A* with a mass of 1 kg has an initial velocity $V_0 = 10\boldsymbol{i} + 6\boldsymbol{j}$ m/sec. After particle *A* strikes particle *B*, the velocity becomes $V = 16\boldsymbol{i} - 3\boldsymbol{j} + 4\boldsymbol{k}$ m/sec. If the time of encounter is 10 msec, what average force was exerted on the particle *A*? What is the change of linear momentum of particle *B*?

The impulse $\boldsymbol{I}$ acting on *A* is immediately determined by computing the change in linear momentum during the encounter:

$$\boldsymbol{I}_A = (1)(16\boldsymbol{i} - 3\boldsymbol{j} + 4\boldsymbol{k}) - (1)(10\boldsymbol{i} + 6\boldsymbol{j})$$
$$= 6\boldsymbol{i} - 9\boldsymbol{j} + 4\boldsymbol{k} \text{ N-sec}$$

Since

$$\int_{t_i}^{t_f} \boldsymbol{F}_A \, dt = (\boldsymbol{F}_{av})_A \Delta t$$

the average force $(\boldsymbol{F}_{av})_A$ becomes

$$(\boldsymbol{F}_{av})_A(0.010) = 6\boldsymbol{i} - 9\boldsymbol{j} + 4\boldsymbol{k}$$

Therefore,

$$(\boldsymbol{F}_{av})_A = 600\boldsymbol{i} - 900\boldsymbol{j} + 400\boldsymbol{k} \text{ N}$$

On the basis of the principle that action equals reaction, an equal but opposite average force must act on the object *B* during the 10-msec time interval. Thus, the impulse on particle *B* is $-\boldsymbol{I}_A$. Equating this impulse to the change in linear momentum, we get

$$\Delta(mV)_B = -\boldsymbol{I}_A = -6\boldsymbol{i} + 9\boldsymbol{j} - 4\boldsymbol{k} \text{ N-sec}$$

During impacts where the exact force variation is unknown, the impulse momentum principle is very useful. We shall examine impacts in more detail in a later section.

Example 12.3

Two bodies, 1 and 2, are connected by an inextensible and weightless cord (Fig. 12.2). Initially, the bodies are at rest. If the dynamic coefficient of friction is μ_d for body 1 on the surface inclined at angle α, compute the velocity of the bodies at any time t before body 1 has reached the end of the incline.

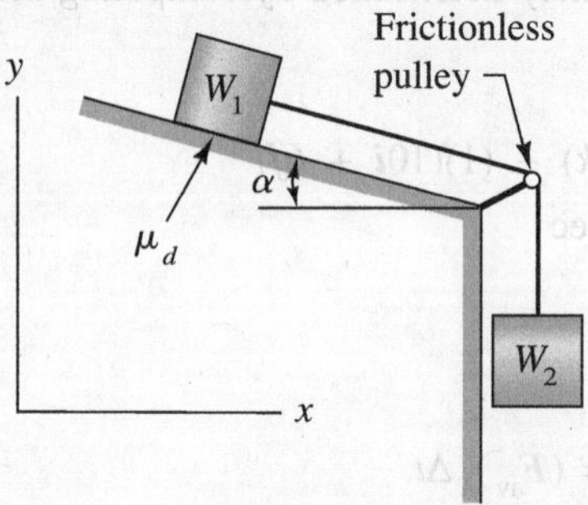

Figure 12.2. Two bodies connected by a cord.

Since only constant forces exist and since a time interval has been specified, we can use momentum considerations advantageously. The free-body diagrams of bodies 1 and 2 are shown in Fig. 12.3. Equilibrium considerations lead to the conclusion that $N_1 = W_1 \cos \alpha$, so the friction force f_1 is

$$f_1 = \mu_d N_1 = \mu_d W_1 \cos \alpha$$

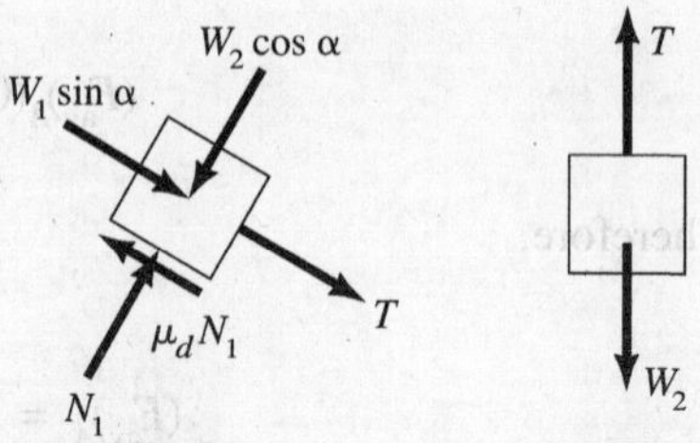

Figure 12.3. Free-body diagrams of W_1 and W_2.

For body 1, take the component of the **linear impulse-momentum** equation along the incline:

$$\int_0^t (-\mu_d W_1 \cos \alpha + W_1 \sin \alpha + T)dt = \frac{W_1}{g}(V - 0)$$

Carrying out the integration, we have

$$(-\mu_d W_1 \cos \alpha + W_1 \sin \alpha + T)t = \frac{W_1}{g} V \qquad \text{(a)}$$

For body 2, we have for the **momentum equation** in the vertical direction:

$$\int_0^t (W_2 - T)dt = \frac{W_2}{g}(V - 0)$$

where, because of the inextensible property of the cable and the frictionless condition of the pulley, the magnitudes of the velocity V and the force T are the same for bodies 1 and 2. Integrating the equation above, we write:

$$(W_2 - T)t = \frac{W_2}{g} V \qquad \text{(b)}$$

Example 12.3 (Continued)

By adding Eqs. (a) and (b), we can eliminate T and solve for the desired unknown V. Thus,

$$(-\mu_d W_1 \cos\alpha + W_1 \sin\alpha + W_2)t = \frac{V}{g}(W_1 + W_2)$$

Therefore,

$$V = \frac{gt}{W_1 + W_2}(W_2 + W_1 \sin\alpha - \mu_d W_1 \cos\alpha) \quad \text{(c)}$$

Note that we have used considerations of linear momentum for a *single* particle each time in solving this problem.

Example 12.4

A conveyor belt is moving from left to right at a constant speed V of 1 ft/sec in Fig. 12.4. Two hoppers drop objects onto the belt at the total rate n of 4 per second. The objects each have a weight W of 2 lb and fall a height h of 1 ft before landing on the conveyor belt. Farther along the belt (not shown) the objects are removed by personnel so that, for steady-state operation, the number N of objects on the belt at any time is 10. If the dynamic coefficient of friction between belt and conveyor bed is .2, estimate the average difference in tension $T_2 - T_1$ of the belt to maintain this operation. The weight of the belt on the conveyor bed is 10 lb.

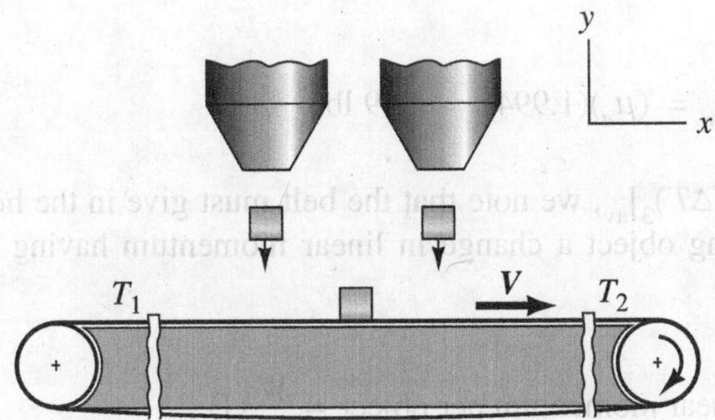

Figure 12.4. Objects falling on moving conveyor.

Example 12.4 (Continued)

We shall superimpose the following effects to get the desired result.

1. A friction force from the bed onto the belt results from the static weight of the ten objects riding on the belt and the weight of the portion of belt on the bed.
2. A friction force from the bed onto the belt results from the force in the y direction needed to change the *vertical* linear momentum of the falling objects from a value corresponding to the free-fall velocity just before impact ($\sqrt{2gh}$) to a value of zero after impact.
3. Finally, the belt must supply a force in the x direction to change the *horizontal* linear momentum of the falling objects from a value of zero to a value corresponding to the speed of the belt.

Thus, we have for the first contribution, which we donate as ΔT_1, the following result:

$$\Delta T_1 = (NW + 10)\mu_d = [(10)(2) + 10](.2) = 6 \text{ lb} \qquad \text{(a)}$$

As for the second contribution, we can only compute an average value $(\Delta T_2)_{av}$ by noting that each impacting object is given a vertical change in linear momentum equal to

$$\begin{aligned}\text{vertical change in linear momentum per object} &= \frac{W}{g}(\sqrt{2gh}) \\ &= \frac{2}{g}\sqrt{(2g)(1)} \\ &= .498 \text{ lb-sec}\end{aligned}$$

where we have assumed a free fall starting with zero velocity at the hopper. For four impacts per second, we have as the total vertical change in linear momentum per second the value (4)(.498) = 1.994 lb-sec. The average vertical force during the 1-sec interval to give the impulse needed for this change in linear momentum is clearly 1.994 lb. Since this result is correct for every second, 1.994 lb is the average normal force that the bed of the conveyor must transmit to the belt for arresting the vertical motion of the falling objects. The desired $[(\Delta T_2]_{av}$ for the belt arising from friction is accordingly given as

$$[(\Delta T)_2]_{av} = (\mu_d)(1.994) = .399 \text{ lb} \qquad \text{(b)}$$

Finally, for the last contribution $[(\Delta T)_3]_{av}$, we note that the belt must give in the horizontal direction for each impacting object a change in linear momentum having the value

$$\begin{aligned}\text{horizontal change in linear momentum per object} &= \frac{W}{g}(1) \\ &= .0621 \text{ lb-sec}\end{aligned}$$

Example 12.4 (Continued)

For four impacts per second we have as the total horizontal change in linear momentum developed by the belt during 1 sec the value (4)(.0621) = .248 lb-sec. The average horizontal force during 1 sec needed for this change in linear momentum is clearly .248 lb. Thus, we have

$$[(\Delta T)_3]_{av} = .248 \text{ lb} \tag{c}$$

The total average difference in tension is then

$$(\Delta T)_{av} = 6 + .399 + .248 = \boxed{6.65 \text{ lb}} \tag{d}$$

12.2 Linear-Momentum Considerations for a System of Particles

In Section 12.1, we considered impulse-momentum relations for a single particle. Although Examples 12.3 and 12.4 involved more than one particle, nevertheless the impulse-momentum considerations were made on one particle at a time. We now wish to set forth impulse-momentum relations for a *system* of particles.

Let us accordingly consider a system of n particles. We may start with Newton's law as developed previously for a system of particles:

$$\boldsymbol{F} = \sum_{j=1}^{n} m_j \frac{d\boldsymbol{V}_j}{dt} \tag{12.3}$$

Since we know that the internal forces cancel, $\boldsymbol{F}$ must be the *total external* force on the system of n particles. Multiplying by dt, as before, and integrating between t_i and t_f, we write:

$$\int_{t_i}^{t_f} \boldsymbol{F}\, dt = \boldsymbol{I}_{\text{ext}} = \left(\sum_{j=1}^{n} m_j \boldsymbol{V}_j\right)_f - \left(\sum_{j=1}^{n} m_j \boldsymbol{V}_j\right)_i \tag{12.4}$$

Thus, we see that *the impulse of the total external force on the system of particles during a time interval equals the sum of the changes of the linear- momentum vectors of the particles during the time interval.*

We now consider an example.

Example 12.5

A 3-ton truck is moving at a speed of 60 mi/hr. [See Fig. 12.5(a).] The driver suddenly applies his brakes at time $t = 0$ so as to lock his wheels in a panic stop. Load A weighing 1 ton breaks loose from its ropes and at time $t = 4$ sec is sliding *relative to the truck* at a speed of 3 ft/sec. What is the speed of the truck at that time? Take μ_d between the tires and pavement to be .4.

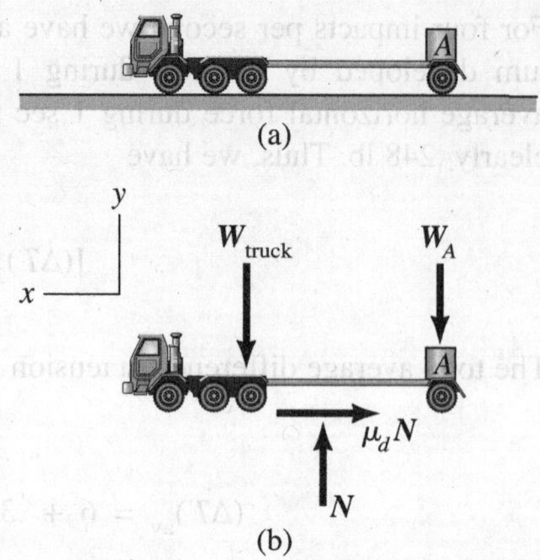

Figure 12.5. Truck undergoing panic stop.

Since we *do not know* the nature of the forces between the truck and load A while the latter is breaking loose, it is easiest to consider the *system* of two particles comprising the truck and the load simultaneously whereby the aforementioned forces become *internal* and are *not* considered. Accordingly, we have shown the system with all the external loads in Fig. 12.5(b). Clearly, $N = (4)(2{,}000) = 8{,}000$ lb and the friction force is $(.4)(8{,}000) = 3{,}200$ lb. We now employ Eq. 12.4 in the x direction as follows:

$$\int_{t_1}^{t_2} F_x\,dt = \left(\sum_j m_j \mathbf{V}_j\right)_2 - \left(\sum_j m_j \mathbf{V}_j\right)_1$$

$$\int_0^4 (-3{,}200)\,dt = \left[\frac{(3)(2{,}000)}{g}V_2 + \frac{(1)(2{,}000)}{g}(V_2+3)\right] - \left[\frac{(4)(2{,}000)}{g}\frac{(60)(5{,}280)}{3{,}600}\right] \qquad \text{(a)}$$

Note that the first quantity inside the first brackets on the right side of Eq. (a) is the momentum of the truck at $t = 4$ sec, and the second quantity inside the same brackets is the momentum of the load at this instant. We may readily solve for V_2:

$$V_2 = 35.7 \text{ ft/sec}$$

Introducing *mass-center* quantities into Eq. 12.4 is easy and sometimes advantageous. You will remember that:

$$M\boldsymbol{r}_c = \sum_{j=1}^{n} m_j \boldsymbol{r}_j \qquad (12.5)$$

Differentiating with respect to time, we get

$$M\boldsymbol{V}_c = \sum_{j=1}^{n} m_j \boldsymbol{V}_j \qquad (12.6)$$

Thus, we see from this equation that *the total linear momentum of a system of particles equals the linear momentum of a particle that has the total mass of the system and that moves with the velocity of the mass center*. Using Eq. 12.6 to replace the right side of Eq. 12.4, we can say:

$$\int_{t_i}^{t_f} \boldsymbol{F}\, dt = \boldsymbol{I}_{\text{ext}} = M(\boldsymbol{V}_c)_f - M(\boldsymbol{V}_c)_i \tag{12.7}$$

Thus, *the total external impulse on a system of particles equals the change in linear momentum of a hypothetical particle having the mass of the entire aggregate and moving with the mass center*.

When the separate motions of the individual particles are reasonably simple, as a result of constraints, and the motion of the mass center is not easily available, then Eq. 12.4 can be employed for linear-momentum considerations as was the case for Example 12.5. On the other hand, when the motions of the particles individually are very complex and the motion of the mass center of the system is reasonably simple, then clearly Eq. 12.7 can be of great value for linear-momentum considerations. Also, as in the case of energy considerations, we note that the single-particle model is really a *special case* of the center-of-mass formulation above, wherein the motion of the center of mass of a body describes sufficiently the motion of the body in question.

Example 12.6

A truck in Fig. 12.6 has two rectangular compartments of identical size for the purpose of transporting water. Each compartment has the dimensions 20 ft × 10 ft × 8 ft. Initially, tank *A* is full and tank *B* is empty. A pump in tank *A* begins to pump water from *A* to *B* at the rate Q_1 of 10 cfs (cubic feet per second) and 10 sec later is delivering water at the rate Q_2 of 30 cfs. If the level of the water in the tanks remains horizontal, what is the average horizontal force needed to restrain the truck from moving during this interval?

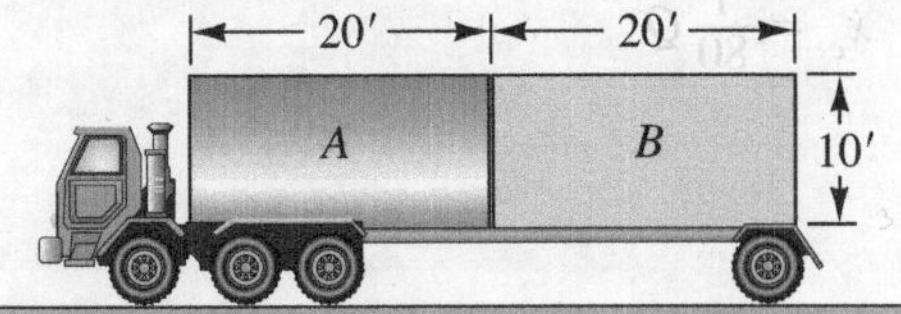

Figure 12.6. Truck with tank compartments.

Example 12.6 (Continued)

In this setup, the mass center of the water in the tanks is moving from left to right and moving *nonuniformly* during the time interval of interest. We show the water in Fig. 12.7 at some time t where the level in tank A has dropped an amount η while, by

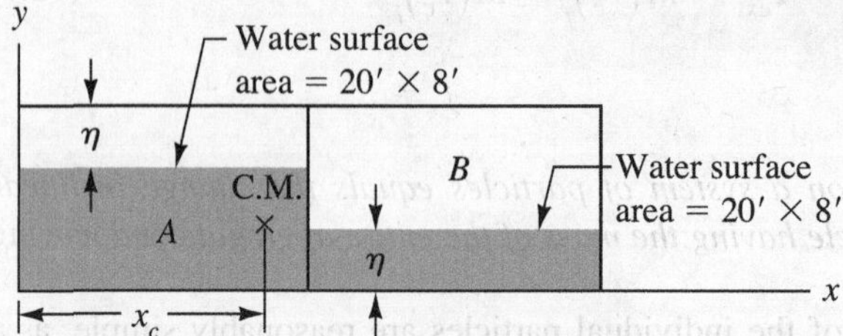

Figure 12.7. Compartments showing flow of water.

conservation of mass, the level in tank B has risen exactly the same amount η. The position x_c of the center of mass at this instant can be readily calculated in terms of η. Thus, using the basic definition of the center of mass, we can say:

$$Mx_c = (M_A)(x_A) + (M_B)(x_B)$$

$$[(20)(8)(10)](\rho)(x_c) = [(20)(8)(10 - \eta)](\rho)(10) + [(20)(8)(\eta)](\rho)(30) \quad \text{(a)}$$

Since we are interested in the time rate of change of x_c so that we can profitably employ Eq. 12.7, we next differentiate with respect to time as follows:

$$[(20)(8)(10)](\rho)(\dot{x}_c) = -[(20)(8)\,\dot{\eta}](\rho)(10) + [(20)(8)(\dot{\eta})](\rho)(30) \quad \text{(b)}$$

But $(20)(8)\dot{\eta}$ is the volume of flow[2] from tank A to tank B at time t. Using Q to represent this volume flow, we get for the equation above:

$$[(20)(8)(10)](\rho)\dot{x}_c = -(\rho)(10)Q + (\rho)(30)Q = (20)(\rho)Q$$

Solving for $\dot{x}_c$, we have

$$\dot{x}_c = \frac{1}{80}Q \quad \text{(c)}$$

[2]Remember that 20 ft × 8 ft is the area of the top water surface in each tank, as shown in Fig. 12.7.

Example 12.6 (Continued)

Now consider the momentum equation in the x direction for the water using the center of mass. We can say from Eq. 12.7:

$$\int_0^{10} F\,dt = \left[(M\dot{x}_c)_2 - (M\dot{x}_c)_1\right]$$

Therefore,

$$\begin{aligned}(F_{av})(10) &= \left[(20)(8)(10)(\rho)\right]\left[(\dot{x}_c)_2 - (\dot{x}_c)_1\right]\\ &= \left[(20)(8)(10)(\rho)\right]\left[\tfrac{1}{80}\left(Q_2 - Q_1\right)\right] \qquad \text{(d)}\end{aligned}$$

where we have used Eq. (c) in the last step. Putting in $Q_2 = 30$ cfs and $Q_1 = 10$ cfs, we then get for the average force during the 10-sec interval of interest on using $\rho = 62.4/g$ slugs/ft^3:

$$F_{av} = 77.5 \text{ lb} \qquad \text{(e)}$$

This is the average horizontal force that the truck exerts on the water. Clearly, this force is also what the ground must exert on the truck in the horizontal direction to prevent motion of the truck during the water transfer operation.

From another viewpoint, this system is not unlike a propulsion system like a jet engine to be studied with the aid of a control volume in your fluids course.

If the total external force on a system of particles is zero, it is clear from the previous discussion that there can be no change in the linear momentum of the system. This is the principle of *conservation of linear momentum*, which means, furthermore, that *with a zero total impulse on an aggregate of particles, there can be no change in the velocity of the mass center*. If at some time t_0 the velocity of the mass center of such a system of particles is zero, then this velocity must remain zero if the impulse on the system of particles is zero. That is, no matter what movements and gyrations the elements of the system may have, they must be such that the center of mass must remain stationary. We reached the same conclusion in Chapter 10, where we found from Newton's law that if the total external force on a system of particles is zero, then the acceleration of the center of mass is zero.[3]

[3]Problems 10.103 and 10.104 are examples of this condition.

12.3 Impulsive Forces

Let us now examine the action involved in the explosion of a bomb that is initially suspended from a wire, as shown in Fig. 12.8. First, consider the situation *directly after* the explosion has been set off. Since very large forces are present from expanding gases, a *fragment* of the bomb receives an appreciable impulse during this short time interval. Also, directly after the explosion, the gravitational forces are no longer counteracted by the supporting wire, so there is an additional impulse acting on the fragment. But since the gravitational force is small compared to forces from the explosion, the gravitational impulse on a fragment can be considered negligibly small for the short period of time under discussion compared to that of the expanding gases acting on the fragment. A plot of the impulsive force (from the explosion) and the force of gravity on a fragment is shown in Fig. 12.9. It is clear from this diagram

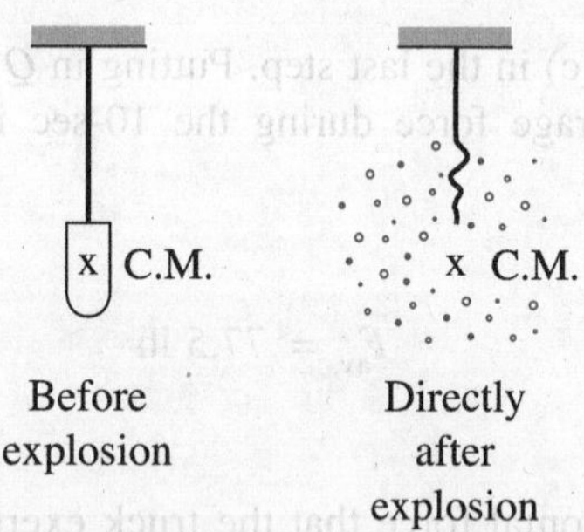

Figure 12.8. Exploding bomb.

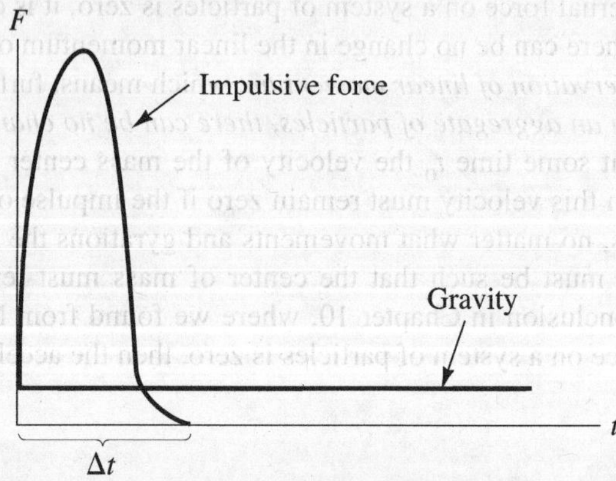

Figure 12.9. Plot of impulsive force and gravity force.

that the impulse from the explosion lasts for a very short time Δt and can be significant, whereas the impulse from gravity during the same short time is by comparison negligible. Forces that act over a very short time but have nevertheless appreciable impulse are called *impulsive forces.* In actions involving very small time intervals, we need only consider impulsive forces. Furthermore, during a very short time Δt an impulsive force acting on a particle can change the velocity of the particle in accordance with the impulse-momentum equation an appreciable amount while the particle undergoes very little change in position during the time Δt.[4] It is simplest in many cases to consider the *change in velocity of a particle from an impulsive force to occur over zero distance.*

Up to now, we have only considered a fragment of the bomb. Now let us consider all the fragments of the bomb taken as a system of particles. Since the explosive action is *internal* to the bomb, the action causes impulses that for any direction have equal and opposite counterparts, and thus *the total impulse on the bomb due to the explosion is zero.* We can thus conclude that *directly after* the explosion *the center of mass of the bomb has not moved appreciably* despite the high velocity of the fragments in all directions, as illustrated in Fig. 12.8. As time progresses beyond the short time interval described above, the gravitational impulse increases and has significant effect. If there were no friction, the center of mass would descend from the position of support as a freely falling particle under this action of gravity.

The following problems will illustrate these ideas.

[4]This idealization can be explained more precisely as follows. For an impulsive force F acting on a body of mass M, we can say from the linear momentum equation

$$\int_0^{\Delta t} F\,dt = F_{AV}\,\Delta t = MV_{\max} \qquad \therefore\ V_{\max} = \left(\frac{F_{AV}}{M}\right)(\Delta t)$$

The maximum movement of the body M during this time interval according to Newton's law on using the above result for V is then

$$x_{\max} = \int_0^{\Delta t} V_{\max}\,dt = \int_0^{\Delta t}\left(\frac{F_{AV}}{M}\,\Delta t\right)dt = \left(\frac{F_{AV}}{M}\right)(\Delta t)^2$$

Note that $V_{\max}$ is proportional to Δt while x is proportional to $(\Delta t)^2$. Clearly for a *very small* interval Δt the value of the movement x of the mass M can be considered *second order* compared to the value of the velocity V. For simplicity, with minimal error, we can say that the *mass M does not move while undergoing a change of velocity in response to an impulsive force.*

Example 12.7

Some top-flight tennis players hit the ball on a service at the instant that the ball is at the top of its trajectory after being released by the free hand. The ball is often given a speed V of 120 mi/hr by the racquet directly after the impact is complete. If the time of duration of the impact process is .005 sec, what is the magnitude of the average force from the racquet on the ball during this time interval? Take the weight of the ball as 1.5 oz.

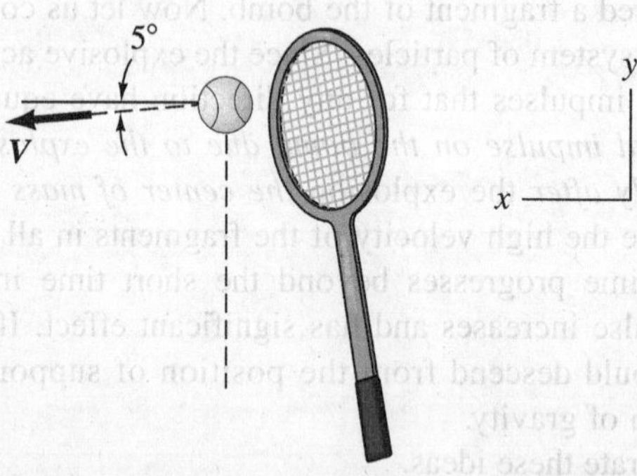

Figure 12.10. Impact of a tennis ball at service.

We have here acting on the ball during a very small time interval an **impulsive** force and the force of gravity. We will ignore the gravity force during the time of impact and we will consider that the ball achieves a post impact velocity while not moving, as explained earlier in the model for impulsive force behavior. As shown in Fig. 12.10, the impulse $\boldsymbol{I}$ generated on the ball by the racquet accordingly is

$$\boldsymbol{I} = \frac{15}{(16)(32.2)}\left[(120)\left(\frac{5{,}280}{3{,}600}\right)\right][\cos 5°\boldsymbol{i} - \sin 5°\boldsymbol{j}]$$

$$= .5124[.996\boldsymbol{i} - .0872\boldsymbol{j}] \text{ lb-sec}$$

Next, we go to the **impulse momentum** equation. Thus

$$(\boldsymbol{F}_{av})(.005) = .5124(.996\boldsymbol{i} - .0872\boldsymbol{j})$$

The magnitude of the average force is finally given as follows:

$$|\boldsymbol{F}_{av}| = 102.6 \text{ lb}$$

After the impact, the ball will have a trajectory determined by gravity, wind forces, and the initial post-impact conditions.

Example 12.8

A 9,000-N idealized cannon with a recoil spring (K = 4,000 N/m) fires a 45-N projectile with a muzzle velocity of 625 m/sec at an angle of 50° (Fig. 12.11). Determine the maximum compression of the spring.

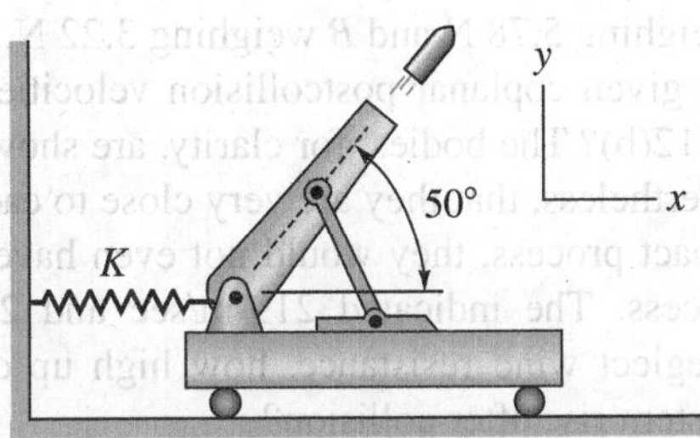

Figure 12.11. Idealized cannon.

The firing of the cannon takes place in a very short time interval. The force on the projectile and the force on the cannon from the explosion are impulsive forces. As a result, the cannon can be considered to achieve a recoil velocity instantaneously without having moved appreciably. Like the exploding bomb, the impulse on the cannon *plus* projectile is zero, as a result of the firing process. Since the linear momentum of the cannon plus projectile is zero just before firing, this linear momentum must be zero directly after firing. Thus, just after firing, we can say for the x direction:

$$(MV_x)_{\text{cannon}} + (MV_x)_{\text{projectile}} = 0 \tag{a}$$

Using V_c for the cannon velocity along the x axis and $V_p = V_c + 625 \cos 50°$ for the projectile velocity along the x axis we get

$$\frac{9{,}000}{g} V_c + \frac{45}{g}\left[(625)(\cos 50°) + V_c\right] = 0$$

Solving for V_c, we get

$$V_c = -2.00 \text{ m/sec} \tag{b}$$

After this initial impulsive action, which results in an instantaneous velocity being imparted to the cannon, the motion of the cannon is then impeded by the spring. We may now use **conservation of mechanical energy** for a particle in this phase of motion of the cannon. Denoting δ as the maximum deflection of the spring, we can say:

$$\frac{1}{2}\frac{9{,}000}{g}(2.00^2) = \frac{1}{2}(4{,}000)(\delta^2)$$

Therefore,

$$\delta = .958 \text{ m}$$

Example 12.9

For target practice, a 9-N rock is thrown into the air and fired on by a pistol. The pistol bullet, of mass 57 g and moving with a speed of 312 m/sec, strikes the rock as it is descending vertically at a speed of 6.25 m/sec. [See Fig. 12.12(a).] Both the velocity of the bullet and the rock are parallel to the *xy* plane. Directly after the bullet hits the rock, the rock breaks up into two pieces, *A* weighing 5.78 N and *B* weighing 3.22 N. What is the velocity of *B* after collision for the given coplanar postcollision velocities of the bullet and the piece *A* shown in Fig. 12.12(b)? The bodies, for clarity, are shown separated in the diagram. Keep in mind, nevertheless, that they are very close to each other at post-impact. In our model of the impact process, they would not even have moved relative to each other during this process. The indicated 219-m/sec and 25-m/sec velocities are in the *xy* plane. If we neglect wind resistance, how high up does the center of mass of the rock and bullet system rise after collision?

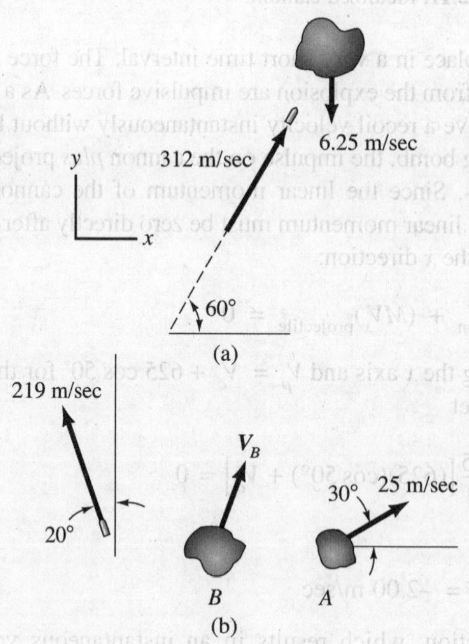

Figure 12.12. Bullet striking a rock.

Linear momentum is conserved during the collision, so we can equate linear momenta directly before and directly after collision. Thus,

$$
\begin{aligned}
&(.057)(312)(.5\boldsymbol{i} + .866\boldsymbol{j}) + \frac{9}{g}(-6.25\boldsymbol{j}) \\
&\quad = (.057)(219)(-\sin 20°\boldsymbol{i} + \cos 20°\boldsymbol{j}) \\
&\qquad + \frac{5.78}{g}25(.866\boldsymbol{i} + .5\boldsymbol{j}) + \frac{3.22}{g}\left[(V_B)_x\boldsymbol{i} + (V_B)_y\boldsymbol{j}\right]
\end{aligned}
$$

Example 12.9 (Continued)

We may solve for the desired quantities $(V_B)_x$ and $(V_B)_y$ to get

$$(V_B)_x = 1.235 \text{ m/sec}$$
$$(V_B)_y = -28.7 \text{ m/sec}$$

We now compute the velocity of the center of mass just before collision. Thus,

$$MV_c = \left(\frac{9}{g} + .057\right)V_c = \frac{9}{g}(-6.25)\boldsymbol{j} + (.057)(312)(.5\boldsymbol{i} + .866\boldsymbol{j})$$

Therefore,

$$V_c = 9.125\boldsymbol{i} + 9.92\boldsymbol{j} \text{ m/sec}$$

Hence, for the center of mass there is an initial velocity upward of 9.92 m/sec just before collision. Directly after collision, since there has been no appreciable external impulse on the system during collision, the center of mass ***still has*** this upward speed. But now considering larger time intervals, we must take into account the action of gravity, which gives the center of mass a downward acceleration of 9.81 m/sec.[2] Thus,

$$\ddot{y}_c = -9.81$$
$$\dot{y}_c = -9.81t + C_1$$
$$y_c = -9.81\frac{t^2}{2} + C_1 t + C_2$$

When $t = 0$, $\dot{y}_c = 9.92$ and we take $y_c = 0$ for convenience. Hence we have

$$\dot{y}_c = -9.81t + 9.92 \qquad \text{(a)}$$
$$y_c = -9.81\frac{t^2}{2} + 9.92t \qquad \text{(b)}$$

Set $\dot{y}_c$ in (a) equal to zero and solve for t. We get

$$t = 1.011 \text{ sec}$$

Substitute this value of t in Eq. (b) and solve for y_c, which now gives the desired maximum elevation of the center of mass after collision. Thus,

$$(y_c)_{max} = 5.01 \text{ m} \qquad \text{(c)}$$

PROBLEMS

12.1. A body weighing 100 lb reaches an incline of 30° while it is moving at 50 ft/sec. If the dynamic coefficient of friction is .3, how long before the body stops?

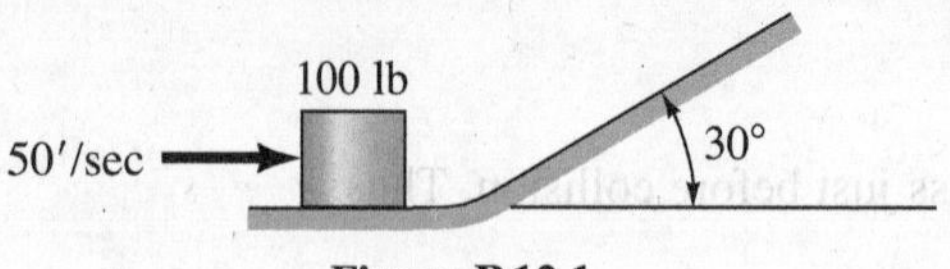

Figure P.12.1.

12.2. A particle of mass 1 kg is initially stationary at the origin of a reference. A force having a known variation with time acts on the particle. That is,

$$\boldsymbol{F}(t) = t^2\boldsymbol{i} + (6t + 10)\boldsymbol{j} + 1.6t^3\boldsymbol{k}\text{ N}$$

where t is in seconds. After 10 sec, what is the velocity of the body?

12.3. A unidirectional force acting on a particle of mass 16 kg is plotted. What is the velocity of the particle at 40 sec? Initially, the particle is at rest.

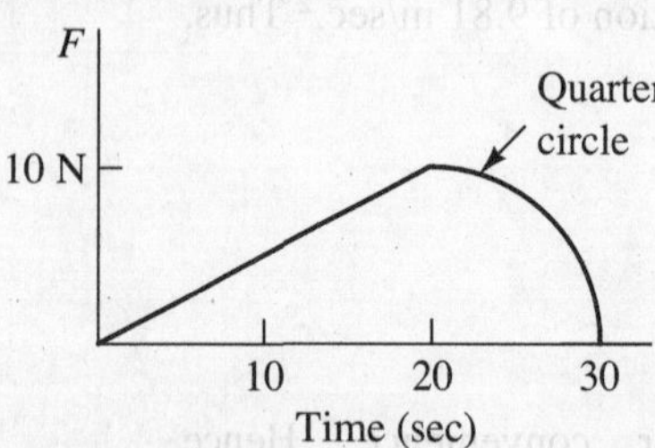

Figure P.12.3.

12.4. A 100-lb block is acted on by a force P, which varies with time as shown. What is the speed of the block after 80 sec? Assume that the block starts from rest and neglect friction. The time axis gives time intervals.

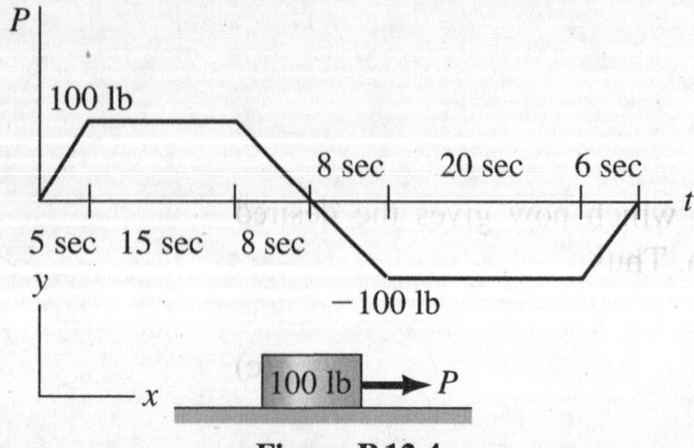

Figure P.12.4.

12.5. If the coefficient of static friction is .5 in Problem 12.4 and the coefficient of dynamic friction is .3, what is the speed of the block after 28 sec?

12.6. A body is dropped from rest. (a) Determine the time required for it to acquire a velocity of 16 m/sec. (b) Determine the time needed to increase its velocity from 16 m/sec to 23 m/sec.

12.7. A body having a mass of 5 lbm is acted on by the following force:

$$\boldsymbol{F} = 8t\boldsymbol{i} + (6 + 3\sqrt{t})\boldsymbol{j} + (16 + 3t^2)\boldsymbol{k}\text{ lb}$$

where t is in seconds. What is the velocity of the body after 5 sec if the initial velocity is

$$\boldsymbol{V}_1 = 6\boldsymbol{i} + 3\boldsymbol{j} - 10\boldsymbol{k}\text{ ft/sec?}$$

12.8. A body with a mass of 16 kg is required to change its velocity from $\boldsymbol{V}_1 = 2\boldsymbol{i} + 4\boldsymbol{j} - 10\boldsymbol{k}$ m/sec to a velocity $\boldsymbol{V}_2 = 10\boldsymbol{i} - 5\boldsymbol{j} + 20\boldsymbol{k}$ m/sec in 10 sec. What average force $\boldsymbol{F}_{av}$ over this time interval will do the job?

12.9. In Problem 12.8, determine the force as a function of time for the case where force varies linearly with time starting with a zero value.

12.10. A hockey puck moves at 30 ft/sec from left to right. The puck is intercepted by a player who whisks it at 80 ft/sec toward goal A, as shown. The puck is also rising from the ice at a rate of 10 ft/sec. What is the impulse on the puck, whose weight is 5 oz?

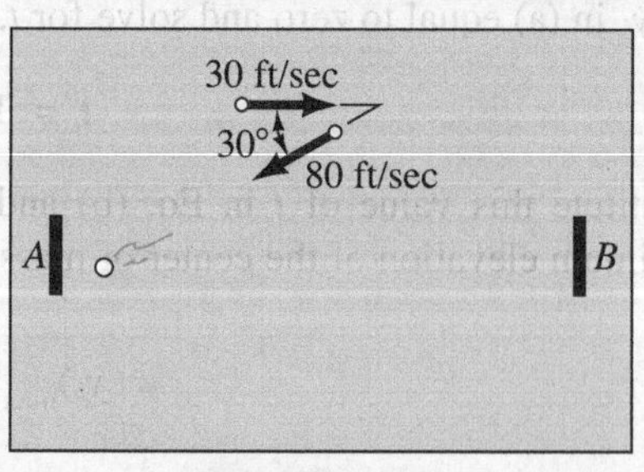

Figure P.12.10.

12.11. Gravel is released from a hopper at the rate of 1 kg/sec. At the exit of the hopper it has a speed of .15 m/s. The belt is moving at a constant speed of 3 m/s. If there is 20 kg of gravel on the conveyor belt at all times and if the belt on the conveyor bed has a weight of 50 N, what is the difference in tension $T_2 - T_1$ for the belt to maintain operation? The dynamic coefficient of friction between bed and belt is 0.4. Assume that the gravel drops 0.2 m from the hopper outlet.

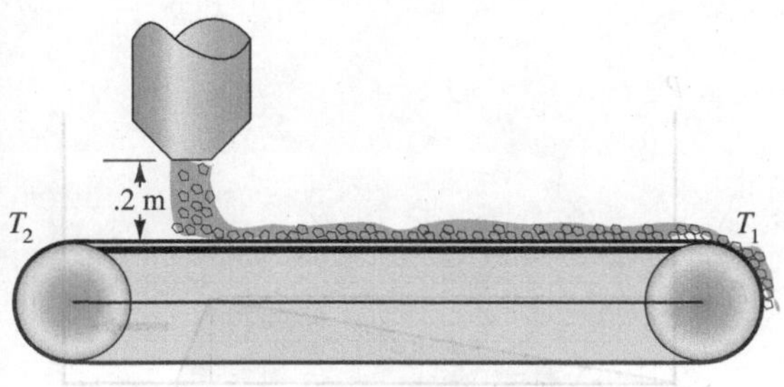

Figure P.12.11.

12.12. Do Problem 12.5 by methods of momentum.

12.13. Do Problem 12.6 by methods of momentum.

12.14. A commuter train made up of two cars is moving at a speed of 80 km/hr. The first car has a mass of 20,000 kg and the second 15,000 kg.

(a) If the brakes are applied simultaneously to both cars, determine the minimum time the cars travel before stopping. The coefficient of static friction between the wheels and rail is .3.

(b) If the brakes on the first car only are applied, determine the time the cars travel before stopping and the force F transmitted between the cars.

12.15. Compute the velocity of the bodies after 10 sec if they start from rest. The cable is inextensible, and the pulleys are frictionless. For the contact surfaces, $\mu_d = .2$.

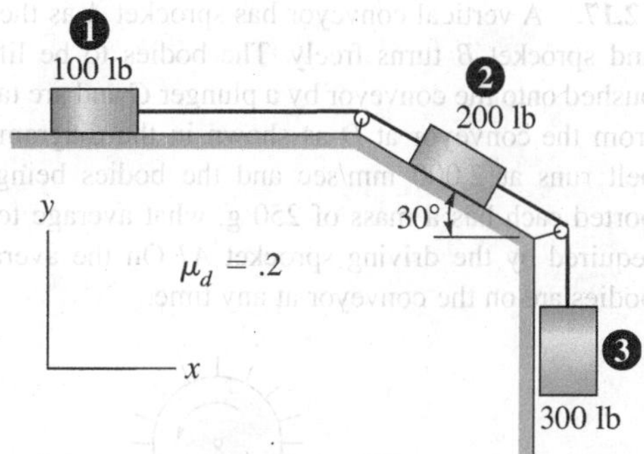

Figure P.12.15.

12.16. Two boxes per second each weighing 100 lb land on a circular conveyor at a speed of 5 ft/sec in the direction of the chute. If there are 6 boxes on the circular belt at any one time, determine the average torque needed to rotate the belt at an angular speed of .2 rad/sec. The dynamic coefficient of friction between the belt and the conveyer bed is .3. What horsepower is needed for operating this belt? Neglect the rotational effect on the boxes themselves as they drop from the chute onto the conveyor. Also, does the radial change in velocity of the boxes affect the torque needed by the conveyor? Neglect any radial slipping of the boxes as they land.

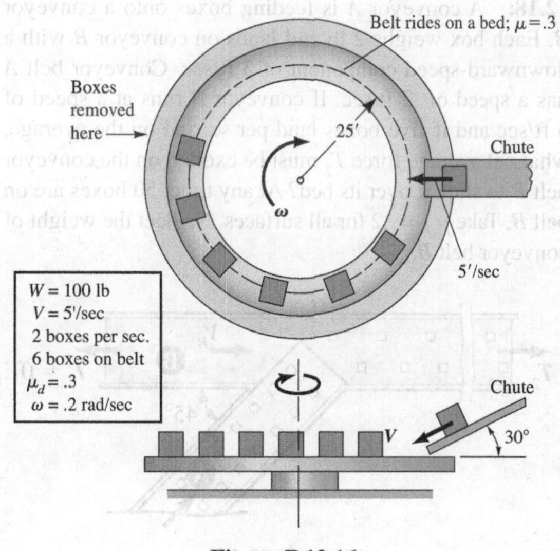

Figure P.12.16.

12.17. A vertical conveyor has sprocket A as the driver, and sprocket B turns freely. The bodies to be lifted are pushed onto the conveyor by a plunger C and are taken off from the conveyor at D as shown in the diagram. If the belt runs at 2,000 mm/sec and the bodies being transported each has a mass of 250 g, what average torque is required by the driving sprocket A? On the average, 40 bodies are on the conveyor at any time.

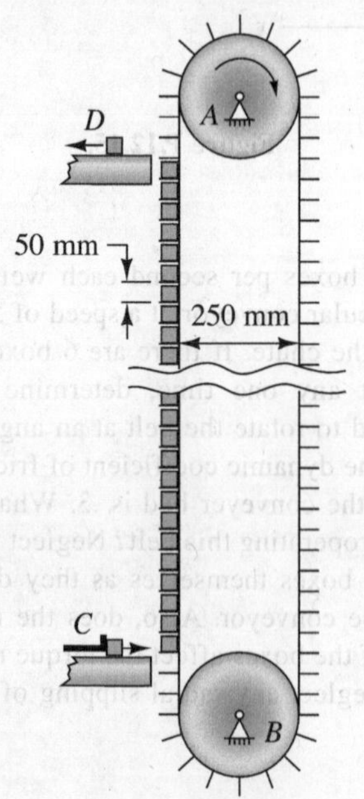

Figure P.12.17.

12.18. A conveyor A is feeding boxes onto a conveyor B. Each box weighs 2 lb and lands on conveyor B with a downward-speed component of 3 ft/sec. Conveyor belt A has a speed of .2 ft/sec. If conveyor B runs at a speed of 5 ft/sec and if five boxes land per second on the average, what net average force T_2 must be exerted on the conveyor belt B to slide it over its bed? At any time, 50 boxes are on belt B. Take μ_d = .2 for all surfaces. Neglect the weight of conveyor belt B.

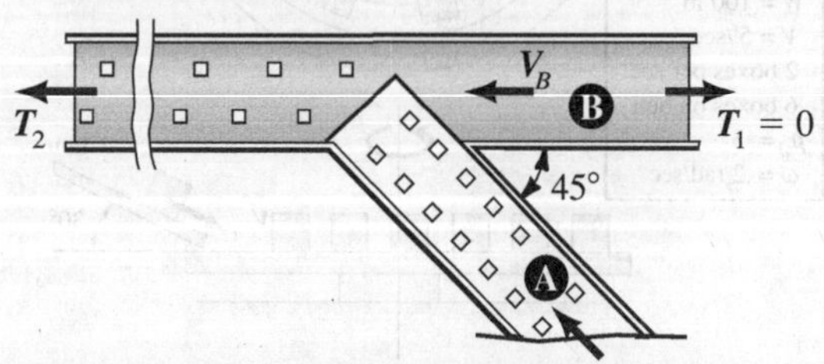

Figure P.12.18.

12.19. An idealized one-dimensional pressure wave (i.e., pressure is a function of one coordinate and time) generated by an explosion travels at a speed V of 1,200 ft/sec, as shown at time t = 0. The peak pressure of this wave is 5 psia. What impulse per square foot is delivered to a wall oriented at right angles to the x axis? The wave is reflected from the wall, and the pressure at the wall is double the incoming pressure at all times. Do the problem for two time intervals corresponding to the interval (a) from when the wave front first touches the wall to when the peak reaches the wall and (b) from when the peak hits the wall to when the end of the wave reaches the wall.

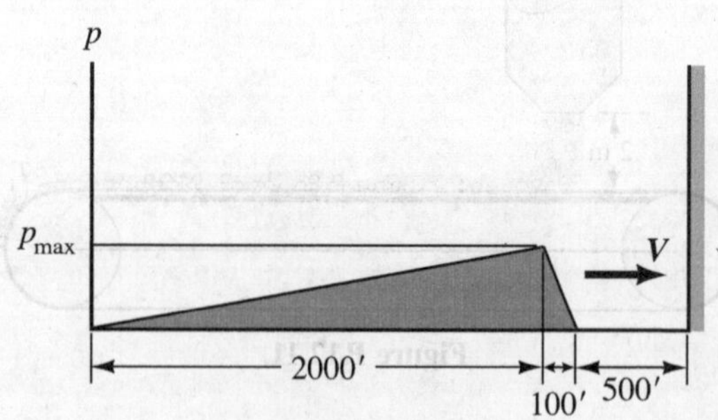

Figure P.12.19.

12.20. Blocks A and B move on frictionless surfaces. The blocks are interconnected with a light bar. Body A weighs 30 lb; the weight of body B is not known. A constant force F of 100 lb is applied at the configuration shown. If a speed of 25 ft/sec is reached by A after 1 sec, what impulse is developed on the vertical wall?

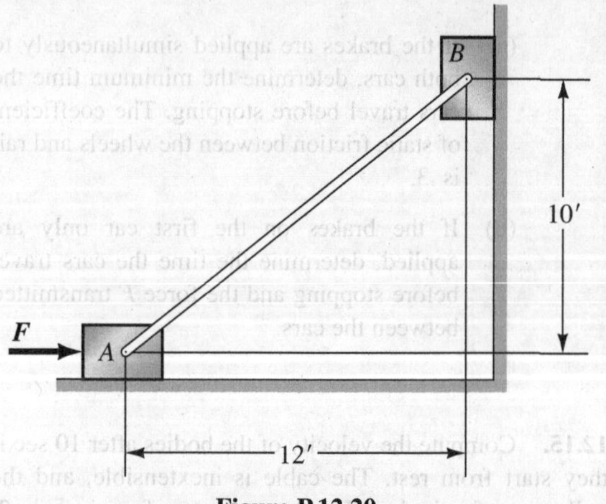

Figure P.12.20.

12.21. In Problem 12.20, compute the impulse on the horizontal surface. A moves 4 ft in 1 sec and $W_B = 20$ lb.

12.22. An antitank airplane fires two 90-N projectiles at a tank at the same time. The muzzle velocity of the guns is 1,000 m/sec relative to the plane. If the plane before firing weighs 65 kN and is moving with a velocity of 320 km/hr, compute the change in its speed when it fires the two projectiles.

12.23. A toboggan has just entered the horizontal part of its run. It carries three people weighing 120 lb, 180 lb, and 150 lb, respectively. Suddenly, a pedestrian weighing 200 lb strays onto the course and is turned end for end by the toboggan, landing safely among the riders. Since the toboggan path is icy, we can neglect friction with the toboggan path for all actions described here. If the toboggan is traveling at a speed of 35 mph just before collision occurs, what is the speed after the collision when the pedestrian has become a rider? The toboggan weighs 30 lb.

Figure P.12.23.

12.24. An 890-N rowboat containing a 668-N man is pushed off the dock by an 800-N man. The speed that is imparted to the boat is .30 m/sec by this push. The man then leaps into the boat from the dock with a speed of .60 m/sec relative to the dock in the direction of motion of the boat. When the two men have settled down in the boat and before rowing commences, what is the speed of the boat? Neglect water resistance.

Figure P.12.24.

12.25. Two vehicles connected with an inextensible cable are rolling along a road. Vehicle B, using a winch, draws A toward it so that the relative speed is 5 ft/sec at $t = 0$ and 10 ft/sec at $t = 20$ sec. Vehicle A weighs 2,000 lb and vehicle B weighs 3,000 lb. Each vehicle has a rolling resistance that is .01 times the vehicle's weight. What is the speed of A relative to the ground at $t = 20$ sec if A is initially moving to the right at a speed of 30 ft/sec?

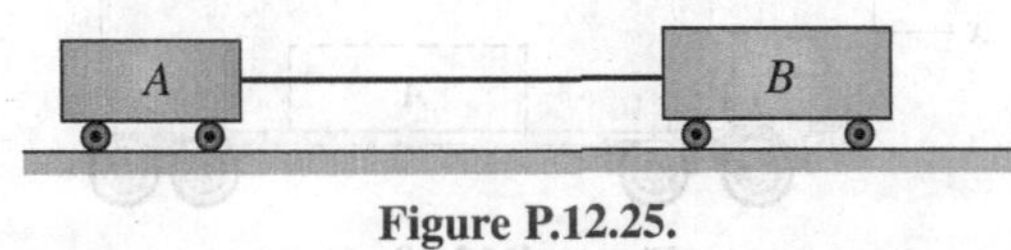

Figure P.12.25.

12.26. Treat Example 12.3 as a two-particle system in the impulse-momentum considerations. Verify the results of Example 12.3 for V. (Be sure to include *all* external forces for the system.)

12.27. Determine the velocity of body A and body B after 3 sec if the system is released from rest. Neglect friction and the inertia of the pulleys.

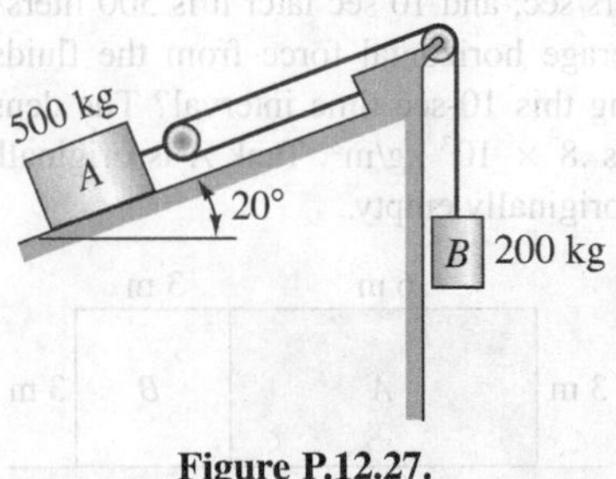

Figure P.12.27.

12.28. Do Problem 12.27 by considering a system of particles. (Be sure to include *all* external forces for the system of bodies A and B.)

12.29. A 40-kN truck is moving at the speed of 40 km/hr carrying a 15-kN load *A*. The load is restrained only by friction with the floor of the truck where there is a dynamic coefficient of friction of .2 and the static coefficient of friction is .3. The driver suddenly jams his brakes on so as to lock all wheels for 1.5 sec. At the end of this interval, the brakes are released. What is the final speed *V* of the truck neglecting wind resistance and rotational inertia of the wheels after load *A* stops slipping? The dynamic coefficient of friction between the tires and the road is .4.

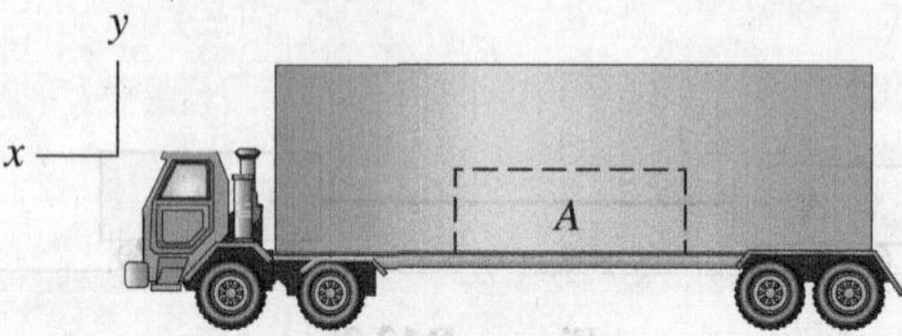

Figure P.12.29.

12.30. A 1,300-kg Jeep is carrying three 100-kg passengers. The Jeep is in four-wheel drive and is under test to see what maximum speed is possible in 5 sec from a start on an icy road surface for which $\mu_s = .1$. Compute V_{max} at $t = 5$ sec.

Figure P.12.30.

12.31. Two adjacent tanks *A* and *B* are shown. Both tanks are rectangular with a width of 4 m. Gasoline from tank *A* is being pumped into tank *B*. When the level of tank *A* is .7 m from the top, the rate of flow *Q* from *A* to *B* is 300 liters/sec, and 10 sec later it is 500 liters/sec. What is the average horizontal force from the fluids onto the tank during this 10-sec time interval? The density of the gasoline is $.8 \times 10^3$ kg/m^3. Tank *A* is originally full and tank *B* is originally empty.

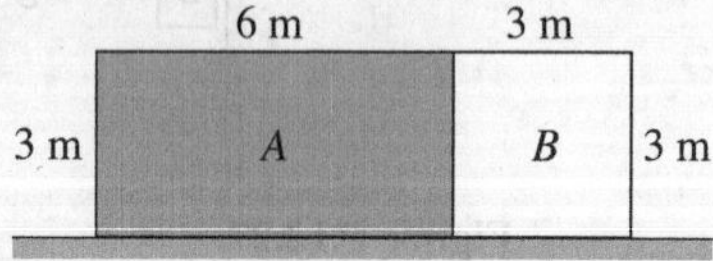

Figure P.12.31.

12.32. Two tanks *A* and *B* are shown. Tank *A* is originally full of water (ρ = 62.4 lbm/ft^3), while tank *B* is empty. Water is pumped from *A* to *B*. If initially 100 cfs of water is being pumped and if this flow increases at the rate of 10 cfs/sec^2 for 30 sec thereafter, what is the average vertical force onto the tanks from the water during this time period, aside from the static dead weight of the water?

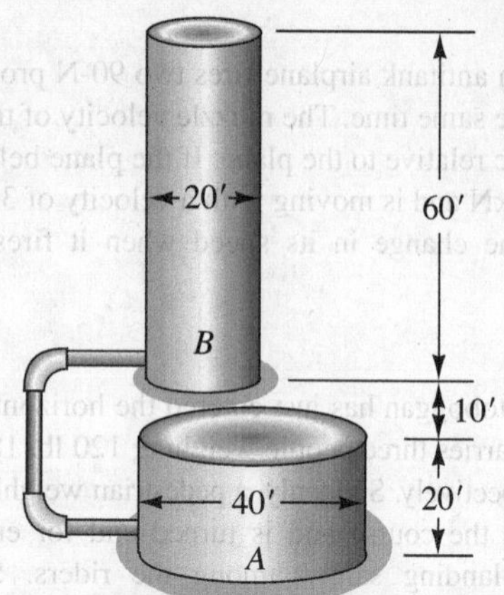

Figure P.12.32

12.33. A device to be detonated is shown in (a) suspended above the ground. Ten seconds after detonation, there are four fragments having the following masses and position vectors relative to reference *XYZ*:

$$m_1 = 5 \text{ kg}$$
$$\boldsymbol{r}_1 = 1{,}000\boldsymbol{i} + 2{,}000\boldsymbol{j} + 900\boldsymbol{k} \text{ m}$$
$$m_2 = 3 \text{ kg}$$
$$\boldsymbol{r}_2 = 800\boldsymbol{i} + 1{,}800\boldsymbol{j} + 2{,}500\boldsymbol{k} \text{ m}$$
$$m_3 = 4 \text{ kg}$$
$$\boldsymbol{r}_3 = 400\boldsymbol{i} + 1{,}000\boldsymbol{j} + 2{,}000\boldsymbol{k} \text{ m}$$
$$m_4 = 6 \text{ kg}$$
$$\boldsymbol{r}_4 = X_4\boldsymbol{i} + Y_4\boldsymbol{j} + Z_4\boldsymbol{k}$$

Find the position $\boldsymbol{r}_4$ if the center of mass of the device is initially at position $\boldsymbol{r}_0$, where

$$\boldsymbol{r}_0 = 600\boldsymbol{i} + 1200\boldsymbol{j} + 2{,}300\boldsymbol{k} \text{ m}$$

Neglect wind resistance.

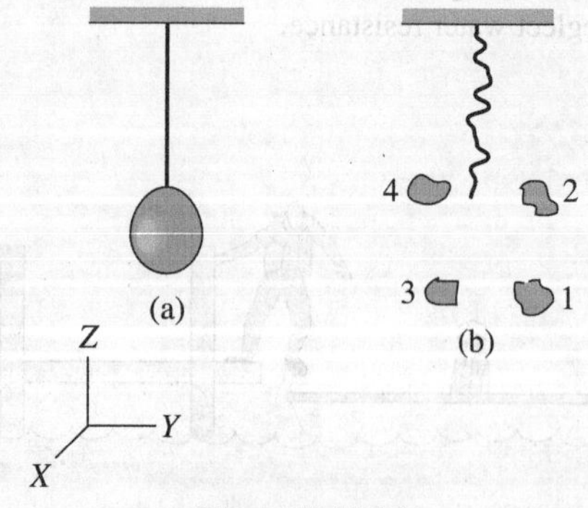

Figure P.12.33.

12.4 Impact

In Section 12.3, we discussed impulsive forces. We shall in this section discuss in detail an action in which impulsive forces are present. This situation occurs when two bodies collide but do not break. The time interval during collision is very small, and comparatively large forces are developed on the bodies during the small time interval. This action is called *impact*. For such actions with such short time intervals, the force of gravity generally causes a negligible impulse. The impact forces on the colliding bodies are always equal and opposite to each other, so the net impulse on the *pair* of bodies during collision is *zero*. This means that the total linear momentum directly after impact (postimpact) equals the total linear momentum directly before impact (preimpact).

We shall consider at this time two types of impact for which certain definitions are needed. We shall call the normal to the *plane of contact* during the collision of two bodies the *line of impact*. If the centers of mass of the two colliding bodies lie along the line of impact, the action is called *central impact* and is shown for the case of two spheres in Fig. 12.13.[5] If, in addition, the velocity vectors of the mass centers approaching the collision are collinear with the line of impact, the action is called *direct central impact*. This action is illustrated by $\mathbf{V}_1$ and $\mathbf{V}_2$ in Fig 12.13. Should one (or both) of the velocities have a line of action not collinear with the line impact—for example, $\mathbf{V}'_1$ and/or $\mathbf{V}'_2$—the action is termed *oblique central impact.*

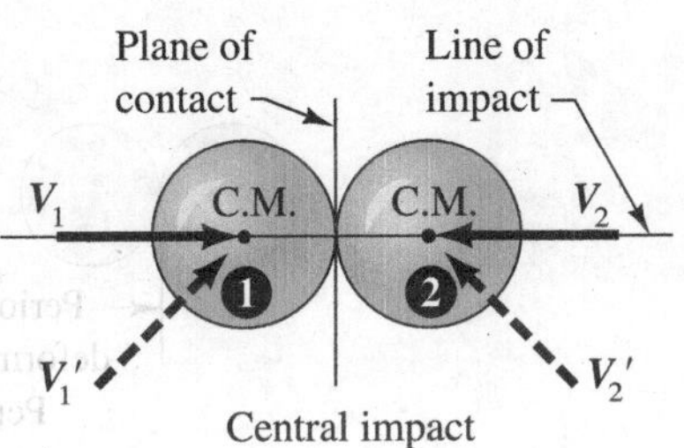

Figure 12.13. Central impact of two spheres.

In either case, linear momentum is conserved during the short time interval from directly before the collision (indicated with the subscript *i*) to directly after the collision (indicated with subscript *f*). That is,

$$(m_1\mathbf{V}_1)_i + (m_2\mathbf{V}_2)_i = (m_1\mathbf{V}_1)_f + (m_2\mathbf{V}_2)_f \tag{12.8}$$

In the *direct-central-impact* case for *smooth* bodies (i.e., bodies with no friction), this equation becomes a single scalar equation since $(\mathbf{V}_1)_f$ and $(\mathbf{V}_2)_f$ are collinear with the line of impact. Usually, the initial velocities are known and the final values are desired, which means that we have for this case one scalar equation involving two unknowns. Clearly, we must know more about the manner of interaction of the bodies, since Eq. 12.8 as it stands is valid for materials of any deformability (e.g., putty or hardened steel) and takes no account of such important considerations. Thus, we cannot consider the bodies undergoing impact only as particles as has been the case thus far, but must, in addition, consider them as deformable bodies of finite size in order to generate enough information to solve the problem at hand.

For the *oblique-impact case*, we can write components of the linear-momentum equation along the line of impact and for smooth (frictionless) bodies, along two other directions at right angles to the line of impact. If we know the initial velocities, then we have six unknown final velocity components and only three equations. Thus, we need even more information to establish fully the final velocities after this more general type of impact. We now consider each of these cases in more detail in order to establish these additional relations.

[5]Noncentral or *eccentric impact* is examined in Chapter 15 for the case of plane motion.

Case 1. Direct Central Impact. Let us first examine the direct-central-impact case. We shall consider the period of collision to be made up of two subintervals of time. The *period of deformation* refers to the duration of the collision, starting from initial contact of the bodies and ending at the instant of maximum deformation. During this period, we shall consider that impulse $\int D\,dt$ acts oppositely on each of the bodies. The second period, covering the time from the maximum deformation condition to the instant at which the bodies just separate,[6] we shall term the *period of restitution*. The impulse acting oppositely on each body during this period we shall indicate as $\int R\,dt$. If the bodies are *perfectly elastic*, they will reestablish their initial shapes during the period of restitution (if we neglect the internal vibrations of the bodies), as shown in Fig. 12.14(a). When the bodies do not reestablish their initial shapes [Fig. 12.14(b)], we say that *plastic deformation* has taken place.

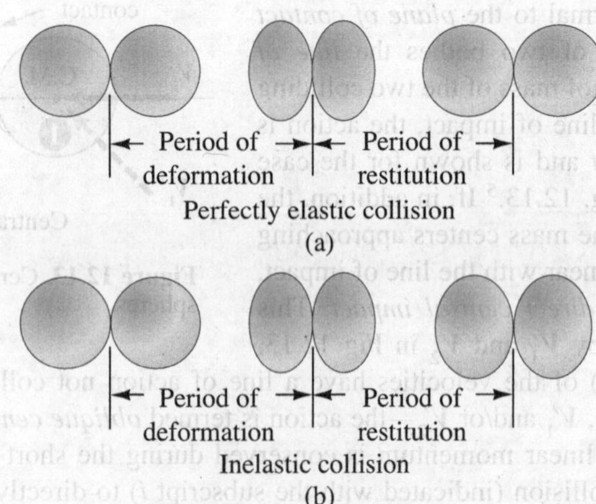

Figure 12.14. Collision process.

The ratio of the impulse during the restitution period $\int R\,dt$ to the impulse during the deformation period $\int D\,dt$ is a number ϵ, which depends mainly on the physical properties of the bodies in collision. We call this number the *coefficient of restitution*. Thus,

$$\epsilon = \frac{\text{impulse during restitution}}{\text{impulse during deformation}} = \frac{\int R\,dt}{\int D\,dt} \tag{12.9}$$

We must strongly point out that the coefficient of restitution depends also on the size, shape, and approach velocities of the bodies before impact. These dependencies result from the fact that plastic deformation is related to the magnitude and nature of the force distributions in the bodies and also to the rate of loading. However, values of ϵ have been established for different materials and can be used for approximate results in the kind of

[6]If they don't separate, the end of the second period occurs when the bodies cease to deform. We call such a process a **plastic** impact.

computations to follow. We shall now formulate the relation between the coefficient of restitution and the initial and final velocities of the bodies undergoing impact.

Let us consider *one* of the bodies during the two phases of the collision. If we call the velocity at the maximum deformation condition $(V)_D$, we can say for mass 1:

$$\int D\,dt = \left[(m_1V_1)_D - (m_1V_1)_i\right] = -m_1\left[(V_1)_i - (V_1)_D\right] \tag{12.10}$$

During the period of restitution, we find that

$$\int R\,dt = -m_1\left[(V_1)_D - (V_1)_f\right] \tag{12.11}$$

Dividing Eq. 12.11 by Eq. 12.10, canceling out m_1, and noting the definition in Eq. 12.9, we can say:

$$\epsilon = \frac{(V_1)_D - (V_1)_f}{(V_1)_i - (V_1)_D} \tag{12.12}$$

A similar analysis for the other mass (2) gives

$$\epsilon = \frac{(V_2)_D - (V_2)_f}{(V_2)_i - (V_2)_D} = \frac{(V_2)_f - (V_2)_D}{(V_2)_D - (V_2)_i} \tag{12.13}$$

In this last expression, we have changed the sign of numerator and denominator. At the intermediate position at the end of deformation and the beginning of restitution the masses have essentially the same velocity. Thus, $(V_1)_D = (V_2)_D$. Since the quotients in Eqs. 12.12 and 12.13 are equal to each other, we can add numerators and denominators to form another equal quotient, as you can demonstrate yourself. Noting the abovementioned equality of the V_D terms, we have the desired result:

$$\epsilon = -\frac{(V_2)_f - (V_1)_f}{(V_2)_i - (V_1)_i} = -\frac{\text{relative velocity of separation}}{\text{relative velocity of approach}} \tag{12.14}$$

This equation involves the coefficient ϵ, which is presumably known or estimated, and the initial and final velocities of the bodies undergoing impact. Thus, with this equation we can solve for the final velocities of the bodies after collision when we use the linear-momentum equation 12.8 for the case of direct central impact.

During a *perfectly elastic* collision, the impulse for the period of restitution equals the impulse for the period of deformation,[7] so the coefficient of restitution is *unity* for this case. For inelastic collisions, the coefficient of restitution is less than unity since the

[7]The impulses are equal because during the period of restitution the body can be considered to undergo identically the reverse of the process corresponding to the deformation period. Thus, from a thermodynamics point of view, we are considering the elastic impact to be a *reversible* process.

impulse is diminished on restitution as a result of the failure of the bodies to resume their original geometries. For a *perfectly plastic* impact, $\epsilon = 0$ [i.e., $(V_2)_f = (V_1)_f$] and the bodies remain in contact. Thus ϵ ranges from 0 to 1.

Case 2. Oblique Central Impact. Let us now consider the case of oblique central impact. The velocity components along the line of impact can be related by the scalar component of the linear-momentum equation 12.8 in this direction and also by Eq. 12.14, where velocity components along the line of impact are used and where the coefficient of restitution may be considered (for smooth bodies) to be the same as for the direct-central-impact case. If we know the initial conditions, we can accordingly solve for those velocity components after impact in the direction of the line of impact. As for the other rectangular components of velocity, we can say that for smooth bodies, these velocity components are unaffected by the collision, since no impulses act in these directions on either body. That is, the velocity components normal to the line of impact for each body are the same immediately after impact as before. Thus, the final velocity components of both bodies can be established, and the motions of the bodies can be determined within the limits of the discussion. The following examples are used to illustrate the use of the preceding formulations.

Note that the mass and materials of the colliding bodies for both direct or oblique central impact can be different from each other.

Example 12.10

Two billiard balls (of the same size and mass) collide with the velocities of approach shown in Fig. 12.15. For a coefficient of restitution of .90, what are the final velocities of the balls directly after they part? What is the loss in kinetic energy?

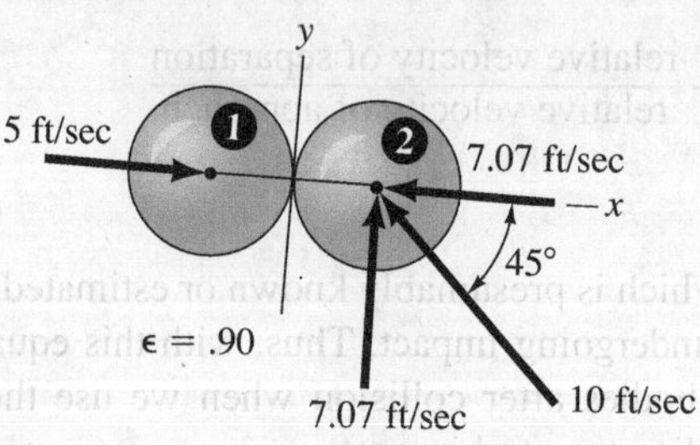

Figure 12.15. Oblique central impact.

A reference is established so that the x axis is along line of impact and the y axis is in the plane of contact such that the reference plane is parallel to the billiard table.

Example 12.10 (Continued)

The approach velocities have been decomposed into components along these axes. The velocity components $(V_1)_y$ and $(V_2)_y$ are unchanged during the action. Along the line of impact, **linear-momentum** considerations lead to

$$5m - 7.07m = m[(V_1)_x]_f + m[(V_2)_x]_f \qquad \text{(a)}$$

Using the **coefficient-of-restitution** relation (Eq. 14.14), we have

$$\epsilon = .90 = -\frac{[(V_2)_x]_f - [(V_1)_x]_f}{-7.07 - 5} \qquad \text{(b)}$$

We thus have two equations, (a) and (b), for the unknown components in the x direction. Simplifying these equations, we have

$$[(V_1)_x]_f + [(V_2)_x]_f = -2.07 \qquad \text{(c)}$$

$$[(V_1)_x]_f + [(V_2)_x]_f = -10.86 \qquad \text{(d)}$$

Adding, we get

$$[(V_1)_x]_f = -6.47 \text{ ft/sec}$$

Solving for $[(V_2)_x]_f$ in Eq. (c), we write

$$[(V_2)_x]_f - 6.47 = -2.07$$

Therefore,

$$[(V_2)_x]_f = 4.40 \text{ ft/sec}$$

The final velocities after collision are then

$$(V_1)_f = -6.47\boldsymbol{i} \text{ ft/sec}$$

$$(V_2)_f = 4.40\boldsymbol{i} + 7.07\boldsymbol{j} \text{ ft/sec}$$

The loss in kinetic energy is given as

$$(\text{KE})_i - (\text{KE})_f = (\tfrac{1}{2}m5^2 + \tfrac{1}{2}m10^2) - [\tfrac{1}{2}m6.47^2 + \tfrac{1}{2}m(7.07^2 + 4.40^2)]$$

$$\Delta\text{KE} = \tfrac{1}{2}m[25 + 100 - (41.9 + 50.0 + 19.33)]$$

$$= 6.89m \text{ ft-lb}$$

Please note that mechanical energy is conserved *only* if ϵ is unity (i.e., a perfectly elastic impact). For all other cases, there is always dissipation of mechanical energy into heat and permanent deformation. However, *all* impacts involve conservation of linear momentum for the system.

Example 12.11

A pile driver is used to force a pile A into the ground (Fig. 12.16) as part of a program to properly prepare the foundation for a tall building. The device consists of a piston C on which a pressure p is developed from steam or air.

The piston is connected to a 1,000-lb hammer B. The assembly is suddenly released and accelerates downward a distance h of 2 ft to impact on pile A weighing 400 lb. If the earth develops a constant resisting force to movement of 25,000 lb, what distance d will the pile move for a drop involving no contribution from p (which is then 0 psig). Take the impact as *plastic*. The weight of the piston and the connecting rod is 100 lb.

We begin by using **conservation of mechanical energy** for the freely falling system to a position just before impact (preimpact). Using the initial configuration as the datum we have

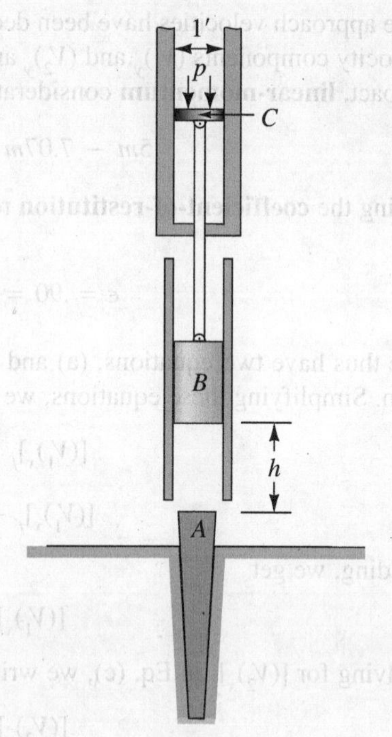

Figure 12.16. Steam-driven pile driver.

$$0 + (1{,}000 + 100)(2) = \frac{1}{2}\left(\frac{1{,}100}{g}\right)V^2 + 0$$

$$\therefore V = \sqrt{2gh} = \sqrt{(2)(32.2)(2)} = 11.35 \text{ ft/sec}$$

Now we get to the impact process. We have **conservation of linear momentum** while the bodies remain hypothetically at the position at which contact is first made. Thus, for plastic impact we can say

$$\frac{1{,}100}{g}(11.35) + 0 = \frac{1{,}500}{g}V \qquad \therefore V = 8.32 \text{ ft/sec}$$

Finally, we come to the post-impact process where we shall use the **work energy equation** for the pile driver and the pile.

$$(1{,}500 - 25{,}000)(d) = 0 - \frac{1}{2}\left(\frac{1{,}500}{g}\right)(8.32^2)$$

where the term on the left side must be negative because the net force on the system (23,500 lb) is in the opposite direction to the motion (see Eq. 13.2). Solving for d we get

$$d = .0686 \text{ ft} = .823 \text{ in}$$

PROBLEMS

12.34. Two cylinders move along a rod in a frictionless manner. Cylinder A has a mass of 10 kg and moves to the right at a speed of 3 m/sec, while cylinder B has a mass of 5 kg and moves to the left at a speed of 2.5 m/sec. What is the speed of cylinder B after impact for a coefficient of restitution ϵ of .8? What is the loss in kinetic energy?

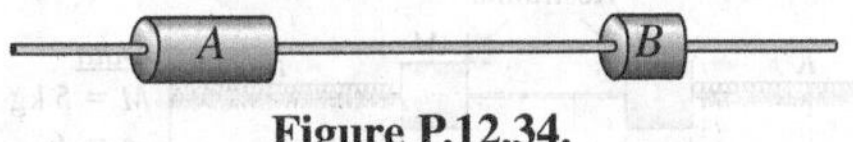

Figure P.12.34.

12.35. In Problem 12.34, what coefficient of restitution is needed for body A to be stationary after impact?

12.36. Two smooth cylinders of identical radius roll toward each other such that their centerlines are perfectly parallel. Cylinder A has a mass of 10 kg, and cylinder B has a mass of 7.5 kg. What is the speed at which cylinder A moves directly after collision for a coefficient of restitution $\epsilon = .75$?

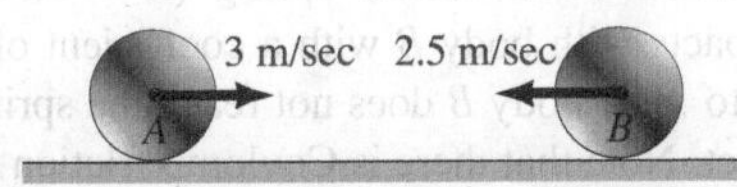

Figure P.12.36.

12.37. Cylinder A, weighing 10 lb, moves toward stationary cylinder B, weighing 40 lb, at the speed of 20 ft/sec. Mass B is attached to a spring having a spring constant K equal to 10 lb/in. If the collision has a coefficient of restitution $\epsilon = .9$, what is the maximum deflection δ of the spring? Assume that there is no friction along the rod and that the spring has negligible mass.

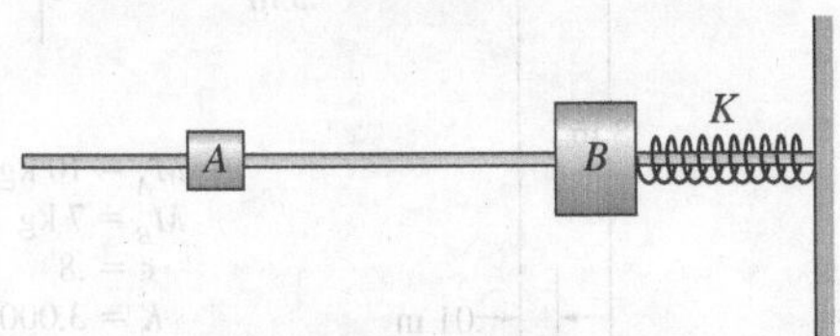

Figure P.12.37.

12.38. Do Problem 12.37 for the case where there is a perfectly plastic impact and the spring is nonlinear such that $5x^{3/2}$ lb of force is required for a deflection of x inches.

12.39. Assume a perfectly plastic impact as the 5-kg body falls from a height of 2.6 m onto a plate of mass 2.5 kg. This plate is mounted on a spring having a spring constant of 1,772 N/m. Neglect the mass of the spring as well as friction, and compute the maximum deflection of the spring after impact.

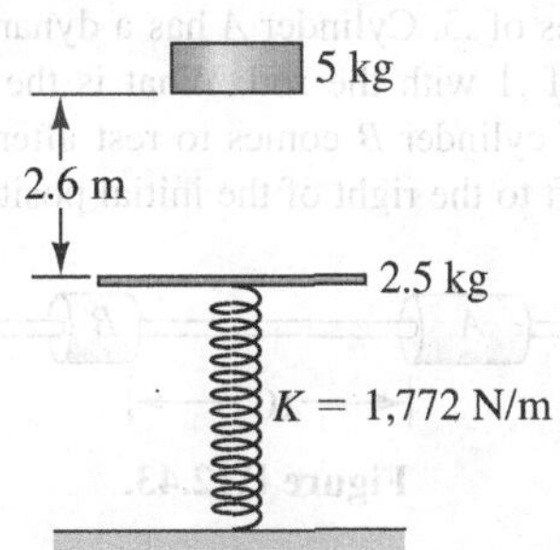

Figure P.12.39.

12.40. Identical spheres B, C, and D lie along a straight line on a frictionless surface. Sphere A, which is identical to the others, moves toward the other spheres at a speed V_A in a direction collinear with the centers of the spheres. For perfectly elastic collisions, what are the final velocities of the bodies?

12.41. In Problem 12.40, (a) What is the final velocity of sphere D if $\epsilon = .80$ for all spheres and $V_A = 50$ ft/sec? (b) Set up a relation for the speed of the $(n + 1)$th sphere in terms of the speed of the nth sphere, again for $\epsilon = .80$ and $V_A = 50$ ft/sec.

12.42. A spherical mass M_1 of 20 lbm is held at an angle θ_1 of 60° before being released. It strikes mass M_2 of 10 lbm with an impact having a coefficient of restitution equal to .75. Mass M_2 is held by a light rod of length 2 ft at the end of which is a torsional spring requiring 500 ft-lb per radian of rotation. The spring has no torque when l_2 is vertical. What is the maximum rotation of l_2 after impact? The length of $l_1 = 18$ in. [*Hint:* The work of a couple C rotating on angle $d\theta$ is $C\,d\theta$. A trial-and-error solution for θ_2 will be necessary.]

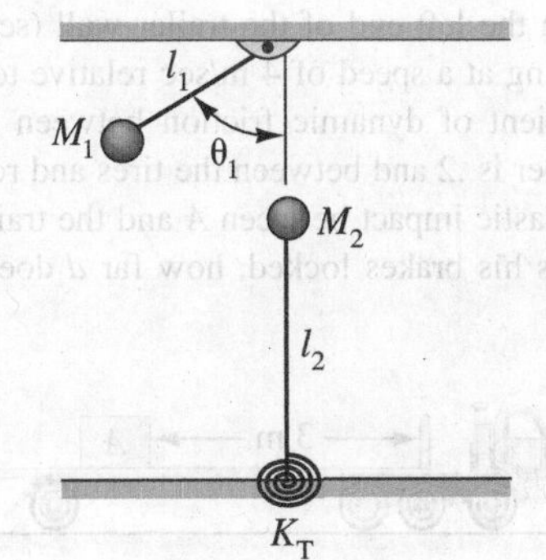

Figure P.12.42.

12.43. Cylinder A, weighing 20 lb, is moving at a speed of 20 ft/sec when it is at a distance 10 ft from cylinder B, which is stationary. Cylinder B weighs 15 lb and has a dynamic coefficient of friction with the rod on which it rides of .3. Cylinder A has a dynamic coefficient of friction of .1 with the rod. What is the coefficient of restitution if cylinder B comes to rest after collision at a distance 12 ft to the right of the initial position?

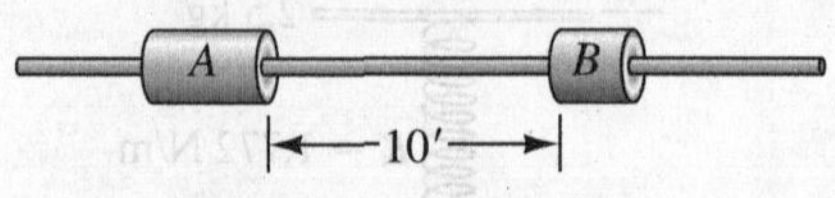

Figure P.12.43.

12.44. A load is being lowered at a speed of 2 m/sec into a barge. The barge weighs 1,000 kN, and the load weighs 100 kN. If the load hits the barge at 2 m/sec and the collision is plastic, what is the maximum depth that the barge is lowered into the water, assuming that the position of loading is such as to maintain the barge in a horizontal position? The width of the barge is 10 m. What are the weaknesses (if any) of your analysis? The density of water is 1,000 kg/m.^{3} [*Hint*: Recall the Archimedes Principle]

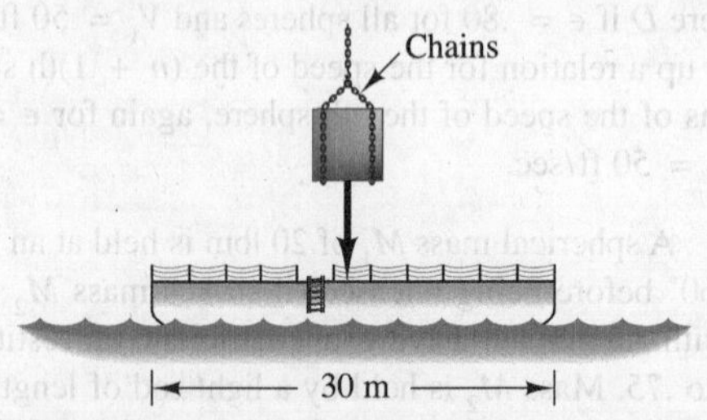

Figure P.12.44.

12.45. A tractor-trailer weighing 50 kN without a load carries a 10-kN load A as shown. The driver jams on his brakes until they lock for a panic stop. The load A breaks loose from its ropes. When the truck has stopped the load is 3 m from the left end of the trailer wall (see diagram) and is moving at a speed of 4 m/sec relative to the truck. The coefficient of dynamic friction between the load A and the trailer is .2 and between the tires and road is .5. If there is a plastic impact between A and the trailer and the driver keeps his brakes locked, how far d does the truck then move?

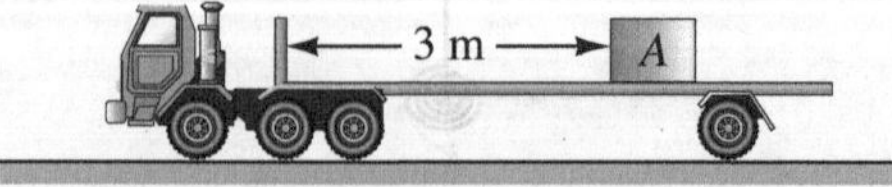

Figure P.12.45.

12.46. Two identical cylinders, each of mass 5 kg, slide on a frictionless rod. Each is fastened to a linear spring (K = 5,000 N/m) whose unstretched length is .65 m. The spring mass is negligible. If the cylinders are released from rest by raising the restraints,

(a) What is their speed just after colliding with a coefficient of restitution of .6?

(b) How close do they come to the walls?

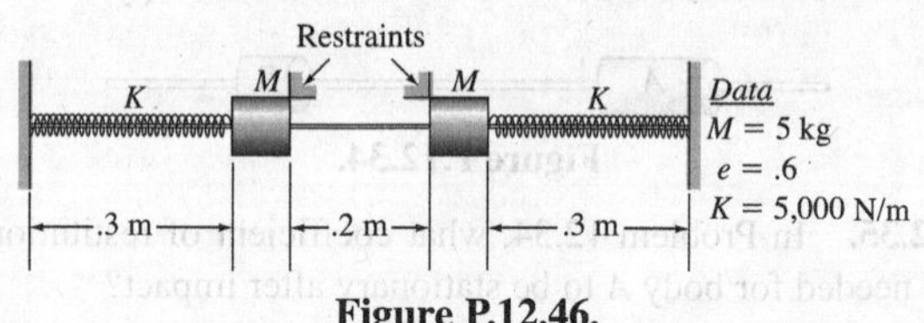

Figure P.12.46.

12.47. A light arm, connected to a mass A, is released from rest at a horizontal orientation. Determine the maximum deflection of the linear spring (K = 3,000 N/m) after A impacts with body B with a coefficient of restitution equal to .8. If body B does not reach the spring, indicate this fact. Note that there is Coulomb friction between the body B and the floor with μ_d = .6. Consider bodies A and B to be small.

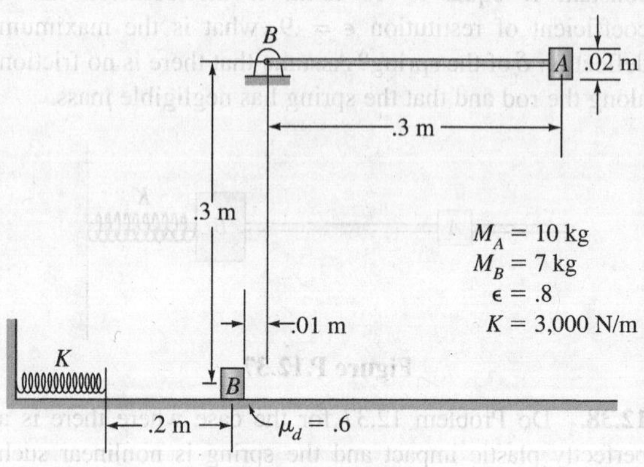

Figure P.12.47.

12.48. Mass M_A slides down the frictionless rod and hits mass M_B, which rests on a linear spring. The coefficient of restitution ϵ for the impact is .8. What is the total maximum deflection δ of the spring?

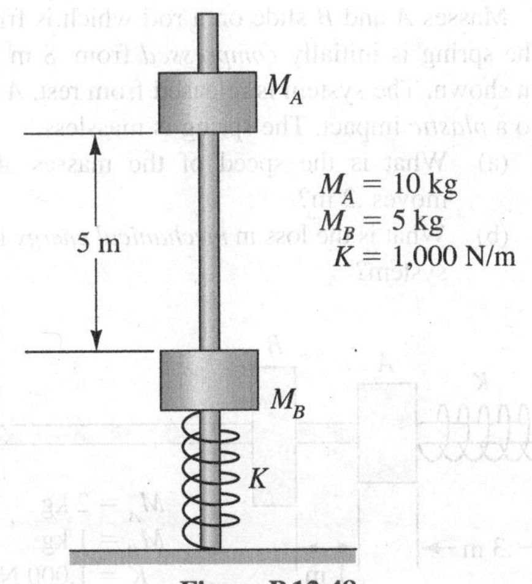

Figure P.12.48.

12.49. A cart A having a mass of 5 kg is released from rest at the position shown. As it rolls along, a constant resisting force of 4 N acts between the wheels and the surface. The cart collides with a block B having a mass of 3 kg. The coefficient of restitution is 0.5. Determine the maximum deflection δ_{max} of the spring having a spring constant of 15 N/m. Also, determine the maximum angle of rotation θ_{max} of the light rod supporting block B. The blocks are small.

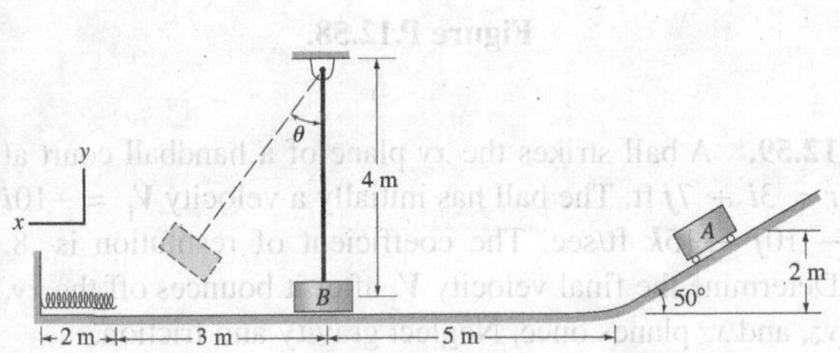

Figure P.12.49.

12.50. A three-seater racing scull is poised for a start. The scull weighs 300 lb, and each occupant weighs about 150 lb. We want to know the speed of the scull after 2 sec. At the sound of the starting gun, each man exerts a 30-lb constant push on the water from each oar in the direction of the axis of the boat. At the 2-sec mark, each man is moving to the left relative to the hull with a speed of 1 ft/sec. Neglect the inertia of the oars as well as water and air friction.

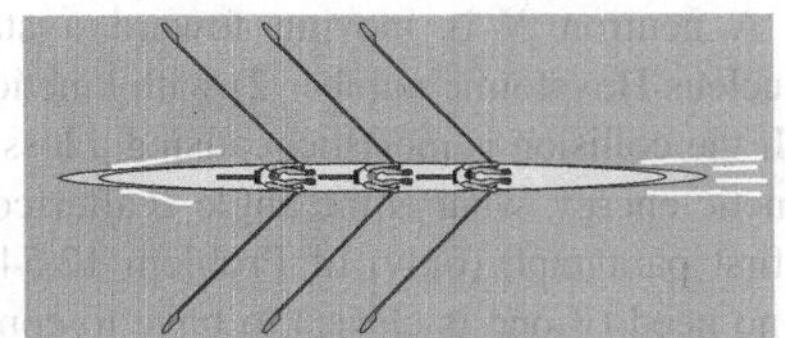
Figure P.12.50.

12.51. Do Example 12.11 for a constant pressure $p = 50$ psig.

12.52. A thin disc A weighing 5 lb translates along a frictionless surface at a speed of 20 ft/sec. The disc strikes a square stationary plate B weighing 10 lb at the center of a side. What are the velocity and direction of motion of the plate and the disc after collision? Assume that the surfaces of the plate and disc are smooth. Take $\epsilon = .7$.

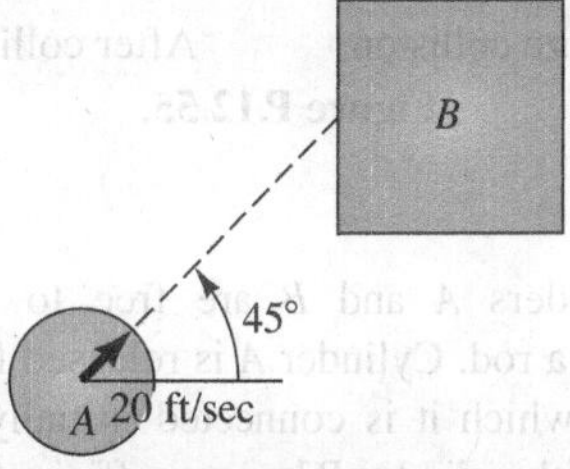

Figure P.12.52.

12.53. In Problem 12.52 at the instant of contact between the bodies a clamping device firmly connects the bodies together so as to form one rigid unit. Find the velocity of the center of mass of the system after impact.

12.54. The theory of collisions of *subatomic* particles is called the theory of *scattering*. The coefficient-of-restitution concept presented for macroscopic bodies in this chapter cannot be used. However, conservation of momentum can be used.

A neutron N shown moving with a speed V_0 strikes a stationary proton P. After collision the velocity of the neutron is V_N and that of the proton is V_P, as shown in the diagram. For an *elastic* collision, prove that $\phi + \theta = \pi/2$. [*Hint:* Use the vector polygon concept and the Pythagorean theorem. Also, take the mass of proton and neutron to be equal.]

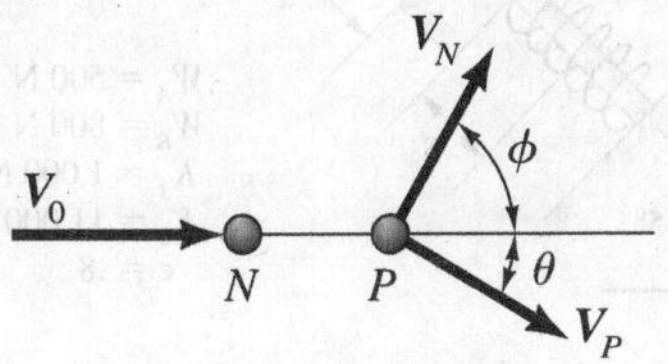

Figure P.12.54.

***12.55.** A neutron N is moving toward a stationary helium nucleus He (atomic number 2) with kinetic energy 10 MeV. If the collision is inelastic, causing a loss of 20% of the kinetic energy, what is the angle θ after collision? See the first paragraph (only) of Problem 12.54. [*Hint:* There is no need (if one is clever) to have to convert the atomic number to kilograms.]

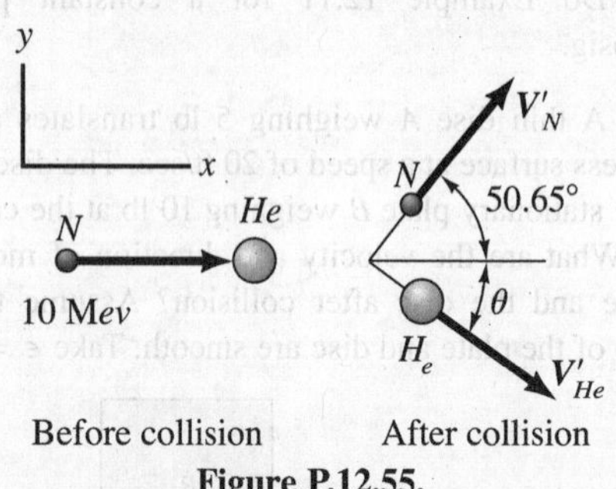

Figure P.12.55.

12.56. Cylinders A and B are free to slide without friction along a rod. Cylinder A is released from rest with spring K_1 to which it is connected initially unstretched. The impact with cylinder B has a coefficient of restitution ϵ equal to .8. Cylinder B is at rest before the impact supported in the position shown by spring K_2. Assume springs are massless.

(a) How much is the lower spring compressed initially?

(b) How much does cylinder B descend after impact before reaching its lowest position?

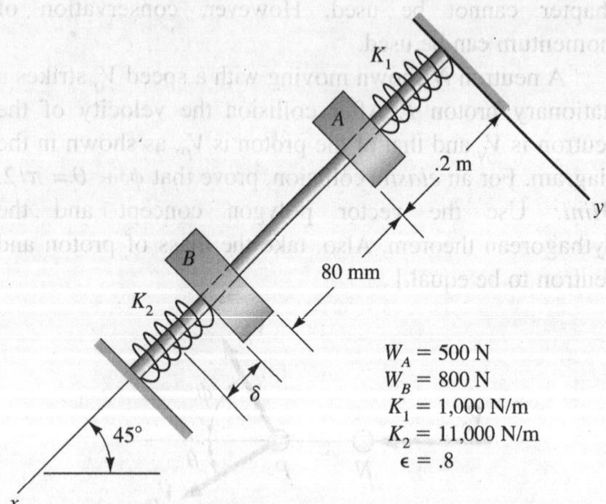

Figure P.12.56.

12.57. Masses A and B slide on a rod which is frictionless. The spring is initially *compressed* from .8 m to the position shown. The system is released from rest. A and B undergo a *plastic* impact. The spring is massless.

(a) What is the speed of the masses after B moves .2 m?

(b) What is the loss in *mechanical energy* for the system?

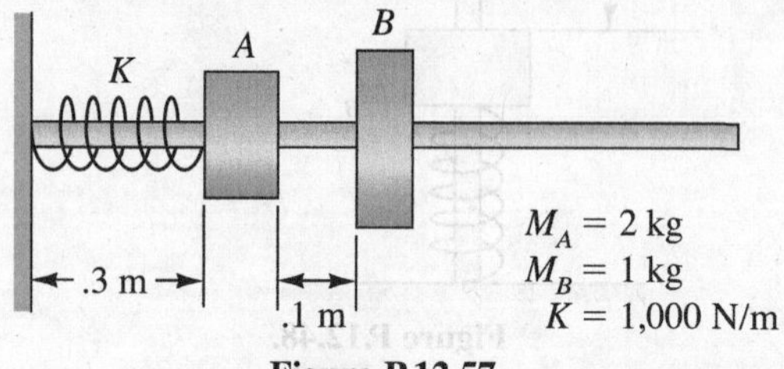

Figure P.12.57.

12.58. A ball is thrown against a floor at an angle of 60° with a speed at impact of 16 m/sec. What is the angle of rebound α if ϵ = .7? Neglect friction.

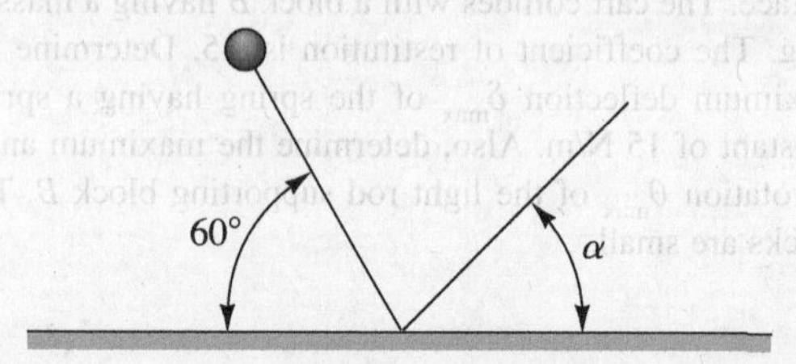

Figure P.12.58.

12.59. A ball strikes the xy plane of a handball court at $\boldsymbol{r} = 3\boldsymbol{i} + 7\boldsymbol{j}$ ft. The ball has initially a velocity $\boldsymbol{V}_1 = -10\boldsymbol{i} - 10\boldsymbol{j} - 15\boldsymbol{k}$ ft/sec. The coefficient of restitution is .8. Determine the final velocity $\boldsymbol{V}_2$ after it bounces off the xy, yz, and xz planes once. Neglect gravity and friction.

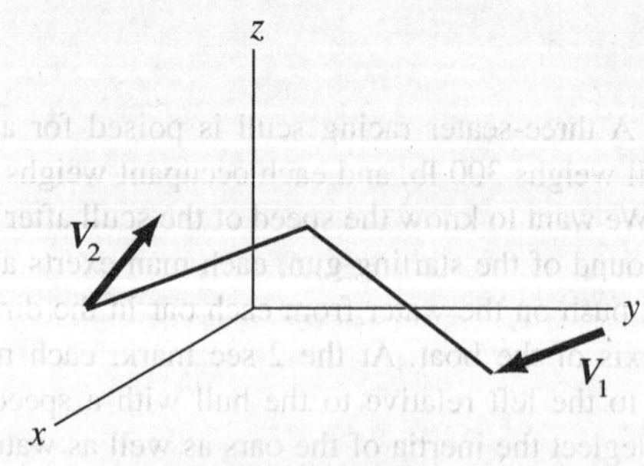

Figure P.12.59.

12.60. A space vehicle in the shape of a cone–cylinder is moving at a speed V m/sec, many times the speed of sound through highly rarefied atmosphere. If each molecule of the gas has a mass m kg and if there are, on the average, n molecules per cubic meter, compute the drag on the cone–cylinder. The cone half-angle is 30°. Take the collision to be perfectly elastic.

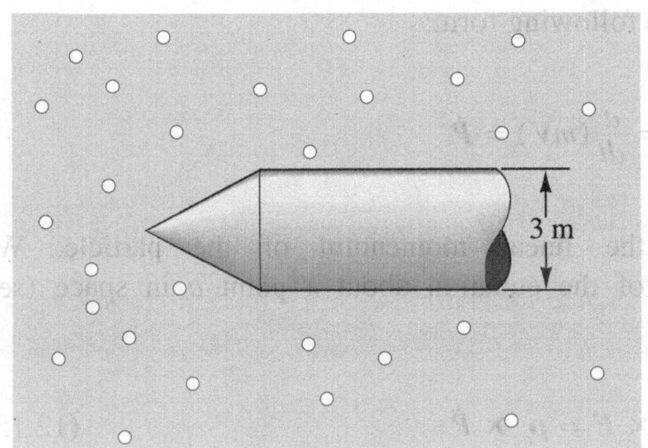

Figure P.12.60.

12.61. Do Problem 12.60 for a case where the collisions are assumed to be inelastic. Assume the coefficient of restitution to be .8.

12.62. A double-wedge airfoil section for a space glider is shown. If the glider moves in highly rarefield atmosphere at a speed V many times greater than the speed of sound, what is the drag per unit length of this airfoil? Assume the collision to be perfectly elastic. There are n molecules per ft^3, each having a mass m in slugs.

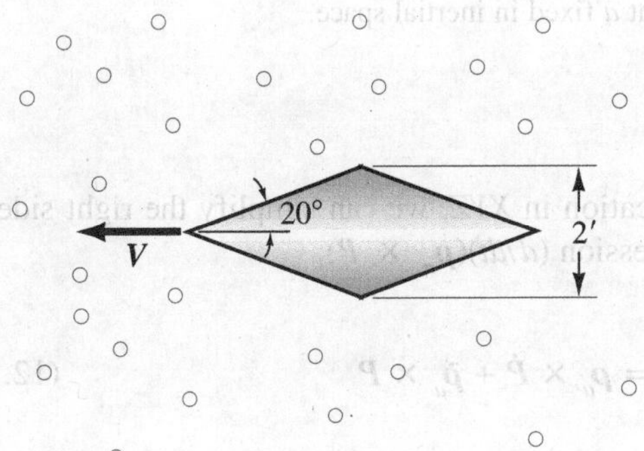

Figure P.12.62.

12.63. Consider a parallel beam of light having an energy flux of S watts/m^2, shining normal to a flat surface that completely absorbs the energy. You learned in physics that an impulse dI is developed on the surface during time dt given by the formula

$$dI = \frac{S}{c}\,dt\,dA$$

where c is the speed of light in vacuo in m/sec. If the surface reflects the light, then we have an impulse dI developed on the surface given as

$$dI = 2\frac{S}{c}\,dt\,dA$$

Compute the force stemming from the reflection of light shining normal to a perfectly reflecting mirror having an area of 1 m^2. The light has an energy flux S of 20 W/m^2. Take the speed $c = 3 \times 10^8$ m/sec. What is the radiation pressure p_{rad} on the mirror?

***12.64.** The Echo satellite when put into orbit is inflated to a 45-m-diameter sphere having a skin made up of a laminate of aluminum over mylar over aluminum. This skin is highly reflectant of light. Because of the small mass of this satellite, it may be affected by small forces such as that stemming from the reflection of light. If a parallel beam of light having an energy density S of .50 W/mm^2 impinges on the Echo satellite, what total force is developed on the satellite from this source? From physics (see Problem 12.63), the radiation pressure, p_{rad}, on a reflecting surface from a beam of light inclined by $\theta°$ from the normal to the surface is

$$p_{\text{rad}} = 2\frac{S}{c}\cos^2\theta$$

The pressure is in the direction of the incident radiation.

Part B: Moment of Momentum

12.5 Moment-of-Momentum Equation for a Single Particle

At this time, we shall introduce another auxiliary statement that follows from Newton's law and that will have great value when extended to the case of a rigid body. We start with Newton's law for a particle in the following form:

$$\boldsymbol{F} = \frac{d}{dt}(m\boldsymbol{V}) = \dot{\boldsymbol{P}}$$

where the symbol $\boldsymbol{P}$ represents the linear momentum of the particle. We next take the moment of each side of the equation about a point a in space (see Fig. 12.24):

$$\boldsymbol{\rho}_a \times \boldsymbol{F} = \boldsymbol{\rho}_a \times \dot{\boldsymbol{P}} \tag{12.15}$$

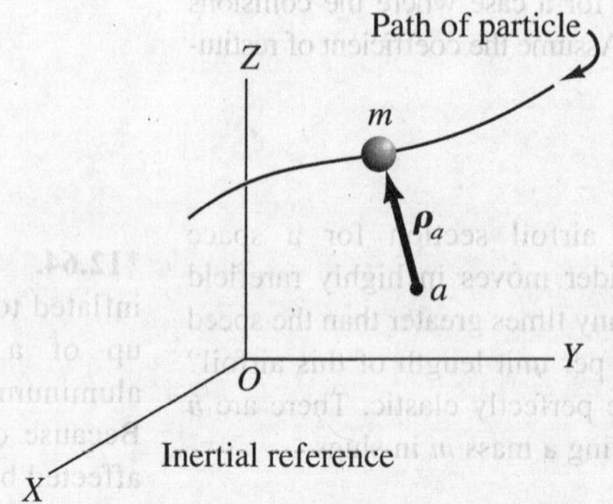

Figure 12.17. Point a fixed in inertial space.

If this point a is positioned at a fixed location in XYZ, we can simplify the right side of Eq. 12.15. Accordingly, examine the expression $(d/dt)(\boldsymbol{\rho}_a \times \boldsymbol{P})$:

$$\frac{d}{dt}(\boldsymbol{\rho}_a \times \boldsymbol{P}) = \boldsymbol{\rho}_a \times \dot{\boldsymbol{P}} + \dot{\boldsymbol{\rho}}_a \times \boldsymbol{P} \tag{12.16}$$

But the expression $\dot{\boldsymbol{\rho}}_a \times \boldsymbol{P}$ can be written as $\dot{\boldsymbol{\rho}}_a \times m\dot{\boldsymbol{r}}$. The vectors $\boldsymbol{\rho}_a$ and $\boldsymbol{r}$ are measured in the same reference from a fixed point a to the particle and from the origin to the particle, respectively (see Fig. 12.18). They are thus different at all times to the extent of a constant vector $\overrightarrow{Oa}$. Note that

$$\boldsymbol{r} = \overrightarrow{Oa} + \boldsymbol{\rho}_a$$

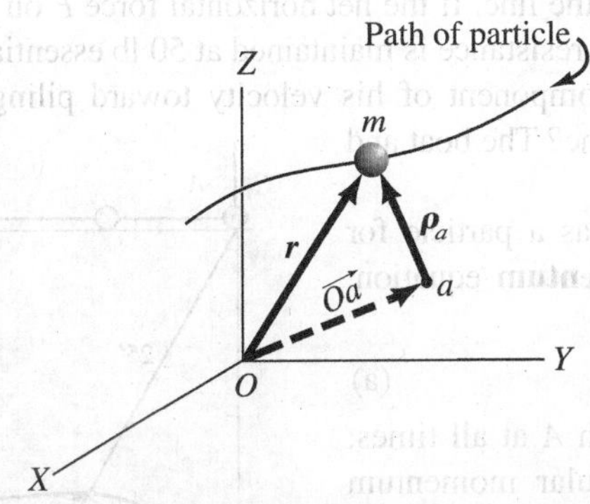

Figure 12.18. Position vectors to m and a.

Therefore,

$$\dot{\boldsymbol{r}} = \dot{\boldsymbol{\rho}}_a$$

Accordingly, the expression $\dot{\boldsymbol{\rho}}_a \times m\dot{\boldsymbol{r}}$ is zero. Thus, Eq. 12.16 becomes

$$\frac{d}{dt}(\boldsymbol{\rho}_a \times \boldsymbol{P}) = \boldsymbol{\rho}_a \times \dot{\boldsymbol{P}} \tag{12.17}$$

and Eq. 12.15 can be written in the form

$$\boldsymbol{\rho}_a \times \boldsymbol{F} = \boldsymbol{M}_a = \frac{d}{dt}(\boldsymbol{\rho}_a \times \boldsymbol{P}) = \dot{\boldsymbol{H}}_a$$

Therefore,

$$\boldsymbol{M}_a = \dot{\boldsymbol{H}}_a \tag{12.18}$$

where $\boldsymbol{H}_a$ *is the moment about point a of the linear momentum vector*. Also, $\boldsymbol{H}$ is termed the *angular momentum vector*. Equation 12.18, then, states that *the moment* $\boldsymbol{M}_a$ *of the resultant force on a particle about a point a, fixed in an inertial reference, equals the time rate of change of the moment about point a of the linear momentum of the particle relative to the inertial reference*. This is the desired alternative form of Newton's law.

The scalar component of Eq. 12.18 along some axis, say the z axis, can be useful. Thus,

$$M_z = \dot{H}_z$$

where M_z is the torque of the total external force about the z axis and H_z is the moment of the momentum (or angular momentum) about the z axis.

Example 12.12

A boat containing a man is moving near a dock (see Fig. 12.19). He throws out a light line and lassos a piling on the dock at A. He starts drawing in on the line so that when he is in the position shown in the diagram, the line is taut and has a length of 25 ft. His speed V_1 is 5 ft/sec in a direction normal to the line. If the net horizontal force F on the boat from tension in the line and from water resistance is maintained at 50 lb essentially in the direction of the line, what is the component of his velocity toward piling A (i.e., V_A) after the man has pulled in 3 ft of line? The boat and the man have a combined weight of 350 lb.

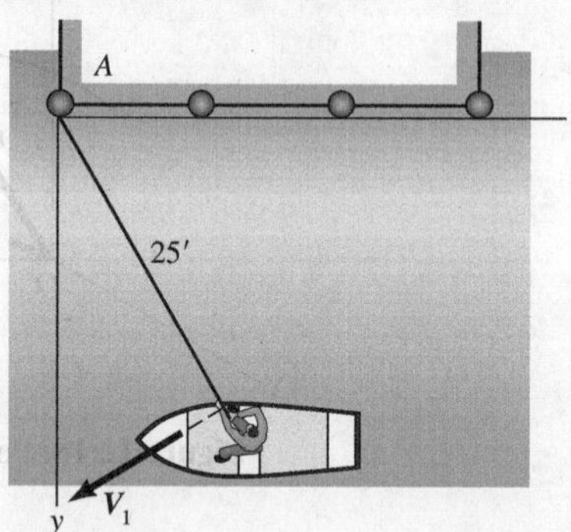

Figure 12.19. Man pulls toward piling.

We may consider the boat and man as a particle for which we can apply the **moment of momentum** equation. Thus,

$$\boldsymbol{M}_A = \dot{\boldsymbol{H}}_z \qquad \text{(a)}$$

Clearly, here $\boldsymbol{M}_A = \boldsymbol{0}$ since $\boldsymbol{F}$ goes through A at all times. Thus, $\boldsymbol{H}_A$ is a constant—that is, the angular momentum about A must be constant. Observing Fig. 12.20, we can say accordingly

$$\boldsymbol{r}_1 \times m\boldsymbol{V}_1 = \boldsymbol{r}_2 \times m\boldsymbol{V}_2$$

Since $\boldsymbol{r}_1$ is perpendicular to $\boldsymbol{V}_1$ and $\boldsymbol{r}_2$ is perpendicular to $(\boldsymbol{V}_2)_t$, we get a simple scalar product from above. Thus

$$(25)(m)(5) = (22)(m)(V_2)_t$$

Therefore,

$$(V_2)_t = 5.68 \text{ ft/sec} \qquad \text{(b)}$$

We need more information to get the desired result V_A toward the piling. We have not yet used the fact that $F = 50$ lb. Accordingly, we now employ the **work–kinetic energy** equation from Chapter 9. Thus,

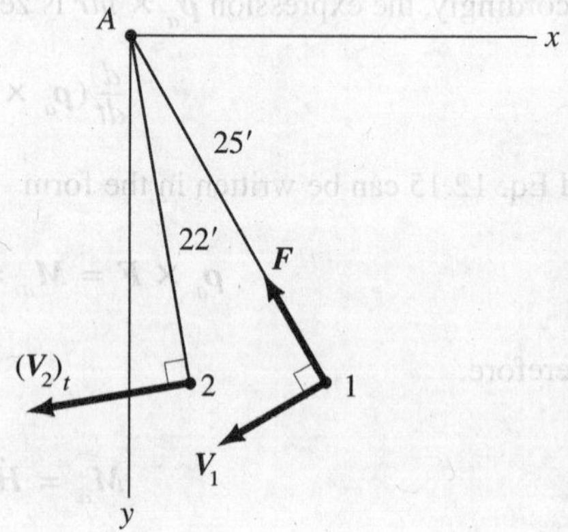

Figure 12.20. Boat at positions 1 and 2.

$$\int_1^2 \boldsymbol{F} \cdot d\boldsymbol{s} = \left(\frac{1}{2}MV^2\right)_2 - \left(\frac{1}{2}MV^2\right)_1$$

$$(50)(3) = \frac{1}{2}\frac{350}{g}(V_2)^2 - \frac{1}{2}\frac{350}{g}(25)$$

Therefore,

$$V_2 = 7.25 \text{ ft/sec} \qquad \text{(c)}$$

Now V_2 is the *total* velocity of the boat at position 2. To get the desired component V_A toward the piling, we can say, using Eqs. (b) and (c):

$$V_2^2 = (V_2)_t^2 + V_A^2$$

$$(7.25)^2 = (5.68)^2 + V_A^2$$

Therefore,

$$V_A = 4.51 \text{ ft/sec}$$

PROBLEMS

12.65. A particle rotates at 30 rad/sec along a frictionless surface at a distance 2 ft from the center. A flexible cord restrains the particle. If this cord is pulled so that the particle moves inward at a velocity of 5 ft/sec, what is the magnitude of the total velocity when the particle is 1 ft from the center?

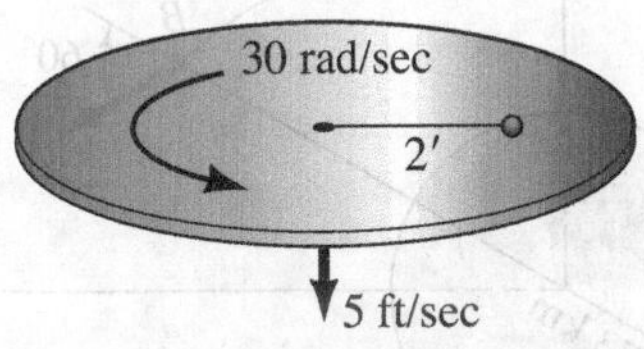

Figure P.12.65.

12.66. A satellite has an apogee of 7,128 km. It is moving at a speed of 36,480 km/hr. What is the transverse velocity of the satellite when $r = 6{,}970$ km?

12.67. A system is shown rotating freely with an angular speed ω of 2 rad/sec. A mass A of 1.5 kg is held against a spring such that the spring is compressed 100 mm. If the device a holding the mass in position is suddenly removed, determine how far toward the vertical axis of the system the mass will move. The spring constant K is .531 N/mm. Neglect all friction and inertia of the bar. The spring is not connected to the mass.

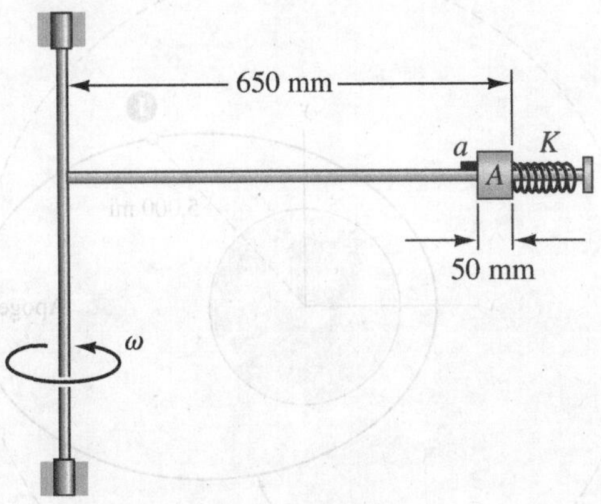

Figure P.12.67.

12.68. Do Problem 12.67 for the case where there is Coulombic friction between the mass A and the horizontal rod with a constant μ_d equal to .4.

12.69. A body A weighing 10 lb is moving initially at a speed of V_1 of 20 ft/sec on a frictionless surface. An elastic cord AO, which has a length l of 20 ft, becomes taut but not stretched at the position shown in the diagram. What is the radial speed toward O of the body when the cord is stretched 2 ft? The cord has an equivalent spring constant of .3 lb/in.

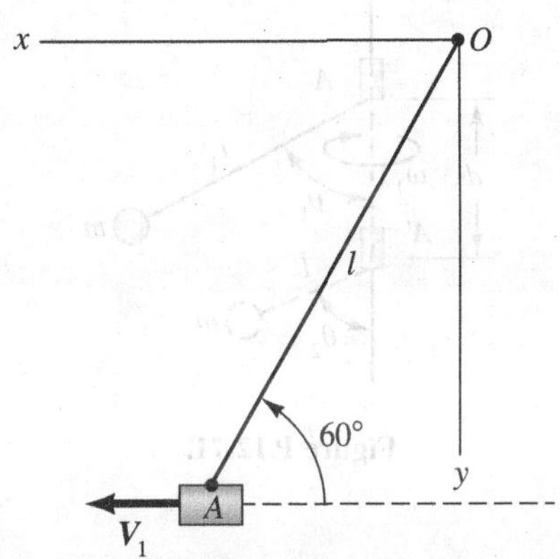

Figure P.12.69.

12.70. A small ball B weighing 2 lb is rotating about a vertical axis at a speed ω_1 of 15 rad/sec. The ball is connected to bearings on the shaft by light inextensible strings having a length l of 2 ft. The angle θ_1 is 30°. What is the angular speed ω_2 of the ball if bearing A is moved up 6 in.?

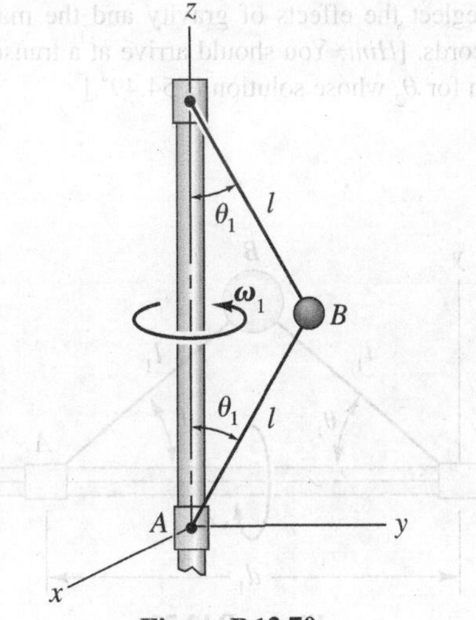

Figure P.12.70.

12.71. A mass m of 1 kg is swinging freely about the z axis at a speed ω_1 of 10 rad/sec. The length l_1 of the string is 250 mm. If the tube A through which the connecting string passes is moved down a distance d of 90 mm, what is ω_2 of the mass? You should get a fourth-order equation for ω_2 which has as the desired root $\omega_2 = 21.05$ rad/sec.

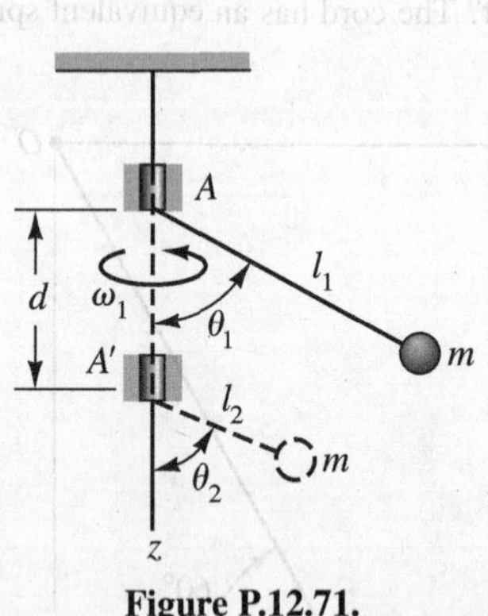

Figure P.12.71.

12.72. A small 2-lb ball B is rotating at angular speed ω_1 of 10 rad/sec about a horizontal shaft. The ball is connected to the bearings with light elastic cords which when unstretched are each 12 in. in length. A force of 15 lb is required to stretch the cord 1 in. The distance d_1 between the bearings is originally 20 in. If bearing A is moved to shorten d by 6 in., what is the angular velocity ω_2 of the ball? Neglect the effects of gravity and the mass of the elastic cords. [*Hint:* You should arrive at a transcendental equation for θ_2 whose solution is 54.49°.]

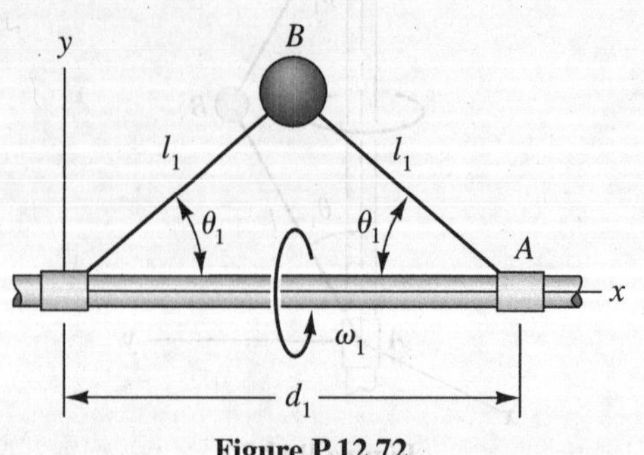

Figure P.12.72.

12.73. A space vehicle is moving at a speed of 37,000 km/hr at position A, which is perigee at a distance of 250 km from the earth's surface. What are the radial and transverse velocity components as well as the distance from the earth's surface at B? The trajectory is in the xy plane.

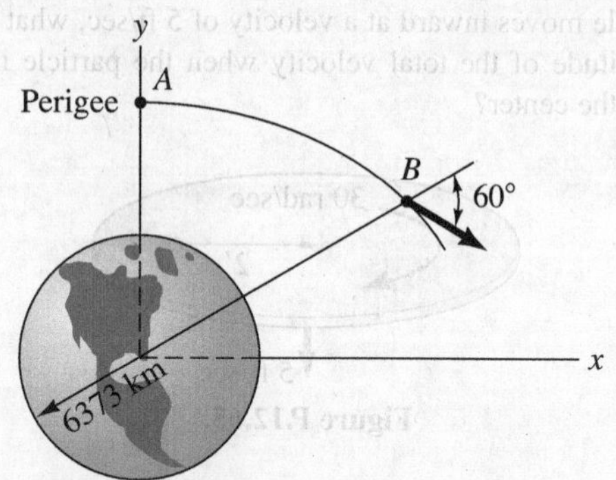

Figure P.12.73.

12.74. A space vehicle is in orbit A around the earth. At position (1) it is 5,000 miles from the center of the earth and has a velocity of 20,000 mi/hr. The transverse velocity at (1) is 15,000 mi/hr. At apogee, it is desired to continue in the circular orbit shown dashed. What change in speed is needed to change orbits when firing at apogee?

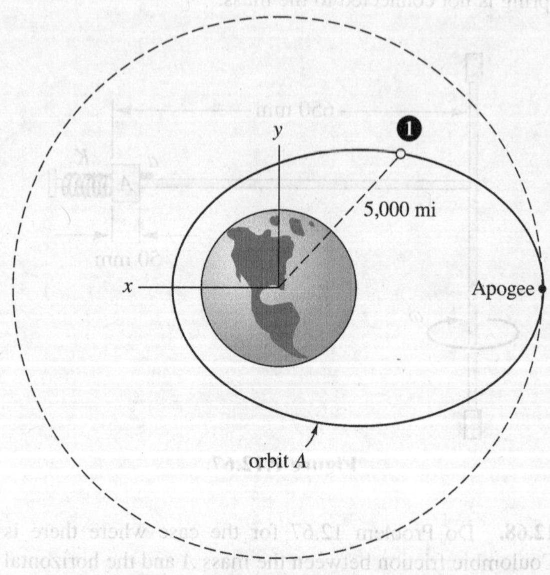

Figure P.12.74.

12.75. Do Problem 10.75 using the principles of conservation of momentum and conservation of mechanical energy.

12.76. In Problem 10.86 find the radial velocity by using the method of conservation of angular momentum and mechanical energy.

12.77. Do Problem 10.82 by the method of conservation of angular momentum and mechanical energy.

12.78. In Problem 10.87, find the height of the bullet above the surface of the planet by the methods of conservation of angular momentum and mechanical energy.

12.79. In Problem 10.119, find the maximum elevation above the earth's surface by the methods of conservation of angular momentum and mechanical energy.

12.80. Do Problem 10.114 by methods of conservation of angular momentum and mechanical energy. [*Hint:* The escape velocity = $\sqrt{2GM/r} = \sqrt{2}\,V_c$.]

12.81. Do Problem 10.113 using the principles of conservation of angular momentum and mechanical energy.

12.82. A space vehicle is in a circular orbit 1,200 km above the surface of the earth. A projectile is shot from this space vehicle at a speed relative to the vehicle of 5,000 km/hr in a radial direction as seen from the vehicle. What are the *apogee* and the *perigee* distances from the center of the earth for the trajectory of the projectile?

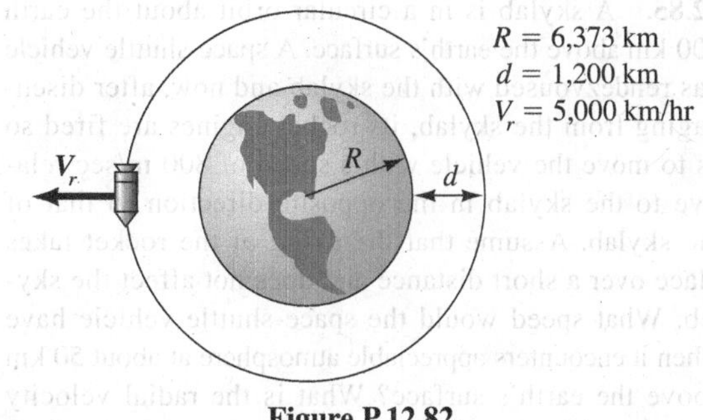

Figure P.12.82.

12.83. A space vehicle is in a circular parking orbit 300 miles above the surface of the earth. If the vehicle is to reach an apogee at location 2 which is 500 miles above the earth's surface, what increase in velocity must the vehicle attain by firing its rockets for a short time at location 1? The radius of the earth is 3,960 miles.

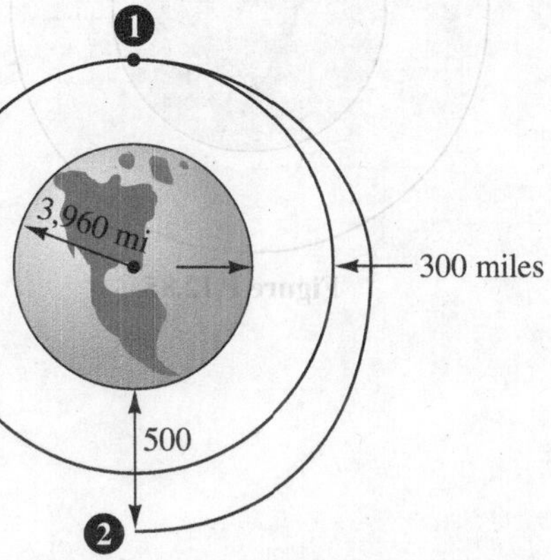

Figure P.12.83.

12.84. A space station is in a circular parking orbit around the earth at a distance of 5,000 mi from the center. A projectile is fired ahead in a direction tangential to the trajectory of the space station with a speed of 5,000 mi/hr relative to the space station. What is the maximum distance from earth reached by the projectile?

12.85. A skylab is in a circular orbit about the earth 500 km above the earth's surface. A space-shuttle vehicle has rendezvoused with the skylab and now, after disengaging from the skylab, its rocket engines are fired so as to move the vehicle with a speed of 800 m/sec relative to the skylab in the opposite direction to that of the skylab. Assume that the firing of the rocket takes place over a short distance and does not affect the skylab. What speed would the space-shuttle vehicle have when it encounters appreciable atmosphere at about 50 km above the earth's surface? What is the radial velocity at this position?

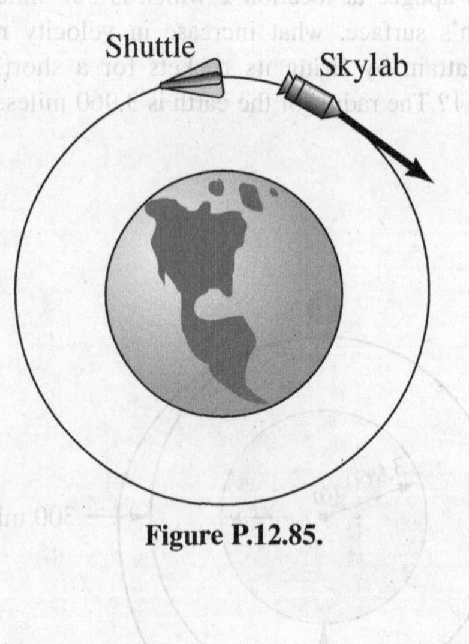

Figure P.12.85.

12.86. A space probe is approaching Mars. When the probe is 50,000 mi from the center of Mars it has a speed V_0 of 10,000 mi/hr with a component $(V_r)_0$ toward the center of Mars of 9,800 mi/hr. How close does the probe come to the surface of Mars? If retrorockets are fired at this lowest position A, what change in speed is needed to alter the trajectory into a circular orbit as shown? The acceleration of gravity at the surface of Mars is 12.40 ft/sec^2, and the radius R of the planet is 2,107 mi.

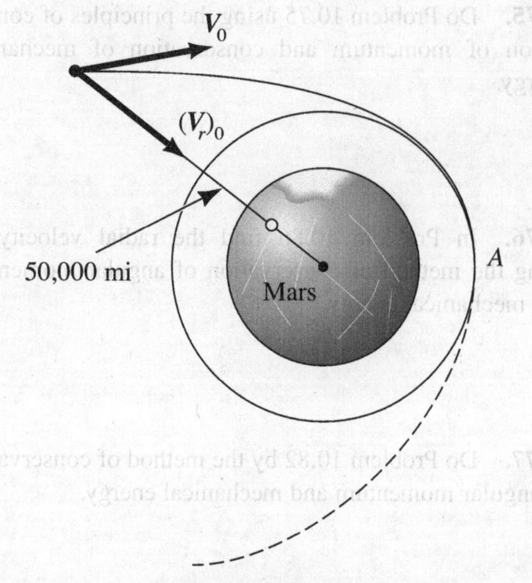

Figure P.12.86.

12.87. In Problem 12.86, a midcourse correction is to be made to get the probe within 1,000 mi from the surface of Mars. If V_0 at r_0 = 50,000 mi is still to be 10,000 mi/hr, what should be the radial velocity component $(V_r)_0$?

12.88. The Apollo command module is in a circular parking orbit about the moon at a distance of 161.0 km above the surface of the moon. The lunar exploratory module is to detach from the command module. The lunar-module rockets are fired briefly to give a velocity V_0 relative to the command module in the opposite direction. If the lunar module is to have a transverse velocity of 1,500 m/sec when it is 80 km from the surface of the moon before rockets are fired again, what must V_0 be? What is the radial velocity at this position? The radius of the moon is 1,733 km, and the acceleration of gravity is 1.700 m/sec^2 at the surface.

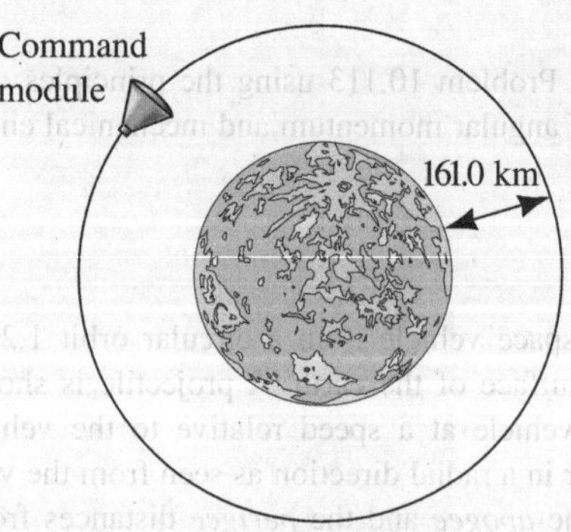

Figure P.12.88.

12.6 Moment-of-Momentum Equations for a System of Particles

We shall now develop the moment-of-momentum equations for an aggregate of particles. The resulting equations will be of vital importance when we apply them to rigid bodies in later chapters. We shall consider a number of cases.

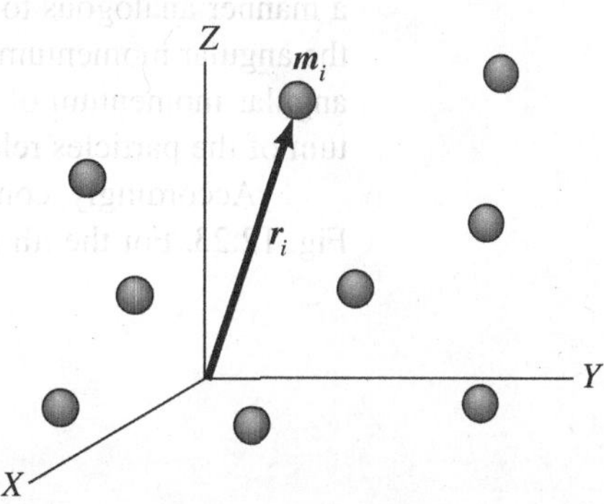

Figure 12.21. System of n particles.

Case 1. Fixed Reference Point in Inertial Space. An aggregate of n particles and an inertial reference are shown in Fig. 12.21. The moment of momentum equation for the ith particle is now written about the origin of this reference:

$$\boldsymbol{r}_i \times \boldsymbol{F}_i + \boldsymbol{r}_i \times \left(\sum_{\substack{j=1 \\ i \neq j}}^{n} \boldsymbol{f}_{ij} \right) = \frac{d}{dt}(\boldsymbol{r}_i \times \boldsymbol{P}_i) \qquad (12.19)$$

where, as usual, $\boldsymbol{f}_{ij}$ is the internal force from the jth particle on the ith particle. We now sum this equation for all n particles:

$$\sum_{i=1}^{n} \boldsymbol{r}_i \times \boldsymbol{F}_i + \sum_{i=1}^{n}\sum_{j=1}^{n} (\boldsymbol{r}_i \times \boldsymbol{f}_{ij}) = \frac{d}{dt}\left[\sum_{i=1}^{n} (\boldsymbol{r}_i \times \boldsymbol{P}_i)\right] = \dot{\boldsymbol{H}}_{\text{total}} \qquad (12.20)$$

where the summation operation has been put after the differentiation on the right side (permissible because of the distributive property of differentiation with respect to addition). For any pair of particles, the internal forces will be equal and opposite and collinear (see Fig. 12.22). Hence, the forces will have a zero moment about the origin. (This result is most easily understood by remembering that, for purposes of taking moments about a point, forces are transmissible.) We can then conclude that the expression

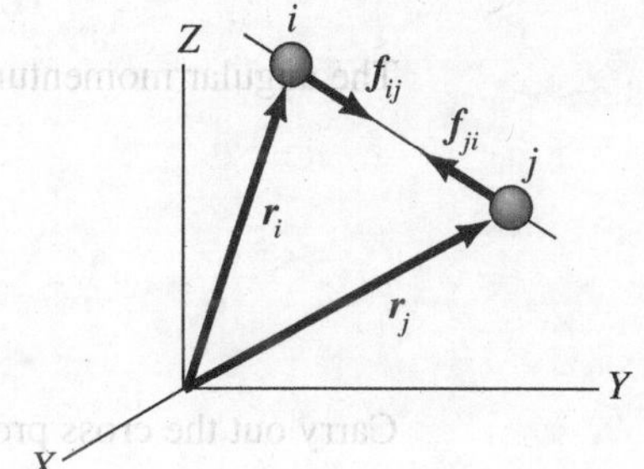

Figure 12.22. Internal equal and opposite forces.

$$\sum_{i=1}^{n}\sum_{j=1}^{n} (\boldsymbol{r}_i \times \boldsymbol{f}_{ij})$$

in this equation is zero. Realizing that $\sum_i \boldsymbol{r}_i \times \boldsymbol{F}_i$ is the total moment of the external forces about the origin, we have as a result for Eq. 12.20:

$$\boldsymbol{M}_a = \dot{\boldsymbol{H}}_a \qquad (12.21)$$

Thus, *the total moment* $\boldsymbol{M}$ *of external forces acting on an aggregate of particles about a point a fixed in an inertial reference* (the point in the development was picked as the origin merely for convenience) *equals the time rate of change of the total moment of the linear momentum relative to the inertial reference, where this moment is taken about the aforementioned point a.*[8]

We may express Eq. 12.21 in a different form by considering the *center of mass*. In a manner analogous to kinetic energy of an aggregate of particles, we can first show that the angular momentum of an aggregate of particles about a fixed point can be given as the angular momentum of the center of mass about the fixed point plus the angular momentum of the particles relative to the center of mass.

Accordingly, consider the center of mass c of an aggregate of particles as shown in Fig. 12.23. For the ith particle, we can say:

$$\boldsymbol{r}_i = \boldsymbol{r}_c + \boldsymbol{\rho}_{ci} \tag{12.22}$$

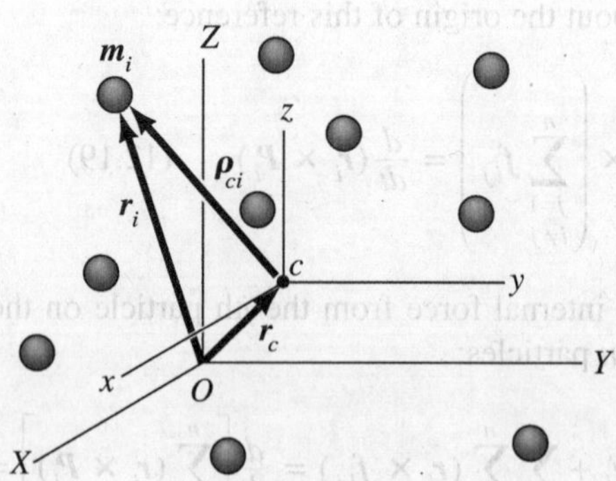

Figure 12.23. c is center of mass of aggregate.

The angular momentum for the aggregate of particles about O is then

$$\begin{aligned} \boldsymbol{H}_0 &= \sum_i (\boldsymbol{r}_c + \boldsymbol{\rho}_{ci}) \times \boldsymbol{P}_i \\ &= \sum_i (\boldsymbol{r}_c + \boldsymbol{\rho}_{ci}) \times \left[(m_i)(\dot{\boldsymbol{r}}_c + \dot{\boldsymbol{\rho}}_{ci})\right] \end{aligned} \tag{12.23}$$

Carry out the cross product and extract $\boldsymbol{r}_c$ from the summations:

$$\boldsymbol{H}_0 = \boldsymbol{r}_c \times M\dot{\boldsymbol{r}}_c + \boldsymbol{r}_c \times \sum_i m_i \dot{\boldsymbol{\rho}}_{ci} + \left(\sum_i m_i \boldsymbol{\rho}_{ci}\right) \times \dot{\boldsymbol{r}}_c + \sum_i \boldsymbol{\rho}_{ci} \times m_i \dot{\boldsymbol{\rho}}_{ci} \tag{12.24}$$

But since c is the center of mass, it follows that

$$\sum_i m_i \boldsymbol{\rho}_{ci} = \boldsymbol{0}$$

$$\sum_i m_i \dot{\boldsymbol{\rho}}_{ci} = \boldsymbol{0}$$

[8]Point a could also be moving with a constant velocity V_0 relative to inertial reference XYZ. However, a would then be fixed in another inertial reference $X'Y'Z'$, which is translating with respect to XYZ at a speed V_0.

Going back to Eq. 12.25, we see that the second and third expressions on the right side are to be deleted and we get then the desired result for $\boldsymbol{H}_0$:

$$\boldsymbol{H}_0 = \boldsymbol{r}_c \times M\dot{\boldsymbol{r}}_c + \sum_i \boldsymbol{\rho}_{ci} \times m_i\dot{\boldsymbol{\rho}}_{ci} = \boldsymbol{r}_c \times M\dot{\boldsymbol{r}}_c + \boldsymbol{H}_c$$

where $\boldsymbol{H}_c$ is the moment about the center of mass of the linear momentum as seen from the center of mass for the aggregate.[9] This may be rewritten and expressed for *any* fixed point a where, using $\boldsymbol{r}_{ac}$ as the position vector from fixed point a to the center of mass c, we have

$$\boldsymbol{H}_a = \boldsymbol{H}_c + \boldsymbol{r}_{ac} \times M\dot{\boldsymbol{r}}_{ac} \tag{12.25}$$

Thus, in a manner analogous to the case of kinetic energy (see Section 11.6), the moment of momentum about point a is the sum of the moment of momentum relative to the center of mass plus the moment of momentum of the center of mass about point a. Note that $\dot{\boldsymbol{r}}_{ac}$ is the velocity of c relative to fixed point a and is thus equal to the velocity $\boldsymbol{V}_c$ of the mass center relative to XYZ. Thus, we can express Eq. (12.25) as

$$\boldsymbol{H}_a = \boldsymbol{H}_c + \boldsymbol{r}_{ac} \times M\boldsymbol{V}_c \tag{12.26}$$

Furthermore, we have for $\dot{\boldsymbol{H}}_a$:

$$\dot{\boldsymbol{H}}_a = \dot{\boldsymbol{H}}_c + \boldsymbol{r}_{ac} \times M\dot{\boldsymbol{V}}_c$$

where we have used the fact that $\dot{\boldsymbol{r}}_{ac} = \boldsymbol{V}_c$ to delete one expression. Note in effect we have put dots over $\boldsymbol{H}_a$, $\boldsymbol{H}_c$, and $\boldsymbol{V}_c$ in Eq. (12.29) to reach to above equation. We may now restate Eq. 12.21 for *a fixed point a* as follows on replacing $\dot{\boldsymbol{H}}_a$ using the above equation. Then, using $\boldsymbol{a}_c$ for $\dot{\boldsymbol{V}}_c$ we have the desired result:

$$\boldsymbol{M}_a = \dot{\boldsymbol{H}}_c + \boldsymbol{r}_{ac} \times M\boldsymbol{a}_c \tag{12.27}$$

Case 2. Reference Point at the Center of Mass. We can use Eq. 12.27 for this purpose. First, we will replace $\boldsymbol{M}_a$ using the left side of Eq. 12.20. But in so doing, we will replace $\boldsymbol{r}_i$ in the first expression by $(\boldsymbol{r}_c + \boldsymbol{\rho}_{ci})$. Note next that Eq. 12.27 calls for stationary point a. We will want a to be the origin O of XYZ and so $\boldsymbol{r}_{ac}$ becomes simply $\boldsymbol{r}_c$. Finally, we replace $\boldsymbol{a}_c$ by $\ddot{\boldsymbol{r}}_c$ in Eq. 12.30 and we have after these steps

$$\sum_i (\boldsymbol{r}_c + \boldsymbol{\rho}_{ci}) \times \boldsymbol{F}_i + \sum_j \sum_i \boldsymbol{r}_i \times \boldsymbol{f}_{ij} = \dot{\boldsymbol{H}}_c + \boldsymbol{r}_c \times M\ddot{\boldsymbol{r}}_c$$

The internal forces $\boldsymbol{f}_{ij}$ give zero contribution in this equation as explained earlier and we have on rearranging the remaining terms in the equation

$$\boldsymbol{r}_c \times \sum_i \boldsymbol{F}_i + \sum_i \boldsymbol{\rho}_{ci} \times \boldsymbol{F}_i = \boldsymbol{r}_c \times M\ddot{\boldsymbol{r}}_c + \dot{\boldsymbol{H}}_c$$

[9] That is, as seen from a reference xyz translating with c relative to XYZ—in other words, as seen by a nonrotating observer moving with c.

From Newton's law for the center of mass, we know that $\sum \boldsymbol{F}_i = M\ddot{\boldsymbol{r}}_c$ and so the first terms on the left and right sides of the equation above cancel. The remaining expression on the left side of the equation is the moment about the center of mass of the external forces. We then get

$$\boldsymbol{M}_c = \dot{\boldsymbol{H}}_c \tag{12.28}$$

We thus get the same formulation for the center of mass as for a fixed point in inertial space. Please note that $\boldsymbol{H}_c$ is the moment about the center of mass of the linear momentum as seen from the center of mass but that the time derivative is as seen from inertial reference XYZ.

Case 3. Point Accelerating Toward or Away from the Mass Center. There is yet a third point of interest to be considered and that is a point a accelerating toward or away from

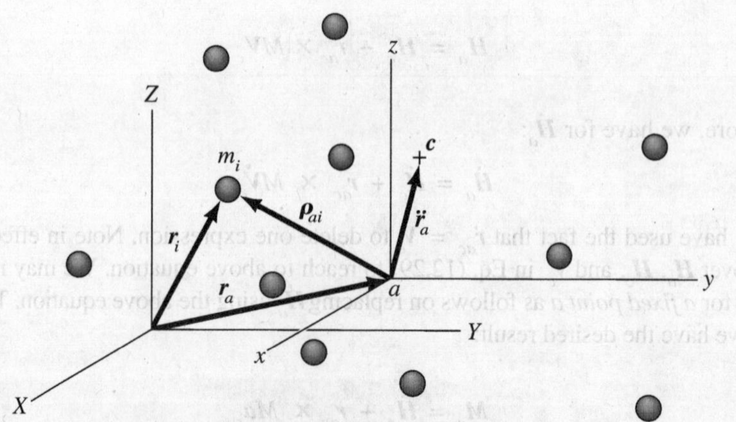

Figure 12.24. Point a accelerates toward or away from c.

the mass center of the aggregate (Fig. 12.24). For such a point, we can again give the same simple equation presented for cases 1 and 2. Thus,

$$\boldsymbol{M}_a = \dot{\boldsymbol{H}}_a \tag{12.25}$$

where $\boldsymbol{H}$ is taken relative to point a (i.e., relative to axes xyz translating with point a). We have asked for the derivation of this equation in Problem 12.96.

The component of the equation $\boldsymbol{M}_a = \dot{\boldsymbol{H}}_a$ for any one of the three cases in say the x direction,

$$M_x = \dot{H}_x$$

can be very useful. Here, M_x is the torque about the x axis, and H_x is the moment of momentum (angular momentum) about the x axis. We now examine such a problem in the following example.

Example 12.13

A heavy chain of length 20 ft lies on a light plate *A* which is freely rotating at an angular speed of 1 rad/sec (see Fig. 12.25). A channel *C* acts as a guide for the chain on the plate, and a stationary pipe acts as a guide for the chain below the plate. What is the speed of the chain after it moves 5 ft starting from rest relative to the platform? Neglect friction, the angular momentum of the plate, and the angular momentum of the vertical section of the chain about its own axis. The chain weight per unit length, *w*, is 10 lb/ft.

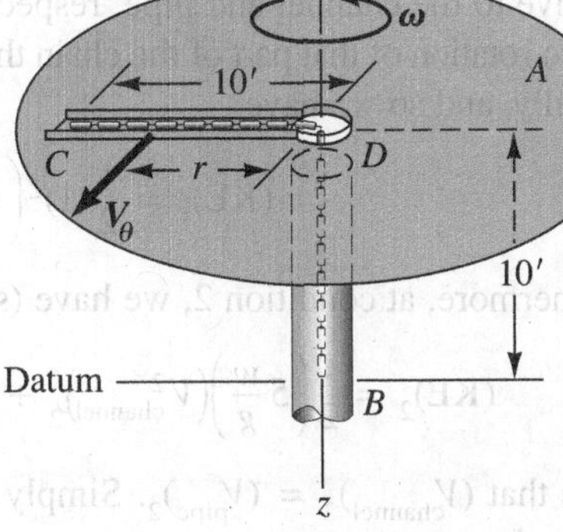

We shall first apply the **moment-of-momentum** equation about point *D* for the chain and plate. Taking the component of this equation along the *z* axis, we can say:

$$M_z = (\dot{H})_z \tag{a}$$

Figure 12.25. Sliding chain.

Clearly $M_z = 0$, and so we have conservation of angular momentum. That is,

$$H_z = \text{constant}$$
$$(H_z)_1 = (H_z)_2 \tag{b}$$

where 1 and 2 refer to the initial condition and the condition after the chain moves 5 ft. We can then say:

$$\int_0^{10} r(V_\theta)_1\left(\frac{w}{g}\,dr\right) = \int_0^{5} r(V_\theta)_2\left(\frac{w}{g}\,dr\right)$$

$$\int_0^{10} r(\omega_1 r)\left(\frac{w}{g}\,dr\right) = \int_0^{5} r(\omega_2 r)\left(\frac{w}{g}\,dr\right)$$

$$(1)\left(\frac{w}{g}\right)\int_0^{10} r^2\,dr = (\omega_2)\left(\frac{w}{g}\right)\int_0^{5} r^2\,dr$$

Therefore,

$$\omega_2 = 8 \text{ rad/sec} \tag{c}$$

To find the speed of movement of the chain, we must next go to *energy* considerations. Because only conservative forces are acting here, we may employ the **conservation-of-mechanical-energy** principle. In so doing, we shall use as a datum the end of the chain *B* at the initial condition (see Fig. 12.25). We can then say:

$$(\text{PE})_1 = (10)(w)(10) + (10)(w)(5) = 1{,}500 \text{ ft-lb}$$

Observing Fig. 12.35, we can say for condition 2:

$$(\text{PE})_2 = (5)(w)(10) + (10)(w)(5) - (5)(w)(2.5)$$
$$= 875 \text{ ft-lb}$$

As for kinetic energy, we have

$$(\text{KE})_1 = \frac{1}{2}\left(10\frac{w}{g}\right)\left(V^2_{\text{channel}}\right)_1 + \frac{1}{2}\left(10\frac{w}{g}\right)\left(V^2_{\text{pipe}}\right)_1 + \frac{1}{2}\int_0^{10}(r\omega_1)^2\,\frac{w}{g}\,dr$$

Example 12.13 (Continued)

where the first two expressions on the right side give the kinetic energy from the motion relative to the channel and pipe, respectively. The last expression is the kinetic energy due to rotation of that part of the chain that is in the channel. Clearly, $V_{channel} = V_{pipe} = 0$ initially, and so we have

$$(KE)_1 = \frac{1}{2}(1)^2\left(\frac{10}{g}\right)\int_0^{10} r^2\, dr = 51.8 \text{ ft-lb}$$

Furthermore, at condition 2, we have (see Fig. 12.35)

$$(KE)_2 = \frac{1}{2}\left(5\frac{w}{g}\right)\left(V^2_{channel}\right)_2 + \frac{1}{2}\left(15\frac{w}{g}\right)\left(V^2_{pipe}\right)_2 + \frac{1}{2}\int_0^5 (r\omega_2)^2 \frac{w}{g}\, dr$$

Note that $(V_{channel})_2 = (V_{pipe})_2$. Simply calling this quantity V_2, we have

$$(KE)_2 = \frac{1}{2}(20)\left(\frac{w}{g}\right)\left(V_2^2\right) + \frac{1}{2}(8^2)\left(\frac{w}{g}\right)\int_0^5 r^2\, dr$$

$$= 3.11\, V_2^2 + 414$$

We can now state

$$(PE)_1 + (KE)_1 = (PE)_2 + (KE)_2$$

$$1{,}500 + 51.8 = 875 + (3.11\, V_2^2 + 414)$$

Therefore,

$$V_2 = 9.19 \text{ ft/sec}$$

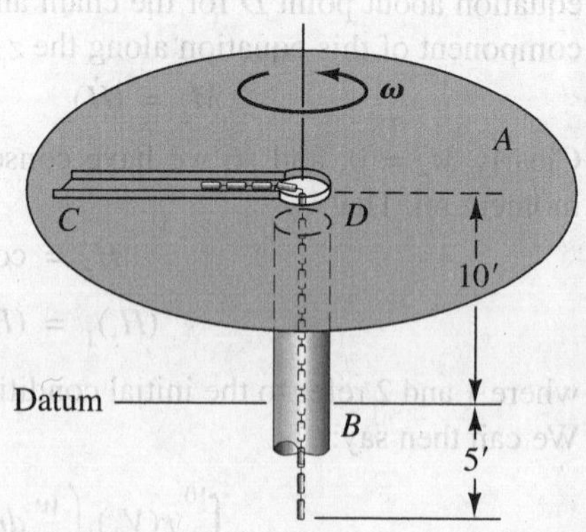

Figure 12.26. Chain after motion of 5 ft.

We can conclude that the chain is moving at a speed of 9.19 ft/sec along the channel and down the stationary pipe and that the plate A is rotating at an angular speed of 8 rad/sec.

Much time will be spent later in the text in applying $\boldsymbol{M}_a = \dot{\boldsymbol{H}}_a$ to a rigid body. There, the rigid body is considered to be made up of an infinite number of contiguous elements. Summations then give way to integration, and so on. The final equations of this section accordingly are among the most important in mechanics.

In the homework assignments, we have included, as in Chapter 11, several very simple rigid-body problems to illustrate the use of the equation $\boldsymbol{M}_a = \dot{\boldsymbol{H}}_a$ and to give an early introduction to rigid-body mechanics.[10] We now illustrate such a problem.

[10]The instructor may wish not to get into rigid-body dynamics at this time. This approach presents no loss in continuity.

Example 12.14

A uniform cylinder of radius 400 mm and mass 100 kg is acted on at its center by a force of 500 N (see Fig. 12.27). What is the friction force f? Take $\mu_s = .2$.

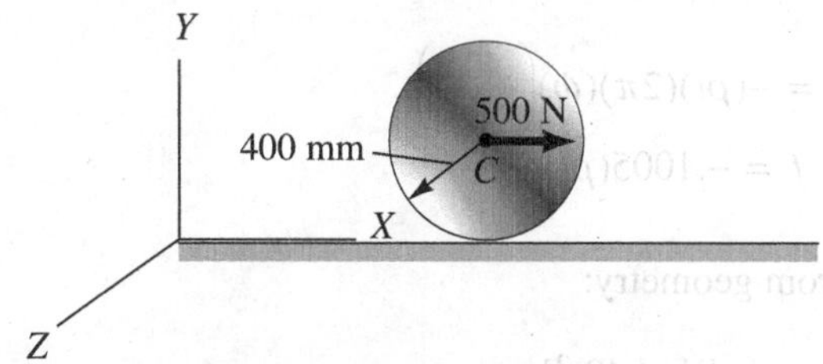

Figure 12.27. Rolling cylinder.

We have shown a free-body diagram of the cylinder in Fig. 12.28. A reference xyz with origin at C translates with the center of mass. We first apply **Newton's law** relative to inertial reference XYZ.[11] Thus, for the X direction we have for the center of mass C:

$$500 - f = 100\ddot{X}_c \qquad \text{(a)}$$

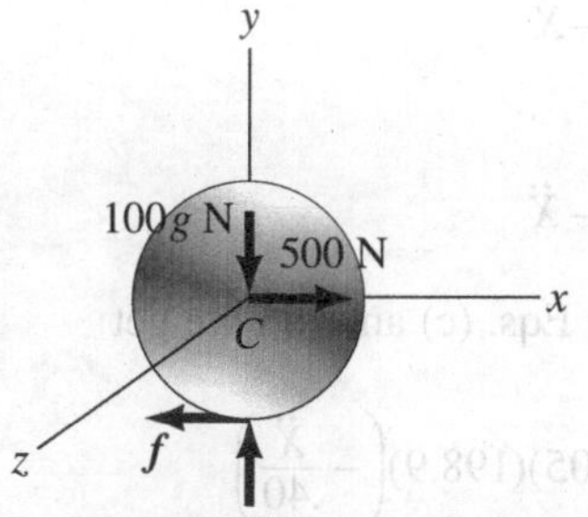

Figure 12.28. Free body.

Next, we write the **moment-of-momentum** equation about the z axis, which goes through the *center of mass*. Thus, noting that we have simple circular motion relative to the z axis for all particles of the cylinder and observing Fig. 12.29:

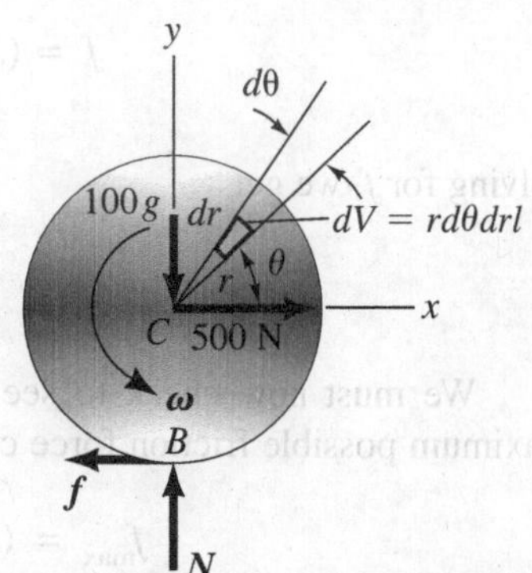

Figure 12.29. Element dV in cylinder. Unknown ω shown as positive.

$$M_z = \frac{d}{dt}(H_z)$$

$$-(f)(.40) = \frac{d}{dt}\left[\int_0^{2\pi}\int_0^{.40} \underbrace{(\rho r\, dr\, d\theta\, l)}_{dm}(r)\underbrace{(r\omega)}_{V_\theta}\right]$$

[11]The reader is cautioned that although the diagram of this problem looks like those solved by energy methods in Section 11.7 of the previous chapter, this is not such a problem since distance is not involved. Instead, we deal with the linear acceleration and the angular acceleration.

Example 12.14 (Continued)

where ρ is the mass density of the cylinder and l is the thickness of the cylinder. Evaluating the integral and differentiating with respect to time as seen from XYZ, we get

$$.40f = -(\rho l)(2\pi)(\dot{\omega})\left(\frac{.40^4}{4}\right)$$

$$f = -.1005(\rho l)\dot{\omega} \qquad \text{(b)}$$

We can determine ρl as follows from geometry:

$$M = 100 = (\rho l)[\pi(.40)^2]$$

$$\rho l = 198.9 \text{ kg/m}^2 \qquad \text{(c)}$$

We have two equations (a) and (b) with three unknowns, f, $\ddot{X}_c$, and $\dot{\omega}$. We now need another independent equation. This equation can be found from **kinematics**. Thus, assuming a *no-slipping* condition, we have pure instantaneous rotation about point B. From your work in physics (we shall later prove this) we can say for point C of the cylinder:

$$(.40)\omega = -\dot{X}$$

Therefore,

$$(.40)\dot{\omega} = -\ddot{X} \qquad \text{(d)}$$

Substituting for ρl and $\dot{\omega}$ in Eq. (b) using Eqs. (c) and (d), we get

$$f = (-.1005)(198.9)\left(-\frac{\ddot{X}}{.40}\right) \qquad \text{(e)}$$

Now solve for $\ddot{X}$ from Eq. (a) and substitute into Eq. (e):

$$f = (.1005)(198.9)\left(\frac{5 - .01f}{.40}\right)$$

Solving for f, we get

$$\boxed{f = 166.6 \text{ N}}$$

We must now check to see whether our no-slipping assumption is valid. The maximum possible friction force clearly is

$$f_{max} = (100)(9.81)(.2) = 196.2 \text{ N}$$

which is greater than the actual friction force, so that the no-slip assumption is consistent with our results.

PROBLEMS

12.89. A system of particles is shown at time t moving in the xy plane. The following data apply:

$$m_1 = 1 \text{ kg}, \quad V_1 = 5i + 5j \text{ m/sec}$$
$$m_2 = 0.7 \text{ kg}, \quad V_2 = -4i + 3j \text{ m/sec}$$
$$m_3 = 2 \text{ kg}, \quad V_3 = -4j \text{ m/sec}$$
$$m_4 = 1.5 \text{ kg}, \quad V_4 = 3i - 4j \text{ m/sec}$$

(a) What is the total linear momentum of the system?
(b) What is the linear momentum of the center of mass?
(c) What is the total moment of momentum of the system about the origin and about point (2, 6)?

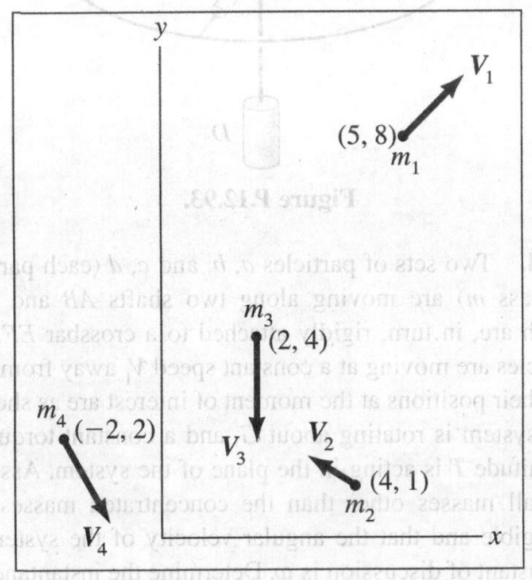

Figure P.12.89.

12.90. A system of particles at time t has the following velocities and masses:

$$V_1 = 20 \text{ ft/sec}, \quad m_1 = 1 \text{ lbm}$$
$$V_2 = 18 \text{ ft/sec}, \quad m_2 = 3 \text{ lbm}$$
$$V_3 = 15 \text{ ft/sec}, \quad m_3 = 2 \text{ lbm}$$
$$V_4 = 5 \text{ ft/sec}, \quad m_4 = 1 \text{ lbm}$$

Determine (a) the total linear momentum of the system, (b) the angular momentum of the system about the origin, and (c) the angular momentum of the system about point a.

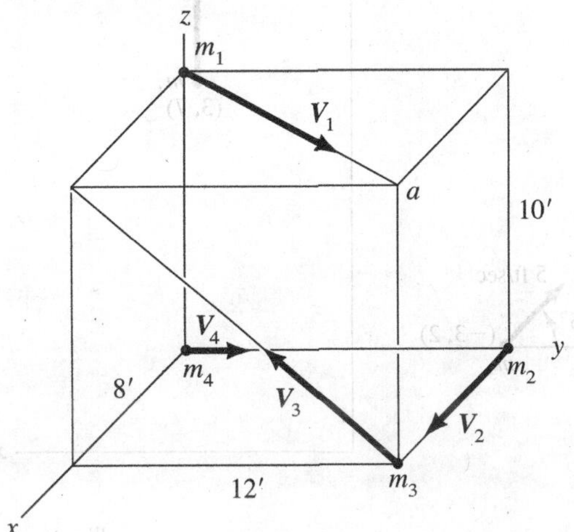

Figure P.12.90.

12.91. A system of particles at time t_1 has masses $m_1 = 2$ lbm, $m_2 = 1$ lbm, $m_3 = 3$ lbm and locations and velocities as shown in (a). The same system of masses is shown in (b) at time t_2. What is the total linear impulse on the system during this time interval? What is the total angular impulse $\int \boldsymbol{M}\, dt$ during this time interval about the origin?

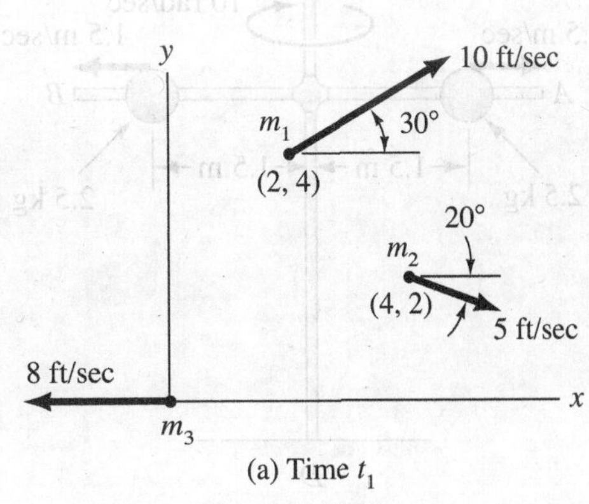

(a) Time t_1

Figure P.12.91-a.

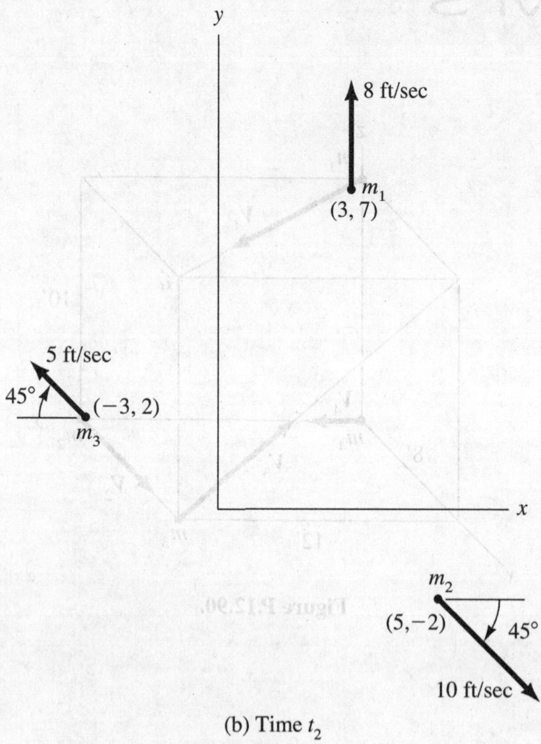

Figure P.12.91-b.

12.92. Two masses slide along bar *AB* at a constant speed of 1.5 m/sec. Bar *AB* rotates freely about axis *CD*. Consider only the mass of the sliding bodies to determine the angular acceleration of *AB* when the bodies are 1.5 m from *CD* if the angular velocity at that instant is 10 rad/sec.

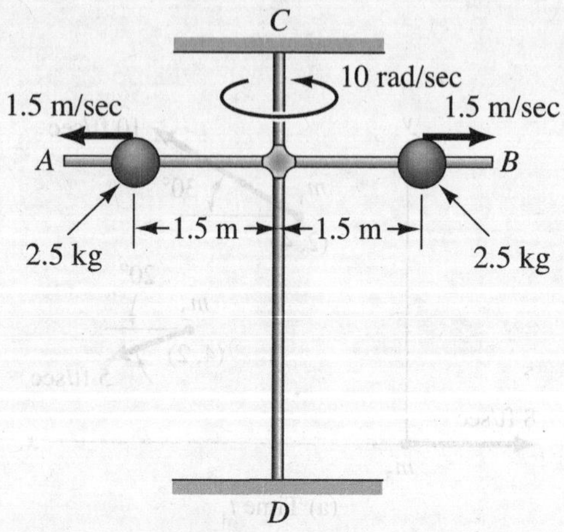

Figure P.12.92.

12.93. A mechanical system is composed of three identical bodies *A*, *B*, and *C* each of mass 3 lbm moving along frictionless rods 120° apart on a wheel. Each of these bodies is connected with an inextensible cord to the freely hanging weight *D*. The connection of the cords to *D* is such that no torque can be transmitted to *D*. Initially, the three masses *A*, *B*, and *C* are held at a distance of 2 ft from the centerline while the wheel rotates at 3 rad/sec. What is the angular speed of the wheel and the velocity of descent of *D* if, after release of the radial bodies, body *D* moves 1 ft? Assume that body *D* is initially stationary (i.e., is not rotating). Body *D* weighs 100 lb.

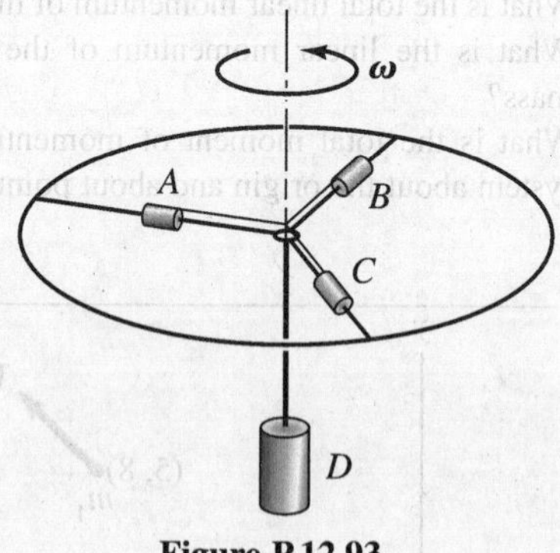

Figure P.12.93.

12.94. Two sets of particles *a*, *b*, and *c*, *d* (each particle of mass *m*) are moving along two shafts *AB* and *CD*, which are, in turn, rigidly attached to a crossbar *EF*. All particles are moving at a constant speed V_1 away from *EF*, and their positions at the moment of interest are as shown. The system is rotating about *G*, and a constant torque of magnitude *T* is acting in the plane of the system. Assume that all masses other than the concentrated masses are negligible and that the angular velocity of the system at the instant of discussion is ω. Determine the instantaneous angular acceleration in terms of m, T, ω, s_1, and s_2.

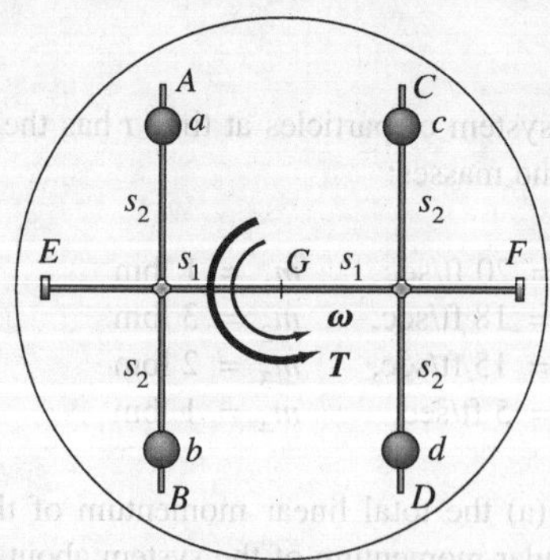

Figure P.12.94.

12.95. A uniform rod with a mass of 7 kg/m lies flat on a frictionless surface. A force of 250 N acts on the rod as shown in the diagram. What is the angular acceleration of the rod? What is the acceleration of the mass center?

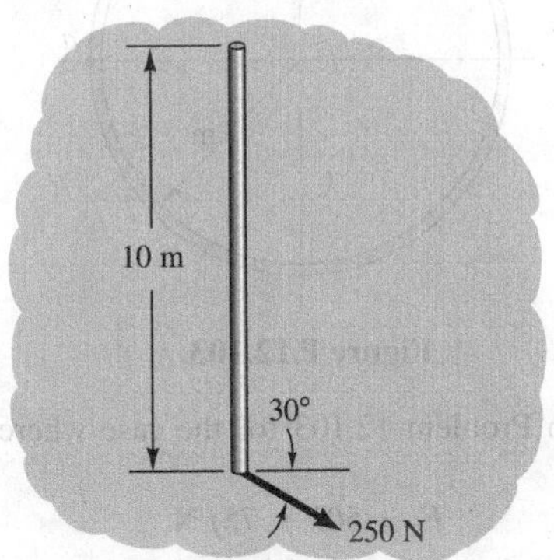

Figure P.12.95

***12.96.** Consider an aggregate of particles with C as the mass center and point A accelerating toward or away from C. Start with the expression for $\dot{\boldsymbol{H}}$ about O given as

$$\boldsymbol{M}_0 = \dot{\boldsymbol{H}}_0 = \frac{d}{dt}\left[\sum_i \boldsymbol{r}_i \times m_i \dot{\boldsymbol{r}}_i\right]$$

$$= \sum_i \boldsymbol{r}_i \times \dot{\boldsymbol{P}}_i = \sum_i (\boldsymbol{r}_A \times \boldsymbol{\rho}_{Ai}) \times \dot{\boldsymbol{P}}$$

Formulate $\boldsymbol{M}_0$ in terms of $\boldsymbol{F}_i$ and use Newton's law to eliminate terms. Next show from the resulting equation that

$$\boldsymbol{M}_A = \left(\sum_i m_i \boldsymbol{\rho}_{Ai}\right) \times \ddot{\boldsymbol{r}}_A + \sum_i \boldsymbol{\rho}_{Ai} \times m_i \ddot{\boldsymbol{\rho}}_{Ai}$$

Replace $\sum_i m_i \boldsymbol{\rho}_{Ai}$ by $M\boldsymbol{\rho}_{Ac}$. Explain why it follows then that

$$\boldsymbol{M}_A = \dot{\boldsymbol{H}}_A$$

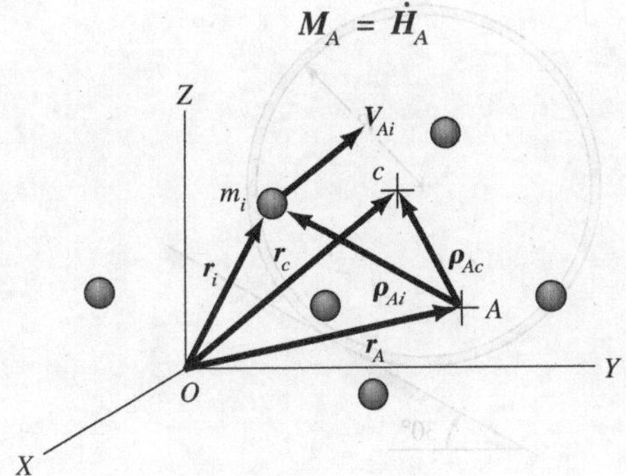

Figure P.12.96.

12.97. A uniform cylinder of radius 1 m rolls without slipping down a 30° incline. What is the angular acceleration of the cylinder if it has a mass of 50 kg?

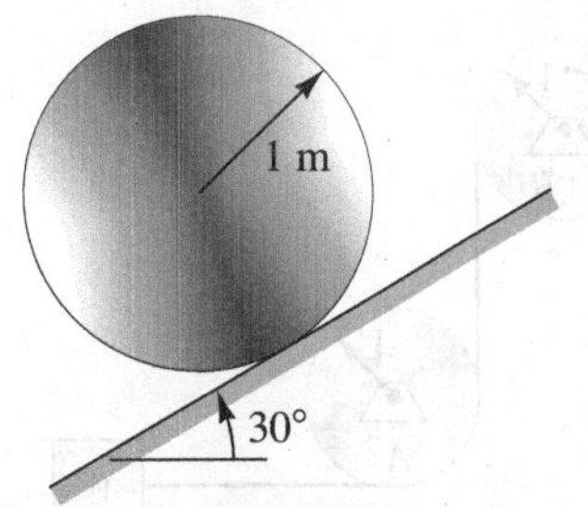

Figure P.12.97.

12.98. A cylinder of length 3 m and mass 45 kg is acted on by a torque $T = (11.25t + 21t^2)$ N-m (where t is in seconds) about its geometric axis. What is the angular speed after 10 sec? The cylinder is at rest when the torque is applied.

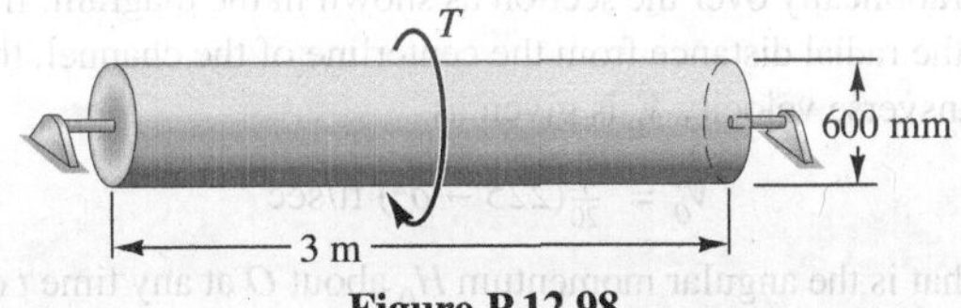

Figure P.12.98.

12.99. A constant torque T of 800 N-m is applied to a uniform cylinder of radius 400 mm and mass 50 kg. A 1.500-kN weight is attached to the cylinder with a light cable. What is the acceleration of W?

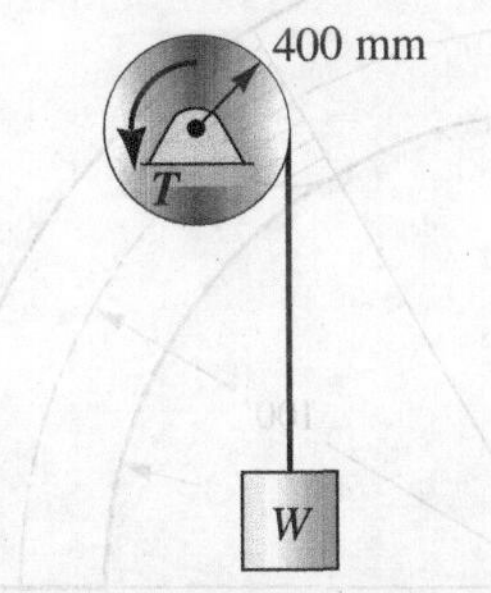

Figure P.12.99.

***12.100.** In Problem 12.99, the torque T is $T = (300 + .2t^2)$ N-m, where t is in seconds. When $t = 0$, the system is at rest. Determine the acceleration of W at the instants when it has zero velocity for $t > 0$.

12.101. A constant torque T of 500 in.-lb is applied to a uniform cylinder of radius 1 ft. A light inextensible cable is wrapped partly around an identical cylinder and is then connected to a block W weighing 100 lb. What is the acceleration of W if the cable does not slip on the cylinders? Take $\mu_d = .3$ for the block. For the cylinders, $W_A = W_B = 100$ lb.

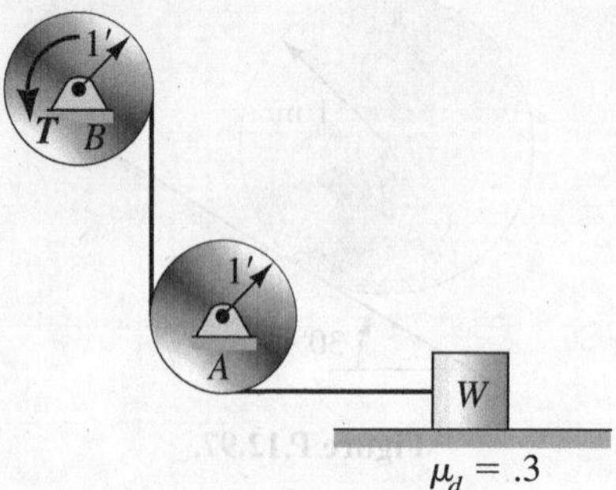

Figure P.12.101.

12.102. A canal with a rectangular cross section is shown having a width of 30 ft and a depth of 5 ft. The velocity of the water is assumed to be zero at the banks and to vary parabolically over the section as shown in the diagram. If δ is the radial distance from the centerline of the channel, the transverse velocity V_θ is given as

$$V_\theta = \frac{1}{20}(225 - \delta^2) \text{ ft/sec}$$

What is the angular momentum H_0 about O at any time t of the water in the circular portion of the canal (i.e., between the x and y axes)? The radial component V_r is zero.

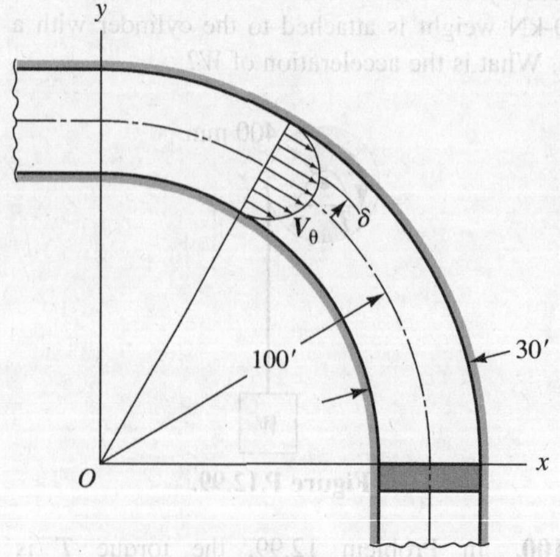

Figure P.12.102.

12.103. A hoop with mass per unit length 6.5 kg/m lies flat on a frictionless surface. A 500-N force is suddenly applied. What is the angular acceleration of the hoop? What is the acceleration of the mass center?

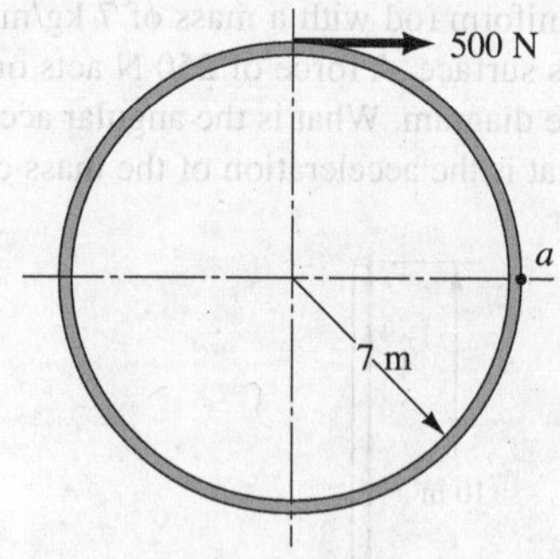

Figure P.12.103.

12.104. Do Problem 12.103 for the case where a force given as:

$$F = 50i + 75j \text{ N}$$

is applied at point a instead of the 500-N force.

12.105. A cylinder weighing 50 lb lies on a frictionless surface. Two forces are applied simultaneously as shown in the diagram. What is the angular acceleration of the cylinder? What is the acceleration of the mass center?

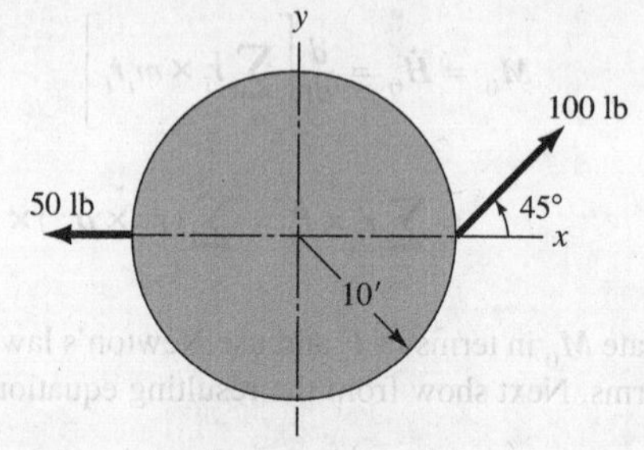

Figure P.12.105.

12.106. A thin uniform hoop rolls without slipping down a 30° incline. The hoop material weighs 5 lb/ft and has a radius R of 4 ft. What is the angular acceleration of the hoop?

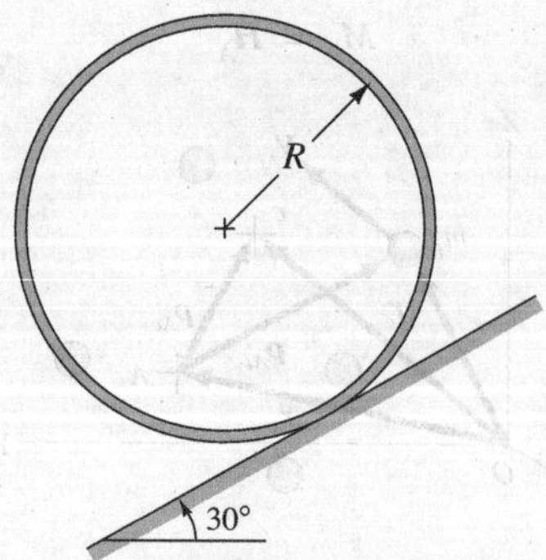

Figure P.12.106.

12.7 Closure

One of the topics studied in this chapter is the impact of bodies under certain restricted conditions. For such problems, we can consider the bodies as particles before and after impact, but during impact the bodies act as deformable media for which a particle model is not meaningful or sufficient. By making an elementary picture of the action, we introduce the coefficient of restitution to yield additional information we need to determine velocities after impact. This is an empirical approach, so our analyses are limited to simple problems. To handle more complex problems or to do the simple ones more precisely, we would have to make a more rational investigation of the deformation actions taking place during impact—that is, a continuum approach to part of the problem would be required. However, we cannot make a careful study of the deformation aspects in this text since the subject of high-speed deformation of solids is a difficult one that is still under careful study by engineers and physicists.

Note in the last two chapters we started with Newton's law $\boldsymbol{F} = M\boldsymbol{a}$ and performed the following operations:

1. Took the dot product of both sides using position vector $\boldsymbol{r}$.
2. Multiplied both sides by dt and integrated.
3. Took the cross product of both sides using position vector $\boldsymbol{r}$.

These steps permitted a surprisingly large number of very useful formulations and concepts that have occupied us for some considerable time. These were the energy methods, the linear-momentum methods, and the moment-of-momentum methods. It should now be clear that Newton's law requires considerable study to fully explore its use.

In our study of moment of momentum for a system of particles, we set forth one of the key equations of mechanics, $\boldsymbol{M}_A = \dot{\boldsymbol{H}}_A$, and we introduced in the examples several considerations whose more careful and complete study will occupy a good portion of the remainder of the text. Thus, in Example 12.14 we have "in miniature," as it were, the major elements involved in the study of much of rigid-body dynamics. Recall that we employed Newton's law for the mass center and the moment-of-momentum equation about the mass center to reach the desired results. In so doing, however, we had to make use of certain elementary kinematical ideas from our earlier work in physics. Accordingly, to prepare ourselves for rigid-body dynamics in Chapters 14 and 15, we shall devote ourselves in Chapter 13 to a rather careful examination of the general kinematics of a rigid body.

Although we shall be much concerned in Chapter 13 with the kinematics of rigid bodies, we shall not cease to consider particles. You will see that an understanding of rigid-body kinematics will permit us to formulate very powerful relations for the general relative motions of a particle involving references that move in any arbitrary manner with respect to each other.

PROBLEMS

12.107. A disc is rotated in the horizontal plane with a constant angular speed ω of 30 rad/sec. A body A with a mass of .4 lbm is moved in a frictionless slot at a uniform speed of 1 ft/sec relative to the platform by a force F as shown. What is the linear momentum of the body relative to the ground reference XY when $r = 2$ ft and $\theta = 45°$? What is the impulse developed on the body as it goes from $r = 2$ ft to $r = 1$ ft? Neglect the mass of the disc.

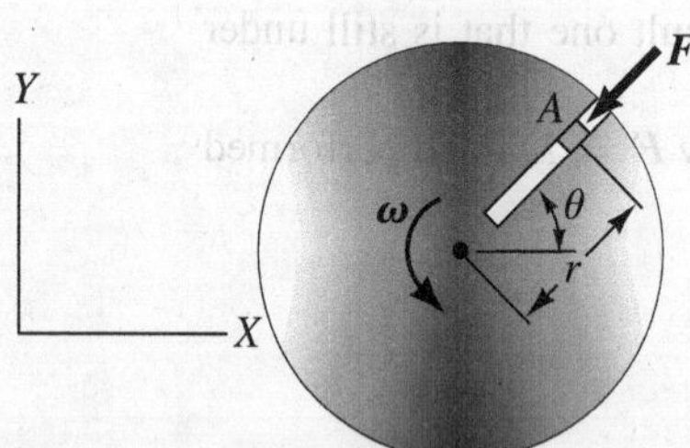

Figure P.12.107.

12.108. Three bodies are towed by a force $F = (100 + 50e^{-t})$ lb as shown. If $W_1 = 30$ lb, $W_2 = 60$ lb, and $W_3 = 50$ lb, what is the speed 5 sec after the application of the given force? The dynamic coefficient of friction is .3 for all surfaces.

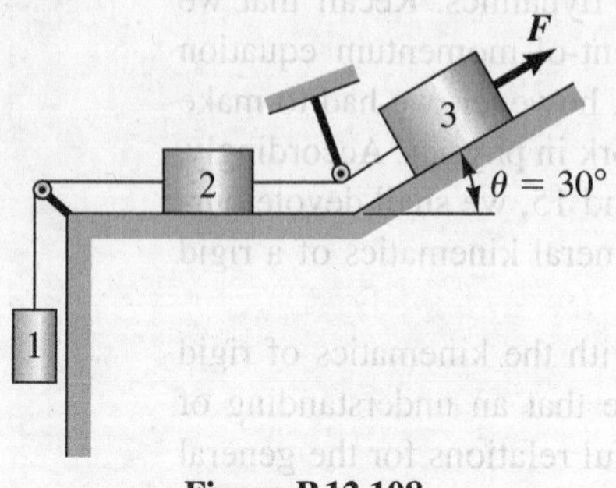

Figure P.12.108.

12.109. A space vehicle is in a circular parking orbit around the earth at a distance of 100 km from the earth's surface. What increase in speed must be given to the vehicle by firing its rockets so as to attain a radial velocity of 2,000 km/hr at an elevation of 200 km?

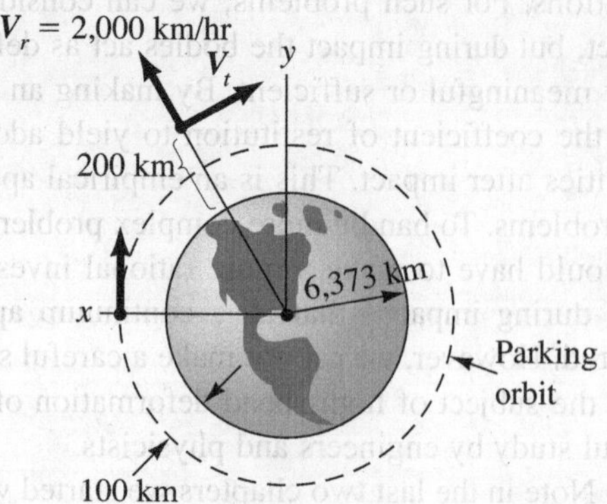

Figure P.12.109.

12.110. A space ship is in a circular parking orbit around the earth at 200 miles above the earth's surface. At space headquarters, they wish to get the vehicle to a position 10,000 miles from the center of the earth with a velocity at this position of 25,000 mi/hr. The command is given to fire rockets directly to the rear for a specified short time interval. What is the change in speed needed for this maneuver? What are the radial and tangential velocity components of the 25,000 mi/hr velocity vector?

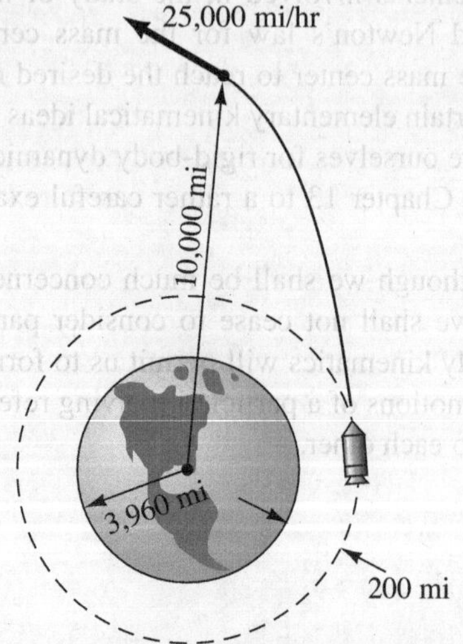

Figure P.12.110.

12.111. A small elastic ball is dropped from a height of 5 m onto a rigid cylindrical body having a radius of 1.5 m. At what position on the x axis does the ball land after the collision with the cylinder?

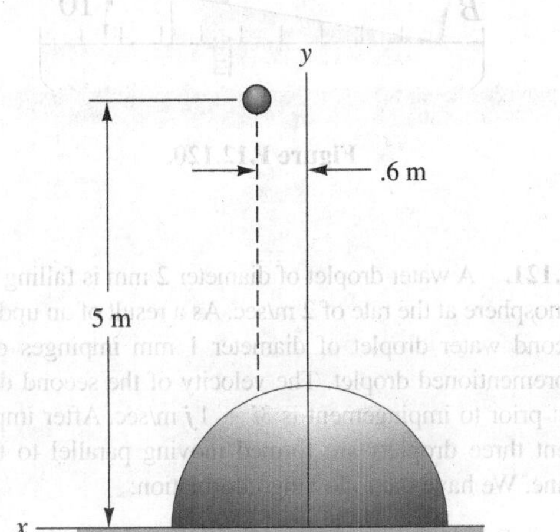

Figure P.12.111.

12.112. Do Problem 12.111 for an inelastic impact with $\epsilon = .6$.

12.113. A small elastic sphere is dropped from position (2, 3, 30) ft onto a hard spherical body having a radius of 5 ft positioned so that the z axis of the reference shown is along a diameter. For a perfectly elastic collision, give the speed of the small sphere directly after impact.

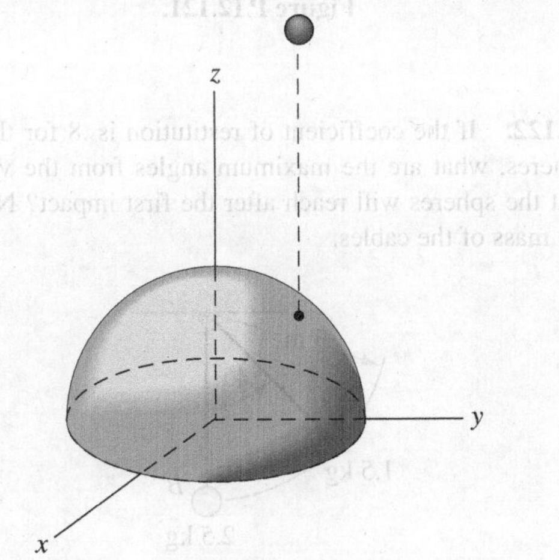

Figure P.12.113.

12.114. Do Problem 12.113 for an inelastic impact with $\epsilon = .6$.

12.115. A bullet hits a smooth, hard, massive two-dimensional body whose boundary has been shown as a parabola. If the bullet strikes 1.5 m above the x axis and if the collision is perfectly elastic, what is the maximum height reached by the bullet as it ricochets? Neglect air resistance and take the velocity of the bullet on impact as 700 m/sec with a direction that is parallel to the x axis.

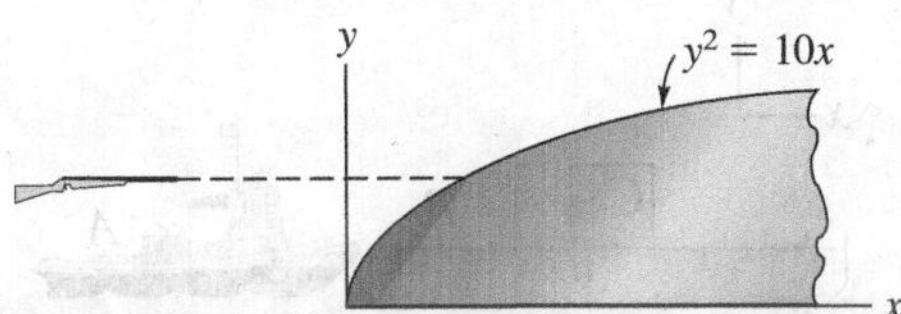

Figure P.12.115.

12.116. In Problem 12.115, assume an inelastic impact with $\epsilon = .6$. At what position along x does the bullet strike the parabola after the impact?

12.117. A space vehicle is in a circular "parking" orbit (1) around the earth 200 km above the earth's surface. It is to transfer to another circular orbit (2) 500 km above the earth's surface. The transfer to the second orbit is done in two stages.

1. Fire rockets so the vehicle has an *apogee* equal to the radius of the second circular orbit. What change of speed is required for this maneuver?
2. At *apogee* rockets are fired again to get into the second circular orbit. What is this second change of speed?

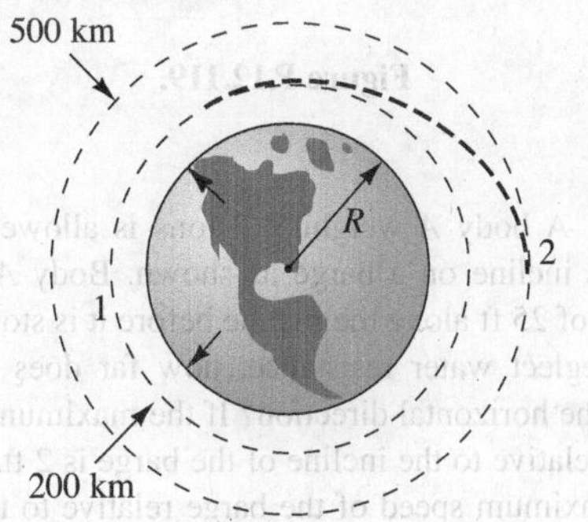

Figure P.12.117.

***12.118.** A tugboat weighing 100 tons is moving toward a stationary barge weighing 200 tons and carrying a load C weighing 50 tons. The tug is moving at 5 knots and its propellers are developing a thrust of 5,000 lb when it contacts the barge. As a result of the soft padding at the nose of the tug, consider that there is plastic impact. If the load C is not tied in any way to the barge and has a dynamic coefficient of friction of .1 with the slippery deck of the barge, what is the speed V of the barge 2 sec after the tug first contacts the barge? The load C slips during a 1-sec interval starting at the beginning of the contact.

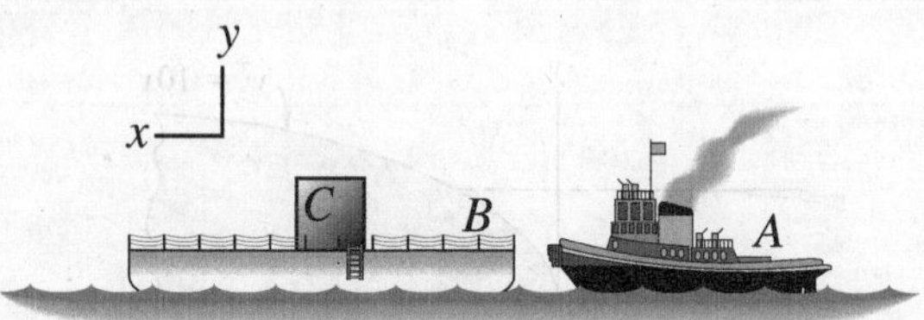

Figure P.12.118.

12.119. A hopper drops small cylinders each weighing 10 N onto a conveyor belt which is moving at a speed of 3 m/sec. At the top, the cylinders are dropped off as shown. If at any time t there are 14 cylinders on the belt and if 10 cylinders are dropped per second from a hopper from a height of 300 mm above the belt, what average torque is needed to operate the conveyor? The weight of the belt that is on the conveyor bed is 100 N. The coefficient of friction μ_d between the belt and the bed is .3. The radius of the driving cylinder is 300 mm. Neglect bearing friction.

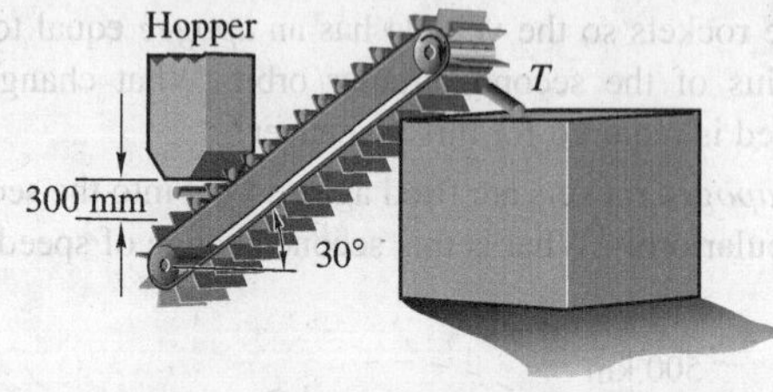

Figure P.12.119.

12.120. A body A weighing 2 tons is allowed to slide down an incline on a barge as shown. Body A moves a distance of 25 ft along the incline before it is stopped at B. If we neglect water resistance, how far does the barge shift in the horizontal direction? If the maximum speed of body A relative to the incline of the barge is 2 ft/sec, what is the maximum speed of the barge relative to the water? The weight of the barge is 20 tons.

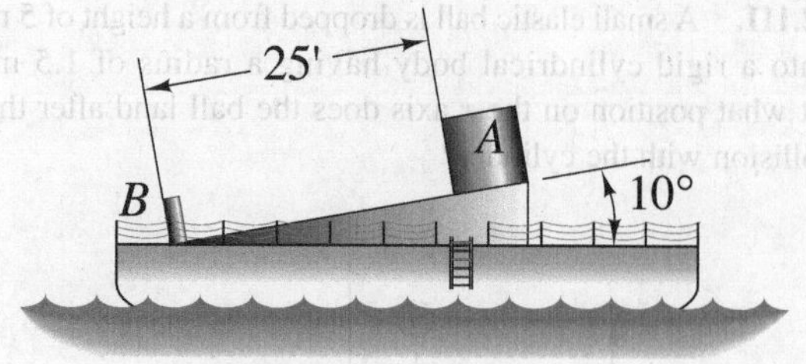

Figure P.12.120.

12.121. A water droplet of diameter 2 mm is falling in the atmosphere at the rate of 2 m/sec. As a result of an updraft, a second water droplet of diameter 1 mm impinges on the aforementioned droplet. The velocity of the second droplet just prior to impingement is $3i + 1j$ m/sec. After impingement three droplets are formed moving parallel to the xy plane. We have the following information:

$$D_1 = .6 \text{ mm}, \quad V_1 = 2 \text{ m/sec}, \quad \theta_1 = 45°$$
$$D_2 = 1.2 \text{ mm}, \quad V_2 = 1 \text{ m/sec}, \quad \theta_2 = 30°$$

Find D_3, V_3, and θ_3.

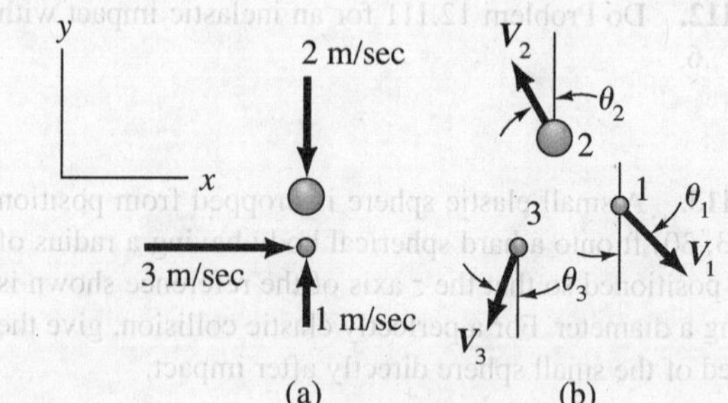

Figure P.12.121.

12.122. If the coefficient of restitution is .8 for the two spheres, what are the maximum angles from the vertical that the spheres will reach after the first impact? Neglect the mass of the cables.

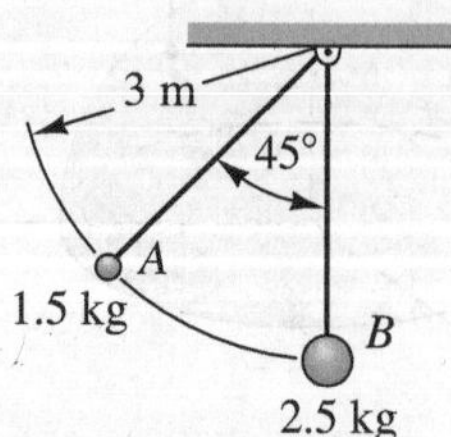

Figure P.12.122.

12.123. Thin discs A and B slide along a frictionless surface. Each disc has a radius of 25 mm. Disc A has a mass of 85 g, whereas disc B has a mass of 227 g. What are the speeds of the discs after collision for $\epsilon = .7$? Assume that the discs slide on a frictionless surface.

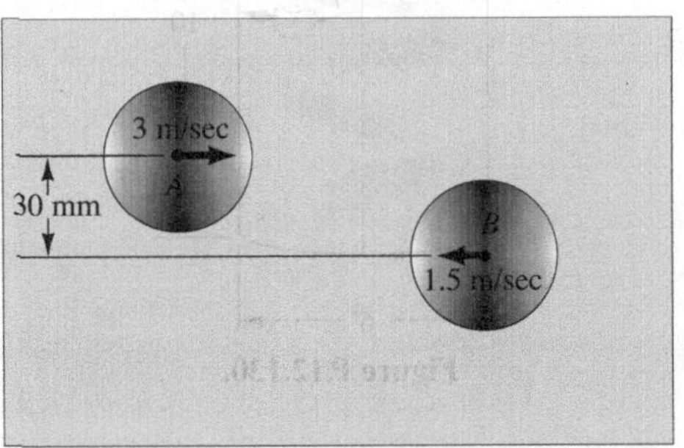

Figure P.12.123.

12.124. A BB is shot at the hard, rigid surface. The speed of the pellet is 300 ft/sec as it strikes the surface. If the direction of the velocity for the pellet is given by the following unit vector:

$$\boldsymbol{\epsilon} = -.6\boldsymbol{i} - .8\boldsymbol{k}$$

what is the final velocity vector of the pellet for a collision having $\epsilon = .7$?

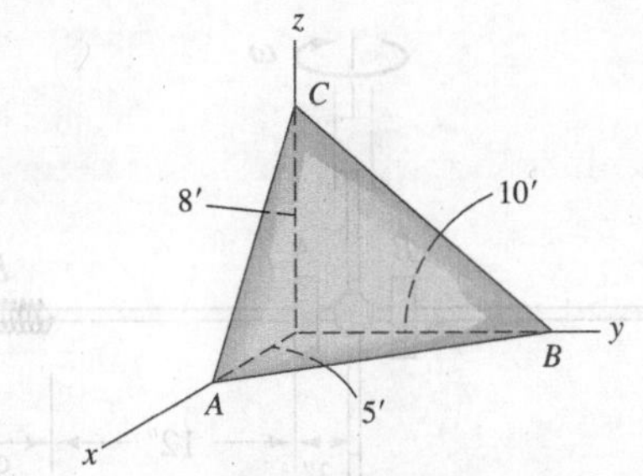

Figure P.12.124.

***12.125.** A chain of wrought iron, with length of 7 m and a mass of 100 kg, is held so that it just touches the support AB. If the chain is released, determine the total impulse during 2 sec in the vertical direction experienced by the support if the impact is plastic (i.e., the chain does not bounce up) and if we move the support so that the links land on the platform and not on each other? [*Hint:* Note that any chain *resting* on AB delivers a vertical impulse. Also check to see if the entire chain lands on AB before 2 sec.]

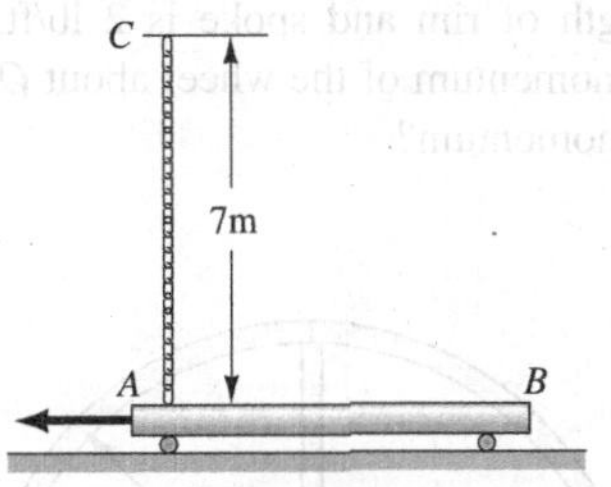

Figure P.12.125.

12.126. Two trucks are shown moving up a 10° incline. Truck A weighs 26.7 kN and is developing a 13.30-kN driving force on the road. Truck B weighs 17.8 kN and is connected with an inextensible cable to truck A. By operating a winch b, truck B approaches truck A with a constant acceleration of .3 m/sec.^{2} If at time $t = 0$ both trucks have a speed of 10 m/sec, what are their speeds at time $t = 15$ sec?

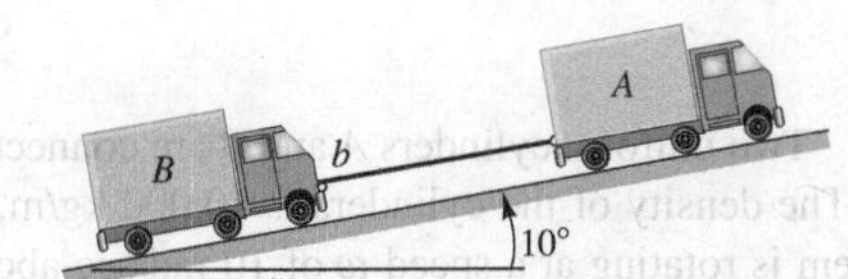

Figure P.12.126.

12.127. Compute the angular momentum about O of a uniform rod, of length $L = 3$ m and mass per unit length m of 7.5 kg/m, at the instant when it is vertical and has an angular speed ω of 3 rad/sec.

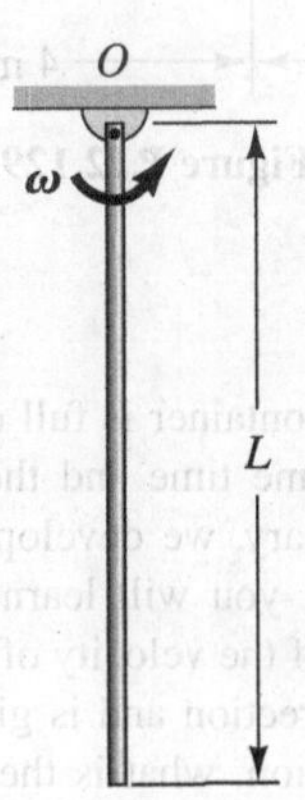

Figure P.12.127.

12.128. A wheel consisting of a thin rim and four thin spokes is shown rotating about its axis at a speed ω of 2 rad/sec. The radius of the wheel R is 2 ft and the weight per unit length of rim and spoke is 2 lb/ft. What is the moment of momentum of the wheel about O? What is the total linear momentum?

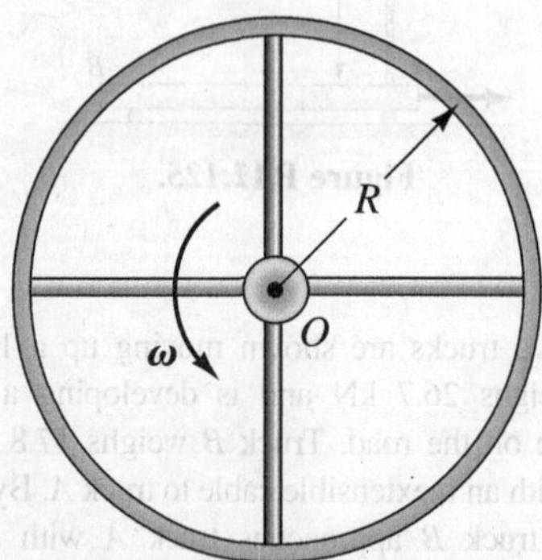

Figure P.12.128.

12.129. Two uniform cylinders A and B are connected as shown. The density of the cylinders is 10,000 kg/m,^{3} and the system is rotating at a speed ω of 10 rad/sec about its geometric axis. What is the angular momentum of the body?

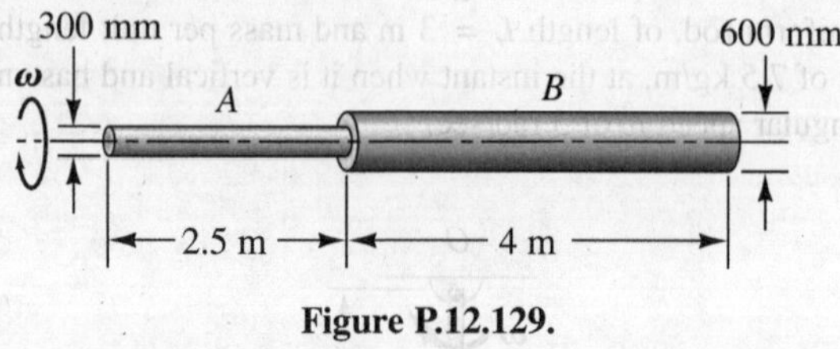

Figure P.12.129.

12.130. A closed container is full of water. By rotating the container for some time and then suddenly holding the container stationary, we develop a rotational motion of the water, which, you will learn in fluid mechanics, resembles a vortex. If the velocity of the fluid elements is zero in the radial direction and is given as $10/r$ ft/sec in the transverse direction, what is the angular momentum of the water?

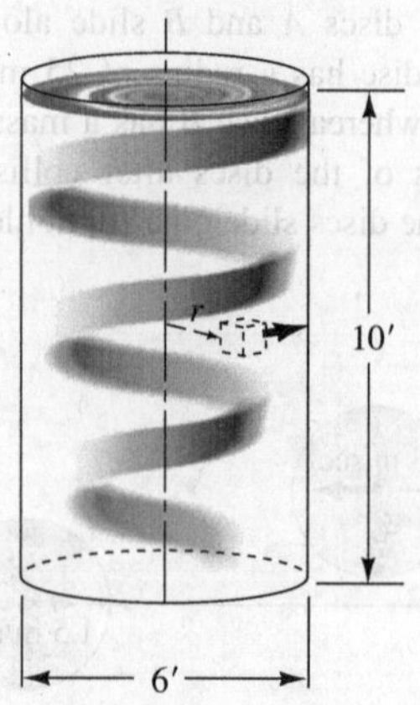

Figure P.12.130.

12.131. Identical thin masses A and B slide on a light horizontal rod that is attached to a freely turning light vertical shaft. When the masses are in the position shown in the diagram, the system rotates at a speed ω of 5 rad/sec. The masses are released suddenly from this position and move out toward the identical springs, which have a spring constant K = 800 lb/in. Set up the equation for the compression δ of the spring once all motions of the bodies relative to the rod have damped out. The mass of each body is 10 lbm. Neglect the mass of the rods and coulombic friction. Show that δ = .08361 in. satisfies your equation.

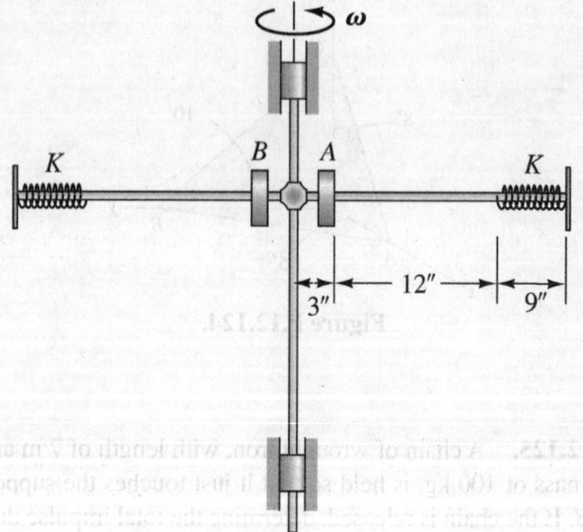

Figure P.12.131.

12.132. A spacecraft has a burnout velocity V_0 of 8,300 m/sec at an elevation of 80 km above the earth's surface. The launch angle α is 15°. What is the maximum elevation h from the earth's surface for the spacecraft?

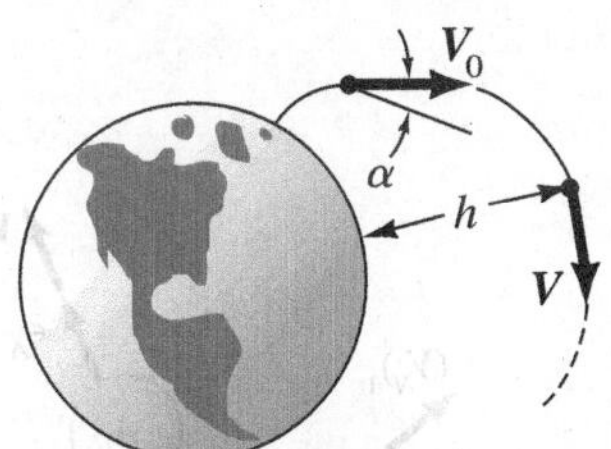

Figure P.12.132.

12.133. A set of particles, each having a mass of 1/2 slug, rotates about axis A–A. The masses are moving out radially at a constant speed of 5 ft/sec at the same time that they are rotating about the A–A, axis. When they are 1 ft from A–A, the angular velocity is 5 rad/sec and at that instant a torque is applied in the direction of motion which varies with time t in seconds as

$$\text{torque} = (6t^2 + 10t) \text{ lb-ft}$$

What is the angular velocity when the masses have moved out radially at constant speed to 2 ft?

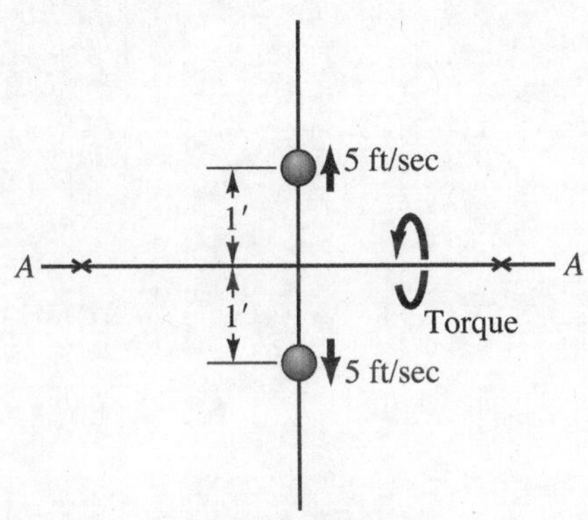

Figure P.12.133.

12.134. A torpedo boat weighing 100,000 lb moves at 40 knots (1 knot = 6,080 ft/hr) away from an engagement. To go even faster, all four 50-caliber machine guns are ordered to fire simultaneously toward the rear. Each weapon fires at a muzzle velocity of 3,000 ft/sec and fires 500 rounds per minute. Each slug weighs 2 oz. How much is the average force on the boat increased by this action? Neglect the rate of change of the total mass of boat.

12.135. A device to be detonated with a small charge is suspended in space [see (a)]. Directly after detonation, four fragments are formed moving away from the point of suspension. The following information is known about these fragments:

$$m_1 = 1 \text{ lbm}$$
$$V_1 = 200\boldsymbol{i} - 100\boldsymbol{j} \text{ ft/sec}$$
$$m_2 = 2 \text{ lbm}$$
$$V_2 = 125\boldsymbol{i} + 180\boldsymbol{j} - 100\boldsymbol{k} \text{ ft/sec}$$
$$m_3 = 1.6 \text{ lbm}$$
$$V_3 = -200\boldsymbol{i} + 150\boldsymbol{j} + 180\boldsymbol{k} \text{ ft/sec}$$
$$m_4 = 3.2 \text{ lbm}$$

What is the velocity V_4?

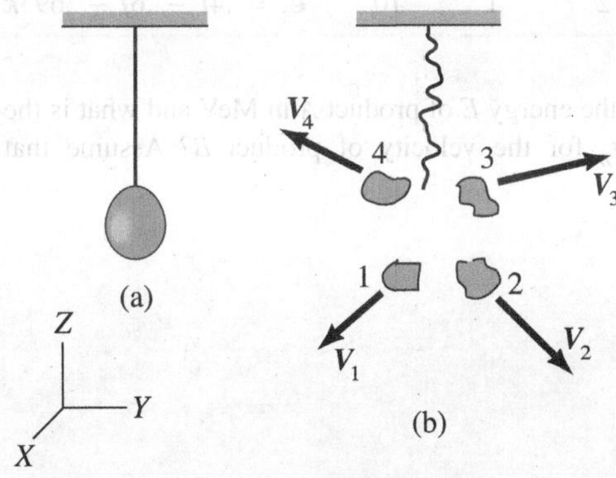

Figure P.12.135.

12.136. A hawk is a predatory bird which often attacks smaller birds in flight. A hawk having a mass of 1.3 kg is swooping down on a sparrow having a mass of 150 g. Just before seizing the sparrow with its claws, the hawk is moving downward with a speed V_H of 20 km/hr. The sparrow is moving horizontally at a speed V_S of 15 km/hr. Directly after seizure, what is the speed of the hawk and its prey? What is the loss in kinetic energy in joules?

12.137. The principal mode of propulsion of an octopus is to take in water through the mouth and then after closing the inlet to eject the water to the rear. If a 5-lb octopus after taking in 1 lb of water is moving at a speed of 3 ft/sec, what is its speed directly after ejecting the water? The water is ejected at an average speed to the rear of 10 ft/sec relative to the initial speed of the octopus. What horsepower is being developed by the octopus in the above action if it occurs in 1 sec?

***12.138.** In the *fission* process in a nuclear reactor, a ^{235}U nucleus first absorbs or captures a neutron [see (a)]. A short time later, the ^{235}U nucleus breaks up into fission products plus neutrons, which may subsequently be captured by other ^{235}U nuclei and maintain a *chain reaction.* Energy is released in each fission. In (b) we have shown the results of a possible fission. The following information is known for this fission:

	Mass No.	Kinetic Energy (MeV)	Direction of V
Product A	138	E	$\epsilon_A = .3i - .2j + .98k$
Product B	96	90	$\epsilon_B = l_B i + m_B j + n_B k$
Neutron 1	1	10	$\epsilon_1 = .6i + .8j$
Neutron 2	1	10	$\epsilon_2 = .4i - .6j - .693k$

What is the energy E of product A in MeV and what is the vector ϵ_B for the velocity of product B? Assume that before fission the nucleus of ^{235}U plus captured neutrons is stationary: [*Hint:* You do not have to actually convert MeV to joules or atomic number to kilograms to carry out the problem.]

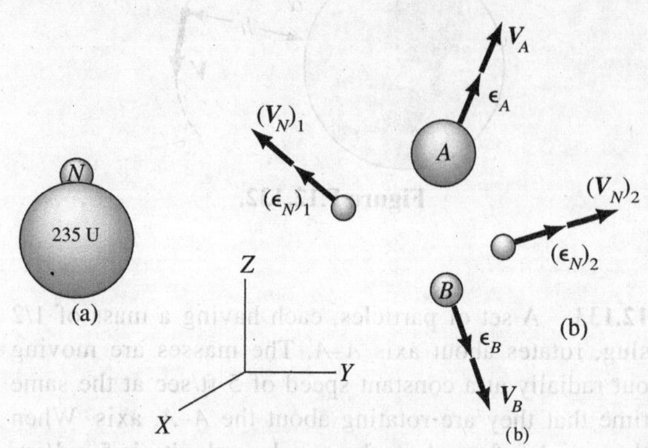

Figure P.12.138.

CHAPTER 13

Kinematics of Rigid Bodies: Relative Motion

13.1 Introduction

In Chapter 9, we studied the kinematics of a particle. During virtually all of this study only a single reference was used. However, at the end of that chapter, we briefly introduced the use of two references—for the case of *simple relative motion* involving two references *translating* with respect to each other.

One of the things we shall do in this chapter is to generalize the formulations for multireference analysis. There are two reasons for doing this. First, we shall be able to analyze complicated motions in a more simple systematic way by using several references. Second, the motion of a particle is often known relative to a moving body (such as an airplane), to which we can fix a reference xyz, while the motion of the plane (and hence xyz) is known relative to an *inertial reference* XYZ (such as the ground). Now **Newton's law** in the form $F = ma$, is valid *only* for an inertial reference. Hence, to use **Newton's law** for the particle we must express the acceleration of the particle relative to the inertial reference directly. Accordingly, for practical reasons we must become involved in multireference systems.

A reference is a rigid body, and, before we can set forth multireference considerations, we must first study the kinematics of a rigid body. In so doing, we will also set the stage for our main effort in the remaining portion of the text involving the dynamics of rigid bodies.

13.2 Translation and Rotation of Rigid Bodies

For purposes of dynamics, a rigid body is considered to be composed of a continuous distribution of particles having fixed distances between each other. We shall profitably define once again two simple types of motion of a rigid body:

Translation. As pointed out in Chapter 9, if a body moves so that all the particles have at time t the same velocity relative to some reference, the body is said to be in *translation* relative to this reference at this time. The velocity of a translating body can vary with

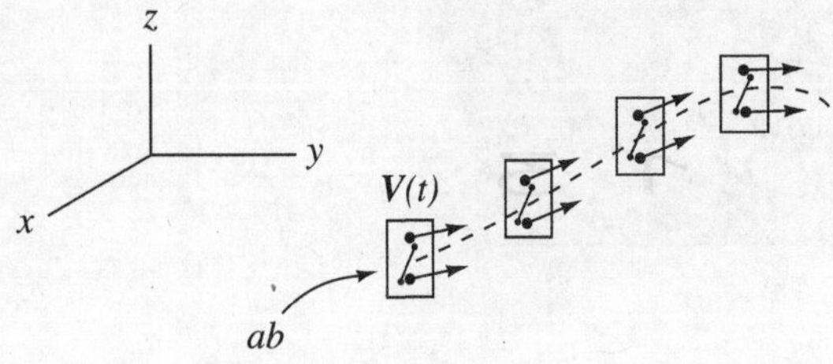

Figure 13.1. Translation of a body.

time and so can be represented as $V(t)$. Accordingly, translational motion does not necessarily mean motion along a straight line. For example, the body shown in Fig. 13.1 is in translation over the interval indicated because at each instant, each particle in the body has a common velocity. A characteristic of translational motion is that a straight line between two points of the body such as ab in Fig. 13.1 always retains an orientation parallel to its *original direction* during the motion.

Rotation. If a rigid body moves so that along some straight line all the particles of the body, or a hypothetical extension of the body, have zero velocity relative to some reference, the body is said to be in *rotation* relative to this reference. The line of stationary particles is called the *axis of rotation*.

We shall now consider how we measure the rotation of a body. A single revolution is defined as the amount of rotation in either a clockwise or a counterclockwise direction about the axis of rotation that brings the body back to its original position. Partial revolutions can conveniently be measured by observing *any* line segment such as AB in the body (Fig. 13.2) from a viewpoint M-M directed along the axis of rotation. In Fig. 13.3, we have shown this view of AB at the beginning of the partial rotation as seen along the axis of rotation, as well as the view $A'B'$ at the end of the partial rotation. The angle β that these lines form will be the same for the initial and final projections viewed along the axis of rotation of *any* line segment so examined in the partial rotation of the rigid body. Accordingly, the angle β so formed during a partial rotation is the measure of rotation.

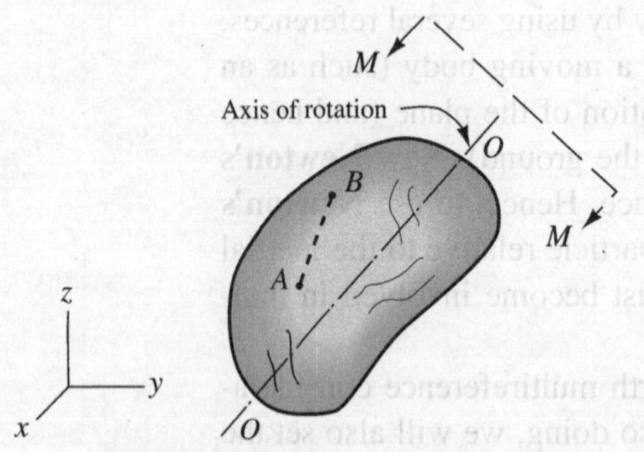

Figure 13.2. Rotation of a body.

In Chapter 1, we pointed out that finite rotations, although they have a magnitude and a direction along the axis of rotation, are not vectors. The superposition of rotations is not commutative, and therefore rotations do not add according to the parallelogram law, which, you will recall, is a requirement of all vector quantities. However, we can show that as rotations become *infinitesimal*, they satisfy in the limit the commutative law of addition, so that infinitesimal rotations $d\beta$ are vector quantities. Therefore, the *angular velocity* is a vector quantity having a magnitude $d\beta/dt$ with an orientation parallel to the axis of rotation and a sense in accordance with the right-hand-screw rule. We shall employ $\boldsymbol{\omega}$ to represent the angular velocity vector. Note that this definition does not prescribe the line of action of this vector, for the line of action may be considered at positions other than the axis of rotation. The line of action depends on the situation at hand (as will be discussed in later sections).

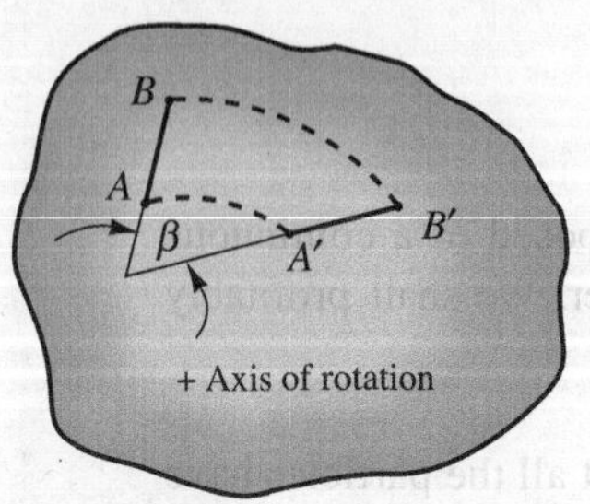

Figure 13.3. Measure of a partial rotation.

13.3 Chasles' Theorem

We have just considered two simple motions of a body, translation and rotation. We shall now demonstrate that at each instant, the motion of any rigid body can be thought of as the superposition of both a translational motion and a rotational motion.

Consider for simplicity a body moving in a plane. Positions of the body are shown tinted at times t and $(t + \Delta t)$ in Fig. 13.4. Let us select any point B of the body. Imagine that the body is displaced without rotation from its position at time t to the position at time $(t + \Delta t)$ so that point B reaches its correct final position B'. The displacement vector for this translation is shown at $\Delta \boldsymbol{R}_B$. To reach the correct orientation for $(t + \Delta t)$, we must now rotate the body an angle $\Delta\phi$ about an axis of rotation which is normal to the plane and which passes through point B'.

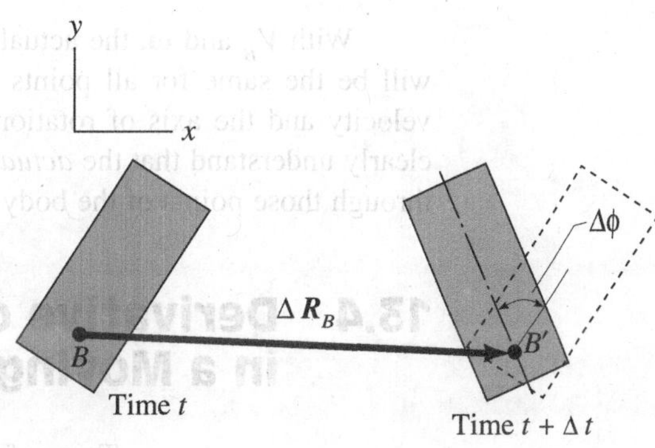

Figure 13.4. Translation and rotation of a rigid body.

What changes would occur had we chosen some other point C for such a procedure? Consider Fig. 13.5, where we have included an alternative procedure by translating the body so that point C reaches the correct final position C'. Next, we must rotate the body an amount $\Delta\phi$ about an axis of rotation which is normal to the plane and which passes through C' in order to get to the final orientation of the body. Thus, we have indicated two routes. We conclude from the diagram that the displacement $\Delta \boldsymbol{R}_C$ differs from $\Delta \boldsymbol{R}_B$, but there is no difference in the amount of rotation $\Delta\phi$. Thus, in general, *$\Delta \boldsymbol{R}$ and the axis of rotation will depend on the point chosen, while the amount of rotation $\Delta\phi$ will be the same for all such points.*

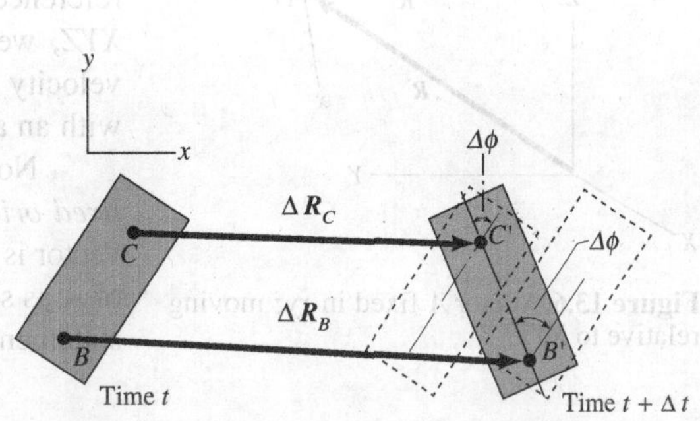

Figure 13.5. Translation and rotation of a rigid body using points B and C.

Consider now the ratios $\Delta R/\Delta t$ and $\Delta\phi/\Delta t$. These quantities can be regarded as an average translational velocity and an average rotational speed, respectively, of the body, which we could superpose to get from the initial position to the final position in the time Δt. Thus, $\Delta R/\Delta t$ and $\Delta\phi/\Delta t$ represent an average measure of the motion during the time interval Δt. *If we go to the limit by letting $\Delta t \to 0$, we have instantaneous translational and angular velocities which, when superposed, give the instantaneous motion of the body.* The displacement vector of the chosen point B in the previous discussion represents the translation of the body during the time Δt. Furthermore, the chosen point B undergoes no other motion during Δt other than that occurring during translation. Thus, we can conclude that, in the limit, the *translational velocity* used for the body corresponds to the *actual instantaneous* velocity of the chosen point B at time t. The angular velocity $\boldsymbol{\omega}$ to be used in the movement of the body, as described above, is the same vector for *all* points B chosen. Accordingly, $\boldsymbol{\omega}$ is the *instantaneous angular* velocity of the body.

We have thus far considered the movement of the body along a plane surface. The same conclusions can be reached for the general motion of an arbitrary rigid body in space. We can then make the following statements for the description of the general motion of a rigid body relative to some reference at time t. These statements comprise **Chasles' theorem.**

1. Select any point B in the body. Assume that all particles of the body have at the time t a velocity equal to V_B, the actual velocity of the point B.
2. Superpose a pure rotational velocity $\boldsymbol{\omega}$ about an axis of rotation going through point B.

With V_B and $\boldsymbol{\omega}$, the actual instantaneous motion of the body is determined, and $\boldsymbol{\omega}$ will be the same for all points B which might be chosen. Thus, only the translational velocity and the axis of rotation change when different points B are chosen. However, clearly understand that the *actual instantaneous axis of rotation* at time t is the one going through those points of the body having zero velocity at time t.

13.4 Derivative of a Vector Fixed in a Moving Reference

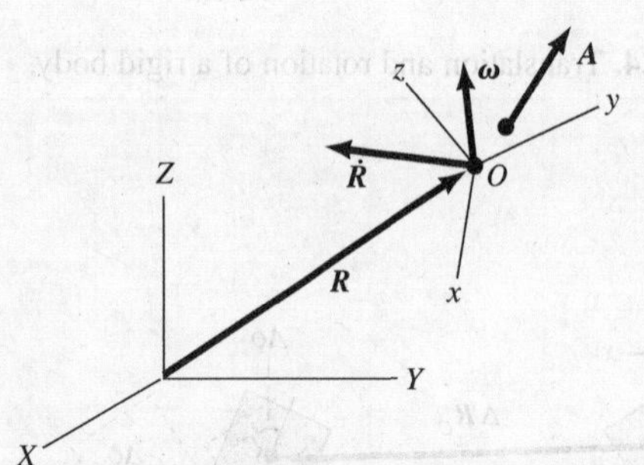

Figure 13.6. Vector A fixed in xyz moving relative to XYZ.

Two references XYZ and xyz move arbitrarily relative to each other in Fig. 13.6. Assume we are observing xyz from XYZ. Since a reference is a rigid system, we can apply **Chasles' theorem** to reference xyz. Thus, to fully describe the motion of xyz relative to XYZ, we choose the origin O, and we superpose a translation velocity $\dot{\boldsymbol{R}}$, equal to the velocity of O, onto a rotational velocity $\boldsymbol{\omega}$ with an axis of rotation through O.

Now suppose that we have a vector $\boldsymbol{A}$ of *fixed length* and of *fixed orientation* as seen from reference xyz. We say that such a vector is "fixed" in reference xyz. Clearly, the time rate of change of $\boldsymbol{A}$ as seen from reference xyz must be zero. We can express this statement mathematically as

$$\left(\frac{d\boldsymbol{A}}{dt}\right)_{xyz} = \boldsymbol{0}$$

However, as seen from XYZ, the time rate of change $\boldsymbol{A}$ will *not* necessarily be zero. To evaluate $(d\boldsymbol{A}/dt)_{XYZ}$, we make use of **Chasles' theorem** in the following manner:

1. Consider the *translational* motion $\dot{\boldsymbol{R}}$. This motion does not alter the direction of $\boldsymbol{A}$ as seen from XYZ. Also, the magnitude of $\boldsymbol{A}$ is fixed; thus, vector $\boldsymbol{A}$ cannot change as a result of this motion.[1]
2. We next consider *solely* a pure rotation about a stationary axis collinear with $\boldsymbol{\omega}$ and passing through point O.

To best observe this rotation, we shall employ at O a *stationary* reference $X'Y'Z'$ positioned so that Z' coincides with the axis of rotation. This reference is shown in Fig. 13.7. Now the vector $\boldsymbol{A}$ is rotating at this instant about the Z' axis. We have shown

[1]The *line of action* of A, however, will change as seen from XYZ. But a change of line of action does not signify a change in the vector, as pointed out in Chapter 1 on the discussion of equality of vectors.

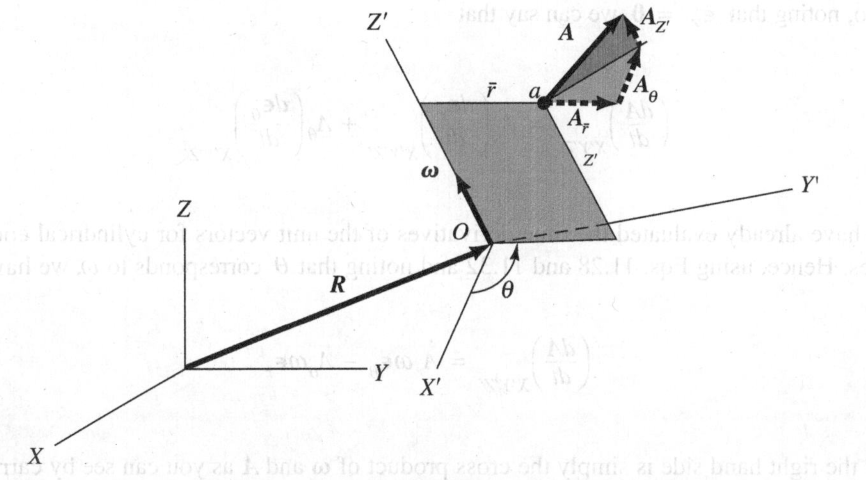

Figure 13.7. Cylindrical components for vector *A*.

cylindrical coordinates to the end of $\boldsymbol{A}$ (i.e., at point a); and have shown cylindrical components A_r, A_θ, and $A_{Z'}$. In Fig. 13.8, we have shown point a with unit vectors $\boldsymbol{\epsilon}_r$, $\boldsymbol{\epsilon}_\theta$, and $\boldsymbol{\epsilon}_{Z'}$, for cylindrical coordinates at this point. We can accordingly express $\boldsymbol{A}$ as

$$\boldsymbol{A} = A_{\bar{r}}\boldsymbol{\epsilon}_{\bar{r}} + A_\theta\boldsymbol{\epsilon}_\theta + A_{Z'}\boldsymbol{\epsilon}_{Z'}$$

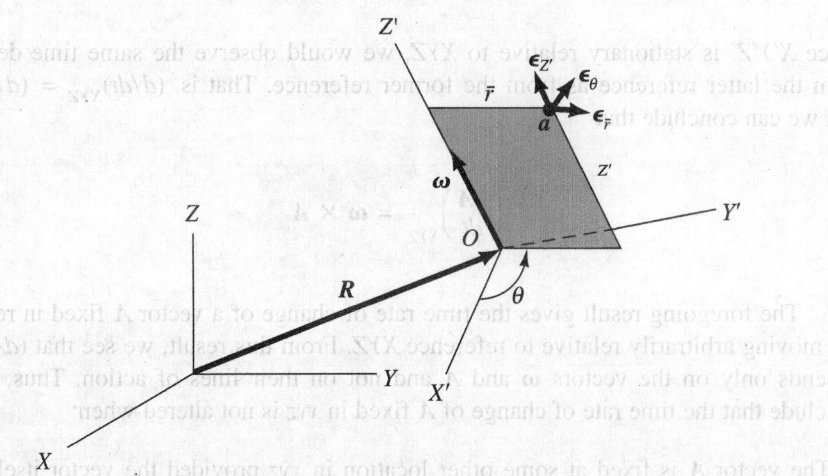

Figure 13.8. Unit vectors for cylindrical coordinates.

Clearly, as $\boldsymbol{A}$ rotates about Z', the values of the cylindrical scalar components of $\boldsymbol{A}$ for $X'Y'Z'$, namely A_r, A_θ, and $A_{Z'}$, do not change. Hence, as seen from $X'Y'Z'$, $\dot{A}_r = \dot{A}_\theta = \dot{A}_Z = 0$.

Also, noting that $\dot{\boldsymbol{\epsilon}}_{Z'} = \mathbf{0}$, we can say that

$$\left(\frac{dA}{dt}\right)_{X'Y'Z'} = A_{\bar{r}}\left(\frac{d\boldsymbol{\epsilon}_{\bar{r}}}{dt}\right)_{X'Y'Z'} + A_\theta\left(\frac{d\boldsymbol{\epsilon}_\theta}{dt}\right)_{X'Y'Z'}$$

We have already evaluated the time derivatives of the unit vectors for cylindrical coordinates. Hence, using Eqs. 11.28 and 11.32 and noting that $\dot{\theta}$ corresponds to ω, we have

$$\left(\frac{dA}{dt}\right)_{X'Y'Z'} = A_{\bar{r}}\omega\boldsymbol{\epsilon}_\theta - A_\theta\omega\boldsymbol{\epsilon}_{\bar{r}}$$

But the right hand side is simply the cross product of $\boldsymbol{\omega}$ and $\boldsymbol{A}$ as you can see by carrying out the cross product with cylindrical components. Thus,

$$\begin{aligned}\boldsymbol{\omega} \times \boldsymbol{A} &= \omega\boldsymbol{\epsilon}_{Z'} \times \left(A_{\bar{r}}\boldsymbol{\epsilon}_{\bar{r}} + A_\theta\boldsymbol{\epsilon}_\theta + A_{Z'}\boldsymbol{\epsilon}_{Z'}\right)\\ &= \omega A_{\bar{r}}\boldsymbol{\epsilon}_\theta - \omega A_\theta\boldsymbol{\epsilon}_{\bar{r}}\end{aligned}$$

We conclude that

$$\left(\frac{dA}{dt}\right)_{X'Y'Z'} = \boldsymbol{\omega} \times \boldsymbol{A}$$

Since $X'Y'Z'$ is stationary relative to XYZ, we would observe the same time derivative from the latter reference as from the former reference. That is, $(d/dt)_{XYZ} = (d/dt)_{X'Y'Z'}$ and we can conclude that

$$\left(\frac{dA}{dt}\right)_{XYZ} = \boldsymbol{\omega} \times \boldsymbol{A} \tag{13.1}$$

The foregoing result gives the time rate of change of a vector $\boldsymbol{A}$ fixed in reference xyz moving arbitrarily relative to reference XYZ. From this result, we see that $(dA/dt)_{XYZ}$ depends only on the vectors $\boldsymbol{\omega}$ and $\boldsymbol{A}$ and not on their lines of action. Thus, we can conclude that the time rate of change of $\boldsymbol{A}$ fixed in xyz is not altered when:

1. The vector $\boldsymbol{A}$ is fixed at some other location in xyz provided the vector itself is not changed.
2. The actual axis of rotation of the xyz system is shifted to a new parallel position.

We can differentiate the terms in Eq. 13.1 a second time. We thus get

$$\left(\frac{d^2A}{dt^2}\right)_{XYZ} = \left(\frac{d\boldsymbol{\omega}}{dt}\right)_{XYZ} \times \boldsymbol{A} + \boldsymbol{\omega} \times \left(\frac{dA}{dt}\right)_{XYZ} \tag{13.2}$$

Using Eq. 13.1 to replace $(d\mathbf{A}/dt)_{XYZ}$ and using $\dot{\boldsymbol{\omega}}$ to replace $(d\boldsymbol{\omega}/dt)_{XYZ}$, since the reference being used for this derivative is clear,[2] we get

$$\left(\frac{d^2A}{dt^2}\right)_{XYZ} = \dot{\boldsymbol{\omega}} \times \boldsymbol{A} + \boldsymbol{\omega} \times (\boldsymbol{\omega} \times \boldsymbol{A}) \tag{13.3}$$

You can compute higher-order derivatives by continuing the process. We suggest that only Eq. 13.1 be remembered and that all subsequent higher-order derivatives be evaluated when needed.

In this discussion thus far, we have considered a vector $\boldsymbol{A}$ fixed in a reference xyz. But a reference xyz is a rigid system and can be considered a *rigid body*. Thus, the words "*fixed in a reference xyz*" in the previous discussion can be replaced by the words "*fixed in a rigid body*." The angular velocity $\boldsymbol{\omega}$ used in Eq. 13.1 is then the angular velocity of the rigid body in which $\boldsymbol{A}$ is fixed. We shall illustrate this condition in the following examples, which you are urged to study very carefully. An understanding of these examples is vital for attaining a good working grasp of rigid-body kinematics.

As an aid in carrying out computations involving the triple cross product, we wish to point out that the product

$$\omega_1 \boldsymbol{k} \times (\omega_1 \boldsymbol{k} \times C\boldsymbol{j}) = -\omega_1^2 C\boldsymbol{j}$$

That is, the product is minus the product of the scalars and has a direction corresponding to the last unit vector, $\boldsymbol{j}$. Remembering this will greatly facilitate our computations.[3]

Additionally, consider a situation where the angular velocity of body A relative to body B is given as $\boldsymbol{\omega}_1$, while the angular velocity of body B relative to the ground is $\boldsymbol{\omega}_2$. What is the *total* angular velocity $\boldsymbol{\omega}_T$ of body A relative to the ground? In such a case, we must remember that the angular velocity $\boldsymbol{\omega}_1$ of body A *relative* to body B is actually the *difference* between the total angular velocity $\boldsymbol{\omega}_T$ of body A as seen from the ground and the angular velocity $\boldsymbol{\omega}_2$ of body B as seen from the ground. Thus,

$$\boldsymbol{\omega}_1 = \boldsymbol{\omega}_T - \boldsymbol{\omega}_2$$

Solving for $\boldsymbol{\omega}_T$, we get

$$\boldsymbol{\omega}_T = \boldsymbol{\omega}_1 + \boldsymbol{\omega}_2$$

We see from above that to get the total angular velocity $\boldsymbol{\omega}_T$, we simply add the various relative angular velocities just as we would with any pair of vectors.

[2]When it is clear from the discussion what reference is involved for a time derivative, we shall use the dot to indicate a time derivative.

[3]Of course, if the $\boldsymbol{j}$ vector were a $\boldsymbol{k}$ vector, then clearly we would arrive at a null value for the triple vector product.

Example 13.1

A disc C is mounted on a shaft AB in Fig. 13.9. The shaft and disc rotate with a constant angular speed ω_2 of 10 rad/sec relative to the platform to which bearings A and B are attached. Meanwhile, the platform rotates at a constant angular speed ω_1 of 5 rad/sec relative to the ground in a direction parallel to the Z axis of the ground reference XYZ. What is the angular velocity vector $\boldsymbol{\omega}$ for the disc C relative to XYZ? What are $(d\boldsymbol{\omega}/dt)_{XYZ}$ and $(d^2\boldsymbol{\omega}/dt^2)_{XYZ}$?

The total angular velocity $\boldsymbol{\omega}$ of the disc relative to the ground is easily given at all times as follows:

$$\boldsymbol{\omega} = \boldsymbol{\omega}_1 + \boldsymbol{\omega}_2 \text{ rad/sec} \tag{a}$$

At the instant of interest as depicted by Fig. 13.9, we have for $\boldsymbol{\omega}$:

$$\boldsymbol{\omega} = 5\boldsymbol{k} + 10\boldsymbol{j} \text{ rad/sec}$$

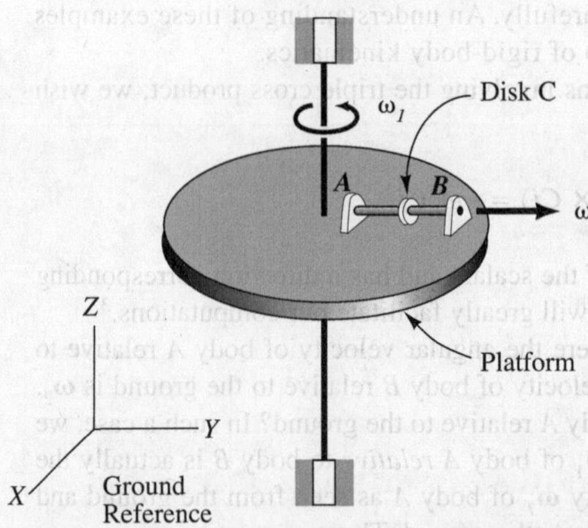

Figure 13.9. Rotating disc on rotating platform.

To get the first time derivative of $\boldsymbol{\omega}$, we go back to Eq. (a), which is always valid and hence can be differentiated with respect to time. Using a dot to represent the time derivative as seen from XYZ, we have

$$\dot{\boldsymbol{\omega}} = \dot{\boldsymbol{\omega}}_1 + \dot{\boldsymbol{\omega}}_2 \tag{b}$$

Consider now the vector $\boldsymbol{\omega}_2$. Note that this vector is constrained in direction to be always collinear with the axis AB of the bearings of the shaft. This clearly is a physical requirement. Also, since ω_2 is of constant value, we may think of the vector $\boldsymbol{\omega}_2$ as *fixed* to the platform along AB. Therefore, since the platform has an angular velocity of $\boldsymbol{\omega}_1$ relative to XYZ, we can say:

$$\dot{\boldsymbol{\omega}}_2 = \boldsymbol{\omega}_1 \times \boldsymbol{\omega}_2 \tag{c}$$

Example 13.1 (Continued)

As for $\dot{\boldsymbol{\omega}}_1$, namely the other vector in Eq. (b), we note that as seen from *XYZ*, $\boldsymbol{\omega}_1$ is a constant vector and so at all times $\dot{\boldsymbol{\omega}}_1 = \mathbf{0}$. Hence Eq. (b) can be written as follows:

$$\dot{\boldsymbol{\omega}} = \boldsymbol{\omega}_1 \times \boldsymbol{\omega}_2 \tag{d}$$

This equation is valid at all times and so can be differentiated again. At the instant of interest as depicted by Fig. 13.9, we have for $\dot{\boldsymbol{\omega}}$:

$$\dot{\boldsymbol{\omega}} = 5\boldsymbol{k} \times 10\boldsymbol{j} = -50\boldsymbol{i} \text{ rad/sec}^2 \tag{e}$$

To get $\ddot{\boldsymbol{\omega}}$, we now differentiate (d) with respect to time. We have

$$\begin{aligned} \ddot{\boldsymbol{\omega}} &= \dot{\boldsymbol{\omega}}_1 \times \boldsymbol{\omega}_2 + \boldsymbol{\omega}_1 \times \dot{\boldsymbol{\omega}}_2 \\ &= \mathbf{0} + \boldsymbol{\omega}_1 \times (\boldsymbol{\omega}_1 \times \boldsymbol{\omega}_2) \end{aligned} \tag{f}$$

where we have used the fact that $\dot{\boldsymbol{\omega}}_1 = \mathbf{0}$ at all times as well as Eq. (c) for $\dot{\boldsymbol{\omega}}_2$. At the instant of interest, we have

$$\ddot{\boldsymbol{\omega}} = 5\boldsymbol{k} \times (5\boldsymbol{k} \times 10\boldsymbol{j}) = -250\boldsymbol{j} \text{ rad/sec}^3$$

Example 13.2

In Example 13.1, consider a position vector $\boldsymbol{\rho}$ between two points on the rotating disc (see Fig. 13.10). The length of $\boldsymbol{\rho}$ is 100 mm and, at the instant of interest, is in the vertical direction. What are the first and second time derivatives of $\boldsymbol{\rho}$ at this instant as seen from the ground reference?

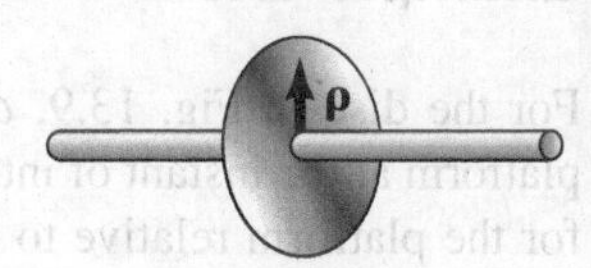

Figure 13.10. Displacement vector ρ in disc.

It should be obvious that the vector $\boldsymbol{\rho}$ is fixed to the disc which has at all times an angular velocity relative to *XYZ* equal to $\boldsymbol{\omega}_1 + \boldsymbol{\omega}_2$. Hence, at all times we can say:

$$\dot{\boldsymbol{\rho}} = (\boldsymbol{\omega}_1 + \boldsymbol{\omega}_2) \times \boldsymbol{\rho} \tag{a}$$

At the instant of interest, we have noting that $\boldsymbol{\rho} = 100\boldsymbol{k}$

$$\dot{\boldsymbol{\rho}} = (5\boldsymbol{k} + 10\boldsymbol{j}) \times 100\boldsymbol{k} = 1{,}000\boldsymbol{i} \text{ mm/sec} \tag{b}$$

To get the second derivative of $\boldsymbol{\rho}$, go back to Eq. (a) and differentiate:

$$\ddot{\boldsymbol{\rho}} = (\dot{\boldsymbol{\omega}}_1 + \dot{\boldsymbol{\omega}}_2) \times \boldsymbol{\rho} + (\boldsymbol{\omega}_1 + \boldsymbol{\omega}_2) \times \dot{\boldsymbol{\rho}}$$

Example 13.2 (Continued)

Noting that $\dot{\boldsymbol{\omega}}_1 = \mathbf{0}$ at all times and, as discussed in Example 13.1, that $\boldsymbol{\omega}_2$ is fixed in the platform, we can say:

$$\ddot{\boldsymbol{\rho}} = (\mathbf{0} + \boldsymbol{\omega}_1 \times \boldsymbol{\omega}_2) \times \boldsymbol{\rho} + (\boldsymbol{\omega}_1 + \boldsymbol{\omega}_2) \times \dot{\boldsymbol{\rho}} \qquad \text{(c)}$$

At the instant of interest we have, on noting Eq. (b):

$$\ddot{\boldsymbol{\rho}} = (5\boldsymbol{k} \times 10\boldsymbol{j}) \times 100\boldsymbol{k} + (5\boldsymbol{k} + 10\boldsymbol{j}) \times 1{,}000\boldsymbol{i} \text{ mm/sec}^2$$

$$\ddot{\boldsymbol{\rho}} = 10\boldsymbol{j} - 10\boldsymbol{k} \text{ m/sec}^2$$

Although we shall later formally examine the case of the time derivative of vector $\boldsymbol{A}$ as seen from XYZ when $\boldsymbol{A}$ is *not fixed* in a body or a reference xyz, we can handle such cases less formally with what we already know. We illustrate this in the following example.

Example 13.3

For the disc in Fig. 13.9, $\omega_2 = 6$ rad/sec and $\dot{\omega}_2 = 2$ rad/sec^2, both relative to the platform at the instant of interest. At this instant, $\omega_1 = 2$ rad/sec and $\dot{\omega}_1 = -3$ rad/sec^2 for the platform relative to the ground. Find the angular acceleration vector $\dot{\boldsymbol{\omega}}$ for the disc relative to the ground at the instant of interest.

The angular velocity of the disc relative to the ground at all times is

$$\boldsymbol{\omega} = \boldsymbol{\omega}_1 + \boldsymbol{\omega}_2 \qquad \text{(a)}$$

For $\dot{\boldsymbol{\omega}}$, we can then say

$$\dot{\boldsymbol{\omega}} = \dot{\boldsymbol{\omega}}_1 + \dot{\boldsymbol{\omega}}_2 \qquad \text{(b)}$$

It is apparent on inspecting Fig. 13.11 that at all times $\boldsymbol{\omega}_1$ is vertical, and so we can say:

$$\dot{\boldsymbol{\omega}}_1 = \frac{d}{dt}_{XYZ}(\omega_1 \boldsymbol{k}) = \dot{\omega}_1 \boldsymbol{k} \qquad \text{(c)}$$

Example 13.3 (Continued)

However, $\boldsymbol{\omega}_2$ is changing direction and, most importantly, is changing magnitude. Because of the latter, $\boldsymbol{\omega}_2$ cannot be considered fixed in a reference or a rigid body for purposes of computing $\dot{\boldsymbol{\omega}}_2$. To get around this difficulty, we fix a unit vector $\boldsymbol{j}'$ *onto the platform* to be collinear with the centerline of the shaft *AB* as shown in Fig. 13.11. We know the angular velocity of this unit vector; it is $\boldsymbol{\omega}_1$ at all times. We can then express $\boldsymbol{\omega}_2$ in the following manner, which is valid at all times:

$$\boldsymbol{\omega}_2 = \omega_2 \boldsymbol{j}' \qquad \text{(d)}$$

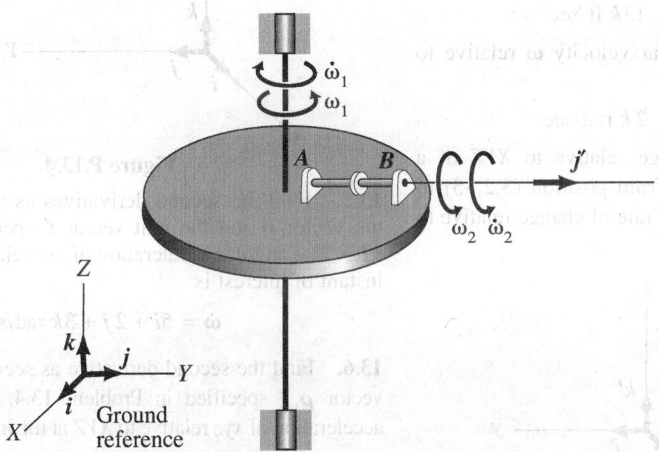

Figure 13.11. Unit vector $\boldsymbol{j}'$ fixed to platform.

We can differentiate the above with respect to time as follows:

$$\dot{\boldsymbol{\omega}}_2 = \dot{\omega}_2 \boldsymbol{j}' + \omega_2 \dot{\boldsymbol{j}}'$$

But $\boldsymbol{j}'$ is *fixed* to the platform which has angular velocity $\boldsymbol{\omega}_1$ relative to *XYZ* at all times. Hence, we have for the above,

$$\dot{\boldsymbol{\omega}}_2 = \dot{\omega}_2 \boldsymbol{j}' + \omega_2(\boldsymbol{\omega}_1 \times \boldsymbol{j}') \qquad \text{(e)}$$

Thus, Eq. (b) then can be given as

$$\dot{\boldsymbol{\omega}} = \dot{\omega}_1 \boldsymbol{k} + \dot{\omega}_2 \boldsymbol{j}' + \omega_2(\boldsymbol{\omega}_1 \times \boldsymbol{j}')$$

This expression is valid at all times and could be differentiated again. At the instant of interest, we can say, noting that $\boldsymbol{j}' = \boldsymbol{j}$ at this instant,

$$\dot{\boldsymbol{\omega}} = -3\boldsymbol{k} + 2\boldsymbol{j} + 6(2\boldsymbol{k} \times \boldsymbol{j})$$

$$\dot{\boldsymbol{\omega}} = -12\boldsymbol{i} + 2\boldsymbol{j} - 3\boldsymbol{k} \text{ rad/sec}^2$$

PROBLEMS

13.1. Is the motion of the cabin of a ferris wheel rotational or translational if the wheel moves at uniform speed and the occupants cause no disturbances? Why?

13.2. A cylinder rolls without slipping down an inclined surface. What is the actual axis of rotation at any instant? Why? How is this axis moving?

13.3. A reference xyz is moving such that the origin O has at time t a velocity relative to reference XYZ given as

$$\boldsymbol{V}_0 = 6\boldsymbol{i} + 12\boldsymbol{j} + 13\boldsymbol{k} \text{ ft/sec}$$

The xyz reference has an angular velocity $\boldsymbol{\omega}$ relative to XYZ at time t given as

$$\boldsymbol{\omega} = 10\boldsymbol{i} + 12\boldsymbol{j} + 2\boldsymbol{k} \text{ rad/sec}$$

What is the time rate of change relative to XYZ of a directed line segment $\boldsymbol{\rho}$ going from position (3,2,–5) to (–2,4,6) in xyz? What is the time rate of change relative to XYZ of position vectors $\boldsymbol{i}'$ and $\boldsymbol{k}'$?

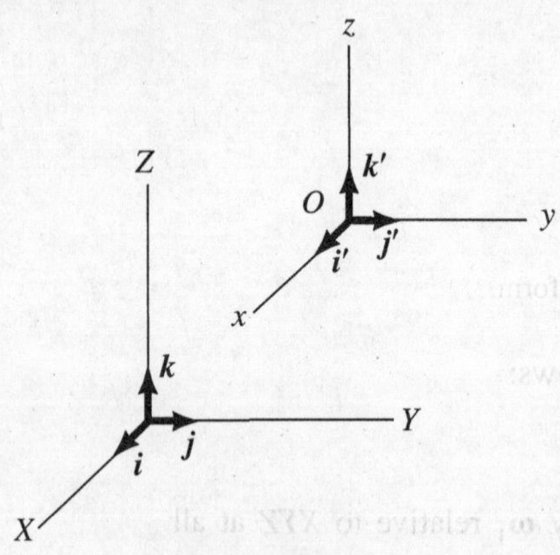

Figure P.13.3.

13.4. A reference xyz is moving relative to XYZ with a velocity of the origin given at time t as

$$\boldsymbol{V}_0 = 6\boldsymbol{i} + 4\boldsymbol{j} + 6\boldsymbol{k} \text{ m/sec}$$

The angular velocity of reference xyz relative to XYZ is

$$\boldsymbol{\omega} = 3\boldsymbol{i} + 14\boldsymbol{j} + 2\boldsymbol{k} \text{ rad/sec}$$

What is the time rate of change as seen from XYZ of a directed line segment $\boldsymbol{\rho}_{1,2}$ in xyz going from position 1 to position 2 where the position vectors in xyz for these points are, respectively,

$$\boldsymbol{\rho}_1 = 2\boldsymbol{i}' + 3\boldsymbol{j}' \text{ m}$$

$$\boldsymbol{\rho}_2 = 3\boldsymbol{i}' - 4\boldsymbol{j}' + 2\boldsymbol{k}' \text{ m}$$

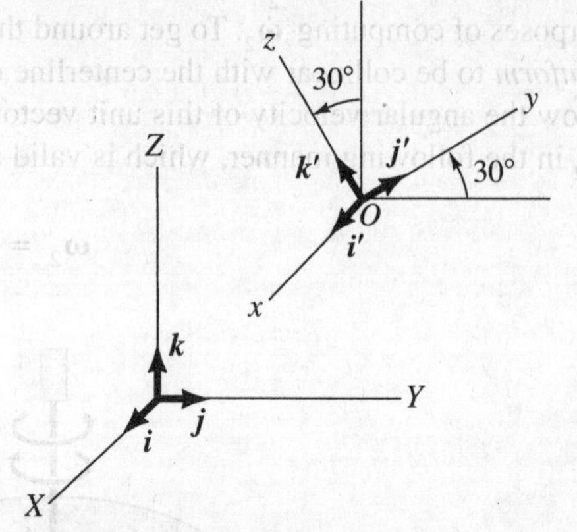

Figure P.13.4.

13.5. Find the second derivatives as seen from XYZ of the vector $\boldsymbol{\rho}$ and the unit vector $\boldsymbol{i}'$ specified in Problem 13.3. The angular acceleration of xyz relative to XYZ at the instant of interest is

$$\dot{\boldsymbol{\omega}} = 5\boldsymbol{i} + 2\boldsymbol{j} + 3\boldsymbol{k} \text{ rad/sec}^2$$

13.6. Find the second derivative as seen from XYZ of the vector $\boldsymbol{\rho}_{1,2}$ specified in Problem 13.4. Take the angular acceleration of xyz relative to XYZ at the instant of interest as

$$\dot{\boldsymbol{\omega}} = 15\boldsymbol{i} - 2\boldsymbol{k} \text{ rad/sec}^2$$

13.7. A platform is rotating with a constant speed ω_1 of 10 rad/sec relative to the ground. A shaft is mounted on the platform and rotates relative to the platform at a speed ω_2 of 5 rad/sec. What is the angular velocity of the shaft relative to the ground? What are the first and second time derivatives of the angular velocity of the shaft relative to the ground?

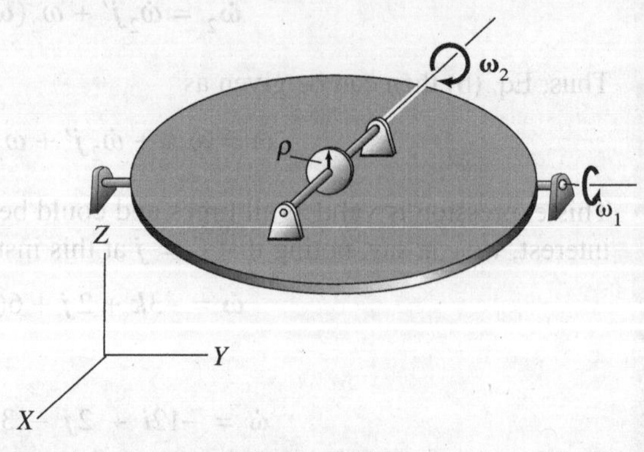

Figure P.13.7.

13.8. In Problem 13.7, what are the first and second time derivatives of a directed line segment $\boldsymbol{\rho}$ in the disc at the instant that the system has the geometry shown? The vector $\boldsymbol{\rho}$ is of length 10 mm.

13.9. A tank is maneuvering its gun into position. At the instant of interest, the turret A is rotating at an angular speed $\dot{\theta}$ of 2 rad/sec relative to the tank and is in position $\theta = 20°$. Also, at this instant, the gun is rotating at an angular speed $\dot{\phi}$ of 1 rad/sec relative to the turret and forms an angle $\phi = 30°$ with the horizontal plane. What are $\boldsymbol{\omega}$, $\dot{\boldsymbol{\omega}}$, and $\ddot{\boldsymbol{\omega}}$ of the gun relative to the ground?

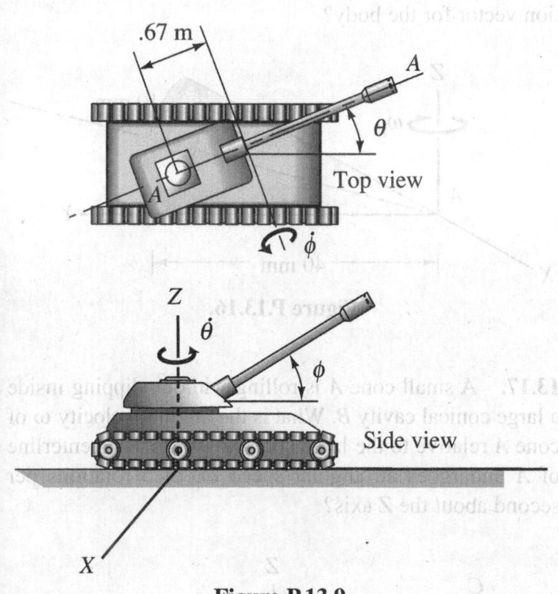

Figure P.13.9.

13.10. In Problem 13.9, determine $\boldsymbol{\omega}$ and $\dot{\boldsymbol{\omega}}$ assuming that the tank is also rotating about the vertical axis at a rate of .2 rad/sec relative to the ground in a clockwise direction as viewed from above.

13.11. A particle is made to move at constant speed V equal to 10 m/sec along a straight groove on a plate B. The plate rotates at a constant angular speed ω_2 equal to 3 rad/sec relative to a platform C while the platform rotates with a constant angular speed ω_1 of 5 rad/sec relative to the ground reference XYZ. Find the first and second derivatives of V as seen from the ground reference.

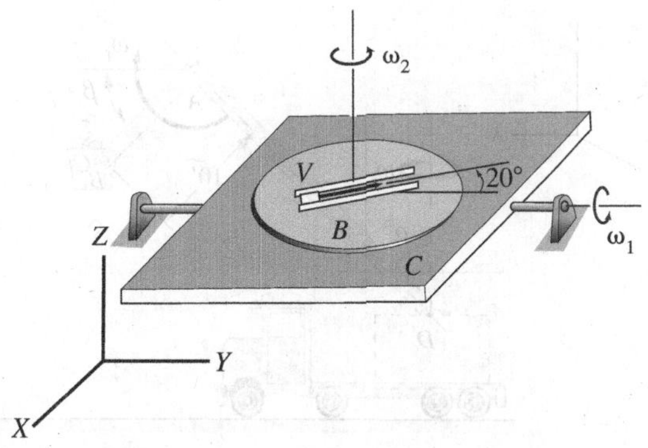

Figure P.13.11.

13.12. A jet fighter plane has just taken off and is retracting its landing gear. At the end of its run on the ground, the plane is moving at a speed of 200 km/hr. If the diameter of the tires is 460 mm and if we neglect the loss of angular speed of the wheels due to wind friction after the plane is in the air, what is the angular speed $\boldsymbol{\omega}$ and the angular acceleration $\dot{\boldsymbol{\omega}}$ of the left wheel (under the wing) at the instant shown in the diagram? Take $\omega_2 = .4$ rad/sec and $\dot{\omega}_2$ is .2 rad/sec² at the instant of interest.

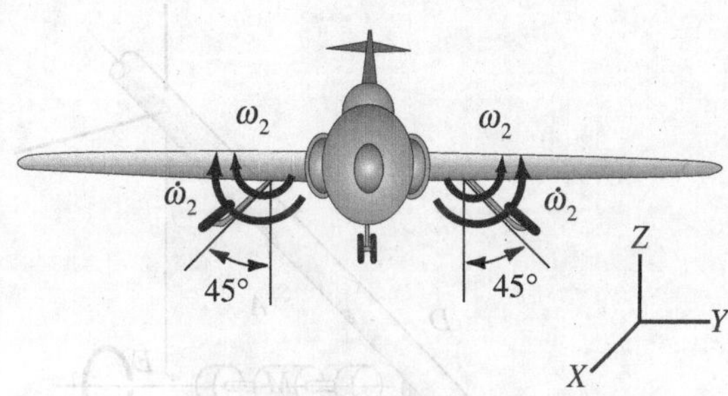

Figure P.13.12.

13.13. A truck is carrying a cockpit for a worker who repairs overhead road fixtures. At the instant shown in the diagram, the base D is rotating with constant speed ω_2 of 1 rad/sec relative to the truck. Arm AB is rotating at constant angular speed ω_1 of 2 rad/sec relative to DA. Cockpit C is rotating relative to AB so as to always keep the man upright. What are $\boldsymbol{\omega}$, $\dot{\boldsymbol{\omega}}$, and $\ddot{\boldsymbol{\omega}}$ of arm AB relative to the ground at the instant of interest? The truck is stationary.

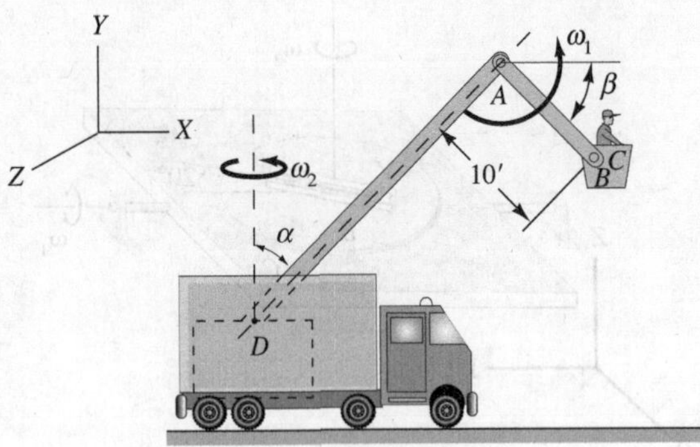

Figure P.13.13.

13.14. An electric motor M is mounted on a plate A which is welded to a shaft D. The motor has a constant angular speed ω_2 relative to plate A of 1,750 rpm. Plate A at the instant of interest is in a vertical position as shown and is rotating with an angular speed ω_1 equal to 100 rpm and a rate of change of angular speed $\dot{\omega}_1$ equal to 30 rpm/sec—all relative to the ground. The normal projection of the centerline of the motor shaft onto the plate A is at an angle of 45° with the edge of the plate FE. Compute the first and second time derivatives of $\boldsymbol{\omega}$, the angular velocity of the motor, as seen from the ground.

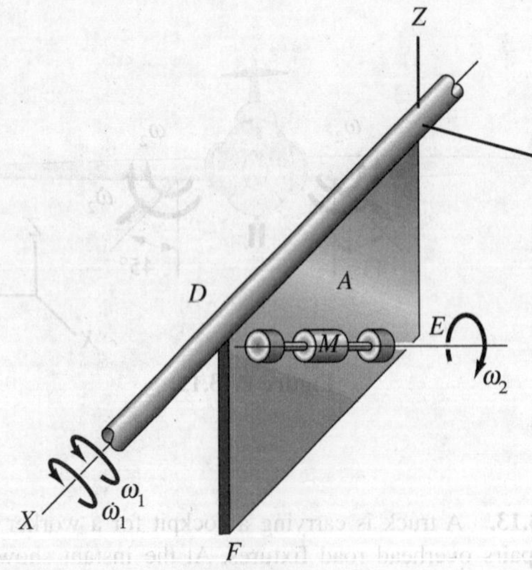

Figure P.13.14.

13.15. A racing car is moving at a constant speed of 200 mi/hr when the driver turns his front wheels at an increasing rate, $\dot{\omega}_1$, of .02 rad/sec^2. If ω_1 = .0168 rad/sec at the instant of interest, what are $\boldsymbol{\omega}$ and $\dot{\boldsymbol{\omega}}$ of the front wheels at this instant? The diameter of the tires is 30 in.

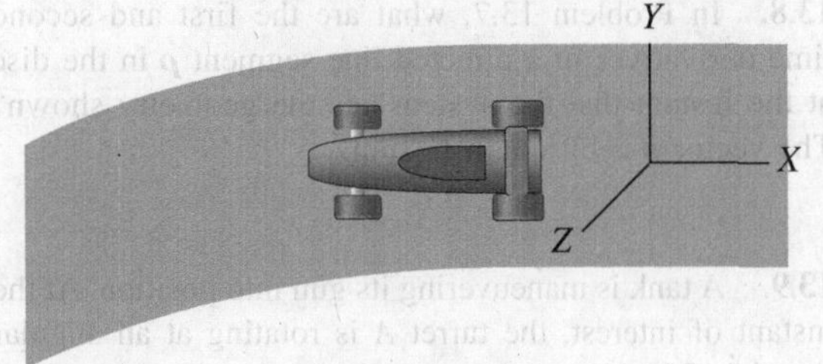

Figure P.13.13.

13.16. A cone is rolling without slipping such that its centerline rotates at the rate ω_1 of 5 revolutions per second about the Z axis. What is the angular velocity $\boldsymbol{\omega}$ of the body relative to the ground? What is the angular acceleration vector for the body?

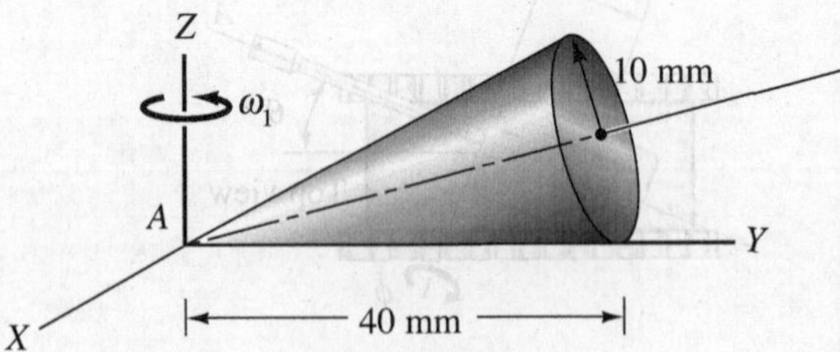

Figure P.13.16.

13.17. A small cone A is rolling without slipping inside a large conical cavity B. What is the angular velocity $\boldsymbol{\omega}$ of cone A relative to the large cone cavity B if the centerline of A undergoes an angular speed ω_1 of 5 rotations per second about the Z axis?

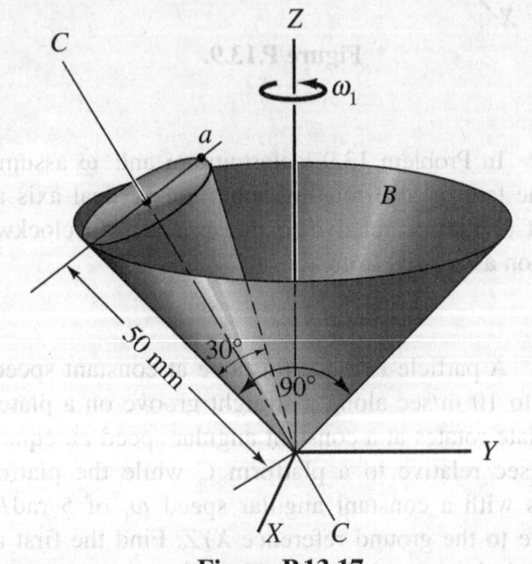

Figure P.13.17.

13.18. An amusement park ride consists of a stationary vertical tower with arms that can swing outward from the tower and at the same time can rotate about the tower. At the ends of the arms, cockpits containing passengers can rotate relative to the arms. Consider the case where cockpit A rotates at angular speed ω_2 relative to arm BC, which rotates at angular speed ω_1 relative to the tower. If θ is fixed at 90°, what are the total angular velocity and the angular acceleration of the cockpit relative to the ground? Use $\omega_1 = .2$ rad/sec and $\omega_2 = .6$ rad/sec.

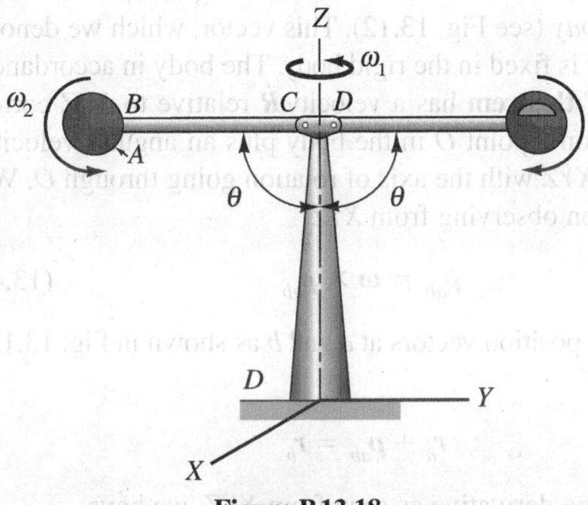

Figure P.13.18.

13.19. In Problem 13.18, find $\dot{\boldsymbol{\omega}}$ of the cockpit for the case where $\dot{\theta} = \omega_3 = .8$ rad/sec at the instant that $\theta = 90°$.

13.20. Mass A is connected to an inextensible wire. Supports C and D are moving as shown.

(a) What is the velocity vector of mass A?

(b) If cylinder G is free to rotate and there is no slipping, what is its angular velocity?

The following data apply:

$h = 2$ m	$(V_x)_2 = .24$ m/s
$L = 3$ m	$(V_y)_2 = .21$ m/s
$l = 2$ m	$R = 1$ m
$(V_x)_1 = .5$ m/s	$\alpha = 45°$
$(V_y)_1 = .6$ m/s	

The last four problems of this set are designed for those students who have studied Example 13.3.

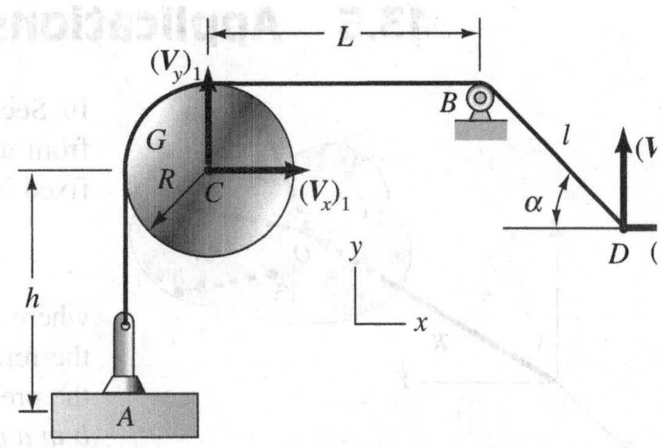

Figure P.13.20.

13.21. In Problem 13.18, find $\dot{\boldsymbol{\omega}}$ of the cockpit A for the case where $\dot{\omega}_1 = .2$ rad/sec² and $\dot{\omega}_2 = .3$ rad/sec².

13.22. In Problem 13.13, find $\dot{\boldsymbol{\omega}}$ of beam AB relative to the ground if at the instant shown the following data apply:

$$\omega_1 = .3 \text{ rad/sec}$$
$$\dot{\omega}_1 = .2 \text{ rad/sec}^2$$
$$\omega_2 = .6 \text{ rad/sec}$$
$$\dot{\omega}_2 = -.1 \text{ rad/sec}^2$$

13.23. In Problem 13.9, find the angular acceleration vector $\dot{\boldsymbol{\omega}}$ for the gun barrel, if, for the instant shown in the diagram, the following data apply:

$\dot{\phi} = .30$ rad/sec,	$\theta = 20°$
$\ddot{\phi} = .26$ rad/sec²,	$\phi = 30°$
$\dot{\theta} = .17$ rad/sec	
$\ddot{\theta} = -.34$ rad/sec²	

13.24. In Problem 13.11, find $\dot{\mathbf{V}}$ if at the instant shown in the diagram:

$$\omega_1 = 5 \text{ rad/sec}$$
$$\dot{\omega}_1 = 10 \text{ rad/sec}^2$$
$$\omega_2 = 2 \text{ rad/sec}$$
$$\dot{\omega}_2 = 3 \text{ rad/sec}^2$$
$$V = 10 \text{ m/sec}$$
$$\dot{V} = 5 \text{ m/sec}^2$$

13.5 Applications of the Fixed-Vector Concept

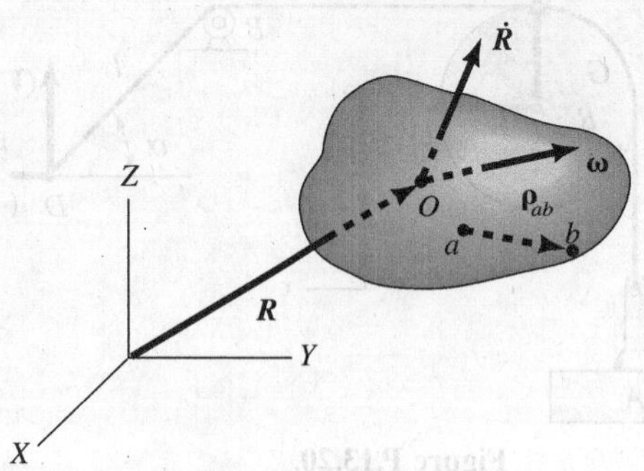

Figure 13.12. ρ_{ab} fixed in rigid body.

In Section 13.4, we considered the time derivative, as seen from a reference *XYZ*, of a vector *A* fixed in a rigid body or fixed in reference *xyz*. The result was a simple formula:

$$\dot{A} = \omega \times A$$

where ω is the angular velocity relative to *XYZ* of the body or the reference in which *A* is fixed. In this section, we shall use the preceding formula for a *vector connecting two points a and b in a rigid body* (see Fig. 13.12). This vector, which we denote as ρ_{ab}, clearly is fixed in the rigid body. The body in accordance with **Chasles' theorem** has a velocity $\dot{R}$ relative to *XYZ* corresponding to some point *O* in the body plus an angular velocity ω relative to *XYZ* with the axis of rotation going through *O*. We can then say on observing from *XYZ*:

$$\dot{\rho}_{ab} = \omega \times \rho_{ab} \tag{13.4}$$

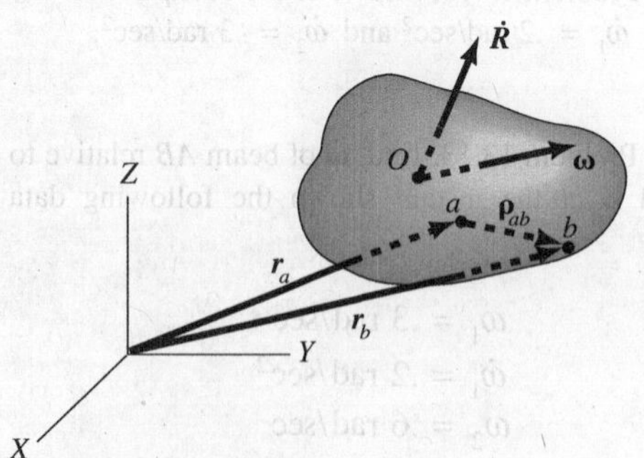

Figure 13.13. Insert position vectors.

Now consider position vectors at *a* and *b* as shown in Fig. 13.13. We can say:

$$r_a + \rho_{ab} = r_b$$

Taking the time derivative as seen from *XYZ*, we have

$$\left(\frac{dr_a}{dt}\right)_{XYZ} + \left(\frac{d\rho_{ab}}{dt}\right)_{XYZ} = \left(\frac{dr_b}{dt}\right)_{XYZ}$$

This equation can be written as

$$\left(\frac{d\rho_{ab}}{dt}\right)_{XYZ} = V_b - V_a \tag{13.5}$$

Since $(d\rho_{ab}/dt)_{XYZ}$ is the *difference* between the velocity of point *b* and that at point *a* as noted above, we can say that $(d\rho_{ab}/dt)_{XYZ}$ is the velocity of point *b relative* to point *a*.[4] Next, using Eq. 13.4 to replace $(d\rho_{ab}/dt)_{XYZ}$, we have, on rearranging terms, a very useful equation:

$$V_b = V_a + \omega \times \rho_{ab} \tag{13.6}$$

In using the foregoing equation, we must be sure that we get the sequence of subscripts correct on ρ since a change in ordering brings about a change in sign (i.e., $\rho_{ab} = -\rho_{ba}$). This equation is a statement of the physically obvious result that *the velocity of particle b of a rigid body as seen from XYZ equals the velocity of any other particle a of this body as seen from XYZ plus the velocity of particle b relative to particle a.*

[4]That is, $(d\rho_{ab}/dt)_{XYZ}$ is the velocity of *b* as seen by an observer translating relative to *XYZ* with point *a*, i.e., as seen by a nonrotating observer moving with *a*.

Differentiating Eq. 13.6 again, we can get a relation involving the acceleration vectors of two points on a rigid body:

$$\boldsymbol{a}_b = \boldsymbol{a}_a + \left(\frac{d\boldsymbol{\omega}}{dt}\right)_{XYZ} \times \boldsymbol{\rho}_{ab} + \boldsymbol{\omega} \times \left(\frac{d\boldsymbol{\rho}_{ab}}{dt}\right)_{XYZ}$$

Hence, we have on using Eq. 13.4 in the last expression

$$\boldsymbol{a}_b = \boldsymbol{a}_a + \dot{\boldsymbol{\omega}} \times \boldsymbol{\rho}_{ab} + \boldsymbol{\omega} \times (\boldsymbol{\omega} \times \boldsymbol{\rho}_{ab}) \tag{13.7}$$

We have thus formulated relations between *the motions of two points of a rigid body as seen from a single reference.* Such relations can be very useful in the study of machine elements.

Before going to the examples, let us now consider the case of a circular cylinder rolling *without slipping* (see Fig. 13.14). The point of contact A of the cylinder with the ground has instantaneously *zero velocity* and hence we have pure instantaneous rotation at any time t about an instantaneous axis of rotation at the line of contact. The velocity of any point B of the cylinder can then easily be found by using Eq. 13.6 for points B and A. Thus:

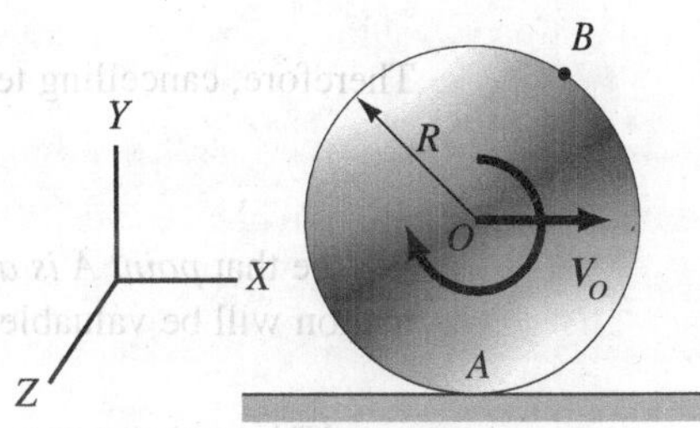

Figure 13.14. Cylinder rolling without slipping on a flat surface.

$$\boldsymbol{V}_B = \boldsymbol{V}_A + \boldsymbol{\omega} \times \boldsymbol{\rho}_{AB}$$

Therefore,

$$\boldsymbol{V}_B = \boldsymbol{0} + \boldsymbol{\omega} \times \boldsymbol{\rho}_{AB} = \boldsymbol{\omega} \times \boldsymbol{\rho}_{AB}$$

From the above equation it is clear that for computing the velocity of any point on the cylinder we can think of the cylinder as *hinged* at the point of contact. In particular for point O, the center of the cylinder, we get from above:

$$\boldsymbol{V}_0 = -\omega R\boldsymbol{i}$$

If the velocity V_0 is known, clearly the angular velocity has a magnitude of V_0/R.

Another way of relating V and ω is to realize that the distance s that O moves must equal the length of circumference coming into contact with the ground. That is, measuring θ from the X axis to the Y axis:

$$s = -R\theta$$

Differentiating we get:

$$V_0 = -R\dot{\theta} = -R\omega$$

thus reproducing the previous result. Differentiating again, we get

$$a_0 = -R\ddot{\theta} = -R\alpha \tag{13.8}$$

relating now the acceleration of O and the angular acceleration α. Clearly, the acceleration vector for O must be parallel to the ground. Again, for computing $\boldsymbol{a}_0$, we have a simple situation.

Next, let us determine the acceleration vector for the *point of contact A* of the cylinder. Thus, we can say for points *A* and *O*:

$$\boldsymbol{a}_O = \boldsymbol{a}_A + \dot{\boldsymbol{\omega}} \times \boldsymbol{\rho}_{AO} + \boldsymbol{\omega} \times (\boldsymbol{\omega} \times \boldsymbol{\rho}_{AO})$$

Therefore,

$$-R\ddot{\theta}\boldsymbol{i} = \boldsymbol{a}_A + \ddot{\theta}\boldsymbol{k} \times R\boldsymbol{j} + \dot{\theta}\boldsymbol{k} \times (\dot{\theta}\boldsymbol{k} \times R\boldsymbol{j}) \tag{13.9}$$

Carrying out the products:

$$-R\ddot{\theta}\boldsymbol{i} = \boldsymbol{a}_A - R\ddot{\theta}\boldsymbol{i} - R\dot{\theta}^2\boldsymbol{j}$$

Therefore, cancelling terms, we get

$$\boldsymbol{a}_A = R\dot{\theta}^2\boldsymbol{j} \tag{13.10}$$

We see that *point A is accelerating upward toward the center of the cylinder.*[5] This information will be valuable for us in Chapter 16 when we study rigid-body dynamics.

[5]This conclusion must apply also to a sphere rolling without slipping on a flat surface. As for acceleration of other points of the cylinder, we do not have a simple formula but must insert data for these points into the acceleration formula valid for two points of a rigid body.

Example 13.4

Wheel *D* rotates at an angular speed ω_1 of 2 rad/sec counterclockwise in Fig. 13.15. Find the angular speed ω_E of gear *E* relative to the ground at the instant shown in the diagram.

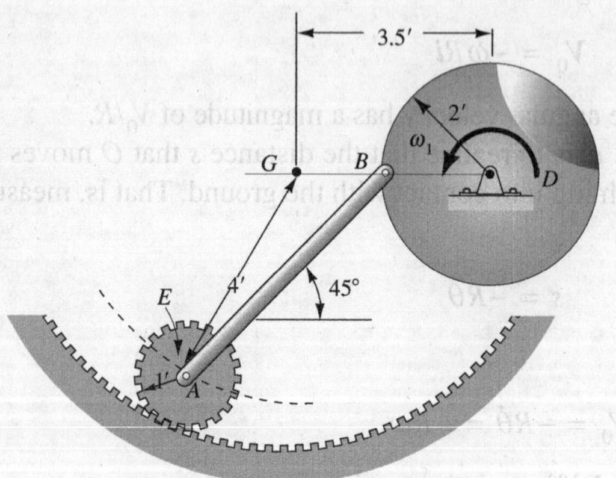

Figure 13.15. Two-dimensional device.

We have information about two points of one of the rigid bodies, namely *AB*, of the device. At *B*, the velocity must be downward with the value of $(\omega_1)(r_D) = 4$ ft/sec

Example 13.4 (Continued)

as shown in Fig. 13.16. Furthermore, since point A must travel a circular path of radius GA we know that A has velocity V_A with a direction at right angles to GA. Accordingly, since the angle between GA and the vertical is $(90° - 45° - \alpha) = (45° - \alpha)$ as can readily be seen on inspecting Fig. 13.16, then the angle between V_A and the horizontal must also be $(45° - \alpha)$ because of the *mutual perpendicularity* of the sides of these angles. If we can determine velocity V_A, we can get the desired angular speed of gear A immediately.

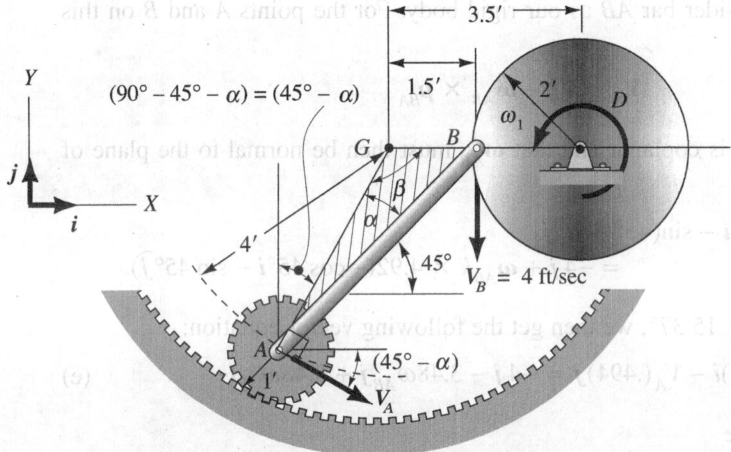

Figure 13.16. Velocity vectors for two points of a rigid body shown.

Before examining rigid body AB, we have some geometrical steps to take. Considering triangle GAB in Fig. 13.16, we can first solve for α using the law of sines as follows:

$$\frac{GA}{\sin(\measuredangle GBA)} = \frac{GB}{\sin\alpha}$$

Therefore, since $\measuredangle GBA = 45°$

$$\frac{4}{\sin 45°} = \frac{1.5}{\sin\alpha} \qquad \text{(a)}$$

Solving for α, we get

$$\alpha = 15.37° \qquad \text{(b)}$$

The angle β is then easily evaluated considering the angles in the triangle GBA. Thus,

$$\beta = 180° - \alpha - \measuredangle GBA$$
$$= 180° - 15.37° - 45° = 119.6° \qquad \text{(c)}$$

Example 13.4 (Continued)

Finally, we can determine AB of the triangle, again using the law of sines. Thus,

$$\frac{AB}{\sin\beta} = \frac{GA}{\sin 45°}$$

$$\frac{AB}{\sin 119.6°} = \frac{4}{.707}$$

Solving for AB, we get

$$AB = 4.92 \text{ ft} \tag{d}$$

We now can consider bar AB as our rigid body. For the points A and B on this body, we can say:

$$\boldsymbol{V}_A = \boldsymbol{V}_B + \boldsymbol{\omega}_{AB} \times \boldsymbol{\rho}_{BA}$$

Noting that the motion is coplanar and that $\boldsymbol{\omega}_{AB}$ must then be normal to the plane of motion, we have[6]

$$V_A[\cos(45° - \alpha)\boldsymbol{i} - \sin(45° - \alpha)\boldsymbol{j}]$$
$$= -4\boldsymbol{j} + \omega_{AB}\boldsymbol{k} \times 4.92(-\cos 45°\boldsymbol{i} - \sin 45°\boldsymbol{j})$$

Inserting the value $\alpha = 15.37°$, we then get the following vector equation:

$$V_A(.869)\boldsymbol{i} - V_A(.494)\boldsymbol{j} = -4\boldsymbol{j} - 3.48\omega_{AB}\boldsymbol{j} + 3.48\omega_{AB}\boldsymbol{i} \tag{e}$$

The scalar equations are

$$.869V_A = 3.48\omega_{AB}$$
$$-.494V_A = -4 - 3.48\omega_{AB} \tag{f}$$

Solving, we get[7]

$$V_A = -10.66 \text{ ft/sec}$$
$$\omega_{AB} = -2.66 \text{ rad/sec} \tag{g}$$

Thus, point A moves in a direction *opposite* to that shown in Fig. 13.16. We now can readily evaluate ω_E, which clearly must have a value of

$$\omega_E = \frac{V_A}{r_E} = \frac{10.66}{1} = \boxed{10.66 \text{ rad/sec}}$$

in the counterclockwise direction.

[6]Our practice will be to consider unknown angular velocities as *positive*. The sign for the unknown angular velocity coming out of the computations will then correspond to the *actual convention* sign for the angular velocity.

[7]By having assumed ω_{AB} as positive and thus *counterclockwise* for the reference xy employed, we conclude from the presence of the minus sign that the assumption is wrong and that ω_{AB} must be *clockwise* for the reference used. It is significant to note that as a result of the initial positive assumption, the result $\omega_{AB} = -2.66$ rad/sec gives at the same time the *correct convention sign* for the actual angular velocity for the reference used.

Example 13.5

In the device in Fig. 13.17, find the angular velocities and angular accelerations of both bars.

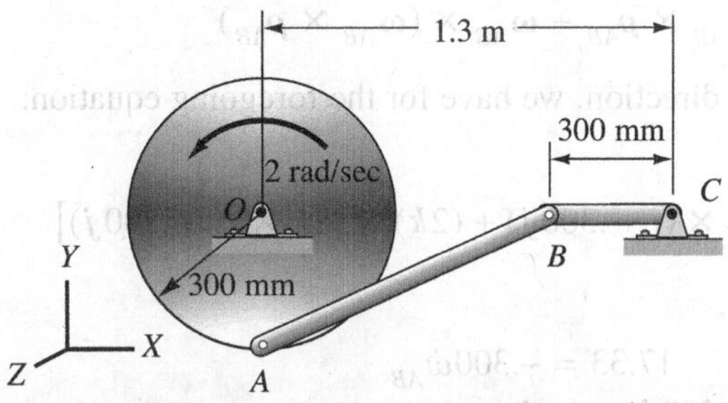

Figure 13.17. Two-dimensional device.

We shall consider points A and B of bar AB. Note first that at the instant shown:

$$\begin{aligned} \boldsymbol{V}_B &= -(.300)(\omega_{BC})\boldsymbol{j} \text{ m/sec} && \text{(a)} \\ \boldsymbol{V}_A &= (2)(.300)\boldsymbol{i} \\ &= .600\boldsymbol{i} \text{ m/sec} && \text{(b)} \end{aligned}$$

Noting that $\boldsymbol{\omega}_{AB}$ must be oriented in the Z direction because we have plane motion in the XY plane, we have for Eq. 13.6:

$$\begin{aligned} \boldsymbol{V}_B &= \boldsymbol{V}_A + \boldsymbol{\omega}_{AB} \times \boldsymbol{\rho}_{AB} \\ -.300\omega_{BC}\boldsymbol{j} &= .600\boldsymbol{i} + (\omega_{AB}\boldsymbol{k}) \times (\boldsymbol{i} + .300\boldsymbol{j}) && \text{(c)} \\ -.300\omega_{BC}\boldsymbol{j} &= .600\boldsymbol{i} + \omega_{AB}\boldsymbol{j} - .300\omega_{AB}\boldsymbol{i} \end{aligned}$$

Note we have assumed ω_{BC} and ω_{AB} as positive and thus counterclockwise. The scalar equations are:

$$\begin{aligned} .600 &= .300\omega_{AB} \\ -.300\omega_{BC} &= \omega_{AB} && \text{(d)} \end{aligned}$$

We then get

$$\begin{aligned} \omega_{AB} &= 2 \text{ rad/sec} \\ \omega_{BC} &= -6.67 \text{ rad/sec} && \text{(e)} \end{aligned}$$

Therefore, ω_{AB} is counterclockwise while ω_{BC} must be clockwise.

Let us now turn to the angular acceleration considerations for the bars. We consider separately now points A and B of bar AB. Thus,

$$\begin{aligned} \boldsymbol{a}_A &= (r\omega^2)\boldsymbol{j} = (.300)(2^2)\boldsymbol{j} = 1.200\boldsymbol{j} \text{ m/sec}^2 \\ \boldsymbol{a}_B &= \rho_{BC}\omega_{BC}^2\boldsymbol{i} + \rho_{BC}\dot{\omega}_{BC}(-\boldsymbol{j}) \\ &= (.300)(-6.67^2)\boldsymbol{i} - .300\dot{\omega}_{BC}\boldsymbol{j} \\ &= 13.33\boldsymbol{i} - .300\dot{\omega}_{BC}\boldsymbol{j} \end{aligned}$$

Example 13.5 (Continued)

Again, we have assumed $\dot{\boldsymbol{\omega}}_{BC}$ positive and thus counterclockwise. Considering bar AB, we can say for Eq. 13.7:

$$\boldsymbol{a}_B = \boldsymbol{a}_A + \dot{\boldsymbol{\omega}}_{AB} \times \boldsymbol{\rho}_{AB} + \boldsymbol{\omega}_{AB} \times (\boldsymbol{\omega}_{AB} \times \boldsymbol{\rho}_{AB}) \tag{f}$$

Noting that $\dot{\boldsymbol{\omega}}_{AB}$ must be in the Z direction, we have for the foregoing equation:

$$\begin{aligned} &13.33\boldsymbol{i} - .300\dot{\omega}_{BC}\boldsymbol{j} \\ &\quad = 1.200\boldsymbol{j} + \dot{\omega}_{AB}\boldsymbol{k} \times (\boldsymbol{i} + .300\boldsymbol{j}) + (2\boldsymbol{k}) \times [2\boldsymbol{k} \times (\boldsymbol{i} + .300\boldsymbol{j})] \end{aligned} \tag{g}$$

The scalar equations are

$$17.33 = -.300\dot{\omega}_{AB}$$
$$-.300\dot{\omega}_{BC} = \dot{\omega}_{AB}$$

We get

$$\dot{\omega}_{AB} = -57.8 \text{ rad/sec}^2$$
$$\dot{\omega}_{BC} = 192.6 \text{ rad/sec}^2$$

Clearly, for the reference used, $\dot{\omega}_{AB}$ must be clockwise and $\dot{\omega}_{BC}$ must be counterclockwise.

Example 13.6

(a) In Example 13.5, find the *instantaneous axis of rotation* for the rod AB.

The intersection of the instantaneous axis of rotation with the xy plane will be a point E in a hypothetical rigid-body extension of bar AB having zero velocity at the instant of interest. We can accordingly say:

$$\boldsymbol{V}_E = \boldsymbol{V}_A + \boldsymbol{\omega}_{AB} \times \boldsymbol{\rho}_{AE}$$

Therefore,

$$\boldsymbol{0} = .60\boldsymbol{i} + (2\boldsymbol{k}) \times (\Delta x\boldsymbol{i} + \Delta y\boldsymbol{j}) \tag{a}$$

where Δx and Δy are the components of the directed line segment from point A to the center of rotation E. The scalar equations are:

$$0 = .60 - 2\Delta y$$
$$0 = 2\Delta x$$

Clearly, $\Delta y = .3$ and $\Delta x = 0$. Thus, the center of rotation is point O.

Example 13.6 (Continued)

We could have easily deduced this result by inspection in this case. The velocity of each point of bar AB must be at *right angles* to a line from the center of rotation to the point. The velocity of point A is in the horizontal direction and the velocity of point B is in the vertical direction. Clearly, as seen from Fig. 13.18, point O is the only point from which lines to points A and B are normal to the velocities at these points.

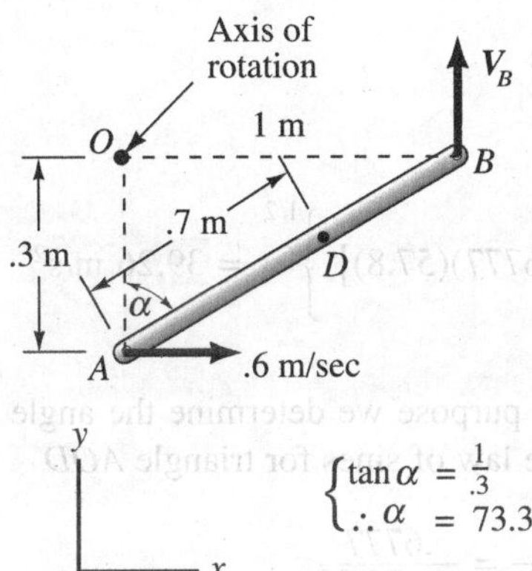

Figure 13.18. Instantaneous axis of rotation of AB.

(b) Now using the instantaneous axis of rotation, find the magnitudes of the velocity and acceleration of point D (Fig. 13.18) using data from the previous example.

In Fig. 13.19, we show the velocity vector normal to line OD. Using the law of cosines for triangle AOD, we can find OD which is a key distance for this example. Thus noting from Fig. 13.18 that $\alpha = 73.3°$, we have

$$\overline{OD} = \left[.7^2 + .3^2 - (2)(.7)(.3)(\cos 73.3°)\right]^{1/2} = .6777 \text{ m}$$

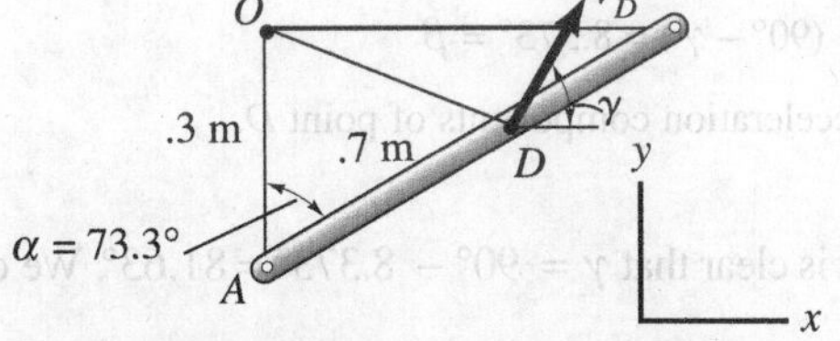

Figure 13.19. Velocity vector for point D.

We then say from rotational motion about the instantaneous center of rotation O,

$$V_D = (.6777)(\omega_{AB}) = (.6777)(2) = 1.355 \text{ m/s}$$

Example 13.6 (Continued)

For the acceleration, we have (see Fig. 13.20)

$$a_D = \left[(a_D)_c^2 + (a_D)_t^2\right]^{1/2}$$

where $(a_D)_c$ and $(a_D)_t$, respectively, are the centripetal and tangential components of acceleration at point D. Noting that r for point D is .6777 m, we get for the above

$$\therefore a_D = \left\{\left(\frac{V_D^2}{r}\right)^2 + \left[(r)(\dot{\omega}_{AB})\right]^2\right\}^{1/2}$$

$$= \left\{\left(\frac{1.355^2}{.6777}\right)^2 + [(.6777)(57.8)]^2\right\}^{1/2} = 39.26 \text{ m/s}^2 \qquad \text{(b)}$$

We now get the vectors V_D and a_D. For this purpose we determine the angle β of the tinted triangle in Fig. 13.20 by first using the law of sines for triangle AOD

$$\frac{.7}{\sin(90° - \beta)} = \frac{.6777}{\sin 73.3°}$$

$$\therefore \beta = 8.373°$$

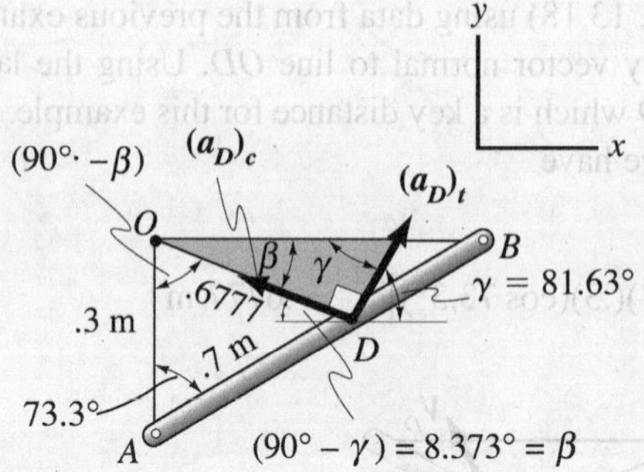

Figure 13.20. Acceleration components of point D.

Hence, looking at the tinted triangle it is clear that $\gamma = 90° - 8.373° = 81.63°$. We can now give V_D (see Fig. 13.19).

$$V_D = V_D(\cos\gamma \boldsymbol{i} + \sin\gamma \boldsymbol{j}) = 1.355(\cos 81.63\boldsymbol{i} + \sin 81.63\boldsymbol{j})$$

$$V_D = .1972\boldsymbol{i} + 1.341\boldsymbol{j} \text{ m/s}$$

Example 13.6 (Continued)

For the acceleration vector, we refer back to Eq. (b) for components of $\boldsymbol{a}_D$. Noting Fig. 13.20, we have

$$\boldsymbol{a}_D = r(\dot{\omega}_{AB})[\cos\gamma\boldsymbol{i} + \sin\gamma\boldsymbol{j}] + \frac{V_D^2}{r}[-\cos\beta\boldsymbol{i} + \sin\beta\boldsymbol{j})$$
$$= (.6777)(-57.8)[\cos 81.63°\boldsymbol{i} + \sin 81.63°\boldsymbol{j}]$$
$$+ \frac{1.355^2}{.6777}[-\cos 8.373°\boldsymbol{i} + \sin 8.373°\boldsymbol{j}]$$

$$\boldsymbol{a}_D = -8.38\boldsymbol{i} - 38.36\boldsymbol{j} \text{ m/s}^2$$

*Example 13.7

A disk E is rotating about a fixed axis HG at a constant angular speed ω_1 of 5 rad/sec in Fig. 13.21. A bar CD is held by the wheel at D by a ball-joint connection and is guided along a rod AB cantilevered at A and B by a collar at C having a second ball-joint connection with CD, as shown in the diagram. Compute the velocity of C.

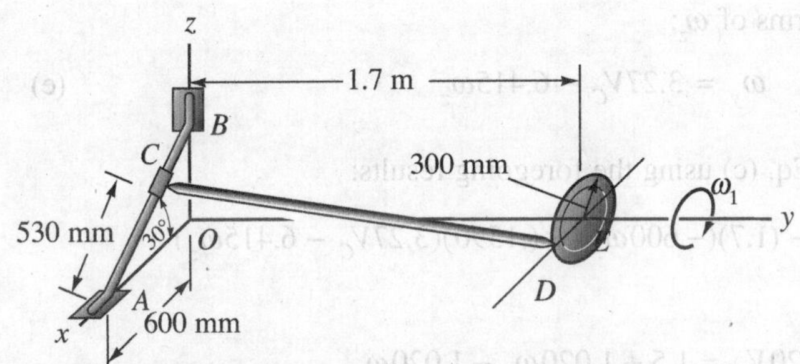

Figure 13.21. Three-dimensional device.

We shall need the vector $\boldsymbol{\rho}_{DC}$. Thus,

$$\boldsymbol{\rho}_{DC} = \boldsymbol{r}_C - \boldsymbol{r}_D$$
$$= [(.600 - .530\cos 30°)\boldsymbol{i} + .530\sin 30°\boldsymbol{k}] - (1.7\boldsymbol{j} + .300\boldsymbol{i})$$
$$= -.1590\boldsymbol{i} - 1.7\boldsymbol{j} + .265\boldsymbol{k} \text{ m}$$

Example 13.7 (Continued)

Now employ Eq. 13.6 for rod *CD*. Thus,

$$V_C = V_D + \boldsymbol{\omega}_{CD} \times \boldsymbol{\rho}_{DC}$$

Therefore, assuming *C* is going from *B* to *A*

$$V_C(\cos 30°\boldsymbol{i} - \sin 30°\boldsymbol{k})$$
$$= (5)(.30)\boldsymbol{k} + (\omega_x\boldsymbol{i} + \omega_y\boldsymbol{j} + \omega_z\boldsymbol{k}) \times (-.1590\boldsymbol{i} - 1.7\boldsymbol{j} + .265\boldsymbol{k})$$
$$V_C(.866\boldsymbol{i} - .500\boldsymbol{k}) = 1.50\boldsymbol{k} - 1.7\omega_x\boldsymbol{k} - .265\omega_x\boldsymbol{j} + .1590\omega_y\boldsymbol{k}$$
$$+ .265\omega_y\boldsymbol{i} - .1590\omega_z\boldsymbol{j} + 1.7\omega_z\boldsymbol{i})$$

The scalar equations are:

$$.866V_C = .265\omega_y + 1.7\omega_z \qquad \text{(a)}$$
$$0 = -.265\omega_x - .1590\omega_z \qquad \text{(b)}$$
$$-.500V_C = 1.50 - 1.7\omega_x + .1590\omega_y \qquad \text{(c)}$$

From these equations, we cannot solve for ω_x, ω_y, and ω_z because the spin of *CD* about its own axis (allowed by the ball joints) can have *any value* without affecting the velocity of slider *C*. However, we can determine V_C, as we shall now demonstrate.

In Eq. (b), solve for ω_x in terms of ω_z:

$$\omega_x = -.600\omega_z \qquad \text{(d)}$$

In Eq. (a), solve for ω_y in terms of ω_z:

$$\omega_y = 3.27V_C - 6.415\omega_z \qquad \text{(e)}$$

Substitute for ω_x and ω_y in Eq. (c) using the foregoing results:

$$-.500V_C = 1.50 - (1.7)(-.600\omega_z) + (.1590)(3.27V_C - 6.415\omega_z)$$

Therefore,

$$-1.020V_C = 1.5 + 1.020\omega_z - 1.020\omega_z$$
$$V_C = -1.471 \text{ m/sec}$$

Hence,

$$\boldsymbol{V}_c = -1.471(\cos 30°\boldsymbol{i} - \sin 30°\boldsymbol{k})$$

$$\boldsymbol{V}_c = -1.274\boldsymbol{i} + .7355\boldsymbol{k} \text{ m/sec}$$

Clearly, contrary to our assumption *C* is going from *A* to *B*.

Before going on to the next section, we wish to point out a simple relation that will be of use in the remainder of the chapter. Suppose that you have a moving particle whose position vector $\boldsymbol{r}$ has a magnitude that is constant (see Fig. 13.22). This position vector, however, has an angular velocity $\boldsymbol{\omega}$ relative to xyz. We wish to know the velocity of the particle P relative to xyz.

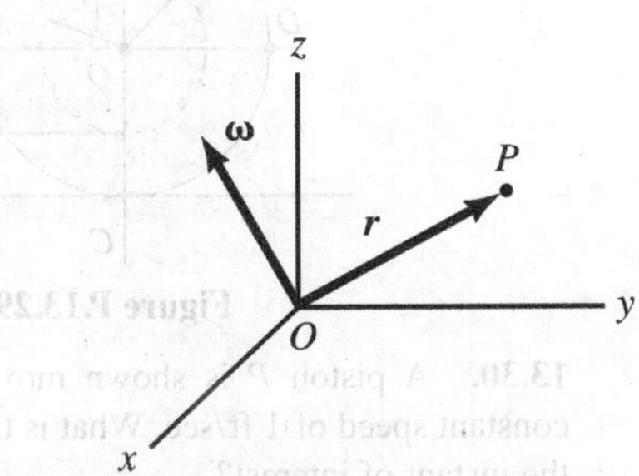

Figure 13.22. Position vector $\boldsymbol{r}$ has constant magnitude but rotates relative to xyz with angular velocity $\boldsymbol{\omega}$.

We could imagine for this purpose that particle P is part of a rigid body attached to xyz at O and rotating with angular velocity $\boldsymbol{\omega}$. This situation is shown in Fig. 13.23. Using Eq. 13.6, we can then say:

$$\boldsymbol{V}_P = \boldsymbol{V}_O + \boldsymbol{\omega} \times \boldsymbol{\rho}_{OP}$$

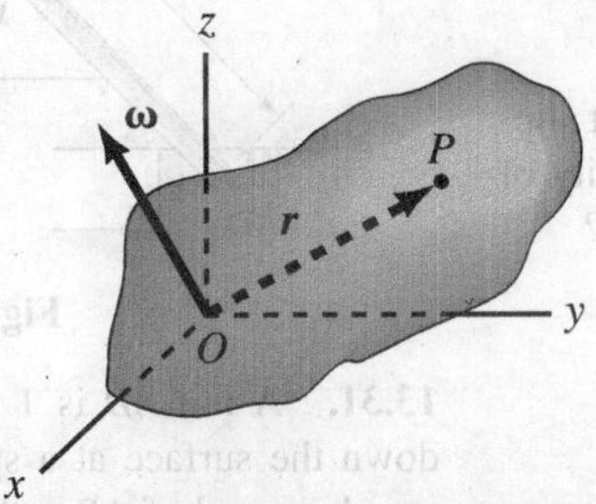

Figure 13.23. P now considered as a point in a rigid body attached at O and having angular velocity $\boldsymbol{\omega}$.

But $\boldsymbol{V}_O = \boldsymbol{0}$ and $\boldsymbol{\rho}_{OP}$ is simply $\boldsymbol{r}$. Hence, we have

$$\boldsymbol{V}_P = \boldsymbol{\omega} \times \boldsymbol{r}$$

We thus have a simple formula for the velocity of a particle moving at a fixed distance from the origin of xyz. This velocity is simply the cross product of the angular velocity $\boldsymbol{\omega}$ of the position vector about xyz and the position vector $\boldsymbol{r}$.

PROBLEMS

13.25. A body is spinning about an axis having direction cosines $l = .5$, $m = .5$, and $n = .707$. The angular speed is 50 rad/sec. What is the velocity of a point in the body having a position vector $\mathbf{r} = 6\mathbf{i} + 4\mathbf{j}$ ft?

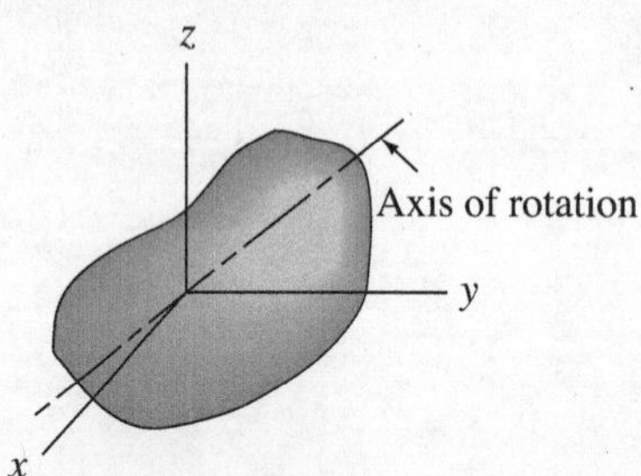

Figure P.13.25.

13.26. In Problem 13.25, what is the relative velocity between a point in the body at position $x = 10$ m, $y = 6$ m, $z = 3$ m and a point in the body at position $x = 2$m, $y = -3$m, $z = 0$ m?

13.27. If the body in Problem 13.25 is given an additional angular velocity $\boldsymbol{\omega}_2 = 6\mathbf{j} + 10\mathbf{k}$ rad/sec, what is the direction of the axis of rotation? Compute the velocity at $\mathbf{r} = 10\mathbf{j} + 3\mathbf{k}$ ft if the actual axis of rotation goes through the origin.

13.28. A wheel is rolling along at 17 m/sec without slipping. What is the angular speed? What is the velocity of point B on the rim of the wheel at the instant shown?

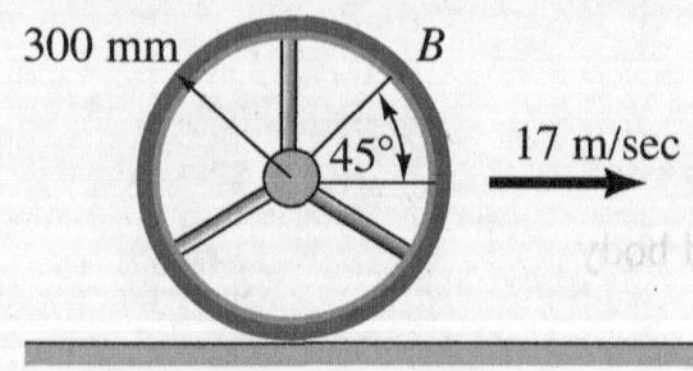

Figure P.13.28.

13.29. A flexible cord is wrapped around a spool and is pulled at a velocity of 10 ft/sec relative to the ground. If there is no slipping at C, what is the velocity of points O and D at the instant shown?

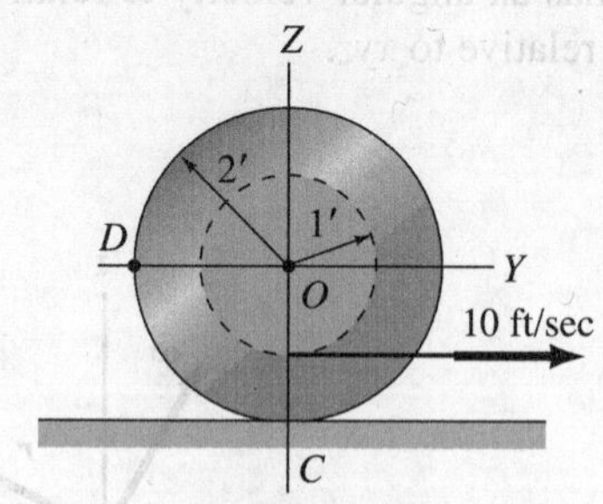

Figure P.13.29.

13.30. A piston P is shown moving downward at the constant speed of 1 ft/sec. What is the speed of slider A at the instant of interest?

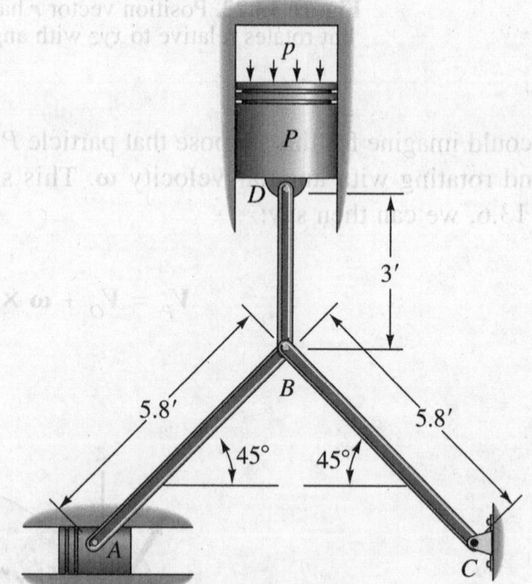

Figure P.13.30.

13.31. A rod AB is 1 m in length. If the end A slides down the surface at a speed V_A of 3 m/sec, what is the angular speed of AB at the instant shown?

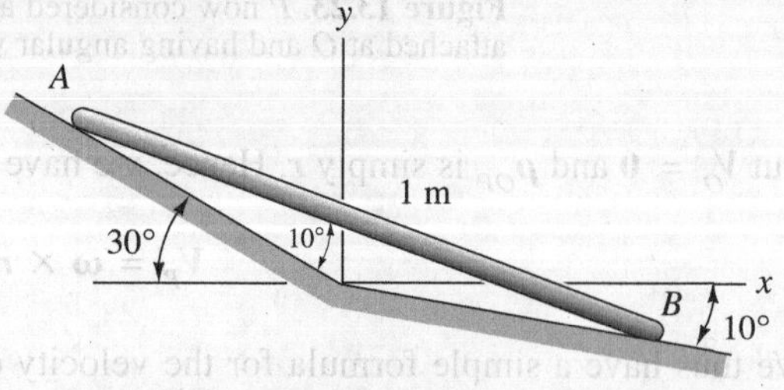

Figure P.13.31.

13.32. A plate moves along a horizontal surface. Components of the velocity for three corners are:

$$(V_A)_x = 2\text{m/sec}$$
$$(V_B)_y = -3\text{m/sec}$$
$$(V_C)_y = 5\text{m/sec}$$

What is the angular speed of the plate and what is the velocity of corner D?

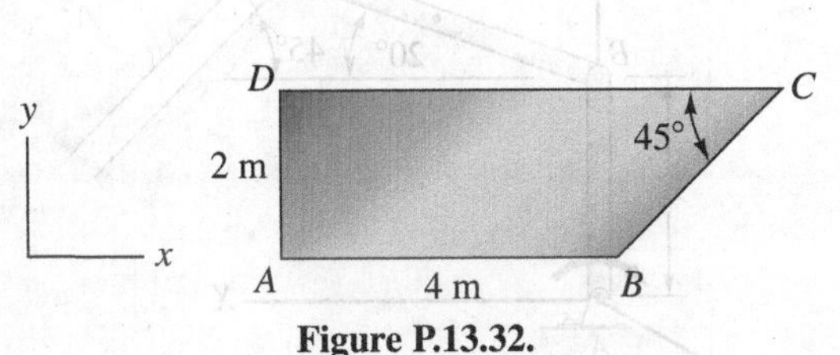

Figure P.13.32.

13.33. Rod DC has an angular speed ω_1 of 5 rad/sec at the configuration shown. What is the angular speed of bar AB?

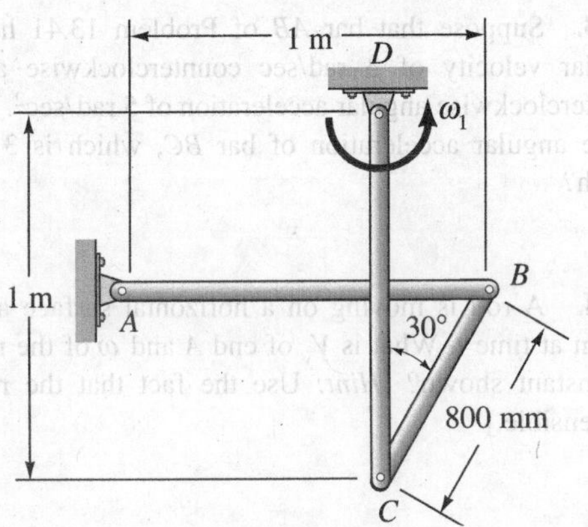

Figure P.13.33.

13.34. A system of meshing gears includes gear A, which is held stationary. Rod $A'C'$ rotates with a speed ω_1 of 5 rad/sec. What is the angular speed of gear C? The gears have the following diameters:

$$D_A = 600 \text{ mm}$$
$$D_B = 350 \text{ mm}$$
$$D_C = 200 \text{ mm}$$

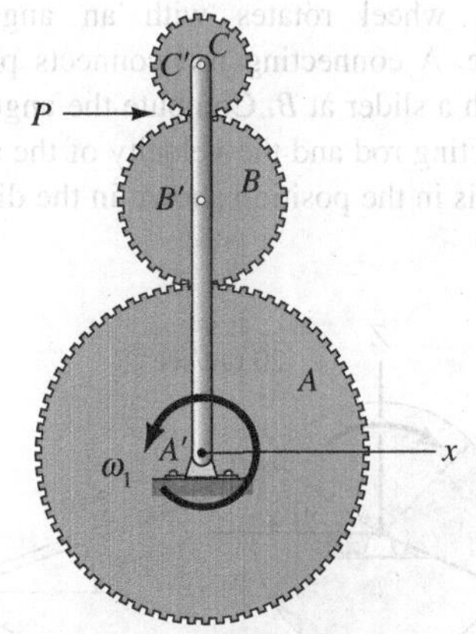

Figure P.13.34.

13.35. In Problem 15.34 take $\omega_1 = 10$ rad/sec. If gear C is to translate, what angular speed should gear A have?

13.36. A bar moves in the plane of the page so that end A has a velocity of 7 m/sec and decelerates at a rate of 3.3 m/sec^2 What are the velocity and acceleration of point C when BA is at 30° to the horizontal?

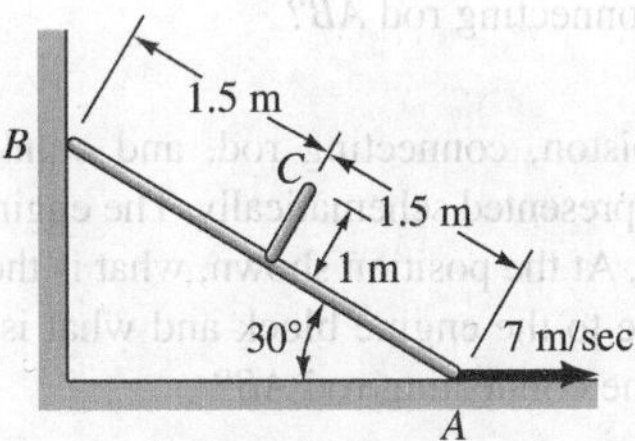

Figure P.13.36.

13.37. Bar AB is rotating at a constant speed of 5 rad/sec clockwise in a device. What is the angular velocity of bar BD and body EFC? Determine the velocity of point D [*Hint:* What is the direction of the velocity of point G?]

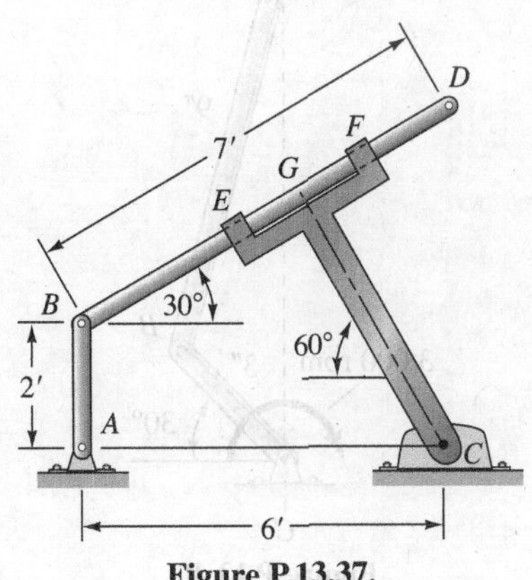

Figure P.13.37.

13.38. A wheel rotates with an angular speed of 20 rad/sec. A connecting rod connects points A on the wheel with a slider at B. Compute the angular velocity of the connecting rod and the velocity of the slider when the apparatus is in the position shown in the diagram.

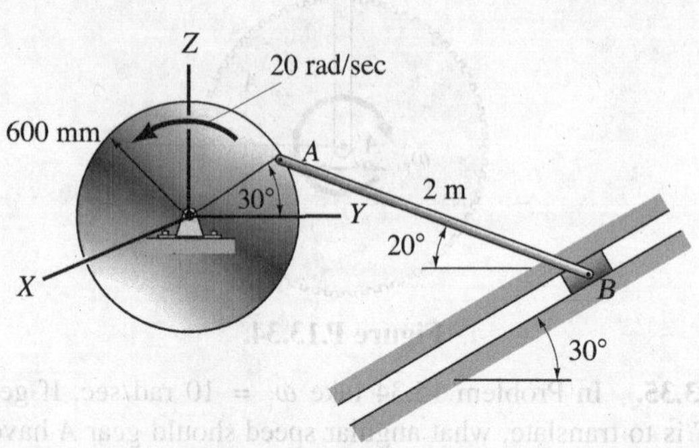

Figure P.13.38.

13.39. In Problem 13.38, if $V_B = 14.30$ m/sec and $\omega_{AB} = 9.33$ rad/sec, where is the instantaneous axis of rotation of connecting rod AB?

13.40. A piston, connecting rod, and crankshaft of an engine are represented schematically. The engine is rotating at 3,000 rpm. At the position shown, what is the velocity of pin A relative to the engine block and what is the angular velocity of the connecting rod AB?

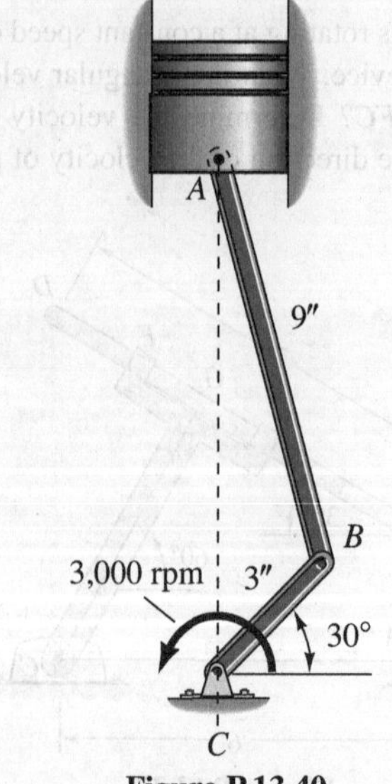

Figure P.13.40.

13.41. Member AB is rotating at a constant speed of 4 rad/sec in a counterclockwise direction. What is the angular velocity of bar BC for the position shown in the diagram? What is the velocity of point D at the center of bar BC? Bar BC is 3 ft in length.

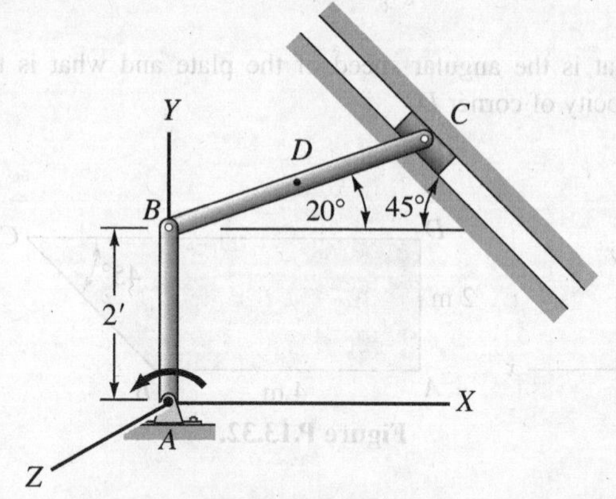

Figure P.13.41.

13.42. In Problem 13.41, determine in the simplest manner the instantaneous axis of rotation for bar BC.

13.43. Suppose that bar AB of Problem 13.41 has an angular velocity of 3 rad/sec counterclockwise and a counterclockwise angular acceleration of 5 rad/sec^2. What is the angular acceleration of bar BC, which is 3 ft in length?

13.44. A rod is moving on a horizontal surface and is shown at time t. What is V_y of end A and ω of the rod at the instant shown? [*Hint:* Use the fact that the rod is inextensible.]

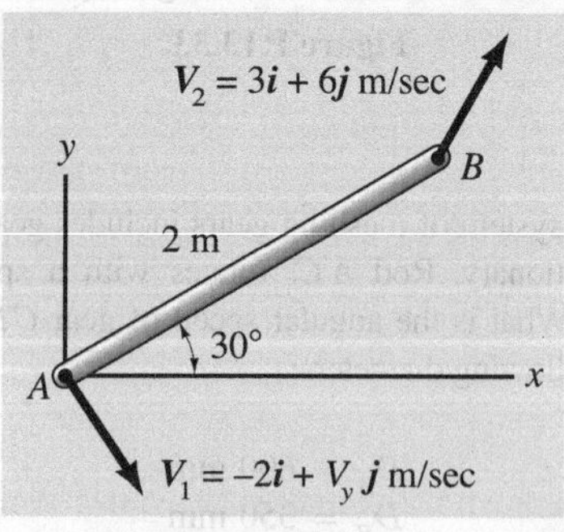

Figure P.13.44.

13.45. A plate $ABCD$ moves on a horizontal surface. At time t corners A and B have the following velocities:

$$V_A = 3i + 2j \text{ m/sec}$$
$$V_B = (V_B)_x i + 5j \text{ m/sec}$$

Find the location of the instantaneous axis of rotation.

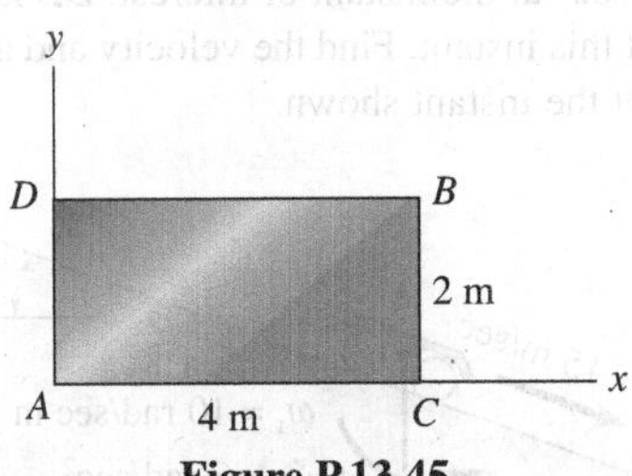

Figure P.13.45.

13.46. Find the velocity and acceleration relative to the ground of pin B on the wheel. The wheel rolls without slipping. Also, find the angular velocity of the slotted bar in which the pin B of the wheel slides when θ of the bar is 30°.

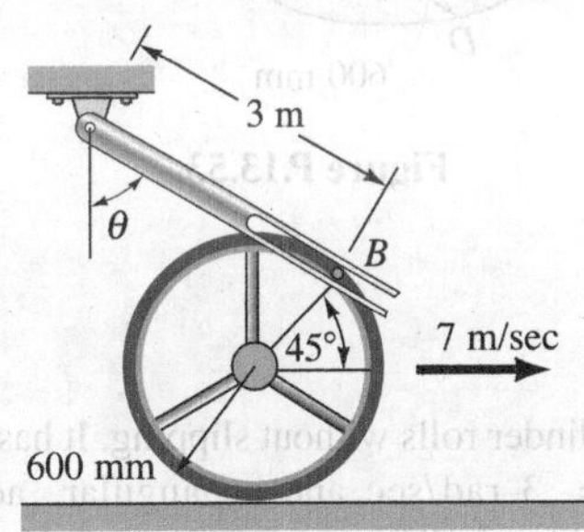

Figure P.13.46.

13.47. If $\omega_1 = 5$ rad/sec and $\dot{\omega}_1 = 3$ rad/sec² for bar CD, compute the angular velocity and angular acceleration of the gear D relative to the ground. Solve the problem using Eqs. 15.6 and 15.7, and then check the result by considering simple circular motion of point D.

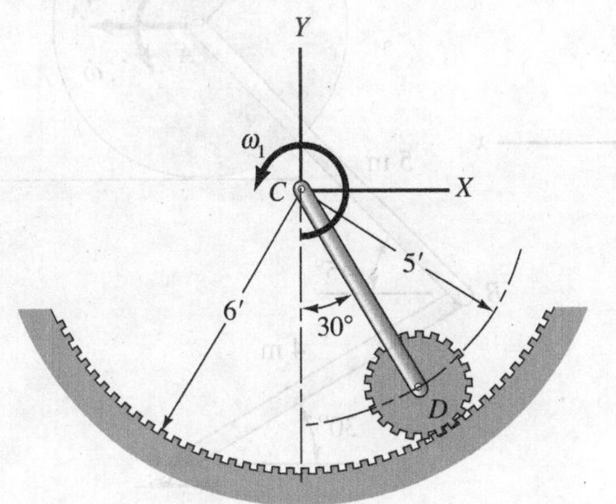

Figure P.13.47.

13.48. A mechanism with two sliders is shown. Slider A at the instant of interest has a speed of 3 m/sec and is accelerating at the rate of 1.7 m/sec². If member AB is 2.5 m in length, what are the angular velocity and angular acceleration for this member?

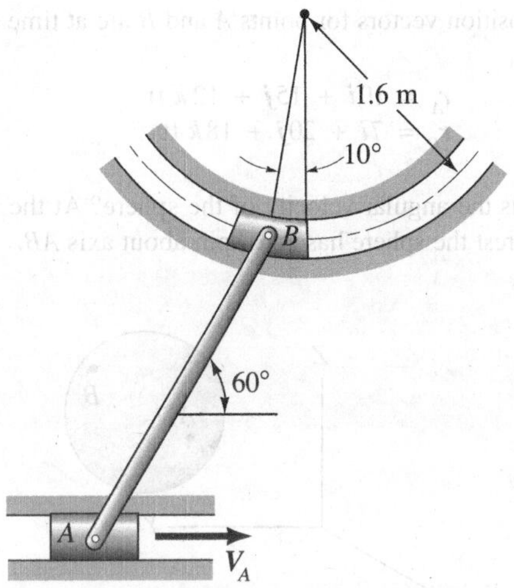

Figure P.13.48.

13.49. In Problem 13.48 find the instantaneous center of rotation of bar AB if V_A is 2.7 m/sec.

13.50. The velocity of corner A of the block is known to be at time t:

$$V_A = 10i + 4j - 3k \text{ m/sec}$$

The angular speed about edge AD is 2 rad/sec, and the angular speeds about the diagonals AF and HE are known to be 3 rad/sec and 6 rad/sec, respectively. What is the velocity of corner B at this instant?

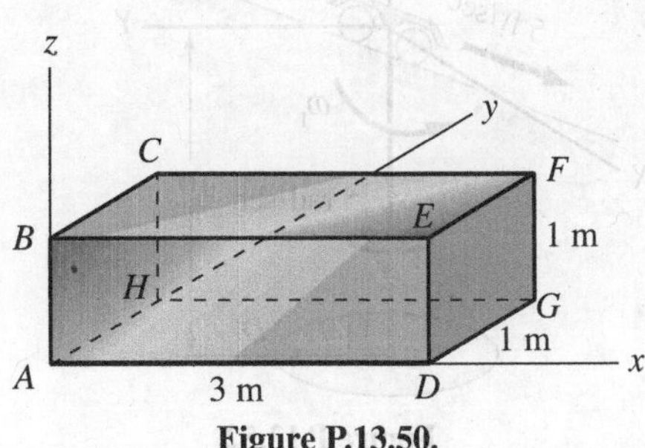

Figure P.13.50.

13.51. A rigid sphere is moving in space. The velocities for two points A and B on the surface have the values at time t:

$$V_A = 6i + 3j + 2k \text{ ft/sec}$$
$$V_B = (V_B)_x i + 6j - 4k \text{ ft/sec}$$

The position vectors for points A and B are at time t:

$$r_A = 10i + 15j + 12k \text{ ft}$$
$$r_B = 7i + 20j + 18k \text{ ft}$$

What is the angular velocity of the sphere? At the instant of interest the sphere has zero spin about axis AB.

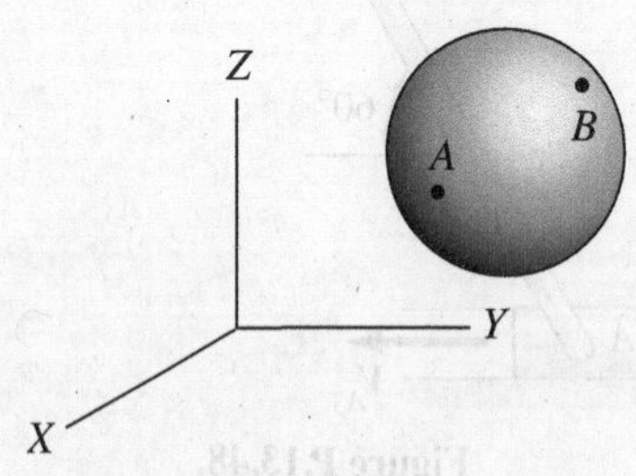

Figure P.13.51.

13.52. A conveyor element moves down the incline at a speed of 5 ft/sec. A shaft and platform move with the conveyor element but have a spin of .5 rad/sec about the centerline AB. Also, the shaft swings in the YZ plane at a speed ω_1 of 1 rad/sec. What is the velocity and acceleration of point D on the platform at the instant it is in the YZ plane, as shown in the diagram? Note that at the instant of interest AB is vertical.

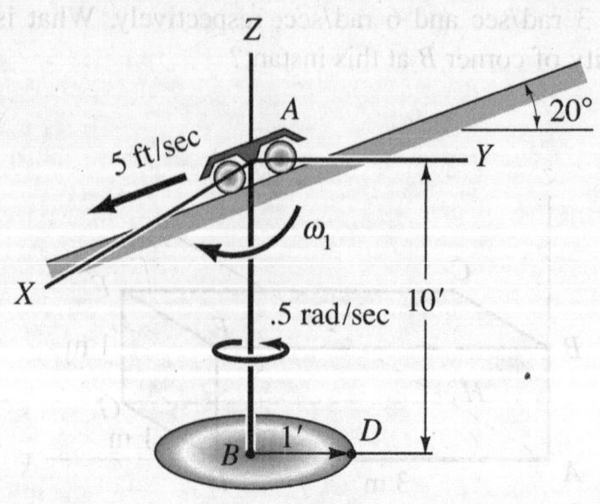

Figure P.13.52

13.53. A conveyor element moves down an incline at a speed of 15 m/sec. A plate hangs down from the conveyor element and, at the instant of interest shown in the diagram, is spinning about AB at the rate of 5 rad/sec. Also, the axis AB swings in the YZ plane at the rate ω_1 of 10 rad/sec and $\dot{\omega}_1 = 3$ rad/sec² at the instant of interest. DB is parallel to the X axis at this instant. Find the velocity and acceleration of point D at the instant shown.

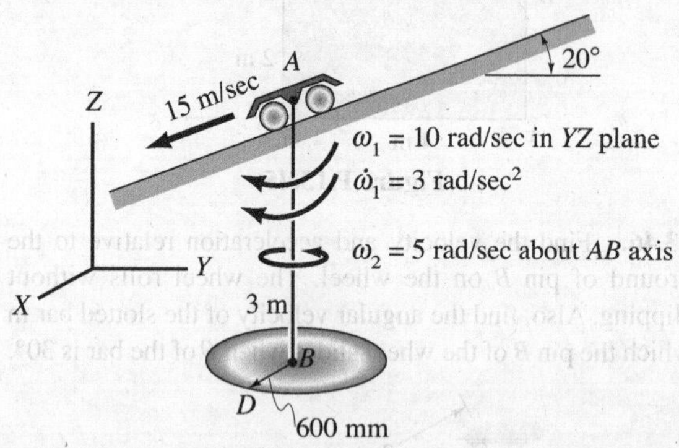

Figure P.13.53.

13.54. A cylinder rolls without slipping. It has an angular velocity $\omega = .3$ rad/sec and an angular acceleration $\dot{\omega} = .014$ rad/sec². What are the angular velocity and angular acceleration of member AB?

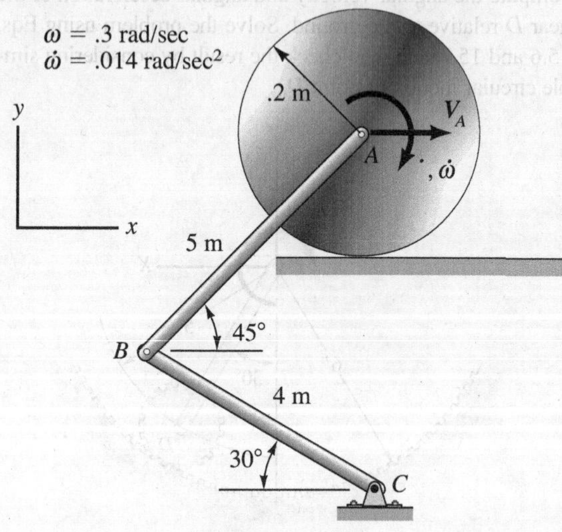

Figure P.13.54.

13.55. Slider A moves in a parabolic slot with speed $\dot{s} = 3$ m/s and $\ddot{s} = 1$ m/s^2 at the instant shown in the diagram. Cylinder E is connected to A by rod AB.

(a) Find the angular velocity of cylinder E at the time of interest.

(b) Also, find the angular acceleration of cylinder E and rod AB at this instant.

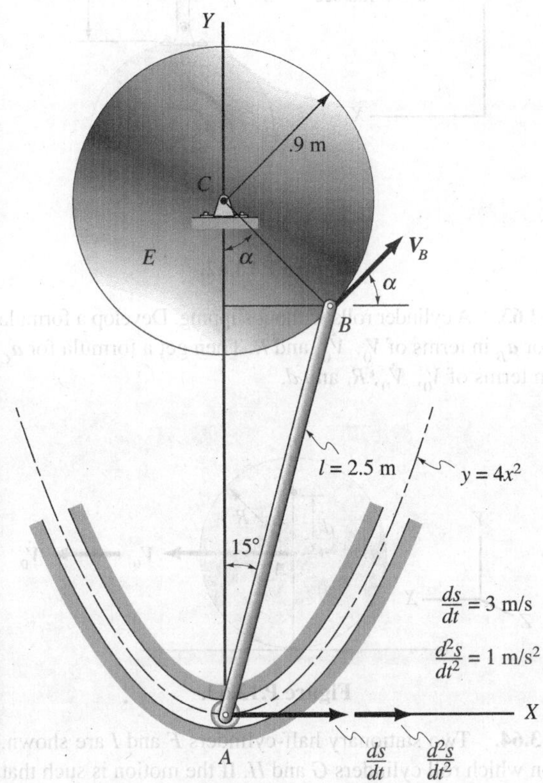

Figure P.13.55.

13.56. Find ω_A and $\dot{\omega}_A$ at the instant shown. The following data apply:

$R_A = .3$ m $R_B = .2$ m $CD = 5$ m $V_B = .2$ m/s

Disc A rolls without slipping.

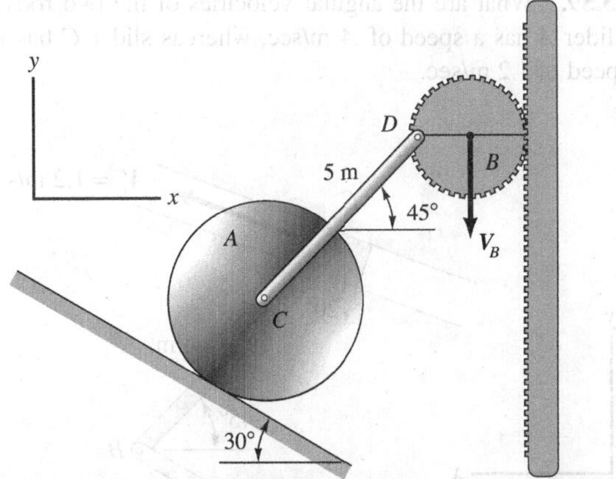

Figure P.13.56.

13.57. Find the velocity and acceleration of the center of A.

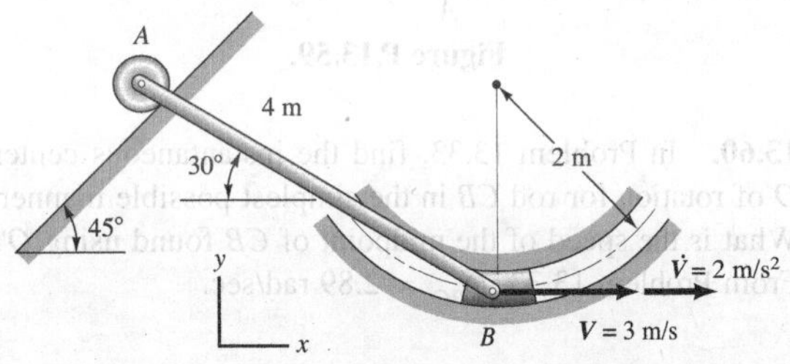

Figure P.13.57.

13.58. What is the angular velocity of rod AD? What is the magnitude of the velocity of point C of rod AD? Rod BC is vertical at the instant of interest.

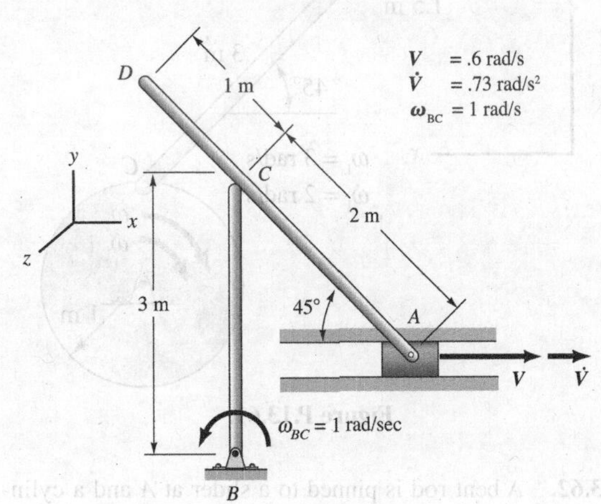

Figure P.13.58.

13.59. What are the angular velocities of the two rods? Slider A has a speed of .4 m/sec, whereas slider C has a speed of 1.2 m/sec.

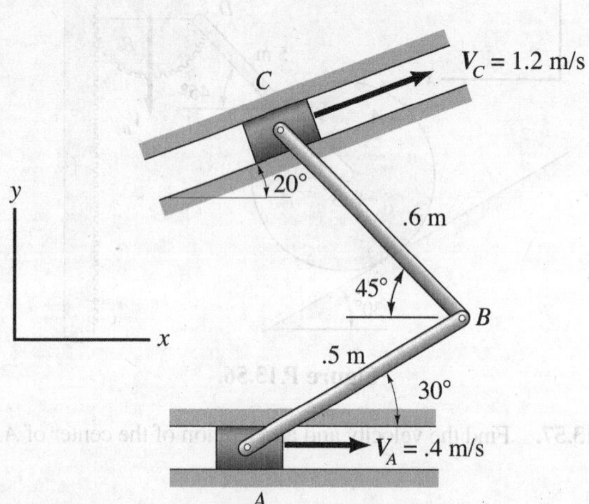

Figure P.13.59.

13.60. In Problem 13.33, find the instantaneous center O of rotation for rod CB in the simplest possible manner. What is the speed of the midpoint of CB found using O? From Problem 13.33, $\omega_{BC} = 2.89$ rad/sec.

13.61. Find ω_E and $\dot{\omega}_E$ at the instant shown.

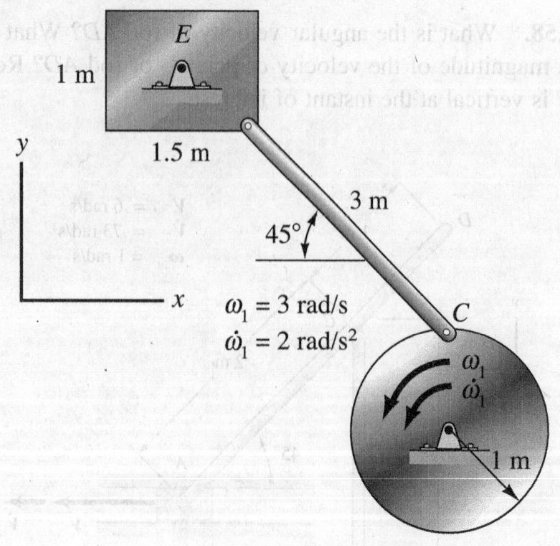

Figure P.13.61.

13.62. A bent rod is pinned to a slider at A and a cylinder at B. Find the velocity and acceleration of the slider at the instant depicted in the diagram.

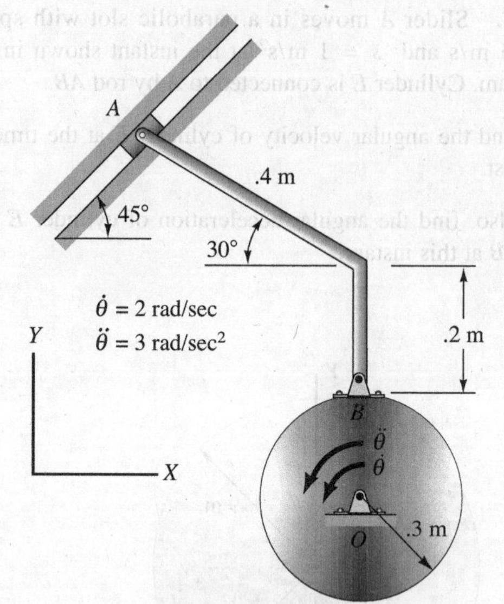

Figure P.13.62.

13.63. A cylinder rolls without slipping. Develop a formula for $\boldsymbol{a}_B$ in terms of V_0, $\dot{V}_0$, and R. Then get a formula for $\boldsymbol{a}_C$ in terms of V_0, $\dot{V}_0$, R, and d.

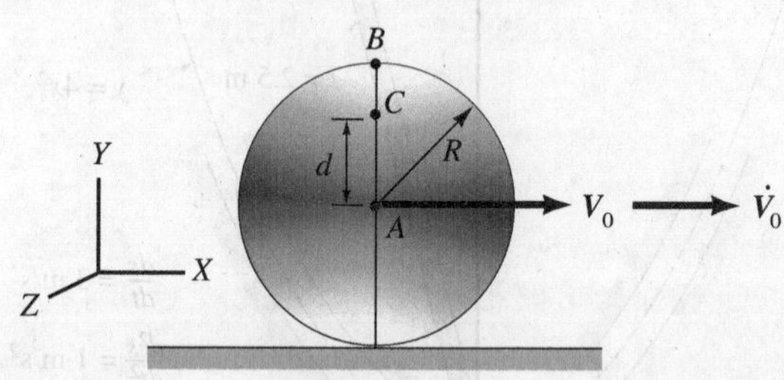

Figure P.13.63.

13.64. Two stationary half-cylinders F and I are shown, on which roll cylinders G and H. If the motion is such that line BA has an angular speed of 2 rad/sec clockwise, what is the angular speed and the angular acceleration of cylinder H relative to the ground? The cylinders roll without slipping.

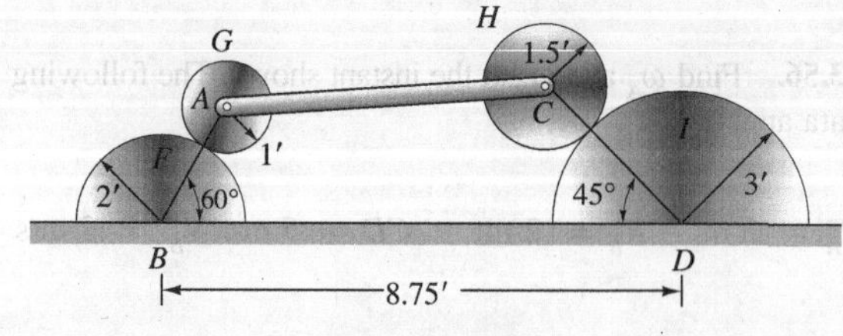

Figure P.13.64.

13.65. In Problem 13.64, assume that cylinder G is rotating at a speed of 5 rad/sec clockwise as seen from the ground. What is the speed and rate of change of speed of point C relative to the ground? Assume that no slipping occurs.

13.66. A wheel D of radius $R_1 = 6$ in. rotates at a speed $\omega_1 = 5$ rad/sec as shown. A second wheel C is connected to wheel D by connecting rod AB. What is the angular speed of wheel C at the instant shown? The radius $R_2 = 12$ in. The wheels are separated by a distance $d = 2$ ft. At A and at B there are ball-and-socket connections.

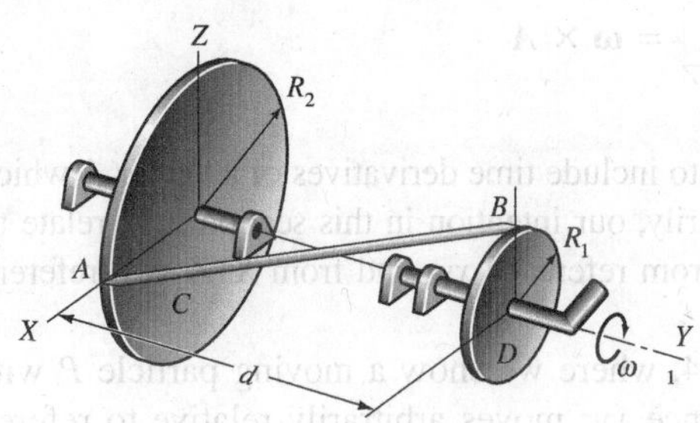

Figure P.13.66.

***13.67.** A bar AB can slide along members CD and FG of a rigid structure. If A is moving at a speed of 300 mm/sec along CD toward D and is at this instant a distance of 300 mm from C, what is the speed of B along FG? At A and B there are ball-and-socket-joint connections.

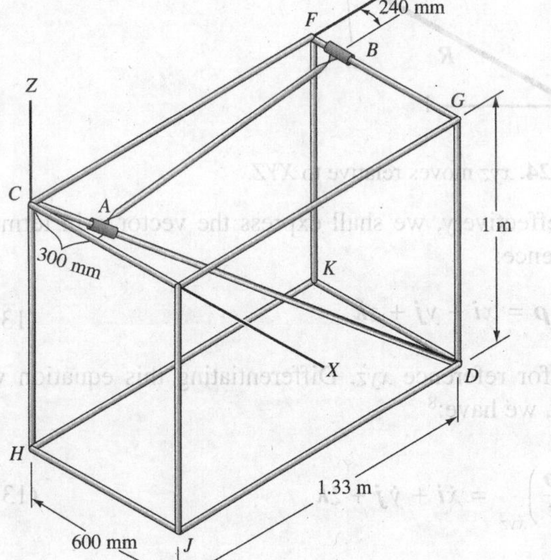

Figure P.13.67.

13.68. Member AB connects two sliders A and B. If $V_B = 5$ m/s and $\dot{V}_B = 3$ m/s², what are ω_{AB} and $\dot{\omega}_{AB}$ at the configuration shown?

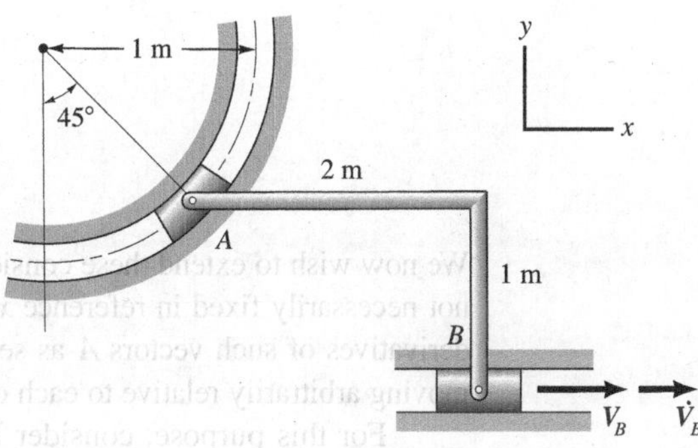

Figure P.13.68.

13.69. Find ω_{AB} and $\dot{\omega}_{AB}$. Cylinder D rolls without slipping with angular motion given as

$$\omega_D = .2 \text{ rad/s and } \dot{\omega}_D = .3 \text{ rad/s}^2$$

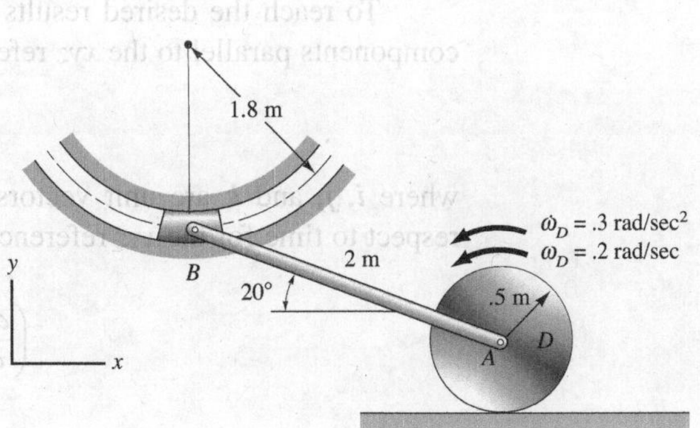

Figure P.13.69.

13.6 General Relationship Between Time Derivatives of a Vector for Different References

In Section 13.4, we considered the time derivatives of a vector $\boldsymbol{A}$ "fixed" in a reference xyz moving arbitrarily relative to XYZ. Our conclusions were:

$$\left(\frac{d\boldsymbol{A}}{dt}\right)_{xyz} = \mathbf{0}$$

$$\left(\frac{d\boldsymbol{A}}{dt}\right)_{XYZ} = \boldsymbol{\omega} \times \boldsymbol{A}$$

We now wish to extend these considerations to include time derivatives of a vector $\boldsymbol{A}$ which is not necessarily fixed in reference xyz. Primarily, our intention in this section is to relate time derivatives of such vectors $\boldsymbol{A}$ as seen both from reference xyz and from XYZ, two references moving arbitrarily relative to each other.

For this purpose, consider Fig. 13.24, where we show a moving particle P with a position vector $\boldsymbol{\rho}$ in reference xyz. Reference xyz moves arbitrarily relative to reference XYZ with translational velocity $\dot{\boldsymbol{R}}$ and angular velocity $\boldsymbol{\omega}$ in accordance with **Chasles' theorem.** We shall now form a relation between $(d\boldsymbol{\rho}/dt)_{xyz}$ and $(d\boldsymbol{\rho}/dt)_{XYZ}$. We shall then extend this result so as to relate the time derivative of any vector $\boldsymbol{A}$ as seen from any two references.

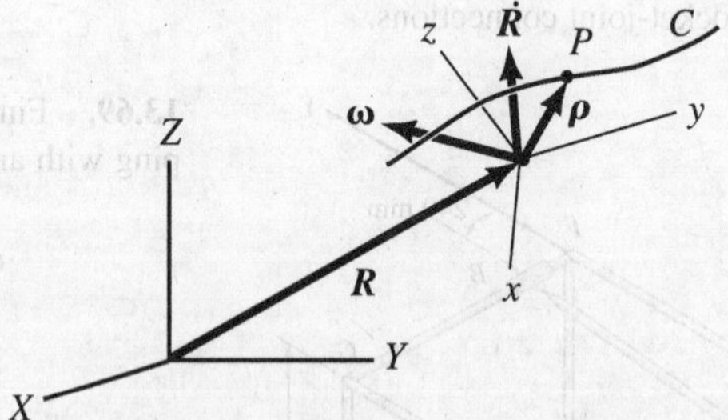

Figure 13.24. xyz moves relative to XYZ.

To reach the desired results effectively, we shall express the vector $\boldsymbol{\rho}$ in terms of components parallel to the xyz reference:

$$\boldsymbol{\rho} = x\boldsymbol{i} + y\boldsymbol{j} + z\boldsymbol{k} \tag{13.11}$$

where $\boldsymbol{i}$, $\boldsymbol{j}$, and $\boldsymbol{k}$ are unit vectors for reference xyz. Differentiating this equation with respect to time for the xyz reference, we have:[8]

$$\left(\frac{d\boldsymbol{\rho}}{dt}\right)_{xyz} = \dot{x}\boldsymbol{i} + \dot{y}\boldsymbol{j} + \dot{z}\boldsymbol{k} \tag{13.12}$$

[8]Note that $\dot{x}$, $\dot{y}$, and $\dot{z}$ are time derivatives of scalars and accordingly there is no identification with any reference as far as the time derivative operation is concerned.

If we next take the derivative of $\boldsymbol{\rho}$ with respect to time for the *XYZ* reference, we must remember that $\boldsymbol{i}, \boldsymbol{j}$, and $\boldsymbol{k}$ of Eq. 13.11 generally will each be a function of time, since these vectors will generally have some rotational motion relative to the *XYZ* reference. Thus, if dots are used for the time derivatives:

$$\left(\frac{d\boldsymbol{\rho}}{dt}\right)_{XYZ} = (\dot{x}\boldsymbol{i} + \dot{y}\boldsymbol{j} + \dot{z}\boldsymbol{k}) + (x\dot{\boldsymbol{i}} + y\dot{\boldsymbol{j}} + z\dot{\boldsymbol{k}}) \tag{13.13}$$

The unit vector $\boldsymbol{i}$ is a vector *fixed* in reference *xyz*, and accordingly $\dot{\boldsymbol{i}}$ equals $\boldsymbol{\omega} \times \boldsymbol{i}$. The same conclusions apply to $\boldsymbol{j}$ and $\boldsymbol{k}$. The last expression in parentheses can then be stated as

$$\begin{aligned}(x\dot{\boldsymbol{i}} + y\dot{\boldsymbol{j}} + z\dot{\boldsymbol{k}}) &= x(\boldsymbol{\omega} \times \boldsymbol{i}) + y(\boldsymbol{\omega} \times \boldsymbol{j}) + z(\boldsymbol{\omega} \times \boldsymbol{k}) \\ &= \boldsymbol{\omega} \times (x\boldsymbol{i}) + \boldsymbol{\omega} \times (y\boldsymbol{j}) + \boldsymbol{\omega} \times (z\boldsymbol{k}) \\ &= \boldsymbol{\omega} \times (x\boldsymbol{i} + y\boldsymbol{j} + z\boldsymbol{k}) = \boldsymbol{\omega} \times \boldsymbol{\rho}\end{aligned} \tag{13.14}$$

In Eq. 13.13 we can replace $(\dot{x}\boldsymbol{i} + \dot{y}\boldsymbol{j} + \dot{z}\boldsymbol{k})$ by $(d\boldsymbol{\rho}/dt)_{xyz}$, in accordance with Eq. 13.12, an $(x\dot{\boldsymbol{i}} + y\dot{\boldsymbol{j}} + z\dot{\boldsymbol{k}})$ by $\boldsymbol{\omega} \times \boldsymbol{\rho}$, in accordance with Eq. 13.14. Hence,

$$\left(\frac{d\boldsymbol{\rho}}{dt}\right)_{XYZ} = \left(\frac{d\boldsymbol{\rho}}{dt}\right)_{xyz} + \boldsymbol{\omega} \times \boldsymbol{\rho} \tag{13.15}$$

We can generalize the preceding result for any vector $\boldsymbol{A}$:

$$\left(\frac{d\boldsymbol{A}}{dt}\right)_{XYZ} = \left(\frac{d\boldsymbol{A}}{dt}\right)_{xyz} + \boldsymbol{\omega} \times \boldsymbol{A} \tag{13.16}$$

where, you must remember, $\boldsymbol{\omega}$ without subscripts will always be the *angular velocity of the xyz reference relative to the XYZ reference.* Note that Eq. 13.1 is a special case of Eq. 13.16 since for $\boldsymbol{A}$ fixed in *xyz*, $(d\boldsymbol{A}/dt)_{xyz} = \boldsymbol{0}$. We shall have much use for this relationship in succeeding sections.

13.7 Relationship Between Velocities of a Particle for Different References

We shall now define the velocity of a particle again in the presence of several references:

The velocity of a particle relative to a reference is the derivative as seen from this reference of the position vector of the particle in the reference.

In Fig. 13.24, the velocities of the particle P relative to the XYZ and the xyz references are, respectively,[9]

$$\boldsymbol{V}_{XYZ} = \left(\frac{d\boldsymbol{r}}{dt}\right)_{XYZ}, \quad \boldsymbol{V}_{xyz} = \left(\frac{d\boldsymbol{\rho}}{dt}\right)_{xyz} \tag{13.17}$$

Since a vector can always be decomposed into *any* set of orthogonal components, $\boldsymbol{V}_{XYZ}$ can be expressed in components parallel to the xyz reference at any time t while $\boldsymbol{V}_{xyz}$ may be expressed in components parallel to the XYZ reference at any time t.

Now, we shall relate these velocities by first noting that

$$\boldsymbol{r} = \boldsymbol{R} + \boldsymbol{\rho} \tag{13.18}$$

Differentiating with respect to time for the XYZ reference, we have

$$\left(\frac{d\boldsymbol{r}}{dt}\right)_{XYZ} \equiv \boldsymbol{V}_{XYZ} = \left(\frac{d\boldsymbol{R}}{dt}\right)_{XYZ} + \left(\frac{d\boldsymbol{\rho}}{dt}\right)_{XYZ} \tag{13.19}$$

The term $(d\boldsymbol{R}/dt)_{XYZ}$ is clearly the velocity of the origin of the xyz reference relative to the XYZ reference, according to our definitions, and we denote this velocity as $\dot{\boldsymbol{R}}$. The term $(d\boldsymbol{\rho}/dt)_{XYZ}$ can be replaced, by use of Eq. 13.15, in which $(d\boldsymbol{\rho}/dt)_{xyz}$ is the velocity of the particle relative to the xyz reference. Denoting $(d\boldsymbol{\rho}/dt)_{xyz}$ simply as $\boldsymbol{V}_{xyz}$, we find that the foregoing equation then becomes the desired relation:

$$\boldsymbol{V}_{XYZ} = \boldsymbol{V}_{xyz} + \dot{\boldsymbol{R}} + \boldsymbol{\omega} \times \boldsymbol{\rho} \tag{13.20}$$

We reiterate the understanding that $\boldsymbol{\omega}$ *without* subscripts represents the angular velocity of xyz relative to XYZ. This $\boldsymbol{\omega}$ always goes into the last expression of Eq. 13.20.

Note that in Sections 13.4 and 13.5 we considered the motion of *two* particles in a rigid body as seen from a *single* reference. Now we are considering the motion of a *single* particle as seen from *two* references.

The multireference approach can be very useful. For instance, we could know the motion of a particle relative to some device, such as a rocket, to which we attach a reference xyz. Furthermore, from telemetering devices, we know the translational and rotational motion (**Chasles' theorem**) of the rocket (and hence xyz) relative to an inertial reference XYZ. It is often important to know the motion of the aforementioned particle relative directly to the inertial reference. The multireference approach clearly is invaluable for such problems.

We now illustrate the use of Eq. 13.20. We shall proceed in a particular methodical way which we encourage you to follow in your homework problems. In these problems, we remind you the dot over a vector generally represents a time derviative as seen from XYZ.

[9]Generally, we have employed $\boldsymbol{r}$ as a position vector and $\boldsymbol{\rho}$ as a displacement vector. With two references, we shall often use $\boldsymbol{\rho}$ to denote a position vector for one of the references.

Example 13.8

An airplane moving at 200 ft/sec is undergoing a roll of 2 rad/min (Fig. 13.25). When the plane is horizontal, an antenna is moving out at a speed of 8 ft/sec relative to the plane and is at a position of 10 ft from the centerline of the plane. If we assume that the axis of roll corresponds to the centerline, what is the velocity of the antenna end relative to the ground when the plane is horizontal?

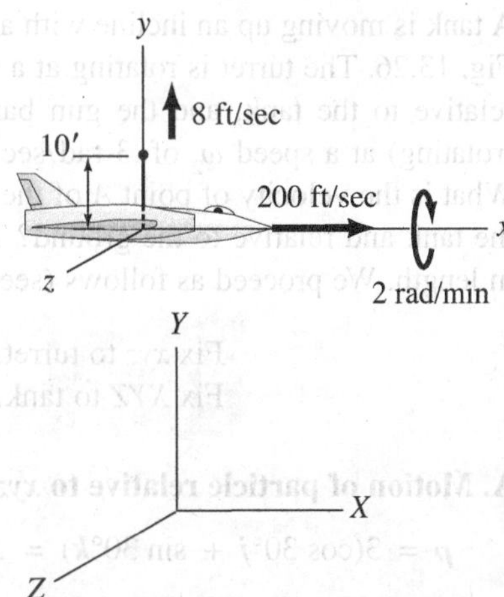

Figure 13.25. *xyz* fixed to plane; *XYZ* fixed to ground.

A *stationary reference XYZ on the ground* is shown in the diagram. A moving reference *xyz is fixed to the plane* with the *x* axis along the axis of roll and the *y* axis collinear with the antenna. We announce this formally as follows:

Fix *xyz* to plane.
Fix *XYZ* to ground.

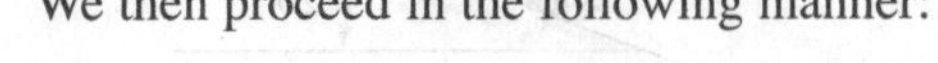

We then proceed in the following manner:

A. Motion of particle (antenna end) relative to *xyz*[10]

$$\boldsymbol{\rho} = 10\boldsymbol{j}\text{ ft}$$
$$\boldsymbol{V}_{xyz} = 8\boldsymbol{j}\text{ ft/sec}$$

B. Motion of *xyz* (moving reference) relative to *XYZ* (fixed reference)

$$\dot{\boldsymbol{R}} = 200\boldsymbol{i}\text{ ft/sec}$$
$$\boldsymbol{\omega} = -\tfrac{2}{60}\boldsymbol{i} = -\tfrac{1}{30}\boldsymbol{i}\text{ rad/sec}$$

We now employ Eq. 13.20 to get

$$\boldsymbol{V}_{XYZ} = \boldsymbol{V}_{xyz} = \dot{\boldsymbol{R}} + \boldsymbol{\omega} \times \boldsymbol{\rho}$$
$$= 8\boldsymbol{j} + 200\boldsymbol{i} + \left(-\frac{\boldsymbol{i}}{30}\right) \times (10\boldsymbol{j})$$

$$\boldsymbol{V}_{XYZ} = 200\boldsymbol{i} + 8\boldsymbol{j} - \tfrac{1}{3}\boldsymbol{k}\text{ ft/sec}$$

[10]Note that since the corresponding axes of the references are parallel to each other at the instant of interest, the unit vectors ***i***, ***j***, and ***k*** apply to either reference at the instant of interest. We will arrange *xyz* and *XYZ* this way whenever possible.

Note from the preceding example that in Part *A*, we are using the dynamics of a particle as presented in Chapters 11-14, while in Part *B* we are implementing **Chasles' theorem** as presented in this chapter. Your author based on long experience urges the student to work in this methodical manner.

Example 13.9

A tank is moving up an incline with a speed of 10 km/hr in Fig. 13.26. The turret is rotating at a speed ω_1 of 2 rad/sec relative to the tank, and the gun barrel is being lowered (rotating) at a speed ω_2 of .3 rad/sec relative to the turret. What is the velocity of point *A* of the gun barrel relative to the tank and relative to the ground? The gun barrel is 3 m in length. We proceed as follows (see Fig. 13.27).

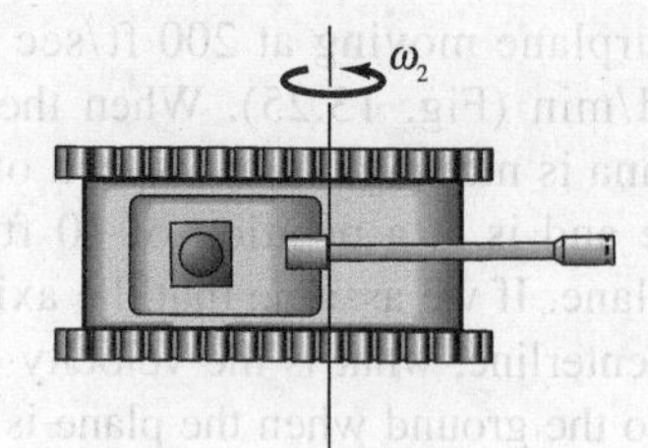

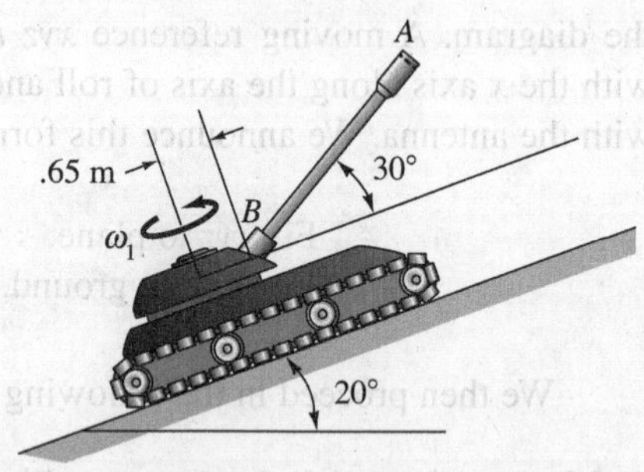

Figure 13.26. Tank with turret and gun barrel in motion.

Fix *xyz* to turret.
Fix *XYZ* to tank.

A. Motion of particle relative to *xyz*

$$\boldsymbol{\rho} = 3(\cos 30°\boldsymbol{j} + \sin 30°\boldsymbol{k}) = 2.60\boldsymbol{j} + 1.50\boldsymbol{k} \text{ m}$$

Since $\boldsymbol{\rho}$ is fixed in the gun barrel, which has an angular velocity $\boldsymbol{\omega}_2$ relative to *xyz*, we have

$$V_{xyz} = \left(\frac{d\boldsymbol{\rho}}{dt}\right)_{xyz} = \boldsymbol{\omega}_2 \times \boldsymbol{\rho} = (-3\boldsymbol{i}) \times (2.60\boldsymbol{j} + 1.5\boldsymbol{k})$$

$$= -.780\boldsymbol{k} + .45\boldsymbol{j} \text{ m/sec}$$

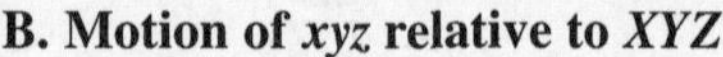

B. Motion of *xyz* relative to *XYZ*

$$\boldsymbol{R} = .65\boldsymbol{j}$$

Since $\boldsymbol{R}$ is fixed in the turret, which is rotating with angular speed $\boldsymbol{\omega}_1$ relative to *XYZ*, we have

$$\dot{\boldsymbol{R}} = \boldsymbol{\omega}_1 \times \boldsymbol{R} = 2\boldsymbol{k} \times .65\boldsymbol{j} = -1.3\boldsymbol{i} \text{ m/sec}$$

$$\boldsymbol{\omega} = \boldsymbol{\omega}_1 = 2\boldsymbol{k} \text{ rad/sec}$$

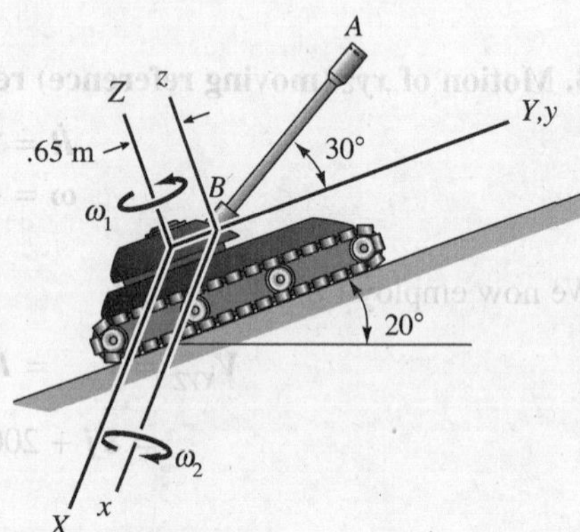

Figure 13.27. *xyz* fixed to turret; *XYZ* fixed to tank.

We can now substitute into the basic equation relating V_{xyz} to V_{XYZ}. That is,

$$V_{XYZ} = V_{xyz} + \dot{\boldsymbol{R}} + \boldsymbol{\omega} \times \boldsymbol{\rho}$$

$$= (-.780\boldsymbol{k} + .45\boldsymbol{j}) - 1.3\boldsymbol{i} + (2\boldsymbol{k}) \times (2.60\boldsymbol{j} + 1.50\boldsymbol{k})$$

$$V_{XYZ} = -6.5\boldsymbol{i} + .45\boldsymbol{j} - .780\boldsymbol{k} \text{ m/sec}$$

This result is the desired velocity of *A* relative to the tank. Since the tank is moving with a speed of (10)(1,000)/(3,600) = 2.78 m/sec relative to the ground, we can say that *A* has a velocity relative to the ground given as

$$V_{\text{ground}} = V_{XYZ} + 2.78\boldsymbol{j}$$

$$V_{\text{ground}} = -6.5\boldsymbol{i} + 3.23\boldsymbol{j} - .780\boldsymbol{k} \text{ m/sec}$$

Example 13.10

A gunboat in heavy seas is firing its main battery (see Fig. 13.28). The gun barrel has an angular velocity ω_1 relative to the turret, while the turret has an angular velocity ω_2 relative to the ship. If we wish to have the velocity components of the emerging shell to be zero in the stationary X and Z directions at a certain specific time t, what should ω_1 and ω_2 be at this instant? At this instant, the ship has a translational velocity given as

$$V_{ship} = .02\boldsymbol{i} + .016\boldsymbol{k} \text{ m/s}$$

Take the inclination of the barrel to be $\theta = 30°$. Determine also the velocity of the gun barrel tip A.

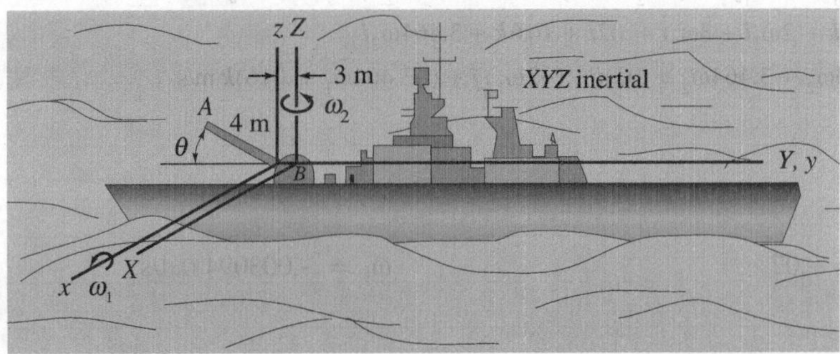

Figure 13.28. A gunboat in heavy seas firing its main battery.

We proceed to solve this problem by the following positioning of axes shown on Fig. 13.28.

> Fix xyz to turret.
> Fix XYZ to the ground (inertial reference).

We can now proceed with the detailed analysis of the problem.

A. Motion of A relative to xyz

$$\boldsymbol{\rho} = -(4)(.866)\boldsymbol{j} + (4)(.5)\boldsymbol{k} = -3.464\boldsymbol{j} + 2\boldsymbol{k} \text{ m}$$

$$V_{xyz} = \boldsymbol{\omega}_1 \times \boldsymbol{\rho} = \omega_1\boldsymbol{i} \times (-3.464\boldsymbol{j} + 2\boldsymbol{k}) = -3.464\omega_1\boldsymbol{k} - 2\omega_1\boldsymbol{j} \text{ m/s}$$

Example 13.10 (Continued)

B. Motion of *xyz* relative to *XYZ*

$$\boldsymbol{R} = -3\boldsymbol{j}\text{ m}$$
$$\dot{\boldsymbol{R}} = \omega_2\boldsymbol{k} \times (-3\boldsymbol{j}) + (.02\boldsymbol{i} + .016\boldsymbol{k}) = (3\omega_2 + .02)\boldsymbol{i} + .016\boldsymbol{k}\text{ m/s}$$
$$\boldsymbol{\omega} = \omega_2\boldsymbol{k}\text{ rad/s}$$

We can now proceed with the calculations.

$$\begin{aligned} \boldsymbol{V}_{XYZ} &= \boldsymbol{V}_{xyz} + \dot{\boldsymbol{R}} + \boldsymbol{\omega} \times \boldsymbol{\rho} \\ &= (-3.464\omega_1\boldsymbol{k} - 2\omega_1\boldsymbol{j}) + (3\omega_2 + .02)\boldsymbol{i} + .016\boldsymbol{k} + (\omega_2\boldsymbol{k}) \times (-3.464\boldsymbol{j} + 2\boldsymbol{k} \\ &= -3.464\omega_1\boldsymbol{k} - 2\omega_1\boldsymbol{i} + 3\omega_2\boldsymbol{i} + .02\boldsymbol{i} + .016\boldsymbol{k} + 3.464\omega_2\boldsymbol{i} \\ \therefore \boldsymbol{V}_{XYZ} &= (3\omega_2 + 3.464\omega_2 + .02)\boldsymbol{i} + (-2\omega_1)\boldsymbol{j} + (-3.464\omega_1 + .016)\boldsymbol{k}\text{ m/s} \end{aligned}$$

Let $(V_{XYZ})_X = \mathbf{0}$

$\therefore 6.464\omega_2 = -.02$ $\qquad$ $\omega_2 = -.003094$ rad/s

Let $(V_{XYZ})_Z = \mathbf{0}$

$\therefore -3.464\omega_1 = -.016$ $\qquad$ $\omega_1 = .004619$ rad/s

Finally, we can give $\boldsymbol{V}_{XYZ}$ as $\qquad$ $\boldsymbol{V}_{XYZ} = -2\omega_1\boldsymbol{j} = -.009238\boldsymbol{j}$ m/s

In some of the homework problem diagrams, in the remainder of the chapter, a set of axes *xyz* has been shown as a suggestion for use by the student. This has been done to help clarify the geometry of the diagram. Also, if the student chooses to use these axes he/she will be able to compare more easily his/her solution with that of the author as presented in the instructor's manual. However, (and note this carefully) the student must decide independently as to how to *fix this reference* and to state clearly as we have done in the examples *how this reference has been fixed.*

Also, we strongly urge the student to make careful clear diagrams and to follow the orderly progression of steps (**A. Motion of particle etc., etc.** followed by **B. Motion of *xyz* etc., etc.**).

PROBLEMS

13.70. A space laboratory, in order to simulate gravity, rotates relative to inertial reference *XYZ* at a rate ω_1. For occupant *A* to feel comfortable, what should ω_1 be? Clearly, at the center room *B*, there is close to zero gravity for zero-*g* experiments. A conveyor along one of the spokes transports items from the living quarters at the periphery to the zero-gravity laboratory at the center. In particular, a particle *D* has a velocity toward *B* of 5 m/sec relative to the space station. What is its velocity relative to the inertial reference *XYZ*?

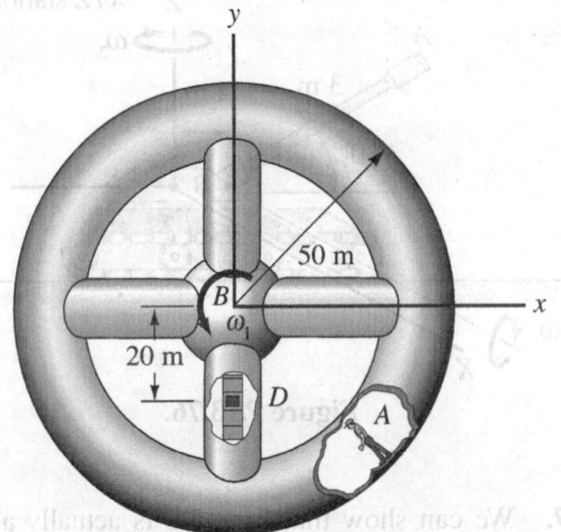

Figure P.13.70.

13.71. Bodies *a* and *b* slide away from each other each with a constant velocity of 5 ft/sec along the axis *C–C* mounted on a platform. The platform rotates relative to the ground reference *XYZ* at an angular velocity of 10 rad/sec about axis *E–E* and has an angular acceleration of 5 rad/sec^2 relative to the ground reference *XYZ* at the time when the bodies are at a distance r = 3 ft from *E–E*. Determine the velocity of particle *b* relative to the ground reference.

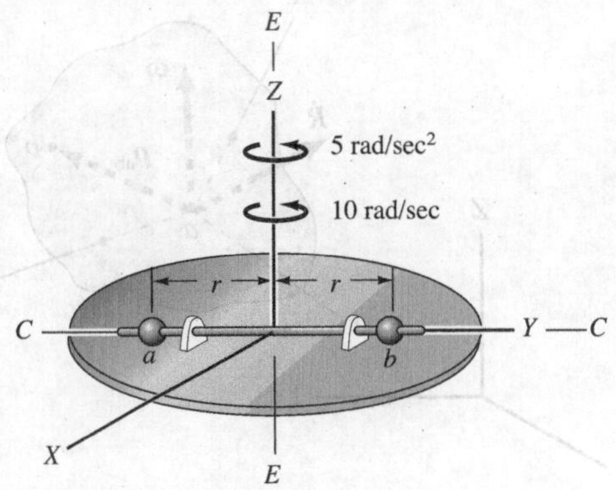

Figure P.13.71.

13.72. A particle rotates at a constant angular speed of 10 rad/sec on a platform, while the platform rotates with a constant angular speed of 50 rad/sec about axis *A–A*. What is the velocity of the particle *P* at the instant the platform is in the *XY* plane and the radius vector to the particle forms an angle of 30° with the *Y* axis as shown?

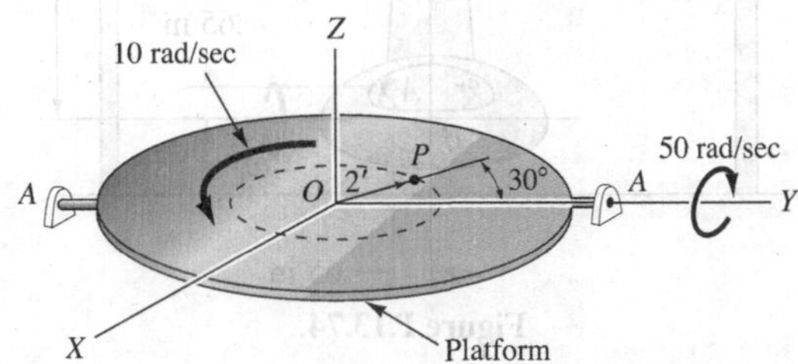

Figure P.13.72.

13.73. A platform *A* is rotating with constant angular speed ω_1 of 1 rad/sec. A second platform *B* rides on *A*, contains a row of test tubes, and has a constant angular speed ω_2 of .2 rad/sec relative to the platform *A*. A third platform *C* is in no way connected with platforms *A* and *B*. *E* on platform *C* is positioned above *A* and *B* and carries dispensers of chemicals which are electrically operated at proper times to dispense drops into the test tubes held by *B* below. What should the angular speed ω_3 be for platform *E* at instant shown if it is to dispense a drop of chemical having a zero tangential velocity relative to the test tube below?

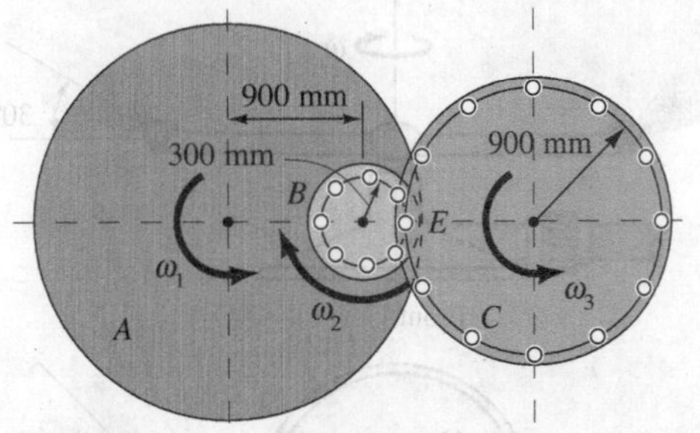

Figure P.13.73.

13.74. In an amusement park ride, the cockpit containing two occupants can rotate at an angular speed ω_1 relative to the main arm *OB*. The arm can rotate with angular speed ω_2 relative to the ground. For the position shown in the diagram and for ω_1 = 2 rad/sec and ω_2 = .2 rad/sec, find the velocity of point *A* (corresponding to the position of the eyes of an occupant) relative to the ground.

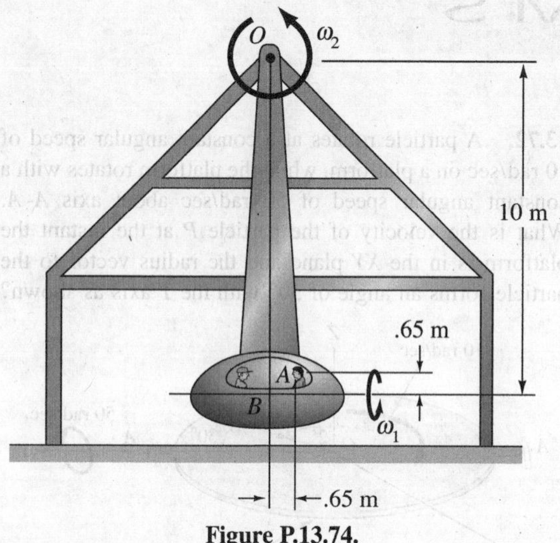

Figure P.13.74.

13.75. A water sprinkler has .4 cfs (cubic ft/sec) of water fed into the base. The sprinkler turns at the rate ω_1 of 1 rad/sec. What is the speed of the jet of water relative to the ground at the exits? The outlet area of the nozzle cross section is .75 in^2. [*Hint:* The volume of flow through a cross section is VA, where V is the velocity and A is the area of the cross section.]

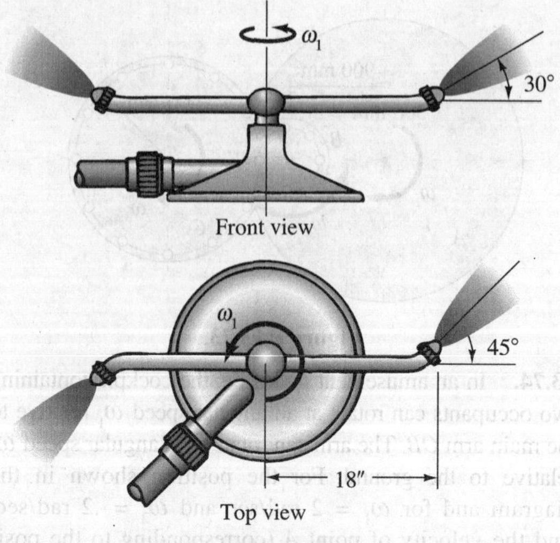

Figure P.13.75.

13.76. A tank is moving over rough terrain while firing its main gun at a fixed target. The barrel and turret of the gun partly compensate for the motion of the tank proper by giving the barrel an angular velocity $\boldsymbol{\omega}_1$ relative to the turret and, simultaneously, by giving the turret an angular velocity $\boldsymbol{\omega}_2$ relative to the tank proper such that any instant the velocity of end A of the barrel has zero velocity in the X and Z directions relative to the ground reference. What should these angular velocities be for the following translational motion of the tank:

$$V_{\text{TANK}} = 10\mathbf{i} + 4\mathbf{k} \text{ m/s}$$

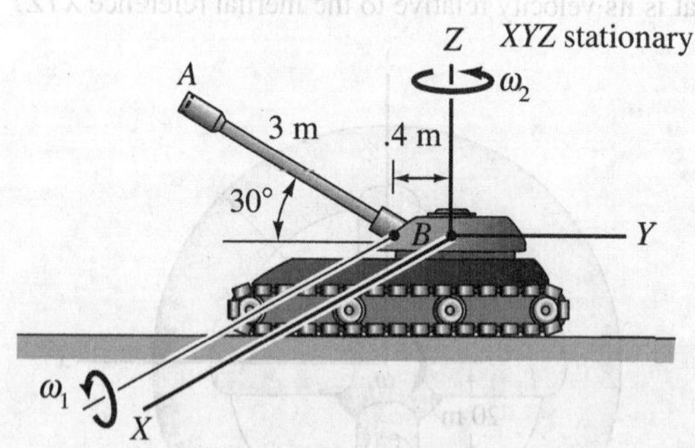

Figure P.13.76.

13.77. We can show that Eq. 13.6 is actually a special case of Eq. 13.20. For this purpose, consider a rigid body moving relative to XYZ. Choose two points a and b in the body. The body has a translational velocity corresponding to the velocity of point a and a rotational velocity $\boldsymbol{\omega}$ as shown in the diagram. Now embed a reference xyz into the body with origin at point a. Next, use this diagram and consider point b to show that Eq. 13.20 can be reformulated to be identical to Eq. 13.6.

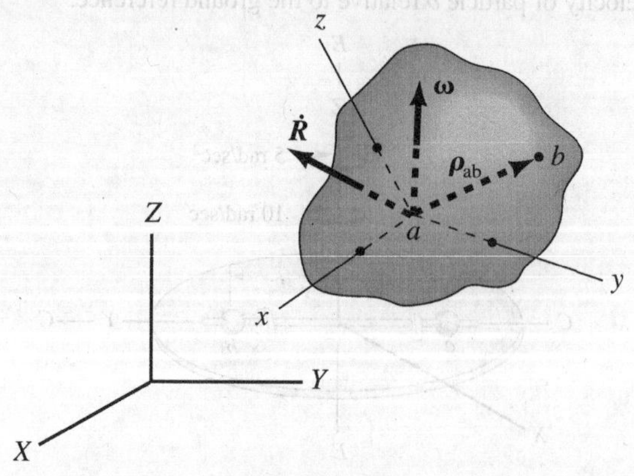

Figure P.13.77.

13.78. A simple-impulse type of turbomachine called a *Pelton* water wheel has a single jet of water issuing out of a nozzle and impinging on the system of buckets attached to a wheel. The runner, which is the assembly of buckets and wheel, has a radius of r to the center of the buckets. The shape of the bucket is also shown where a horizontal midsection of the bucket has been taken. Note that the jet is split in two parts by the bucket and is rotated relative to the bucket in the horizontal plane as measured by β. If we neglect gravity and friction, the speed of the water relative to the bucket is unchanged during the action. Suppose that 8 liters of water per second flow through the nozzle, whose cross-sectional area at the exit is 2,000 mm². If $r = 1$ m, what should ω_1 be (in rpm) for the water on average to have zero velocity relative to the ground in the Y direction when it comes off the bucket? Take $\beta = 10°$. (Why is it desirable to have the exit velocity equal to zero in the Y direction?) See the hint of Problem 13.75.

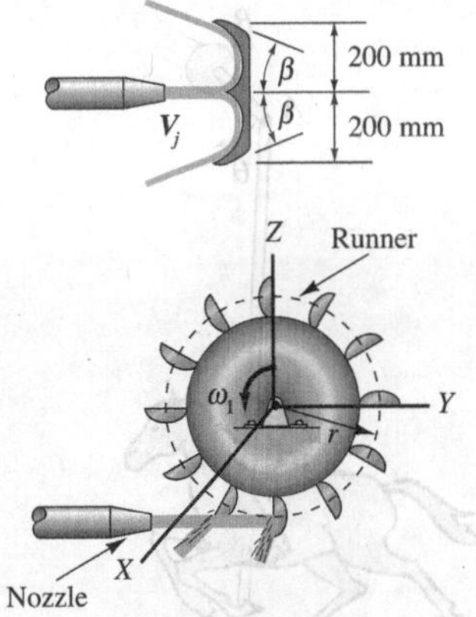

Figure P.13.78.

13.79. A propeller-driven airplane is moving at a speed of 130 km/hr. Also, it is undergoing a yaw rotation of 1/4 rad/sec and is simultaneously undergoing a loop rotation of 1/4 rad/sec. The propeller is rotating at the rate of 100 rpm with a sense in the positive Y direction. What is the velocity of the tip of the propeller a relative to the ground at the instant that the plane is horizontal as shown? The propeller is 3 m in total length and at the instant of interest the blade is in a vertical position.

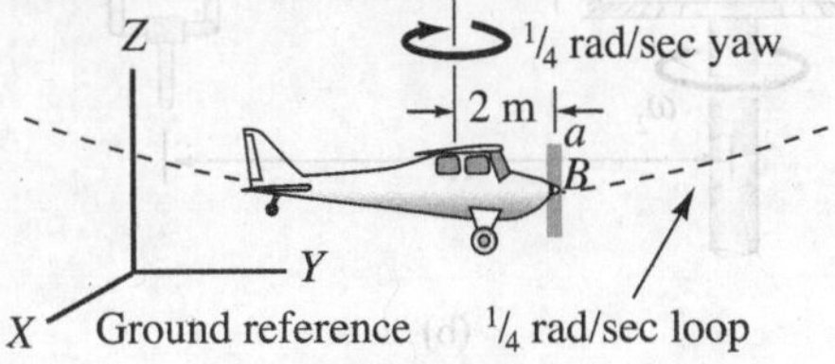

Figure P.13.79.

13.80. A crane moves to the right at a speed of 5 km/hr. The boom OB, which is 15 m long, is being raised at an angular speed ω_2 relative to the cab of .4 rad/sec, while the cab is rotating at an angular speed ω_1 of .2 rad/sec relative to the base. What is the velocity of pin B relative to the ground at the instant when OB is at an angle of 35° with the ground? The axis of rotation O of the boom is 1 m from the axis of rotation A–A of the cab, as shown in the diagram.

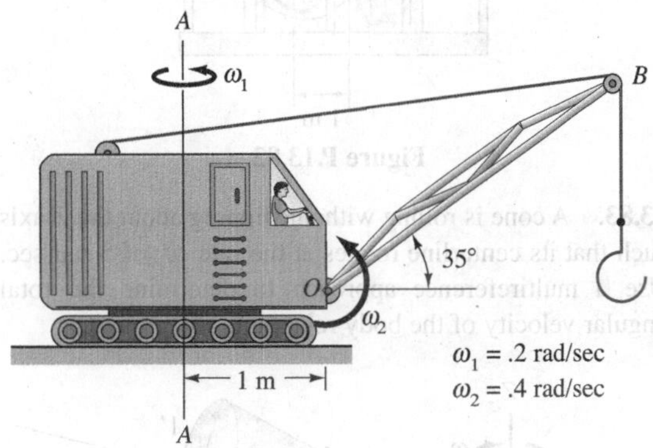

Figure P.13.80.

13.81. A power shovel main arm AC rotates with angular speed ω_1 of .3 rad/sec relative to the cab. Arm ED rotates at a speed ω_2 of .4 rad/sec relative to the main arm AC. The cab rotates about axis A–A at a speed ω_3 of .15 rad/sec relative to the tracks which are stationary. What is the velocity of point D, the center of the shovel, at the instant of interest shown in the diagram? AB has a length of 5 m and BD has a length of 4 m.

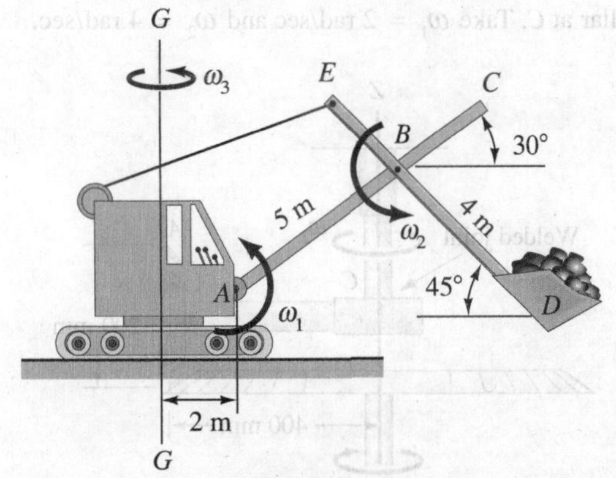

Figure P.13.81.

13.82. An antiaircraft gun is shown in action. The values of ω_1 and ω_2 are .3 rad/sec and .6 rad/sec, respectively. At the instant shown, what is the velocity of a projectile *normal* to the direction of the gun barrel when it just leaves the gun barrel as seen from the ground?

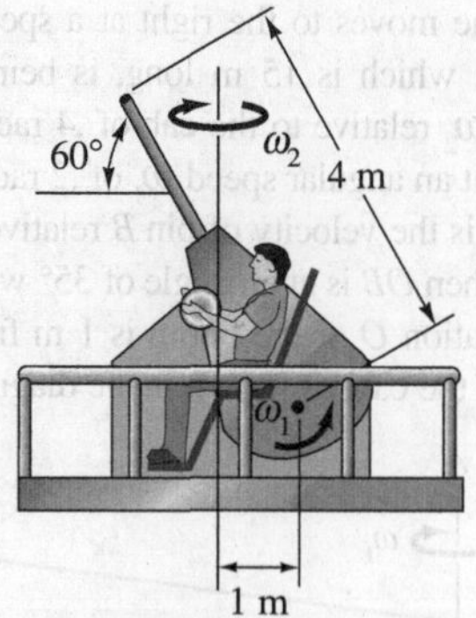

Figure P.13.82.

13.83. A cone is rolling without slipping about the Z axis such that its centerline rotates at the rate ω_1 of 5 rad/sec. Use a multireference approach to determine the total angular velocity of the body relative to the ground.

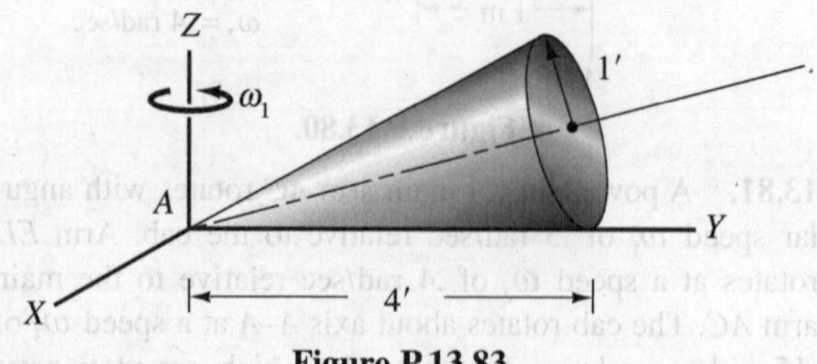

Figure P.13.83.

13.84. Find the velocity of gear tooth A relative to the ground reference XYZ. Note that ω_1 and ω_2 are both relative to the ground. Bevel gear A is free to rotate in the collar at C. Take $\omega_1 = 2$ rad/sec and $\omega_2 = 4$ rad/sec.

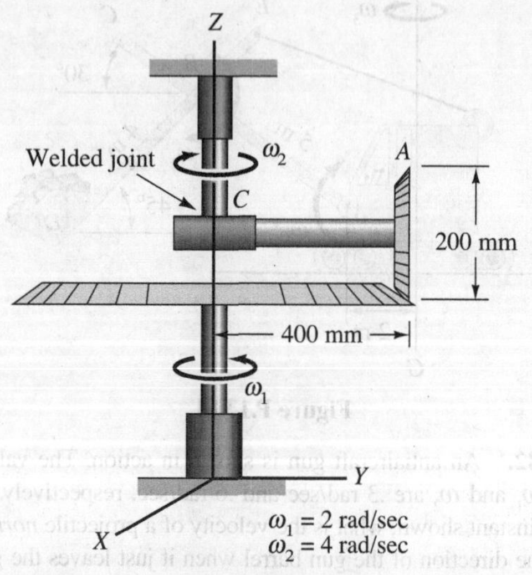

Figure P.13.84.

13.85. In a merry-go-round, the main platform rotates at the rate ω_1 of 10 revolutions per minute. A set of 45° bevel gears causes B to rotate at an angular speed $\dot{\theta}$ relative to the platform. The horse is mounted on AB, which slides in a slot at C and is moved at A by shaft B, as indicated in the diagram, where part of the merry-go-round is shown. If $AB = 1$ ft and $AC = 15$ ft, compute the velocity of point C relative to the platform. Then, compute the velocity of point C relative to the ground. Take $\theta = 45°$ at the instant of interest. What is the angular velocity of the horse relative to the platform and relative to the ground at the instant of interest?

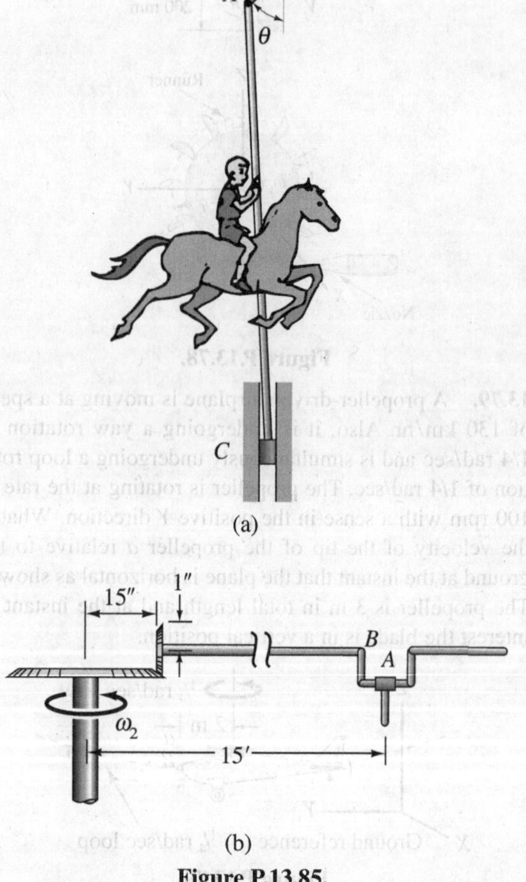

Figure P.13.85.

13.86. Rod BO rotates at a constant angular speed $\dot{\theta}$ of 5 rad/sec clockwise. A collar A on the rod is pinned to a slider C, which moves in the groove shown in the diagram. When $\theta = 60°$, compute the speed of the collar A relative to the ground. What is the speed of collar A relative to the rod?

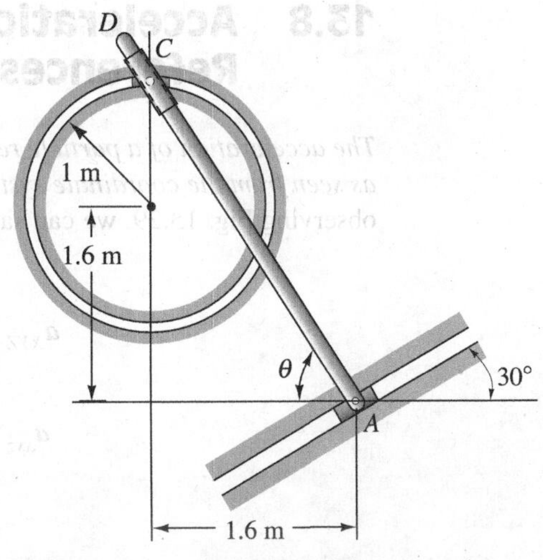

Figure P.13.88.

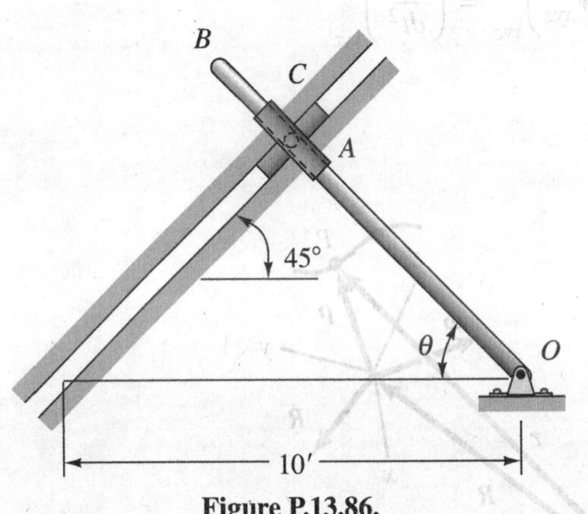

Figure P.13.86.

13.89. In Problem 13.88, assume, in addition to the rotation of bar AD, that pin A is moving at a speed of 1.6 m/sec up the grooved incline.

13.90. Rod AC is connected to a gear D and is guided by a bearing B. Bearing B can rotate only in the plane of the gears. If the angular speed of AC is 5 rad/sec clockwise, what is the angular speed of gear D relative to the ground? The diameter of D is 2 ft.

13.87. Work Problem 13.86 assuming that pin O is on rollers moving to the right at a speed of 3 ft/sec relative to the ground. In addition, OB rotates at a constant angular speed $\dot{\theta}$ of 5 rad/sec clockwise all at the instant of interest.

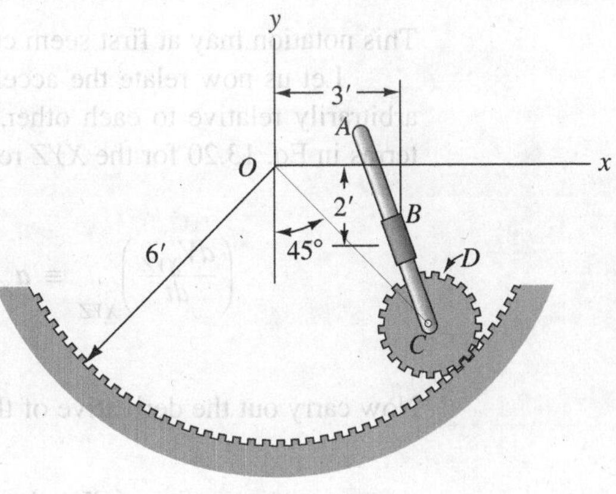

13.88. Rod AD rotates at a constant speed $\dot{\theta}$ of 2 rad/sec. Collar C on the rod DA is constrained to move in the circular groove shown in the diagram. When the rod is at the position shown, compute the speed of collar C relative to the ground. What is the speed of collar C relative to the rod AD? Point A is stationary.

Figure P.13.90.

13.8 Acceleration of a Particle for Different References

The acceleration of a particle relative to a coordinate system is simply the time derivative, as seen from the coordinate system, of the velocity relative to the coordinate system. Thus, observing Fig. 13.29, we can say:

$$a_{XYZ} = \left(\frac{d}{dt}V_{XYZ}\right)_{XYZ} = \left(\frac{d^2 r}{dt^2}\right)_{XYZ}$$
$$a_{xyz} = \left(\frac{d}{dt}V_{xyz}\right)_{xyz} = \left(\frac{d^2 \rho}{dt^2}\right)_{xyz} \tag{13.21}$$

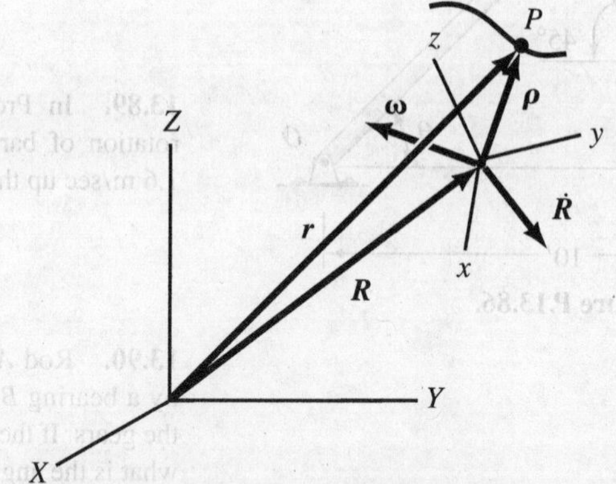

Figure 13.29. *xyz* moves arbitrarily relative to *XYZ*.

This notation may at first seem cumbersome to you, but it will soon be simplified.

Let us now relate the acceleration vectors of a particle for two references moving arbitrarily relative to each other. We do this by differentiating with respect to time the terms in Eq. 13.20 for the *XYZ* reference. Thus,

$$\left(\frac{dV_{XYZ}}{dt}\right)_{XYZ} \equiv a_{XYZ} = \left(\frac{dV_{xyz}}{dt}\right)_{XYZ} + \ddot{R} + \left[\frac{d}{dt}(\omega \times \rho)\right]_{XYZ} \tag{13.22}$$

Now carry out the derivative of the cross product using the product rule.

$$a_{XYZ} = \left(\frac{dV_{xyz}}{dt}\right)_{XYZ} + \ddot{R} + \omega \times \left(\frac{d\rho}{dt}\right)_{XYZ} + \left(\frac{d\omega}{dt}\right)_{XYZ} \times \rho \tag{13.23}$$

To introduce more physically meaningful terms, we can replace

$$\left(\frac{d\boldsymbol{V}_{xyz}}{dt}\right)_{XYZ} \text{ and } \left(\frac{d\boldsymbol{\rho}}{dt}\right)_{XYZ}$$

using Eq. 13.16 in the following way:

$$\left(\frac{d\boldsymbol{V}_{xyz}}{dt}\right)_{XYZ} = \left(\frac{d\boldsymbol{V}_{xyz}}{dt}\right)_{xyz} + \boldsymbol{\omega} \times \boldsymbol{V}_{xyz}$$

$$\left(\frac{d\boldsymbol{\rho}}{dt}\right)_{XYZ} = \left(\frac{d\boldsymbol{\rho}}{dt}\right)_{xyz} + \boldsymbol{\omega} \times \boldsymbol{\rho}$$

Substituting into Eq. 13.23, we get

$$\boldsymbol{a}_{XYZ} = \left(\frac{d\boldsymbol{V}_{xyz}}{dt}\right)_{xyz} + \boldsymbol{\omega} \times \boldsymbol{V}_{xyz} + \ddot{\boldsymbol{R}} + \boldsymbol{\omega} \times \left(\frac{d\boldsymbol{\rho}}{dt}\right)_{xyz} + \boldsymbol{\omega} \times (\boldsymbol{\omega} \times \boldsymbol{\rho}) + \left(\frac{d\boldsymbol{\omega}}{dt}\right)_{XYZ} \times \boldsymbol{\rho}$$

You will note that $(d\boldsymbol{V}_{xyz}/dt)_{xyz}$ is $\boldsymbol{a}_{xyz}$; that $(d\boldsymbol{\rho}/dt)_{xyz}$ is $\boldsymbol{V}_{xyz}$; and that $(d\boldsymbol{\omega}/dt)_{XYZ}$ is $\dot{\boldsymbol{\omega}}$. Hence, rearranging terms, we have

$$\boldsymbol{a}_{XYZ} = \boldsymbol{a}_{xyz} + \ddot{\boldsymbol{R}} + 2\boldsymbol{\omega} \times \boldsymbol{V}_{xyz} + \dot{\boldsymbol{\omega}} \times \boldsymbol{\rho} + \boldsymbol{\omega} \times (\boldsymbol{\omega} \times \boldsymbol{\rho}) \tag{13.24}$$

where $\boldsymbol{\omega}$ and $\dot{\boldsymbol{\omega}}$ are the angular velocity and acceleration, respectively, of the *xyz* reference relative to the *XYZ* reference. The vector $2(\boldsymbol{\omega} \times \boldsymbol{V}_{xyz})$ is called the *Coriolis acceleration vector*; we shall examine its interesting effects in Section 13.10.

Although Eq. 13.24 may seem somewhat terrifying at first, you will find that, by using it, problems that would otherwise be tremendously difficult can readily be carried out in a systematic manner. *You should keep in mind when solving problems that any of the methods developed in Chapter 9 can be used for determining the motion of the particle relative to the xyz reference or for determining the motion of the origin of xyz relative to the XYZ reference.* We shall now examine several problems, in which we shall use the notation, $\boldsymbol{\omega}_1$, $\boldsymbol{\omega}_2$, etc., to denote the various angular velocities involved. The notation, $\boldsymbol{\omega}$ (i.e., without subscripts), however, will we repeat be reserved to represent the angular velocity of the *xyz* reference relative to the *XYZ* reference.

Example 13.11

A stationary truck is carrying a cockpit for a worker who repairs overhead fixtures. At the instant shown in Fig. 13.30, the base D is rotating at angular speed ω_2 of .1 rad/sec with $\dot{\omega}_2 = .2$ rad/sec^2 relative to the truck. Arm AB is rotating at angular speed ω_1 of .2 rad/sec with $\dot{\omega}_1 = .8$ rad/sec^2 relative to DA. Cockpit C is rotating relative to AB so as to always keep the man upright. What are the velocity and acceleration vectors of the man relative to the ground if $\alpha = 45°$ and $\beta = 30°$ at the instant of interest? Take $DA = 13$ m.

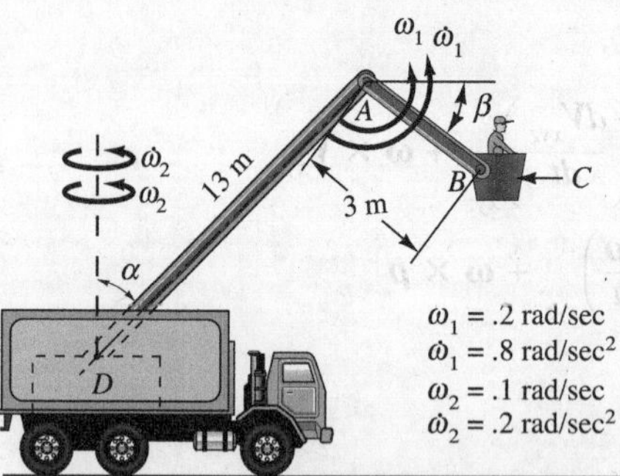

Figure 13.30. Truck with moving cockpit.

Because of the rotation of the cockpit C relative to arm AB to keep the man vertical, clearly, each particle in that body including the man has the same motion as point B of arm AB. Therefore, we shall concentrate our attention on this point.

Fix xyz to arm DA.
Fix XYZ to truck.

This situation is shown in Fig. 13.31.

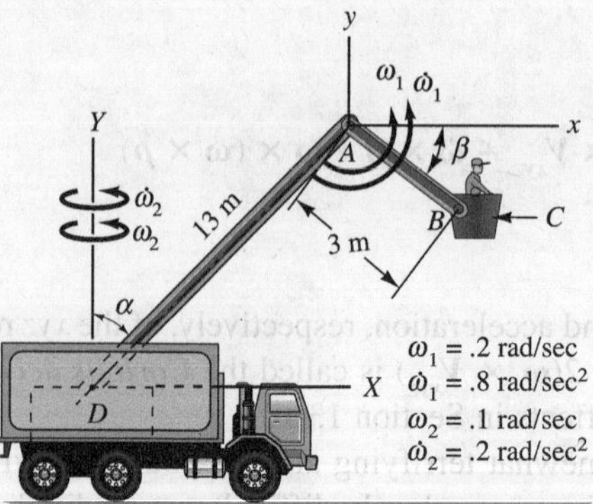

Figure 13.31. xyz fixed to DA; XYZ fixed to truck.

A. Motion of B relative to xyz

$$\boldsymbol{\rho} = 3(\cos\beta\boldsymbol{i} - \sin\beta\boldsymbol{j}) = 2.60\boldsymbol{i} - 1.5\boldsymbol{j}\text{ m}$$

Example 13.11 (Continued)

Since $\boldsymbol{\rho}$ is fixed in *AB*, which has angular velocity $\boldsymbol{\omega}_1$ relative to *xyz*, we have

$$\begin{aligned} \boldsymbol{V}_{xyz} &= \boldsymbol{\omega}_1 \times \boldsymbol{\rho} = (.2\boldsymbol{k}) \times (2.60\boldsymbol{i} - 1.5\boldsymbol{j}) \\ &= .520\boldsymbol{j} + .3\boldsymbol{i} \text{ m/sec} \\ \boldsymbol{a}_{xyz} &= \left(\frac{d\boldsymbol{\omega}_1}{dt}\right)_{xyz} \times \boldsymbol{\rho} + \boldsymbol{\omega}_1 \times \left(\frac{d\boldsymbol{\rho}}{dt}\right)_{xyz} \end{aligned}$$

As seen from *xyz*, only the value of $\boldsymbol{\omega}_1$ and not its direction is changing. Also note that $(d\boldsymbol{\rho}/dt)_{xyz} = \boldsymbol{V}_{xyz}$. Hence,

$$\begin{aligned} \boldsymbol{a}_{xyz} &= (.8\boldsymbol{k}) \times (2.60\boldsymbol{i} - 1.5\boldsymbol{j}) + (.2\boldsymbol{k}) \times (.520\boldsymbol{j} + .3\boldsymbol{i}) \\ &= 1.09\boldsymbol{i} + 2.14\boldsymbol{j} \text{ m/sec}^2 \end{aligned}$$

B. Motion of *xyz* relative to *XYZ*

$$\boldsymbol{R} = 13(.707\boldsymbol{i} + .707\boldsymbol{j}) = 9.19\boldsymbol{i} + 9.19\boldsymbol{j} \text{ m}$$

Since $\boldsymbol{R}$ is fixed in *DA*, and since *DA* rotates with angular velocity $\boldsymbol{\omega}_2$ relative to *XYZ*, we have

$$\begin{aligned} \dot{\boldsymbol{R}} &= \boldsymbol{\omega}_2 \times \boldsymbol{R} = (.1\boldsymbol{j}) \times (9.19\boldsymbol{i} + 9.19\boldsymbol{j}) \\ &= -.919\boldsymbol{k} \text{ m/sec} \\ \ddot{\boldsymbol{R}} &= \dot{\boldsymbol{\omega}}_2 \times \boldsymbol{R} + \boldsymbol{\omega}_2 \times \dot{\boldsymbol{R}} \\ &= (.2\boldsymbol{j}) \times (9.19\boldsymbol{i} + 9.19\boldsymbol{j}) + (.1\boldsymbol{j}) \times (-.919\boldsymbol{k}) \\ &= -1.838\boldsymbol{k} - .0919\boldsymbol{i} \text{ m/sec}^2 \\ \boldsymbol{\omega} &= \boldsymbol{\omega}_2 = .1\boldsymbol{j} \text{ rad/sec} \\ \dot{\boldsymbol{\omega}} &= \dot{\boldsymbol{\omega}}_2 = .2\boldsymbol{j} \text{ rad/sec}^2 \end{aligned}$$

Hence,

$$\begin{aligned} \boldsymbol{V}_{XYZ} &= \boldsymbol{V}_{xyz} + \dot{\boldsymbol{R}} + \boldsymbol{\omega} \times \boldsymbol{\rho} \\ &= .520\boldsymbol{j} + .3\boldsymbol{i} - .919\boldsymbol{k} + (.1\boldsymbol{j}) \times (2.60\boldsymbol{i} - 1.5\boldsymbol{j}) \end{aligned}$$

$$\boldsymbol{V}_{XYZ} = .3\boldsymbol{i} + .520\boldsymbol{j} - 1.179\boldsymbol{k} \text{ m/sec}$$

$$\begin{aligned} \boldsymbol{a}_{XYZ} &= \boldsymbol{a}_{xyz} + \ddot{\boldsymbol{R}} + 2\boldsymbol{\omega} \times \boldsymbol{V}_{xyz} + \dot{\boldsymbol{\omega}} \times \boldsymbol{\rho} + \boldsymbol{\omega} \times (\boldsymbol{\omega} \times \boldsymbol{\rho}) \\ &= 1.09\boldsymbol{i} + 2.14\boldsymbol{j} - 1.838\boldsymbol{k} - .0919\boldsymbol{i} \\ &\quad + 2(.1\boldsymbol{j}) \times (.520\boldsymbol{j} + .3\boldsymbol{i}) + (.2\boldsymbol{j}) \times (2.60\boldsymbol{i} - 1.5\boldsymbol{j}) \\ &\quad + (.1\boldsymbol{j}) \times [(.1\boldsymbol{j}) \times (2.60\boldsymbol{i} - 1.5\boldsymbol{j})] \end{aligned}$$

$$\boldsymbol{a}_{XYZ} = .978\boldsymbol{i} + 2.14\boldsymbol{j} - 2.42\boldsymbol{k} \text{ m/sec}^2$$

Notice that the essential aspects of the analysis come in the consideration of parts A and B of the problem, while the remaining portion involves direct substitution and vector algebraic operations.

Example 13.12

A wheel rotates with an angular speed ω_2 of 5 rad/sec on a platform which rotates with a speed ω_1 of 10 rad/sec relative to the ground as shown in Fig. 13.32. A valve gate *A* moves down the spoke of the wheel, and when the spoke is vertical the valve gate has a speed of 20 ft/sec, an acceleration of 10 ft/sec^2 along the spoke, and is 1 ft from the shaft centerline of the wheel. Compute the velocity and acceleration of the valve gate relative to the ground at this instant.

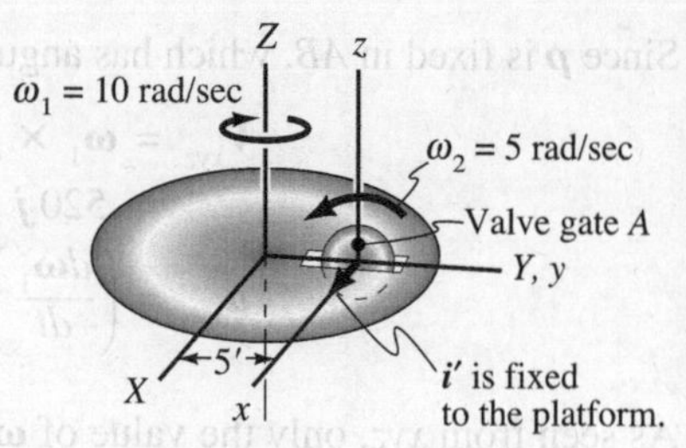

Figure 13.32. *xyz* fixed to wheel; *XYZ* fixed to ground.

Fix xyz to wheel.
Fix XYZ to ground.

A. Motion of particle relative to *xyz*

$$\boldsymbol{\rho} = \boldsymbol{k}\text{ ft}$$
$$\boldsymbol{V}_{xyz} = -20\boldsymbol{k}\text{ ft/sec}$$
$$\boldsymbol{a}_{xyz} = -10\boldsymbol{k}\text{ ft/sec}^2$$

B. Motion of *xyz* relative to *XYZ*

$$\boldsymbol{R} = 5\boldsymbol{j}\text{ ft}$$

Since $\boldsymbol{R}$ is fixed to the platform:

$$\dot{\boldsymbol{R}} = \boldsymbol{\omega}_1 \times \boldsymbol{R} = (-10\boldsymbol{k}) \times (5\boldsymbol{j}) = 50\boldsymbol{i}\text{ ft/sec}$$
$$\ddot{\boldsymbol{R}} = \dot{\boldsymbol{\omega}}_1 \times \boldsymbol{R} + \boldsymbol{\omega}_1 \times \dot{\boldsymbol{R}}$$
$$= \boldsymbol{0} + (-10\boldsymbol{k}) \times (50\boldsymbol{i}) = -500\boldsymbol{j}\text{ ft/sec}^2$$
$$\boldsymbol{\omega} = \boldsymbol{\omega}_2 + \boldsymbol{\omega}_1 = 5\boldsymbol{i} - 10\boldsymbol{k}\text{ rad/sec}$$
$$\dot{\boldsymbol{\omega}} = \dot{\boldsymbol{\omega}}_2 + \dot{\boldsymbol{\omega}}_1$$

Note that $\boldsymbol{\omega}_2$ is of constant magnitude but, because of the bearings of the wheel, $\boldsymbol{\omega}_2$ must rotate with the platform. In short, we can say that $\boldsymbol{\omega}_2$ is *fixed* to the platform and so $\dot{\boldsymbol{\omega}}_2 = \boldsymbol{\omega}_1 \times \boldsymbol{\omega}_2$. Hence,

$$\dot{\boldsymbol{\omega}} = \boldsymbol{\omega}_1 \times \boldsymbol{\omega}_2 + \boldsymbol{0}$$
$$= (-10\boldsymbol{k}) \times (5\boldsymbol{i}) = -50\boldsymbol{j}\text{ rad/sec}^2$$

We then have

$$\boldsymbol{V}_{XYZ} = \boldsymbol{V}_{xyz} + \dot{\boldsymbol{R}} + \boldsymbol{\omega} \times \boldsymbol{\rho}$$
$$= -20\boldsymbol{k} + 50\boldsymbol{i} + (5\boldsymbol{i} - 10\boldsymbol{k}) \times \boldsymbol{k}$$

$$\boldsymbol{V}_{XYZ} = 50\boldsymbol{i} - 5\boldsymbol{j} - 20\boldsymbol{k}\text{ ft/sec}$$

Example 13.12 (Continued)

Also,

$$\begin{aligned}\boldsymbol{a}_{XYZ} &= \boldsymbol{a}_{xyz} + \ddot{\boldsymbol{R}} + 2\boldsymbol{\omega} \times \boldsymbol{V}_{xyz} + \dot{\boldsymbol{\omega}} \times \boldsymbol{\rho} + \boldsymbol{\omega} \times (\boldsymbol{\omega} \times \boldsymbol{\rho}) \\ &= -10\boldsymbol{k} - 500\boldsymbol{j} + 2(5\boldsymbol{i} - 10\boldsymbol{k}) \times (-20\boldsymbol{k}) + (-50\boldsymbol{j}) \times \boldsymbol{k} \\ &\quad + (5\boldsymbol{i} - 10\boldsymbol{k}) \times [(5\boldsymbol{i} - 10\boldsymbol{k}) \times \boldsymbol{k}]\end{aligned}$$

$$\boldsymbol{a}_{XYZ} = -100\boldsymbol{i} - 300\boldsymbol{j} - 35\boldsymbol{k} \text{ ft/sec}^2$$

Example 13.13

In Example 13.12, the wheel accelerates at the instant under discussion with $\dot{\omega}_2 = 5$ rad/sec², and the platform accelerates with $\dot{\omega}_1 = 10$ rad/sec² (see Fig. 13.32). Find the velocity and acceleration of the valve gate *A*.

If we review the contents of parts A and B of Example 13.12, it will be clear that only $\ddot{\boldsymbol{R}}$ and $\dot{\boldsymbol{\omega}}$ are affected by the fact that $\dot{\omega}_1 = 10$ rad/sec² and $\dot{\omega}_2 = 5$ rad/sec². In this regard, consider $\boldsymbol{\omega}_2$. It is no longer of constant value and cannot be considered as *fixed* in the platform. However we can express $\boldsymbol{\omega}_2$ as $\omega_2 \boldsymbol{i}'$ *at all times*, wherein $\boldsymbol{i}'$ is *fixed* in the platform as shown in Fig. 13.32. Thus, we can say for $\boldsymbol{\omega}$:

$$\boldsymbol{\omega} = \omega_2 \boldsymbol{i}' + \boldsymbol{\omega}_1$$

Therefore,

$$\begin{aligned}\dot{\boldsymbol{\omega}} &= \dot{\omega}_2 \boldsymbol{i}' + \omega_2 \dot{\boldsymbol{i}}' + \dot{\boldsymbol{\omega}}_1 \\ &= 5\boldsymbol{i}' + 5(\boldsymbol{\omega}_1 \times \boldsymbol{i}') - 10\boldsymbol{k} \\ &= 5\boldsymbol{i}' + 5(-10\boldsymbol{k}) \times \boldsymbol{i}' - 10\boldsymbol{k}\end{aligned}$$

At the instant of interest, $\boldsymbol{i}' = \boldsymbol{i}$. Hence,

$$\dot{\boldsymbol{\omega}} = 5\boldsymbol{i} - 50\boldsymbol{j} - 10\boldsymbol{k} \text{ rad/sec}^2$$

Hence, we use the above $\dot{\boldsymbol{\omega}}$ in part B of Example 13.12 to compute $\boldsymbol{V}_{XYZ}$ and $\boldsymbol{a}_{XYZ}$. The computation of $\ddot{\boldsymbol{R}}$ is straightforward and so we can compute $\boldsymbol{a}_{XYZ}$ accordingly. We leave the details to the reader.

An understanding of Examples 13.11, 13.12, and 13.13, involving two angular velocities of component parts is sufficient for most of the homework problems of this section covering a wide range of applications. In the next example, we have three angular velocities to deal with. We urge you to examine it carefully if time allows. It is an interesting problem, and comprehension of the three different analyses given will ensure a strong grasp of multireference kinematics.

Example 13.14[11]

To simulate the flight conditions of a space vehicle, engineers have developed the *centrifuge*, shown diagrammatically in Fig. 13.33. A main *arm*, 40 ft long, rotates about the *A–A* axis. The pilot sits in a *cockpit*, which can rotate about axis *C–C*. The *seat* for the pilot can rotate inside the cockpit about an axis shown as *B–B*. These rotations are controlled by a computer that is set to simulate certain maneuvers corresponding to the entry and exit from the earth's atmosphere, malfunctions of the control system, and so on. When a pilot sits in the cockpit, his/her head has a position which is 3 ft from the seat as shown in Fig. 13.33. At the instant of interest the main arm is rotating at 10 rpm and accelerating at 5 rpm^2. The cockpit is rotating at a constant speed about *C–C* relative to the main arm at 10 rpm. Finally, the seat is rotating at a constant speed of 5 rpm relative to the cockpit about axis *B–B*. How many *g*'s of acceleration relative to the ground is the pilot's head subject to?[12] Note that the three axes, *A–A*, *C–C*, and *B–B*, are orthogonal to each other at time *t*.

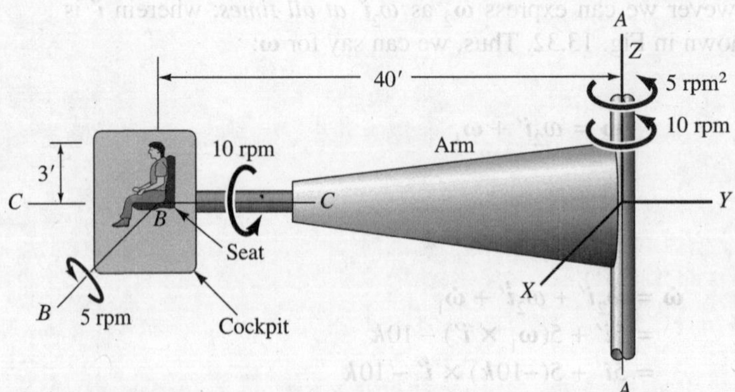

Figure 13.33. Centrifuge for simulating flight conditions.

[11]Example 13.14 was given as two homework problems in both the first and second editions of this text. They were so instructive that for subsequent editions the author decided to move the problems into the main text.

[12]A *g* of acceleration is an amount of acceleration equal to that of gravity (32.2 ft/sec^2 or 9.81 m/sec^2). Thus, a 4*g* acceleration is equivalent to an acceleration of 128.8 ft/sec^2.

Example 13.14 (Continued)

In Fig. 13.34 the arm of the centrifuge rotates relative to the ground at an angular velocity of ω_1. The cockpit meanwhile rotates relative to the arm with angular speed ω_2. Finally, the seat rotates relative to the cockpit at an angular speed ω_3. For constant ω_2, we see that, because of bearings in the arm, the vector $\boldsymbol{\omega}_2$ is "fixed" in the arm. Also, for constant ω_3, because of bearings in the cockpit, the vector $\boldsymbol{\omega}_3$ is "fixed" in the cockpit. Before we examine the acceleration of the pilot's head, note that at the instant of interest:

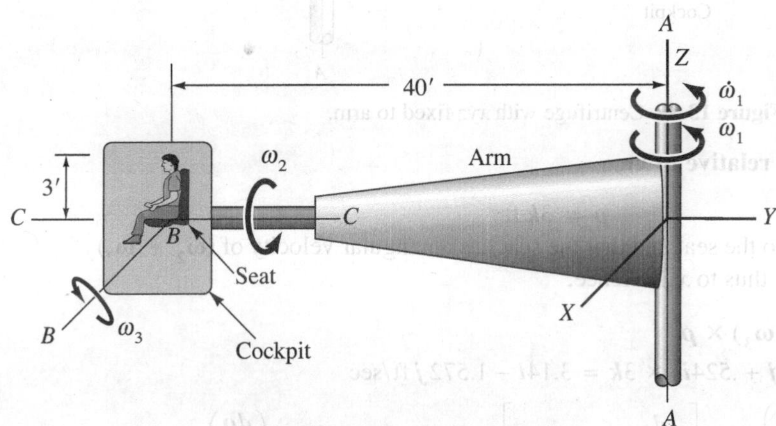

Figure 13.34. Centrifuge listing $\boldsymbol{\omega}$'s.

$$\omega_1 = \omega_2 = 10 \text{ rpm} = 1.048 \text{ rad/sec}$$
$$\dot{\omega}_1 = 5 \text{ rpm}^2 = .00873 \text{ rad/sec}^2$$
$$\omega_3 = 5 \text{ rpm} = .524 \text{ rad/sec}$$

We shall do this problem using three different kinds of moving references *xyz*.

ANALYSIS I

Fix *xyz* to arm.
Fix *XYZ* to ground.

Note in Fig. 13.35 that *xyz* and the arm to which it is fixed are shown dark. Note also that the axes *xyz* and *XYZ* are parallel to each other at the instant of interest.

Example 13.14 (Continued)

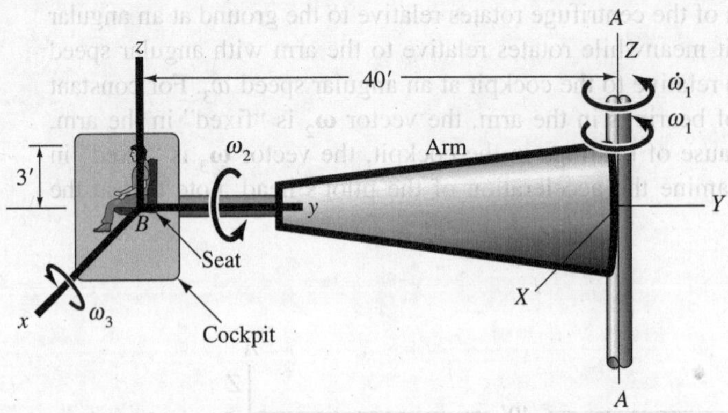

Figure 13.35. Centrifuge with *xyz* fixed to arm.

A. Motion of particle relative to *xyz*

$$\boldsymbol{\rho} = 3\boldsymbol{k} \text{ ft}$$

Note that $\boldsymbol{\rho}$ is "fixed" to the seat and that the seat has an angular velocity of $(\boldsymbol{\omega}_2 + \boldsymbol{\omega}_3)$ relative to the arm and thus to *xyz*. Hence,

$$\begin{aligned}\boldsymbol{V}_{xyz} &= (\boldsymbol{\omega}_2 + \boldsymbol{\omega}_3) \times \boldsymbol{\rho} \\ &= (1.048\boldsymbol{j} + .524\boldsymbol{i}) \times 3\boldsymbol{k} = 3.14\boldsymbol{i} - 1.572\boldsymbol{j} \text{ ft/sec}\end{aligned}$$

$$\boldsymbol{a}_{xyz} = \left(\frac{d\boldsymbol{V}_{xyz}}{dt}\right)_{xyz} = \left[\frac{d}{dt}_{xyz}(\boldsymbol{\omega}_2 + \boldsymbol{\omega}_3)\right] \times \boldsymbol{\rho} + (\boldsymbol{\omega}_2 + \boldsymbol{\omega}_3) \times \left(\frac{d\boldsymbol{\rho}}{dt}\right)_{xyz}$$

Clearly, relative to the arm, and thus to *xyz*, $\boldsymbol{\omega}_2$ is constant. And $\boldsymbol{\omega}_3$ is fixed in the cockpit that has an angular velocity of $\boldsymbol{\omega}_2$ relative to *xyz*. Thus, we have

$$\begin{aligned}\boldsymbol{a}_{xyz} &= (0 + \boldsymbol{\omega}_2 \times \boldsymbol{\omega}_3) \times \boldsymbol{\rho} + (\boldsymbol{\omega}_2 + \boldsymbol{\omega}_3) \times \boldsymbol{V}_{xyz} \\ &= (1.048\boldsymbol{j} \times .524\boldsymbol{i}) \times 3\boldsymbol{k} + (1.048\boldsymbol{j} + .524\boldsymbol{i}) \times (3.14\boldsymbol{i} - 1.572\boldsymbol{j}) \\ &= -4.12\boldsymbol{k} \text{ ft/sec}^2\end{aligned}$$

B. Motion of *xyz* relative to *XYZ*

$$\boldsymbol{R} = -40\boldsymbol{j} \text{ ft}$$

Note that $\boldsymbol{R}$ is fixed in the arm, which has an angular velocity $\boldsymbol{\omega}_1$ relative to *XYZ*. Hence,

$$\begin{aligned}\dot{\boldsymbol{R}} &= \boldsymbol{\omega}_1 \times \boldsymbol{R} = 1.048\boldsymbol{k} \times (-40\boldsymbol{j}) = 41.9\boldsymbol{i} \text{ ft/sec} \\ \ddot{\boldsymbol{R}} &= \boldsymbol{\omega}_1 \times \dot{\boldsymbol{R}} + \dot{\boldsymbol{\omega}}_1 \times \boldsymbol{R} \\ &= 1.048\boldsymbol{k} \times 41.9\boldsymbol{i} + .00873\boldsymbol{k} \times (-40\boldsymbol{j}) \\ &= 43.9\boldsymbol{j} + .349\boldsymbol{i} \text{ ft/sec}^2 \\ \boldsymbol{\omega} &= \boldsymbol{\omega}_1 = 1.048\boldsymbol{k} \text{ rad/sec} \\ \dot{\boldsymbol{\omega}} &= \dot{\boldsymbol{\omega}}_1 = .00873\boldsymbol{k} \text{ rad/sec}^2\end{aligned}$$

Example 13.14 (Continued)

We can now substitute into the following equation:

$$\boldsymbol{a}_{XYZ} = \boldsymbol{a}_{xyz} + \ddot{\boldsymbol{R}} + 2\boldsymbol{\omega} \times \boldsymbol{V}_{xyz} + \dot{\boldsymbol{\omega}} \times \boldsymbol{\rho} + \boldsymbol{\omega} \times (\boldsymbol{\omega} \times \boldsymbol{\rho})$$

Therefore,

$$\boldsymbol{a}_{XYZ} = 3.64\boldsymbol{i} + 50.5\boldsymbol{j} - 4.12\boldsymbol{k} \text{ ft/sec}^2$$

$$|\boldsymbol{a}_{XYZ}| = \frac{\sqrt{3.64^2 + 50.5^2 + 4.12^2}}{32.2} = 1.578\, g$$

ANALYSIS II

Fix xyz to cockpit.
Fix XYZ to ground.

This situation is shown in Fig. 13.36.

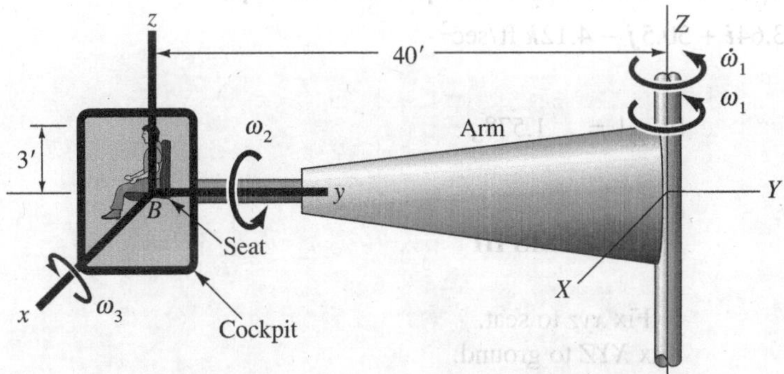

Figure 13.36. Centrifuge with xyz fixed to cockpit.

A. Motion of particle relative to xyz

$$\boldsymbol{\rho} = 3\boldsymbol{k} \text{ ft}$$

Note that $\boldsymbol{\rho}$ is fixed to the seat, which has an angular velocity of $\boldsymbol{\omega}_3$ relative to the cockpit and thus relative to xyz. Hence,

$$\boldsymbol{V}_{xyz} = \boldsymbol{\omega}_3 \times \boldsymbol{\rho} = .524\boldsymbol{i} \times 3\boldsymbol{k} = -1.572\boldsymbol{j} \text{ ft/sec}$$

$$\boldsymbol{a}_{xyz} = \left(\frac{d\boldsymbol{\omega}_3}{dt}\right)_{xyz} \times \boldsymbol{\rho} + \boldsymbol{\omega}_3 \times \left(\frac{d\boldsymbol{\rho}}{dt}\right)_{xyz}$$

But $\boldsymbol{\omega}_3$ is constant as seen from the cockpit and thus from xyz. Hence,

$$\boldsymbol{a}_{xyz} = \boldsymbol{0} \times \boldsymbol{\rho} + \boldsymbol{\omega}_3 \times \boldsymbol{V}_{xyz} = .524\boldsymbol{i} \times (-1.572\boldsymbol{j})$$
$$= -.824\boldsymbol{k} \text{ ft/sec}^2$$

Example 13.14 (Continued)

B. Motion of *xyz* relative to *XYZ*. The origin of *xyz* in this analysis has the same motion as the origin of *xyz* in the previous analysis. Thus, we use the results of analysis I for $\boldsymbol{R}$ and its time derivatives.

$$\boldsymbol{R} = -40\boldsymbol{j} \text{ ft}$$
$$\dot{\boldsymbol{R}} = 41.9\boldsymbol{i} \text{ ft/sec}$$
$$\ddot{\boldsymbol{R}} = 43.9\boldsymbol{j} + .349\boldsymbol{i} \text{ ft/sec}^2$$
$$\boldsymbol{\omega} = \boldsymbol{\omega}_1 + \boldsymbol{\omega}_2 = 1.048\boldsymbol{j} + 1.048\boldsymbol{k} \text{ rad/sec}$$
$$\dot{\boldsymbol{\omega}} = \dot{\boldsymbol{\omega}}_1 + \dot{\boldsymbol{\omega}}_2$$

We are given $\dot{\boldsymbol{\omega}}_1$ about the Z axis and $\boldsymbol{\omega}_2$ is fixed in the arm, which is rotating with angular velocity $\boldsymbol{\omega}_1$ relative to the XYZ reference. Hence,

$$\dot{\boldsymbol{\omega}} = \dot{\boldsymbol{\omega}}_1 + \boldsymbol{\omega}_1 \times \boldsymbol{\omega}_2 = .00873\boldsymbol{k} + (1.048\boldsymbol{k} \times 1.048\boldsymbol{j})$$
$$= -1.098\boldsymbol{i} + .00873\boldsymbol{k} \text{ rad/sec}^2$$

We can now substitute into the key equation, 13.24:

$$\boldsymbol{a}_{XYZ} = \boldsymbol{a}_{xyz} + \ddot{\boldsymbol{R}} + 2\boldsymbol{\omega} \times \boldsymbol{V}_{xyz} + \dot{\boldsymbol{\omega}} \times \boldsymbol{\rho} + \boldsymbol{\omega} \times (\boldsymbol{\omega} \times \boldsymbol{\rho})$$
$$= 3.64\boldsymbol{i} + 50.5\boldsymbol{j} - 4.12\boldsymbol{k} \text{ ft/sec}^2$$

$$|\boldsymbol{a}_{XYZ}| = \boxed{1.578g}$$

ANALYSIS III

Fix *xyz* to seat.
Fix *XYZ* to ground.

This situation is shown in Fig. 13.37.

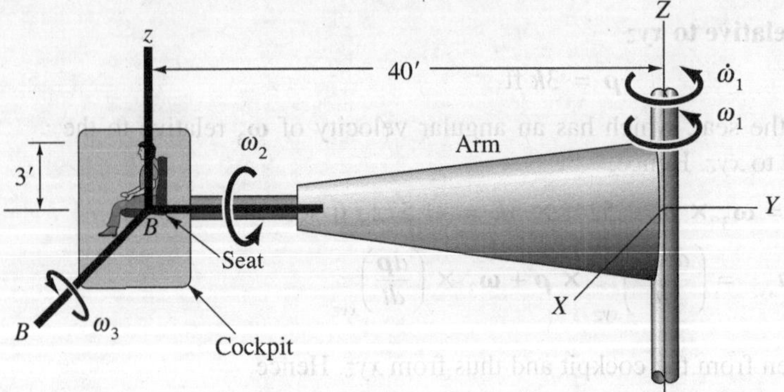

Figure 13.37. Centrifuge with *xyz* fixed to seat.

Example 13.14 (Continued)

A. Motion of particle relative to *xyz*

$$\boldsymbol{\rho} = 3\boldsymbol{k} \text{ ft}$$

Since the particle is fixed to the seat and is thus fixed in *xyz*, we can say:

$$\boldsymbol{V}_{xyz} = \boldsymbol{0}$$
$$\boldsymbol{a}_{xyz} = \boldsymbol{0}$$

B. Motion of *xyz* relative to *XYZ*. Again, the origin of *xyz* has identically the same motion as in the previous analyses. Thus, we have the same results as before for $\boldsymbol{R}$ and its derivatives.

$$\boldsymbol{R} = -40\boldsymbol{j} \text{ ft}$$
$$\dot{\boldsymbol{R}} = 41.9\boldsymbol{i} \text{ ft/sec}$$
$$\ddot{\boldsymbol{R}} = 43.9\boldsymbol{j} + .349\boldsymbol{i} \text{ ft/sec}^2$$
$$\boldsymbol{\omega} = \boldsymbol{\omega}_1 + \boldsymbol{\omega}_2 + \boldsymbol{\omega}_3 = 1.048\boldsymbol{k} + 1.048\boldsymbol{j} + .524\boldsymbol{i}$$
$$\dot{\boldsymbol{\omega}} = \dot{\boldsymbol{\omega}}_1 + \dot{\boldsymbol{\omega}}_2 + \dot{\boldsymbol{\omega}}_3$$

Note that $\dot{\boldsymbol{\omega}}_1$ is given. Also, $\boldsymbol{\omega}_2$ is fixed in the arm, which rotates with angular speed $\boldsymbol{\omega}_1$ relative to *XYZ*. Finally, $\boldsymbol{\omega}_3$ is fixed in the cockpit, which has an angular velocity $\boldsymbol{\omega}_2 + \boldsymbol{\omega}_1$ relative to *XYZ*. Thus,

$$\begin{aligned}\dot{\boldsymbol{\omega}} &= \dot{\boldsymbol{\omega}}_2 + \boldsymbol{\omega}_1 \times \boldsymbol{\omega}_2 + (\boldsymbol{\omega}_1 + \boldsymbol{\omega}_2) \times \boldsymbol{\omega}_3 \\ &= .00873\boldsymbol{k} + (1.048\boldsymbol{k} \times 1.048\boldsymbol{j}) + (1.048\boldsymbol{k} + 1.048\boldsymbol{j}) \times (.524\boldsymbol{i}) \\ &= -1.098\boldsymbol{i} + .549\boldsymbol{j} - .540\boldsymbol{k}\end{aligned}$$

We now go to the basic equation, 13.24.

$$\boldsymbol{a}_{XYZ} = \boldsymbol{a}_{xyz} + \ddot{\boldsymbol{R}} + 2\boldsymbol{\omega} \times \boldsymbol{V}_{xyz} + \dot{\boldsymbol{\omega}} \times \boldsymbol{\rho} + \boldsymbol{\omega} \times (\boldsymbol{\omega} \times \boldsymbol{\rho})$$

Substituting, we get

$$\boldsymbol{a}_{XYZ} = 3.64\boldsymbol{i} + 50.5\boldsymbol{j} - 4.12\boldsymbol{k} \text{ ft/sec}^2$$

$$|\boldsymbol{a}_{XYZ}| = 1.578g$$

In the final example of this series, we have a case where it is advantageous to use cylindrical coordinates in parts of the problem and then later to convert to rectangular coordinates.

Example 13.15

A submersible (see Fig. 13.38) is moving relative to the ground reference XYZ so as to have the following motion at the instant of interest for point A fixed to shaft CD which in turn is fixed to the submersible:

$$\boldsymbol{V} = 3\boldsymbol{i} + .6\boldsymbol{j} \text{ m/s}$$
$$\boldsymbol{a} = 2\boldsymbol{i} + 3\boldsymbol{j} - .5\boldsymbol{k} \text{ m/s}^2$$

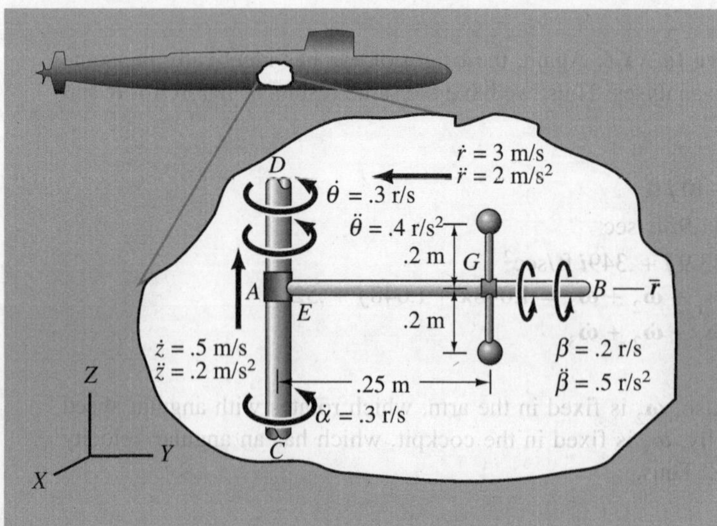

Figure 13.38. Rotating device inside a moving submersible. CD is fixed to the submersible.

At the instant of interest, the vessel has an angular speed of rotation $\dot{\alpha} = .3$ rad/s about the centerline of CD as seen from the ground reference. A horizontal rod EB has the following angular motion about CD:

$$\dot{\theta} = .3 \text{ rad/s} \qquad \ddot{\theta} = .4 \text{ rad/s}^2$$

Two spheres, each of mass 1 kg, are mounted on a rod turning about EB with the following angular motion

$$\dot{\beta} = .2 \text{ rad/s} \qquad \ddot{\beta} = .5 \text{ rad/s}^2$$

Also, the rod and the attached spherical masses advance toward CD at the following rate:

$$\dot{r} = -3 \text{ m/s} \qquad \ddot{r} = -2 \text{ m/s}^2$$

at a time when $r = .25$ m. Finally, at the instant of interest, the horizontal rod EB moves up along vertical rod CD with a speed of .5 m/s and a rate of change of speed of .2 m/s². What force must rod EB exert at point G at this instant *due only to the motion of the two spheres*?

Example 13.15 (Continued)

We will first consider the motion of the *center of mass* of the rotating spheres which clearly must be *G*. We now proceed to get the acceleration of *G* relative to *XYZ* (see Fig. 13.39).

Fix *xyz* to the vessel at *A*.
Fix *XYZ* to the ground.

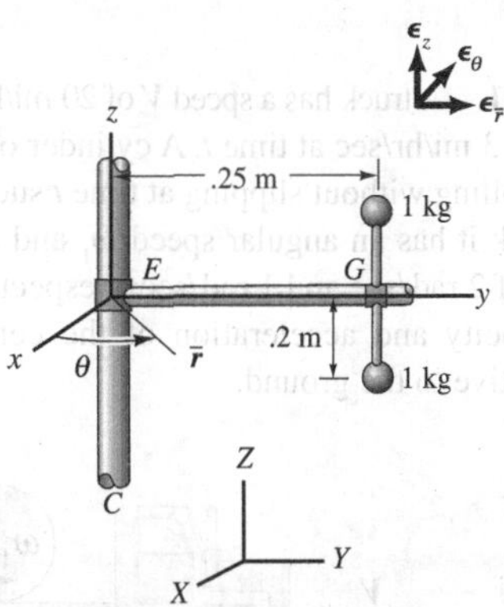

Figure 13.39. Reference *xyz* fixed to the vessel.

A. Motion of G relative to *xyz* (using cylindrical coordinates). Use Figs. 13.38 and 13.39.

$$\boldsymbol{\rho} = .25\boldsymbol{\epsilon}_{\bar{r}} = .25\boldsymbol{j}\text{ m}$$

$$\boldsymbol{V}_{xyz} = \dot{\bar{r}}\boldsymbol{\epsilon}_{\bar{r}} + \bar{r}\dot{\theta}\boldsymbol{\epsilon}_\theta + \dot{z}\boldsymbol{\epsilon}_z = -3\boldsymbol{\epsilon}_{\bar{r}} + (.25)(.3)\boldsymbol{\epsilon}_\theta + .5\boldsymbol{\epsilon}_z$$

$$= -3\boldsymbol{j} - .075\boldsymbol{i} + .5\boldsymbol{k} = -.075\boldsymbol{i} - 3\boldsymbol{j} + .5\boldsymbol{k}\text{ m/s}$$

$$\boldsymbol{a}_{xyz} = (\ddot{\bar{r}} - \bar{r}\dot{\theta}^2)\boldsymbol{\epsilon}_{\bar{r}} + (\bar{r}\ddot{\theta} + 2\dot{\bar{r}}\dot{\theta})\boldsymbol{\epsilon}_\theta + \ddot{z}\boldsymbol{\epsilon}_z$$

$$= \left[-2 - (.25)(.3)^2\right]\boldsymbol{\epsilon}_{\bar{r}} + \left[(.25)(.4) + (2)(-3)(.3)\right]\boldsymbol{\epsilon}_\theta + .2\boldsymbol{\epsilon}_z$$

$$= -2.023\boldsymbol{\epsilon}_{\bar{r}} - 1.7\boldsymbol{\epsilon}_\theta + .2\boldsymbol{\epsilon}_z = 1.7\boldsymbol{i} - 2.023\boldsymbol{j} + .2\boldsymbol{k}\text{ m/s}^2$$

B. Motion of *xyz* relative to *XYZ*

$$\dot{\boldsymbol{R}} = 3\boldsymbol{i} + .6\boldsymbol{j}\text{ m/s}$$
$$\ddot{\boldsymbol{R}} = 2\boldsymbol{i} + 3\boldsymbol{j} - .5\boldsymbol{k}\text{ m/s}^2$$
$$\boldsymbol{\omega} = .3\boldsymbol{k}\text{ rad/s}$$
$$\dot{\boldsymbol{\omega}} = \boldsymbol{0}\text{ rad/s}^2$$

We may now express $\boldsymbol{a}_{XYZ}$ for point *G*. Thus

$$\boldsymbol{a}_{XYZ} = \boldsymbol{a}_{xyz} + \ddot{\boldsymbol{R}} + 2\boldsymbol{\omega} \times \boldsymbol{V}_{xyz} + \dot{\boldsymbol{\omega}} \times \boldsymbol{\rho} + \boldsymbol{\omega} \times (\boldsymbol{\omega} \times \boldsymbol{\rho})$$

$$= (1.7\boldsymbol{i} - 2.023\boldsymbol{j} + .2\boldsymbol{k}) + (2\boldsymbol{i} + 3\boldsymbol{j} - .5\boldsymbol{k}) + 2(.3\boldsymbol{k})$$
$$\times (-.075\boldsymbol{i} - 3\boldsymbol{j} + .5\boldsymbol{k}) + \boldsymbol{0} \times \boldsymbol{\rho} + (.3\boldsymbol{k}) \times (.3\boldsymbol{k} \times .25\boldsymbol{j})$$
$$= 5.5\boldsymbol{i} + .9095\boldsymbol{j} - .3\boldsymbol{k}\text{ m/s}^2$$

Now we apply **Newton's law** to the mass center *G* at the instant of interest. Denoting the force from the rod *AB* onto *G* as $\boldsymbol{F}_{\text{ROD}}$, we get

$$\boldsymbol{F}_{\text{ROD}} - 2\text{mg}\boldsymbol{k} = 5.5\boldsymbol{i} + .9095\boldsymbol{j} - .3\boldsymbol{k}$$
$$\therefore \boldsymbol{F}_{\text{ROD}} = (2)(1)(9.81)\boldsymbol{k} + 5.5\boldsymbol{i} + .9095\boldsymbol{j} - .3\boldsymbol{k}$$

$$\boldsymbol{F}_{\text{ROD}} = 5.5\boldsymbol{i} + .9095\boldsymbol{j} + 19.32\boldsymbol{k}\text{ N}$$

This is our desired result.

PROBLEMS

13.91. A truck has a speed V of 20 mi/hr and an acceleration $\dot{V}$ of 3 mi/hr/sec at time t. A cylinder of radius equal to 2 ft is rolling without slipping at time t such that relative to the truck it has an angular speed ω_1 and angular acceleration $\dot{\omega}_1$ of 2 rad/sec and 1 rad/sec^2, respectively. Determine the velocity and acceleration of the center of the cylinder relative to the ground.

Figure P.13.91.

13.92. A wheel rotates with an angular speed ω_2 of 5 rad/sec relative to a platform, which rotates with a speed ω_1 of 10 rad/sec relative to the ground as shown. A collar moves down the spoke of the wheel, and, when the spoke is vertical, the collar has a speed of 20 ft/sec, an acceleration of 10 ft/sec^2 along the spoke, and is positioned 1 ft from the shaft centerline of the wheel. Compute the velocity and acceleration of the collar relative to the ground at this instant. Fix xyz to platform and use cylindrical coordinates.

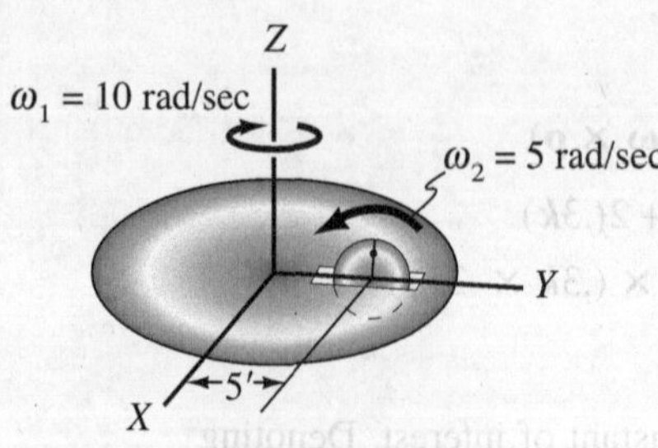

Figure P.15.92.

13.93. In Problem 13.71, determine the acceleration of the particle at the instant of interest.

13.94. In Problem 13.72, find the acceleration of the particle P relative to the ground reference.

13.95. In Problem 13.74, find the acceleration of point A relative to the ground.

13.96. In Problem 13.79, find the acceleration of the tip of the propeller relative to the ground reference. Take the yaw rotation to be zero and the loop rotation radius r to be 500 m.

13.97. In Problem 13.80, find the acceleration of point B relative to the ground.

13.98. In Problem 13.80, find the acceleration of point B relative to the ground for the following data at the instant of interest shown in the diagram.

$$\omega_1 = .2 \text{ rad/sec}$$
$$\dot{\omega}_1 = -.1 \text{ rad/sec}^2$$
$$\omega_2 = .4 \text{ rad/sec}$$
$$\dot{\omega}_2 = .3 \text{ rad/sec}^2$$

13.99. In Problem 13.82, determine the acceleration of the top tip of the gun relative to the ground.

13.100. In Problem 13.74, find the acceleration of point A relative to the ground for the configuration shown. Take $\omega_1 = 2$ rad/sec $\dot{\omega}_1 = 3$ rad/sec^2, $\omega_2 = .1$ rad/sec, and $\dot{\omega}_2 = 2$ rad/sec^2. How many g's of acceleration is this point subject to?

13.101. In Problem 13.81, find the acceleration of D relative to the ground. [*Hint:* Use two position vectors to get $\boldsymbol{\rho}$.]

13.102. Find the acceleration of gear tooth A relative to the ground in Problem 13.84.

13.103. In Problem 13.92, the wheel accelerates at the instant under discussion with 5 rad/sec^2 relative to the platform, and the platform increases its angular speed at 10 rad/sec^2 relative to the ground. Find the velocity and acceleration of the collar relative to the ground.

13.104. As with the velocity equation 13.20, we can easily show that Eq. 13.7, relating accelerations between two points on a rigid body, is actually a special case of Eq. 13.24. Thus, consider the diagram showing a rigid body moving arbitrarily relative to *XYZ*. Choose two points *a* and *b* in the body and embed a reference *xyz* in the body with the origin at *a*. Now express the acceleration of point *b* as seen from the two references. Show how this equation can be reformulated as Eq. 13.7.

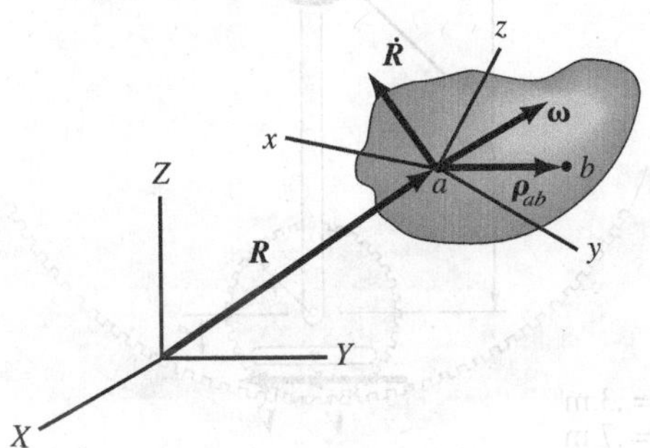

Figure P.13.104.

13.105. Solve Problem 13.81 for the following data:

$$\omega_1 = .3 \text{ rad/sec}$$
$$\dot{\omega}_1 = .2 \text{ rad/sec}^2$$
$$\omega_2 = .40 \text{ rad/sec}$$
$$\dot{\omega}_2 = .10 \text{ rad/sec}^2$$
$$\omega_3 = .15 \text{ rad/sec}$$
$$\dot{\omega}_3 = -.2 \text{ rad/sec}^2$$

13.106. In Problem 13.82, find the component of acceleration of the projectile relative to the ground which is normal to the gun barrel at the instant that the projectile just leaves the barrel. Use the following data:

$$\omega_1 = 3 \text{ rad/sec}$$
$$\dot{\omega}_1 = 2 \text{ rad/sec}^2$$
$$\omega_2 = -.6 \text{ rad/sec}$$
$$\dot{\omega}_2 = -.4 \text{ rad/sec}^2$$

13.107. In Problem 13.88, find the magnitude of the acceleration of collar *C* relative to the ground for the following data at the instant shown:

$$\dot{\theta} = 2 \text{ rad/sec}$$
$$\ddot{\theta} = 4 \text{ rad/sec}^2$$

13.108. A truck is moving at a constant speed $V = 1.7$ m/sec at time *t*. The truck loading compartment has at this instant a constant angular speed $\dot{\theta}$ of .1 rad/sec at an angle $\theta = 45°$. A cylinder of radius 300 mm rolls relative to the compartment at a speed ω_1 of 1 rad/sec, accelerating at a rate $\dot{\omega}_1$ of .5 rad/sec² at time *t*. What are the velocity and acceleration of the center of the cylinder relative to the ground at time *t*? The distance *d* at time *t* is 5 m.

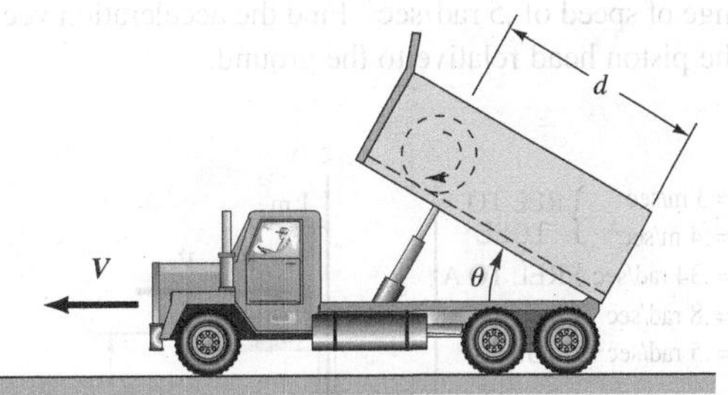

Figure P.13.108.

***13.109.** In Example 13.14, suppose at the instant of interest that there is an angular acceleration $\dot{\omega}_2 = .3$ rad/sec² of the cockpit relative to the arm and that there is an angular acceleration $\dot{\omega}_3 = .2$ rad/sec² of the seat relative to the cockpit. Find the number of *g*'s to which the pilot's head is subjected. Follow analysis I in the example. [*Hint:* The angular velocity $\boldsymbol{\omega}_3$ can always be expressed as $\omega_3\hat{\mathbf{c}}$, where $\hat{\mathbf{c}}$ is a unit vector *fixed to the cockpit* having a direction along the *x* axis at the instant of interest.]

13.110. A ferris wheel is out of control. At the instant shown, it has an angular speed ω_1 equal to .2 rad/sec and a rate of change of angular speed $\dot{\omega}_1$ of .04 rad/sec² relative to the ground. At this instant a "chair" shown in the diagram has an angular speed ω_2 relative to the ferris wheel equal to .25 rad/sec and a rate change of speed $\dot{\omega}_2$, again relative to the ferris wheel, equal to .03 rad/sec². In the figure, we have shown details of the passenger at this instant. Note that the hinge of the seat is at *A*. How many *g*'s of acceleration is the passenger's head subject to?

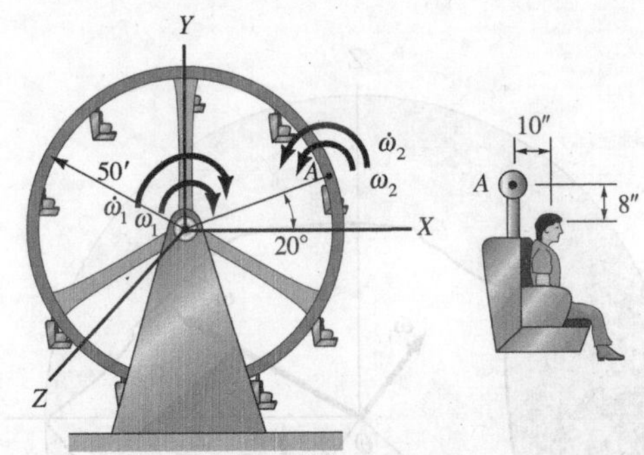

Figure P.13.110.

13.111. A shaft *BC* rotates relative to platform *A* at a speed $\omega_1 = .34$ rad/sec. A rod is welded to *BC* and is vertical at the instant of interest. A tube is fixed to the vertical rod in which a small piston head is moving relative to the tube at a speed *V* of 3 m/sec with a rate of change of speed $\dot{V}$ of .4 m/sec². The platform *A* has an angular velocity relative to the ground given as $\omega_2 = .8$ rad/sec with a rate of change of speed of .5 rad/sec². Find the acceleration vector of the piston head relative to the ground.

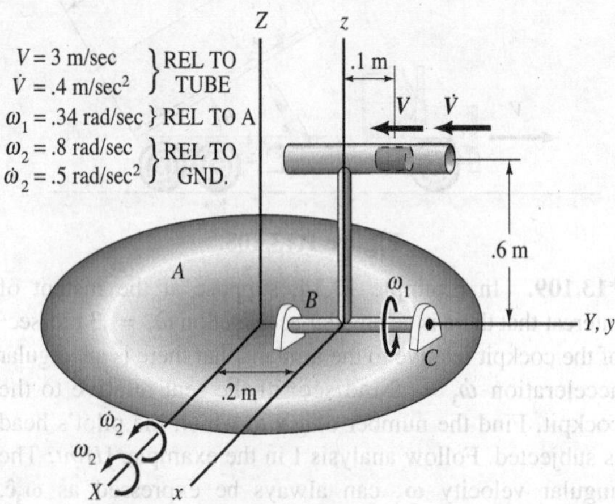

Figure P.13.111.

13.112. A communications satellite has the following motion relative to an inertial reference *XYZ*.

$$\boldsymbol{\omega}_1 = 3\boldsymbol{i} + 4\boldsymbol{j} + 10\boldsymbol{k} \text{ rad/s} \quad \dot{\boldsymbol{\omega}}_1 = 2\boldsymbol{i} + 3\boldsymbol{k} \text{ rad/s}^2$$
$$\boldsymbol{V}_0 = \boldsymbol{a}_0 = \boldsymbol{0}$$

A wheel at *A* is rotating relative to the satellite at a constant speed $\omega_2 = 5$ rad/s. What is the acceleration of point *D* on the wheel relative to *XYZ* at the instant shown? The following additional data apply:

$$OA = 1 \text{ m} \quad AD = .2 \text{ m}$$

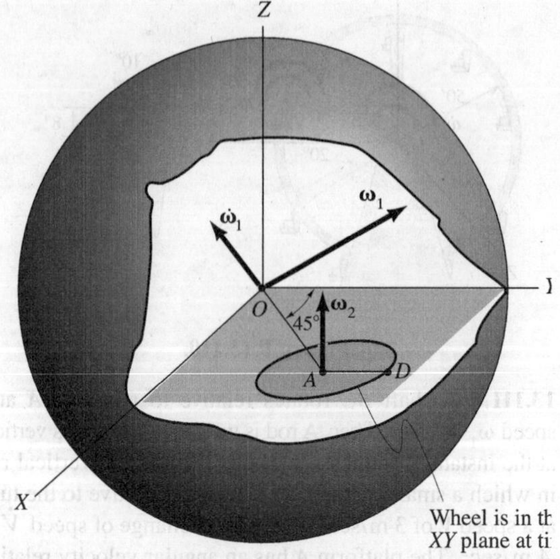

Figure P.13.112.

13.113. A particle moves in a slot of a gear with speed $V = 2$ m/s and a rate of change of speed $\dot{V} = 1.2$ m/s² both relative to the gear. Find the acceleration vector for the particle at the configuration shown relative to the ground reference *XYZ*.

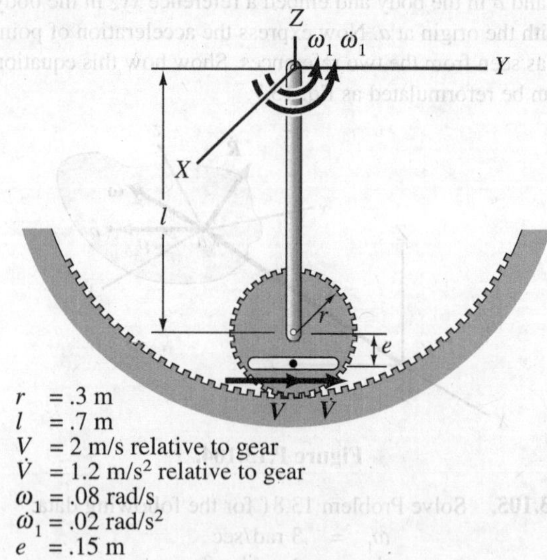

$r = .3$ m
$l = .7$ m
$V = 2$ m/s relative to gear
$\dot{V} = 1.2$ m/s² relative to gear
$\omega_1 = .08$ rad/s
$\dot{\omega}_1 = .02$ rad/s²
$e = .15$ m

Figure P.13.113.

13.114. A submarine is undergoing an evasive maneuver. At the instant of interest, it has a speed $V_s = 10$ m/s and an acceleration $a_s = 15$ m/s² at its center of mass. It also has an angular velocity about its center of mass *C* of $\omega_1 = .5$ rad/s and an angular acceleration $\dot{\omega}_1 = .02$ rad/s². Inside is a part of an inertial guidance system that consists of a wheel spinning with speed $\omega_2 = 20$ rad/s about a vertical axis of the ship. Along a spoke shown at the instant of interest, a particle is moving toward the center with r, $\dot{r}$ and $\ddot{r}$ as given in the diagram. What is the acceleration of the particle relative to inertial reference *XYZ*? Just write out the formulations for $\boldsymbol{a}_{XYZ}$ but do not carry out the cross products.

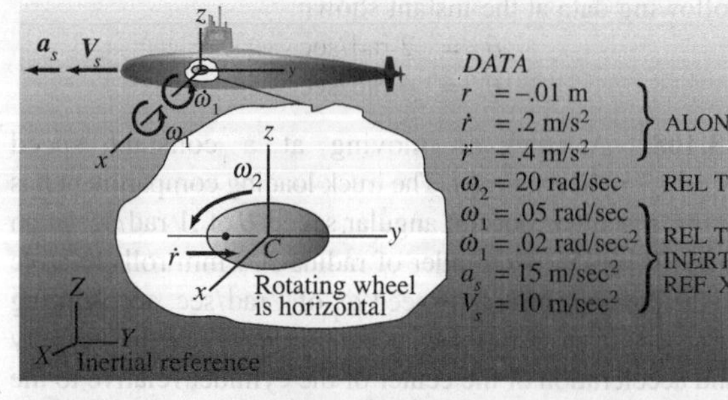

Figure P.13.114.

13.115. Find V and a of B relative to XYZ. The single blade portion AB is rotating as shown relative to the helicopter with speed ω_2 about an axis parallel to the X axis at the instant of interest while the entire blade system is rotating about the vertical axis with speed ω_1.

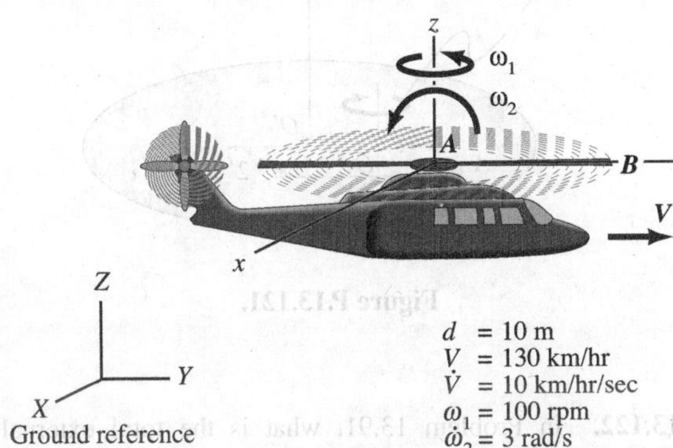

Figure P.13.115.

13.116. An F-16 fighter plane is moving at a constant speed of 800 km/hr while undergoing a loop as shown in the diagram. At the instant of interest, it has an angular roll velocity ω_2 of 5 rad/min relative to the ground. A solenoid is activated at this instant and moves the moving portion of a gate valve downward at a speed of 3 m/s with an acceleration of 5 m/s² all relative to the plane. What is the total external force on the moving portion of the gate valve if it is 1 m above the axis of roll at the instant of interest? The moving portion of the valve has a weight of 10 N.

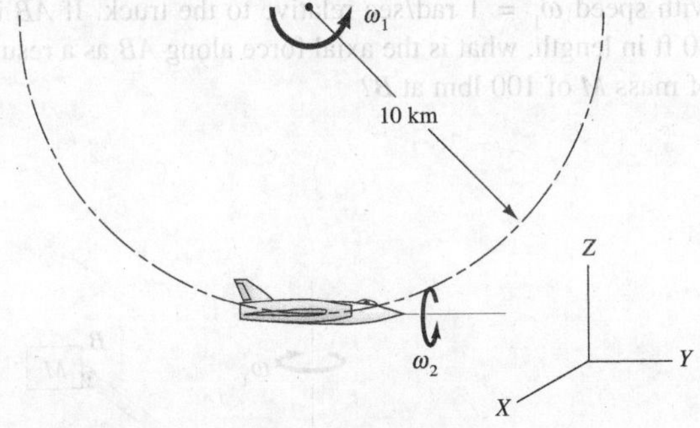

Figure P.13.116.

13.117. A robot moves a body held by its "jaws" G as shown in the diagram. What is the velocity and acceleration of point A at the instant shown relative to the ground? Arm EH is welded to the vertical shaft MN. Arm HKG is one rigid member which rotates about EH. How do you want to fix xyz?

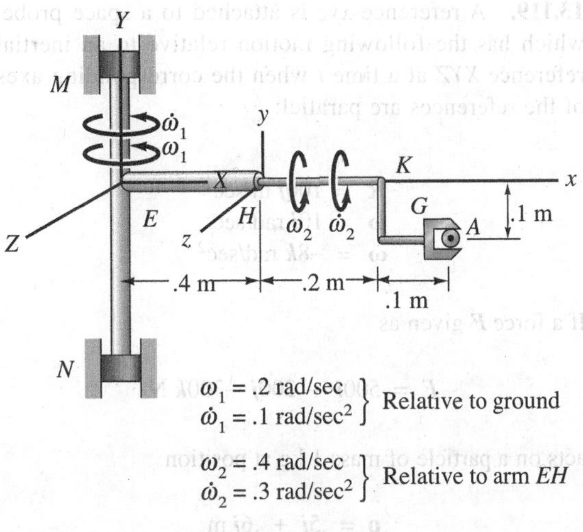

Figure P.13.117.

13.118. The turret of the main gun of a destroyer has at time t an angular velocity $\omega_1 = 2$ rad/s and a rate of change of angular velocity $\dot{\omega}_1 = 3$ rad/s² both relative to the ship. The gun barrel has $\omega_2 = .5$ rad/s and $\dot{\omega}_2 = .3$ rad/s² relative to the turret.

(a) Find the acceleration of the tip A of the gun at time t relative to the destroyer.

(b) If the destroyer has a translational acceleration relative to land equal to

$$\boldsymbol{a}_{\text{Destroyer}} = .05\boldsymbol{i} + .26\boldsymbol{j} - 2.2\boldsymbol{k}\ \text{m/s}^2$$

what is the acceleration of A relative to the land at time t?

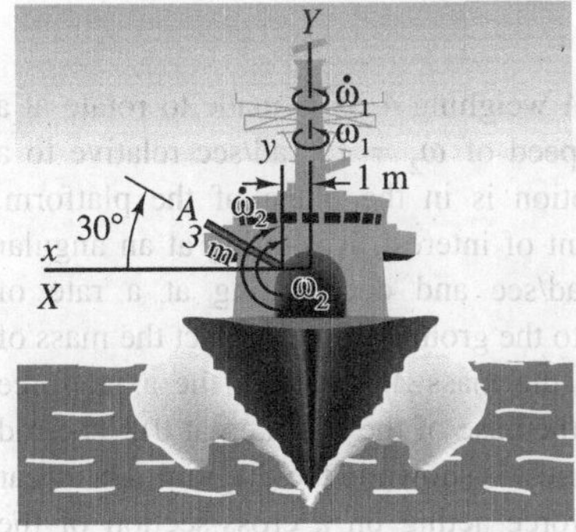

Figure P.13.118.

13.119. A reference xyz is attached to a space probe, which has the following motion relative to an inertial reference XYZ at a time t when the corresponding axes of the references are parallel:

$$\ddot{\boldsymbol{R}} = 100\boldsymbol{j} \text{ m/sec}^2$$
$$\boldsymbol{\omega} = 10\boldsymbol{i} \text{ rad/sec}$$
$$\dot{\boldsymbol{\omega}} = -8\boldsymbol{k} \text{ rad/sec}^2$$

If a force $\boldsymbol{F}$ given as

$$\boldsymbol{F} = 500\boldsymbol{i} + 200\boldsymbol{j} - 300\boldsymbol{k} \text{ N}$$

acts on a particle of mass 1 kg at position

$$\boldsymbol{\rho} = .5\boldsymbol{i} + .6\boldsymbol{j} \text{ m}$$

what is the acceleration vector relative to the probe? The particle has a velocity $\boldsymbol{V}$ relative to xyz of

$$\boldsymbol{V} = 10\boldsymbol{i} + 20\boldsymbol{j} \text{ m/sec}$$

13.120. In the space probe of Problem 13.119, what must the velocity vector $\boldsymbol{V}_{xyz}$ of the particle be to have the acceleration

$$\boldsymbol{a}_{xyz} = 495.2\boldsymbol{i} + 100\boldsymbol{j} \text{ m/sec}^2$$

if all other conditions are the same? Is there a component of $\boldsymbol{V}_{xyz}$ that can have any value for this problem?

13.121. A mass A weighing 4 oz is made to rotate at a constant angular speed of $\omega_2 = 15$ rad/sec relative to a platform. This motion is in the plane of the platform, which, at the instant of interest, is rotating at an angular speed $\omega_1 = 10$ rad/sec and decelerating at a rate of 5 rad/sec² relative to the ground. If we neglect the mass of the rod supporting the mass A, what are the axial force and shear force at the base of the rod (i.e., at 0)? The rod at the instant of interest is shown in the diagram. The shear force is the total force acting on a cross-section of the member in a direction *tangent* to the section.

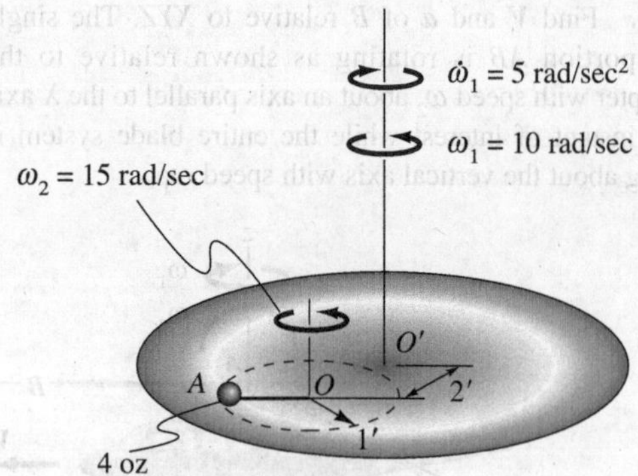

Figure P.13.121.

13.122. In Problem 13.91, what is the total external force acting on the cylinder for the case when

$$V = 5 \text{ ft/sec}$$
$$\dot{V} = -2 \text{ ft/sec}^2$$
$$\omega_1 = 2 \text{ rad/sec}$$
$$\dot{\omega}_1 = 1 \text{ rad/sec}^2?$$

The mass of the cylinder is 100 lbm.

13.123. A truck is moving at constant speed V of 10 mi/hr. A crane AB is at time t at $\theta = 45°$ with $\dot{\theta} = 1$ rad/sec and $\ddot{\theta} = .2$ rad/sec². Also at time t, the base of AB rotates with speed $\omega_1 = 1$ rad/sec relative to the truck. If AB is 30 ft in length, what is the axial force along AB as a result of mass M of 100 lbm at B?

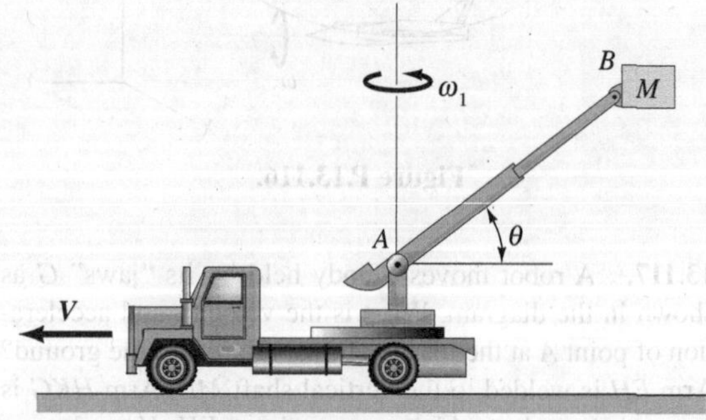

Figure P.13.123.

13.124. An exploratory probe shot from the earth is returning to the earth. On entering the earth's atmosphere, it has a constant angular velocity component ω_1 of 10 rad/sec about an axis normal to the page and a constant component ω_2 of 50 rad/sec about the vertical axis. The velocity of the probe at the time of interest is 1,300 m/sec vertically downward with a deceleration of 160 m/sec². A small sphere is rotating at ω_3 = 5 rad/sec inside the probe, as shown. At the time of interest, the probe is oriented so that the trajectory of the sphere in the probe is in the plane of the page and the arm is vertical. What are the axial force in the arm and the bending moment at its base (neglect the mass of the arm) at this instant of time, if the sphere has a mass of 300 g? (The bending moment is the couple moment acting on the cross section of the beam.)

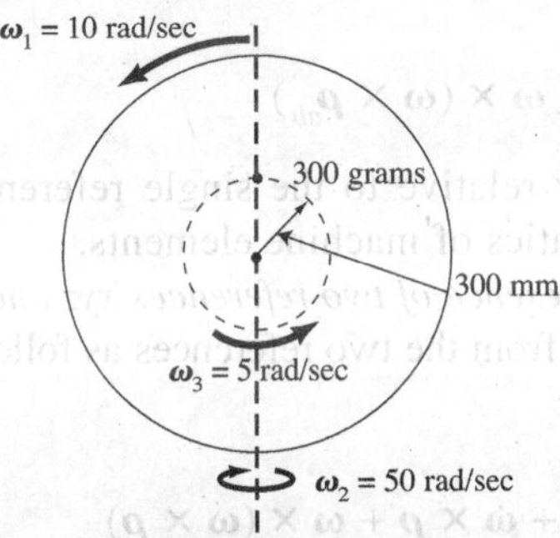

Figure P.13.124.

13.125. A river flows at 2 ft/sec average velocity in the Northern Hemisphere at a latitude of 40° in the north–south direction. What is the Coriolis acceleration of the water relative to the center of the earth? What is the Coriolis force on 1 lbm of water?

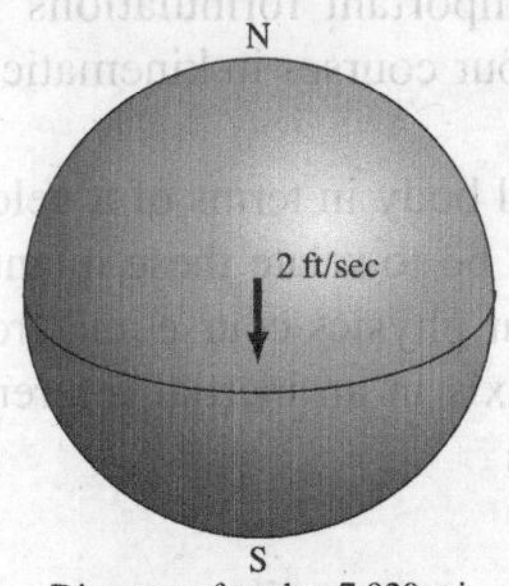

Figure P.13.125.

13.126. A clutch assembly is shown. Rods *AB* are pinned to a disc at *B*, which rotates at an angular speed ω_1 = 1 rad/sec and $\dot{\omega}_1$ = 2 rad/sec² at time *t*. These rods extend through a rod *EF*, which rotates with the rods and at the same time is moving to the left with a speed *V* of 1 m/sec. At the instant shown, corresponding to time *t*, what is the axial force on the member *AB* as a result of the motion of particle *A* having a mass of .6 kg?

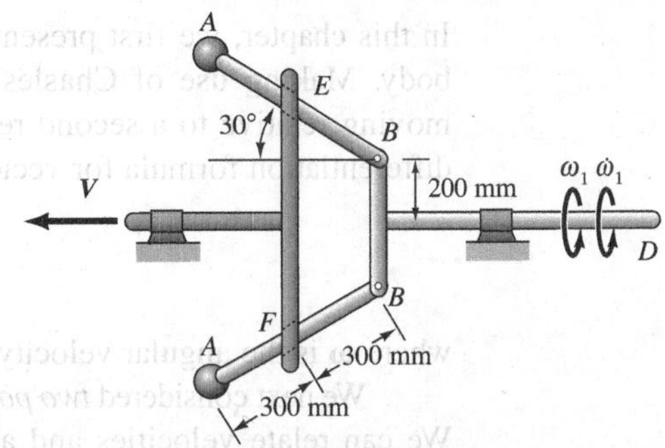

Figure P.13.126.

13.127. A flyball governor is shown. The weights *C* and *D* each have a mass of 200 g. At the instant of interest, θ = 45° and the system is rotating about axis *AB* at a speed ω_1 of 2 rad/sec. At this instant, collar *B* is moving upward at a speed of .5 m/sec. If we neglect the mass of the members, find the axial forces in the members at the instant of interest. What is the total shear force F_s on the members? (The shear force is the force component tangent to the cross section of the member.)

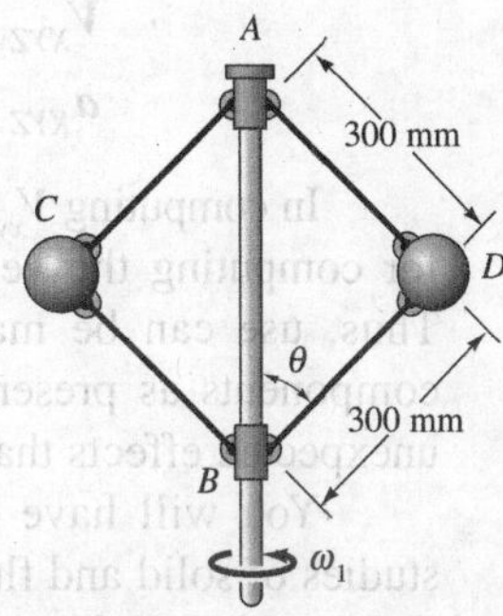

Figure P.13.127.

13.128. A man throws a ball weighing 3 oz from one side of a rotating platform to a man diametrically opposite, as shown. What is the Coriolis acceleration and force on the ball? Relative to the platform, in what direction does the ball tend to go as a result of the Coriolis force?

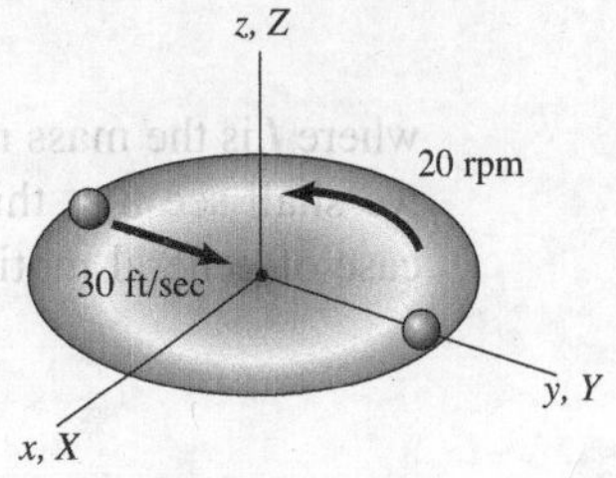

Figure P.13.128.

13.9 Closure

In this chapter, we first presented Chasles' theorem for describing the motion of a rigid body. Making use of Chasles' theorem for describing the motion of a reference *xyz* moving relative to a second reference *XYZ*, we presented next a simple but much used differentiation formula for vectors *A fixed* in the reference *xyz* or a rigid body. Thus,

$$\left(\frac{dA}{dt}\right)_{XYZ} = \boldsymbol{\omega} \times \boldsymbol{A}$$

where $\boldsymbol{\omega}$ is the angular velocity of *xyz* or the rigid body relative to *XYZ*.

We next considered *two points fixed in a rigid body in the presence of a single reference.* We can relate velocities and accelerations of the points relative to the aforementioned reference as follows:

$$\boldsymbol{V}_b = \boldsymbol{V}_a + \boldsymbol{\omega} \times \boldsymbol{\rho}_{ab}$$
$$\boldsymbol{a}_b = \boldsymbol{a}_a + \dot{\boldsymbol{\omega}} \times \boldsymbol{\rho}_{ab} + \boldsymbol{\omega} \times (\boldsymbol{\omega} \times \boldsymbol{\rho}_{ab})$$

where $\boldsymbol{\omega}$ is the angular velocity of the body relative to the single reference. These relations can be valuable in studies of kinematics of machine elements.

We then considered *one particle in the presence of two references xyz and XYZ.* We expressed the velocity and acceleration as seen from the two references as follows:

$$\boldsymbol{V}_{XYZ} = \boldsymbol{V}_{xyz} + \dot{\boldsymbol{R}} + \boldsymbol{\omega} \times \boldsymbol{\rho}$$
$$\boldsymbol{a}_{XYZ} = \boldsymbol{a}_{xyz} + \ddot{\boldsymbol{R}} + 2\boldsymbol{\omega} \times \boldsymbol{V}_{xyz} + \dot{\boldsymbol{\omega}} \times \boldsymbol{\rho} + \boldsymbol{\omega} \times (\boldsymbol{\omega} \times \boldsymbol{\rho})$$

In computing $\boldsymbol{V}_{xyz}$, $\boldsymbol{a}_{xyz}$, $\dot{\boldsymbol{R}}$, and $\ddot{\boldsymbol{R}}$, we use the various techniques presented in Chapter 9 for computing the velocity and acceleration of a particle relative to a given reference. Thus, use can be made of Cartesian components, path components, and cylindrical components as presented in that chapter. We then explored some interesting and often unexpected effects that occur when we use a noninertial reference.

You will have occasion to use these two important formulations in your basic studies of solid and fluid mechanics as well as in your courses in kinematics of machines and machine design.

Now that we can express the motion of a rigid body in terms of a velocity vector $\dot{\boldsymbol{R}}$ and an angular velocity vector $\boldsymbol{\omega}$, our next job will be to relate these quantities with the forces acting on the body. You may recall from your physics course and from the end of Chapter 12 that for a body rotating about a fixed axis in an inertial reference, we could relate the torque T and the angular acceleration α as

$$T = I\alpha$$

where I is the mass moment of inertia of the body about the axis of rotation. In Chapter 14, we shall see that this motion is a special case of plane motion, which itself is a special case of general motion.

PROBLEMS

13.129. A light plane is circling an airport at constant elevation. The radius R of the path = 3 km and the speed of the plane is 120 km/hr. The propeller of the plane is rotating at 100 rpm relative to the plane in a clockwise sense as seen by the pilot. What are $\boldsymbol{\omega}$, $\dot{\boldsymbol{\omega}}$, and $\ddot{\boldsymbol{\omega}}$ of the propeller as seen from the ground at the instant shown in the diagram?

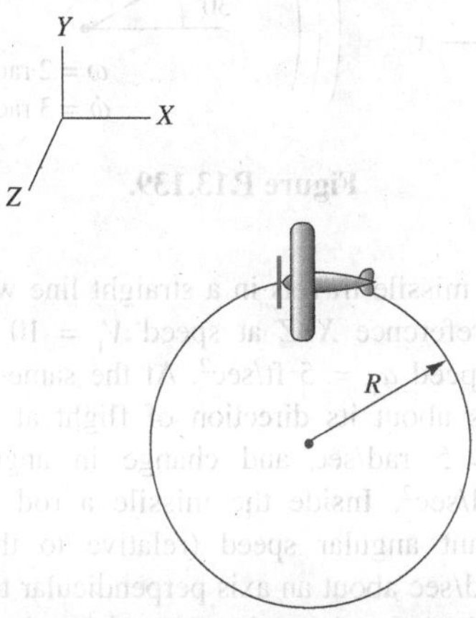

Figure P.13.129.

***13.130.** In Problem 13.7, find the angular acceleration $\dot{\boldsymbol{\omega}}$ of the disc for the configuration shown in the diagram, if, at the instant shown, the following data apply:

$$\omega_1 = 3 \text{ rad/sec}$$
$$\dot{\omega}_1 = 2 \text{ rad/sec}^2$$
$$\omega_2 = -10 \text{ rad/sec}$$
$$\dot{\omega}_2 = -4 \text{ rad/sec}^2$$

13.131. A slider A has at the instant of interest a speed V_A of 3 m/sec with a deceleration of 2 m/sec². Compute the angular velocity and angular acceleration of bar AB at the instant of interest. What is the position of the instantaneous axis of rotation of bar AB?

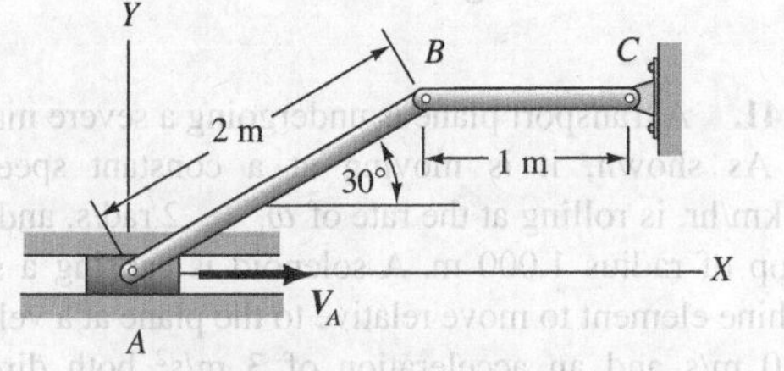

Figure P.13.131.

13.132. A cylinder C rolls without slipping on a half-cylinder D. Rod BA is 7 m long and is connected at A to a slider which at the instant of interest is moving in a groove at the speed V of 3 m/sec and increasing its speed at the rate of 2 m/sec². What is the angular speed and the angular acceleration of cylinder C relative to the ground?

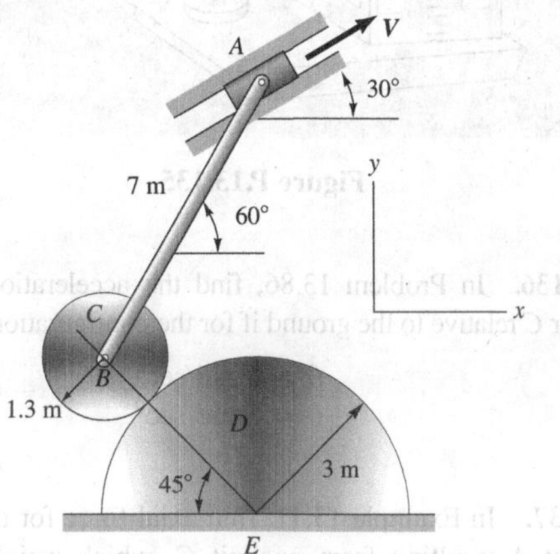

Figure P.13.132.

13.133. A wheel is rotating with a constant angular speed ω_1 of 10 rad/sec relative to a platform, which in turn is rotating with a constant angular speed ω_2 of 5 rad/sec relative to the ground. Find the velocity and acceleration relative to the ground at a point b on the wheel at the instant when it is directly vertically above point a.

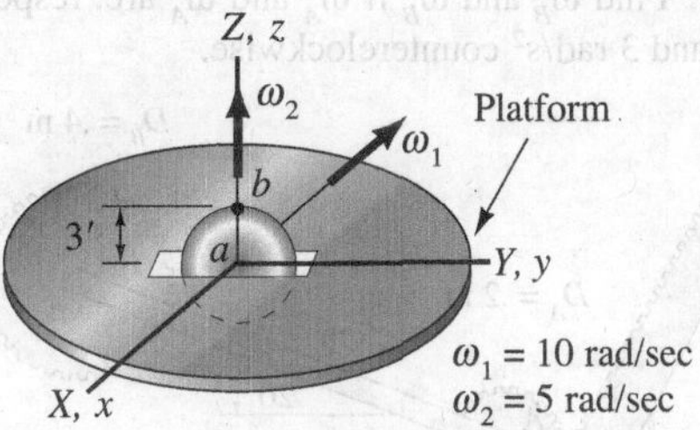

Figure P.13.133.

***13.134.** Solve Problem 13.133 for the case where ω_1 is increasing in value at the rate of 5 rad/sec² and where ω_2 is increasing in value at the rate of 10 rad/sec².

***13.135.** A barge is shown with a derrick arrangement. The main beam AB is 40 ft in length. The whole system at the instant of interest is rotating with a speed ω_1 of 1 rad/sec and an acceleration $\dot{\omega}_1$ of 2 rad/sec² relative to the barge. Also, at this instant $\theta = 45°$, $\dot{\theta} = 2$ rad/sec, and $\ddot{\theta} = 1$ rad/sec². What are the velocity and acceleration of point B relative to the barge?

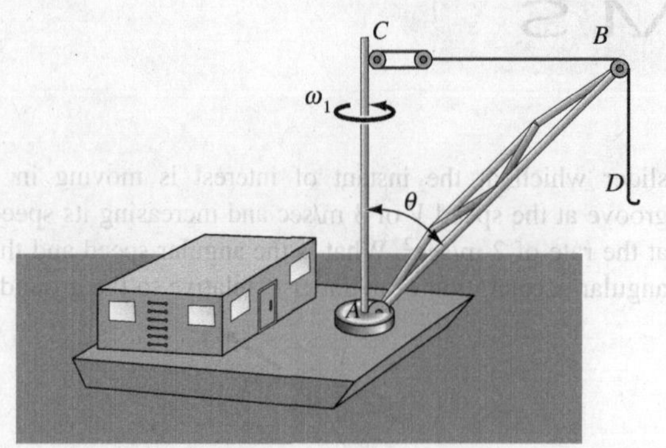

Figure P.13.135.

***13.136.** In Problem 13.86, find the acceleration of the collar C relative to the ground if for the configuration shown:

$$\dot{\theta} = 5 \text{ rad/sec}$$
$$\ddot{\theta} = 8 \text{ rad/sec}^2$$

13.137. In Example 15.11, find axial force for the beam AB at A resulting from cockpit C, which weighs (with occupant) $136g$ N. The following data apply:

$$\beta = 20°$$
$$\alpha = 60°$$
$$\omega_1 = .2 \text{ rad/sec}$$
$$\omega_2 = .1 \text{ rad/sec}$$

13.138. Find ω_B and $\dot{\omega}_B$ if ω_A and $\dot{\omega}_A$ are, respectively, 2 rad/s and 3 rad/s² counterclockwise.

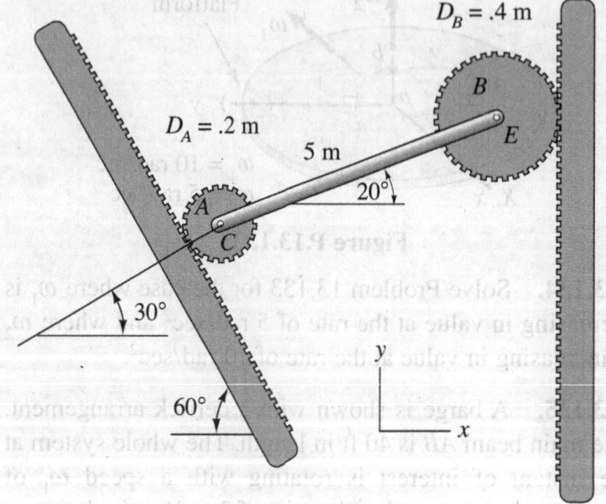

Figure P.13.138.

13.139. Find ω_{AC} and $\dot{\omega}_{AC}$ at the instant shown.

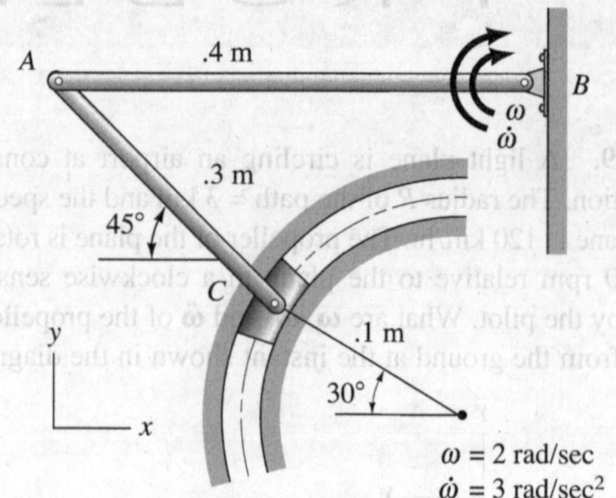

Figure P.13.139.

13.140. A missile travels in a straight line with respect to inertial reference XYZ at speed $V_1 = 10$ ft/sec and change of speed $a_1 = 5$ ft/sec². At the same instant the missile rolls about its direction of flight at an angular speed $\omega_1 = 5$ rad/sec and change in angular speed $\omega_1 = 5$ rad/sec². Inside the missile a rod is rotating at a constant angular speed (relative to the missile) $\omega_2 = 10$ rad/sec about an axis perpendicular to the page. Find the velocity and acceleration of the tip of the rod relative to XYZ at the instant shown.

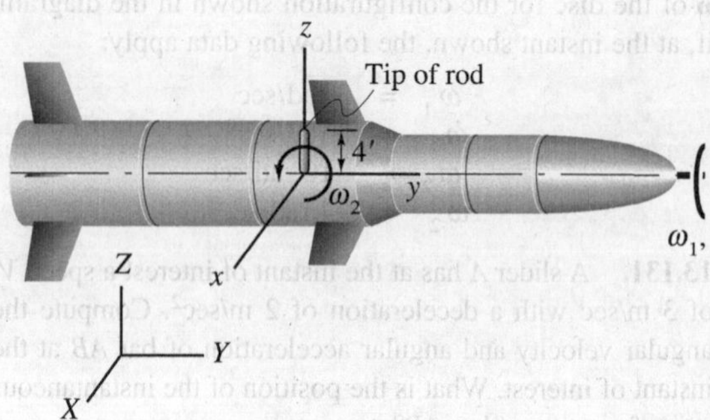

Figure P.13.140.

13.141. A transport plane is undergoing a severe maneuver. As shown, it is moving at a constant speed of 400 km/hr, is rolling at the rate of $\omega_1 = .2$ rad/s, and is in a loop of radius 1,000 m. A solenoid is causing a small machine element to move relative to the plane at a velocity of 10 m/s and an acceleration of 3 m/s² both directed downward. If the mass of the machine element is 10 kg, what is the force on it from the plane at this instant? The machine element is at position A at the time of interest.

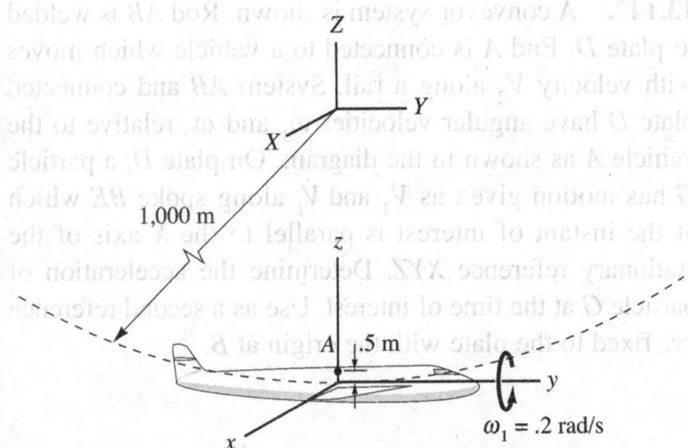

Figure P.13.141.

13.142. A rod moves in the plane of the paper in such a way that end *A* has a speed of 3 m/sec. What is the velocity of point *B* of the rod when the rod is inclined at 45° to the horizontal? *B* is at the upper support.

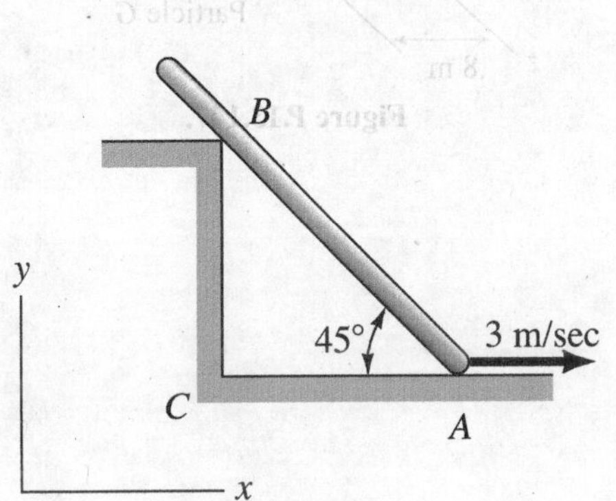

Figure P.13.142.

13.143. A vehicle on a monorail has a speed $\dot{S}$ = 10 m/s and an acceleration $\ddot{S}$ = 4 m/s² relative to the ground reference *XYZ* when it reaches point *A*. Inside the vehicle, a 3 kg mass slides along a rod which at the time of interest is parallel to the *X* axis. This rod rotates about a vertical axis with ω = 1 rad/s and $\dot{\omega}$ = 2 rad/s² relative to the vehicle at the time of interest. Also at this time, the radial distance *d* of the mass is .2 m and its radial velocity v = .4 m/s inward. What is the dynamic force on the mass at this instant?

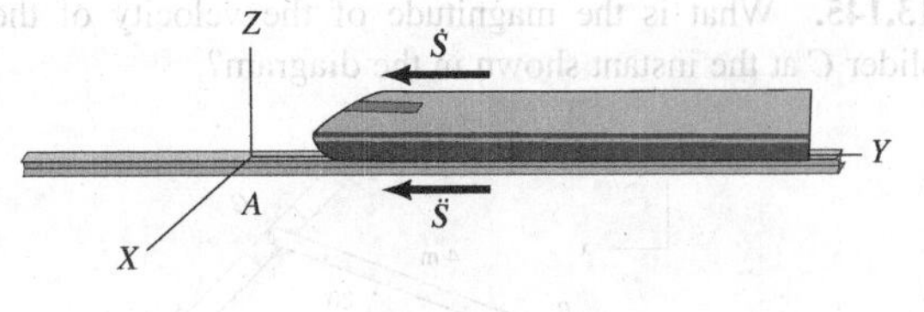

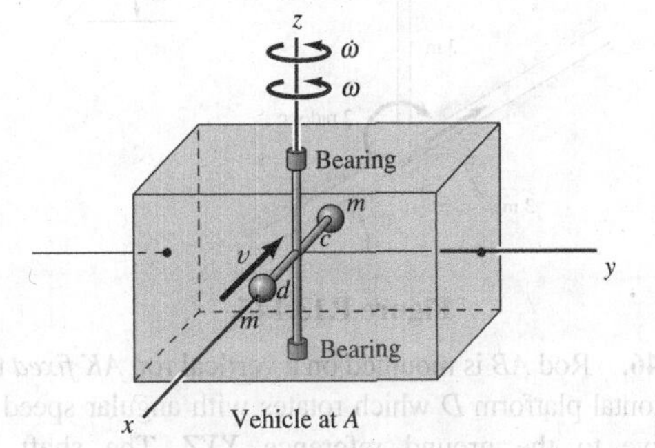

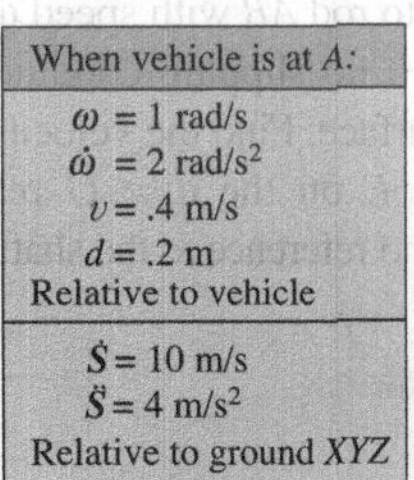

When vehicle is at *A*:

ω = 1 rad/s
$\dot{\omega}$ = 2 rad/s²
v = .4 m/s
d = .2 m
Relative to vehicle

$\dot{S}$ = 10 m/s
$\ddot{S}$ = 4 m/s²
Relative to ground *XYZ*

Figure P.13.143.

13.144. Cylinder *A* rolls without slipping. What are ω_{BC}, $\dot{\omega}_{BC}$, and $\dot{\omega}_{CD}$ At the instant shown, V_A = 5 m/s and $\dot{V}_A$ = 3 m/s². Use an intuitive approach only as a check over a formal approach.

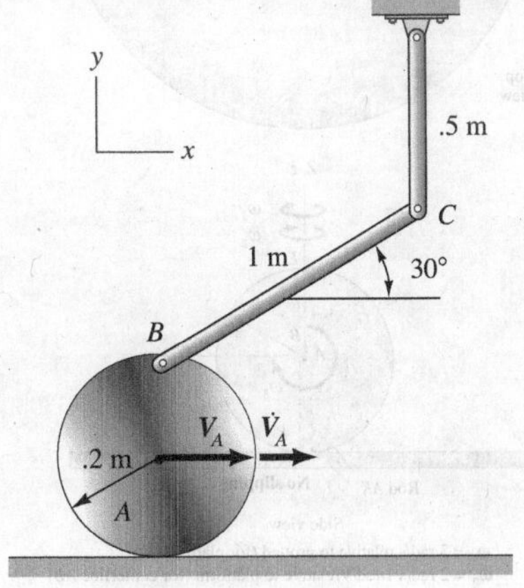

Figure P.13.144.

13.145. What is the magnitude of the velocity of the slider C at the instant shown in the diagram?

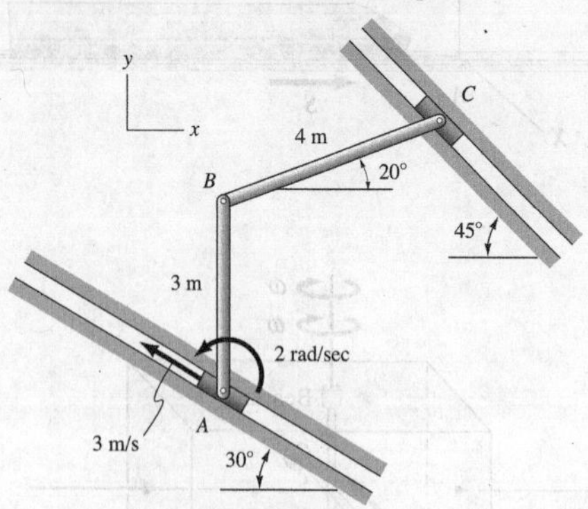

Figure P.13.145.

13.146. Rod AB is mounted on a vertical rod AK *fixed* to a horizontal platform D which rotates with angular speed ω_1 relative to the ground reference XYZ. The shaft AB meanwhile rotates relative to the platform with speed ω_2. Disc G rotates relative to rod AB with speed ω_3 whose value can be determined by a no slipping condition for the disc and platform contact surface. Find the velocity and acceleration vectors for point E on the disc G relative to XYZ. [Suggestion: Fix a second reference to the shaft AB as shown.]

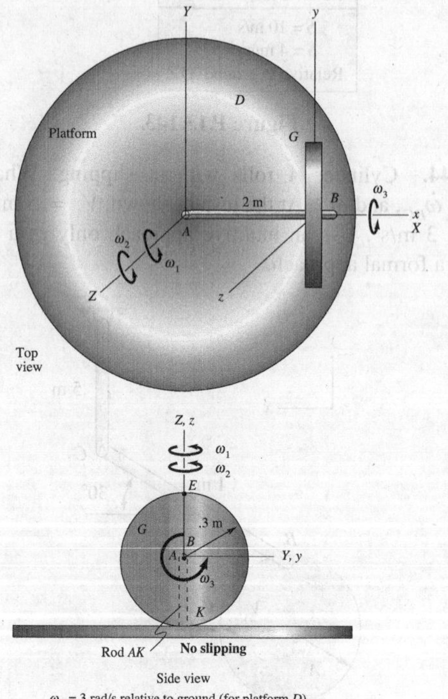

Figure P.13.146.

13.147. A conveyor system is shown. Rod AB is welded to plate D. End A is connected to a vehicle which moves with velocity V_2 along a rail. System AB and connected plate D have angular velocities ω_1 and ω_2 relative to the vehicle A as shown in the diagram. On plate D, a particle G has motion given as V_1 and $\dot{V}_1$ along spoke BE which at the instant of interest is parallel to the X axis of the stationary reference XYZ. Determine the acceleration of particle G at the time of interest. Use as a second reference xyz fixed to the plate with the origin at B.

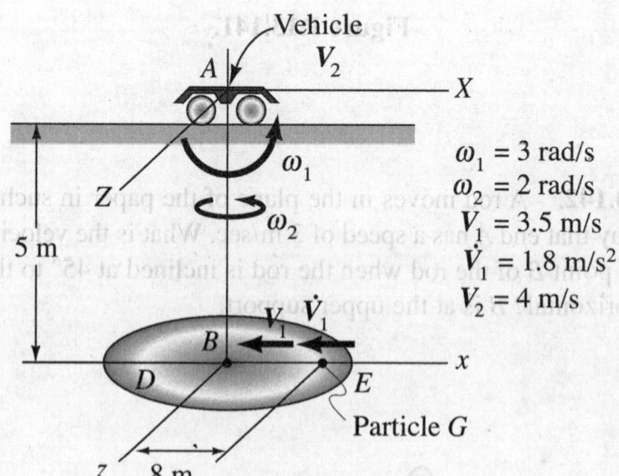

Figure P.13.147.

CHAPTER 14

Kinetics of Plane Motion of Rigid Bodies

14.1 Introduction

In **kinematics** we learned that the motion of a rigid body at any time t can be considered to be a superposition of a translational motion and a rotational motion. The translational motion may have the actual instantaneous velocity of any point of the body, and the angular velocity of the rotation, $\boldsymbol{\omega}$, then has its axis of rotation through the chosen point. A convenient point is, of course, the center of mass of the rigid body. The translatory motion can then be found from particle dynamics. You will recall that the motion of the center of mass of any aggregate of particles (this includes a rigid body) is related to the total external force by the equation

$$\boldsymbol{F} = M\dot{\boldsymbol{V}}_c \tag{14.1}$$

where M is the total mass of the aggregate. Integrating this equation, we get the motion of the center of mass. To ascertain fully the motion of the body, we must next find $\boldsymbol{\omega}$. As we saw in Chapter 12,

$$\boldsymbol{M}_A = \dot{\boldsymbol{H}}_A \tag{14.2}$$

for any system of particles where the point A about which moments of force and linear momentum are to be taken can be (1) the mass center, (2) a point fixed in an inertial reference, or (3) a point accelerating toward or away from the mass center. For these points, we shall later show that the angular velocity vector $\boldsymbol{\omega}$ is involved in the equation above when it is applied to rigid bodies. Also, the inertia tensor will be involved. After we find the motion of the mass center from Eq. 14.1 and the angular velocity $\boldsymbol{\omega}$ from Eq. 14.2, we get the instantaneous motion by letting the entire body have the velocity $\boldsymbol{V}_c$ plus the angular velocity $\boldsymbol{\omega}$, with the axis of rotation going through the center of mass.[1]

[1]After the general development and after looking at the solution of three-dimensional problems, one may wish to come back to Chapter 14 to study plane motion dynamics in detail. This approach is entirely optional.

14.2 Moment-of-Momentum Equations

Consider now a rigid body wherein each particle of the body moves parallel to a plane. Such a body is said to be in *plane motion* relative to this plane. We shall consider that axes *XY* are in the aforementioned plane in the ensuing discussion. The *Z* axis is then normal to the velocity vector of each point in the body. Furthermore, we consider only the situation where *XYZ* is an *inertial reference*. A body undergoing plane motion relative to *XYZ* as described above is shown in Fig. 14.1.

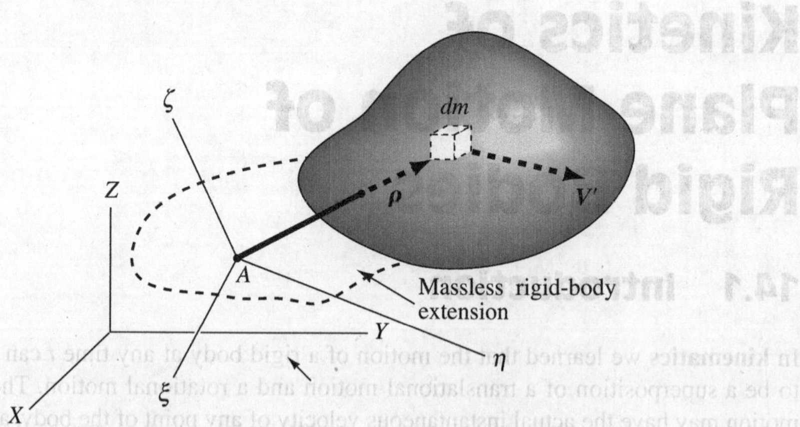

Figure 14.1. Body undergoing plane motion parallel to *XY* plane. $\xi\eta\zeta$ translates with *A*.

Choose some point *A* which is part of this body or a hypothetical massless rigid body extension of this body. An element *dm* of the body is shown at a position $\boldsymbol{\rho}$ from *A*. The velocity $\mathbf{V}'$ of *dm* relative to *A* is simply the velocity of *dm* relative to any reference $\xi\eta\zeta$, which *translates* with *A* relative to *XYZ*. Similarly, the linear momentum of *dm* relative to *A* (i.e., $\mathbf{V}'\,dm$) is the linear momentum of *dm* relative to $\xi\eta\zeta$ translating with *A*. We can now give the moment of this momentum (i.e., the angular momentum) $d\mathbf{H}_A$ about *A* as

$$d\mathbf{H}_A = \boldsymbol{\rho} \times \mathbf{V}'\,dm = \boldsymbol{\rho} \times \left(\frac{d\boldsymbol{\rho}}{dt}\right)_{\xi\eta\zeta} dm$$

But since *A* is fixed in the body (or in a hypothetical massless extension of the body) and dm is a part of the body having mass, the vector $\boldsymbol{\rho}$ must be *fixed* in the body and, accordingly,

$$\left(\frac{d\boldsymbol{\rho}}{dt}\right)_{\xi\eta\zeta} = \boldsymbol{\omega} \times \boldsymbol{\rho}$$

where $\boldsymbol{\omega}$ is the angular velocity of the body relative to $\xi\eta\zeta$. However, since $\xi\eta\zeta$ translates relative to *XYZ*, $\boldsymbol{\omega}$ is the angular velocity of the body relative to *XYZ* as well. Hence, we can say:

$$d\mathbf{H}_A = \boldsymbol{\rho} \times (\boldsymbol{\omega} \times \boldsymbol{\rho})dm \tag{14.3}$$

Note that the angular velocity $\boldsymbol{\omega}$ for the plane motion relative to the XY plane must have a direction *normal* to the XY plane.

Having helped us reach Eq. (14.3), we no longer need reference $\xi\eta\zeta$ and so we now dispense with it. Instead we *fix* reference xyz to the body at point A such that the z axis is *normal* to the plane of motion while the other two axes have arbitrary orientations normal to z (see Fig. 14.2).

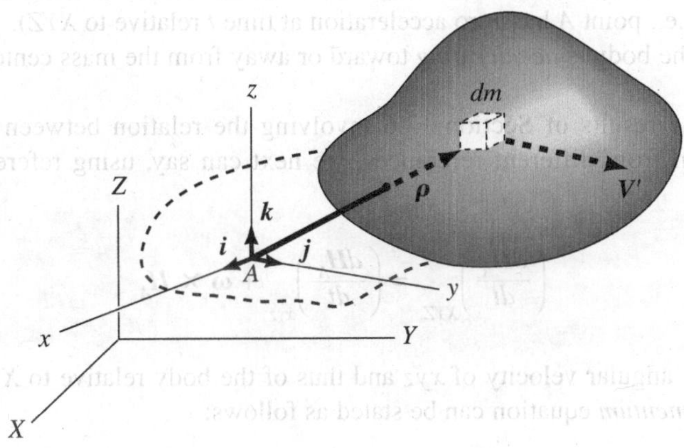

Figure 14.2. Reference xyz fixed to body at A; reference XYZ is inertial.

Note that the z axis will remain normal to XY as the body moves because of the plane motion restriction. Next, we evaluate Eq. 14.3 in terms of components relative to xyz as follows:

$$(dH_A)_x\boldsymbol{i} + (dH_A)_y\boldsymbol{j} + (dH_A)_z\boldsymbol{k} = (x\boldsymbol{i} + y\boldsymbol{j} + z\boldsymbol{k}) \times \left[(\omega\boldsymbol{k}) \times (x\boldsymbol{i} + y\boldsymbol{j} + z\boldsymbol{k})\right]dm$$

The scalar equations resulting from the foregoing vector equations are

$$\begin{aligned}(dH_A)_x &= -\omega xz\,dm\\(dH_A)_y &= -\omega yz\,dm\\(dH_A)_z &= \omega(x^2 + y^2)\,dm\end{aligned}$$

Integrating over the entire body, we get[2]

$$\begin{aligned}(H_A)_x &= -\iiint_M \omega xz\,dm = -\omega\iiint_M xz\,dm = -\omega I_{xz}\\(H_A)_y &= -\iiint_M \omega yz\,dm = -\omega\iiint_M yz\,dm = -\omega I_{yz}\\(H_A)_z &= \iiint_M \omega(x^2 + y^2)\,dm = \omega\iiint_M (x^2 + y^2)\,dm = \omega I_{zz}\end{aligned} \tag{14.4}$$

[2]Note that the massless extension of the rigid body has zero density and hence does not contribute to the integration.

We now have the angular momentum components for reference xyz at A.[3] Note that, because xyz is *fixed* to the body, the inertia terms I_{xz}, I_{yz}, and I_{zz} must be *constants*.

In order to employ the moment-of-momentum equation, $\boldsymbol{M}_A = \dot{\boldsymbol{H}}_A$, at time t, we next restrict point A of the body to be any one of the following three cases.[4]

1. Point A of the body is the *center of mass* of the body.
2. Point A of the body is *fixed* or moving with *constant velocity* at time t in inertial reference XYZ (i.e., point A has zero acceleration at time t relative to XYZ).
3. Point A of the body is *accelerating* toward or away from the mass center at time t.

Using the results of Section 13.6 involving the relation between derivatives of vectors as seen from different references, we next can say, using references XYZ and xyz:

$$\left(\frac{d\boldsymbol{H}_A}{dt}\right)_{XYZ} = \left(\frac{d\boldsymbol{H}_A}{dt}\right)_{xyz} + \boldsymbol{\omega} \times \boldsymbol{H}_A$$

where $\boldsymbol{\omega}$ is the angular velocity of xyz and thus of the body relative to XYZ. Hence, the *moment-of-momentum* equation can be stated as follows:

$$\boldsymbol{M}_A = \left(\frac{d\boldsymbol{H}_A}{dt}\right)_{xyz} + \boldsymbol{\omega} \times \boldsymbol{H}_A \tag{14.5}$$

Using Eqs. 14.4 for the components of $\boldsymbol{H}_A$, we get for this equation:

$$\boldsymbol{M}_A = \frac{d}{dt_{xyz}}(-\omega I_{xz}\boldsymbol{i} - \omega I_{yz}\boldsymbol{j} + \omega I_{zz}\boldsymbol{k}) + \omega \boldsymbol{k} \times (-\omega I_{xz}\boldsymbol{i} - \omega I_{yz}\boldsymbol{j} + \omega I_{zz}\boldsymbol{k})$$

Noting that $\boldsymbol{i}$, $\boldsymbol{j}$, and $\boldsymbol{k}$ are constant vectors as seen from xyz, as are the inertia terms, we get

$$\boldsymbol{M}_A = -\dot{\omega} I_{xz}\boldsymbol{i} - \dot{\omega} I_{yz}\boldsymbol{j} + \dot{\omega} I_{zz}\boldsymbol{k} - \omega^2 I_{xz}\boldsymbol{j} + \omega^2 I_{yz}\boldsymbol{i}$$

The scalar forms of the equation above are then

$$(M_A)_x = -I_{xz}\dot{\omega} + I_{yz}\omega^2 \tag{14.6a}$$

$$(M_A)_y = -I_{yz}\dot{\omega} - I_{xz}\omega^2 \tag{14.6b}$$

$$(M_A)_z = I_{zz}\dot{\omega} \tag{14.6c}$$

[3]We now see the motivation for presenting earlier the definitions of mass moments and products of inertia. Clearly the inertia tensor enters prominently in the evaluation of the angular motion of a rigid body.

[4]The "body" here includes the hypothetical, massless, rigid-body extension as well as the actual body.

It is important to note emphatically that the angular velocity as given by ω (and later by $\dot{\theta}$) is always taken *relative to the inertial reference XYZ*, whereas the moments of forces (as given by $(M_A)_x$, $(M_A)_y$, and $(M_A)_z$ as well as the inertia tensor components are always taken about the axes *xyz fixed to the body at A* (Eqs. (14.6). Eqs. (14.6) are the *general angular momentum equations for plane motion*. The last equation is probably familiar to you from your work in physics. There you expressed it as

$$T = I\alpha \tag{14.7}$$

or as

$$T = I\ddot{\theta} \tag{14.8}$$

We shall now consider special cases of plane motion, starting with the most simple case and going toward the most general case. However, please remember that for *all* plane motions relative to an inertial reference, the moment of the forces about the z axis at A *always* equals $I_{zz}\dot{\omega}$. The other two equations of 14.6 may get simplified for various special plane motions.

Also, to use Eqs. 14.6, we must remember that point A is *part of the body* (because we took $\boldsymbol{\rho}$ to be fixed in the body so we could use $d\boldsymbol{\rho}/dt = \boldsymbol{\omega} \times \boldsymbol{\rho}$) and is also one of the *three acceptable points* presented in Chapter 12. Furthermore, the *xyz* axes are *fixed to the body* to render the inertia tensor components constant.

Finally, we note from the derivation that ω, θ, and their derivatives are measured from the inertial reference *XYZ*.

14.3 Pure Rotation of a Body of Revolution About Its Axis of Revolution

A uniform body of revolution is shown in Fig. 14.3. If the body undergoes pure rotation about the axis of revolution fixed in inertial space reference *XYZ*, we then have plane motion parallel to any plane for which the axis of revolution is a normal. A reference *xyz* is *fixed* to the body such that the z axis is collinear with the axis of revolution. Since all points along the axis of revolution are fixed in inertial space *XYZ*, we can choose for the origin of reference *xyz* any point A along this axis. The x and y axes forming a right handed triad then have arbitrary orientation. For simplicity, we choose *xyz* collinear with axes *XYZ* at time t. Clearly, the plane *zy* is a plane of symmetry for this body, and the x axis is normal to this plane of symmetry. From our work in Chapter 8, recall[5] that, as a consequence, $I_{xy} = I_{xz} = 0$. Similarly, with y normal to a plane of symmetry, *xz*, we

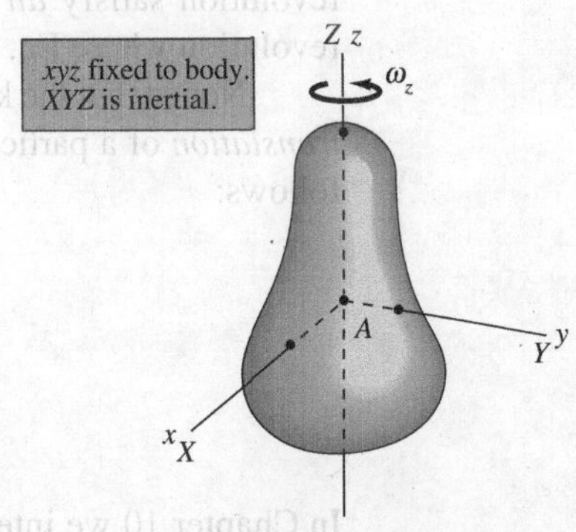

Figure 14.3. Rigid uniform body of revolution at time *t*.

[5]We pointed out in Chapter 8 that if an axis, such as the x axis, is normal to a plane of symmetry, then the products of inertia with x as a subscript must be zero. This is similarly true for other axes normal to a plane of symmetry.

conclude that $I_{yx} = I_{yz} = 0$. Hence, xyz are principal axes. Returning to Eq. 14.6, we find that only one equation of the set has nonzero moment, and that is the familiar equation

$$M_z = I_{zz}\dot{\omega}_z \tag{14.9}$$

The other pair of equations from 14.6 yield

$$\begin{aligned} M_x &= 0 \\ M_y &= 0 \end{aligned} \tag{14.10}$$

Now we turn to *Newton's law*. For this we must use the inertial reference XYZ. Note that at the instant t shown in Fig. 14.3, axes xyz and axes XYZ have been taken as collinear. This means that at this instant the forces on the body needed in Newton's law such as F_X can be denoted as F_x since the directions of X and x are the same at this instant and it is only the direction that is significant here. Since the center of mass of the body is stationary at all times (it is on the axis of rotation), we can accordingly say from *Newton's law*:

$$\begin{aligned} \sum F_x &= 0 \\ \sum F_y &= 0 \\ \sum F_z &= 0 \end{aligned} \tag{14.11}$$

Thus, the applied forces at any time t, the supporting forces, and the weight of the body of revolution satisfy *all* the equations of equilibrium *except* for motion about the axis of revolution where Eq. 14.9 applies.

Notice that the key equation (14.9) has the *same form* as Newton's law for *rectilinear translation* of a particle along an axis, say the x axis. We write both equations together as follows:

$$M_z = I_{zz}\ddot{\theta} \tag{14.12a}$$

$$F_X = M\ddot{X} \tag{14.12b}$$

In Chapter 10 we integrated Eq. 14.12b for various kinds of force functions: time functions, velocity functions, and position functions. The same techniques used then to integrate Eq. 14.12b can now be used to integrate Eq. 14.12a, where the moment functions can also be time functions, angular velocity functions, and angular position functions.

We illustrate these possibilities in the following examples. The first example involves a torque which in part is a function of angular position θ.

Example 14.1

A stepped cylinder having a radius of gyration $k = .40$ m and a mass of 200 kg is shown in Fig. 14.4. The cylinder supports a weight W of mass 100 kg with an inextensible cord and is restrained by a linear spring whose constant K is 2 N/mm. What is the angular acceleration of the stepped cylinder when it has rotated 10° after it is released from a state of rest? The spring is initially unstretched. What are the supporting forces at this time?

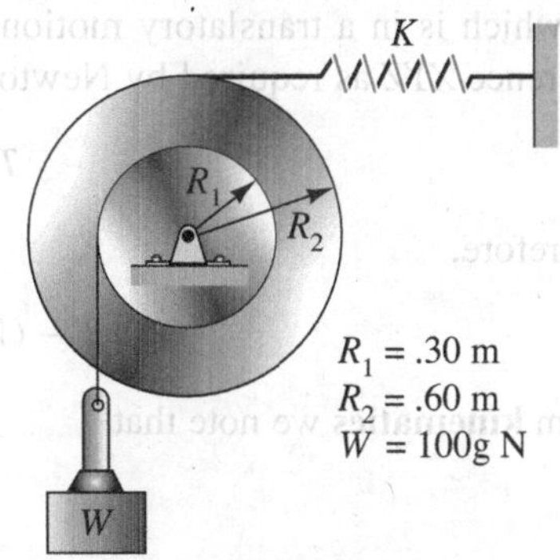

Figure 14.4. Stepped cylinder.

We have shown free-body diagrams of the stepped cylinder and the weight W in Fig. 14.5. A tension T from the cord is shown acting both on the weight W and the stepped cylinder. We have here for the stepped cylinder a body of revolution rotating about its axis of symmetry along which we have chosen point A. Axes xyz are fixed to the body at A with z along the axis of rotation. Furthermore we have shown inertial axes XYZ at A collinear with xyz at the time t.

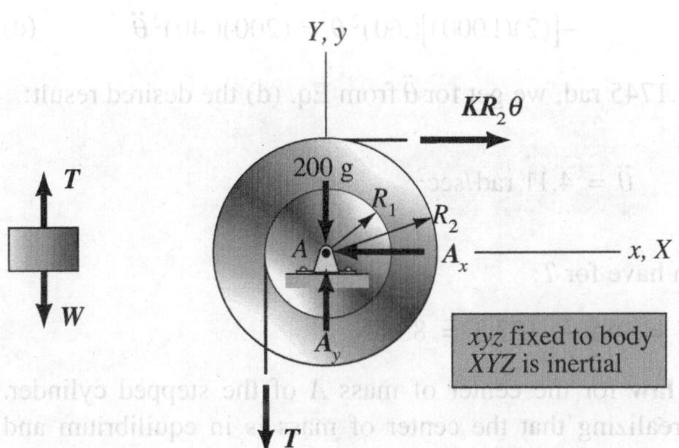

Figure 14.5. Free-body diagrams of components at time t.

We can accordingly apply the **moment-of-momentum** equation about the centerline of the cylinder:

$$TR_1 - KR_2^2\theta = I\ddot{\theta} = (Mk^2)\ddot{\theta}$$

$$T(.30) - [(2)(1{,}000)](.60)^2\theta = (200)(.40)^2\ddot{\theta} \qquad \text{(a)}$$

where as indicated earlier $\dot{\theta}$ is the angular velocity of the body relative to stationary axes XYZ and where θ is the rotation of the cylinder in radians from a position corresponding to the unstretched condition of the spring. Now considering the weight

Example 14.1 (Continued)

W, which is in a translatory motion, we can say, from **Newton's law** using inertial reference XYZ as required by Newton's law

$$T - W = M\ddot{Y}$$

Therefore,

$$T - (100)(9.81) = 100\ddot{Y} \qquad \text{(b)}$$

From **kinematics** we note that

$$R_1\ddot{\theta} = -\ddot{Y}$$

Therefore,

$$.30\ddot{\theta} = -\ddot{Y} \qquad \text{(c)}$$

Substituting for T in Eq. (a) using Eq. (b) and for $\ddot{Y}$ using Eq. (c), we then have

$$\left[(100)(9.81) + (100)(-.30\ddot{\theta})\right](.30)$$
$$-[(2)(1{,}000)](.60)^2\,\theta = (200)(.40)^2\,\ddot{\theta} \qquad \text{(d)}$$

When $\theta = (10°)(2\pi/360°) = .1745$ rad, we get for $\ddot{\theta}$ from Eq. (d) the desired result:

$$\ddot{\theta} = 4.11 \text{ rad/sec}^2$$

From Eqs. (b) and (c), we then have for T:

$$T = 981 - 123.3 = 858 \text{ N}$$

We next use **Newton's law** for the center of mass A of the stepped cylinder. Thus, considering Fig. 14.5, realizing that the center of mass is in equilibrium and assuming collinear orientation of the two sets of axes at time t, we can say $A_X \equiv A_x$ and $A_Y \equiv A_y$ at time t since only direction is involved here. Thus summing forces,

$$-858 - 200g + A_y = 0$$

$$A_y = 2{,}820 \text{ N}$$

$$-A_x + [(2)(1{,}000)](.60)(.1745) = 0$$

$$A_x = 209 \text{ N}$$

It should be clear on examining Fig. 14.4 that the motion of the cylinder, after W is released from rest, will be rotational oscillation. This motion ensues because the spring develops a restoring torque much as the spring in the classic spring–mass system (Fig. 14.6) supplies a restoring force. We shall study torsional oscillation or vibration in Chapter 14 when we consider vibrations. The key concepts and mathematical techniques for both motions you will find to be identical.

The torque in the next example is, in part, a function of time.

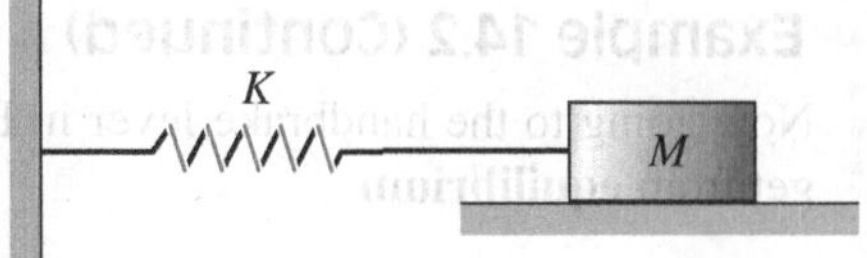

Figure 14.6. Classic spring–mass system.

Example 14.2

A cylinder A is rotating at a speed ω of 1,750 rpm (see Fig. 14.7) when the light handbrake system is applied using force F = $(10t + 300)$ N with t in seconds. If the cylinder has a radius of gyration of 200 mm and a mass of 500 kg, how long a time does it take to halve the speed of the cylinder? The dynamic coefficient of friction between the belt and the cylinder is .3.

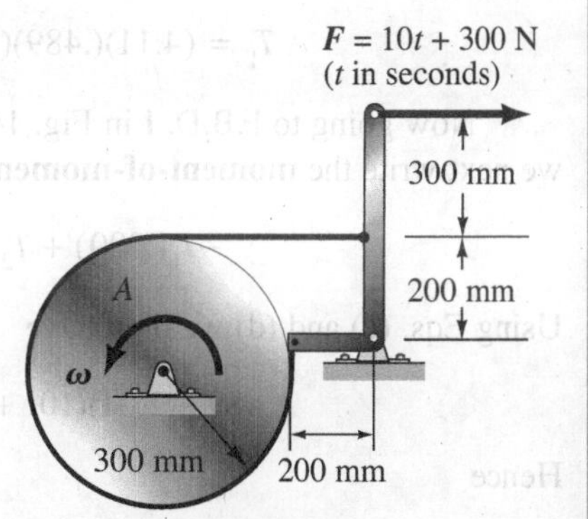

Figure 14.7. Cylinder and handbrake system.

We start by showing the free body of the cylinder and of the brake lever in Fig. 14.8. From the belt formula of Chapter 6, we can say for the belt tensions on the cylinder

$$\frac{T_1}{T_2} = e^{\mu_d \beta} = e^{(.3)\left(\frac{3}{4}\right)(2\pi)} = 4.11$$

$$\therefore T_1 = 4.11T_2 \qquad \text{(a)}$$

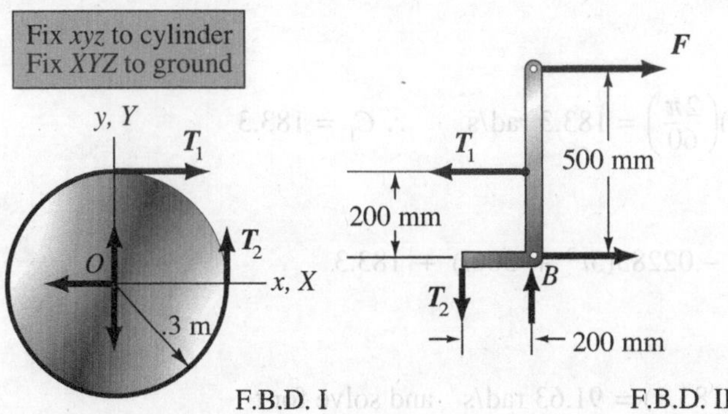

Figure 14.8. Free-body diagrams of cylinder and handbrake lever.

Example 14.2 (Continued)

Now going to the handbrake lever in F.B.D. II and taking moments about point B, we get from **equilibrium**

$$-F(.500) + T_1(.200) + T_2(.200) = 0$$

Inserting $F = 10t + 300$ N, we get for $(T_1 + T_2)$

$$T_1 + T_2 = 2.5(10t + 300) \qquad \text{(b)}$$

Substitute for T_1 using Eq. (a).

$$4.11T_2 + T_2 = 2.5(10t + 300)$$

$$\therefore T_2 = .489(10t + 300) \qquad \text{(c)}$$

Also from (a)

$$T_1 = (4.11)(.489)(10t + 300) = 2.01(10t + 300) \qquad \text{(d)}$$

Now going to F.B.D. I in Fig. 14.8 and using axes xyz fixed to the cylinder at O, we next write the **moment-of-momentum** equation.

$$-T_1(.300) + T_2(.300) = (500)(.200)^2(\ddot{\theta})$$

Using Eqs. (c) and (d) we then have

$$(.489 - 2.01)(10t + 300)(.300) = (500)(.200)^2\ddot{\theta}$$

Hence

$$\ddot{\theta} = -.02283(10t + 300)$$

Integrating

$$\dot{\theta} = -.02283\left(10\frac{t^2}{2} + 300t\right) + C_1 \qquad \text{(e)}$$

When $t = 0$,

$$\dot{\theta} = (1{,}750)\left(\frac{2\pi}{60}\right) = 183.3 \text{ rad/s} \qquad \therefore C_1 = 183.3$$

Thus

$$\dot{\theta} = -.02283(5t^2 + 300t) + 183.3$$

Set

$$\dot{\theta} = \left(\frac{1}{2}\right)(183.3) = 91.63 \text{ rad/s} \quad \text{and solve for } t$$

$$91.63 = -.02283(5t^2 + 300t) + 183.3$$

Example 14.2 (Continued)

Rearranging, we get

$$t^2 + 60t - 803 = 0$$

Using the quadratic formula, we have

$$t = \frac{-60 \pm \sqrt{60^2 + (4)(803)}}{2}$$

$$t = 11.27 \text{ sec}$$

14.4 Pure Rotation of a Body with Two Orthogonal Planes of Symmetry

Consider next a uniform body having *two* orthogonal planes of symmetry. Such a body is shown in Fig. 14.9, where in (a) we have shown the aforementioned planes of symmetry and in (b) we have shown a view along the intersection of the planes of symmetry. We shall consider pure rotation of such a body about a stationary axis collinear with the intersection of the planes of symmetry, which we take as the z axis. The origin A can be taken anywhere along the axis of rotation, and the x and y axes are fixed in the planes of symmetry, as shown in the diagram. XYZ is taken collinear with xyz at time t. We leave it to the reader to show that the identical equations apply to this case as to the previous case of a body of revolution.

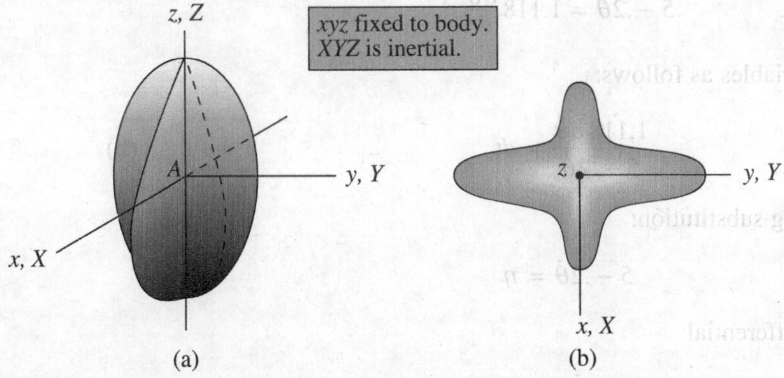

Figure 14.9. Body with two orthogonal planes of symmetry at time t.

The torque in the next example is a function of angular speed.[6]

[6]From now on it will be understood that in the diagrams xyz will be fixed to the body and that XYZ will be fixed to the ground and hence to be an inertial reference. This will avoid cluttering the diagrams unnecessarily.

Example 14.3

A thin-walled shaft is shown in Fig. 14.10. On it are welded identical plates A and B, each having a mass of 10 kg. Also welded onto the shaft at right angles to A and B are two identical plates C and D, each having a mass of 6 kg. The thin-walled shaft is of diameter 100 mm and has a mass of 15 kg. The wind resistance to rotation of this system is given for small angular velocities as $.2\dot{\theta}$ N-m, with $\dot{\theta}$ in rad/sec. Starting from rest, what is the time required for the system to reach 100 rpm if a torque T of 5 N-m is applied? What are the forces on the bearings G and E when this speed is reached?

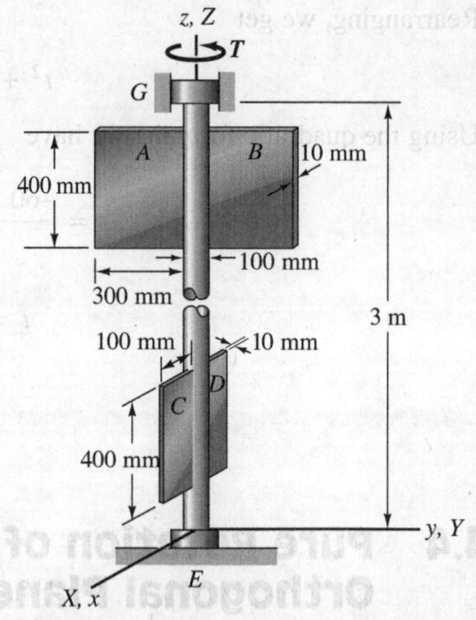

Figure 14.10. Device with rotational resistance.

We have here a body with two orthogonal planes of symmetry. The body is rotating about the axis of symmetry which we take as the z axis for axes xyz fixed to the body at E. As usual we position xyz at time t to be collinear with inertial reference XYZ.

The **moment-of-momentum** equation about the z axis is given as follows using moment of inertia formulas for plates (see inside covers) along with the parallel-axis formula and noting for the thin-walled shaft that we use $I_{zz} = Mr^2$ where, as an approximation, we take the outside radius for r.

$$5 - .2\dot{\theta} = \Big\{(15)(.05)^2 + 2\big[\tfrac{1}{12}(10)(.30^2 + .01^2) + 10(.20)^2\big] + 2\big[\tfrac{1}{12}(6)(.10^2 + .01^2) + (6)(.10)^2\big]\Big\}\ddot{\theta} \quad \text{(a)}$$

This becomes

$$5 - .2\dot{\theta} = 1.118\frac{d\dot{\theta}}{dt}$$

We can separate the variables as follows:

$$\frac{1.118\,d\dot{\theta}}{5 - .2\dot{\theta}} = dt \quad \text{(b)}$$

Now make the following substitution:

$$5 - .2\dot{\theta} = \eta$$

Therefore, taking the differential

$$-.2d\dot{\theta} = d\eta$$

Hence, we have for Eq. (b):

$$-5.59\frac{d\eta}{\eta} = dt$$

Example 14.3 (Continued)

Integrate to get

$$-5.59 \ln \eta = t + C_1$$

Hence, replacing η

$$-5.59 \ln(5 - .2\dot{\theta}) = t + C_1 \tag{c}$$

When $t = 0$, $\dot{\theta} = 0$, and we have for C_1:

$$-5.59 \ln(5) = C_1$$

Therefore,

$$C_1 = -8.99$$

Hence, Eq. (c) becomes

$$-5.59 \ln(5 - .2\dot{\theta}) = t - 8.99$$

Let $\dot{\theta} = (100/60)(2\pi) = 10.47$ rad/sec in the above equation. The desired time t for this speed to be reached then is

$$t = 3.03 \text{ sec}$$

We now consider the supporting forces for the system. For reasons set forth in Section 14.3 we know that

$$M_x = 0$$

$$M_y = 0$$

for the other **moment of momentum** equations. Also from **Newton's law**, while noting that xyz and XYZ are collinear at time t,

$$\sum F_x = 0$$

$$\sum F_y = 0$$

$$\sum F_z = 0$$

for the center of mass. Clearly, the dead weights of bodies (in the z direction) give rise to a constant supporting force of $[2(10 + 6) + 15]g = 461$ N at bearing E. All other forces are zero.

$E_z = 461$ N
All other support forces are zero

14.5 Pure Rotation of Slablike Bodies

We now consider bodies that have a *single* plane of symmetry, such as is shown in Fig. 14.11. Such bodies we shall call *slablike* bodies. We have oriented the body in Fig. 14.8 so that the plane of symmetry is parallel to the XY plane. We shall now consider the pure rotation of such a body about a fixed axis normal to the XY plane and going through a point A in the plane of symmetry of the body. We fix a reference xyz to the body at point A with xy in the plane of symmetry and z along the axis of rotation.

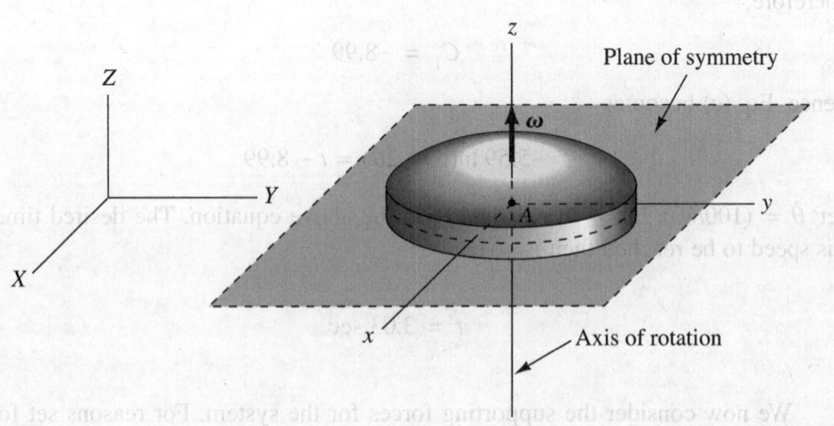

Figure 14.11. Slablike body undergoing pure rotation.

The angular velocity $\boldsymbol{\omega}$ is then along the z axis. Since z is normal to the plane of symmetry, it is clear immediately that $I_{zx} = I_{zy} = 0$. And so the *moment of momentum* equations become for this case:

$$\begin{aligned} M_x &= 0 \\ M_y &= 0 \qquad (14.13) \\ M_z &= I_{zz}\dot{\omega}_z \end{aligned}$$

If the center of mass is not at a position along the axis of rotation, then we no longer have equilibrium conditions for the center of mass. It will be undergoing circular motion. However, through *Newton's law* we can relate the external forces on the body to the acceleration of the mass center. We may then have to use the *kinematics* of rigid-body motion to yield enough equations to solve the problem. We now illustrate this case.

Example 14.4

A uniform rod of weight W and length L supported by a pin connection at A and a wire at B is shown in Fig. 14.12. What is the force on pin A at the instant that the wire is released? What is the force at A when the rod has rotated 45°?

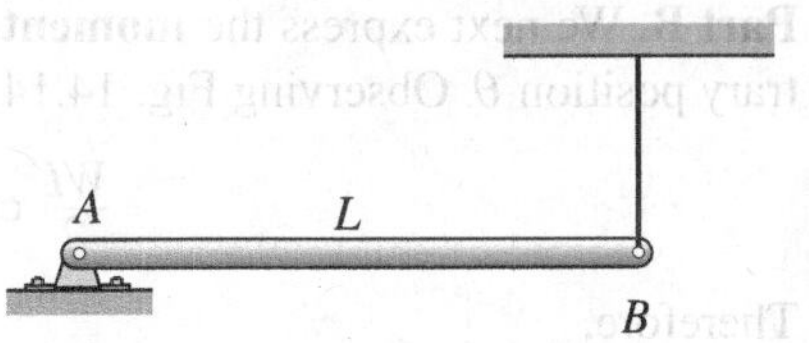

Figure 14.12. Rod supported by wire.

Part A. A free-body diagram of the rod is shown in Fig. 14.13 at the instant that the wire is released at B. We fix xyz to the body at A. XYZ is stationary. The **moment-of-momentum** equation about the axis of rotation at A, on using the formula for I of a rod about a transverse axis at the end, yields

$$\frac{WL}{2} = I\ddot{\theta} = \frac{1}{3}\left(\frac{W}{g}\right)L^2\ddot{\theta}$$

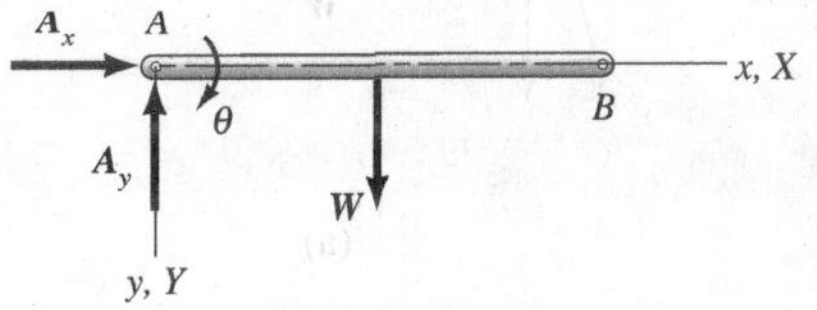

Figure 14.13. Wire suddenly cut.

Therefore,

$$\ddot{\theta} = \frac{3}{2}\frac{g}{L} \quad \text{at time } t = 0 \qquad \text{(a)}$$

Using simple **kinematics** of plane circular motion, we can determine the acceleration of the mass center at $t = 0$. Thus, using the inertial reference

$$\ddot{X} = 0, \qquad \ddot{Y} = \frac{L}{2}\ddot{\theta} = \frac{3}{4}g \qquad \text{(b)}$$

where we have used Eq. (a) in the last step. Next express **Newton's law** for the mass center using $A_y \equiv A_Y, A_x \equiv A_X$:[7]

$$\frac{W}{g}\ddot{X} = A_x, \qquad \frac{W}{g}\ddot{Y} = W - A_y$$

Accordingly, at time $t = 0$ we have, on noting Eqs. (b):

$$A_x = 0, \qquad A_y = \tfrac{1}{4}W \qquad \text{(c)}$$

Thus, we see that at the instant of releasing the wire there is an upward force of $\frac{1}{4}W$ on the left support.

[7]Note that we can replace A_X by A_x, etc., because of the common direction of X and x as well as the other corresponding axes at time t. However, we cannot replace $\ddot{X}$ by $\ddot{x}$ and $\ddot{Y}$ by $\ddot{y}$. The reason for this is that while the orientations of the respective axes are the same at time t, the fact remains that because of the rotational motion of xyz relative to XYZ, the velocities and accelerations of a particle relative to xyz and XYZ will be different requiring us to use only the XYZ axes when dealing with derivatives of the particle coordinates in Newton's law.

Example 14.4 (Continued)

Part B. We next express the **moment-of-momentum** equation for the rod at any arbitrary position θ. Observing Fig. 14.14, we get

$$\frac{WL}{2}\cos\theta = \frac{1}{3}\frac{W}{g}L^2\ddot{\theta}$$

Therefore,

$$\ddot{\theta} = \frac{3}{2}\frac{g}{L}\cos\theta \tag{d}$$

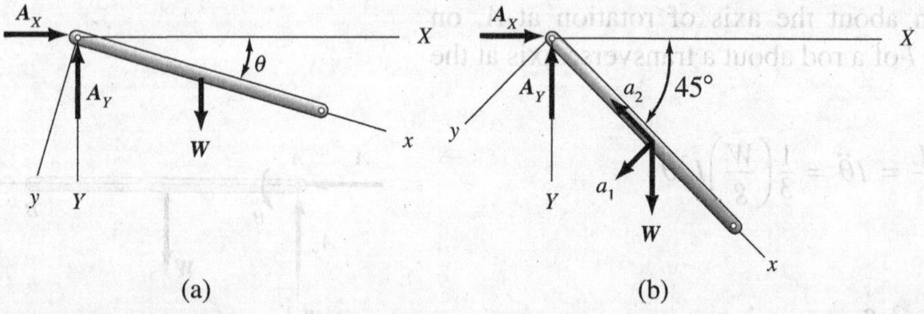

Figure 14.14. (a) Rod at position θ; (b) rod at $\theta = 45°$.

Consequently, at $\theta = 45°$ we have

$$\ddot{\theta} = (1.5)(.707)\frac{g}{L} = 1.060\frac{g}{L} \tag{e}$$

We shall also need $\dot{\theta}$, and accordingly we now rewrite Eq. (d) as follows:

$$\ddot{\theta} \equiv \left(\frac{d\dot{\theta}}{d\theta}\right)\left(\frac{d\theta}{dt}\right) \equiv \left(\frac{d\dot{\theta}}{d\theta}\right)(\dot{\theta}) = \frac{3}{2}\frac{g}{L}\cos\theta \tag{f}$$

Separating variables, we get

$$\dot{\theta}\,d\dot{\theta} = \frac{3}{2}\frac{g}{L}\cos\theta\,d\theta$$

which we integrate to get

$$\frac{\dot{\theta}^2}{2} = \frac{3}{2}\frac{g}{L}\sin\theta + C$$

When $\theta = 0$, $\dot{\theta} = 0$; accordingly, $C = 0$. We then have

$$\dot{\theta}^2 = 3\frac{g}{L}\sin\theta \tag{g}$$

At the instant of interest, $\theta = 45°$ and we get for $\dot{\theta}^2$:

$$\dot{\theta}^2 = 3\frac{g}{L}(.707) = 2.12\frac{g}{L} \tag{h}$$

For $\theta = 45°$, we can now give the acceleration component a_1 of the center of mass directed normal to the rod and component a_2 directed along the rod [see Fig. 14.14(b)]. From **kinematics** we can say, using Eqs. (e) and (h):

Example 14.4 (Continued)

$$a_1 = \frac{L}{2}\ddot{\theta} = \frac{L}{2}\left(1.060\frac{g}{L}\right) = .530g$$

$$a_2 = \frac{L}{2}(\dot{\theta})^2 = \frac{L}{2}\left(2.12\frac{g}{L}\right) = 1.060g \qquad \text{(i)}$$

Now, employing **Newton's law** for the mass center, we have on noting that xy is *no longer collinear* with XY.

$$A_X = \frac{W}{g}(-a_1 \sin 45° - a_2 \cos 45°)$$

Therefore, using Eqs. (i)

$$A_X = -1.124W$$

Also, from **Newton's law**

$$-A_Y + W = \frac{W}{g}(-a_2 \sin 45° + a_1 \cos 45°)$$

Therefore, again using Eq. (i)

$$A_Y = 1.375W \qquad \text{(j)}$$

Consider next the case of a body undergoing pure rotation about an axis which, for some point A in the body (or massless hypothetical extension of the body), is a *principal* axis (see Fig. 14.15). For a reference xyz fixed at A with z collinear with the axis of rotation, it is clear that $I_{zx} = I_{zy} = 0$, and hence the moment of momentum equations simplify to the exact same forms as presented here for the rotation of slablike bodies.

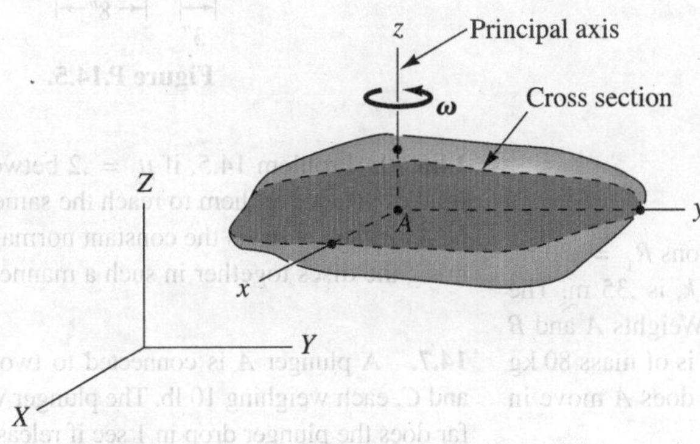

Figure 14.15. Axis z is a principal axis for point A.

PROBLEMS

14.1. A shaft and disc of steel having a density of 7,626 kg/m^3 are subjected to a constant torque T of 67.5 N-m, as shown. After 1 min, what is the angular velocity of the system? How many revolutions have occurred during this interval? Neglect friction of the bearings. Use $\frac{1}{2}Mr^2 = I$ for disc.

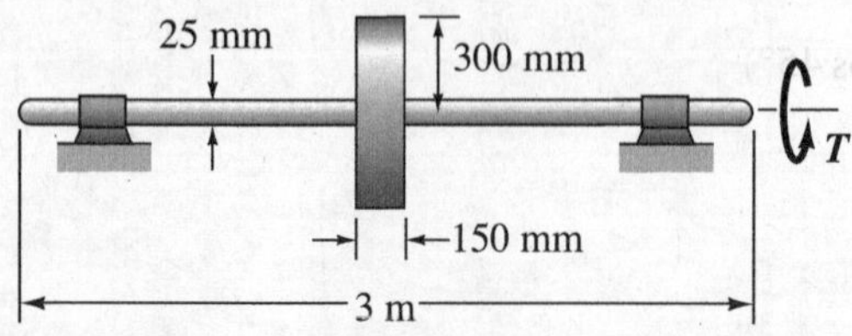

Figure P.14.1.

14.2. In Problem 14.1 include wind and bearing friction losses by assuming that they are proportional to angular speed $\dot{\theta}$. If the disc will halve its speed after 5 min from a speed of 300 rpm when there is no external applied torque, what is the resisting torque at 600 rpm?

14.3. A 400-lb flywheel is shown. A 100-lb block is held by a light cable wrapped around the hub of the flywheel, the diameter of which is 1 ft. If the initially stationary weight descends 3 ft in 5 sec, what is the radius of gyration of the flywheel?

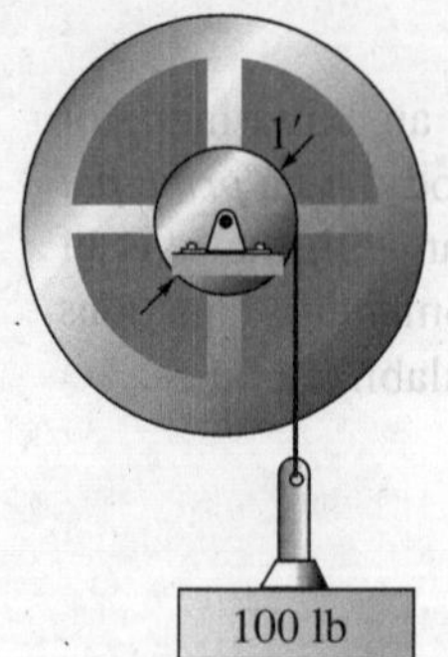

Figure P.14.3.

14.4 A stepped cylinder has the dimensions $R_1 = .30$ m, $R_2 = .65$ m, and the radius of gyration, k, is .35 m. The mass of the stepped cylinder is 100 kg. Weights A and B are connected to the cylinder. If weight B is of mass 80 kg and weight A is of mass 50 kg, how far does A move in 5 sec? In which direction does it move?

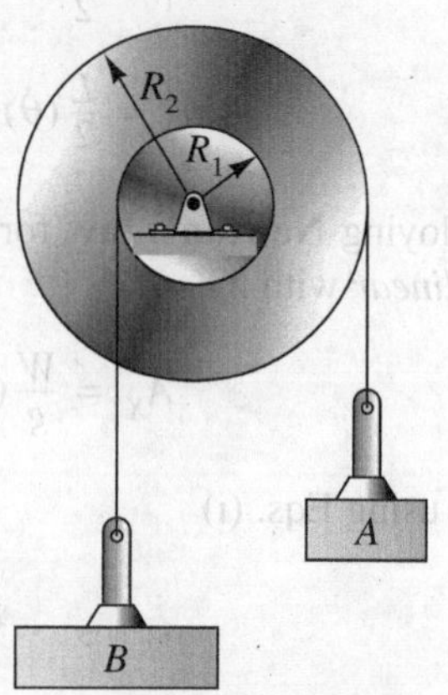

Figure P.14.4.

14.5. Two discs E and F of diameter 1 ft rotate in frictionless bearings. Disc F weighs 100 lb and rotates with angular speed ω_1 of 10 rad/sec, whereas disc E weighs 30 lb and rotates with angular speed ω_2 of 5 rad/sec. Neglecting the angular momentum of the shafts, what is the total angular momentum of the system relative to the ground? Use Eq. 14.4 to compute H from first principles. Consider that the discs are forced together along the axis of rotation. What is the common angular velocity when friction has reduced relative motion between the discs to zero?

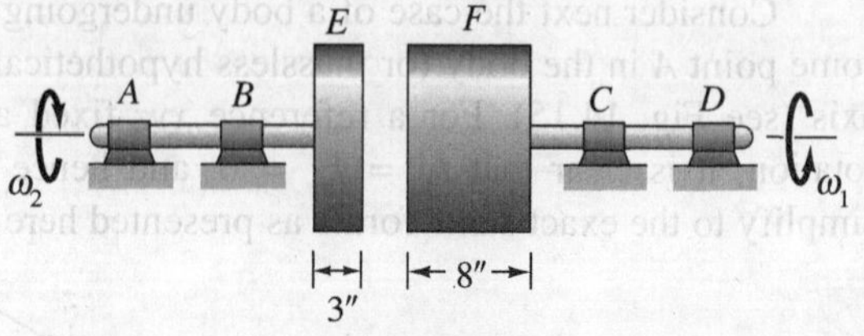

Figure P.14.5.

14.6 In Problem 14.5, if $\mu_s = .2$ between the discs, and it takes 30 sec for them to reach the same angular speed of 6.54 rad/sec, what is the constant normal force required to bring the discs together in such a manner?

14.7. A plunger A is connected to two identical gears B and C, each weighing 10 lb. The plunger weighs 40 lb. How far does the plunger drop in 1 sec if released from rest?

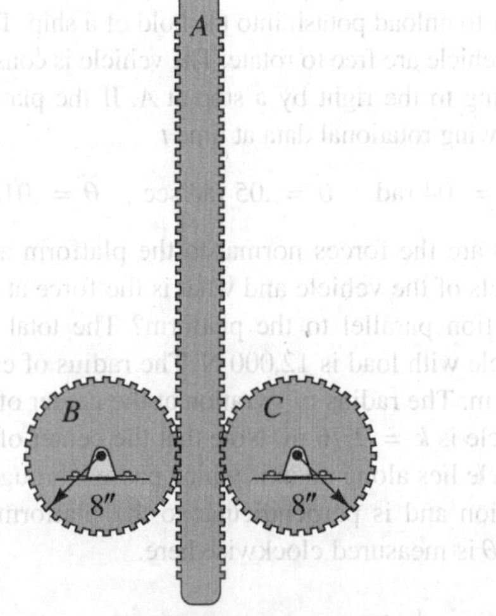

Figure P.14.7.

14.8. A pulley A and its rotating accessories have a mass of 1,000 kg and a radius of gyration of .25 m. A simple hand brake is applied as shown using a force P. If the dynamic coefficient of friction between belt and pulley is .2, what must force P be to change ω from 1,750 rpm to 300 rpm in 60 sec?

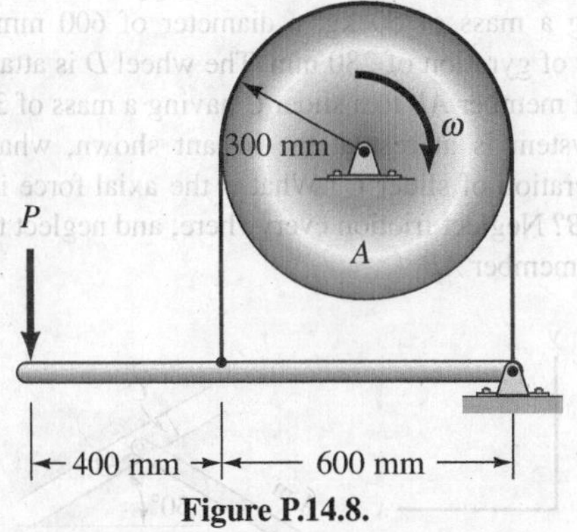

Figure P.14.8.

14.9. A flywheel is shown. There is a viscous damping torque due to wind and bearing friction which is known to be $-.04\omega$ N-m, where ω is in rad/sec. If a torque $T = 100$ N-m is applied, what is the speed in 5 min after starting from rest? The mass of the wheel is 500 kg, and the radius of gyration is .50 m.

Figure P.14.9.

14.10. Two cylinders and a rod are oriented in the vertical plane. The rod is guided by bearings (not shown) to move vertically. The following are the weights of the three bodies:

$$W_A = 1{,}000 \text{ N} \quad W_B = 300 \text{ N} \quad W_C = 200 \text{ N}$$

What is the acceleration of the rod? There is rolling without slipping.

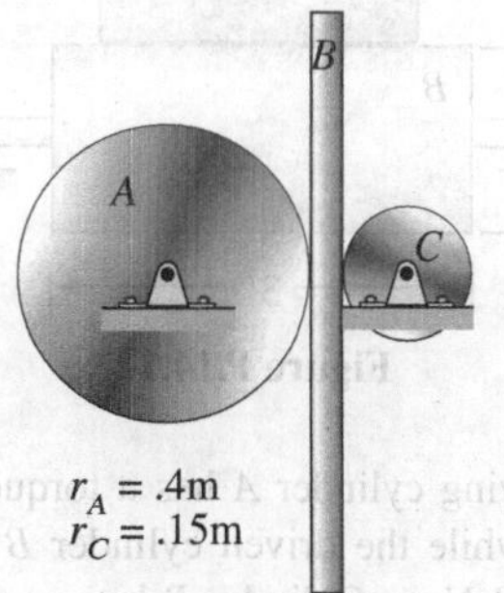

Figure P.14.10.

14.11. A bar A weighing 80 lb is supported at one end by rollers that move with the bar and a stationary cylinder B weighing 68 lb that rotates freely. A 100 lb force is applied to the end of the bar. What is the friction force between the bar and the cylinder as a function of x? Indicate the ranges for non-slipping and slipping conditions. Take $\mu_s = .5$ and $\mu_d = .3$.

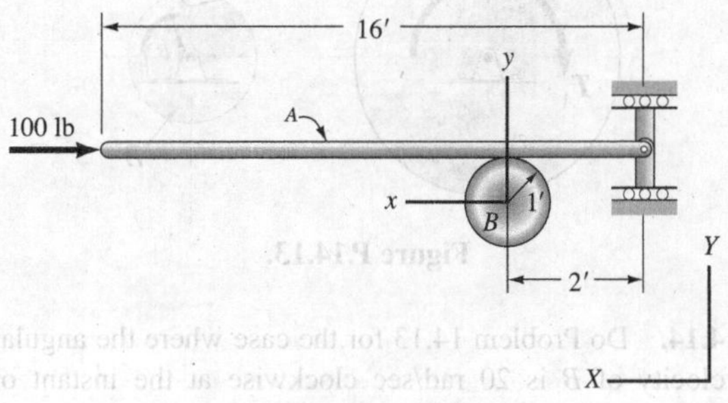

Figure P.14.11.

14.12. Cylinder A has an angular speed ω_1 of 3 rad/sec when it is lowered onto cylinder B, which has an angular speed ω_2 of 5 rad/sec before contact is made. What are the final angular velocities of the cylinders resulting from friction at the surfaces of contact? The mass of A is 500 lbm and of B is 400 lbm. If $\mu_d = .3$ for the contact surface of the cylinders and if the normal force transmitted from A to B is 10 lb, how long does it take for the cylinders to reach a constant speed?

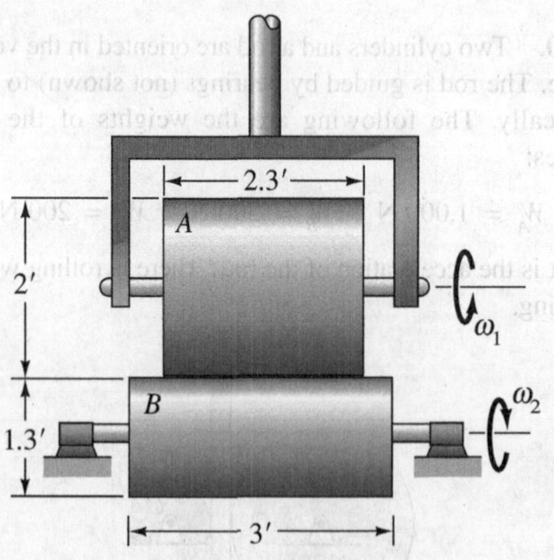

Figure P.14.12.

14.13. A driving cylinder A has a torque T_A of 30 N-m applied to it while the driven cylinder B has a resisting torque T_B of 10 N-m. Cylinder B has a mass of 15 kg, a radius of gyration of 100 mm, and a diameter of 400 mm. Cylinder A has a mass of 50 kg, a radius of gyration of 200 mm, and a diameter of 800 mm. Rod CD is a light rod connecting the cylinders. What is the angular acceleration of cylinder A at the instant shown if the system is stationary at this instant? Rod CD is 1 m long.

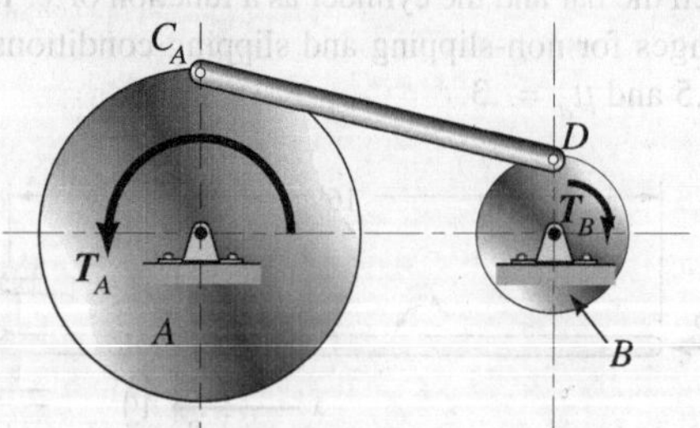

Figure P.14.13.

14.14. Do Problem 14.13 for the case where the angular velocity of B is 20 rad/sec clockwise at the instant of interest.

14.15. A loading vehicle is on a platform that can rotate in order to unload potash into the hold of a ship. The wheels of the vehicle are free to rotate. The vehicle is constrained from moving to the right by a stop at A. If the platform has the following rotational data at time t

$$\theta = .04 \text{ rad} \qquad \dot{\theta} = .05 \text{ rad/sec} \qquad \ddot{\theta} = .012 \text{ rad/sec}^2$$

what are the forces normal to the platform acting on the wheels of the vehicle and what is the force at the stop in a direction parallel to the platform? The total mass of the vehicle with load is 12,000 N. The radius of each wheel is 0.15 m. The radius of gyration at the center of mass of the vehicle is $k = 2.76$ m. Note that the center of mass of the vehicle lies along an axis which passes through the axis of rotation and is perpendicular to the platform. Also, note that θ is measured clockwise here.

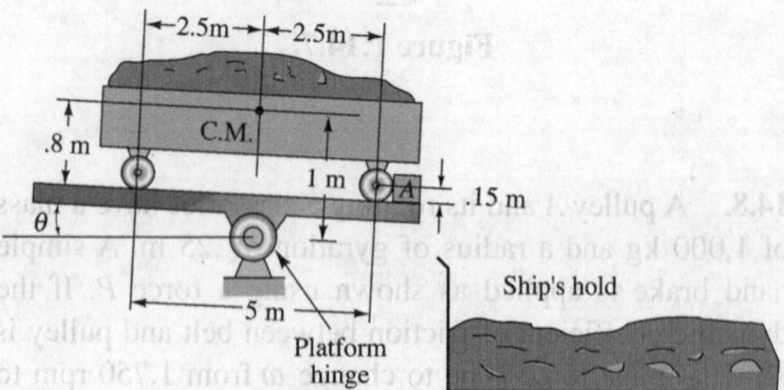

Figure P.14.15.

14.16. A torque T of 100 N-m is applied to a wheel D having a mass of 50 kg, a diameter of 600 mm, and a radius of gyration of 280 mm. The wheel D is attached by a light member AB to a slider C having a mass of 30 kg. If the system is at rest at the instant shown, what is the acceleration of slider C? What is the axial force in member AB? Neglect friction everywhere, and neglect the inertia of member AB.

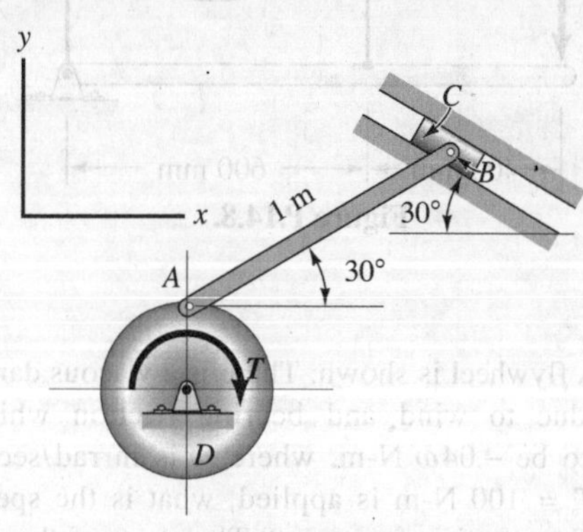

Figure P.14.16.

14.17. Do Problem 14.16 for the case where at the instant of interest $\omega_D = 2$ rad/sec counterclockwise.

14.18. A torque T of 50 N-m is applied to the device shown. The bent rods are of mass per unit length 5 kg/m. Neglecting the inertia of the shaft, how many rotations does the system make in 10 sec? Are there forces coming onto the bearings other than from the dead weights of the system?

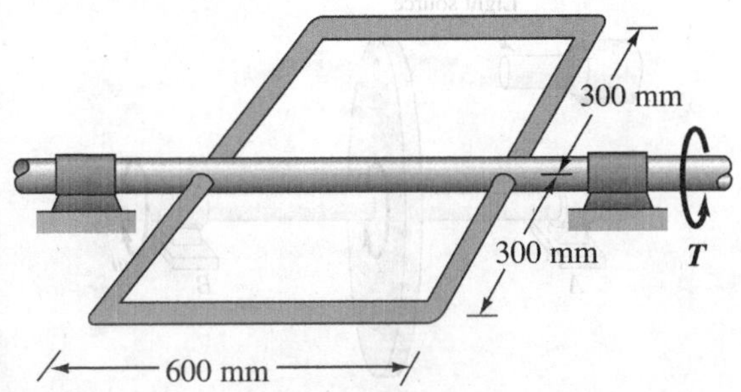

Figure P.14.18.

14.19. An idealized torque-versus-angular-speed curve for a shunt, direct-current motor is shown as curve A. The motor drives a pump which has a resisting torque-versus-speed curve shown in the diagram as curve B. Find the angular speed of the system as a function of time, after starting, over the range of speeds given in the diagram. Take the moment of inertia of motor, connecting shaft, and pump to be I.

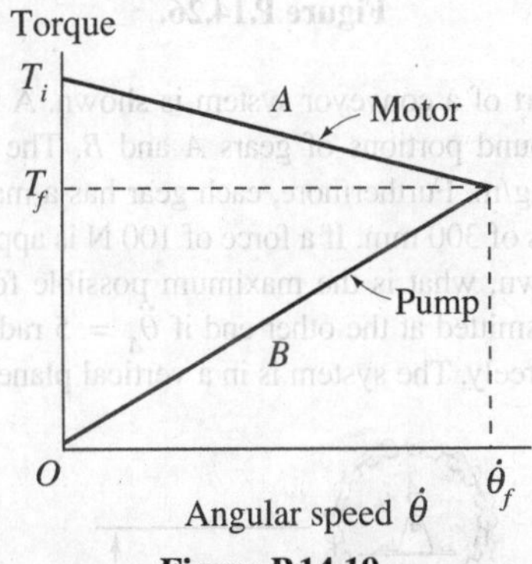

Figure P.14.19.

14.20. Rods of length L have been welded onto a rigid drum A. The system is rotating at a speed ω of 5,000 rpm. By this time, you may have studied stress in a rod in your strength of materials class. In any case, the stress is the normal force per unit area of cross section of the rod. If the cross-sectional area of the rod is 2 in^2 and the mass per unit length is 5 lbm/ft, what is the normal stress τ_{rr} on a section at any position r? The length L of the rods is 2 ft. What is τ_{rr} at $r = 1.5$ ft? Consider the upper rod when it is vertical.

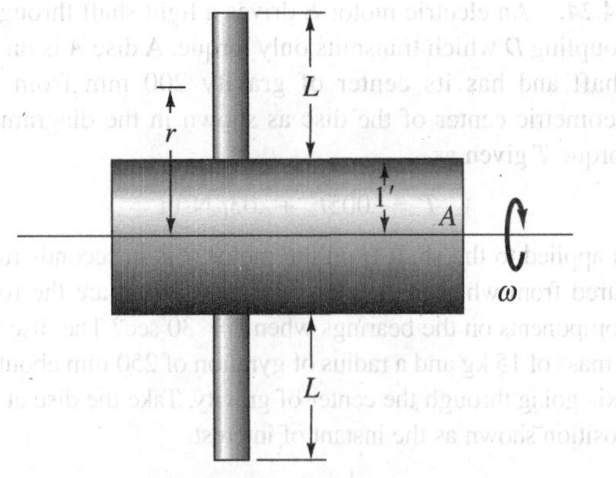

Figure P.14.20.

***14.21.** In Problem 14.20, consider that the mass per unit length varies linearly from 5 lbm/ft at $r = 1$ ft (at the bottom of the rod) to 6 lbm/ft at $r = 3$ ft (at the top of the rod). Find τ_{rr} at any position r and then compute τ_{rr} for $r = 1.5$ ft.

14.22. A plate weighing 3 lb/ft^2 is supported at A and B. What are the force components at B at the instant support A is removed?

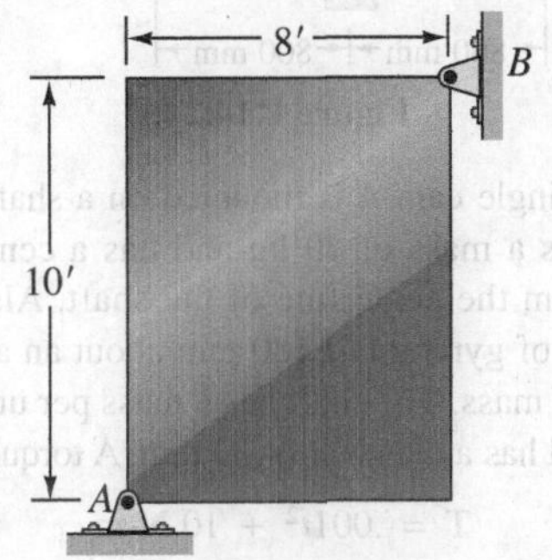

Figure P.14.22.

14.23. When the uniform rigid bar is horizontal, the spring at C is compressed 3 in. If the bar weighs 50 lb, what is the force at B when support A is removed suddenly? The spring constant is 50 lb/in.

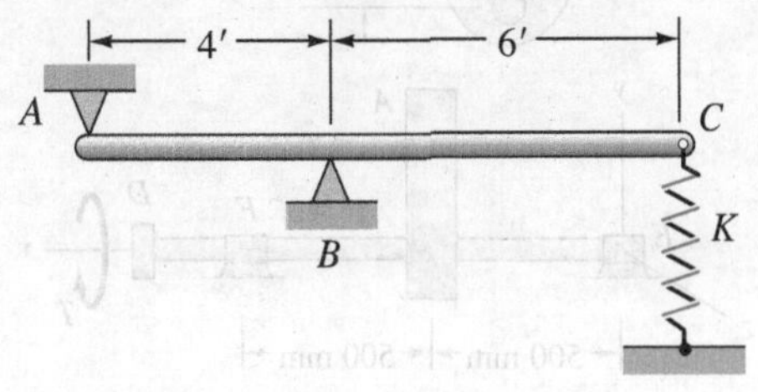

Figure P.14.23.

14.24. An electric motor E drives a light shaft through a coupling D which transmits only torque. A disc A is on the shaft and has its center of gravity 200 mm from the geometric center of the disc as shown in the diagram. A torque T given as

$$T = .005t^2 + .03t \text{ N-m}$$

is applied to the shaft from the motor (t is in seconds measured from when the system is at rest). What are the force components on the bearings when $t = 30$ sec? The disc has a mass of 15 kg and a radius of gyration of 250 mm about an axis going through the center of gravity. Take the disc at the position shown as the instant of interest.

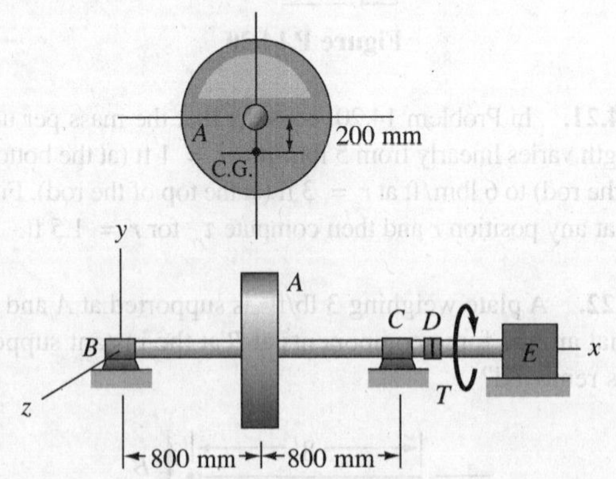

Figure P.14.24.

14.25. A single cam A is mounted on a shaft as shown. The cam has a mass of 10 kg and has a center of mass 300 mm from the centerline of the shaft. Also, the cam has a radius of gyration of 180 mm about an axis through the center of mass. The shaft has a mass per unit length of 10 kg/m and has a diameter of 30 mm. A torque T given as

$$T = .001t^2 + 10 \text{ N-m}$$

is applied at coupling D (t is in seconds). What are the force components in the bearings after 25 sec if the cam has the position shown in the diagram at this instant?

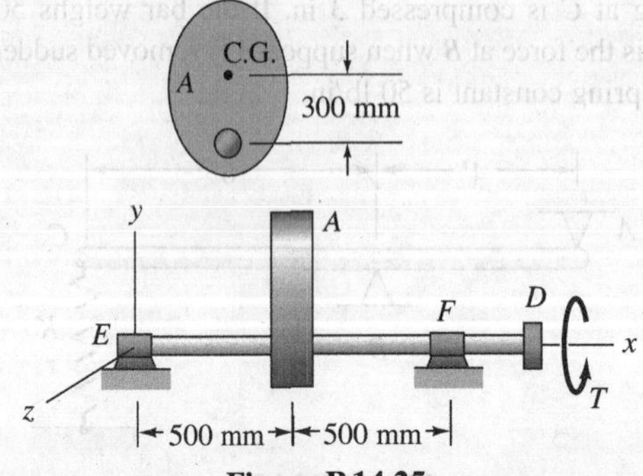

Figure P.14.25.

14.26. A circular plate is rotating at a speed of 10,000 rpm. A hole has been cut out of the plate so that the beam of light is allowed through for very short intervals of time. Such a device is called a *chopper*. If the plate weighed 20 lb originally and the material removed for the hole weighed 3 oz, what are the forces in the bearings A and B from the circular plate for the instant shown?

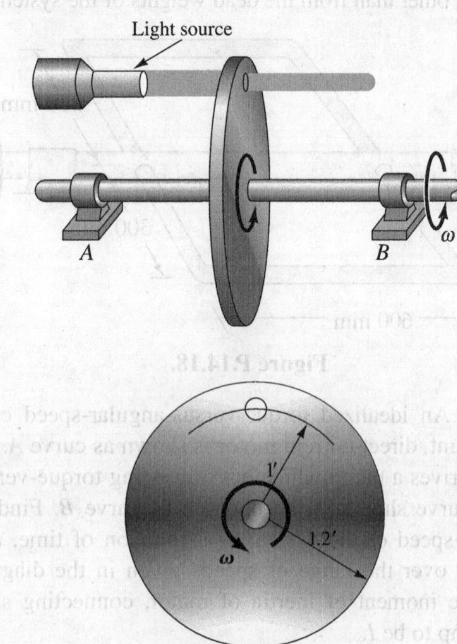

Figure P.14.26.

14.27. Part of a conveyor system is shown. A link belt is meshed around portions of gears A and B. The belt has a mass of 5 kg/m. Furthermore, each gear has a mass of 3 kg and a radius of 300 mm. If a force of 100 N is applied at one end as shown, what is the maximum possible force T that can be transmitted at the other end if $\ddot{\theta}_A = 5$ rad/sec^2? The gears turn freely. The system is in a vertical plane.

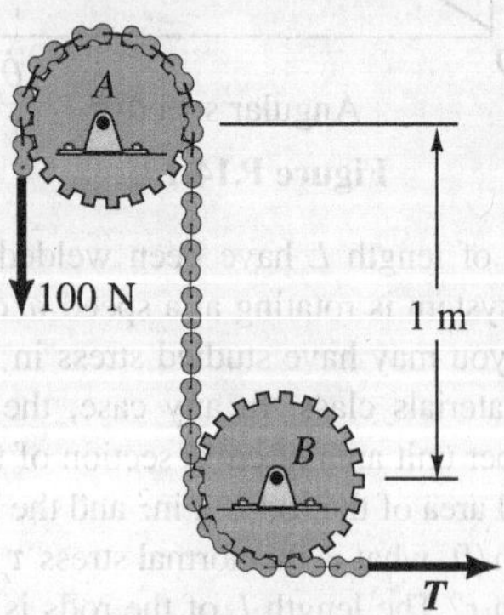

Figure P.14.27.

14.28. A hollow cylinder A of mass 100 kg can rotate over a stationary solid cylinder B having a mass of 70 kg. The surface of contact is lubricated so that there is a resisting torque between the bodies given as $0.2\,\dot{\theta}_A$ N-m with $\dot{\theta}_A$ in radians per second. The outer cylinder is connected to a device at C which supplies a force equal to $-50\ddot{Y}$ N. Starting from a counter-clockwise angular speed of 1 rad/sec, what is the angular speed of cylinder A after the force at C starts to move downward and moves 0.7 m?

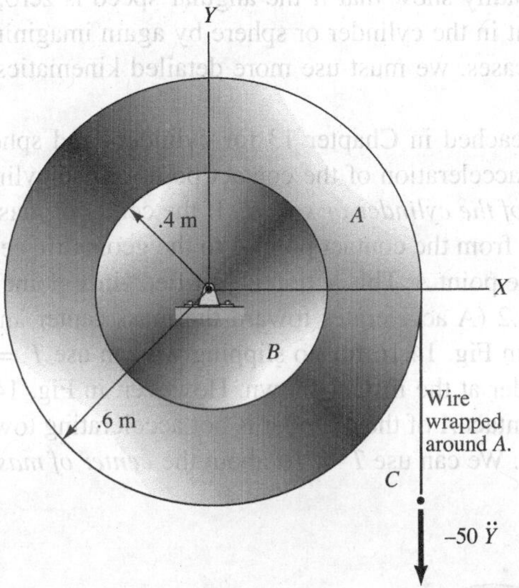

Figure P.14.28.

14.29. A box C weighing 150 lb rests on a conveyor belt. The driving drum B has a mass of 100 lbm and a radius of gyration of 4 in. The driven drum A has a mass of 70 lbm and a radius of gyration of 3 in. The belt weighs 3 lb/ft. Supporting the belt on the top side is a set of 20 rollers each with a mass of 3 lbm, a diameter of 2 in., and a radius of gyration of .8 in. If a torque T of 50 ft-lb is developed on the driving drum, what distance does C travel in 1 sec starting from rest? Assume that no slipping occurs.

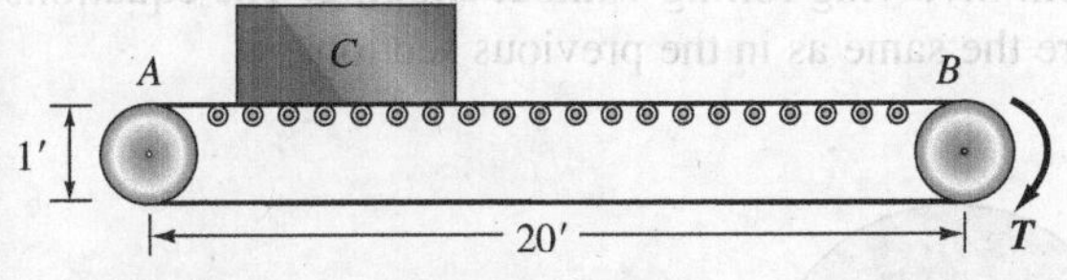

Figure P.14.29.

14.30. A uniform slender member is supported by a hinge at A. A force P is suddenly applied at an angle α with the horizontal. What value should P have and at what distance d should it be applied to result in zero reactive forces at A at the configuration shown if $\alpha = 45°$? The weight of the member AB is W. What is the angular acceleration of the bar for these conditions at the instant of interest?

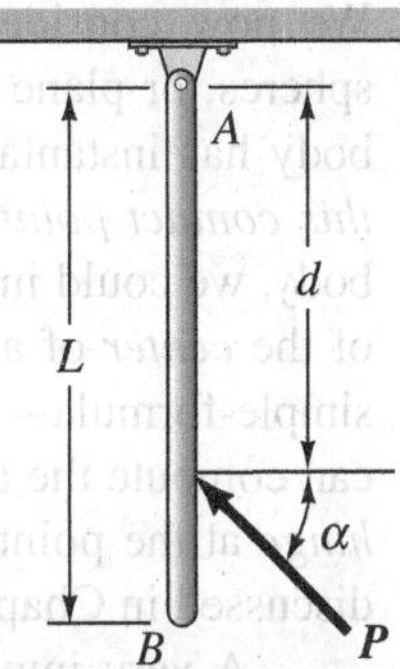

Figure P.14.30.

***14.31.** A rod AB is welded to a rod CD, which in turn is welded to a shaft as shown. The shaft has the following angular motion at time t:

$$\omega = 10 \text{ rad/sec}$$
$$\dot{\omega} = 40 \text{ rad/sec}^2$$

What are the shear force, axial force, and bending moment along CD at time t as a function of r? The rods have a mass per unit length of 5 kg/m. Neglect gravity.

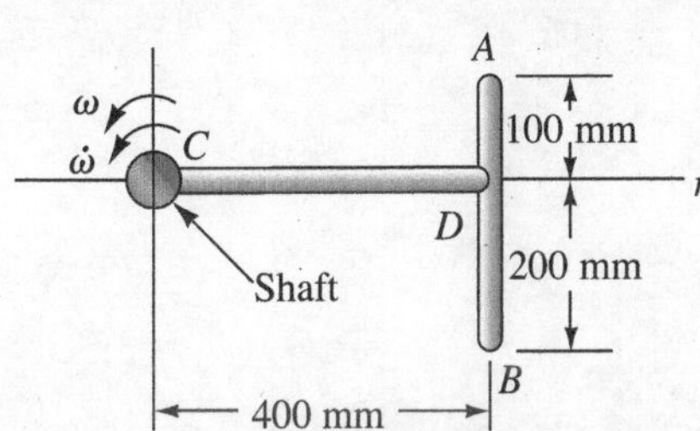

Figure P.14.31.

14.32. A four-bar linkage is shown (the ground is the fourth linkage). Each member is 300 mm long and has a mass per unit length of 10 kg/m. A torque T of 5 N-m is applied to each of bars AB and DC. What is the angular acceleration of bars AB and CD?

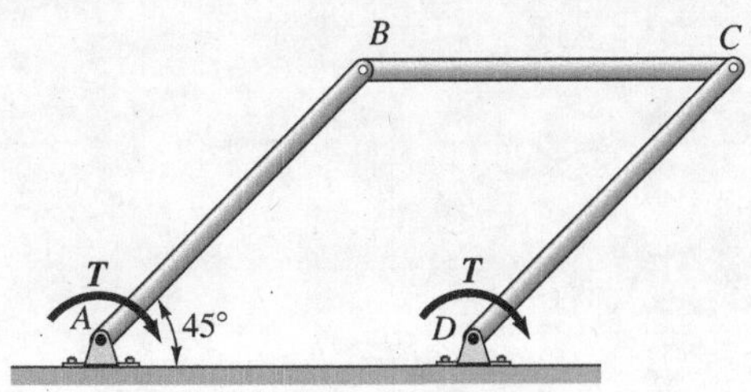

Figure P.14.32.

14.6 Rolling Slablike Bodies

We now consider the rolling without slipping of slablike bodies such as cylinders, spheres, or plane gears. As we have indicated in Chapter 13, the point of contact of the body has instantaneously *zero velocity*, and we have *pure instantaneous rotation about this contact point*. We pointed out that for getting velocities of points on such a rolling body, we could imagine that there is a *hinge* at the point of contact. Also, the acceleration of the *center* of a rolling without slipping sphere or cylinder can be computed using the simple formula—$R\ddot{\theta}$. Finally, you can readily show that if the angular speed is zero, we can compute the acceleration of any point in the cylinder or sphere by again imagining a *hinge* at the point of contact. For other cases, we must use more detailed kinematics, as discussed in Chapter 13.

A very important conclusion we reached in Chapter 13 for cylinders and spheres was that for rolling without slipping the acceleration of the contact point on the cylinder or sphere is *toward the geometric center of the cylinder or sphere*. If the center of mass of the body lies anywhere along the line AO from the contact point A to the geometric center O, then clearly we can use Eq. 14.6 for the point A. This action is justified since point A is then an example of case 3 in Section 14.2 (A accelerates toward the mass center and is part of the cylinder). Thus, for the body in Fig. 14.16 for no slipping we can use $T = I\alpha$ about the point of contact A of the cylinder at the instant shown. However, in Fig. 14.17 we cannot do this because the point of contact A of the cylinder is not accelerating toward the center of mass as in the previous case. We can use $T = I\alpha$ about the *center of mass* in the latter case.

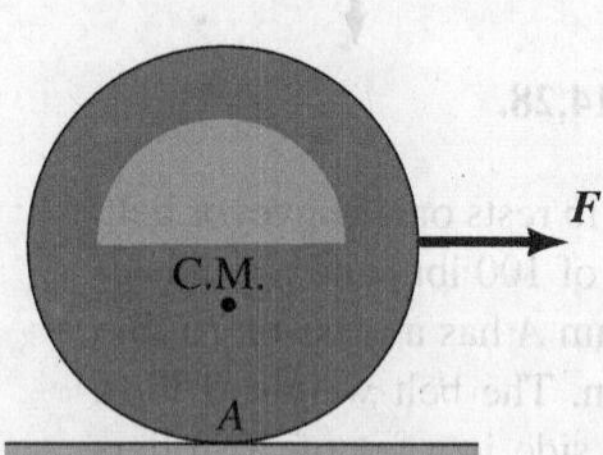

Figure 14.16. Point A accelerates toward center of mass.

We shall now examine a problem involving rolling without slipping. The equations of motion, you can readily deduce, are the same as in the previous section.

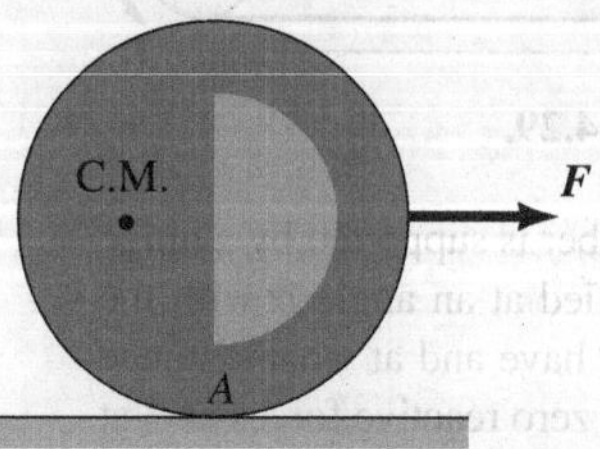

Figure 14.17. Point A does not accelerate toward center of mass.

Example 14.5

A steam roller is shown going up a 5° incline in Fig. 14.18. Wheels A have a radius of gyration of 1.5 ft and a weight each of 500 lb, whereas roller B has a radius of gyration of 1 ft and a weight of 5,000 lb. The vehicle, minus the wheels and roller but including the operator, has a weight of 7,000 lb with a center of mass positioned as shown in the diagram. The steam roller is to accelerate at the rate of 1 ft/sec^2. In part A of the problem, we are to determine the torque T_{eng} from the engine onto the drive wheels.

Figure 14.18. Steam roller moving up incline.

Part A. In Fig. 14.19, we have shown free-body diagrams of the drive wheels and the roller. Note we have combined the two drive wheels into a single 1,000-lb wheel. In each case, the point of contact on the wheel accelerates toward the mass center of the wheel and we can put to good use the **moment-of-momentum** equation (14.9) for the points of contact on the cylinders. Accordingly, we fix xyz to cylinder A and

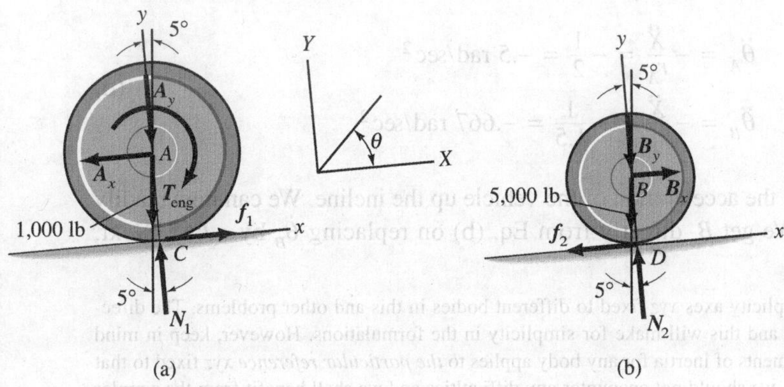

Figure 14.19. Free-body diagrams of driving wheels and roller.

Example 14.5 (Continued)

we fix another reference xyz to cylinder B at their respective points of contact as has been shown.[8] Hence, for cylinder A we have

$$(A_x + 1{,}000 \sin 5°)(2) - T_{eng} = \frac{1{,}000}{g}(1.5^2 + 2^2)\ddot{\theta}_A \quad \text{(a)}$$

where we have employed the parallel-axis theorem in computing the moment of inertia about the line of contact at C. Similarly, for the roller, we have

$$(5{,}000 \sin 5° - B_x)(1.5) = \frac{5{,}000}{g}(1^2 + 1.5^2)\ddot{\theta}_B \quad \text{(b)}$$

We have here two equations and no fewer than five unknowns. By considering the free body of the vehicle minus wheels shown diagrammatically in Fig. 14.20, we can say from **Newton's law** noting once again that $A_X \equiv A_x$, etc., because of the parallel orientation of axes XYZ with axes xyz of Fig. 14.19(a) and Fig. 14.19(b) at time t

$$A_x - B_x - 7{,}000 \sin 5° = \frac{7{,}000}{g}(1) \quad \text{(c)}$$

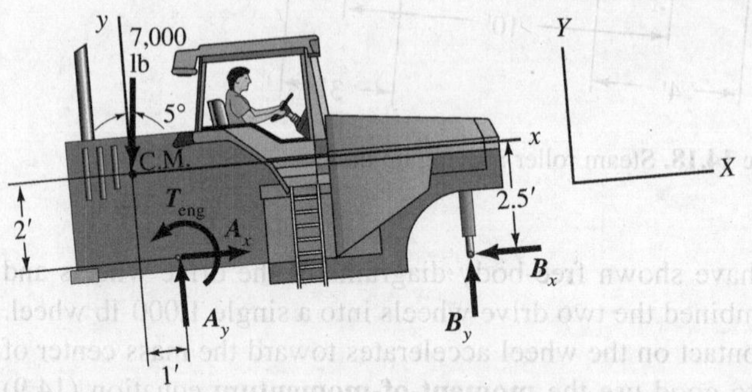

Figure 14.20. Free-body diagram of vehicle without wheels and roller.

Finally, from **kinematics** we can say:

$$\ddot{\theta}_A = -\frac{\ddot{X}}{r_A} = -\frac{1}{2} = -.5 \text{ rad/sec}^2$$

$$\ddot{\theta}_B = -\frac{\ddot{X}}{r_B} = -\frac{1}{1.5} = -.667 \text{ rad/sec}^2 \quad \text{(d)}$$

where $\ddot{X} = 1$ ft/sec² is the acceleration of the vehicle up the incline. We can now readily solve the equations. We get B_x directly from Eq. (b) on replacing $\ddot{\theta}_B$ by $-.667$. Next,

[8]We shall use for simplicity axes xyz fixed to different bodies in this and other problems. The directions will thus be the same and this will make for simplicity in the formulations. However, keep in mind that the computation of moments of inertia for any body applies to *the particular reference xyz* fixed to that body. Keeping this in mind, we should not encounter any difficulties and we shall benefit from the simpler notation.

Example 14.5 (Continued)

we get A_x from Eq. (c). Finally, going back to Eq. (a), we can solve for T_{eng}. The results are:

$$T_{eng} = 3{,}250 \text{ ft-lb}$$
$$A_x = 1{,}487 \text{ lb}$$
$$B_x = 660 \text{ lb}$$

Part B. Determine next the normal forces N_1 and N_2 at the wheels and roller, respectively.

We can express **Newton's law** for the wheels and roller in the direction normal to the incline by using the free-body diagrams of Fig. 14.19. Thus,

$$N_1 - A_y - 1{,}000 \cos 5° = 0 \tag{e}$$

$$N_2 - B_y - 5{,}000 \cos 5° = 0 \tag{f}$$

Next, we consider the free body of the vehicle without the wheels and roller (Fig. 14.20). **Newton's law** in the y direction for the center of mass then becomes

$$A_y + B_y - 7{,}000 \cos 5° = 0 \tag{g}$$

The **moment-of-momentum** equation about the center of mass of the vehicle without wheels and roller is

$$A_x(2) - B_x(2.5) + A_y(1) + B_y(11) + T_{eng} = 0 \tag{h}$$

We have four equations in four unknowns. Solve for A_y in Eq. (g), and substitute into Eq. (h). Inserting known values for A_x, B_x, and T_{eng}, we have

$$(1{,}487)(2) - (660)(2.5) + 7{,}000 \cos 5° - B_y + B_y(11) + 3{,}250 = 0$$

Therefore,

$$B_y = 1{,}155 \text{ lb}$$

Now from Eq. (g) we get A_y:

$$A_y = 7{,}000 \cos 5° + 1{,}155 = 8{,}128 \text{ lb}$$

Finally, from Eqs. (e) and (f) we get N_1 and N_2.

$$N_1 = 8{,}128 + 1{,}000 \cos 5° = 9{,}120 \text{ lb}$$

$$N_2 = -1{,}155 + 5{,}000 \cos 5° = 3{,}830 \text{ lb}$$

Hence, on each wheel we have a normal force of 4,560 lb, and for the roller we have a normal force of 3,830 lb.

*Example 14.6

A gear A weighing 100 N is connected to a stepped cylinder B (see Fig. 14.21) by a light rod DC. The stepped cylinder weighs 1 kN and has a radius of gyration of 250 mm along its centerline. The gear A has a radius of gyration of 120 mm along its centerline. A force F = 1,500 N is applied to the gear at D. What is the compressive force in member DC if, at the instant that F is applied, the system is stationary?

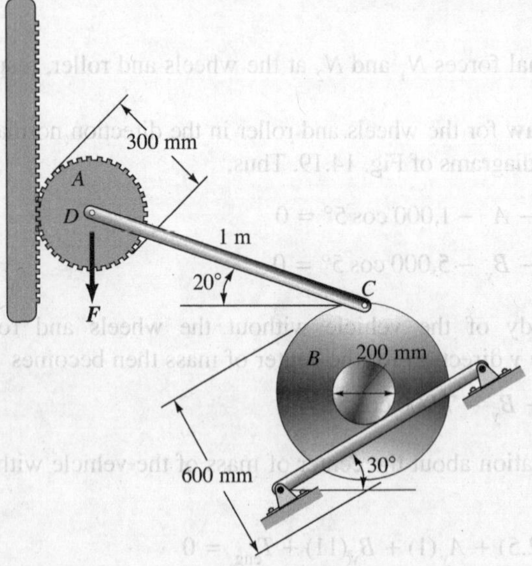

Figure 14.21. Stepped cylinder connected to a gear by a light rod.

Noting that DC is a two-force compressive member, we draw the free-body diagrams for the gear and the stepped cylinder in Fig. 14.22.

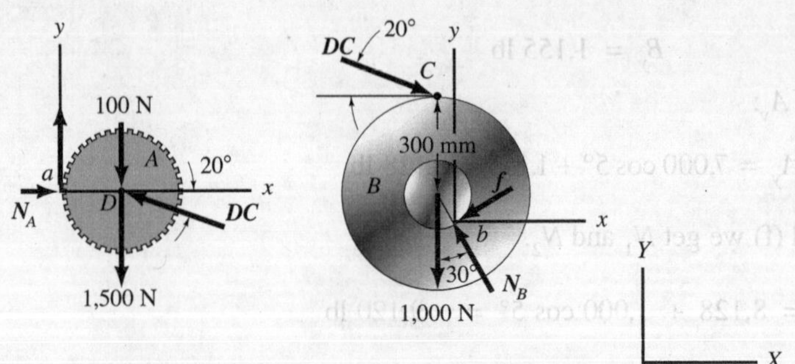

Figure 14.22. Free-body diagram of gear and cylinder.

The **moment-of-momentum** equations about the contact points for both bodies (points a and b, respectively) are

Example 14.6 (Continued)

$$-(100 + 1{,}500)(.15) + DC(\sin 20°)(.15) = \left(\frac{100}{9.81}\right)[(.12)^2 + (.15)^2]\ddot{\theta}_A$$

$$(DC \sin 20° + 1{,}000)(\sin 30°)(.10) - (DC \cos 20°)(.30 + .10 \cos 30°)$$
$$= \frac{1{,}000}{9.81}[(.25)^2 + (.10)^2]\ddot{\theta}_B$$

These equations simplify to the following pair:

$$.0513DC - 240 = .376\ddot{\theta}_A \quad \text{(a)}$$

$$-.3462DC + 50 = 7.39\ddot{\theta}_B \quad \text{(b)}$$

Clearly, we need an equation from **kinematics** at this time. Considering rod DC, we can say:[9]

$$\boldsymbol{a}_c = \boldsymbol{a}_D + \dot{\boldsymbol{\omega}}_{DC} \times \boldsymbol{\rho}_{DC} + \boldsymbol{\omega}_{DC} \times (\boldsymbol{\omega}_{DC} \times \boldsymbol{\rho}_{DC})$$

$$\ddot{\theta}_B \boldsymbol{k} \times [(.30 + .10 \cos 30°)\boldsymbol{j} - (.10 \sin 30°)\boldsymbol{i}]$$
$$= \ddot{Y}_D \boldsymbol{j} + \dot{\omega}_{DC} \boldsymbol{k} \times (\cos 20° \boldsymbol{i} - \sin 20° \boldsymbol{j}) + \boldsymbol{0}$$

The scalar equations are

$$-.3866\ddot{\theta}_B = .342\dot{\omega}_{DC} \quad \text{(c)}$$

$$-.05\ddot{\theta}_B = \ddot{Y}_D + .940\dot{\omega}_{DC} \quad \text{(d)}$$

Also, from **kinematics** we can say, considering gear A:

$$\ddot{Y}_D = .15\ddot{\theta}_A \quad \text{(e)}$$

Multiply Eq. (c) by .940/.342 and rewrite Eq. (d) below it with $\ddot{Y}_D$ replaced by using Eq. (e):

$$-1.063\ddot{\theta}_B = .940\dot{\omega}_{DC}$$

$$-.05\ddot{\theta}_B - .15\ddot{\theta}_A = .940\dot{\omega}_{DC}$$

Subtracting, we get

$$-1.013\ddot{\theta}_B + .15\ddot{\theta}_A = 0$$

Therefore,

$$\ddot{\theta}_A = 6.75\ddot{\theta}_B \quad \text{(f)}$$

Solving Eq. (a), (b), and (f) simultaneously gives us for DC the result

$$DC = 1{,}511 \text{ N (compression)}$$

Also, note that $\ddot{Y}$ comes out negative indicating that D accelerates downward.

[9]Note that because cylinder B has zero angular velocity, we can imagine it to be hinged at b for computing $\boldsymbol{a}_C$. Also, we do not know the sign of $\ddot{Y}_D$ and so we leave it as positive and thus let the mechanics yield the correct sign at the end of the calculations.

14.7 General Plane Motion of a Slablike Body

We now consider *general plane motion* of slablike bodies. The motion to be studied will be parallel to the plane of symmetry. Accordingly, we use the center of mass. The angular velocity vector $\boldsymbol{\omega}$ will be normal to the plane of symmetry, and, in accordance with Chasles' theorem, will be taken to pass through the center of mass. The translational velocity vector V_c will be parallel to the plane of symmetry. We fix a reference xyz at the center of mass of the body such that the xy plane coincides with the plane of symmetry as shown in Fig. 14.23. As usual, take the inertial reference parallel to xyz at time t. Note that the actual instantaneous axis of rotation is also shown. For the same

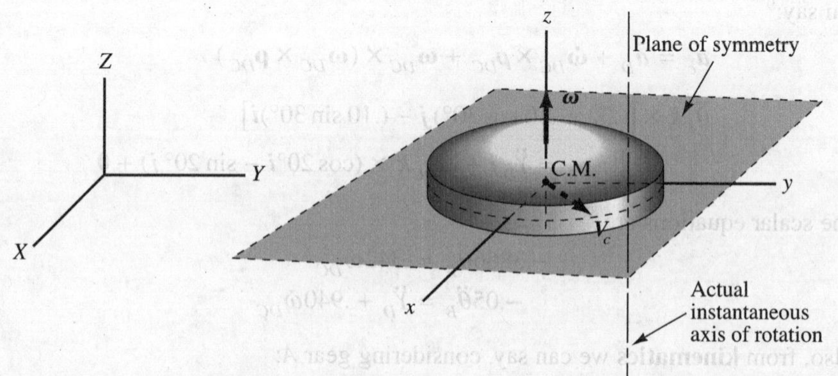

Figure 14.23. Slablike body undergoing general plane motion parallel to XY.

reasons used in Section 14.5 for slablike bodies, the *moment-of-momentum* equations become

$$M_x = 0 \tag{14.14a}$$

$$M_y = 0 \tag{14.14b}$$

$$M_z = I_{zz}\dot{\omega} \tag{14.14c}$$

Furthermore, considering the center of mass, we must have *equilibrium* in the z direction.

$$\sum F_z = 0 \tag{14.15}$$

while the full form of *Newton's law* holds in the x and y directions.

As in Section 14.5, we can generalize the results of this section to include the plane motion of a body having at the center of mass a *principal* axis z normal to the plane of motion XY. Clearly, for principal axes xyz at A, we get the same equations of motion as for the slablike body.

Example 14.7

Find the acceleration of block B shown in Fig. 14.24. The system is in a vertical plane and is released from rest. The cylinders roll without slipping along the vertical walls and along body B. Neglect friction along the guide rod. The 150 N-m torque M_A is applied to cylinder A.

We first draw free-body diagrams of the three bodies comprising the system as shown in Fig. 14.25 where it will be noticed that we have deleted the horizontal forces since they play no role in this problem. As usual, XYZ is our inertial reference. Points a, b, d, and e in F.B.D. I and F.B.D. III, respectively, are contact points in the respective bodies where, we repeat, there is rolling without slipping. Furthermore, it should be clear that points a and b are accelerating toward respective mass centers. Accordingly we fix xyz to the cylinders at these points.

We may immediately write the **moment-of-momentum** equations for the two cylinders about their respective points of contact a and b. Thus

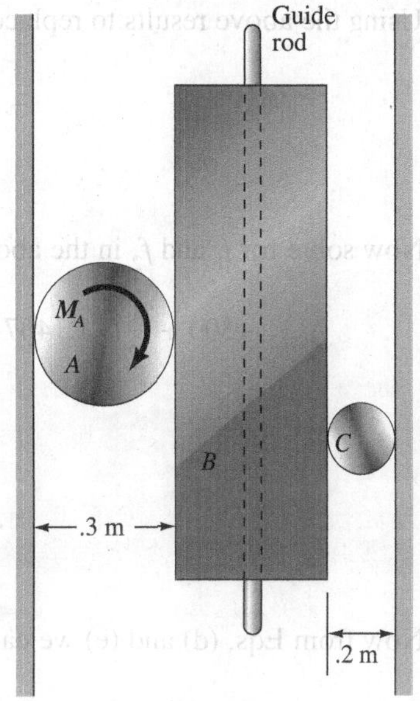

Guide rod

M_A

A

B

C

.3 m

.2 m

Data

$W_A = 100$ N
$W_B = 300$ N
$W_C = 50$ N
$M_A = 150$ N-m

Figure 14.24. A block and two cylinders in a vertical plane.

F.B.D. I

$$-(f_1)(.3) - (100)(.15) - 150 = \left[\frac{1}{2}\frac{100}{g}(.15)^2 + \frac{100}{g}(.15)^2\right]\ddot{\theta}_A$$

$$\therefore -.3f_1 - 165 = .344\ddot{\theta}_A \qquad \text{(a)}$$

F.B.D. III

$$(f_2)(.2) + (50)(.1) = \left[\frac{1}{2}\frac{50}{g}(.1)^2 + \frac{50}{g}(.1)^2\right]\ddot{\theta}_C$$

$$.2f_2 + 5 = .07645\ddot{\theta}_C \qquad \text{(b)}$$

Next going to F.B.D. II we employ **Newton's law** since we have simple translation. Referring now to the inertial reference XYZ we have

F.B.D. II

$$-300 + f_1 + f_2 = \frac{300}{g}\ddot{Y}_B \qquad \text{(c)}$$

Since the three bodies are interconnected by nonslip rolling conditions we must next consider the **kinematics** of the system. Thus

$$.3\ddot{\theta}_A = \ddot{Y}_B$$
$$.2\ddot{\theta}_C = -\ddot{Y}_B$$

Example 14.7 (Continued)

Using the above results to replace $\ddot{\theta}_A$ and $\ddot{\theta}_C$ in Eqs. (a) and (b) we get

$$-.3f_1 - 165 = 1.1467\ddot{Y}_B \quad \text{(d)}$$
$$.2f_2 + 5 = -.3823\ddot{Y}_B \quad \text{(e)}$$

Now solve for f_1 and f_2 in the above equations and substitute into Eq. (c). We get

$$-300 + \frac{1}{.3}(-1.1467\ddot{Y}_B - 165) + \frac{1}{.2}(-.3823\ddot{Y}_B - 5) = \frac{300}{g}\ddot{Y}_B$$
$$36.31\ddot{Y}_B = -875$$

$$\therefore \quad \ddot{Y}_B = -24.10 \text{ m/s}^2$$

Now from Eqs. (d) and (e) we can determine f_1 and f_2.

$$f_1 = -457.9 \text{ N} \qquad f_2 = 21.07 \text{ N}$$

Thus, cylinder A forces body B downward while cylinder C resists this motion.

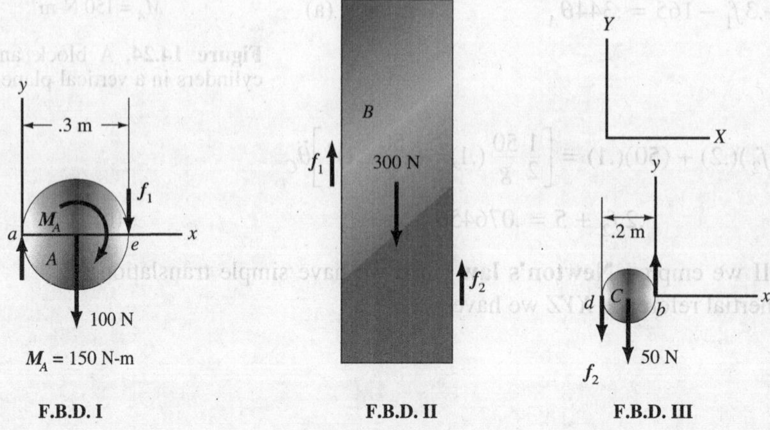

Figure 14.25. Free-body diagrams of the system elements with horizontal forces deleted.

Notice, unless we want to determine the friction forces at the walls there is no need to use **Newton's law** for the cylinders. Also note that we could not use the **moment-of-momentum** equations for points c and d of the cylinders even though there is rolling without slipping there. The reason for this, as you must know, is that these points *do not accelerate toward or away from the mass centers of the cylinders.*

Example 14.8

A stepped cylinder having a weight of 450 N and a radius of gyration k of 300 mm is shown in Fig. 14.26(a). The radii R_1 and R_2 are, respectively, 300 mm and 600 mm. A total pull T equal to 180 N is exerted on the ropes attached to the inner cylinder. What is the ensuing motion? The coefficients of static and dynamic friction between the cylinder and the ground are, respectively, .1 and .08.

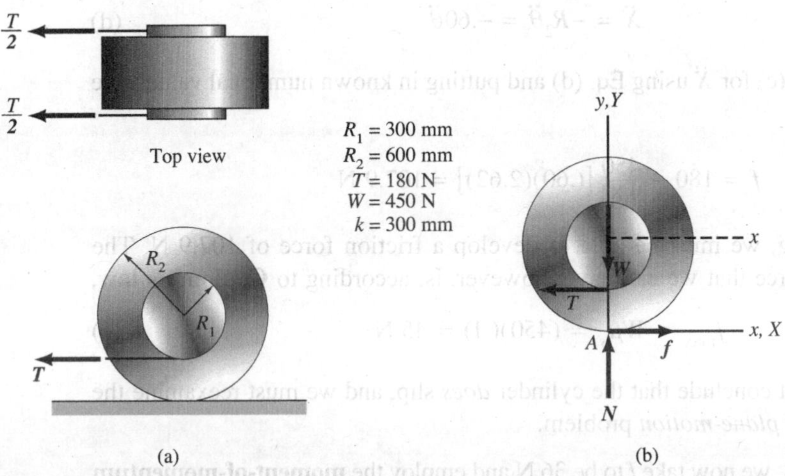

Figure 14.26. (a) Stepped cylinder; (b) free-body diagram of cylinder. *XY* is stationary.

A free-body diagram of the cylinder is shown in Fig. 14.26(b). Let us assume first that there is *no slipping* at the contact surface. Of course, we will have to later check this supposition. We have then pure instantaneous rotation about contact point A. Fix xyz to the body at A. XYZ as usual is stationary. We can then say for the **moment-of-momentum** equation about the axis of contact:

$$T(R_2 - R_1) = \left(\frac{W}{g}k^2 + \frac{W}{g}R_2^2\right)\ddot{\theta} \qquad \text{(a)}$$

wherein we have used the parallel-axis theorem for moment of inertia. Inserting numerical values, we can solve directly for $\ddot{\theta}$ at the instant that the force T is applied. Thus,

$$(180)(.30) = \left[\frac{450}{g}(.30)^2 + \frac{450}{g}(.60)^2\right]\ddot{\theta}$$

Therefore,

$$\ddot{\theta} = 2.62 \text{ rad/sec}^2 \qquad \text{(b)}$$

Example 14.8 (Continued)

We must now check our assumption of no slipping. Employ **Newton's law** for the mass center. In the X direction we get

$$-180 + f = \frac{W}{g}\ddot{X} \tag{c}$$

Using **kinematics** we note that

$$\ddot{X} = -R_2\ddot{\theta} = -.60\ddot{\theta} \tag{d}$$

Substituting into Eq. (c) for $\ddot{X}$ using Eq. (d) and putting in known numerical values, we can solve for f:

$$f = 180 - \frac{450}{9.81}[(.60)(2.62)] = 107.9 \text{ N}$$

Thus, for no slipping, we must be able to develop a friction force of 107.9 N. The maximum friction force that we can have, however, is, according to **Coulomb's law**,

$$f_{max} = W\mu_s = (450)(.1) = 45 \text{ N} \tag{e}$$

Accordingly, we must conclude that the cylinder *does* slip, and we must reexamine the problem as a *general plane-motion* problem.

Using $\mu_d = .08$, we now take f to be 36 N and employ the **moment-of-momentum** equation for the **center of mass** with xyz now fixed at the **center of mass**. We then have (Fig. 14.26(b))

$$fR_2 - TR_1 = \frac{W}{g}k^2\ddot{\theta} \tag{f}$$

Inserting numerical values, we get for $\ddot{\theta}$:

$$\ddot{\theta} = -7.85 \text{ rad/sec}^2 \tag{g}$$

Now, using **Newton's law** in the X direction for the mass center, we get

$$-T + f = \frac{W}{g}\ddot{X} \tag{h}$$

Inserting numerical values, we get for $\ddot{X}$:

$$\ddot{X} = -3.14 \text{ m/sec}^2 \tag{i}$$

Thus, the cylinder has a linear acceleration of 3.14 m/sec^2 to the left and an angular acceleration of 7.85 rad/sec^2 in the clockwise direction. Equations (g) and (i) are valid at all times, so we can integrate them if we like to get θ and X at any time t.

Example 14.9

A 4.905-kN flywheel rotating at a speed ω of 200 rpm (see Fig. 14.27) breaks away from the steam engine that drives it and falls on the floor. If the coefficient of dynamic friction between the floor and the flywheel surface is .4, at what speed will the flywheel axis move after 2 sec? At what speed will it hit the wall A? The radius of gyration of the flywheel is 1 m and its diameter is 2.30 m. Do not consider effects of bouncing in your analysis. Neglect rolling resistance (Section 7.7) and wind friction losses.

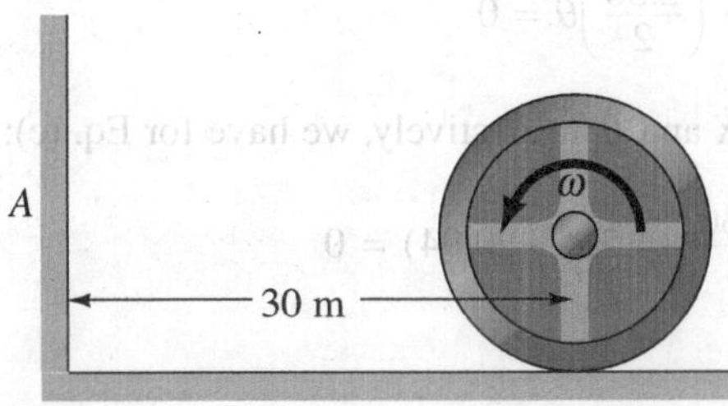

Figure 14.27. Runaway flywheel at initial position.

We assume slipping occurs when the flywheel first touches the floor (see Fig. 14.28). **Newton's law** for the center of mass of the flywheel is

$$(.4)N = \left(\frac{4{,}905}{9.81}\right)\ddot{X}$$

Therefore,

$$\ddot{X} = 3.92 \text{ m/sec}^2$$

Integrate twice:

$$\dot{X} = 3.92t + C_1 \qquad \text{(a)}$$

$$X = 1.962t^2 + C_1 t + C_2 \qquad \text{(b)}$$

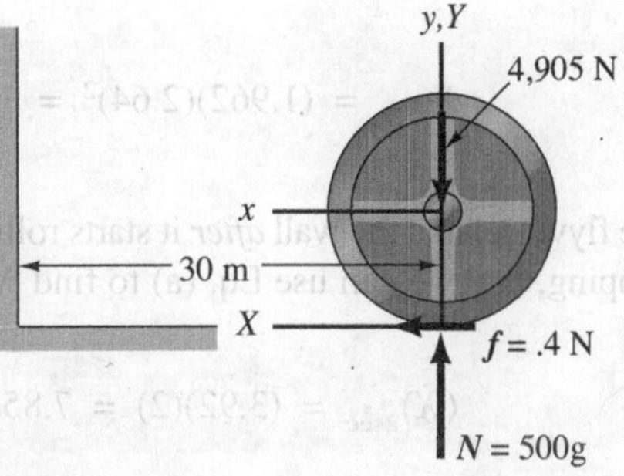

Figure 14.28. xy fixed at initial position.

At $t = 0$, $\dot{X} = 0$ and $X = 0$. Hence, $C_1 = 0$ and $C_2 = 0$. The **moment-of-momentum** equation for axes fixed to the body at the center of mass is next given.

$$(.4)(N)\left(\frac{2.30}{2}\right) = \left(\frac{4{,}905}{9.81}\right)(1)^2\ddot{\theta}$$

Therefore,

$$\ddot{\theta} = 4.51 \text{ rad/sec}^2$$

Integrate twice:

$$\dot{\theta} = 4.51t + C_3 \qquad \text{(a)}$$

$$\theta = 2.26t^2 + C_3 t + C_4 \qquad \text{(b)}$$

Example 14.9 (Continued)

When $t = 0$, $\theta = 0$, and $\dot{\theta} = -(200)(2\pi/60) = -20.94$ rad/sec. Hence, $C_3 = -20.94$ and $C_4 = 0$.

We now ask when does the slipping stop? Clearly, it stops when there is *zero velocity* of the point of contact of the cylinder.[10] From **kinematics** we have for this condition:

$$\dot{X} + \left(\frac{2.30}{2}\right)\dot{\theta} = 0 \qquad \text{(e)}$$

Substituting from Eq. (a) and (c) for $\dot{X}$ and $\dot{\theta}$, respectively, we have for Eq. (e):

$$3.92t + \left(\frac{2.30}{2}\right)(4.51t - 20.94) = 0$$

Therefore,

$$t = 2.64 \text{ sec}$$

Since we get a time here greater than zero, we can be assured that the initial slipping assumption is valid. The position $X_{N.S.}$ at the time of initial no-slipping is deduced from Eq. (b). Thus,

$$X_{N.S.} = (1.962)(2.64)^2 = 13.67 \text{ m}$$

Accordingly, the flywheel hits the wall *after* it starts rolling without slipping. At $t = 2$ sec, there is still slipping, and we can use Eq. (a) to find $\dot{X}$ at this instant. Thus,

$$(\dot{X})_{2\text{sec}} = (3.92)(2) = 7.85 \text{ m/sec}$$

The speed, once there is no further slipping, is constant, and so the speed at the wall is found by using $t = 2.64$ sec in Eq. (a). Thus,

$$(\dot{X})_{\text{wall}} = (3.92)(2.64) = 10.35 \text{ m/sec}$$

[10]Note that the friction will accelerate the center of mass of the flywheel (which starts out with a zero velocity) while at the same time friction will decrease the angular speed of the flywheel (which starts out at its maximum angular speed). Thus the contact point of the flywheel will be subject to two opposing speeds, one of which is increasing and the other of which is decreasing. When there occurs a cancellation of these speed, we have rolling without slipping and there ceases to be coulombic friction present. And so, neglecting the other resistances to motion there ceases to be any change of speed of the flywheel.

PROBLEMS

14.33. A stepped cylinder is released from a rest configuration where the spring is stretched 200 mm. A constant force F of 360 N acts on the cylinder, as shown. The cylinder has a mass of 146 kg and has a radius of gyration of 1 m. What is the friction force at the instant the stepped cylinder is released? Take $\mu_s = .3$ for the coefficient of friction. The spring constant K is 270 N/m.

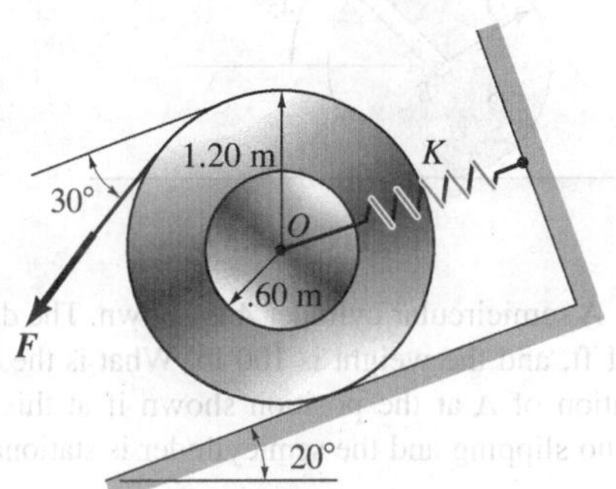

Figure P.14.33.

14.34. A stepped cylinder is held on an incline with an inextensible cord wrapped around the inner cylinder and an outside agent (not shown). If the tension T on the cord at the instant that the cylinder is released by the outside agent from the position shown is 100 lb, what is the initial angular acceleration? What is the acceleration of the mass center? Use the following data:

$$W = 300 \text{ lb}$$
$$k = 3 \text{ ft}$$
$$R_1 = 2 \text{ ft}$$
$$R_2 = 4 \text{ ft}$$
$$\mu_s = .1$$

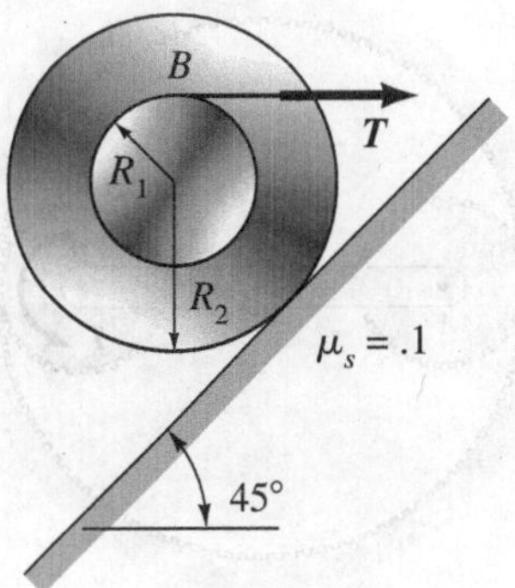

Figure P.14.34.

14.35. The cylinder shown is acted on by a 100-lb force. At the contact point A, there is viscous friction such that the friction force is given as

$$f = .05V_A$$

where V_A is the velocity of the cylinder at the contact point in ft/sec. The weight of the cylinder is 30 lb, and the radius of gyration k is 1 ft. Set up a third-order differential equation for finding the position of O as a function of time.

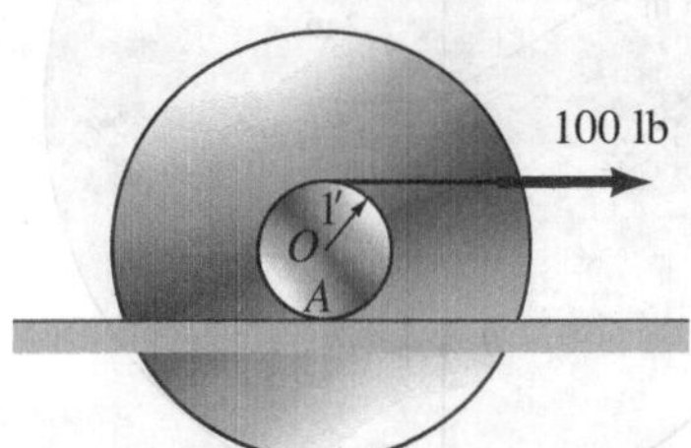

Figure P.14.35.

14.36. The cylinder shown weighs 445 N and has a radius of gyration of .27 m. What is the minimum coefficient of friction at A that will prevent the body from moving? Using half of this coefficient of friction, how far d does point O move in 1.2 sec if the cylinder is released from rest?

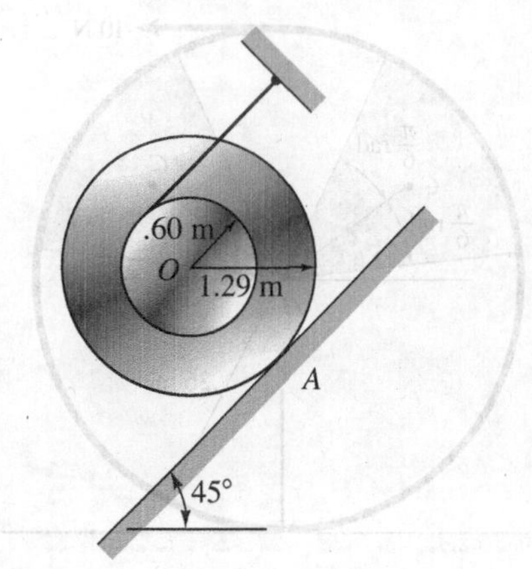

Figure P.14.36.

14.37. The velocities of two points of a cylinder, namely A and B, are

$$V_A = 6 \text{ m/s} \qquad V_B = 2 \text{ m/s}$$

What is the velocity of point D? If the cylinder has a mass of 4.2 kg, what is the angular acceleration for a dynamic coefficient of friction $\mu_d = .35$?

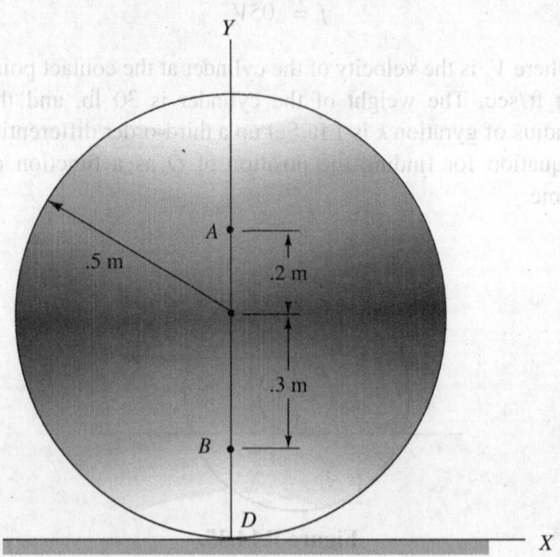

Figure P.14.37.

14.38. A thin ring having a mass of 0.4 kg is released from rest and rolls without slipping under the action of a 10-N force. Two identical metal sectors each of mass 0.83 kg are attached to the ring. What is the angular acceleration of the ring? Each sector has a radius of gyration at its centroid equal to $k = 0.18$ m.

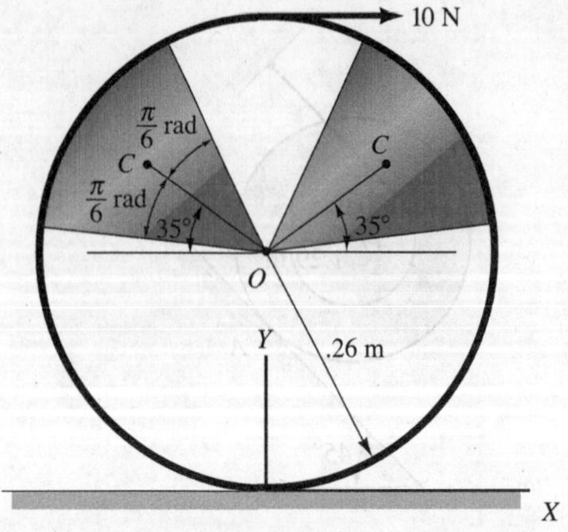

Figure P.14.38.

14.39. A light rod AB connects a plate C with a cylinder D which may roll without slipping. A torque T of value 50 ft-lb is applied to plate C. What is the angular acceleration of cylinder D when the torque is applied? The plate weighs 100 lb and the cylinder weighs 200 lb.

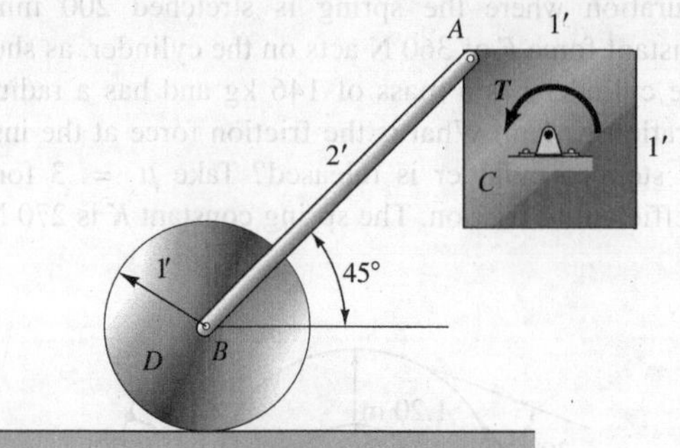

Figure P.14.39.

14.40. A semicircular cylinder A is shown. The diameter of A is 1 ft, and the weight is 100 lb. What is the angular acceleration of A at the position shown if at this instant there is no slipping and the semicylinder is stationary?

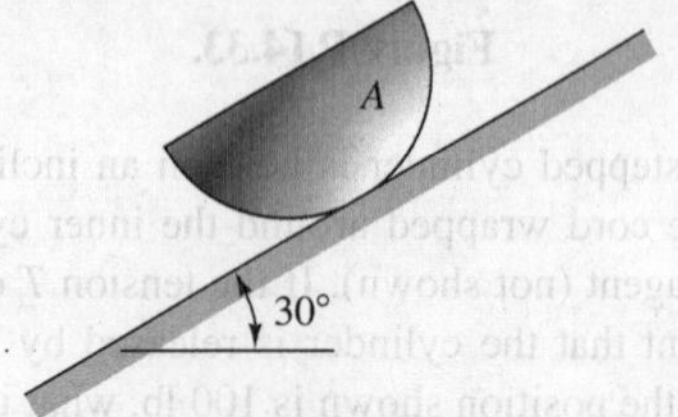

Figure P.14.40.

14.41. A 20 kg bar AB connects two gears G and H. These gears each have a mass of 5 kg and a diameter of 300 mm. A torque T of 5 N-m is applied to gear H. What is the angular speed of H after 20 sec if the system starts from rest? The system is in a horizontal plane. Bar AB is 2 m long.

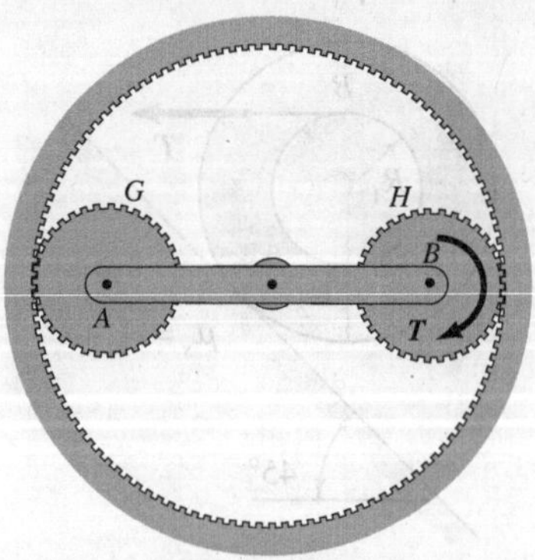

Figure P.14.41.

14.42. A bar C weighing 445 N rolls on cylinders A and B, each weighing 223 N. What is the acceleration of bar C when the 90-N force is applied as shown? There is no slipping.

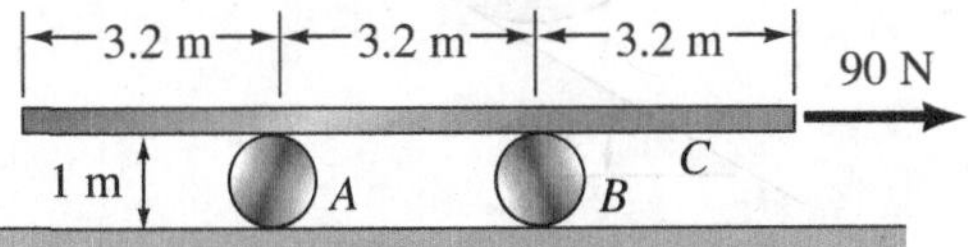

Figure P.14.42.

14.43. In Problem 14.42, at what position of the bar relative to the wheels does slipping first occur after the force is applied? Take $\mu_s = .2$ for the bottom contact surface and $\mu_s = .1$ for the contact surface between bar and cylinders. From Problem 14.42, $\ddot{x}_c = 1.442$ m/sec^2.

14.44. A platform B, of weight 30 lb and carrying block A of weight 100 lb, rides on gears D and E as shown. If each gear weighs 30 lb, what distance will platform B move in .1 sec after the application of a 100-lb force as shown?

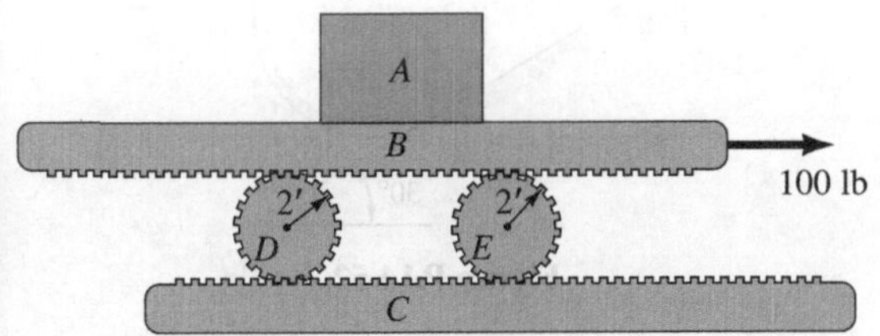

Figure P.14.44.

14.45. A crude cart is shown. A horizontal force P of 100 lb is applied to the cart. The coefficient of static friction between wheels and ground is .6. If $D = 3$ ft, what is the acceleration of the cart to the right? The wheels weigh 50 lb each. Neglect friction in the axle bearings. The total weight of cart with load is 322 lb. Treat the wheels as simple solid cylinders.

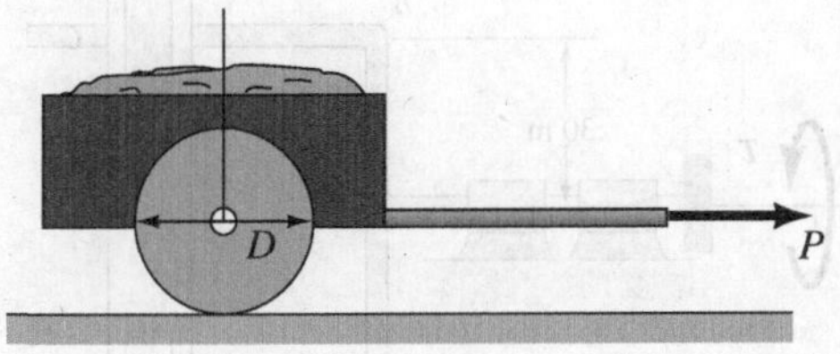

Figure P.14.45.

14.46. What minimum force component P is required to cause the cart in Problem 14.45 to move so that the wheels slip rather than roll without slipping?

14.47. A pulley system is shown. Sheave A has a mass of 25 kg and has a radius of gyration of 250 mm. Sheave B has a mass of 15 kg and has a radius of gyration of 150 mm. If released from rest, what is the acceleration of the 50-kg block? There is no slipping.

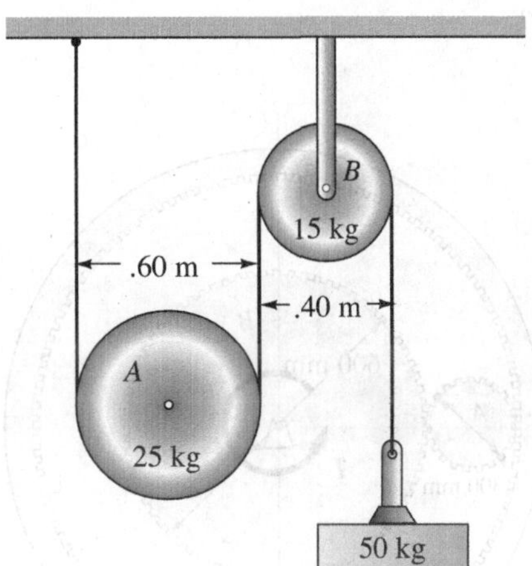

Figure P.14.47.

14.48. A steam locomotive drive system is shown. Each drive wheel weighs 5 kN and has a radius of gyration of 400 mm. At the instant shown, a pressure $p = .50$ N/mm^2 above atmospheric acts on the piston to drive the train backward. If the train is moving at 1 m/sec backward at the instant shown, what is its acceleration? Members AB and BC are to be considered stiff but light in comparison to other parts of the engine. Also, the piston assembly can be considered light. Only the driving car is in action in this problem. It has one driving system, on each side of the locomotive as shown below. It has two additional wheels of the size and mass described above on each side of the locomotive plus additional small wheels whose rotational inertia we shall neglect. The drive train minus its eight large wheels has a weight of 150 kN. Assume no slipping, and neglect friction in the piston assembly.

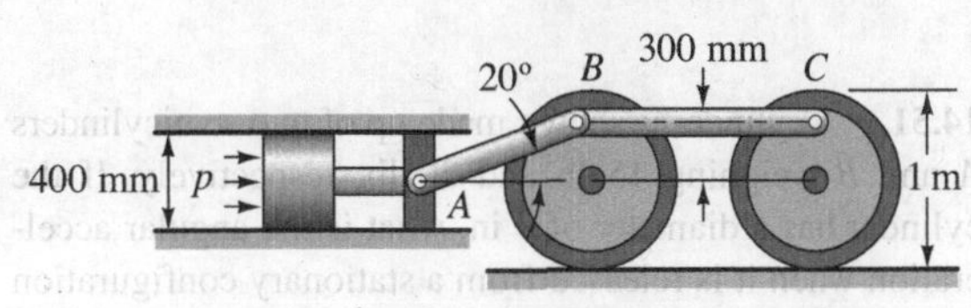

Figure P.14.48.

14.49. A system of interconnected gears is shown. Gear B rotates about a fixed axis, and gear D is stationary. If a torque T of 2.5 N-m is applied to gear B at the configuration shown, what is the angular acceleration of gear A? Gear A has a mass of 1.36 kg while gear B has a mass of 4.55 kg. The system is in a vertical orientation relative to the ground. What vertical force is transmitted to stationary gear D?

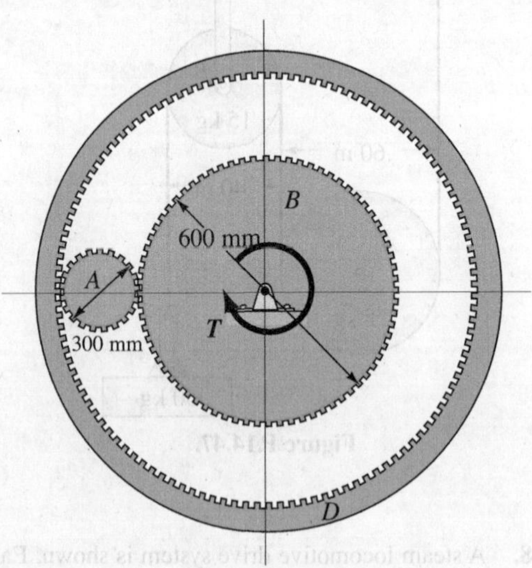

Figure P.14.49.

14.50. A solid semicircular cylinder of weight W and radius R is released from rest from the position shown. What is the friction force at that instant?

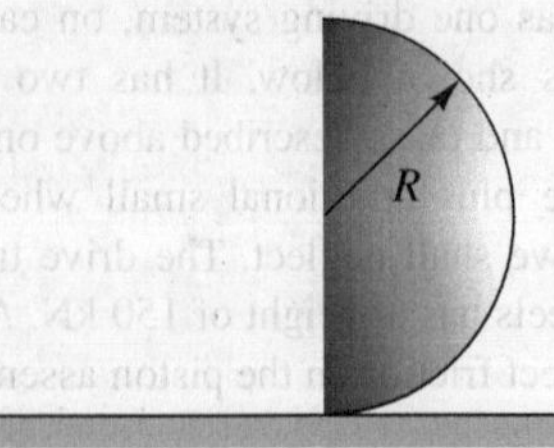

Figure P.14.50.

14.51. A cylinder is shown made up of two semicylinders A and B weighing 15 lb and 30 lb, respectively. If the cylinder has a diameter of 3 in, what is the angular acceleration when it is released from a stationary configuration at the position shown? Assume no slipping.

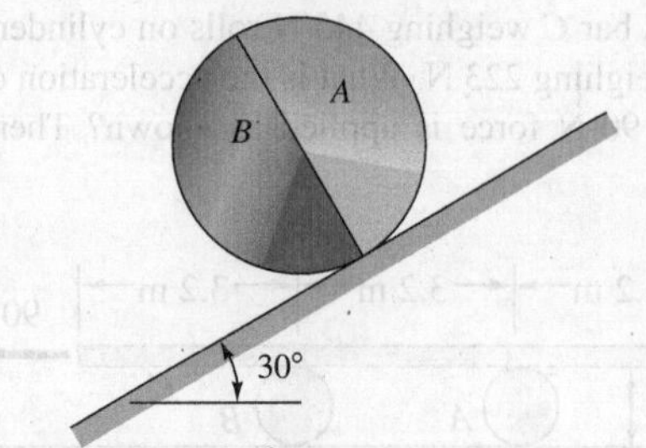

Figure P.14.51.

14.52. A thin-walled cylinder is shown held in position by a cord AB. The cylinder has a mass of 10 kg and has an outside diameter of 600 mm. What are the normal and friction forces at the contact point C at the instant that cord AB is cut? Assume that no slipping occurs.

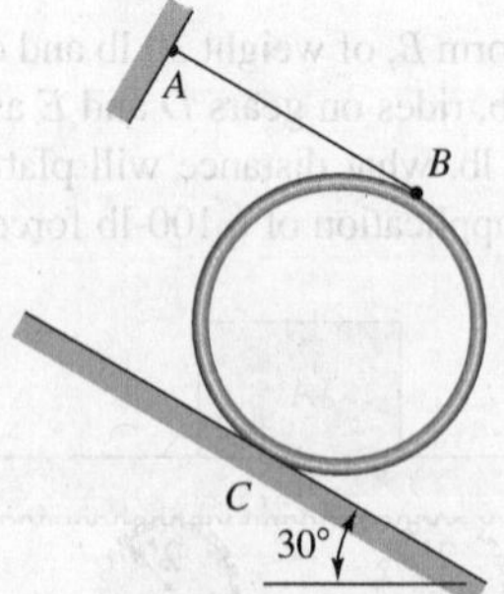

Figure P.14.52.

14.53. A bent rod $CBEF$ is welded to a shaft. At the ends C and F are identical gears G and H, each of mass 3 kg and radius of gyration 70 mm. The gears mesh with a large stationary gear D. A torque T of 50 N-m is applied to the shaft. What is the angular speed of the shaft after 10 sec if the system is initially at rest? The bent rod has a mass per unit length of 5 kg/m. Will there be forces on the bearings of the shaft other than those from gravity? Why?

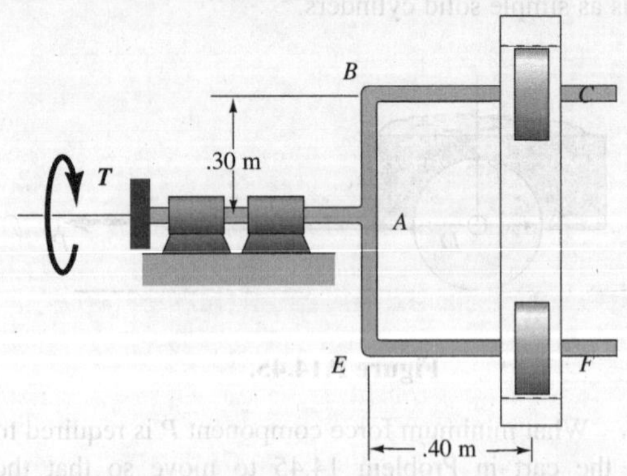

Figure P.14.53. *(Continued on next page)*

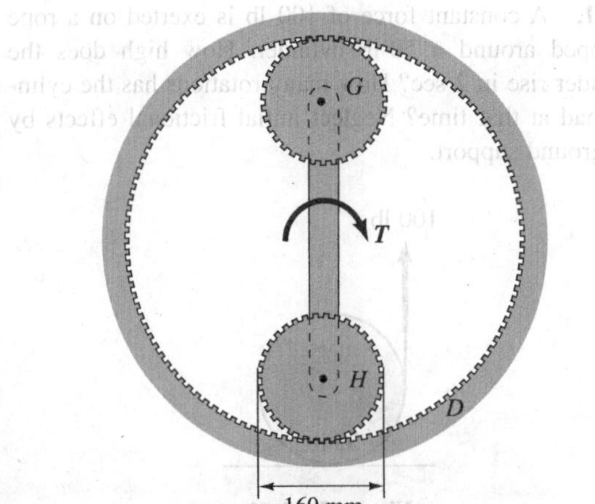

Figure P.14.53.

14.54. A tractor and driver has a mass of 1,350 kg. If a total torque T of 300 N-m is developed on the two drive wheels by the motor, what is the acceleration of the tractor? The large drive wheels each have a mass of 90 kg, a diameter of 1 m, and a radius of gyration of 400 mm. The small wheels each have a mass 20 kg and have a diameter of 300 mm with a radius of gyration of 100 mm.

Figure P.14.54.

14.55. A block B weighing 100 lb rides on two identical cylinders C and D weighing 50 lb each as shown. On top of block B is a block A weighing 100 lb. Block A is prevented from moving to the left by a wall. If we neglect friction between A and B and between A and the wall and we consider no slipping at the contact surfaces of the cylinders, what is the angular speed of the cylinders after 2 sec for $P = 80$ lb?

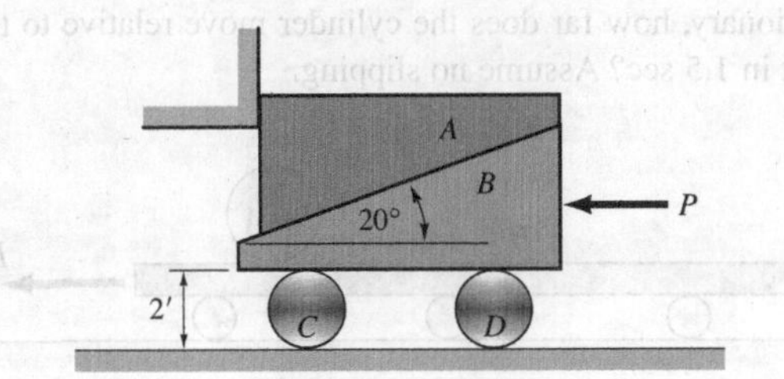

Figure P.15.55.

14.56. A cable is wrapped around two pulleys A and B. A force T is applied to the end of the cables at G. Each pulley weighs 5 lb and has a radius of gyration of 4 in. The diameter of the pulleys is 12 in. A body C weighing 100 lb is supported by pulley B. Suspended from body C is a body D weighing 25 lb. Body D is lowered from body C so as to accelerate at the rate of 5 ft/sec^2 relative to body C. What force T is then needed to pull the cable downward at G at the increasing rate of 5 ft/sec^2?

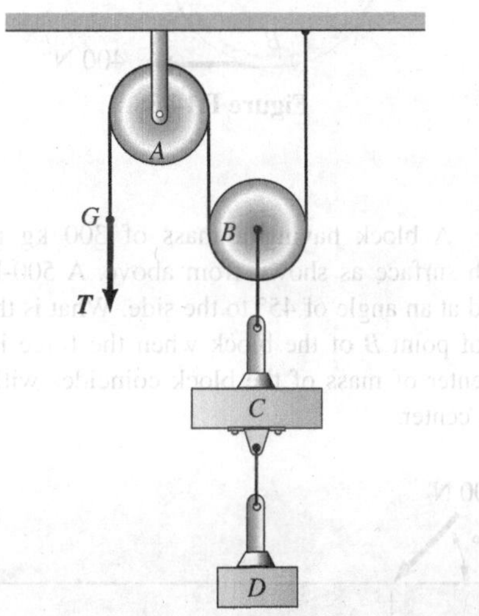

Figure P.14.56.

14.57. A cylinder A is acted on by a torque T of 1,000 N-m. The cylinder has a mass of 75 kg and a radius of gyration of 400 mm. A light rod CD connects cylinder A with a second cylinder B having a mass of 50 kg and a radius of gyration of 200 mm. What is the force in member CD when torque T is applied? The system is stationary at the instant the torque is applied. Assume no slipping of cylinder C along the incline.

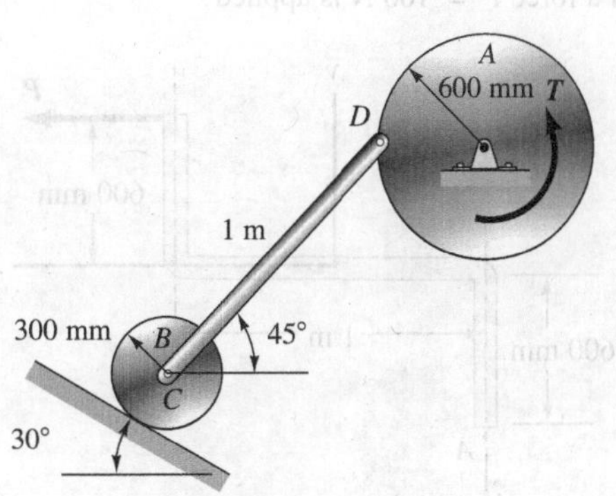

Figure P.14.57.

14.58. A ring rests on a smooth surface as shown from above. The ring has a mean radius of 2 m and a mass of 30 kg. A force of 400 N is applied to the ring at *B*. What is the acceleration of point A on the ring?

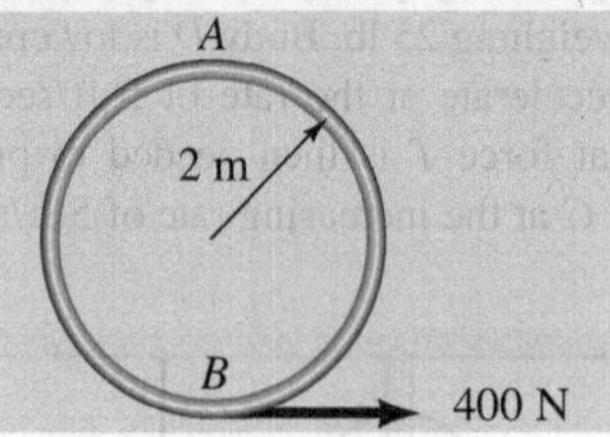

Figure P.14.58.

14.59. A block having a mass of 300 kg rests on a smooth surface as shown from above. A 500-N force is applied at an angle of 45° to the side. What is the acceleration of point *B* of the block when the force is applied? The center of mass of the block coincides with the geometric center.

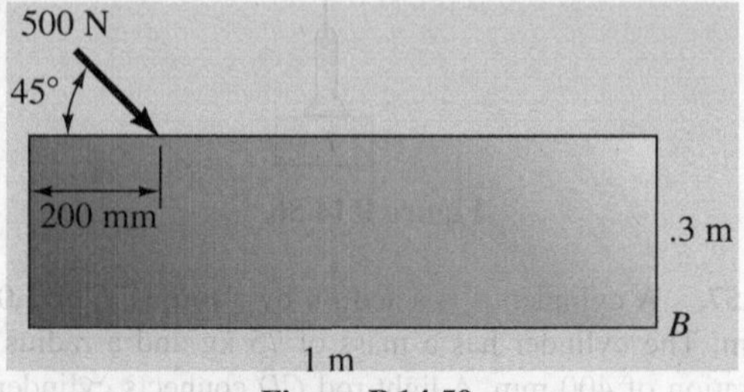

Figure P.14.59.

14.60. A bent rod rests on a smooth surface. The rod has a mass of 20 kg. What is the acceleration of point *A* when a force $P = 100$ N is applied?

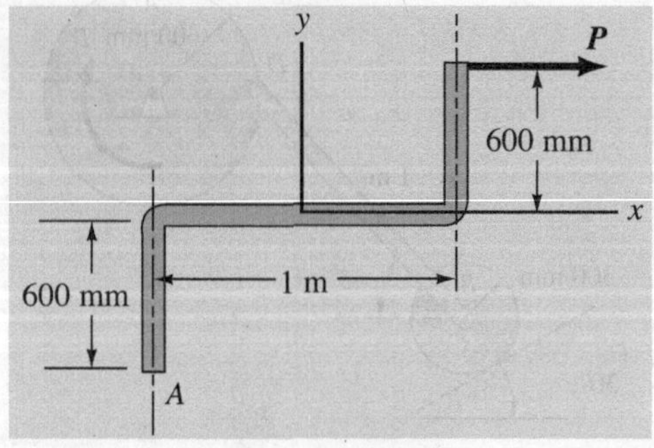

Figure P.14.60.

14.61. A constant force of 100 lb is exerted on a rope wrapped around a 50-lb cylinder. How high does the cylinder rise in 2 sec? How many rotations has the cylinder had at that time? Neglect initial frictional effects by the ground support.

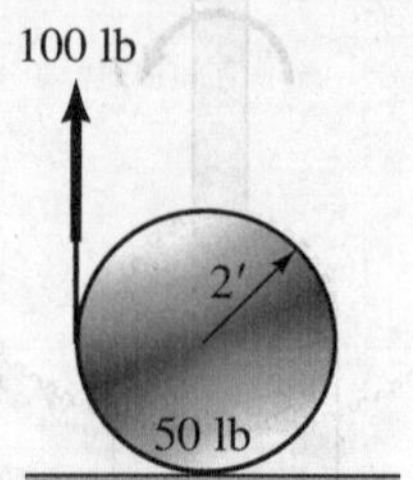

Figure P.14.61.

14.62. A bar of weight *W* equal to 100 lb is at rest on a horizontal surface *S* at the instant that a force *P* equal to 60 lb is applied. Show that the center of rotation at the instant that the force *P* is applied is 3.27 ft from the left end. The coefficient of friction μ_s equals .2. The length of the bar is 10 ft. Assume that the normal force on the surface is uniform.

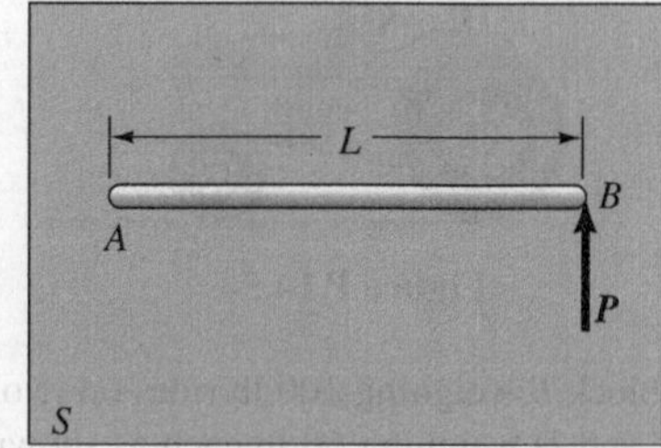

Figure P.14.62.

14.63. A cart *B* is given a constant acceleration of 5 m/sec^2. On the cart is a cylinder *A* having a mass of 5 kg and a diameter of 600 mm. If the system is initially stationary, how far does the cylinder move relative to the cart in 1.5 sec? Assume no slipping.

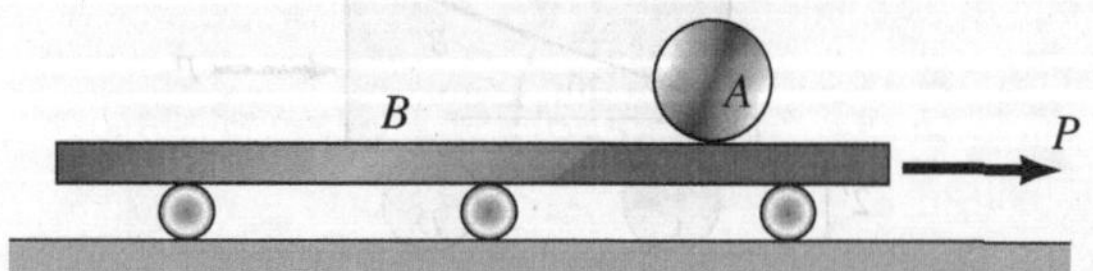

Figure P.14.63.

14.64. A gear A meshes with a stationary pinion B. The hub H of gear A moves along frictionless guide rails DE. The system is in a vertical plane. What are the vertical and angular speeds of the gear after 2 sec if the system is initially at rest? The gear has a mass of 20 kg.

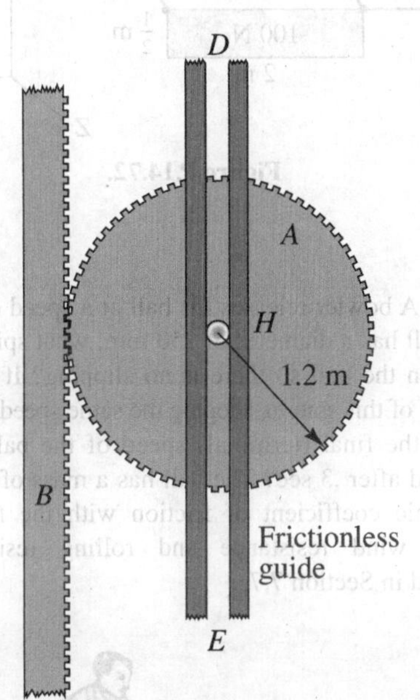

Figure P.14.64.

14.65. In Problem 13.43, compute the forces at pin B and pin C. The mass of rod BC is 25 lbm and the mass of rod AB is 18 lbm. The following data stem from the solution of Problem 13.41:

$$\omega_{BC} = 3.35 \text{ rad/sec} \quad \dot{\omega}_{BC} = 39.7 \text{ rad/sec}^2$$
$$a_c = 116.6 \text{ m/sec}^2$$

Neglect the mass of the slider at C, but do not consider it to be frictionless. A strain gauge informs us that there is a torque of 20 ft-lb acting on rod AB at A.

14.66. A circular disc is shown with a circular hole. It rests on a frictionless surface and the view shown is from above. A force F = .04 lb acts on the disc. The thickness of the disc is 2 in and the specific weight is 350 lb/ft^3. What is the initial linear acceleration of the center of mass and the angular acceleration of the disc?

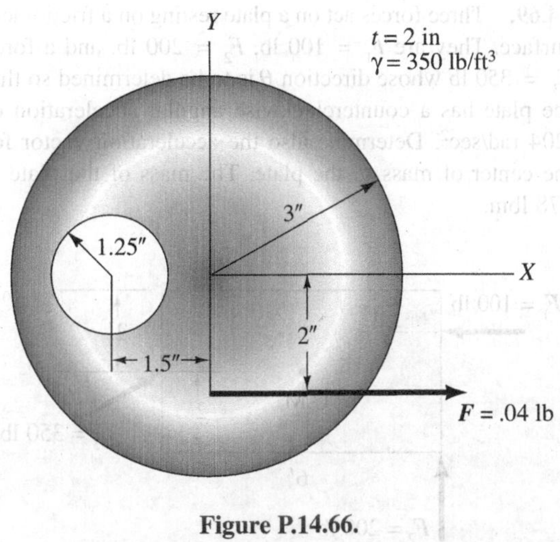

Figure P.14.66.

14.67. In the preceding problem the disc is in the vertical plane and is held up by a horizontal surface where the coefficients of friction are μ_s = .005 and μ_d = .003. What are the initial linear acceleration of the center of mass and the initial angular acceleration of the disc? Start by assuming no slipping.

14.68. A cylinder A slides off the flatbed of a truck (Fig. P.14.68.) onto the road with zero angular velocity. The mass of the cylinder is 100 kg; its radius is 1 m; and its radius of gyration about the axis through the center of mass is 0.75 m. The coefficient of friction μ_d between the cylinder and the pavement is 0.6. If marks on the pavement from the cylinder while it is sliding extend over a distance along the road of 3 m, and if the axis of the cylinder remains perpendicular to the sides of the road during the action, what was the approximate speed of the truck when the cylinder slid off. Neglect the speed of the cylinder relative to the truck when it slides off. [*Hint:* What do the pavement marks signify?]

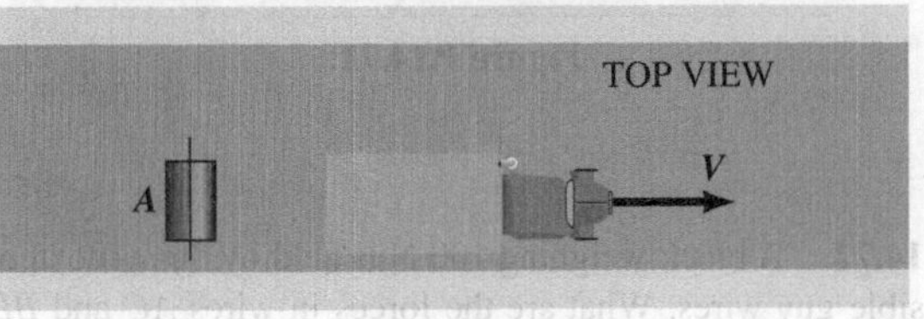

Figure P.14.68.

14.69. Three forces act on a plate resting on a frictionless surface. They are $F_1 = 100$ lb, $F_2 = 200$ lb, and a force $F_3 = 350$ lb whose direction θ is to be determined so that the plate has a counterclockwise angular acceleration of .204 rad/sec^2. Determine also the acceleration vector for the center of mass of the plate. The mass of the plate is 178 lbm.

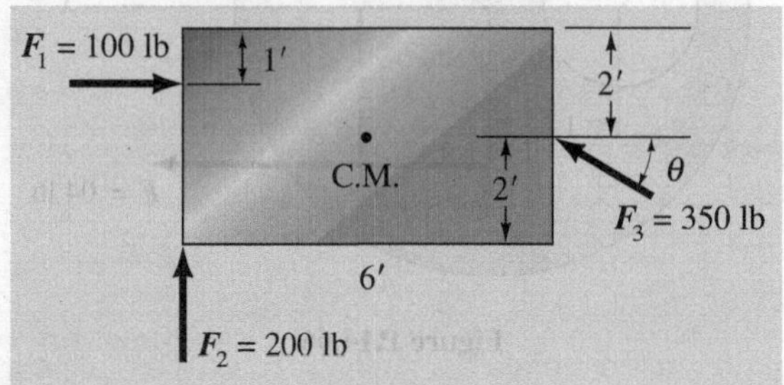

Figure P.14.69.

14.70. In Problem 14.63, what force P is needed to uniformly accelerate the cart so that the cylinder A moves 1 m in 2 sec relative to the cart. Cart B has a mass of 10 kg. Neglect the inertia of the small rollers supporting the cart, and assume there is no slipping.

14.71. A wedge B is shown with a cylinder A of mass 20 kg and diameter 500 mm on the incline. The wedge is given a constant acceleration of 20 m/sec^2 to the right. How far d does the cylinder move in sec relative to the incline if there is no slipping? The system starts from rest.

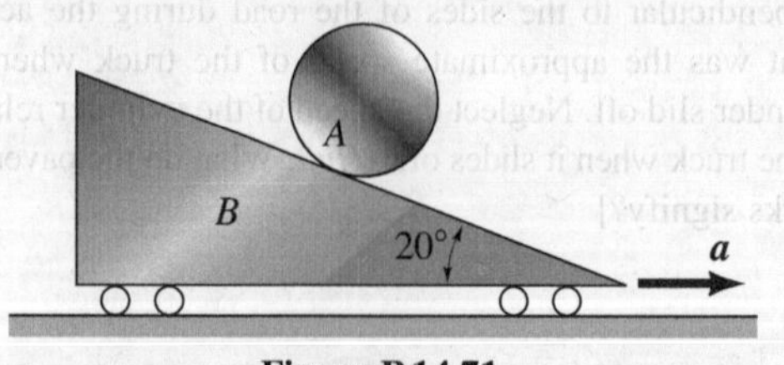

Figure P.14.71.

14.72. A block weighing 100 N is held by three inextensible guy wires. What are the forces in wires AC and BD at the instant that wire EC is cut?

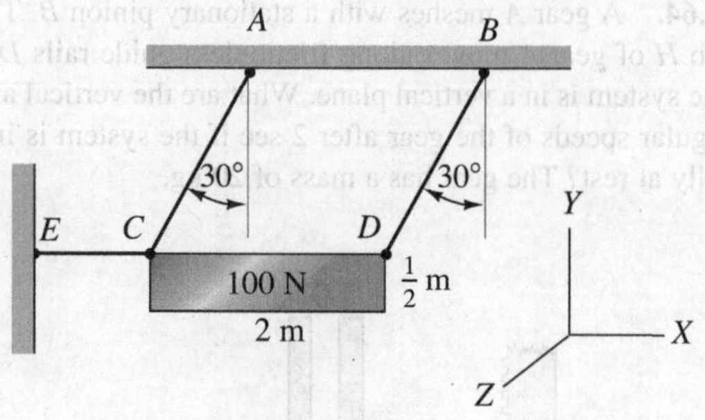

Figure P.14.72.

14.73. A bowler releases his ball at a speed of 3 m/sec. If the ball has a diameter of 250 mm, what spin ω should be put on the ball so there is no slipping? If he puts on only half of this spin ω, keeping the same speed of 3 m/sec, what is the final (terminal) speed of the ball? What is the speed after .3 sec? The ball has a mass of 1.8 kg and a dynamic coefficient of friction with the floor of .1. Neglect wind resistance and rolling resistance (as discussed in Section 7.7).

Figure P.14.73.

14.74. In Problem 14.73, suppose ω is 1.4 times the value ω = 24 rad/sec needed for no slipping. What is the terminal speed of the ball? What is the speed of the ball after .3 sec?

14.75. A tug is pushing on the side of a barge which is loaded with sand and which has a total weight of 50 tons. The tug generates a 3-kN force which is always normal to the barge. If the barge rotates 5° in 20 sec, what is the moment of inertia of the barge at the center of mass which we take at the geometric center of the barge? The distance d for this maneuver is 5 m. What is the acceleration of the center of mass of the barge at t = 35 sec? Neglect the resistance of the water.

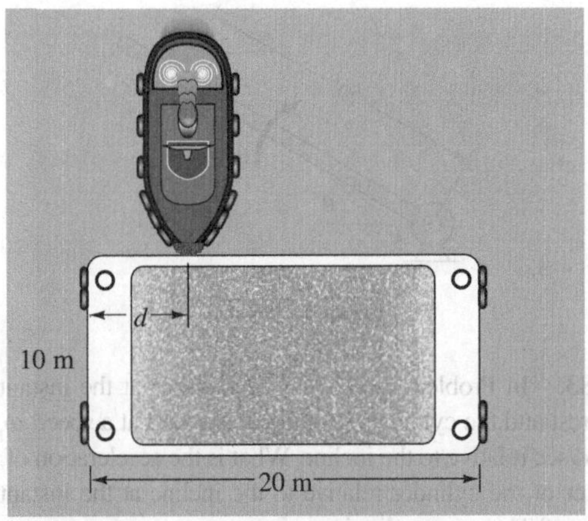

Figure P.14.75.

14.76. A plate A having a mass of 100 lbm is supported by a rod at one end and by a linear spring having a spring constant $K = 190$ lb/in at the other. If the rod BC is suddenly released at B, what is the angular acceleration of the plate? Also, what is the acceleration of the plate mass center? The plate is oriented as shown at the instant of cutting the left support.

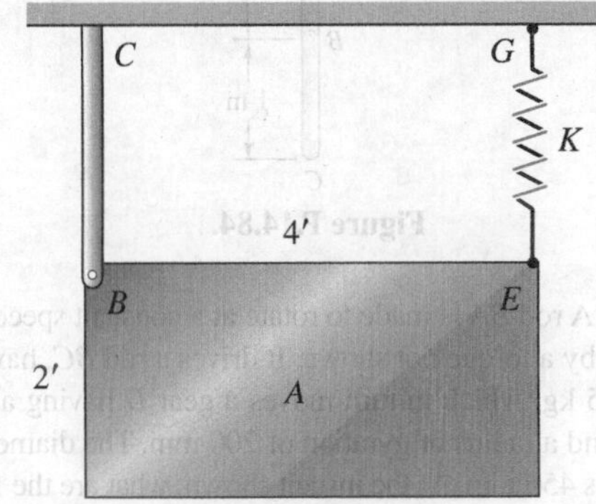

Figure P.14.76.

14.77. The great English liner, the *Queen Elizabeth* (QE II), is the last transatlantic luxury ship left. All the others have been either scrapped or made into pleasure excursion ships. The QE II is 700 ft long and weighs 60,000 tons.

(a) At a top speed of 40 knots, with the engines producing 110,000 horsepower, what is the thrust coming from the propellers?

(b) In a harbor, two tugboats are turning an initially stationary QE II as shown in the diagram. Determine an approximate value for the angular acceleration of the ship. Consider the ship to be a long uniform rod. Include an additional one fourth of the mass of the ship to account for the water which must be moved to accommodate the movement of the ship. Each tug develops a force of 5,000 lb.

(c) If the tugs remain perpendicular to the QE II, and if we assume constant angular acceleration, how many minutes are required to turn the ship 10 degrees?

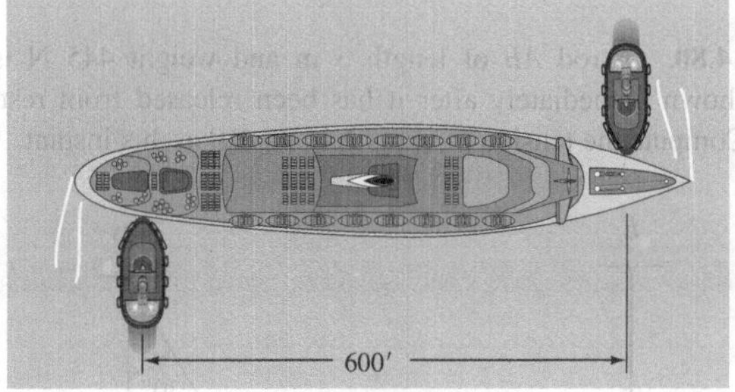

Figure P.14.77.

14.78. A cable supports cylinder A of mass 40 kg and then wraps around a light cylinder B and finally supports cylinder C of mass 20 kg. If the system is released from rest, what are the accelerations of the centers of A and C? There is no slipping.

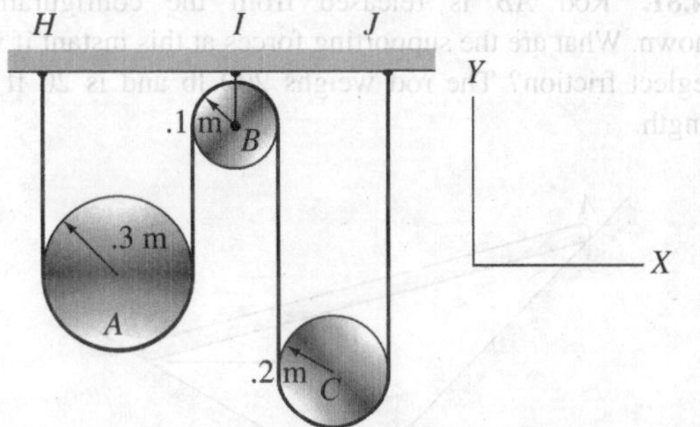

Figure P.14.78.

14.79. A cylinder of mass 20 kg can rotate inside a block B whose mass is 35 kg. There is a constant resisting torque for this rotation given as 200 N-m. A horizontal 1,000-N force is applied to a cable firmly wrapped around the cylinder. What is the velocity of the block after moving 0.5 m? What is the angular velocity of the cylinder when the block reaches this position? The coefficient of dynamic friction between block B and the floor is 0.3. The system is shown in a vertical orientation.

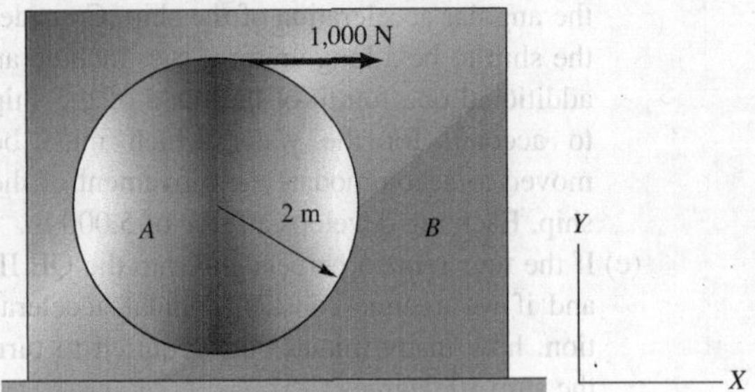

Figure P.14.79.

14.80. A rod AB of length 3 m and weight 445 N is shown immediately after it has been released from rest. Compute the tension in wires EA and DB at this instant.

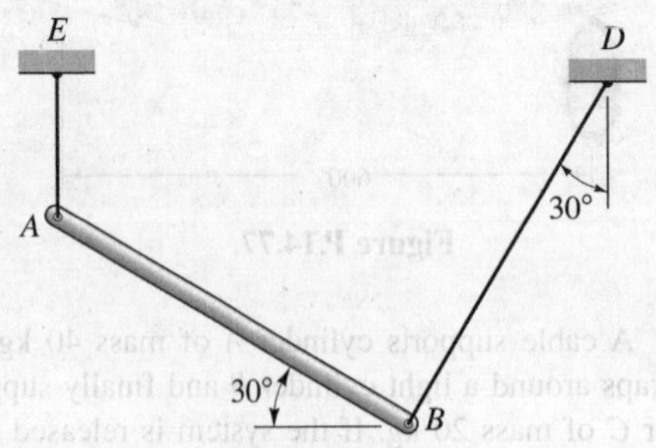

Figure P.14.80.

14.81. Rod AB is released from the configuration shown. What are the supporting forces at this instant if we neglect friction? The rod weighs 200 lb and is 20 ft in length.

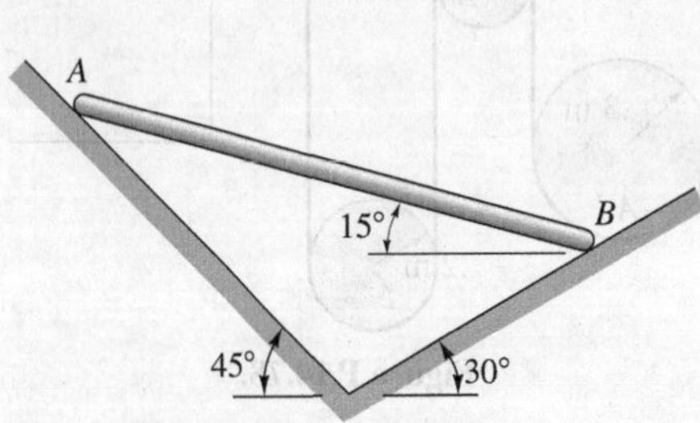

Figure P.14.81.

14.82. A cylinder weighing 100 lb with a radius of 1 ft is held fixed on an incline that is rotating at $\frac{1}{2}$ rad/sec. The cylinder is released when the incline is at position θ equal to 30°. If the cylinder is 20 ft from the bottom A at the instant of release, what is the initial acceleration of the center of the cylinder relative to the incline? There is no slipping.

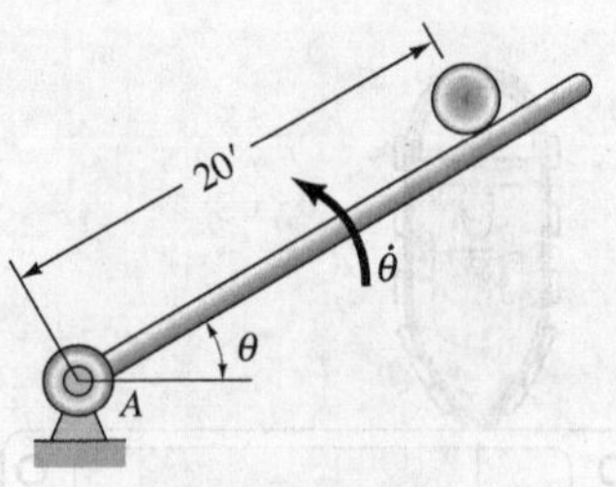

Figure P.14.82.

14.83. In Problem 14.82, $\ddot{\theta}$ = 2 rad/sec² at the instant of interest and the cylinder is rolling downward at a speed ω_1 of 3 rad/sec relative to the incline. What is the acceleration of the center of the cylinder relative to the incline at the instant of interest? There is no slipping.

14.84. Two identical bars, each having a mass of 9 kg, hang freely from the vertical. A force of 45 N is applied at the center of the upper bar AB. What are the angular accelerations of the bars?

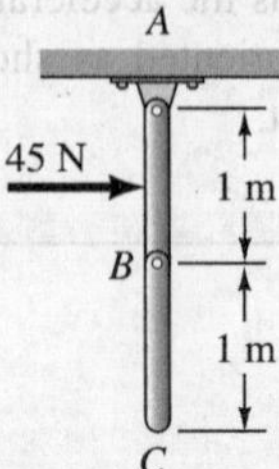

Figure P.14.84.

14.85. A rod BA is made to rotate at a constant speed ω of 100 rpm by a torque not shown. It drives a rod BC, having a mass of 5 kg, which in turn moves a gear D having a mass of 3 kg and a radius of gyration of 200 mm. The diameter of the gear is 450 mm. At the instant shown, what are the forces transmitted by the pins at C and at B?

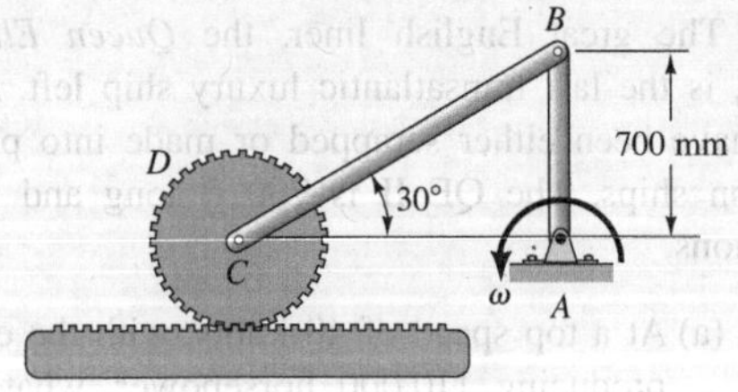

Figure P.14.85.

14.86. A 1-m rod AB weighing 10 kg is suspended at one end by a cord BC and at the other end rides on an inclined surface on small wheels. Initially the wheels are being held at the position shown. At the instant the wheels are released, what is the angular acceleration of rod AB? Neglect friction on the inclined surface.

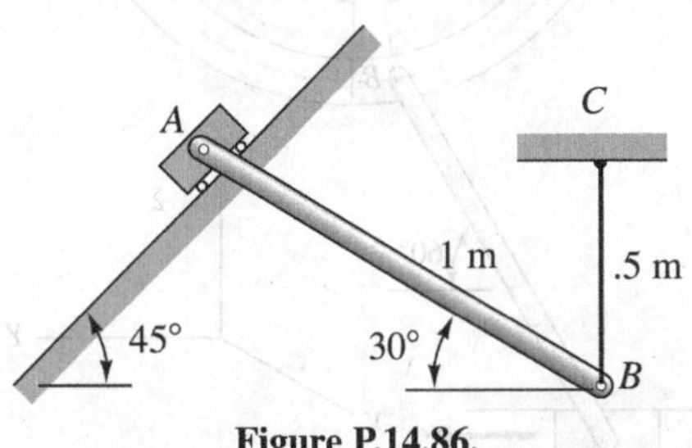

Figure P.14.86.

14.87. A truck is carrying a 10,000 N crate A held by two steel cables each under a tensile force of 800 N. At corner B there is a stop. The truck undergoes a crash with a constant deceleration of 4.6 g's. Cable FG breaks and the crate starts to rotate about corner B where we have the stop. What are the angular acceleration and the forces at the stop at the instant of the break? The center of mass of the crate is at the geometrical center of the crate and the radius of gyration about the center of mass is 2.3 m. Consider that when the rotational acceleration first occurs, the crate has yet to move and that cable DE still has the original tension, whereas cable FG has snapped.

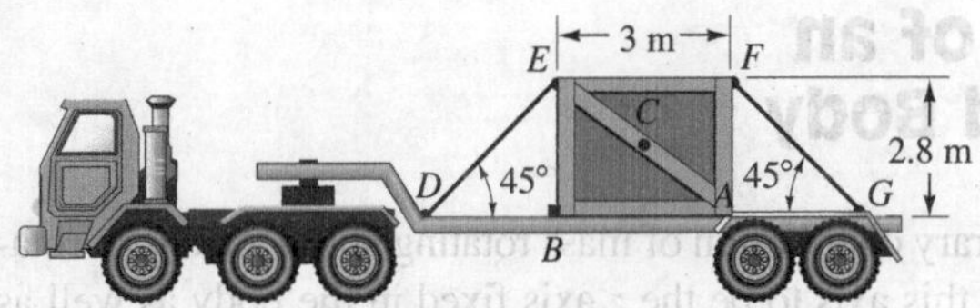

Figure P.14.87.

14.88. An object with a cylindrical hub, thin spokes, and a thin rim rests on a frictionless horizontal surface. The hub weighs 100 N and the spokes and rim each weigh 15 N/m. What is the acceleration of center A when the indicated force is applied? What is the angular acceleration of the system? Finally, what is the acceleration of point B? Use calculus when considering the spokes and the rim. For the hub, the angular momentum about an axis through the center of mass is given as

$$\int_V \mathbf{r} \times dm\mathbf{V} = \frac{1}{2} MR^2 \omega \mathbf{k}$$

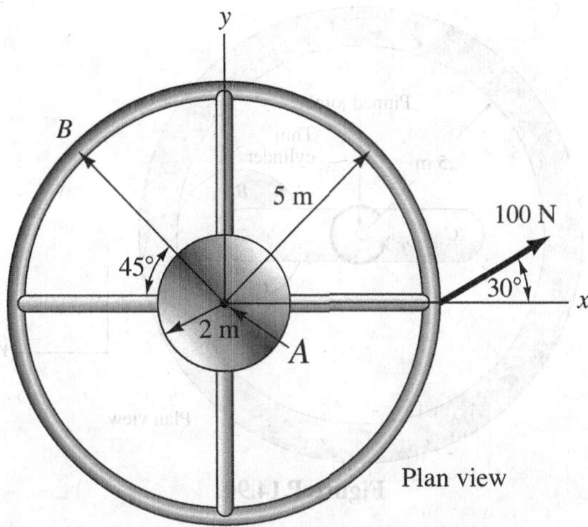

Figure P.14.88.

14.89. We are looking down from above at a baseball player swinging his bat in a horizontal plane in the process of hitting a baseball at point B. The player is holding the bat such that the resultant of the force from his hands is at point A. At what distance d should the ball hit the bat to render as zero the normal force component from the batter's hands onto the centerline of the bat? This point B is called the *center of percussion*. Because this kind of hit feels effortless to the batter it is often called the "sweet spot" by athletes in both baseball and tennis. The radius of gyration about the center of mass is k_z.

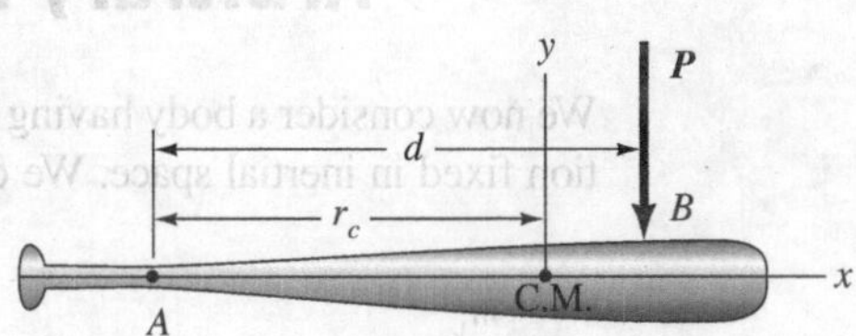

Figure P.14.89.

14.90. A torque T = 10 N-m is applied to body C. If there is no slipping, how many rotations does cylinder B make in 1 sec if the system starts from rest? A is stationary at all times. The system starts from rest. We are observing the system from above. The following data apply.

$$M_C = 50 \text{ kg} \qquad M_B = 30 \text{ kg}$$
$$k_C = .2 \text{ m} \qquad k_B = .1 \text{ m}$$

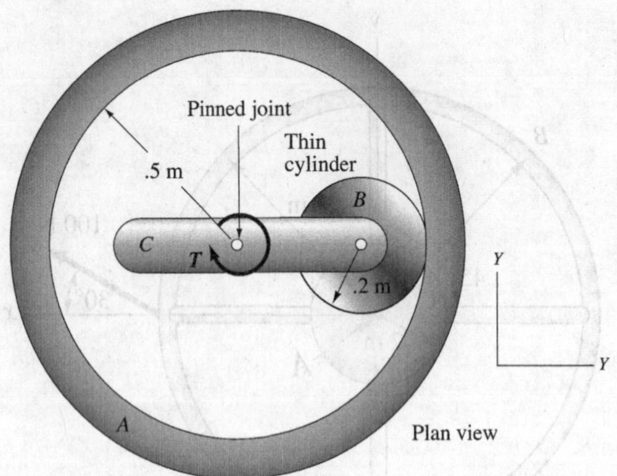

Figure P.14.90.

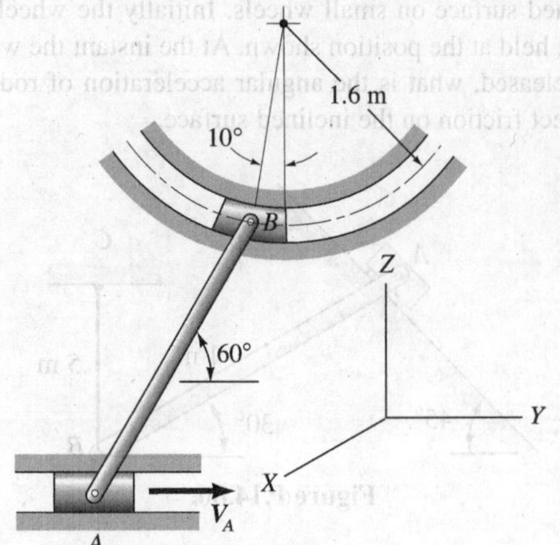

Figure P.14.91.

14.91. In Problem 13.48, we determined the following results from kinematics:

$$\omega_{AB} = -.609 \text{ rad/s} \qquad V_B = 4.38 \text{ m/s}$$
$$\dot{\omega}_{AB} = 14.71 \text{ rad/sec}^2 \quad \dot{V}_B = -33.28 \text{ m/s}^2$$

The mass of rod *AB* is 10 kg. If we neglect the masses of the sliders, what are the forces coming onto the end pins of the rod? The horizontal slot in which slider *A* is moving is frictionless.

14.92. In the preceding problem, the slider at *A* no longer moves in the slot without friction and we do not know the friction force there. However, we have a strain gage mounted on rod *AB* giving data indicating a 200 N compressive axial force at *A*. Using the data of the previous problem, compute the force components at the ends of the rod.

14.8 Pure Rotation of an Arbitrary Rigid Body

We now consider a body having an arbitrary distribution of mass rotating about an axis of rotation fixed in inertial space. We consider this axis to be the z axis fixed in the body as well as being an inertial coordinate axis Z. We can take the origin of xyz anywhere along the z axis since all such points are fixed in inertial space. The *moment-of-momentum* equations to be used will now be the general equations 14.6 since I_{zx} and I_{zy} will generally not equal zero. If the center of mass is along the z axis, then it obviously has no acceleration, and so we can then apply the rules of statics to the center of mass. For other cases we shall often need to use *Newton's law* for the center of mass. In this regard it will be helpful to note from the definition of the center of mass that for a system of rigid bodies such as is shown in Fig. 14.29

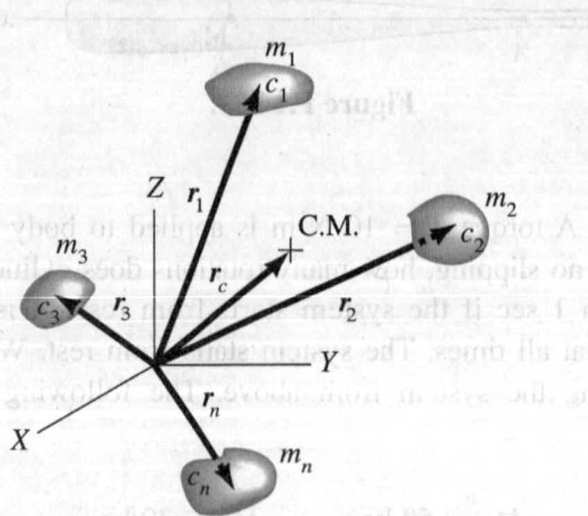

Figure 14.29. *n* rigid bodies having a total mass *M*.

$$M\mathbf{r}_c = \sum_{i=1}^{n} m_i \mathbf{r}_i \tag{14.16}$$

where m_i is the mass of the ith rigid body, r_i is the position vector to the center of mass of the ith rigid body, M is the total mass, and r_c is the position vector to the center of mass of the system. We can then say on differentiating:

$$M\dot{r}_c = \sum_{i=1}^{n} m_i \dot{r}_i \tag{14.17}$$

$$M\ddot{r}_c = \sum_{i=1}^{n} m_i \ddot{r}_i \tag{14.18}$$

In *Newton's law* for the mass center of a system of rigid bodies, we conclude that we can use the centers of mass of the component parts of the system as given on the right side of Eq. 14.18 rather than the center of mass of the total mass.

Example 14.10

A shaft has protruding arms each of which weighs 40 N/m (see Fig. 14.30). A torque T gives the shaft an angular acceleration $\dot{\omega}$ of 2 rad/sec². At the instant shown in the diagram, ω is 5 rad/sec. If the shaft without arms weighs 180 N, compute the vertical

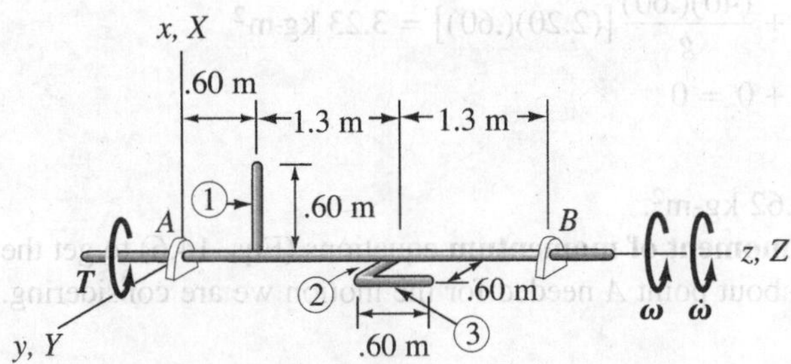

Figure 14.30. Rotating shaft with arms.

and horizontal forces at bearings A and B (see Fig. 14.31). Note that we have numbered the various arms for convenient identification.

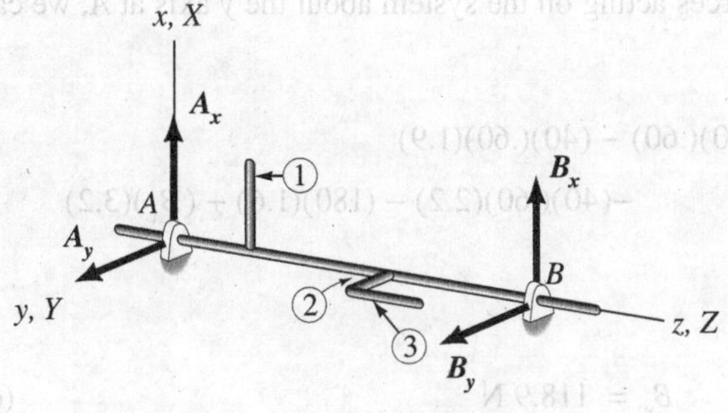

Figure 14.31. Supporting forces.

Example 14.10 (Continued)

We first fix a reference *xyz* to the shaft at *A*. Also at *A* we fix an inertial reference *XYZ* to the ground. We can directly use Eqs. 14.6a and 14.6b about point *A*. For this reason, we shall compute the required products of inertia of the shaft system for reference *xyz*. Accordingly, using the parallel-axis theorem, we have:

$$(I_{xz})_{\text{arm}(1)} = 0 + \frac{(40)(.60)}{g}[(.6)(.3)] = .440 \text{ kg-m}^2$$

$$(I_{xz})_{\text{arm}(2)} = 0 + \frac{(40)(.60)}{g}[(1.9)(0)] = 0$$

$$(I_{xz})_{\text{arm}(3)} = 0 + \frac{(40)(.60)}{g}[(2.20)(0)] = 0$$

Hence, for the system, $I_{xz} = .440 \text{ kg-m}^2$.

We next consider I_{zy}. Accordingly, we have

$$(I_{zy})_{\text{arm}(1)} = 0 + \frac{(40)(.60)}{g}[(.60)(0)] = 0$$

$$(I_{zy})_{\text{arm}(2)} = 0 + \frac{(40)(.60)}{g}[(1.9)(0.30)] = 1.394 \text{ kg-m}^2$$

$$(I_{zy})_{\text{arm}(3)} = 0 + \frac{(40)(.60)}{g}[(2.20)(.60)] = 3.23 \text{ kg-m}^2$$

$$(I_{zy})_{\text{shaft}} = 0 + 0 = 0$$

Hence, for the system, $I_{zy} = 4.62 \text{ kg-m}^2$.

We can now employ the **moment of momentum** equations (Eqs. 14.6) to get the required moments M_x and M_y about point *A* needed for the motion we are considering. Thus, we have

$$M_x = -(2)(.440) + (5^2)(4.62) = 114.7 \text{ N-m} \qquad \text{(a)}$$

$$M_y = -(2)(4.62) - (5^2)(.440) = -20.2 \text{ N-m} \qquad \text{(b)}$$

Summing moments of all the forces acting on the system about the *y* axis at *A*, we can say (see Fig. 14.31):

$$M_y = -20.2 = -(40)(.60)(.60) - (40)(.60)(1.9)$$
$$-(40)(.60)(2.2) - (180)(1.6) + (B_x)(3.2)$$

Therefore, we require

$$B_x = 118.9 \text{ N} \qquad \text{(c)}$$

Example 14.10 (Continued)

Summing moments about the x axis at A, we can say

$$M_x = 114.7 = -B_y(3.2)$$

Therefore, we require

$$B_y = -35.8 \text{ N} \qquad \text{(d)}$$

We next use **Newton's law** considering the three arms to be three particles at their mass centers as has been shown in Fig. 14.32. In the x direction at time t we have,

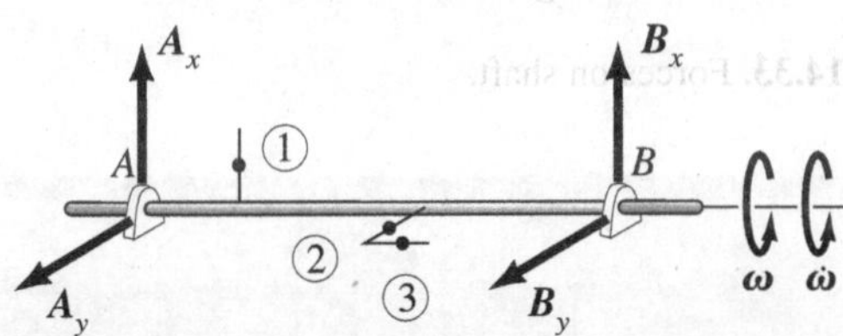

Figure 14.32. Arms replaced by mass centers.

using Eq. 14.18 and noting that each of the aforementioned particles has circular motion:

$$118.9 + A_x - 180 - (3)[(40)(.60)] = -\frac{(40)(.60)}{g}(.30)(\omega^2)$$
$$-\frac{(40)(.60)}{g}(.30)(\dot{\omega}) - \frac{(40)(.60)}{g}(.60)(\dot{\omega})$$

Setting $\omega = 5$ rad/sec and $\dot{\omega} = 2$ rad/sec^2, we get

$$A_x = 110.3 \text{ N} \qquad \text{(e)}$$

In the y direction, we can say similarly at time t

$$A_y - 35.8 =$$
$$\frac{(40)(.60)}{g}(.30)(\dot{\omega}) - \frac{(40)(.60)}{g}(.30)(\omega^2) - \frac{(40)(.60)}{g}(.60)(\omega^2)$$

Example 14.10 (Continued)

Therefore,

$$A_y = -17.78 \text{ N} \tag{f}$$

The forces acting on the shaft are shown in Fig. 14.33. The reactions to these forces are then the desired forces on the bearings. In the z direction it should be clear that there is no force on the bearings.

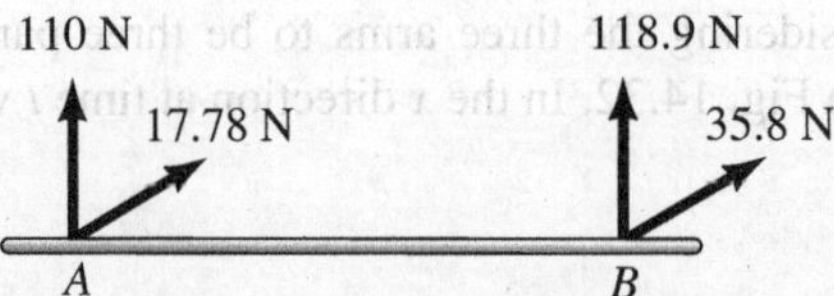

Figure 14.33. Forces on shaft.

If, in the last example, we had ignored the constant forces of gravity, we would have determined forces at bearings A and B that are due entirely to the motion of the body. Forces computed in this way are called *dynamic forces*. If the body were rotating with constant speed ω, these forces would clearly have constant values in the x and y directions. Since the xy axes are rotating with the body relative to the ground reference XYZ, such dynamic forces must also rotate relative to the ground about the axis of rotation with the speed ω of the body. This means that, in any *fixed* direction normal to the shaft at a bearing, there will be a *sinusoidal force variation* with a frequency corresponding to the angular rotation of the shaft. Such forces can induce vibrations of large amplitude in the structure or support if a natural frequency or multiple of a natural frequency is reached in these bodies. When a shaft creates rotating forces on the bearings by virtue of its own rotation, the shaft is said to be unbalanced. We shall set up criteria for balancing a rotating body in the next section.

PROBLEMS

14.93. A shaft shown supported by bearings A and B is rotating at a speed ω of 3 rad/sec. Identical blocks C and D weighing 30 lb each are attached to the shaft by light structural members. What are the bearing reactions in the x and y directions if we neglect the weight of the shaft?

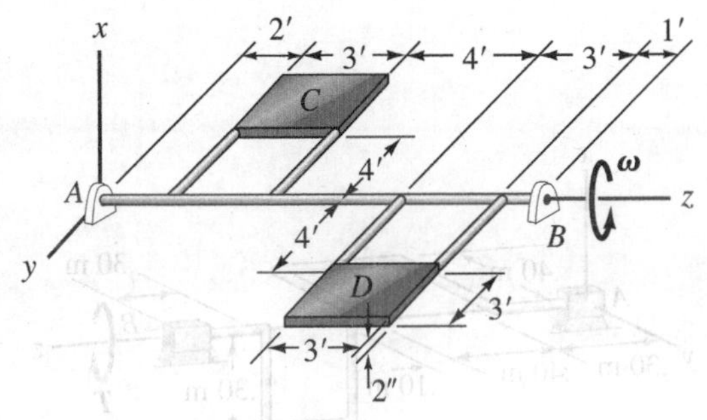

Figure P.14.93.

14.94. Shaft AB is rotating at a constant speed ω of 20 rad/sec. Two rods having a weight of 10 N each are welded to the shaft and support a disc D weighing 30 N. What are the supporting forces at the instant shown?

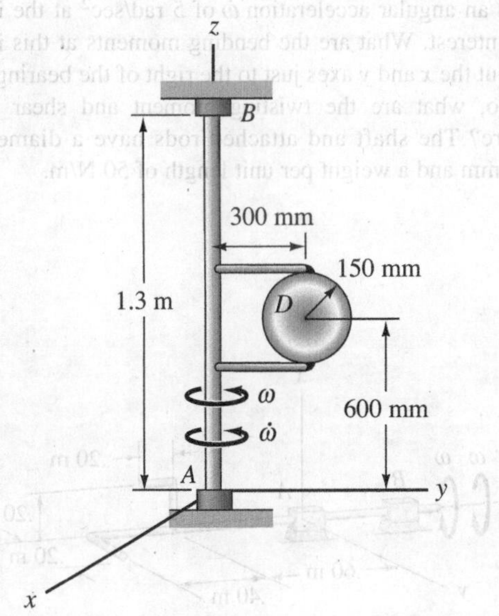

Figure P.14.94.

14.95. Do Problem 14.94 for the case in which $\omega = 20$ rad/sec and $\dot{\omega} = 38$ rad/sec^2 at the instant of interest as shown.

14.96. A uniform wooden panel is shown supported by bearings A and B. A 100-lb weight is connected with an inextensible cable to the panel at point G over a light pulley D. If the system is released from rest at the configuration shown, what is the angular acceleration of the panel, and what are the forces at the bearings? The panel weighs 60 lb.

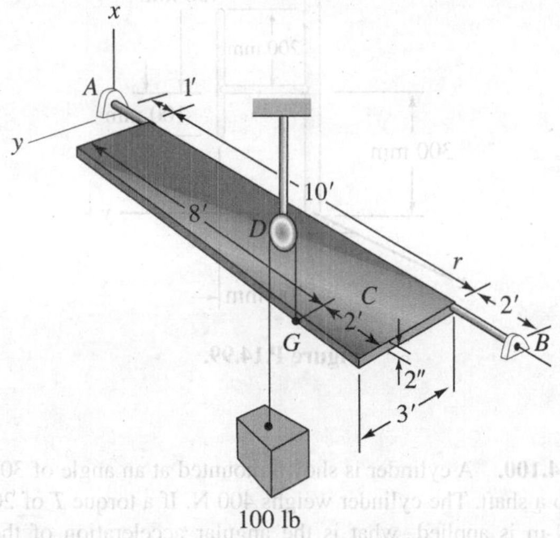

Figure P.14.96.

14.97. Do Problem 14.96 when there is a frictional torque at the bearings of 10 ft-lb and the pulley has a radius of 1 ft and a moment of inertia of 10m-ft^2.

14.98. A thin rectangular plate weighing 50 N is rotating about its diameter at a speed ω of 25 rad/sec. What are the supporting forces in the x and y directions at the instant shown when the plate is parallel to the yz plane?

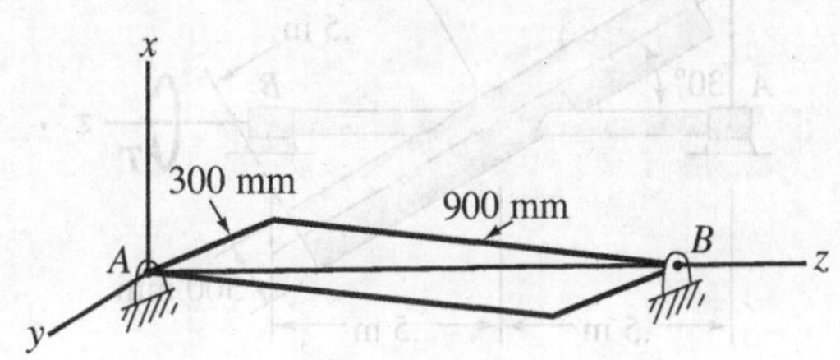

Figure P.14.98.

14.99. A shaft is shown rotating at a speed of 20 rad/sec. What are the supporting forces at the bearings? The rods welded to the shaft weight 40 N/m. The shaft weighs 80 N.

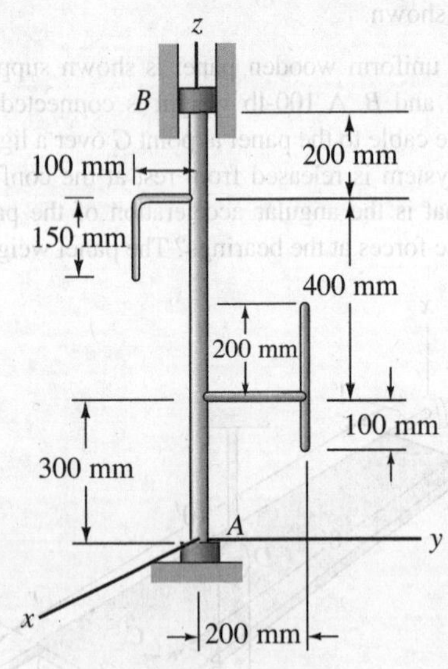

Figure P.14.99.

14.100. A cylinder is shown mounted at an angle of 30° to a shaft. The cylinder weighs 400 N. If a torque T of 20 N-m is applied, what is the angular acceleration of the system? What are the supporting forces in the x and y directions at the configuration shown wherein the system is stationary? Neglect the mass of the shaft. The centerline of the cylinder is in the xz plane at the instant shown.

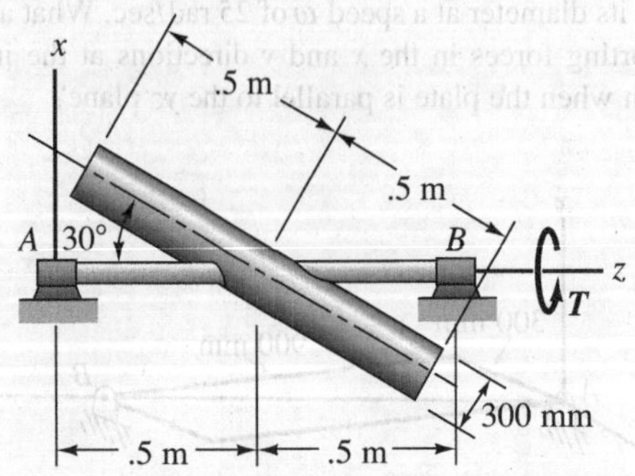

Figure P.14.100.

14.101. A bent shaft has applied to it a torque T including gravity given as

$$T = 10 + 5t \text{ N-m}$$

where t is in seconds. What are the supporting forces at the bearings in the x and y directions when $t = 3$ sec? The shaft is made from a rod 20 mm in diameter and weighing 70 N/m. At $t = 3$ sec, the position of the shaft is as shown.

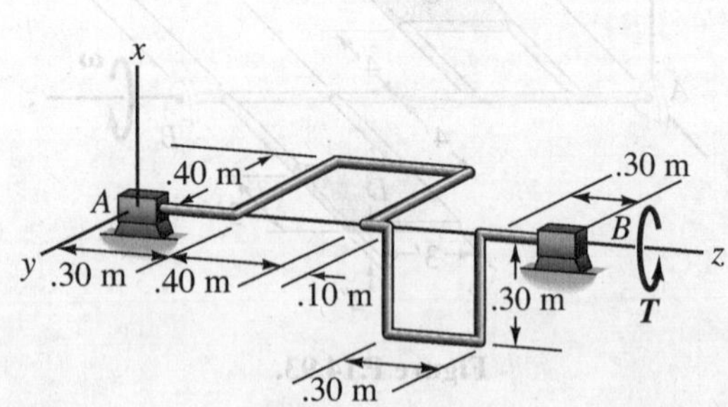

Figure P.14.101.

***14.102.** A shaft has an angular velocity ω of 10 rad/sec and an angular acceleration $\dot{\omega}$ of 5 rad/sec² at the instant of interest. What are the bending moments at this instant about the x and y axes just to the right of the bearing at A? Also, what are the twisting moment and shear forces there? The shaft and attached rods have a diameter of 20 mm and a weight per unit length of 50 N/m.

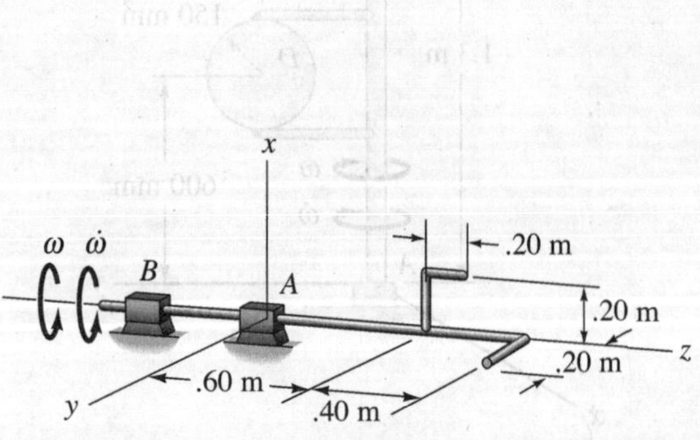

Figure P.14.102.

14.103. Balance the system in planes A and B at a distance 1 ft from centerline. Use two weights.

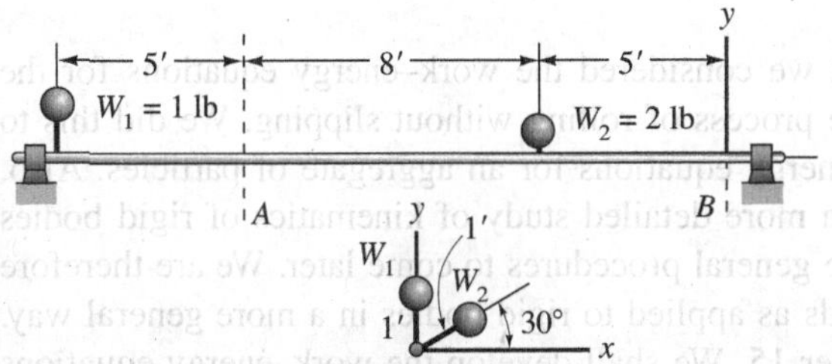

Figure P.14.103.

14.104. Balance the system in Problem 14.103 by using a weight in plane A of 1½ lb and a weight in plane B of 1 lb. You may choose suitable radii in these planes.

14.105. Balance the shaft of Problem 14.94 using rods weighing 50 N/m and welded to the shaft normal to the centerline just next to bearings A and B. Determine the lengths of these rods and their orientations relative to the xy axes.

14.106. A disc and a cylinder are mounted on a shaft. The disc has been mounted eccentrically so that the center of mass is ½ in from the centerline of the shaft. Balance the shaft using balancing planes 5 ft in from bearing A and 3 ft in from bearing B, respectively. The balancing masses each weigh 3 lb and can be regarded as particles. Give the proper position of the balancing masses in these planes.

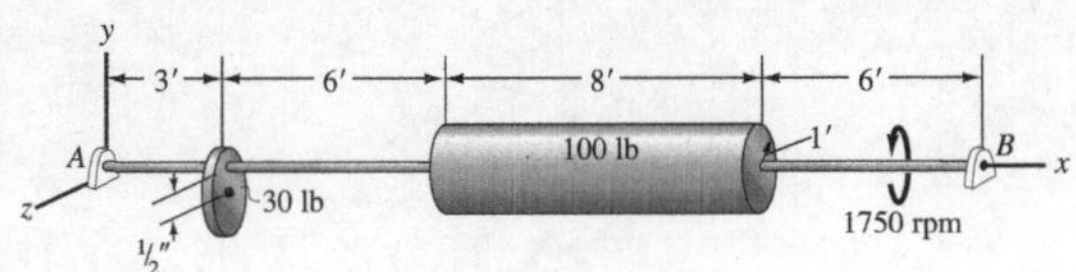

Figure P.14.106.

14.107. Balance the shaft described in Problem 14.106 by removing a small chunk of metal from each of the end faces of the 100-lb cylinder at a position 10 in from the shaft centerline. What are the weights of these chunks and what are their orientations?

14.108. Balance the shaft in Problem 14.99 using balancing planes just next to bearings A and B. At bearing B use a small balancing sphere of weight 30 N and at bearing A use a rod having a weight per unit length of 35 N/m.

14.109. A disc is shown mounted off-center at B on a shaft CD that rotates with angular speed ω. The diameter of the shaft is 2 in. The disc weighs 50 lb and has a diameter of 6 ft. Balance the system using two rods, each weighing 10 lb/ft and having a diameter of 2 in. The rods are to be attached normal to the shaft at position 1 ft in from bearing C and 2 ft. in from bearing D. Determine the lengths of these rods and their inclination.

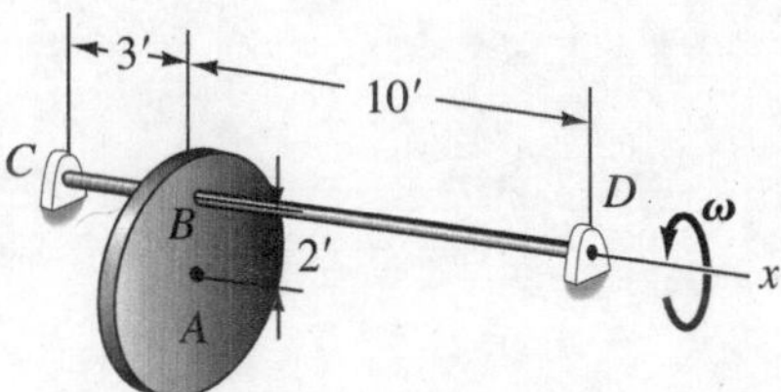

Figure P.14.109.

14.110. Balance the shaft shown for Problem 14.101. Use a balancing plane just next to bearing A and one just next to bearing B. Attach a circular plate normal to the shaft at each bearing, and cut a hole with diameter 60 mm at the proper position in the plate to balance the system. The plates are 30 mm thick and have a specific weight of 8×10^{-5} N/mm^3.

14.9 Closure

In this chapter, we have developed the moment-of-momentum equations for plane motion of a rigid body. We applied this equation to various cases of plane motion starting from the simplest case and going to the most difficult case. Many problems of engineering interest can be taken as plane-motion problems; the results of this chapter are hence quite important.

Recall next that in Chapter 12 we considered the work–energy equations for the plane motion of simple bodies in the process of rolling without slipping. We did this to help illustrate the use of the work–energy equations for an aggregate of particles. Also, this undertaking served to motivate a more detailed study of kinematics of rigid bodies and to set forth in miniature the more general procedures to come later. We are therefore now ready to examine energy methods as applied to rigid bodies in a more general way. This will be done in Part A of Chapter 15. We shall develop the work–energy equations for three-dimensional motion and apply them to all kinds of motions, including plane motions. The student should not have difficulty in going directly to the general case; indeed, a better understanding of the subject should result.

In Part B of Chapter 15, we shall consider the impulse-momentum equations for rigid bodies. This will be an extension of the useful impulse-momentum methods discussed in Chapter 12 for particles and aggregates of particles. Again, we shall be able to go to the general case and then apply the results to three- and two-dimensional motions (plane motions).

PROBLEMS

14.111. A circular plate A is used in electric meters to damp out rotations of a shaft by rotating in a bath of oil. The plate and its shaft have a mass of 300 g and a radius of gyration of 100 mm. If the shaft and plate very thin down from 30 rpm to 20 rpm in 5 sec, what angular acceleration can be developed by a .005 N-m torque when ω of the shaft is 10 rpm? Assume that the damping torque is proportional to the angular speed.

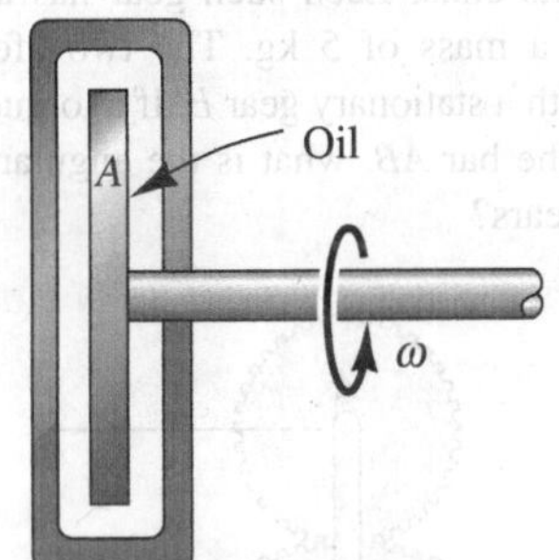

Figure P.14.111.

14.112. The dynamic coefficient of friction for contact surfaces E and G is .2 and for A is .3. If a force P of 250 lb is applied, what will be the tension in the cord HB? Start by assuming no slipping at A. Check your assumption at the end of the calculation.

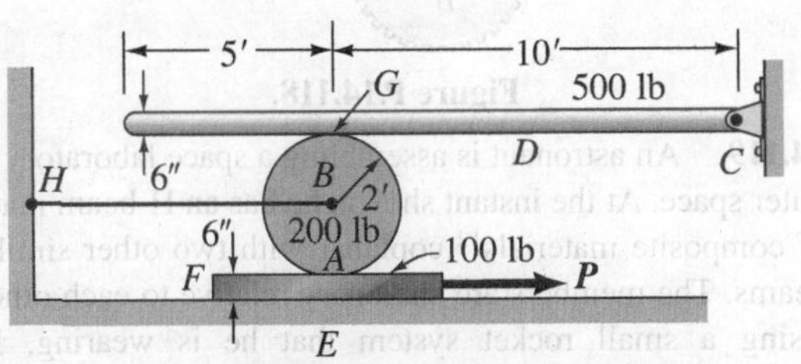

Figure P.14.112.

14.113. A torque T of 15 N-m is applied to a rod AB as shown. At B there is a pin which slides in a frictionless slot in a disc E whose mass is 10 kg and whose radius of gyration is 300 mm. If the system is at rest at the instant shown, what is the angular acceleration of the rod and the disc? The rod has a mass of 18 kg.

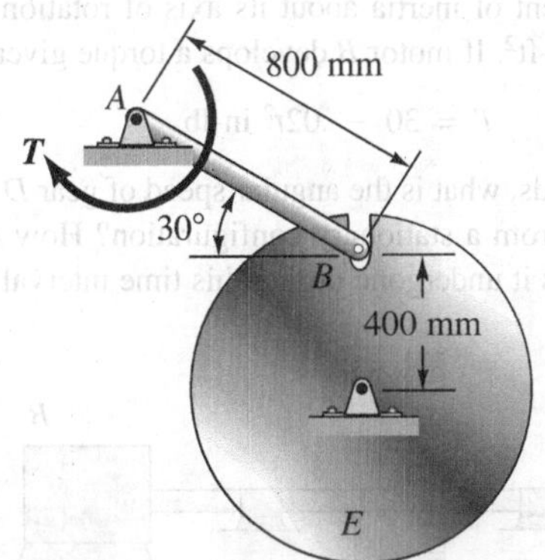

Figure P.14.113.

14.114. A platform A has a torque T applied about its axis of rotation. The platform has a mass of 1,000 kg and a radius of gyration of 2 m. A block B rests on the platform but is prevented from sliding off by very thin stops C and D. The block has a mass of 1 kg and has dimensions 200 mm × 200 mm × 200 mm. The center of mass of the block is at its geometric center. If a torque $T = 20t^2 + 50t$ N-m is applied with t in seconds, when and how does the block first tip? (Because of the small size and mass of B, consider the system to be a slablike body.)

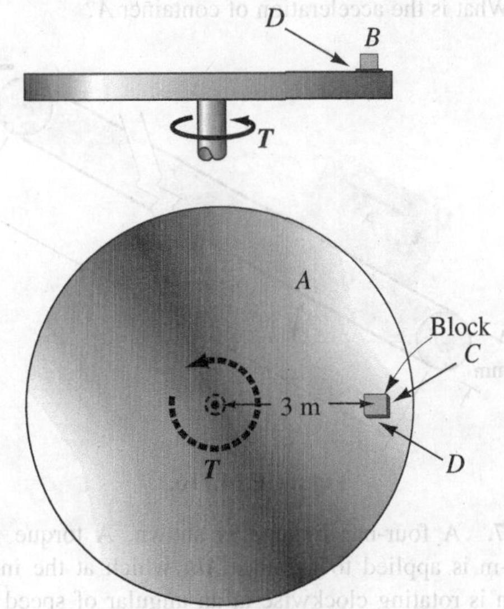

Figure P.14.114.

14.115. A motor B drives a gear C which connects with gear D to driven device A. The top system of shaft and gear has a moment of inertia I_1 about the axis of rotation of 3 lbm-ft^2, whereas the bottom system of device A and gear D has a moment of inertia about its axis of rotation of I_2 equal to 1 lbm-ft^2. If motor B develops a torque given as

$$T = 30 - .02t^2 \text{ in-lb}$$

with t in seconds, what is the angular speed of gear D 6 sec after starting from a stationary configuration? How many revolutions has it undergone during this time interval?

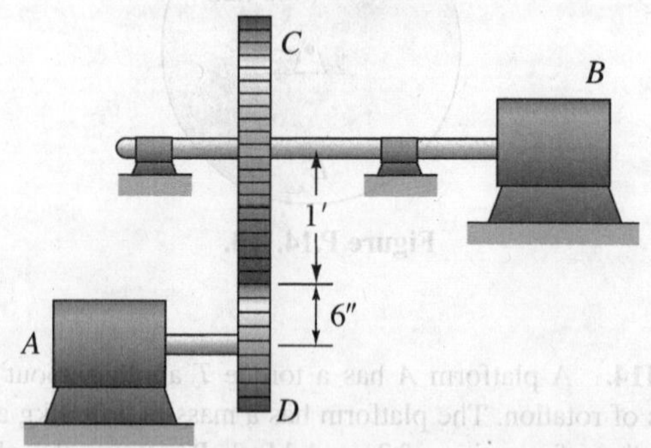

Figure P.14.115.

14.116. A 50-kg container A is being transported by a conveyor as shown. A torque T of 200 N-m is applied to the driving drum. Both driving and driven drums each have a mass of 10 kg and a radius of gyration of 130 mm. The belt has a mass per unit length of 3 kg/m and a dynamic coefficient of friction of .3 with the conveyor bed. What is the acceleration of container A?

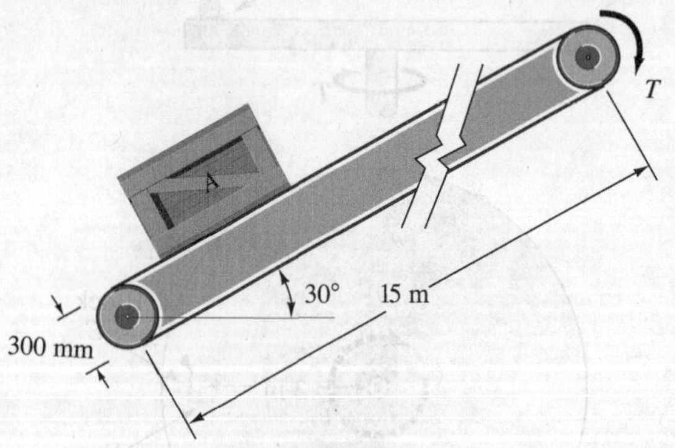

Figure P.14.116.

14.117. A four-bar linkage is shown. A torque T of 10 N-m is applied to member AB, which at the instant shown is rotating clockwise at an angular of speed of 3 rad/sec. Bars AB and CD are 300 mm long. Bar BC is a circular arc of radius 400 mm and length 450 mm. All bars have a mass of 10 kg per meter. What is the angular acceleration of bar AB at the instant shown?

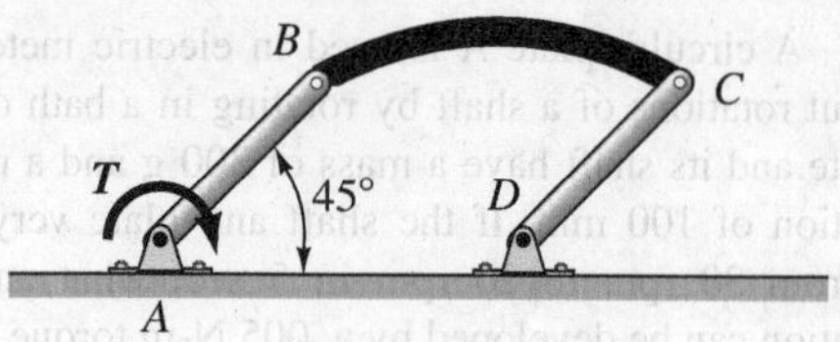

Figure P.14.117.

14.118. Bar AB having a mass of 20 kg is connected to two gears at its ends. Each such gear has a diameter of 300 mm and a mass of 5 kg. The two aforementioned gears mesh with a stationary gear E. If a torque of 100 N-m is applied to the bar AB, what is the angular acceleration of the small gears?

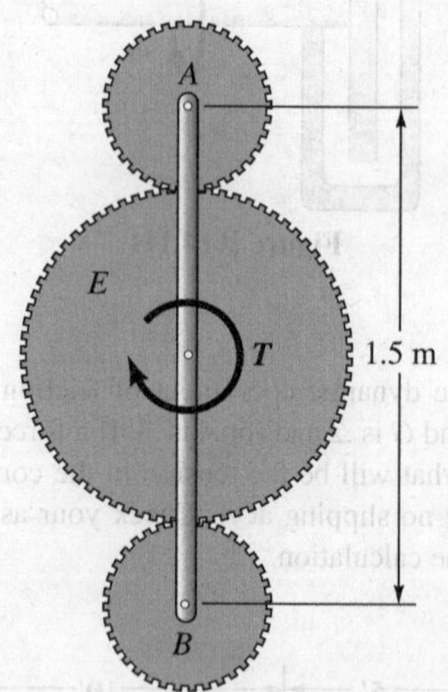

Figure P.14.118.

14.119. An astronaut is assembling a space laboratory in outer space. At the instant shown, he has an H-beam made of composite materials[12] coplanar with two other similar beams. The members are stationary relative to each other. Using a small rocket system that he is wearing, he develops a force F of 200 N always normal to the beam A and in the plane of the beams. At what position d should he exert this force to have the beam A parallel to the tops of beams C and D in 2 sec? What is the acceleration of the center of mass of A at time $t = 1$ sec? Beam A has a mass 200 kg and is 10 m long. Consider beam A to be a long slender rod.

[12]Made up of plastics and various kinds of fibers.

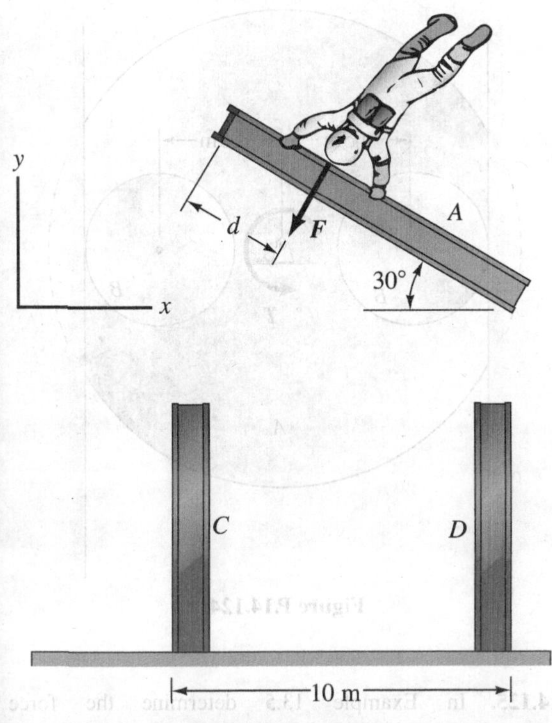

Figure P.14.119.

14.120. Rod *AB* of length 20 ft and weight 200 lb is released from rest at the configuration shown. *AC* is a weightless wire, and the incline at *B* is frictionless. What is the tension in the wire *AC* at this instant?

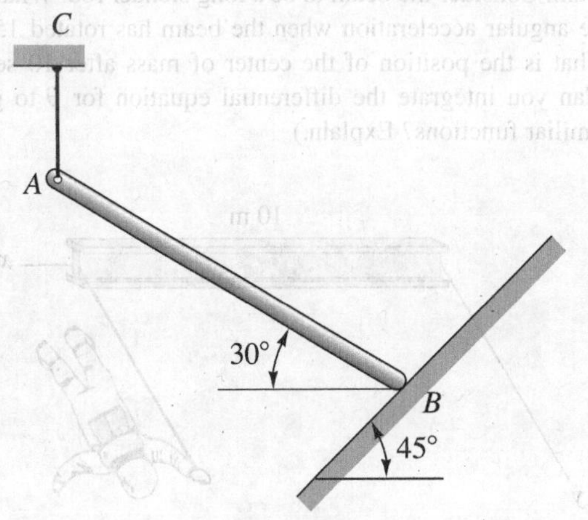

Figure P.14.120.

14.121. Rod *AB* of length 1 m and mass 10 kg is pinned to a disc *D* having a mass of 20 kg and a diameter of .5 m. A torque T = 15 N-m is applied to the disc. What are the angular accelerations of the rod *AB* and disc at this instant?

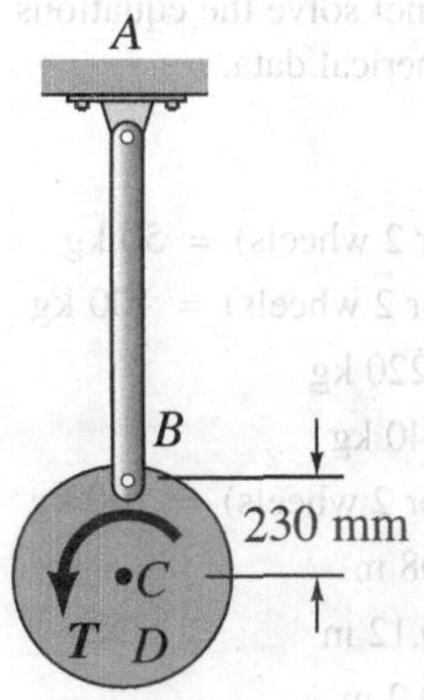

Figure P.14.121.

14.122. A four-bar linkage is shown. Bar *AB* has the following angular motion at the instant shown.

$$\omega_1 = 2 \text{ rad/sec}$$
$$\dot{\omega}_1 = 3 \text{ rad/sec}^2$$

What are the supporting forces at *D* at this instant for the following data?

mass of *AB* = 4 kg
mass of *BC* = 3 kg
mass of *CD* = 6 kg

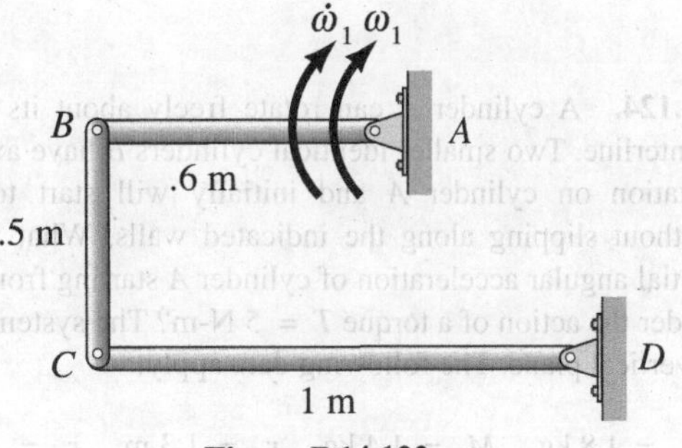

Figure P.14.122.

14.123. Starting from rest, find the acceleration of device A when the 10,000 N force is applied. Proceed as follows:

(a) Draw and label four free-body diagrams.
(b) Write a system of equations whose number equals the number of unknowns.
(c) Do not solve the equations but do put in any numerical data.

Data:

M_J (for 2 wheels) = 50 kg
M_E (for 2 wheels) = 100 kg
M_A = 220 kg
M_C = 40 kg
M_E (for 2 wheels) = 100 kg
k_J = .08 m
k_E = 0.12 m
D_J = 0.2 m
D_E = 0.3 m

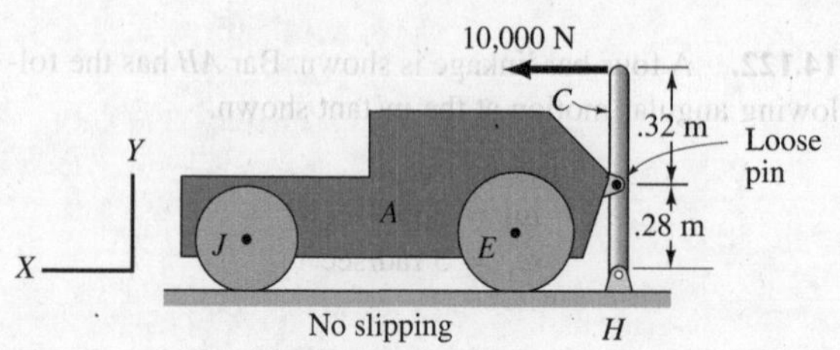

Figure P.14.123.

14.124. A cylinder A can rotate freely about its fixed centerline. Two smaller identical cylinders B have axes of rotation on cylinder A and initially will start to roll without slipping along the indicated walls. What is the initial angular acceleration of cylinder A starting from rest under the action of a torque $T = 5$ N-m? The system is in a vertical plane. The following data apply:

$M_A = 1.8$ kg $\quad M_B = 1.4$ kg $\quad r_A = 1.3$ m $\quad r_B = .04$ m

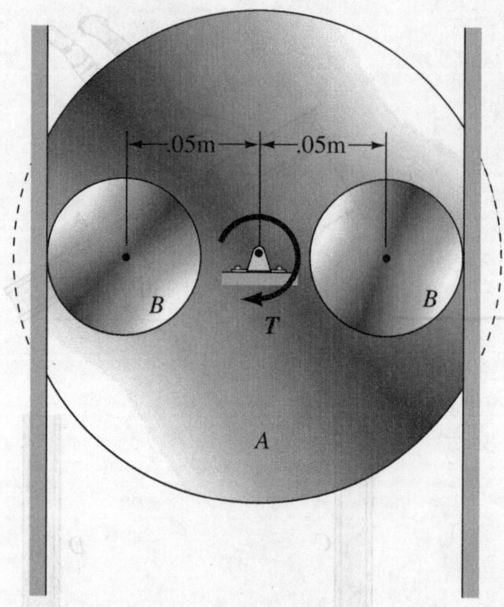

Figure P.14.124.

14.125. In Example 13.5 determine the force components at pins A and B. The mass of rod AB is 8 kg and the mass of BC is 5 kg. Use the kinematic results of the example.

14.126. A 10-m I-beam having a mass of 400 kg is being pulled by an astronaut with his space propulsion rig as shown. The force is 40 N and is always in the same direction. Initially, the beam is stationary relative to the astronaut, and the connecting cord is at right angles to the beam. Consider the beam to be a long slender rod. What is the angular acceleration when the beam has rotated 15°? What is the position of the center of mass after 10 sec? (Can you integrate the differential equation for θ to get familiar functions? Explain.)

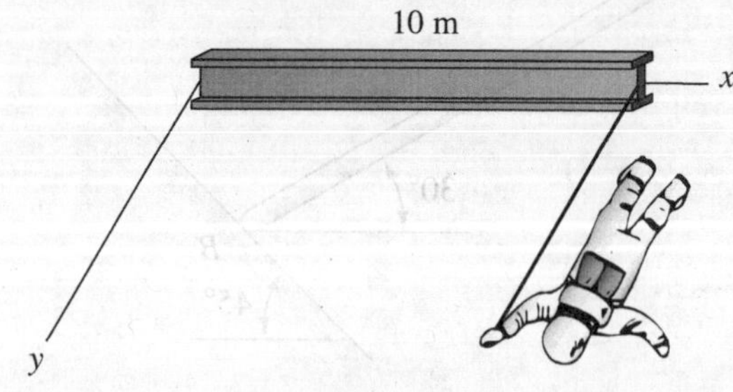

Figure P.14.126.

14.127. A rectangular box having a mass of 20 kg is being transported on a conveyor belt. The center of gravity of the box is 150 mm above the conveyor belt, as shown. What is the maximum starting torque T for which the box will not tip? The belt has a mass per unit length of 2 kg/m, and the driving and driven drums have a mass of 5 kg each and a radius of gyration of 130 mm. The dynamic coefficient of friction between the belt and conveyor bed is .2.

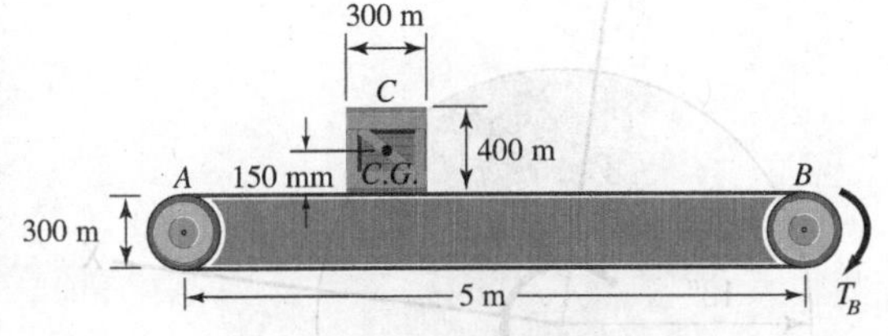

Figure P.14.127.

14.128. A ring is shown supported by wire AB and a smooth surface. The ring has a mass of 10 kg and a mean radius of 2 m. A body D having a mass of 3 kg is fixed to the ring as shown. If the wire is severed, what is the acceleration of body D?

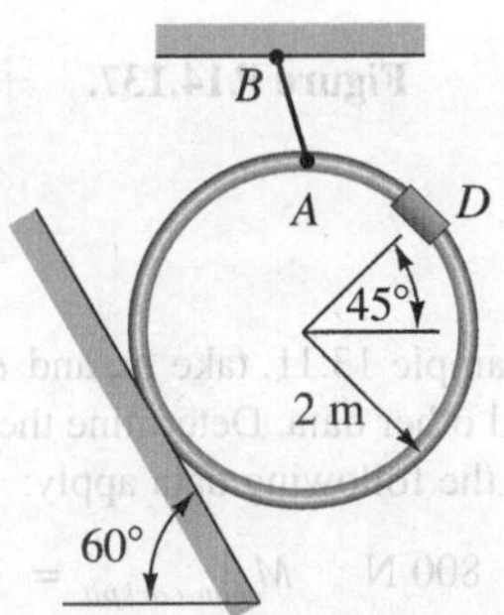

Figure P.14.128.

14.129. If the rod shown is released from rest at the configuration shown, what are the supporting forces at A and B at that instant? The rod weighs 100 lb and is 10 ft long. The static coefficient of friction is .2 for all surface contacts.

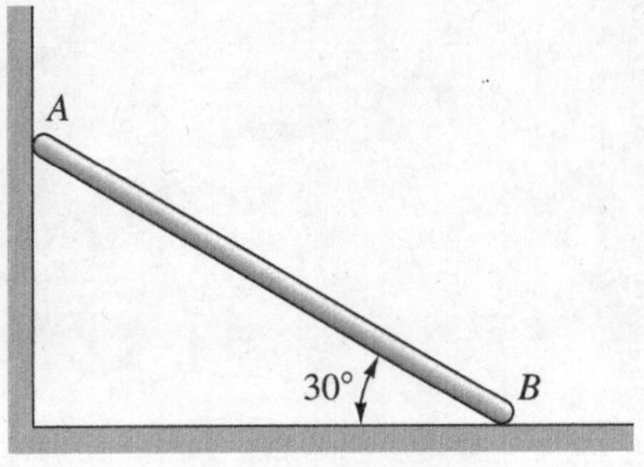

Figure P.14.129.

14.130. Do Problem 14.129 for the case where end A is moving downward at a speed of 10 ft/sec at the instant shown and where $\mu_d = .2$.

14.131. In Problem 14.129, find by inspection the instantaneous axis of rotation for the rod. What are the magnitude and direction of the acceleration vector for the axis of rotation at the instant the rod is released? We know from Problem 14.129 that $\dot{\omega} = 3.107$ rad/sec^2 and $\boldsymbol{a}_c = 7.77\boldsymbol{i} - 13.45\boldsymbol{j}$ ft/sec^2 for the center of mass.

14.132. Identical bars AB and BC are pinned as shown with frictionless pins. Each bar is 2.3 m in length and has a mass of 9 kg. A force of 450 N is exerted at C when the bars are inclined at 60°. What is the angular acceleration of the bars?

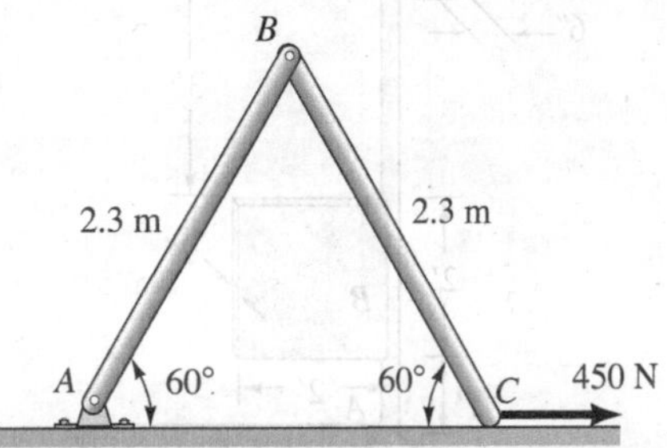

Figure P.14.132.

14.133. A compressor is shown. Member AB is rotating at a constant speed ω_1 of 100 rpm. Member BC has a mass of 2 kg and piston C has a mass of 1 kg. The pressure p on the piston is 10,000 Pa. At the instant shown, what are the forces transmitted by pins B and C?

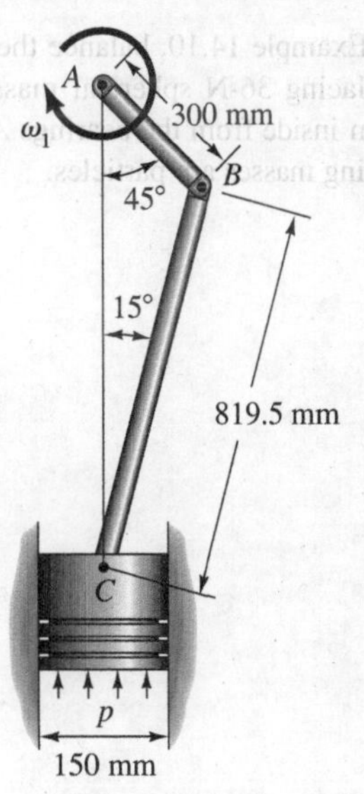

Figure P.14.133.

***14.134.** A thin vertical shaft rotates with angular speed ω of 5 rad/sec in bearings A and D as shown in the diagram. A uniform plate B weighing 50 lb is attached to the shaft as is a disc C weighing 30 lb. What are the bearing reactions at the configuration shown? The shaft weighs 20 lb and the thickness of disc and plate is 2 in.

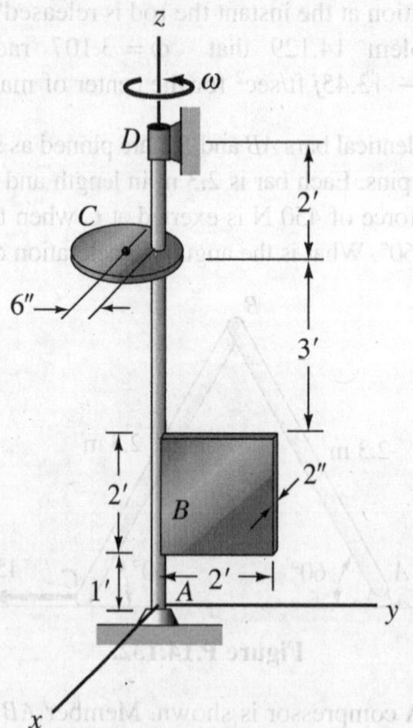

Figure P.14.134.

***14.135.** Do Problem 14.100 for the case where $\omega = 5$ rad/sec at the instant of interest.

***14.136.** In Example 14.10, balance the rotating system by properly placing 36-N spherical masses in balancing planes 300 mm inside from the bearings A and B. Assume that the balancing masses are particles.

14.137. A 25-lb cylinder which is spinning at a rate ω_0 of 500 rpm is placed on an 8 degree incline. The coefficients of friction are $\mu_s = .4$ and $\mu_d = .3$. How far does the cylinder move before there is rolling without slipping? How much time elapses before the cylinder stops moving instantaneously?

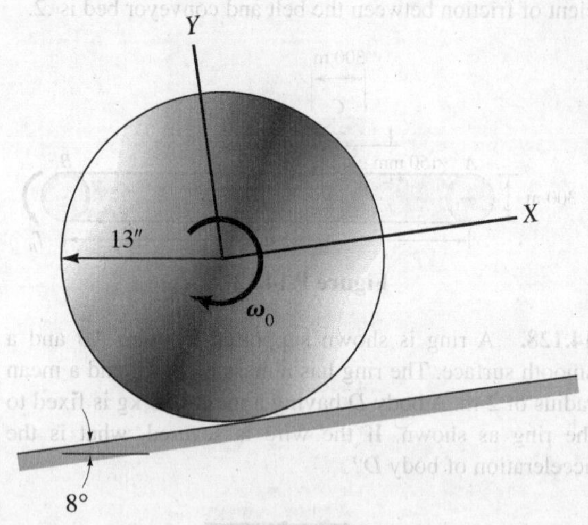

Figure P.14.137.

14.138. In Example 13.11, take ω_2 and $\dot{\omega}_2$ both equal to zero but keep all other data. Determine the force system at A knowing that the following data apply:

$$W_{AB} = 800 \text{ N} \qquad M_{man\text{-}cockpit} = 135 \text{ kg}$$

Note that the man and cockpit are translating and can be considered as a particle of mass 135 kg.

CHAPTER 15

Energy and Impulse–Momentum Methods for Rigid Bodies

15.1 Introduction

Let us pause to reflect on where we have been thus far in dynamics and where we are about to go. In Chapter 10, you will recall, we worked directly with *Newton's law* and integrated it several times to consider the motion of a *particle*. Then, in Chapter 11 and 12, we formulated certain useful integrated forms from *Newton's law* and thereby presented the *energy methods* and the *linear impulse-momentum* methods also for a *particle*. At the end of Chapter 12, we derived the important *angular momentum* equation, $M_A = \dot{H}_A$. In Chapter 14, we returned to Newton's law and along with the angular momentum equation, $M_A = \dot{H}_A$, carried out integrations to solve *plane motion* problems of rigid bodies. In the present chapter, we shall come back to *energy methods* and *linear impulse-momentum methods*—this time for the *general motion* of rigid bodies. In addition, we shall use a certain integrated form of the angular momentum equation $M_A = \dot{H}_A$, namely the *angular impulse-momentum equation*. These equations at times will be applied to a single rigid body. At other times, we shall apply them to several interconnected rigid bodies considered as a whole. When we do the latter, we say we are dealing with a *system* of rigid bodies. We shall consider energy methods first.

Part A: Energy Methods

15.2 Kinetic Energy of a Rigid Body

First, we shall derive a convenient expression for the kinetic energy of a rigid body. We have already found (Section 11.6) that the kinetic energy of an aggregate of particles relative to any reference is the sum of two parts, which we list again as:

1. The kinetic energy of a hypothetical particle that has a mass equal to the total mass of the system and a motion corresponding to that of the mass center of the system, plus
2. The kinetic energy of the particles relative to the mass center.

Mathematically,

$$\text{KE} = \tfrac{1}{2} M|\dot{\boldsymbol{r}}_c|^2 + \tfrac{1}{2}\sum_{i=1}^{n} m_i|\dot{\boldsymbol{\rho}}_i|^2 \tag{15.1}$$

where $\boldsymbol{\rho}_i$ is the position vector from the mass center to the ith particle.

Let us now consider the foregoing equation as applied to a rigid body which is a special "aggregate of particles" (Fig. 15.1). In such a case, the velocity of any particle relative to the mass center becomes

$$\dot{\boldsymbol{\rho}}_i = \boldsymbol{\omega} \times \boldsymbol{\rho}_i \tag{15.2}$$

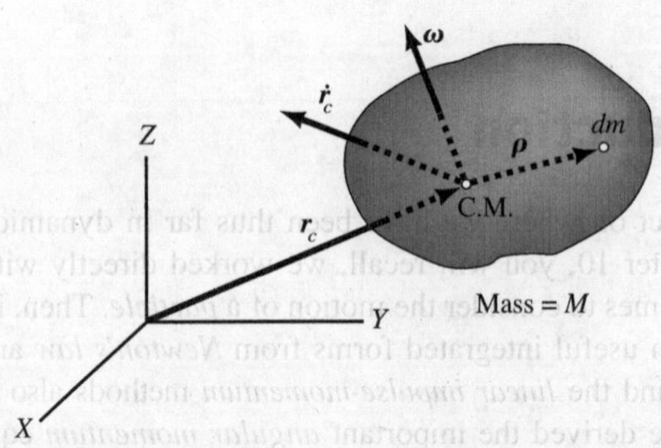

Figure 15.1. Rigid body.

where $\boldsymbol{\omega}$ is the angular velocity of the body relative to reference XYZ in which we are computing the kinetic energy. For the rigid body, the discrete particles of mass m_i become a continuum of infinitesimal particles each of mass dm, and the summation in Eq. 15.1 then becomes an integration. Thus, we can say for the rigid body, replacing $|\dot{\boldsymbol{r}}_c|^2$ by V_c^2.

$$\text{KE} = \tfrac{1}{2} MV_c^2 + \tfrac{1}{2}\iiint_M |\boldsymbol{\omega} \times \boldsymbol{\rho}|^2 \, dm \tag{15.3}$$

where $\boldsymbol{\rho}$ represents the position vector from the center of mass to any element of mass dm. Let us now choose a set of orthogonal directions xyz at the center of mass, so we can carry out the preceding integration in terms of the scalar components of $\boldsymbol{\omega}$ and $\boldsymbol{\rho}$. This step is illustrated in Fig. 15.2. We first express the integral in Eq. 15.3 in the following manner:

$$\iiint_M |\boldsymbol{\omega} \times \boldsymbol{\rho}|^2 \, dm = \iiint_M (\boldsymbol{\omega} \times \boldsymbol{\rho}) \cdot (\boldsymbol{\omega} \times \boldsymbol{\rho}) \, dm \tag{15.4}$$

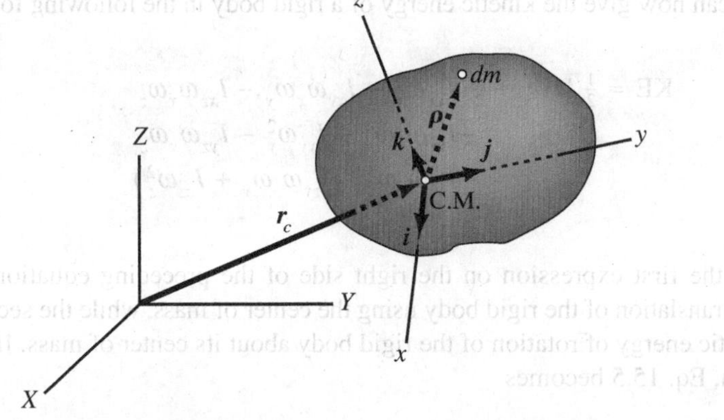

Figure 15.2. Fix *xyz* at center of mass.

Inserting the scalar components, we get:

$$\iiint_M |\boldsymbol{\omega} \times \boldsymbol{\rho}|^2\, dm = \iiint_M \Big\{\Big[\left(\omega_x \boldsymbol{i} + \omega_y \boldsymbol{j} + \omega_z \boldsymbol{k}\right) \times (x\boldsymbol{i} + y\boldsymbol{j} + z\boldsymbol{k})\Big]$$

$$\bullet \Big[\left(\omega_x \boldsymbol{i} + \omega_y \boldsymbol{j} + \omega_z \boldsymbol{k}\right) \times (x\boldsymbol{i} + y\boldsymbol{j} + z\boldsymbol{k})\Big]\Big\}\, dm$$

Carrying out first the cross products and then the dot product in the integrand and collecting terms, we form the following relation on extracting the ω's from the integrals.

$$\begin{aligned}\iiint_M |\boldsymbol{\omega} \times \boldsymbol{\rho}|^2\, dm = &\left[\iiint_M (z^2 + y^2)\, dm\right]\omega_x^2 - \left[\iiint_M xy\, dm\right]\omega_x\omega_y - \left[\iiint_M xz\, dm\right]\omega_x\omega_z - \\ &\left[\iiint_M yx\, dm\right]\omega_y\omega_x + \left[\iiint_M (x^2 + z^2)\, dm\right]\omega_y^2 - \left[\iiint_M yz\, dm\right]\omega_y\omega_z - \\ &\left[\iiint_M zx\, dm\right]\omega_z\omega_x - \left[\iiint_M zy\, dm\right]\omega_z\omega_y + \left[\iiint_M (x^2 + y^2)\, dm\right]\omega_z^2\end{aligned}$$

You will recognize that the integrals are the moments and products of inertia for the *xyz* reference. Thus,[1]

$$\begin{aligned}\iiint_M |\boldsymbol{\omega} \times \boldsymbol{\rho}|^2\, dm = &\; I_{xx}\omega_x^2 - I_{xy}\omega_x\omega_y - I_{xz}\omega_x\omega_z \\ &- I_{yx}\omega_y\omega_x + I_{yy}\omega_y^2 - I_{yz}\omega_y\omega_z \\ &- I_{zx}\omega_z\omega_x - I_{zy}\omega_z\omega_y + I_{zz}\omega_z^2\end{aligned}$$

[1]Note that we have deliberatley used a matrix like array for ease in remembering the formulation.

We can now give the kinetic energy of a rigid body in the following form:

$$\begin{aligned}\text{KE} = \tfrac{1}{2}MV_c^2 + \tfrac{1}{2}(I_{xx}\omega_x^2 - I_{xy}\omega_x\omega_y - I_{xz}\omega_x\omega_z \\ -I_{yx}\omega_y\omega_x + I_{yy}\omega_y^2 - I_{yz}\omega_y\omega_z \\ -I_{zx}\omega_z\omega_x - I_{zy}\omega_z\omega_y + I_{zz}\omega_z^2)\end{aligned} \tag{15.5}$$

Note that the first expression on the right side of the preceding equation is the kinetic energy of translation of the rigid body using the center of mass, while the second expression is the kinetic energy of rotation of the rigid body about its center of mass. If principal axes are chosen, Eq. 15.5 becomes

$$\text{KE} = \tfrac{1}{2}MV_c^2 + \tfrac{1}{2}(I_{xx}\omega_x^2 + I_{yy}\omega_y^2 + I_{zz}\omega_z^2) \tag{15.6}$$

Note that for this condition the kinetic energy terms for rotation have the same form as the kinetic energy term that is due to translation, with the moment of inertia corresponding to mass and angular velocity corresponding to linear velocity.

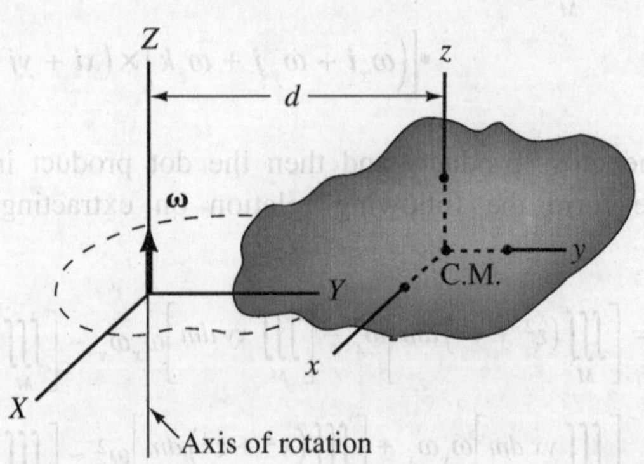

Figure 15.3. Rigid body undergoing pure rotation about Z axis.

As a special case, we shall consider the calculation of kinetic energy of any rigid body undergoing *pure rotation* relative to XYZ with angular velocity $\boldsymbol{\omega}$ about an actual axis of rotation[2] going through part of the body or through a rigid hypothetical massless extension of the body. We have shown this situation in Fig. 15.3, where the Z axis is chosen to be collinear with the $\boldsymbol{\omega}$ vector and the axis of rotation. The reference xyz at the center of mass is chosen parallel to XYZ. Clearly, $\omega_x = \omega_y = 0$ and $\omega_z = \omega$ so Eq. 15.5 becomes

$$\text{KE} = \tfrac{1}{2}MV_c^2 + \tfrac{1}{2}I_{zz}\omega^2$$

[2]An actual axis of rotation in XYZ is a line along which the velocity relative to XYZ is zero.

where I_{zz} is about an axis which goes through the mass center parallel to Z. Note that $V_c = \omega d$, where d is the distance between the axis of rotation Z and the z axis at the center of mass. We then have

$$\begin{aligned} \text{KE} &= \tfrac{1}{2}(Md^2)\omega^2 + \tfrac{1}{2}I_{zz}\omega^2 \\ &= \tfrac{1}{2}(I_{zz} + Md^2)\omega^2 \end{aligned}$$

But the bracketed expression is the moment of inertia of the body about the axis of rotation Z. Denoting this moment of inertia simply as I, we get for the kinetic energy:

$$\text{KE} = \tfrac{1}{2}I\omega^2 \tag{15.7}$$

This simple expression for pure rotation is completely analogous to the kinetic energy of a body in pure translation.

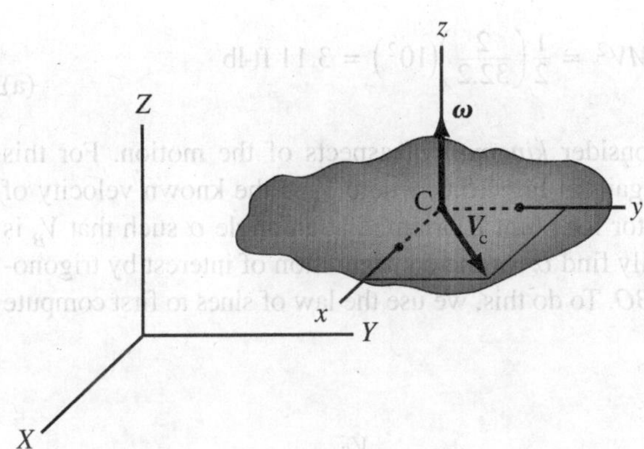

5igure 15.4. Plane motion relative to the *XY* plane.

For a body undergoing *general plane motion* (see Fig. 15.4) parallel to the *XY* plane, where *xyz* are taken at the center of mass and oriented parallel to *XYZ*, we get from Eq. 15.5:

$$\text{KE} = \tfrac{1}{2}MV_c^2 + \tfrac{1}{2}I_{zz}\omega_z^2 \tag{15.8}$$

We now illustrate the calculation of the kinetic energy in the following example.

Example 15.1

Compute the kinetic energy of the crank system in the configuration shown in Fig. 15.5. Piston *A* weighs 2 lb, rod *AB* is 2 ft long and weighs 5 lb, and flywheel *D* weighs 100 lb with a radius of gyration of 1.2 ft. The radius *r* is 1 ft. At the instant of interest, piston *A* is moving to the right at a speed *V* of 10 ft/sec.

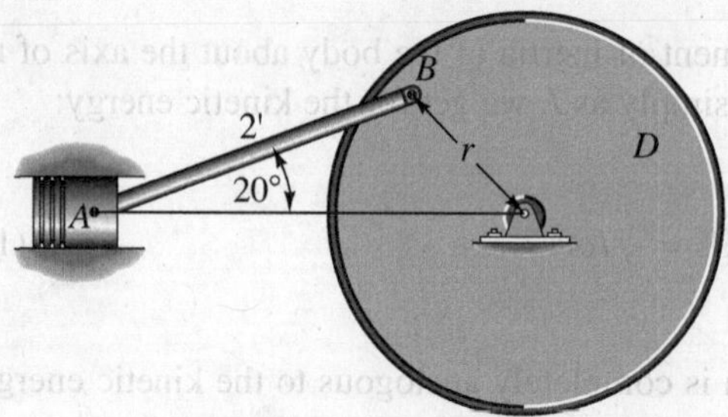

Figure 15.5. Crank system.

We have here a translatory motion (piston *A*), a plane motion (rod *AB*), and a pure rotation (flywheel *D*). Thus, for piston *A* we have for the kinetic energy:

$$(\text{KE})_A = \frac{1}{2}MV^2 = \frac{1}{2}\left(\frac{2}{32.2}\right)(10^2) = 3.11 \text{ ft-lb} \tag{a}$$

For the rod *AB*, we must first consider *kinematical* aspects of the motion. For this purpose we have shown rod *AB* again in Fig. 15.6, where V_A is the known velocity of point *A* and V_B is the velocity vector for point *B* oriented at an angle α such that V_B is perpendicular to *OB*. We can readily find α for the configuration of interest by trigonometric considerations of triangle *ABO*. To do this, we use the law of sines to first compute the angle β:

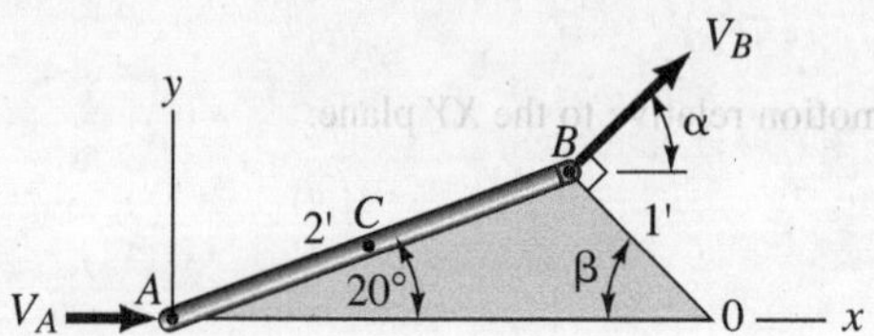

Figure 15.6. Kinematics of rod *AB*.

$$\frac{2}{\sin\beta} = \frac{1}{\sin 20°}$$

Therefore,

$$\beta = 43.2° \tag{b}$$

Example 15.1 (Continued)

Because V_B is at right angles to OB, we have for the angle α:

$$\alpha = 90° - \beta = 46.8° \tag{c}$$

From **kinematics** of a rigid body we can now say:

$$V_B = V_A + (\omega_{AB}\boldsymbol{k}) \times \boldsymbol{\rho}_{AB}$$

Hence,

$$\begin{aligned} V_B(\cos\alpha\boldsymbol{i} + \sin\alpha\boldsymbol{j}) &= 10\boldsymbol{i} + \omega_{AB}\boldsymbol{k} \times (2\cos 20°\boldsymbol{i} + 2\sin 20°\boldsymbol{j}) \\ \therefore\ V_B(.648\boldsymbol{i} + .729\boldsymbol{j}) &= 10\boldsymbol{i} + 1.879\omega_{AB}\boldsymbol{j} - .684\omega_{AB}\boldsymbol{i} \end{aligned} \tag{d}$$

From this we solve for ω_{AB} and V_B. Thus,

$$\begin{aligned} V_B &= 10.53 \text{ ft/sec} \\ \omega_{AB} &= 4.09 \text{ rad/sec} \end{aligned} \tag{e}$$

To get the velocity of the mass center C of AB, we proceed as follows:

$$\begin{aligned} \boldsymbol{V}_C &= \boldsymbol{V}_A + (\omega_{AB}\boldsymbol{k}) \times \boldsymbol{\rho}_{AC} \\ &= 10\boldsymbol{i} + 4.09\boldsymbol{k} \times (.940\boldsymbol{i} + .342\boldsymbol{j}) \\ &= 10\boldsymbol{i} + 3.84\boldsymbol{j} - 1.399\boldsymbol{i} = 8.60\boldsymbol{i} + 3.84\boldsymbol{j} \text{ ft/sec} \end{aligned} \tag{f}$$

We can now calculate $(\text{KE})_{AB}$, the kinetic energy of the rod:

$$\begin{aligned} (\text{KE})_{AB} &= \frac{1}{2}M_{AB}V_c^2 + \frac{1}{2}I_{zz}\omega_{AB}^2 \\ &= \frac{1}{2}\left(\frac{5}{32.2}\right)(8.60^2 + 3.84^2) + \frac{1}{2}\left(\frac{1}{12}\frac{5}{32.2}2^2\right)(4.09^2) \\ &= 7.32 \text{ ft-lb} \end{aligned} \tag{g}$$

Finally, we consider the flywheel D. The angular speed ω_D can easily be computed using V_B of Eq. (e). Thus,

$$\omega_D = \frac{V_B}{r} = \frac{10.53}{1} = 10.53 \text{ rad/sec} \tag{h}$$

Accordingly, we get for $(\text{KE})_D$:

$$(\text{KE})_D = \left[\frac{1}{2}\left(\frac{100}{32.2}\right)(1.2^2)\right](10.53^2) = 248 \text{ ft-lb} \tag{i}$$

The total kinetic energy of the system can now be given as

$$\begin{aligned} \text{KE} &= (\text{KE})_A + (\text{KE})_{AB} + (\text{KE})_D \\ &= 3.11 + 7.32 + 248 = \boxed{258 \text{ ft-lb}} \end{aligned} \tag{j}$$

15.3 Work–Energy Relations

We presented in Chapter 11 the work–energy relation for a *single* particle m_i in a *system* of n particles (see Fig. 15.7) to reach the following equation:

$$\int_1^2 \boldsymbol{F}_1 \bullet d\boldsymbol{r}_i + \int_1^2 \sum_{\substack{j=1 \\ j \neq i}}^{n} \boldsymbol{f}_{ij} \bullet d\boldsymbol{r}_i = \tfrac{1}{2}(m_i V_i^2)_2 - \tfrac{1}{2}(m_i V_i^2)_1 = (\Delta \text{KE})_i \tag{15.9}$$

where $\boldsymbol{f}_{ij}$ is the force from particle j onto particle i and is an internal force. (Note that since a particle cannot exert a force on itself, $\boldsymbol{f}_{ii} = \boldsymbol{0}$.) Now consider that the particle m_i is part of a rigid body, as shown in Fig. 15.8. From **Newton's third law** we can say that

$$\boldsymbol{f}_{ij} = -\boldsymbol{f}_{ji} \tag{15.10}$$

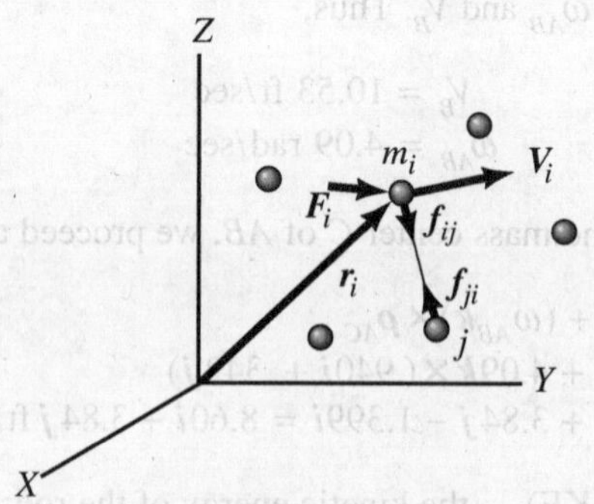

Figure 15.7. System of particles.

It might be intuitively obvious to the reader that for any motion of a rigid body the totality of internal forces $\boldsymbol{f}_{ij}$ can do no work. If not, read the following proof to verify this claim.

Suppose the rigid body moves an infinitesimal amount. We employ *Chasles' theorem*, whereby we give the entire body a displacement $d\boldsymbol{r}$ corresponding to the actual displacement of particle m_i (see Fig. 15.8). The total work done by $\boldsymbol{f}_{ij}$ and $\boldsymbol{f}_{ji}$ is clearly zero for this displacement as a result of Eq. 15.10. In addition, we will have a rotation $d\boldsymbol{\phi}$ about an axis of rotation going through m_i.

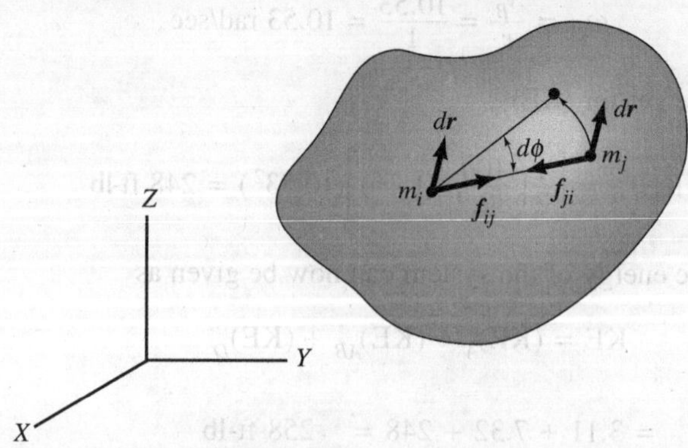

Figure 15.8. Particles of a rigid body.

We can decompose $d\boldsymbol{\phi}$ into orthogonal components, such that one component $d\phi_1$ is along the line between m_i and m_j (and thus collinear with $\boldsymbol{f}_{ij}$) and two components are at right angles to this line (see Fig. 15.9). Clearly, the work done by the forces $\boldsymbol{f}_{ij}$ and $\boldsymbol{f}_{ji}$ for $d\boldsymbol{\phi}_1$ is zero. Also, the movement of m_j for the other components of $d\boldsymbol{\phi}$ is at right angles to $\boldsymbol{f}_{ji}$, and again there is no work done. Consequently, the work done by $\boldsymbol{f}_{ij}$ and $\boldsymbol{f}_{ji}$ is zero during the total infinitesimal movement. And since a finite movement is a sum of such infinitesimal movements, the work done for a finite movement of m_i and m_j is zero. But a rigid body consists of *pairs* of interacting particles such as m_i and m_j. Hence, on summing Eq. 15.9 for all particles of a rigid body, we can conclude that the *work done by forces internal to a rigid body for any rigid-body movement is always zero.*

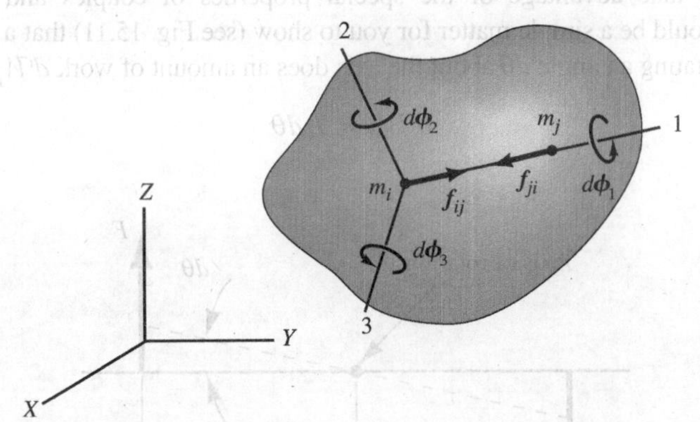

Figure 15.9. Rectangular components of $d\boldsymbol{\phi}$.

We must clearly point out here that although the internal forces *in* a rigid body can do no work, forces *between rigid bodies* of a *system* of rigid bodies *can* do a net amount of work even though Newton's third law applies and even though these forces are *internal to the system.* We shall say more about this later when we discuss systems of rigid bodies.

We accordingly compute the work done on a rigid body in moving from configuration I to configuration II by summing the work terms for all the *external* forces. Thus, for the body shown in Fig. 15.10, we can express the work between I and II in the following manner:

$$(\text{work})_{\text{I,II}} = \underset{\text{path 1}}{\int_{\text{I}}^{\text{II}}} \boldsymbol{F}_1 \bullet d\boldsymbol{s}_1 + \underset{\text{path 2}}{\int_{\text{I}}^{\text{II}}} \boldsymbol{F}_2 \bullet d\boldsymbol{s}_2 + \cdots + \underset{\text{path } n}{\int_{\text{I}}^{\text{II}}} \boldsymbol{F}_n \bullet d\boldsymbol{s}_n \tag{15.11}$$

In this equation, we must remember, the dot products of *nonconservative* forces are to be integrated over the *actual paths* along which the *points of application of the forces on the rigid body move*. We must take into account the variations of direction and magnitude of these nonconservative forces along their paths. For conservative forces we can use the concept of potential energy.

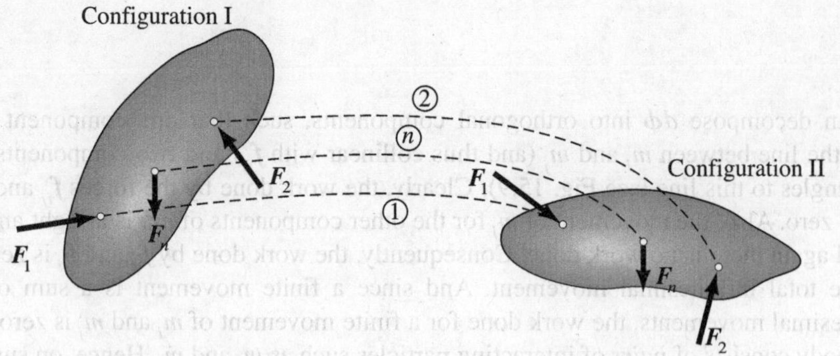

Figure 15.10. Rigid body moves from configuration I to configuration II.

Although we can treat *couples* as sets of discrete forces in the foregoing manner, it is often useful to take advantage of the special properties of couples and to treat them separately. It should be a simple matter for you to show (see Fig. 15.11) that a torque T about an axis upon rotating an angle $d\theta$ about the axis does an amount of work $d\mathcal{W}_K$ given as

$$d\mathcal{W}_K = T\, d\theta$$

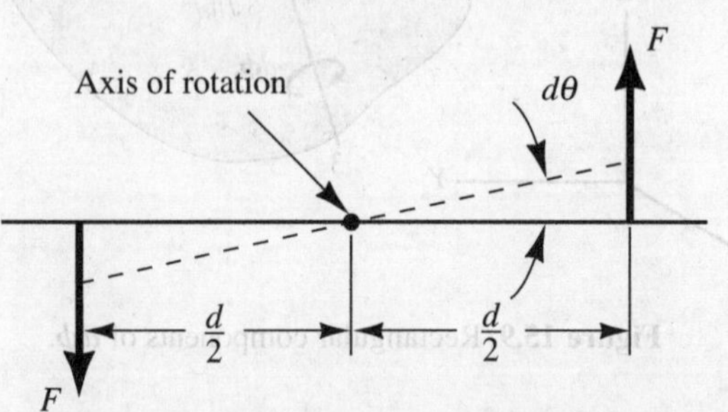

Figure 15.11. Work of torque T about an axis rotating an angle $d\theta$ about the axis.

Dividing and multiplying by dt, we can say further that

$$d\mathcal{W}_K = T\frac{d\theta}{dt}\, dt = T\dot{\theta}\, dt$$

Integrating, we get

$$\mathcal{W}_K = \int_{t_1}^{t_2} T\dot{\theta}\, dt \tag{15.12}$$

In this case the torque T and angular speed $\dot{\theta}$ are about the same axis. The generalization of Eq. 15.12 for any moment $\boldsymbol{M}$ and any angular velocity $\boldsymbol{\omega}$ (see Fig. 15.12) then is

$$\mathcal{W}_K = \int_{t_1}^{t_2} \boldsymbol{M} \bullet \boldsymbol{\omega}\, dt \tag{15.13}$$

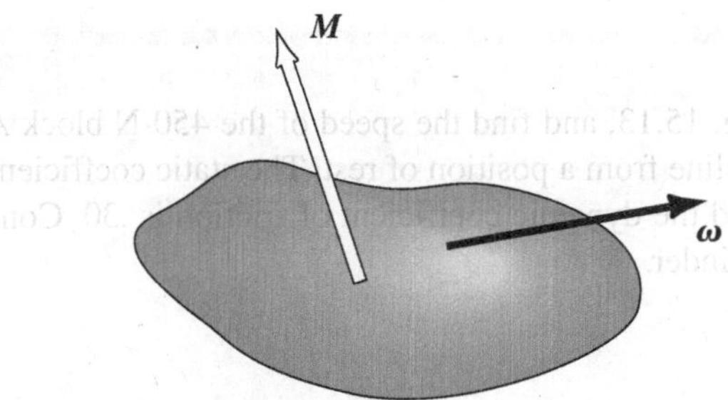

Figure 15.12. $d\mathcal{W}_K = \boldsymbol{M} \cdot \boldsymbol{\omega}\, dt$.

We thus have formulations for finding the work done by external forces and couples on a rigid body. For the conservative forces, we know from Chapter 11 that we can use for work a quantity that is minus the change in potential energy from I to II without having to specify the path taken.

Using this information for computing work, we can then say for any rigid body:

$$\mathcal{W}_K \text{ from I to II} = \Delta\text{KE} \tag{15.14}$$

where $\mathcal{W}_K$ is the work by *external* forces.

If there are only *conservative* external forces present, we can also say for the rigid body:

$$(\text{PE})_\text{I} + (\text{KE})_\text{I} = (\text{PE})_\text{II} + (\text{KE})_\text{II} \tag{15.15}$$

If both conservative *and* nonconservative external forces are present, we can say:

$$\text{nonconservative } \mathcal{W}_K \text{ from I to II} = \Delta\text{PE} + \Delta\text{KE} \tag{15.16}$$

These three equations parallel the three we developed for a particle in Chapter 11.

The foregoing equations are expressed for a *single* rigid body. For a *system* of *interconnected rigid* bodies, we distinguish between two types of forces internal to the system. They are

1. Forces internal to any rigid body of the system.
2. Forces *between* rigid bodies of the system.

For a system of bodies, as in the case of a single rigid body, forces of category 1 can do no work. However, if the forces *between two bodies* of 2 system do not move the same distance over the same path, then there may be a net amount of work done on the system by these internal forces. We must include such work contributions when employing Eqs. 15.14–15.16 *for a system of interconnected rigid bodes.*[3] Example 15.4 is an example of this situation.

[3]Recall from Chapter 13, that Eq. 13.16 and hence Eqs. 13.14 and 13.15 are valid for *any* aggregate of particles provided we include the work of internal forces both conservative and nonconservative.

Example 15.2

Neglect the weight of the cable in Fig. 15.13, and find the speed of the 450-N block *A* after it has moved 1.7 m along the incline from a position of rest. The static coefficient of friction along the incline is .32, and the dynamic coefficient of friction is .30. Consider the pulley *B* to be a uniform cylinder.

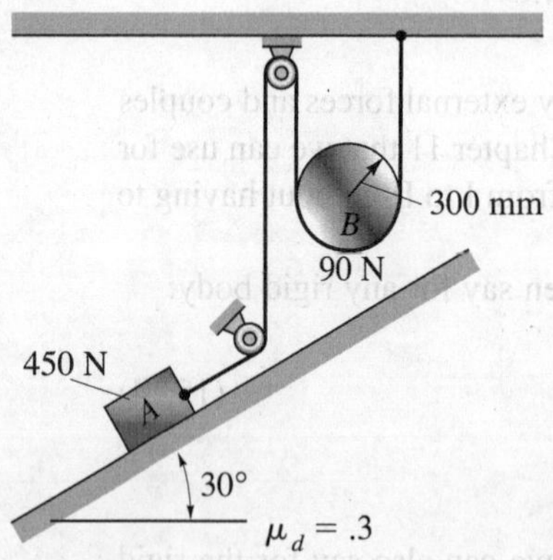

Figure 15.13. Pulley system.

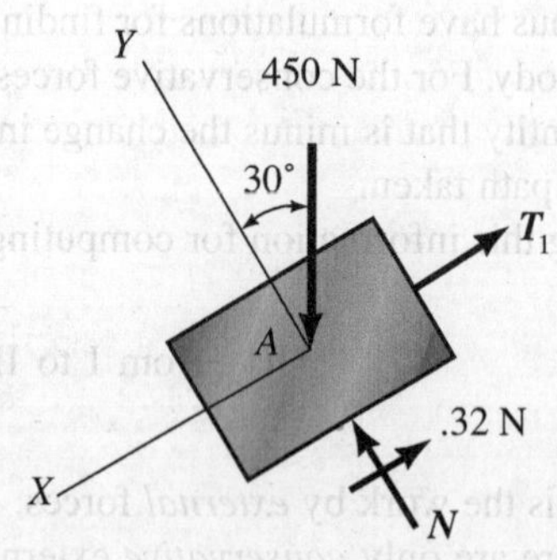

Figure 15.14. Free body of block.

We must first decide which way the block moves along the incline. To overcome friction and move down the incline, the block must create a force in the downward direction of the cable exceeding 90/2 = 45 N. Considering the block *A* alone (see Fig. 15.14), we can readily decide that the maximum force T_1 to allow *A* to start sliding downward is

$$\begin{aligned}(T_1)_{\max} &= -(.32)\text{ N} + 450 \sin 30^\circ \\ &= -(.32)(450)\cos 30^\circ + 450 \sin 30^\circ = 100.3 \text{ N}\end{aligned}$$

Clearly, the block goes down the incline.

We now use the **work–kinetic energy** equation separately for each body. Thus, for the block we have, using now the dynamic coefficient of friction

$$(450 \sin 30^\circ)(1.7) - (450)(\cos 30^\circ)(.30)(1.7) - T_1(1.7) = \frac{1}{2}\frac{450}{g}V_A^2$$

Therefore,

$$T_1 = 108.1 - 13.49V_A^2 \qquad \text{(a)}$$

Example 15.2 (Continued)

We now consider the cylinder for which the free-body diagram is displayed in Fig. 15.15. Note that the cylinder is in effect rolling without slipping along the right supporting cable. Hence, T_2 does no work as explained in Chapter 11.[4] Thus, the **work–kinetic energy** equation is as follows using the formula $\frac{1}{2}Mr^2$ for I of the cylinder:

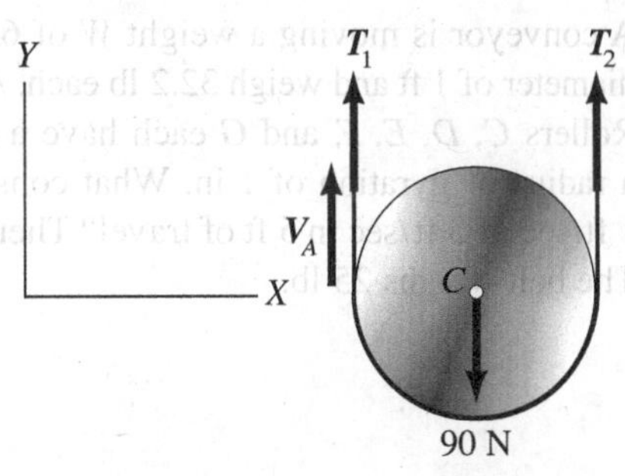

Figure 15.15. Free body of cylinder.

$$T_1(1.7) - 90\left(\frac{1.7}{2}\right) = \frac{1}{2}\frac{90}{g}V_C^2 + \frac{1}{2}\left[\frac{1}{2}\left(\frac{90}{g}\right)(.30)^2\right]\omega^2$$

Therefore,

$$T_1 = 45 + 2.70V_C^2 + .1214\omega^2 \qquad \text{(b)}$$

From **kinematics** we can conclude on inspection that

$$V_C = \frac{1}{2}V_A \qquad \text{(c)}$$

$$\omega = -\frac{V_A}{.60} \qquad \text{(d)}$$

Subtracting Eq. (b) from Eq. (a) to eliminate T_1, and then substituting from Eq. (c) and Eq. (d) for V_C and ω, we then get for V_A:

$$V_A = 2.09 \text{ m/sec} \qquad \text{(e)}$$

F

This problem can readily be solved as a system, i.e, without disconnecting the bodies. You are asked to do this in Problem 15.25.

[4]Recall that the point of contact of the cylinder has zero velocity, and hence the friction force (in this case, T_2) transmits no power to the cylinder.

In the previous example, we considered the problem to be composed of two *discrete* bodies. We proceeded by expressing equations for each body separately. In the following example, we will consider a *system* of bodies expressing equations for the whole system directly. In this problem, the forces between any two bodies of the system have the *same velocity*; consequently, from Newton's third law, we can conclude that these forces contribute zero net work. We shall illustrate a case where this condition is not so in Example 15.5.

Example 15.3

A conveyor is moving a weight W of 64.4 lb in Fig. 15.16. Cylinders A and B have a diameter of 1 ft and weigh 32.2 lb each. Also, they each have a radius of gyration of .4 ft. Rollers C, D, E, F, and G each have a diameter of 3 in., weigh 10 lb each, and have a radius of gyration of 1 in. What constant torque will increase the speed of W from 1 ft/sec of 3 ft/sec in 5 ft of travel? There is no slipping at any of the rollers and drums. The belt weighs 25 lb.

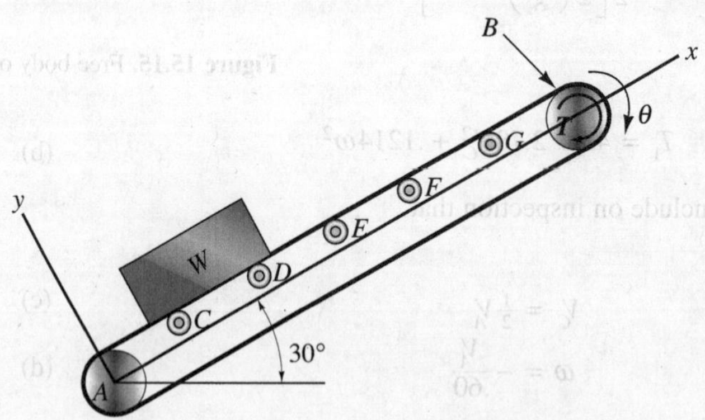

Figure 15.16. Conveyor moving weight W.

We shall use the **work–kinetic energy** relation specified in Eq. 15.14 for this problem. Only external forces and torques do work for the system; the interactive forces between bodies do work in amounts that clearly cancel each other because of the condition of no slipping. Hence,

$$T\theta - W(5)(\sin 30°) = (\text{KE})_2 - (\text{KE})_1 \tag{a}$$

where T is the applied torque. The general expression for the kinetic energy is

$$\text{KE} = 5\left(\frac{1}{2}I_{\text{roll}}\omega_{\text{roll}}^2\right) + 2\left(\frac{1}{2}I_{\text{cyl}}\omega_{\text{cyl}}^2\right) + \frac{1}{2}M_{\text{belt}}V_{\text{belt}}^2 + \frac{1}{2}\frac{W}{g}V_{\text{weight}}^2 \tag{b}$$

From **kinematics** we can say:

$$\omega_{\text{roll}} = -\frac{V_{\text{belt}}}{\left(\frac{1}{2}\right)\left(\frac{3}{12}\right)} = -8V_{\text{belt}}$$

$$\omega_{\text{cyl}} = -\frac{V_{\text{belt}}}{\left(\frac{1}{2}\right)(1)} = -2V_{\text{belt}}$$

$$V_{\text{weight}} = V_{\text{belt}}$$

Example 15.3 (Continued)

Using these results, we can give the kinetic energy at the end and at the beginning of the interval of interest as

$$\begin{aligned}(KE)_2 &= 5\left\{\left(\frac{1}{2}\right)\left(\frac{10}{g}\right)\left(\frac{1}{12}\right)^2[(8)(3)]^2\right\} + 2\left\{\left(\frac{1}{2}\right)\left(\frac{32.2}{g}\right)(.4)^2[(2)(3)]^2\right\}\\ &\quad + \frac{1}{2}\left(\frac{25}{g}\right)(3)^2 + \frac{1}{2}\left(\frac{64.4}{g}\right)(3^2)\\ &= 21.4 \text{ ft-lb}\end{aligned}$$

$$\begin{aligned}(KE)_2 &= 5\left\{\left(\frac{1}{2}\right)\left(\frac{10}{g}\right)\left(\frac{1}{12}\right)^2[(8)(1)]^2\right\} + 2\left\{\left(\frac{1}{2}\right)\left(\frac{32.2}{g}\right)(.4)^2[(2)(1)]^2\right\}\\ &\quad + \frac{1}{2}\left(\frac{25}{g}\right)(1^2) + \frac{1}{2}\left(\frac{64.4}{g}\right)(1^2)\\ &= 2.37 \text{ ft-lb}\end{aligned}$$

Substituting these results into Eq. (a), we get

$$T\theta - (64.4)(5)(\sin 30°) = 21.4 - 2.37 \qquad \text{(c)}$$

From **kinematics** again we can say for θ, on considering the rotation of a cylinder and the distance traveled by the belt:

$$(r_{\text{cyl}})(\theta) = 5$$

Therefore,

$$\theta = \frac{5}{\frac{1}{2}} = 10 \text{ rad}$$

Substituting back into Eq. (c), we can then solve for the desired torque T:

$$T = \tfrac{1}{10}[21.4 - 2.37 + (64.4)(5)(\sin 30°)]$$

$$T = 18.00 \text{ ft-lb}$$

In the next example, we have a case of internal forces between bodies that satisfy Newton's third law but do not have identical velocities.

Example 15.4[5]

A diesel-powered electric train moves up a 7° grade in Fig. 15.15. If a torque of 750 N-m is developed at each of its six pairs of drive wheels, what is the increase of speed of the train after it moves 100 m? Initially, the train has a speed of 16 km/hr. The train weighs 90 kN. The drive wheels have a diameter of 600 mm. Neglect the rotational energy of the drive wheels.

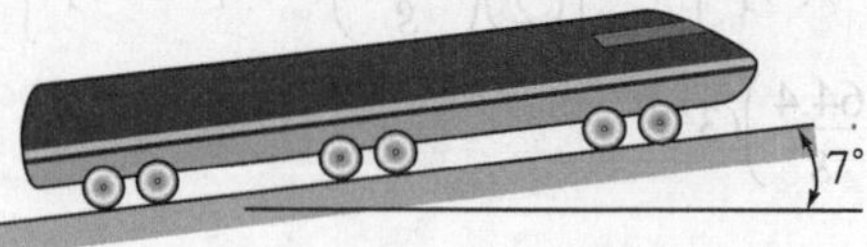

Figure 15.15. Diesel–electric train.

We shall consider the train as a *system of rigid bodies* including the 6 pairs of wheels and the body. We have shown the train in Fig. 15.18 with the external forces, W, N, and f. In addition, we have shown certain internal torques M.[6] The torques shown act on the *rotors* of the motors, and, as the train moves, these torques rotate and accordingly do work. The *reactions* to these torques are equal and opposite to M according to **Newton's third law** and act on the *stators* or the motors (i.e., the field coils). The stators are stationary, and so the reactions to M do *no* work as the train moves. Thus, we have an example of equal and opposite internal forces between bodies of a system performing a nonzero net amount of work. We now employ Eq. 15.16. Thus,

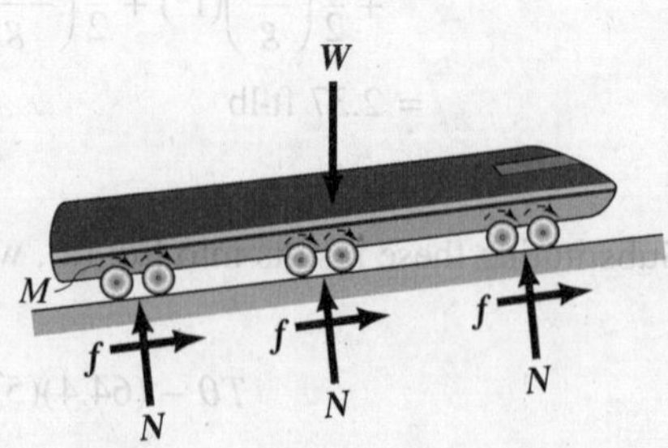

Figure 15.18. External and internal forces and torques.

$$\Delta PE + \Delta KE = \mathcal{W}_K \qquad \text{(a)}$$

Using the initial configuration as the datum, we have[7]

$$\begin{aligned}&[(90{,}000)(100 \sin 7° - 0)]\\&\quad + \left\{\frac{1}{2}\frac{90{,}000}{g}V^2 - \frac{1}{2}\frac{90{,}000}{g}\left[\frac{(16)(1{,}000)}{3{,}600}\right]^2\right\}\\&\quad = (6)(750)(\theta)\end{aligned} \qquad \text{(b)}$$

[5]This problem was undertaken in Chapter 11 as a system of particles. Here, we consider it from the viewpoint of a system of interconnected rigid bodies.

[6]Figure 15.18, accordingly, is *not* a free-body diagram.

[7]Recall that for a *rolling* body with *no slipping*, the friction force does *no work*. If we were considering the *center of mass* of the train, then *f would move* with the center of mass and then do work as we shall see in Example 15.5.

Example 15.4 (Continued)

where θ is the clockwise rotation of the rotor in radians. Assuming direct drive from rotor to wheel, we can compute θ as follows for the 100-m distance over which the train moves:

$$\theta = \underbrace{\left(\frac{100}{2\pi r}\right)}_{\text{rev}} \underbrace{(2\pi)}_{\text{rad/rev}} = \frac{100}{.30} \text{ rad} \qquad \text{(c)}$$

Substituting into Eq. (b) and solving for V, we get

$$V = 10.38 \text{ m/sec}$$

Hence,

$$\Delta V = \frac{(10.38)(3{,}600)}{1000} - 16 = \boxed{21.4 \text{ km/hr}}$$

In Chapter 11, we also developed a work–energy equation involving the *mass center* of any system of particles. You will recall that

$$\int_1^2 \boldsymbol{F} \bullet d\boldsymbol{r}_c = \tfrac{1}{2}(MV_c^2)_2 - \tfrac{1}{2}(MV_c^2)_1 \qquad (15.17\text{a})$$

where $\boldsymbol{F}$ is the *total external force* (only!) which hypothetically *moves with the center of mass*. This equation applies to a rigid body. Note that an external torque makes no work contribution here since equal and opposite forces each having identical motion (that of the mass center) can do no net amount of work. For a system of *interconnected rigid bodies*, we can say:

$$\int_1^2 \boldsymbol{F} \bullet d\boldsymbol{r}_c = \left[\sum_i \tfrac{1}{2} M_i (V_c)_i^2\right]_2 - \left[\sum_i \tfrac{1}{2} M_i (V_c)_i^2\right]_1 \qquad (15.17\text{b})$$

The force $\boldsymbol{F}$ includes *only external forces* (internal forces between interconnecting bodies are equal and opposite and must move with the mass center of the system; hence they contribute no work to the left side of the foregoing equation). On the other hand, external friction forces on wheels rolling without slipping must move with the mass center of the system in this formulation and thus can *do work*, in contrast to the previous approach, in which the mass center is not used. On the right side, we have summed the kinetic energies of each of the mass centers of the constituent bodies of the system.[8] We now illustrate the use of Eq. 15.17b.

[8]We discussed this topic in Section 16.8.

Example 15.5

A vehicle for traversing swamplands is shown in Fig. 15.19. The vehicle has four-wheel drive and weighs 22.5 kN. Each wheel weighs 2 kN has a diameter of 2.5 m and has a radius of gyration of 180 mm. If each wheel gets a torque of 100 N-m, what is the speed after 20 m of travel starting from rest? Also, determine the friction force from the ground on each wheel. The weight of the vehicle includes that of passengers and baggage. Neglect rolling resistance since the vehicle in this problem is moving on a hard surface. Consider rolling without slipping.

Figure 15.19. A vehicle used in swampland.

We have shown the free-body diagram of the system in Fig. 15.20. We will first use the **system of particles** approach. This includes internal torques from the vehicle frame onto the wheels and the reaction torques from the wheels onto the frame. The former will do internal work because these torques rotate with the wheels. The reaction torques on the frame do not rotate and obviously do no work. Also, because of the no slipping condition, the friction forces from the ground onto the tires do no work as has been explained at length in Chapter 11. Hence, we can say

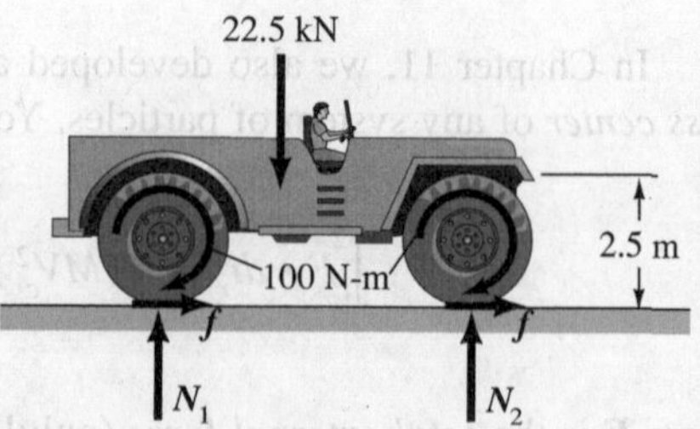

Figure 15.20. External forces and internal torques on swampland vehicle. Query: Is this a freebody diagram?

$$(4)(T)(\theta) = (\text{KE})_2 - (\text{KE})_1$$

Recalling that θ = distance/radius, we get

$$\therefore (4)(100)\left(\frac{20}{1.25}\right) = \left[\left(\frac{4}{2}\right)\left(\frac{2{,}000}{9.81}\right)(.1800)^2\left(\frac{V}{1.25}\right)^2\right] + \frac{1}{2}\left(\frac{22{,}500}{9.81}\right)(V^2)$$

$$\therefore \quad V = 2.354 \text{ m/s}$$

To get the friction forces f (why are they the same for each wheel?) we use the **center of mass approach**. Thus

$$W_k = (\Delta\text{KE})_{CM}$$

$$\therefore (4)(f)(20) = \frac{1}{2}\left[\frac{22{,}500}{9.81}\right](2.354)^2$$

$$\therefore \quad f = 79.4 \text{ N}$$

We thus have the desired information.

PROBLEMS

In the following problems, neglect friction unless otherwise instructed.

15.1. A uniform solid cylinder of radius 2 ft and weight 200 lb rolls without slipping down a 45° incline and drags the 100-lb block *B* with it. What is the kinetic energy of the system if block *B* is moving at a speed of 10 ft/sec? Neglect the mass of connecting agents between the bodies.

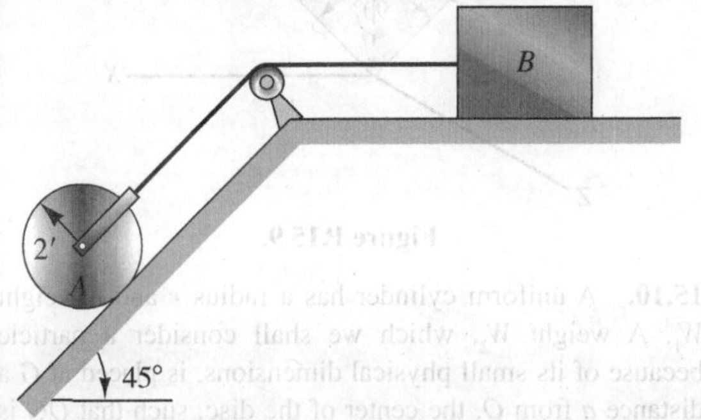

Figure P.15.1.

15.2. A steam roller with driver weighs 5 tons. Wheel A weighs 1 ton and has a radius of gyration of .8 ft. Drive wheels B have a total mass of 1,000 lb and a radius of gyration of 1.6 ft. If the steam roller is coasting at a speed of 5 ft/sec with motor disconnected, what is the total kinetic energy of the system?

Figure P.15.2.

15.3. A thin disc weighing 450 N is suspended from an overhead conveyor moving at a speed of 10 m/sec. If the disc rotates at a speed of 5 rad/sec in the plane of the page (i.e., *ZY* plane), compute the kinetic energy of the disc relative to the ground.

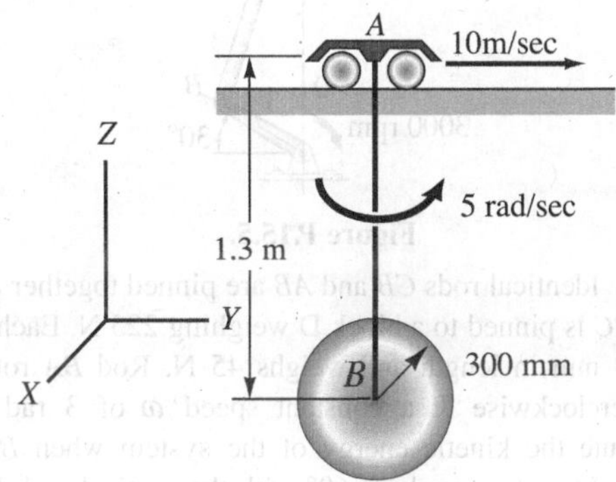

Figure P.15.3.

15.4. Two slender rods *CD* and *EA* are pinned together at *B*. Rod *EA* is rotating at a speed *ω* equal to 2 rad/sec. Rod *CD* rides in a vertical slot at *D*. For the configuration shown in the diagram, compute the kinetic energy of the rods. Rod *CD* weighs 50 N and rod *EA* weighs 80 N.

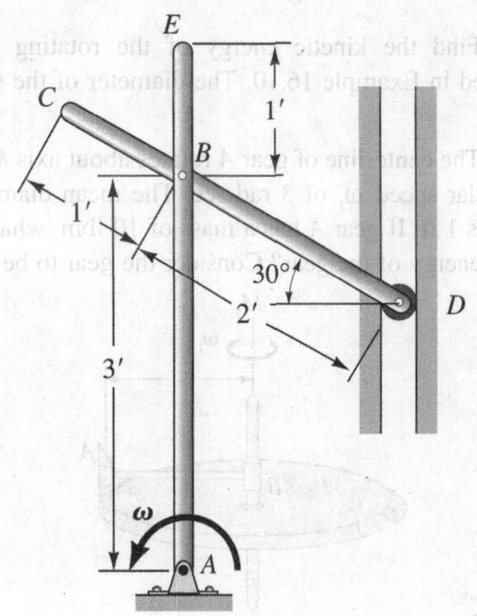

Figure P.15.4.

15.5. Consider the connecting rod AB to be a slender rod weighing 2 lb, and compute its kinetic energy for the data given.

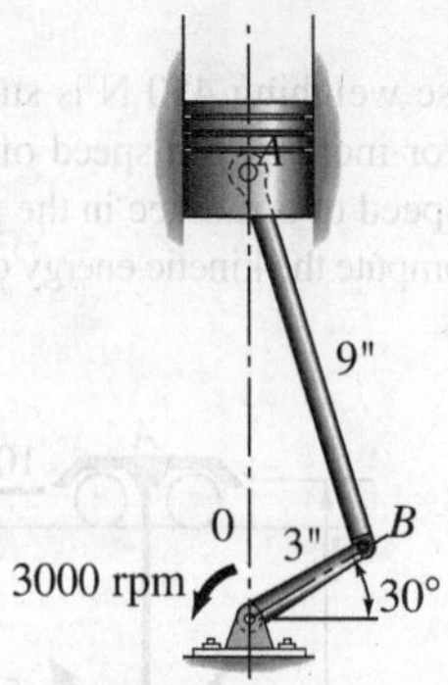

Figure P.15.5.

15.6. Identical rods CB and AB are pinned together at B. Rod BC is pinned to a block D weighing 225 N. Each rod is 600 mm in length and weighs 45 N. Rod BA rotates counterclockwise at a constant speed ω of 3 rad/sec. Compute the kinetic energy of the system when BA is oriented (a) at an angle of 60° with the vertical and (b) at an angle of 90° with the vertical (the latter position is shown dashed in the diagram).

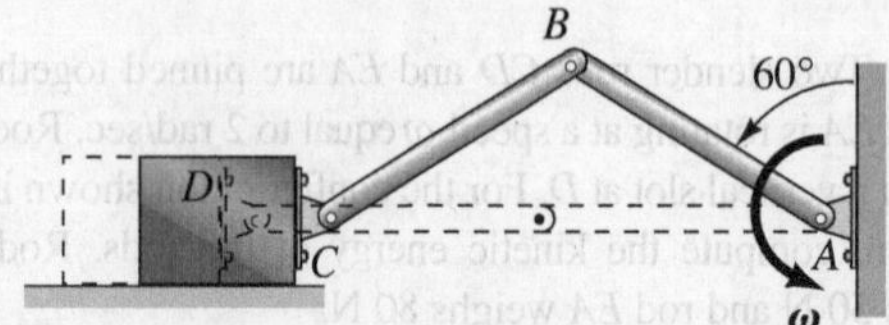

Figure P.15.6.

15.7. Find the kinetic energy of the rotating system described in Example 16.10. The diameter of the shaft is 50 mm.

15.8. The centerline of gear A rotates about axis M–M at an angular speed ω_1 of 3 rad/sec. The mean diameter of gear A is 1 ft. If gear A has a mass of 10 lbm, what is the kinetic energy of the gear? Consider the gear to be a disc.

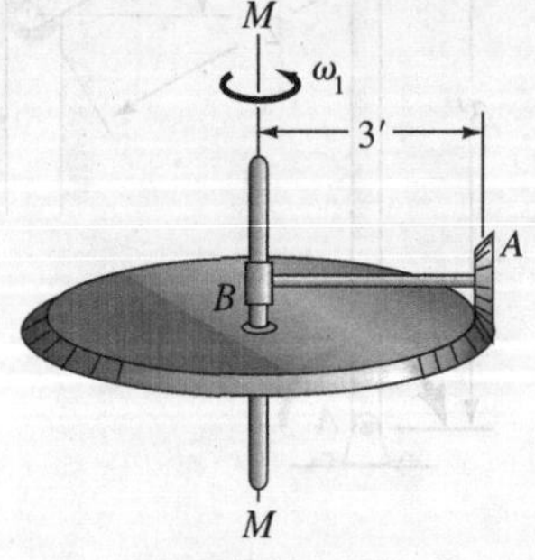

Figure P.15.8.

15.9. A cone B weighing 20 lb rolls without slipping inside a conical cavity C. The cone has a length of 10 ft. The centerline of the cone rotates with an angular speed ω_1 of 5 rad/sec about the Y axis. Compute the kinetic energy of the cone.

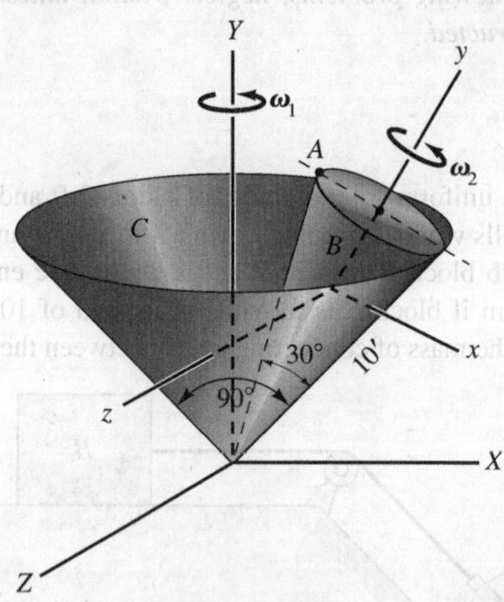

Figure P.15.9.

15.10. A uniform cylinder has a radius r and a weight W_1. A weight W_2, which we shall consider a particle because of its small physical dimensions, is placed at G a distance a from O, the center of the disc, such that OG is vertical. What is the angular velocity of the cylinder when, after it is released from rest, the point G reaches its lowest elevation, as shown at the right? The cylinder rolls without slipping.

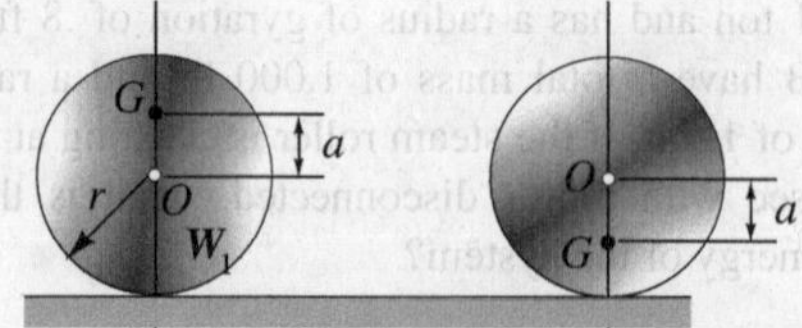

Figure P.15.10.

15.11. A homogeneous solid cylinder of radius 300 mm is shown with a fine wire held fixed at A and wrapped around the cylinder. If the cylinder is released from rest, what will its velocity be when it has dropped 3 m?

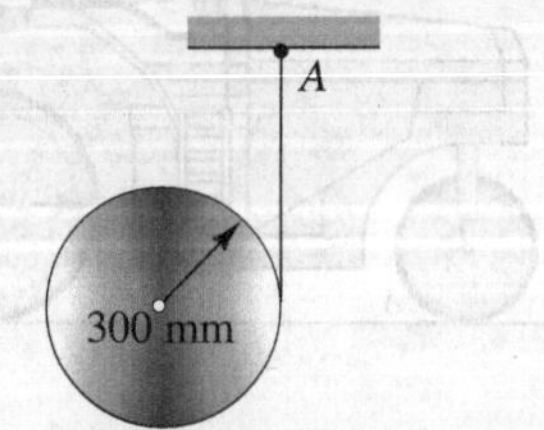

Figure P.15.11.

15.12. Three identical bars, each of length l and weight W, are connected to each other and to a wall with smooth pins at A, B, C, and D. A spring having spring constant K is connected to the center of bar BC at E and to a pin at F, which is free to slide in the slot. Compute the angular speed $\dot{\theta}$ as a function of time if the system is released from rest when AB and DC are at right angles to the wall. The spring is unstretched at the outset of the motion. Neglect friction.

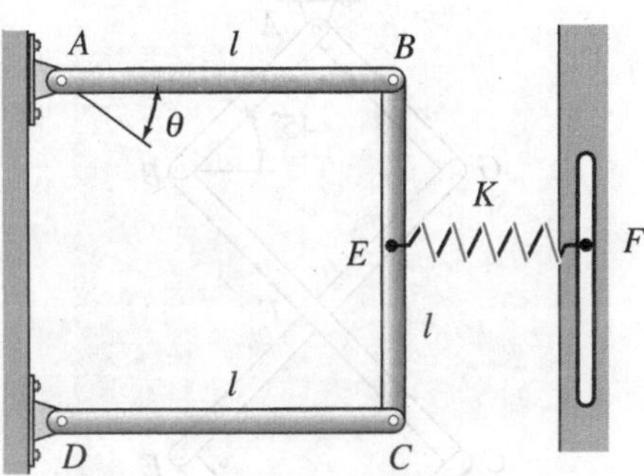

Figure P.15.12.

15.13. A 3-m rod AB weighing 225 N is guided at A by a slot and at B by a smooth horizontal surface. Neglect the mass of the slider at A, and find the speed of B when A has moved 1 m along the slot after starting from a rest configuration shown in the diagram.

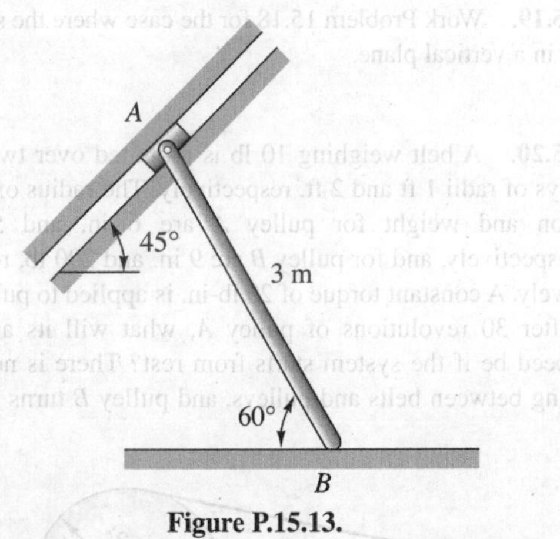

Figure P.15.13.

15.14. A stepped cylinder has radii of 600 mm for the smaller radius and 1.3 m for the larger radius. A rectangular block A weighing 225 N is welded to the cylinder at B. The spring constant K is .18 N/mm. If the system is released from a configuration of rest, what is the angular speed of the cylinder after it has rotated 90°? The radius of gyration for the stepped cylinder is 1 m and its mass is 36 kg. The spring is unstretched in the position shown.

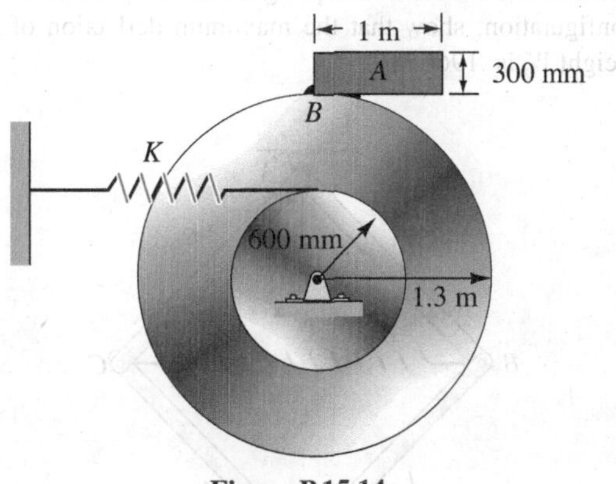

Figure P.15.14.

15.15. A cylinder of diameter 2 ft is composed of two semicylinders C and D weighing 50 lb and 80 lb, respectively. Bodies A and B, weighing 20 lb and 50 lb, respectively, are connected by a light, flexible cable that runs over the cylinder. If the system is released from rest for the configuration shown, what is the speed of B when the cylinder has rotated 90°? Assume no slipping.

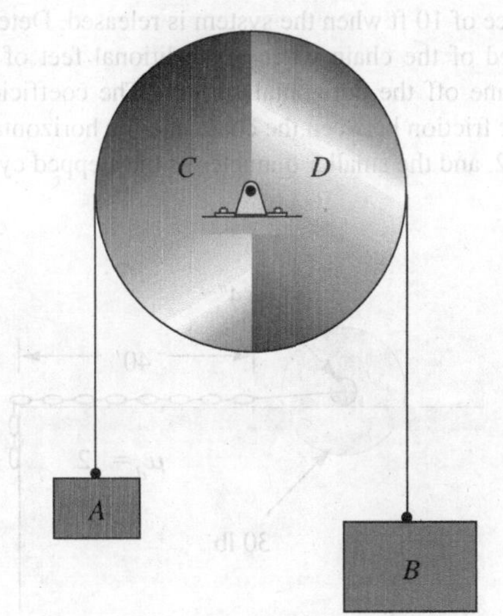

Figure P.15.15.

15.16. Four identical rods, each of length $l = 1.3$ m and weight 90 N, are connected at the frictionless pins A, B, C, and D. A compression spring of spring constant $K = 5.3$ N/mm connects pins B and C, and a weight W_2 of 450 N is supported at pin D. The system is released from a configuration where $\theta = 45°$. If the spring is not compressed at that configuration, show that the maximum deflection of the weight W_2 is .1966 m.

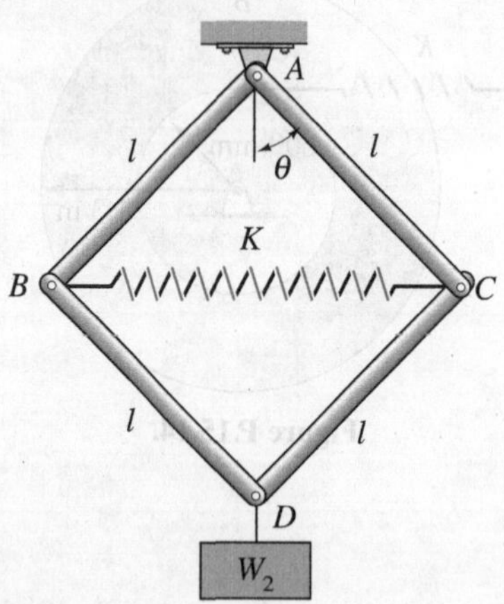

Figure P.15.16.

15.15. A stepped cylinder weighing 30 lb with a radius of gyration of 1 ft is connected to a 50-ft chain weighing 100 lb. The chain hangs down from the horizontal surface a distance of 10 ft when the system is released. Determine the speed of the chain when 30 additional feet of chain have come off the horizontal surface. The coefficient of dynamic friction between the chain and the horizontal surface is .2, and the smaller diameter of the stepped cylinder is 4 in.

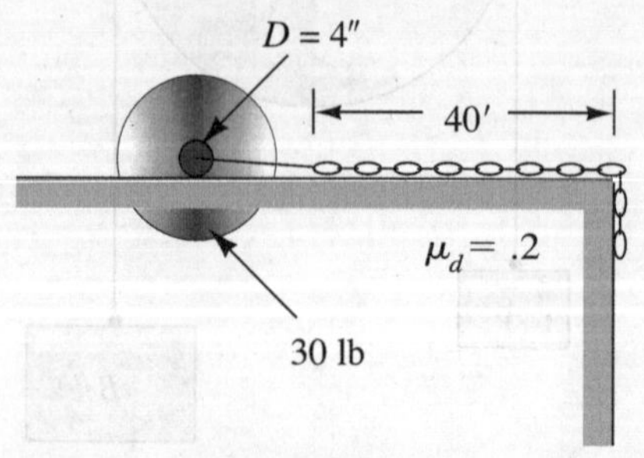

Figure P.15.15.

15.18. The linkage system rests on a frictionless plane. The lengths AB, BF, etc., are each 300 mm, and the bars, all of the same stock, weigh 67.5 N/m. A force F of 450 N is applied at D. What is the speed of D after it moves 300 mm? The system is stationary in the configuration shown. The view of the linkage system is from above.

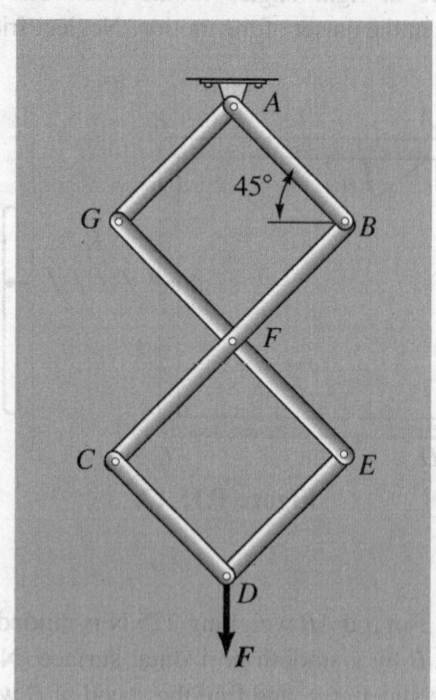

Figure P.15.18.

15.19. Work Problem 15.18 for the case where the system is in a vertical plane.

15.20. A belt weighing 10 lb is mounted over two pulleys of radii 1 ft and 2 ft, respectively. The radius of gyration and weight for pulley A are 6 in. and 50 lb, respectively, and for pulley B are 9 in. and 200 lb, respectively. A constant torque of 20 lb-in. is applied to pulley A. After 30 revolutions of pulley A, what will its angular speed be if the system starts from rest? There is no slipping between belts and pulleys, and pulley B turns freely.

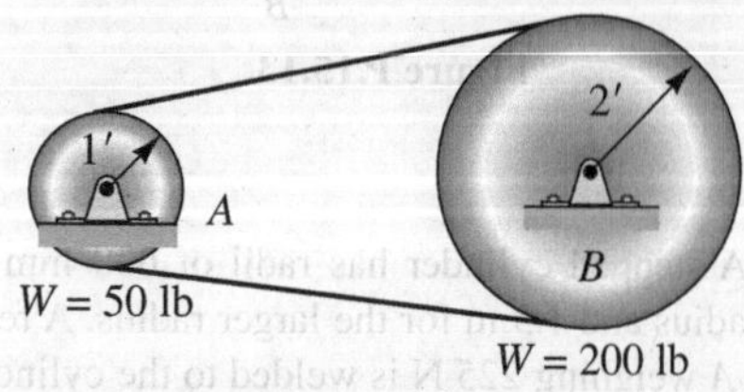

Figure P.15.20.

15.21. Two identical members, AB and BC, are pinned together at B. Also member BC is pinned to the wall at C. Each member weighs 32.2 lb and is 20 ft long. A spring having a spring constant $K = 20$ lb/ft is connected to the centers of the members. A force $P = 100$ lb is applied to member AB at A. If initially the members are inclined 45° to the ground and the spring is unstretched, what is $\dot{\beta}$ after A has moved 2 ft? System is in a vertical plane.

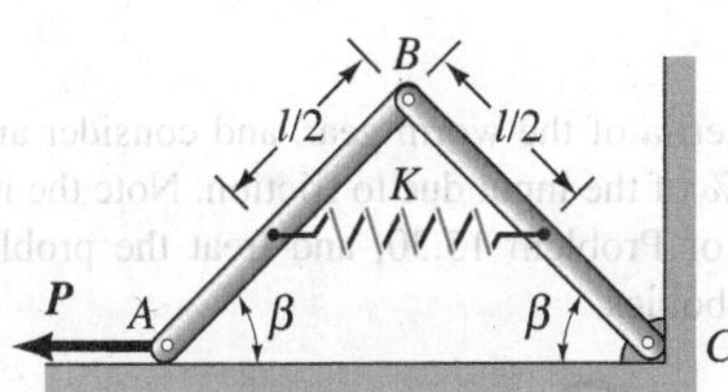

Figure P.15.21.

15.22. A flexible cord of total length 50 ft and weighing 50 lb is pinned to a wall at A and is wrapped around a cylinder having a radius of 4 ft and weighing 30 lb. A 50-lb force is applied to the end of the cord. What is the speed of the cylinder after the end of the cord has moved 10 ft? The system starts from rest in the configuration shown in the diagram. Neglect potential energy considerations arising from the sag of the upper cord.

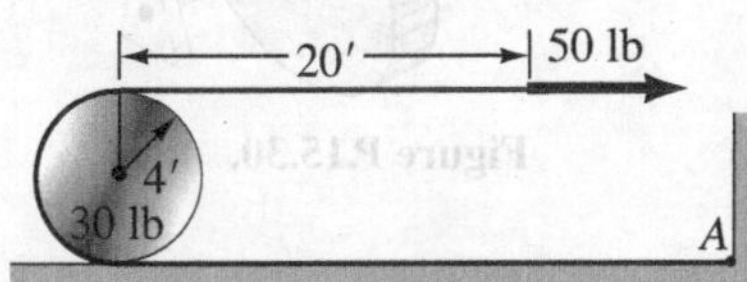

Figure P.15.22.

15.23. In Problem 15.5, suppose that an average pressure of 20 psig (above atmosphere) exists in the cylinder. What is the rpm after the crankshaft has rotated 60° from position shown? The crank rod OB weighs 1 lb and has a radius of gyration about O of 2 in. The diameter of the piston is 4 in., and its weight is 8 oz. The crankshaft is rotating at 3,000 rpm at the position shown. Take the center of mass of the crank rod at the midpoint of OB.

15.24. A right circular cone of weight 32.2 lb, height 4 ft, and cone angle 20° is allowed to roll without slipping on a plane surface inclined at an angle of 30° to the horizontal. The cone is started from rest when the line of contact is parallel to the X axis. What is the angular speed of the centerline of the cone when it has its maximum kinetic energy?

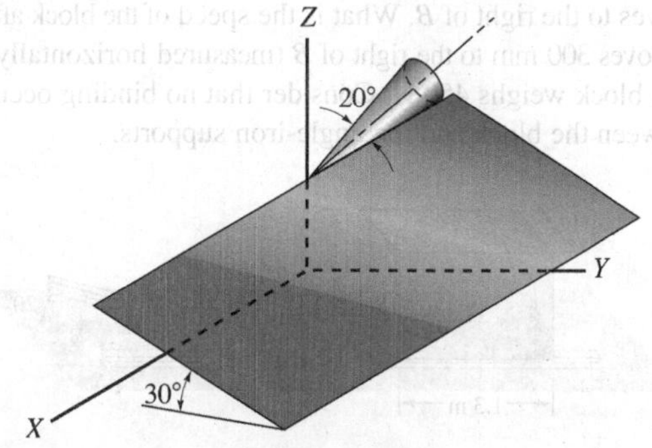

Figure P.15.24.

15.25. Work Example 15.2 by considering the system to be the block, pulley, and cable.

15.26. A weight W_1 is held with alight flexible wire. The wire runs over a stationary semicylinder of radius R equal to 1 ft. A pulley weighing 32.2 lb and having a radius of gyration of unity rides on the wire and supports a weight W_2 of 16.1 lb. If W_1 weighs 128.8 lb and the dynamic coefficient of friction for the semicylinder and wire is .2 what is the drop in the weight W_1 for an increase in speed of 5 ft/sec of weight W_1 starting from rest? The diameter d of the small pulley is 1 ft.

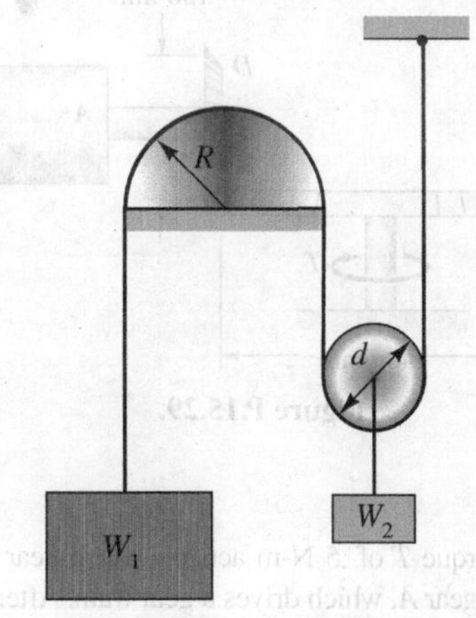

Figure P.15.26.

15.27. A solid uniform block A moves along two frictionless angle-iron supports at a speed of 6 m/sec. One of the supports is inclined at an angle of 20° from the horizontal at B and causes the block to rotate about its front lower edge as it moves to the right of B. What is the speed of the block after it moves 300 mm to the right of B (measured horizontally)? The block weighs 450 N. Consider that no binding occurs between the block and the angle-iron supports.

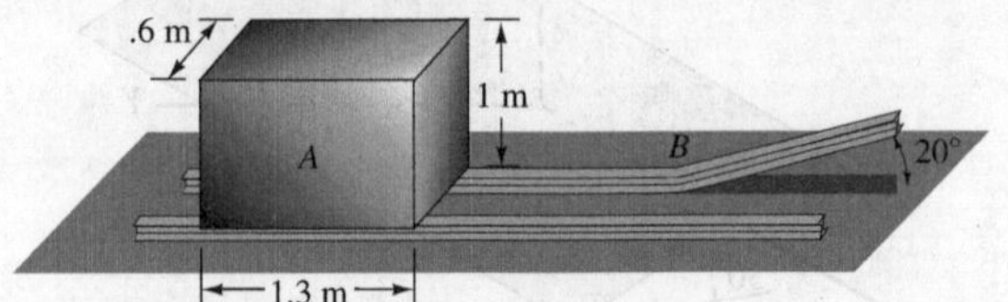

Figure P.15.27.

15.28. In Problem 15.27, will the block reach an instantaneous zero velocity and then slide back or will it tip over onto face A?

15.29. A torque $T = .30$ N-m is applied to a bevel gear B. Bevel gear D meshes with gear B and drives a pump A. Gear B has a radius of gyration of 150 mm and a weight of 50 N, whereas gear D and the impeller of pump A have a combined radius of gyration of 50 mm and a weight of 100 N. Show that the *work of the contact forces between gears B and D is zero.* Next, find how many revolutions of gear B are needed to get the pump up to 200 rpm from rest. Treat the problem as a system of bodies.

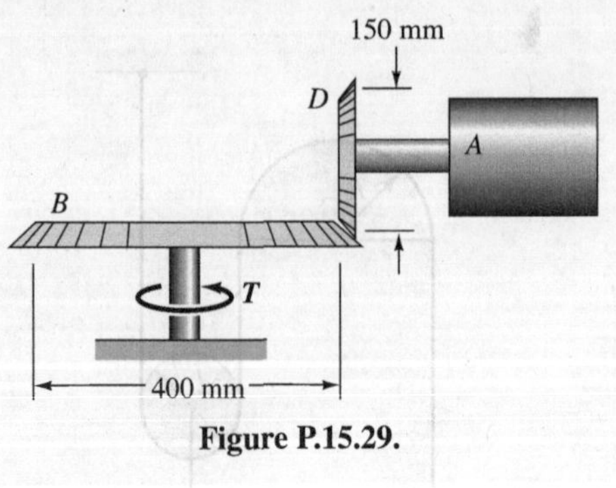

Figure P.15.29.

15.30. A torque T of .5 N-m acts on worm gear E, which meshes with gear A, which drives a gear train. After five revolutions of gear H, what is its angular speed? One revolution of worm gear E corresponds to .2 revolution of gear A. Use the following data:

	k (mm)	W (N)	D (mm)
A	30	10	100
B	30	10	100
C	100	40	300
H	200	100	500

Neglect inertia of the worm gear, and consider an energy loss of 10% of the input due to friction. Note the italicized statement of Problem 15.30, and treat the problem as a system of bodies.

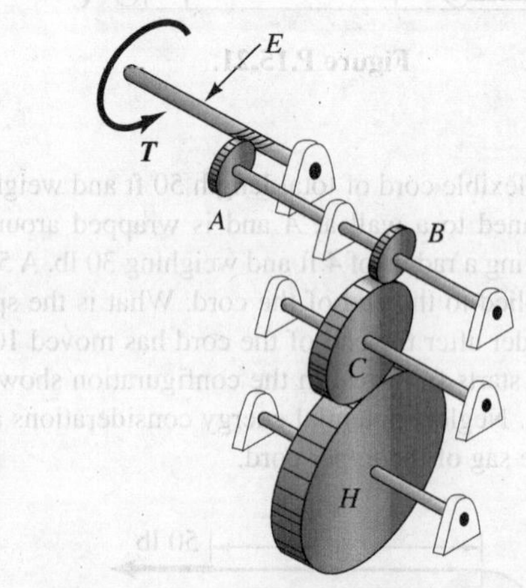

Figure P.15.30.

15.31. A force F of 450 N acts on block A weighing 435 N. Block A rides on identical uniform cylinders B and C, each weighing 290 N and having a radius of 300 mm. If there is no slipping, what is the speed of A after it moves 1 m?

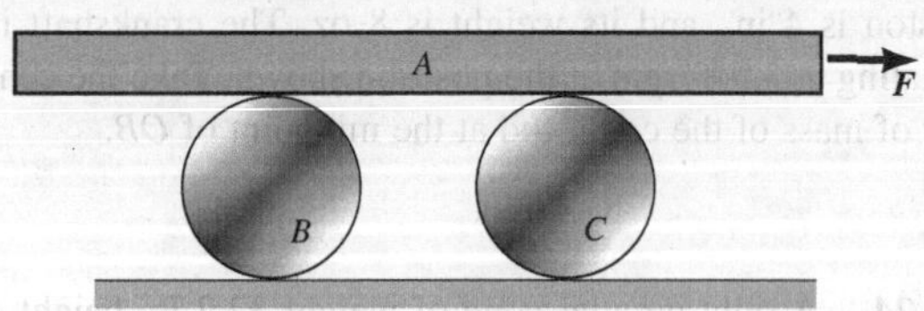

Figure P.15.31.

15.32. Work Example 15.5 using the center-of-mass approach for the whole system. What are the friction forces on the wheels from the ground?

15.33. An electric train (one car) uses its motors as electric generators for braking action. Suppose that this train is moving down a 15° incline at a speed initially of 10 m/sec and, during the next 100 m, the generators develop 1.5 kW-hr of energy. What is the speed of the train at the end of this interval? The train with passengers weighs 200 kN. Each of the eight wheels weighs 900 N and has a radius of gyration of 250 mm and a diameter of 600 mm. Neglect wind resistance, and consider that there is no slipping. The efficiency of the generators for developing power is 90%. [*Hint:* One watt is 1 N-m/sec.] Do not use center of mass approach.

15.34. Work Problem 15.33 using the center of mass approach for the whole train. Also, find the average friction force from the rail onto the wheels. Consider that each wheel is attached to a generator.

15.35. A windlass has a rotating part which weighs 75 lb and has a radius of gyration of 1 ft. When the suspended weight of 20 lb is dropping at a speed of 20 ft/sec, a 100-lb force is applied to the lever at *A*. This action applies the brake shoe at *B*, where there is a coefficient of friction of .5. How far will the 20-lb weight drop before stopping?

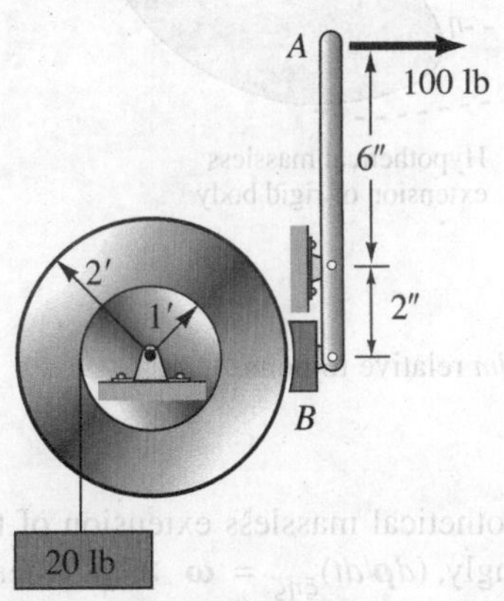

Figure P.15.35.

15.36. A square-threaded screw has a diameter of 50 mm and is inclined 45° to the horizontal. The pitch of the thread is 5 mm, and it is single-threaded. A body *A* weighing 290 N and having a radius of gyration of 300 mm screws onto the shaft. A torque *T* of 45 N-m is applied to *A* as shown. What is the angular speed of *A* after three revolutions starting from a rest configuration? Neglect friction.

Figure P.15.36.

15.37. A uniform block *A* weighing 64.4 lb is pulled by a force *P* of 50 lb as shown. The block moves along the rails on small, light wheels. One rail descends at an angle of 15° at point *B*. If the force *P* always remains horizontal, what is the speed of the block after it has moved 5 ft in the horizontal direction? The block is stationary at the position shown. Assume that the block does not tilt forward.

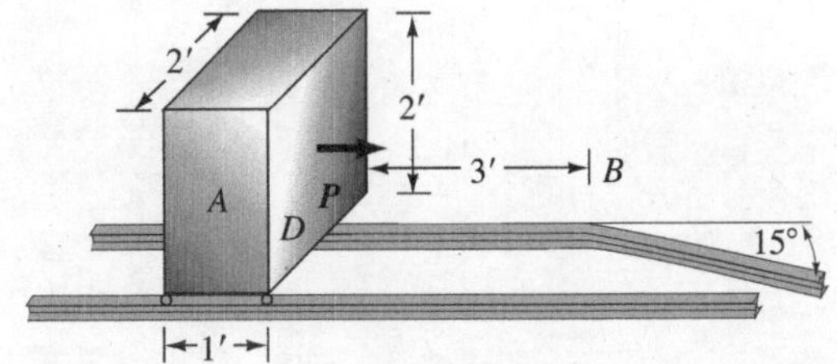

Figure P.15.37.

15.38. A solid uniform rod *AB* connects two light slider bearings *A* and *B*, which move in a frictionless manner along the indicated guide rods. The rod *AB* has a mass of 150 lbm and a diameter of 2 in. Smooth ball-joint connections exist between the rod and the bearings. If the rod is released from rest at the configuration shown, what is the speed of the bearing *A* when it has dropped 2 ft?

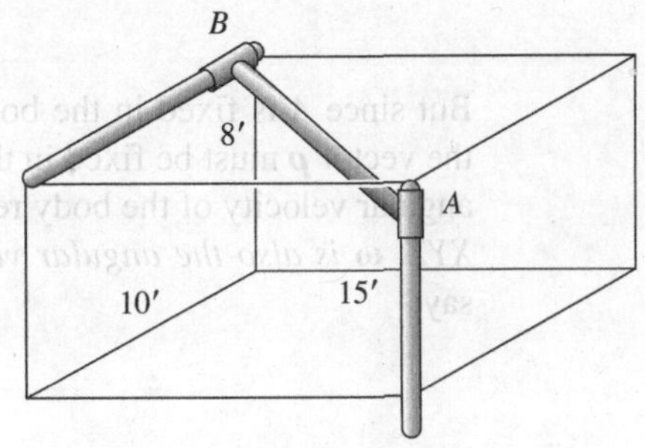

Figure P.15.38.

Part B: Impulse-Momentum Methods

15.4 Angular Momentum of a Rigid Body About Any Point in the Body

As we go to three dimensions, we will need formulations for linear momentum and angular momentum of rigid bodies. The linear momentum is simply $\iiint \boldsymbol{V}\,dm = M\boldsymbol{V}_c$. We shall now formulate an expression for the more complicated angular momentum $\boldsymbol{H}$ of a rigid body about a point. For this purpose, we choose a point A in a rigid body or hypothetical massless extension of the rigid body as shown in Fig. 15.21. An element of mass dm at a position $\boldsymbol{\rho}$ from A is shown. The velocity $\boldsymbol{V}'$ of dm relative to A is simply the velocity of dm relative to reference $\xi\eta\zeta$ which *translates* with A relative to XYZ. Similarly, the linear momentum of dm relative to A is the linear momentum of dm relative to a reference $\xi\eta\zeta$ translating with A. We can now give the angular momentum $d\boldsymbol{H}_A$ for element dm about A as

$$d\boldsymbol{H}_A = \boldsymbol{\rho} \times \boldsymbol{V}'\,dm = \boldsymbol{\rho} \times \left(\frac{d\boldsymbol{\rho}}{dt}\right)_{\xi\eta\zeta} dm \qquad (15.18a)$$

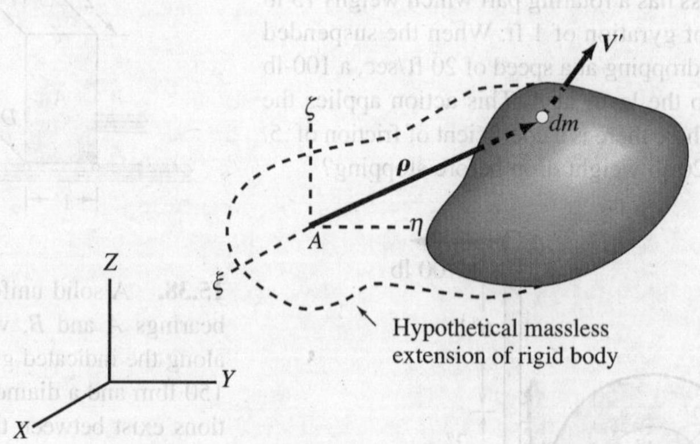

Figure 15.21. Velocity of *dm* relative to point *A*.

But since A is fixed in the body (or in the hypothetical massless extension of the body), the vector $\boldsymbol{\rho}$ must be fixed in the body. Accordingly, $(d\boldsymbol{\rho}/dt)_{\xi\eta\zeta} = \boldsymbol{\omega} \times \boldsymbol{\rho}$, where $\boldsymbol{\omega}$ is the angular velocity of the body relative to $\xi\eta\zeta$. However, since $\xi\eta\zeta$ translates with respect to XYZ, *$\boldsymbol{\omega}$ is also the angular velocity of the body relative to XYZ as well*. Hence, we can say:

$$d\boldsymbol{H}_A = \boldsymbol{\rho} \times (\boldsymbol{\omega} \times \boldsymbol{\rho})\,dm \qquad (15.18b)$$

We shall find it convenient to express Eq. 15.18 in terms of orthogonal components. For this purpose, imagine an arbitrary reference xyz fixed to the body or rigid-body extension

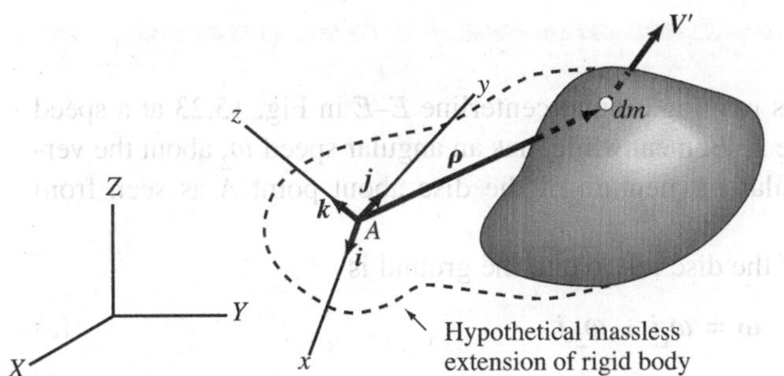

Figure 15.22. Reference *xyz* at *A*.

of the body having the origin at *A* and any arbitrary orientation relative to *XYZ*,[9] as shown in Fig. 15.22. We next decompose each of the vectors in Eq. 15.18b into rectangular components in the ***i***, ***j***, and ***k*** directions associated with the *x*, *y*, and *z* axes, respectively. Thus,

$$d\boldsymbol{H}_A = (dH_A)_x\boldsymbol{i} + (dH_A)_y\boldsymbol{j} + (dH_A)_z\boldsymbol{k} \tag{15.19a}$$

$$\boldsymbol{\rho} = x\boldsymbol{i} + y\boldsymbol{j} + z\boldsymbol{k} \tag{15.19b}$$

$$\boldsymbol{\omega} = \omega_x\boldsymbol{i} + \omega_y\boldsymbol{j} + \omega_z\boldsymbol{k} \tag{15.19c}$$

We then have for Eq. 15.18b:

$$(dH_A)_x\boldsymbol{i} + (dH_A)_y\boldsymbol{j} + (dH_A)_z\boldsymbol{k} = (x\boldsymbol{i} + y\boldsymbol{j} + z\boldsymbol{k}) \times [(\omega_x\boldsymbol{i} + \omega_y\boldsymbol{j} + \omega_z\boldsymbol{k}) \times (x\boldsymbol{i} + y\boldsymbol{j} + z\boldsymbol{k})]\,dm \tag{15.20}$$

Carrying out the cross products and collecting terms, we have

$$(dH_A)_x = \omega_x(y^2 + z^2)\,dm - \omega_y xy\,dm - \omega_z xz\,dm \tag{15.21a}$$

$$(dH_A)_y = -\omega_x yx\,dm + \omega_y(x^2 + z^2)\,dm - \omega_z yz\,dm \tag{15.21b}$$

$$(dH_A)_z = -\omega_x zx\,dm - \omega_y zy\,dm + \omega_z(x^2 + y^2)\,dm \tag{15.21c}$$

If we integrate these relations for all the mass elements *dm* of the rigid body, we see that the components of the inertia tensor for point *A* appear:

$$(H_A)_x = I_{xx}\omega_x - I_{xy}\omega_y - I_{xz}\omega_z \tag{15.22a}$$

$$(H_A)_y = -I_{yx}\omega_x + I_{yy}\omega_y - I_{yz}\omega_z \tag{15.22b}$$

$$(H_A)_z = -I_{zx}\omega_x - I_{zy}\omega_y + I_{zz}\omega_z \tag{15.22c}$$

We thus have components of the angular momentum vector $\boldsymbol{H}_A$ for a rigid body about point *A* in terms of an arbitrary set of directions *x*, *y*, and *z* at point *A*.

We now illustrate the calculation of $\boldsymbol{H}_A$ in the following example.

[9]At this time we can forget about the axes ξ, η, and ζ. They only become necessary when we ask the question: What is the velocity or linear momentum of a particle relative to point *A*. To repeat, the velocity or linear momentum of a particle relative to point *A* is the velocity or momentum relative to a reference $\xi\eta\zeta$ translating with point *A* as seen from *XYZ* or, in other words, relative to a nonrotating observer moving with *A*.

Example 15.6

A disc B has a mass M and is rotating around centerline E–E in Fig. 15.23 at a speed ω_1 relative to E–E. Centerline E–E, meanwhile, has an angular speed ω_2 about the vertical axis. Compute the angular momentum of the disc about point A as seen from ground reference XYZ.

The angular velocity of the disc relative to the ground is

$$\boldsymbol{\omega} = \omega_1 \boldsymbol{i} + \omega_2 \boldsymbol{j} \tag{a}$$

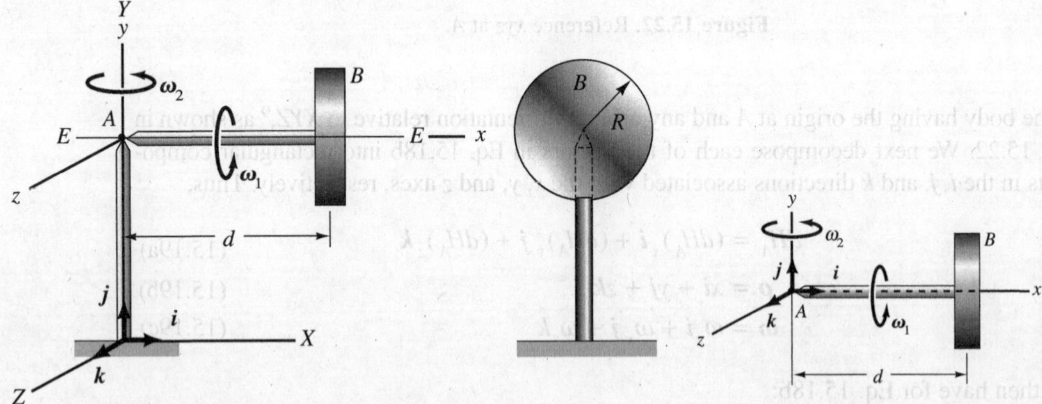

Figure 15.23. Rotating disc. **Figure 15.24.** xyz axes at A.

Consider a set of axes xyz with the origin at A and fixed to the body having cylinder B. At the instant of interest, the xyz axes are parallel to the inertial reference XYZ and we can say (see Fig. 15.24):

$$\omega_x = \omega_1, \qquad \omega_y = \omega_2, \qquad \omega_z = 0 \tag{b}$$

The inertia tensor for the disc taken at A is next presented.

$$\begin{aligned} I_{xx} &= \tfrac{1}{2} MR^2 & I_{xy} &= 0 & I_{xz} &= 0 \\ I_{yx} &= 0 & I_{yy} &= \tfrac{1}{4} MR^2 + Md^2 & I_{yz} &= 0 \\ I_{zx} &= 0 & I_{zy} &= 0 & I_{zz} &= \tfrac{1}{4} MR^2 + Md^2 \end{aligned} \tag{c}$$

Note that the product-of-inertia terms are zero because the xy and the xz planes are planes of symmetry. Clearly, moments of inertia for the xyz axes are principal moments of inertia. Now going to Eq. 15.22, we have

$$\begin{aligned} (H_A)_x &= \frac{MR^2}{2}\omega_1 \\ (H_A)_y &= \left(\tfrac{1}{4} MR^2 + Md^2\right)\omega_2 \\ (H_A)_z &= 0 \end{aligned} \tag{d}$$

As seen in Example 15.5, when *xyz* are *principal* axes, $\boldsymbol{H}_A$ simplifies to

$$\boldsymbol{H}_A = I_{xx}\omega_x\boldsymbol{i} + I_{yy}\omega_y\boldsymbol{j} + I_{zz}\omega_z\boldsymbol{k} \tag{15.23}$$

which is analogous to the linear momentum vector $\boldsymbol{P}$. That is,

$$\boldsymbol{P} = MV_x\boldsymbol{i} + MV_y\boldsymbol{j} + MV_z\boldsymbol{k}$$

Note that mass plays the same role as does I, and $\boldsymbol{V}$ plays the same role as does $\boldsymbol{\omega}$.

In Chapter 14, we considered with some care the plane motion of a slablike body, the motion being parallel to the plane of symmetry of the body. We have shown such a case in Fig. 15.25. A reference xyz is shown fixed to the body at A with xy at the midplane of the body and with the z axis oriented normal to the plane of motion. Recall now that for a slablike body the plane xy must be a plane of symmetry or be a principal plane, and consequently that $I_{zx} = I_{zy} = 0$. Also, the only nonzero component of $\boldsymbol{\omega}$ is ω_z. Going back to Eq. 15.22, we see that only $(H_A)_z$ is nonzero, with the result

$$(H_A)_z = (I_{zz})_A\omega_z \tag{15.24}$$

Because of the importance of plane motion of slablike bodies, we shall often use the foregoing simple formula.

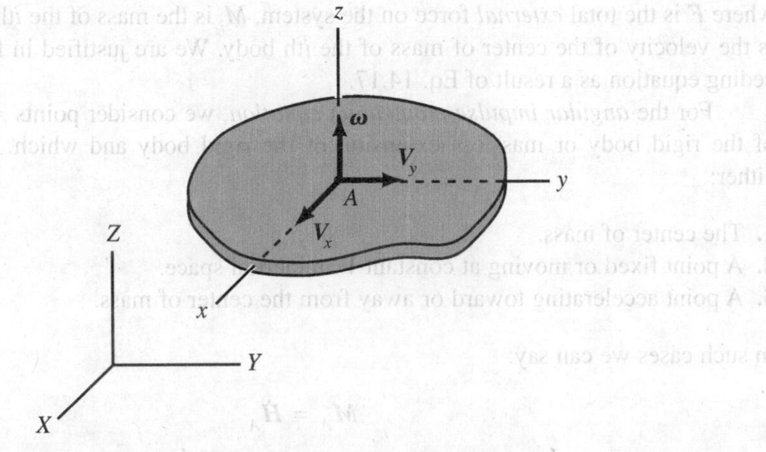

Figure 15.25. Slablike body in plane motion.

We leave it for you to show that the Eq. 15.24 also applies to a body of revolution rotating about its axis of symmetry in inertial space, where z is taken along this axis. Also, Eq. 15.24 is valid for a body having two orthogonal planes of symmetry rotating about an axis corresponding to the intersection of these planes of symmetry in inertial space, where z is taken along this axis. We are now ready to relate linear and angular momenta with force system causing the motion.

15.5 Impulse-Momentum Equations

You will recall that **Newton's law** for the center of mass of any body is

$$\boldsymbol{F} = M\frac{d\boldsymbol{V}_c}{dt} = \frac{d}{dt}(M\boldsymbol{V}_c)$$

where $\boldsymbol{F}$ is the total *external* force on the body. The corresponding *linear impulse-momentum* equation can then be given as

$$\int_{t_1}^{t_2} \boldsymbol{F}\,dt = \boldsymbol{I}_{\text{lin}} = (M\boldsymbol{V}_C)_2 - (M\boldsymbol{V}_C)_1 \qquad (15.25)$$

where $\boldsymbol{I}_{\text{lin}}$ is the *linear impulse*. For a system of n rigid bodies we have, for the foregoing equation:

$$\int_{t_1}^{t_2} \boldsymbol{F}\,dt = \boldsymbol{I}_{\text{lin}} = \left[\sum_{i=1}^{n} M_i(\boldsymbol{V}_C)_i\right]_2 - \left[\sum_{i=1}^{n} M_i(\boldsymbol{V}_C)_i\right]_1 \qquad (15.26)$$

where $\boldsymbol{F}$ is the total *external* force on the system, M_i is the mass of the ith body, and $(\boldsymbol{V}_C)_i$ is the velocity of the center of mass of the ith body. We are justified in forming the preceding equation as a result of Eq. 14.17.

For the *angular impulse-momentum equation*, we consider points A which are part of the rigid body or massless extension of the rigid body and which, in addition, are either:

1. The center of mass.
2. A point fixed or moving at constant $\boldsymbol{V}$ in inertial space.
3. A point accelerating toward or away from the center of mass.

In such cases we can say:

$$\boldsymbol{M}_A = \dot{\boldsymbol{H}}_A$$

where $\boldsymbol{H}_A$ is given by Eq. 15.22. Integrating with respect to time, we then get the desired *angular impulse-momentum equation*:

$$\int_{t_1}^{t_2} \boldsymbol{M}_A\,dt = \boldsymbol{I}_{\text{ang}} = (\boldsymbol{H}_2)_A - (\boldsymbol{H}_1)_A \qquad (15.27)$$

where $\boldsymbol{I}_{\text{ang}}$ is the *angular impulse*. We now illustrate the use of the *linear impulse-* and *angular impulse-momentum equations*.

Example 15.7

A thin bent rod is sliding along a smooth surface (Fig. 15.26). The center of mass has the velocity

$$V_C = 10\boldsymbol{i} + 15\boldsymbol{j} \text{ m/sec}$$

and the angular speed ω is 5 rad/sec counterclockwise. At the configuration shown, the rod is given two simultaneous impacts as a result of a collision. These impacts have the following impulse values:

$$\int_{t_1}^{t_2} F_1\,dt = 5 \text{ N-sec}$$

$$\int_{t_1}^{t_2} F_2\,dt = 3 \text{ N-sec}$$

What is the angular speed of the rod and the linear velocity of the mass center, directly after the impact? The rod weighs 35 N/m.

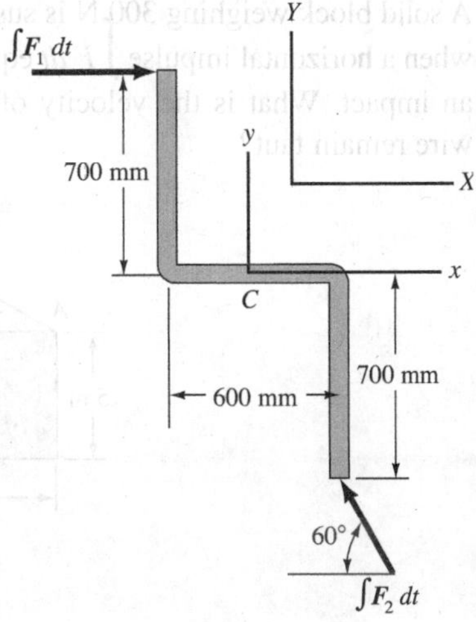

Figure 15.26. Bent rod slides on smooth horizontal surface.

The velocity of the mass center after the impact can easily be determined using the **linear impulse-momentum equation** (Eq. 15.25). Thus, we have

$$5\boldsymbol{i} + 3\sin 60°\boldsymbol{j} - 3\cos 60°\boldsymbol{i} = (.7 + .7 + .6)\left(\frac{35}{g}\right)(\boldsymbol{V}_2 - 10\boldsymbol{i} - 15\boldsymbol{j})$$

Solving for $\boldsymbol{V}_2$:

$$\boldsymbol{V}_2 = 10.49\boldsymbol{i} + 15.36\boldsymbol{j} \text{ m/sec} \tag{a}$$

For the angular velocity, we use the **angular impulse momentum equation** (Eq. 15.27) simplified for the case of plane motion of a slablike body. Again using the center of mass at which we fix *xyz*, we have for Eq. (15.27):

$$\int_1^2 \boldsymbol{M}\,dt = (I_{zz}\omega_2 - I_{zz}\omega_1)\boldsymbol{k} \tag{b}$$

Putting in numerical data and canceling $\boldsymbol{k}$, we get

$$-(5)(.70) + (3)(\sin 60°)(.30) - (3)(\cos 60°)(.70) = I_{zz}(\omega_2 - 5) \tag{c}$$

We next compute I_{zz} at C:

$$\begin{aligned} I_{zz} &= \frac{1}{12}\left[\frac{35}{g}(.60)\right](.60)^2 \\ &\quad + 2\left[\frac{1}{12}\left(\frac{35}{g}\right)(.70)(.70)^2 + \left(\frac{35}{g}\right)(.70)(.30^2 + .35^2)\right] \\ &= 1.330 \text{ kg-m}^2 \end{aligned} \tag{d}$$

Going back to Eq. (c), we can now give ω_2:

$$\omega_2 = 2.16 \text{ rad/sec}$$

Example 15.8

A solid block weighing 300 N is suspended from a wire (see Fig.15.27) and is stationary when a horizontal impulse $\int F\,dt$ equal to 100 N-sec is applied to the body as a result of an impact. What is the velocity of corner A of the block just after impact: Does the wire remain taut?

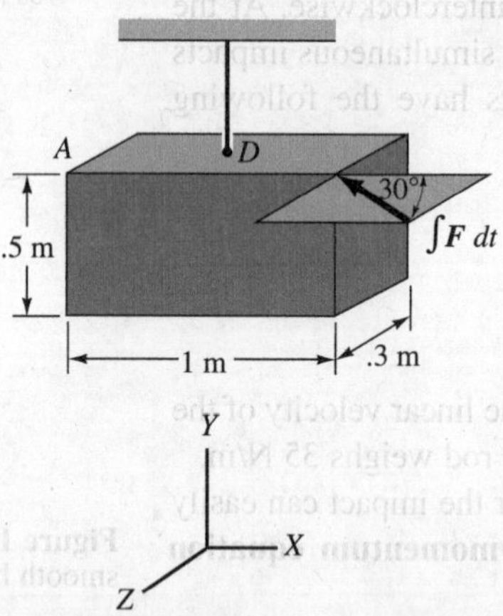

Figure 15.27. Stationary block under impact.

For the **linear momentum equation** we can say for the center of mass (see Fig. 15.28) in the z, x, and y directions:

$$-100 \sin 30° = \frac{300}{g}\left[(V_c)_z - 0\right]$$

$$-100 \cos 30° = \frac{300}{g}\left[(V_c)_x - 0\right] \qquad \text{(a)}$$

$$0 = \frac{300}{g}\left[(V_c)_y - 0\right]$$

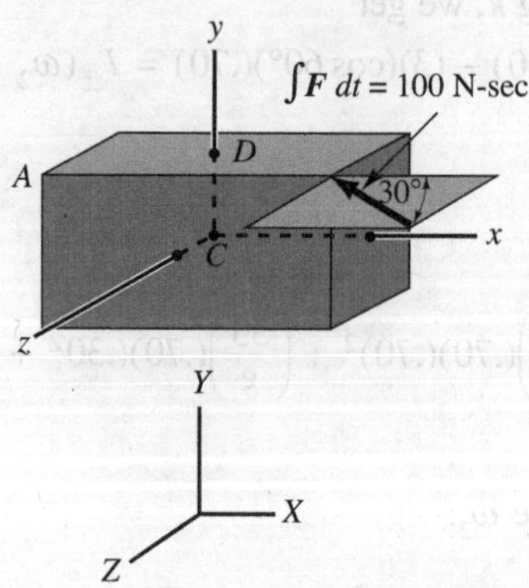

Figure 15.28. xyz fixed at center of mass.

Example 15.8 (Continued)

From these equations, we get

$$V_C = -2.83\boldsymbol{i} - 1.635\boldsymbol{k} \text{ m/sec} \tag{b}$$

For the **angular momentum equation** about C, we can say, on noting that xyz are principal axes:

$$\begin{aligned} -(100)(\sin 30°)(.25) &= I_{xx}\omega_x - 0 \\ (100)(\sin 30°)(.5) - 100(\cos 30°)\left(\frac{.3}{2}\right) &= I_{yy}\omega_y - 0 \\ (100)(\cos 30°)(.25) &= I_{zz}\omega_z - 0 \end{aligned} \tag{c}$$

Note that

$$\begin{aligned} I_{xx} &= \frac{1}{12}\left(\frac{300}{g}\right)(.3^2 + .5^2) = .866 \text{ kg-m}^2 \\ I_{yy} &= \frac{1}{12}\left(\frac{300}{g}\right)(1^2 + .3^2) = 2.78 \text{ kg-m}^2 \\ I_{zz} &= \frac{1}{12}\left(\frac{300}{g}\right)(1^2 + .5^2) = 3.19 \text{ kg-m}^2 \end{aligned} \tag{d}$$

We then get for $\boldsymbol{\omega}$ as seen from XYZ from Eq. (c):

$$\boldsymbol{\omega} = -14.43\boldsymbol{i} + 4.32\boldsymbol{j} + 6.79\boldsymbol{k} \text{ rad/sec}$$

Hence, for the velocity of point A, we have

$$\begin{aligned} V_A &= V_C + \boldsymbol{\omega} \times \boldsymbol{\rho}_{CA} \\ &= -2.83\boldsymbol{i} - 1.635\boldsymbol{k} + (-14.43\boldsymbol{i} + 4.32\boldsymbol{j} + 6.79\boldsymbol{k}) \\ &\quad \times (-.5\boldsymbol{i} + .25\boldsymbol{j} + .15\boldsymbol{k}) \end{aligned}$$

$$V_A = -3.88\boldsymbol{i} - 1.230\boldsymbol{j} - 3.08\boldsymbol{k} \text{ m/sec}$$

Finally, to decide if wire remains taut, find the velocity of point D after impact.

$$\begin{aligned} V_D &= V_C + \boldsymbol{\omega} \times \boldsymbol{\rho}_{CD} \\ &= -2.83\boldsymbol{i} - 1.635\boldsymbol{k} + (-14.43\boldsymbol{i} + 4.32\boldsymbol{j} + 6.79\boldsymbol{k}) \times (.25\boldsymbol{j}) \\ &= -4.53\boldsymbol{i} - 5.24\boldsymbol{k} \text{ m/sec} \end{aligned}$$

Since there is zero velocity component in the y direction stemming from the given impulse, we can conclude that the wire remains taut.

In the following example we consider a problem involving a system of interconnected bodies.

Example 15.9

A tractor weighs 2,000 lb, including the driver (Fig. 15.29). The large driver wheels each weigh 200 lb with a radius of 2 ft and a radius of gyration of 1.8 ft. The small wheels weigh 40 lb each, with a radius of 1 ft and a radius of gyration of 10 in. The tractor is pulling a bale of cotton weighing 300 lb. The coefficient of friction between the bale and the ground is .2. What torque is needed on the drive wheels from the motor for the tractor to go from 5 ft/sec to 10 ft/sec in 25 sec? Assume that the tires do not slip.

Figure 15.29. Tractor pulling bale of cotton.

We have shown a free-body diagram of the system in Fig. 15.30. Noting on inspection that $N_1 = 300 \cos 5°$, we can give the **linear momentum equation** for the system in the X direction as

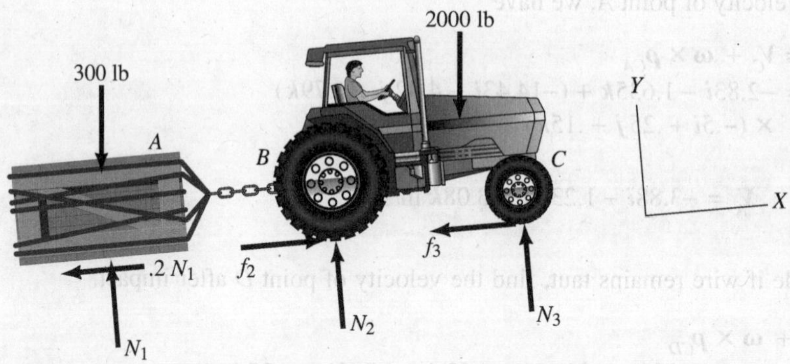

Figure 15.30. Free-body diagram of system.

$$\left[f_2 - f_3 - (.2)(300)\cos 5° - 300 \sin 5° - 2{,}000 \sin 5°\right](25) = \frac{2{,}300}{g}(10 - 5)$$

Therefore,

$$f_2 - f_3 = 275 \quad \text{(a)}$$

Example 15.9 (Continued)

We next consider free-body diagrams of the wheels in Fig. 15.31. The **impulse-angular momentum equation** for the drive wheels then can be given about the center of mass as

$$\left[-T + f_2(2)\right](25) = \frac{400}{g}(1.8)^2\left[(\omega_B)_2 - (\omega_B)_1\right]$$

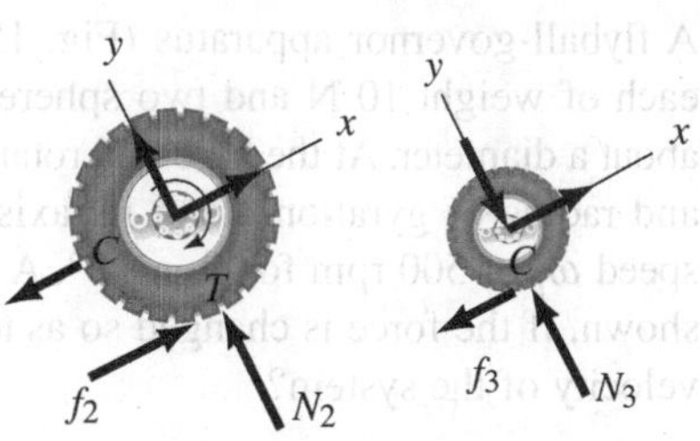

Figure 15.31. Free-body diagrams of wheels xy axes are fixed to wheels.

Noting from **kinematics** that $\omega_B = -V/2$, we have

$$-T + 2f_2 = -4.025 \qquad \text{(b)}$$

The **impulse-angular momentum equation** about the center of mass for the front wheels is then

$$-(f_3)(1)(25) = \frac{80}{g}\left(\frac{10}{12}\right)^2\left[(\omega_C)_2 - (\omega_C)_1\right]$$

Noting from **kinematics** again that $\omega_C = -V/1$, we get from the equation above:

$$f_3 = .345 \text{ lb} \qquad \text{(c)}$$

From Eq. (a) we may now solve for f_2. Thus,

$$f_2 = .345 + 275 = 275.3 \text{ lb}$$

Finally, from Eq. (b) we get the desired torque T:

$$T = 4.025 + (2)(275.3)$$

$$T = 555 \text{ ft-lb}$$

In Example 15.8, there was no obvious convenient stationary point or stationary axis which could be considered as part of a rigid-body extension of *all* the bodies at any time. Therefore, in order to use the formulas for $\boldsymbol{H}$ given by Eq. 15.22, we considered rigid bodies *separately*. In the following example, we have a case where there is a stationary axis present which can be considered as part of (or a hypothetical rigid-body extension of) all bodies in the system at the instants of interest. And for this reason, we shall consider the angular momentum equation for the entire system using this stationary axis. Also, if the torque about such a common axis for a system of bodies is zero, then the angular momentum of the system about the aforestated axis must be *conserved*. In the example to follow, we shall also illustrate conservation of angular momentum about an axis for such a case.

Example 15.10

A flyball-governor apparatus (Fig. 15.32) consists of four identical arms (solid rods) each of weight 10 N and two spheres of weight 18 N and radius of gyration 30 mm about a diameter. At the base and rotating with the system is a cylinder B of weight 20 N and radius of gyration along its axis of 50 mm. Initially, the system is rotating at a speed ω_1 of 500 rpm for $\theta = 45°$. A force F at the base B maintains the configuration shown. If the force is changed so as to decrease θ from 45° to 30°, what is the angular velocity of the system?

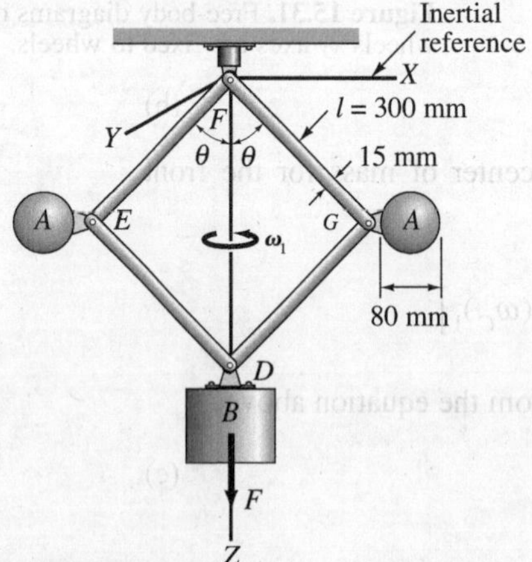

Figure 15.32. Flyball governor apparatus; Y axis is normal to page.

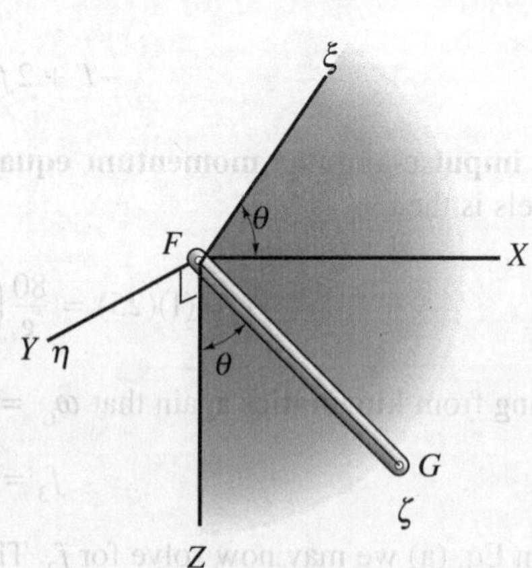

Figure 15.33. $\xi\eta\zeta$ are principal axes of FG at F.

Clearly, there is zero torque from external forces about the stationary axis FD which we take as a Z axis at all times. Hence, we have conservation of angular momentum about this axis at all times. And, since the axis is an axis of rotation for all bodies of the system,[10] we can use Eq. 15.22 for computing H about FD for all bodies in the system. As a first step we shall need I_{ZZ} for the members of the system.

Consider first member FG, which is shown in Fig. 15.33. The axes $\xi\eta\zeta$ are principal axes of inertia for the rod at F. The η axis is collinear with the Y axis, and these are normal to the page. The axes XYZ are reached by $\xi\eta\zeta$ by rotating $\xi\eta\zeta$ about the η axis an angle θ. Using the transformation equations for I_{ZZ}, we can say:

$$(I_{ZZ})_{FG} = I_{\xi\xi}\left[\cos\left(\frac{\pi}{2}+\theta\right)\right]^2 + I_{\eta\eta}\left(\cos\frac{\pi}{2}\right)^2 + I_{\zeta\zeta}(\cos\theta)^2$$

$$= \left[\frac{1}{3}\frac{10}{g}(.30)^2\right]\sin^2\theta + 0 + \left[\frac{1}{2}\frac{10}{g}(.0075)^2\right]\cos^2\theta \quad \text{(a)}$$

For the sphere we have, using the parallel axis theorem

$$(I_{ZZ})_{\text{sphere}} = \frac{18}{g}(.03)^2 + \frac{18}{g}[(.30)\sin\theta + .04]^2 \quad \text{(b)}$$

[10]That is, at any time t the stationary axis FD either is part of a rigid body directly or is part of a hypothetical extension of a rigid body for all bodies of the system.

Example 15.10 (Continued)

Finally, for cylinder *B* we have

$$(I_{ZZ})_{\text{cyl}} = \frac{20}{g}(.05)^2 \qquad \text{(c)}$$

Conservation of angular momentum about the *Z* axis then prescribes the following:

$$\left[4(I_{ZZ})_{FG} + 2(I_{ZZ})_{\text{sphere}} + (I_{ZZ})_{\text{cyl}}\right]_{\theta=45^\circ} \frac{(500)(2\pi)}{60}$$
$$= \left[4(I_{ZZ})_{FG} + 2(I_{ZZ})_{\text{sphere}} + (I_{ZZ})_{\text{cyl}}\right]_{\theta=30^\circ} \omega_2$$

Substituting from Eqs. (a), (b), and (c), we have

$$\left(4\left[\frac{1}{3}\frac{10}{g}(.30)^2(.707)^2 + \frac{1}{2}\frac{10}{g}(.0075)^2(.707)^2\right] + 2\left\{\frac{18}{g}(.03)^2 + \frac{18}{g}[(.30)(.707) + .04]^2\right\} + \frac{20}{g}(.05)^2\right)\frac{(500)(2\pi)}{60}$$
$$= \left(4\left[\frac{1}{3}\frac{10}{g}(.30)^2(.5)^2 + \frac{1}{2}\frac{10}{g}(.0075)^2(.866)^2\right] + 2\left\{\frac{18}{g}(.03)^2 + \frac{18}{g}[(.30)(.5) + .04]^2\right\} + \frac{20}{g}(.05)^2\right)\omega_2$$

Therefore,

$$\omega_2 = 92.4 \text{ rad/sec} = 883 \text{ rpm}$$

Before closing the section, we note that we have worked with *fixed points* or axes in inertial space and with the *mass center*. What about a point *accelerating toward the mass center*? A common example of such a point is the point of contact *A* of a cylinder rolling without slipping on a circular arc with the center of mass of the cylinder coinciding with the geometric center of the cylinder. We can then say:

$$\boldsymbol{M}_A = \dot{\boldsymbol{H}}_A$$

The question then arises: Can we form the familiar angular momentum equation from above about point *A*? In other words, is the following equation valid for plane motion?

$$\int_{t_1}^{t_2} M_A\, dt = I_A\omega_2 - I_A\omega_1 \qquad (17.28)$$

The reason we might hesitate to do this is that the point of contact *continually changes* during a time interval when the cylinder is rolling. However, we have asked you to prove in Problem 15.60 that for rolling without slipping along a circular or straight path, the equation above is still valid.[11] At times it can be very useful.

[11]The statement is actually valid for point *A* when there is rolling of the cylinder without slipping on a *general* path.

PROBLEMS

15.39. A uniform cylinder C of radius of 1 ft and thickness 3 in. rolls without slipping at its center plane on the stationary platform B such that the centerline of CD makes 2 revolutions per second relative to the platform. What is the angular momentum vector for the cylinder about the center of mass of the cylinder? The cylinder weighs 64.4 lb.

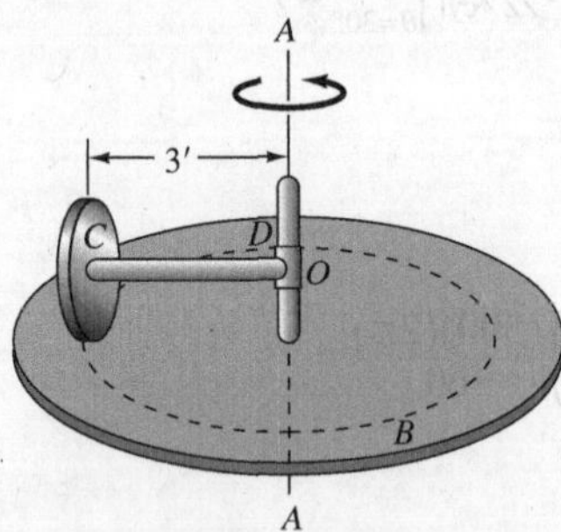

Figure P.15.39.

15.40. In Problem 15.39, find the angular momentum of the disc about the stationary point O along the vertical axis A–A.

15.41. A platform rotates at an angular speed of ω_1, while a cylinder or radius r and length a mounted on the platform rotates relative to the platform at an angular speed of ω_2. When the axis of the cylinder is collinear with the stationary Y axis, what is that angular momentum vector of the cylinder about the center of mass of the cylinder? The mass of the cylinder is M.

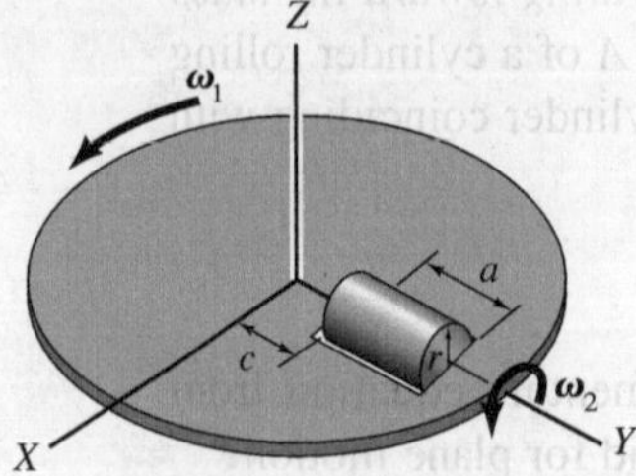

Figure P.15.41.

15.42. A disc A rotates relative to an inclined shaft CD at the ratio ω_2 of 3 rad/sec while shaft CD rotates about vertical axis FE at the rate ω_1 of 4 rad/sec relative to the ground. What is the angular momentum of the disc about its mass center as seen from the shaft CD? What is the angular momentum of the disc about its mass center as seen from the ground? The disc weighs 290 N.

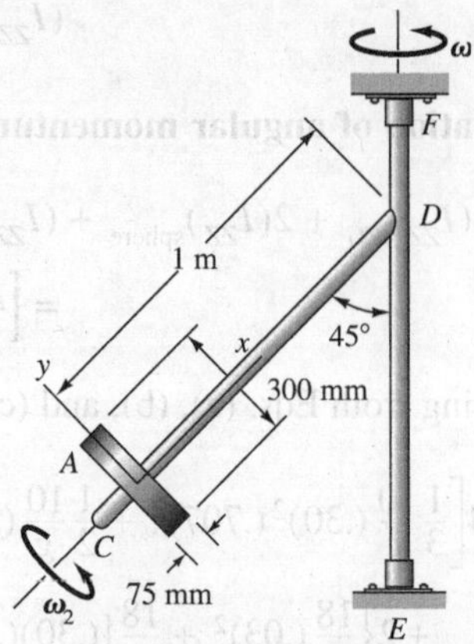

Figure P.15.42.

15.43. Work Problem 14.5 by methods of momentum.

15.44. A flywheel having a mass of 1,000 kg is brought to speed of 200 rpm in 360 sec by an electric motor developing a torque of 60 N-m. What is the radius of gyration of the wheel?

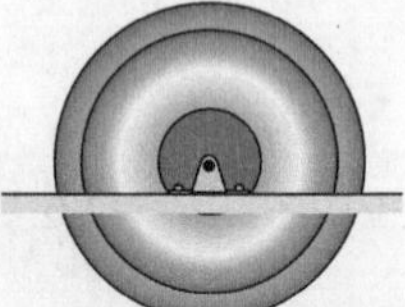

Figure P.15.44.

15.45. Work Problem Example 14.2 by method of angular momentum.

15.46. Work Problem 14.8 by method of angular momentum.

15.47. A light plane is coming in for a landing at a speed of 100 km/hr. The wheels have zero rotation just before touching the runway. If $\frac{1}{10}$ the weight of the plane is maintained by the upward force of the runway for the first second, what is the approximate length of the skid mark left by the wheel on the runway? The wheels each weigh 100 N and have a radius of gyration of 180 mm and a diameter of 450 mm. The plane weighs with load 8,000 N. The coefficient of friction between the tire and runway is .3.

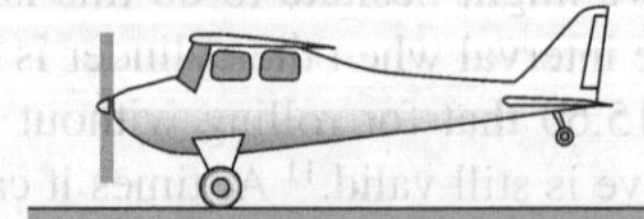

Figure P.15.47.

15.48. Work Problem 15.47 for the case where the upward force from the ground on the plane during the first second after touchdown is

$$N = 8{,}000\,t^2 \text{ N}$$

where t is in seconds after touchdown.

15.49. A circular conveyor carries cylinders a from position A through a heat treatment furnace. The cylinders are dropped onto the conveyor at A from a stationary position above and picked up at B. The conveyor is to turn at an average speed ω of 2 rpm. The cylinders are dropped onto the conveyor at the rate of 9 per minute. If the resisting torque due to friction is 1 N-m, what average torque T is needed to maintain the prescribed angular motion? Each cylinder weighs 300 N and has a radius of gyration of 150 mm about its axis.

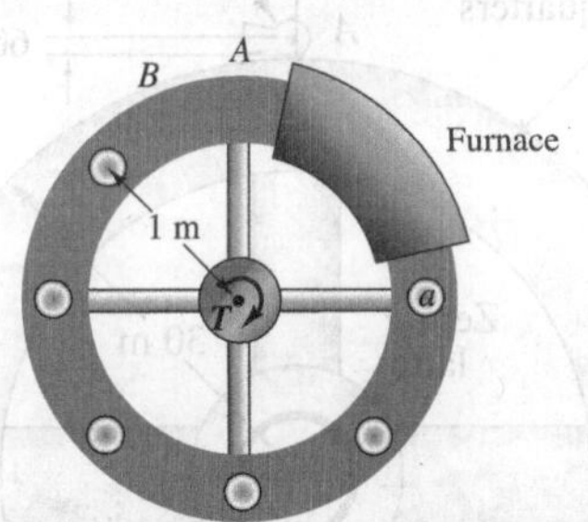

Figure P.15.49.

15.50. A circular *towing tank* has a main arm A which has a mass of 1,000 kg and a radius of gyration of 1 m. On the arm rides the model support B, having a mass of 200 kg and having a radius of gyration about the vertical axis at its mass center of 600 mm. If a torque T of 50 N-m is developed on A when B is at position $r = 1.8$ m, what will be the angular speed 5 sec later if B moves out at a constant radial speed of .1m/sec. The initial angular speed of the arm A is 2 rpm. Neglect the drag of the model.

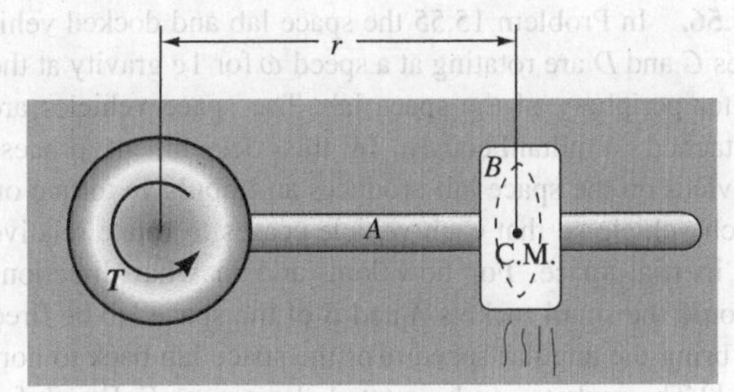

Figure P.15.50.

15.51. A steam roller with driver weighs 5 tons. Wheel A weighs 2 tons and has a radius of gyration of .8 ft. Drive wheels B have a total weight of 1 ton and a radius of gyration of 1.5 ft. If a total torque of 400 ft-lb is developed by the engine on the drive wheels, what is the speed of the steam roller after 10 sec starting from rest? There is no slipping.

Figure P.15.51.

15.52. An electric motor D drives gears C, B, and device A. The diameters of gears C and B are 6 in. and 16 in., respectively. The mass of A is 200 lbm. The combined mass of the motor armature and gear C is 50 lbm, while the radius of gyration of this combination is 8 in. Also, the mass of B is 20 lbm. If a constant counterclockwise torque of 60 lb-ft is developed on the armature of the motor, what is the speed of A in 2 sec after starting from rest? Neglect the inertia of the small wheels under A.

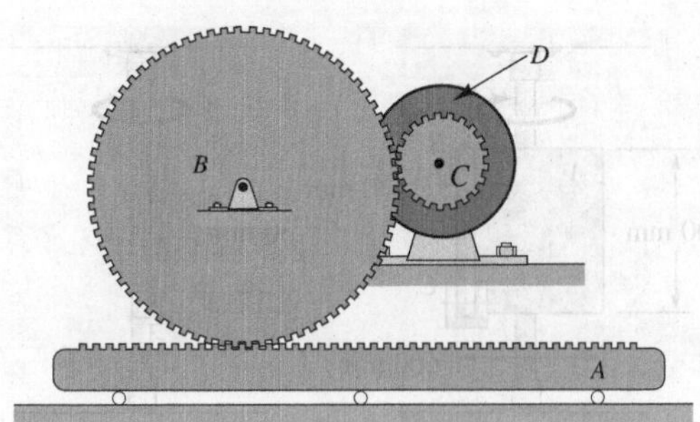

Figure P.15.52.

15.53. A conveyor is moving a weigh W of 64.4 lb. Cylinders A and B have a diameter of 1 ft and weigh 32.2 lb each. Also, they each have a radius of gyration of .8 ft. Rollers C, D, E, F and G each have a diameter of 3 in., weigh 10 lb each, and have a radius of gyration of 2 in. What constant torque T will increase the speed of W from 3 ft/sec to 5 ft/sec in 3 seconds? The belt weighs 50 lb.

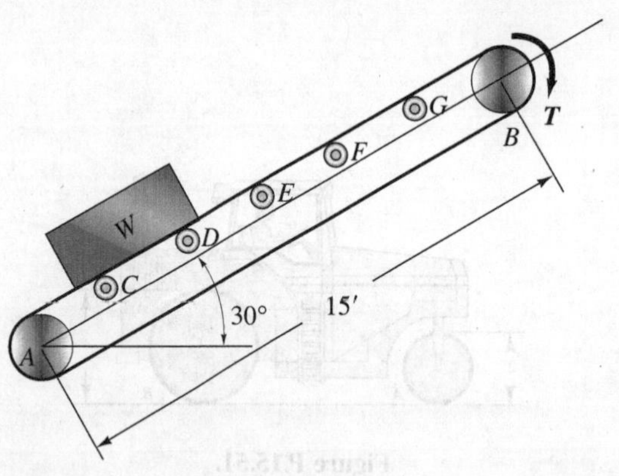

Figure P.15.53.

15.54. A rectangular block A is rotating freely at a speed ω of 200 rpm about a light hollow shaft. Attached to A is a circular rod B which can rotate out from the block A about a hinge at C. This rod weighs 20 N. When the system is rotating at the speed ω of 200 rpm, the rod B is vertical as shown. If the catch at the upper end of B releases so that B falls to a horizontal orientation (shown as dashed in the diagram), what is the new angular velocity? The block A weighs 60 N. Neglect the inertia of the shaft.

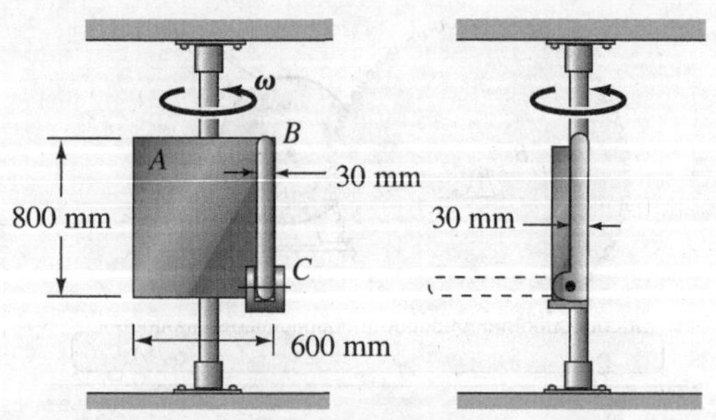

Figure P.15.54.

15.55. A space laboratory is in orbit and has an angular velocity ω of .5 rad/sec relative to inertial space so as to have a partial "gravitational" force for the living quarters on the outside ring. The space lab has a mass of 4,500 kg and a radius of gyration about its axis equal to 20 m. Two space ships C and D, each of mass 1,000 kg, are shown docked so as also to get the benefit of "gravity." the centers of mass of each vehicle is .7 m from the wall of the space lab. The radius of gyration of each vehicle about an axis at the mass center parallel to the axis of the space lab is .5 m. Small rocket engines A and B are turned on to develop a thrust each of 250 N. How long should they be on to increase the angular speed so as to have 1g of gravity at the outer radius of the space lab?

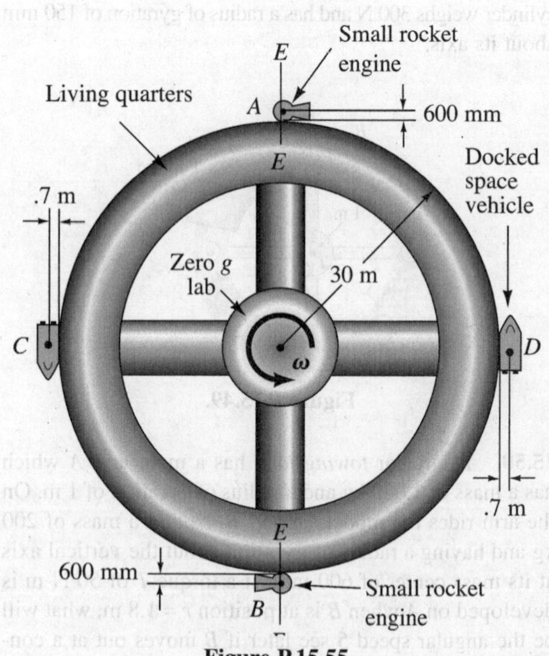

Figure P.15.55.

15.56. In Problem 15.55 the space lab and docked vehicles C and D are rotating at a speed ω for 1g gravity at the outer periphery of the space lab. The space vehicles are detached simultanFeously. In this detachment process devices on the space lab produces an impulsive torque on each vehicle so that each vehicle ceases to rotate relative to inertial space. For how long and in what directions should the small rockets A and B of the space lab be fired to bring the angular speed ω of the space lab back to normal? The rockets can be rotated about axes E–E and, for this maneuver, they are developing a thrust of 100 N each.

15.57. A turbine is rotating freely with a speed ω of 6,000 rpm. A blade breaks off at its base at the position shown. What is the velocity of the center of mass of the blade just after the fracture? Does the blade have an angular velocity just after fracture assuming that no impulsive torques or forces occur at the fracture? Explain.

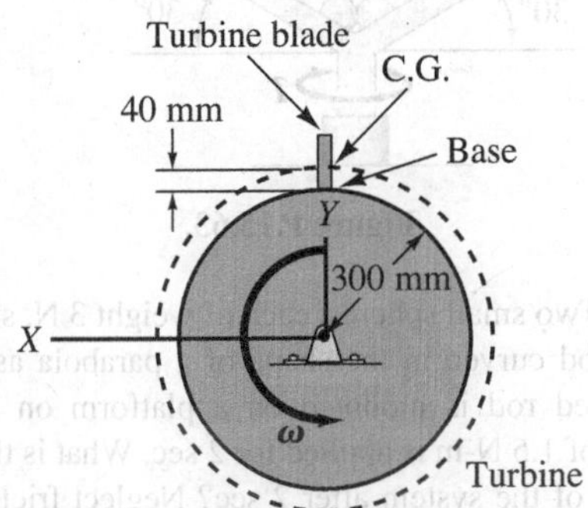

Figure P.15.57.

15.58. Two rods are welded to a drum which has an angular velocity of exactly 2,000 rpm. The rod A breaks at the base at the position shown. If we neglect wind friction, how high up does the center of mass of the rod go? What is the angular orientation of the rod at the instant that the center of mass reaches its apex? Assume that there are no impulsive torques or forces at fracture. The Y axis is vertical. Neglect friction. Use exact value of 2,000 rpm throughout.

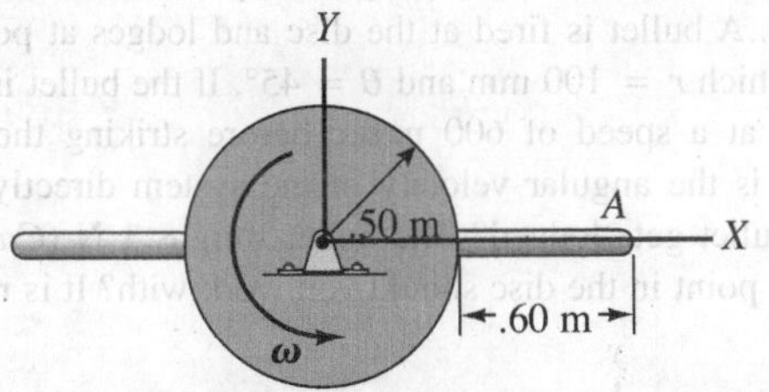

Figure P.15.58.

15.59. A stepped cylinder is release d on a 45° incline where the dynamic coefficient of friction at the contact is .2 and the static coefficient of friction is .22. What is the angular speed of the stepped cylinder after 4 sec? The stepped cylinder has a weight of 500 N and a radius of gyration about its axis of 250 mm. Be sure to check to see if the cylinder moves at all!

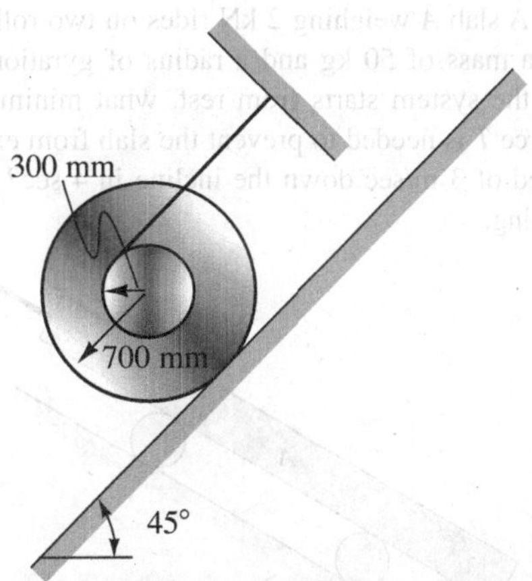

Figure P.15.59.

15.60. Prove that you can apply the angular momentum equation about the contact point A on a cylinder rolling without slipping on a circular (and hence including a straight) path. A force P always normal to OA acts on the cylinder as do a couple moment T and weight W. Specifically, prove that

$$\int_{t_1}^{t_2} (2Pr + W \sin \theta\, r + T)\, dt = M(k^2 + r^2)(\omega_2 - \omega_1) \tag{a}$$

where ω is the total angular speed of the cylinder. [*Hint:* Express the *angular momentum* equation about C and then, from *Newton's law* using the cylindrical component in the transverse direction, show on integrating that

$$\int_{t_1}^{t_2} (f + P + W \sin \theta)\, dt = -M(R - r)(\dot{\theta}_2 - \dot{\theta}_1)$$

where f is the friction force at the point of contact. Now let xy rotate with line OC. From *kinematics* first show that $R\theta = -r\phi$, where ϕ is the rotation of the cylinder relative to xy. Then, show that $\omega = -[(R - r)/r]\dot{\theta}$. From these three considerations, you should readily be able to derive Eq. (a).]

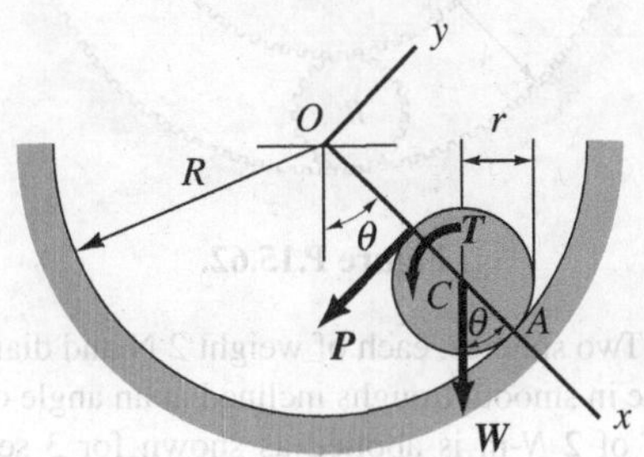

Figure P.15.60.

15.61 A slab A weighing 2 kN rides on two rollers each having a mass of 50 kg and a radius of gyration of 200 mm. If the system starts from rest, what minimum constant force T is needed to prevent the slab from exceeding the speed of 3 m/sec down the incline in 4 sec? There is no slipping.

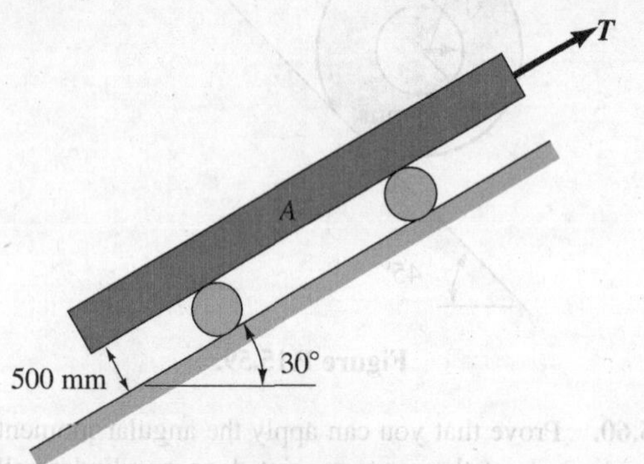

Figure P.15.61.

15.62. A main gear A rotates about a fixed axis and meshes with four identical floating gears B. The floating gears, in turn, mesh with the stationary gear F. If a torque T of 200 N-m is applied to the main gear A, what is its angular speed in 5 sec? The following data apply:

$$M_A = 100 \text{ kg}, \quad k_A = 250 \text{ mm}$$
$$M_B = 20 \text{ kg}, \quad k_B = 40 \text{ mm}$$

The system is horizontal. Read the first sentence of Problem 15.60.

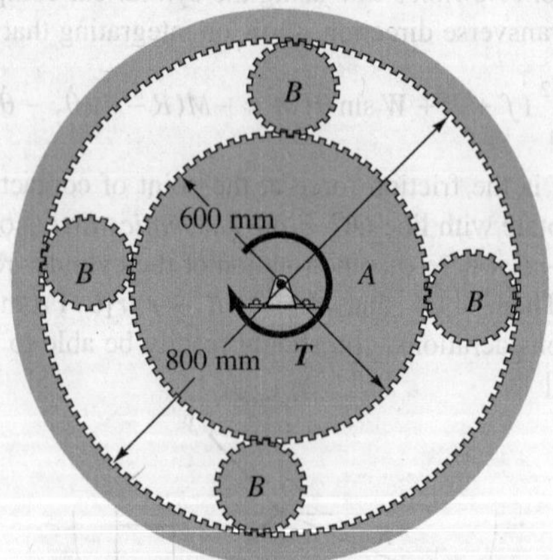

Figure P.15.62.

15.63. Two spheres, each of weight 2 N and diameter 30 mm, slide in smooth troughs inclined at an angle of 30°. A torque T of 2 N-m is applied as shown for 3 sec and is then zero. How far d up the inclines do the spheres move? The support system exclusive of the spheres has a weight of 10 N and a radius of gyration for the axis of rotation of 100 mm. Neglect friction and wind-resistance losses. Treat the spheres as particles.

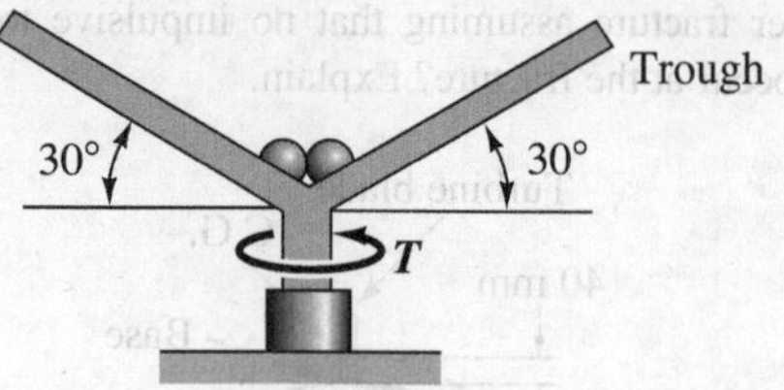

Figure P.15.63.

15.64. Two small spheres, each of weight 3 N, slide on a smooth rod curved in the shape of a parabola as shown. The curved rod is mounted on a platform on which a torque T of 1.5 N-m is applied for 2 sec. What is the angular speed of the system after 2 sec? Neglect friction. The curved rod and the platform have a mass of 5 kg and a radius of gyration of 200 mm.

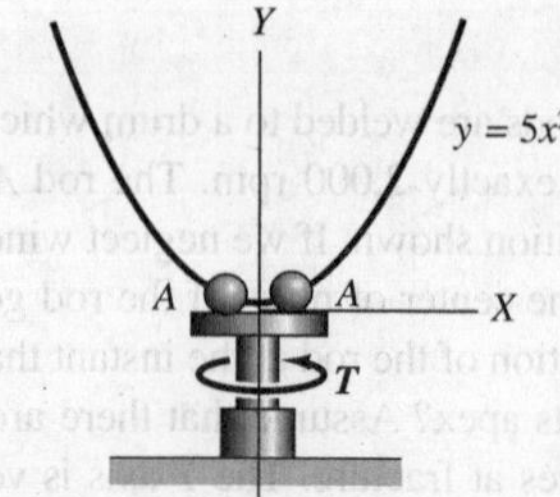

Figure P.15.64.

15.65. A disc of mass 3 kg is suspended between two wires. A bullet is fired at the disc and lodges at point A, for which $r = 100$ mm and $\theta = 45°$. If the bullet is traveling at a speed of 600 m/sec before striking the disc, what is the angular velocity of the system directly after the bullet gets lodged? The bullet weighs 1 N (*Caution:* What point in the disc should you work with? It is not 0!)

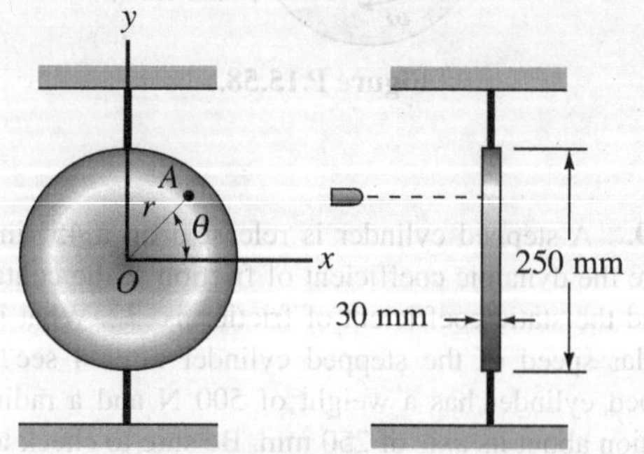

Figure P.15.65.

15.6 Impulsive Forces and Torques: Eccentric Impact

In Chapter 12, we introduced the concept of an *impulsive force*. Recall that an impulsive force F acts over a very short time interval Δt but has a very high value during this interval such that the impulse $\int_0^{\Delta t} F\,dt$ is significant. The impulse of other ordinary forces (not having very high peaks during Δt) is usually neglected for the short interval Δt. The same concept applies to torques, so that we have *impulsive torques*. The impulsive force and impulsive torque concepts are most valuable for the consideration of impact of bodies. Here, the collision forces and torques are impulsive while the other forces, such as gravity forces, have negligible impulse during collision.

In Chapter 12, we considered the case of *central impact* between bodies. Recall that for such problems the mass centers of the colliding bodies lie along the line of impact.[12] At this time, we shall consider the *eccentric impact of slablike* bodies undergoing plane motion such as shown in Fig. 15.34. For eccentric impact, at *least one of the mass centers does not lie along line of impact*. The bodies in Fig. 15.32 have just begun contact whereby point A of one body has just touched point B of the other body. The velocity of point A just before contact (preimpact) is given as $(V_A)_i$, while the velocity just before contact for point B is $(V_B)_i$. (The i stands for "initial," as in earlier work.) We shall consider only *smooth bodies,* so that the impulsive forces acting on the body at the point of contact are *collinear* with the line of impact. As a result of the impulsive forces, there is a *period of deformation*, as in our earlier studies, and a *period of restitution*. In the period of deformation, the bodies are deforming, while in the period of restitution there is a complete or partial recovery of the original geometries. At the end of the period of deformation, the points A and B have the *same velocity* and we denote this velocity as V_D. Directly after impact (post impact), the velocities of points A and B are denoted as $(V_A)_f$ and $V_B)_f$, respectively, where the subscript f is used to connote the final velocity resulting solely from the impact process. We shall be able to use the *linear impulse-momentum equation* and the *angular impulse-momentum* equation to relate the velocities, both linear and angular, for preimpact and postimpact states. These equations do not take into account the nature of the material of the colliding bodies, and so additional information is needed for solving these problems.

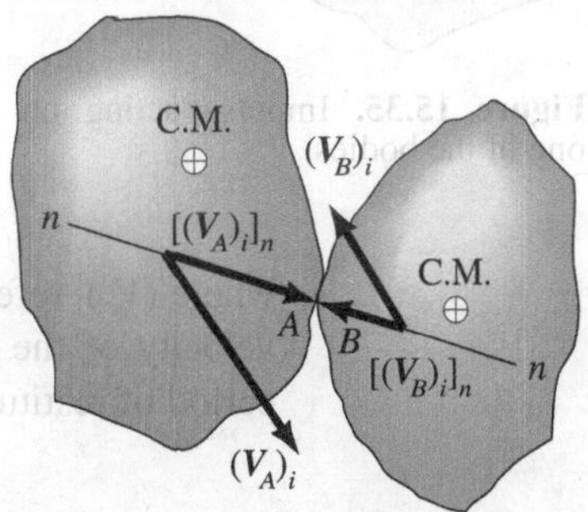

Figure 15.34. Eccentric impact between two bodies.

For this reason, we use the ratio between the impulse on each body during the period of restitution, $\int R\,dt$, and the impulse on each body during the period of deformation, $\int D\,dt$. As in central impact, the ratio is a number ϵ, called the *coefficient of restitution,* which depends primarily on the materials of the bodies in collision. Thus,

$$\epsilon = \frac{\int R\,dt}{\int D\,dt} \tag{15.29}$$

[12]The line of impact is normal to the plane of contact between the bodies.

We shall now show that the components along the line of impact *n–n* of V_A and V_B, taken at pre- and postimpact, are related to ϵ by the very same relation that we had for central impact. That is,

$$\epsilon = -\frac{[(V_B)_f]_n - [(V_A)_f]_n}{[(V_B)_i]_n - [(V_A)_i]_n} \tag{15.30}$$

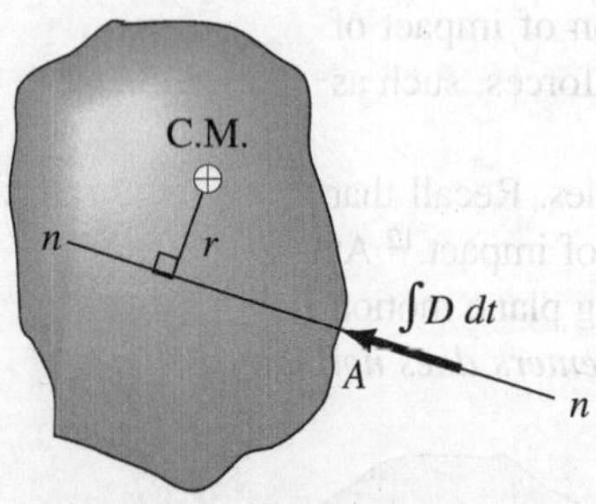

Figure 15.35. Impulse acting on one of the bodies.

Recall that the numerator represents the relative velocity of separation along *n* of the points of contact, whereas the denominator represents the relative velocity of approach along *n* of these points.

We shall first consider the case where the bodies are in *no way constrained* in their plane of motion; we can then neglect all impulses except that coming from the impact. We now consider the body having contact point *A*. In Fig. 15.35 we have shown this body with impulse $\int D\,dt$ acting. Using the component of the *linear momentum equation* along line of impact *n–n*, we can say for the center of mass;

$$\int D\,dt = M[(V_C)_D]_n - M[(V_C)_i]_n \tag{15.31}$$

where $(V_C)_i$ refers to the preimpact velocity of the center of mass and where $(V_C)_D$ is the velocity of the center of mass at the end of the deformation period. Similarly, for the period of restitution, we have

$$\int R\,dt = M[(V_C)_f]_n - M[(V_C)_D]_n \tag{15.32}$$

For angular momentum, we can say for the deformation period using *r* as the distance from the center of mass to *n–n*:

$$r\int D\,dt = I\omega_D - I\omega_i \tag{15.33}$$

Similarly, for the period of restitution:

$$r\int R\,dt = I\omega_f - I\omega_D \tag{15.34}$$

Now, substitute the right sides of Eqs. 15.31 and 15.32 into Eq. 15.29. We get on cancellation of *M*:

$$\epsilon = \frac{[(V_C)_f]_n - [(V_C)_D]_n}{[(V_C)_D]_n - [(V_C)_i]_n} = \frac{[(V_C)_D]_n - [(V_C)_f]_n}{[(V_C)_i]_n - [(V_C)_D]_n} \tag{15.35}$$

Next, substitute for the impulses in Eq. 15.29 using Eqs. 15.33 and 15.34. We get on canceling only *I*:

$$\epsilon = \frac{r\omega_f - r\omega_D}{r\omega_D - r\omega_i} = \frac{r\omega_D - r\omega_f}{r\omega_i - r\omega_D} \tag{15.36}$$

Adding the numerators and denominators of Eqs. 15.35 and 15.36, we can then say on rearranging the terms:

$$\epsilon = \frac{\left\{\left[(V_C)_D\right]_n + r\omega_D\right\} - \left\{\left[(V_C)_f\right]_n + r\omega_f\right\}}{\left\{\left[(V_C)_i\right]_n + r\omega_i\right\} - \left\{\left[(V_C)_D\right]_n + r\omega_D\right\}} \tag{15.37}$$

We pause now to consider the *kinematics* of the motion. We can relate the velocities of points A and C on the body (see Fig. 15.36) as follows:

$$\boldsymbol{V}_A = \boldsymbol{V}_C + \boldsymbol{\omega} \times \boldsymbol{\rho}_{CA} \tag{15.38}$$

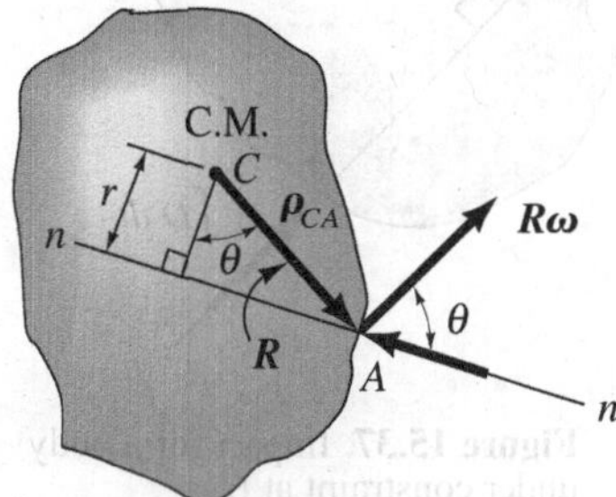

Figure 15.36. Slab with $|\boldsymbol{\rho}_{CA}|$ as R.

Note that the magnitude of $\boldsymbol{\rho}_{CA}$ is R as shown in the diagram. Since $\boldsymbol{\omega}$ is normal to the plane of symmetry of the body and thus to $\boldsymbol{\rho}_{CA}$, the value of the last term in Eq. 15.38 is $R\omega$ with a direction normal to R as has been shown in Fig. 15.36. The components of the vectors in Eq. 15.38 in direction n can then be given as follows:

$$(V_A)_n = (V_C)_n + \omega R \cos\theta \tag{15.39}$$

Since $R \cos\theta = r$ (see Fig. 15.36), we conclude that

$$(V_A)_n = (V_C)_n + r\omega \tag{15.40}$$

With the preceding result applied to the initial condition (i), the final condition (f), and the intermediate condition (D), we can now go back to Eq. 15.37 and replace the expressions inside the braces ({ }) by the left side of Eq. 15.40 as follows:

$$\epsilon = \frac{\left[(V_A)_D\right]_n - \left[(V_A)_f\right]_n}{\left[(V_A)_i\right]_n - \left[(V_A)_D\right]_n} \tag{15.41}$$

A similar process for the body having contact point B will yield the preceding equation with subscript B replacing subscript A:

$$\epsilon = \frac{\left[(V_B)_D\right]_n - \left[(V_B)_f\right]_n}{\left[(V_B)_i\right]_n - \left[(V_B)_D\right]_n} = \frac{\left[(V_B)_f\right]_n - \left[(V_B)_D\right]_n}{\left[(V_B)_D\right]_n - \left[(V_B)_i\right]_n} \tag{15.42}$$

Now add the numerators and denominators of the right side of Eq. 15.41 and the extreme right side of Eq. 15.42. Noting that

$$\left[(V_A)_D\right]_n = \left[(V_B)_D\right]_n \tag{15.43}$$

we get Eq. 15.30, thus demonstrating the validity of that equation.

Let us next consider the case where one or both bodies undergoing impact is constrained to *rotate about a fixed axis*. We have shown such a body in Fig. 15.37 where O is the axis of rotation and point A is the contact point. If an impulse is developed at the point of contact A (we have shown the impulse during the period of deformation), then clearly there will be an impulsive force at O, as shown in the diagram. We shall employ the

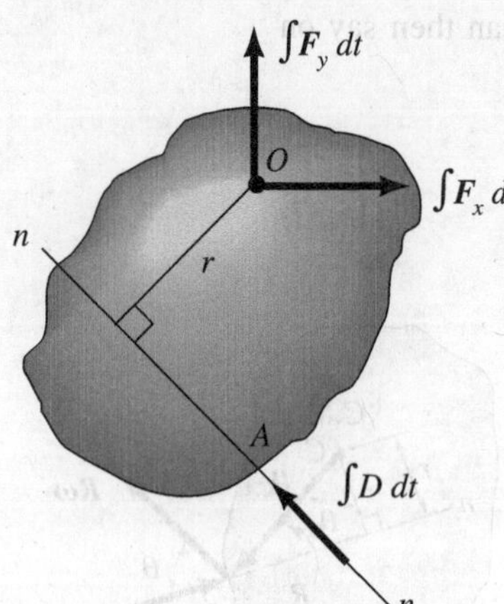

Figure 15.37. Impact for a body under constraint at O.

angular impulse-momentum equation about the fixed point O. Thus, we have for the period of deformation and the period of restitution:

$$r\int D\,dt = I_0\omega_D - I_0\omega_i$$
$$r\int R\,dt = I_0\omega_f - I_0\omega_D$$

Now solve for the impulses in the equation above, and substitute into Eq. 15.29. Canceling I_0, the moment of inertia about the axis of rotation at O, we get

$$\epsilon = \frac{r\omega_f - r\omega_D}{r\omega_D - r\omega_i} = \frac{r\omega_D - r\omega_f}{r\omega_i - r\omega_D} \tag{15.44}$$

In Fig. 15.38 we see that

$$V_A = R\omega$$

Therefore,

$$(V_A)_n = R\omega\cos\theta = r\omega$$

Using the above result in Eq. 15.44, we get

$$\epsilon = \frac{\left[(V_A)_D\right]_n - \left[(V_A)_f\right]_n}{\left[(V_A)_i\right]_n - \left[(V_A)_D\right]_n} \tag{15.45}$$

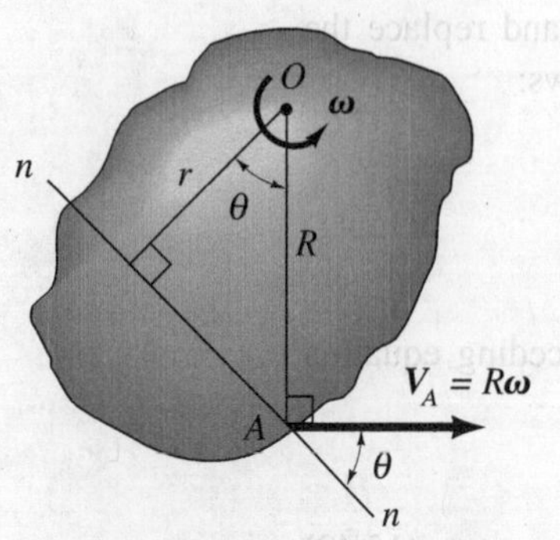

Figure 15.38. Velocity of point A is $R\omega$.

But this expression is identical to Eq. 15.41. And by considering the second body, which is either free or constrained to rotate, we get an equation corresponding to Eq. 15.42. We can then conclude that Eq. 15.30 is valid for impact where one or both bodies are constrained to rotate about a fixed axis.

In a typical impact problem, the motion of the bodies preimpact is given and the motion of the bodies postimpact is desired. Thus, there could be four unknowns—two velocities of the mass centers of the bodies plus two angular velocities. The required equations for solving the problem are formed from linear and angular momentum considerations of the bodies taken separately or taken as a system. Only impulsive forces are taken into consideration during the time interval spanning the impact. If the bodies are considered separately, we simply use the formulations of Section 15.5, remembering to observe Newton's third law at the point of impact between the two bodies. Furthermore, we must use the coefficient of restitution Eq. 15.30. Generally, kinematic considerations are also needed to solve the problem. When there are no other impulsive forces other than those occurring at the point of impact, it might be profitable to consider the bodies as one system. Then, clearly, as a result of Newton's third law, we must have *conservation of linear momentum* relative to an inertial reference, and also we must have *conservation of angular momentum* about *any one axis* fixed in inertial space.

We now illustrate these remarks in the following series of examples.

Example 15.11

A rectangular plate A weighing 20 N has two identical rods weighing 10 N each attached to it (see Fig. 15.39). The plate moves on a plane smooth surface at a speed of 5 m/sec. Moving oppositely at 10 m/sec is disc B, weighing 10 N. A perfectly elastic collision ($\epsilon = 1$) takes place at G. What is the speed of the center of mass of the plate just after collision (postimpact)? Solve the problem two ways: consider the bodies separately and consider the bodies as a system.

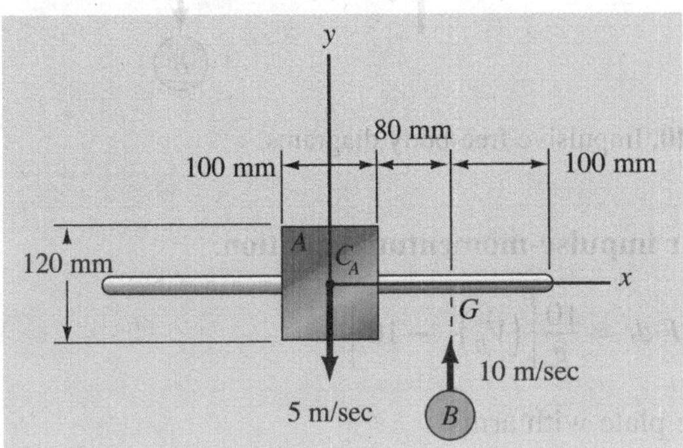

Figure 15.39. Colliding bodies.

Solution 1. We can consider the disc to be a particle since this body will only translate. Let point G be the point of contact on the rod. We then have, for Eq. 15.30:

$$\epsilon = 1 = -\frac{(V_G)_f - (V_B)_f}{-5 - 10}$$

Therefore,

$$(V_G)_f - (V_B)_f = 15 \tag{a}$$

Now, consider linear and angular momentum for each of the bodies. For this purpose we have shown the bodies in Fig. 15.40 with only impulsive forces acting. We might call such a diagram an "impulsive free-body diagram." We then see that C_A must move in the plus or minus y direction after impact. We can then say for body A, using **linear impulse-momentum** and **angular impulse-momentum equations** (the latter about the center of mass):

$$\int F\,dt = \frac{40}{g}\left[\left(V_{C_A}\right)_f - (-5)\right] \tag{b}$$

$$\begin{aligned}(.13)\int F\,dt &= \left[\frac{1}{12}\left(\frac{20}{g}\right)(.10^2 + .12^2)\right.\\ &\quad \left. + 2\left(\frac{1}{12}\right)\left(\frac{10}{g}\right)(.18)^2 + 2\left(\frac{10}{g}\right)(.14)^2\right](\omega_A - 0) \\ &= .0496\omega_A\end{aligned} \tag{c}$$

Example 15.11 (Continued)

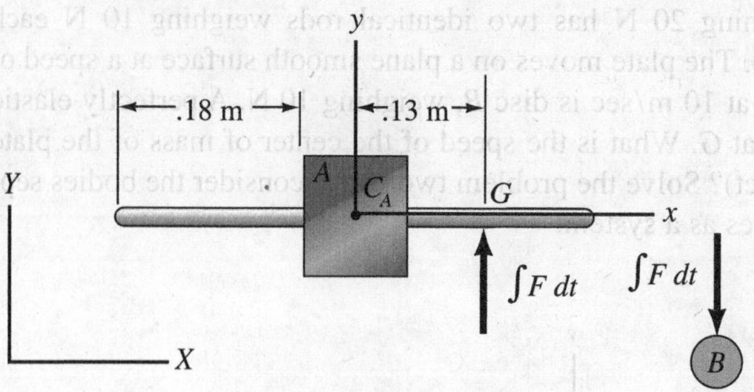

Figure 15.40. Impulsive free-body diagrams.

For body B we have for the **linear impulse-momentum equation**:

$$-\int F\,dt = \frac{10}{g}\left[\left(V_B\right)_f - 10\right] \tag{d}$$

From **kinematics** we have for the plate with arms:

$$(V_G)_f = (V_{C_A})_f + (\omega_A)(.13) \tag{e}$$

We have now a complete set of equations which we can proceed to solve. We get:

$$(V_{C_A})_f = -.305 \text{ m/sec}$$

Solution 2. Equation (a) above from the coefficient of restitution and Eq. (e) above from kinematics also apply to solution 2. Substituting for $(V_G)_f$ in Eq. (a) of solution 1 using Eq. (e) of solution 1, we get for solution 2:

$$(V_{C_A})_f + .13\omega_A - (V_B)_f = 15 \tag{a}$$

Conservation of linear momentum for the system leads to the requirement in the y direction that

$$\frac{40}{g}(-5) + \frac{10}{g}(10) = \frac{40}{g}(V_{C_A})_f + \frac{10}{g}(V_B)_f$$

Therefore,

$$4(V_{C_A})_f + (V_B)_f = -10 \tag{b}$$

Also **angular momentum is conserved** about any fixed z axis. We choose the axis at the position corresponding to C_A at the time of impact. Noting that I for the plate and arms is .0496 kg-m^2 from solution 1 (see Eq. (c)) and noting that B can be considered as a particle, we have

Example 15.11 (Continued)

$$\underbrace{\frac{10}{g}(10)(.13)}_{\substack{\text{angular}\\ \text{momentum}\\ \text{preimpact}}} = \underbrace{(.0496)\omega_A + \frac{10}{g}(V_B)_f(.13)}_{\substack{\text{angular}\\ \text{momentum}\\ \text{postimpact}}}$$

Therefore,

$$\omega_A = -2.67(V_B)_f + 26.7 \qquad \text{(c)}$$

Solving Eqs. (a), (b), and (c) simultaneously we get:

$$\left(V_{C_A}\right)_f = -.305 \text{ m/sec}$$

We leave it to you to demonstrate there has been *conservation of mechanical energy* during this perfectly elastic impact. We could have used this fact in place of Eq. (a) in solution 1.

Note that there was some saving of time and labor in using the system approach throughout for the preceding problem wherein a rigid body, namely the plate and its arms, collided with a body, the disc, which could be considered as a particle. In problems involving two colliding rigid bodies neither of which can be considered a particle, we must consider the bodies separately since the system approach does not yield a sufficient number of independent equations as you can yourself demonstrate.

In the preceding example, the bodies were not constrained except to move in a plane. If one or both colliding bodies is pinned, the procedure for solving the problem may be a little different than what was shown in Example 15.11. Note that there will be unknown supporting impulsive forces at the pin of any pinned body. If we are not interested in the supporting impulsive force for a pinned body, we only consider *angular momentum* about the pin for that pinned body; in this way the undesired unknown supporting impulsive forces at the pin do not enter the calculations. Other than this one factor, the calculations are the same as in the previous example.

You will recall from momentum considerations of particles that we considered the collision of a comparatively small body with a very massive one. We could not use the conservation of momentum equation for the collision between such bodies since the velocity change of the massive body went to zero as the mass (mathematically speaking) went to infinity, thus producing an indeterminacy in our idealized formulations. We shall next illustrate the procedure for the collision of a very massive body with a much smaller one. You will note that we cannot consider a system approach for linear or angular momentum conservation for the same reasons set forth in Chapter 12.

Example 15.12

A 20-lb rod AB is dropped onto a massive body (Fig. 15.41). What is the angular velocity of the rod postimpact for the following conditions:

A. Smooth floor; elastic impact.
B. Rough floor (no slipping); plastic impact.

In either case, the velocity of end B preimpact is

$$V_B = \sqrt{2gh} = \sqrt{(2)(32.2)(2)} = 11.35 \text{ ft/sec}$$

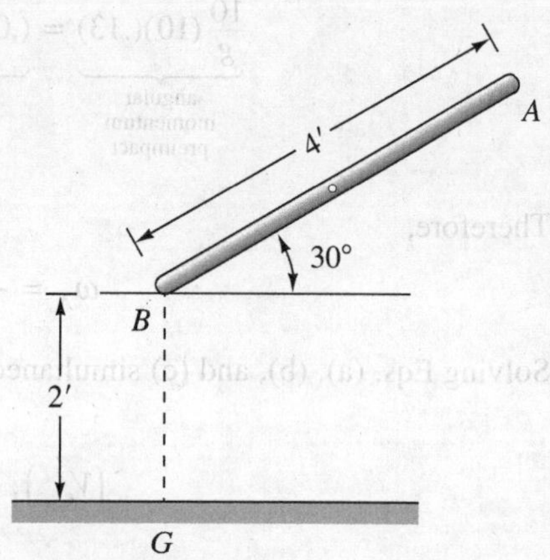

Figure 15.41. Falling rod on a massive body.

We now consider each case separately.

Case A. Equation 15.30 can be used here. Thus,

$$\epsilon = 1 = -\frac{\left[(V_B)_f\right]_n - 0}{-11.35 - 0}$$

Therefore,

$$\left[(V_B)_f\right]_n = 11.35 \text{ ft/sec} \qquad \text{(a)}$$

Next, considering rod AB in Fig. 15.42 we have for **linear impulse-** and **angular impulse-momentum** considerations (the latter about an axis at the center of mass).

$$\int F\,dt = \frac{20}{g}\left[(V_C)_f - (-11.35)\right] \qquad \text{(b)}$$

$$-(2)(\cos 30°)\int F\,dt = \frac{1}{12}\left(\frac{20}{g}\right)(4^2)\omega_f \qquad \text{(c)}$$

From **kinematics**, we have[13]

$$(\boldsymbol{V}_B)_f = (\boldsymbol{V}_C)_f + \boldsymbol{\omega}_f \times \boldsymbol{\rho}_{CB}$$

$$\left[(V_B)_x\right]_f \boldsymbol{i} + 11.35\boldsymbol{j} = (V_C)_f\,\boldsymbol{j} + \omega_f \boldsymbol{k} \times (2)(-.866\boldsymbol{i} - .5\boldsymbol{j})$$

$$\left[(V_B)_x\right]_f \boldsymbol{i} + 11.35\boldsymbol{j} = (V_C)_f\,\boldsymbol{j} - 1.732\omega_f \boldsymbol{j} + \omega_f \boldsymbol{i}$$

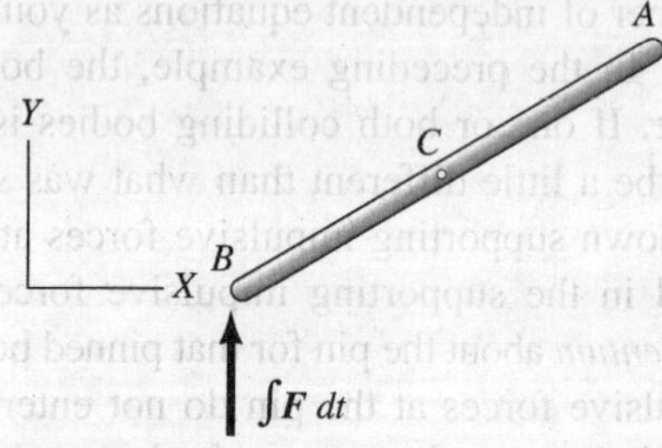

Figure 15.42. Impulsive free-body diagram.

Hence, the scalar equations are

$$11.35 = (V_C)_f - 1.732\omega_f \qquad \text{(d)}$$

$$\left[(V_B)_x\right]_f = \omega_f \qquad \text{(e)}$$

[13]Note that since the initial velocity of C is vertical and since the impulsive force is vertical, the final velocity of C must be vertical. This fact is used in the kinematics.

Example 15.12 (Continued)

We now have a complete set of equations considering $\int F\,dt$ as an unknown.

Solving, we have for the desired unknowns:

$$\omega_f = -9.07 \text{ rad/sec}$$
$$[(V_B)_x]_f = -9.07 \text{ ft/sec}$$

You can demonstrate that energy has been conserved in this action. We could have used this fact in lieu of Eq. (a) for this problem.

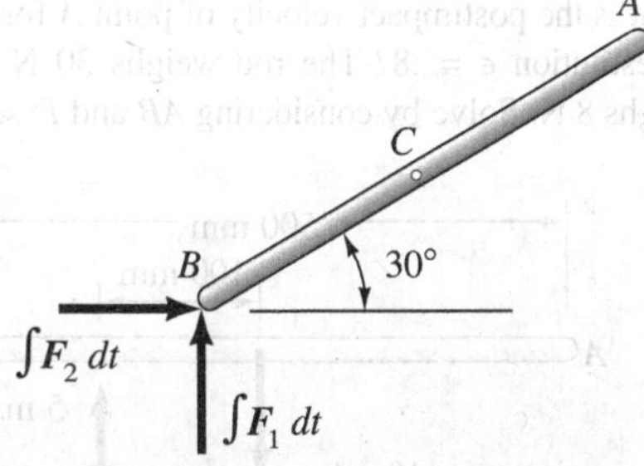

Figure 15.43. Impulsive free-body diagram.

Case B. Here we have no slipping on the rough surface and zero vertical movement of point *B*. Accordingly, we have shown rod *BA* with vertical and horizontal impulses in Fig. 15.43. The **linear momentum equations** for the center of mass then are

$$\int F_1\,dt = \frac{20}{g}\left\{[(V_C)_y]_f - (-11.35)\right\} \qquad \text{(f)}$$

$$\int F_2\,dt = \frac{20}{g}\left\{[(V_C)_x]_f - 0\right\} \qquad \text{(g)}$$

The **angular impulse-momentum equation** about the center of mass is

$$-2(\cos 30°)\int F_1\,dt + 2(\sin 30°)\int F_2\,dt = \frac{1}{12}\left(\frac{20}{g}\right)(4^2)\omega_f \qquad \text{(h)}$$

From **kinematics,** noting that at postimpact there is *pure rotation* about point *B*, we have

$$[(V_C)_y]_f = 2(\cos 30°)(\omega_f) = 1.732\omega_f \qquad \text{(i)}$$

$$[(V_C)_x]_f = -2(\sin 30°)(\omega_f) = -\omega_f \qquad \text{(j)}$$

Substitute for $\int F_1\,dt$ and $\int F_2\,dt$ from Eqs. (f) and (g) into Eq. (h). We get on then employing Eqs. (i) and (j):

$$-(2)(.866)\left(\frac{20}{g}\right)(1.732\omega_f + 11.35) + (2)(.5)\left(\frac{20}{g}\right)(-\omega_f) = \frac{1}{12}\left(\frac{20}{g}\right)(4^2)\omega_f$$

Therefore,

$$\omega_f = -3.69 \text{ rad/sec}$$

PROBLEMS

15.66. A rod AB slides on a smooth surface at a speed of 10 m/sec and hits a disc D moving at a speed of 5 m/sec. What is the postimpact velocity of point A for a coefficient of restitution $\epsilon = .8$? The rod weighs 30 N and the disc weighs 8 N. Solve by considering AB and D separately.

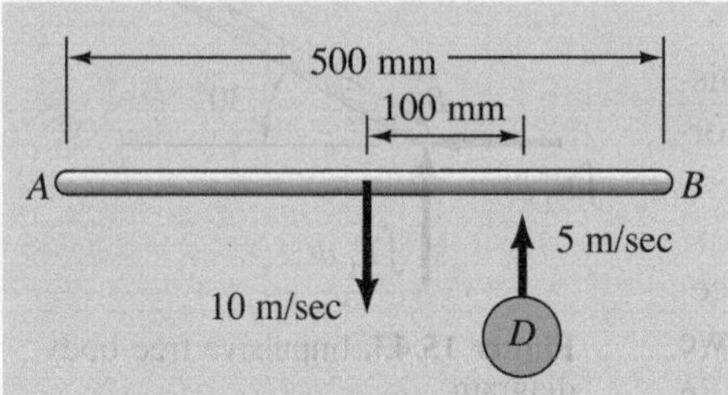

Figure P.15.66.

15.67. Solve Problem 15.66 by considering the two bodies as a system.

15.68. Rods AB and HD translate on a horizontal frictionless surface. When they collide at G we have a coefficient of restitution of .7. Rod HD weighs 70 N and rod AB weighs 40 N. What is the postimpact angular speed of AB?

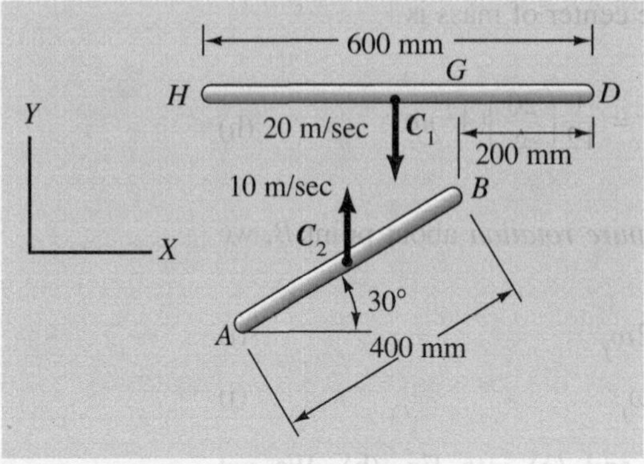

Figure P.15.68.

15.69. Two slender rods 20 lb each with small 45° protuberances are shown on a smooth horizontal surface. The rods are identical except for the position of the protuberances. Rod A moves to the left at a speed of 10 ft/sec while rod B is stationary. If the protuberance surfaces are smooth, and an impact having $\epsilon = .8$ occurs, what should d be in order for the postimpact angular speed of A to be 5 rad/sec? Neglect the protuberance for determination of moments of inertia and centers of mass.

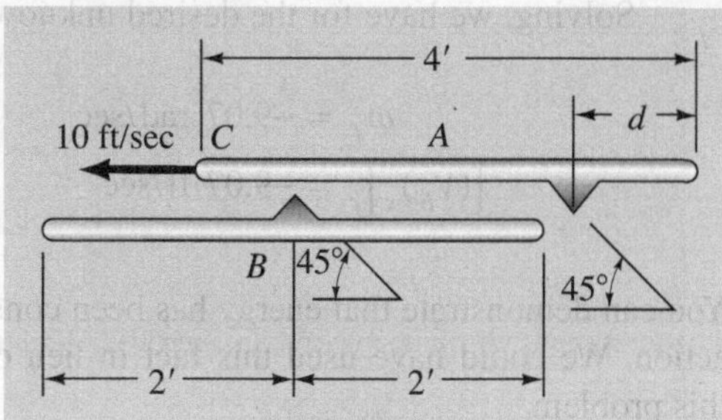

Figure P.15.69.

15.70. A 22-N sphere moving at a speed of 10 m/sec hits the end of a 1-m rod having a mass of 10 kg. The coefficient of restitution for the impact is .9. What is the postimpact angular velocity of the rod if it is stationary just before impact? The rod is pinned at O.

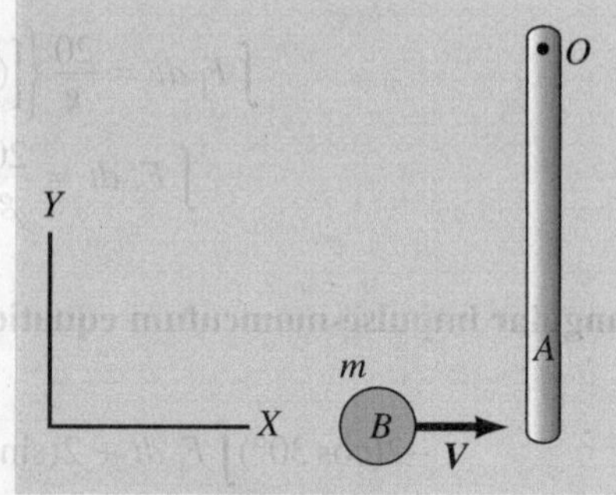

Figure P.15.70.

15.71. A rod A weighing 50 N rotates freely about a hinge at a speed of 10 rad/sec on a frictionless surface just prior to hitting a disc B weighing 20 N and moving at a speed of 5 m/sec. If the coefficient of restitution is .8, what is the postimpact angular speed of A? The contact surface between the rod and the disc is smooth.

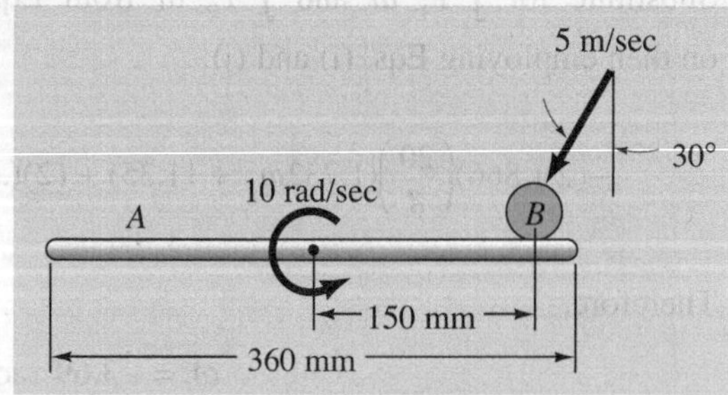

Figure P.15.71.

15.72. A stiff bent rod is dropped so that end A strikes a heavy table D. If the impact is plastic and there is no sliding at A, what is the postimpact speed of end B? The rod weighs per unit length 30 N/m.

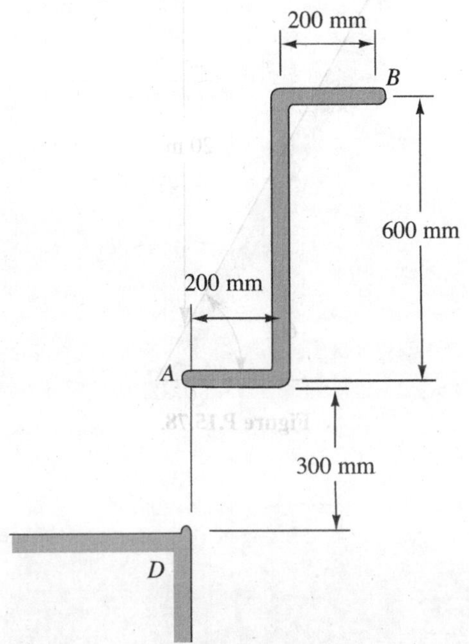

Figure P.15.72.

15.73. Solve Problem 15.72 for an elastic impact at A. Take the surface of contact to be smooth. Demonstrate that energy has been conserved.

15.74. A drumstick and a wooden learner's drum are shown in (a) of the diagram. At the beginning of a drum roll, the drumstick is horizontal. The action of the drummer's hand is simulated by force components F_x and F_y, which make A a stationary axis of rotation. Also, there is a constant couple M of .5 N-m. If the action starts from a stationary position shown, what is the frequency of the drum roll for perfectly elastic collision between the drumstick and the learner's drum? The mass of the drumstick is 200 g. Idealize the drumstick as a uniform slender rod.

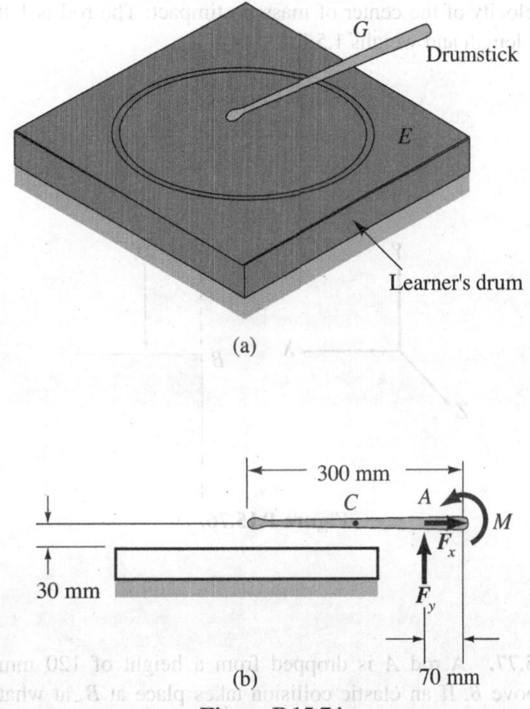

Figure P.15.74.

15.75. A block of ice 1 ft × 1 ft × $\frac{1}{2}$ ft slides along a surface at a speed of 10 ft/sec. The block strikes stop D in a plastic impact. Does the block turn over after the impact? What is the highest angular speed reached? Take $\gamma = 62.4$ lb/ft².

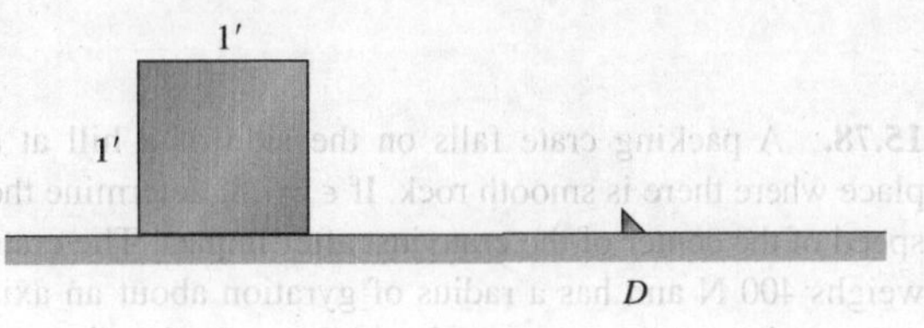

Figure P.15.75.

15.76. A horizontal rigid rod is dropped from a height 10 ft above a heavy table. The end of the rod collides with the table. If the coefficient of impact ϵ between the end of the rod and the corner of the table is .6, what is the postimpact angular velocity of the rod? Also, what is the velocity of the center of mass postimpact: The rod is 1 ft in length and weighs 1.5 lb.

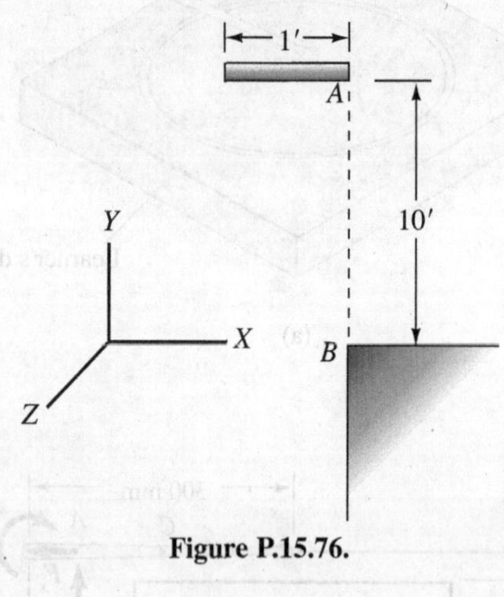

Figure P.15.76.

15.77. A rod A is dropped from a height of 120 mm above B. If an elastic collision takes place at B, at what time Δt later does a second collision with support D take place? The rod weighs 40 N.

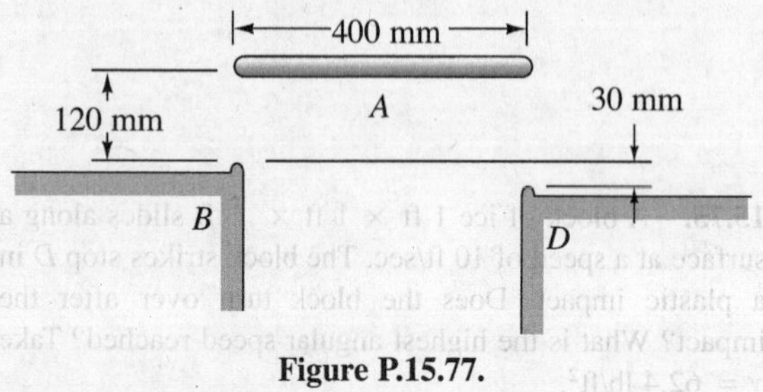

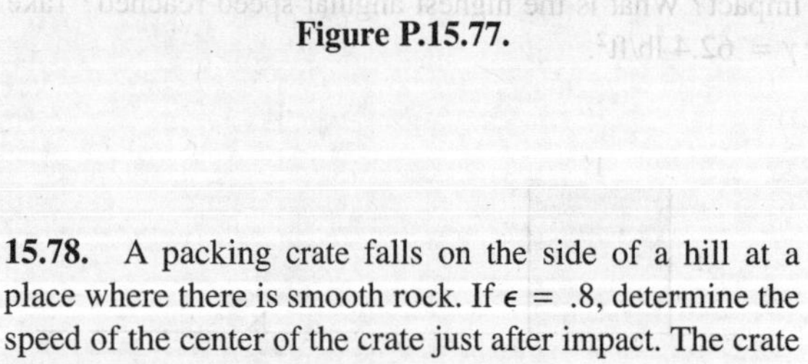

Figure P.15.77.

15.78. A packing crate falls on the side of a hill at a place where there is smooth rock. If $\epsilon = .8$, determine the speed of the center of the crate just after impact. The crate weighs 400 N and has a radius of gyration about an axis through the center and normal to the face in the diagram of $\sqrt{2}$ m.

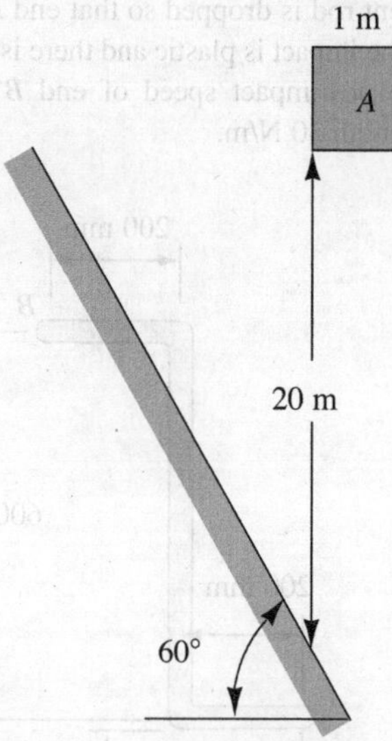

Figure P.15.78.

15.79. An arrow moving essentially horizontally at a speed of 20 m/sec impinges on a stationary wooden block which can rotate freely about light rods DE and GH. The arrowhead weighs 1.5 N, and the shaft, which is 250 mm long, weighs .7 N. If the block weighs 10 N, what is the angular velocity of the block just after the arrowhead becomes stuck in the wood? The arrowhead sinks into the wood at point A such that the shaft sticks out 250 mm from the surface of the wood.

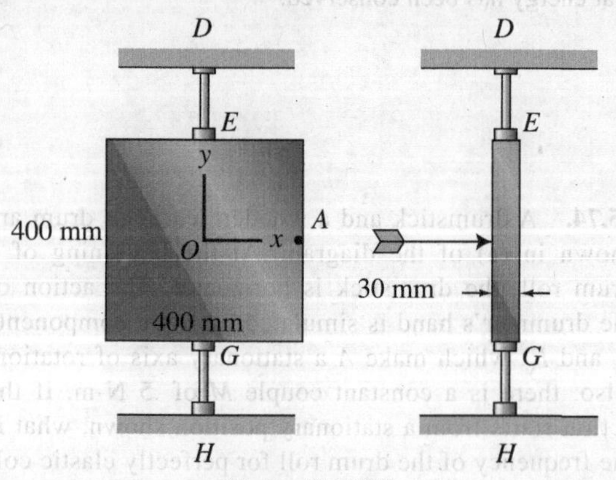

Figure P.15.79.

15.7 Closure

Let us now pause for an overview of the text up to this point. Recall the following:

(a) In Chapter 11, we studied energy methods for single particles and systems of particles with a short introduction to coplanar rigid body motions.
(b) Then in Chapter 12, we presented linear and angular momentum principles for a single particle and systems of particles with again a short introduction to plane motion of rigid bodies.

These methods are derived from Newton's law. In short, they are integrated forms of Newton's law which bring in useful concepts and greater ease in solving many problems. Next, after studying *Chasles' theorem* and kinematics of rigid bodies in Chapter 13, we went to Chapter 14.

(c) In Chapter 14, we studied the dynamics of plane motion for rigid bodies using Newton's law and the equation $\boldsymbol{M}_A = \dot{\boldsymbol{H}}_A$ directly rather than integrated of forms of these equations, namely the energy and momentum equations.

Now in Chapter 15, we went through a partial recycle of the above steps. That is, we went back to Chapters 11 and 12 this time for three-dimensional formulations and focused on rigid bodies and systems of rigid bodies. Naturally, there is an overlap with those earlier chapters plus an expansion of viewpoints to include a general formulation and use of $\boldsymbol{H}$ and a look at eccentric coplanar impact of more complex bodies. Again, with the aforementioned integrated forms of *Newton's law*, we were able to solve some interesting problems.

We are now at a stage comparable to what we were in just preceding Chapter 14.

PROBLEMS

15.80. Gear E rotates at an angular speed ω of 5 rad/sec and drives four smaller "floating" gears A, B, C, and D, which roll within stationary gear F. What is the kinetic energy of the system if gear E weighs 50 lb and each of the small gear weighs 10 lb?

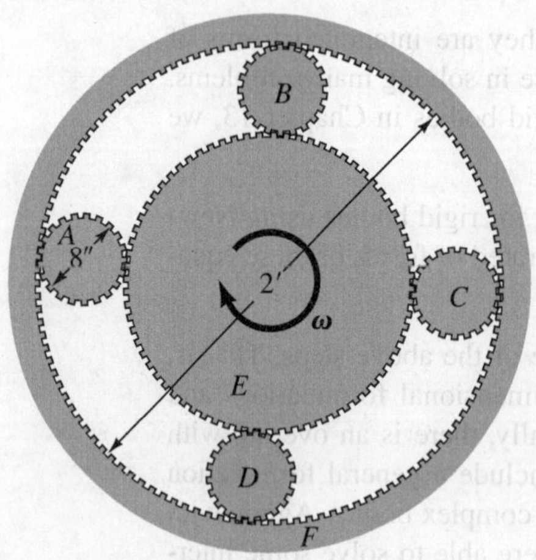

Figure P.15.80.

15.81. A homogeneous rectangular parallelepiped weighing 200 N rotates at 20 rad/sec about a main diagonal held by bearings A and B, which are mounted on a vehicle moving at a speed 20 m/sec. What is the kinetic energy of the rectangular parallelepiped relative to the ground?

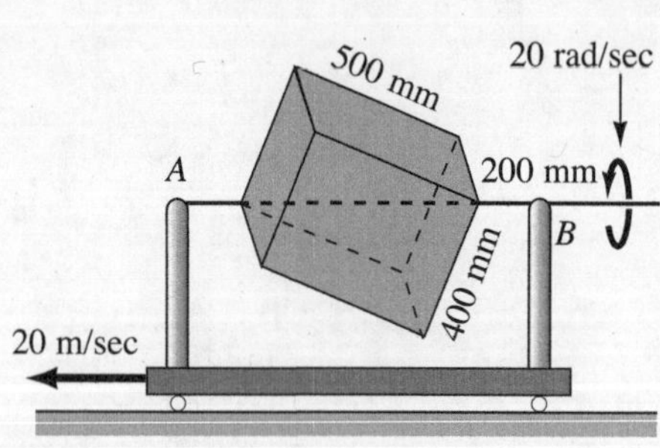

Figure P.15.81.

15.82. A rod AB is held at a position $\theta = 45°$ and released. HA is a light wire to which the rod is attached. When AB is vertical, it strikes a stop E and undergoes an elastic collision. What position d of the stop will result in a zero postimpact velocity of point A? The rod has a mass $M = 10$ kg.

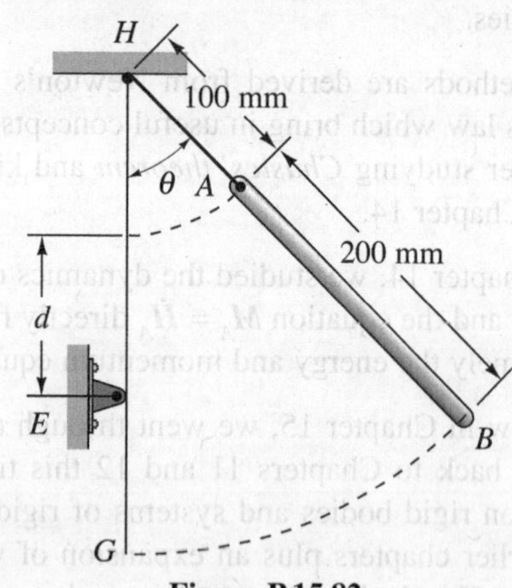

Figure P.15.82.

15.83. A stepped cylinder weighs 100 lb and has a radius of gyration of 4 ft. A 50-lb block A is welded to the cylinder. If the spring is unstretched in the configuration shown and has a spring constant K of .5 lb/in., what is the angular speed of the cylinder after it rotates 90°? Assume that the cylinder rolls without slipping.

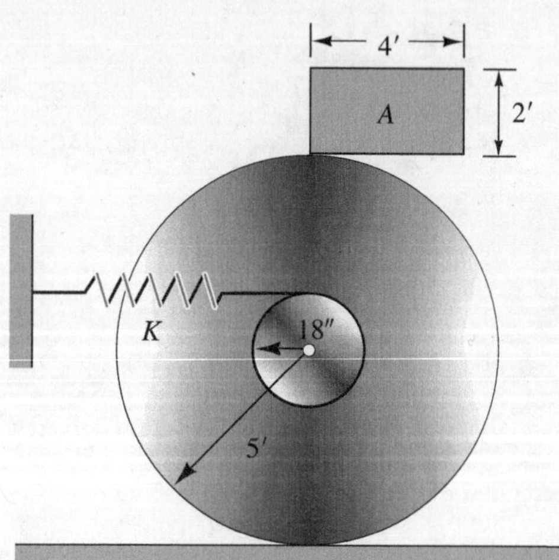

Figure P.15.83.

15.84. A tractor with driver weighs 3,000 lb. If a torque of 200 ft-lb is developed on each of the drive wheels by the motor, what is the speed of the tractor after it moves 10 ft? The large drive wheels each weigh 200 lb and have a diameter of 3 ft and a radius of gyration of 1 ft. The small wheels each weigh 40 lb and have a diameter of 1 ft with a radius of gyration of .4 ft. Do not use the center-of-mass approach for the whole system.

Figure P.15.84.

15.85. Work Problem 15.84 using the center of mass of the whole system. Find the friction forces on each wheel.

15.86. What is the angular momentum vector about the center of mass of a homogeneous rectangular parallelepiped rotating with an angular velocity of 10 rad/sec about a main diagonal? The sides of the rectangular parallelepiped are 1 ft, 2 ft, and 4 ft, as shown, and the weight is 4,000 lb.

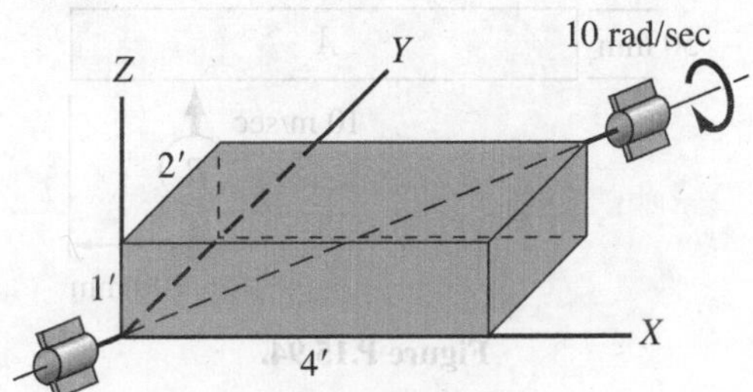

Figure P.15.86.

15.87. A force P of 200 N is applied to a cart. The cart minus the four wheels weighs 150 N. Each wheel has a weight of 50 N, a diameter of 400 mm, and a radius of gyration of 150 mm. The load A weighs 500 N. If the cart starts from rest, what is its speed in 20 sec? The wheels roll without slipping.

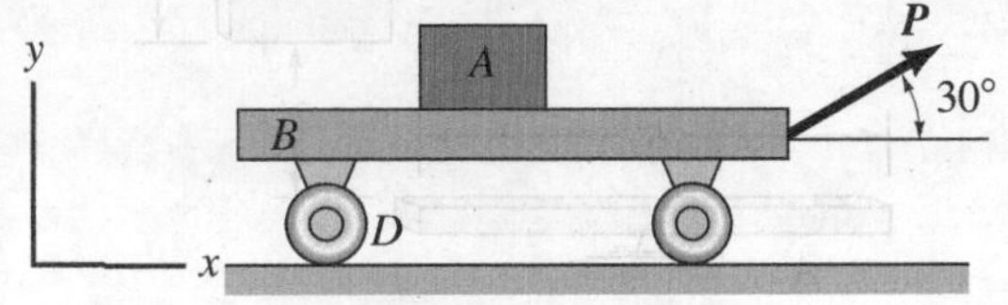

Figure P.15.87.

***15.88.** A rod weighing 90 N is guided by two slider bearings A and B. Smooth ball joints connect the rod to the bearings. A force F of 45 N acts on bearing A. What is the speed of A after it has moved 200 mm? The system is stationary for the configuration shown. Neglect friction.

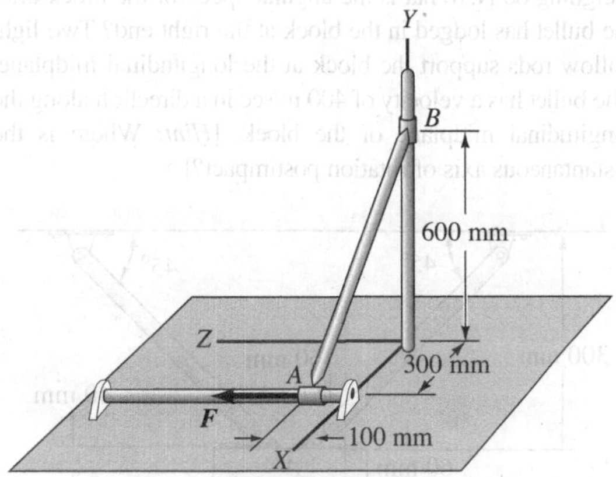

Figure P.15.88.

***15.89.** A device with thin walls contains water. The device is supported on a platform on which a torque T of .5 N-m is applied for 2 sec and is zero thereafter. What is the angular velocity ω of the system at the end of 2 sec assuming that, as a result of low viscosity, the water has no rotation relative to the ground in the vertical tubes. The system of tubes and platform have a weight of 50 N and a radius of gyration of 200 mm. [*Hint:* The pressure in the liquid is equal at all times to the specific weight γ times the distance d below the free surface of the liquid.] Note that for water γ = 9,810 N/m^3. Note: water heights in the tubes change with angular speed.

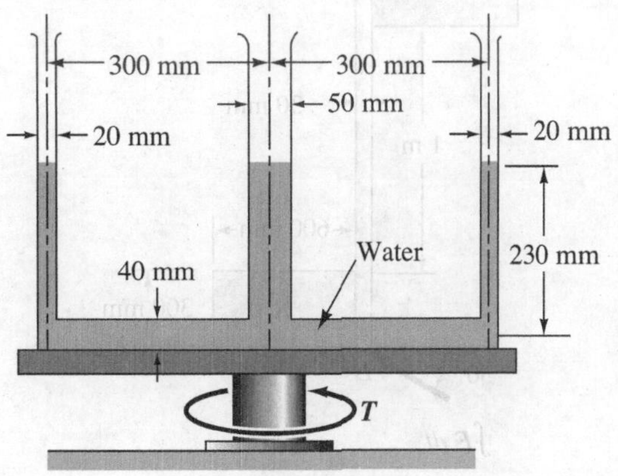

Figure P.15.89.

***15.90.** In Problem 15.89 what is the final angular speed of the system when viscosity has had its full effect and there is no movement of the water relative to the container walls? Disregard other frictional effects.

15.91. A bullet weighing $\frac{1}{2}$ N is fired at a wooden block weighing 60 N. What is the angular speed of the block after the bullet has lodged in the block at the right end? Two light hollow rods support the block at the longitudinal midplane. The bullet has a velocity of 400 m/sec in a direction along the longitudinal midplane of the block. [*Hint:* Where is the instantaneous axis of rotation postimpact?]

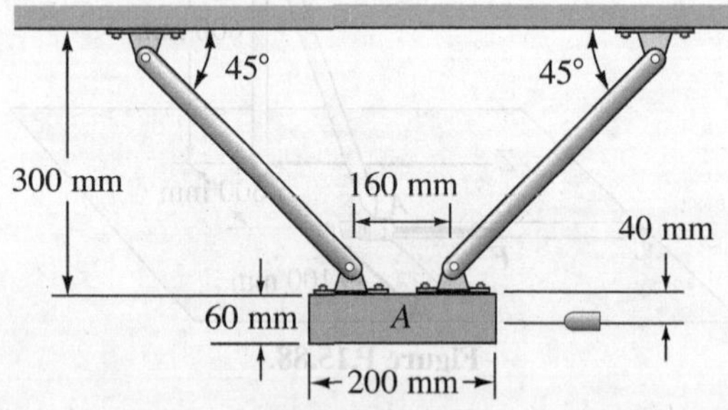

Figure P.15.91.

15.92. A device consists of two identical rectangular plates each weighing 100 N welded to a rod weighing 50 N. The device rests horizontally on a smooth surface. As a result of a collision, a horizontal impulse of 30 N-sec is delivered to point D. What is the velocity of corner E on plate A postimpact?

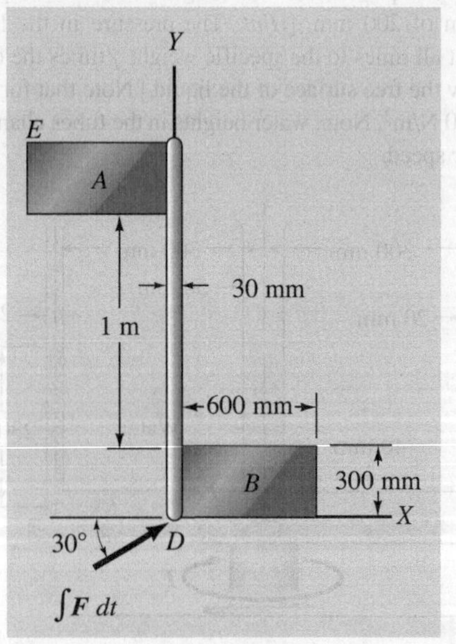

Figure P.15.92.

15.93 A cylinder A of weight 150 N and radius of gyration 100 mm is placed onto a conveyor belt which is moving at the constant speed of $V_B = 10$ m/sec. Give the speed of the axis of the cylinder at time $t = 5$ sec. The coefficient of friction between the cylinder and the belt is .5.

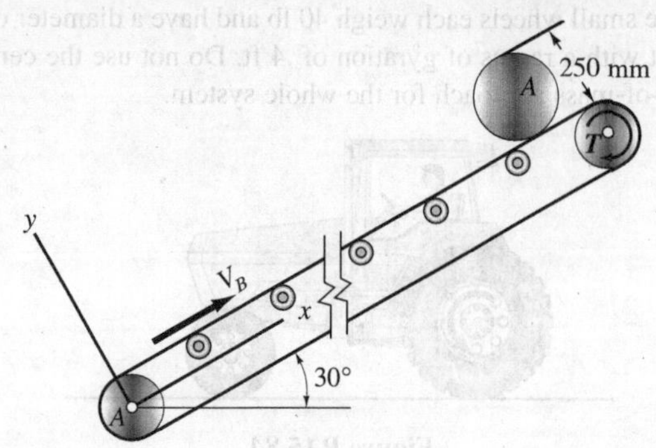

Figure P.15.93.

15.94 A rectangular rod A rests on a smooth horizontal surface. A disc B moves toward the rod at a speed of 10/m sec. What is the postimpact angular velocity of the rod for a coefficient of restitution of .6? The rod weighs 50 N, and the disc B weighs 8 N.

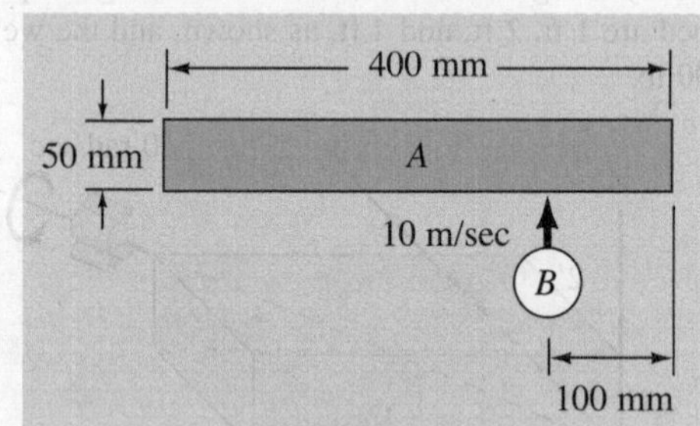

Figure P.15.94.

15.95. A block B weighs 10 lb is dropped onto the end of a rectangular rod A weighing 20 lb. What is the angular speed of rod A postimpact for $\epsilon = .7$? The rod is pinned at its center as shown.

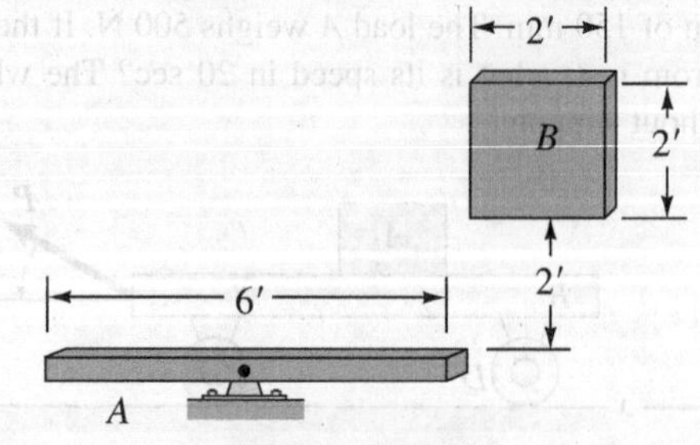

Figure P.15.95.

CHAPTER 16

Vibrations

16.1 Introduction

You will recall that in Chapter 10 we said we would defer a more general examination of particle motion about a fixed point until the very end of the text. We do this to take full advantage of any course in differential equations that you might be taking simultaneously with this course. Accordingly, we shall now continue the work begun in Chapter 10.

16.2 Free Vibration

Let us begin by reiterating what we have done earlier leading to the study of vibrations. Recall that we examined the case of a particle in rectilinear translation acted on either by a constant force, a force given as a function of time, a force that is a function of speed, or, finally, a force that is a function of position. In each case we could separate the variables and effect a quadrature to arrive at the desired algebraic equations, including constants of integration. In particular we considered, as a special case of a force given as a function of position, the linear restoring force resulting from the action (or equivalent action) of a linear spring. Thus, for the spring-mass system shown in Fig. 16.1, the differential equation of motion was shown to be

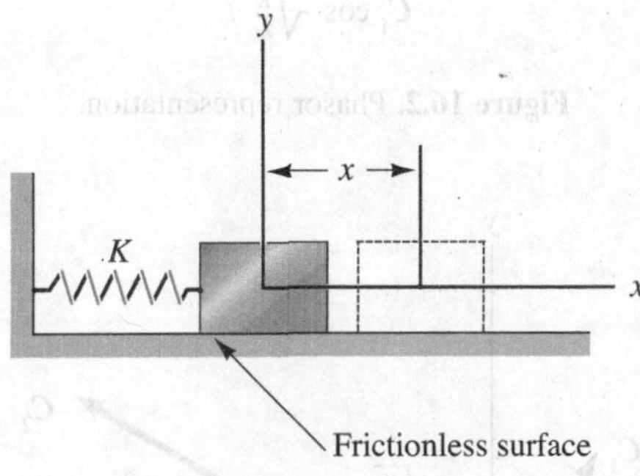

Figure 16.1. Spring–mass system.

$$\frac{d^2x}{dt^2} + \frac{K}{m}x = 0 \tag{16.1}$$

where K is the spring constant and where x is measured from the static equilibrium position of the mass. You will now recognize this equation from your studies in mathematics as a second-order, linear differential equation with constant coefficients.

Instead of rearranging the equation to effect a quadrature, as we did in the previous case,[1] we shall take a more general viewpoint toward the solving of differential equations.

[1]Recall that this can be done by replacing d^2x/dt^2 by $(dV/dx)(dx/dt)$, which is simply $V(dV/dx)$.

To solve a differential equation, we must find a function of time, $x(t)$, which when substituted into the equation satisfies the equation (i.e., reduces it to an identity $0 = 0$). We can either guess at $x(t)$ or use a formal procedure. You have learned in your differential equations course that the most general solution of the above equation will consist of a linear combination of two functions that cannot be written as multiples of each other (i.e, the functions are linearly independent). There will also be two arbitrary constants of integration. Thus, $C_1 \cos \sqrt{K/m}\,t$ and $C_2 \sin \sqrt{K/m}\,t$ will satisfy the equation, as we can readily demonstrate by substitution, and are independent in the manner described. We can therefore say

$$x = C_1 \cos \sqrt{\frac{K}{m}}t + C_2 \sin \sqrt{\frac{K}{m}}t \tag{16.2}$$

where C_1 and C_2 are the aforementioned constants of integration to be determined from the initial conditions.

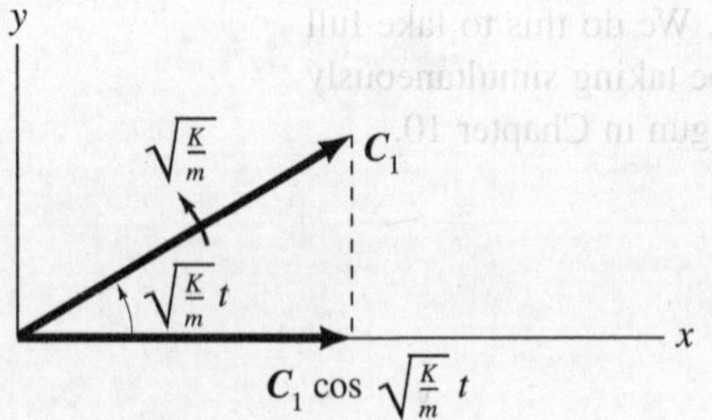

Figure 16.2. Phasor representation.

We can conveniently represent each of the above functions by employing rotating vectors of magnitudes that correspond to the coefficients of the functions. This representation is shown in Fig. 16.2, where, if the vector C_1 rotates counterclockwise with an angular velocity of $\sqrt{K/m}$ radians per unit time and if C_1 lies along the x axis at time $t = 0$, then the projection of this vector along the x axis represents one of the functions of Eq. 16.2, namely $C_1 \cos \sqrt{K/m}\,t$. Vectors used in this manner are called *phasors*.

Consider now the function $C_2 \sin \sqrt{K/m}\,t$, which we can replace by $C_2 \cos (\sqrt{K/m}\,t - \pi/2)$, as we learned in elementary trigonometry. The phasor representation for this function, therefore, would be a vector of magnitude C_2 that rotates with angular velocity $\sqrt{K/m}$ and that is out of phase by $\pi/2$ with the phasor C_1 (Fig. 16.3). Thus, the projection of C_2 on the x axis is the other function of Eq. 16.2. Clearly, because vectors C_1 and C_2 rotate at the same angular speed, we can represent the combined contribution by simply summing the vectors and considering the projection of the resulting single vector along the x axis. This summation is shown in Fig. 16.4 where vector C_3 replaces the vectors C_1 and C_2. Now we can say:

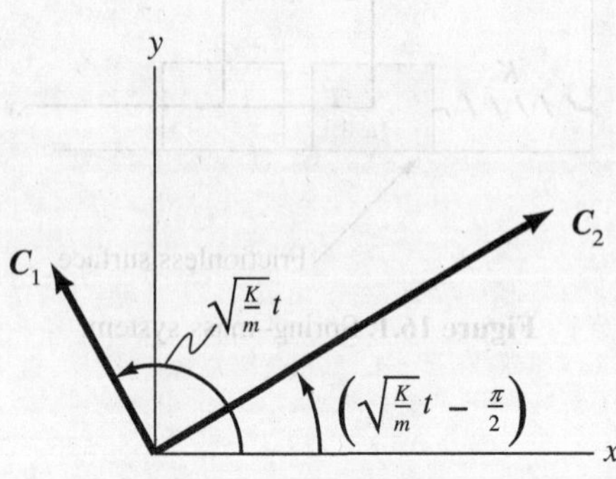

Figure 16.3. Phasors $\pi/2$ out of phase.

$$C_3 = \sqrt{C_1^2 + C_2^2}, \qquad \beta = \tan^{-1}\frac{C_2}{C_1} \tag{16.3}$$

Because C_1 and C_2 are arbitrary constants, C_3 and β are also arbitrary constants. Consequently, we can replace the solution given by Eq. 16.2 by another equivalent form:

$$x = C_3 \cos\left[\sqrt{\frac{K}{m}}t - \beta\right] \tag{16.4}$$

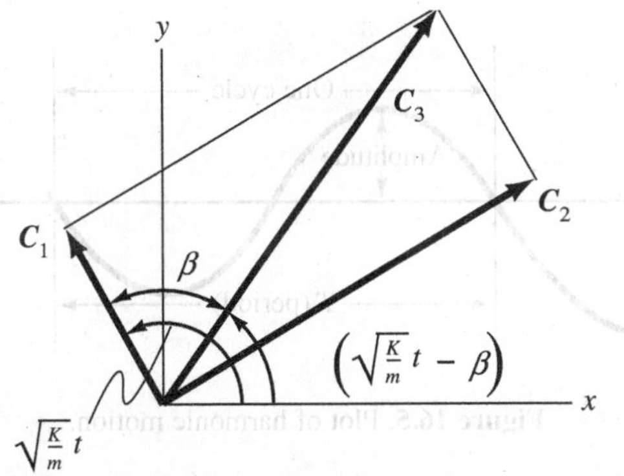

Figure 16.4. Vector sum of phasors.

From this form, you probably recognize that the motion of the body is *harmonic motion*. In studying this type of motion, we shall use the following definitions:

Cycle. The cycle is that portion of a motion (or series of events in the more general usage) which, when repeated, forms the motion. On the phasor diagrams, a cycle would be the motion associated with one revolution of the rotating vector.

Frequency. The number of cycles per unit time is the frequency. The frequency is equal to $\sqrt{K/m}/2\pi$ for the above motion, because $\sqrt{K/m}$ has units of radians per unit time. Often $\sqrt{K/m}$ is termed the *natural frequency* of the system in radians per unit time or, when divided by 2π, in cycles per unit time. The *natural frequency* is denoted generally in the following ways:

$$\omega_n = \sqrt{K/m} \text{ rad/sec}$$

$$f_n = \frac{1}{2\pi}\sqrt{K/m} \text{ cycles/sec}$$

Period. The period, T, is the time of one cycle, and is therefore the reciprocal of frequency. That is,

$$T = \frac{2\pi}{\sqrt{K/m}} \tag{16.5}$$

Amplitude. The largest displacement attained by the body during a cycle is the amplitude. In this case, the amplitude corresponds to the coefficient C_3.

Phase angle. The phase angle is the angle between the phasor and the x axis when $t = 0$ (i.e., the angle β).

A plot of the motion as a function of time is presented in Fig. 16.5, where certain of these various quantities are shown graphically.

It is usually easier to use the earlier form of solution, Eq. 16.2, rather than Eq. 16.4 in satisfying initial conditions. Accordingly, the position and velocity can be given as

$$x = C_1 \cos\sqrt{\frac{K}{m}}t + C_2 \sin\sqrt{\frac{K}{m}}t$$

$$V = -C_1\sqrt{\frac{K}{m}} \sin\sqrt{\frac{K}{m}}t + C_2\sqrt{\frac{K}{m}} \cos\sqrt{\frac{K}{m}}t$$

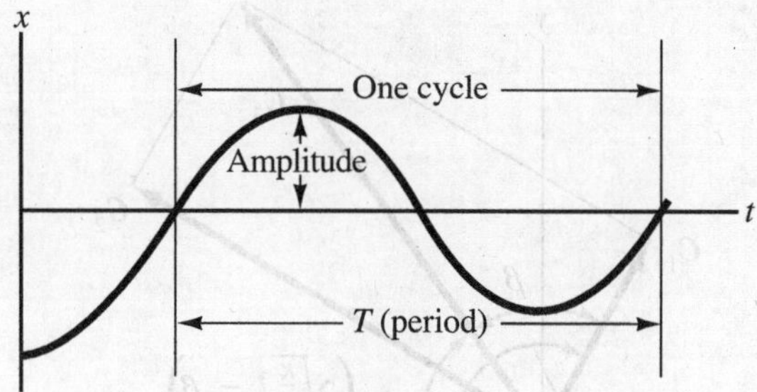

Figure 16.5. Plot of harmonic motion.

The initial conditions to be applied to these equations are:

$$\text{when } t = 0 \qquad x = x_0,\ V = V_0$$

Substituting, we get

$$x_0 = C_1, \qquad V_0 = C_2\sqrt{\frac{K}{m}}$$

Therefore, the motion is given as

$$x = x_0 \cos\sqrt{\frac{K}{m}}t + \frac{V_0}{\sqrt{K/m}}\sin\sqrt{\frac{K}{m}}t \tag{16.6a}$$

$$V = -x_0\sqrt{\frac{K}{m}}\sin\sqrt{\frac{K}{m}}t + V_0\cos\sqrt{\frac{K}{m}}t \tag{16.6b}$$

We can generalize from these results by noting that any agent supplying a linear restoring force for all rectilinear motions of a mass can take the place of the spring in the preceding computations. We must remember, however, that to behave this way the agent must have negligible mass. Thus, we can associate with such agents an *equivalent spring constant* K_e, which we can ascertain if we know the static deflection δ permitted by the agent on application of some known force F. We can then say:

$$K_e = \frac{F}{\delta}$$

Once we determine the equivalent spring constant, we immediately know that the natural frequency of the system is $(1/2\pi)\sqrt{K_e/m}$ cycles per unit time. This natural frequency is the number of cycles the system will repeat in a unit time if some initial disturbance is imposed on the mass. Note that this natural frequency depends only on the "stiffness" of the system and on the mass of the system and is not dependent on the amplitude of the motion.[2]

We shall now consider several problems in which we can apply what we have just learned about harmonic motion.

[2]Actually, when the amplitude gets comparatively large, the spring ceases to be linear, and the motion does depend on the amplitude. Our results do not apply for such a condition.

Example 16.1

A mass weighing 45 N is placed on the spring shown in Fig. 16.6 and is released very slowly, extending the spring a distance of 50 mm. What is the natural frequency of the system? If the mass is given a velocity instantaneously of 1.60 m/sec down from the equilibrium position, what is the equation for displacement as a function of time?

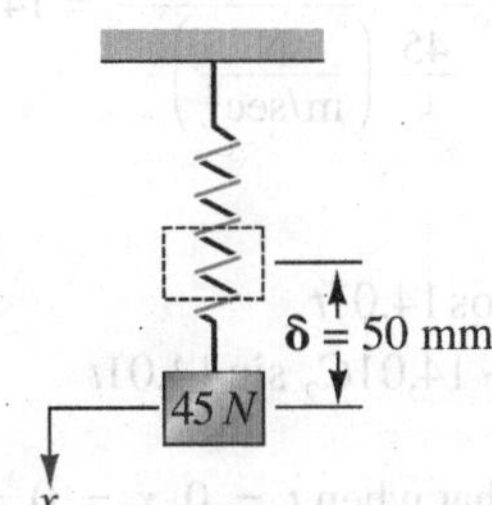

Figure 16.6. x measured from static deflection position.

The spring constant is immediately available by the equation

$$K = \frac{F}{\delta} = \frac{45 \text{ N}}{50 \text{ mm}} = .9 \text{ N/mm}$$

The equation of motion for the mass can be written for a reference whose origin is at the static equilibrium position shown in the diagram. Thus,

$$m\frac{d^2x}{dt^2} = W - K(x + \delta)$$

where δ is the distance from the unextended position of the spring to the origin of the reference. However, from our initial equation, $\delta = F/K = W/K$. Therefore, we have

$$m\frac{d^2x}{dt^2} = W - K(x + \frac{W}{K}) = -Kx$$

and the equation becomes identical to Eq. 16.1:

$$\frac{d^2x}{dt^2} + \frac{K}{m}x = 0 \qquad \text{(a)}$$

Thus, the motion will be an oscillation about the position of static equilibrium, which is an extended position of the spring. Measuring x from the *static equilibrium position* and considering the spring force as $-Kx$, we can thus disregard the weight on the body in writing Newton's law for the body to reach Eq. (a) most directly.

Example 16.1 (Continued)

Accordingly, we can use the results stemming from our main discussion. Employing the notation ω_n as the natural frequency in units of radians per unit time, we have

$$\omega_n = \sqrt{\frac{K}{m}} = \sqrt{\frac{\left(\frac{.9\ \text{N}}{\text{mm}}\right)\left(\frac{1{,}000\ \text{mm}}{\text{m}}\right)}{\frac{45}{g}\ \text{kg}}} = \sqrt{\frac{\left(\frac{.9\ \text{N}}{\text{mm}}\right)\left(\frac{1{,}000\ \text{mm}}{\text{m}}\right)}{\frac{45}{g}\left(\frac{\text{N}}{\text{m/sec}^2}\right)}} = 14.01\ \text{rad/sec}$$

The motion is now given by the equations

$$x = C_1 \sin 14.01t + C_2 \cos 14.01t$$
$$\dot{x} = 14.01C_1 \cos 14.01t - 14.01C_2 \sin 14.01t$$

From the specified initial conditions, we know that when $t = 0, x = 0$, and $x = 1.6$ m/sec. Therefore, the constants of integration are

$$C_2 = 0, \qquad C_1 = \frac{1.60}{14.01} = .1142$$

The desired equation, then, is

$$x = .1142 \sin 14.01t \text{ m}$$

Example 16.2

A body weighing 22 N is positioned in Fig. 16.7 on the end of a slender cantilever beam whose mass we can neglect in considering the motion of the body at its end.

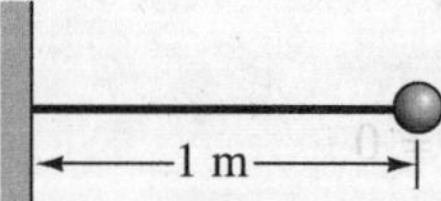

Figure 16.7. Slender cantilever beam with weight at end.

If we know the geometry and the composition of the cantilever beam, and if the deflection involved is small, we can compute from strength of materials the deflection of the end of the beam that results from a vertical load there. This deflection is directly proportional to the load. In this case, suppose that we have computed a deflection of

Example 16.2 (Continued)

12.5 mm for a force of 4 N (see Fig. 16.8). What would be the natural frequency of the body weighing 22 N for small oscillations in the vertical direction?

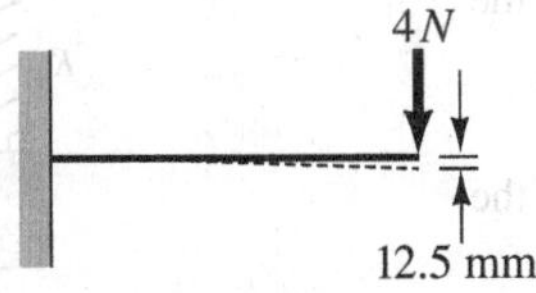

Figure 16.8. Beam acts as linear spring.

Because the motion is restricted to small amplitudes, we can consider the mass to be in rectilinear motion in the vertical direction in the same manner as the mass on the spring in the previous case. The beam now supplies the linear restoring force. The formulations of this section are once again applicable. The equivalent spring constant is found to be

$$K_e = \frac{F}{\delta} = \frac{4}{12.5} = .32 \text{ N/mm}$$

The natural frequency for vibration of the 22-N weight at the end of the cantilever is then

$$\omega_n = \sqrt{\frac{(.32)(1,000)}{22/9.81}} = \boxed{11.94 \text{ rad/sec}}$$

Before starting out on the exercises, we wish to point out results of Problem 16.2 that will be of use to you. In that problem, you are asked to show that the equivalent spring constant for springs in *parallel* and subject to the same deflection [see Fig. P.16.2(a)] is simply the sum of the spring constants of the springs. That is,

$$K_e = K_1 + K_2 \tag{16.7}$$

For springs in series [Fig. P.16.2(b)] we have

$$\frac{1}{K_e} = \frac{1}{K_1} + \frac{1}{K_2} \tag{16.8}$$

Also, we wish to point out that, for small values of θ, we can approximate sin θ by θ and cos θ by unity. To justify this, expand sin θ and cos θ as power series and retain first terms.

PROBLEMS

16.1. If a 5-kg mass causes an elongation of 50 mm when suspended from the end of a spring, determine the natural frequency of the spring–mass system.

16.2. (a) Show that the spring constant is doubled if the length of the spring is halved.

(b) Show that two springs having spring constants K_1 and K_2 have a combined spring constant of $K_1 + K_2$ when connected in parallel, and have a combined spring constant whose reciprocal is $1/K_1 + 1/K_2$ when combined in series.

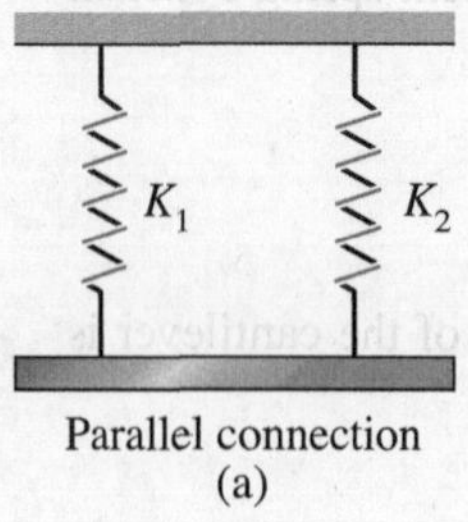

Parallel connection
(a)

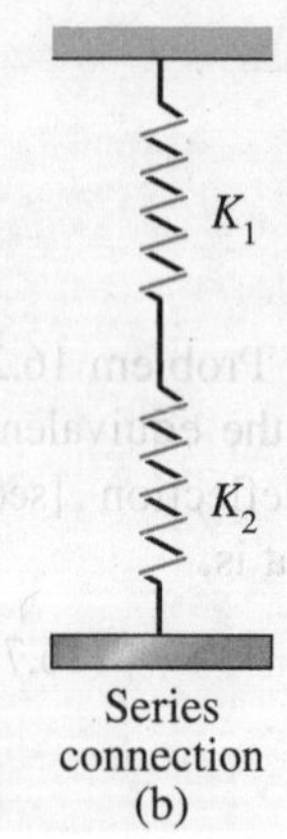

Series connection
(b)

Figure P.16.2.

16.3. A mass M of 2 kg rides on a vertical frictionless guide rod. With only one spring K_1, the natural frequency of the system is 2 rad/sec. If we want to increase the natural frequency threefold, what must the spring constant K_2 of a second spring be?

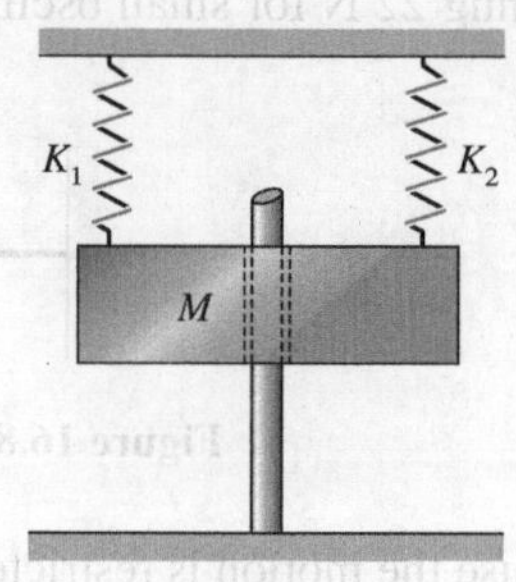

Figure P.16.3.

16.4. A mass M Of 100 g rides on a frictionless guide rod. If the natural frequency with spring K_1 attached is 5 rad/sec, what must K_2 be to increase the natural frequency to 8 rad/sec?

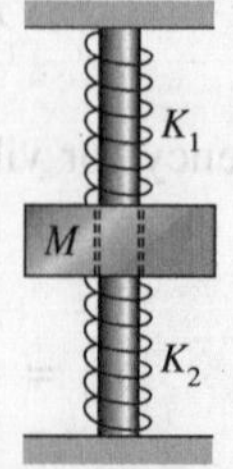

Figure P.16.4.

16.5. For small oscillations, what is the natural frequency of the system in terms of a, b, K, and W? (Neglect the mass of the rod.)

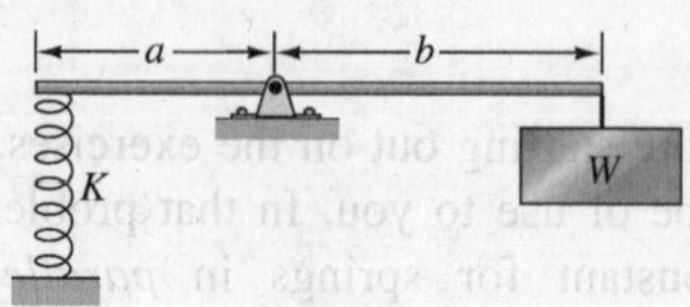

Figure P.16.5.

16.6. A rod is supported on two rotating grooved wheels. The contact surfaces have a coefficient of friction of μ_d. Explain how the rod will oscillate in the horizontal direction if it is disturbed in that direction. Compute the natural frequency of the system.

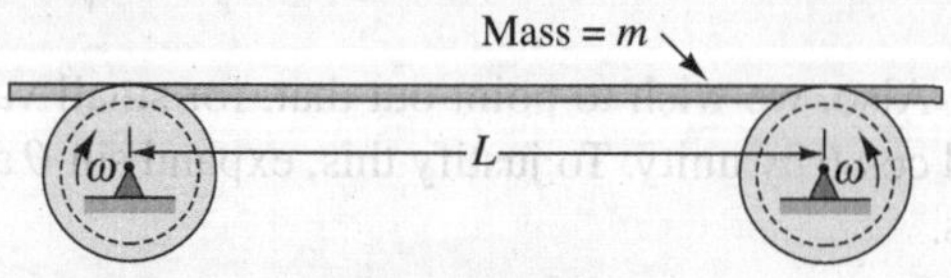

Figure P.16.6.

16.7. A mass is held so it just makes contact with a spring. If the mass is released suddenly from this position, give the amplitude, frequency, and the center position of the motion. First use the *undeformed position* to measure x. Then, do the problem using x' from the static equilibrium position.

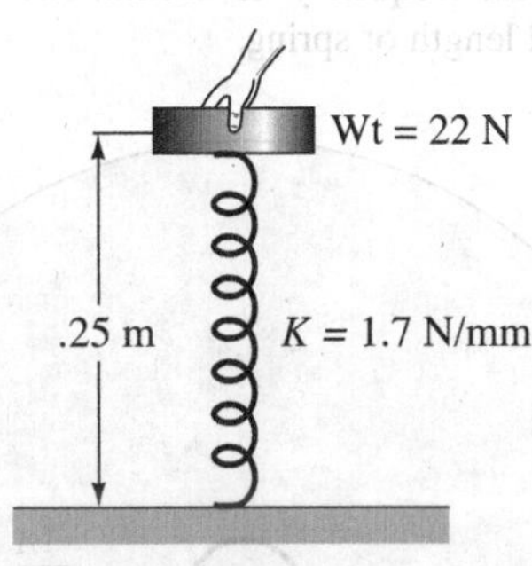

Figure P.16.7.

16.8. A *hydrometer* is a device to measure the *specific gravity* of liquids. The hydrometer weighs .36 N, and the diameter of the cylindrical portion above the base is 6 mm. If the hydrometer is disturbed in the vertical direction, what is the frequency of vibration in cycles/sec as it bobs up and down? Recall from *Archimedes' principle* that the buoyant force equals the weight of the water displaced. Water has a density of 1,000 kg/m^3.

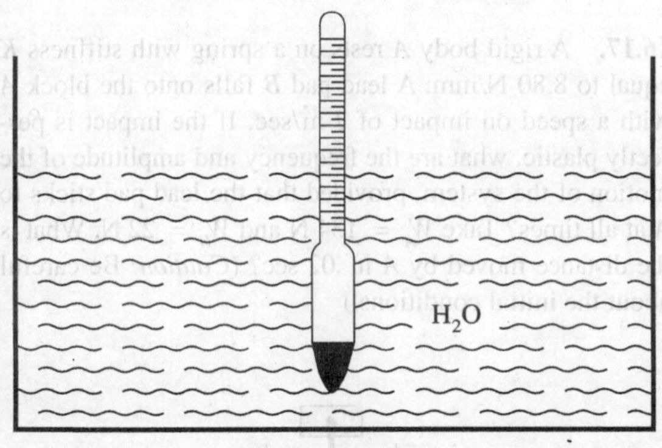

Figure P.16.8.

***16.9.** The hydrometer of Problem 16.8 is used to test the specific gravity of battery acid in a car battery. What is the period of oscillation in this case? [*Hint:* Note if hydrometer goes down, the battery acid surface will have to rise a certain amount simultaneously. The battery acid has a density of 1,100 kg/m^3.]

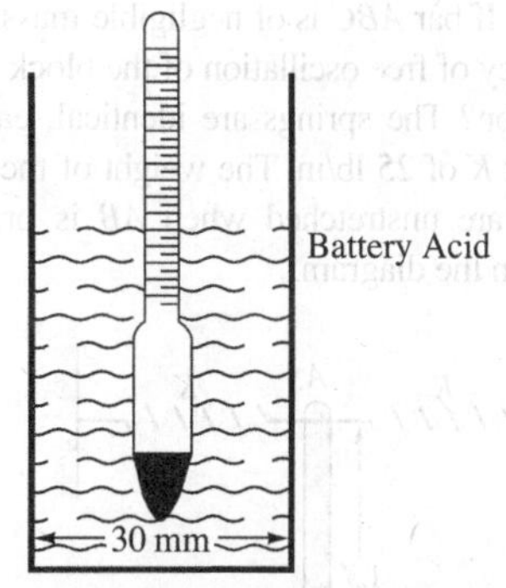

Figure P.16.9.

16.10. A 30-kg block is suspended using two light wires. What is the frequency in cycles/sec at which the block will swing back and forth in the x direction if it is slightly disturbed in this direction? [*Hint:* For small θ, $\sin\theta \approx \theta$ and $\cos\theta \approx 1$.]

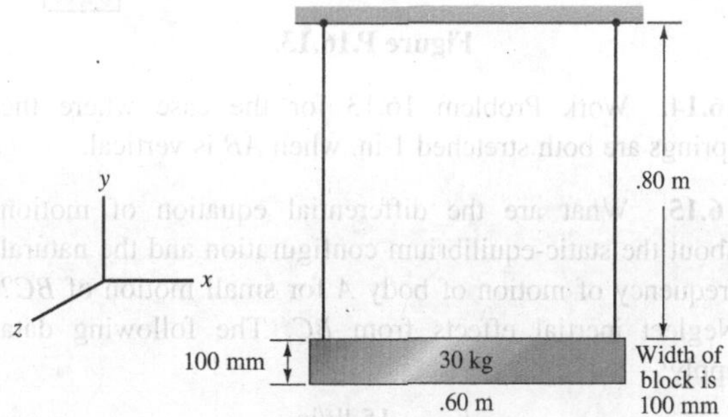

Figure P.16.10.

16.11. In Problem 16.10, what is the period of small oscillations for a small disturbance that causes the block to move in the z direction?

16.12. What is the natural frequency of motion for block A for small oscillation? Consider BC to have negligible mass and body A to be a particle. When body A Is attached to the rod, the static deflection is 25 mm. The spring constant K_1 is 1.75 N/mm. Body A weighs 110 N. What is K_2?

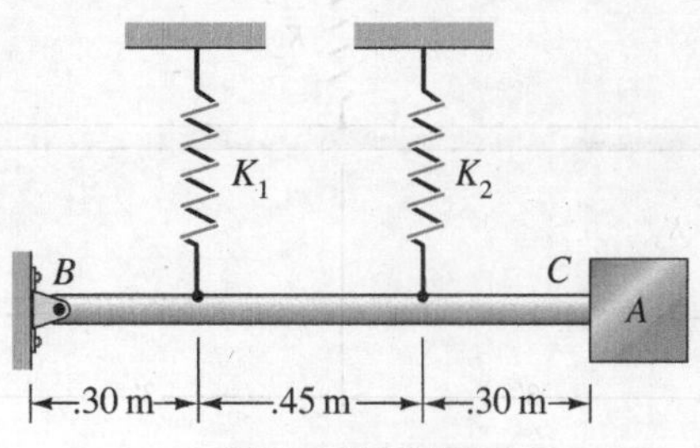

Figure P.16.12.

16.13. If bar ABC is of negligible mass, what is the natural frequency of free oscillation of the block for small amplitude of motion? The springs are identical, each having a spring constant K of 25 lb/in. The weight of the block is 10 lb. The springs are unstretched when AB is oriented vertically as shown in the diagram.

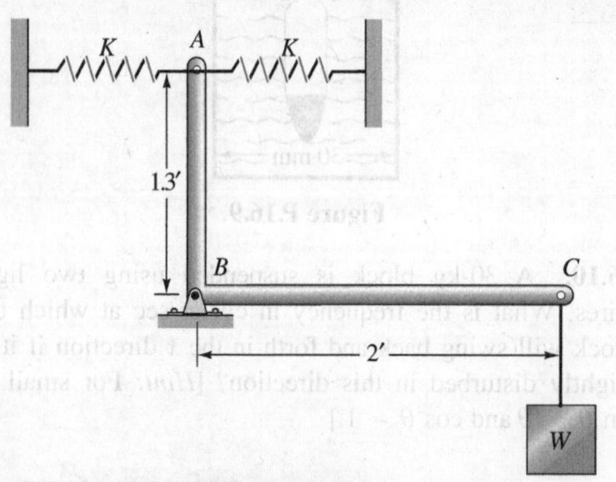

Figure P.16.13.

16.14. Work Problem 16.13 for the case where the springs are both stretched 1 in. when AB is vertical.

16.15. What are the differential equation of motion about the static-equilibrium configuration and the natural frequency of motion of body A for small motion of BC? Neglect inertial effects from BC. The following data apply:

$$K_1 = 15 \text{ lb/in.}$$
$$K_2 = 20 \text{ lb/in.}$$
$$K_3 = 30 \text{ lb/in.}$$
$$W_A = 30 \text{ lb}$$

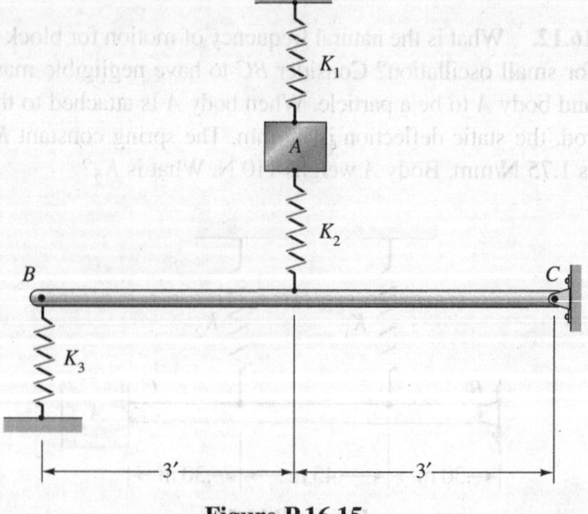

Figure P.16.15.

16.16. A horizontal platform is rotating with a uniform angular speed of ω rad/sec. On the platform is a rod CD on which slides a cylinder A having weight W. The cylinder is connected to C through a linear spring having a spring constant K. What is the equation of motion for A relative to the platform after it has been disturbed? What is the natural frequency of oscillation? Take r_0 as the unstretched length of spring.

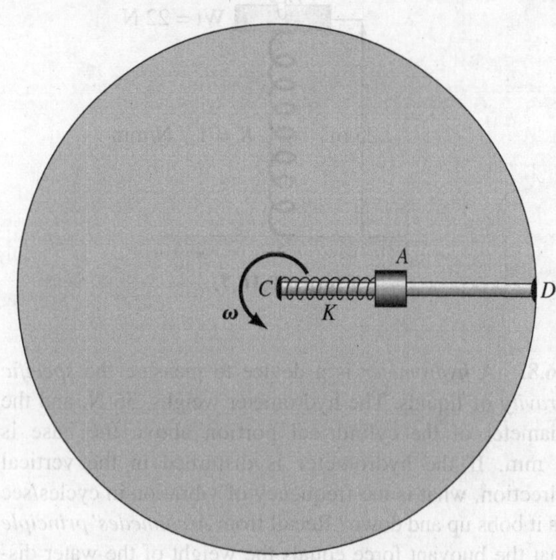

Figure P.16.16.

16.17. A rigid body A rests on a spring with stiffness K equal to 8.80 N/mm. A lead pad B falls onto the block A with a speed on impact of 7 m/sec. If the impact is perfectly plastic, what are the frequency and amplitude of the motion of the system, provided that the lead pad sticks to A at all times? Take $W_A = 134$ N and $W_B = 22$ N. What is the distance moved by A in .02 sec? (*Caution:* Be careful about the initial conditions.)

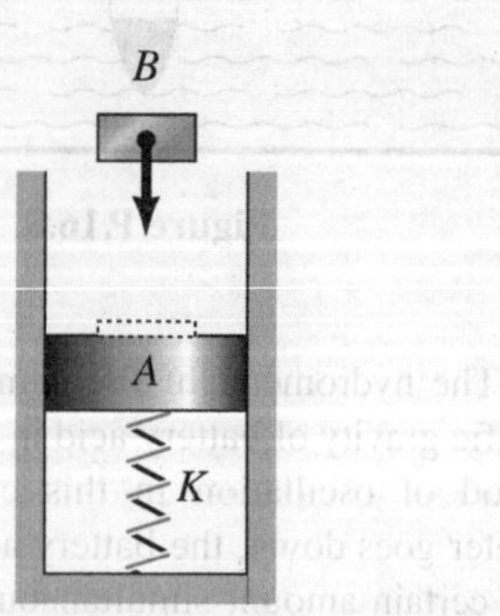

Figure P.16.17.

16.18. A small sphere of weight 5 lb is held by taut elastic cords on a frictionless plane. If 50 lb of force is needed to cause an elongation of 1 in. for each cord, what is the natural frequency of small oscillation of the weight in a transverse direction? Also, determine the natural frequency of the weight in a direction along the cord for small oscillations. Neglect the mass of the cord. The tension in the cord in the configuration shown is 100 lb.

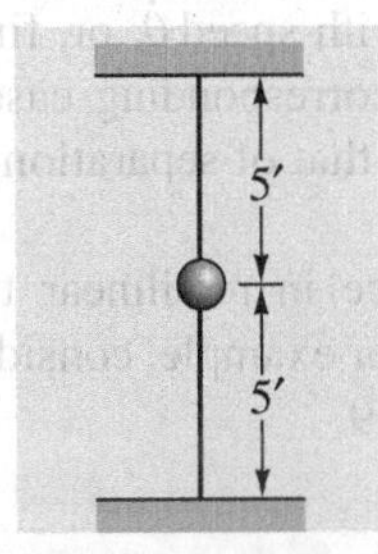

Figure P.16.18.

16.19. Body A weighs 445 N and is connected to a spring having a spring constant K_1 of 3.50 N/mm. At the right of A is a second spring having a spring constant K_2 of 8.80 N/mm. Body A is moved 150 mm to the left from the configuration of static equilibrium shown in the diagram, and it is released from rest. What is the period of oscillation for the body? [*Hint:* Work with half a cycle.]

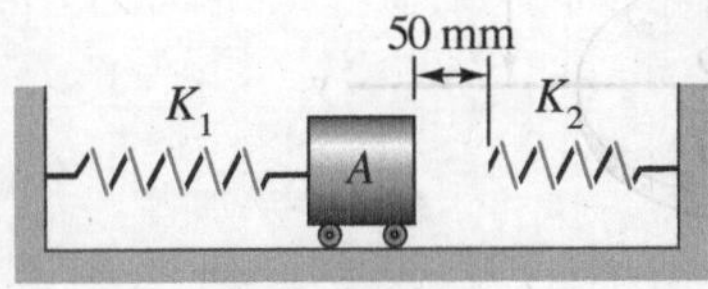

Figure P.16.19.

16.20. Body A, weighing 50 lb, has a speed of 20 ft/sec to the left. If there is no friction, what is the period of oscillation of the body for the following data:

$$K_1 = 20 \text{ lb/in.}$$
$$K_2 = 10 \text{ lb/in.}$$

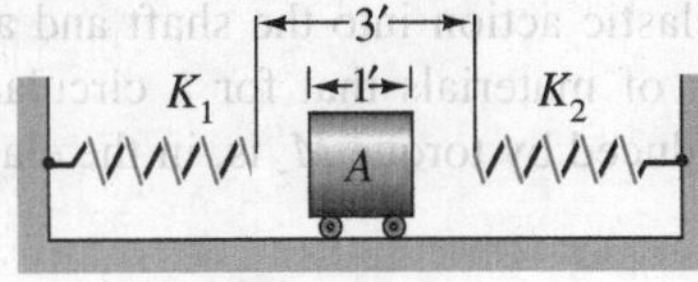

Figure P.16.20.

16.21. A spherical body A of mass 2 kg is attached by a light rod to a shaft BC which is inclined by an angle of 30°. For small, rotational oscillations about BC, what is the natural frequency of the system? [*Hint:* Recall that the moment about an axis n is $(\boldsymbol{r} \times \boldsymbol{F}) \cdot \hat{\boldsymbol{n}}$.]

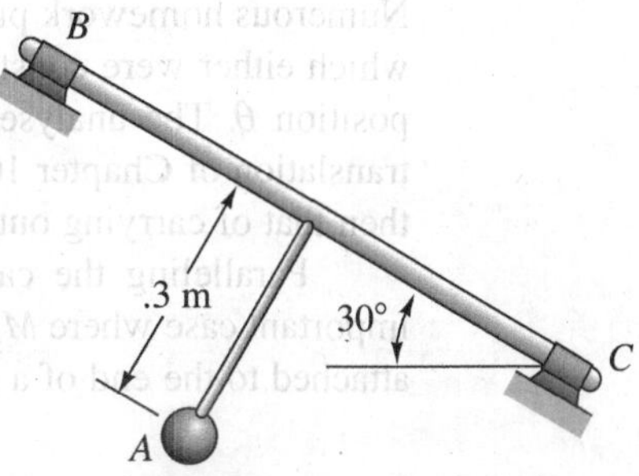

Figure P.16.21.

***16.22.** A cube A, .25 m on a side, has a specific gravity of 1.10 and is attached to a cone having a specific gravity of .8. What is the equation for up-and-down motion of the system? Neglect the mass and buoyant force for rod CD. For very small oscillations, what is the approximate natural frequency?

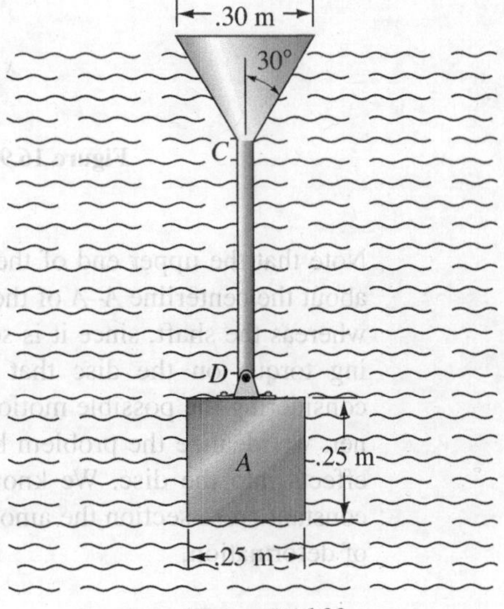

Figure P.16.22.

16.3 Torsional Vibration

We showed in Section 16.2 that, for a body constrained to rotate about an axis fixed in inertial space, the angular momentum equation about the fixed axis is

$$I_{zz}\dot{\omega}_z = I_{zz}\ddot{\theta} = M_z \tag{16.9}$$

Numerous homework problems involved the determination of $\dot{\theta}$ and θ for applied torques which either were constant, varied with time, varied with speed $\dot{\theta}$, or, finally varied with position θ. The analyses paralleled very closely the corresponding cases for rectilinear translation of Chapter 10. Primarily, the approach was that of separation of variables and then that of carrying out one or more quadratures.

Paralleling the case of the linear restoring force in rectilinear translation is the important case where M_z is a *linear restoring torque*. For example, consider a circular disc attached to the end of a light shaft as shown in Fig. 16.9.

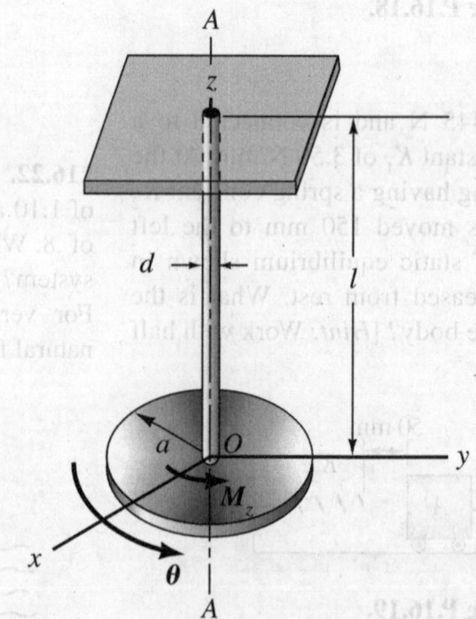

Figure 16.9. Shaft–disc analog of spring–mass system.

Note that the upper end of the shaft is fixed. If the disc is twisted by an external agent about the centerline *A–A* of the shaft, then the disc will rotate essentially as a rigid body, whereas the shaft, since it is so much thinner and longer, will twist and supply a restoring torque on the disc that tries to bring the disc back to its initial position. In considering the possible motions of such a system disturbed in the aforementioned manner, we idealize the problem by *lumping* all elastic action into the shaft and all inertial effects into the disc. We know from strength of materials that for a circular shaft of constant cross section the amount of twist θ induced by torque M_z is, in the elastic range of deformation,

$$\theta = \frac{M_z L}{GJ} \tag{16.10}$$

where G is the shear modulus of the shaft material, J is the polar moment of area of the shaft cross section, and L is the length of the shaft. We can set forth the concept of a torsional spring constant K_t given as

$$K_t = \frac{M_z}{\theta} \tag{16.11}$$

For the case at hand, we have

$$K_t = \frac{GJ}{L} \tag{16.12}$$

Thus, the thin shaft has the same role in this discussion as the light linear spring of Section 16.2. Employing Eq. 16.11 for M_z and using the proper sign to ensure that we have a restoring action, we can express Eq. 16.9 as follows:

$$\ddot{\theta} + \frac{K_t}{I_{zz}}\theta = 0 \tag{16.13}$$

Notice that this equation is identical in form to Eq. 16.1. Accordingly, all the conclusions developed in that discussion apply with the appropriate changes in notation. Thus, the disc, once disturbed by being given an angular motion, will have a *torsional* natural oscillation frequency of $(\omega_n)_t = \sqrt{K_t/I_{zz}}$ rad/sec. The equation of motion for the disc is

$$\theta = C_1 \cos\sqrt{\frac{K_t}{I_{zz}}}t + C_2 \sin\sqrt{\frac{K_t}{I_{zz}}}t \tag{16.14}$$

where C_1 and C_2 are constants of integration to be determined from initial conditions. Thus, for $\theta = \theta_0$ and $\dot{\theta} = \dot{\theta}_0$ at $t = 0$ we have

$$\theta = \theta_0 \cos\sqrt{\frac{K_t}{I_{zz}}}t + \frac{\dot{\theta}_0}{\sqrt{K_t/I_{zz}}} \sin\sqrt{\frac{K_t}{I_{zz}}}t \tag{16.15}$$

In the example just presented, the linear restoring torque stemmed from a long thin shaft. There could be other agents that can develop a linear restoring torque on a system otherwise free to rotate about an axis fixed in inertial space. We then talk about an *equivalent* torsional spring constant. We shall illustrate such cases in the following examples.

Example 16.3

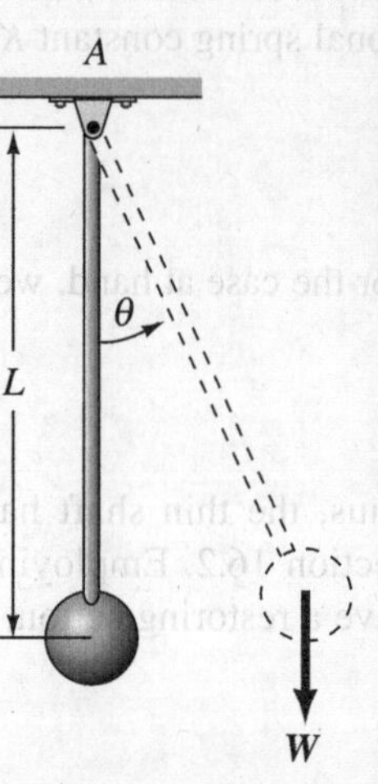

Figure 16.10. Pendulum.

What are the equation of motion and the natural frequency of oscillation for small amplitude of a simple plane pendulum shown in Fig. 16.10? The pendulum rod may be considered massless.

Because the pendulum bob is small compared to the radius of curvature of its possible trajectory of motion, we may consider it as a particle. The pendulum has one degree of freedom, and we can use θ as the independent coordinate.[3] Notice from the diagram that there is a restoring torque about point A developed by gravity given as

$$M_x = -WL\sin\theta \tag{a}$$

where W is the weight of the bob. If the amplitude of the motion θ is very small, we can replace $\sin\theta$ by θ and so for this case we have a linear restoring torque given as

$$M_x = -WL\theta \tag{b}$$

We then have an equivalent torsional spring constant for the system

$$K_t = WL \tag{c}$$

The equation of possible *small-amplitude* motions for the pendulum is given as

$$-WL\theta = (ML^2)\ddot{\theta} \tag{d}$$

where we have used the **moment-of-momentum equation** about the fixed point A. Rearranging terms, we get

$$\ddot{\theta} + \frac{WL}{ML^2}\theta = 0 \tag{e}$$

Noting that $W = Mg$, we have

$$\ddot{\theta} + \frac{g}{L}\theta = 0 \tag{f}$$

Accordingly, the natural frequency of oscillation is

$$\omega_n = \sqrt{\frac{g}{L}} \text{ rad/sec} \tag{g}$$

The equation of motion for this system is

$$\theta = C_1 \cos\sqrt{\frac{g}{L}}t + C_2 \sin\sqrt{\frac{g}{L}}t \tag{h}$$

where C_1 and C_2 are computed from known conditions at some time t_0.

[3]One degree of freedom means that one independent coordinate locates the system.

Example 16.4

A stepped disc is shown in Fig. 16.11 supporting a weight W_1 while being constrained by a linear spring having a spring constant K. The mass of the stepped disc is M and the radius of gyration about its geometric axis is k. What is the equation of motion for the system if the disc is rotated a small angle θ_1 counterclockwise from its static-equilibrium configuration and then suddenly released from rest? Assume the cord holding W_1 is weightless and perfectly flexible.

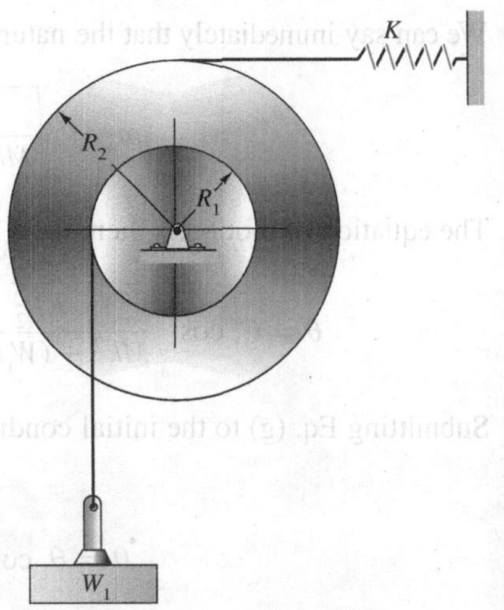

Figure 16.11. Stepped disc.

If we measure θ from the static-equilibrium position as shown in Fig. 16.12(a) the spring is stretched an amount $R_2(\theta + \theta_0)$ wherein θ_0 is the amount of rotation induced by the weight W_1 to reach the static-equilibrium configuration. Consequently, applying the **angular momentum equation** to the stepped disc about the axis of rotation, we get

$$R_1T - KR_2^2(\theta+\theta_0) = Mk^2\ddot{\theta} \qquad \text{(a)}$$

Next consider the suspended weight W_1. Clearly we have only translation for this body, for which **Newton's law** gives us

$$T - W_1 = -\frac{W_1}{g}R_1\ddot{\theta} \qquad \text{(b)}$$

where we have made the assumption that the **cord** is always taut and is inextensible and have considered the **kinematics** of the motion. We may replace T in Eq. (a) using Eq. (b) as follows:

$$R_1W_1 - \frac{W_1}{g}R_1^2\ddot{\theta} - KR_2^2(\theta+\theta_0) = Mk^2\ddot{\theta} \qquad \text{(c)}$$

Rearranging terms, we get

$$\left(Mk^2 + \frac{W_1}{g}R_1^2\right)\ddot{\theta} + KR_2^2\theta = R_1W_1 - KR_2^2\theta_0 \qquad \text{(d)}$$

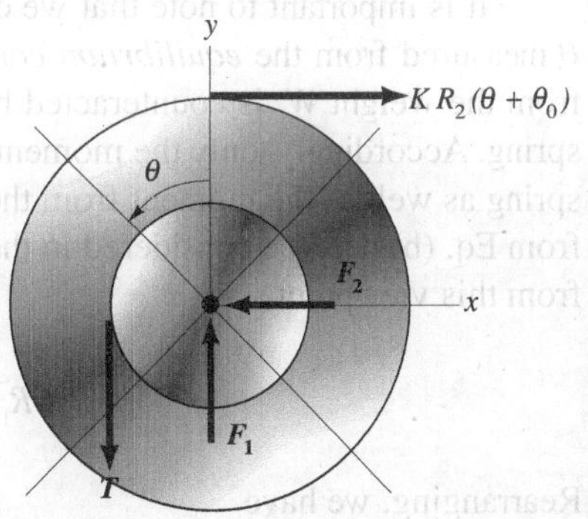

(a)

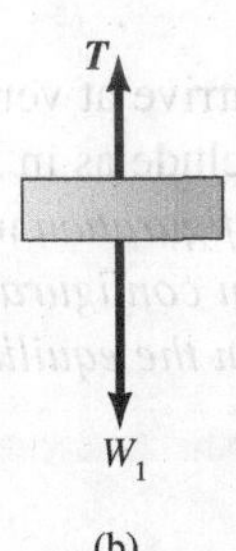

(b)

Figure 16.12. Free-body diagrams.

Considering the **static-equilibrium** configuration of the system, we see on summing moments about the axis of rotation that the right side of the equation above is zero. Accordingly, we have for Eq. (d):

$$\ddot{\theta} + \frac{KR_2^2}{Mk^2 + (W_1/g)R_1^2}\theta = 0 \qquad \text{(e)}$$

Example 16.4 (Continued)

We can say immediately that the natural torsional frequency of the system is

$$\omega_n = \sqrt{\frac{KR_2^2}{Mk^2 + (W_1/g)R_1^2}} \text{ rad/sec} \tag{f}$$

The equation of motion is then

$$\theta = C_1 \cos \sqrt{\frac{KR_2^2}{Mk^2 + (W_1/g)R_1^2}}\,t + C_2 \sin \sqrt{\frac{KR_2^2}{Mk^2 + (W_1/g)R_1^2}}\,t \tag{g}$$

Submitting Eq. (g) to the initial conditions to determine C_1 and C_2, we get

$$\theta = \theta_1 \cos \sqrt{\frac{KR_2^2}{Mk^2 + (W_1/g)R_1^2}}\,t \tag{h}$$

It is important to note that we could have reached Eq. (e) more directly by using θ measured from the *equilibrium configuration*. That is, use the fact that the moment from the weight W_1 is counteracted by the static moment from the stretch $R_2\theta_0$ of the spring. Accordingly, only the moment from the force $-R_2K\theta$ from further stretch of the spring as well as the moment from the inertial force $-(W_1/g)R_1\ddot{\theta}$ of the hanging weight from Eq. (b) need be considered in the angular-momentum equation (a). Thus, we have from this viewpoint:

$$-\frac{W_1}{g}R_1^2\ddot{\theta} - KR_2^2\theta = Mk^2\ddot{\theta} \tag{i}$$

Rearranging, we have

$$\ddot{\theta} + \frac{KR_2^2}{Mk^2 + (W_1/g)R_1^2}\theta = 0 \tag{j}$$

Accordingly, we arrive at very same differential equation (e) in a more direct manner. We can again conclude as in Example 16.1 that *when the coordinate is measured from an equilibrium configuration we can forget about contributions of torques that are present for the equilibrium configuration and include only new torques developed when there is a departure from the equilibrium configuration.*

Before you start on the problems, we wish to point out that shafts directly connected to each other (see the shafts on the right side of the disc in Fig. P.16.23) are analogous to springs in *series* as far as the equivalent torsional spring constant is concerned. On the other hand, shafts on opposite sides of the disc are analogous to springs in *parallel* as far as the equivalent torsional spring constant is concerned. You should have no trouble justifying these observations.

PROBLEMS

16.23. Compute the equivalent torsional spring constant of the shaft on the disc.

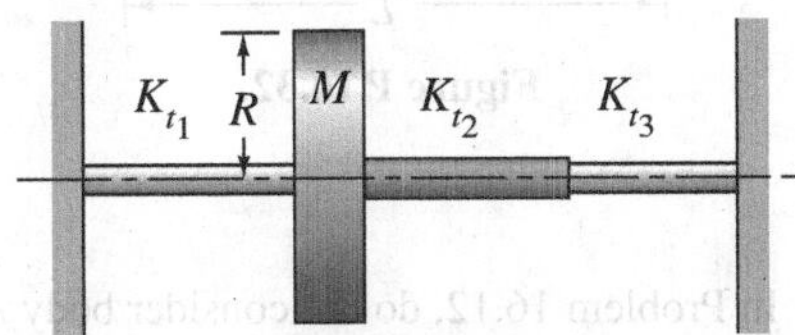

Figure P.16.23.

16.24. What is the equivalent torsional spring constant on the disc from the shafts? The modulus of elasticity G for the shafts is 10×10^{10} N/m^2. What is the natural frequency of the system? If the disc is twisted 10° and then released, what will its angular position be in 1 sec? Neglect the mass of the shafts. The disc weighs 143 N.

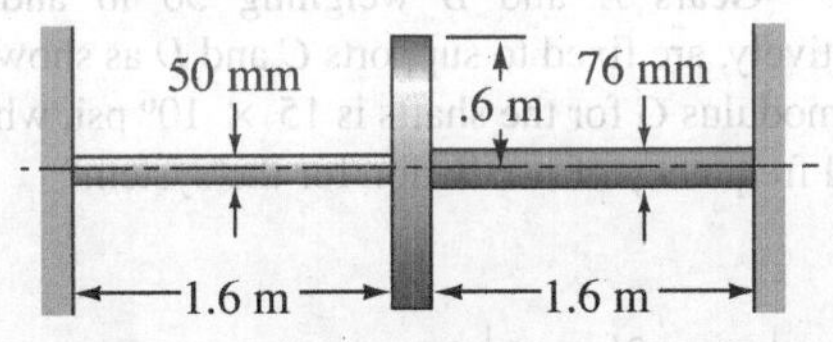

Figure P.16.24.

16.25. A small pendulum is mounted in a rocket that is accelerating upward at the rate of $3g$. What is the natural frequency of rotation of the pendulum if the bob has a mass of $50g$? Neglect the weight of the rod. Consider small oscillations.

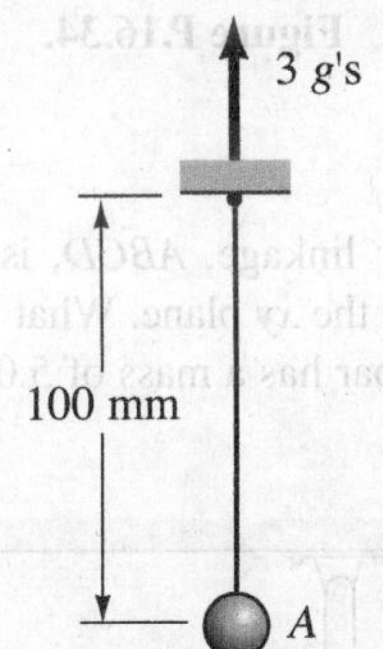

Figure P.16.25.

16.26. Work Problem 16.25 for the case where the rocket is decelerating at $.6g$ in the vertical direction.

16.27. What is the natural frequency for small oscillations of the compound pendulum supported at A?

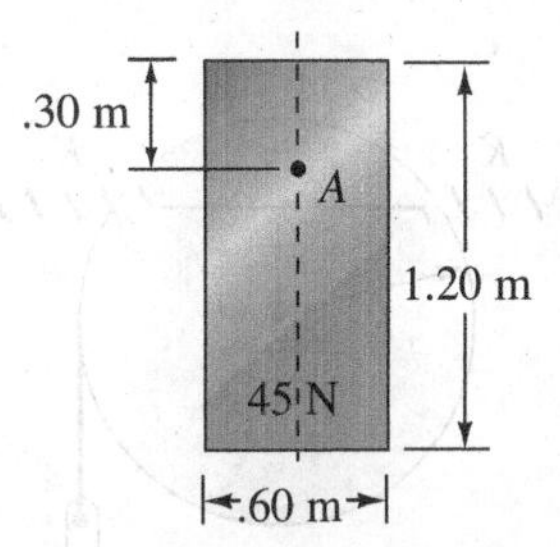

Figure P.16.27.

16.28. A slender rod weighing 140 N is held by a frictionless pin at A and by a spring having a spring constant of 8.80 N/mm at B

(a) What is the natural frequency of oscillation for small vibrations?

(b) If point B of the rod is depressed 25 mm at $t = 0$ from the static-equilibrium position, what will its position be when $t = .02$ sec?

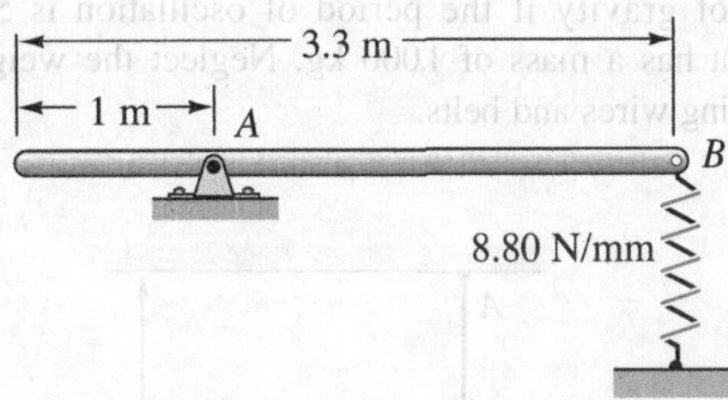

Figure P.16.28.

16.29. What is the natural frequency of the pendulum shown for small oscillations? Take into account the inertia of the rod whose mass is m. Also, consider the bob to be a sphere of diameter D and mass M, rather than a particle. The length of the rod is l.

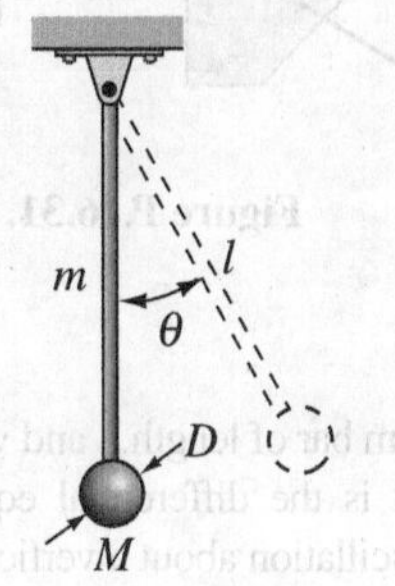

Figure P.16.29.

16.30. A cylinder of mass M and radius R is connected to identical springs and rotates without friction about O. For small oscillations, what is the natural frequency? The cord supporting W_1 is wrapped around the cylinder.

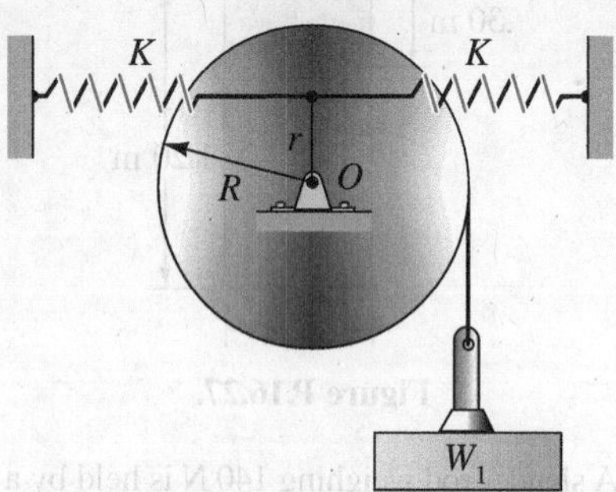

Figure P.16.30.

16.31. The author's 22-ft Columbia sailboat is suspended by straps from a crane. The boat is made to swing freely about support A in the xy plane (the plane of the page). What is the radius of gyration about the z axis at the center of gravity if the period of oscillation is 5 sec? The boat has a mass of 1,000 kg. Neglect the weight of supporting wires and belts.

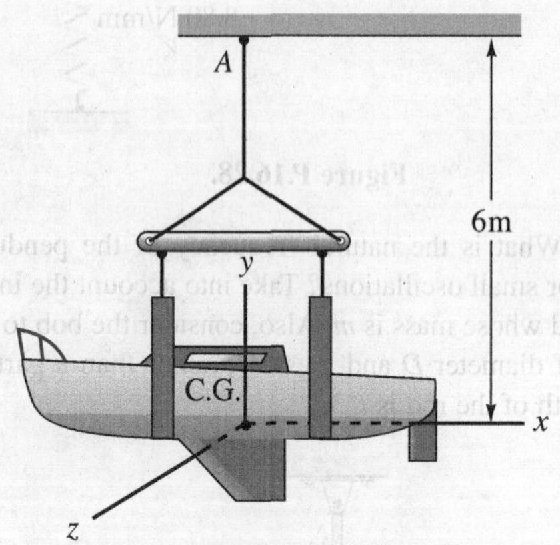

Figure P.16.31.

16.32. A uniform bar of length L and weight W is suspended by strings. What is the differential equation of motion for small torsional oscillation about a vertical axis at the center of mass at C? What is the natural frequency?

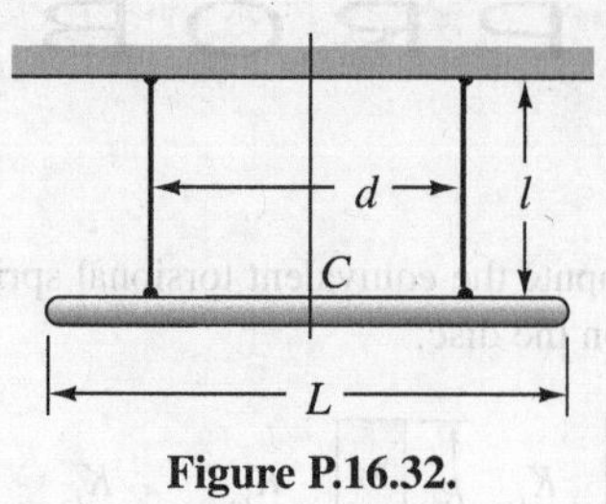

Figure P.16.32.

16.33. In Problem 16.12, do not consider body A to be a particle, and compute the natural frequency of the system for small vibrations. Take the dimension of A to be that of a 150-mm cube. If ω_n for the particle approach is 16.81 rad/sec, what percentage error is incurred using the particle approach?

16.34. Gears A and B weighing 50 lb and 80 lb, respectively, are fixed to supports C and D as shown. If the shear modulus G for the shafts is 15×10^6 psi, what is the natural frequency of oscillation for the system?

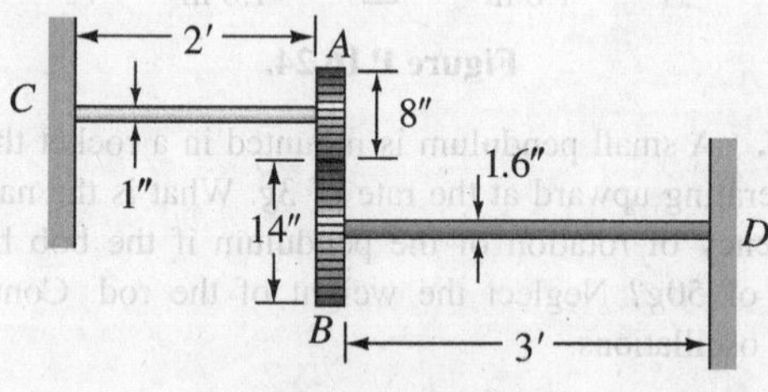

Figure P.16.34.

16.35. A four-bar linkage, $ABCD$, is disturbed slightly so as to oscillate in the xy plane. What is the frequency of oscillation if each bar has a mass of 5.0 g/mm?

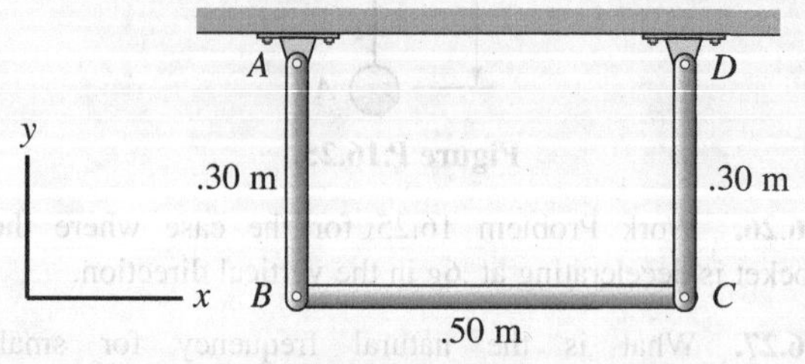

Figure P.16.35.

16.36. A plate A weighing 1 kN is attached to a rod CD. If at the instant that the rod CD is torsionally unstrained, the plate has an angular speed of 2 rad/sec about the centerline of CD, what is the amplitude of twist developed by the rod? Take $G = 6.90 \times 10^{10}$ N/m^2 for the rod.

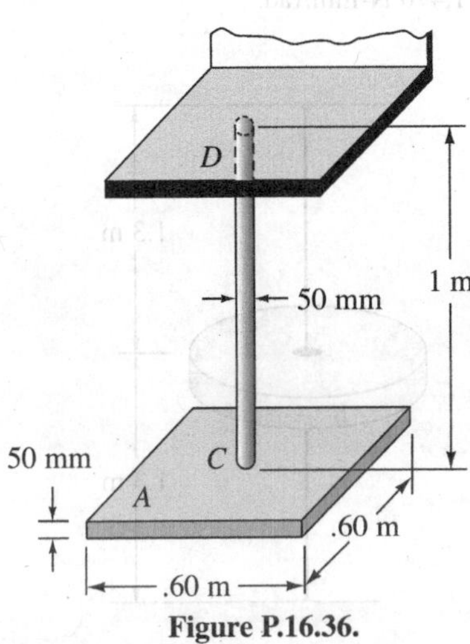

Figure P.16.36.

16.37. A block A having a uniform density of 300 lb/ft^3 is suspended by a fixed shaft of length 3 ft as shown. If the area of the top surface of the block is 5 ft^2, what are the values of a and b for extreme values of natural torsional frequency of the system? The shear modulus G for the shaft is 15×10^6 psi. Compute the natural frequency for the extreme cases.

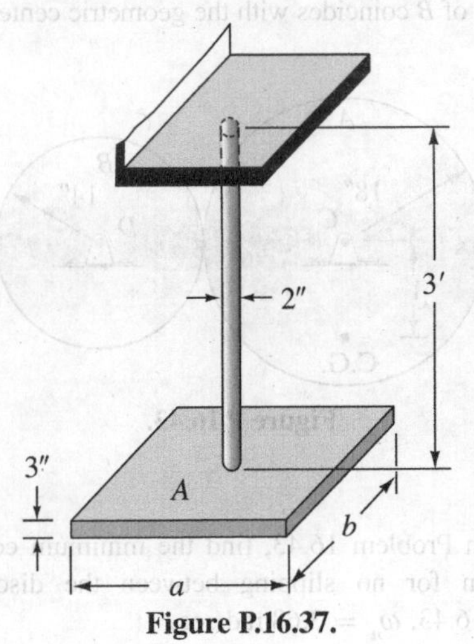

Figure P.16.37.

16.38. A rod of weight W and length L is restrained in the vertical position by two identical springs having spring constant K. A vertical load P acts on top of the rod. What value of P, in terms of W, L, and K, will cause the rod to have a natural frequency of oscillation about A approaching zero for small oscillations? What does this signify physically? [*Hint:* For small θ we may take $\cos\theta = 1$.]

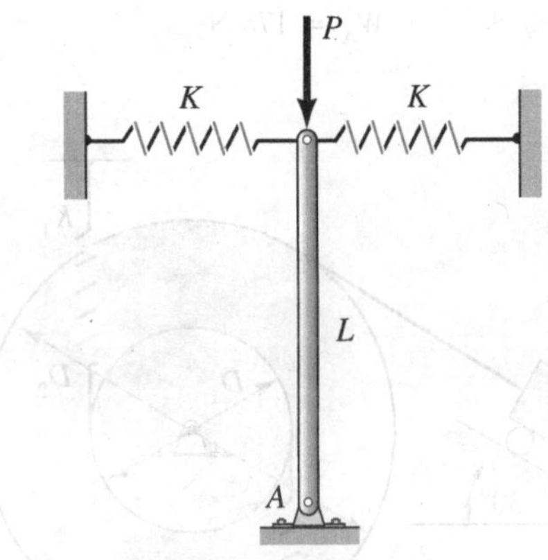

Figure P.16.38.

16.39. A 1-m rod weighing 60 N is maintained in a vertical position by two identical springs having each a spring constant of 50 N/mm. What vertical force P will cause the natural frequency of the rod about A to approach zero value for small oscillations? [*Hint:* For small θ we can say that $\cos\theta = 1$.]

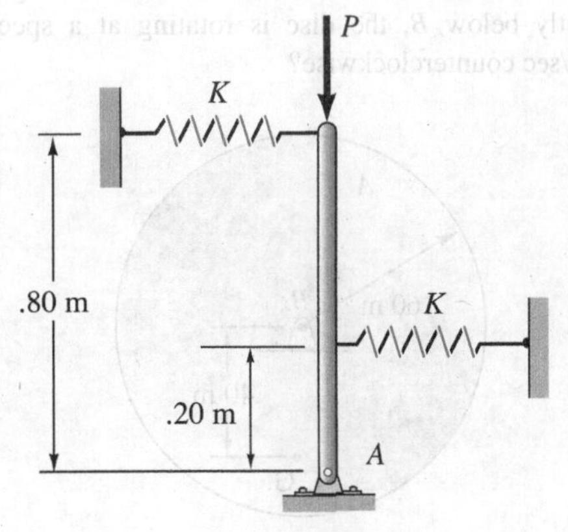

Figure P.16.39.

16.40. What is the natural frequency of torsional vibration for the stepped cylinder? The mass of the cylinder is 45 kg, and the radius of gyration is .46 m. The following data also apply:

$$D_1 = .30 \text{ m}$$
$$D_2 = .60 \text{ m}$$
$$K_1 = .875 \text{ N/mm}$$
$$K_2 = 1.8 \text{ N/mm}$$
$$W_A = 178 \text{ N}$$

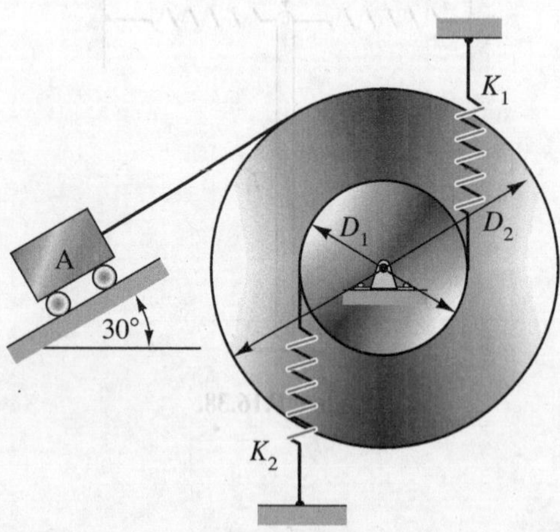

Figure P.16.40.

16.41. A disc A weighs 445 N and has a radius of gyration of .45 m about its axis of symmetry. Note that the center of gravity does not coincide with the geometric center. What are the amplitude of oscillation and frequency of oscillation if, at the instant that the center of gravity is directly below B, the disc is rotating at a speed of .01 rad/sec counterclockwise?

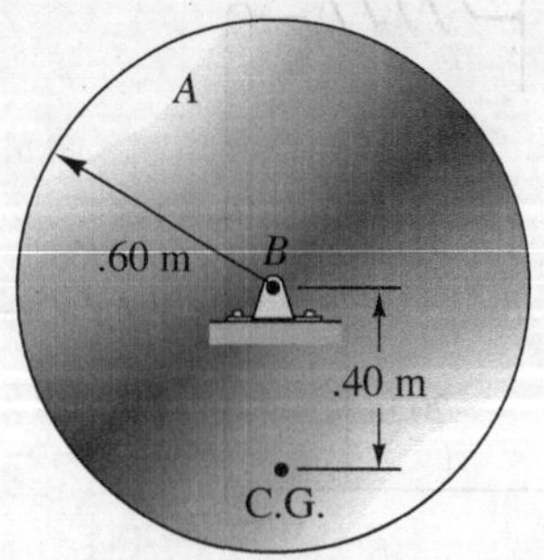

Figure P.16.41.

16.42. A disc B is suspended by a flexible wire. The tension in the top wire is 4,450 N. If the disc is observed to have a period of lateral oscillation of .2 sec for very small amplitude and a period of torsional oscillation of 5 sec, what is the radius of gyration of the disc about its geometric axis? The torsional spring constant for each of the wires is 1,470 N-mm/rad.

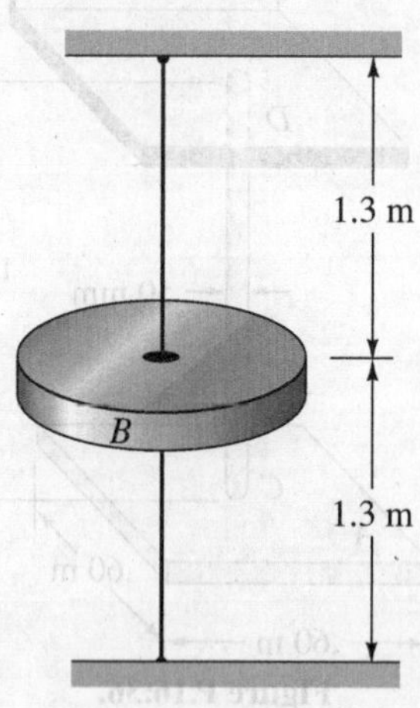

Figure P.16.42.

16.43. Two discs are forced together such that, at the point of contact, a normal force of 50 lb is transmitted from one disc to the other. Disc A weighs 200 lb and has a radius of gyration of 1.4 ft about C, whereas disc B weighs 50 lb and has a radius of gyration about D of 1 ft. What is the natural frequency of oscillation for the system, if disc A is rotated 10° counterclockwise and then released? The center of gravity of B coincides with the geometric center.

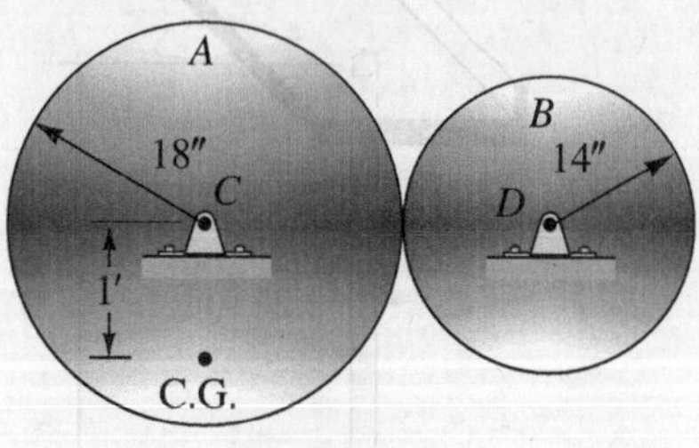

Figure P.16.43.

16.44. In Problem 16.43, find the minimum coefficient of friction for no slipping between the discs. From Problem 16.43, $\omega_n = 3.68$ rad/sec.

16.4 Simple Harmonic Motion

A body is considered to be in simple harmonic motion, if it moves along a straight line in such a way that its acceleration is always proportional to its distance from a fixed point and is directed towards the fixed point.

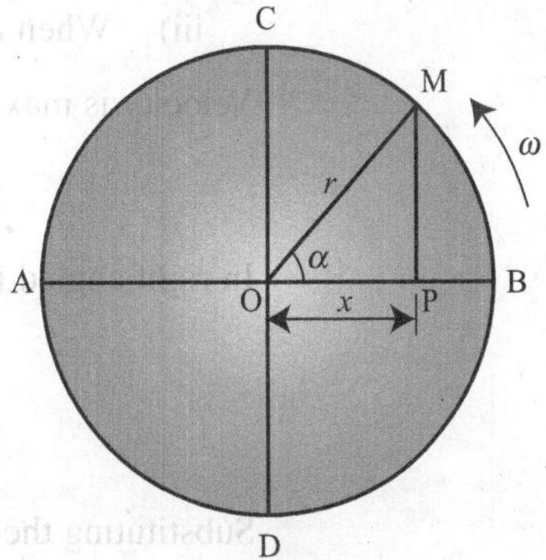

Figure 16.13

Let ω = Constant Angular velocity of the body
O = Center of the circle
AB = Fixed diameter of the circle
M = Position of the body
r = Radius of the circle
P = Projection of M on diameter AB.

When the body moves along the circumference of the circle from B to C, the projection of the body on the diameter AB i.e., Point P moves from B to O and when the body moves from C to A the projection of the body on the diameter moves from O to A. The motion of P along the diameter AB is known as simple harmonic motion.

Let $\angle AOP = \alpha$

t = Time taken by the body when moving from B to M

Let the distance of the point P from the center of the circle is equal to OP or x.

$$OP = x = OM \cos\alpha$$
$$= r \cos\alpha$$

But α is angular displacement and is equal to the product of angular velocity (ω) and time (t)

$$\alpha = \omega t,\ x = r\cos\omega t \tag{16.16}$$

Equation 16.16 gives the displacement of point P from the center of the circle. Differentiating Eq. (16.16) w.r.t. t.

$$\frac{dx}{dt} = \frac{d}{dt}(r\cos\omega t) = -r\omega\sin\omega t$$

$$\text{But } \frac{dx}{dt} = \text{Velocity}$$

$$V = -r\omega \times \sin\omega t \tag{16.17}$$

$$= -r\omega\sin\alpha \tag{16.18}$$

Equation (16.17) gives the velocity of point P in terms of angular velocity and Eq. (16.18) gives velocity in terms of angular displacement. The presence of negative sign in Eq. (16.17) and Eq. (16.18) indicates that with the increase of x the velocity decreases.

The values of velocity for different values of α are given as:

i) When $\alpha = 0$, $V = -r\omega \sin 0 = 0$
ii) When $\alpha = 90°$, $V = -r\omega \sin 90 = -r\omega$
iii) When $\alpha = 180$, $V = 0$

Velocity is maximum when point P is at 0 and

$$V_{\max} = -\omega r \tag{16.19}$$

In right angled triangle OMP

$$\text{Sin}\alpha = \frac{\text{PM}}{\text{OM}} = \frac{\sqrt{\text{OM}^2 - \text{OP}^2}}{\text{OM}} = \frac{\sqrt{r^2 - x^2}}{r}$$

Substituting the value of Sinα in Eq. (16.18)

$$V = -r\omega \frac{\sqrt{r^2 - x^2}}{r} = -\omega\sqrt{r^2 - x^2} \tag{16.20}$$

Equation (16.20) gives the velocity in terms of x and radius of circle.

Acceleration of the point P

Differentiating Equation (16.17) w.r.t t we can get acceleration of the point P.

Acceleration $f = \frac{dv}{dt} = \frac{d}{dt}(-r\omega \sin \omega t)$

$$= -r\omega^2 \cos \omega t \ (or) - r\omega^2 \cos \alpha$$

$$= -\omega^2 r \cos \omega t \tag{16.21}$$

$$= -\omega^2 x \tag{16.22}$$

From Eq. (7) it can be inferred that acceleration of particle P is directly proportional to the distance of P from the center. The negative sign indicates that acceleration is directed towards the center of the circle. Hence the point P moves with simple harmonic motion.

i) When $\alpha = 0$, $f = -r\omega^2 \cos 0 = -r\omega^2$
ii) When $\alpha = 90$, $f = -r\omega^2 \cos 90 = 0$
iii) When $\alpha = 180°$, $f = -r\omega^2 \cos 180 = r\omega^2$

Acceleration is observed to be maximum when $\alpha = 0°$ or $180°$

$$f_{\max} = \pm r\omega^2 \tag{16.23}$$

Terms applied to SHM

1. **Amplitude:** The maximum possible displacement of point P from its mean position. Radius of circle will be the amplitude.
2. **Oscillation:** Change in position of P from B to A and back from A to B is called an oscillation.

3. **Period of SHM:** It is the time required for one complete oscillation of point P. It is equivalent to the time taken by point M to make one complete revolution.

Let T = Time period of SHM.
ω = Constant angular velocity of the body M.
T = Time required for the body M to complete one revolution

$$= \frac{\text{Angular displacement for one revolution}}{\text{Constant angular velocity}}$$

$$= \frac{2\pi}{\omega} \tag{16.24}$$

4. **Frequency:** It is defined as the number of cycles per second and is mathematically given as

$$\frac{1}{\mathrm{T}} = \frac{1}{2\pi/\omega} = \frac{\omega}{2\pi} \text{ cycles/sec or Hz}$$

Example 16.5

Find the velocity and acceleration after 1 sec from the extreme position of a body, moving with simple harmonic motion with an amplitude of 0.8 m and period of complete oscillation of 4 sec.

Sol. Given:

Given data amplitude, a = 0.8 m
Period of oscillation, T = 4 sec

We know that, $T = \dfrac{2\pi}{\omega}$

$$\omega = \frac{2\pi}{\mathrm{T}} = \frac{2\pi}{4} = \frac{\pi}{2} \text{ rad/sec}$$

$$= 1.57 \text{ rad/sec}$$

Since the time in the present problem is measured from the extreme position

$$V = -a\omega \sin \omega t$$

$$= -0.8 \times 1.57 \sin(1.57 \times 1)$$

$$= -1.256 \text{ m/sec}$$

Acceleration is given by, $f = -\omega^2 a \cos \omega t$

$$= -1.57^2 \times 0.8 \times \cos(1.57 \times 1)$$

$$= 0 \text{ m/sec}^2$$

Example 16.6

A body is moving with SHM and has velocity of 16 m/sec and 6 m/sec at a distance of 3 m and 4 m, respectively, from the center. Find the amplitude and time period of the body.

Sol. Given:

$$V_1 = 16 \text{ m/sec at 3 m from center}$$
$$x_1 = 3 \text{ m}$$
$$V_2 = 4 \text{ m/sec at 4 m from center}$$
$$x_2 = 4 \text{ m}$$

Let a = Amplitude of the body

T = Time period of the body

From Eq. 16.24

$$T = \frac{2\pi}{\omega}$$
$$\omega = \frac{2\pi}{T}$$

From Eq. 16.20

$$V = -\omega\sqrt{a^2 - x^2}$$
$$V_1 = -\omega\sqrt{a^2 - x_1^2}$$

Substituting $16 = -\omega\sqrt{a^2 - 3^2}$

and $V_2 = -\omega\sqrt{a^2 - x_2^2}$

$$6 = -\omega\sqrt{a^2 - 4^2}$$
$$\frac{16}{6} = \frac{\sqrt{a^2 - 3^2}}{\sqrt{a^2 - 4^2}} \qquad \text{(a)}$$

squaring on both sides

$$7.111 = \frac{a^2 - 3^2}{a^2 - 4^2}$$
$$7.111\,a^2 - 113.8 - a^2 + 9 = 0$$
$$6.111\,a^2 - 104.8 = 0$$
$$a = \frac{\sqrt{104.8}}{6.111} = 4.14 \text{ m}$$

Substituting the value of a in Eq. (a)

Example 16.6 (Continued)

$$16 = -\omega\sqrt{4.14 - 3^2}$$

$$\omega = -\frac{16}{2.85} = 5.61 \text{ rad/sec (Numerically)}$$

$$T = \frac{2\pi}{\omega} = \frac{2\pi}{5.61} = 1.12 \text{ sec}$$

Example 16.7

In a mechanism, a cross-head moves in a straight guide with SHM. At distances of 250 mm and 400 mm from its mean position, it has velocity of 12 m/sec and 6 m/sec, respectively. Find the amplitude, maximum velocity, and period of vibration. If the cross-head weight is 0.5 kg(f), calculate the maximum force on it in the direction of motion.

Sol. Given:

$$x_1 = 250 \text{ mm}, \qquad V_1 = 12 \text{ m/sec}$$
$$= 0.25 \text{ m}$$
$$x_2 = 400 \text{ mm}, \qquad V_2 = 6 \text{ m/sec}$$
$$= 0.4 \text{ m}$$

Let amplitude = a

Using the relation,

$$V = -\omega\sqrt{a^2 - x^2}$$

$$V_1 = -\omega\sqrt{a^2 - x_1^2}$$

$$12 = -\omega\sqrt{a^2 - .250^2} \qquad \text{(a)}$$

$$V^2 = -\omega\sqrt{a^2 - x_2^2}$$

$$6 = -\omega\sqrt{a^2 - 0.4^2} \qquad \text{(b)}$$

$$\frac{\text{a}}{\text{b}} \qquad \frac{12}{6} = \frac{\sqrt{a^2 - 0.25^2}}{\sqrt{a^2 - 0.4^2}}$$

Squaring on both sides,

$$\Rightarrow 4 = \frac{a^2 - 0.25^2}{a^2 - 0.4^2} = \frac{a^2 - 0.625}{a^2 - 0.16}$$

$$\Rightarrow 4a^2 - 0.64 - a^2 + 0.0625 = 0$$

Example 16.7 (Continued)

$$3a^2 = 0.5175$$

$$a = \sqrt{\frac{0.5175}{3}} = 0.439 \text{ m}$$

From Eq. (a) $12 = -\omega\sqrt{0.439^2 - 0.25^2}$

$$\omega = \frac{12}{\sqrt{0.439^2 - 0.25^2}} = -33.3 \text{ rad/sec}$$

$$= 33.3 \text{ rad/sec (Numerically)}$$

Maximum velocity, $V_{max} = -a\omega$

$$= -0.439 \times (-33.3)$$

$$= 14.6 \text{ m/sec.}$$

Period of vibration, $T = \dfrac{2\pi}{\omega} = \dfrac{2\pi}{33.3} = 0.189 \text{ sec}$

Maximum force on cross head

$$W = 0.5\ kg(f)$$

$$m = \frac{w}{g} = \frac{0.5}{9.81}$$

Maximum force $= m \times f_{max} = \dfrac{0.5}{9.81} \times f_{max}$

$$f_{max} = \pm\omega^2 a$$

$$= \pm 33.3^2 \times 0.439$$

$$= 486.8 \text{ m/sec}^2.$$

$$\text{Maximum force} = \frac{0.5}{9.81} \times 486.8 = 24.81 \text{ kg(f)}$$

PROBLEMS

16.45. A particle is moving with simple harmonic motion. When the particle is 1.75 m from the mid-path its velocity is 6 m/sec and when 4 m from the mid-path its velocity is 2 m/sec. Find the angular velocity, periodic time, and its maximum acceleration.

16.46. The time period of a simple harmonic motion is 3 sec, and the particle oscillates through a distance of 100 mm on each side of the mean position. Find the maximum velocity and maximum acceleration of the particle.

16.47. The piston of an engine moves with SHM. The crank rotates at 80 rpm and the stroke is 160 cm. Find the velocity and acceleration of the piston, when it is at a distance of 40 cm from the center.

16.48. A body is moving with SHM and has an amplitude of 3.5 m and period of complete oscillation as 2.5 sec. Find the time required by the body in passing between two points which are at a distance of 4.5 and 2.5 m from the center and are on the same side.

16.49. A particle is moving with SHM and performs 6 complete oscillations per minute. If the particle is 4 cm from the center of oscillation determine the amplitude, velocity of the particle, and maximum acceleration. The velocity of the particle at a distance of 5 cm from the center of oscillation is 0.6 times the maximum velocity.

16.5 Simple Pendulum

Simple pendulum consists of a light, inelastic, and flexible string with a heavy bob suspended at the lower end and the upper end fixed at O.

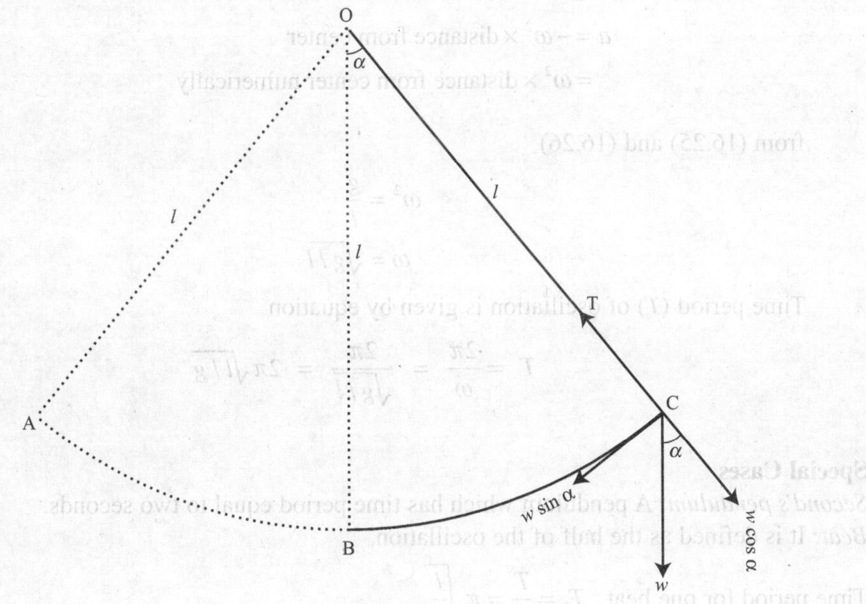

Figure 16.14

The bob is allowed to oscillate freely in a vertical plane.

Let l = Length of string = OA = OB = OC

m = Mass of bob

W = Weight of bob = $m \times g$

The pendulum is in equilibrium, when the bob is at B. But upon displacement to a new position at C and release, it starts oscillating between A and C with B as mean position.

The forces acting on the bob when it is at C are

(i) weight of the bob, $w = mg$ acting vertically downwards.

(ii) Tension T in the string.

$W\cos\alpha$ and $W\sin\alpha$ are the resolved components of W along the string and at right angles to the string, respectively.

Since the bob is in equilibrium condition at C, T = $W\cos\alpha$

The component $W\sin\alpha$ acting towards B will be unbalanced and will give rise to an acceleration (f) in the direction CB.

$$\text{Acceleration, } a = \frac{\text{Force}}{\text{Mass}} = \frac{w \sin\alpha}{m} = \frac{mg\ \sin\alpha}{m} = g\ \sin\alpha$$

If α is very small, $\sin\alpha = \alpha$

$$a = g\alpha = g\frac{\text{Length of arc CB}}{l} \tag{16.25}$$

Both g and l are constant.

Acceleration is proportional to CB i.e., the distance from B.

Acceleration is acting towards A. Because of this the bob moves with simple harmonic motion about the mean position B.

Acceleration of a body undergoing SHM is given by equation

$$a = -\omega^2 \times \text{distance from center}$$

$$= \omega^2 \times \text{distance from center numerically} \tag{16.26}$$

from (16.25) and (16.26)

$$\omega^2 = \frac{g}{l}$$

$$\omega = \sqrt{g/l}$$

Time period (T) of oscillation is given by equation

$$T = \frac{2\pi}{\omega} = \frac{2\pi}{\sqrt{g/l}} = 2\pi\sqrt{l/g}$$

Special Cases

Second's pendulum: A pendulum which has time period equal to two seconds.

Beat: It is defined as the half of the oscillation.

$$\text{Time period for one beat, } T_b = \frac{T}{2} = \pi\sqrt{\frac{l}{g}}$$

Gain or Loss of Oscillations due to charge in g or l

Time period for n beats $T_n = n\pi\sqrt{\frac{l}{g}}$

$$n = \frac{T_n}{n\sqrt{\frac{l}{g}}} = \frac{T_n}{\pi}\sqrt{\frac{g}{l}}$$

Taking log on both sides; $\log T_n - \log\pi + \frac{1}{2}[\log g - \log l]$

Taking differentials, $\frac{dn}{n} = 0 - 0 + \frac{1}{2}\left[\frac{dg}{g} - \frac{dl}{l}\right]$

$$\frac{dn}{n} = \frac{1}{2}\left[\frac{dg}{g} - \frac{dl}{l}\right]$$

If l is constant and g is variable

$$\frac{dn}{n} = \frac{1}{2}\frac{dg}{g}$$

If g is constant and l is variable.

$$\frac{dn}{n} = \frac{1}{2}\left[0 - \frac{dl}{l}\right] = \frac{-dl}{2l}$$

Example 16.8

A pendulum has a string of length 99.39 cm. How much length of the pendulum must be shortened to keep the correct time of the pendulum if it losses 4 sec a day.

Sol. Given:

No. of Seconds or beats, the pendulum looses in a day

$$dn = -4\,\text{sec}$$

No. of beats in a day,

$$n = 24 \times 60 \times 60\,\text{sec}$$

Let dl be the length by which the pendulum should be shortened to keep the correct time.

$$\frac{dn}{n} = \frac{-dl}{2l}$$

$$\frac{-4}{24 \times 60 \times 60} = \frac{-dl}{2 \times 99.39}$$

$$dl = \frac{4 \times 2 \times 99.39}{24 \times 60 \times 60} = 0.00920\text{ cm}$$

Example 16.9

A clock with a second's pendulum is running correct at a place where the acceleration due to gravity is 9.81 cm/sec^2. Find the length of the pendulum. What would be the loss or gain in a day if the clock is taken to a place where acceleration due to gravity is 9.82 cm/sec^2.

Sol. Given:

For a second's pendulum time period, $T = 2$ sec.

Acceleration due to gravity is given as 9.81 cm/sec^2

$$T = 2\pi\sqrt{\frac{l}{g}}$$

$$2 = 2\pi\sqrt{\frac{l}{9.81}} \qquad \Rightarrow \ l = 99.39 \text{ cm}$$

Number of beats in a day, $n = 24 \times 60 \times 60 = 86{,}400$

Acceleration due to gravity at new place

$$g_1 = 9.82 \text{ cm/sec}^2$$

Change in gravity, $dg = g_1 - g = 982 - 981 = 1$ cm/sec^2

We know $\dfrac{dn}{n} = \dfrac{dg}{2g}$

$$\frac{dn}{86,400} = \frac{1}{2 \times 9.81}$$

$$dn = 44.03 \text{ sec}$$

The clock will gain 44.03 sec a day.

16.6 Compound Pendulum

It consists of a rigid body suspended from a fixed axis that does not pass through the body's center of mass. It is free to swing from side to side about the pivot point.

Using the relation

$$\tau = I\,\alpha$$

I – Moment of inertia around the axis through P

$$-Mgl\ \sin\alpha = I\frac{d^2\alpha}{dt^2}$$

If we assume that α is small then sinα = α

$$\frac{d^2\alpha}{dt^2} = -\frac{\text{Mgl}\alpha}{\text{I}} = \text{W}^2\theta$$

$$\Rightarrow \text{W} = \sqrt{\frac{\text{Mgl}}{\text{I}}}$$

Time period is given by $T = \dfrac{2\pi}{\omega} = \sqrt{\dfrac{I}{mgl}}$

If we compare it with the Time period of simple pendulum Eq (6.26) reduces to

$$\text{T} = 2\pi\sqrt{\frac{\text{L}}{\text{g}}}$$

where $L = \dfrac{I}{Ml}$

Hence, it can be concluded that a compound pendulum behaves like a simple pendulum with effective length L.

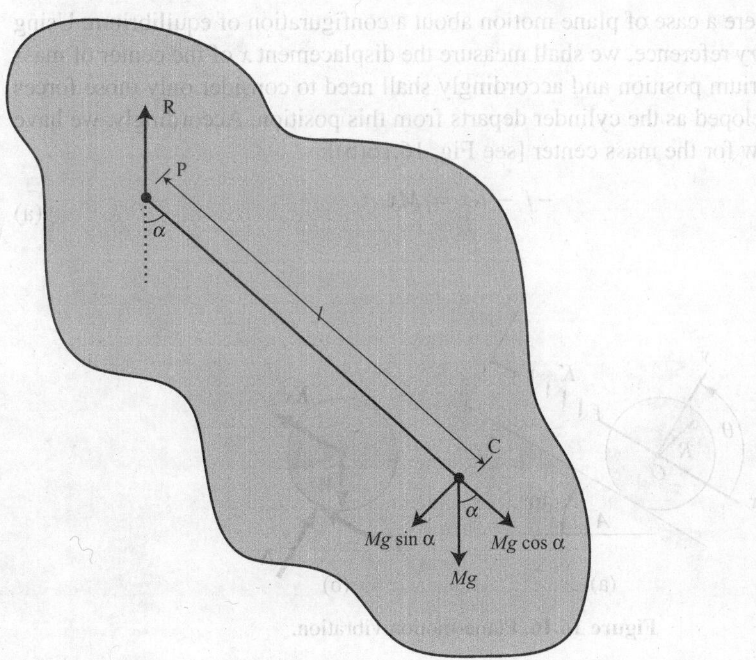

Figure 16.15

*16.7 Examples of Other Free-Oscillating Motions

In the previous sections, we examined the rectilinear translation of a rigid body under the action of a linear restoring force as well as the pure rotation of a rigid body under the action of a linear restoring torque. In this section, we shall first examine a body with one degree of freedom undergoing *plane motion* governed by a differential equation of motion of the form given in the previous section. The dependent variable for such a case varies harmonically with time, and we have a *vibratory plane motion*. Consider the following example.

Example 16.10

Shown in Fig. 16.16(a) on an inclined plane is a uniform cylinder maintained in a position of equilibrium by a linear spring having a spring constant K. If the cylinder rolls without slipping, what is the equation of motion when it is disturbed from its equilibrium position?

We have here a case of plane motion about a configuration of equilibrium. Using xyz as a *stationary* reference, we shall measure the displacement x of the center of mass from the equilibrium position and accordingly shall need to consider only those forces and torques developed as the cylinder departs from this position. Accordingly, we have for **Newton's law** for the mass center [see Fig. 16.16(b)]:

$$-f - Kx = M\ddot{x} \tag{a}$$

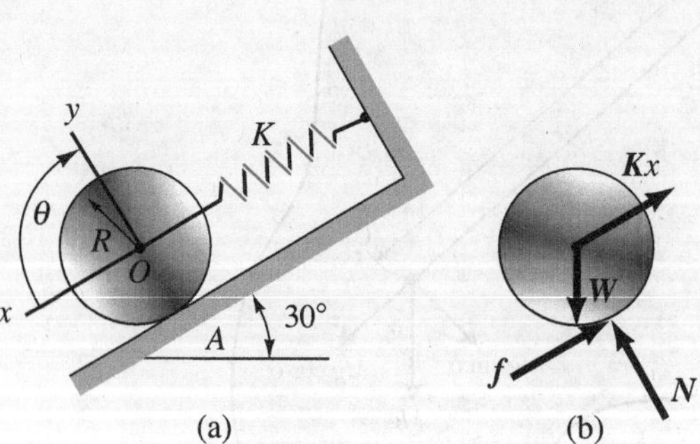

Figure 16.16. Plane-motion vibration.

Example 16.10 (Continued)

Now employ the **angular-momentum** equation about the geometric axis of the cylinder at O. Using θ to measure the rotation of the cylinder about this axis from the equilibrium configuration, we get

$$-fR = \tfrac{1}{2} MR^2\ddot{\theta} \qquad \text{(b)}$$

Noting from **kinematics** that $\ddot{x} = -R\,\ddot{\theta}$ as a result of the no-slipping condition, we have for Eq. (b):

$$fR = \frac{1}{2} MR^2\left(\frac{\ddot{x}}{R}\right)$$

Therefore,

$$f = \tfrac{1}{2} M\ddot{x} \qquad \text{(c)}$$

Substituting for f in Eq. (a) using this result, we have

$$M\ddot{x} = -\tfrac{1}{2} M\ddot{x} - Kx$$

Therefore,

$$\ddot{x} + \frac{2}{3}\frac{K}{M}x = 0 \qquad \text{(d)}$$

We could also have arrived at the differential equation above by noting that we have instantaneous pure rotation about the line of contact A as a result of the no-slipping condition. Thus, the **angular-momentum** equation can be used as follows about the point of contact on the cylinder:

$$\left(\tfrac{1}{2} MR^2 + MR^2\right)\ddot{\theta} = KxR \qquad \text{(e)}$$

Noting as before that $\ddot{\theta} = -\ddot{x}/R$, we get

$$-\frac{3}{2} MR^2 \frac{\ddot{x}}{R} = KxR$$

Therefore, as before

$$\ddot{x} + \frac{2}{3}\frac{K}{M}x = 0 \qquad \text{(f)}$$

Example 16.10 (Continued)

We may solve the differential equation to give us

$$x = x_0 \cos\sqrt{\frac{2}{3}\frac{K}{M}}t + \frac{\dot{x}_0}{\sqrt{\frac{2}{3}K/M}}\sin\sqrt{\frac{2}{3}\frac{K}{M}}t \tag{g}$$

where x_0 and $\dot{x}_0$ are the initial position and speed of the center of mass, respectively. Since $\theta = -x/R$ (we have here only one degree of freedom as a result of the no-slipping condition), we have for θ from Eq. (g):

$$\theta = \frac{-x_0}{R}\cos\sqrt{\frac{2}{3}\frac{K}{M}}t - \frac{\dot{x}_0}{R\sqrt{\frac{2}{3}K/M}}\sin\sqrt{\frac{2}{3}\frac{K}{M}}t \tag{h}$$

*16.8 Energy Methods

Up to now, the procedure has been primarily to work with *Newton's law* or the *angular-momentum equation* in reaching the differential equation of interest. There is an alternative approach to the handling of free vibration problems that may be very useful in dealing with simple systems and in setting up approximate calculations for more complex systems. Suppose we know for a one-degree-of-freedom system that only linear restoring forces and torques do work during possible motions of the system. Then, the agents developing such forces are *conservative* force agents and may be considered to store potential energy. You will recall from Section 11.2 that the *total mechanical energy* for such systems is *conserved*. Thus, we have

$$\text{PE} + \text{KE} = \text{constant} \tag{16.27}$$

Also, we know from our present study that the system must oscillate harmonically when disturbed and then allowed to move freely with only the linear restoring agents doing work. Thus, if κ is the independent coordinate measured from the static-equilibrium configuration, we have

$$\kappa = A\sin(\omega_n t + \beta) \tag{16.28}$$

Hence,

$$\dot{\kappa} = A\omega_n \cos(\omega_n t + \beta) \tag{16.29}$$

Now at the instant when $\kappa = 0$, we are at the static-equilibrium position and the potential energy of the system is a minimum. Since the total mechanical energy must be conserved at all times once such a motion is under way, it is also clear that the kinetic energy must be at a maximum at that instant. If we take the lowest potential energy as zero, then we have for the total mechanical energy simply the maximum kinetic energy. Also, when the body is undergoing a change in direction of its motion at the outer extreme position, the kinetic energy is zero instantaneously, and accordingly the potential energy must be a maximum and equal to the total mechanical energy of the system. Thus, we can equate the maximum potential energy with the maximum kinetic energy.

$$(\text{KE})_{\text{max}} = (\text{PE})_{\text{max}} \tag{16.30}$$

In computing the $(\text{KE})_{\text{max}}$, we will involve $(\dot{\kappa})_{\text{max}}$ and hence $A\omega_n$, whereas for the $(\text{PE})_{\text{max}}$ we will involve $(\kappa)_{\text{max}}$ and hence A. In this way we can set up quickly an equation for ω_n, the natural frequency of the system. For example, if we have the simple linear spring-mass system of Fig. 16.1, we can say:

$$(\text{PE}) = \tfrac{1}{2} Kx^2$$

Therefore,

$$(\text{PE})_{\text{max}} = \tfrac{1}{2} K(x_{\text{max}})^2 = \tfrac{1}{2} KA^2$$

where we have made use of our knowledge that $x = A \sin(\omega_n t + \beta)$. And, noting that $\dot{x} = A\omega_n \cos(\omega_n t + \beta)$, we have

$$(\text{KE})_{\text{max}} = \tfrac{1}{2} M(\dot{x}_{\text{max}})^2 = \tfrac{1}{2} M(A\omega_n)^2$$

Now, equating these expressions, we get

$$\tfrac{1}{2} KA^2 = \tfrac{1}{2} M(A\omega_n)^2$$

Therefore,

$$\omega_n = \sqrt{\frac{K}{M}}$$

which is the expected result. We next illustrate this approach in a more complex problem.

Example 16.11

A cylinder of radius r and weight W rolls without slipping along a circular path of radius R as shown in Fig. 16.17. Compute the natural frequency of oscillation for small oscillation.

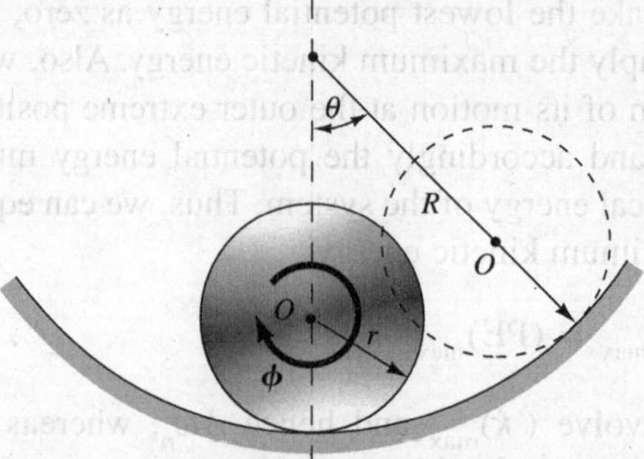

Figure 16.17. Cylinder rolls without slipping.

This system has one degree of freedom. We can use ϕ, the angle of rotation of the cylinder about its axis of symmetry, as the independent coordinate, or we may use θ as shown in the diagram. To relate these variables for no slipping we may conclude, on observing the motion of point O, that for small rotation:

$$(R - r)\theta = r\phi$$

Therefore,

$$\theta = \frac{r}{R - r}\phi \tag{a}$$

The only force that does work during the possible motions of the system is the force of gravity W. The torque developed by W about the point of contact for a given θ is easily determined after examining Fig. 16.18 to be

$$\text{torque} = Wr\sin\theta = Wr\sin\left(\frac{r}{R - r}\phi\right) \tag{b}$$

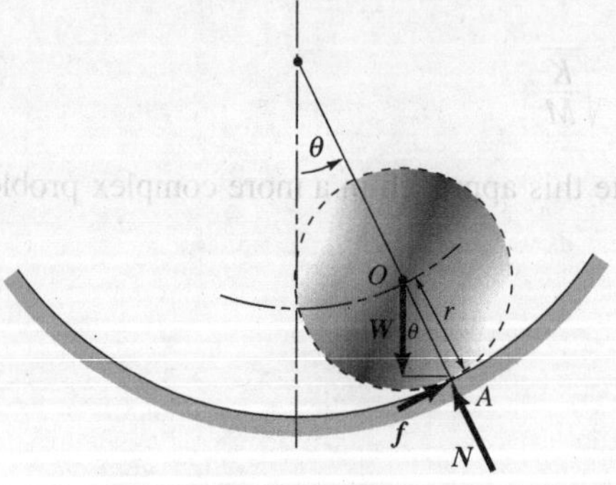

Figure 16.18. Free-body diagram of cylinder.

Example 16.11 (Continued)

This is a restoring torque, and because we limit ourselves to *small oscillations* it becomes $W[r^2\phi/(R-r)]$, which is clearly a linear restoring torque. Because the force doing work on the cylinder is *conservative*, and because it results in a *linear restoring torque*, we can employ the energy formulation of this section.

The motion may be considered to be given as follows:

$$\theta = C\sin(\omega_n t + \beta) \tag{c}$$

or, using Eq. (a),

$$\phi = \frac{R-r}{r} C\sin(\omega_n t + \beta) \tag{d}$$

Expressing the maximum potential and kinetic energies and using C for θ_{max} and the lowest position of O as the datum, we have:

$$\begin{aligned}(PE)_{max} &= W(R-r)(1-\cos\dot{\theta}_{max}) \\ &= W(R-r)(1-\cos C)\end{aligned} \tag{e}$$

$$\begin{aligned}(KE)_{max} &= \frac{1}{2}\frac{W}{g}(R-r)^2\dot{\theta}^2_{max} + \frac{1}{4}\frac{W}{g}r^2\dot{\phi}^2_{max} \\ &= \frac{1}{2}\frac{W}{g}(R-r)^2(C\omega_n)^2 + \frac{1}{4}\frac{W}{g}r^2\left(\frac{R-r}{r}C\omega_n\right)^2\end{aligned} \tag{f}$$

We have used Eq. (d) in the last expression of Eq. (f). Expanding cos C in a power series and retaining the first two terms $(1-C^2/2)$, we then get, on equating the right sides of the above equations:

$$W(R-r)\frac{C^2}{2} = \frac{W}{2g}\left[(R-r)^2 + \frac{(R-r)^2}{2}\right]\omega_n^2 C^2$$

Therefore,

$$\omega_n = \sqrt{\frac{g}{\frac{3}{2}(R-r)}} \tag{g}$$

PROBLEMS

16.50. A cylinder of diameter 1 m is shown. The center of gravity of the cylinder is .30 m from the geometric center, and the radius of gyration is .60 m at the center of mass. What is the natural frequency of oscillation for small vibrations without slipping? The cylinder weighs 220 N. Work the problem by two methods.

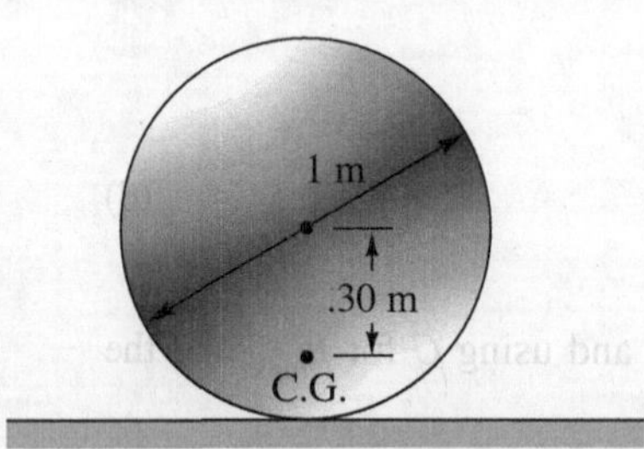

Figure P.16.50.

16.51 A stepped cylinder is maintained along the incline by a spring having a spring constant K. What is the formulation for the natural frequency of oscillation for the system? What is the maximum friction force? Take the weight of the cylinder as W and the radius of gyration about the geometric centerline O as k. The initial conditions are $\theta = \theta_0$ at $t = 0$, and $\dot{\theta} = \dot{\theta}_0$ at $t = 0$. There is no slipping.

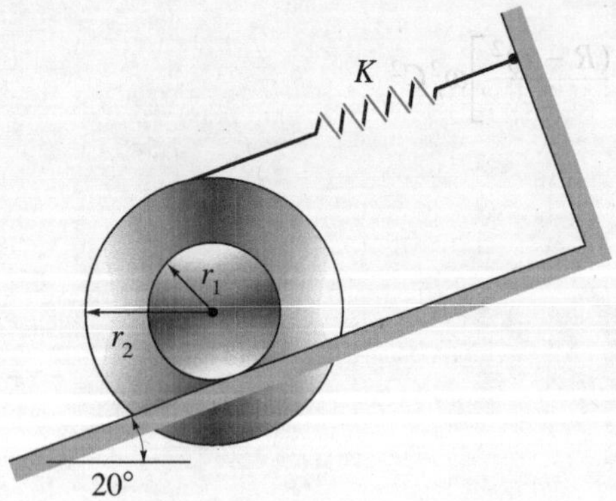

Figure P.16.51.

16.52 Two masses are attached to a light rod. The rod rides on a frictionless horizontal rail. $M_1 = 45$ kg and $M_2 = 14$ kg. What is the natural frequency of oscillation of the system if a small impulsive torque is applied to the system when it is in a rest configuration? Consider the masses as particles. [*Hint:* Consider motion about the center of mass of the system.]

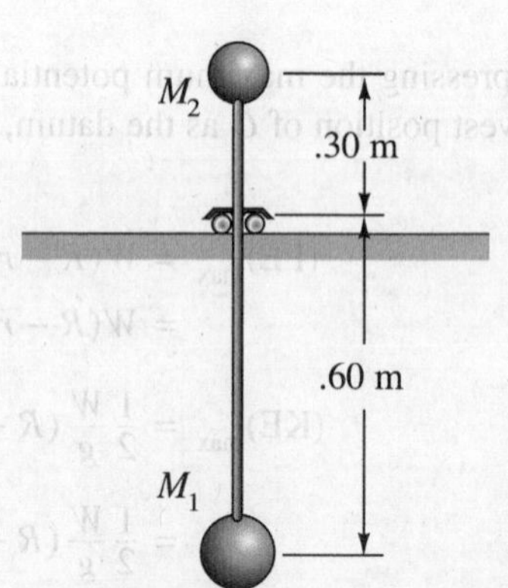

Figure P.16.52.

16.53. Work Problem 16.30 by energy methods.

16.54. Work Problem 16.13 by energy methods.

16.55. Work Problem 16.40 by energy methods.

16.56. A *manometer* used for measuring pressures is shown. If the mercury has a length L in the tube, what is the formulation for the natural frequency of movement of the mercury?

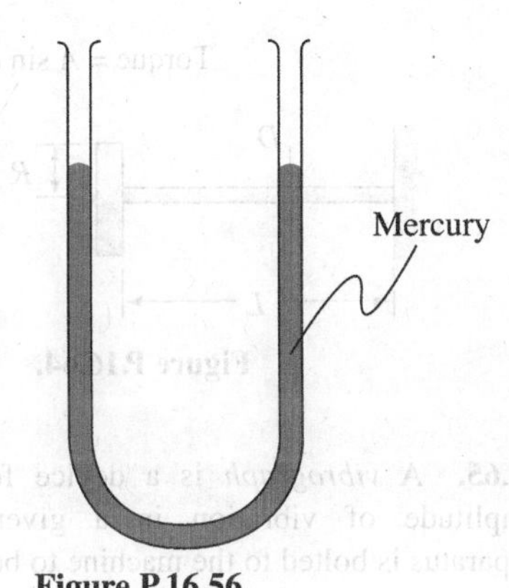

Figure P.16.56.

16.57. An *inclined manometer* is often used for more accurate pressure measurements. If the mercury in the tube has a length L, what is the natural frequency of oscillation of the mercury in the tube?

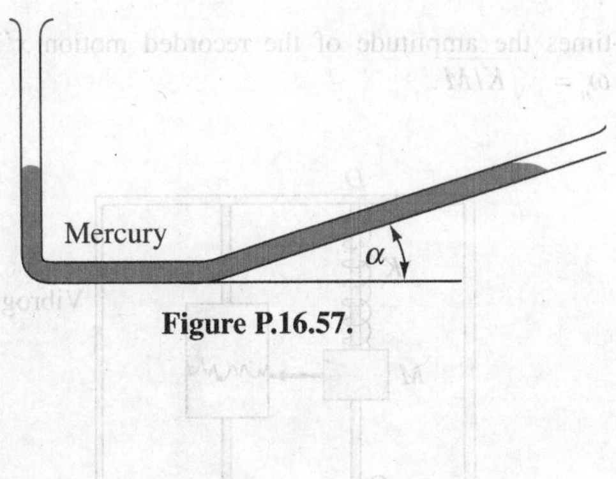

Figure P.16.57.

16.58. A *differential* manometer is used for measuring high pressures. What is the natural frequency of oscillation of the mercury?

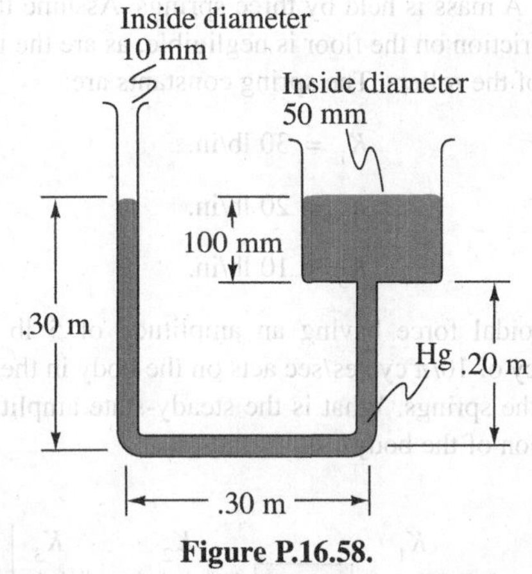

Figure P.16.58.

16.59. A stepped cylinder rides on a circular path. For small oscillations, what is the natural frequency? Take the radius of gyration about the geometric axis O as k and the weight of the cylinder as W.

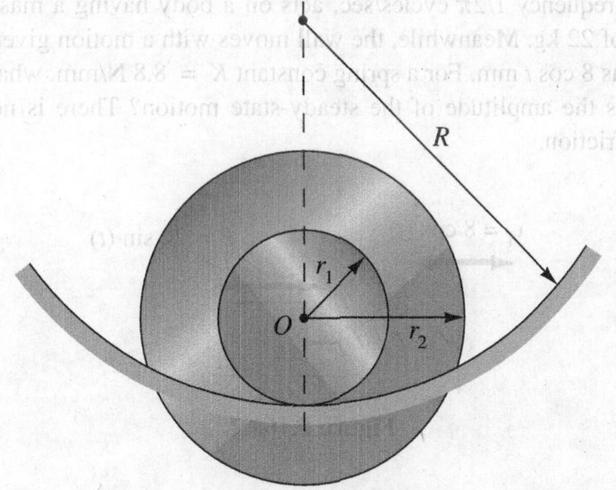

Figure P.16.59.

16.60. A mass is held by three springs. Assume that the rolling friction on the floor is negligible, as are the inertial effects of the rollers. The spring constants are:

$K_1 = 30$ lb/in.

$K_2 = 20$ lb/in.

$K_3 = 10$ lb/in.

A sinusoidal force having an amplitude of 5 lb and a frequency of $10/\pi$ cycles/sec acts on the body in the direction of the springs. What is the steady-state amplitude of the motion of the body?

Figure P.16.60.

16.61. In Problem 16.60, the initial conditions are:

(a) The initial position of the body is 3 in. to the right of the static-equilibrium position.

(b) The initial velocity is zero.

(c) At $t = 0$, the sinusoidal disturbing force has a value of 5 lb in the positive direction.

Find the position of the body after 3 sec.

16.62 A sinusoidal force, with amplitude F of 22 N and frequency $1/2\pi$ cycles/sec, acts on a body having a mass of 22 kg. Meanwhile, the wall moves with a motion given as 8 cos t mm. For a spring constant $K = 8.8$ N/mm, what is the amplitude of the steady-state motion? There is no friction.

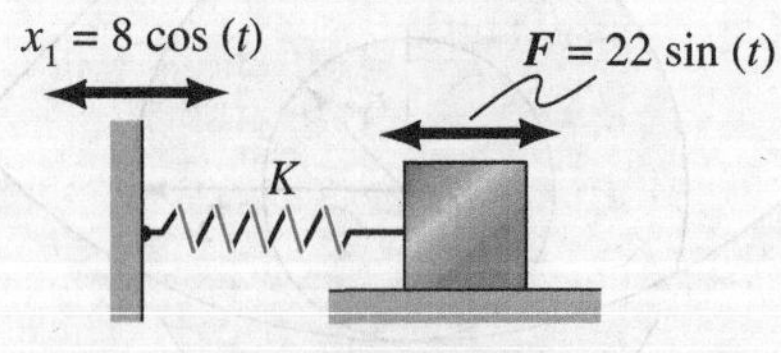

Figure P.16.62.

16.63. In Problem 16.62, suppose that the disturbing force F is 22 sin $(t + \pi/4)$. What is the amplitude of the steady-state motion?

16.64. A torque $T = A \sin \omega t$ is applied to the disc. Express the solution for the transient torsional motion and the steady-state torsional motion in terms of arbitrary constants of integration. Take the shear modulus of elasticity of the shaft as G [*Hint:* Recall that K_t for a shaft is GJ/L.]

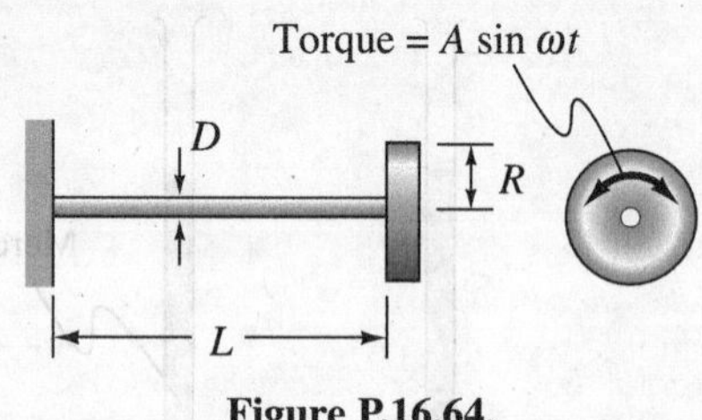

Figure P.16.64.

16.65. A *vibrograph* is a device for measuring the amplitude of vibration in a given direction. The apparatus is bolted to the machine to be tested. A seismic mass M in the vibrograph rides along a rod CD under constraint of a linear spring of spring constant K. If the machine being tested has a harmonic motion x of frequency ω in the direction of C–D, then M will have a steady-state oscillatory frequency also of frequency ω. The motion of M relative to the vibrograph is given as x' and is recorded on the rotating drum. Show that the amplitude of motion of the machine is

$$\left| \frac{(\omega/\omega_n)^2 - 1}{(\omega/\omega_n)^2} \right|$$

times the amplitude of the recorded motion x', where $\omega_n = \sqrt{K/M}$.

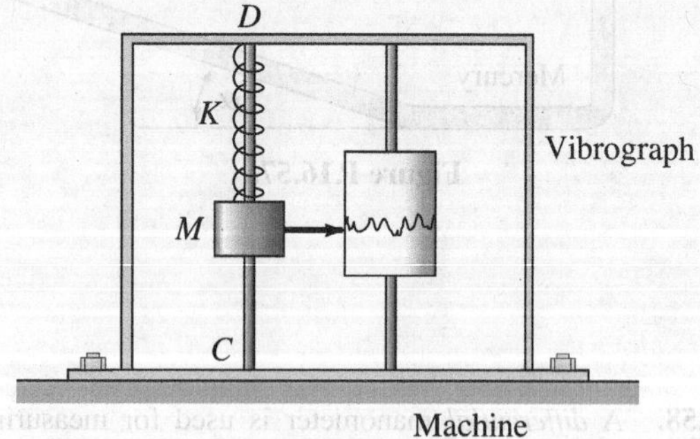

Figure P.16.65.

16.66. A vibrograph is attached rigidly to a diesel engine for which we want to know the vibration amplitude. If the seismic spring-mass system has a natural frequency of 10 cycles/sec, and if the seismic mass vibrates relative to the vibrograph with an amplitude of 1.27 mm when the diesel is turning over at 1,000 rpm, what is the amplitude of vibration of the diesel in the direction of the vibrograph? The seismic mass weighs 4.5 N. See Problem 16.60 before doing this problem.

16.67. Explain how you could devise an instrument to measure torsional vibrations of a shaft in a manner analogous to the way the vibrograph measures linear vibrations of a machine. Such instruments are in wide use and are called *torsiographs*. What would be the relation of the amplitude of oscillations as picked up by your apparatus to that of the shaft being measured? See Problem 16.60 before doing this problem.

16.68. A trailer of weight W moves over a washboard road at a constant speed V to the right. The road is approximated by a sinusoid of amplitude A and wavelength L. If the wheel B is small, the center of the wheel will have a motion x closely resembling the aforementioned sinusoid. If the trailer is connected to the wheel through a linear spring of stiffness K, formulate the steady-state equation of motion x' for the trailer. List all assumptions. What speed causes resonance?

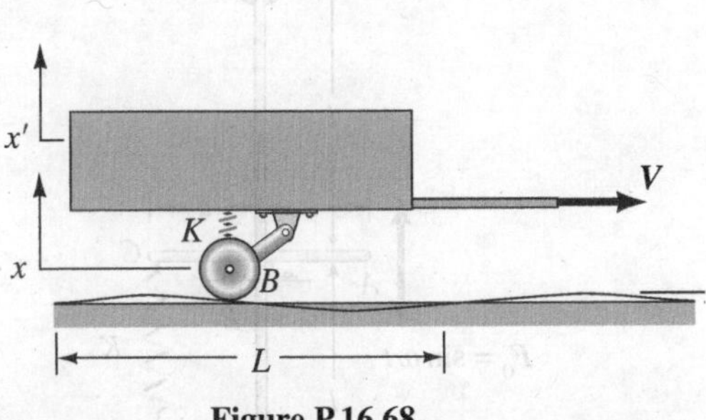

Figure P.16.68.

16.69. In Problem 16.68, compute the amplitude of motion of the trailer for the following data:

$$W = 5.34 \text{ kN}$$
$$V = 16 \text{ km/hr}$$
$$K = 43.8 \text{ N/mm}$$
$$L = 10 \text{ m}$$
$$A = 100 \text{ mm}$$

What is the resonance speed V_{res} for this case? From Problem 16.68, we have

$$x'_p = \frac{A}{|1-(2\pi V/L)^2(W/gK)|}\sin\frac{2\pi Vt}{L}$$

16.70. A cantilever beam of length L has an electric motor A weighing 100 N fastened to the end. The tip of the cantilever beam descends 12 mm when the motor is attached. If the *center of mass* of the armature of the motor is a distance 2 mm from the axis of rotation of the motor, what is the amplitude of vibration of the motor when it is rotating at 1,750 rpm? The armature weighs 40 N. Neglect the mass of the beam.

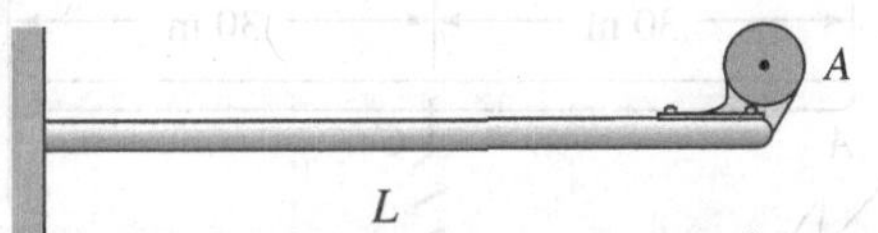

Figure P.16.70.

16.71. Suppose that a 2-N block is glued to the top of the motor in Problem 16.70, where the maximum strength of the bond is ½ N. At what minimum angular speed ω of the motor will the block fly off?

16.72. An important reason for mounting rotating and reciprocating machinery on springs is to decrease the transmission of vibration to the foundation supporting the machine. Show that the amplitude of force transmitted to the ground, F_{TR}, for such cases is

$$F_0\left|\frac{1}{1-(\omega/\omega_n)^2}\right|$$

where F_0 is the disturbing force from the machine. The factor $|1/[1-(\omega/\omega_n)^2]|$ is called the *relative transmission factor*. Show that, unless the springs are soft, $(\omega_n < \omega/\sqrt{2})$, the use of springs actually increases the transmission of vibratory forces to the foundation.

16.73. In Example 16.7, what is the amplitude of the force transmitted to the foundation? What must K of the spring system be to decrease the amplitude by one-half? See Problem 16.72.

16.74. A machine weighing W N contains a reciprocating mass of weight w N having a vertical motion relative to the machine given approximately as $x' = A \sin \omega t$. The machine is mounted on springs having a total spring constant K. This machine is guided so that it can move only in the vertical direction. What is the differential equation of motion for this machine? What is the formula for the amplitude of the machine for steady-state operation?

16.75. A mass M of .5 kg is suspended from a stiff rod AB via a spring whose spring constant K is 100 N/m. The end of rod AB is given a vertical sinusoidal motion $\delta_A = 2 \sin 14t$ mm, with t in seconds. What is the maximum force on the rod at C long after the motion has started?

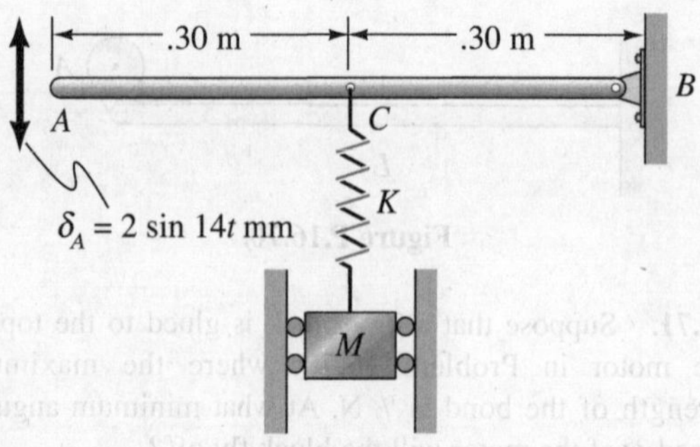

Figure P.16.75.

16.76. In Problem 16.75, what range of frequencies of the motion δ_A must be excluded to keep the maximum force at C less than 7 N? Consider only steady-state motion. [*Hint:* Note that below resonance disturbance and motion are in phase whereas above resonance they are 180° out of phase. Therefore, x_p in Eq. 16.27 will be positive below resonance and negative above resonance.]

16.77. A bob B of weight W is suspended from a vehicle A which is made to have a motion $x_A = \delta \sin \omega t$. If δ is very small, what should ω be so that bob B has an amplitude of motion equal to 1.5δ?

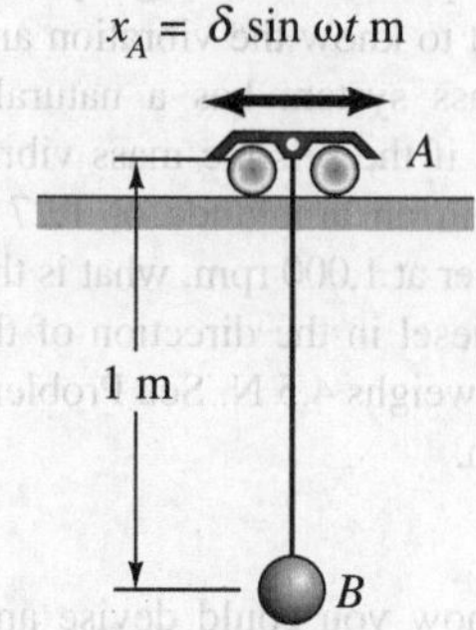

Figure P.16.77.

16.78. wo spheres each of mass M = 2 kg are welded to a light rod that is pinned at B. A second light rod AC is welded to the first rod. At A we apply a disturbance $F_o \sin \omega t$. At the other end C, there is a restraining spring which is unstretched when AC is horizontal. If the amplitude of steady-state rotation of the system is to be kept below .02 rad, what ranges of frequencies ω are permitted? The following data apply:

l = 300 mm
K = 7.0 N/mm
F_o = 10 N
a = 100 mm

See the hint in Problem 16.76.

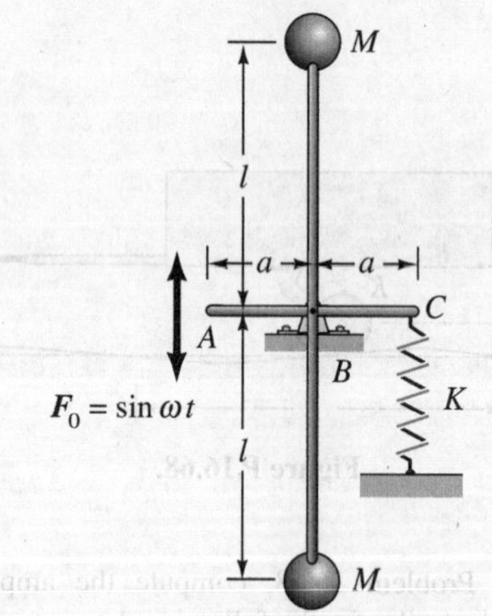

Figure P.16.78.

16.79. In Problem 16.78, what is the angle of rotation of the system 10 sec after the application of the sinusoidal load? The system is stationary at time t = 0. Take ω = 13 rad/sec.

16.80. A rod of length L and weight W is suspended from a light support at A. This support is given a movement $x_A = \delta \sin \omega t$, where δ is very small compared to L. At what frequency, ω, should A be moved if the amplitude of motion of tip, B, is to be 1.5δ?

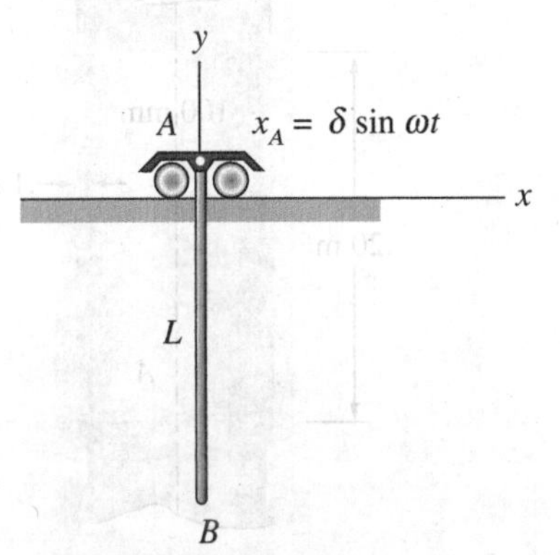

Figure P.16.80.

16.81. A body of weight W N is suspended between two springs. Two identical dashpots are shown. Each dashpot resists motion of the block at the rate of c N/m/sec. What is the equation of motion for the block? What is c for critical damping?

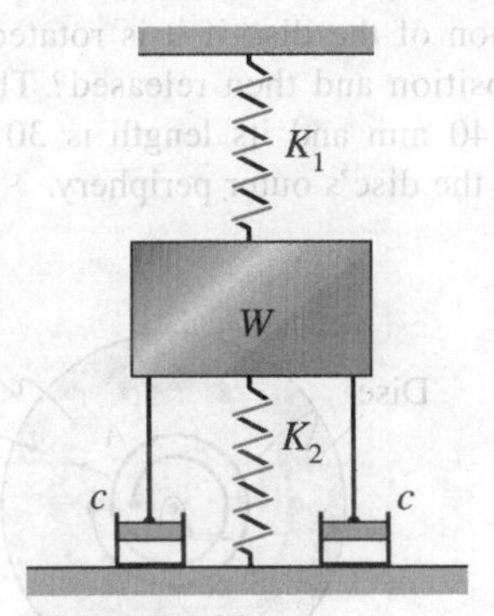

Figure P.16.81.

16.82. In Problem 16.81, the following data apply:

$$W = 445 \text{ N}$$
$$K_1 = 8.8 \text{ N/mm}$$
$$K_2 = 14.0 \text{ N/mm}$$
$$c = 825 \text{ N/m/sec}$$

Is the system underdamped, overdamped, or critically damped? If the weight W is released 150 mm above its static-equilibrium configuration, what are the speed and position of the block after .1 sec? What force is transmitted to the foundation at that instant?

16.83. The damping constant c for the body is $\frac{1}{2}$lb/ft/sec. If, at its equilibrium position, the body is suddenly given a velocity of 10 ft/sec to the right, what will the frequency of its motion be? What is the position of the mass at $t = 5$ sec?

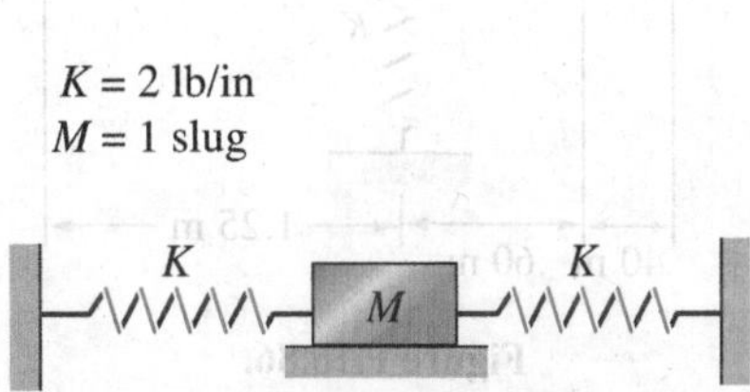

Figure P.16.83.

16.84. The damping in Problem 16.83 is increased so that it is twice the critical damping. If the mass is released from a position 3 in. to the right of equilibrium, how far from the equilibrium position is it in 5 sec? Theoretically, does it ever reach the equilibrium position?

16.85. A plot of a free damped vibration is shown. Show that ln (x_1/x_2), where x_1 and x_2 are two succeeding peaks, can be given as $(c/4m)\tau$. The expression ln (x_1/x_2) is called the *logarithmic decrement* and is used in vibration work.

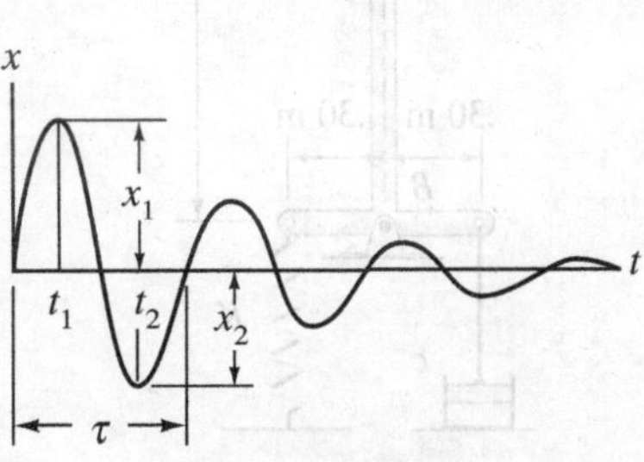

Figure P.16.85.

16.86. A rod of length $2\frac{1}{4}$ m and weight 200 N is shown in the static-equilibrium position supported by a spring of stiffness $K = 14$ N/mm. The rod is connected to a dashpot having a damping force c of 69 N/m/sec. If an impulsive torque gives the rod an angular speed clockwise of $\frac{1}{2}$ rad/sec at the position shown, what is the position of point A at $t = .2$ sec?

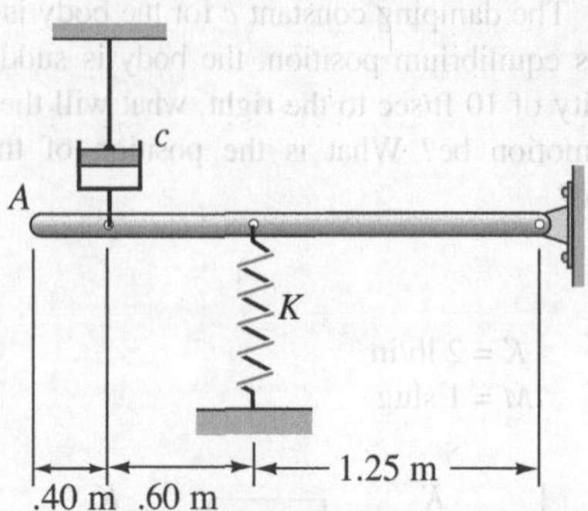

Figure P.16.86.

16.87. A spherical ball of weight 134 N is welded to a vertical light rod which in turn is welded at B to a horizontal rod. A spring of stiffness $K = 8.8$ N/mm and a damper c having a value 179 N/m/sec are connected to the horizontal rod. If A is displaced 75 mm to the right, how long does it take for it to return to its vertical configuration?

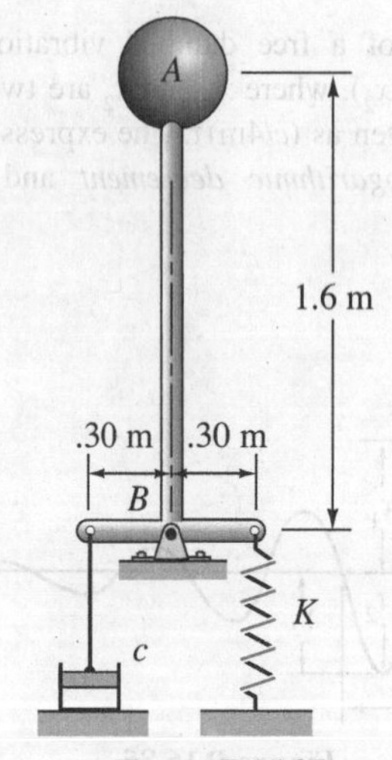

Figure P.16.87.

16.88. Cylinder A of weight 200 N slides down a vertical cylindrical chute. A film of oil of thickness .1 mm separates the cylinder from the chute. If the air pressure before and behind the cylinder is maintained at the same value of 15 psig, what is the maximum velocity that the cylinder can attain by gravity? The coefficient of viscosity of the oil is .00800 N-sec/m^2.

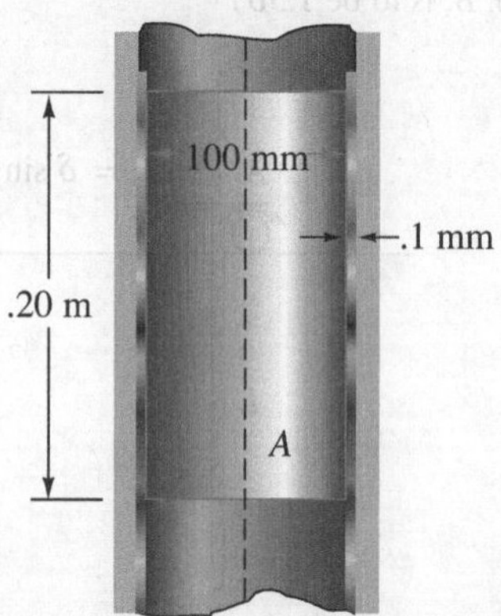

Figure P.16.88.

16.89. A disc A with a mass of 5 kg is constrained during rotation about its axis by a torsional spring having a constant K_T equal to 2×10^{-5} N-m/rad. The disc is in a journal having a diameter 2 mm larger than the disc. Oil having a viscosity .0085 N-sec/m^2 fills the outer space between disc and journal. If we assume a linear profile for the oil film, what is the frequency of oscillation of the disc if it is rotated from its equilibrium position and then released? The diameter of the disc is 40 mm and its length is 30 mm. The oil acts only on the disc's outer periphery.

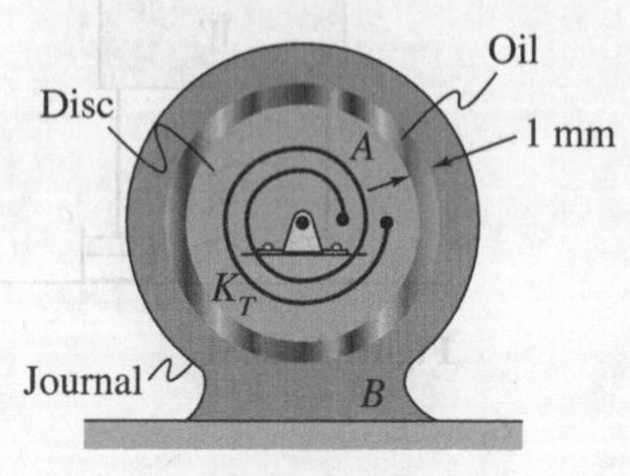

Figure P.16.89.

16.90. A block W weighing 60 N is released from rest at a configuration 100 mm above its equilibrium position. It rides on a film of oil whose thickness is .1 mm and whose coefficient of viscosity is .00950 N-sec/m^2. If K is 50 N/m, how far down the incline will the block move? The block is .20 m on each edge.

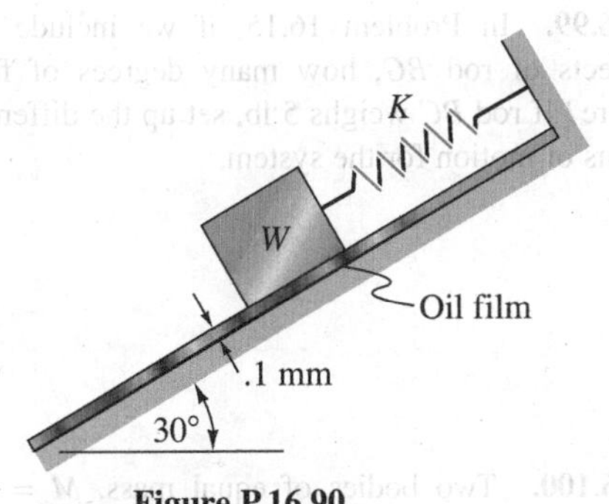

Figure P.16.90.

*16.91. In Problem 16.90, set up two simultaneous equations to determine the spring constant K so that, after W is released, W comes back to its equilibrium position with no oscillation. As a short project, solve for K using a computer.

16.92. A disc B of diameter 100 mm rotates in a stationary housing filled with oil of viscosity .00600 N-sec/m^2. The disc and its shaft have a mass of 30 g and a radius of gyration of 20 mm. The shaft and disc connect to a device that supplies a linear restoring torque of 5 N-mm/rad. Use a linear velocity profile for the oil and find how long each oscillation of the disc about its axis takes.

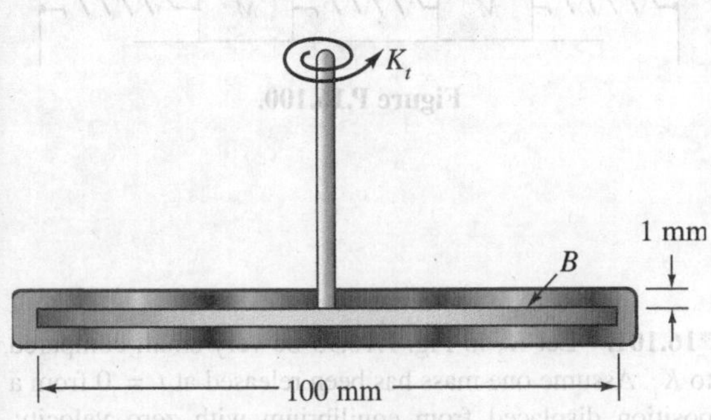

Figure P.16.92.

16.93. Examine the case of the spring-mass system with viscous damping for a sinusoidal forcing function given as $F_0 \sin \omega t$. Go through the steps in the text leading up to Eq. 16.41 for this case.

16.94. A force $F = 35 \sin 2t$ N acts on a block having a weight of 285 N. A spring having stiffness K of 550 N/m and a dashpot having a damping factor c of 68 N-sec/m are connected to the body. What is the amplitude of steady-state motion for the body, and what is the maximum force transmitted to the wall?

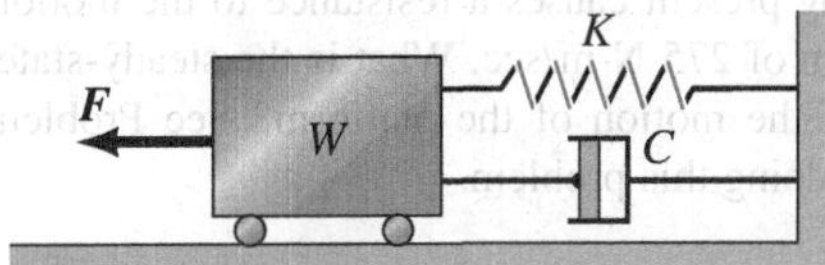

Figure P.16.94.

16.95. In Example 16.11, compute K for an amplitude of steady-state vibration of $\frac{1}{2}$ in.

16.96. A block of mass M rests on two springs, the total spring constant of which is K. Also, there is a dashpot of constant c. A small sphere of mass m is attached to M and is made to rotate at a speed of ω. The distance from the center of rotation to the sphere is r. Derive the equation of motion for the block first by considering the motions of M and m separately as single masses. Show that you could reach the same equation of motion by lumping the masses M and m into one body of mass $(M + m)$ on which a sinusoidal disturbance equal to $mr\omega^2 \sin \omega t$ (from the rotating sphere) is applied.

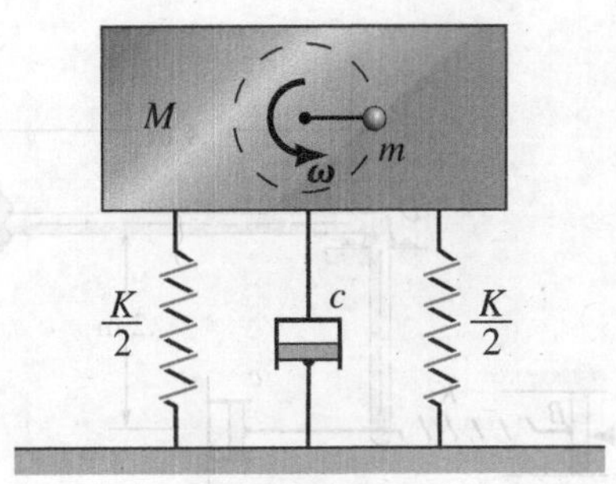

Figure P.16.96.

16.97. A platform weighing 222 N deflects the spring 50 mm when placed carefully on the spring. A motor weighing 22 N is then clamped on top of the platform and rotates an eccentric mass m which weighs 1 N. The mass m is displaced 150 mm from the axis of rotation and rotates at an angular speed of 28 rad/sec. The viscous damping present causes a resistance to the motion of the platform of 275 N-m/sec. What is the steady-state amplitude of the motion of the platform? See Problem 16.96 before doing this problem.

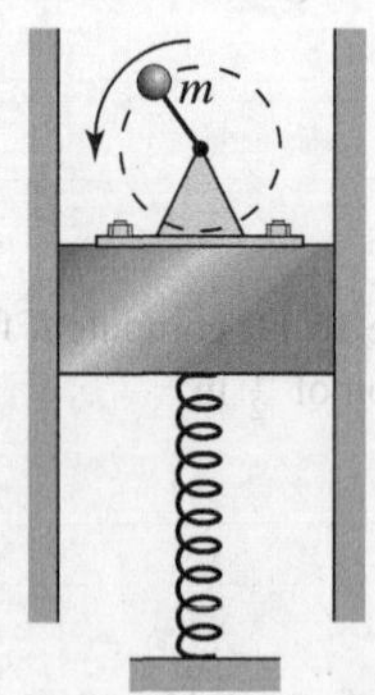

Figure P.16.97.

16.98. A body weighing 143 N is connected by a light rod to a spring of stiffness K equal to 2.6 N/mm and to a dashpot having a damping factor c. Point B has a given motion x' of 30.5 sin t mm with t in seconds. If the center of A is to have an amplitude of steady-state motion of 20 mm, what must c be?

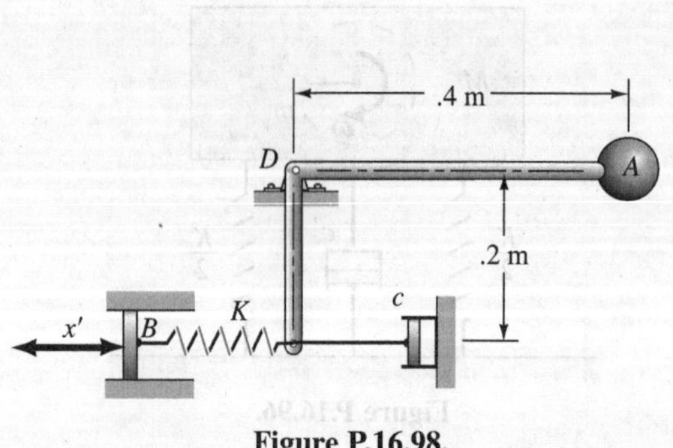

Figure P.16.98.

***16.99.** In Problem 16.15, if we include the inertial effects of rod BC, how many degrees of freedom are there? If rod BC weighs 5 lb, set up the differential equations of motion for the system.

***16.100.** Two bodies of equal mass, M = 1 slug, are attached to walls by springs having equal spring constants K_1 = 5 lb/in. and are connected to each other by a spring having a spring constant K_2 = 1 lb/in. If the mass on the left is released from a position $(x_1)_0$ = 3 in. at t = 0 with zero velocity and the mass at the right is stationary at x_2 = 0 at this instant, what is the position of each mass at the time t = 5 sec? The coordinates x_1 and x_2 are measured from the static-equilibrium positions of the body.

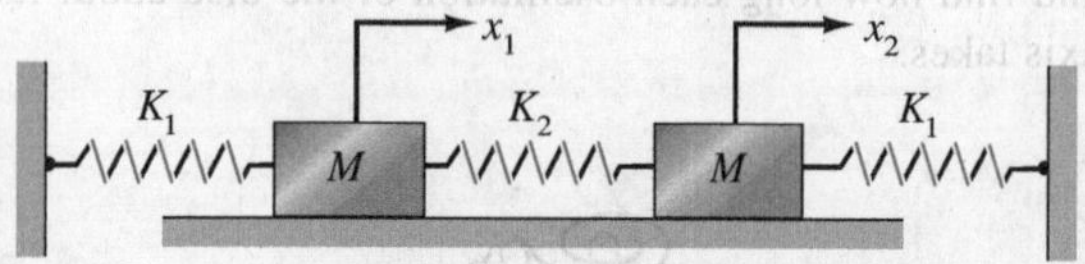

Figure P.16.100.

***16.101.** Let K_2 in Fig. P.16.95 be very small compared to K_1. Assume one mass has been released at t = 0 from a position displaced from equilibrium with zero velocity, while the other mass is released from the initial stationary position at that instant with zero velocity. Show that one mass will have a maximum velocity while the other will have a minimum velocity and that there will be a continual transfer of kinetic energy from one mass to the other at a frequency equal to the beat frequency of the natural frequencies of the system. [*Hint*: Study the phasors $A\cos\sqrt{\frac{K_1}{M}}t$ and $A\cos\sqrt{\frac{K_1}{M}+\frac{2K_2}{M}}t$.]

16.9 Closure

This introductory study of vibration brings to a close the present study of particle and rigid-body mechanics. As you progress to the study of deformable media in your courses in solid and fluid mechanics you will find that particle mechanics and, to a lesser extent rigid-body mechanics, will form cornerstones for these disciplines. And in your studies involving the design of machines and the performance of vehicles you will find rigid-body mechanics indispensable.

It should be realized, however, that we have by no means said the last word on particle and rigid-body mechanics. More advanced studies will emphasize the variational approach introduced in statics. With the use of the calculus of variations, such topics as Hamilton's principle, Lagrange's equation,[5] and Hamilton-Jacobi theory will be presented and you will then see a greater unity between mechanics and other areas of physics such as electromagnetic theory and wave mechanics. Also the special theory of relativity will most surely be considered.

Finally, in your studies of modern physics you will come to more fully understand the limitations of classical mechanics when you are introduced to quantum mechanics.

[5]Some of these topics are covered in the author's text written with C.L. Dym, "Energy and Finite Elements Methods in Structural Mechanics" Taylor and Francis.

PROBLEMS

16.102. If $K_1 = 2K_2 = 1.8K_3$, what should K_3 be for a period of free vibration of .2 sec? The mass M is 3 kg.

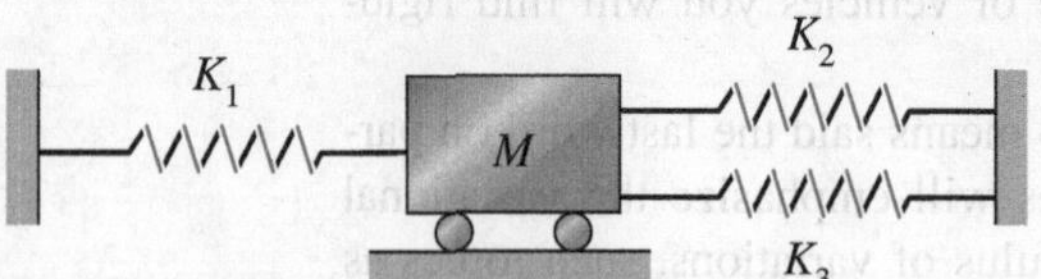

Figure P.16.102.

16.103. In Problem 16.60, determine the natural frequency of the system. If the mass is deflected 2 in. and then released, determine the displacement from equilibrium after 3 sec. Finally, determine the *total* distance traveled during this time.

Figure P.16.103.

16.104. Find the natural frequency of motion of body A for small rotation of rod BD when we neglect the inertial effects of rod BD. The spring constant K_2 is .9 N/mm and the spring constant K_1 is 1.8 N/mm. The weight of block A is 178 N. Neglect friction everywhere. Rod BD weighs 44 N.

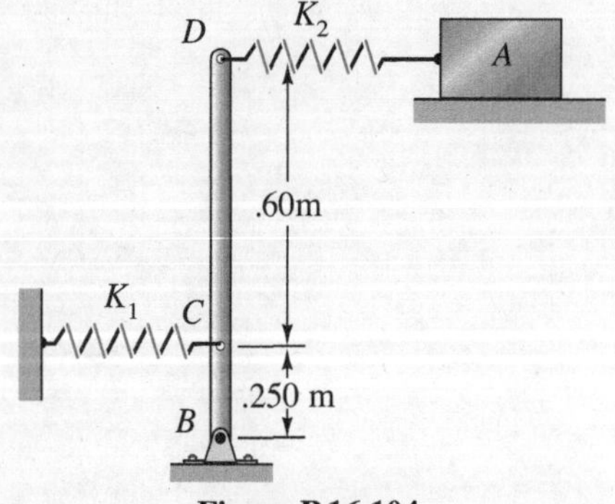

Figure P.16.104.

16.105. What is the equivalent spring constant for small oscillations about the shaft AB? Neglect all mass except the block at B, which weighs 100 lb. The shear modulus of elasticity G for the shaft is 15×10^6 psi. What is the natural frequency of the system for torsional oscillation of small amplitude?

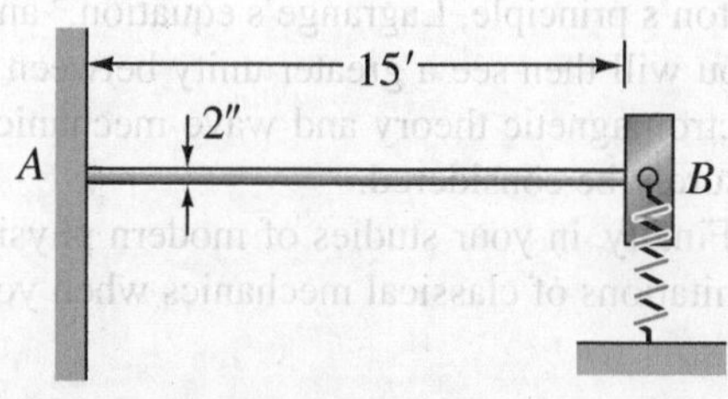

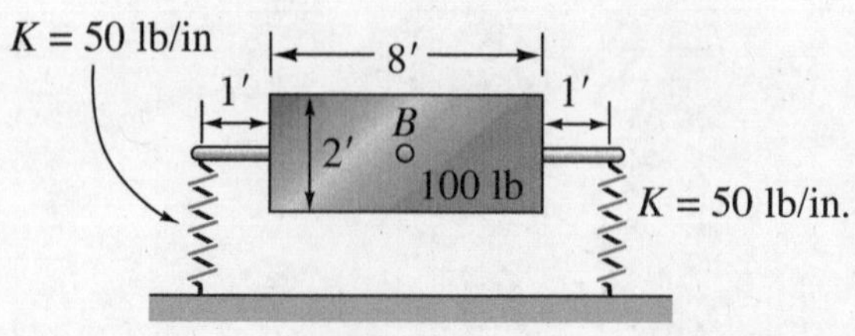

Figure P.16.105.

16.106. What is the radius of gyration of the speedboat about a vertical axis going through the center of gravity, if it is noted that the boat will swing about this vertical axis one time per second? The mass of the boat is 500 kg.

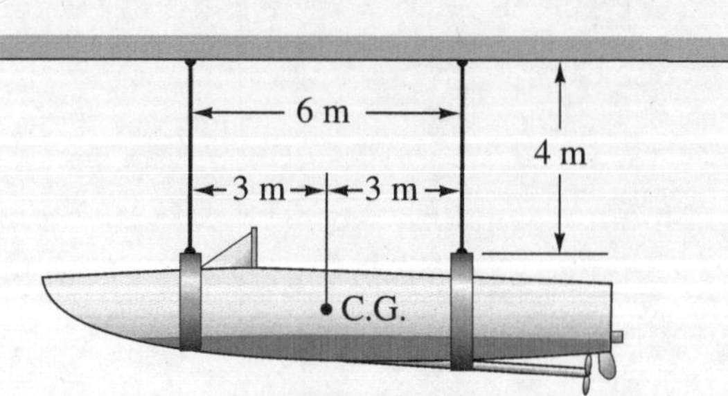

Figure P.16.106.

16.107. A rod of length L and mass M is suspended from a frictionless roller. If a small impulsive torque is applied to the rod when it is in a state of rest, what is the natural frequency of oscillation about this state of rest?

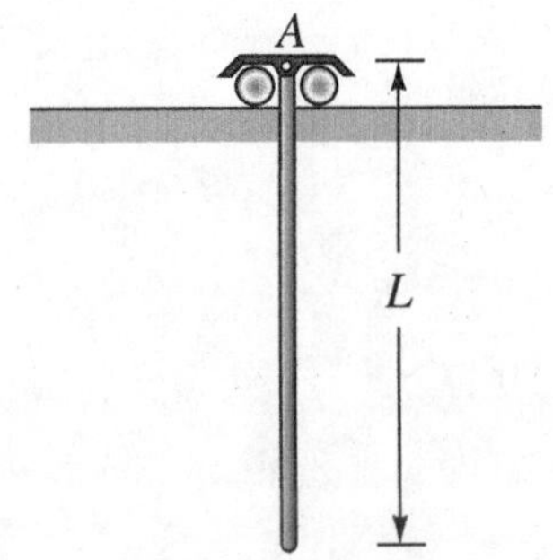

Figure P.16.107.

***16.108.** Solve Problem 16.12 by energy methods.

16.109. A block is acted on by a force F given as

$$F = 90 + 22 \sin 80t \text{ N}$$

and is found to oscillate, after transients have died out, with an amplitude of .5 mm about a position 50 mm to the left of the static-equilibrium position corresponding to the condition when no force is present. What is the weight of the body?

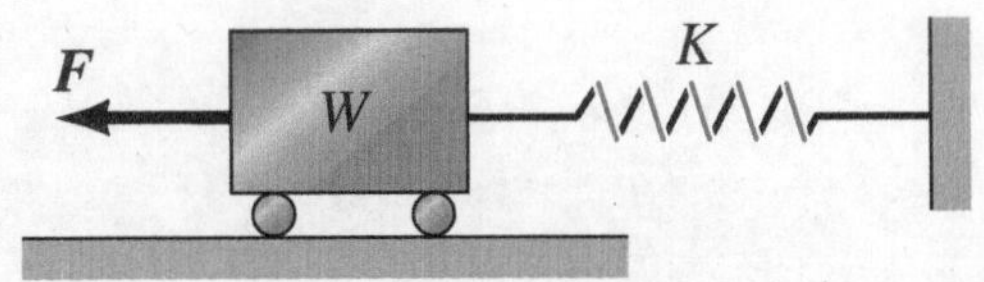

Figure P.16.109.

16.110. A light stiff rod ACB is made to rotate about C such that θ varies sinusoidally with $\omega = 20$ rad/sec. What should the amplitude of rotation θ_0 be to cause a steady-state amplitude of vibration of M to be 20 mm? The following data apply:

$$M = 3 \text{ kg}$$
$$K = 2.5 \text{ N/mm}$$
$$l = 1 \text{ m}$$

What is the one essential difference between the motions of the two masses?

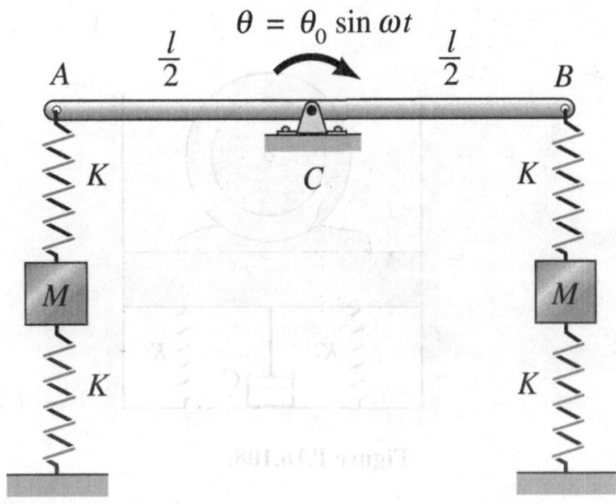

Figure P.16.110.

16.111. In Problem 16.105, what is the angle θ after 5 sec? System is stationary at the outset and AB is horizontal. Take $\omega = 39$ rad/sec and θ_0 as .02 rad.

16.112. A body rests on a conveyor moving with a speed of 5 ft/sec. If the damping constant is 2 lb/ft/sec, determine the equilibrium force in the spring. If the body is displaced 3 in. to the left from the equilibrium position, what is the time for the mass to pass through the equilibrium position again?

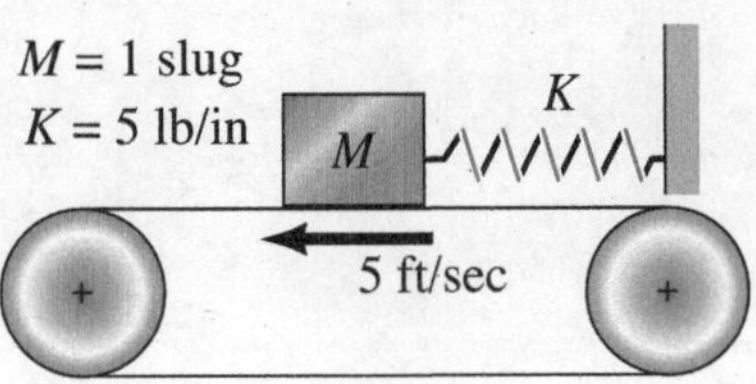

Figure P.16.112.

***16.113.** A motor is mounted on two springs of stiffness 8.8 N/mm each and a dashpot having a coefficient c of 96 N-sec/m. The motor weighs 222 N. The armature of the motor weighs 89 N with a center of mass 5 mm from the geometric centerline. If the machine rotates at 1,750 rpm, what is the amplitude of motion in the vertical direction of the motor? Determine the maximum force transmitted to the ground.

***16.114.** In Problem 16.108, what is the resonant condition for the system? What is the amplitude of motion for this case? To what value must c be changed if the amplitude at this motor speed is to be halved?

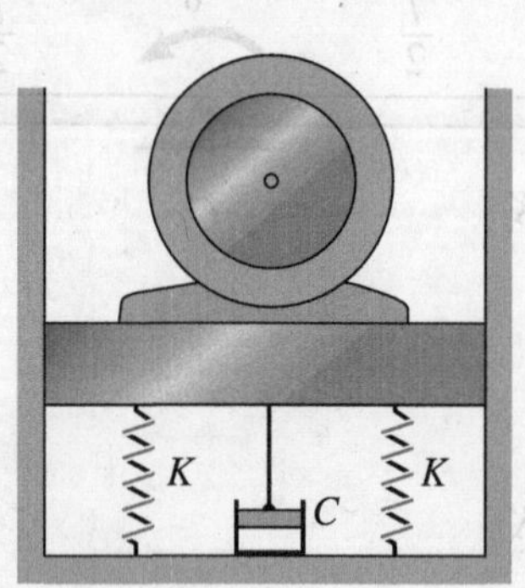

Figure P.16.108.

APPENDIX I

Integration Formulas

1. $\int \frac{x\,dx}{a+bx} = \frac{1}{b^2}[a + bx - a\ln(a+bx)]$

2. $\int \frac{dx}{a^2 - x^2} = \frac{1}{2a}\ln\left(\frac{a+x}{a-x}\right)$

3. $\int \sqrt{x^2 \pm a^2}\,dx = \frac{1}{2}\left[x\sqrt{x^2 \pm a^2} \pm a^2 \ln\left(x + \sqrt{x^2 \pm a^2}\right)\right]$

4. $\int \sqrt{a^2 - x^2}\,dx = \frac{1}{2}\left(x\sqrt{a^2 - x^2} + a^2 \sin^{-1}\frac{x}{a}\right)$

5. $\int x\sqrt{a^2 - x^2}\,dx = -\frac{1}{3}\sqrt{(a^2 - x^2)^3}$

6. $\int x\sqrt{a + bx}\,dx = -\frac{2(2a - 3bx)\sqrt{(a+bx)^3}}{15b^2}$

7. $\int x^2\sqrt{a^2 - x^2}\,dx = -\frac{x}{4}\sqrt{(a^2 - x^2)^3} + \frac{a^2}{8}\left(x\sqrt{a^2 - x^2} + a^2 \sin^{-1}\frac{x}{a}\right)$

8. $\int x^2\sqrt{a^2 \pm x^2}\,dx = \frac{x}{4}\sqrt{(x^2 \pm a^2)^3} \mp \frac{a^2}{8}x\sqrt{x^2 \pm a^2} - \frac{a^4}{8}\ln\left(x + \sqrt{x^2 \pm a^2}\right)$

9. $\int \frac{dx}{\sqrt{a^2 - x^2}} = \sin^{-1}\frac{x}{a}$

10. $\int \frac{dx}{\sqrt{x^2 + a^2}} = \ln\left(x + \sqrt{x^2 + a^2}\right) = \sinh^{-1}\frac{x}{a}$

11. $\int x^m e^{ax}\,dx = \frac{x^m e^{ax}}{a} - \frac{m}{a}\int x^{m-1}e^{ax}\,dx$

12. $\int x^m \ln x\,dx = x^{m+1}\left(\frac{\ln x}{m+1} - \frac{1}{(m+1)^2}\right)$

13. $\int \sin^2\theta\,d\theta = \frac{1}{2}\theta - \frac{1}{4}\sin 2\theta$

14. $\int \cos^2 \theta \, d\theta = \frac{1}{2}\theta + \frac{1}{4}\sin 2\theta$

15. $\int \sin^3 \theta \, d\theta = -\frac{1}{3}\cos\theta(\sin^2\theta + 2)$

16. $\int \cos^m \theta \sin\theta \, d\theta = -\dfrac{\cos^{m+1}\theta}{m+1}$

17. $\int \sin^m \theta \cos\theta \, d\theta = \dfrac{\sin^{m+1}\theta}{m+1}$

18. $\int \sin^m \theta \, d\theta = -\dfrac{\sin^{m-1}\theta\cos\theta}{m} + \dfrac{m-1}{m}\int \sin^{m-2}\theta \, d\theta$

19. $\int \theta^2 \sin\theta \, d\theta = 2\theta\sin\theta - (\theta^2 - 2)\cos\theta$

20. $\int \theta^2 \cos\theta \, d\theta = 2\theta\cos\theta + (\theta^2 - 2)\sin\theta$

21. $\int \theta \sin^2\theta \, d\theta = \frac{1}{4}\left[\sin\theta(\sin\theta - 2\theta\cos\theta) + \theta^2\right]$

22. $\int \sin m\theta \cos m\theta \, d\theta = -\dfrac{1}{4m}\cos 2m\theta$

23. $\displaystyle\int \frac{d\theta}{(a + b\cos\theta)^2}$

$$= \frac{1}{(a^2 - b^2)}\left(\frac{-b\sin\theta}{a + b\cos\theta} + \frac{2a}{\sqrt{a^2 - b^2}}\tan^{-1}\frac{\sqrt{a^2 - b^2}\tan\frac{\theta}{2}}{a + b}\right)$$

24. $\int \theta\sin\theta \, d\theta = \sin\theta - \theta\cos\theta$

25. $\int \theta\cos\theta \, d\theta = \cos\theta + \theta\sin\theta$

APPENDIX II

Computation of Principal Moments of Inertia

We now turn to the problem of computing the principal moments of inertia and the directions of the principal axes for the case where we do not have planes of symmetry. It is unfortunate that a careful study of this important calculation is beyond the level of this text. However, we shall present enough material to permit the computation of the principal moments of inertia and the directions of their respective axes.

The procedure that we shall outline is that of extremizing the mass moment of inertia at a point where the inertia-tensor components are known for a reference xyz. This will be done by varying the direction cosines l, m, and n of an axis k so as to extremize I_{kk} as given by Eq. 8.13. We accordingly set the differential of I_{kk} equal to zero as follows:

$$\begin{aligned} dI_{kk} = {} & 2lI_{xx}\,dl + 2mI_{yy}\,dm + 2nI_{zz}\,dn \\ & -2lI_{xy}\,dm - 2mI_{xy}\,dl - 2lI_{xz}\,dn \\ & -2nI_{xz}\,dl - 2mI_{yz}\,dn - 2nI_{yz}\,dm = 0 \end{aligned} \tag{II.1}$$

Collecting terms and canceling the factor 2, we get

$$\begin{aligned} (lI_{xx} - mI_{xy} - nI_{xz})dl + (-lI_{xy} + mI_{yy} - nI_{yz})dm & \\ + (-lI_{xz} - mI_{yz} + nI_{zz})dn = 0 & \end{aligned} \tag{II.2}$$

If the differentials dl, dm, and dn were *independent* we could set their respective coefficients equal to zero to satisfy the equation. However, they are not independent because the equation

$$l^2 + m^2 + n^2 = 1 \tag{II.3}$$

must at all times be satisfied. Accordingly, the differentials of the direction cosines must be related as follows[1]:

$$l\,dl + m\,dm + n\,dn = 0 \tag{II.4}$$

[1]We are thus extremizing I_{kk} in the presence of a constraining equation.

We can of course consider any two differentials as independent. The third is then established in accordance with the equation above.

We shall now introduce the *Lagrange multiplier* λ to facilitate the extremizing process. This constant is an arbitrary constant at this stage of the calculation. Multiplying Eq. II.4 by λ and subtracting Eq. II.4 from Eq. II.2 we get when collecting terms:

$$\left[(I_{xx}-\lambda)l - I_{xy}m - I_{xz}n\right]dl + \left[-I_{xy}l + (I_{yy}-\lambda)m - I_{yz}n\right]dm + \left[-I_{xz}l - I_{yz}m + (I_{zz}-\lambda)n\right]dn = 0 \tag{II.5}$$

Let us next consider that *m* and *n* are independent variables and consider the value of λ so chosen that the coefficient of *dl* is zero. That is,

$$(I_{xx}-\lambda)l - I_{xy}m - I_{xz}n = 0 \tag{II.6}$$

With the first term Eq. II.5 disposed of in this way, we are left with differentials *dm* and *dn*, which are independent. Accordingly, we can set their respective coefficients equal to zero in order to satisfy the equation. Hence, we have in addition to Eq. II.6 the following equations:

$$\begin{aligned} -I_{xy}l + (I_{yy}-\lambda)m - I_{yz}n &= 0 \\ -I_{xz}l - I_{yz}m + (I_{zz}-\lambda)n &= 0 \end{aligned} \tag{II.7}$$

A necessary condition for the solution of a set of direction cosines *l, m,* and *n,* from Eqs. II.6 and II.7, which does not violate Eq. II.3[2] is that the determinant of the coefficients of these variables be zero. Thus:

$$\begin{vmatrix} (I_{xx}-\lambda) & -I_{xy} & -I_{xz} \\ -I_{xy} & (I_{yy}-\lambda) & -I_{yz} \\ -I_{xz} & -I_{yz} & (I_{zz}-\lambda) \end{vmatrix} = 0 \tag{II.8}$$

This results in a cubic equation for which we can show there are three real roots for λ. Substituting these roots into any two of Eqs. II.6 and II.7 plus Eq. II.3, we can determine three direction cosines for each root. These are the direction cosines for the principal axes measured relative to *xyz*. We could get the principal moments of inertia next by substituting a set of these direction cosines into Eq. 9.13 and solving for I_{kk}. However, that is not necessary, since it can be shown that the three Lagrange multipliers *are* the principal moments of inertia.

[2]This precludes the possibility of a trivial solution $l = m = n = 0$.

ANSWERS

2.2. $F = 38.5$ N @ 66.6° from x axis.

2.4. $L = 2.75$ km.

2.6. $B = 17.32$ N $\alpha = 60°$

2.8. $F_A = 100$ N @ −120° with horizontal.
$F_B = 76.53$ N @ −67.5° with horizontal.
$F_C = 76.53$ N @ −22.5° with horizontal.

2.10. $F = 846$ N @ 17.68° with horizontal.

2.12. $T_{AC} = 767.2$ N $\alpha = 36.8°$

2.14. $F = 1{,}206$ N.

2.16. $F = 137.5$ N @ 43.34° from x direction.

2.18. $F = 242$ lb @ 3.07° from x direction.

2.22. $F_{slot} = 36.4$ lbf; $F_{vertical} = 81.5$ lbf.

2.24. $F_{BC} = .630F$ $F_{vert} = .590F$.

2.26. 2,690 lb 803.8 lb.

2.28. $F_{AC} = 707$ N $\alpha = 90°$.

2.30. $\boldsymbol{F} = 215.9\boldsymbol{i} + 1968\boldsymbol{j} + 3151\boldsymbol{k}$ N.

2.32. $F = 37.4$ lb l = .267 m = .535 n = −.802

2.34. $F_1 + F_2 = 918.6\boldsymbol{i} - 1581\boldsymbol{j} - 835.8\boldsymbol{k}$ N.

2.36. $F_x = 400$ N $F_y = -1007$ N.

2.40. $\boldsymbol{F} = 25.7\boldsymbol{i} + 24.7\boldsymbol{j} + 16\boldsymbol{k}$ lb.

2.42. $\boldsymbol{A} = \pm 5\sqrt{2}\,\boldsymbol{i} \mp 5\sqrt{2}\,\boldsymbol{k}$.

2.44. $\boldsymbol{f} = .465\boldsymbol{i} + .814\boldsymbol{j} + .349\boldsymbol{k}$.
$\boldsymbol{F} = 46.5\boldsymbol{i} + 81.4\boldsymbol{j} + 34.9\boldsymbol{k}$ N.

2.46. −164 −.465 −10.5.

2.48. $\boldsymbol{D} = 10\boldsymbol{i} - .769\boldsymbol{j} - 3.77\boldsymbol{k}$.

2.52 .418 ft.

2.54. 37.05 N.

2.56. $A = 2.5$ N $\alpha = 45.7°$

2.58. -75 ft^2 95.94°.

2.60. −28.83 N.

2.62. 47.51°.

2.64. $(18\boldsymbol{i} + 20\boldsymbol{j} - 42\boldsymbol{k})$ 47.

2.68. $.804\boldsymbol{i} + .465\boldsymbol{j} + .372\boldsymbol{k}$

2.70. 640 m^2.

2.74. −29.6.

2.76. (a) $-43\boldsymbol{i} + 49\boldsymbol{j} + 2\boldsymbol{k}$.
(b) −136.
(c) −136.

2.78. $L_{DT} = 373$ miles $L_{TC} = 689$ miles
162 miles longer.

2.80. $F_x = 57.1$ N $F_y = 342.8$ N $F_z = 971.4$ N.

2.82. $F = 231.3$ N.

2.84. 32.4 ft-lb.

2.86. 112.4° −190.72 lb.

2.88. 35.7 kN.

2.90. 19.87 m.

2.92. $\boldsymbol{F} = -80.1 \times 10^{-13}\boldsymbol{k}$ N.

2.94. 251 ft^2.

2.96. 18.21m $(l, m, n) = (.5145, .6860, .5145)$.

2.98. $T_{AB} = 500$ N $T_{AC} = 866$ N $\alpha = 30°$.

2.100 15.90°.

3.2. 4i - 16j - 3k ft.

3.4. $6\boldsymbol{i} + 7.16\boldsymbol{j} + 7.598\boldsymbol{k}$ m.

3.6. −1067.4 N-m 1959.3 N-m.

3.8. $\pm 2\sqrt{2z}\,\boldsymbol{j} + z\boldsymbol{k}$.

3.10. −18,026 N-m −6,824.8 N-m.

3.12. 13.8 m.

3.14. $-257.5\boldsymbol{k}$ N-m.

3.16. 8944 ft-lb 4472 ft-lb 0 ft-lb.

3.18. $180\boldsymbol{i} - 50\boldsymbol{k}$ kN-m.
$30\boldsymbol{i} + 75\boldsymbol{j} - 50\boldsymbol{k}$ kN-m.
$-1{,}551.8\boldsymbol{i} + 75\boldsymbol{j} + 389.4\boldsymbol{k}$ kN-m.

3.20. $-84\boldsymbol{i} + 94\boldsymbol{j} - 46\boldsymbol{k}$ N-m.

3.22. $\boldsymbol{M}_A = \frac{10}{\sqrt{3}}a\boldsymbol{k} - \frac{10}{\sqrt{3}}a\boldsymbol{j}$ lb-ft.
$\boldsymbol{M}_D = \boldsymbol{0}$ lb-ft.
$\boldsymbol{M}_I = -\frac{10}{\sqrt{3}}a\boldsymbol{i} + \frac{10}{\sqrt{3}}a\boldsymbol{j}$ lb-ft
etc.

3.24. $(7.277F_{BA} + 8.575F_{CD} - 10{,}000)\boldsymbol{i} +$
$(2.911F_{BA} - 5.145F_{CD})\boldsymbol{j} +$
$(-2.911F_{BA} + 5.145F_{CD})\boldsymbol{k}$ ft-lb.

3.26. $1.3296F$ ft-lb.

3.28. 22.295 N-m $\boldsymbol{M}_D = -30\boldsymbol{i} + 100\boldsymbol{j} - 36\boldsymbol{k}$ N-m.

3.30. −146.2 lb 77.00 lb-ft.

3.32. 5,769 lb-ft.

3.34. −113.5 kN-m −117.5 kN-m.

3.38. 300 lb-ft 600 lb-ft 1,200 lb-ft.

3.40. 175 N.

3.42. $-7{,}270\boldsymbol{k}$ lb-ft $10{,}405\boldsymbol{k}$ lb-ft.

3.44. $\boldsymbol{M}_A = \boldsymbol{M}_P = -261\boldsymbol{i} - 261\boldsymbol{j}$ N-m.

3.46. 420 ft-lb.

3.48. $\boldsymbol{C} = 100\boldsymbol{i} + 50\boldsymbol{j} - 224\boldsymbol{k}$ lb-ft.

3.50. −1,857 N-m.

3.52 $\boldsymbol{M}_P = 48\boldsymbol{i} - 36\boldsymbol{j} - 225.6\boldsymbol{k}$ N-m.
$M_E = -146.9$ N-m.

3.54. $M = 3{,}635$ lb-in @ 68.2° to horizontal.

3.56. $\boldsymbol{C} = 35.35\boldsymbol{i} + 22.36\boldsymbol{j} + 80.07\boldsymbol{k}$ N-m.

3.58. 120 lb-ft 160 lb-ft 60 lb.

3.60. 1,750 N-m.

3.62. 408.4 N-m.

3.64. 1,179 ft-lb.

3.66. $-452\boldsymbol{k}$ lb-ft $-1{,}202\boldsymbol{k}$ lb-ft.

3.68 $10{,}600\boldsymbol{i} + 4{,}500\boldsymbol{j} - 5{,}000\boldsymbol{k}$ m.
12,554 m.

3.70. −27.2 N-m.

3.72. 277.6 N-m.

4.2. $F = 15$ kN $C = 30$ kN-m C.C.W.

4.4. $F = 150$ N $C = 187.5$ N-m.

4.6. $M_A = 8\boldsymbol{k}$ kN-m $F_A = F_B = -10\boldsymbol{j}$ kN.
$\mathbf{M}_B = 22.72\boldsymbol{k}$ kN-m.

4.8. Move 200 lb force 5 ft to left.

4.10. $\boldsymbol{F}_A = -200\boldsymbol{j} - 150\boldsymbol{k}$ N.
$\boldsymbol{C}_A = 34.0\boldsymbol{i} + 26.0\boldsymbol{k}$ N-m.

4.12. $\boldsymbol{F} = 20\boldsymbol{i} - 60\boldsymbol{j} + 30\boldsymbol{k}$ N.
$\boldsymbol{C}_A = 900\boldsymbol{i} + 680\boldsymbol{j} + 760\boldsymbol{k}$ N-m.

4.14. 12.375 m from origin.

4.16. $\boldsymbol{F} = -44.567\boldsymbol{i} - 33.425\boldsymbol{j} - 22.283\boldsymbol{k}$ N.
$\boldsymbol{C} = -31.007\boldsymbol{i} - 160.46\boldsymbol{j} + 302.71\boldsymbol{k}$ kN-m.

4.18. $\boldsymbol{F}_R = -204\boldsymbol{i} + 604\boldsymbol{j} + 408\boldsymbol{k}$ lb.
$\boldsymbol{C}_A = 8{,}160\boldsymbol{i} + 2{,}032\boldsymbol{j} + 3{,}264\boldsymbol{k}$ ft-lb.
$\boldsymbol{C}_B = 2{,}032\boldsymbol{j} - 816\boldsymbol{k}$ ft-lb.

4.20. $\boldsymbol{F}_R = 400\boldsymbol{i} - 1{,}900\boldsymbol{j} + 600\boldsymbol{k}$ N.
$\boldsymbol{C}_R = 24{,}200\boldsymbol{i} + 2{,}000\boldsymbol{j} - 3{,}300\boldsymbol{k}$ N-m.

4.22. $\boldsymbol{F}_R = -50\boldsymbol{k}$ kN $\boldsymbol{C}_R = 7.5\boldsymbol{i} + 50\boldsymbol{j}$ kN-m.

4.24. $F_1 = 26.9$ lb $F_2 = 17.6$ lb $F_3 = 35.5$ lb.

4.26. $\boldsymbol{F} = 2{,}100\boldsymbol{i} - 500\boldsymbol{j}$ lb $x = -.05$ ft.

4.28. $\boldsymbol{F}_R = 400\boldsymbol{i} - 1600\boldsymbol{j}$ N
$\boldsymbol{C}_R = -1{,}800\boldsymbol{i} - 2{,}400\boldsymbol{j} - 12{,}600\boldsymbol{k}$ N-m.

4.30. $\boldsymbol{F}_R = 39.44\boldsymbol{i} + 74.7\boldsymbol{j}$ N.
$x = .669$ m.

4.32. $\boldsymbol{F}_R = -100\boldsymbol{k}$ N $x = 2.5$ m $y = 2.2$ m.
$\boldsymbol{F}_R = \boldsymbol{0}$ $\boldsymbol{C}_R = 280\boldsymbol{i} + 450\boldsymbol{j}$ N-m.

4.34. $\boldsymbol{F}_R = \boldsymbol{0}$ $\boldsymbol{C}_k = 860\boldsymbol{i} + 900\boldsymbol{j}$ N-cm.

4.36. $\boldsymbol{F}_R = 353.5\boldsymbol{i} - 653.5\boldsymbol{j}$ N $x = 9.2739$ m.

4.38. $\boldsymbol{F}_R = 43{,}000$ lb $x = 23.209$ ft.

4.42. $\boldsymbol{F} = 3{,}164\boldsymbol{k}$ lb.

4.44. $x = 10.37$ in. $y = z = 0$.

4.48. $x = .289$ $y = .400$.

4.50. 10.84 m.

4.52. $x = 1.292$ ft $y = 7.63$ ft.

4.54. $x = .210$ ft $y = .0756$ ft.

4.56. $y = 3.808$ m.

4.58. $\boldsymbol{F} = -37.5\boldsymbol{k}$ kN $x = .844$ m $y = 1.067$ m
(origin at front left lower corner of the load).
original weight 90 kN.
lost weight = 52.5 kN.

4.60. $\boldsymbol{F}_R = -111{,}780\boldsymbol{k}$ lb $x = -3.78$ ft
(in front of trailer C.G.).
$\boldsymbol{F}_R = -86{,}386\boldsymbol{k}$ lb $x = -7.329$ ft.

4.62. $W = .02739\gamma$. $x_c = .2042$ m $y_c = .1$ m
$z_c = -.1968$ m.

4.64. $F_R = 2.315 \times 10^6$ lb $x = 20$ ft $y = 12.44$ ft.

4.66. $F_R = 262$ N.

4.68. 4,025 N-m 596.3 N-m.

4.70. $p_a = 7.8644 \times 10^4$ *Pa*.

4.72. $F = 61{,}740$ lb 10.386 ft.

4.74. $F = 8761$ lb 2.615 ft.

4.76. $F = 1.697 \times 10^{10}$ N 24.04 m from base along inclined surface.

4.78. $F = 666 \times 10^8$ N.
@ $\frac{40}{3}$ m from base.

4.80. $x = 21.2$ ft $y = 4.67$ ft.

4.82. $x_c = .07441$ m.

4.84. $\boldsymbol{F} = -400\boldsymbol{j}$ kN $\boldsymbol{M}_{\text{outside lane}} = -900\boldsymbol{k}$ kN-m
$\boldsymbol{M}_{\text{inside lane}} = -300\boldsymbol{k}$ kN-m.

4.86. 1,415 kN 1.915 m from front of load.

4.88. $\boldsymbol{F}_R = 4.698\boldsymbol{i} - 10.09\boldsymbol{j}$ kN.
$\boldsymbol{M}_R = -3\boldsymbol{i} + 3.2316\boldsymbol{k}$ kN-m.

4.90. $r' = 17.57x^2 - 7.027x + .7026$.

4.92. $l = 3.51$ m.

4.94. $F = 2{,}200$ lb $x = 4.136$ ft.

4.96. −217.2 N-m.

4.98. $\boldsymbol{F}_R = 500\boldsymbol{i} - 1{,}220\boldsymbol{j}$ N
$\boldsymbol{C}_R = 240\boldsymbol{i} - 2{,}190\boldsymbol{k}$ N-m.

4.100. $\boldsymbol{F}_R = 58{,}900$ lb $x = 3.919$ from bottom left corner.

4.102. $x = 1.20$ ft $y = 1.125$ ft $z = \frac{t}{2}$ ft.

4.104. $y = 1.299$ ft.

4.106. −1,550 lb $x = 20.04$ ft.

4.108. $x = 10.94$ ft $y = 12.67$ ft.

4.110. 485.397 kN @ 1.636 m below top of gate.
132.381 kN @ 1 m below top of gate.

4.112. $x = 4.98$ m $y = 2.65$ m.

5.18. $T_{BC} = 26.79$ lb $T_{BA} = 37.9$ lb.

5.20. $T_{AB} = 111.8$ lb $T_{BD} = 109.9$ lb $T_{BC} = 63.7$ lb.

5.22. 27 lb 36.35 lb 4.653 in.

5.24. 671 N, 447 N, 1,000 N.

5.26. 409.8 N-m.

5.28. $A_x = 4{,}233$ N, $A_y = 3{,}140$ N, $B_x = 4{,}233$ N,
$B_y = -1{,}840$ N.

5.30. $A_x = 2{,}000$ N, $A_y = 6{,}970$ N, $M_A = 62{,}360$ N-m
$B = 530.4$ N.

5.32. $A_x = 8.00$ kN $A_y = -12.50$ kN $B_x = -8.00$ kN
$B_y = 22.5$ kN.

5.34. $F_{BD} = 462$ lb $F_{BC} = 215$ lb $F_{AB} = 375$ lb.

5.36. $T_A = 125.0$ N $f = 125$ N.

5.38. $T = 117.4$ N $\alpha = 79.8°$

5.40. $A_x = 433$ N, $A_y = 790$ N $M_A = 1{,}167$ N-m.

5.42. $A_x = 200$ N, $A_y = 3{,}464$ N $M_A = 1{,}192$ N-m.

5.44. $A_x = 9{,}736$ N $A_y = -2{,}670$ N $B_x = -9{,}736$ N $B_y = 6{,}670$ N.

5.46. $\alpha = 17.46°$.

5.48. 58.7 ft-lb.

5.50. $\beta = \tan^{-1}\left[\dfrac{\dfrac{W_2}{\tan\alpha_1} - \dfrac{W_1}{\tan\alpha_2}}{W_1 + W_2}\right]$.

5.52. $A_x = 408$ N $A_y = -258$N $G_x = -749.5$ N $G_y = 583$ N.

5.54. $T = 20$ N-m.

5.56. $T_{AB} = 40.5$ kN $T_{BC} = 39.3$ kN.

5.58. 72 lb 447 lb.

5.60. 4 m.

5.62. $F = \dfrac{W}{2}\dfrac{r_1 - r_2}{r_2}$.

5.64. $R_x = 0$ $R_y = 58{,}900$ lb $M_B = 142{,}500$ ft-lb.

5.66. $R_x = 0$, $R_y = 4{,}953$ lb $M = 61{,}082.5$ ft-lb.

5.68. $R_C = 1{,}006$ lb $M_C = 10{,}108$ ft-lb. $R_D = 294$ lb $M_D = 2{,}770$ ft-lb.

5.70. 270 lb 253 lb.

5.72. $W_{max} = 2.5$ ton $B_y = 10.8$ ton $A_y = 1.7$ ton $C_y = 1.7$ ton $M_c = 34$ ft-ton $D_y = 1.7$ ton $M_D = 85$ ft-ton.

5.74. 17.55 ton (2 supports) 24.9 ton M = 70 ton-ft.

5.76. $F_1 = 293$ N $F_2 = 52.37$ N $F_3 = 254.3$ N.

5.78. $\boldsymbol{A} = -500\boldsymbol{j} + 1{,}500\boldsymbol{k}$ lb
$\boldsymbol{M}_A = 26{,}260\boldsymbol{i} - 9{,}100\boldsymbol{k}$ ft-lb
$\boldsymbol{B} = -500\boldsymbol{j}$ lb
$\boldsymbol{M}_B = -9{,}100\boldsymbol{k}$ ft-lb.

5.80. $F = 57.14$ lb $A_y = 160.1$ lb $B_y = 93.0$ lb $B_z = 22.6$ lb $A_z = 18.08$ lb.

5.82. $A_x = 0$ $A_y = 50$ N $A_z = 100$ N
$M_x = -200$ N-m.
$M_y = -220$ N-m $M_z = 110$ N-m.

5.84. $B_y = 3{,}000$ lb $B_z = 26{,}500$ lb $C_z = 17{,}000$ lb.

5.86. 3 ft $A_z = 50$ N $B_z = 50$ N.

5.88. $P = 33.7$ lb $A_z = 25$ lb $A_y = -7.50$ lb.

5.92. 11,394 lb.

5.94. 65,300 lb.

5.96. $F_{EC} = 4.36$ kN $A_x = .800$ kN $A_y = 196.7$ kN.

5.98. $B_x = 165.9$ N $B_y = 206.3$ N $F_{CB} = 208.8$ N.

5.100. $C_x = 576$ lb $C_y = 1{,}326$ lb $M_c = 8009.6$ N-m.

5.102. $T_1 = 209$ N $T_2 = 202$ N $F = 286$ N.

5.104. $A = 10{,}640$ lb $C = 6{,}520$ lb $D = 6{,}520$ lb $G = H = 7{,}500$ lb $F = 10{,}640$ lb.

5.106. 1154 N

5.108. 9.418×10^6 *Pa.*

5.110. $\tau = (6.833 - 6.50z - \frac{z^3}{3}) \times 10^4$ *Pa.*

5.112. $\tau_{zz} = 281.7$ psi

5.114. $G_x = 750$ lb $G_y = 650$ lb.

5.116. $F_x = A_x = 806$ lb $F_y = A_y = 500$ lb.

5.118. $R_1 = 41$ kN $R_2 = 88$ kN $R_3 = 141$ kN.

5.120. $F_{CD} = 1{,}694$ lb $F_{EF} = 4{,}931$ lb.

5.122. $A_x = 13.15$ lb $A_y = 142.6$ lb.

5.124. $T_E = 1{,}688$ N $T_D = 2{,}840$ N.

5.126. $T = 30.67$ lb.

5.128. $C_x = -322.3$ lb $C_y = -161.3$ lb $C_z = -583.6$ lb $D_x = -322.7$ lb $D_z = -398.9$ lb.

5.130. $T = 2{,}121$ lb $H = 3{,}000$ lb $V = 3{,}000$ lb
$T = 1{,}148$ lb $H = 2{,}121$ lb $V = 5{,}121$ lb.

5.132. $T_{DB} = 353$ lb $T_{AC} = 289$ lb $T_{EC} = 394$ lb.

5.134. $T_R = 1.928$ N $T_P = 709$ N $T_A = 2{,}129$ N
$T_D = 1{,}012$ N $l = .0828$ $m = .728$ $n = .618$.

5.136. $F_{AB} = 8{,}000$ lb $E_x = 5{,}000$ lb $E_y = 6{,}928$ lb $M_E = 5{,}000$ ft-lb.

5.138. $A_x = -3{,}984$ N $A_y = -7{,}800$ N $B_x = 3{,}984$ N $B_y = 4{,}800$ N.

5.140. $A_x = -925$ N $A_y = 150$ N.

5.142. 63.75 kN-m.

5.144. $F = 996.7$ N.

5.146. $A_y = B_y = 18{,}170$ N $A_x = 63{,}660$ N.

5.148. $C_x = 600$ lb $C_y = 0$ $C_z = -320$ lb $F_D = 1{,}033$ lb $F_A = 312.5$ lb.

5.150. $p = 208$ psi gage.

5.152. $\boldsymbol{A} = 433\boldsymbol{i} - 533\boldsymbol{j} - 19.00\boldsymbol{k}$ lb $E = 240$ lb.
$\boldsymbol{H} = -433\boldsymbol{i} + 433\boldsymbol{j} - 520\boldsymbol{k}$ lb $K = 0$.

5.154. $A_x = -28.13$ lb $A_y = -98$ lb $B_x = 70$ lb $B_y = 0$.

5.156. $f = 15.08$ kN.

5.158. $A_x = 0$ $A_y = 1{,}938$ lb $C_y = B_y = 3{,}062$ lb $D_y = 1{,}938$ lb. Ram force = 9,798 lb.

5.160. $T_{EF} = 1{,}174$ N.

5.162. $A_x = 1.399 \times 10^4$ N $A_y = 5.549 \times 10^3$ N. $M_A = 1.355 \times 10^4$ N-m $B = 2{,}887$ N.

6.4. 600 lb.

6.6. 151.0 lb 139.0 lb.

6.8. $\mu_s = 0.322$.

6.10. $F = 52.76$ N.

6.14. $(\mu_s)_{max} = 0.249$.

6.16. at start $F = 4{,}142$ lb
for movement $F = 4{,}414$.

6.18. $P = 733.6$ N.

6.20. $f = 115.5$ N
$\mu_{min} = .578$.

6.24. C moves 1.5 m to left.

6.26. $T = 123.8$ ft-lb.

6.28. 750 N (clockwise)
50 N (counterclockwise).

6.30. $F = 196.7$ N.

6.34. 32.68 lb $< W_1 <$ 67.32 lb.
$W_1 = 50$ lb (no friction).

6.36. $\mu_{min} = 0.225$.

6.38. $\alpha_{min} = 28.0°$.

6.40. 4.76 ft.

6.40. 1.691 m.

6.44. 251.5 lb.

6.46. $W_2 = 7.245$ N $\quad \mu_{min} = 0.0279$.

6.51. $T = \frac{P\mu}{3}\left[\frac{D_2^3 - D_1^3}{D_2^2 - D_1^2}\right]$.

6.53. $T = \frac{PD\mu_d}{4}$.

6.55. $T = 320$ N.

6.57. $T = 188.6$ N-m.

6.59. 832 N.

6.61. 267 lb.

6.63. 136.8 N.

6.65. $\mu_s = 0.260$.

6.67. $\mu_s = 0.048$.

6.69. $P = 202$ N $\quad \mu_5 = 0.556$.

6.71. $x = 277$ mm.

6.73. $F_x = 256$ N $\quad F_y = 150.4$ N.

6.75. $F_{min} = 2{,}210$ N $\quad \mu_s = 0.4$.

6.81. 133.6 in.-lb 155.6 in.-lb.

6.83. $W = 6{,}100$ lb.

6.87. 0.2387°.

6.89. $\mu_s = .00308$.

6.91. $F = 128.2$ N.

6.93. $\theta = 30.96°$ $\quad T = 292$ lb.

6.95. 7.11 N.

6.97. 3,333 N.

6.99. $P = 13.22$ lb C.W. $\quad P = 163.2$ lb C.C.W.

6.101. 40.7 N-m.

6.103. 94.7 mm.

6.105. 35.55° above horizontal.

6.109. $\theta = 41.9°$.

6.111. 256 N.

6.113. 1.002×10^4 N.

6.115. $W_2 = 429.18$ N.

6.117. $W_c = 810$ N.

6.119. $T = 877$ N-m.

7.4. $M_x = 125$ ft^3 $\quad M_y = 200$ ft^3.

7.8. $x = 14.30$ ft $\quad y = 5.74$ ft.

7.12. $x_c = 0.1691$ m $\quad y_c = 0.363$ m.

7.14. $x_c = 2.02$ mm $\quad y_c = 83.6$ mm.

7.16. $x_c = 0.1217$ m $\quad y_c = 0.4478$ m.

7.18. $x_c = 7.23$ in. $\quad y_c = 2.55$ in. $\quad M_z' = 226.1$ in.^{3}

7.20. $x_c = y_c = 242.22$ mm.

7.22. $x_c = 1.7172$ m $\quad y_c = 0.1722$ m.

7.24. $x_c = 2.889$ m $\quad y_c = 2.667$ m.

7.26. $y_c = 315.7$ mm.

7.28. $x_c = 40.1$ mm $\quad y_c = 19.51$ mm.

7.30. $x = 12.92$ ft.

7.32. $x_c = \frac{a}{2} \quad y_c = \frac{2}{3}b \quad z_c = \frac{4}{3}c$.

7.34. $z = \frac{3}{8}a$.

7.36. $\boldsymbol{r}_c = 1.179\boldsymbol{i} + .955\boldsymbol{j} + .284\,\boldsymbol{k}$ m.

7.38. $x_c = 2.571$ m.

7.40. $x = 8.22$ m $\quad y = z = 0$.

7.42. $x_c = y_c = 0 \quad z_c = 9.78$ in.

7.44. center of volume $x = -9.75$ mm
$y = 122.55$ mm $\quad z = 0$
center of mass and gravity
$x = -32.76$ mm $\quad y = 447$ mm $\quad z = 0$.

7.46. $x = 41.95$ mm $\quad y = 493$ mm $\quad z = 608$ mm.

7.48. $\boldsymbol{r}_c = .742\boldsymbol{j} - .1720\,\boldsymbol{k}$ m.

7.50. $y_c = \frac{2}{3}\left(\frac{b}{\pi}\right)$

7.52. $A = 1{,}751$ in.^{2} $\quad V = 5{,}242$ in.^{3}

7.54. $A = .8624$ m^2 $\quad V = .06329$ m^3

7.56. $\boldsymbol{r}_c = 3.39\boldsymbol{i} + 3.44\boldsymbol{j} - 3\boldsymbol{k}$ in.

7.58. $I_{xx} = \frac{ab^3\pi}{4}$.

7.60. $I_{xx} = 1.113$ units^4 $\quad I_{yy} = .535$ units^4.
$I_{xy} = .750$ units^4.

7.62. $I_{yy} = \frac{5}{2}\pi^2 - 8$ ft^4.

7.66. $I_{xx} = 666.7$ ft^4 $\quad I_{yy} = 1{,}257$ ft^4.
$I_{xy} = 791.7$ ft^4.

7.68. $I_{xx} = 4{,}540$ ft^4 $\quad I_{yy} = 11{,}250$ ft^4.

7.70. $I_{xx} = 6.646$ ft^4 $\quad I_{yy} = 15.39$ ft^4
$I_{xy} = 9.52$ ft^4.

7.76. $I_{xx} = 1{,}512$ ft^4 $\quad I_{xy} = 559$ ft^4.

7.78. $I_{xx} = \frac{5\sqrt{3}}{16}R^4 \quad I_{yy} = \sqrt{3}R^4\left(\frac{5}{16}\right) \quad I_{xy} = 0$.

7.80. $I_{x_cx_c} = 8.463 \times 10^7$ mm^4 $\quad I_{y_cy_c} = 1.8869 \times 10^8$ mm^4.

7.82. $I_{x'x'} = 140.6$ ft^4 $\quad I_{y'y'} = 439$ ft^4 $\quad I_{x'y'} = 213$ ft^4.

7.84. $I_{xx} = 91.92$ ft^4 $\quad I_{yy} = 1{,}455$ ft^4 $\quad I_{xy} = -96.52$ ft^4.
$J_p = 1{,}547$ ft^4.

7.88. 109.2 mm^4 $\quad$ 2,367 mm^4.

7.90. 1,299 mm^4 $\quad$ 79.1 mm^4.

7.94. $I_1 = .003781$ m^4 $\quad I_2 = .0002557$ m^4.

7.96. $x = 3.16$ mm $\quad y = 1.323$ mm.
$I_1 = 2.11 \times 10^4$ mm^4 $\quad I_2 = 8.0675 \times 10^3$ mm^4.

7.98. $y_c = 1.671$ m.

7.100. $I_{xx} = 3.29 \times 10^7$ mm^4 $\quad I_{x_cx_c} = 9.85 \times 10^6$ mm^4
$I_{yy} = 2.89 \times 10^6$ mm^4 $\quad I_{y_cy_c} = 1.607 \times 10^6$ mm^4
$I_{xy} = 6.90 \times 10^6$ mm^4 $\quad I_{x_cy_c} = 1.457c \times 10^6$ mm^4.

7.102. $x_c = \frac{5}{6}h \quad y_c = \frac{20}{21\pi}a \quad z_c = 0.$

7.104. 6.75° 96.75°.

7.106. $A = 151.24 \text{ m}^3 \quad V = 113.0 \text{ m}^3.$

7.108. $x_c = 1.500$ in. $\quad I_{x_cx_c} = 83.250 \text{ in.}^4$
$I_{y_cy_c} = 19.250 \text{ in}^4 \quad I_{x_cy_c} = 0.$

7.110. $12{,}708 \text{ m}^4.$

7.112. $M_z = 10.616s \text{ in.}^3$
$M_x = 185.7 - 0.465y^2$
$M_x = (.1886)[316.8 + (4 + 3\eta)\sqrt{(\eta + 2)^3}.$

8.2. $4{,}333 \text{ lbm-ft}^2 \quad 15{,}333 \text{ lbm-ft}^2.$

8.4. $I_{yy} = \frac{1}{12}M(a^2 + b^2) \quad I_{zz} = \frac{1}{12}M(b^2 + l^2)$
$I_{xx} = \frac{1}{12}M(a^2 + l^2).$

8.6. $\frac{1}{2}Mr^2.$

8.8. $2.67 \times 10^7 \text{ grams-mm}^2$
$3.95 \times 10^8 \text{ grams-mm}^2.$

8.10. $1.723 \times 10^6 \text{ grams-mm}^2.$

8.12. $3{,}959 \text{ kg-mm}^2.$

8.14. $I_{xx} = 1.004 \times 10^5 \text{ mm}^4$
$I_{yy} = 1{,}875 \text{ mm}^4$
$I_{xy} = 1.172 \times 10^4 \text{ mm}^4$
$(I_{xx})_M = 205 \text{ kg-mm}^2$
$(I_{yy})_M = 3.82 \text{ kg-mm}^2$
$(I_{xy})_M = 23.9 \text{ kg-mm}^2.$

8.16. $37.0 \text{ kg-mm}^2 \quad 166.8 \text{ kg-mm}^2 \quad 203.8 \text{ kg-mm}^2.$

8.20. $I_{y'y'} = 6.21 \text{ slug-ft}^2 \quad I_{z'z'} = I_{x'x'} = 29.0 \text{ slug-ft}^2$
$I_{xx} = 119.1 \text{ slug-ft}^2 \quad I_{yy} = 18.63 \text{ slug-ft}^2$
$I_{zz} = 106.6 \text{ slug-ft}^2.$

8.22. $I_{xx} + I_{yy} + I_{zz} = \frac{M}{6}(a^2 + b^2 + c^2)$
$+ 2M(x^2 + y^2 + z^2).$

8.24. $I_{x''x''} = 3{,}870 \text{ kg-m}^2$
$I_{y''y''} = 4{,}615 \text{ kg-m}^2.$

8.26. $1.258 \times 10^{-2} \text{ kg-m}^2 \quad 1.239 \times 10^{-2} \text{ kg-m}^2$
$3.91 \times 10^{-3} \text{ kg-m}^2.$

8.28. $1.650 \text{ kg-m}^2 \quad 1.438 \text{ kg-m}^2 \quad .513 \text{ kg-m}^2.$

8.30. $k = 2.54$ ft $\quad I = 28{,}400 \text{ lbm-ft}^2.$

8.32. $I_{xx} = 8{,}093 \text{ lbm-in.}^2 \quad I_{yy} = 1{,}603 \text{ lbm-in.}^2$
$I_{xy} = 1{,}282 \text{ lbm-in.}^2$

8.34. $I_{cc} = 37.81 \text{ kg-m}^2.$

8.36. $3{,}026 \text{ lbm-ft}^2.$

8.38. $.524 \text{ kg-m}^2.$

8.40. $-.00696 \text{ kg-m}^2.$

8.44. $362 \text{ kg-mm}^2.$

8.46. $(I_{xx})_M = 534 \text{ kg-mm}^2$
$(I_{yy})_M = 16.59 \times 10^2 \text{ kg-mm}^2$
$(I_{zz})_M = 21.9 \times 10^2 \text{ kg-mm}^2$
$(I_{xy})_M = 5.68 \times 10^2 \text{ kg-mm}^2$
$(I_{xz})_M = (I_{yz})_M = 0.$

8.48. $.0493 \text{ kg-m}^2 \quad .00187 \text{ kg-m}^2 \quad .0503 \text{ kg-m}^2.$

8.50. $1{,}075 \text{ kg-m}^2 \quad 92.35 \text{ kg-m}^2.$

8.52. $I_{yy} = 7.58\rho_0 \text{ kg-m}^2 \quad I_{yz} = -1.789\rho_0 \text{ kg-m}^2.$

9.2. $\dot{x} = 15.30$ ft/sec $\quad \dot{y} = 5.351$ ft/sec.

9.4. $\boldsymbol{a} = .6765\boldsymbol{i} - 1.2075\boldsymbol{j} - .3177\boldsymbol{k} \text{ ft/sec}^2.$

9.6. $a(3) = 12 \text{ m/sec}^2 \quad d = 60.94$ m.

9.8. 2.0177 ft/sec $\quad 1.522 \text{ ft/sec}^2.$

9.10. $\vec{V} = 2\boldsymbol{i} - 1.296\boldsymbol{j}$ m/sec
$\boldsymbol{a} = .6\boldsymbol{i} + 9.21\boldsymbol{j} \text{ m/sec}^2$
$a_n = 8.05 \text{ m/sec}^2.$

9.12. $$V = V_n\boldsymbol{i} + \left[-V_A + V_c \sin\alpha + \frac{V_c^2(\cos^2\alpha)(t)}{\{h^2 + [V_c(\cos\alpha)(t)]^2\}^{1/2}}\right]\boldsymbol{j}.$$

9.14. $x^4 + 149.9(y - x) = 0 \quad d = 5.312$ ft.

9.16. $9.06 < d < 25.16$ ft.

9.18. $\delta = 9.148$ m.

9.20. $d = 89.97$ ft.

9.22. 4.22 ft.

9.24. $\bar{x} = 9.25 \times 10^4 \text{ m} \quad \bar{y} = .856 \times 10^4 \text{ m}.$

9.26. $d = 280 \text{ m} \quad \Delta\tau = 2.04$ sec.

9.28. $$400\cos\alpha\left[\sin\beta - \frac{\cos\beta}{4}\right] = -21.72.$$
$12{,}000 = [400\cos\alpha\cos\beta +$
$$1.178]\left(\frac{400\sin\alpha}{4.905}\right).$$

9.30. $\boldsymbol{a}_1 = 12.5\boldsymbol{j} \text{ ft/sec}^2 \quad \boldsymbol{a}_2 = 21.65\boldsymbol{i} - 12.5\boldsymbol{j} \text{ ft/sec}^2$
$\boldsymbol{a}_3 = \boldsymbol{O}.$

9.32. $a_n = .460 \text{ m/sec}^2 \quad V = 73.63$ km/hr.

9.34. $\omega = 17.13$ rpm.

9.36. $\boldsymbol{\epsilon}_n = .995\boldsymbol{i} - .0995\boldsymbol{j} \quad R = 101.5$ m.

9.38. $|\boldsymbol{a}| = 10.616 \text{ ft/sec}^2.$

9.40. $V = 891.4$ km/hr.

9.42. $7.68 \text{ ft/sec}^2 \quad 49.4 \text{ ft/sec}^2.$

9.44. $x = -8.94$ ft $\quad y = -2.24$ ft.

9.46. $\boldsymbol{a} = .02\boldsymbol{\epsilon}_t + .01504\boldsymbol{\epsilon}_n \text{ m/sec}^2.$

9.48. $\dot{\boldsymbol{a}} = 5\boldsymbol{\epsilon}_t + 6\boldsymbol{\epsilon}_n \text{ m/sec}^3.$

9.50. $\dot{\theta} = 1.063$ rad/sec
$\ddot{\theta} = .275 \text{ rad/sec}^2.$

9.52. $\Omega = 53.3$ rad/sec $\quad \dot{\Omega} = 72 \text{ rad/sec}^2.$

9.54. $\boldsymbol{V} = 21.5\boldsymbol{\epsilon}_{\bar{r}} + 358\boldsymbol{\epsilon}_\theta$ mm/sec
$\boldsymbol{a} = -517\boldsymbol{\epsilon}_{\bar{r}} + 127.4\,\boldsymbol{\epsilon}_\theta \text{ mm/sec}^2.$

9.56. $\boldsymbol{V} = -3\boldsymbol{\epsilon}_{\bar{r}} + 3\boldsymbol{\epsilon}_\theta$ m/sec
$\boldsymbol{a} = -16.60\boldsymbol{\epsilon}_{\bar{r}} - 28.8\boldsymbol{\epsilon}_\theta \text{ m/sec}^2.$

9.58. $\boldsymbol{V}(2) = .7524\boldsymbol{\epsilon}_{\bar{r}} + 1.139\boldsymbol{\epsilon}_\theta + 24\boldsymbol{\epsilon}_z$ m/sec
$\boldsymbol{a}(2) = -1.064\boldsymbol{\epsilon}_{\bar{r}} + 2.364\boldsymbol{\epsilon}_\theta + 12\boldsymbol{\epsilon}_z \text{ m/sec}^2.$

9.60. $\boldsymbol{r} = 11.66\boldsymbol{\epsilon}_{\bar{r}} + (t^3 + 10)\boldsymbol{\epsilon}_z$ ft
$\boldsymbol{V} = 11.66\boldsymbol{\epsilon}_{\bar{r}} + 3t^2\boldsymbol{\epsilon}_z$ ft/sec
$\boldsymbol{a} = 6t\boldsymbol{\epsilon}_z$ ft/sec^2.

9.62. $\boldsymbol{V} = 1.061\boldsymbol{\epsilon}_{\bar{r}} + .04242\boldsymbol{\epsilon}_\theta + 1.061\boldsymbol{\epsilon}_z$ m/sec
$\boldsymbol{a} = -5.268\boldsymbol{\epsilon}_{\bar{r}} + .4329\boldsymbol{\epsilon}_\theta + 5.345\boldsymbol{\epsilon}_z$ m/sec.

9.64. $\boldsymbol{a} = -4\boldsymbol{\epsilon}_{\bar{r}} + 3\boldsymbol{\epsilon}_\theta + .01989\boldsymbol{\epsilon}_z$ ft/sec^2.

9.66. $\boldsymbol{V} = 1.500\boldsymbol{\epsilon}_{\bar{r}} + .600\boldsymbol{\epsilon}_\theta + 2.60\boldsymbol{\epsilon}_z$ m/sec
$\boldsymbol{a} = -1.200\boldsymbol{\epsilon}_{\bar{r}} + 6\boldsymbol{\epsilon}_\theta - 9.81\boldsymbol{\epsilon}_z$ m/sec^2.

9.68. $\boldsymbol{V} = -.01062\boldsymbol{\epsilon}_{\bar{r}} + 12.27\boldsymbol{\epsilon}_\theta$ ft/sec
$\boldsymbol{a} = -77.28\boldsymbol{\epsilon}_{\bar{r}} + 1.093\boldsymbol{\epsilon}_\theta$ ft/sec^2.

9.70. $\bar{r} = .382 + 2.48\,z^2$ ft
$\dot{Z} = .00694$ ft/sec
$\dot{Y} = .0239$ ft/sec.

9.72. Particle 1 $2.397\boldsymbol{i} - .397\boldsymbol{j}$ m/sec
Particle 2 $.397\boldsymbol{i} + 2.603\boldsymbol{j}$ m/sec, etc.

9.74. $\boldsymbol{F} = -10{,}680\boldsymbol{i} + 7.36\boldsymbol{j}$ N.

9.76. $\omega = 27.4$ rad/sec $\quad \Omega = 553$ rad/sec.

9.78. $\boldsymbol{V}_{xyz} = -10\boldsymbol{j} - 12\boldsymbol{k}$ ft/sec
$\boldsymbol{a}_{xyz} = 5\boldsymbol{j} - 34.2\boldsymbol{k}$ ft/sec^2.

9.80. Hit at 3.946 m to right of center.

9.82. Hit at 77 ft from bow.

9.84. $\boldsymbol{F} = -.0652\boldsymbol{k}$ N.

9.86. $\alpha = 18.55°$.

9.88. 6.76 sec $\quad \beta = 80.53°$.

9.90. $\boldsymbol{F} = -5.73\boldsymbol{i} + 46.1\boldsymbol{j}$ N.

9.92. $\boldsymbol{V}_{xyz} = 36.1\boldsymbol{i} + 19.44\boldsymbol{j} - 99.2\boldsymbol{k}$ m/sec
$\boldsymbol{a}_{xyz} = -1094\boldsymbol{i} + 4.44\boldsymbol{j} - 8.85\boldsymbol{k}$ m/sec.^{2}

9.94. $\boldsymbol{F} = 163.1\boldsymbol{i} + 244.6\boldsymbol{j} + 1.009 \times 10^4\boldsymbol{k}$ N.

9.96. $\boldsymbol{V} = 136\boldsymbol{i} + 80\boldsymbol{j} + 5\boldsymbol{k}$ ft/sec
$\boldsymbol{r} = 1{,}523\boldsymbol{i} + 1{,}004\boldsymbol{j} + 106\boldsymbol{k}$ ft.

9.98. Hit at midpoint of freighter.

9.100. $y = 34.19$ km.

9.102. $\boldsymbol{V} = 1.742\boldsymbol{i} + .8360\boldsymbol{j}$ m/sec
$\boldsymbol{a} = 6.338\boldsymbol{i} + 10.322\boldsymbol{j}$ m/sec.

9.104. 128.5 ft/sec.

9.106. $A = 635.8$ m^2.

9.108. .00255 g's.

9.110. $\boldsymbol{V} = \boldsymbol{\epsilon}_{\bar{r}} + 10\boldsymbol{\epsilon}_\theta + 2\boldsymbol{\epsilon}_z$ ft/sec
$\boldsymbol{a} = -20\boldsymbol{\epsilon}_{\bar{r}} + 4\boldsymbol{\epsilon}_\theta$ ft/sec^2 $\quad \boldsymbol{a} \bullet \boldsymbol{\epsilon} = -15.2$ ft/sec^2.

9.112. $a = 7.21$ ft/sec^2.

9.114. $\bar{x} = 17.49$ m
$\bar{y} = 15.29$ m
$\dot{x} = 10$ m/sec $\quad \dot{y} = .1623$ m/sec.

9.116. $d = 982.2$ m $\quad \delta = .463$ m.

9.118. $\alpha = 27.62°$.

9.120. $\boldsymbol{F}_{\max} = .01962\boldsymbol{i} - .02943\boldsymbol{j} + .304\boldsymbol{k}$ N.

9.122. $\boldsymbol{V} = 149\boldsymbol{i} + 200\boldsymbol{j} + 60\boldsymbol{k}$ m/sec.
Tensile force on rod is 45 N.

9.124. $\boldsymbol{\epsilon}_0 = .2931\boldsymbol{i} + .1466\boldsymbol{j} + .5863\boldsymbol{k}$.

10.2. $\boldsymbol{r} = 10\boldsymbol{i} + 8\boldsymbol{j} + 7\boldsymbol{k}$ m.

10.4. 99.4 ft $\quad$ 2.485 sec.

10.6. 6.26 sec.

10.10. $d = .01021$ m.

10.12. 162.8 m.

10.14. 10.37 m.

10.16. 62.90 m/sec.

10.18. $V_A = 14.12$ km/hr $\quad V_B = 6.833$ km/hr.

10.20. 4 m/sec.

10.22. 435 m/sec $\quad$ 16,490 m.

10.24. 96.8 ft/sec $\quad$ 64.1 ft.

10.26. 82.82 lb $\quad$ 13.22 ft/sec^2.

10.28. 17.22 ft $\quad$ distance = 17.22 ft.

10.30. 77.53 m/sec^2.

10.32. $d = 13.86$ mm $\quad l = 20$ mm.

10.34. 22.4 m/sec $\quad d = 108.1$ m.

10.36. $A = 48.83$ m^2.

10.38. $d = 2.4754$ ft $\quad V = -2.72$ ft/sec.

10.40. $t = .522$ sec.

10.42. $t = 8.58$ hours.

10.44. $V = 1.9365$ ft/sec.

10.46. $t = 2.10$ sec $\quad h = 92.7$ ft.

10.48. $h = 77{,}280$ ft.

10.50. $F = 2411$ lb.

10.54. $\dot{\theta} = 31.29$ rpm.

10.56. $d = g/\omega^2$.

10.58. 350 N.

10.60. $\omega = 9.64$ rad/sec.

10.62. $\boldsymbol{\epsilon}_n = -.594\boldsymbol{i} + .396\boldsymbol{j} + .701\boldsymbol{k}$.

10.66. $V_E = 35{,}921$ km/hour $\quad V_C = 25{,}397$ km/hour.

10.68. $\ddot{\theta} = 1.131$ rad/sec^2 $\quad T = .1978$ N.

10.70. $\Delta t = 2.07$ hours.

10.74. $V_{\max} = 15{,}094$ km/hour $\quad \tau = 5.36$ hours.

10.76. $V = 194{,}500$ mph.

10.78. $\tau = 548$ hours.

10.80. 484 mph.

10.82. $h = 56.1$ miles.

10.86. $V_r = 6210$ km/hour $\quad \epsilon = .326$.

10.90. $\omega = 4.43$ rad/sec.

10.92. $a_t = -8.51$ m/sec^2.

10.94. $(F_n)_{\text{TOTAL}} = \dfrac{250}{g}\left[\dfrac{(1+a^4\sinh^2 ax)^{3/2}}{a^3\cosh ax}\right]^{-1}(100) + 250\cos\,[\tan^{-1}(a^2\sinh ax)]$.

10.96. From S = 15.66 mm to 138.0 mm.

10.100. $(V_1 - V_C)_{at\ t=2} = -487\boldsymbol{i} + 15{,}450\boldsymbol{j} + 84.2\boldsymbol{k}$ ft/sec
$(V_2 - V_C)_{at\ t=2} = -581\boldsymbol{i} + 15{,}450\boldsymbol{j} + 319\boldsymbol{k}$ ft/sec.

10.102. $F = -30.822$ kN.

10.104. 2.01m.

10.106. .1171 m/sec.

10.108. $\Delta t = 48.9$ sec.

10.112. 6620 miles 1.123 hours 24,290 mph.

10.114. $(\Delta V)_1 = 1131$ km/hour $(\Delta V)_2 = 1088$ km/hour.

10.116. $F = 199.5$ N.

10.118. $\Delta V = 408$ km/hour.

10.120. 224 lb.

10.122. $-Mg + K(r_0 - r) = M(\ddot{r} - r\dot{\theta}^2)$
$g\theta = r\ddot{\theta} + 2\dot{r}\dot{\theta}$.

10.124. $a_c = 1.4\,(6t + 15)\boldsymbol{i} - 14\boldsymbol{k}$ ft/sec^2
$(x_c) - (x_c)_0 = 2498$ ft
$(y_c) - (y_c)_0 = 13.05$ ft.
$(z_c) - (z_c)_0 = -686$ ft.

10.126. $V = 0$.

10.128. $V = .484$ ft/sec.

10.130. 3130 ft/sec.

10.132. $V = 4.706$ m/sec.

10.134. 38.0 ft.

10.136. $\dot{s}_B = .309$ m/sec.

10.138. .5933 g's.

11.2. $V = 34.8$ ft/sec.

11.4. $V = 3.98$ m/sec.

11.6. $V = 6.04$ ft/sec.

11.8. $F = 19.6$ kN.

11.10. -45.2 m/sec^2.

11.12. $V = 6.002$ ft/sec.

11.14. $KE = 1334$ ft-lb.

11.16. $F = 1800$ lb.

11.18. .211 m/sec.

11.20. $\omega = 8.74$ rad/sec.

11.22. $d = 1.835$ m.

11.24. $d = .312$ m.

11.26. $V = 18.75$ m/sec.

11.28. $W = 121.0$ kN.

11.30. .288 ft.

11.32. 24.7 HP 16.57 kW.

11.34. Power $= 932\, a_0 V + \kappa V^3$
for $a_0 = 0$ Power $= \kappa V^3$.

11.36. $\Delta t = 6.44$ sec.

11.38. 8333 N.

11.40. $V = 3.40$ ft/sec.

11.42. $\delta = 106.1$ mm.

11.46. $d = -5z \cos x + xy + 2y^2 z +$ const.

11.48. $V = 6.397$ ft/sec.

11.50. $V = .389$ m/sec.

11.52. $\dot{V} = 107.3$ ft/sec^2.

11.54. $h = 23.33$ ft.

11.56. $V = 34.80$ ft/sec.

11.58. $\delta = 1.518$ ft.

11.60. $V = 55.8$ ft/sec.

11.62. $V = 2.853$ m/sec.

11.64. $\delta = 2.398$ mm.

11.66. $d = 149.12$ m $F_{AB} = 58{,}955$ N $F_{CB} = 29{,}477$ N.

11.68. $V = \left[\frac{q}{L}\left(L^2 - a^2\right)\right]^{1/2}$.

11.70. $(\mathcal{W}_K)_{\text{internal}} = -\frac{1}{2}\frac{\mathcal{W}_1}{g}V_0^{\,2} + 5.7\left(\mathcal{W}_1 + \mathcal{W}_2\right)$.

11.72. $KE = 61.02$ N-m.

11.74. $V = 6.78$ ft/sec.

11.78. $KE = .1069$ ft-lb ERROR $= 1.952\%$.

11.80. $V = 6.138$ m/sec.

11.82. $d = 1.835$ m.

11.84. $F = 1.863$ lb $f = .621$ lb.

11.86. $(V_C)_f = 20.72$ ft/sec $(V_C)_{f=0} = 25.4$ ft/sec.

11.88. $F_C = 31.2$ lb.

11.90. $\omega = 6.83$ rad/sec.

11.92. $d = 3.823$ m $\begin{cases} \text{f} = 0 \text{ horizontal surface} \\ \text{f} = 83.35 \text{ N on incline} \end{cases}$.

11.94. $V = 1.414$ m/sec $f_{\text{total}} = 915$ N.

11.96. $\omega = 3.17$ rad/sec.

11.98. $V_A = 12.94$ ft/sec $a = 4.29$ ft/sec^2.

11.100. $\omega = .499$ rad/sec.

11.102. 1.157 HP Power Input $= 6.70$ HP.

11.104. .558 HP.

11.106. $\theta = 48.0°$.

11.110. 5471 HP.

11.112. 15,900 HP.

11.114. $V = 1.716$ m/sec 58.5 N.

11.116. $\omega = 11.90$ rad/sec $f = 130.0$ N.

11.118. 114.9 ft/sec.

11.120. $\mathcal{W} = 17.63$ lb $F_n = 36.33$ N.

11.122. $V = 9.17$ m/sec $f = 85.7$ N.

11.124. $V = 2.32$ ft/sec.

11.126. $V = .4937$ m/s $f = 186.1$ N.

12.2. $V = 333\boldsymbol{i} + 400\boldsymbol{j} + 4{,}000\boldsymbol{k}$ m/sec.

12.4. $V = 177.1$ ft/sec to left.

12.6. $t = 1.631$ sec $\Delta t = .7136$ sec.

12.8. $F_{AV} = 12.8\boldsymbol{i} - 14.40\boldsymbol{j} + 48.0\boldsymbol{k}$ N.

12.10. $I = -.964\boldsymbol{i} - .388\boldsymbol{j} + .97\boldsymbol{k}$ lb-sec.

12.12. $t = .646$ sec.

12.14. $t = 7.55$ sec $t = 13.21$ sec $F = 25{,}233$ N.

12.16. 1.961 HP.

12.18. $T_2 = 21.70$ lb.

12.20. $T_x = 76.7$ lb-sec.

12.22. $\Delta V = 10.00$ km/hour.
12.24. $V = .402$ m/sec.
12.28. $V_A = 5.183$ m/sec $V_B = 10.37$ m/sec.
12.30. 4.905 m/sec.
12.32. $F_\omega = 965$ lb.
12.34. $(V_B)_f = 4.10$ m/sec $\Delta KE = 18.15$ N-m.
12.36. $(V_A)_f = -1.130$ m/sec.
12.38. $\delta = 5.610$ in.
12.42. $\theta = 11.81°$.
12.44. $\delta = .750$ m.
12.46. 4.648 m/s $\delta = .36$ m.
12.48. $\delta = .7990$ m.
12.50. 14.86 ft/sec.
12.52. $(V_A)_f = -1.88\boldsymbol{i} + 14.14\boldsymbol{j}$
$(V_B)_f = 8.01\boldsymbol{j}$.
12.56. $\Delta = .05142$ m $\delta = .059$ m.
12.58. 50.5°.
12.60. Drag $= 3.53\ mnV^2$.
12.62. Drag $= .468\ mnV^2$.
12.64. $F_x = \dfrac{4}{3}\dfrac{\pi s R^2}{c}$ 3.53 N.
12.66. 37,306 km/hour.
12.68. $d = .205$ m.
12.70. $\omega_2 = 8.292$ rad/sec.
12.72. $\omega_2 = 5.107$ rad/sec.
12.74. 4071 mi/hour.
12.76. 6210 km/hour.
12.78. $h = 183$ km.
12.80. $(\Delta V)_1 = 1135$ km/hour
$(\Delta V)_2 = 1085$ km/hour.
12.82. 6354 km 9373 km.
12.84. $r_{max} = 32{,}534$ miles.
12.86. $r_A = 8900$ miles $\Delta V = 7285$ mph.
12.88. $V_0 = 741$ km/hour
$V_r = 826$ km/hour.
12.90. $\boldsymbol{P} = 2.02\boldsymbol{i} - .0435\boldsymbol{j} + .596\boldsymbol{k}$ slug-ft/sec
$\boldsymbol{H}_0 = 1.975\boldsymbol{i} - 1.326\boldsymbol{j} - 25.8\boldsymbol{k}$ slug-ft²/sec
$\boldsymbol{H}_a = -5.61\boldsymbol{i} - 16.76\boldsymbol{j} - 1.212\boldsymbol{k}$ slug-ft²/sec.
12.92. $\dot{\omega} = -20.0$ rad/sec².
12.94. $\dot{\omega} = \dfrac{T - 8s_2Vm\omega}{4m(s_1^{\,2} + s_2^{\,2})}$.
12.98. $\omega = 593$ cycles/sec $= 3725$ rad/sec.
12.100. 21.0 m/sec².
12.102. $H = 34.4 \times 10^6$ slug-ft²/sec.
12.104. $\dot{\omega} = .0375$ rad/sec² $\boldsymbol{V}_C = .1749\boldsymbol{i} + .262\boldsymbol{j}$ m/sec².
12.106. $\dot{\omega} = 2.01$ rad/sec².
12.108. $V = 27.7$ ft/sec.
12.110. $(V_2)_t = 12{,}975$ mi/hour
$(V_2)_r = 21{,}370$ mi/hour
$\Delta V = 31{,}190 - 17{,}260 = 13{,}930$ mi/hour.

12.112. $x = 4.464$ m.
12.114. $V_f = 34.35$ ft/sec.
12.116. $x = -20{,}566$ m.
12.118. $V = 2.778$ ft/sec $= 1.645$ knots.
12.120. $\delta = 2.238$ ft $V_B = .1791$ ft/sec.
12.122. $\theta_A = 5.47°$ $\theta_B = 29.9°$.
12.124. $\boldsymbol{V}_f = 161\boldsymbol{i} + 170.8\boldsymbol{j} - 26.1\boldsymbol{k}$ ft/sec.
12.126. $(V_A)_f = 26.6$ m/sec. $(V_B)_f = 31.1$ m/sec.
12.128. $H_0 = 7.57$ slug-ft²/sec $P_0 = 0$.
12.130. $H_0 = 5480$ slug-ft²/sec.
12.132. $h = 3000$ km $r_{max} = 9371$ km.
12.134. $F = 2329$ lb.
12.136. $V = 18$ km/hour $\Delta KE = 3.24$ joules.
12.138. $E = 58.96$ Mev $l_B = -.325$ $m_B = .18746$
$n_B = -.97417$.

13.4. -10.66i + 7.31j - 35.18k m/s.
13.6. $-521\boldsymbol{i} + 108.7\boldsymbol{j} + 65.2\boldsymbol{k}$ m/s.
13.8. $\dot{\boldsymbol{\rho}} = 10\boldsymbol{i} + 5\boldsymbol{j}$ mm/sec
$\ddot{\boldsymbol{\rho}} = -1250\boldsymbol{k}$ mm/sec².
13.10. $\boldsymbol{\omega} = .940\boldsymbol{i} + .342\boldsymbol{j} + 1.8\boldsymbol{k}$ rad/sec.
$\dot{\boldsymbol{\omega}} = -.616\boldsymbol{i} + 1.692\boldsymbol{j}$ rad/sec².
13.12. $\boldsymbol{\omega} = -.4\boldsymbol{i} + 170.8\boldsymbol{j} - 170.8\boldsymbol{k}$ rad/sec.
$\dot{\boldsymbol{\omega}} = -.2\boldsymbol{i} - 68.32\boldsymbol{j} - 68.32\boldsymbol{k}$ rad/sec².
13.14. $\dot{\boldsymbol{\omega}} = 3.14\boldsymbol{i} + 1356\boldsymbol{j}$ rad/sec²
$\ddot{\boldsymbol{\omega}} = 407\boldsymbol{j} + 14{,}200\boldsymbol{k}$ rad/sec³.
13.16. $\boldsymbol{\omega} = -121.6\boldsymbol{j}$ rad/sec.
$\dot{\boldsymbol{\omega}} = 3823\boldsymbol{i}$ rad/sec².
13.18. $\dot{\boldsymbol{\omega}} = .12\boldsymbol{j}$ rad/sec²
$\ddot{\boldsymbol{\omega}} = -.024\boldsymbol{i}$ rad/sec³.
13.20. $\boldsymbol{V}_A = .5\boldsymbol{i} + .1212\boldsymbol{j}$ m/s
$\omega = .4788$ rad/sec.
13.22. $\dot{\boldsymbol{\omega}} = .18\boldsymbol{i} - .1\boldsymbol{j} + .2\boldsymbol{k}$ rad/sec².
13.24. $\dot{\boldsymbol{V}} = -20.5\boldsymbol{i} - 2.14\boldsymbol{j} + 17.10\boldsymbol{k}$ m/sec².
13.26. $\boldsymbol{V} = -243\boldsymbol{i} + 207\boldsymbol{j} + 25\boldsymbol{k}$ m/sec.
13.28. $\omega = 56.7$ rad/sec $\boldsymbol{V}_B = 29\boldsymbol{j} - 12.03\boldsymbol{k}$ m/sec.
13.30. $V_A = -2$ ft/sec.
13.32. $\omega = 4$ rad/sec $\boldsymbol{V}_D = -6\boldsymbol{i} - 19\boldsymbol{j}$ m/sec.
13.34. $\omega_C = -10$ rad/sec.
13.36. $\boldsymbol{V}_C = -.548\boldsymbol{i} - .373\boldsymbol{j}$ m/sec
$\boldsymbol{a}_C = -43.4\boldsymbol{i} - 63.7\boldsymbol{j}$ m/sec².
13.38. $\omega_{AB} = -9.33$ rad/sec $\boldsymbol{V}_B = -12.38\boldsymbol{j} - 7.149\boldsymbol{k}$ m/sec.
13.40. $\omega_{BA} = -54.7$ rad/sec $V_A = 79.8$ ft/sec.
13.44. $V_y = 14.66$ m/sec $\omega = -5$ rad/sec.
13.46. $\boldsymbol{V}_B = 11.95\boldsymbol{i} - 4.949\boldsymbol{j}$ m/sec
$\boldsymbol{a}_B = -57.77\boldsymbol{i} - 57.77\boldsymbol{j}$ m/sec²
$\theta = 2.625$ rad/sec.
13.48. $\omega_{AB} = -.609$ rad/sec $\dot{\omega}_{AB} = 14.71$ rad/sec².
13.50. $\boldsymbol{V}_B = 5.03\boldsymbol{i} + 2\boldsymbol{j} - 3\boldsymbol{k}$ m/sec.

13.52. $V_D = -.5\boldsymbol{i} - 14.7\boldsymbol{j} - 2.710\boldsymbol{k}$ ft/sec
$a_D = -1.25\boldsymbol{j} + 10\boldsymbol{k}$ ft/sec^2.

13.54. $\boldsymbol{\omega}_{AB} = -0.0176\boldsymbol{k}$ rad/sec
$\dot{\boldsymbol{\omega}} = -4.34 \times 10^{-4}\boldsymbol{k}$ rad/sec^2.

13.56. $\omega_A = 3.643$ rad/sec
$\dot{\omega}_A = -6.45$ rad/sec^2.

13.58. $\omega = 1.273$ rad/sec $\quad V_C = 2.163$ m/sec.

13.60. $V = 1.156$ m/sec.

13.62. $V_A = 5.478$ m/sec $\quad a_A = -929.6$ m/sec^2.

13.64. $\omega_H = 3.93$ rad/sec $\quad \dot{\omega}_H = 20.7$ rad/sec^2.

13.66. $\omega_C = 5.00$ rad/sec.

13.68. $\omega_{AB} = -5$ rad/sec $\quad \dot{\omega}_{AB} = -360.8$ rad/sec^2.

13.70. $\boldsymbol{V}_{XYZ} = 8.86\boldsymbol{i} + 5\boldsymbol{j}$ m/sec.

13.72. $\boldsymbol{V}_{XYZ} = -17.32\boldsymbol{i} - 10\boldsymbol{j} - 50\boldsymbol{k}$ ft/sec.

13.74. $\boldsymbol{V}_{XYZ} = 1.30\boldsymbol{i} + 1.8\boldsymbol{j} + .130\boldsymbol{k}$ m/sec.

13.76. $\omega_1 = 15.40$ rad/sec $\quad \omega_2 = -31.37$ rad/sec.

13.78. $\omega_1 = 1.985$ rad/sec.

13.80. $\boldsymbol{V}_{XYZ} = -2.05\boldsymbol{i} + 4.92\boldsymbol{j} - 2.66\boldsymbol{k}$ m/sec.

13.82. $\boldsymbol{V}_{\perp \text{ barrel}} = -.908\boldsymbol{i} - .5253\boldsymbol{j} - .6\boldsymbol{k}$ m/sec.

13.84. $\boldsymbol{V}_{XYZ} = 4.00\boldsymbol{i}$ m/sec.

13.86. $V_{XYZ} = 37.9$ ft/sec rel. to ground
$V_{xyz} = 9.813$ ft/sec rel. to rod.

13.88. $V_{xyz} = 3.756$ m/sec $\quad V_{XYZ} = 7.165$ m/sec.

13.90. $\omega_D = 9.028$ rad/sec.

13.92. $\boldsymbol{V}_{XYZ} = 50\boldsymbol{i} - 5\boldsymbol{j} - 20\boldsymbol{k}$ ft/sec.
$\boldsymbol{a}_{XYZ} = -100\boldsymbol{i} - 300\boldsymbol{j} - 35\boldsymbol{k}$ ft/sec^2.

13.94. $\boldsymbol{a}_{XYZ} = 2600\boldsymbol{i} - 173.2\boldsymbol{j} - 1732\boldsymbol{k}$ ft/sec^2.

13.96. $\boldsymbol{a}_{XYZ} = -161.9\boldsymbol{k}$ m/sec^2.

13.98. $\boldsymbol{a}_{XYZ} = -5.08\boldsymbol{i} + 2.314\boldsymbol{j} + 2.70\boldsymbol{k}$ m/sec^2.

13.100. $\boldsymbol{a}_{XYZ} = 1.950\boldsymbol{i} - 18.69\boldsymbol{j} - 1.207\boldsymbol{k}$ m/sec^2.
$|\boldsymbol{a}_{XYZ}| = 1.919$ g's.

13.102. $\boldsymbol{a}_{XYZ} = -25.6\boldsymbol{j} - 57.6\boldsymbol{k}$ m/sec^2.

13.106. $\boldsymbol{a}_{\perp \text{gun barrel}} = -.403\boldsymbol{i} - .233\boldsymbol{j} - 1.647\boldsymbol{k}$.

13.108. $\boldsymbol{V}_{XYZ} = -1.113\boldsymbol{i} + .1202\boldsymbol{j}$ m/sec
$\boldsymbol{a}_{XYZ} = .0969\boldsymbol{i} - .1859\boldsymbol{j}$ m/sec^2.

13.110. $\boldsymbol{a}_{XYZ} = -1.204\boldsymbol{i} - 2.570\boldsymbol{j}$ ft/sec^2.
$|\boldsymbol{a}_{XYZ}| = .08814$ g's.

13.112. $\boldsymbol{a}_{XYZ} = -73.22\boldsymbol{i} - 112.63\boldsymbol{j} + 67.3\boldsymbol{k}$ m/sec^2.

13.114. $\boldsymbol{a}_{XYZ} = -.4\boldsymbol{j} + (-15\boldsymbol{j}) + 2(5\boldsymbol{i} + 20\boldsymbol{k}) \times (.2\boldsymbol{j}) + (.02\boldsymbol{i} - 10\boldsymbol{j}) \times (-.01\boldsymbol{j}) + (.5\boldsymbol{i} + 20\boldsymbol{k}) \times [(.5\boldsymbol{i} + 20\boldsymbol{j}) \times (-.01\boldsymbol{j})]$ m/sec^2.

13.116. $\boldsymbol{F} = -.51\boldsymbol{i} + .1359\boldsymbol{j} - .07176\boldsymbol{k}$ N.

13.118. $\boldsymbol{a}_{XYZ} = -15.50\boldsymbol{i} + .405\boldsymbol{j} - 7.80\boldsymbol{k}$ m/sec^2
$(\boldsymbol{a}_A)_{XYZ} = -1.05\boldsymbol{i} + .665\boldsymbol{j} - 2.2\boldsymbol{k}$ m/sec^2.

13.122. $\boldsymbol{F} = 12.42\boldsymbol{j}$ lb.

13.124. Axial force = 30.7 N compression
Bending moment = 27 N-m.

13.126. $F_{\text{axial}} = 1.791$ N.

13.128. $-125.7\boldsymbol{i}$ ft/sec^2 $\quad$ .732 lb in plus x direction.

13.130. $\dot{\boldsymbol{\omega}} = -4\boldsymbol{i} + 2\boldsymbol{j} + 30\boldsymbol{k}$ rad/sec^2.

13.132. $\omega_C = 2.07$ rad/sec $\quad \dot{\omega}_C = -1.799$ rad/sec^2.

13.134. $\boldsymbol{a}_{XYZ} = -300\boldsymbol{i} + 15\boldsymbol{j} - 300\boldsymbol{k}$ ft/sec^2.

13.136. 162.2 ft/sec^2.

13.138. $\omega_B = -.5077$ rad/sec
$\dot{\omega}_B = -1.0115$ rad/sec^2.

13.140. $\boldsymbol{V}_{XYZ} = 20\boldsymbol{i} - 30\boldsymbol{j}$ ft/sec
$\boldsymbol{a}_{XYZ} = 20\boldsymbol{i} + 5\boldsymbol{j} - 500\boldsymbol{k}$ ft/sec^2.

13.142. $\boldsymbol{V}_B = 1.500\boldsymbol{i} - 1.500\boldsymbol{j}$ m/sec.

13.144. $V_C = 10$ m/sec $\quad \omega_{CD} = 20$ rad/sec $\quad \omega_{BC} = 0$.

13.146. $\boldsymbol{V}_{XYZ} = 2\boldsymbol{j}$ m/sec
$\boldsymbol{a}_{XYZ} = -13.33\boldsymbol{k}$ m/sec^2.

14.2. 2.12 N-m.

14.4. 16.70 m down.

14.6. $F = .672$ lb.

14.8. $P = 678$ N.

14.10. $a_B = -3.27$ m/sec^2.

14.12. $t = .285$ sec.

14.14. $\ddot{\theta}_A = -.8657$ rad/sec^2.

14.16. $F_{AB} = 163.4$ N.

14.18. 552 Rev.

14.20. $\tau_{rr} = 71{,}833$ psi.

14.22. $F_x = 87.7$ lb $\quad F_y = 169.7$ lb.

14.24. $B_y = C_y = 2244$ N $\quad B_z = C_z = 5.27$ N.

14.26. $A_y = B_y = 3200$ lb.

14.28. $\dot{\theta}_A = -.9583$ rad/sec.

14.30. $\ddot{\theta} = \dfrac{2y}{L}$

14.32. $\ddot{\theta} = -49.97$ rad/sec^2.

14.34. $a_0 = -12.13$ ft/sec^2 $\quad \ddot{\theta} = -1.035$ rad/sec^2.

14.36. -2.08 m.

14.38. $\ddot{\theta} = -14.96$ rad/sec^2.

14.40. 28.5 rad/sec^2.

14.42. 1.442 m/sec^2.

14.44. .1056 ft.

14.46. $P = 1437$ lb.

14.48. 3.43 m/sec^2.

14.50. $f = .283\mathcal{W}$.

14.52. $f = 24.53$ N $\quad N = 85.0$ N.

14.54. .405 m/sec^2.

14.56. $T = 68.53$ lb.

14.58. $\boldsymbol{a} = \boldsymbol{0}$.

14.60. $\boldsymbol{a} = -2.512\boldsymbol{i} + 6.26\boldsymbol{j}$ m/sec^2.

14.64. 13.08 m/sec $\quad \dot{\theta} = 10.90$ rad/sec.

14.66. $\ddot{X} = .1361$ ft/sec^2 $\quad \ddot{Y} = 0 \quad \ddot{\theta} = .6932$ rad/sec^2.

14.68. $V_0 = 7.732$ m/sec.

14.70. $P = 8.75$ N.

14.72. $BD = 49.5$ N $\quad AC = 37.0$ N.

14.74. $\dot{X} = -3.343$ m/sec
$(\dot{X})_{.3} = -3.294$ m/sec.
14.76. $\ddot{\theta} = 19.32$ rad/sec^2 $\ddot{Y} = -16.100$ m/sec^2.
14.78. $\ddot{Y}_A = -2.18$ m/sec^2 $\ddot{Y}_C = 2.18$ m/sec^2.
14.80. $T_B = 246$ N $T_A = 201$ N.
14.82. $\dot{\omega} = 7.40$ rad/sec^2 $\ddot{X} = -7.40$ ft/sec^2.
14.84. $\ddot{\theta}_{BC} = -6.43$ rad/sec^2 $\ddot{\theta}_{AB} = 4.29$ rad/sec^2.
14.86. $\ddot{\theta}_{AB} = 6.54$ rad/sec^2.
14.88. $\boldsymbol{a}_A = 1.1308\boldsymbol{i} + .6529\boldsymbol{j}$ m/sec^2
$\dot{\omega} = .17125$ rad/sec^2
$\boldsymbol{a}_B = .52549\boldsymbol{i} + .04752\boldsymbol{j}$ m/sec^2.
14.90. .222 Rev.
14.92. $A_x = -32$ N
$A_y = -212.4$ N
$B_x = -176.6$ N
$B_y = -26.40$ N.
14.94. $B_x = 0$
$A_x = 0$
$B_y = -235$ N
$A_y = -254$ N
$A_z = 50$ N.
14.96. $\dot{\omega} = -6.26$ rad/sec^2 $B_x = 6.90$ lb $B_y = 0$
$A_x = 28.9$ lb $A_y = 0$.
14.98. $B_x = 25$N
$B_y = 60.2$ N
$A_x = 25$ N
$A_y = -60.2$ N
14.100. $B_x = 200$ N
$B_y = -21.93$ N
$A_x = 200$ N
$A_y = -21.93$ N.
14.102. $M_x = -9.171$ N-m
$V_x = -48.9$ N
$M_y = 20.9$ N-m
$V_y = 8.66$ N
$M_z = -.3415$ N-m.
14.104. $r_3 = 1.260$ ft $\theta_3 = 249.5°$
$r_4 = 1.098$ ft $\theta_4 = 192.2°$.
14.106. $y_D = -.667$ in.
$z_D = 0$
$y_C = 5.67$ in.
$z_C = 0$.
14.108. $\theta_A = 270°$ $L_A = .318$ m.
14.110. $\theta_B = 223.1°$ $r_B = 1.727$ m
$\theta_4 = 254.3°$ $r_4 = 2.21$ m.
14.112. $T = 169.6$ lb.
14.114. $t = 9.174$ sec.
14.116. 16.90 m/sec^2.
14.118. -41 rad/sec^2.
14.120. 60.4 lb.
14.122. $D_x = 6.96$ N $D_y = 31.23$ N.
14.124. $\ddot{\theta}_A = -3.26$ rad/sec^2.
14.126. $\ddot{\theta} = .0580$ rad/sec^2
$X = 5$ m $Y = 5$ m.
14.128. $\boldsymbol{a}_D = 4.06\boldsymbol{i} - 7.20\boldsymbol{j}$ m/sec^2.
14.130. $N_A = 17.13$ lb
$N_B = 38.9$ lb
$F_A = 3.43$ lb
$F_B = 7.78$ lb.
14.132. $\ddot{\theta} = -18.34$ rad/sec^2.
14.134. $A_x = 0$ $A_y = -21.83$ lb
$A_z = 100$ lb.
14.136. $\theta_C = 226.2°$ $r_C = .256$ m
$\theta_D = 266.8°$ $r_D = .416$ m.
14.138. $A_x = -192.7$ N $A_y = 2500$ N
$M_A = 5647.5$ N-m.

15.2. 4627 ft-lb.
15.4. 24.5 N-m.
15.6. 45.8 N-m 4.954 N-m.
15.8. $KE = 18.95$ ft-lb.
15.10. $\omega = \left[\dfrac{8\mathcal{W}_2 a_g}{3\mathcal{W}_1 r^2 + 2\mathcal{W}_2 (r-a)^3}\right]^{\frac{1}{2}}$.
15.12. $V_B = -.575$ m/sec.
15.14. $\omega = 2.79$ rad/sec.
15.18. $V_D = 5.40$ m/sec.
15.20. $\omega_A = 20$ rad/sec $\omega_B = 10.0$ rad/sec.
15.22. $V_C = 11.93$ ft/sec.
15.24. $\omega_z = 2.79$ rad/sec.
15.26. $\Delta = .985$ ft.
15.30. $\dot{\theta}_H = 35.30$ rad/sec.
15.32. $V = 2.354$ m/s $f = 79.4$ N.
15.34. $V = 4.60$ m/sec $f = 7480$ N.
15.36. 25.3 rad/sec.
15.38. $V_A = 13.61$ ft/sec.
15.40. $(H_0)_x = 0$ $(H_0)_y = 12\pi \dfrac{\text{slug-ft}^2}{\text{sec}}$
$(H_0)_z = 74\pi \dfrac{\text{slug-ft}^2}{\text{sec}}$.
15.42. $(H_c)_x = 1.941$ kg-m^2/sec $(H_c)_y = .510$ kg-m^2/sec
$(H_c)_z = 0$.
15.44. $h = 1.016$ m.
15.46. $T_1 = 1130$ N $P = 678.1$ N.
15.48. $L = 8.47$ m.
15.50. $\omega = 2.732$ RPM.
15.52. $V_A = -27.3$ ft/sec.
15.54. $\omega = 9.09$ rad/sec.
15.56. $t = .0467$ sec.
15.58. $y_{\max} = 1431$ m 116.1°.
15.62. $\omega_A = -108.7$ rad/sec.
15.64. $\omega = 9.90$ rad/sec.

15.66. $V_A = -11.04$ m/sec.
15.68. $(\omega_2)_f = -165.2$ rad/sec $\quad (\omega_1)_f = 24.22$ rad/sec.
15.70. $\omega_f = 7.64$ rad/sec.
15.72. $V_B = 1.542\boldsymbol{i} - 1.028\boldsymbol{j}$ m/sec.
15.74. $f = 15.05$ per sec.
15.76. -15.24 ft/sec $\quad \omega_f = 60.9$ rad/sec.
15.78. $[(V_C)_y]_f = -12.39$ m/sec.
$[(V_C)_x]_f = 12.85$ m/sec.
15.80. 15.52 ft-lb.
15.82. $d = .1246$ m.
15.84. $V = 7.29$ ft/sec.
15.86. $(H_C)_x = -14{,}550\,\dfrac{\text{lbm-ft}^2}{\text{sec}}$

$(H_C)_y = -24{,}700\,\dfrac{\text{lbm-ft}^2}{\text{sec}}$

$(H_C)_z = -14{,}530\,\dfrac{\text{lbm-ft}^2}{\text{sec}}$

15.88. 2.32 m/sec.
15.90. $\omega = 4.11$ rad/sec.
15.92. $V_E = -.3117\boldsymbol{i} - .434\boldsymbol{j}$ m/sec.
15.94. $\omega = 14.78$ rad/sec.

16.4. $K_2 = 3.90$ N-m.
16.6. $\omega_n = \sqrt{\dfrac{2\mu g}{L}}$ rad/sec
16.8. .438 cycles/sec.
16.10. .557 cycles/sec.
16.12. 19.81 rad/sec; 8.34 N/mm.
16.14. 28.6 rad/sec.
16.16. $\ddot{r} + \left(\dfrac{K}{m} - \omega^2\right)r = \dfrac{Kr_0}{m}$

$\omega_n = \sqrt{\dfrac{K}{M} - \omega^2}$

16.18. 16.05 rad/sec; $\quad$ 87.9 rad/sec.
16.20. 1.420 sec.
16.22. $\ddot{x} + .326x + .01230x^2 + .0001550x^3 = 0$
$\omega = .0909$ cycles/sec.
16.24. $(K_t)_{eq} = 2.43 \times 10^5$ N-m/rad
$\omega_n = 304$ rad/sec
$\theta = -7.42°$.
16.26. $\omega_n = 1$ cycle/sec.
16.28. 7.88 cycles/sec
$\theta = -.342°$.
16.30. $\omega_n = \sqrt{\dfrac{2Kr^2}{R^2\left[(M/2) - (W_1/g)\right]}}$

16.32. $\dfrac{1}{12}\dfrac{W}{g}L^2\,\ddot{\theta} + \dfrac{Wd^2}{4l}\,\theta = 0$

$\omega_n\sqrt{\dfrac{3d^2g}{lL^2}}$ rad/sec.

16.34. 235 rad/sec.
16.36. 1.377°.
16.38. $P = [(2KL - W/2)]$
16.40. 2.32 rad/sec.
16.42. $k = .518$ m.
16.44. $\mu = 0.809$.
16.51. $\omega_n = \sqrt{\dfrac{K(r_1+r_2)^2}{(W/g)(k^2+r_1^2)}}$

$f_{\max} = \dfrac{\left[I_{00}\omega^2 - K(r_1+r_2)(r_2)\right]A}{r_1}$

16.53. $\omega_n = \sqrt{\dfrac{2Kr^2}{\frac{1}{2}MR^2 + (W_1/g)R^2}}$.
16.55. 2.32 rad/sec.
16.57. $\omega_n = \sqrt{\dfrac{(3+\sin\alpha)g}{L}}$.
16.59. $\omega_n = \sqrt{\dfrac{g}{(R-r_1)\left[1+(k^2/r_1^2)\right]}}$.
16.61. $x = .004185$ ft.
16.63. .00992 m.
16.69. 110.7 m $\quad V = 51.4$ km/hr.
16.71. $\omega = 15.95$ rad/sec.
16.73. 17.89 N $\quad K = 553 \times 10^3$ N/m.
16.75. 9.805 N.
16.77. 1.808 rad/sec.
16.79. $-11.28°$.
16.81. $m\ddot{x} + 2c\dot{x} + (K_1 + K_2)x = 0$

$c_{cr} = \sqrt{\dfrac{W}{g}(K_1 + K_2)}$

16.83. $x(5) = -.211$ in. $\quad f \cong 6.92$ rad/sec.
16.87. $t = .321$ sec.
16.89. $\omega = .0225$ cycle/sec.
16.91. $e^{-.311\tau}\left[\left(-.05 - \dfrac{.0155}{\alpha}\right)e^{\alpha\tau} + \left(-.05 + \dfrac{0.155}{\alpha}\right)e^{\alpha\tau}\right] = 0$

$e^{-.311\tau}\left[(-.05\alpha - .01555)e^{\alpha\tau} - (-.05\alpha + .01555)e^{-\alpha\tau}\right]$

$-.311e^{-.311\tau}\left[\left(-.05 - \dfrac{.01555}{\alpha}\right)e^{\alpha\tau} + \left(-.05 + \dfrac{.0155}{\alpha}\right)e^{\alpha\tau}\right] = 0$

16.95. 8.73 lb/in.
16.97. .000705 m.

16.99. 2 degrees of freedom $\ddot{x} + 451x - 773\theta = 0$
$\ddot{\theta} - 386x + 8120\theta = 0.$

16.102. $K_3 = 799$ N/m.

16.103. 1.972 in.; 80.03 in.

16.104. 2.35 rad/sec.

16.105. 48.2 rad/sec.

16.106. $k = .748$ m.

16.107. $\omega_n \sqrt{\frac{6g}{L}}$

16.107. $\omega = 19.80$ rad/sec.

16.109. $W = 64.6$ N.

16.110. $\theta_0 = .0608$ rad.

16.111. .003014 rad.

16.112. .221 sec.

16.113. $A = .001454$ m
$(F_{tr})_{max} = 337$ N.

16.114. 23.6 rad/sec
.01115 m
$c = 192.0$ N-sec/m.

Index

SI UNIT PREFIXES

Multiplication Factor	*Prefix*	*Symbol*	*Pronunciation*	*Term*
1 000 000 000 000 = 10^{12}	tera	T	as in terrace	one trillion
1 000 000 000 = 10^9	giga	G	jig 'a(a as in about)	one billion
1 000 000 = 10^6	mega	M	as in megaphone	one million
1 000 = 10^3	kilo	k	as in kilowatt	one thousand
100 = 10^2	hecto	h	heck 'toe	one hundred
10 = 10	deka	da	deck 'a (a as in about)	ten
0.1 = 10^{-1}	deci	d	as in decimal	one tenth
0.01 = 10^{-2}	centi	c	as in sentiment	one hundredth
0.001 = 10^{-3}	milli	m	as in military	one thousandth
0.000 001 = 10^{-6}	micro	μ	as in microphone	one millionth
0.000 000 001 = 10^{-9}	nano	n	nan 'oh (an as in ant)	one billionth
0.000 000 000 001 = 10^{-12}	pico	p	peek 'oh	one trillionth

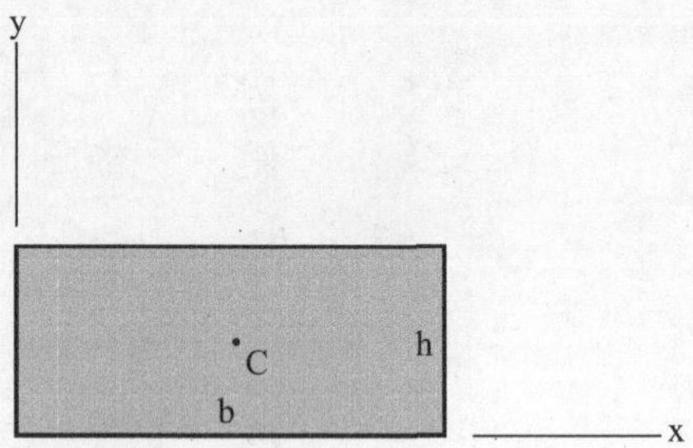

Rectangle

$$A = bh \qquad (I_{xx})_c = \frac{1}{12}bh^3$$

$$x_c = \frac{b}{2} \qquad (I_{yy})_c = \frac{1}{12}hb^3$$

$$y_c = \frac{h}{2} \qquad (I_{xy})_c = 0$$

$$J_c = \frac{1}{12}bh(b^2 + h^2)$$

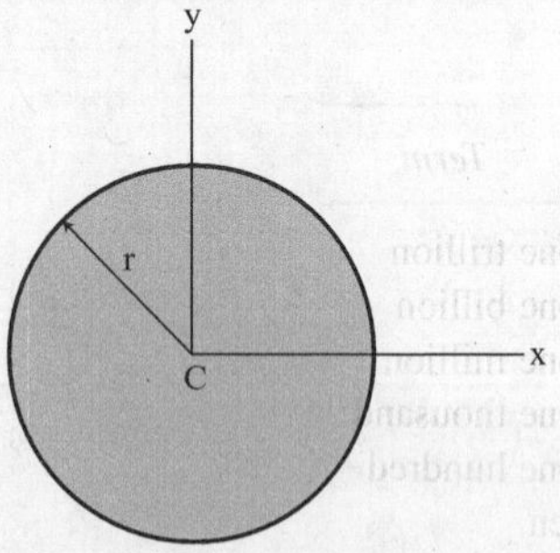

Circle

$$A = \pi r^2 \qquad (I_{xx})_c = \frac{1}{4}\pi r^4$$

$$x_c = 0 \qquad (I_{yy})_c = \frac{1}{4}\pi r^4$$

$$y_c = 0 \qquad J_c = \frac{1}{2}\pi r^4$$

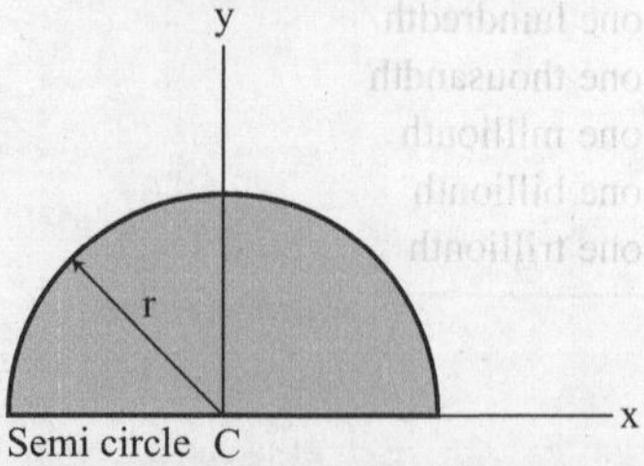

Semi circle

$$A = \frac{\pi r^2}{2} \qquad (I_{xx})_c = .00686d^4$$

$$x_c = 0 \qquad (I_{yy})_c = \frac{1}{8}\pi r^4$$

$$y_c = .424r$$

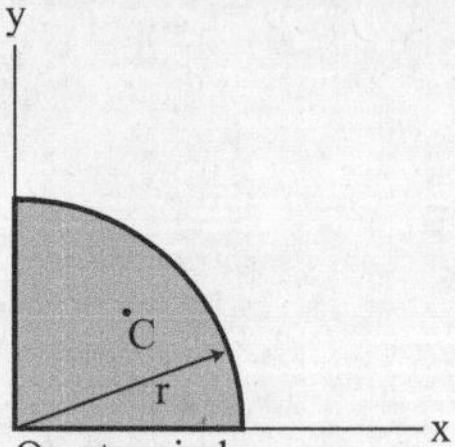

Quarter circle

$$A = \frac{\pi r^2}{4} \qquad (I_{xx})_c = .0549r^4$$

$$x_c = \frac{4r}{3\pi} \qquad (I_{yy})_c = .0549r^4$$

$$y_c = \frac{4r}{3\pi}$$

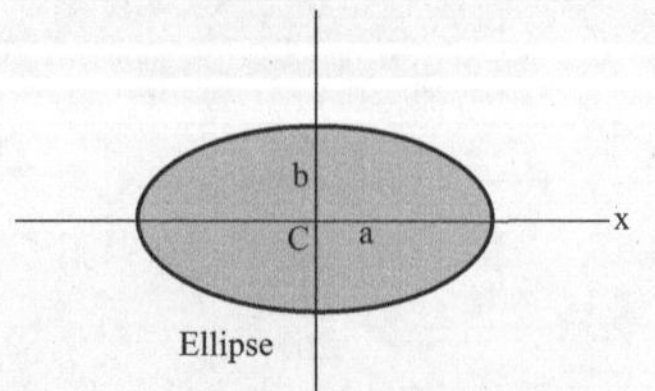

Ellipse

$$A = \pi ab \qquad (I_{xx})_c = \frac{\pi ab^3}{4}$$

$$(I_{yy})_c = \frac{\pi ba^3}{4}$$

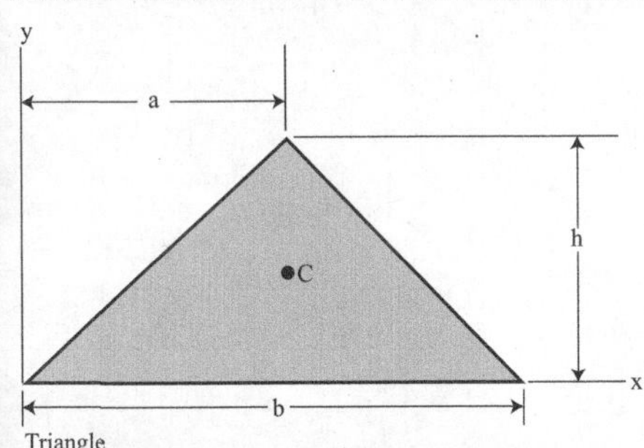

Triangle

$$A = \frac{1}{2}bh \qquad (I_{xx})_c = \frac{bh^5}{36}$$

$$x_c = \frac{1}{3}(a+b) \qquad (I_{yy})_c = \frac{bh}{36}(b^2 - ab + a^2)$$

$$y_c = \frac{1}{3}h \qquad (I_{xy})_c = \frac{bh^2}{72}(2a-b)$$

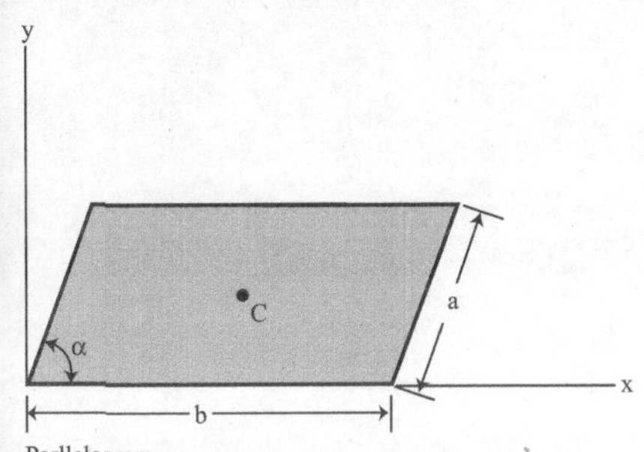

Parllelogram

$$A = ab\sin\alpha \qquad (I_{xx})_c = \frac{a^3 2}{12}\sin^3\alpha$$

$$x_c = \frac{1}{2}(b + a\cos\alpha) \qquad (I_{yy})_c = \frac{ab}{12}\sin\alpha(b^2 + a^2\cos^2\alpha)$$

$$y_c = \frac{1}{2}(a\sin\alpha) \qquad (I_{xy})_c = \frac{a^3 b}{12}\sin^2\alpha\cos\alpha$$

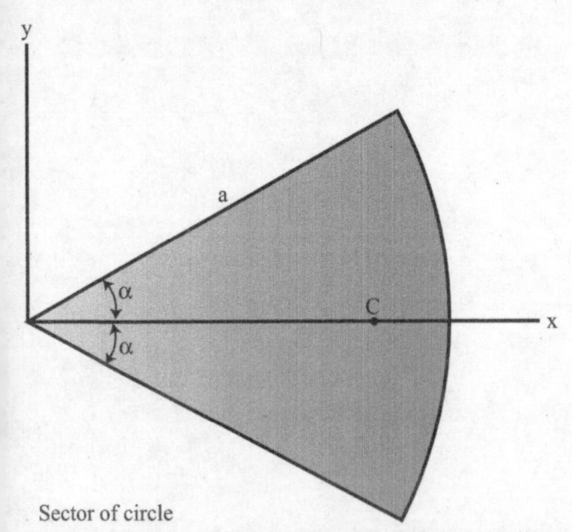

Sector of circle

$$A = a^2\alpha \qquad (I_{xx})_c = \frac{a^4}{4}\left(\alpha - \frac{\sin 2\alpha}{2}\right)$$

$$x_c = \frac{2}{3}\frac{a}{\alpha}\sin\alpha \qquad I_{yy} = \frac{a^4}{4}\left(\alpha + \frac{\sin 2\alpha}{2}\right)$$

$$y_c = 0 \qquad (I_{xy})_c = 0$$

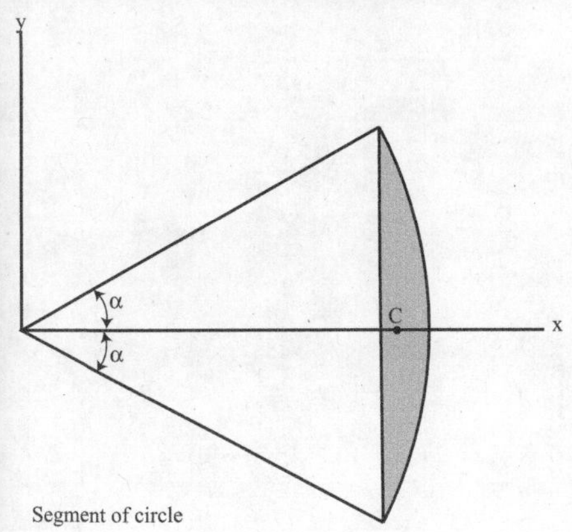

Segment of circle

$$A = a^2\left(\alpha - \frac{\sin 2\alpha}{2}\right) \qquad (I_{xx})\frac{a^4}{4}\left(\alpha - \frac{\sin 2\alpha}{2}\right)\left[1 - \frac{2\sin^3\alpha\cos\alpha}{3\left(\alpha - \frac{\sin 2\alpha}{2}\right)}\right]$$

$$x_c = \frac{2}{3}a\frac{\sin^3\alpha}{\alpha - \frac{\sin 2\alpha}{2}} \qquad I_{yy} = \frac{a^4}{4}\left(\alpha - \frac{\sin 2\alpha}{2}\right)\left[1 + \frac{2\sin^3\alpha\cos\alpha}{\alpha - \frac{\sin^2\alpha}{2}}\right]$$

$$y_c = 0 \qquad (I_{xy})_c = 0$$